Botany: An Ecological Approach

William A. Jensen
University of California, Berkeley

Frank B. Salisbury
Utah State University

Wadsworth Publishing Company, Inc.
Belmont, California

ISBN: 0–534–00092–4
L. C. Cat. Card No.: 70–183575
Printed in the United States of America

1 2 3 4 5 6 7 8 9 10–76 75 74 73 72

Biological drawings by Darwen & Vally Hennings.
Graphs & diagrams by Carleton Brown.

Cover Photo by Eliot Porter.

CHAPTER 3 *Fig. 3–A:* Courtesy of W. M. Laetsch, University of California, Berkeley. *Fig. 3–2: From Vascular Plants: Form and Function* by Frank B. Salisbury and Robert V. Parke, p. 33. Wadsworth Publishing Company, Inc., 1970. Used by permission. *Fig. 3–3:* The photograph was taken at the California Institute of Technology and supplied by Jack Griffith and James Bonner. *Fig. 3–5:* After Frey-Wyssling, *Ber. deut. botan. Ges.* 55, 1937, p. 132, redrawn from p. 206, *Principles of Plant Physiology* by James Bonner and Arthur W. Galston. W. H. Freeman and Company. Copyright © 1952. Used by permission. *Fig. 3–12:* From *Plant Physiology* by Frank B. Salisbury and Cleon Ross, p. 213. Wadsworth Publishing Company, Inc., 1969. Used by permission. *Fig. 3–14:* From *Plant Physiology* by Frank B. Salisbury and Cleon Ross, p. 238. Wadsworth Publishing Company, Inc., 1969. Used by permission. *Fig. 3–17:* From *Plant Physiology* by Frank B. Salisbury and Cleon Ross, p. 245. Wadsworth Publishing Company, Inc., 1969. Used by permission.

CHAPTER 4 *Fig. 4–A:* Courtesy of Daniel Branton, Department of Botany, University of California, Berkeley. *Fig. 4–1:* Courtesy of Roderick Park, Department of Botany, University of California, Berkeley. *Fig. 4–2:* From *Plant Physiology* by Frank B. Salisbury and Cleon Ross, p. 25. Wadsworth Publishing Company, Inc., 1969. Used by permission. *Fig. 4–3:* From *Vascular Plants: Form and Function* by Frank Salisbury and Robert V. Parke, p. 49. Wadsworth Publishing Company, Inc., 1970. Used by permission. *Fig. 4–4:* From *Plant Physiology* by Frank B. Salisbury and Cleon Ross, p. 157. Wadsworth Publishing Company, Inc., 1969. Used by permission. *Fig. 4–5:* From Emanual Epstein in *Plant Biochemistry* by James Bonner and J. E. Varner, p. 458. Academic Press, Inc., 1965. Used by permission. *Fig. 4–7:* From *Plant Physiology* by Frank B. Salisbury and Cleon Ross, p. 40. Wadsworth Publishing Company, Inc., 1969. Used by permission.

CHAPTER 5 *Fig. 5–H:* Supplied by Provost R. Gaurth Hansen, Utah State University. *Fig. 5–1:* Redrawn from data of John H. Wolcott and Cleon Ross, 1967, *Plant Physiol.*, 42:275. *Fig. 5–2:* From *Plant Physiology* by Frank B. Salisbury and Cleon Ross, p. 224. Wadsworth Publishing Company, Inc., 1969. Used by permission. *Fig. 5–3:* From *Plant Physiology* by Frank B. Salisbury and Cleon Ross, p. 554. Wadsworth Publishing Company, Inc., 1969. Used by permission.

CHAPTER 6 *Fig. 6–A and 6–G:* Courtesy of W. M. Laetsch, Department of Botany, University of California, Berkeley. *Fig. 6–B:* Courtesy of Paul Kurgens, Department of Botany and Plant Pathology, Colorado State University. *Fig. 6–C:* Courtesy of Mary Allen, Department of Botany, University of California, Berkeley. *Fig. 6–D:* Courtesy of Daniel Branton, Department of Botany, University of California, Berkeley. *Fig. 6–E:* Courtesy of Roderick Park, Department of Botany, University of California, Berkeley. *Fig. 6–F:* Courtesy of P. Walne and H. J. Arnott (top) and H. J. Arnott and R. M. Brown, Jr. (bottom). Cell Research Institute, University of Texas, Austin. *Fig. 6–1:* Courtesy of David Stetler, Dartmouth College. *Fig. 6–3:* Adapted from the bean leaf data of S. E. Balegh and Orlin Biddulph, 1970. *Plant Physiol.*, 46:1. *Fig. 6–6:* From "The Path of Carbon in Photosynthesis" by James A. Bassham, *Scientific American*, 206:88. Copyright © 1962 by Scientific American, Inc. All rights reserved. Used by permission. *Fig. 6–8:* Adapted from James A. Bassham, 1964, *Ann. Rev. Plant Physiol.*, 15:101. *Fig. 6–10:* Adapted from data of R. C. Dougherty et al., 1966, J. *Amer. Chen. Soc.*, 88:5037. *Fig. 6–12:* Courtesy of E. L. Vigil, University of Chicago.

CHAPTER 7 *Fig. 7–1:* From *Plant Physiology* by Frank B. Salisbury and Cleon Ross, p. 146. Wadsworth Publishing Company, Inc., 1969. Used by permission. *Fig. 7–3:* From *Plant Physiology* by Frank B. Salisbury and Cleon Ross, p. 319. Wadsworth Publishing Company, Inc., 1969. Used by permission. *Fig. 7–4:* From *Plant Physiology* by Frank B. Salisbury and Cleon Ross, p. 305. Wadsworth Publishing Company, Inc., 1969. Used by permission. *Fig. 7–5:* From *Plant Physiology* by Frank B. Salisbury and Cleon Ross, p. 308. Wadsworth Publishing Company, Inc., 1969. Used by permission. *Fig. 7–6:* From *Plant Physiology* by Frank B. Salisbury and Cleon Ross, p. 310. Wadsworth Publishing Company, Inc. Used by permission. *Fig. 7–8:* From *Plant Physiology* by Frank B. Salisbury and Cleon Ross, p. 314. Wadsworth Publishing Company, Inc., 1969. Used by permission.

CHAPTER 8 *Essay 8–1:* From *The Double Helix* by James D. Watson. Atheneum Publishers, © 1968. Used by permission. *Fig. 8–1:* Courtesy of William J. Wiebe, Department of Microbiology, University of Georgia, Athens. *Fig. 8–4.* From *The Plant Cell*, Second Edition, by William A. Jensen. Wadsworth Publishing Company, Inc., 1970. Used by permission. *Fig. 8–7:* From *The Plant Cell*, Second Edition, by William A. Jensen. Wadsworth Publishing Company, Inc., 1970. Used by permission. *Fig. 8–12:* From *The Plant Cell*, Second Edition, by William A. Jensen. Wadsworth Publishing Company, Inc., 1970. Used by permission. *Fig. 8–19:* Courtesy of Hilton Mollenhauer, Kettering Research Laboratory, Yellow Springs, Ohio. *Fig. 8–20:* Courtesy of M. H. F. Wilkins, Medical Research Council Biophysics Unit, King's College, London.

CHAPTER 9 *Fig. 9–2:* Courtesy of James Cronshaw and Katherine Esau, Department of Biological Sciences, University of California, Santa Barbara. *Fig. 9–3:* Courtesy of Allen Ichida and Melvin S. Fuller, Department of Botany, University of Georgia, Athens. *Fig. 9–5:* Courtesy of Paul Kurgens, Department of Botany and Plant Pathology, Colorado State University, Fort Collins. Diagram from *Cell and Molecular Biology* by E. J. Du Praw, by permission of Academic Press, New York, 1968. *Fig. 9–10:* After Taylor, *Amer. Nat.*, 91:1957. *Fig. 9–11:* From *Cell and Molecular Biology* by E. J. Du Praw, by permission of Academic Press, New York, 1968. *Fig. 9–12:* Courtesy of Anddrew S. Bajer, University of Oregon.

CHAPTER 10 *Fig. 10–A:* Courtesy of W. M. Laetsch, Department of Botany, University of California, Berkeley.

CHAPTER 11 *Fig. 11–A:* Studio portrait by Maull and Fox, 1856. *Fig. 11–3:* Courtesy of Sherwin Carlquist, from *Hawaii, A Natural History*, Natural History Press, New York, 1971.

CHAPTER 12 *Essay 12–1:* From *Nature and Man's Fate* by Garrett Hardin. Copyright © 1959 by Garrett Hardin. Reprinted by permission of Holt, Rinehart and Winston, Inc., and the author.

CHAPTER 13 *Fig. 13–A:* Courtesy of W. M. Laetsch, Director, Botanical Garden, University of California, Berkeley; photograph taken by Albert Blaker, Scientific Photography Laboratory, University of California, Berkeley. *Fig. 13–1:* From Woodworth, Jugenheimer, and Lenz, Agronomy Journal, 44:60 (1952). *Fig. 13–2:* From *Processes of Organic Evolution* by Ledyard Stebbins, © 1966. Used by permission of Prentice-Hall, Inc., Englewood Cliffs, New Jersey. *Fig. 13–3:* From *Biology of Plants* by P. Raven and H. Curtis. Worth Publishing Co., 1970. *Fig. 13–4:* From *Cell Heredity* by Ruth Sager and Francis Ryan, p. 52. © 1961 by John Wiley and Sons, Inc. Used by permission. *Fig. 13–6:* From *Processes of Organic Evolution* by G. Ledyard Stebbins, © 1966. Used by permission of Prentice-Hall, Inc., Englewood Cliffs, New Jersey.

CHAPTER 14 *Figs. 14–A, 14–B, 14–C, 14–D, and 14–F:* From *Prehistoric Animals* by J. Augusta and Z. Burian. Paul Hamlyn Books, Feltham, England, © 1960. *Fig. 14–E:* Courtesy

1964. Copyright 1964 by the American Association for the Advancement of Science. Used by permission. *Fig. 31–5:* Data from Folke Skoog and C. O. Miller, 1957. *Society of Experimental Biology Symposia,* 11:118–131. Used by permission.

CHAPTER 32 *Fig. 32–2:* From H. A. Evans and Alvin Nason, *Plant Physiol.,* 28:233–254. Used by permission. *Fig. 32–3:* Courtesy of Ralph Baker, Colorado State University. The technician is Marilyn Young. *Fig. 32–4:* Modified from Perry R. Stout, *Proc. 9th Ann. Calif. Fertilizer Conf.,* pp. 21–23, 1961. Used by permission.

CHAPTER 33 *Fig. 33–A:* Courtesy of A. S. Foster, University of California, Berkeley. *Fig. 33–4:* From *Plant Physiology* by Bernard S. Meyer and Donald B. Anderson, p. 239. D. Van Nostrand Company, Inc., 1939. Used by permission of the author. *Fig. 33–7:* Specimen was supplied by Herman Wiebe. *Fig. 33–8:* From "Sap Pressure in Vascular Plants" by P. F. Scholander, et al. in *Science,* 148:339–346, April 16, 1965. Copyright 1965 by the American Association for the Advancement of Science. Used by permission. *Fig. 33–11:* From "Movement of Organic Substances in Trees" by Martin H. Zimmerman in *Science,* 133:73–79, January 13, 1961. Copyright 1961 by the American Association for the Advancement of Science. Used by permission. *Fig. 33–13:* Courtesy of Orlin Biddulph, Washington State University. All rights reserved.

CHAPTER 35 *Fig. 35–1:* From Burke K. Zimmerman and Winslow R. Briggs, 1963, *Plant Physiol.,* 38:248. Used by permission. *Fig. 35–2:* Absorption spectra for Pr and Pfr purified from oats are data of Warren L. Butler, et al., 1965. In T. W. Goodwin (ed.), *Chemistry and Biochemistry of Plant Pigments,* Academic Press, New York, p. 203. Action spectra shown in the middle graph were redrawn from data of M. W. Parker, et al., 1949, *Amer. J. Bot.,* 36:194. Action spectra for promotion of bean hypocotyl hook opening shown in the lower graph were redrawn from R. B. Withrow, et al., 1957, *Plant Physiol.,* 32:453. Action spectra for stimulation and subsequent inhibition of lettuce seeds (Grand Rapids Variety) were redrawn from data of Sterling B. Hendricks, 1960, *Comparative Biochemistry of Photoreactive Systems,* Academic Press, New York, p. 307. All data were used by permission. *Fig. 35–3:* Courtesy of Harry Borthwick, United States Dept. of Agriculture, Agricultural Research Service, Beltsville, Maryland. *Fig. 35–4:* Adapted from Hans Mohr and M. Wehrung, 1960, *Planta,* 55:438. Used by permission. *Fig. 35–5:* Data from James A. Lackhart, 1961, *Amer. J. Bot.,* 48:387. Used by permission. *Fig. 35–6:* From James A. Lochart, 1963, *Adv. Frontiers of Plant Sciences,* 7:1, after G. Meyer, 1959, *Acta Botanica Neerlandica,* 8:189. Used by permission. *Fig. 35–10:* From *The Physiological Clock* by Erwin Bünning, p. 62. Springer-Verlag, 1964. Used by permission. *Fig. 35–11:* Data after R. Zimmer in *The Physiological Clock* by Erwin Bünning, p. 61. Springer-Verlag, 1964. Used by permission. *Fig. 35–13:* Curves are hypothetical. Data from "Endogenous Circadian Rhythm in Cytoplasm of Acetabularia: Influence of the Nucleus," by E. Schweiger, et al. in *Science,* 146:658–659, October 30, 1964. Copyright 1964 by the American Association for the Advancement of Science. Used by permission.

CHAPTER 36 *Fig. 36–O:* From "Hormonal Induction of Flowering in Conifers" by Richard P. Pharis and John N. Owens in *Yale Scientific Magazine,* November 1966. Used by permission. *Fig. 36–1:* Modified from *The Flowering Process* by Frank B. Salisbury, p. 206. Pergamon Press, 1963. Used by permission. *Fig. 36–2:* From *The Flowering Process* by Frank B. Salisbury. Pergamon Press, 1963. Used by permission. *Fig. 36–3:* From *The Biology of Flowering* by Frank B. Salisbury. The Natural History Press, New York, 1971. *Fig. 36–6:* (*left*) From *Plant Physiol.,* 30:327–334, 1955. *Fig. 36–8:* From *The Flowering Process* by Frank B. Salisbury. Pergamon Press, 1963. Used by permission. *Fig. 36–9:* all of these experiments were performed by David D. Gibby, *Plant Physiol.,* 47:784–789 (1971). *Fig. 36–10:* Data for Xanthium from Frank B. Salisbury and James Bonner, 1956, *Plant Physiol.,* 310:141–147; data for Hyoscyamus from H. Claes and A. Lang, 1947, *Z. Natur for schung,* 26:56–63. *Fig. 36–11:* Data for both figures from various publications of Karl C. Hamner. Figures from *The Flowering Process* by Frank B. Salisbury. Pergamon Press, 1963. Data and figures used by permission. *Fig. 36–12:* Data from Frank B. Salisbury, 1965, *Planta,* 66:1–26.

CHAPTER 37 *Facing photograph for Chapter 37:* Photograph by National Aeronautics and Space Administration.

CHAPTER 38 *Essay 38–1:* Modified from R. M. Klein and D. T. Klein, *Discovering Plants,* Natural History Press, Garden City, New York, 1968. Used by permission. *Figs. 38–A, 38–C, 38–D, 38–E, 38–F, 38–G, 38–H, and 38–K:* Courtesy of the United States Forest Service. *Fig. 38–3:* From *Communities and Ecosystems* by Robert H. Whittaker, pp. 155–156. The Macmillan Company, 1970. Used by permission of the author. *Fig. 38–7A:* Rainfall data from *Climate and Man,* The 1941 Yearbook of Agriculture.

CHAPTER 39 *Fig. 39–2:* Curves from U.S. Dept. of Commerce, Weather Bureau chart supplied by A. Richardson, who also furnished the data. See also Lois M. Cox, 1967, *Utah Science,* 28(2):53.

CHAPTER 40 *Essay 40–1:* Prepared largely from material supplied by R. Philip Hammond, Oak Ridge National Laboratory, Tennessee. *Fig. 40–4A:* Modified from *Fundamentals of Ecology* by Eugene P. Odum. W. B. Saunders Company, 1959. Used by permission. *Fig. 40–5:* Modified from *Fundamentals of Ecology* by Eugene Odum. W. B. Saunders Company, 1959. Used by permission. *Fig. 40–7:* Data of R. Kindeman, 1942, *Ecology,* 23:399–418. *Fig. 40–11:* From *Biological Science* by William T. Keeton, p. 715. W. W. Norton & Company, 1967. Used by permission. *Fig. 40–13B:* For more details, see Frank B. Salisbury, 1964, *Ecology,* 45:1–9.

of Wayne Fry, Department of Paleontology, University of California, Berkeley. *Fig. 14–2:* Courtesy of J. William Schopf, Department of Geology, Paleobiological Laboratory, University of California, Los Angeles. *Fig. 14–3:* From J. William Schopf, *Biological Reviews*, 45:319 (1970). Used by permission of Cambridge University Press. *Fig. 14–5:* Modified from "The Breakup of Pangaea" by Robert S. Dietz and John C. Holden, *Scientific American*, 223:32. Copyright © 1970 by Scientific American, Inc. All rights reserved. Used by permission.

CHAPTER 15 *Fig. 15–A:* Courtesy of Roderick Park, Department of Botany, University of California, Berkeley. *Figs. 15–B, 15–C, 15–D:* From *Viruses in Plant Hosts* by Katherine Esau. (Madison: The University of Wisconsin Press, © 1968 by the Regents of the University of Wisconsin.) Used by permission. *Fig. 15–1:* From *Nonseed Plants: Form and Function*, Second Edition, by William T. Doyle. Wadsworth Publishing Company, Inc., 1970. Used by permission. *Fig. 15–2:* Modified from *Nonseed Plants: Form and Function*, Second Edition, by William T. Doyle. Wadsworth Publishing Company, Inc., 1970. *Fig. 15–3:* From *Nonseed Plants: Form and Function*, Second Edition, By William T. Doyle. Wadsworth Publishing Company, Inc., 1970. Used by permission.

CHAPTER 16 *Fig. 16–H:* From C. C. Remsen et al. in *Journal of Bacteriology*, Vol. 95, 1968. Used by permission. *Fig. 16–1:* Adapted from *Plant Pathology* by George N. Agrios. Academic Press, Inc., 1969. *Fig. 16–3:* From *Principles of Microbial Ecology* by Thomas D. Brock. © 1966. Used by permission of Prentice-Hall, Inc., Englewood Cliffs, New Jersey. *Fig. 16–4:* Modified from *Biology of Microoganisms* by Thomas P. Brock. © 1970. Used by permission of Prentice-Hall, Inc., Englewood Cliffs, New Jersey. *Fig. 16–6:* Modified from *Biology of Microorganisms* by Thomas D. Brock. © 1970. Used by permission of Prentice-Hall, Inc., Englewood Cliffs, New Jersey. *Fig. 16–9:* Modified from *An Evolutionary Survey of the Plant Kingdom* by Robert F. Scagel, et al. Wadsworth Publishing Company, Inc., 1965. *Fig. 16–10:* Courtesy of William J. Liebe, Department of Microbiology, University of Georgia, Athens. *Fig. 16–12:* From *An Evolutionary Survey of the Plant Kingdom* by Robert F. Scagel, et al. Wadsworth Publishing Company, Inc., 1965. Used by permission. *Fig. 16–13:* From *Nonseed Plants: Form and Function*, Second Edition, by William T. Doyle. Wadsworth Publishing Company, Inc., 1970. Used by permission. *Fig. 16–14:* From *The Microbial World*, Third Edition, by R. Y. Stanier, et al. © 1970. Used by permission of Prentice-Hall, Inc., Englewood Cliffs, New Jersey. *Fig. 16–17:* A. from "A Thermophilic Acidophilic Mycoplasma Isolated from a Coal Refuse Pile," by G. Darland, et al., *Science*, Vol. 170, p. 1416, Fig. la. Copyright 1970 by the American Association for the Advancement of Science. Used by permission. C. From *Biology of Microorganisms* by Thomas D. Brock. © 1970. Used by permission of Prentice-Hall, Inc., Englewood Cliffs, New Jersey.

CHAPTER 17 *Figs. 17–A and 17–D:* (Chlamydomonas) Courtesy of Norma Lang, University of California, Davis. *Fig. 17–C:* (Spirulina) Courtesy of John West, University of California, Berkeley. *Fig. 17–6:* From *Concepts of Ecology* by Edward J. Kormondy. © 1969. Used by permission of Prentice-Hall, Inc., Englewood Cliffs, New Jersey. *Fig. 17–8:* Courtesy of Daniel Branton, University of California, Berkeley. *Figs. 17–10A, B, and C:* Courtesy of Paul Krugens, Colorado State University, Fort Collins. *Figs. 17–11B and C:* Courtesy of M. L. Chiappino and B. E. Volcani, University of California, San Diego.

CHAPTER 18 *Fig. 18–2:* From *An Evolutionary Survey of the Plant Kingdom* by Robert F. Scagel, et al. Wadsworth Publishing Company, Inc., 1965. Used by permission. *Fig. 18–10:* Courtesy of John West, Department of Botany, University of California, Berkeley. *Fig. 18–11:* Courtesy of Paul Kurgens, Colorado State University, Fort Collins. *Fig. 18–12:* Modified from *Plant Diversity: An Evolutionary Approach* by Robert F. Scagel, et al. Wadsworth Publishing Company, Inc., 1969. *Fig. 18–14:* Courtesy of Paul Kurgens, Colorado State University, Fort Collins. *Fig. 18–16:* From *Plant Diversity: An Evolutionary Approach* by Robert F. Scagel, et al. Wadsworth Publishing Company, Inc., 1969. Used by permission.

CHAPTER 19 *Fig. 19–B:* Courtesy of Ralph Emerson, Department of Botany, University of California, Berkeley. *Figs. 19–C, 19–D, and 19–E:* Courtesy of Melvin S. Fuller, Department of Botany, University of Georgia, Athens. *Fig. 19–1:* Modified from *The Biology of Fungi* by C. T. Ingold. Hutchinson Educational Ltd., London, 1961. Used by permission. *Fig. 19–2:* Modified from *The Biology of Fungi* by C. T. Ingold. Hutchinson Educational Ltd., London, 1961. Used by permission. *Fig. 19–3:* Modified from *The Biology of Fungi* by C. T. Ingold. Hutchinson Educational Ltd., London, 1961. Used by permission. *Fig. 19–4:* Modified from *The Biology of Fungi* by C. T. Ingold. Hutchinson Educational Ltd., London, 1961. Used by permission. *Fig. 19–5:* Modified from *The Biology of Fungi* by C. T. Ingold. Hutchinson Educational Ltd., London, 1961. Used by permission. *Fig. 19–7:* From *Plant Diversity: An Evolutionary Approach* by Robert F. Scagel, et al. Wadsworth Publishing Company, Inc., 1969. Used by permission. *Fig. 19–8:* Modified from *The Biology of Fungi* by C. T. Ingold. Hutchinson Educational Ltd., London, 1961. Used by permission. *Fig. 19–9:* From *Plant Pathology* by George N. Agrios. Academic Press, Inc., 1969. Used by permission. *Fig. 19–11:* A. Modified from *The Biology of Fungi* by C. T. Ingold. Hutchinson Educational Ltd., London, 1961. Used by permission. B. From *Plant Diversity: An Evolutionary Approach* by Robert F. Scagel, et al. Wadsworth Publishing Company, Inc., 1969. Used by permission. *Fig. 19–12:* From *Introduction to the Fungi* by John Webster. Cambridge University Press, 1970. Used by permission.

CHAPTER 20 *Figs. 20–A, 20–B, 20–C, 20–D, 20–E, 20–F, 20–G, 20–H, and 20–I:* Courtesy of Victor Duran, Berkeley, California. *Figs. 20–J, 20–K, 20–L, 20–M, 20–N, and 20–O:* Courtesy of O'Neill Ray Collins, Department of Botany, University of California, Berkeley. *Fig. 20–2:* From *The Biology of Fungi* by C. T. Ingold. Hutchinson Educational Ltd., London, 1961. Used by permission. *Fig. 20–5:* From *Plant Diversity: An Evolutionary Approach* by Robert F. Scagel, et al. Wadsworth Publishing Company, Inc., 1969. Used by permission. *Fig. 20–11:* Courtesy of Victor Duran, Berkeley, California. *Fig. 20–12:* From *The Biology of Fungi* by C. T. Ingold. Hutchinson Educational Ltd., London, 1961. Used by permission. *Fig. 20–14:* A. From *The Biology of Fungi* by C. T. Ingold. Hutchinson Educational Ltd., London, 1961. Used by permission. B. Courtesy of J. Motta. *Fig. 20–16:* Courtesy of O. R. Collins, University of California, Berkeley. *Fig. 20–20:* From *Plant Pathology* by George N. Agrios. Academic Press, Inc., 1969. Used by permission.

CHAPTER 21 *Figs. 21–B, 21–C, 21–D, 21–E, 21–F, and 21–G:* Courtesy of Victor Duran, Berkeley, California. *Fig. 21–2:* From *Plant Diversity: An Evolutionary Approach* by Robert F. Scagel, et al. Wadsworth Publishing Company, Inc., 1969. Used by permission. *Fig. 21–5:* Courtesy of Robert Knaupt, Department of Botany, University of California, Berkeley. *Fig. 21–7:* Courtesy of Ralph Emerson, University of California, Berkeley.

CHAPTER 22 *Fig. 22–A:* Taken by Barry Juniper, Oxford University. *Fig. 22–4:* Courtesy of Gordon McBride, Department of Botany, University of Michigan, Ann Arbor.

CHAPTER 23 *Fig. 23–A:* Courtesy of Howard Arnott, University of Texas, Austin. *Figs. 23–B: (right) and 24–C:* From *Nonseed Plants: Form and Function*, Second Edition, by William T. Doyle. Wadsworth Publishing Company, Inc., 1970. Used by permission. *Fig. 23–1:* Courtesy of P. V. Glob, Director, The Danish National Museum. *Fig. 23–6:* From *Nonseed Plants: Form and Function*, Second Edition, by William T. Doyle. Wadsworth Publishing Company, Inc., 1970. Used by permission. *Fig. 23–8:* From *The Biology of Higher Cryptograms* by William T. Doyle. The Macmillan Co., London, 1970. Used by permission. *Fig. 23–9:* Modified from *Spore Liberation* by C. T. Ingold. Oxford University Press, 1965. *Fig. 23–11:* From *Nonseed Plants: Form and Function*, Second Edition, by William T. Doyle. Wadsworth Publishing Company, Inc., 1970. Used by permission. *Fig. 23–12:* From *Plant Diversity: An Evolutionary Approach* by Robert F. Scagel, et al. Wadsworth Publishing Company, Inc., 1969. Used by permission.

CHAPTER 24 *Fig. 24–C:* Courtesy of W. M. Laetsch, University of California, Berkeley. *Fig. 24–E:* Courtesy of E. P. Healey, University of California, Irvine. *Fig. 24–G:* Courtesy of J. Cronshaw, University of California, Santa Barbara. *Fig. 24–12:* From *Plant Cytology* by F. A. L. Clowes and B. E. Juniper. Blackwell Scientific Publications, Ltd., Oxford, England. © 1968. Used by permission. *Fig. 24–14:* From *Plant Cytology* by F. A. L. Clowes and B. E. Juniper. Blackwell Scientific Publications, Ltd., Oxford, England. © 1968. Used by permission. *Fig. 24–16:* Courtesy of Elizabeth Frank. *Fig. 24–19:* Courtesy of Mozingo, University of California, Berkeley. Work supported by the NSF Grants # GK–6428 and GK–3845. *Fig. 24–20:* From *Plant Physiology* by Frank B. Salisbury and Cleon Ross. Wadsworth Publishing Company, Inc., 1969. Used by permission. *Fig. 24–24:* From G. Avery, *Amer. Jour. Bot.*, 20, 1933. *Fig. 24–26:* From *Plant Anatomy* by Katherine Esau. © 1965 by John Wiley and Sons, Inc. Used by permission.

CHAPTER 25 *Figs. 25–A, 25–B, and 25–C:* From *Evolution and Plants of the Past* by Harlan P. Banks. Wadsworth Publishing Company, Inc., 1970. Used by permission. *Fig. 25–H:* From *Nonseed Plants: Form and Function*, Second Edition, by William T. Doyle. Wadsworth Publishing Company, Inc., 1970. Used by permission. *Fig. 25–2:* From *Evolution and Plants of the Past* by Harlan P. Banks. Wadsworth Publishing Company, Inc., 1970. Used by permission. *Fig. 25–4:* Redrawn by permission from *Comparative Morphology of Vascular Plants* by Adriance S. Foster and Ernest M. Gifford, Jr. W. H. Freeman and Company. Copyright © 1959. *Fig. 25–5:* Redrawn by permission from Comparative Morphology of Vascular Plants by Adriance S. Foster and Ernest M. Gifford, Jr. W. H. Freeman and Company. Copyright © 1959. *Fig. 25–6:* Redrawn by permission from *Comparative Morphology of Vascular Plants* by Adriance S. Foster and Ernest M. Gifford, Jr. W. H. Freeman and Company. Copyright © 1959. *Fig. 25–7:* From *Plant Diversity: An Evolutionary Approach* by Robert F. Scagel, et al. Wadsworth Publishing Company, Inc., 1969. Used by permission.

CHAPTER 26 *Fig. 26–13:* Courtesy of Diana Myles, Department of Botany, University of California, Berkeley. *Fig. 26–14:* Redrawn by permission from *Comparative morphology of Vascular Plants* by Adriance S. Foster and Ernest M. Gifford, Jr. W. H. Freeman and Company. Copyright © 1959. *Fig. 26–17:* Courtesy of Diana Myles, Department of Botany, University of California, Berkeley.

CHAPTER 27 *Fig. 27–4:* Redrawn by permission from *Comparative Morphology of Vascular Plants* by Adriance S. Foster and Ernest M. Gifford, Jr. W. H. Freeman and Company. Copyright © 1959. *Fig. 27–5:* From W. N. Stewart and T. Delevoryas, *Bot. Rev.*, 22, 1956. *Fig. 27–6:* Redrawn by permission from *Comparative Morphology of Vascular Plants* by Adriance S. Foster and Ernest M. Gifford, Jr. W. H. Freeman and Company. Copyright © 1959. *Fig. 27–13:* Redrawn by permission from *Comparative Morphology of Vascular Plants* by Adriance S. Foster and Ernest M. Gifford, Jr. W. H. Freeman and Company. Copyright © 1959. *Fig. 27–14:* Redrawn by permission from *Comparative Morphology of Vascular Plants* by Adriance S. Foster and Ernest M. Gifford, Jr. W. H. Freeman and Company. Copyright © 1959. *Fig. 27–15:* Redrawn by permission from *Comparative Morphology of Vascular Plants* by Adriance S. Foster and Ernest M. Gifford, Jr. W. H. Freeman and Company. Copyright © 1959. *Fig. 27–16:* Courtesy of Charles B. Beck, University of Michigan, Ann Arbor.

CHAPTER 28 *Fig. 28–1:* Modified from *The History of Life* by A. Lee McAlester. Prentice-Hall, Inc., Englewood Cliffs, New Jersey, 1968. *Fig. 28–9:* From *An Evolutionary Survey of the Plant Kingdom* by Robert F. Scagel, et. al. Wadsworth Publishing Company, Inc., 1965. Used by permission. *Fig. 28–10:* From *An Evolutionary Survey of the Plant Kingdom* by Robert F. Scagel, et al. Wadsworth Publishing Company, Inc., 1965. Used by permission. *Fig. 28–11:* From *An Evolutionary Survey of the Plant Kingdom* by Robert F. Scagel, et al. Wadsworth Publishing Company, Inc., 1965. Used by permission. *Fig. 28–12:* From *An Evolutionary Survey of the Plant Kingdom* by Robert F. Scagel, et al. Wadsworth Publishing Company, Inc., 1965. Used by permission. *Fig. 28–13:* From *An Evolutionary Survey of the Plant Kingdom* by Robert F. Scagel, et al. Wadsworth Publishing Company, Inc., 1965. Used by permission. *Fig. 28–17:* Courtesy of Patricia Schulz, Rosary College, River Forest, Illinois. *Fig. 28–20:* Courtesy of Patricia Schulz, Rosary College, River Forest, Illinois. *Fig. 28–22:* From *Plant Diversity: An Evolutionary Approach* by Robert F. Scagel, et al. Wadsworth Publishing Company, Inc., 1969. Used by permission. *Fig. 28–24:* From *Plant Diversity: An Evolutionary Approach* by Robert F. Scagel, et al. Wadsworth Publishing Company, Inc., 1969. Used by permission. *Fig. 28–26:* From *Plant Diversity: An Evolutionary Approach* by Robert F. Scagel, et al. Wadsworth Publishing Company, Inc., 1969. Used by permission. *Fig. 28–27:* From *An Evolutionary Survey of the Plant Kingdom* by Robert F. Scagel, et al. Wadsworth Publishing Company, Inc., 1969. Used by permission. *Fig. 28–27:* From *Plant Diversity: An Evolutionary Approach* by Robert F. Scagel, et al. Wadsworth Publishing Company, Inc., 1969. Used by permission.

CHAPTER 29 *Fig. 29–2:* From *Plant Classification* by Lyman Benson. D. C. Heath and Company, 1957. Used by permission of the publisher. *Fig. 29–I:* Modified by Lyman Benson from Bessey's original chart. Original publication: Charles E. Bessey, *Annals of the Missouri Botanical Garden*, 2:118, 1915. *Fig. 29–J:* Based upon the chart of Bessey but modified by Frederic and Edith Clements in *Rocky Mountain Flowers*. H. W. Wilson Company, 1945. The chart was originally prepared in full color and copyrighted by the National Geographic Society. *Fig. 29–K:* From *The Evolution and Classification of Flowering Plants* by Arthur Cronquist. Houghton Mifflin Company, 1968. Used by permission.

CHAPTER 30 *Fig. 30–2:* Data from *The Experimental Control of Plant Growth* by F. W. Went. The Ronald Press Company. Copyright © 1957. Used by permission. *Fig. 30–4:* From *Plant Physiology* by Frank B. Salisbury and Cleon Ross, p. 445. Wadsworth Publishing Company, Inc., 1969. Used by permission. *Fig. 30–5:* From *Vascular Plants: Form and Function* by Frank B. Salisbury and Robert V. Parke, p. 118. Wadsworth Publishing Company, Inc., 1970. Used by permission. *Fig. 30–6:* From *Phytohormones* by F. W. Went and K. V. Thimann, p. 41. The Macmillan Company, 1937. Used by permission of the author. *Fig. 30–8:* Redrawn from Walter A. Shropshire and R. B. Withrow, 1958, *Plant Physiol.*, 33:360–365. Used by permission. *Fig. 30–9:* From *Plant Physiology* by Frank B. Salisbury and Cleon Ross, p. 490. Wadsworth Publishing Company, Inc., 1969. Used by permission. *Fig. 30–14:* From *Growth and Organization in Plants* by F. C. Steward, p. 223. Addison-Wesley Publishing Company, 1968. Used by permission. *Fig. 30–15:* Adapted from K. Mothes and L. Engelbrecht, *Phytochem.*, 1:58, 1961. Used by permission of Pergamon Press.

CHAPTER 31 *Fig. 31–1:* From *Plant Morphogenesis* by E. W. Sinnott, p. 32. McGraw-Hill Book Company, Inc., 1960. Used by permission. *Fig. 31–2:* From "Growth and Development of Cultured Plant Cells" by F. C. Steward, et al. in *Science*, Vol. 143, pp. 20–27, January 3,

Acknowledgments

The authors of some of the essays and end papers deserve special recognition. They contributed not only their time but a part of themselves in writing for us. The contributors are: Harlan P. Banks, Charles Beck, Lyman Benson, James Bonner, Erwin Bunning, Edward Cantino, Sherwin Carlquist, Lincoln Constance, Wade Dewey, Adriance Foster, David Gibby, Karl Hammer, Garrett Hardin, Sterling Hendricks, Ralph Johnson, Leonard Machlis, John Raper, A. W. Roach, G. Ledyard Stebbins, A. H. Sturtevant, Kenneth Thimann, Johannes van Overbeek, E. Dean Vest, James Watson, Frits Went, Ralph Wetmore, Robert Whittaker.

In particular, we would like to thank Lee Kavaljian for the valuable editorial assistance he provided at various stages of the manuscript preparation. We also owe special thanks to Patricia Schulz for her willing assistance in the final days of proofreading and putting the book together. Among the host of people who generously gave of their time and ideas, we especially thank: Mary Ashton, Herbert Baker, Orlin Biddulph, David Bierhorst, W. D. Billings, James Bonner, Daniel Branton, Lincoln Constance, Eleanor Crump, Adriance Foster, Jerry Hill, Arthur Holmgren, James Jensen, Stefan Kirchanski, Watson Laetsch, James Lockhart, Raymond T. Lynn, Jerry Motta, Robert Ordnuff, A. W. Roach, Richard Shaw, L. K. Shumway, Paul Silva, Janet Stein, Erich Steiner, Sanford Tepfer, Warren Wagner, David Walker, Hermann Wiebe, Robert Whittaker.

We are indebted to Patricia Hagius, Nancy Williams, Marilyn Salisbury, Alice Denny, Steven L. Kimball, and Joan Jensen for their help in typing and indexing.

Once again, as authors, we are delightedly impressed by our colleagues' willingness to supply us with photographs—all of which are acknowledged beginning on the copyright page. We want to warmly thank all of these cooperative individuals for their time and effort.

Preface

Botany is the scientific study of plants. But it is also men and ideas, facts and concepts, the search for truth, and the challenges and frustrations of unsolved problems. In short, it is a living science that is worthy of man's most serious intellectual efforts.

The first man who planted a seed with the hope that it would yield food—and wondered what he could do to help it grow—was a botanist. With the concept of cultivating plants came a dependence on plants that slowly developed into a modern social and economic order. But this dependence also brought the very real threat of famine, if the crop failed. It also brought another, rather ironic peril—a surplus of food that allows the human population to increase at such rates that no amount of cultivation will ultimately be able to feed man unless these rates are slowed.

But botany is more than the growing of crops and the fighting of crop disease. It is the joy of seeing spring flowers and knowing why they are growing in a particular spot and why they are blooming now. It is seeing plants, knowing their names, knowing where they come from, and knowing what they are used for. Above all, it is an increased awareness of the living world of plants that surrounds us and upon which we are dependent.

Ecology is the scientific study of plants and animals and their interrelationships with their environment. But it is also a unifying concept in biology that relates activities within an organism and between organisms. It is the science that answers the question: why does that plant grow there? It is also the science that relates growth to decay and photosynthesis to respiration. And it is now being called upon to unravel—and perhaps to remedy—the train of events initiated in the earth's ecosystem by man's activities. Obviously, ecology is related to botany, and we have therefore used it as the unifying theme in this book. But this is a botany book with an ecological approach—not an ecology text.

Anyone writing a book on a general topic like botany is faced with two problems—giving enough facts and ideas so that the field can be understood and, at the same time, presenting a broad overview so that these facts can be pieced together into a meaningful whole. We have tried to do this by devoting Part 1 (consisting of two chapters) to a general introduction to the major areas of botany. We start this introduction by taking a walk in an acre of woodland, which leads us down through the layers of organization to the cellular and molecular levels. After this general introduction, we work our way through the cell in more detail (Part 2) and then cover genetics and evolution (Part 3). Using this as a base, we begin a second examination of the plant kingdom, covering the viruses, bacteria, algae, and fungi in Part 4 and the land plants in Part 5. The physiology and development of vascular plants is discussed in Part 6. Finally, in Part 7 we deal with land-plant ecology, which in a sense completes the cycle and returns us to the first chapter. By using this approach, we hope that the general aspects of botany will be apparent at the beginning so that later the detailed material will be more interesting and more meaningful. Throughout all these chapters, the thread of ecology serves as our unifying factor.

We also believe that a textbook like this should clearly indicate which material is basic to an understanding of botany and which material is essentially enrichment material that extends both the level and the interest of the primary text. Our response to this need has been to organize the chapters into four main sections: photo essays, core text, essays, and end papers. The core text focuses on primary aspects of botany, and the enrichment materials (photo essays, essays, and end papers) provide both a more detailed examination of certain topics mentioned in the text and interesting discussions that relate botany to the world around us.

Most chapters begin with a photo essay. Thus a photo essay may be a pictorial introduction to marine algae or ferns, or it may be a look into a herbarium or a biochemistry laboratory. In either case, it sets the scene for the chapter.

The chapter proper—what we think of as the core text—is written in an easy-to-read, narrative style and has an introduction calculated to arouse and hold your interest. In the core itself we stress the basic and essential elements of the subject being discussed. Wherever possible an ecological approach has been used; for example, in the two chapters on algae, the environment in which they live is related to their morphology and life cycles. We have also stressed the relation of form and function; our discussions of photosynthesis and chloroplasts and mitochondria and respiration are good examples. Finally, wherever it has been possible in the core text, we have introduced information and examples relevant to current problems and man's role in the ecosystem.

Within the chapters—but separate from the text—are the essays, which cover a wide range of topics and serve several functions. Sometimes, they expand upon certain topics covered in the core text by presenting additional material in greater detail. They also introduce interesting and relevant asides, such as recipes for making yogurt and sushi, directions for growing bonsai trees, an account of Lord Kevin's disagreement with Darwin about the age of the earth, and a discussion of enzyme detergents. Another important function of the essays is related to the way in which science operates. We know that botany is more than facts, and we have tried to emphasize this by examining the evolution of important concepts and the men behind them. But we have gone one step further—we have let botanists speak for themselves! We wrote to several dozen friends and colleagues asking them to tell us about their work and about themselves. In return, we received a fascinating collection of letters that provide an insight into not only how an idea is developed in the laboratory but also why these scientists became botanists.

Many chapters have end papers that are either an extension of the core material or additional background material related to the chapter itself. For example, Chapter 9 covers mitosis and meiosis and has an end paper on the ultrastructure of cell division; Chapter 6 discusses photosynthesis and chloroplasts and has an end paper on light; Chapter 19, which includes the water molds, has an end paper on sex hormones in fungi; and Chapter 29, which details angiosperm systematics, has an end paper on the evolution of systems of classification. End papers allow the student to pursue some topics in greater depth, and they can also be used as topics for class discussions, since they often present both sides of a controversial problem. You can take or leave the essays and end papers—but we bet that you'll both read and enjoy them!

Well, here it is—our approach to a botany text!

Contents

Part 1 Introduction

1 Ecological Botany 2

Photo Essay: An Acre of Woodland 2

2 What Is Life? 18

Photo Essay: From Kilometers to Microns 18
End Paper: Metric Measurement 28

Part 2 The Cell

3 Cell Structure 31

Essay: Seeing the Unseen World 32
End Paper: The Chemistry of the Cell: Carbohydrates 42

4 Cell Membranes 51

End Paper: Fats and Amino Acids 60
End Paper: Colloids: Sols, Gels, and Hydration 64

5 Cell Energetics: Enzymes and Metabolism 66

Photo Essay: The Enzymology Laboratory 66
Essay: The Universes of Protein Variety 73
End Paper: Assay and Purification of Alpha Amylase 77

6 Photosynthesis: Chloroplasts 78

Photo Essay: Chloroplasts—The Green Machines 78
Essay: Chemistry of the Photosynthetic Pigments 97
End Paper: Photochemistry and the Nature of Light 98

7 Respiration: Mitochondria 101

8 Cell Biosynthesis 112

Photo Essay: The Nucleus—Control Center of the Cell 112
Essay: The Double Helix by James Watson 119
Essay: X-ray Diffraction 129
Essay: The Nobel Prize 131
End Paper: Autoradiography 134

9 Chromosomes, Mitosis, and Meiosis 136

Photo Essay: From Little Acorns Grow 136
End Paper: The Details of Mitosis 147

10 Cell Biosynthesis and Control 153

Essay: Turning Genes On: The Most Exciting Period of My Scientific Life by James Bonner 157

Part 3 Genetics and Evolution

11 Darwin and Evolution 161

Essay: Observations on the Evolution of Island Species by Sherwin Carlquist 165
Essay: Darwin, Kelvin, and Time 168
End Paper: Darwin and Wallace 168

12 Genetics: The Basis of Variation 172

Photo Essay: The Art and Science of Plant Breeding 172
Essay: Mendel and the Selection of Experimental Plants 177
Essay: The Priest and the Professor by Garrett Hardin 181
Essay: The Importance of the Individual in the Advance of Science—A Contrary View by A. H. Sturtevant 184
Essay: On Making a Green Revolution by Wade Dewey 186
End Paper: Hybrid Vigor and Plant Breeding 190
End Paper: Sex Determination 192

13 Genetics, Evolution, and Systematics 195

Essay: Experiments with the Origin of Species by G. Ledyard Stebbins 197
Essay: Linnaeus and the Origins of Systematics 209

14 Evolution of the Plant Kingdom 214

Photo Essay: Plants of the Past 214
End Paper: The Making of Fossils 239

Part 4 Viruses, Bacteria, Algae, and Fungi

15 Viruses 240

Photo Essay: TMV 240
End Paper: Viruses and Plant Diseases 249

16 Bacteria 250

Photo Essay: Bacteria—Unseen Friends and Foes 250
Essay: Robert Koch—Genius Recognized 262
Essay: Whiter than White—Enzyme Detergents 267
Essay: Bacteria and Yogurt 270
End Paper: Bacteria and Food 273

17 Algae of the Soil and Freshwater 274

Photo Essay: The Living Waters 274
Essay: The Relationship of Bacteria and the Blue-Green Algae 288
End Paper: Cyanophyta, Euglenophyta, Chrysophyta, Xanthophyta, and Chlorophyta 290

18 Algae of the Oceans 298

Photo Essay: Life between the Tides 298
Essay: Norimaki—Sushi 309
End Paper: Phaeophyta, Rhodophyta, Pyrrophyta, and Algal Evolution 317

19 The Fungi: Water Molds, Bread Molds, and Mildews 326

Photo Essay: The Fate of Frogs 326
Essay: The Best Laid Plans of Mice and Men . . . by Edward Cantino 336
End Paper: Sex Hormones in Fungi by John Raper and Leonard Machlis 344

20 The Fungi: Yeasts, Mushrooms, Rusts, Imperfect Fungi, and Slime Molds 348

Photo Essay: Mushrooms and Toadstools 348
End Paper: The Biology of Rusts 368

21 Fungal Ecology 372

Photo Essay: Lichens 372

Part 5 Land Plants: Natural History

22 The Invasion of the Land 385

23 The Bryophytes 390

Photo Essay: Mosses and Liverworts 390
End Paper: The Evolution of Bryophytes 405

24 The Anatomy and Development of Vascular Plants 406

Photo Essay: The Architecture of Life 406
Essay: Meristems and Idioplasts by Adriance Foster 418

25 The Lower Vascular Plants 444

Photo Essay: The Invaders 444
Essay: Chipping Away at Early Land Plants by Harlan Banks 451

26 The Ferns 458

Photo Essay: Fronds, Fiddleheads, and Ferns 458
End Paper: The Evolution of Ferns 472

27 The Gymnosperms 474

Photo Essay: Remembrance of Things Past 474
Essay: Two Lines or Three? 485
End Paper: The Discovery of the Progymnosperms by Charles Beck 493

28 The Angiosperms 497

Essay: The Study of Fertilization in Angiosperms 515

29 The Angiosperms: Taxonomy and Major Groups 524

Photo Essay: The Herbarium 524
Essay: The Role of the Taxonomist in Biology by Lincoln Constance 528
Essay: Hybrid Oaks: An Exercise in Discovering and Solving Problems by Lyman Benson 531
End Paper: The Development of an Evolutionary System of Classification 544

Part 6 Land Plants: Physiology and Development

30 Development: Plant Hormones 551

Essay: Model Building and an Analysis of Growth 554
Essay: Why a Biologist? Some Reminiscences by Frits Went 557
Essay: Coconuts and Cytokinins by Johannes van Overbeek 571

31 Development: Differentiation 574

Photo Essay: Around and Around We Go—Symmetries and Phyllotaxes 574
Essay: What Happens Where and Why—The Experimental Study of Development by Ralph Wetmore 583
Essay: A Search for the Root-Forming Hormone by Kenneth Thimann 587
Essay: Bonsai by Ralph Johnson 594
Essay: Vegetative or Asexual Propagation 598

32 Transfers between Plants and the Environment 601

Essay: Nitrate, Soil, and Men by A. W. Roach 607
End Paper: Energy Exchange: Some Mathematical Concepts 615

33 Transport in Plants 621

34 Plant Response to the Environment: Temperature and Dormancy 635

35 Light and the Biological Clock 643

Essay: The Discovery of Phytochrome by Sterling Hendricks 648
Essay: Discovering and Studying the Biological Clock by Erwin Bunning 654
End Paper: The Nature of the Biological Clock 658

36 Environmental Control of Reproduction 663

Essay: Induction of Reproduction in Conifers 667
Essay: Discovering the Photoperiodic Sensitivity of the Cocklebur by Karl Hamner 671
End Paper: Photoperiodism and the Biological Clock 676

Part 7 Ecology

37 Descriptive Ecology 681

Essay: An Hypothesis Rejected: The Natural Distribution of Vegetation by Robert Whittaker 689

38 The Principal Biomes of the World 694

Photo Essay: Here, There, and Everywhere—Some Representative Biomes 694
Essay: Making a Terrarium 711

39 Physiological Ecology 712

Photo Essay: The Desert and the Alpine Tundra 712
Essay: The Art of the Scientific Paper 720
Essay: Shadscale and Fungus: Desert Partners by E. Dean Vest 725

40 Functional Ecology: The Ecosystem 728

Photo Essay: Time and Again: The Ecology of Plant Succession 728
Essay: Man, Energy, and Ecosystems 734

Index 749

1
Ecological Botany

An Acre of Woodland

A woodland, where moisture is abundant, is a marvelous place to be introduced to plants. In an acre of woodland, which might include a stream or a pond, we can find representatives of most major plant groups. Of course, not all woodlands are the same, and the types of trees and other plants present will vary, depending upon climate, soil, past history, and many other factors.

A For this introduction to the world of plants we have chosen an eastern beech-maple forest in Michigan. Such forests, covering extensive areas of the United States, should be familiar to many of you. This photo shows the most common trees found in our forest: the smooth-barked beech (Fagus grandifolia) *and the sugar maple* (Acer saccharum). *These trees lose their leaves in the fall and are therefore called deciduous. During the summer the floor of our forest is heavily shaded, except for flecks of sun.*

B But our woodland acre consists of more than trees, and among the other plants found growing here are the various shrubs. These form an intermediate stratum in the forest, as this flowering dogwood (Cornus florida) *shows. Shrubs are particularly common near edges or openings in the forest, along roads, and near streams and ponds.*

C A shrub commonly found in deciduous forests in the East and in streamside forests in the West is this chokecherry (Prunus virginiana), *whose flower and foliage are shown in this photo.*

D On the forest floor we find still another layer of plants. Many of these, unlike the trees and shrubs, live for only one year, growing from seed to maturity and death in a few weeks or months. A common species found in our particular woodland is the wild strawberry (Fragaria virginiana). *The strawberry is also an example of a large group of plants known as dicotyledons or dicots for short. Among the distinguishing characteristics of these plants are their relatively broad leaves with net veins and seeds with two special food-storage leaves.*

E We can also find various species of grasses, such as this Panicum, *on the forest floor. The veins in the leaves of these plants are parallel and therefore to help identify them as members of the second great group of flowering plants known as the monocotyledons or monocots.*

F Some of the flowers growing on the floor of the forest will be found only at the edges of our woodland, while others can grow in the deep shade cast by the trees and shrubs. This is true of the sweet william (Phlox divaricata), *common throughout the eastern United States.*

B

C

D

E

F

G *In our woodland some species of plants are easily recognized because they are so common, while others are rarely seen—and then only by those who know where to look. Both types are shown in this photograph. The rare plant on the right is the beautiful lady's slipper orchid* (Cypripedium calceolus). *On the left is the common poison ivy* (Rhus toxicodendron), *with its characteristic pattern of leaves in a group of three. Anyone who frequently walks in woodlands should learn to recognize this unpleasant member of the flora!*

H *Many woodlands contain swampy regions, and our acre is no exception. A variety of plants grow around and in such wet areas, and a few live floating on the surface of the water. One such plant is duckweed* (Lemna sp.), *a minute flowering plant, far smaller than the maple seeds shown floating next to them.*

I *Flowering plants are not the only inhabitants of our woodland. Ferns such as this bracken fern* (Pteridium sp.) *are important members of the flora in many forests. These plants, which lack flowers and seeds, must reproduce through the production of spores.*

J *Another spore-producing plant found in our forest is this horsetail or scouring rush* (Equisetum arvense). *This plant has two kinds of branches: the highly branched, vegetative kind on the left and the reproductive kind on the right. The reproductive branch has a cone at its tip and leaves greatly reduced in size.*

G

H

I

J

K Mosses are another flowerless group of plants frequently found in woodland acres across the continent. They can grow in many parts of the forest, but here we find them on the surface of a rock. The thick clumps of moss are the dark patches, which appear cracked and dead because the weather has been dry for some time. Following the next rainstorm, they will become green again and begin to grow. Also growing on the rock are numerous, much lighter colored plants called lichens. These are an unusual combination of two quite different plants, growing together in such close association that they actually form a new group of plants.

L Life in our woodland acre is not simply those plants that we see growing from the surface of the soil. The soil itself is a living mat of organisms that are critical to the life of the forest. Algae, bacteria, and fungi all inhabit the soil, but most of the time we are unaware that they are there. Under certain conditions some of these organisms, like the fungi, become readily apparent. We see them because they form fruiting bodies that rise from the forest floor in the shape of the familiar mushroom. Each species produces a distinct type of fruiting body, not all of which resemble the common toadstool. Here, we see a small, spherical fruiting body of a group of fungi known as the puffballs, so-named because when mature they release their spores in a puff when someone steps on them!

K

L

We human beings are curious creatures. We search for understanding of our universe, our world, and ourselves. Botany is one small but important part of that search. It is concerned with the plants upon which the existence of man—and all other animals—depends. Although this quest has been conducted casually from the beginnings of mankind, during the last 250 years it has evolved into an intensive study. Because of its intensity, its formalized methods, its organization of knowledge, and its special tools, we now call our search a science, the science of botany.

The first two chapters will introduce this search and some of its results. We will survey the entire field, partially to introduce some important concepts to be used as building blocks in subsequent chapters. For example, the orientation of this book is ecological; yet ecology is such an all-encompassing science that its detailed discussion must be postponed until basic terms and ideas have been presented. Although some new terms are defined immediately (most are discussed and defined in later chapters), we have generally avoided this practice, believing that it is distracting and that most terms will already be somewhat familiar.

An Acre of Woodland

Let us begin our search of the plant world by visiting a patch of woodland, perhaps an acre or two, preferably with a small clearing and a stream meandering through one corner. This is a good place to begin not only because so much of the plant world is represented but also because woodlands, whether they are natural or planted by man, are readily accessible to almost everyone. Much of our present civilization developed in forested areas, and man seems to enjoy the presence of trees in his environment. Often, he clears the forest for his grassland crops, wheat and corn, but when he breaks the treeless prairie sod to plant these same crops or to build cities, he also plants trees around his dwellings and in parks. Even in the desert, he plants trees and waters them, making his own acre of woodland. Our mental visit to a representative woodland will emphasize (but not exclusively) the kind of forest that one might encounter in much of the eastern United States or western Europe—the great deciduous forest.

In the forest we are in a world dominated by plants, where trees are the overwhelming aspect. In much of the temperate zone, the trees are deciduous, providing an appearance of dense foliage during the warm months and barren twigs the rest of the year. There will probably be maples, elms, and numerous others such as oaks, sycamores, ashes, willows, and birches. Wherever Western man has extended his civilization, he has planted these species, as well as the conifers (spruces, firs, and pines) of western Europe. Of course, in his travels he has encountered and put to use a host of other deciduous and evergreen species: cottonwoods, aspens, Douglas firs, sugar pines, loblolly pines, sequoias, and eucalyptuses. Many western or southern woodlands might consist primarily of these species. In the tropics and subtropics, our woodland might contain such broadleaved, evergreen trees as the pepper tree, the camphor tree, and many of the acacias. Most of the trees fall readily into two groups: the broadleaved, deciduous trees that are really flowering plants and the needleleaved, evergreen trees without true flowers. There are a few interesting exceptions that we shall encounter later.

Next to the trees, the shrubs are most conspicuous. Again, the variety is tremendous, probably even greater than the variety of trees. In temperate climates the dogwoods, serviceberries, willows, raspberries, thimbleberries, and most others will be deciduous. A few such as the holly, the Oregon grape (a Western plant hardly large enough to be called a shrub), and the rhododendrons are broadleaved evergreens, and the yew is a needleleaved, evergreen shrub. If the trees form a dense canopy in our acre of woodland, the shrubs may be limited to the edge of the clearing or perhaps the stream bank, but if there are few trees, the shrubs may form a dense thicket that is almost impenetrable.

To see the herbs (smaller, nonwoody plants) in their splendor, we must make a visit some time during spring and summer. The numbers of kinds of herbs may well exceed those of both the trees and the shrubs. It would not be uncommon to find 20 to 50 or more varieties within our acre.

Even in a short visit, we can see that the herbs fall into two general categories. There are the narrowleaved, parallel-veined grasses in the group of flowering plants called monocots. Lilies, iris, jack-in-the-pulpit (although its leaves are neither narrow nor parallel-veined), and orchids, as well as the grasses, are also monocots. Then there are the broadleaved, net-veined herbs in the group called dicots. Buttercups, roses, sunflowers, geraniums, mints, phlox, snapdragons, and a legion of other familiar and exotic flowering plants are dicots. Most of the deciduous trees and shrubs are also dicots, but palm trees are monocots. A few of the grasses and other herbs will be found in the shade of the trees and shrubs, but the clearing will probably be the area most densely populated with these plants. The grasses might dominate in the clearing, while dicots may be more common in dense shade.

The complexity of the environment is impressive, even within our acre. The high light intensities in the clearing allow more plant growth and tend to increase the temperature and decrease the relative humidity. Under the forest canopy, light intensities may change within minutes from extremely low levels to the full intensity of sunlight, as the shifting rays of the sun penetrate the canopy and provide moving sun flecks on the forest floor. Clearly, the plants that grow there must be adapted to an environment different from that in the clearing.

Let us examine our acre a bit more closely, discovering several nonflowering plants (neither monocots nor dicots). Depending upon the region, we may find one or more species of ferns, usually with their finely divided leaves and minute spores in small clusters (rather than seed-producing flowers) on the underside of the leaves. Some (e.g., the brackens) will probably be in the clearing, while others will be in the shade. If our area has a rocky cliff, we may find delicate ferns growing in the crevices of the rock, either in the sun or in the shade. Near the stream or perhaps in the clearing, we might find horsetails with their jointed stems and small cones at the tips. In a few natural woodlands there may be club mosses (*Selaginella* and *Lycopodium*). It will require some rather careful observation, perhaps even reference to later chapters in this book, to distinguish the club mosses from the true mosses.

The true mosses are more familiar to us. You will find them growing on the moist soil itself, in moist and protected places on rocks, on rocks in the stream or on its banks, or even on the bark of some of the trees.

But be careful. You will also find lichens and patches of algae growing on many of these same objects, and they superficially resemble the mosses. The lichens are usually in drier habitats such as the dry parts of the rock or the south side of the tree trunk, while the mosses and algae might be confined to the north side or the crotches formed where the trunk branches. The true mosses can be easily distinguished from the lichens or the algae by their minute and delicate, leaflike structures, which are not possessed by the latter. (Some lichens, however, have a form that is not completely unlike a leaf.) The mosses have a fruiting body that is unique and diagnostic, when it is present.

If we look carefully and if our woodland is frequently moist and cool, we may find liverworts as well as mosses. Their plant body is reminiscent of some of the lichens, but they tend to occur directly on moist soil and are a "truer" green than the pale blue-greens, or even reds or blacks, of the lichens. In any case, the liverwort fruiting structure clearly distinguishes this plant, although it is not usually seen.

Looking up at the tree canopy, we may see mistletoe growing on one or more of the deciduous or coniferous trees. With investigation, we discover that this is a flowering plant and a separate organism from its host tree. Although it is yellow-green and capable of photosynthesis, it penetrates its host with peglike organs called haustoria to obtain water and nutrients. The mistletoe is a parasite, and with careful searching, we could find others in our woodland. We might also find plants that must obtain their nutrients not from a live host, as the mistletoe does, but from dead organic matter. A few of these such as the coralroot and the Indian pipe are flowering plants; most are fungi. Plants that depend upon dead organic matter for their nutrients are called saprophytes.

For the next level of our investigation, it will be convenient and sometimes essential to have a magnifying glass. Kneeling on the forest floor, we see that the layer of fallen leaves contains numerous molds. Using our magnifying glass, we can probably find small fungi of this type on decaying wood and even on the bark of some trees. If it is midsummer and the rains have been plentiful, the forest floor may be populated with a variety of eye-catching mushrooms. (Nonbotanists may call the poisonous ones toadstools.) Most will have the familiar form, but some (the puffballs) may be almost round like an amorphous ball of putty, and others (the stinkhorns) may have a phallic shape. A few (e.g., the so-called field mushrooms) may be delicious and safe to eat, but many others such as *Amanita* (the deadly white angel) may be extremely poisonous. As we study these fungi more carefully, digging in the soil and searching for their origins, we discover that below ground they are a mass of filaments called hyphae, many of which are attached to the roots of the trees and other plants in an arrangement that benefits both organisms. Nutrients from the soil are made more readily available to the trees by fungal action, and the fungi obtain their nutrient energy sources (e.g., sugars) from the trees. The fungal hyphae associated with the roots form a combination called mycorrhiza, and it was only relatively recently recognized that these hyphae develop into the intricate, mushroom fruiting bodies.

On decaying logs and stumps, we may find other fungi that, while not completely unlike mushrooms, typically consist of a shelflike structure growing out from a vertical surface. These are the reproductive structures of the bracket fungi, the hyphae of which are important in speeding the decay of dead wood.

If we really look carefully at the moist, decaying wood and debris in our forest, we may be able to find slime molds—those interesting organisms who spend half their life as animal-like plasmodia (masses of protoplasm) creeping slowly over wet decaying wood and who finally form an intricate, beautiful, and plantlike fruiting body.

In stagnating pools by the stream, we may find the delicate, fine hyphae of a saprophytic water mold radiating from the carcass of a dead fly. With further searching, we can also find parasitic fungi growing on virtually all of the higher plants of our acre. For example, we may find one stage of the wheat rust growing on a barberry bush, or we may find damping-off fungi growing in the delicate new stem of a seedling at the soil level. The number and activity of plant fungal parasites alone could provide scientific study for a lifetime.

Now, let's get out the microscope (and the necessary associated equipment) and really look closely. We will find all kinds of algae growing in and on the soil. Some of these will be blue-green algae without distinct nuclei in their cells; others will be green algae with distinct nuclei. Probably, if we look carefully, we will also be able to find diatoms—those microscopic organisms that live in intricately carved glass shells. Fungal hyphae should also be apparent, and there will be a vast number of minute organisms, again without organized nuclei and usually lacking green color. These are the bacteria. Many are tiny balls, sometimes forming chains like beads; others are rod-shaped, also sometimes in chains; and a few are spiral with whiplike appendages that provide locomotion. The bacteria not only occur in the soil but are virtually everywhere in our acre—on the trees, in the layer of decaying leaf litter, and even floating in the air itself. They are almost all saprophytes or parasites, and they play an absolutely essential role in decaying the organic matter produced by the forest.

Let's go to the stream. Probably, if the current is not too rapid, we will find aquatic higher plants such as sedges, rushes, cattails, watercress, and the monkey flower (in the West). If our acre includes a pond, we may find water lilies or the floating duckweed. Partially exposed rocks in the stream will probably be covered with mosses. A few of the submerged green plants may be higher plants, but the slimy, filamentous masses of green are algae. If we check the water itself, we will find floating algae and bacteria, particularly if the water is somewhat stagnant.

A really detailed study of the organisms in our acre would show that some harbor viruses or even rickettsiae or mycoplasma—those organisms larger than viruses but smaller than bacteria.

Because this is a botany book—and to simplify our study—we have ignored the animals. But of course we should not. The animals are interacting strongly with the plants in a great variety of ways. They are consuming the plants for food, and they are breathing the oxygen produced by the plants. The carbon dioxide produced by the animals (and the plants, especially when they decay) is used in turn by the living green plants. The excreta produced by the animals provides fertilizer for the plants.

If we investigated the animals in detail, we would find an even greater variety than we did for the plants. In the stream or in a pond, we would find single-celled animals, many of whom consume bacteria, called protozoa. We might find a hydra, barely visible to the eye, with its tentacles pulling food into the body opening at their center. Rotifers could be seen in the water with the aid of a microscope. Flatworms (the planaria) may occur in the stream. Roundworms, the nematodes, are almost as universally present in the soil and on the plants and in some of the animals as are the bacteria. Snails and slugs may be creeping in the stream, across the moist earth, or on the plants themselves. The soil will contain a tremendous variety of arthropods, especially insects and mites, and also spiders, centipedes, millipedes, and a few crustaceans in the water. The vertebrates will almost surely be represented in our acre by fish in the stream, frogs and salamanders, snakes and lizards, possibly turtles, and certainly birds and rodents. There may be other mammals such as deer or bear, and during our investigation, we can be certain that one unique mammal will be present—man, the botanist.

Incidentally, we should note that a few important marine groups of animals and plants will not be represented in our woodland (e.g., the starfish, the sponge, the lamp shell, the brown algae, and probably the red algae, which do have a very few freshwater species).

Man, the Curious Animal

Man must be recognized as a natural part of our woodland acre; yet, among the myriad life forms, he is distinguished in a special way from all the others. His curiosity and the ability to formulate questions in response to it far transcend that of any other creature. He has a contemplative, reflective intelligence capable of forming abstract ideas and concepts. Man the botanist formulates increasingly complex and often important questions about the plant world. Man the scientist has developed the scientific method, a fairly systematic approach to answering questions about his world and his universe. It involves special tools such as the microscope, but perhaps more importantly, it involves special ways of thinking, that is, special in the degree to which they have been formalized and refined. Actually, they are merely extensions of the way man always thinks when he is seeking answers. He gains experience by observation and description; he then formulates questions and devises means to test tentative answers both by performing experiments and by comparing possible answers with other known facts; and after performing his experiments, he carries out a logical thinking process, which may be either inductive (from limited facts to broad generalizations, especially important in science) or deductive (from general concepts to particular conclusions). These conclusions are then used to formulate further questions or hypotheses that will initiate another turn in this cyclical process of learning by experience and logic. Finally, the results of a study are published. More than any other way, the scientific method differs from everyday learning in that results *must* be published—to allow others to build upon.

We will develop our introduction to botanical science by considering a number of the questions that might grow logically out of our walk in the woods and by briefly summarizing some of the known answers to these questions and some of the areas of current interest and activity. Hopefully, this book will make it clear to the reader (if it is not already) that botanical science is not a collection of cut and dried, musty facts to memorize. There are such facts, and a lot of them, but the real excitement of modern science, botanical or any other, is in seeking answers to the remaining questions. Challenging and important areas remain almost without a theory (e.g., the question of how an organism develops from a fertilized egg, the secret of establishing proper balances in the world of living things, the biological clock, the operation of the human brain). Seeking answers to these questions is providing a few lucky individuals with some of the most exciting activity ever afforded to a human being. Some of the answers, when found, will probably thoroughly upset the world in which we live—for better or for worse. But let us begin with a prosaic question.

Organizing Diversity: Taxonomy

How can we organize the tremendous diversity encountered in nature? The diversity of our woodland acre has two aspects. First, there is a fantastic variety in the kinds of organisms. Second, there is great diversity in the kinds of woodlands (which should be apparent from our frustrated attempts to satisfy all of our readers, regardless of where they live). A major means of ordering a diverse set of things or phenomena is through classification—grouping things together according to their similarities. Classification of organisms is called taxonomy or systematics. Plant communities are also classified as a part of the science of ecology. Organizing things may allow us to better ask the important questions, so let us get on with the task!

This search for order in the plant kingdom must extend far beyond our walk in the forest. In fact, it has been going on for hundreds of years everywhere on the earth's surface—in the tropics, in the desert, in the ocean, on the prairie, in a multitude of forest types, in the arctic tundra, in the canyons, and on the flanks and tops of the mountains. And almost everywhere, except perhaps in lands of eternal snow or rapidly shifting sands, numerous kinds of plants have been found, collected, pressed and dried, and brought back to the taxonomists' laboratories (called herbaria). Hundreds of years ago (certainly as early as the Romans and the Greeks) this search was motivated largely by the desire to find medicinal plants. By the early 1700s, however, the search also began to be strongly motivated by man's desire to understand the plant world surrounding him.

A milestone in this effort was achieved by a young physician-botanist, Carl von Linné, who latinized his Swedish name to Carolus Linnaeus. He became obsessed with the desire to formulate a scheme that could classify all higher plants, and he eventually championed two innovations that have guided our taxonomic efforts ever since. First, he based the relationships in his classification scheme on floral structure, organizing according to the numbers of pistils and stamens (female and male flower parts). Second, he adhered uniformly to a system of naming called the binomial system. This brought order into what was then a chaotic and confused situation in which some plants might have as many as six or eight different Latin names. Linnaeus gave each plant a single species designation; then he grouped related species together to form genera, the name of which precedes the species designation. (If a species such as *Ginkgo biloba* has no apparent close relatives, then it constitutes the sole member of its genus.) In many cases (especially those originally named by Linnaeus), the so-called scientific name is the same as the Latin name originally provided by the Romans, but as plants unknown to the Romans were discovered, they were often given names unknown in Latin but always latinized according to the rules of Latin grammar and the convention accepted by botanists. Scientific names are always written in *italics* or underlined, and the genus is always capitalized.

Linnaeus' scheme of classification was rather arbitrary, as was the system implied in our discussion of the acre of woodland where we spoke of such categories as trees, shrubs, herbs, saprophytes, and parasites. Following the 1859 publication of Charles Darwin's *Origin of Species*, biologists became convinced that all organisms were related through evolutionary descent, and it then became reasonable to try to establish a natural or evolutionary system of classification. Within such a system, organizational relationship implies a real genetic relationship. Members of a genus, for example, may have changed relatively little in their descent from a common ancestor, while members of different families have changed much more.

Taxonomy, then, was the mother of botany. This is reasonable and natural, since the first thing we must do in studying our world is describe it. It would seem that by now, after more than 220 years, we ought to be done with this task. But many organisms remain to be named and described; species names are still being changed; and at higher levels, we encounter comparable degrees of uncertainty. We shall see in Chapters 14, 28, and 29 that botanists cannot yet agree upon a thoroughly satisfactory classification scheme; that is, we lack enough information to provide agreement about evolutionary relationships.

The names of the primary organizational levels used by botanical taxonomists are generally agreed upon. They are established in the Botanical Congresses, held once every five years. The principal taxa (a taxon is a unit of classification) are usually listed in the form shown in the following example of the cocklebur, which Linnaeus named *Xanthium strumarium* L.

Kingdom:	Plantae
Division:	Tracheophyta
Class:	Angiospermae
Subclass:	Dicotyledonae
Order:	Asterales
Family:	Compositae
Genus (plural is genera):	*Xanthium*
Species designation:	*strumarium*
Author:	L.
Abbrev. version:	*X. strumarium*

These are the main groups, but various authors have added other levels such as subdivisions, superclasses, or series. For example, since we have mentioned dicots, that subclass is included. Subspecies or races are also often used; so are hybrids, varieties, strains, and cultivars, the latter three to designate useful hybrids or selections formed and propagated by man.

Life Cycles

What criteria can we use to organize the diversity of organisms? Of primary importance are the organs and principles of reproduction. We shall be discussing these in Chapters 15 through 29, but the concepts are so basic that they deserve to be introduced now.

Sexual reproduction leads to the concept of a life cycle. In virtually all plants this life cycle includes a

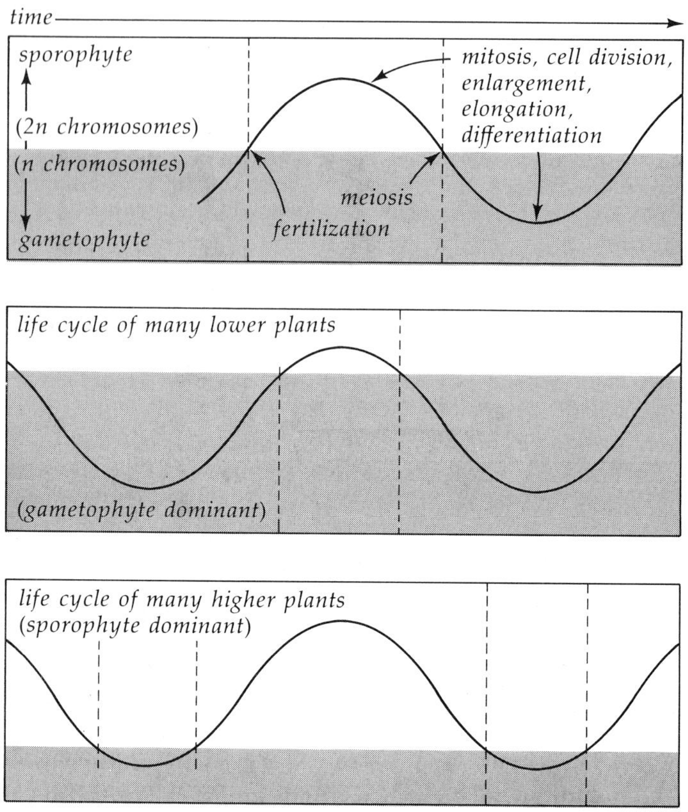

1–1 *A generalized life-cycle diagram.*

very distinct alternation of generations, in which one type of organism produces another type, which then produces the first type to complete the cycle. A generalized life cycle is shown in Figure 1–1. One type of organism is initiated by the union of two sex cells called gametes (in many cases, a sperm and an egg). If each of the gametes contributes a quantity of genetic material equal to n, then the resulting cell (the zygote) will contain a quantity of genetic material equal to $2n$. (We usually use n to designate the number of chromosomes in the gametes.) This union of gametes producing a zygote is called fertilization. The zygote may then divide by the process of cell division called mitosis, in which the genetic material is duplicated each time the nucleus divides so that the nuclei continue to contain $2n$ chromosomes. To produce gametes again, this number must be cut in half. This occurs by the process of cell division called meiosis, in which the pairs of chromosomes are separated, and the resulting cells each contain only n chromosomes.

In the plant kingdom the cells produced by meiosis (having n chromosomes) are almost never themselves the gametes; they are spores. The spores then grow by one to several mitotic divisions to produce the second kind of organism, which in some cases (e.g., the mosses and ferns) is fairly complex. In other cases only a few cell divisions occur, and the organism produced can only be seen with the help of a microscope, having only 3 to 8 cells or nuclei. In any case, this organism with n chromosomes, called the gametophyte, finally produces the gametes that can unite to form a zygote, completing the life cycle. The $2n$ organism developing from the zygote is called the sporophyte, since it produces spores. Because of its $2n$ condition, it is said to be diploid, while the n gametophyte is said to be haploid.

In Figure 1–1 the horizontal line represents the division between the sporophyte and the gametophyte. Movement from left to right along the curve proceeds with the passage of time. Crossing the line from below is fertilization; from above, meiosis. Many later figures in the book illustrate the various stages in the life cycles of plants.

In all multicellular animals and most protozoa (single-celled animals), the results of meiosis are the gametes, and the organisms themselves are always diploid. In plants, however, there is a great variety in the extent to which the gametophyte develops, and it is never eliminated entirely. In most of the green algae, the gametophyte is the predominant generation, and the sporophyte may be reduced to a single cell, the zygote. Often, the zygote forms a resting stage during dry or cold conditions, but its first division following this is a meiotic one, producing spores that then develop into the gametophyte. In some of the green algae, both the sporophyte and the gametophyte form fairly complex multicellular organisms; in some of the other groups of algae, this is the rule. In a few cases the sporophyte and the gametophyte are identical in appearance (e.g., *Ulva* or sea lettuce, a green alga), and in some (most brown algae, especially kelps) the sporophyte is the predominant generation.

On land the most conspicuous part of the moss plant is the green, "leafy" gametophyte. The fruiting bodies that form and are dependent for their nourishment upon this gametophyte represent the sporophyte. In the ferns the gametophyte and the sporophyte are independent of each other. The sporophyte is the larger leafy fern plant familiar to most of us, and the gametophyte is a small (smaller than a dime), flat, often heart-shaped organism. In some of the club mosses (*Selaginella*) the sporophyte is predominant, and the gametophyte is almost seedlike. In the gymnosperms (conifers, etc.) the sporophyte is the conspicuous plant, a shrub or tree, and the gametophytes are minute organisms that become the pollen (male) or that develop into much of the seed (female). They are totally dependent upon the sporophyte. In the flowering plants the female gametophyte has been further reduced so that it consists only of a tiny, typically 7-celled, 8-nucleate structure, also contained within and completely dependent upon the sporophyte.

In Figure 1–1 the relation between sporophyte and gametophyte is implied by the position of the horizontal line. It is high in most green algae with their dominant gametophytes, intermediate in many specialized algae and in mosses and ferns, and low in higher plants and animals. In a general way, the evolutionary trends in the plant kingdom seem to progress from a predominant gametophyte generation with a minimal sporophyte to a large, predominant sporophyte and a greatly reduced gametophyte. Details of this fascinating story will be considered in later chapters.

Communities, Ecosystems, and Biomes

Can we organize the diversity of vegetation types occurring on the earth's surface? Clearly, the acre of woodland was only one of many kinds of communities that might be found. The variations in environments on this planet are most extensive, and plant distribution tends to correspond. Nevertheless, it is not difficult to recognize certain broad groups of vegetation, and the existence of names in our language for these groups implies that they have been recognized for a long time.

We need some hierarchy of names comparable to the divisions, classes, orders, etc., used by the taxonomist. Several systems of terms have been proposed, but no completely satisfactory system has been accepted by all ecologists. Three principal terms and their relatives are useful, although they are far less precise than the ones used by taxonomists. First, there is the plant community, which includes those plants living together in any unit the ecologist wishes to consider. We would, for example, refer to the plants in our acre of woodland as a plant community. If we wanted, we could refer to the mosses, lichens, algae, and other things growing on the bark of one tree as a community, or we might consider the entire eastern forest as a community. All real communities include animals as well as plants, so biotic community is a more realistic term.

Second, if we consider not only the plants and animals but also the physical environment, we are dealing with an ecosystem, a concept often discussed today. When one speaks of ecosystems, one usually thinks of the exchanges of energy and matter taking place within a community and its environment. The ecosystem includes the community; indeed, the community is the biological phase of an ecosystem.

Third, it is possible to group similar biotic communities on a continental or even worldwide scale into biomes. If only plants are being considered, we speak of formations. A biome consists of all the communities or ecosystems having a similar dominant plant growth form and occurring in broadly similar environments. There are, for example, grassland, deciduous forest, shrubland, plankton (single-celled plants living in the water), and tidal zone algal biomes. Some authors suggest that when biotic communities are grouped on a worldwide rather than continental scale, we should speak of biome-types. We shall consider the world's biome-types in some detail in Chapter 38, but for an insight into the diversity of plant communities and their classification, we will now introduce seven major biome-types.

Aquatic systems. Much remains to be learned about the biotic communities or biomes of the ocean. There are the plankton communities of the open oceans, which are relatively sterile since there are so few organisms, the deepwater communities of the continental shelves with their red algae, the rich and interesting communities of the tidal zones, and such special but localized communities as the Sargasso Sea northeast of the West Indies in the Atlantic Ocean (see Chapter 18). In addition to the marine communities, there are those found in freshwater lakes and streams, which will vary widely in response to the nutrients and oxygen in the water, its temperature, light intensity, and a number of other factors. (Any detailed discussion of these aquatic communities must await our investigation of the algae in Chapters 15, 17, and 18.)

Swamp and marsh. Only a few higher plants occur in the truly submerged, aquatic communities just discussed. In bogs and marshlands, however, higher plants (grasses or grasslike plants) are the dominant aspect of the community. Again, there is much variety, depending upon the environment. There are striking differences between saltwater and freshwater marshes. Higher plants must have a remarkable physiology to survive and grow in salt water, but a number of them do, particularly the mangrove trees that form swamp forests along thousands of kilometers of coastline in the tropics. Temperate-zone, freshwater marshes are populated by cattails, rushes, sedges, grasses, and a number of other species (mostly monocots). Many swamps in warm, temperate climates in the southeastern United States are dominated by bald cypress trees.

Forests. We have already examined a representative forest. There are many types, however, besides our temperate deciduous or mixed deciduous–coniferous forest. There are tropical rain forests where rain falls nearly every day, and there are tropical seasonal forests where a number of months during the year are quite dry. There are the coniferous forests of the North (the taiga), which are similar to but significantly different from the forest we discussed. In any case, most forests require considerable available moisture.

Reduced forests—woodland or shrubland. Where rainfall is limited, particularly where the relatively limited quantity of moisture falls only during a part of the year, and where other conditions are suitable, reduced forests occur. Usually, these are not forests in which the familiar trees have been reduced in size; rather, they consist of trees that are genetically smaller, sometimes almost shrubs. In climates such as those near the shores of the Mediterranean Sea, southern California, and parts of Australia, shrublands appear. Here the dominant plants are commonly broadleaved, evergreen shrubs or small trees (dicots) with hard, leathery leaves. Often, the trees are thorny, and they may form such a dense growth that penetration by man is difficult or impossible. Because of the dense growth and the long dry period each year, fires are common. (We have all read about the disastrous fires in southern California where expensive homes are built in this vegetation type.) Tropical communities of thorny shrubs or trees are called thorn scrub or thorn woodland.

Reduced forests, with the crowns of small trees not touching or overlapping and usually with well-developed undergrowth, are called woodlands. In the Great Basin of western North America, we find the pygmy conifer woodlands, consisting of small juniper and pinyon pine trees. These forests are typically very open and quite different from the mediterranean shrub types. The pygmy conifer forests occur in areas where rainfall is too limited for other forests but where the low temperatures during winter and the fairly even distribution of moisture throughout the year provide more available water than in the nearby deserts.

Grasslands. Over vast, relatively flat expanses of the earth's surface, both in the tropics and in the temperate zone, grasses dominate the landscape. In the temperate zone they occur where a moderate amount of precipitation is evenly distributed throughout the year, sometimes tending to be highest in the summer. Since these are regions of fertile soils and high productivity, they provide the area for most agricultural crops. Few now remain in their original condition. Nevertheless, the crops grown on most of these acres are grass crops. In the United States corn is grown in the region that once supported tallgrass prairie, and wheat is grown in the shortgrass regions. Tropical grasslands, typically with widely scattered groves of trees, are called savannas.

Tundras. In the extreme northern regions and on the tops of high mountains, temperatures are too low for the growth of trees, and a reduced herbaceous vegetation covers the landscape. In temperate climates the plants must be adapted to a very short growing season, sometimes as short as four or five weeks (but typically 8 to 11 weeks). Frost can occur at any time during the year, even at the height of the growing season. Daily temperatures seldom reach, let alone exceed, what we would consider to be comfortable room temperature. Total moisture may be relatively low, although the low temperatures retard evaporation so that moisture is seldom limiting for plant growth. The tundra is a severe but starkly beautiful landscape.

Deserts. Where moisture is extremely limiting and daytime temperatures are frequently high, plant growth is strongly restricted, and deserts result. There are the warm deserts of the tropics and temperate zones and the cold deserts of temperate and arctic zones. Here grow cacti, saltbrush, greasewood, sage, and a host of others. Typically, plants are shrubby and spaced some distance apart, but there are few, if any, extensive areas of the earth's surface completely devoid of plants.

In studying the plants of the world and how they function, it is often interesting to examine extremes. The desert and the tundra offer excellent examples, and we shall again discuss them in later chapters.

About 10% of the earth's land area is occupied by agricultural lands. These constitute artificial ecosystems of tremendous importance to mankind—his survival depends upon them. As indicated, much of the agricultural land resembles a simplified grassland biome, and large forest regions have been replaced by grassland agriculture.

Let us return to our woodland acre. We now see it as an ecosystem, consisting of numerous individuals that can be classified into recognizable groups. The acre itself is similar to many others and thus is part of a biome. But there is more here than just something to be named and classified. This is a highly dynamic system of things. The plants are taking energy from the sun to grow and to reproduce. Both plants and animals are transforming the elements into a host of complex molecules. They are consuming and being consumed. They are reacting delicately to the continually changing factors of the environment. They are interacting with each other in a myriad of complex ways. They are reproducing and dying. In short, our ecosystem acre is throbbing and pulsing with *life*.

2

What Is Life?

From Kilometers to Microns

There are many ways to study our biological surroundings. We can examine whole biomes, or we can analyze a cell part. And there are as many areas of study between these extremes as there are dimensions and men. We would like to take you into the world of the cell by starting with a view of the world meaningful in terms of kilometers. Then, by steps of ten, we shall shrink the scope of our attention until we reach a microscopic world measurable only in units as small as microns.

A *Let us start our trip with an aerial view of a portion of the western United States. This is Tony Grove Lake east of Logan, Utah. The long line represents 1 km (0.62 miles), while the shorter line indicates 100 m (109 yd). At this size scale we can see relatively large units of vegetation, and a number of biotic communities are clearly visible —the lake itself is an aquatic community. Even the evergreens to the left are part of a coniferous biome, which would require a much larger scale to be seen in its entirety.*

B *Moving in, we now see Tony Grove Lake from the forested cliffs immediately above and behind the lake. The longer line is 100 m and the shorter line, 10 m (32.8 ft). At this scale, which is comparable in size with our acre of woodland in Chapter 1, only small stands of vegetation are visible.*

C *Ten times closer—the longer line is now 10 m and the shorter one, 1 m (39.4 in.)—individual shrubs are readily apparent, and large animals* (Homo sapiens) *are just visible.*

D *Small plants can now be seen as individuals, but shrubs and large animals still dominate this scale, where the longer line is 1 m and the shorter one, 1 dm (3.94 in.).*

A

B

C

D

E *Here, we see a dogtooth violet* (Erythronium grandiflorum) *in the microenvironment of the forest floor. The longer line is now 1 dm and the shorter one, 1 cm (0.394 in.). At this level the macrostructure of the individual herbaceous plant becomes apparent, while only small portions of shrubs and large animals remain in view.*

F *Coming 10 times closer to the leaf of the dogtooth violet allows us to barely discern the epidermal cells of the leaf (the specks on the leaf are mostly pollen grains). Here, the longer line is 1 cm and the shorter line, 1 mm (0.0394 in.).*

G *Now, we have reached a scale where we must continue our investigation with a microscope. When our leaf is sectioned and viewed with a light microscope, the cells composing it become clearly visible, and even the larger cell parts such as the nucleus and the chloroplasts can be seen. The longer line is now 1 mm and the shorter one, 100 μ (0.0039 in.).*

H *By using higher magnifications of the microscope, we can see the internal structure of the cells much more clearly. Now, the longer line is 100 μ and the shorter one, 10 μ (0.00039 in.).*

I *If we exchange our light microscope for an electron microscope, we can then see the internal structure of chloroplasts, mitochondria, and other organelles. The longer line is now only 10 μ and the shorter one, 1 μ (0.00004 in.).*

E

F

Watson and Crick's discovery gave us some insight into not only how the gene might reproduce itself but also how one gene might differ from another. The four nucleotides can be arranged in innumerable ways exactly as dots and dashes and spaces between letters and between words (again a total of four symbols) can be arranged in an infinite variety. Given enough symbols, it is possible to translate all accumulated human knowledge and all that could possibly be accumulated into the Morse code. A single gene may contain a thousand or more nucleotides; hence it seemed reasonable that each gene might be a message in itself. For example, one gene might transmit the message: "Synthesize an enzyme that will break down sucrose into fructose and glucose."

How is the message of the gene in its nucleotide language translated into the amino acid language of the enzyme? An understanding of this came during the 1960s. The message of the gene is first copied in the nucleus onto another nucleic acid called ribonucleic acid (RNA). The double helix unwinds a bit, exposing the bonded parts of the component chains so that the RNA can form in close association with the DNA. (Incidentally, all of these events are themselves controlled by enzymes.) This transcription occurs as the nucleotides of the forming RNA are complementarily bonded to the nucleotides of the DNA. The result is a substance called messenger-RNA (m-RNA), an appropriate designation because this molecule carries the genetic message from the nucleus into the cytoplasm of the cell. In the cytoplasm the m-RNA acts in conjunction with minute submicroscopic cellular particles called ribosomes.

The ribosomes translate the message. Amino acids become activated and attached to another form of RNA called transfer-RNA (t-RNA). These t-RNA molecules with their specific amino acids arrive in their kinetic ramblings at the site of the ribosomes with associated m-RNA molecules. Each molecule of t-RNA recognizes a specific spot on the m-RNA by complementary bonding between three of its nucleotides (a codon) and three of the m-RNA nucleotides. Thus the translation occurs by special t-RNA molecules, and the principle of nucleotide recognition results from complementary bonding. When the amino acids attached to their t-RNA molecules have been suitably aligned along the m-RNA strand, enzymes present at the ribosome hook them together, forming a connecting link called a peptide bond between each two amino acids. The result is the chain of amino acids constituting a protein. The order of arrangement of the amino acids in the protein is determined by the order of arrangement of the nucleotides in the DNA, and a portion of the amino acid arrangement constituting the active site confers upon the protein its enzymatic activity.

We have therefore answered our question about the origin of the enzymes. The DNA genes with the help of at least two kinds of RNA, ribosomes, and a number of enzymes control the synthesis of other enzymes. At the cellular level, we now have a fairly coherent story to tell about the nature of life. *Life is genes perpetuating themselves in cells by making enzymes that make precursors for more genes.* An extremely important generalization is that this story appears to apply to all organisms—bacteria, earthworms, spinach plants, elephants, pigeons, pink bread mold, cocklebur plants, redwoods, and even the botanist or molecular biologist who contemplates it all.

But it hardly answers our question about diversity; rather it gives us an insight into the *uniformity* of life—*all* organisms have genes, enzymes, and the other paraphernalia. Nevertheless, it does open the way to understanding diversity in terms of uniformity. We need only assume that there is a vast number of diverse enzymes in diverse organisms. These might account for the diversity of both form and response to the environment. Some enzymes may be more sensitive to temperature than others (in ways that we completely fail to understand at the moment), or some may confer upon their possessors an ability to survive with less moisture. Unfortunately, we have not yet found time to study in earnest the diversity of enzymes in different species, although preliminary probes have been made. Inevitably, the whole picture presented here brings up another question, perhaps the most fundamental of all.

Development

How do the functions of the genes and enzymes within cells account for the growth, development, and continuing function of multicellular organisms? It has been conclusively demonstrated that the genes are completely reproduced at every cell division so that *all* cells in an organism have a complete set of genes, the same set originally possessed by the zygote. But, clearly, all the genes are not operating in all the cells all the time. Genes that determine flower structure operate only in certain cells during development of the flower. Genes that determine the formation of the root cap operate only in developing root cap cells. Furthermore, the system is essentially foolproof. Occasionally, monsters are produced, but the deviations from the plan of the organism are always relatively minor. An eye never haphazardly (or otherwise) appears on one's navel!

How do specific genes manage to function only at specific times and specific places during the development of an organism? Apparently, there is some kind of program that controls development. Perhaps, as a computer might be programmed to operate a factory or to fly a guided missile, some similar system in the zygote is in charge of organismal development. If it is, we remain in the dark about what it is or how it works. This may well be the most important, completely unanswered part of our question, what is life. This is the future, and contemplating it makes one tingle with excitement—tempered by a little healthy fear. Understanding development implies being able to control it, but will man's mind ever be equal to the responsibility of controlling life itself?

The Functioning Ecosystem

Having pursued the nature of life to the misty borderlines of the unknown, we can return to our acre of woodland and the original question of diversity of species and communities. If the diversity of a community is due to the diversity of its species and if the diversity of species is a diversity of enzymes and some poorly understood developmental program, then we are entitled to study the *functions of ecosystems* in terms of the functions of their component organisms and their environments. Our acre of woodland was much more than a mere collection of species. It was a collection of individuals that were interacting with each other and with their environments in a great number of intricate and involved ways.

How do ecosystems function? Throughout this book, we will have many opportunities to consider this question, but we will postpone a fairly comprehensive discussion of it until the final chapters. As an introduction, we will briefly explain six approaches to ecosystem function.

Energy transfer. We have already indicated that the very term ecosystem implies studies of energy transfer in biotic communities. The source of energy is the sun. Its rays warm the ecosystem, causing evaporation and other processes. A small portion of this radiation is converted photosynthetically to the chemical bond energy of the organic compounds that constitute the organisms in the ecosystem. All this photosynthetically converted energy is originally contained in the green plants. They constitute the producers in the system and form the base of an imaginary energy pyramid. The primary consumers are the herbivores that feed directly upon the plants. Of course, most of these are grazing animals, including insects, but a few such as the mistletoe might be plant parasites. Secondary consumers are the carnivores that eat the primary consumers. Again, they are largely animals, but plant parasites on the plant parasites (e.g., a fungus growth on the mistletoe) is not an impossible thought. There might well be tertiary and higher orders of consumers, and one consumer will seldom be limited to a single food source. Thus the energy pyramid becomes an energy web, as the intricate net of producers and consumers develops. Finally, at the top of the pyramid, or perhaps lying evenly over the entire structure, are the decomposers. These are organisms that decay the dead remains of all the others. It is difficult to define the decomposers accurately. The primary consumers are in a sense decomposing the producers. Usually, however, when we speak of the decomposers, we are thinking of the bacteria and fungi that decay the dead remains of other organisms.

Productivity. How productive are the original producers; that is, how efficient are they at photosynthetically converting the sun's energy to chemical compounds? And then, how efficient are the consumers at converting the foods they eat into their own substance? These questions form a subtopic under energy transfer that has received considerable attention from ecologists.

Cycling of elements. Much study of ecosystems has been concerned with the transfer of elements from the soil or the air through the producers, consumers, etc., and back to the soil. This is an important phase of understanding ecosystems.

Transfer of compounds. Some organisms produce certain compounds that are harmful or beneficial to other organisms. They may be excreted, released upon death and decay, or eaten by consumers. These substances are an important phase in the functioning of an ecosystem. Indeed, they tend to make us philosophical about ecosystems. Is the ecosystem some kind of superorganism? Certainly, the functions of a single organism transcend the functions of its cells. A growing stem, not any one of its cells, bends toward the light. Do ecosystem functions transcend the functions of their components? The transfer of compounds between organisms in the ecosystem reminds us of the metabolic processes that go on within an organism. This is especially true when we remember that most consumers must obtain in their diets not only raw energy sources (such as sugars) but also very specific compounds (such as vitamins) that they are unable to synthesize in their own metabolism. In Chapter 40 we conclude that the superorganism view of ecosystems, although valuable, is really an imperfect analogy.

Inter- and intraspecies interactions. The members of a population (organisms all belonging to a single species) interact with each other. These interactions have led to intensive studies of population problems. Most of this work has been done with animals, but the same principles certainly apply, at least in a modified form, to plant populations. Individuals compete not only with members of their own species but also with other species. Sometimes, the interaction between individuals of different species is very intimate, in which case it is called symbiosis. If the symbiosis benefits both members of the partnership (as the mycorrhizae growing on the forest tree roots), then it is called mutualism. (Many books define symbiosis as we have defined mutualism.) If one benefits to the other's detriment, the symbiosis is called parasitism (the mistletoe). If the two associate closely to the benefit of one organism but to no apparent benefit or harm of the other, then the symbiosis is called commensalism (as an orchid growing on a tree in the tropics). Grazing is when an animal consumes plants for food. Predation is when an animal actively hunts and consumes another. It should be obvious from this brief summary that we are still dealing with the details of energy transfers in ecosystems.

Ecosystem functions. When all of these things are considered together, we arrive at a picture of balances and feedbacks in ecosystems that represents the sum total of ecosystem functions. We become aware that disturbing one aspect of an ecosystem almost invariably results in changes in other aspects. We further discover that there is a tendency in ecosystems for feedback reactions to maintain the status quo. This is called homeostasis. For example, if a given component of an ecosystem is reduced by heavy grazing or some other factor, once the pressure is removed, the ecosystem component will tend to return to its original state. Incidentally, in agriculture man consciously applies practices designed to upset the balances of natural ecosystems so that the productivity of the newly established ecosystem will be channeled toward production of food for himself. Organisms that compete with man for food (pests: weeds, insects, pathogens) are removed by cultivation or with chemicals. It is tricky business, but we are now dependent upon making it work.

Not all ecosystems tend to be balanced and to exhibit homeostasis. Some communities are gradually changing with time. Following a severe disturbance of the environment such as a fire or an avalanche, the community does not restore itself to its original state in one step. Only a few individuals are well suited to the newly altered environment; they will invade first. As they grow and develop, they tend to modify the environment somewhat, making it less unsuitable for other species. A few of these will begin to move in, further altering the environment. For example: soils develop; moisture is better retained; and temperatures are altered. In cases where the available species are limited and well known, it is possible to predict with some accuracy the progress of such a succession of communities. The final community is called the climax, and the process of change is called succession. In a sense, this seems to be the epitome of ecosystem function. We may think of it as being comparable to the development of an organism, although it is much easier to understand and probably does not depend upon some mysterious developmental program.

Man is an overwhelming component of most present ecosystems. He has managed to perturb, either positively or negatively, most of the ecosystem functions just described. There is much talk about his negative effects, and we will often refer to the problems of pollution and the destruction of natural ecosystems. On the other hand, by applying the principles that he learns, man can benefit both himself and the ecosystems by controlling their functions in a sort of mutualistic symbiosis (although not as closely associated in space as this might imply). He can do this not only with natural ecosystems but with the artificial agricultural ones upon which he depends for his nourishment and hence his life. Surely, such a mutually beneficial interaction between man and the ecosystems of his planet should be his guiding and directing goal.

Origins

How do we account for the origins of life itself and all of its diverse manifestations? So far, we have been considering things as they are, but inevitably, we must ask where they came from and how they became what they are. The current answer is evolutionary theory. The concepts developed in this story are a great unifying concept in biology. The theory attempts first to describe and then to explain the changes in organisms and communities that have occurred over an immense span of time. Charles Darwin and Alfred R. Wallace provided the key with their theory of natural selection, presented to the world in 1858 at a meeting of the Linnaean Society. Darwin had been working on the idea for at least 20 years, and when he heard that Wallace had arrived at the same conclusion, he decided he could wait no longer to present his theory. Darwin and Wallace noticed that the potential for increasing populations always exists; any two parents of any species (plant or animal) can produce more than two offspring—often many more. Yet the space for growth of a population is not infinite; for man, the earth's surface provides the ultimate limitation on population size. This situation must finally lead to a struggle for existence. (Thomas Malthus had emphasized earlier the struggle for available food, and his writings provided the clue for both Darwin and Wallace.) Darwin and Wallace also noted that a great variety exists within every population, and they reasoned that some members of the population must be better able to survive (compete for food, for mates, etc.) than others. The result is a *survival of the fittest.* Hence population pressures create a continual struggle for existence in which the fittest tend to survive and reproduce—this is natural selection.

To account for evolution, natural selection must be able to act upon a continual source of new variety within the population. Otherwise, natural selection will tend to keep a population well adapted to its environment, but long-term evolutionary changes will not occur. Darwin was at a loss to explain the origin of this variability. He finally fell back on the concept of the inheritance of acquired characteristics, a concept that has been rather conclusively disproven. At the turn of the century, the science of genetics began to develop, and it was soon discovered that the genes are subject to occasional change—mutation. We now think in terms of changes in the order of nucleotides in the DNA chains and of breakages and other changes in the chromosomes. Modern evolutionary theory holds that the ultimate origin of variability upon which natural selection acts is genetic mutation. Even here, however, we may be faced with problems. Now that we understand the order of complexity of genes and enzymes, we cannot be absolutely certain that random mutations of genes occurring for about four billion years can account for all the complexity observable in organisms. But the fact of evolution seems indisputable, and we will not hesitate to build upon this foundation in discussing many other topics in this book.

End Paper 2–1 Metric Measurement

Except for a few countries, including the United States, most of the world keeps track of things by the metric system of measurement, a contribution of the French revolution. As everyone knows, the system has a decimal base whose advantages are familiar to Americans because of our decimal money system. Nevertheless, we have not been troubled enough by the English system of weights and measures to bother changing. Indeed, as long as one is dealing with only a few of the elements of this system (quarts and gallons, miles, yards, feet, inches, pounds and ounces), things go fairly well—although even at this level, addition, subtraction, multiplication, and division would be much easier with the metric system.

In science we are almost constantly involved (especially in some fields) in making manipulations with bits of data obtained by measurement. It is simply too cumbersome to bother with a nondecimal system. The metric system is a must. Furthermore, when we are discussing tiny objects such as the cell and its component parts, there are no suitable units in the English system, unless one wants to use decimal fractions of an inch. It is imperative, then, that the student familiarize himself with metric measurements. This is not a difficult task because another advantage of the metric system is that it can be easily learned. Of course, getting a real feeling for any system of measurement requires some experience with that system. To try to provide some of that experience, metric measurements are used extensively and almost exclusively in this book. In this end paper we will provide an introduction to the metric system, especially those linear measurements that are used in describing cells.

Start with a meter (m). Originally, the French scientists who devised the system wanted the meter to be one ten-millionth of the distance between the equator and the pole—a distance that could theoretically be measured any time, even if standards were lost. Unfortunately, they were not able to measure this distance exactly, and the meter turned out to be a standard metal bar maintained in a Paris vault. More recently, it has been possible to define the meter in terms of a specific wavelength of light so that it could be reproduced any time, even if the standard bar were destroyed. (A meter is a little more than a yard—39.37 inches.) Multiply the meter by a thousand to get the kilometer (km); one mile = 1.62 kilometers. Divide the meter into ten equal lengths, and we obtain decimeters (dm); deci meaning 1/10. Divide the meter into 100 equal lengths, and we have centimeters (cm); centi = 1/100. Divide it into 1000 intervals, and we have millimeters (mm); milli = 1/1,000 = 10^{-3}. (An inch is equal to 2.54 cm or 25.4 mm.) Divide a meter into one-million intervals, and we obtain micrometers or microns (μ); micro = 1/1,000,000 = 10^{-6}. Divide by one billion, and we have nanometers (nm); nano = 1/1,000,000,000 = 10^{-9}. In the past they were often called millimicrons (mμ). Divide the nanometer by 10 to get angstroms (A), the smallest unit normally used in our discussion of the universe.

In microscopy the micron is the most frequently used unit, since it is convenient to measure cells in terms of microns. (To keep our system more consistent, we are beginning to use the term micrometer instead of micron.) In measuring wavelengths of light, it is convenient to use nanometers and angstroms. In discussions of atomic and molecular structures, angstroms are used almost exclusively.

All of this is very logical, but a person should have some real gut feeling for it. Certainly, he should memorize and understand the prefixes: kilo, deci, centi, milli, micro, and nano. He might further think of some common items and their dimensions in the metric system: The fingernail on the little finger of an adult male is about 1 cm wide—the thumbnail of an adult female may be about the same. About 20 pages of this book make a pile 1 mm thick. A thin cell wall is about 1 μ thick, and a typical cell in a meristematic region might be about 50 μ in diameter. The wavelength of green light is about 500 nm. The DNA double helix is about 20 A thick.

The main thing is don't panic! The metric system is not a great mystery devised to further confuse the poor student. Rather, it has greatly simplified the measurement of things. Depending upon your abilities as a memorizer, you should be able to learn the entire metric system in 10 to 30 minutes of intensive study. On the other hand, it would certainly take days and possibly weeks to learn the entire English system—if anybody wanted to learn it all.

Let's quickly summarize the rest of the metric system. Volume bears a direct relationship to the linear measurements. Hence a cubic decimeter is called a liter (l). A thousandth of a liter is a milliliter (ml), and this would be equal to one cubic centimeter (cm^3). Actually, the original standards were not perfect; thus a milliliter and a cubic centimeter are not exactly equal. For our purposes, however, it is important to realize that they are essentially equal. Of course, there are microliters (μl) and possibly even nanoliters (nl)—although we have never heard of these being used.

The system of weights is also tied to linear measure and volume measure. Thus a cubic centimeter of water (a milliliter) weighs almost exactly 1 gram (g). Of course, water expands and contracts with changes in temperature; so this has to be defined at a given temperature (4°C). A thousand grams is a kilogram (kg), and a thousand kilograms is a metric ton (not or ginally considered part of the system). A thousandth of a gram is a milligram (mg), and we are now to the level of accuracy where micrograms (μg) are often used. It is not completely uncommon to encounter nanograms (ng) or even picograms (pg); pico = 10^{-12}. (A plant nucleus may have 10 to 100 pg of DNA.)

For temperatures, the Celsius or centigrade scale is used with the metric system. Water freezes at 0°C and boils at 100°C, provided the pressure is exactly one atmosphere (atm)—the average atmospheric pressure at sea level. For time, seconds, minutes, and hours are used, although they do not form a decimal system. The second is divided into decimal fractions using the terms milli, micro, and nano.

Now see if you can't learn this in 10 minutes: A meter is about one ten-millionth of the distance between the equator and the pole. It divides into deci-, centi-, milli-, micro-, and nanometers (and once more to the angstrom) and multiplies into kilometers. A cubic decimeter is almost exactly equal to a liter, which divides into milli- and microliters. A milliliter of water weighs almost exactly one gram, which multiplies to kilograms and divides into milli-, micro-, and nanograms. Water freezes at 0°C and boils (at sea level) at 100°C. Wasn't that easy? The hard part is trying to learn the cumbersome English equivalents for this simple system.

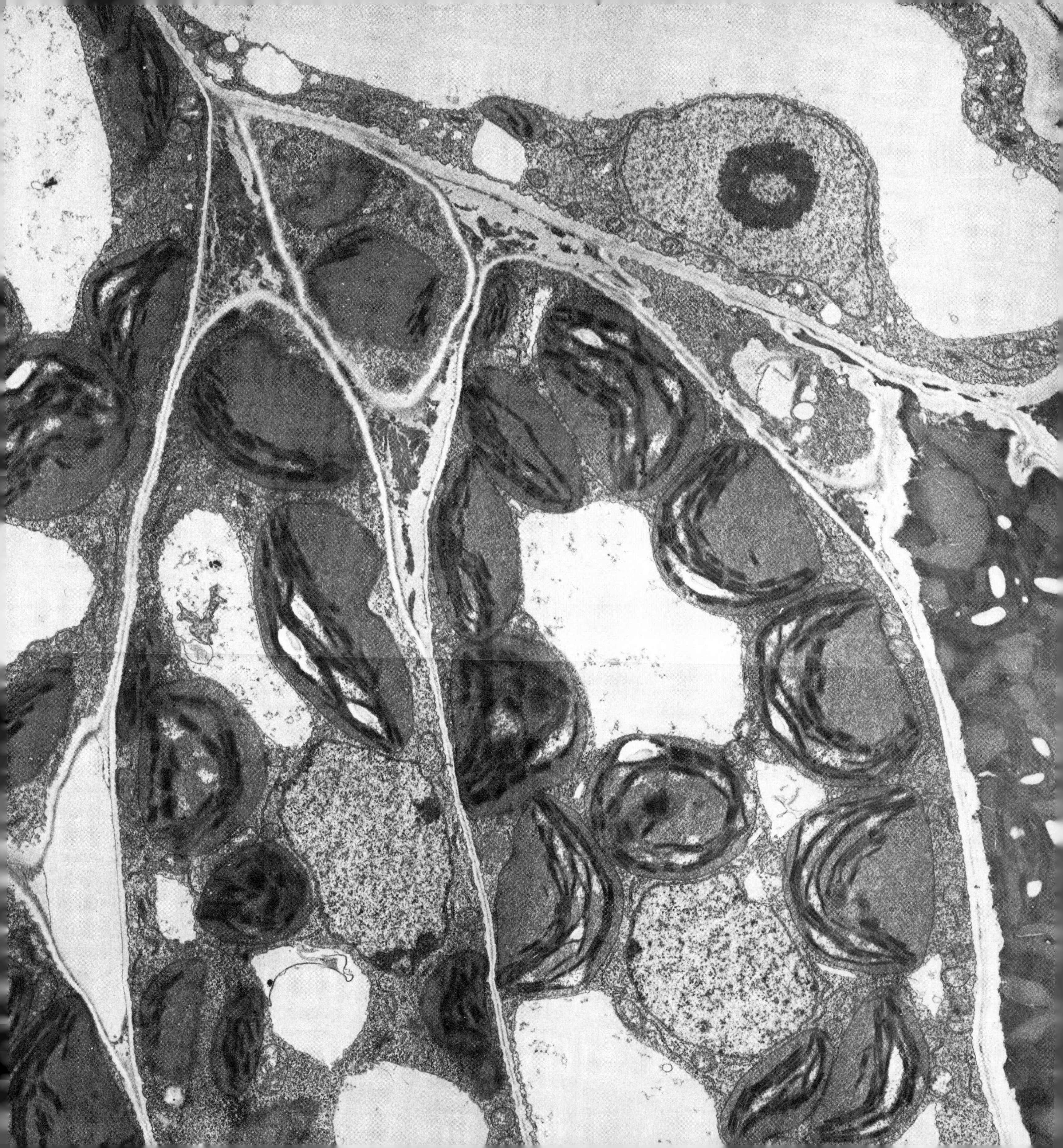

3

Cell Structure

Matthias Schleiden and Theodor Schwann are usually credited with the formulation of the cell theory, which states that *all plants and animals are composed of cells.* In their own words:

> Although plants present so great a variety of external form, yet they are no less remarkable for the simplicity of their internal structure. This extraordinary diversity in figure is produced solely by different modes of function of simple elementary structures, which, though they present various modifications, are yet throughout essentially the same, namely cells.
>
> *Theodor Schwann, 1839.*

> Each cell leads a double life: an independent one, pertaining to its own development alone; and another incidental, insofar as it has become an integral part of a plant.
>
> *Matthias Schleiden, 1838.*

But they were merely summarizing the state of the field rather than postulating a startling new concept. In 1824, the French biologist Dutrochet had clearly stated the same thing, and in 1833, the great Scottish botanist Robert Brown made the same essential assumption from his observations of plants, which included the recognition of the nucleus as a universal cell part. Cells themselves had first been seen and named by Robert Hooke in 1665. In the intervening years cells had become increasingly well known, although many of the most elementary facts concerning them were disputed.

Finally, in the mid-1800s, the design of the microscope improved to the point where objects in the micron range (see End Paper 2–1) could be seen. This is about the limit set by the wavelengths of light. Equally important, techniques were developed to study cells with these improved microscopes. Procedures had to be invented that permitted the preservation of the cells and enabled the material to be sectioned. Finally, staining methods had to be discovered that allowed parts of the cell to be seen (see Essay 3–1).

3–A An epidermal cell (top) and several palisade parenchyma cells from the leaf of a sugarcane plant. These cells illustrate both the similarities and the differences that exist between plant cells. Both types of cells contain the same basic organelles, yet they differ in number, size, and degree of development, resulting in cells with different appearances and different functions.

Essay 3–1 Seeing the Unseen World

Our knowledge of the cell is directly related to the means of seeing it. The Greeks and Romans had no way to magnify objects, so their world was limited to things large enough and close enough to be seen with the naked eye. Not until the mid-1300s did monks in northern Italy develop the art of lens grinding. Their aim was to produce spectacles to aid failing eyesight. Slowly, the art spread across Europe. If an object looks larger when viewed through one lens, what happens when two are used, one above the other? The simple experiment necessary to answer this question must have been done many times in the years following the discovery of lens grinding. This quite probably led to the development of the compound microscope by Hans and Zacharias Jansen in Holland about 1590. These Dutch lens grinders and spectacle makers mounted two lenses in a tube and found that they obtained an increase in magnification over a single lens. Their microscope looked much like a telescope; in fact, the discovery of the microscope and the telescope are completely intertwined.

The invention of the microscope is credited to the Jansens, but the history of the modern compound microscope, that is, one with two sets of lenses, begins with the remarkable Robert Hooke. He was a contemporary of Antony van Leeuwenhoek (1632–1723), a Dutch lens grinder of great skill. Although Leeuwenhoek made many valuable observations with his fine lenses, he did not invent the compound microscope, as is sometimes implied. Hooke, an Englishman of broad interests and great intellect, was a biologist, a physicist, a chemist, an artist, and an architect. In 1665 he published the Micrographia. *This book included his observations made with a microscope he designed and illustrated in his work. The microscope Hooke designed and the modern microscope have the same principle of operation (Fig. 3–0). In both, light passes through the object to be examined and into the first lens of the microscope. Since this lens is close to the* object *being viewed, it is termed the objective. The objective lens magnifies the image of the object a fixed number of times, depending upon the way the lens is ground. This magnified image then passes through a second lens, which is near the eye, and therefore called the eyepiece or ocular. The ocular magnifies the image of the object even more and, at the same time, converges the light so that the eye can focus on the image of the object. The amount of magnification that is achieved by a microscope is a product of the magnifying power of the objective times the magnifying power of the ocular.*

There is an absolute limit to what one can see with the light microscope. Physical theory makes it clear that objects smaller than about half the wavelength of the light used in an attempt to observe them simply cannot be seen. They are literally too small to cast a shadow. The shortest wavelengths of light that can be used in a microscope with special quartz lenses are in the ultraviolet part of the spectrum and are about 0.25 μ long. This means that an ultramicroscope using photographic film to see the images (the eye cannot see ultraviolet light) might detect objects as small as 0.12 μ long. An ordinary light microscope uses light with average wavelengths of about 0.5 μ (green light); so the ultimate limit of visibility in such a microscope is about 0.25 μ. No matter how carefully the craftsman grinds his lenses, objects smaller than this simply cannot be seen with light. Only the development of the electron microscope overcame this difficulty.

In the electron microscope a beam of electrons replaces the beam of light, and magnetic lenses replace those of glass, resulting in an increase of magnification to the 1000s of times and in the ability to see objects in the angstrom range.

More is needed than simply better microscopes—either light or electron—if biological material is to be successfully examined. What is needed is a set of techniques (called microtechniques) to prepare the material for examination.

Microtechnique is as much an art as it is a science. Professor Adriance Foster of the University of California, Berkeley, likes to tell of the student who came to him at the end of a laboratory and complimented him on the beauty of the microscope slides that were used. The only question the student had was that it seemed incredible anyone could paint so small a picture! This is not, of course, the art we mean, but the fact is that good microscope preparations often rely more on the skill of the operator than on the application of well-understood scientific principles.

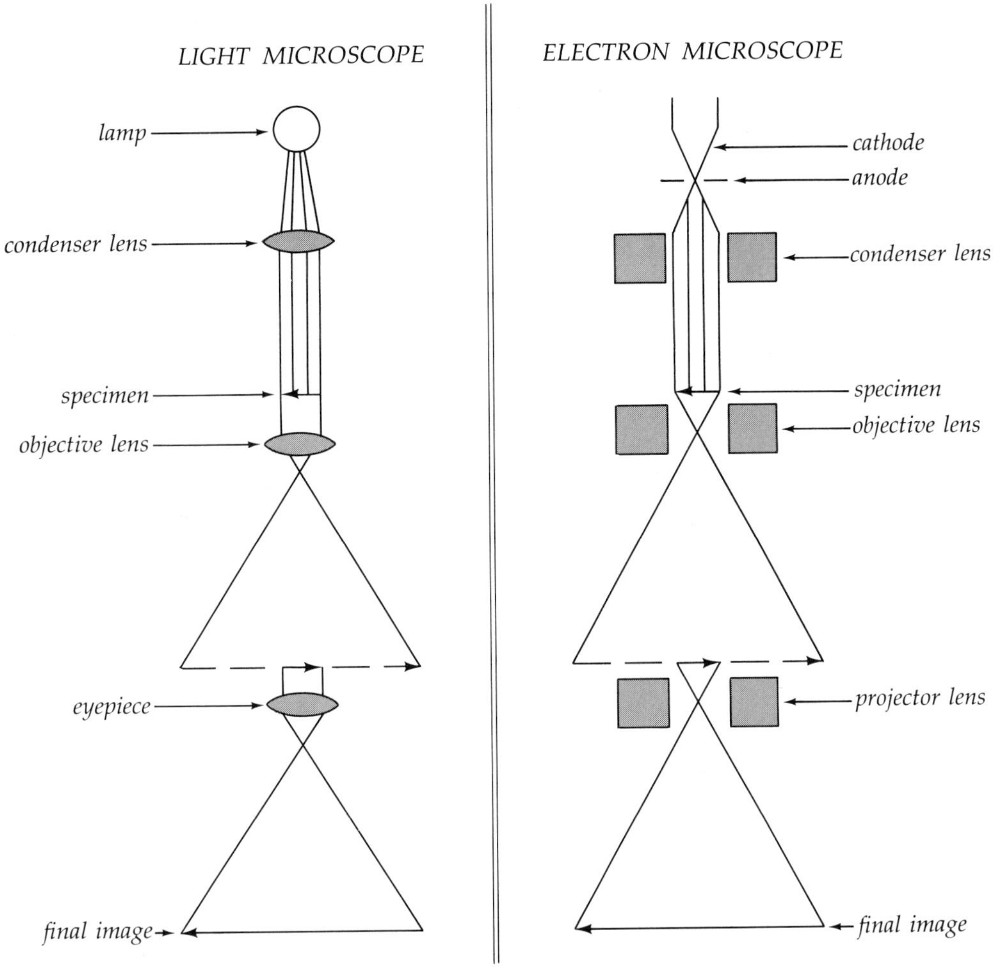

3–0 Comparison between the components of the light and electron microscopes.

To prepare specimens to be viewed with either the light or electron microscope, the same basic steps must be taken.

(1) The tissue must be fixed—killed and preserved. The term fix denotes the fact that soluble compounds of the cell are stabilized or fixed in place. If they are not, gross distortion of the cell will occur, and accurate observation will be impossible. For light microscopy, compounds such as alcohol, formalin, and acetic acid, which are protein precipitants, are used. The more rigorous demands of the electron microscope have led to the use of gluteraldehyde, osmium, and potassium permanganate. (Interestingly, gluteraldehyde is now being widely used with the light microscope.)

(2) The water in the tissue must now be removed and replaced with an organic solvent. This step is called dehydration and is a necessary preparation for Step 3. Usually, alcohol in increasingly high concentrations is used to replace the water. Finally, the alcohol is replaced by a solvent miscible with the embedding medium.

(3) A compound is now introduced into the tissue, which is first liquid and then transformed to a solid, that will hold the tissue firm during sectioning. For the light microscope, this is paraffin wax, which is added hot and then allowed to cool and solidify. Plastic is used for electron microscope preparations and is solidified through heat polymerization. Plastic is used for the electron microscope preparations because it is harder and therefore thinner sections can be cut.

(4) The tissue, now firmly infiltrated and embedded, is cut on a glorified sausage cutter known as a microtome. In a regular microtome, the tissue is held in a chuck and moved up and down against a stationary knife. The tissue is advanced a given amount each time the tissue block

makes a cycle. The sections adhere to one another, end to end, and a ribbon is formed. For the electron microscope, thinner sections must be cut, and an ultramicrotome must be used. The operation is the same, but the action is much finer. Instead of a steel knife, glass or diamond knives, which are sharper, are used. The sections are floated on a water surface as they are cut.

(5) The paraffin sections are next mounted on glass slides, while the plastic sections are picked up on fine metal grids or screens. Light will pass through glass but electrons will not. Hence, in the electron microscope, only that part of the section located over the hole in the grid will be seen.

(6) Little will be seen in any case without first staining the tissue. In the case of the light microscope, the paraffin is removed and the tissue stained with various dyes. These, of course, absorb light and provide contrast to the various parts of the cell. Usually, more than one dye is used—the combination of fast green, which colors the wall and cytoplasm green, and safranin, which stains lignified walls, nucleoli, and chromosomes red. Stains for the electron microscope involve such heavy metals as lead, uranium, and osmium, since these deflect the electrons bombarding the tissue and increase the contrast of the final image.

For the light microscope, the final preparations can be critically examined through the instrument directly with the eye. But human eyes cannot see electrons; so we are forced to use indirect means to observe the specimens. The electrons may be used to excite a phosphorescent screen on which the image can be seen or used to expose a photographic negative. Photographs provide the best possible image and are used for all critical observations.

All of these steps have an effect on the final image. The most important perhaps is fixation. For example, in the case of the electron microscope, potassium permanganate fixation results in beautiful, sharp membranes and clear pictures, but the chemical removes many proteins and all ribosomes from the cell. Gluteraldehyde-osmium, on the other hand, preserves the ribosomes and proteins better, but because of this, the membranes are more difficult to see.

Using these techniques, a series of brilliant biologists observed the principal parts of the cell in the latter part of the nineteenth century. By then, the role of the chromosomes in mitosis and meiosis was understood. During the early twentieth century, the concept of the gene as a particle on a chromosome was fully developed.

Studying Cells

How do cells work? Many of the early microscopists attempted to find the answer. One of these was the French botanist Francois Raspail, who in 1825 established the field of microscopic histochemistry. Its goal is to localize and identify substances and enzyme activities within cells. In 1830 Raspail published his major work at his own expense after it was rejected by a committee consisting of a physiologist, a chemist, and a botanist. The reason for the rejection, according to Raspail, was that the physiologist was ignorant of chemistry, the chemist of microscopy, and the botanist of both!

The problem was that the microscopists needed chemical data, which were not yet available, to answer the questions they were able to ask. As a result, the emphasis shifted from microscopy to organic chemistry and biochemistry. By grinding up cells and analyzing the resulting suspensions, chemists and biochemists learned about the compounds of which cells are made. In the 1800s and early 1900s, great progress was made in our understanding of carbohydrates, proteins, and fats. In the 1870s, in a grim German medieval castle converted to a biochemistry laboratory, the Swiss scientist Friedrich Miescher isolated and characterized deoxyribonucleic acid (DNA), preparing the way for our recognition much later that this substance was the genetic material.

As before, new instruments were required for another major leap in understanding. To bridge the gap between developing ideas of cellular structure and cellular function, three instruments played an important role.

Centrifuge. In the late 1930s this instrument came into use in separating cellular components. In the centrifuge suspensions of tissues could be spun so fast that even particles only slightly denser than the medium in which they were suspended would settle out in the resulting centrifugal fields. If tissues were carefully homogenized so that the subcellular particles were not disrupted, nuclei could be spun out intact, as could chloroplasts, mitochondria, ribosomes, and a few other cellular particles. The biochemist and cell biologist could then investigate the chemical composition and metabolic activities of these isolated cellular components. A later development was the ultracentrifuge, which produces such high velocities that even protein molecules can be spun down. A beam of light passed through the spinning glass tubes detects the moving front of protein by changes in the index of refraction of the medium. Such an instrument is called an analytical ultracentrifuge.

Electron microscope. In the late 1940s the electron microscope became available to biologists. By the 1960s techniques for its use had been developed to the point where this instrument could begin to contribute a vast amount of data, illuminating our understanding of the cell and leading to a sizable revolution in our understanding of cell structure and function. Its operating principle is analogous to the light microscope, except that a beam of electrons is used instead of a beam of light, and the beam is focused by magnetic fields rather than glass lenses (Fig. 3–1). An electron beam has an extremely short wavelength (a fraction of an angstrom); hence there are virtually no resolution limits analogous to those limiting use of the light microscope. The limits are imposed by the state of our technology in building electron microscopes (this can be improved) and, more importantly, by the electron beam that is so energetic it tends to destroy particles in the size range of small and medium-sized molecules.

There is also a problem in obtaining sections thin enough to transmit the electron beam, but this has been solved by developing ultramicrotomes with glass or diamond knives (sections cut as thin as 100 to 500 A compared to 40,000 to 50,000 A with an ordinary microtome) and by embedding the sample tissue in plastics instead of the softer paraffins or other waxes. Contrast is a problem as with the light microscope, but this is solved by using special heavy metal stains (the metal

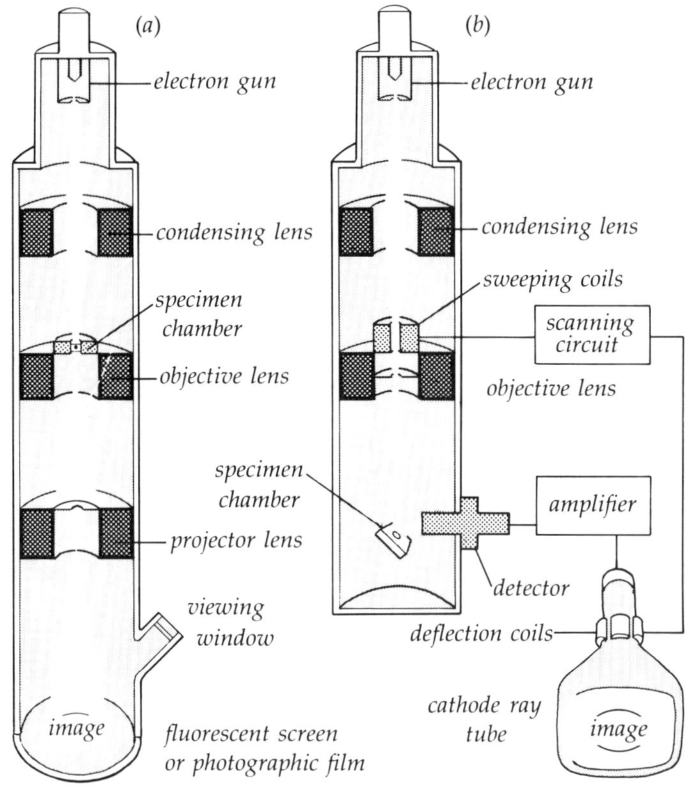

3–1 The principle of operation of both conventional and scanning electron microscopes.

atoms strongly deflect the electron beam) and by employing important special techniques such as shadowing and freeze etching.

If the surface of the specimen is rough, it can be shadowed by evaporating heavy metal atoms from a point source held to one side of the line of view. This will produce apparent shadows, accentuating the surface texture. In freeze etching tissue is frozen and then fractured rather than sliced cleanly. Metal is shadowed onto the surface of the fractured tissue, and the tissue is then dissolved away with suitable chemicals. The surface of the shadowed metal will be an image of the surface of the fractured tissue and can be observed with the electron microscope.

The newest variation of the electron microscope involves the use of whole pieces of tissue and bouncing electrons from the surface rather than having them pass through a section of the tissue. This is the scanning electron microscope in which a beam of electrons bombards the surface of the object. The scattered electrons, together with secondary electrons emitted by the specimen itself, are then detected and amplified to form an image on a cathode tube. This image has great depth of focus and a three-dimensional appearance. Pictures taken with the scanning electron microscope have great beauty and are useful in understanding the spatial relationships of the ultrastructural form. (We have presented several examples in this book.)

Phase-contrast microscope. The third instrument is an adaptation of the light microscope. In the phase-contrast microscope substances within the cell that appear quite transparent in ordinary light are made visible because of slight differences in density and a consequent slight difference in index of refraction. The advantage of this instrument is that cells can be observed in an intact state without staining and, if suitable tissues are used, without sectioning. For example, it has been possible to observe mitochondria in living cells and to study their rates of formation and change—features that are not apparent in fixed samples.

With the development of these instruments and their creative application, it is possible for us to begin to relate cellular structure to cellular function. In this and the next few chapters, we shall present a general examination of what is known about life as functioning cells. First, we shall briefly consider the parts of cells as revealed by light and electron microscopes. Second, we shall examine membranes and the cell wall, emphasizing the structure-function relationship. Third, we shall discuss the regulation of water movement and of ion and other solute passage into and out of cells. Fourth, we shall examine cellular energy transfer—energy capture by chloroplasts (photosynthesis), energy release by mitochondria (respiration), and energy utilization (synthetic metabolism), all under the control of enzymes. Fifth, we shall discuss how these cellular functions relate to each other and how they are controlled by the DNA and RNA molecules, particularly in the nucleus and in the ribosomes. Finally (in Chapter 24 and in other chapters), after we have outlined and examined many of the various kinds of plants and their morphologies, we shall consider how cells specialize to form the complex tissues, organs, and organ systems that make up organisms.

3–2 A summary of the generalized green-plant cell structure

Cell wall (in living cells, usually penetrated by strands of cytoplasm called plasmodesmata)

Protoplast (contents of the cell exclusive of the cell wall)

Cytoplasm

(1) Outer cytoplasmic membrane (plasmalemma)
(2) Vacuolar membrane (tonoplast)
(3) Endoplasmic reticulum (ER)
(4) Mitochondria
(5) Ribosomes
(6) Lysosomes
(7) Microbodies (or peroxisomes)
(8) Dictyosomes
(9) Plastids

(a) Proplastids
(b) Leucoplasts
(c) Chloroplasts
(d) Chromoplasts

Nucleus

(1) Nuclear envelope
(2) Nuclear sap
(3) Chromatin
(chromosomes during cell division)
(4) Nucleolus

Vacuole(s)

Ergastic substances

(1) Crystals
(2) Tannins
(3) Fats and oils
(4) Starch grains
(5) Protein bodies

Cell Structure

There is really no such thing as a "typical" cell; hence not all of the features we will now describe can be illustrated in one or two photographs or drawings. Table 3–2 summarizes the known parts of a plant cell in tabular form.

Surrounding virtually all cells in almost all plants is the cell wall. Walls of adjoining cells are separated by a middle lamella. The wall itself may be a primary one only (e.g., in young cells, storage cells, etc.), or it may consist of a primary layer as well as a secondary layer laid down during cell maturation (e.g., in the mature conducting cells of wood). Since only plants have cell walls, the wall is highly characteristic of these organisms. The most obvious function of the wall is to provide rigidity or support for the cell, but less obvious is the fact that it must do so while allowing ready access of liquids to the cell itself. Hence the wall is not only rigid, but it is also highly porous and permeable to water solutions.

There is also the interesting relationship between cell wall and plasma membranes found in transfer walls. These are found in situations where large amounts of material are moving from one cell to another. In such cases a thick, highly convoluted but porous wall is found. This wall is freely permeable to molecules, and its elaborate shape increases tremendously the surface area of the plasma membrane.

The mature or maturing plant cell typically includes a large vacuole filled with water and dissolved materials. Substances dissolved in a solvent are called solutes. In mature cells the vacuole may occupy most of the cell (90% or more). We shall see that the vacuole, by taking up water through osmosis, provides pressure within the cell and thus cell and tissue turgidity. The importance of this is apparent when you think of a wilted plant in which pressure in the vacuoles has decreased considerably. The large, central vacuole is also almost unique to plant cells.

The passage of water and solutes is regulated by several membranes in the cell. Around the cell but within the wall (around the protoplast) is the plasma membrane or plasmalemma. This membrane is in intimate contact with the cell wall, sometimes penetrating its porosity. (Incidentally, the wall and the membrane are often confused. If you read about the walls around animal cells—even in some textbooks—translate it as "membranes." Almost by definition, animals do not have cell walls, although traces of cellulose have been detected in human skin!)

The cell wall is penetrated by plasmodesmata—minute strands connecting adjacent cells with membranes and cytoplasm. Around the vacuole is another membrane (different in certain ways from the plasma membrane) called the vacuolar membrane or the tonoplast. Many of the cellular organelles described below are surrounded by membranes, and sometimes these membranes differ in important ways from the plasma membrane or the tonoplast.

It used to be popular in biology classes to discuss the protoplasm. Because the term seemed to imply certain mystical properties of life not supported by presently available facts, it has fallen somewhat into disuse. Nevertheless, the term seems valuable if it is defined structurally as consisting of the cytoplasm and the nucleus. The cytoplasm would be all of the cell except the wall, the vacuole, and the nucleus. Although the idea seems more arbitrary now than it used to, we can still think, at least to a first approximation, that the protoplasm (cytoplasm plus nucleus) is the "living" part of the cell, although some enzymes probably exist in the cell wall and possibly the vacuole.

The cytoplasm is highly complex. In many plant cells it contains chloroplasts, which are the organelles responsible for photosynthesis. These are themselves highly complex and unique to plant cells, as we shall see in Chapter 6. They may be isolated in the centrifuge for studies on photosynthesis. Other plastids also occur in plant cells. These may contain oils, pigments, or proteins, but common plastids contain starch grains. Minute bodies were observed in the cytoplasm as early as 1900 and called mitochondria. With the development of the electron microscope, these proved to have an intricate double membrane structure, and when isolated in the ultracentrifuge, it was found that the metabolic steps of respiration, as well as other chemical activities, occur in conjunction with these organelles.

A network of membranes that became apparent in early electron microscope preparations has subsequently been shown to be a double membrane system comparable to a thin-walled, highly collapsed, and folded sack. This endoplasmic reticulum (ER) is attached to the nuclear membrane and may be continuous with the tonoplast. At first the existence of the ER was disputed by some because the cytoplasm of many plant cells is highly fluid and exhibits active streaming. This controversy has now subsided, and it is agreed that the ER is a universal feature of the cell.

In many plant cells the ER has numerous granules or particles attached to one side. This is called rough ER. In other cells having smooth ER, the granules do not appear. The particles have been identified and separated by ultracentrifugation. They are called ribosomes. Sometimes, they occur free (not attached to the ER) in the cytoplasm, and they are even known to occur in the nucleus and other cellular organelles. Their function is protein synthesis. Sometimes, a number of them (perhaps 2 to 5 or more) are apparently held together in a group by a long molecule of RNA. Such a group is referred to as a polyribosome or polysome.

Stacks of membranes with convoluted margins and surrounded by vesicles (spherical bodies) are visible in electron micrographs. These are called dictyosomes. There is excellent evidence that the vesicles migrate from the dictyosome and fuse with the plasma membrane, with the contents contributing to the growth of the wall. This is one function of dictyosomes, but there are probably others.

When mitochondria were separated by centrifugation, they were always contaminated by particles of similar size and density. Although these were membrane-enclosed, they did not have the complex inner membranes of the mitochondria. They are called lysosomes, and their apparent function is to digest other cellular particles. They contain numerous hydrolytic enzymes capable of this digestion, and they often contain fragments of the other cellular particles. There is evidence that lysosomes originate from an ER-dictyosome complex.

In recent years another class of cellular particles has been observed. The function of these microbodies is currently being intensively investigated. They apparently take part in certain special aspects of cellular biochemistry such as fat metabolism or photorespiration in plants. Other bodies called spherosomes have also been observed, but their function remains a mystery.

Microtubules are observed in many electron micrographs, often associated with the plasma membrane. These are interesting structures that we shall discuss later. They appear during nuclear division, forming part of the spindle (see Chapter 9).

The nucleus is the control center of the cell. This is a relatively large cell organelle—the average size ranging from 5 to 30 μ. The nucleus is surrounded by a membrane called the nuclear envelope that, when viewed with the electron microscope, is seen to be filled with numerous relatively large pores. Within the nucleus is one or more nucleoli (singular: nucleolus). Also found within the nucleus are the chromosomes. These are composed of chromatin, which is a combination of DNA and protein. During cell division, the chromatin becomes tightly coiled and the chromosomes quite short, but between divisions the chromatin coiling relaxes so that the chromosomes are difficult to resolve in the nucleus.

In examining photographs and drawings of cells and in reading our structural descriptions, it is easy to overlook the extremely dynamic nature of the cell. Indeed, this aspect of cell life should probably be emphasized much more in discussions such as this one. Samuel Wildman and his colleagues at the University of California in Los Angeles have made numerous studies of living cells using the phase microscope. They examined certain plant hair cells, for example, which can be seen in the intact state without the preparations that would lead to their death. Moving pictures were taken; frequently things happened so rapidly in the cell that these could be taken at normal speeds (rather than using time-lapse techniques, which is necessary in photographing such processes as mitosis). The streaming of the cytoplasm was readily apparent in such observations. Furthermore, it was possible, particularly by examining individual frames on the movie film, to study the changes that were taking place in such organelles as the mitochondria. Certainly, we must keep these observations in mind whenever we find ourselves thinking of the living cell as the kind of static entity that might be implied by sections displayed in an electron micrograph.

Some important cell structures still cannot be seen in a meaningful way. These are the molecules of metabolism. There are the protein enzymes, the nucleic acids, the fats, carbohydrates, and numerous small molecules such as sugars and amino acids. We know much about these cellular structures at the molecular level, but what we know has been deduced by applying the powerful but indirect approaches of organic chemistry. For the most part, the molecules have not been seen.

Someday we may actually see atomic structure. In 1969 workers at the California Institute of Technology succeeded in photographing the double helix of DNA (Fig. 3–3). The photograph appears blurry because the background molecules provide a granulation of particles nearly as large as those constituting the double helix itself. Nevertheless, the helix is visible.

We have also been able to see protein molecules, but so far they appear only as rather uninteresting blobs that reveal little or nothing about their function. Only recently have protein molecules been seen in what is close to their true form using high resolution scanning electron microscopes. The electron microscope has the potential power to resolve atoms. If this could ever be achieved, we might be able to determine molecular structures directly. But how can one see an individual molecule in a mass of others, and how can one keep the electron beam from destroying that molecule? Researchers who are intimately involved in the perfection of electron microscopes believe that these problems will be solved.

The Cell Wall

Plants are plants largely because of the cell wall. The wood that is so familiar to us is nothing but a collection of cell walls. The fact that plants are extremely rigid compared to animals is due largely to the presence of cell walls. During the development of an animal from the zygote to the adult, cells often move from one place to another. In plants cell walls are cemented to each other by the middle lamella, and although they may stretch and become thicker, they cannot move much in relation to each other. Organ systems of plants are much simpler than those of animals, and this may be partially due to the more rigid form of growth imposed by cell walls.

Primary cell walls in newly developing cells are quite thin, on the order of 1 to 3 μ in thickness (Fig. 3–4). These primary walls are elastic, increasing their areas many times as the cell grows. When the cell has reached its final size, the secondary wall begins to form (Fig. 3–4). This is not elastic, but it gives the final rigidity to the cell (e.g., as in wood). Although the secondary wall is rigid, it is also highly permeable to water and minerals dissolved in the water.

When the wall is carefully examined with the light microscope, it can occasionally be resolved into a series of layers. These layers are difficult to see, and few details of the fine structure of the wall can be observed. With the electron microscope, much of the ultrastructure can be seen and studied in detail (Fig. 3–4). The material of the wall is arranged in strands called microfibrils. Between these microfibrils there is a matrix, which does not appear in the figure because it was removed during the course of preparing the wall for observation with the electron microscope.

Microfibrils can be identified by physical and chemical means as bundles of long, chain-like cellulose molecules. These in turn are composed of repeating units of glucose molecules. Glucose is a sugar containing 6 carbon atoms, 12 hydrogen atoms, and 6 oxygen atoms. Sugars such as glucose are carbohydrates—a class of compounds extremely important in the structure and function of the cell. To better appreciate these interesting and vital compounds, we conclude this chapter with a short article (see End Paper 3–1) summarizing the essential points or take-home lessons of carbohydrate chemistry.

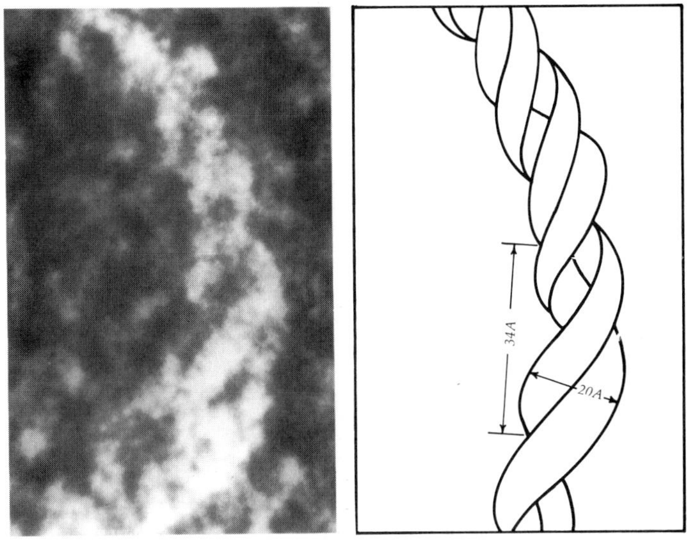

3–3 An electron micrograph of DNA from the pea plant (× about 4.4 million). The DNA was rotary shadowed with tungsten. The actual photograph is on the left, while an interpretive drawing is shown on the right.

The microfibrils of the cell wall form a rather loose network—most lie at least roughly parallel to each other, but with many pores or open spaces between. These allow free passage of solutions and also provide room for the other components of the cell wall.

In cells that are destined to elongate, the microfibrils are laid down around the cellular circumference, much as barrel hoops around a barrel (Fig. 3–5). As the cell elongates, the microfibrils that were first laid down are pulled until they tend to lie parallel to the elongation axis of the cell. As the cell wall is expanding during growth, new microfibrils are continually being laid down around the circumference. The ones laid down after the cell has already elongated considerably will not have a chance to be pulled completely into the axis of elongation. The last ones to be laid down will not be pulled at all. Thus the outer microfibrils will be

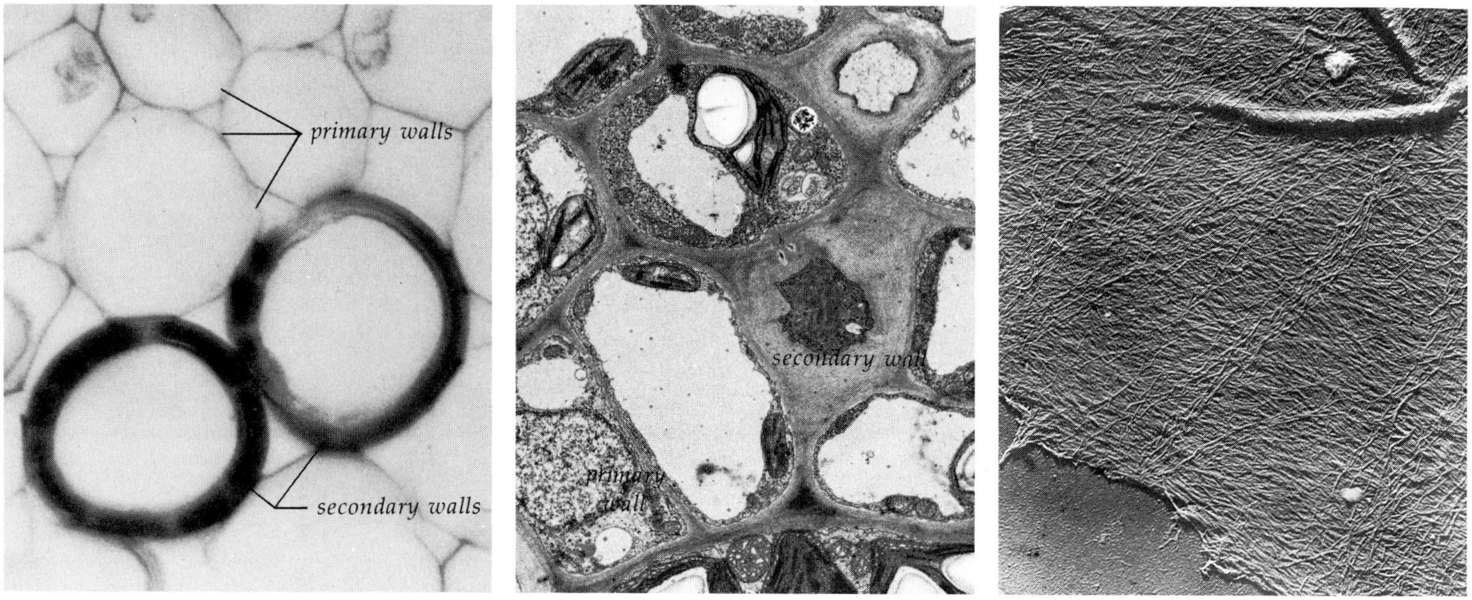

(a) (b) (c)

3–4 *(a) The appearance of primary and secondary walls as seen with the light microscope. (b) The primary and secondary walls as they appear in section when viewed with the electron microscope. (c) The primary wall as it appears in the electron microscope after it has been shadowed with a heavy metal to show the orientation of the microfibrils in the various layers of the wall.*

nearly parallel to the length of the cell, and the inner ones will approach right angles to these. This forms a porous meshwork that, as the microfibrils become cemented to each other, provides much strength.

Other cell wall components are the pectic substances. These occur particularly in the middle lamella but also in the primary and secondary cell walls. They are also polysaccharides, but they consist of long chains of galacturonic acid molecules. The pectic substances have a number of unusual chemical properties that are important in understanding their role in the cell wall. Perhaps the most important of these properties is that the pectic substances can form gels. Such gels can act as cementing substances and are thus ideally suited to the functional role that the pectic substances play in holding cells together. Gels are also permeable to water and solute, and this too is important to the physiology of the cell (see End Paper 4–2).

The pectic substances can be broken down by suitable enzymes called pectinases. For example: in ripening fruits, as the middle lamella is dissolved, the cells begin to come apart, and the fruit gets mushy. This also occurs in certain plant diseases in which fungal organisms produce the pectinases, thereby dissolving their way into the plant.

Besides cellulose and the pectic substances, there are other carbohydrates in all parts of cell walls. The hemicelluloses, which contain pentose residues, are not well understood, but they are presumed to be quite important.

Secondary cell walls contain an extremely important compound called lignin. This substance is not a carbohydrate; rather it contains benzene ring structures in its molecules (Fig. 3–6). The benzene ring is an extremely common structure in organic chemistry, and a whole field (called aromatics) deals with compounds that include the benzene ring. Lignin has been difficult to study because when one attempts to break the bonds and separate the residues for examination, the residues themselves tend to break down. Yet a rather complex chemistry has been developing (which we will not consider here), and lignin structure is beginning to yield to study. The molecule is known to contain complex residues such as coniferyl alcohol. In any case, lignin is important in conferring the rigidity upon the secondary cell walls. When paper is manufactured, lignin must be

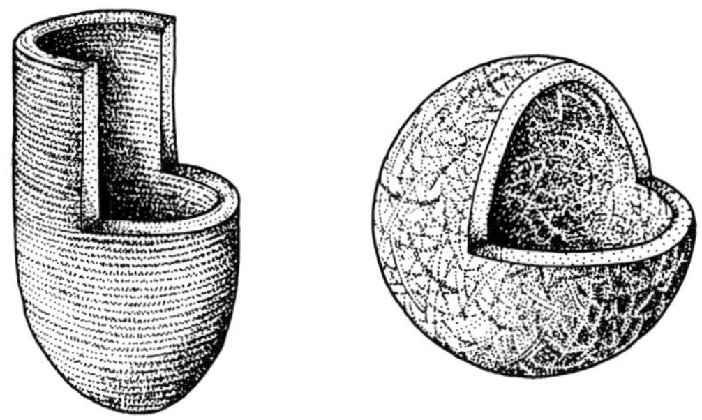

3–5 *A schematic drawing showing the arrangement of long-chain cellulose molecules in the primary walls of two types of cells: transverse orientation, as in an elongating cell (left), and random orientation, as in a cell expanding more or less uniformly (right).*

removed from the pulp, and one of our principal pollution problems concerns what to do with excess lignin. In a sense, this has always been a problem because lignin breaks down in the soil under bacterial influence more slowly than virtually any other natural compound.

In certain fungi chitin occurs in the cell walls. Interestingly enough, this same material is responsible for the hard exoskeletons of insects and other arthropods. Chitin is a polysaccharide of acetyl glucosamine residues. This substance is a derivative of glucose containing nitrogen and two extra carbons. Some cell walls, notably in the grasses and in the horsetails (*Equisetum*), contain quantities of silica, sometimes as crystals. Probably, the majority of primary cell walls also contain protein, which could have enzymatic activity, but so far it has been little studied.

The cell wall is a good place to appreciate how the molecular structures of the wall components account for cell-wall functions. The glucose rings in their chair form confer upon cellulose its specific properties of porosity and strength. The pectins, and probably the hemicelluloses, have properties determined by their molecular structures that make them ideal as cementing substances to hold the cellulose microfibrils together and (via the middle lamella) to cement together the primary walls of adjacent cells. Lignin, chitin, silica, proteins, and other materials confer special properties on special cells, where needed. Finally, within the porosity of the walls, there are free negative charges, possibly due to ionized carboxyl groups of pectic acid or to other materials. Cations (positively charged metal ions such as potassium, magnesium, calcium, and sodium) are electrically attracted to these negative spots and become attached there. One such attached cation might be subsequently removed by another, particularly if the other is present at a higher concentration. This process is called cation exchange, and we shall discuss it in relation to soil formation in Chapter 38. Its significance in the cell wall is not apparent, but it could provide a reservoir for nutrient elements.

The cell wall illustrates beautifully the intimate relationship between structure and function always so characteristic of living things. To really appreciate this relationship, we must have a thorough understanding of the chemistry and molecular structures of the cell-wall constituents. The chemical properties of cellulose, pectin, lignin, and others are conferred upon these substances by the manner in which simpler molecules such as glucose are put together by highly specific enzymes—this makes the cell wall what it is and ultimately determines many of the characteristics of plants.

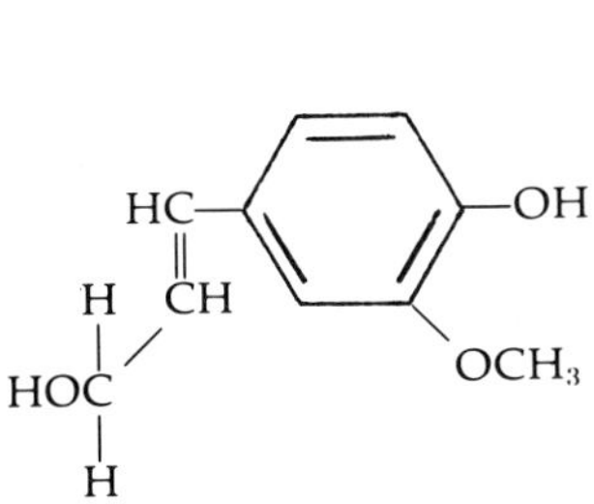

coniferyl alcohol

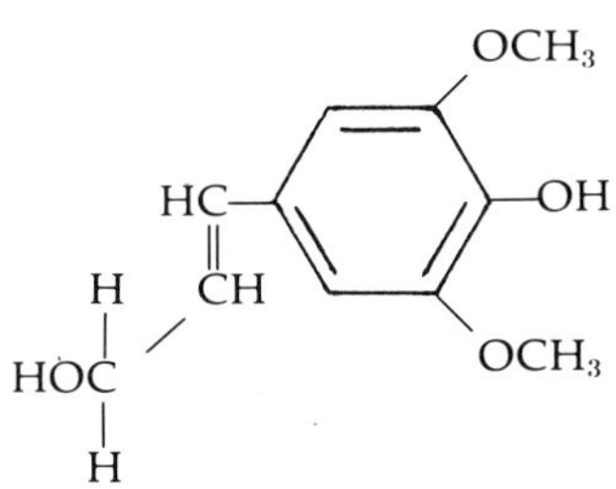

sinapyl alcohol

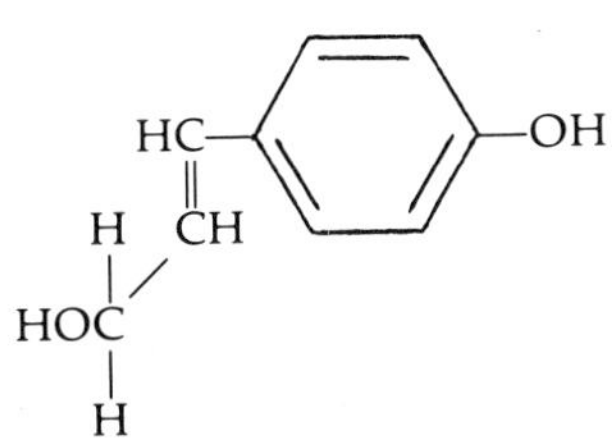

p-coumaryl alcohol

3–6 *Some of the molecular components of lignin. Note the six-membered rings (benzene rings)— molecular structures we did not encounter in the carbohydrates.*

End Paper 3–1 The Chemistry of the Cell: Carbohydrates

An appreciation for the beauty of the structure–function relationship of cells and organisms depends upon a visualization of the molecular level. The place to start in understanding molecules is the atom. These are the smallest particles into which an element can be divided without losing its characteristic properties. The atoms of an element owe their distinctive properties to the number of protons in the nucleus, that is, to the number of positive charges. For example: hydrogen has one positive charge (one proton); carbon has six; nitrogen has seven; oxygen has eight; and sulfur has sixteen. These charges are neutralized by negatively charged electrons circling the nucleus at various levels. If the atom loses one or more electrons, it becomes a positively charged ion; if it gains one or more electrons, it becomes a negatively charged ion. Atoms also have neutrons (no charge) in their nuclei; their number can vary for a given element, producing isotopes . Atoms undergo chemical reactions by losing or gaining electrons from other atoms (forming ionic bonds) or by sharing electrons to form covalent bonds (Fig. 3–7). When atoms thus reacting with each other form bonds, molecules are produced.

In the branch of chemistry called organic, the carbon atom plays a particularly important role. It has four electrons in its outer shell, but there is room for eight. Thus a carbon atom can form four covalent bonds with four hydrogen atoms, sharing an electron with each. The four electrons from the hydrogen atoms complete the complement of eight required for the carbon atom; the hydrogen atoms are also "happy" because they each need one more electron to complete their outer shell, which can handle two electrons. Incidentally, the molecule containing one carbon atom and four hydrogen atoms is methane (CH_4). Two carbon atoms can share one of each of their electrons with each other, forming a single bond. The remaining three electrons of each carbon atom can share with an electron of a hydrogen atom, forming the compound ethane ($H_3C—CH_3$). Two carbon atoms can share two of each of their electrons with each other, forming a so-called double bond. This leaves room for only two hydrogens for each carbon, forming the compound ethylene ($H_2C{=}CH_2$). If three electrons are shared between two carbon atoms, acetylene ($HC{\equiv}CH$) is the result, but triple bonds are not common in organic molecules. Of course, carbon atoms can also share

3–7 Illustration of some principles relating to elements and chemical bonds. The circles at the top represent one way of visualizing atoms. In the neutral atom the number of electrons must always equal the number of protons in the nucleus. The inner shell can hold 2 electrons, and the outer shell can hold 8. Helium (He) is an inert gas because its electron shell is full. The outer shell of hydrogen (H) can accept 1 electron; carbon (C) can accept 4; and oxygen (O) can accept 2. The shaded drawing indicates a more correct way of visualizing the O atom. The dumbbell-shaped spaces are to be visualized as being at right angles to each other. The 6 outer electrons are most likely to be found in these spaces. Sodium (Na) reacts with chlorine (Cl) by transferring its single outer electron to the outer shell of the Cl atom, filling that shell with the required 8 electrons. The result is that the Na atom has one less electron than protons in the nucleus and hence is positive. The Cl atom becomes negative after accepting the electron from the Na atom. Such charged atoms are called ions. The electrical attraction between positive and negative ions is called an ionic bond. In a crystal of sodium chloride (NaCl), Na and Cl ions are not molecules but are "packed" as if they were to fill the space in the corner of a cube, as indicated by the drawing of a NaCl crystal lattice. If electrons are shared, as in methane and formaldehyde, covalent bonds are formed. These are by far the most common in the organic molecules constituting living organisms. A third kind of bond is the hydrogen bond. Hydrogen bonds occur in water because the water molecule is not symmetrical, as indicated by the drawings at the lower right. The two H atoms are attached to the O atom at an angle of 105° rather than 180°. This results in one side of the molecule being slightly positive compared with the other side, which is slightly negative. The positive parts of the water molecule attract the negative parts of other water molecules (and vice versa).

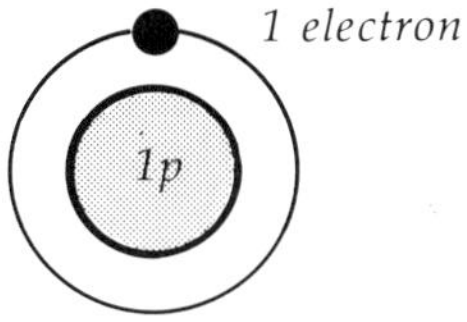

Hydrogen (H)

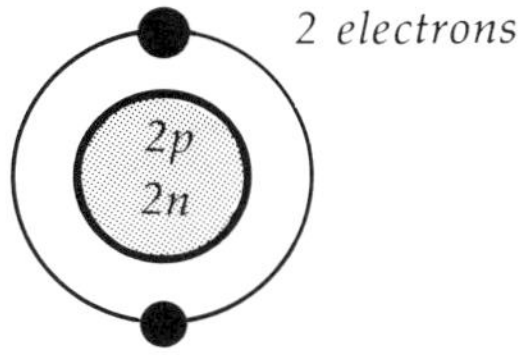

Helium (He)

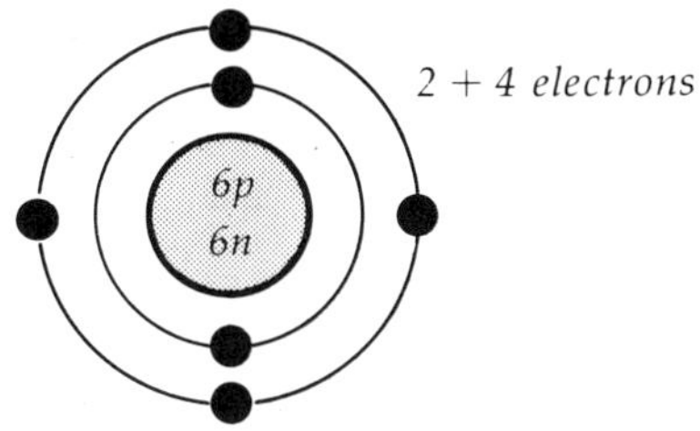

Carbon (C)

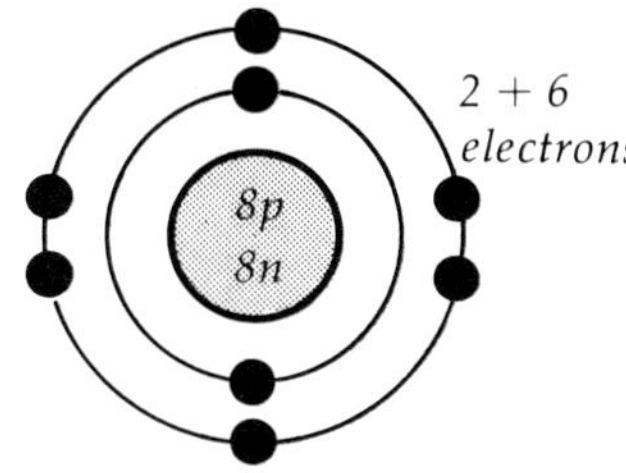

Oxygen (O)

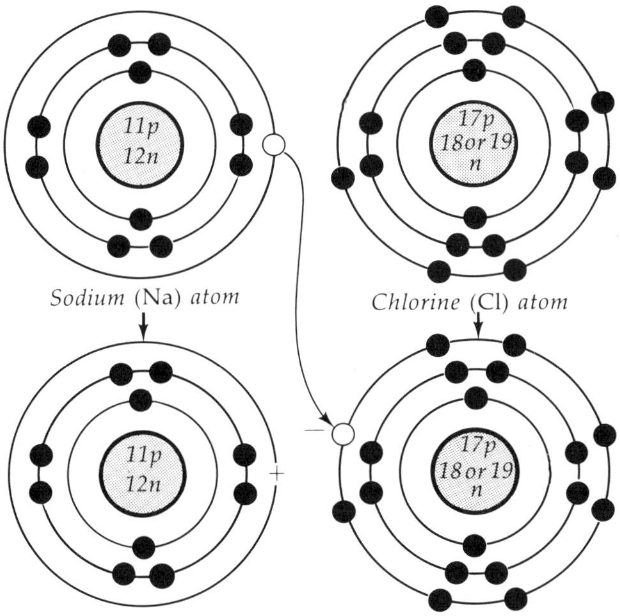

Sodium ion

Chloride ion

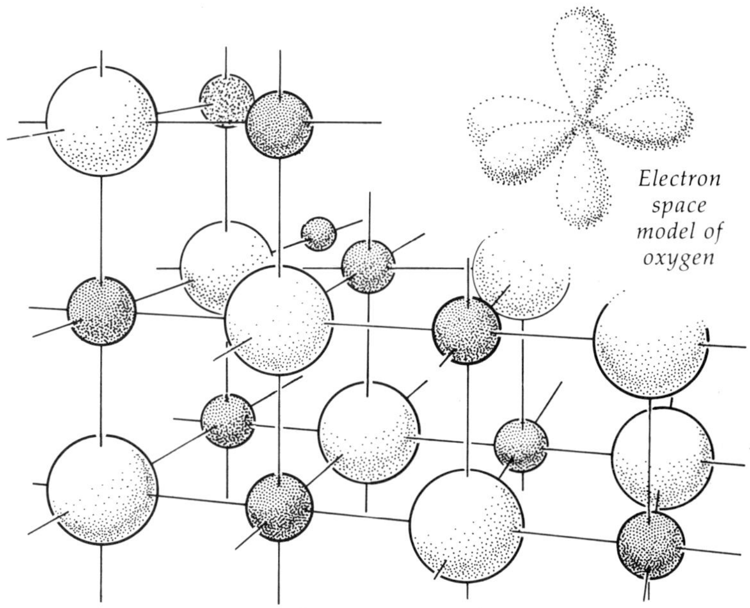

Electron space model of oxygen

Ionic bonds in a crystal of sodium chloride

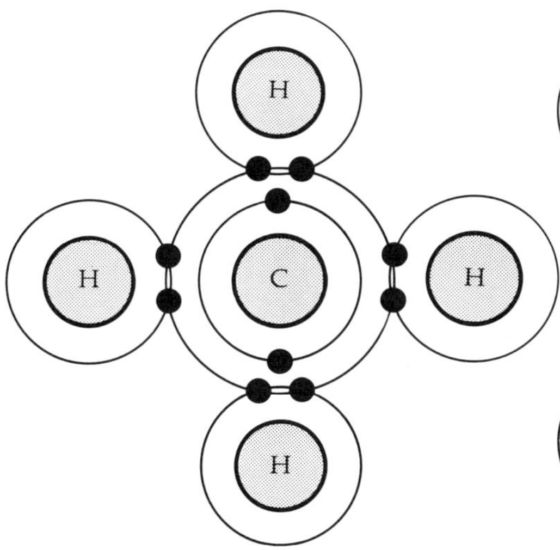

Covalent Bonds: Methane (4 single bonds)

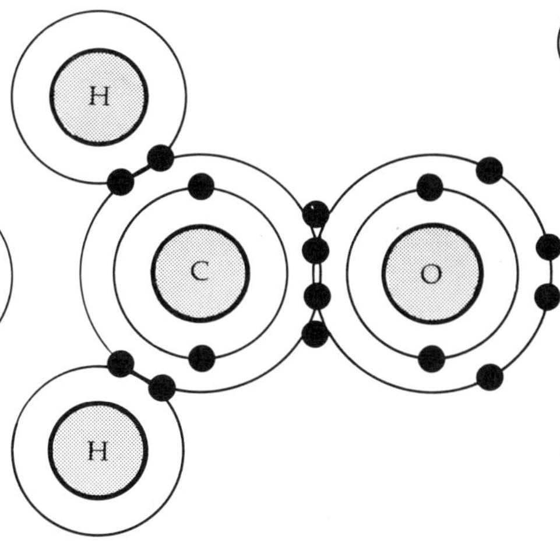

Formaldehyde (double bond)

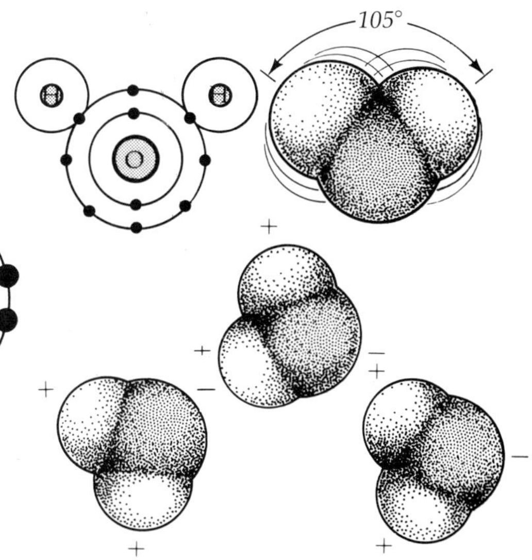

Hydrogen bonding illustrated by water molecules

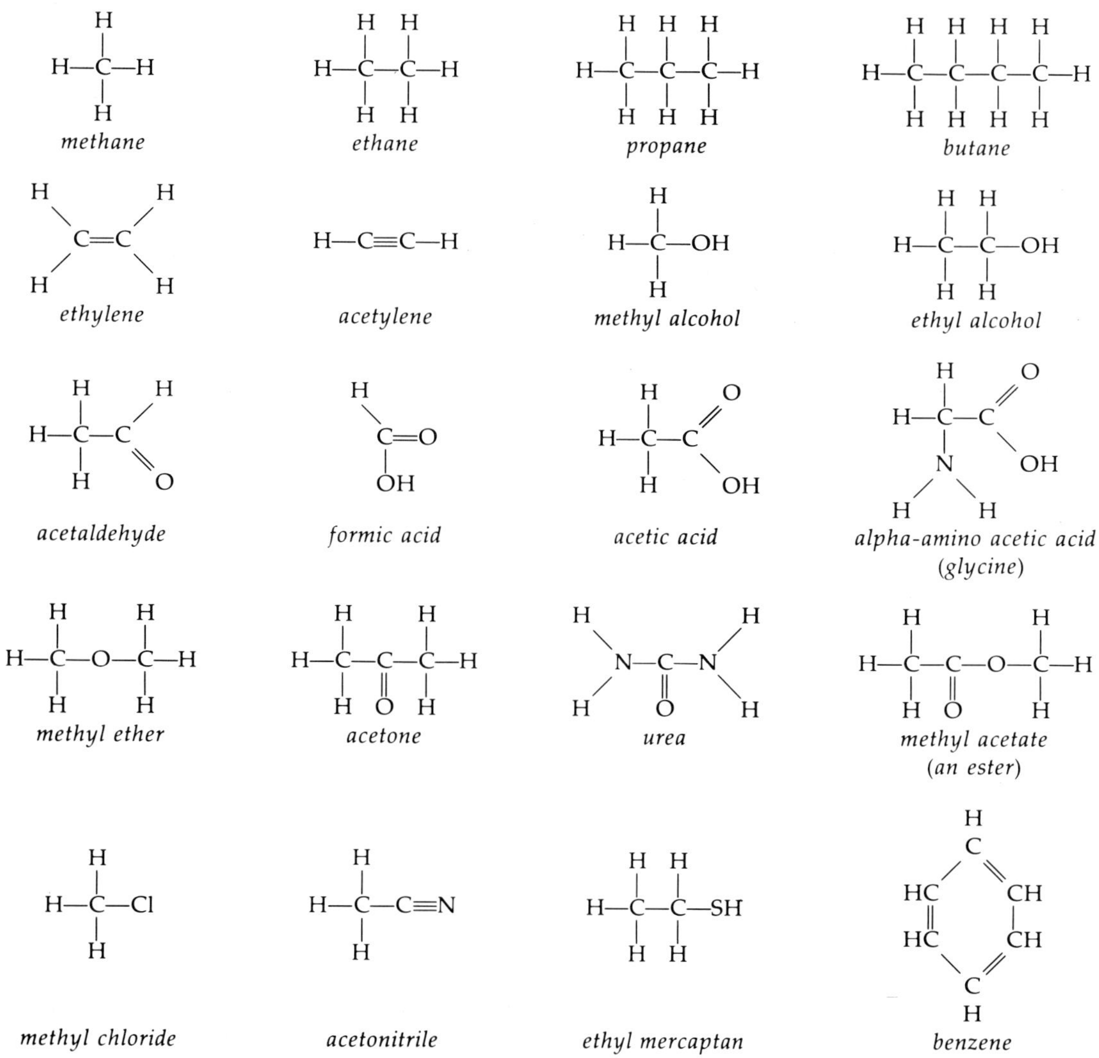

3–8 Some common organic molecules.

electrons with many other atoms essential to the compounds of living things such as oxygen, nitrogen, and sulfur (Fig. 3–8).

In both primary and secondary walls, cellulose is an extremely important, if not the most important, component. As a matter of fact, in terms of quantity, this compound is the most important carbon-containing compound on earth. Fifty percent of lumber consists of cellulose, as does 90% of cotton. To understand cellulose, we must understand something about the chemistry of carbohydrates in general.

(1) Carbohydrates are organic compounds containing approximately two hydrogen atoms and one oxygen atom for every carbon atom. The general formula is $(CH_2O)_n$; that is, for each C, there is an H_2O (water—although it doesn't exist as such), thus suggesting the name carbohydrate (hydrate referring to water).

(2) The basic building blocks of carbohydrates are called monosaccharides, and they contain various numbers of carbon atoms (Fig. 3–9).

Three-carbon atoms = trioses: These compounds are important intermediates in the metabolic pathways of photosynthesis, respiration, etc. They seldom have the $(CH_2O)_3$ formula, although they approach it.

Four-carbon atoms = quadroses: These are uncommon, although one plays a role in photosynthesis. Four-

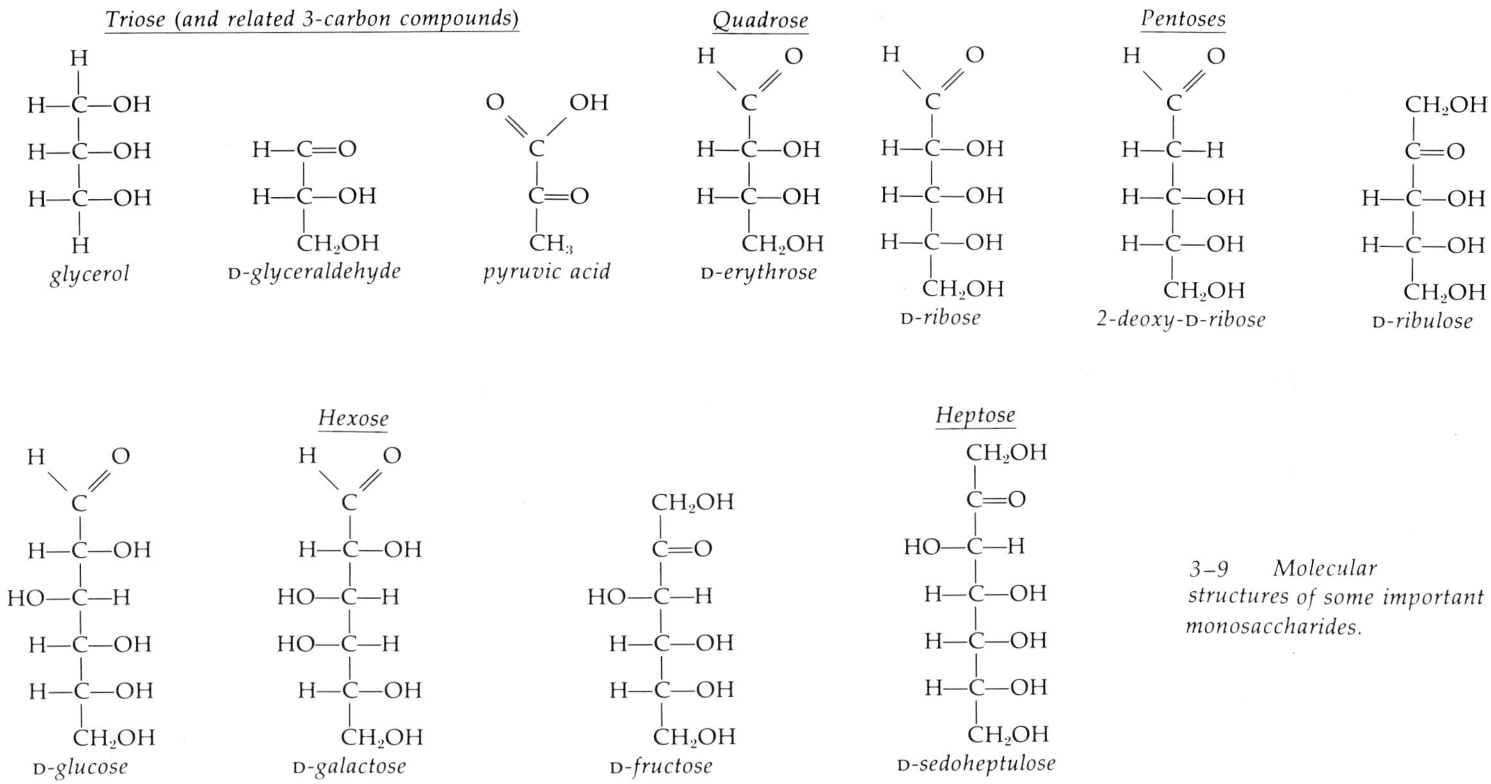

3–9 *Molecular structures of some important monosaccharides.*

carbon acids, rather closely related to the sugars, are important in respiration.

Five-carbon atoms = pentoses: These are involved in respiration and photosynthesis, and they also form an important structural component of the nucleic acids and hence of all life. Certain gums peculiar to specific plants (e.g., both algae and higher plants) consist largely of pentoses as do the hemicelluloses common in all cell walls.

Six-carbon atoms = hexoses: These are the often-discussed sugars important in respiration and photosynthesis and also as the building blocks of starch and cellulose.

Seven-carbon atoms = heptoses: These are uncommon except for one that is an intermediate in photosynthesis.

(3) The carbohydrates exhibit stereoisomerism. If, as shown in Figure 3–10, four different atoms or molecules are attached to a single carbon atom, there are two mirror-image ways to make the attachment. The mirror images are like the right and the left hand. At first they appear identical, but it soon becomes apparent that they cannot be superimposed upon each other. When tartaric acid was synthesized in the laboratory and allowed to crystallize, it was discovered that two kinds of crystals were formed—mirror images of each other. If crystals of one kind were dissolved in water, a beam of polarized light would rotate in one direction (e.g., right) by passing through this solution. A solution of the other kind of crystals caused the beam of polarized light to rotate in the opposite direction (left).

Sugar molecules, then, can differ from each other not only in the number of carbons and the number and position of attachment of hydrogens and oxygens but also in whether these are attached in a right- or a left-handed way. Study the four hexose sugars shown in Figure 3–9. In glucose, for example, the top (no. 1) carbon has only three things attached to it: a hydrogen, an oxygen (by a double bond), and the rest of the sugar molecule. The bottom (no. 6) carbon has only three kinds of things attached to it: two hydrogens, an —OH (hydroxyl—alcohol), and the rest of the molecule. The four carbons in between each have four different things attached to them, and they exhibit stereoisomerism.

The chain of carbons in sugars is really zigzagged because four things attached to a carbon atom will form the

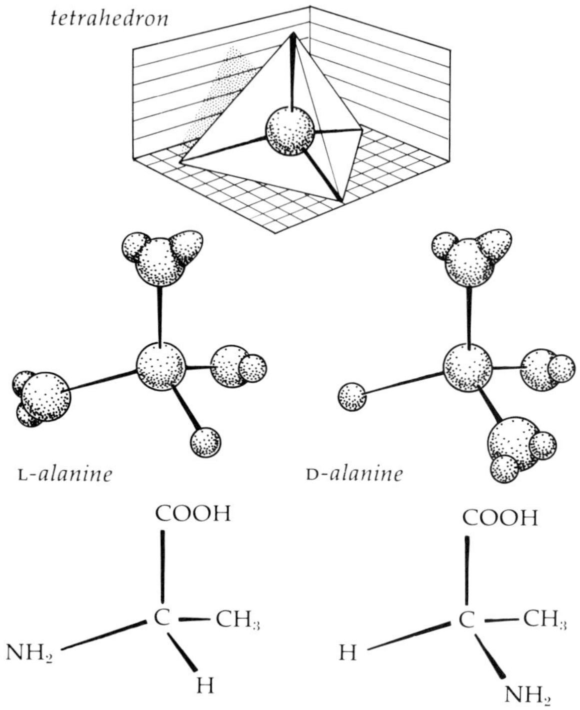

3–10 Tetrahedral models representing nonsuperimposable forms of the amino acid alanine. The top figure is an attempt to show a tetrahedron; the middle figures are expanded molecular models; and the bottom figures indicate positions of the various atoms.

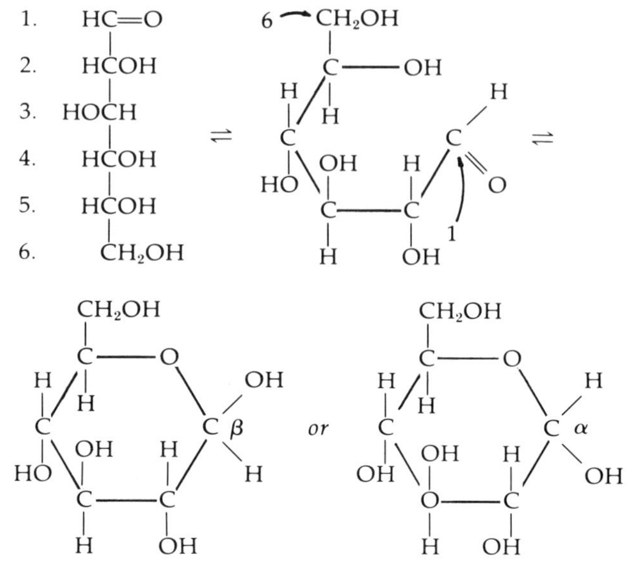

3–11 Formation of the ring structures of sugars, specifically D-glucose.

configuration of a tetrahedron (Fig. 3–10). This zigzag is difficult, if not impossible, to represent conveniently on a two-dimensional sheet of paper; so the carbon chain is written as though it were straight. If the so-called functional group (the —OH) on the bottom asymmetric carbon is on the right, the sugar is said to be of the dextro (meaning right) form. Thus glucose is correctly called D-glucose (sometimes called dextrose). If every atom on the glucose molecule were reversed, giving a complete mirror image of D-glucose, L-glucose would be the result, and the two molecules would exhibit identical chemical properties. But living organisms (because they use stereospecific enzymes) could react only with the D-glucose and not the L-glucose. If only the bottom —OH group were moved to the left side, we would not produce L-glucose but a completely different sugar (L-mannose). Changing the position of only one group alters the chemistry of the sugar, but changing all of them produces a chemically identical, mirror-image stereoisomer (same solubility, density, ordinary chemical reactivity, etc.).

(4) Many sugars in solution form rings rather than straight chains. In glucose the ring forms by the no. 1 carbon reacting through an oxygen with the no. 5 carbon to form a six-membered ring, as shown in Figure 3–11. When this occurs, the no. 1 carbon now has four different things attached to it (instead of only three); thus it becomes asymmetric. This means that there are two possible ways for the ring to form. As indicated in the figure, the —OH group can be either pointing up or pointing down. The two forms are called alpha (α) and beta (β).

(5) Since the chains are really zigzagged not straight, when a ring forms, it is distorted rather than flat. Model building has shown that the ring can be distorted either in the form of a "boat" or in the form of a "chair," as shown in Figure 3–12. When a sugar such as glucose is in solution, the molecules are continuously forming and breaking the ring, and these will occur in both the boat and the chair forms. *Thus all five forms of glucose are present in the cell: open chain, α-boat, β-boat, α-chair, and β-chair.*

(6) The rings may be attached to each other in various ways. This occurs by removal of a water molecule from between two of the sugar molecules, as shown in Figure 3–13.

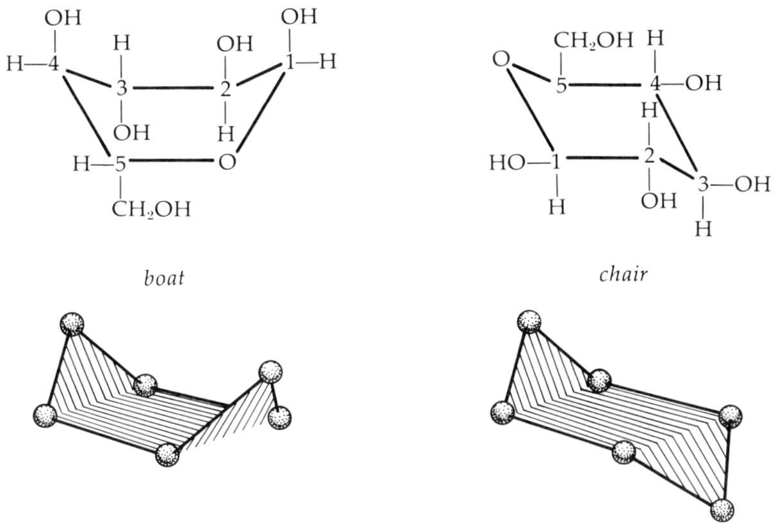

3–12 The boat and chair forms of a hexose-ring sugar molecule like glucose.

If two so-called residues (single sugar molecules) hook together, they form a disaccharide. Two glucose molecules form the disaccharide maltose, and a glucose and a fructose molecule form the disaccharide sucrose (in terms of quantity, probably the most important sugar in the plant kingdom). Three residues form a trisaccharide, and four a tetrasaccharide. When numerous residues hook together, they form polysaccharides.

Cellulose molecules consist of glucose molecules in the β-chair ring form (Fig. 3–14). Since the molecules hook together through the groups that were formed by closing the ring in the β configuration, it is said that the glucose molecules of cellulose are hooked together by the β linkage. A single molecule of cellulose contains 3000 to 10,000 glucose residues in an unbranched chain.

The relevance of all this chemistry to cellular structure and function is that when β chairs of glucose are constructed as models and hooked together in the β linkage to form a long chain, the chain proves to be almost perfectly straight. Furthermore, it can be seen that several straight-chain glucose molecules can fit nicely alongside each other, bonding to each other through the exposed hydrogen and —OH groups with weak bonds called hydrogen bonds. Such bonds, which do not share electrons as the covalent bonds do, form because of the slightly positive charge on the hydrogen atoms and the slightly negative charge on the oxygens. The result of this particular molecular arrangement is that cellulose is completely insoluble in water or organic solvents and that it has an extremely high tensile strength (exceeding that of steel). Since cellulose forms the tubular pathways through which water moves in the plant, it is certainly appropriate that it should be insoluble, and of course, its strength is also essential in providing support for the plant structure.

It is instructive to compare cellulose with starch. Starch also consists of glucose residues, but they are in the boat form and have the α linkage (Fig. 3–15). This produces *coiled* instead of straight chains. (Incidentally, the coil is of just the right dimensions to accommodate iodine molecules and produce the characteristic blue color used in the iodine test for starch.) One form of starch (amylose) consists of unbranched glucose chains (200 to 1000 residues) and is fairly soluble in water. The other form (amylopectin) has highly branched chains and is somewhat less soluble, although it too can be dissolved (or suspended) in hot water. Amylopectin is believed to have from 2000 upward to 200,000 glucose residues in each molecule. The relative insolubility of starch, even though it consists of sugar molecules that are highly soluble in water, is due to the large size of the molecules, especially amylopectin.

How can we account for the differences between starch and cellulose? After all, both are synthesized from the same kind of molecule—ultimately the open chains of glucose. The difference is due to the highly specific nature of the enzymes responsible for the synthesis of these polysaccharides. The enzyme that synthesizes cellulose recognizes only β chairs of glucose; whereas that synthesizing starch recognizes only α boats.

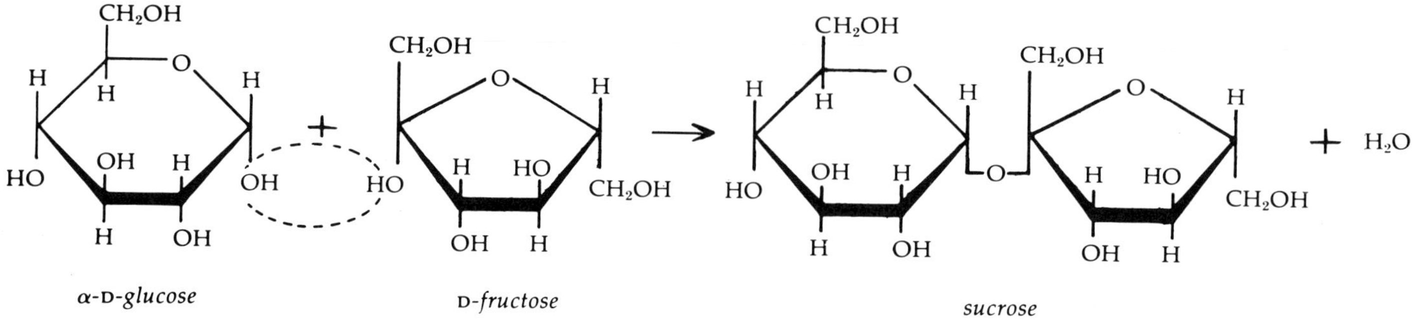

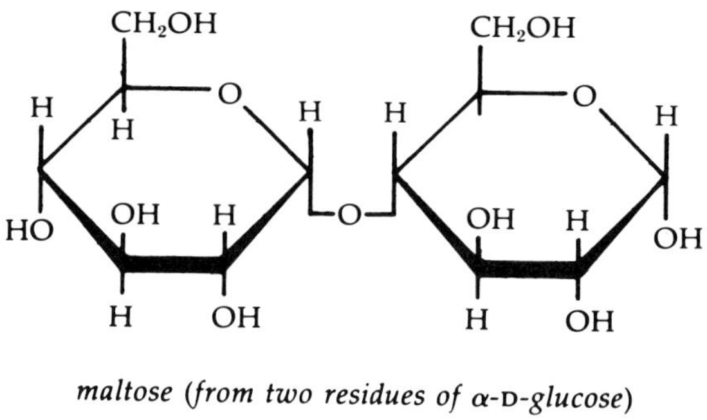

3–13 Formation of a bond between a molecule of glucose and a molecule of fructose, forming the disaccharide sucrose. The disaccharide maltose is also shown.

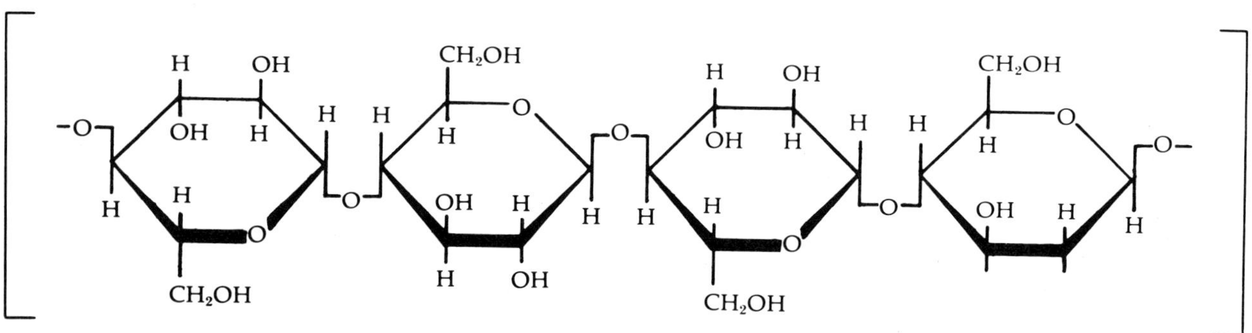

3–14 The molecular arrangement of α-D-glucose residues in cellulose. A segment of 4 residues in the molecular chain is shown.

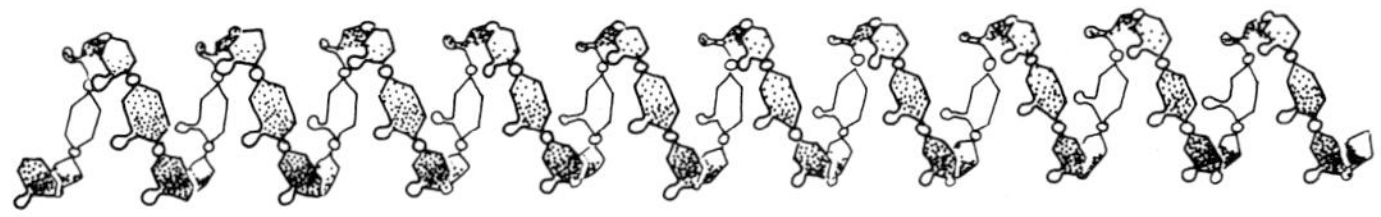

amylose

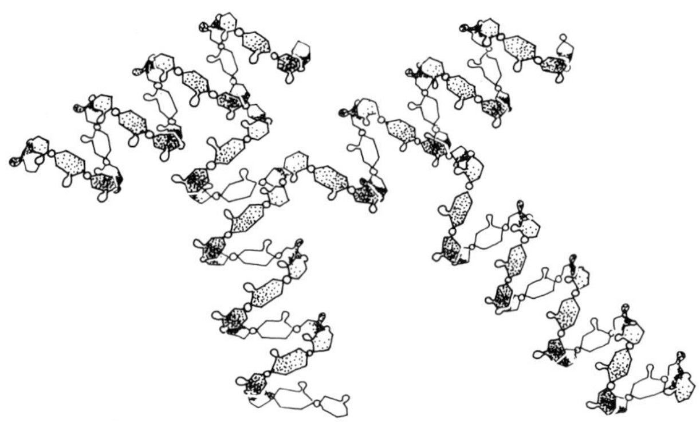

amylopectin

3–15 A schematic representation of the arrangement of glucose residues in the two kinds of starch molecules. Amylose and amylopectin are similar, except that amylopectin is branched.

4

Cell Membranes

4–A *Membranes of a cell as seen in a freeze-etched preparation. This procedure permits a surface view of membranes, thus revealing internal structures unseen in other methods. The large structure in the center of the page is a mitochondrion, while to the right we see a vacuole and to the left (running in an irregular band) the plasma membrane.*

Consider a freshwater algal cell in the stream or pond of our woodland acre. It faces two kinds of problems. First, if the enzymes and cellular organelles inside are to function properly, concentrations of certain minerals such as potassium and calcium must be considerably higher than they are in the water of the stream or pond; also many of the small soluble molecules of metabolism such as sugars and amino acids must be much higher in the cell than in the surrounding water (where they may be virtually nonexistent). Second, as we shall see in our discussion of osmosis, when solutes exist at a high concentration in one part of the system and not in the other (e.g., within the cell but not in the surrounding water), water will tend to diffuse toward the point of higher concentration of solutes. For our algal cell, this could mean swelling and eventually bursting. Fortunately, the rigid plant cell wall solves this problem.

An algal cell in the ocean has opposite problems. There is much sodium chloride dissolved in the water, and this could be damaging to the enzymes and cellular organelles within the cell. It must be kept out. Cells all over the earth are exposed to a great variety of environments in addition to fresh and salt water. The algal and bacterial cells growing in the films of water adhering to soil particles provide an excellent example. Many cells of a higher plant may be exposed almost directly to the air or the soil solution, but many others within the plant are exposed only to the water solutions that have entered the plant and have been somewhat influenced in their composition by this entry.

There are many fascinating situations in which the problems are much more extreme. The Dead Sea and the Great Salt Lake have saturated solutions of salt along with many other dissolved materials; yet algal and bacterial cells grow there, as they do in mineral springs in Yellowstone National Park and many other places. Bacteria are known to exist in strong alkaline or strong acid solutions or even in such unlikely liquids as gasoline or phenol—substances that inactivate most enzymes.

One of the principal functions of membranes is to control the passage of solutes in and out of the cell or cellular organelles, solving at least partially the problem just cited. Even in the ocean, potassium and calcium may build up to levels much higher inside the cell than outside. Sugars and amino acids are maintained within the cell, and sodium chloride and other toxic materials are excluded (depending, naturally, upon the cell we are dealing with). Many cells are killed in sea water, for example, while others are not. In any case, one of the most important interfaces between the organism and its environment (or between a cellular organelle and *its* environment) is the membrane.

The Cell Membrane

How do we go about studying the membrane? Since it is only about 75 A thick (7.5 nm), it can't be seen in the light microscope. But several lines of evidence clearly indicated that it must be there before it was seen with the electron microscope. Three approaches have been taken in studying membranes. First, one may study membrane properties—what penetrates, what doesn't, and how fast? Second, in a few cases it has been possible to study membrane chemistry. We can investigate things that dissolve it, and some membranes have been isolated in quantities large enough for conventional chemical study, namely, the membranes of chloroplasts, mitochondria, and red blood cells (the latter called erythrocyte ghosts). Third, with the electron microscope we can see cross sections of membranes and occasionally a surface view. The technique of freeze etching provides beautiful surface pictures (see Figure 4–A).

Most germane to this work is the concept of permeability. We speak of various substances such as water, cytoplasm, cell walls, and membranes as being *permeable to some other material* (typically water or some solute dissolved in the water). To quantitate the concept of permeability, we measure the solute crossing a unit area of membrane (or other medium) in a measured interval of time. A simple approach is to use a radioactive solute, since radioactivity measurements are fairly simple and extremely accurate when small amounts are used. Let us summarize four highly generalized results of such studies, realizing, of course, that there will always be exceptions at the detail level when one is considering different kinds of cells and different species.

Cells must be alive and metabolizing, or their membranes become highly permeable to virtually everything. If a cell is killed (high temperatures, poisons, etc.), or if its metabolism is inhibited (low temperatures, specific inhibitors, etc.), many of the ions and small molecules in the cell will begin to leak out through the membranes.

Dissolved gases such as oxygen, nitrogen, and carbon dioxide (O_2, N_2, CO_2), and water molecules penetrate cell membranes very rapidly.

Nonionized, water-soluble hydrophilic molecules penetrate cells at rates inversely proportional to their molecular size (i.e., small molecules penetrate fastest). As early as 1867, Moritz Traube in Germany made this observation and suggested the molecular sieve hypothesis or model. This hypothesis says that cell membranes must contain numerous minute holes in the size range of the molecules being studied. Clearly, the smaller the molecule, the greater its expected chances of encountering a pore to penetrate. Incidentally, even the smallest molecules penetrate much slower than water or dissolved gases—a central fact in our discussion of osmosis.

The penetration of lipid materials occurs at rates proportional to their lipid solubility. In 1895 Charles E. Overton in Switzerland observed that regardless of the size of such hydrophobic or lipophilic ("water-fearing" or "fat-loving") materials, penetration is much faster if they are highly lipid-soluble. Overton suggested the lipid model for membranes, which said that the entire membrane was essentially a lipid and that those hydrophobic substances that could dissolve in the membrane would penetrate best.

It occurred to subsequent investigators that the models of Traube and Overton could be nicely reconciled by a membrane that was a lipid matrix with watery holes. Lipid soluble materials would go through the matrix and hydrophilic materials through the holes. Recent work at the U.S. Department of Agriculture laboratories in Beltsville, Maryland, has supported this

balls and the other with black balls. The balls have perfect elasticity and are in motion, colliding and bouncing off each other in the same way that molecular particles do. Obviously, there is a certain probability that some of the black balls will go through the window instead of striking the walls. Of course, white ones will go in the opposite direction. Rates will depend upon the number of balls, their velocities, and the size of the window. As more and more black ones go through the window, the concentration (number per unit volume) decreases in the one chamber and increases in the other. Finally, there will be approximately an equal number of black balls and white balls in both rooms, and the chances that they will go from left to right will be about the same as from right to left. This is equilibrium, and it is *dynamic* because movement between the rooms is still occurring—but at equal rates in opposite directions.

Diffusion, however, is not strictly a function of the concentration gradient. More accurately, it occurs in response to a gradient in free energy of the particles. The concepts of free energy and chemical potential (although not equivalent are essentially so for our purposes) are developed after considerable preparatory instruction in courses in physical chemistry and thermodynamics. We will not give a complete definition of these terms, but we will summarize the way they are influenced by certain factors important in plant cells, thus providing some intuitive insight into the concepts.*

As *concentration* increases, so does chemical potential, as implied in the example of the white and black balls in the two connecting rooms. Both increasing *temperature* and increasing *pressure* increase the average kinetic energy (energy of motion) of molecular-sized particles; so increasing either or both of these two factors increases chemical potential. If we are discussing the chemical potential of water molecules, we see that their chemical potential is *decreased by the presence of substances dissolved in the water.*

Diffusion occurs whenever there is a gradient in chemical potential between one part of a system and another part, provided that there is nothing obstructing movement of molecules between the two parts of the system. Obviously, such a gradient could be

*The term diffusion pressure has been used in botany and plant physiology texts in essentially the same context as free energy or chemical potential.

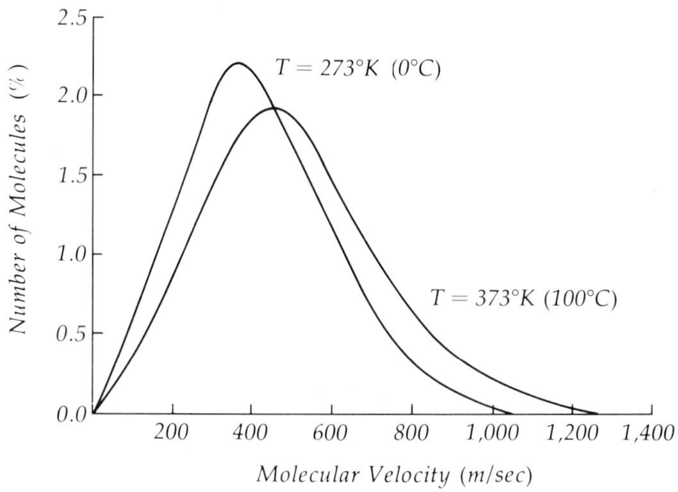

4–2 The Maxwell-Boltzmann distribution of molecular velocities in a gas (oxygen) at two temperatures 100° C apart.

established by differences in concentration, temperature, pressure, or dissolved substances.

Rate of diffusion will depend upon how steep the gradient in chemical potential is between the two parts of the system, that is, upon how great the differences are over some measured distance. The rate will also be influenced by how fast the particles are moving; increasing temperature increases their rate of movement and hence diffusion, and small particles diffuse faster because they are moving faster than larger particles at a given temperature. Rate of diffusion also depends upon the permeability of the medium through which diffusion is taking place—being faster when the medium is highly permeable—and permeability of the medium may in turn be influenced by other factors such as temperature, again illustrating the complexities of interaction between objects and their environments.

In the plant we are interested in diffusion of all kinds of things: ions, molecules, gases, etc. But when we are discussing *osmosis* in the plant, we are concerned only with the diffusion of water molecules. Hence it is convenient to talk about the chemical potential of water, since this is what determines diffusion of water. We will refer to this specifically as the water potential.

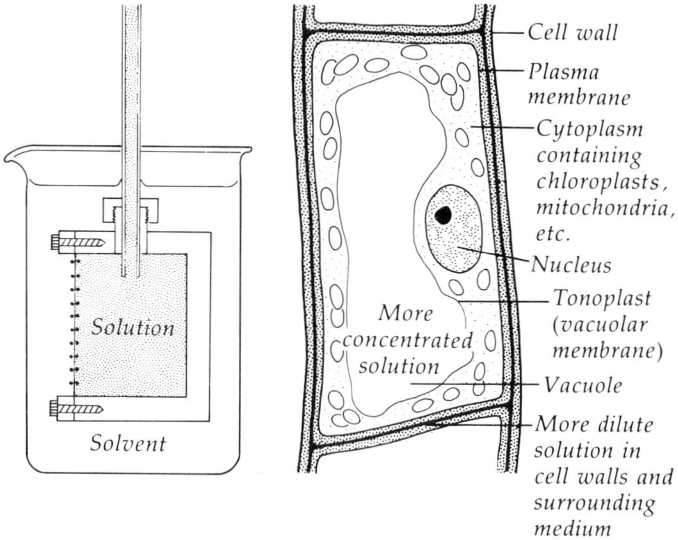

4–3 *A mechanical osmometer (left), and the plant cell as an osmotic system (right).*

Osmosis and Water Potential

The phenomenon of osmosis is a special case of diffusion. The essential feature of an osmotic system is the presence of a *membrane* that will maintain, for a reasonable time interval, concentration and pressure differences on either side. Thus, in the typical osmometer (Fig. 4–3), we might have a solution of sugar or any other solute separated from water by a membrane. As previously discussed, solutes do pass through membranes, but for our purposes we shall begin by assuming they do not.

The presence of solute particles in the water inside the membrane decreases the water potential of the water molecules. This establishes a water potential gradient, with high water potential on the outside of the membrane and low water potential on the inside of the membrane. Water, then, will tend to diffuse into the osmometer. (If *all* the solute is on one side of the membrane, its chemical potential difference across the membrane will be extreme, but since the solute particles cannot pass through the membrane, this gradient cannot be relieved by diffusion.)

As water molecules move into the system, there is a dilution of the solution inside. In many cases, this dilution is minimal and can be ignored. A much more important effect is the buildup of pressure, which may result from the rigidity of the cell wall, if the osmotic system under consideration is a plant cell, or from a column of water exerting pressure due to its weight in the typical laboratory osmometer. (Both instances are illustrated in Fig. 4–3.) As pressure increases, water potential also increases, and at some given pressure, the water potential inside the osmometer becomes equal to the water potential on the outside. When this occurs, there will no longer be a gradient in water potential; so diffusion (osmosis) will stop. Of course, water molecules will still be moving through the membrane because of their kinetic activity, but rate of movement will be the same in both directions.

Many incorrect interpretations of osmosis have been proposed in the past. The most common of these asserts that diffusion occurs in response to differences in water concentration inside and outside the membrane; it states that the solute particles take up space and hence decrease the concentration of the water. In many cases this is simply not true. The concentration of water (number of water molecules per unit volume) does not change in a few instances where the addition of solute does not change the volume of the resulting solution. In most cases, concentration of water is slightly influenced by the addition of solute, but in no case can this influence explain the quantitative aspects of osmosis. Osmosis depends upon the number of solute particles compared to the number of water molecules (the mole fraction), and it is virtually independent of possible volume changes in the water caused by the addition of solute.

Incidentally, it is unclear why osmosis should depend upon the *number* of solute particles but be independent of their nature. A sodium ion, for example, has essentially the same effect as a nonionized sugar molecule and almost the same effect as a protein molecule. This seems incongruous, but it is supported by nearly a century of experimental data.

Reduced to the barest facts, osmosis is easy to understand. It is a special case of diffusion, and diffusion of water molecules occurs down a water potential gradient from points of high water potential to points of low water potential. At equilibrium, water potentials are equal throughout the system. Two factors are important in establishing the water-potential gradient that accounts for osmosis: the difference in concentration of solutes in different parts of the system (this situation being maintained by the membrane), and the pressure differences between different parts of a system (these differences being maintained in plants by the presence of the cell wall). Two factors might be important in osmosis but normally are ignored: water concentration is virtually equal throughout any plant system, and temperature is also fairly uniform or simply ignored (but perhaps shouldn't be). Of course, there is more to understanding osmosis, particularly the application of measurements and quantitative concepts to the process.

As we shall see in Chapter 33, osmosis is extremely important to plants—and to animals, for that matter. It confers turgidity upon plant cells. Water in the stems of plants, especially in the trunks of tall trees, is under tension—it has a negative pressure. As pressure raises water potential, tension lowers it. Hence, if water potentials within the cells weren't even lower, water would move out of the living cells into the stem or trunk. The plant would wilt. But because the cells contain solutes, their water potentials are even lower than those in the stem, and they do not wilt. The tensions in the water of stems is an unavoidable necessity if the plant is to be tall; so osmosis is also essential for tall plants.

Again, we must emphasize the importance of the cell wall in all this. Without that strong enclosure, cells with water potentials as low as those in plants would swell and burst in a pure water medium. This is called lysis, and it happens to blood cells, for example. They are osmotically balanced to match the serum in which they normally occur. The opposite problem is plasmolysis, in which the medium outside the cell has a lower water potential than the cell; hence water moves out of the cell into the medium. When this happens to a plant cell, the protoplast shrinks away from the cell wall. This happens to freshwater algal cells placed in sea water or to bacterial cells placed in salt or sugar solutions of high concentrations (which is why honey or salt pork doesn't spoil). If salt is poured on the soil around the base of a plant, water will move out of the plant into the salty soil, since the water potential there is lower than within the plant. The cells will plasmolyze, and the plant will wilt and often die. Clearly, there is an important relationship between a plant and its osmotic environment.

But some plants (called halophytes) do grow with their roots in salt water. How do they manage? Sometimes they absorb much of the salt but don't seem to be damaged by it, as most plants would be. This way they maintain an internal water potential equal to or below that of the external medium. One question is how their protoplasm avoids damage by the salt. We are discovering that their enzymes can tolerate it. On the other hand, the mangrove tree of the tropics grows with its roots in the ocean, but water in its trunk has much less salt than sea water. Why doesn't the water move out of the trunk into the sea? Apparently because it is under tension within the cells of the wood in the trunk, to the point where the water potential is lower than in sea water so that the plant continues to absorb water from the sea (filtering out much salt). The living cells in the plant (not the dead cells in the wood) do contain more salt than sea water; so they are not under tension.

There is one more important complication to all this. We can extend our concept of water potential beyond the liquids in cells and around them. We can speak of the water potential of the air, for example. Since water potential is almost always (except when relative humidity is at or close to 100%) far lower in air than it is within the plant, water will nearly always tend to evaporate from plants into the air (transpiration—see Chapter 32). Furthermore, we can speak of the water potential of a hydrating surface such as cellulose or protein (see End Paper 4–2). Water molecules tend to become adsorbed to such a surface and to a degree that can be expressed by the water potential. Sometimes plants tend to take up water primarily by this mechanism (hydration) rather than by osmosis. Hydration always plays an important role in keeping water in the cell walls. The forces involved are often extreme, perhaps on the order of a hundred times as great as those that can be developed by osmosis. The forces of hydration are also highly important in soils, where clay particles tend to adsorb water strongly.

Active Transport of Solutes across Membranes

When the special metabolic environment in the cell is upset, death results. So far in our discussion of osmosis, we have ignored the movement of solutes into and out of cells, but now it is time to consider that phase of cell life. The problem, as stated in the preceding section, is that things tend to diffuse in the direction of equilibrium, with diffusion finally resulting in

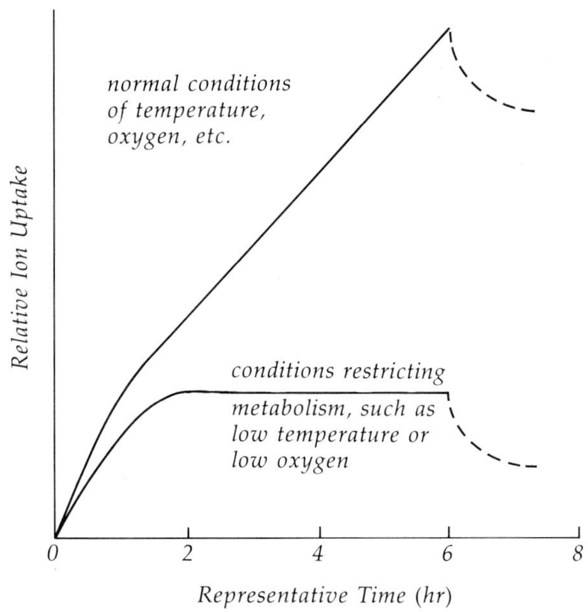

4–4 Passive and active uptake of ions as a function of time. The top, sharply slanting curve is obtained under conditions in which active metabolic accumulation is possible. The lower curve, leveling off after about 2 hours, is obtained when accumulation is largely prevented. The dashed lines indicate the release of ions when tissues are placed in water. About the same quantity is released in either case, these being the ions taken up passively and not adsorbed onto negative sites in the tissue.

equal concentrations (chemical potentials) throughout a system. Thus the environment inside the cell should become like the overwhelmingly large environment outside. This running down toward fatal equilibrium will occur unless something resists the trend. That something is the living membrane.

The phenomenon of control of ion uptake is demonstrated by a simple experiment. Tissue (e.g., a section of carrot, potato, or a seedling root system) is placed in a limited volume of solution, and the uptake of ions (e.g., potassium or radioactive rubidium) is measured as a function of time (Fig. 4–4). If the metabolic processes of tissues are inhibited (e.g., by low temperature, low oxygen, certain metabolic poisons, etc.), uptake reaches equilibrium in a relatively short time (e.g., 30 min.). This is a passive uptake and readily follows the predictions of thermodynamics. If the tissue is subsequently placed in pure water, the ions will diffuse out into the water (except, perhaps, for a few that have taken part in ion exchange in the cell walls). If the tissue is allowed to respire, however, it will accumulate solutes for a long interval of time and to internal concentrations far higher than those of the outside medium. If this tissue is returned to pure water, the accumulated ions are not released, nor are they exchanged if the tissue is placed in another solution.

How does this process of active uptake or accumulation work? In another kind of experiment, uptake of ions is measured for a standard interval of time and as a function of the concentration of ions in the medium (Fig. 4–5). The result is a saturation curve; that is, with increasing concentration in the medium, ion uptake increases but only to a point. Beyond that saturation point, increasing concentration does not result in increased uptake. Such an experimental result suggested the concept of a carrier substance that is part of the membrane. Perhaps the ion is bound to this substance on the outside of the membrane, using cellular metabolic energy for the binding. The carrier with its attached ion would then diffuse across the membrane where the ion is released (perhaps under the control of an enzyme). In such a situation, if binding can occur only on the outside and release only on the inside, we could understand both the lack of exchange and the saturation curve. Saturation would be reached when all of the carrier molecules were combined with ions (saturated). Increasing concentration of ions would have no effect because there would be no more carrier molecules with which they might react. Using actual experimental data, this model of a carrier based upon a saturation curve can be supported mathematically, although we will not discuss the details here. (Actually the mathematics were developed for a perfectly analogous situation concerned with the manner in which enzymes function—see Chapter 5.)

Emmanuel Epstein, then at the United States Department of Agriculture laboratories in Maryland, was among the first to obtain the saturation curve and apply the mathematics. In further experiments, he studied the uptake of one ion (e.g., potassium) in the presence of another (e.g., rubidium). Each of the ions inhibited the uptake of the other as though the two were competing with each other for a site on the same carrier molecules. Epstein found that rubidium would

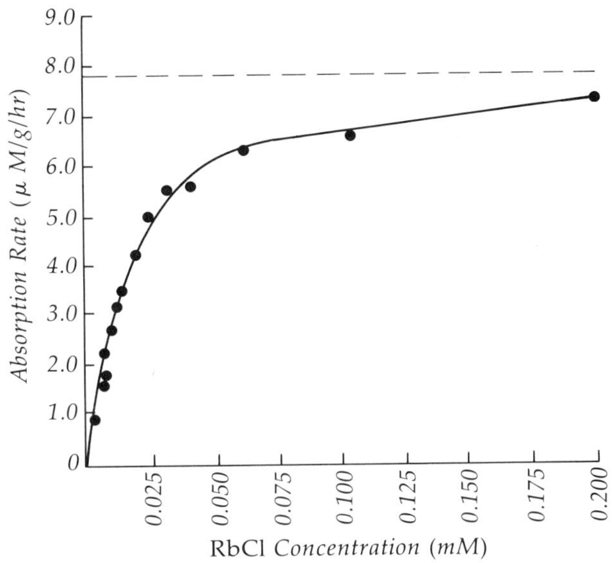

4–5 Ion absorption as a function of concentration in the outside medium, showing rapid uptake with increase in concentration up to the saturating concentration (in this case, about 0.050 millimoles/liter). The data are taken from the absorption rate of rubidium ions by excised barley roots.

compete with potassium and cesium but not with sodium, for example. Epstein's competing groups are summarized in Table 4–6. Organic compounds such as amino acids and sugars are also accumulated by cells. If they were not, they would tend to leak out of the cell to the detriment of cellular metabolism.

There has been recent and exciting work on the carrier concept. Carrier molecules are being isolated, and the entire process is being related to important metabolic steps and chemosmotic events. Accumulation, for example, appears to be closely related to production of the energy-rich compound ATP (see Chapter 5) in chloroplasts and mitochondria. Several enzymes are being localized on cell membranes. Obviously, we have much to learn about the arrangement of protein and other molecules in membranes to account for movement in one direction only. The picture is certainly more complex than implied either by permeability studies or by electron micrographs.

Sometimes, cells are capable of taking up large molecules such as those of proteins or nucleic acids, or even virus particles. Occasionally, enzymes are secreted by roots, and many microorganisms secrete enzymes, as do the cells of animal digestive systems. How do such large molecules get into or out of cells? Apparently, one way for these to enter is the process of pinocytosis. Electron micrographs reveal how particles may be engulfed in an invagination of the cellular membrane. The membrane coalesces inside the cell in such a way that the particle is inside of the cell and surrounded by the membrane. Eventually, the membrane is digested or otherwise destroyed, and the large particle is then free in the cytoplasm (Fig. 4–7). It is easy to see pinocytosis in *Amoeba,* but apparently even some cells of higher plants exhibit pinocytotic invaginations. Certainly, the plasma membrane is not always stretched tightly against the cell wall.

We should note the importance of the vacuole in all this. The vacuole is almost as distinctive to plants as is the cell wall. The tonoplast (vacuolar membrane) does not have properties identical to those of the plasma membrane, as indicated not only by its appearance in electron micrographs (although the two membranes are nearly identical) but also by numerous permeability experiments. The concentrations of salts, pigments, sugars, etc., in the vacuole are often considerably higher than in the cytoplasm. The high concentration in the vacuolar sap maintains cell and consequently plant turgidity by osmosis. Sometimes, the materials contained in the vacuole would actually be harmful to the surrounding cytoplasm. In citrus fruits, for example, the vacuolar sap is so acidic that it could damage the cytoplasm. Occasionally, crystals occur in the vacuole. Calcium oxalate crystals are especially common. Recently, a few enzymes have been

*4–6 Groups of ions that compete with each other for sites on a carrier. For example, rubidium, potassium, and cesium compete with each other but not with sodium.**

Ions that compete with each other	Ions that do not compete with those
Rb^+, K^+, Cs^+	Na^+
Cl^-, Br^-, I^-	NO_3^-
Ca^{++}, St^{++}, Ba^{++}	Mg^{++}
$SO_4^{=}$, $SeO_4^{=}$	NO_3^-, $PO_3^{\equiv}$

*Data for barley roots summarized by E. Epstein, *Ann. Rev. Plant Physiol.* 7:1, 1956.

detected inside of the vacuole, although we don't know what they are doing there. There has been much discussion of how the vacuole forms, but we are still not sure. The best suggestion is that numerous minute vacuoles form in developing cells and then grow and coalesce with each other as the cell grows. Indeed, cell growth is largely a matter of uptake of water and solutes by the vacuole, that is, vacuolar growth.

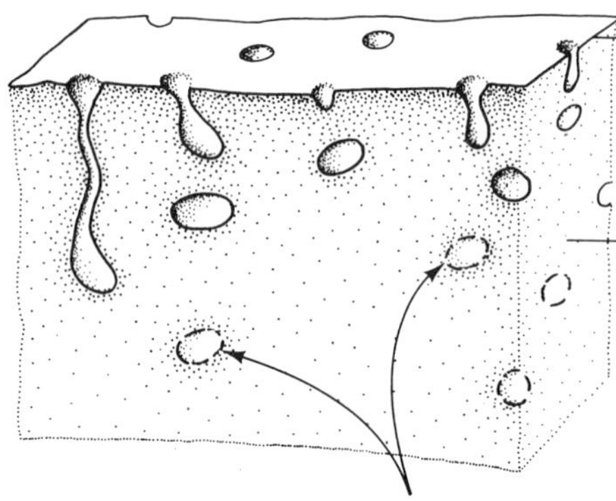

4–7 Pinocytosis, showing the origin and breakdown of pinocytotic vesicles from the plasma membrane.

Some Final Thoughts

Not many years ago, we had a most fuzzy concept of the membrane. Although the picture is still not distinct, it is getting sharper, and it appears amazingly complex! Even the cell wall, whose functions are relatively simple, has a high order of complexity. But consider the membranes with their many convolutions. The term unit membrane implies simplicity, and certainly the distinct lines of electron micrographs are quite simple. Yet the membranes are not simple; they are highly complex. They must contain or secrete enzymes to form the wall, and they must include the complex enzyme carrier systems of accumulation and the pores and matrix that somehow account for the lipid-sieve observations.

The membranes are closely associated with other microstructures, increasing the complexity of our developing picture. Consider the plasmodesmata, the connections through the cell wall. Membranes appear to be continuous from cell to cell; thus we may be misleading ourselves when we think of a single cell (a protoplast) as an individual unit. Perhaps, the biggest mystery of all concerns the ER. It apparently connects the plasma membrane to the nuclear envelope and possibly even to the tonoplast. It contains protein-synthesizing ribosomes. Obviously, it is a part of the cell's membrane system, but just what role does it play?

Although we have learned a great deal about the cell, we still couldn't build one.

End Paper 4–1 Fats and Amino Acids

Fats form a complex group. Different substances called lipids or fats may have little in common regarding their molecular structures, but all are soluble in such solvents as acetone, ether, or benzene. The fats in membranes are the so-called true fats and oils and their derivatives. They are somewhat more complex than the polysaccharides, since they are made up of more units. Yet their chemistry is not difficult to understand.

True fats consist of fatty acids (Fig. 4–8) combined with glycerol. Glycerol is a so-called triple alcohol, consisting of 3 carbons with an —OH group attached to each:

$$\begin{array}{c} H_2COH \\ | \\ HCOH \\ | \\ H_2COH \end{array}$$

The fatty acids consist of long chains of carbon, always with an even number of carbon atoms in the chain. (Fatty acids

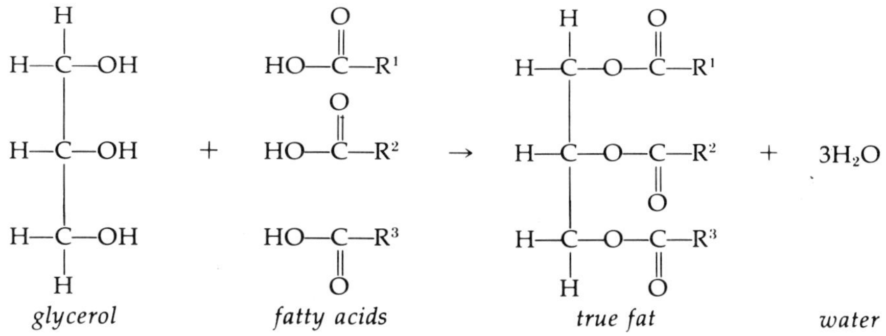

4–8 The formation of true fats by esterification between glycerol (a triple alcohol) and fatty acids.

are built up two carbons at a time.) Usually, there are 12 to 18 carbons in a fatty acid chain. At one end of this chain is a carboxyl (acid) group. The chain may be *saturated,* in which case each carbon has its full complement of hydrogens, and there are no double bonds between carbons—as there are if the fatty acid is unsaturated. There are many kinds of fatty acids, and they vary in the number of carbons and in the degree of saturation (Fig. 4–9). A few rare ones also have branched chains, but most of the common ones have only straight chains.

Fatty acids are esterified to glycerol (Fig. 4–8) by removal of water between a carboxyl and an alcohol. Typically, the three fatty acids esterified to glycerol are different, although two or even all three of them can be alike.

Two special lipids are especially important in membranes. In the phospholipids phosphoric acid is esterified to one of the carbons of glycerol, and frequently something else is esterified in turn to the phosphoric acid. Figure 4–10 illustrates the phospholipids. Incidentally, the phospholipids seem to be especially important in nerve cells, particularly in the brain.

Glycolipids are fats in which a sugar has been attached to one of the carbons of the glycerol. Usually, the sugar is galactose or glucose, but it may be a disaccharide. The glycolipids are especially important in chloroplast membranes.

Most common proteins are made up of 20 amino acids (which may be modified slightly in some proteins to give a larger number). There are five important things to remember about amino acids.

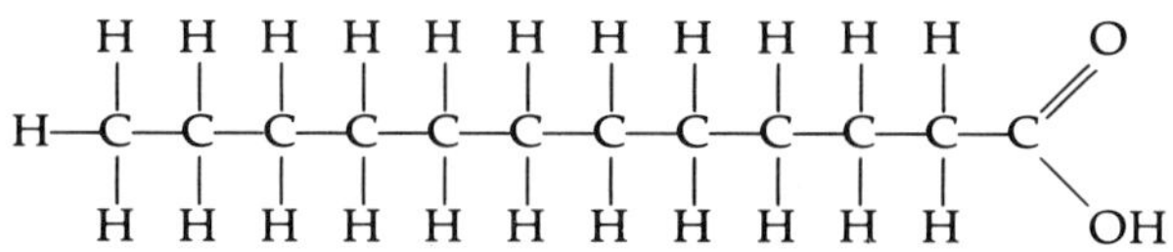

lauric (C_{12})

$CH_3(CH_2)_{16}COOH$

stearic (C_{18})

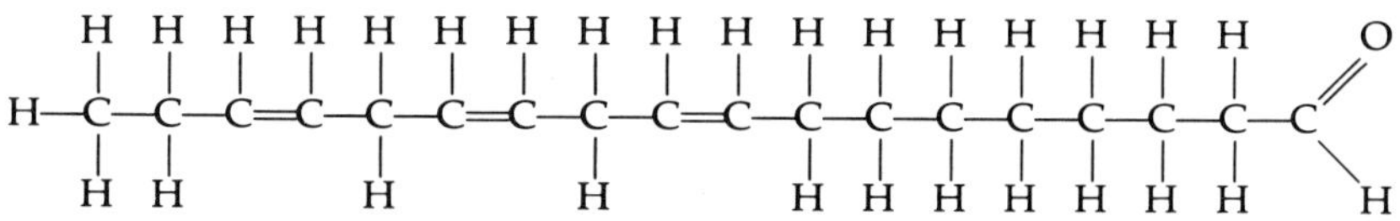

linolenic (C_{18})

4–9 Some representative fatty acids. Lauric and stearic acids are saturated; linolenic is unsaturated, as indicated by the double bonds,

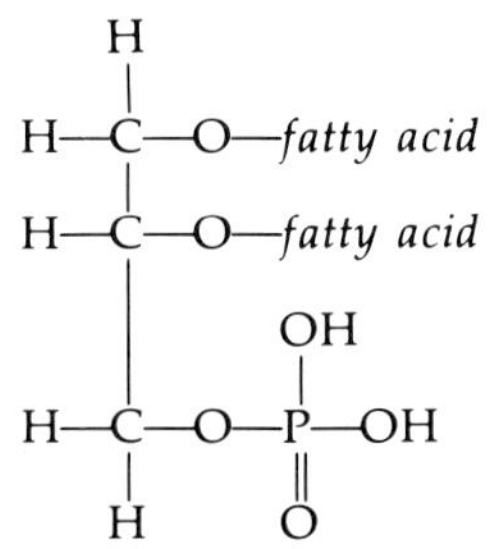

4–10 A generalized molecular structure of phosphatidic acid, a phospholipid. A phosphate group has replaced one of the fatty acids on glycerol.

Amino acids are acids because they have a carboxyl group (—COOH, as in glucuronic acids and fatty acids).

Amino acids are amino compounds because they have an amino group (—NH_2) *on the carbon next to the carboxyl group* (Fig. 4–11). An amino group consists of a nitrogen attached to a carbon with two hydrogens attached to the nitrogen.

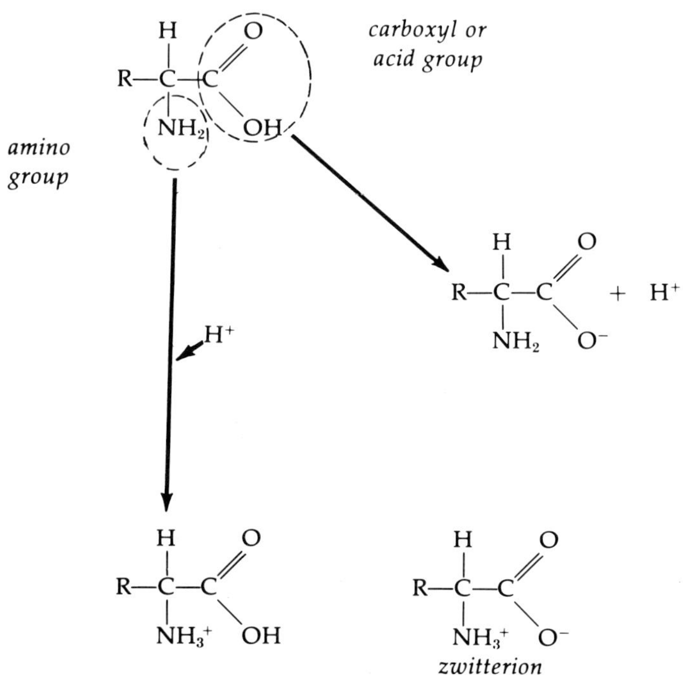

4–11 A generalized molecular structure of amino acids. R stands for characteristic parts of the molecule, as shown in Figure 4–12. The carboxyl is an acid, being able to donate a proton; the amino group is a base, being able to accept a proton.

An amino group can accept a hydrogen ion (H^+) to become —NH_3^+. (A hydrogen ion—a hydrogen atom without its electron—is a proton. The hydrogen atom is so simple that its nucleus consists of only one proton, while other atoms have two to many, along with neutrons. Certain hydrogen isotopes also have neutrons: deuterium, 1 neutron, and tritium, 2 neutrons. Incidentally, tritium is radioactive.)

A carboxyl group (—COOH) can donate a hydrogen ion (a proton) to become —$COOH^-$. By definition, acids are materials that can donate protons, while bases are materials that accept protons. *An amino acid, then, is part acid and part base.* Under proper conditions, the two parts will neutralize each other. This acid-base nature of amino acids is extremely important to life functions. Amino acids have been called double ions, or by the Germans *zwitterions* (mongrel or hybrid ions).

Amino acids differ from each other by differences in the part of the molecule represented by R in Figure 4–11. There is an almost infinite number of possibilities for the R part of the amino acid; hence there are many more amino acids than the 20 commonly found in most proteins.

In examining the amino acids of Figure 4–12, note the following features: Five of the amino acids have only carbons and hydrogens in the R part. Three of these have branched carbon chains. Two of the amino acids have additional amino groups or their derivatives. These are especially basic amino acids. Four of the amino acids have additional carboxyl groups (in two cases, so-called amide derivatives) and are especially acidic. Two of the amino acids have hydroxyl groups, although in one case this forms an ether (which we have not discussed). Two of the amino acids have benzene rings in the R part; one of these has a hydroxyl group. Four of the amino acids have a basic (proton-accepting) nitrogen. Obviously, these are fairly complex molecules. Three of the amino acids contain sulfur, and cystine is a combination of two molecules of cysteine with two hydrogens removed. The 20 amino acids originally synthesized into protein do not include hydroxyproline or cystine. These are additions to the completed protein.

Amino acids form stereoisomers, as noted in the discussion of sugars. The second carbon has four different things attached and is therefore an asymmetric carbon. While the sugars are designated D, the naturally occurring amino acids are all of the L form (L meaning levo or left).

In proteins, amino acids are combined with each other by forming peptide bonds. This happens as the amino group of one amino acid becomes attached to the carboxyl group of another by the removal of a molecule of water, illustrated in Figure 4–13.

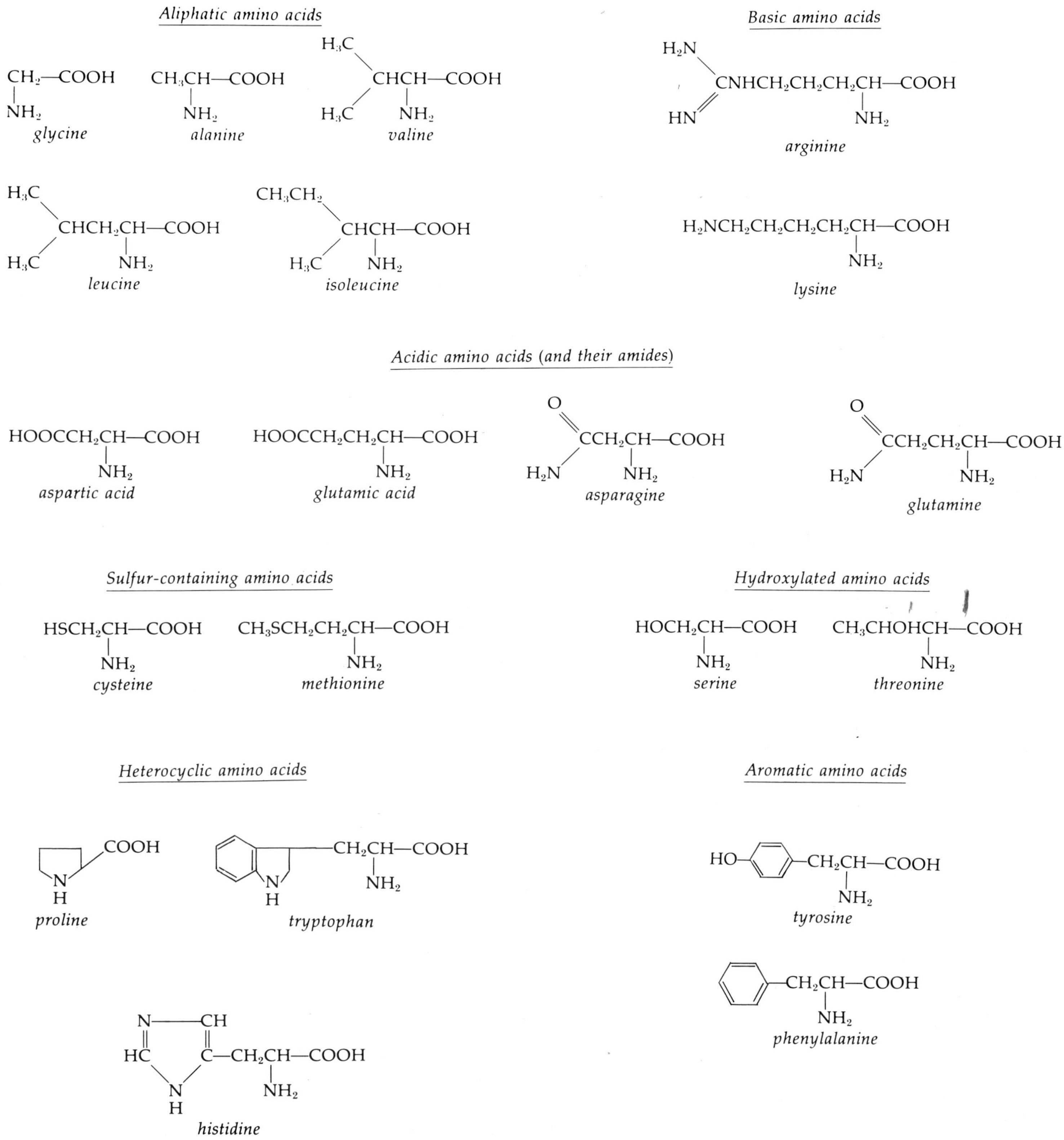

4–12 *Molecular structures of the 20 amino acids.*

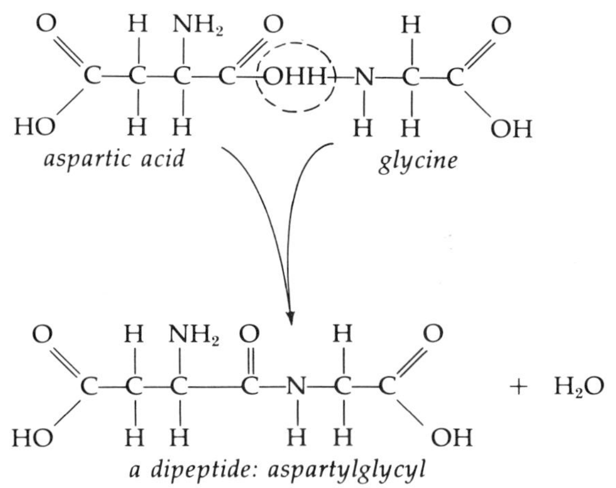

4–13 Formation of a peptide bond between 2 amino acids by removal of a water molecule. Note that the dipeptide in this case has 2 free carboxyl groups and 1 free amino group.

End Paper 4–2 Colloids: Sols, Gels, and Hydration

Protoplasm is characterized by the colloidal state, which exhibits a large quantity of surface or interface compressed into a very small volume such as a cell. Imagine (as in Fig. 4–14) cutting a cubic centimeter of solid material into minute cubes about the size of protein molecules. The original cube had 6 cm^2 of surface, but the total surface of all the minute cubes amounts to about 600 m^2!

Colloids (from the Greek word for glue) are particles larger than the atoms and molecules that form true solutions but small enough to remain suspended in water. Unlike particles of about a micron or more in diameter, colloids do not settle out in response to gravity because they are continually being bombarded by the molecules of water in which they are suspended. (Nor are they large enough to be trapped by filter paper.) Since, at any given temperature, the molecules have an average energy but vary greatly in the energy of individual molecules (some moving very slowly, others more rapidly—their velocities constantly changing as they collide with each other, there is a good chance that a colloidal particle will be struck at any given instant harder on one side than on the other. With proper illumination, colloidal particles can be seen in the light microscope as points of light (not as particles with shape), and these points of light are seen to be continually dancing around with a rapid, erratic motion called Brownian movement (after its discoverer Robert Brown). This is the most direct evidence that we have supporting the kinetic theory of matter.

There are several different kinds of colloidal systems. An emulsion consists of minute droplets of liquid (sometimes larger than the size range just described) suspended in another liquid with which it is immiscible—droplets of fat in water (cream) or droplets of water in fat (butter). Some of the properties of cell membranes are similar to those of emulsions.

Protoplasm exhibits many of the properties of a colloidal sol-gel system. In the sol, colloidal particles (as of the molecules of gelatin, a protein) are suspended in water. The system is virtually as fluid as the water itself—gelatin dessert when it is hot, for example. As the gel forms, the colloidal particles begin to contact each other at various points, resulting in a system of rather tenuous rigidity (as in the cooled and set gelatin dessert). Solutes diffuse through the gel about as fast as they will diffuse through perfectly still water; so we think of gels in terms of the brush-pile theory. The colloidal particles such as the gelatin molecules are thought of as being elongated like sticks. In the sol state these are unable to contact each other due to their kinetic activity. In the gel state they contact at a number of points, like the sticks in a pile of brush, but there can be a free movement of solute particles among them—as a mouse can dart through the pile of brush almost as rapidly as he could run if it were not there.

In some systems such as the ameboid forms of the slime molds discussed in Chapter 20, the protoplasm seems to alternate in the cell from a sol form in which it can flow freely to the more rigid gel form, accounting for movement of the

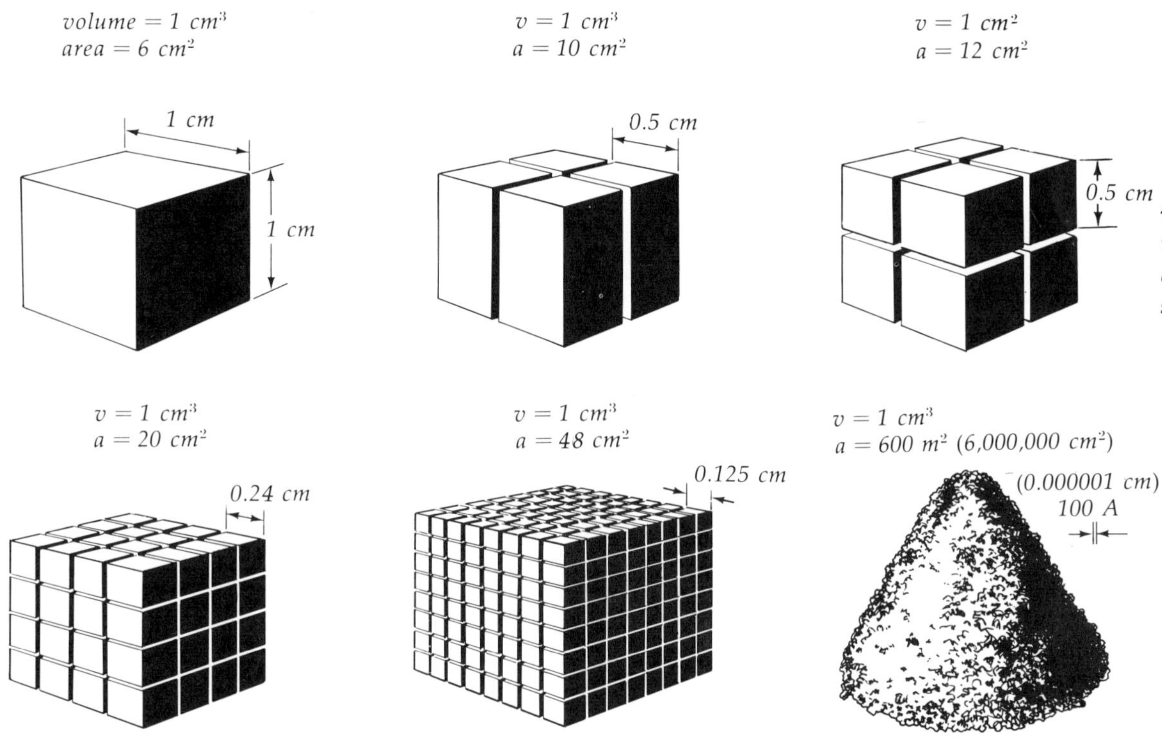

4–14 An illustration of the relationship between surface area and volume as particle size decreases.

organism. In virtually all cells, protoplasm has some of the properties of sols and gels—protein molecules are typically large enough to be in the colloidal range, bestowing these properties upon protoplasm. The middle lamella between cells, consisting largely of pectin, is another example of a gel. (Pectin extracted from green fruits where it occurs in rich supply is used in making jams and jellies.) The cell wall itself is similar in many ways to a water-saturated gel. The micells of cellulose (Chapter 3) would be similar to the sticks in our brush pile. This structure allows free passage of solutions through the cell wall.

The large surface areas of colloidal systems have important consequences for cells. In the protoplasm the activities of enzymes are related to the fact that they are proteins in the colloidal size range, exhibiting large surface areas for the control of metabolic reactions. Proteins and cellulose typically have surfaces covered with many negative charges. This makes them highly attractive toward water (they are highly hydrophilic), and water molecules tend to become adsorbed on the surface in a process known as hydration or imbibition. Many characteristics of living organisms are determined by the fact that they contain these hydrated surfaces. This is especially true in plants because of the cell wall and middle lamella.

In hydration the water molecules are attached by strong forces to the hydrating surfaces. When a colloidal hydrophilic material such as cellulose or various protein gels (e.g., those associated with the brown algae) becomes hydrated, a measurable quantity of heat is given off (the heat of hydration). As the water molecules attach to the surface, they occupy volume, and the entire system tends to swell. If it is confined, extremely high pressures can develop. This is the source of the pressures produced by growing roots or fungal hyphae that are capable of breaking rocks or concrete.

The charges on the surfaces of colloidal material also account for the phenomenon of cation exchange mentioned at the end of Chapter 3. Certain ions attach more tenaciously than others, which they will replace on the colloidal surface when both are present at equal concentrations. Water is "hard" because of the presence of calcium ions that react with soap to make an insoluble ring around the bathtub. A water softener contains an ion exchange colloid that is first "charged" by passing a high concentration of sodium chloride (salt) through it. Sodium ions become attached to the colloid, and excess sodium chloride is then washed out. When the "hard" water is passed through, the calcium ions (Ca^{++}) become attached to the colloid, releasing the sodium ions (Na^{+})—two for each calcium ion. Since sodium ions don't form an insoluble precipitate with soap, the water is said to be "soft."

5

Cell Energetics: Enzymes and Metabolism

The Enzymology Laboratory

In the 1920s ways to isolate and purify enzymes were discovered. This opened a new era in the study of the cell because enzymes are the proteins that control the metabolic reactions in all living cells. They are the key to understanding thousands of problems in plants, animals, and bacteria. Consequently, the procedures developed to study and isolate enzymes occupy a central position in modern laboratory work in cell biology. These methods have been refined to the point where enzymes can be extracted from tissues and then purified until, in some cases, one enzyme is isolated from all the others—sometimes as a pure crystal. In this photo essay we will present just a few of the many important procedures and pieces of equipment used in such purification.

A Leaf tissue is first cut into small pieces, then homogenized in a blender, which spins razor-sharp blades at a high velocity. The glass tubes in the lower left of the photo are plungers that can also be used to homogenize tissue; the large stainless steel vessel above them is used with another kind of blender. In homogenizing the tissue the enzymes are released from the cells and brought into solution.

B Next, the homogenate is filtered to remove cell debris; the enzymes pass through the filter paper with the liquid.

C The enzymes can now be concentrated and purified. The first steps usually involve fractional precipitation, in which various salts, acids or bases, organic solvents, etc., are added to the solution of enzyme molecules obtained by filtering. Addition of these materials causes certain enzymes to precipitate out of solution, while others do not. They can then be separated by filtering or centrifugation. Usually, these procedures must be carried out at low temperatures, and care must be taken to maintain suitable concentrations of added materials, or proteins will be denatured (changed so enzymatic activity is lost). After these initial purification steps (applied to nearly all proteins), other, more sophisticated techniques can be applied. One of these is dialysis, as shown in the photograph. An enzyme is placed in a cellophane tube tied at both ends, and this is hung in a container of water or solution. Small molecules pass through the cellophane into the outside liquid, which may be changed at intervals, while the enzymes, which are large molecules, remain inside.

D Another method of purifying enzymes is through column chromatography. In this procedure a mixture of substances is allowed to pass through a column of absorbent material, shown on the left. The different substances in the mixture will be adsorbed to different degrees and hence separated as they come out at the bottom. The entire apparatus is maintained in a refrigerated chamber, and the samples are collected automatically in a series of test tubes.

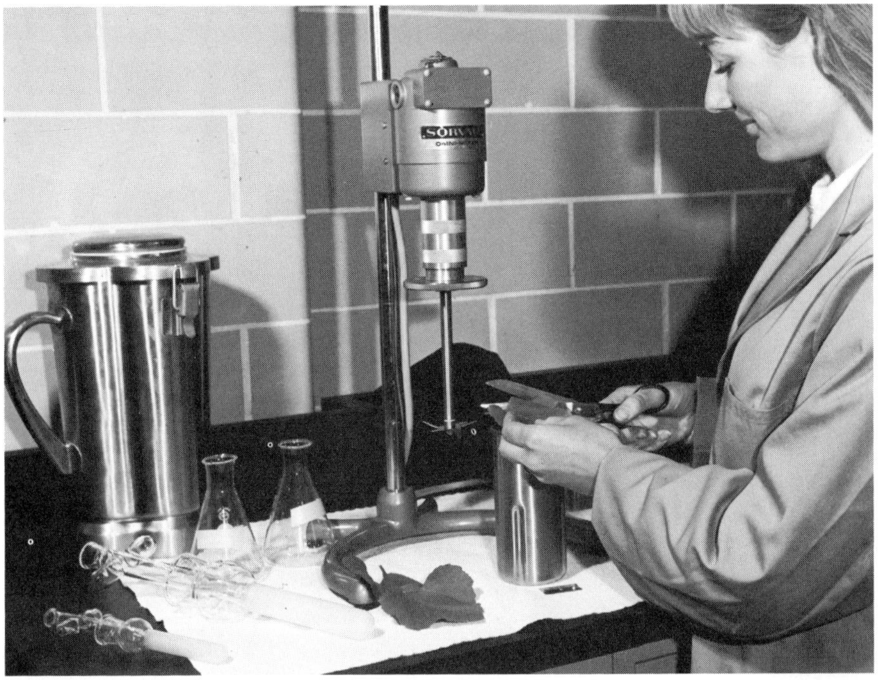

A

SARGENT
MAGNETIC STIRRER
ON
OFF
DESIGNED AND MFD. BY
E. H. SARGENT & CO.

C

B

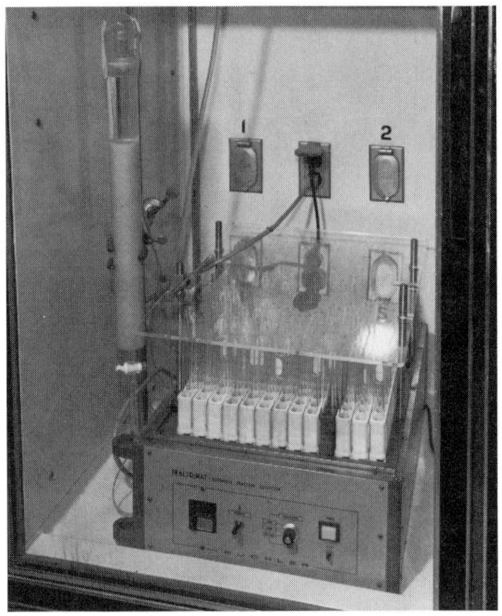
1
2

D

E Proteins can then be separated by allowing them to migrate in an electrical field. Protein molecules with a high charge will migrate more rapidly toward the pole with the opposite charge than will the proteins with a lower charge. Dr. John O. Evans is preparing a gel electrophoresis apparatus in which proteins migrate through the column of polyacrilamide gel in response to the electrical potential applied from one end of the gel to the other.

F Columns of gel stained so that proteins appear as dark bands in the column.

G Pure enzyme crystals of human-liver UDP-glucose pyrophosphorylase.

H Pure enzyme crystals of calf-liver UDP-glucose pyrophosphorylase.

E

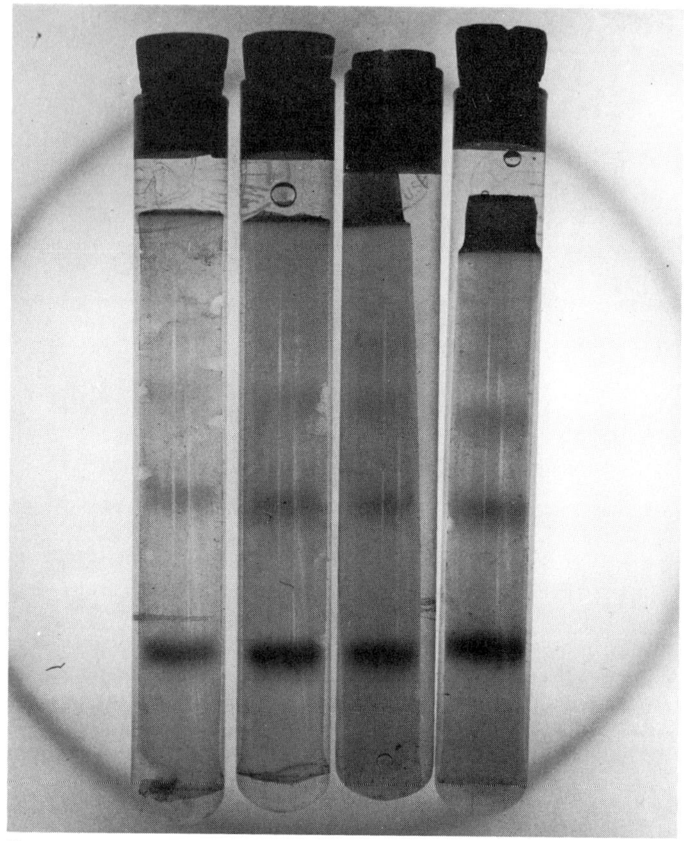

F

G

H

It is high noon on a cloudless day, and our woodland acre is bathed in sunlight. All seems serene and quiet, but there is a fever of activity. Billions of trillions of chemical bonds are breaking or forming at any instant, producing countless new metabolic molecules in the cells of the acre's occupants. Solutions are streaming from cell to cell, and water and air molecules are evaporating, condensing, and diffusing in both directions across the interfaces between the organisms and their environments. The sunlight is driving all this activity, and little of it is haphazard. These are the functions of life that result in growth, development, reproduction—and perhaps even the joy of living. Here is life in all its glory.

We have said that life is functioning cells. But how do they function? By making more of themselves, that is, by making more DNA and all the things required to make more DNA. This is the way we have been looking at life. Now, it is time to view it from a somewhat different angle. *Life is a particular kind of transformation of matter.*

What is matter? The physicist defines it as something that occupies space and has mass. If life is a particular kind of transformation of matter, just how is matter transformed or acted upon? Matter is transformed or acted upon by energy. About all we can say about *energy* is that it is something that can transform or act upon matter. We can observe energy only by observing its effects on matter.

Based upon much experience, we have concluded that matter and energy are conserved; that is, they are neither created nor destroyed—except in atomic reactions, in which one can be converted into the other. Einstein gave us the well-known formula for this atomic conversion: $E = mc^2$, where E is energy, m is mass, and c is the velocity of light. So far as we know, the equation has no direct significance to life functions, although there is one all important indirect relationship: energy from the sun comes from such a nuclear fire. Furthermore, disintegrating (radioactive) atoms are undergoing a transformation of a small portion of their matter to energy. The radioactivity given off is capable of causing mutations in the genetic material of living organisms. Scientists frequently utilize these radioactive atoms as tracers in their studies, particularly in studies of metabolism—the main topic of this and the following two chapters.

Energy and Energy Transfer

Energy has two states: *kinetic* (when it is actually acting upon matter, as a stone rolling down a hill) and *potential* (when it has the power to act upon matter but isn't, as the stone at the top of the hill). Gasoline in the can has potential energy; when it burns or explodes, it releases kinetic energy.

Although, as we shall see, subdivision would be easy, it is convenient to think of five principal forms of energy.

Mechanical energy. Mechanical energy is *matter in motion*—the rolling stone. We include heat in this category because heat (as we saw in the previous chapter) is a function of the kinetic activity of molecules. In considering mechanical energy of objects larger than molecules, it is important to think of the effects of gravity or electrical charge. Opposite electrical charges attract and similar charges repel each other; this attraction or repulsion can result in a movement of matter or an expression of kinetic mechanical energy. The forces of gravity exist between all particles of matter, but these forces are fundamentally much weaker than electrical forces.* Nevertheless, when large quantities of matter are assembled together, as in the earth, gravitational forces become powerful and can have a strong influence upon matter and the release of mechanical kinetic energy. Land plants must always contend with the forces of gravity, and their forms and anatomical structures are strongly influenced by this still mysterious force.

---

*Given two electrons at some distance apart, the electrical force between them (a repulsion) is 4.17×10^{42} times as great as the gravitational force between them (an attraction). In such a situation, the difference is over 42 *orders of magnitude* (powers of ten).

Chemical energy. The potential energy of chemical bonds can be transferred from one molecule to another as the molecules undergo chemical reactions. Sometimes, this chemical potential energy is converted to heat, that is, an increased velocity of movement of the particles that are reacting. In one sense, chemical potential energy is analogous to the potential energy of our stone at the top of the hill. It is sometimes due to the position of an electron in relation to the nucleus of an atom. If the electron falls closer to the nucleus, some of the potential energy is released as kinetic energy.

Electrical energy. Moving electrons exhibit kinetic electrical energy. Electrons that are capable of moving if the conditions are right have potential electrical energy. This potential electrical energy is measured as voltage. It exists, for example, between the electrical contacts in the wall socket. If a conductor of electricity is placed in contact with these two bits of metal in the socket, electrons flow through the conductor, and the potential energy is converted to kinetic. Electrons may also move during chemical reactions from one compound to another. In this case, we speak of *electrochemical transformations* of energy. Such transformations are extremely important to life functions.

Radiant energy. The energy of the electromagnetic spectrum is called radiant. The spectrum extends from the cosmic and X-rays through ultraviolet light, visible light, infrared light, and radio waves. It is always a form of kinetic energy, although matter under certain circumstances has the potential to give off radiant energy (but this energy can also be released in other ways). An atom capable of releasing radiant energy is said to be *excited.*

Atomic energy. This is the energy released by the conversion of matter to energy, as in an atomic explosion.

Transformations of energy from one form to another occur constantly in the whole universe. If we had infinite time at our disposal, we could probably think of an infinite number of examples. An oft-cited example is the waterwheel turning (mechanical energy) an electrical generator, producing a current (electrical energy) to light a bulb (radiant energy) to produce light that acts on the eye (a photochemical transformation), which sends a nervous impulse to the brain (a bioelectrical transformation). At each step, some of the energy is transformed to heat; ultimately all of it might be. In our woodland acre, sunlight (radiant energy) drives photosynthesis (photochemical transformation) to make carbohydrates (chemical energy) that are then eaten by animals and oxidized in respiration (electrochemical transformations) to make the compound ATP (chemical energy), which can then move ions across membranes, synthesize starches or proteins, result in growth, cause muscles to contract, eyes to see, brains to comprehend this sentence, and assist in virtually all other life functions. Again, some energy is converted to heat at each step, and eventually it all may end up as heat. There are four especially important energy transformations of living organisms.

Photochemical. This is when radiant energy is converted to chemical energy, as in photosynthesis.

Electrochemical. This is when chemical bond energy is transferred from one compound to another by transferring electrons. It is a part of both the photosynthetic and the respiratory processes. Often, we call this kind of energy transfer oxidation or reduction. If electrons are *removed* from a substance, it is said that the substance is *oxidized.* (In one kind of oxidation, the electrons are removed by combining with oxygen.) If electrons are *added* to a substance, the substance is said to be *reduced.* Often, reduction occurs by addition of a hydrogen atom with its electron; sometimes only the electron is added, which may be derived from a hydrogen atom or from some other atom. Oxidation and reduction are extremely important aspects of the topics discussed in this and subsequent chapters, so you should familiarize yourself with the concepts right now. *Oxidation is removal of electrons, and reduction is addition of electrons.*

Chemical. This energy transformation occurs when any kind of chemical bond is broken or formed. This happens whenever the bonds of metabolic molecules change, whether electrons are exchanged or not.

Mechanical. Many of the functions of living organisms result from a transformation of chemical-bond energy to mechanical energy. The best example might be the contraction of muscles in animals, but growth and movement in plants, though less spectacular, is an equally valid example.

There are two important laws of energy transfer: The First Law of Thermodynamics states that energy can be neither created nor destroyed. We know that atomic reactions are an exception, but in studying life functions, the law is completely valid.

The Second Law of Thermodynamics can be stated in several different ways. One is that heat cannot be converted completely to work. A practical consequence is that energy transformations are never 100% efficient. Some of the energy always appears in the system as heat. Of course, the efficiency of energy transfer varies considerably for different processes, but it is worth remembering that the energy transformations of cells are often around 30% efficient—that is, 70% of the original energy becomes heat.

The second law is sometimes stated more philosophically. This involves the concept of entropy, which (among other things) is a measure of the disorder of a system. The greater the disorder of a system, the higher the entropy. With this idea in mind, the second law may be stated by saying that *in any transfer of energy, the total entropy of the system (that is, its disorder) increases.* Consider diffusion: A crystal of potassium permanganate placed in the bottom of a beaker of water constitutes a system with a relatively high degree of order (the ions of the crystal are nicely localized and ordered). As the ions dissolve and begin to diffuse by their random, kinetic activities, finally filling the beaker uniformly, the system becomes much more disordered (random) than it was to begin with, and consequently the entropy increases.

Energy can only be observed by its effects upon matter, and by these effects, it can be measured. Heat energy can be measured, for example, in a calorimeter. In such a device, a small quantity of some substance such as glucose may be burned. All of the heat produced by this reaction is absorbed by water surrounding the reaction. The increase in the temperature of the water can then be observed with an accurate thermometer. The calorie (cal) is the amount of energy required to raise the temperature of one gram of water by 1°C (approximately—depending upon the exact temperature to begin with). The so-called large calorie or kilocalorie (kcal) is equal to 1000 of the small calories—and is counted by dieters.

In another energy-measuring device a "black body" is exposed to radiant energy. A black body is one that absorbs all of the radiant energy falling upon it. We can think, for example, of a large spherical container painted black on the inside with a small opening; all of the radiant energy entering the opening will be absorbed by the inside wall or reflected and then absorbed. (Only an insignificant, minute fraction will come back out of the hole.) Although such a black body has been used by physicists, a small blackened thermister (temperature-measuring device) is more commonly used, and a correction is applied for the small amount of energy that is not absorbed. In either case, the amount of energy absorbed may be calculated by first measuring the increase in temperature of the black body.

These, then, are some of the forms of energy and some of the transformations of energy occurring in our acre of woodland—a typical ecosystem in which the energy of sunlight is transformed in millions of ways to drive the metabolic transformations of matter going on in the organisms. Such chemical transformations are collectively called metabolism: the sum total of all of the chemical activities of living organisms. How is metabolism controlled in the living plant? The control is exercised by the enzymes to which we have frequently referred in previous chapters. It is now time to discuss these enzymes in more detail, beginning with the chemistry of proteins—since all enzymes are proteins.

Proteins

Proteins are composed of amino acids, as described in End Paper 4–1. These interesting molecules, with their acidic (carboxyl) and basic (amino) groups, are tied together by peptide bonds to form long chains. Each protein enzyme is an unique entity, differing from all others in at least six important ways.

Primary structure. The order of arrangement of amino acids in the chain making up a protein molecule is referred to (according to one usage) as the primary structure. This order of arrangement has now been determined for several important proteins. It was first achieved for beef insulin, being completed in 1953 after a decade of work by Frederick Sanger and his colleagues in England. The insulin was broken into short chains of amino acids (polypeptides), and their structures were then determined. Finally, these were put together as one might solve a jigsaw puzzle, yielding the structure of the entire insulin molecule with its 51 amino acids (a relatively small protein). Sanger received the Nobel prize for this work in 1958. Sequences can now be determined more rapidly, thanks to the discovery of several enzymes, each of which hydrolyzes only a specific kind of peptide bond.

Secondary structure. Formation of peptide bonds often results in a tendency for the backbone of the proteins to coil into a helix. This was worked out by Linus Pauling in 1950; he received the Nobel prize in 1954 for his efforts.

Tertiary structure. The amino acid chains (themselves often coiled into an α-helix) are typically folded in various ways, being bonded together at points of contact by hydrogen or even disulfide bonds. Disulfide bonds occur when 2 molecules of the amino acid cysteine combine (releasing 2 hydrogen atoms) to form the amino acid cystine (see Fig. 4–12). Secondary and tertiary structures are studied by X-ray diffraction (see Essay 8–1) and by other techniques.

Quaternary structure. Often, 2 or more polypeptide chains (sometimes similar to each other, sometimes different) are attached to each other, frequently through disulfide bonds.

Surface charges. The extra amino, carboxyl, hydroxyl, and sulfur groups can ionize in various ways (or be oxidized or reduced), providing surface charges on the protein molecule and influencing water solubility. Portions of the surface may be hydrophilic (attract water), and others may be hydrophobic (repel water).

Surface geometry. The various hydrocarbon chains and rings (the R parts of the amino acids, p. 63) sticking out add further geometric variety and often lipid solubility. Surface charges and geometry are especially important in giving the protein molecular surfaces the characteristics responsible for enzyme activity.

With all of these ways that proteins can differ from each other, consider the possible variety! Actually, this variety is ultimately determined by the order of arrangement of the amino acids, which in turn determines coiling, tertiary structure, and surface charge and configuration. The number of possible arrangements of amino acids is extremely great, as pointed out in Essay 5–1.

Enzyme Characteristics

Not all proteins are enzymes. Some play structural roles; hair, for example is almost pure protein. Others are stored, particularly in seeds. On the other hand, all enzymes are proteins; so the enzyme chemist is a protein chemist who must deal with the special features of enzymes. The primary problem is that of denaturation. At high temperatures (above 50°C or so) and in the presence of strong acids, bases, heavy metal ions, organic solvents, some detergents, or even air (as in whipped egg white), the physical properties of most proteins change markedly. Usually, they become much less soluble in water, and they also lose their enzymatic activity. This denaturation is due to the unfolding of the amino acid chain, which breaks hydrogen bonds and exposes certain labile or hydrophobic groups to the surrounding environment. New hydrogen bonds form between different molecules, welding the whole into an insoluble mass, and exposure of otherwise protected hydrophobic groups also results in a decrease of solubility. Enzymatic activity is dependent upon the exact structure of the protein molecule (the active site), so changing the structure results in a loss of activity. Usually, denaturation is irreversible,* but this is not always the case. Addition of urea denatures several proteins, but if the urea is removed or diluted greatly, the proteins assume their original shapes and again exhibit their biological activity.

Because of the ease with which proteins may be denatured, the biochemist usually does his preparations at low temperatures (in an ice bucket or a cold room with the thermostat set close to 0°C). He must take many other precautions such as keeping his solutions free of heavy metal ions, adjusting acidity (pH) accurately, and adding special compounds to prevent oxidation of certain groups in the protein.

The photo essay for this chapter illustrates some of the techniques used by protein or enzyme chemists, and End Paper 5–1 presents a detailed outline of an enzyme analysis and an enzyme purification. Both should provide a little of the flavor of this important activity of modern biologists.

One of the most important and striking characteristics of enzymes is their specificity. A given enzyme

*Humpty Dumpty sat on a wall. Humpty Dumpty had a great fall. All the King's horses and all the King's biochemists couldn't restore Humpty Dumpty's denatured proteins again.

will act only upon a single substrate or perhaps upon a group of substrates closely related chemically. The enzyme is usually named according to the substrate upon which it acts. The name is formed by adding *-ase* to the substrate or perhaps to a word or two describing the substrate and the kind of reaction. Thus *hexokinase* acts upon hexoses, changing them in a certain way. *Cytochrome oxidase* is an enzyme that oxidizes a molecule called cytochrome. *Malic acid dehydrogenase* removes hydrogens from malic acid. (A few enzyme names do not follow these rules because names were given before the rules were agreed upon.)

Essay 5–1 The Universes of Protein Variety

We can obtain some idea of the possible variety of protein structures by considering the ways that 20 kinds of amino acids can be arranged in a linear order. This proves to be fairly simple to calculate. If there is only one amino acid in the "chain," there could be 20 different kinds of "chains." If there are two, there will be $20 \times 20 = 400$ kinds of chains. (Each amino acid can combine with each of the other possible 20.) If there are three amino acids, the possible arrangements will be $20 \times 20 \times 20 = 8000$. The power rule for calculating the possible arrangements soon becomes apparent. The possible arrangements are equal to 20^n (20 to the *n*th power), where *n* is equal to the number of amino acids in the chain.

The number of possible arrangements is absolutely beyond comprehension. For example, a small protein, ferredoxin, contains 97 amino acids (and also three to seven iron atoms). There could be 20^{97} (10^{126}) possible proteins of this size. Ten to the 126th power is equal to a 1 followed by 125 zeros. This is more than all of the atomic-sized particles in the universe. Indeed, it is over a billion times the size of the known universe in cubic angstroms! Cytochrome c has 104 amino acids ($20^{104} = 10^{135.5}$ possible kinds of protein that size). Hemoglobin has 574 ($20^{574} = 10^{747}$ possibilities).

When we are considering possible kinds of proteins, we are dealing with information theory. The situation is similar to that of a written language. In our language, we have 26 letters and several symbols (a rounded total of about 30). Thus a sentence as long as Shakespeare's "THIS ABOVE ALL: TO THINE OWN SELF BE TRUE!" *could exist in $30^{45} = 10^{64}$ forms. We could write this sentence by allowing monkeys to pound typewriters randomly or by blowing up a print shop and then searching for the sentence among the bits of type on the street. Let's build small computers instead. Each one occupies a volume of 1 liter, and all are connected. They cover all the earth to a depth of 2 km. Each one changes a character in a sequence of 45 characters and compares the result with the sentence above. It does this a trillion times each second, and no two computers ever duplicate their efforts. To try more sentences, we cover 10^{20} identical planets to a depth of 2 km with such computers, still programming them so that no two ever duplicate the same sequence of characters. Now, we let the computers on all these planets run for 4 billion years, the time required to form all possible 45-character sequences (sentences), using 30 characters: $30^{45} = 1.3 \times 10^{67}$ sequences. One of these will be our Shakespearean sentence. Of course, it could have been produced during the first second of the billion-year period or during the last, but to try all possible combinations would require all computers on all the planets running all the time. Such an example helps one realize that 10^{67} is a truly large number.*

Most of the forms of the Shakespearean sentence would be nothing but nonsense; similarly most of the possible kinds of proteins would probably have no biological significance. In any case, it is apparent that if metabolism requires special molecules, surely the proteins should provide suitable possibilities.

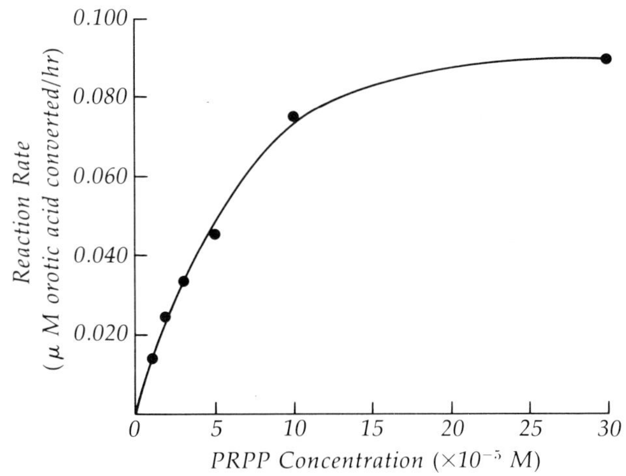

5–1 Rate of an enzymatic reaction as a function of concentration of the substrate. In this case, the substrate is 5-phosphoribosyl-1-pyrophosphate, PRPP, and the enzyme is orotidine-5′-phosphate pyrophosphorylase. The reaction is measured by measuring the micromoles of orotic acid converted by the enzyme each hour. The orotic acid combines with the PRPP. Note that the enzyme is about saturated in this case at 10 to 15 $\times 10^{-5}$ M PRPP. Note also how the shape of this curve resembles that for the curve of active ion uptake or accumulation in Figure 4–5.

Often, a portion of an enzyme molecule is called a prosthetic group. Sometimes, it is a metal ion, and other times it may be a simple to relatively complex organic molecule. In some cases, it is attached tightly to the protein, and in other cases, it may be attached loosely (even removable by dialysis—the procedure originally used to discover some of these prosthetic groups). We will refer to these groups as coenzymes, although some authors suggest that the term coenzyme should be reserved for the organic molecules, while the ions should be called metal activators.

The metal activators turn out to be the essential mineral elements (see Chapter 32). The organic molecules prove in a majority of cases to be identical to or derivatives of molecules that we call vitamins. These vitamins are compounds that animals are unable to synthesize for themselves and must therefore obtain in their diet, and that are essential to metabolism because they are coenzymes. Plants are able to synthesize all of these molecules by their own metabolism.

Rather recently, it was discovered that more than one protein can sometimes catalyze the same reaction. In such a case, the different proteins are referred to as isozymes. Sometimes, these proteins are very closely related. Lactic acid dehydrogenase, for example, consists of a molecule with two kinds of chains. These chains occur in different proportions to each other, producing five different isozymes.

How do enzymes work? This is the real question. Much evidence indicates that enzymes function by the formation of an enzyme-substrate complex. In the previous chapter, we considered the saturation curve for ion accumulation. The same argument applies to enzyme action. (It was developed first for enzyme action.) In the presence of a given amount of enzyme, increasing the concentration of substrate increases the rate of the reaction—but only up to a point (the saturation level, Fig. 5–1). Increasing substrate above this saturation concentration no longer increases the rate of reaction. This is apparently because all of the enzyme molecules have been used up in forming the enzyme-substrate complex, in which case addition of further substrate molecules can have no effect. The concept is illustrated by the following equation:

$$E + S \rightarrow ES \rightarrow E + P,$$

where E is the enzyme, S is the substrate or substrates, ES is the enzyme-substrate complex, and P is the product or products of the reaction.

Apparently formation of the enzyme-substrate complex lowers an energy barrier. As illustrated in Figure 5–2, the products of a reaction must always exist at a lower energy level than the substrate or reactants. This is another consequence of the Second Law of Thermodynamics. The transformation from substrate(s) to product(s) requires an input of energy; that is, the reaction must go over an energy barrier. Due to formation of the enzyme-substrate complex, this energy barrier is lowered so that the reaction can occur more easily.

But how is the enzyme-substrate complex formed? As we mentioned in Chapter 2, it is formed because of an active site on the enzyme molecule—the molecule has a special structural configuration that is an exact mirror image of the substrate molecule, allowing the two to fit together intimately. Apparently, in this fitting together, some kind of tension may be applied to the substrate molecule, causing a bond to break—or perhaps pressure is applied, causing a bond to form. Obviously, understanding the active site is one of the keys to understanding the nature of life. The active site may not depend upon the entire protein molecule. Papain, for example, is a plant enzyme capable of breaking down other proteins. Its activity is unaffected by removing as much as $^2/_3$ of the molecule. Nevertheless, the molecule includes a total of about 300 amino acids; so we might imagine that the exact configuration of the active site depends upon the structural arrangement of as many as 100 amino acids, although this is difficult to prove because the procedures required to remove any of these might upset the active site in some indirect way. In any case, it is obvious that the active site must be small enough to fit the substrate molecule. At the same time, there are so many different enzymes (so many different active sites) and so few amino acids involved in actually contacting a substrate molecule, that we can imagine that many more amino acids must be necessary for the exact configuration of the active site beyond those actually contacting the substrate molecule—but of course we don't know *how* many. The amino acids not contacting a substrate may tend to configure those that do.

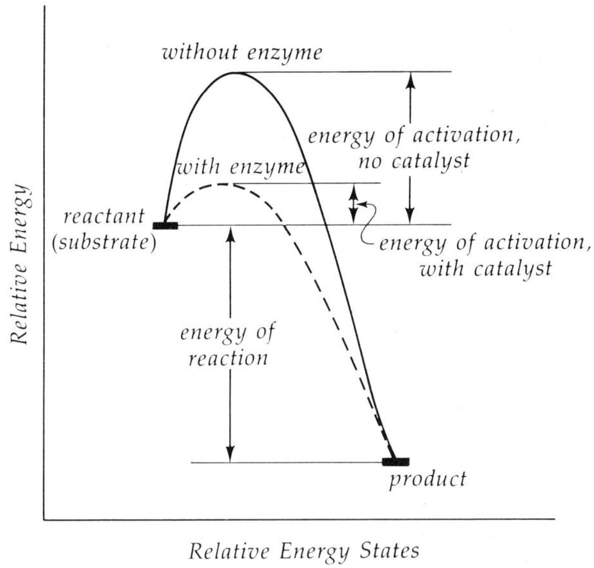

5–2 Energy diagram for a metabolic reaction occurring in the presence and the absence of an enzyme. Reacting substrate molecules must pass over an "energy barrier," even though the product may be at a lower energy level than the substrate. The enzyme lowers the activation energy required, thus increasing the fraction of molecules that can react in a given time.

Enzyme action is highly dependent upon the enzyme's environment. We have discussed the interaction of enzyme and substrate concentration. If either enzyme or substrate are increased in their concentration (below saturation), the rate of the reaction will increase. Furthermore, virtually all enzymatic reactions are reversible to some slight extent. Hence, as the products build up (increasing their free energy levels), the reaction tends to go in the reverse direction, and the increasing product concentration often inhibits the reaction.

Temperature has a strong influence upon the rate of enzymatic reactions. Increasing temperature usually increases the rate of any chemical reaction, an increase of 10°C often about doubling the rate. With enzymatically controlled reactions, this is true at low temperatures, but at high temperatures, increasing temperature tends to decrease the rate, finally halting the reaction completely. Apparently, this is because at temperatures above 30°C or so, many proteins tend to be denatured. Thus, at high temperatures, there are two competing effects: promotion of the chemical reaction by increasing temperature and inhibition by denaturation of the controlling enzyme (Fig. 5–3). Interestingly, there are some plants and animals that grow close to or below the freezing point where enzymatic reactions in most others proceed extremely slowly, if at all. At the other end of the scale, there are certain bacteria growing in the hot springs of Yellowstone National Park almost at the boiling point (about 90°C). So far, we do not understand how these organisms can break the "rules" so nicely.

The rate of a chemical reaction is also affected strongly by the acidity of the solution surrounding the enzyme. This is because the active site is strongly influenced by the presence of positive and negative

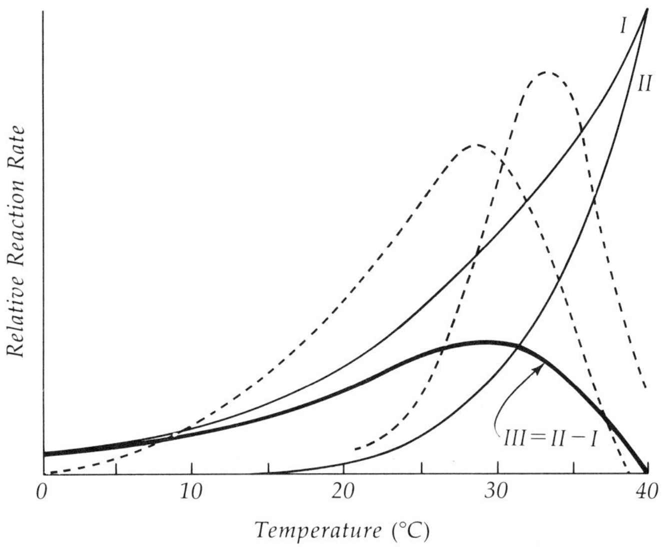

5–3 Effects of temperature upon enzyme reactions. The dotted lines indicate responses of two hypothetical enzymes to temperature. In one case, there is no activity below 0° C, and optimum is reached at about 26 to 27° C—the maximum is about 39° C. In the other case, minimum, optimum, and maximum are somewhat higher. The solid lines indicate how these responses might be understood. Curve I indicates increasing rate of enzymatic reaction with increasing temperature. Curve II indicates more rapidly increasing rate of enzymatic destruction with increasing temperature. The heavy solid line (III) is obtained by subtracting II from I. It could represent the enzymatically controlled reaction.

charges on the protein, and these are, as should be clear from the dual nature of amino acids, highly sensitive to acidity. As the acidity increases, more amino groups become more positive, and fewer carboxyl groups remain negative. Most proteins have a certain acidity at which they exhibit an optimum level of activity. Substances affected by acidity in this manner are said to be amphoteric.

There are various compounds that inhibit enzymatic reactions in different ways. Some of these react with the active site. Poisons such as carbon monoxide, fluoride, arsenate, and others are good examples. There is another class of compounds called competitive inhibitors that have a molecular structure similar but not identical to the substrate molecule. They also form an enzyme-substrate complex, but their structure is such that no reaction is possible. Thus they occupy active sites that could otherwise be occupied by substrate molecules, inhibiting their reaction.

More recently, it was discovered that some enzymes are capable of reacting with certain compounds in such a way that the enzymatic reaction is either promoted or inhibited. In the case of these so-called allosteric enzymes, molecules that inhibit do not react with the active site but react at some other point on the enzyme, thereby changing the configuration of the active site so it is no longer effective. Promoters may also react at some point other than the active site, making the active site more effective—or in a few cases the substrate itself or some close relative may react with the active site, increasing its effectiveness even more. Often, these molecules are of considerable importance in controlling the specific metabolic reactions of living cells.

Some Thoughts about Enzymes and Environment

If ecology is our concern, we are interested in the interactions between organisms and their environment. These interactions are often the responses of enzymes to environment. After all, life is largely a matter of functioning enzymes, and enzymes are notoriously sensitive to environmental influences. A good ecologist might devote much study to the science of enzymology. He should understand the role of energy in driving the transformations of matter as controlled by enzymes. He should continually be aware that most such reactions are driven by heat energy but that high temperatures inactivate most enzymes. Organism responses to temperatures are typically enzyme responses. He might be aware of the reason why certain minerals, vitamins, and energy sources are required in the diets of many organisms: to allow their enzymes to function. The poisons of natural environments are usually enzyme poisons, as are those now being released into the environment with industrial and other wastes. Mercury and lead form heavy metal ions capable of denaturing enzymes. Carbon monoxide inactivates an enzyme essential to respiration. Many pesticides act as competitive inhibitors on specific enzymes. We shall multiply these examples in later chapters, but it should already be clear that much of ecology depends upon energy-driven, enzymatic reactions.

End Paper 5–1 Assay and Purification of Alpha Amylase

Alpha amylase is an enzyme that catalyzes the hydrolysis (breakdown) of starch. It occurs in abundance in many germinating seeds and is present in smaller quantities in most tissue, including leaves. It is sometimes important to measure the quantity of an enzyme and also to purify the enzyme for a detailed study of its properties. The following explanation should provide some insight into how the enzyme chemist proceeds with the characterization of an enzyme.

To begin an amylase assay, three solutions must be prepared. One is a starch solution made by adding 0.15 g of potato starch, 0.60 g of potassium dihydrogen phosphate (KH_2PO_4), and 0.022 g of calcium chloride ($CaCl_2$) to water to make a final volume of 100 ml. This solution is boiled for 1 minute, cooled in an ice bath, and centrifuged at 4000 × g for 10 minutes. The clear supernatant is the starch solution; the material in the bottom of the tube is discarded. An iodine solution is prepared by adding 1.0 ml of a stock iodine solution (6.0 g of KI and 60 mg I in 100 ml of water) to 100 ml of 0.05 N HCl. An acetate buffer solution is prepared from sodium acetate and acetic acid, adding water to a final concentration of 0.05 M acetate adjusted to pH 5.0. The buffer will maintain the proper acidity during extraction.

To prepare for the assay, a mortar and pestle are chilled with cracked ice, the acetate buffer is chilled in an ice bath, and a water bath is set at 25°C. A spectrophotometer (it measures specific wavelengths of light absorbed by a solution) is set for wavelength 620 nm (orange light). About 2.5 g of pure quartz sand is added to the mortar along with about 1 g of the endosperm portions of 3- to 4-day-old barley seedlings (or wheat, rye, oats, or rice) and 5.0 ml of acetate buffer. The tissue is ground until most of the cells are broken, and the homogenate is then poured into a centrifuge tube. The mortar and pestle are rinsed with 5.0 ml of buffer, and the rinsings are added to the homogenate in the centrifuge tube. This extract is centrifuged at 4000 × g for 10 minutes at 0°C. The clear supernatant contains the enzyme and is poured (decanted) into a clean test tube already chilled in the ice bath. The residue of sand and cellular debris is discarded.

For the assay, 0.10 ml of the extract is pipetted into each of four test tubes in the 25°C water bath. To two of these, 1.0 ml of iodine reagent is immediately pipetted followed by 1.0 ml of starch solution and 13.0 ml of distilled water. The iodine solution denatures the enzyme and reacts with the starch to produce a blue color. The intensity of this color is measured with the spectrophotometer; this is a control indicating the amount of starch present in the 1.0 ml aliquot of the starch solution. At time zero, 1.0 ml of starch solution is added to the other two tubes of enzyme. After 5 minutes, 1.0 ml of iodine solution is added followed by 13.0 ml of distilled water. The enzyme will have hydrolyzed some of the starch so that the blue color will be less intense in these tubes. (If there is no blue color, all of the starch was hydrolyzed, so reduce either the time of incubation or the aliquot of the extract taken for assay). The color is read in the spectrophotometer, and the readings for the two tubes are averaged and compared with the averages of the two control (no enzyme) tubes. The difference in optical density is a measure of the starch hydrolyzed by the enzyme, which is a function of enzyme concentration times incubation time. That is:

Starch hydrolyzed = enzyme concentration × time

There are several ways to purify alpha amylase, but the following procedure is especially simple:

Barley seeds are germinated for 3 to 4 days at room temperature. The endosperm portions from about 100 seedlings are ground with about 50 ml of 0.02 M CaCl and about 10 g of quartz sand. The homogenate is centrifuged for 10 minutes at 4000 × g. The supernatant is decanted into a beaker and heated with stirring to 70°C for 30 minutes. Amylase is not denatured by heating to 70°C, but most other proteins are. After heating, the denatured proteins are removed by centrifuging the solution at 4000 × g for 10 minutes. The supernatant solution, which contains the amylase, is cooled to 0°C, and sufficient 95% ethanol is added to bring the final concentration of the solution to 40% with respect to ethanol. Now, 0.5 ml of a 2% glycogen (animal starch) solution is added with stirring to the cold ethanolic solution of the enzyme.

Glycogen is a substrate for the enzyme and binds to it to form an enzyme-glycogen complex that is insoluble in cold 40% ethanol. At 0°C the action of the enzyme is too slow to hydrolyze any significant proportion of the glycogen. The mixture is centrifuged at 10,000 × g for 20 minutes. The small pellet contains the amylase. The supernatant is discarded, and the pellet is taken up in about 2 ml of buffer. The buffer solution of the enzyme is incubated for about 2 hours at 25°C to allow the enzyme to hydrolyze the glycogen. The end products of the hydrolysis (glucose, low molecular-weight oligosaccharides, and limit dextrins) are removed by dialysis (put solution in a cellophane tube—small molecules pass out through pores in the cellophane and amylase protein stays in). At this point, the solution contains alpha amylase as the only protein and small quantities of limit dextrin that are not removed by the dialysis procedure. In a study using this procedure, the alpha amylase was examined by gel electrophoresis and found to contain four closely related kinds of alpha amylase molecules. In the germinating seedling these four kinds of amylase are synthesized by the aleurone layer in response to gibberellins (plant hormones) synthesized in the nodal region of the embryo.

6

Photosynthesis: Chloroplasts

Chloroplasts—The Green Machines

Chloroplasts are among the most important and beautiful parts of a plant cell. Here, the essential process of photosynthesis is carried out and carbohydrates—so critical to all life—are produced. All chloroplasts are surrounded by a double membrane, and they also contain other double membranes known as thylakoids. These thylakoids may all be separate from one another, or they may be grouped together in various combinations. In many of the chloroplasts present in vascular plants, small thylakoids are stacked on top of one another to form grana. Between the thylakoids are regions that contain DNA and ribosomes. In the blue-green algae the thylakoids are not surrounded by an outer membrane but are in direct contact with the cytoplasm.

A *A section of a chloroplast of a tobacco leaf, as viewed with the electron microscope. In this type of chloroplast the grana are clearly visible.*

B *In this section of a chloroplast of a red alga, the thylakoids are separated from each other, and there are no grana.*

C *A thin section of a blue-green* (Gleocapsa alpicola) *algal cell in which the thylakoids clearly run throughout the cytoplasm of the cell.*

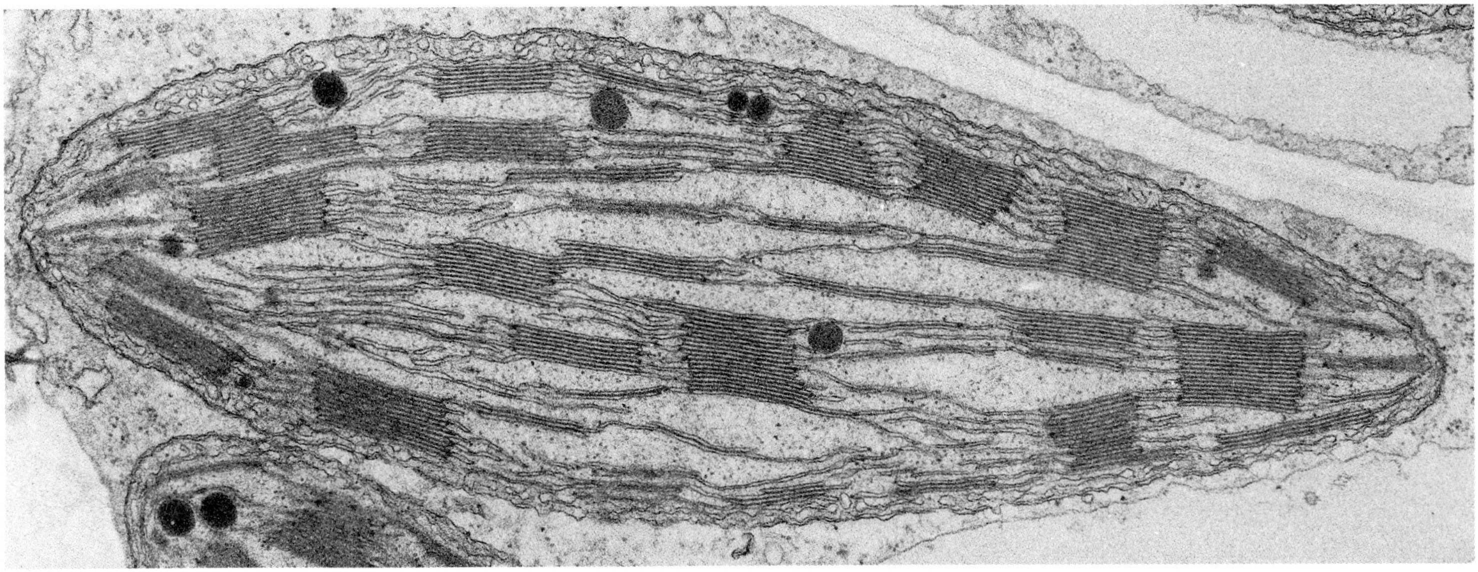

A

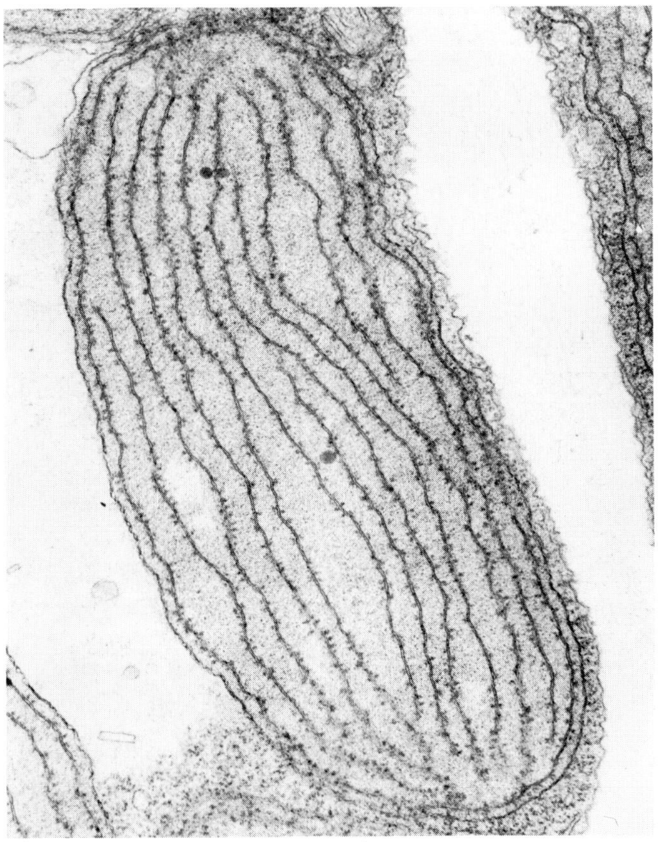

B

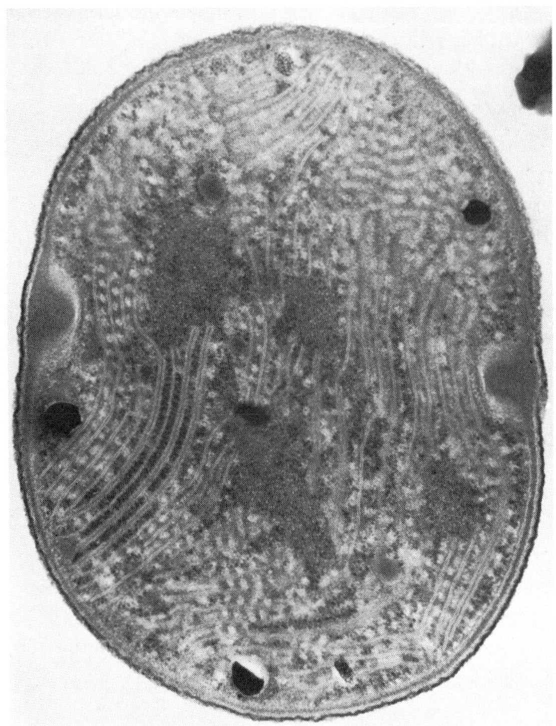

C

D A freeze-etched preparation of chloroplast thylakoids showing the structure of the membranes. The large particles buried within intact membrane are exposed by this method of tissue preparation.

The thylakoid membranes present in the chloroplasts are complex in structure. The pigments and many of the enzymes of the chloroplast are incorporated in these membranes. This creates a membrane that is quite different in appearance from the plasma membrane, and when viewed with such techniques as freeze-etching, it reveals many subunits and particles.

E The pyrenoid in the chloroplast of the green alga Chlorella.

Starch is frequently stored between the thylakoids. In some algae and a few other plants, starch is synthesized in conjunction with a specialized portion of the chloroplast known as the pyrenoid. Some algal chloroplasts contain precise arrangements of orange-red pigments that act as eyespots. Organisms containing such eyespots are sensitive to the direction and intensity of light.

F The eyespot region in the chloroplast of the green alga Chlamydomonas. *The top photo is a surface view, while the bottom is the actual eyespot region.*

G A prolamellar body in the chloroplast of a dark-grown tobacco cell (× 42,000).

Dark-grown chloroplasts do not normally develop elaborate thylakoid systems; instead, they produce a structure known as the prolamellar body. This is a crystal-like arrangement of tubular membranes, which are rearranged when the chloroplast is illuminated to form thylakoids.

H Starch-storing plastids in the pollen of Magnolia (× 32,000).

Many other types of plastids besides chloroplasts are found in plant cells. These include plastids that lack pigments and that are involved with the storage of various food products. There are also plastids rich in pigments other than the chlorophylls, which causes them to appear red or yellow rather than green. Colorless plastids are also found in many plant cells and may give rise to still other types of plastids, including the chloroplasts.

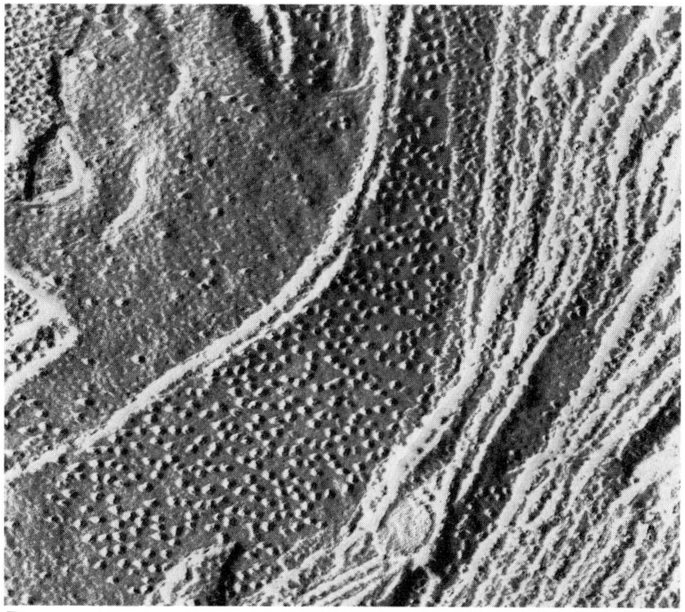

D

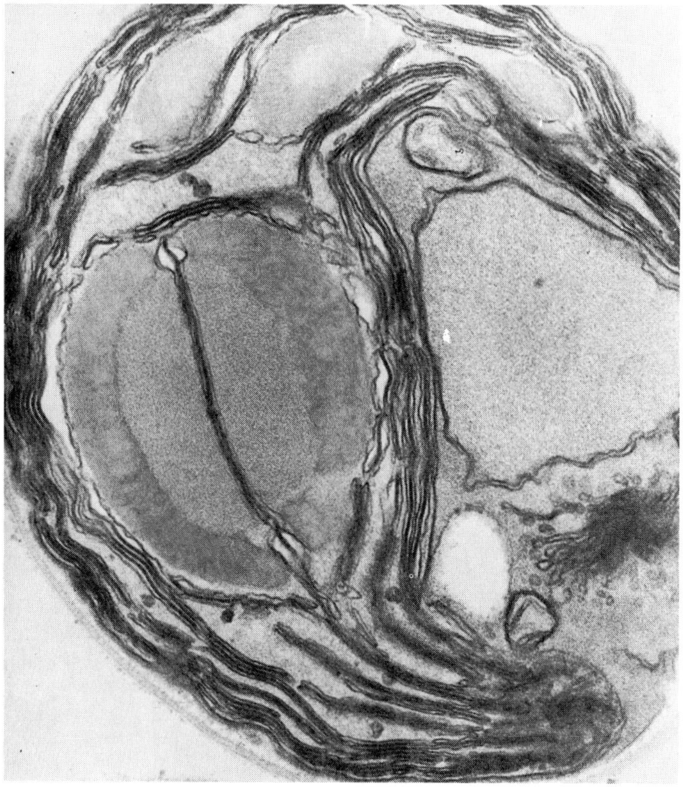

E

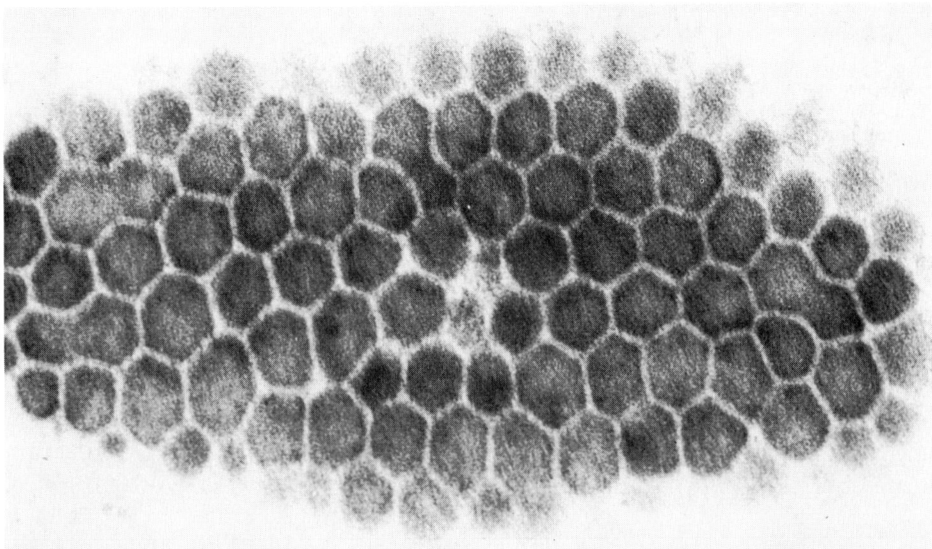

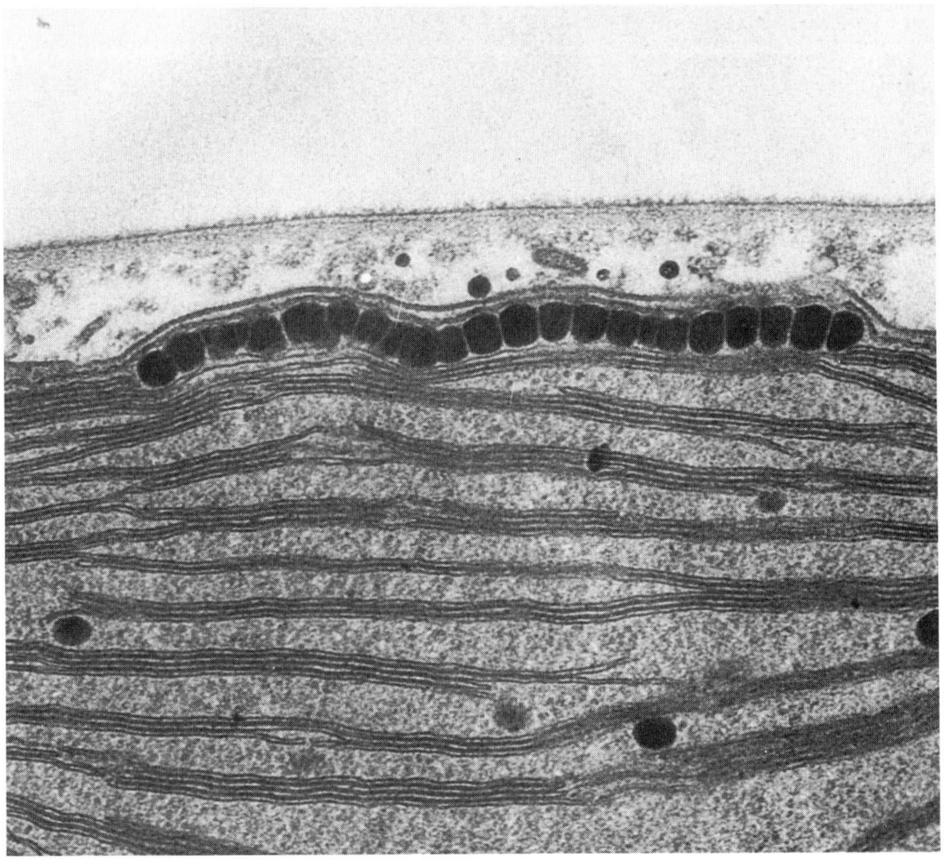

F

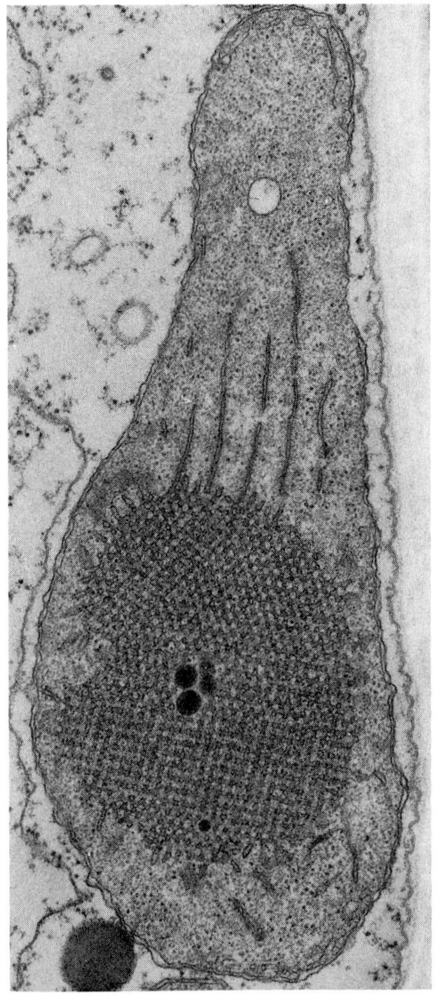

G

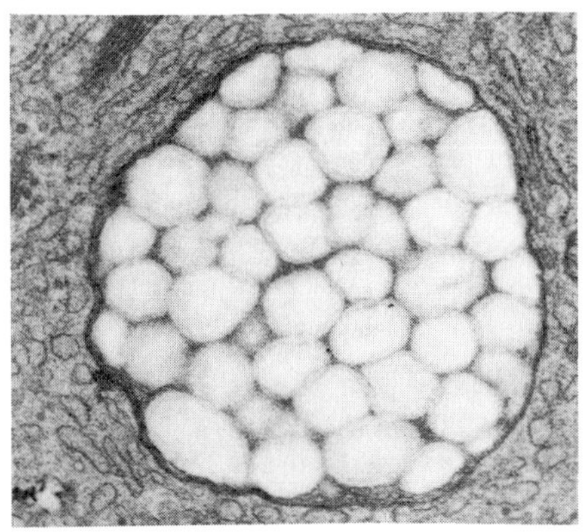

H

It is still high noon, and our woodland acre is still bathed in sunlight. This radiant energy from space is driving the woodland ecosystem, but to do so, it must undergo an energy transformation, since most metabolic reactions are driven by chemical-bond energy rather than by radiant energy. The primary photochemical energy transformation of ecosystems is photosynthesis. There are numerous photochemical reactions in ecosystems (one causes plants to orient their leaves at right angles to the rays of the sun), but these are responses in which the radiant energy is used up in triggering or causing the response. None is left over for other uses. Photosynthesis converts vast amounts of radiant energy striking the earth into the chemical-bond energy upon which virtually all other metabolic reactions ultimately depend.

Furthermore, carbon dioxide molecules are incorporated into the carbon chains that constitute the skeletons of all the organic molecules of metabolism. Photosynthesis not only drives the ecosystem (as it is in turn driven by light) and provides the building blocks from which the ecosystem is constructed, but it also removes the carbon dioxide produced by the respiration of plants and animals and resupplies the air with the oxygen used in respiration. Without photosynthesis, our atmosphere might have vast quantities of carbon dioxide (released in volcanic eruptions) and virtually no oxygen—as do the atmospheres of Venus and to a lesser extent Mars. The process can be summarized by a general equation, which we shall elaborate upon in this chapter:

$$\text{Light energy} + \text{carbon dioxide} + \text{water} \xrightarrow{\text{green plants}} \text{carbohydrates (etc.)} + \text{oxygen}$$

Photosynthesis provides us with an excellent example of the structure-function relationship. It works because suitable pigments and enzymes are properly arranged in chloroplasts. In the early 1950s it was shown that all the essential machinery for photosynthesis is present in the chloroplast. Of course, the cell provides a suitable environment for chloroplast function, and unless this environment is duplicated in the test tube (proper acidity, substrates, etc.), isolated chloroplasts are not capable of photosynthesis.

Chloroplasts

Learning to isolate chloroplasts opened the way for modern studies of photosynthesis. Plant cells must be ruptured without damaging the chloroplasts, which is conveniently done by grinding leaves with sand in a mortar and pestle. Chloroplasts contain solutes that lower their water potential considerably; so if they are released into pure water, they will take up the water by osmosis and burst. Thus they must be ground in a sugar solution having a suitable osmotic potential.

In the early 1940s Robin Hill found that isolated chloroplast fragments (intact chloroplasts were not yet available) would carry out the so-called Hill Reaction. In this reaction chloroplasts are capable of using light energy to transfer the hydrogen atoms from water to some acceptor substance, releasing free oxygen. This is a reduction process (the addition of hydrogen with its electrons to the substrate—or sometimes only the hydrogen's electron, depending upon the substrate). Such isolated chloroplast fragments were not capable, however, of carrying out the other steps of photosynthesis until all of the environmental conditions essential for this process were discovered.

Chloroplasts have been known as important parts of the plant cell since the 1700s. Actually, plastids in general and chloroplasts in particular are as unique to plant cells as are cell walls and vacuoles. There may be on the order of 20 to 100 chloroplasts per leaf cell, and in higher plants they are usually disc shaped. They are from 0.5 to 1.0 μ thick and from 2 to 4 μ in diameter. Many algal cells have only a single chloroplast, and it may assume several different and often highly characteristic shapes.

Some of chloroplast structure had been discerned with the light microscope, but the real ultrastructure of these organelles was only clearly understood with the help of electron micrographs. Chloroplasts seen in electron micrographs are enclosed by a double membrane system, each membrane resembling the unit membrane discussed in Chapter 3. Within the chloroplast, a number of other membranes called thylakoids are readily apparent—a few appear to extend all the way across the chloroplast. Most take part in disc-shaped stacks of membranes called grana, with 40 to

60 grana in each chloroplast. Some of the thylakoids extend from one granum to another, but all of the grana membranes do not extend across the chloroplast. The grana are embedded in a matrix called stroma, which is largely protein. It is not easy to describe chloroplast structure in words, but the structure becomes readily apparent when we examine the photo essay for this chapter.

All chloroplasts do not have the same internal structure. The plastids of the mosses, ferns, and seed plants are generally like that shown in Figure 6–A, but the green and brown algae have plastids consisting of a double-limiting membrane and lamellae, generally lacking grana stacks (Fig. 6–B). In the blue-green algae, no limiting membrane exists, and parallel membranes containing the photosynthetic pigments extend directly into the cytoplasm (Fig. 6–C).

Origin and Development of Plastids

In some algae and mosses, the light microscope can reveal chloroplasts dividing in living cells. Plastid division has also been studied in algae and mosses with the aid of the electron microscope. The chloroplasts of ferns and seed plants have not yet been seen to divide (with the light microscope), and the mode of their origin has long been controversial.

Two means of chloroplast reproduction have been suggested for ferns and seed plants: (1) division of plastids and (2) development from much smaller and simpler bodies called proplastids, which presumably came from division or from some other cell part. For many years, the second alternative was generally accepted because the light microscope had not revealed mature plastids dividing in living cells, as they had in algae and mosses. In addition, the light microscope revealed many small bodies in the cell that could be proplastids.

As data from the electron microscope and biochemical studies have accumulated, the weight of the evidence has shifted to the first suggestion—direct division. Electron microscope evidence includes photographs such as the one shown in Figure 6–1, in which chloroplasts appear to be dividing or to have recently divided. Such observations were first dis-

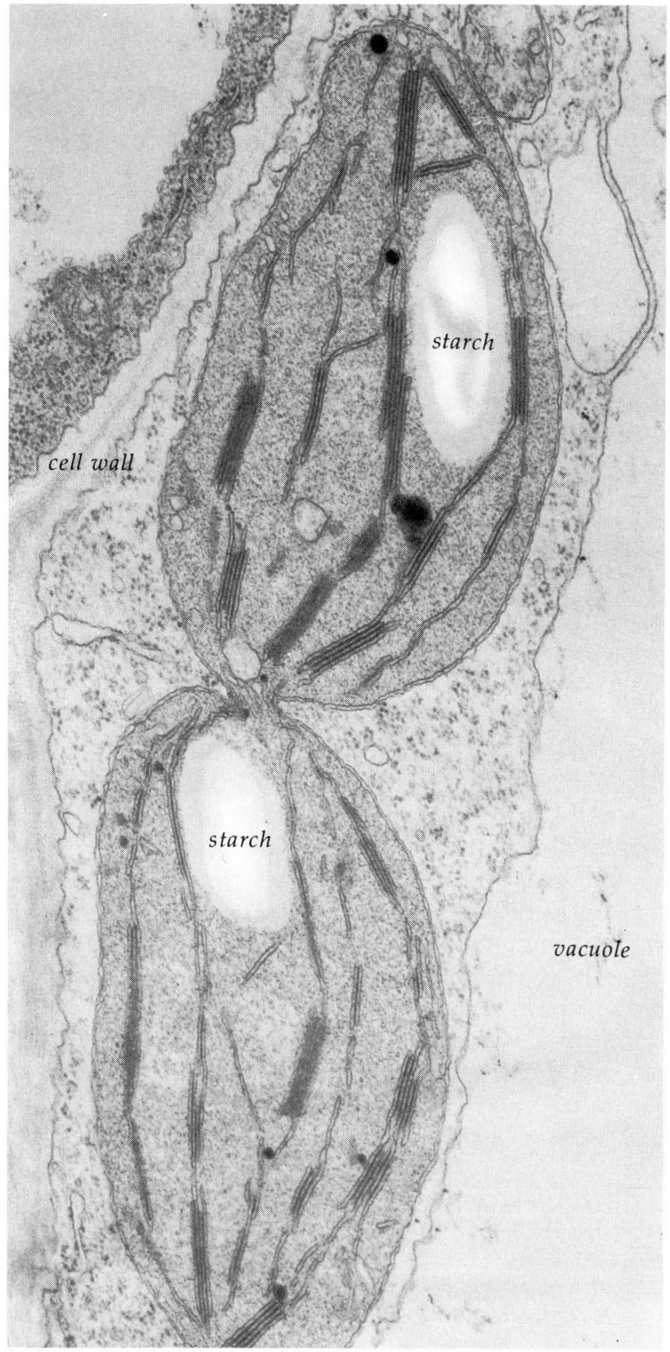

6–1 *A dividing chloroplast of a tobacco cell.*

counted because they might represent curved or twisted plastids that were sectioned in planes, giving the appearance of plastids dividing when, in fact, they were not. Too many pictures of dividing plastids have now been seen with the electron microscope for such simple explanations to hold.

Biochemical data have demonstrated the presence of substantial quantities of DNA, RNA, and ribosomes in plastids; thus they contain all the necessary chemical constituents to be self-reproducing. This does not imply that their division or development is not under the control of the nucleus, but it does mean that in many ways they are independent organelles. Both the DNA and ribosomes found in plastids differ significantly from their counterparts in the nucleus and cytoplasm of the cell. This has led some investigators to suggest that the chloroplasts were originally free-living, blue-green, algal-like cells that became incorporated into early protocells.

In meristematic cells of the root and shoot, the plastids are usually poorly developed, containing few lamellae. The lamellae are formed through the fusion of vesicles produced by invaginations of the inner layer of the double membrane surrounding the chloroplast. Tubular structures are formed and arranged in a lattice. These structures, called prolamellar bodies, are more elaborate in dark-grown plants. When the plant is placed in light, the prolamellar body becomes rearranged into the usual thylakoid structure found in the mature chloroplast (Fig. 6–G). Also contained within the developing chloroplast are membrane-bound vesicles that are continuous with the thylakoids. The materials within these vesicles are apparently utilized during the development of the chloroplast, since they change and become reduced in size in the course of the chloroplast's maturation.

A variety of factors influence plastid development, including such external conditions as light, mineral nutrition, and temperature. Plastid development is also under the control of genes that, as the inheritable units of the cell, ultimately determine the development of the organism. How such control is exercised and how nuclear control is related to the self-determining development of the plastid is an intriguing and unanswered question. It is interesting that a chloroplast contains about the same quantity of DNA as a single cell of the bacterium *Escherichia coli*. This is enough to control synthesis of 2000 to 2500 enzymes—certainly enough to account for all the reactions of photosynthesis.

Chloroplast Pigments

Since chloroplasts are colored bright green, it seemed reasonable to investigate the chemistry of the pigments that bestow this color. The First Law of Photochemistry states that light to be effective in any kind of photochemical reaction must first be absorbed. Substances that absorb light are commonly called pigments; so a reasonable question to ask concerns the pigment that absorbs the light used in photosynthesis. The pigments of chloroplasts were studied and found to fall into two groups: the chlorophylls (of which there are commonly two kinds and sometimes more) and the carotenoids (of which there are several kinds). Do all these pigments absorb light used in photosynthesis, or does only one?

The principal approach to this problem was developed in 1883 by the German plant physiologist T. W. Engelmann. His experiment was elegantly simple. A strand of the green alga *Spirogyra* was observed in the field of a microscope. This was illuminated with a light spectrum produced by a prism and in the presence of a form of oxygen-motile bacteria that tends to move toward points of highest oxygen concentration in a solution. Those colors of light most effective in photosynthesis (hence in producing oxygen) would be expected to cause the greatest accumulation of oxygen-motile bacteria, and this is what happened (Fig. 6–2). Considerable numbers of bacteria congregated along the algal strand in the blue and red parts of the spectrum, but only a few collected in the green part of the spectrum. If the number of bacteria are plotted on a graph as a function of the wavelength of the light used to produce them, a so-called action spectrum is obtained. In general, an action spectrum is obtained when one plots any kind of biological response as a function of wavelength (Fig. 6–3).

Isolated pigments can be studied by measuring an absorption spectrum. To do this, light of some wavelength is first passed through a test tube containing solvent but no pigment, and this light is detected in an instrument called a spectrophotometer. The needle on the instrument is set at zero optical density. Another test tube with the same amount of solvent but containing the pigment is then inserted in place of the so-called blank. The instrument then indicates how much light of the given wavelength is absorbed by the pigment. Another wavelength is then tested, and so on.

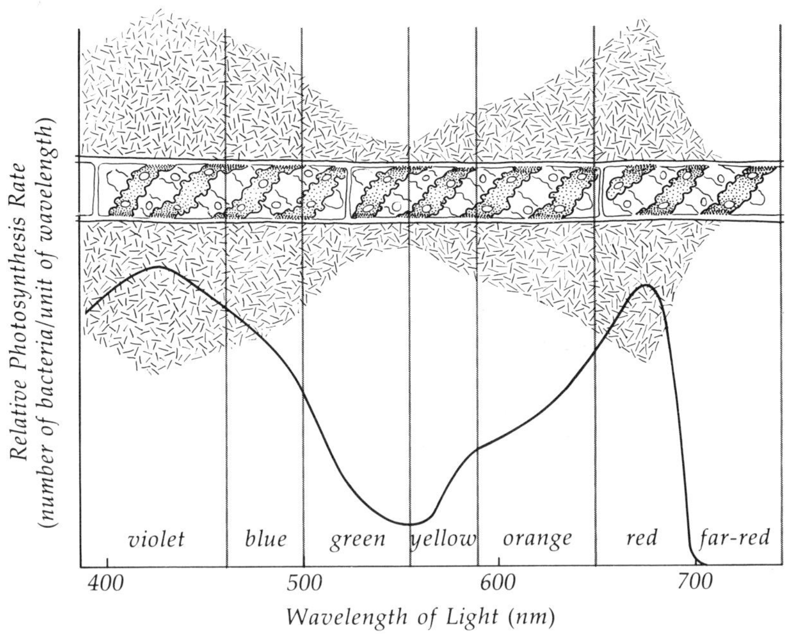

6–2 A schematic drawing showing Engelmann's experiment in which he obtained the action spectrum for photosynthesis in Spirogyra *by illuminating a filament of* Spirogyra *with a spectrum in the presence of oxygen-motile bacteria. The curve represents the results of this and similar experiments, some of which measure photosynthesis by* CO_2 *uptake.*

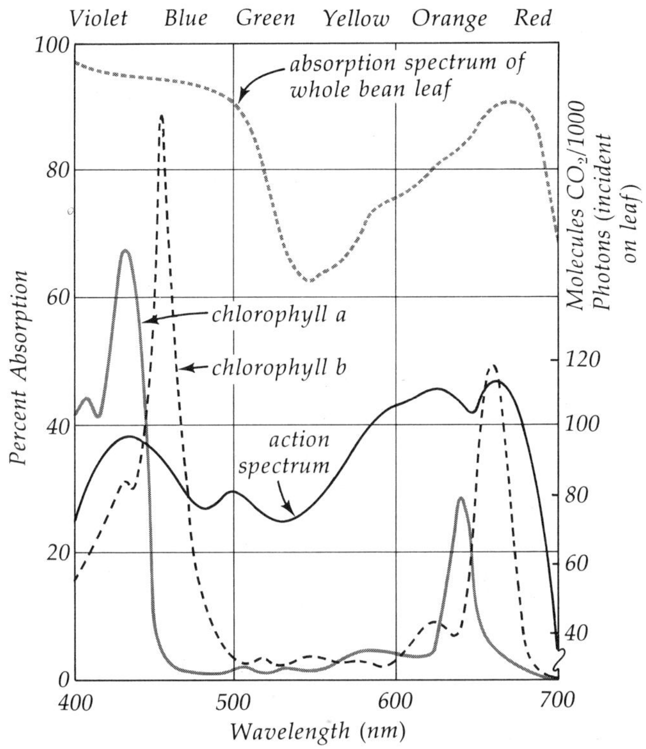

6–3 A detailed, photosynthetic action spectrum of the bean leaf compared with an absorption spectrum of an intact bean leaf, as well as absorption spectra of chlorophyll a and chlorophyll b dissolved in ether. Peaks of the chlorophyll spectra in ether are somewhat different from those in the intact plant.

Since to be effective in a photochemical response light must first be absorbed, we can deduce that the absorption spectrum of a pigment should match the action spectrum for the response being studied—if that pigment absorbs the light used in the response. Thus action spectra such as those of Engelmann should match the absorption spectrum of the pigment effective in photosynthesis. Actually, not many action spectra have been obtained, considering the number of possible plants that might be studied and other variables that should be important. Nevertheless, as is apparent in Figure 6–3, the action spectrum for photosynthesis does roughly* match the absorption spectrum of chlorophyll a and chlorophyll b. Apparently, these chlorophylls are essential to photosynthesis in all photosynthesizing plants that contain them (the vast majority of plants). The chemistry of the photosynthetic pigments is summarized in Essay 6–1.

Action spectra data indicate that some carotenoids also absorb light used in photosynthesis. This is true for the brown pigment *fucoxanthin* that occurs in the brown algae. Since the carotenoids are universally present in chloroplasts, it has long been suspected that they must play a role in photosynthesis, even though in higher plants they may not absorb the light used in the photosynthesis. The presence of carotenoids in the chloroplasts prevents the oxidation of chlorophyll caused by light; thus their most important role in the chloroplast seems to be a protective one. In-

*Absorption spectra of the chlorophylls dissolved in ether (as shown in Fig. 6–3) are somewhat different from absorption spectra in the intact plant.

terestingly, algae growing in snow—or in birdbaths—where light intensities are especially high have unusually large amounts of carotenoids. Indeed, the snow algae appear red because of these substances.

There is another group of pigments present in the red and the blue-green algae. These are the phycobilins (phycoerythrin in red algae and phycocyanin in blue-green algae). They are water-soluble pigments attached to proteins but structurally related to the porphyrins such as chlorophyll (see Essay 6–1). It has been shown that these pigments absorb light used in the photosynthesis of red and blue-green algae, and further that this occurs by a transfer of energy from the phycobilin pigment to chlorophyll. This was illustrated in an interesting experiment in which a mixture of phycobilins and chlorophyll a was illuminated with green light absorbed by the phycobilin but not by the chlorophyll. The result was a fluorescence of the chlorophyll, indicating that the energy was transferred from the one pigment to the other.

The chlorophyll and carotenoid pigments of chloroplasts exist on the thylakoid membranes. Undoubtedly, this structuring is important to the photosynthetic process.

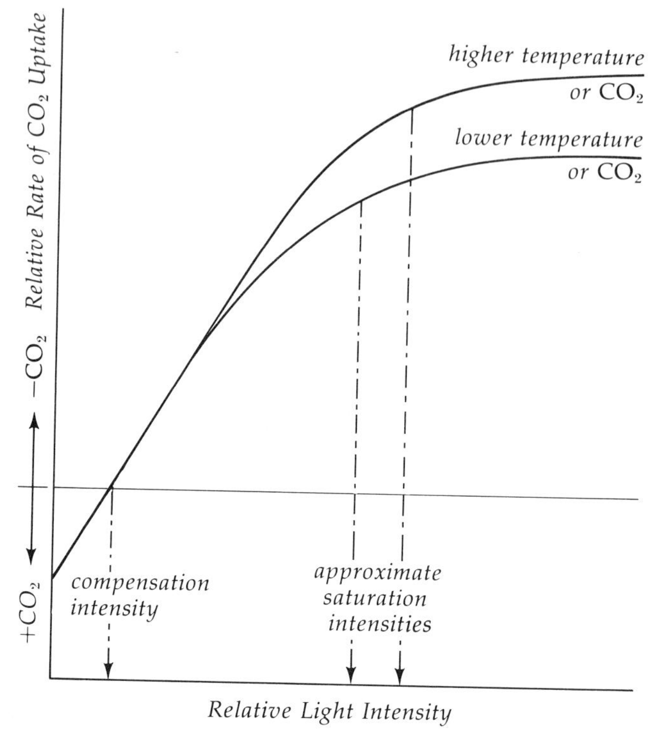

6–4 *Carbon dioxide evolution ($+CO_2$) or uptake ($-CO_2$) in green plants as a function of relative light intensity. The compensation point is where uptake and evolution are equal (photosynthesis equals respiration), and the saturation point is the light intensity above which little increase in photosynthesis occurs in response to increasing intensities. Low temperatures or CO_2 concentrations lower the maximum photosynthesis rates at high light intensities.*

Photosynthesis: Light Reaction

What are the chemical mechanisms of photosynthesis? Finding out would not only be intellectually satisfying—it could also have significant practical consequences. Let us begin with a fairly simple experiment. First, we need a way to quantitatively measure the rate of photosynthesis. This could be done by measuring the amount of carbohydrate produced (e.g., the amount of new plant material formed—often done in ecological studies), the amount of oxygen given off (often done in aquatic environments but not an easy approach in air since there is so much oxygen there already), or the amount of carbon dioxide absorbed. Usually, carbon dioxide uptake is measured. The amount of carbon dioxide in air passing over an illuminated leaf held in a transparent container is measured, usually by the amount of infrared light absorbed by a sample of the air.

In our experiment we will measure the rate of carbon dioxide uptake as a function of light intensity. The results can be plotted as in Figure 6–4 . In total darkness the plant will be giving off carbon dioxide in respiration. As the light intensity gradually increases, less and less carbon dioxide will be given off because more and more will be used in photosynthesis. At some light intensity (the compensation intensity), the amount of carbon dioxide given off in respiration will exactly equal that taken up in photosynthesis. Obviously, if the plant is to grow, it must have average light intensities higher than the compensation point. As light intensities continue to increase above the compensation intensity, more and more carbon dioxide is taken up until finally, at some light intensity, saturation is reached, after which increasing light intensity no longer increases the rate of carbon dioxide uptake.

The experiment may be repeated under different environmental conditions. For example, it may be performed at a high temperature and at a low temperature; or perhaps with high and low available concentrations of carbon dioxide in the air. Figure 6–4 shows what we might expect. At low light intensities, photosynthesis rates are not greatly affected by changing temperatures or carbon dioxide concentrations, but the effect of changing light intensities continues to be important. At saturation, however, temperatures and carbon dioxide concentrations become quite important. Saturation is reached at lower light intensities when temperatures or carbon dioxide concentrations are lower.

The experiment can be interpreted in this way: At low light intensities, light is limiting, while temperature and carbon dioxide concentrations are not; at high light intensities, the light-absorbing system is saturated; now chemical reactions become limiting—and subject to changes in temperature and carbon dioxide levels. Thus photosynthesis consists of a light reaction and a dark reaction. Light is not required for the dark reaction, but of course, under normal conditions the light and the dark reactions proceed in the light. Scientists have now studied both reactions in some detail, and a summary of the results follows. Probably, some of the most intricate and sophisticated biochemistry ever carried out has been done in these studies of photosynthesis.

What happens in the light reaction? Light energy absorbed by chlorophyll is used to split water with the release of oxygen and the formation of two important high-energy compounds: reduced nicotinamide adenine dinucleotide phosphate (NADPH) and adenosine triphosphate (ATP). Let's examine the steps leading to these.

Chlorophyll absorbs light because some change is possible within the molecule that requires exactly the energy of a photon of light. A photon (or quantum) is a packet of radiant energy that cannot be subdivided. Its energy varies depending on the color or wavelength of light; blue-light photons have a shorter wavelength but more energy than red-light photons. (The question of the nature of light and a fuller discussion of its various manifestations are presented in End Paper 6–1.)

When chlorophyll absorbs photons, there is a change in the level or orbit of electrons. One of three things can occur to such an "excited" chlorophyll molecule as the energy is released and the molecule returns to its original or ground energy state: the energy can be converted to heat; it can be released as another photon (fluoresce); or it can be captured in a chemical bond. This last possibility is the primary reaction of photosynthesis.

Within the chloroplast, chlorophyll molecules do not act singly but in groups of several hundred known as photosynthetic units. These units are part of the membrane system of the chloroplast. Photons of light are absorbed by these groups of molecules and passed around until the energy is concentrated in a reaction center. This rather indirect procedure is necessary because about 8 to 10 photons of red light are needed to fix a single carbon dioxide molecule in photosynthesis. Yet, only one photon can act at a time; hence a mechanism to accumulate energy in the reaction center has evolved.

This energy is first used to split water, releasing oxygen (the Hill Reaction) and reducing NADP to NADPH, as shown in the following reaction:

$$\text{Photons} + \text{NADP} + H_2O \xrightarrow{\text{chloroplasts}} \text{NADPH} + H^+ + \tfrac{1}{2}O_2 \uparrow$$

When NADP becomes reduced, its energy content is higher. NADPH is thus an energy-rich compound—energy that can be used to reduce many other important cellular components.

At the same time ATP is formed:

$$\text{Photons} + \text{ADP} + \text{Pi} \xrightarrow{\text{chloroplasts}} \text{ATP}$$

ADP stands for adenosine diphosphate; Pi for inorganic phosphate; and ATP for adenosine triphosphate (Fig. 6–5). The inorganic phosphate has been added to the ADP in conjunction with a complex sequence of electron-transferring enzymes that will be described in our discussion of respiration. ATP contains less than a third as much energy as NADPH, but it is much more important in the cell, since it is used in a legion of energy-requiring metabolic processes such as synthesis of polysaccharides, proteins or fats, and uptake of ions by cells.

In this part of photosynthesis, an electron may be ejected by chlorophyll when a photon is absorbed. This electron is then passed along a series of enzymes—the energy being used along the way to convert ADP + Pi to ATP. Finally, the electron is returned to the chlorophyll molecule, restoring it to a state in which it can absorb another photon. Considering this part of the photosynthetic process, we conclude that *chlorophyll acts in photosynthesis as an electron pump driven by light.*

The light reaction of photosynthesis thus consists of two processes: water is broken down, and then oxygen is produced (as are two energy-rich compounds). It may be summarized by the following equation:

$$\text{Photons} + H_2O + \text{NADP} + \text{ADP} + \text{Pi} \xrightarrow{\text{chloroplasts}} \text{NADPH}^- + H^+ + \text{ATP} + \tfrac{1}{2}O_2$$

The rest of photosynthesis simply includes necessary elaborations for higher plant life, but the light reaction is the real heart of the matter.

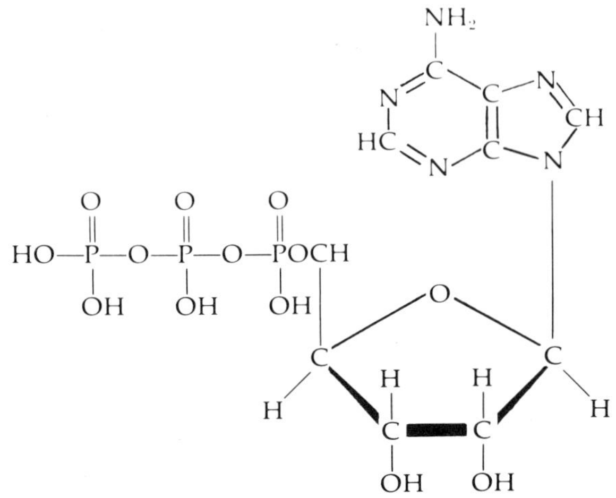

6–5 *The structure of adenosine triphosphate (ATP).*

Photosynthesis: Dark Reactions

Just how is carbon dioxide converted to sugars, cellulose, and all the other products of metabolism? The energy of NADPH and ATP is used to drive a cycle of complex reactions in which, at one stage, CO_2 is joined or "fixed" to a five-carbon compound called ribulose-diphosphate. Later in the cycle, three-carbon compounds are produced (triose phosphates) that can be used in synthesizing more complex compounds required by the cell. Ribulose diphosphate is regenerated, completing the cycle. The cycle involves 10 to 15 separate reactions, the entire cycle being driven by the energy-rich products of the light reaction. The critical point is that the CO_2 molecules are handled one at a time in a cycle.

The development of our knowledge of carbon-dioxide fixation reactions is one of the most impressive triumphs of modern biochemistry. As an example of how science progresses and to provide you with a greater insight into these remarkable and important reactions, let us examine the research that led to the conclusions just summarized and indicate the direction current research is heading.

The generalized formula for carbohydrate $(CH_2O)_n$ is the formula for formaldehyde; so it seemed logical to scientists in the early 1800s that oxygen was removed from carbon dioxide and the carbon added to a molecule of water, producing formaldehyde. Formaldehyde molecules might then be tied together to produce sugars, as in a familiar laboratory procedure. For about a century it was thought that this is exactly what happened. We've already seen that the oxygen comes from the water, not from carbon dioxide. Furthermore, after decades of trying, it had to be admitted that not a trace of formaldehyde could be found as a photosynthetic product. So much for logic!

If logic wouldn't solve the problem, how could we find out what happens in the dark reaction? The coming of the atomic age provided the needed tool. Atomic reactors produced considerable quantities of ^{14}C, the radioactive isotope of carbon. As discussed in End Paper 3–1, an element is an element because of the number of protons in its nucleus. Carbon 12 (^{12}C) has six protons and six neutrons, but carbon 14 (^{14}C) has six protons and eight neutrons. Thus in chemical reactions carbon 14 acts almost identically to carbon 12. The ^{14}C nucleus is not stable, and given a quantity of ^{14}C atoms, we can predict with certainty that half will have undergone a radioactive disintegration after 5,770 years. Of the naturally occurring carbon atoms, only 1.11% are ^{14}C, and the rest are ^{12}C.

Say that we allow a plant to photosynthesize in the presence of naturally occurring carbon dioxide. If we then want to know what happens to this carbon dioxide in the photosynthetic process, we have no way of dis-

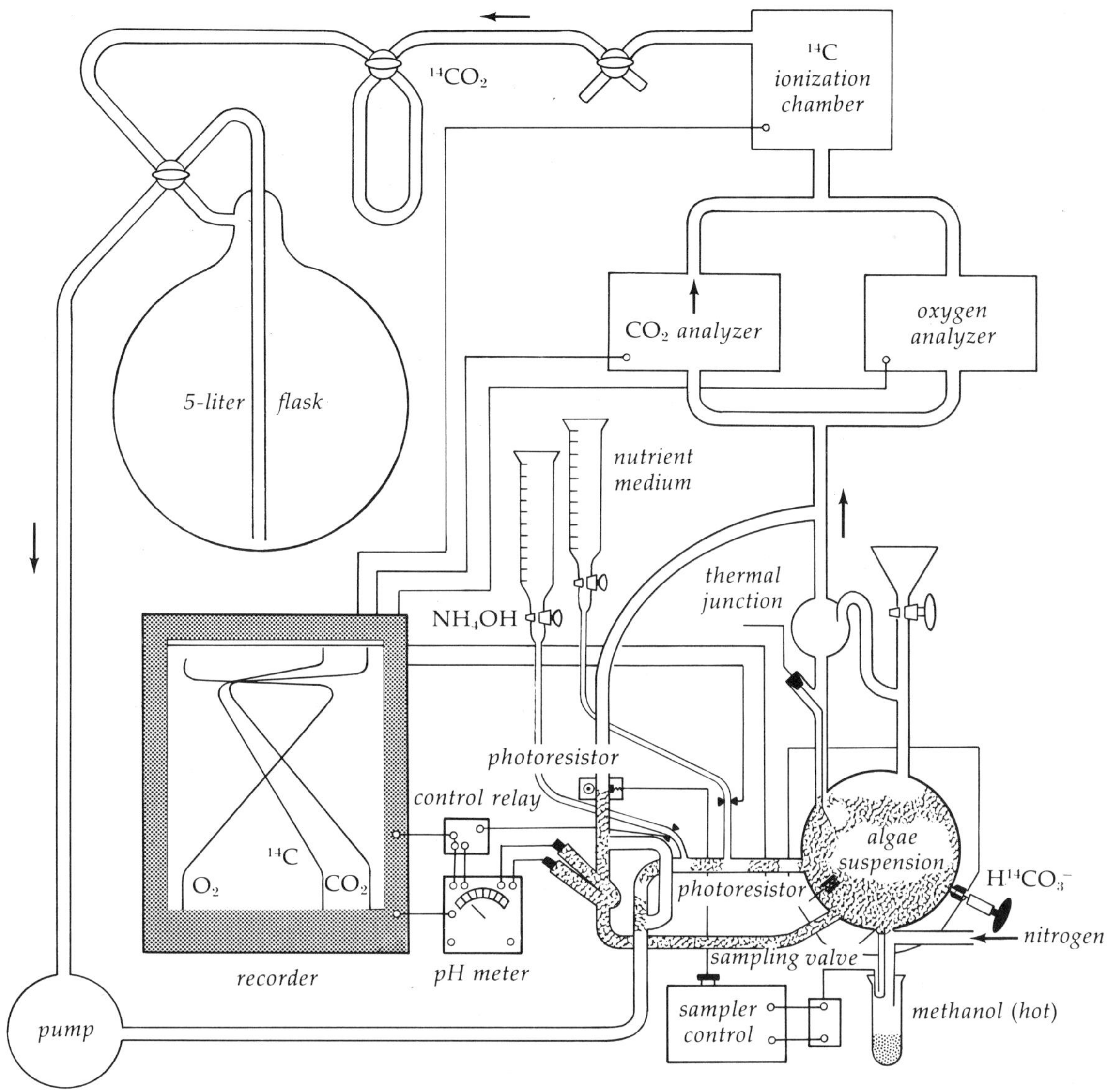

6–6 The University of California system used to expose algae to $^{14}CO_2$ for short time periods under steady state photosynthetic conditions. Total CO_2, O_2, and $^{14}CO_2$ were analyzed and recorded as a function of time. Carbon 14 was injected into the algae suspension with a syringe (lower right) as $NaH^{14}CO_3$. *After short periods of ^{14}C fixation, some of the algae were dropped into boiling methanol to kill them and to extract labeled compounds.*

tinguishing the carbon atoms that were carbon dioxide from all the others already present in the plant tissue. If, however, we allow the plant to photosynthesize in the presence of carbon dioxide containing a percentage of ^{14}C much higher than 1.11%, we will be able to keep track of it during the photosynthetic process. Most of the ^{14}C atoms occurring in the plant tissue after photosynthesis will be those that had recently entered as carbon dioxide.

The experiments were originally done by Melvin Calvin, Andrew Benson, and their colleagues at the University of California in Berkeley. They used suspensions of the single-celled green alga *Chlorella*. Depending upon the acidity of the solution, much of the carbon dioxide in water exists in the form of bicarbonate ions (HCO_3^-). Thus they could add bicarbonate containing radioactive ^{14}C to a suspension of *Chlorella* cells in the light and then stop the reaction after some time interval by injecting a fairly large quantity of boiling alcohol, enough to quickly kill all of the cells and extract the carbohydrate produced in photosynthesis (Fig. 6–6).

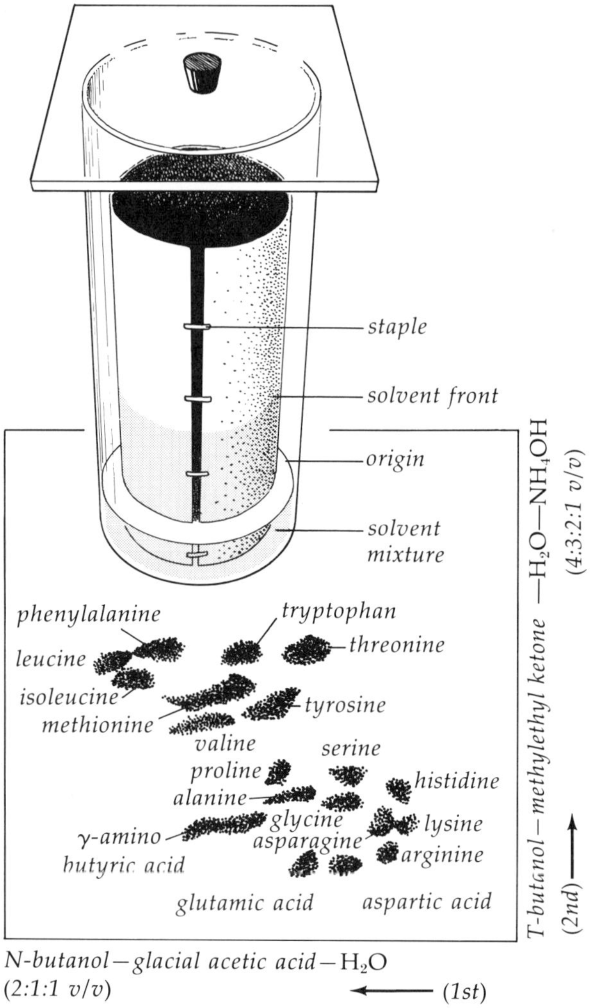

6–7 The apparatus used for paper chromatography. A drop of solution containing a mixture of closely related substances is allowed to evaporate close to the corner of a large sheet of filter paper. The filter paper is then placed in an insulated, vapor-tight cabinet or in a large glass jar. The paper becomes equilibrated with the atmosphere in the jar before it is allowed to drop into the liquid. When the paper is dropped into the solvent mixture, solvents begin to move by capillarity, down the paper or up the paper. The different substances applied in the drop of solution have different solubilities in the solvent, and they also have different tenacities for the filter paper. The result is that they move at different rates in relation to the solvent front, thus becoming separated from each other. Usually, they do not become completely separated after moving in one direction; so the filter paper is dried, turned on its edge, and placed in a different mixture of solvents. In such a two-dimensional chromatogram, the substances are further separated, and if solvents are suitably chosen, all of the unknowns may be separated from each other. To detect radioactive substances, the paper is dried and placed in contact with a piece of X-ray film. It may be left there for hours, days, or even weeks or months. Finally, the film is developed, and wherever there is a spot of radioactivity, there will be an exposed dark spot on the film. This chromatogram illustrates separation of amino acids. The amino acids are made visible (purple) on the paper by treatment with a reagent called ninhydrin. The filter paper may be cut up and the unknown substances washed off and identified by conventional chemical procedures. The position of the spot is a preliminary identification, since a number of chromatograms may be prepared with samples of known compounds.

Identifying the radioactive compounds produced in photosynthesis was a long and difficult task. All of this was done in the late 1940s by using the newly available technique of paper chromatography. In this procedure (certainly one of the most important reasons for the rapid advances in biochemistry since World War II), closely related molecules can be separated from each other, as explained in the caption for Figure 6–7.

When Calvin and Benson allowed *Chlorella* to attain a steady rate of photosynthesis and then introduced radioactive bicarbonate into the solution (stopping the reaction after as short a time as only 7 seconds), as many as 12 radioactive compounds had already been formed. After 60 seconds, numerous radioactive spots appeared on the chromatograms (see Fig. 6–7) including a number of phosphorylated sugars (sugars with a phosphate attached), amino acids, and organic acids. To answer the question of what first happens to the carbon dioxide, the time had to be shortened to below 5 seconds. With 2 second photosynthesis times, although six or seven compounds appeared on the chromatograph, one was considerably more radioactive than the others: phosphoglyceric acid, abbreviated PGA.

Apparently the PGA was the first product in the fixation of carbon dioxide. This compound contained *three* carbon atoms (rather than the one carbon atom of the "logical" formaldehyde), but only one of the three carbon atoms (a terminal one) was radioactive. It appeared that the first step in the dark reaction of photosynthesis was the addition of carbon dioxide to a so-called "two-carbon fragment." A search was instigated for the fragment. Again, Mother Nature proved to be more complex than we might have anticipated. Carbon dioxide was added not to a two-carbon fragment but to the five-carbon sugar ribulose-diphosphate. Adding the carbon dioxide molecule produced a six-carbon phosphorylated sugar that was immediately split to produce two molecules of PGA. All of this occurred in the presence of what must surely be considered one of the most important enzymes of living things: carboxydismutase, also referred to as ribulose-diphosphate carboxylase. This combining of carbon dioxide with ribulose-diphosphate in the presence of carboxydismutase is not an energy-requiring process. It goes on spontaneously when concentrations of carbon dioxide, ribulose-diphosphate, and the enzyme are high enough. As a matter of fact, the enzyme has been found in several microorganisms not capable of photosynthesis. Similar enzymes that will fix carbon dioxide in the dark are known to exist in virtually all organisms, including animals.

Surprisingly, the affinity of CO_2 for this enzyme is rather low; yet photosynthesis proceeds even when concentrations of CO_2 in the atmosphere are very low. Perhaps this is because such large quantities of this enzyme occur in plants (about 16% of the total protein in a spinach leaf, for example).

In the photosynthetic process of higher plants, the energy of NADPH and ATP is used to convert PGA to a close relative—phosphoglyceraldehyde, which is considerably more energy rich than PGA.

Where did the ribulose-diphosphate come from? Calvin, Benson, and their colleagues addressed themselves to this question, studying carefully the products of photosynthesis produced after various intervals of photosynthesizing time. Bit by bit, they constructed the scheme illustrated in Figure 6–8. Note that the ribulose-diphosphate is produced after the PGAldehyde passes through a number of transformations, at several of which energy from NADPH and ATP is required. Furthermore, hexose sugars and other compounds are also produced—one hexose sugar being produced for each six carbon dioxide molecules fixed in the dark reaction or for each six times that the complex cycle is completed. This phase of photosynthesis is now known to occur in the stroma of chloroplasts; whereas the light reaction occurs on the thylakoid membranes.

The entire cycle is called the Calvin-Benson Cycle, and Calvin received a Nobel prize in 1961 for his work on it. It should be noted that the sugars are all phosphorylated rather than occurring as free sugar molecules. Although the steps are not shown, the sugars are often combined with nucleotides (the same building blocks used to make nucleic acids; see Chapter 8), and then in this form, they are combined to produce such disaccharides as sucrose or such polysaccharides as starch or cellulose. Free six-carbon sugars are not the immediate product of photosynthesis, although this is often implied in discussions of photosynthesis. It is also important to note (and this too is unapparent from Fig. 6–8) that amino acids, organic acids, and other compounds are produced very early in the photosynthetic process.

Hatch and Slack Pathway

In the 1960s George O. Burr and his colleagues at the Sugar Cane Research Institute in Hawaii discovered that PGA was not the first photosynthetic product in sugarcane plants—a species that is unusually efficient in photosynthesis. Rather, the products first labeled most heavily were four-carbon acids. Later work, particularly by M. D. Hatch and C. R. Slack in Australia, showed that these acids were formed as carbon dioxide was added to the three-carbon compound phosphoenol pyruvic acid in the presence of the enzyme phosphoenol pyruvate carboxylase to produce the four-carbon acid oxaloacetic acid. The enzyme for this reaction has an affinity for carbon dioxide considerably higher than that of carboxydismutase. In this kind of photosynthesis, now referred to as the Hatch and Slack Pathway, the four-carbon acids were further metabolized according to reactions long familiar to us as part

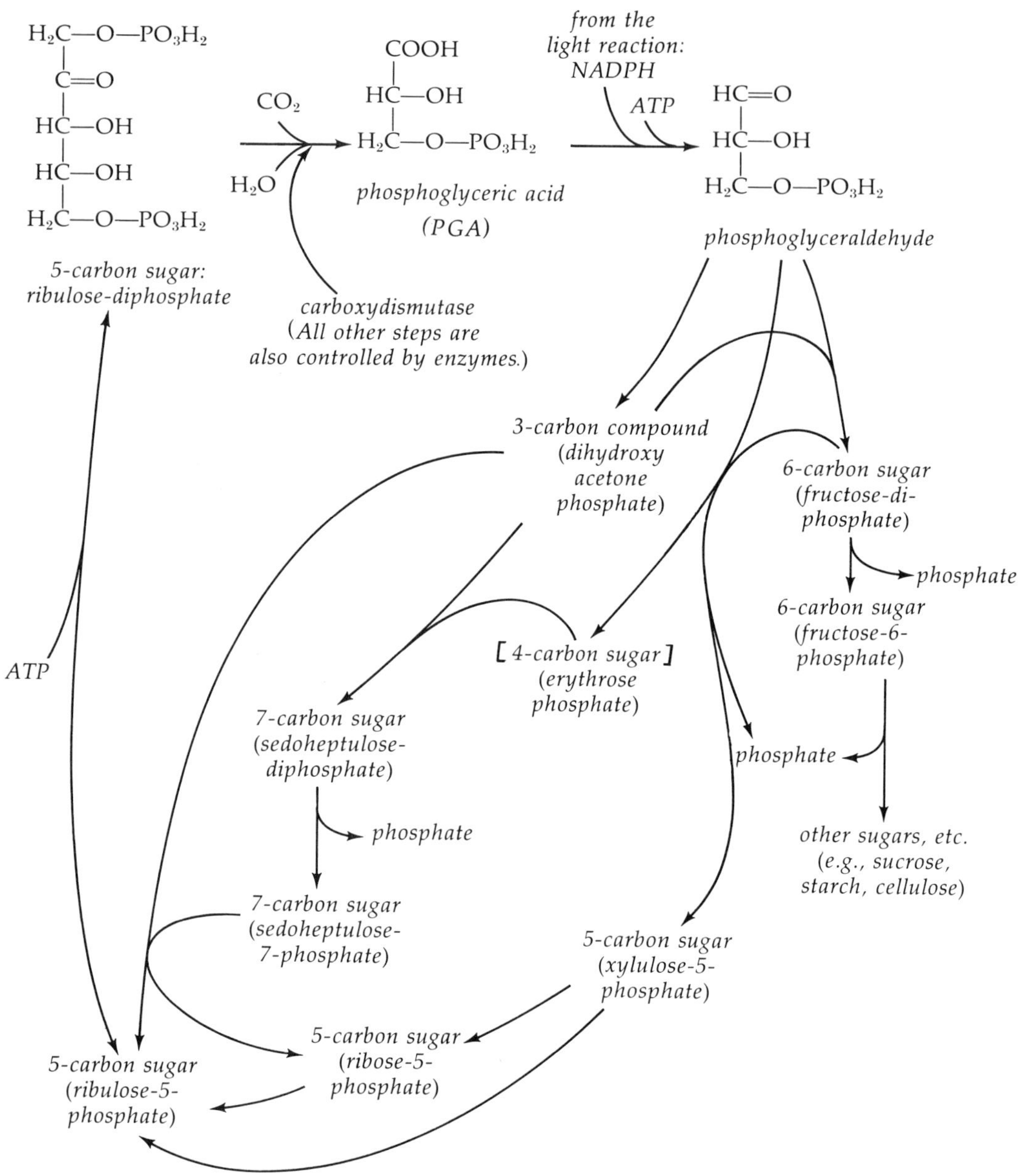

6–8 A schematic outline of the Calvin-Benson Cycle, accounting for uptake of CO_2 in the photosynthesis of many species.

of respiration (see Chapter 7). Again, the products are carbohydrates, amino acids, and many other compounds. Proposed reactions for this pathway (Fig. 6–9) have been worked out only recently.

The Hatch and Slack Pathway occurs in several species of tropical grasses as well as in sugarcane. It occurs in corn (which was originally a tropical grass) and in such unrelated species of dicots as amaranth (*Amaranthus*) and desert saltbush (*Atriplex*). The fact that this unique metabolic pathway occurs in a number of unrelated species poses an interesting problem for the evolutionist.

Recently, plant physiologists were excited by the discovery of photorespiration. Using isotopes of oxygen, it was shown that increasing light intensities increases respiration in some plants. The effect of light

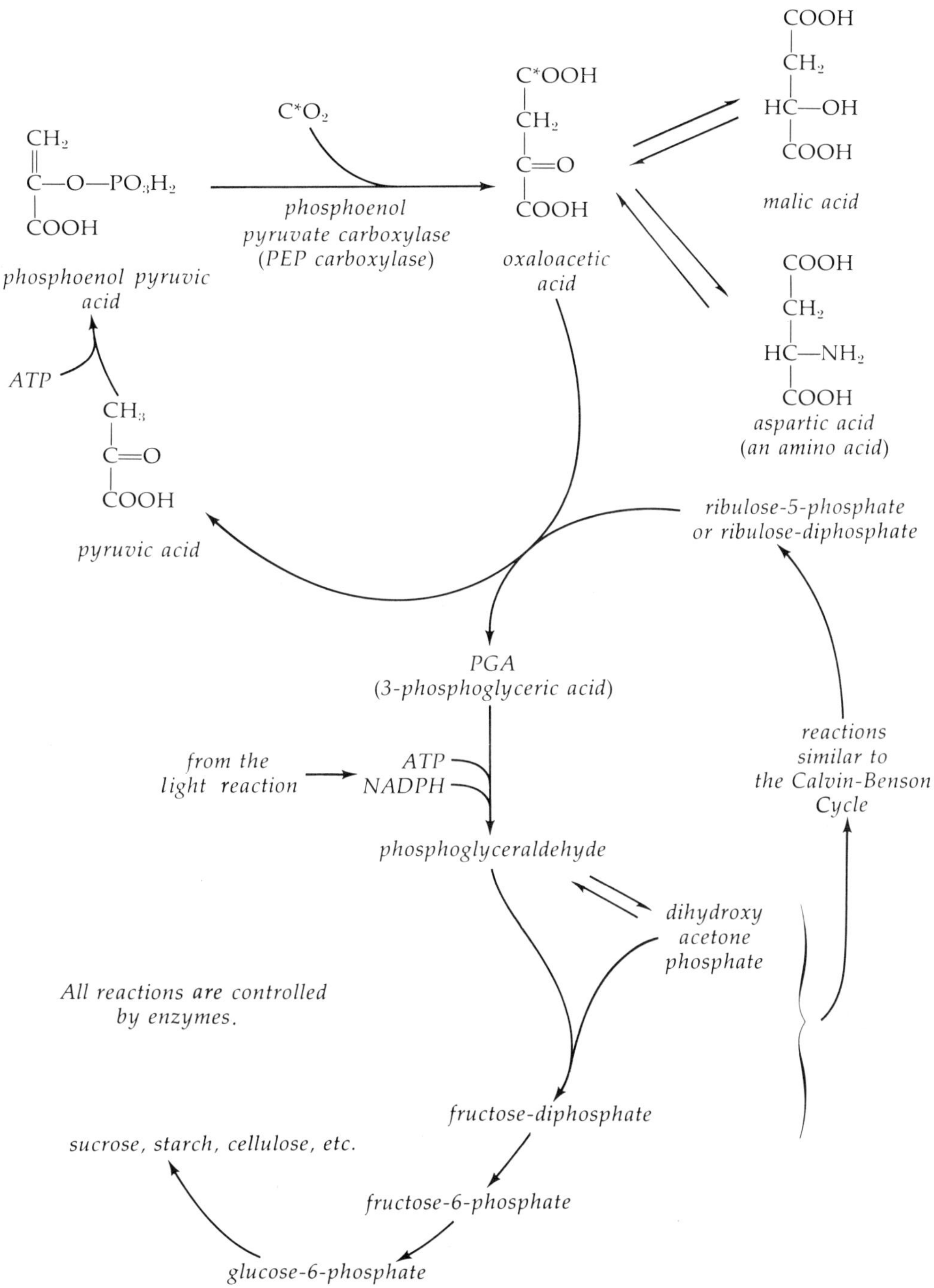

6–9 The Hatch and Slack or C_4-dicarboxylic acid Pathway of carbon dioxide fixation as it occurs in many species, especially tropical grasses. The asterisk indicates that the upper carbon in oxaloacetic acid comes from the carbon of CO_2. In a sense, the cycle may be thought of as an alternative way to add carbon to ribulose-diphosphate. Only the upper few reactions in the diagram are unique; the others are the same as in the Calvin-Benson Cycle.

on respiration had always been a problem. In most early studies, all that could be done was to subtract the CO_2 given off in the dark (respiration) from the CO_2 taken up during photosynthesis in the light, assuming that respiration was not affected by light.

In photorespiration an intermediate of photosynthesis (glycolic acid) is oxidized more rapidly in the light. Interestingly, those plants utilizing the Hatch and Slack Pathway of photosynthesis do not exhibit photorespiration. This difference, as well as the greater affinity of phosphenol pyruvate carboxylase for carbon dioxide, helps us to understand why the Hatch and Slack plants are so much more efficient in their photosynthesis. Clearly, a goal for future plant breeders should be to produce agricultural plants that minimize photorespiration and emphasize Hatch and Slack photosynthesis.

We began our discussion of photosynthesis with a summary equation for the process. We can now quantitate this, using symbols. Note that the asterisks indicate that the oxygen originally comes from the water, and note the 8 to 10 photons taking part in the process. Carbohydrate is indicated by its general equation rather than as some specific compound such as glucose (which is seldom a product). The equation does not indicate that photosynthetic products might also be amino acids or organic acids. Such an equation would have to include nitrogen on the left side of the equation and amino acids on the right, and it would also have to adjust for the fact that the organic and amino acids cannot have the generalized formula for carbohydrate. Here is the somewhat simplified summary equation:

$$2nH_2O^* + nCO_2 + n(8 \text{ to } 10 \text{ photons}) \xrightarrow[\text{(enzymes in cell environment)}]{\text{chloroplasts}} (CH_2O)_n + nO_2^* + nH_2O$$

Starch and Amyloplasts

The sugars formed at the end of photosynthesis may be used directly in metabolism. Typically, sucrose is formed by joining two six-carbon sugars. Sucrose is the form in which most carbohydrate is moved through most plants. Another alternative is to store the product of photosynthesis in the form of starch. Starch is a major, but not the only, carbohydrate food-storage form in the plant.

The formation and storage of starch may take place in the chloroplast itself. In the algae this is the usual case, and the starch is associated with a special structure or area within the chloroplast—the pyrenoid (Fig. 6–E). The chloroplasts of the ferns and seed plants lack pyrenoids and, although there are exceptions, usually do not form large amounts of starch. In these plants the mass of starch is stored in special colorless plastids called amyloplasts, which contain the enzymes necessary for the formation of starch. The starch is deposited in layers; one layer is laid down on top of the previous one until a large starch grain is formed (Fig. 6–H).

The amyloplasts are not the only type of storage plastid. Another type is involved in the storage of protein; it is found in the seeds of many plants and is a ready source of amino acids during periods of rapid growth associated with germination. Still another plastid in this same group is known to store oils. Fats and oils are even better energy sources than the carbohydrates and are frequently found as reserve energy supplies in both plant and animal cells.

Microbodies

For years, single membrane-bound spherical or cup-shaped particles, some 0.5 to 1.5 μ in diameter, have been observed in sections of plant tissue (Fig. 6–11). These particles are darker inside than the surrounding cytoplasm, and some contain crystalline material. They show no elaborate internal structure such as that seen in the plastids or mitochondria. Such particles have been ignored or classed as enlarged dictyosome vesicles, small vacuoles, spherosomes, or lysosomes. We now know that they constitute a separate class of cell organelles called microbodies.

Plant microbodies are found in all types of plant cells, frequently in close association with segments of the ER. They are known to contain oxidative enzymes, and some appear to have specific biochemical roles in plant cells. The best enzymatically characterized microbody is the glyoxysome, found in the cells of germinating seeds in which oils and fats are the principal storage forms. The enzymes present in the glyoxysome convert the oils and fats to carbohydrates that can then be translocated to other parts of the plant. Another microbody, the peroxisome, is known to be present in the leaves of some plants and is believed to be involved

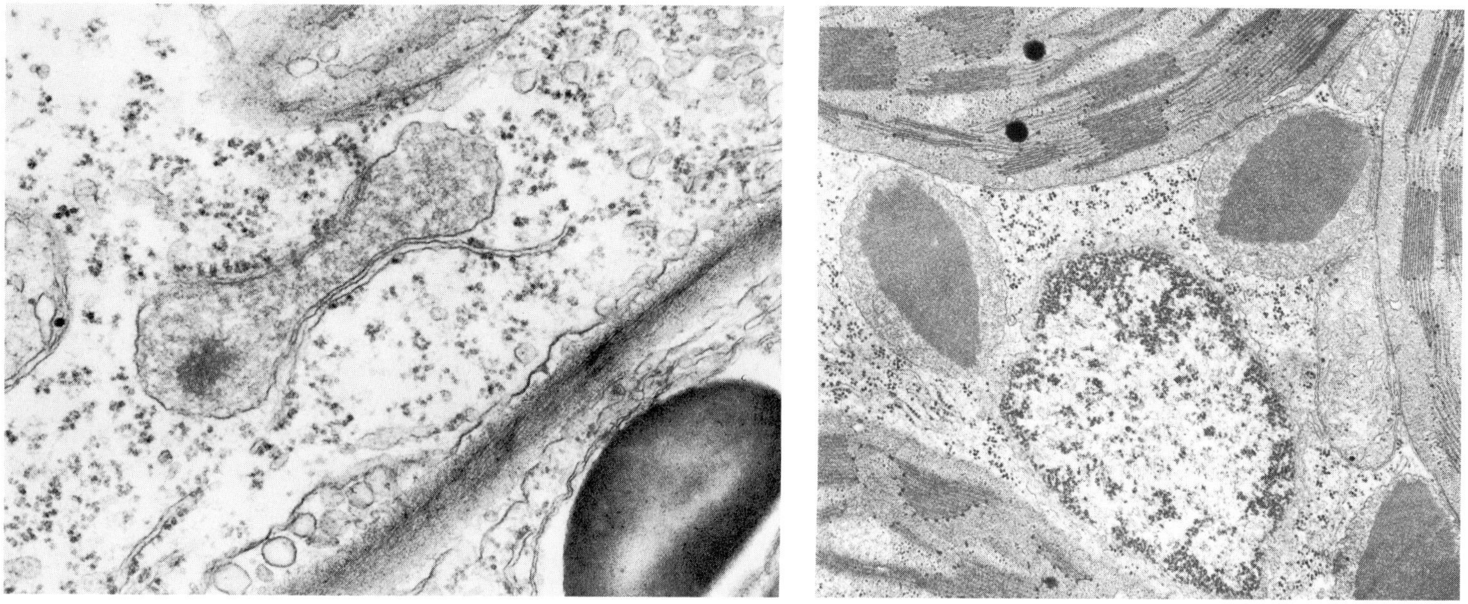

(*a*) (*b*)

6–10 *Microbodies in various plant cells: (a) epidermal cells of* Byrum *capsule, and (b) leaf cell of* Hedychium.

in amino acid synthesis and in photorespiration—the light-dependent respiration mentioned earlier.

Photosynthesis in Nature

In Chapters 37 to 40 we shall describe the role of photosynthesis in the ecosystem. But we should follow our look at the cellular and metabolic details of the process with a short discussion of a few principles of photosynthesis in nature.

As might well be imagined, several environmental factors strongly influence photosynthesis rates. These rates are first limited by penetration of carbon dioxide to the cell. Much of this is through the openings on leaf surfaces called stomates (see Chapter 15). Some of the penetration occurs through the waxy covering of the leaf called the cuticle. There has been considerable argument as to the relative importance of these two kinds of carbon dioxide penetration. Apparently, the importance depends upon the species. In many species penetration of the cuticle is minimal, but in some it is rather high.

As indicated in our discussion of the two photosynthetic pathways, rates can be influenced differently by light intensity, depending upon the species. Species utilizing the Hatch and Slack Pathway photosynthesize more at both low and high light intensities. Furthermore, plants may become adapted to light intensity. So-called sun plants have relatively high compensation points and saturation levels, while shade plants have noticeably lower levels. Also, the leaves of these two kinds of plants become adapted morphologically in ways that we shall discuss in Chapter 15.

As a rule, light quality (the balance of colors in the spectrum) is fairly constant over most of the earth's surface. Since much of the blue and the red parts of the spectrum are filtered out by the canopy of leaves, however, the spectrum reaching a forest floor is much richer in green light. This is also true at great depths in the ocean or in other bodies of water. Water absorbs more red than blue, but floating green algae absorb blue and red more than green. In the oceans red algae grow at greater depths than green. The red algae are able to utilize the green light absorbed by their red-colored phycoerythrin (phycobilin) pigments. These pigments in those plants are clearly ecologically adaptive. Whether a similar situation occurs with plants growing on the forest floor remains to be studied.

(Such plants most probably do not use phycobilin pigments, but they may be adapted in other ways to the light rich in green.)

The temperature range over which plants can photosynthesize is surprisingly large. Certain microorganisms photosynthesize at temperatures as high as 70°C, while many conifers and lichens maintain this process, although weakly, at temperatures several degrees below freezing. The optimum temperatures for photosynthesis probably fall within 20 to 35°C for most species of temperate regions. Typically, maximum photosynthesis occurs at temperatures a few degrees lower than those optimal for most rapid respiration.

In the long run the factor most important in limiting photosynthesis of land plants is availability of water. Unless the plant has adequate water, growth and photosynthesis will not occur.

Since world food supplies depend upon photosynthesis, it is important to understand photosynthetic efficiencies. If about 3 photons of light are required to account for the energy trapped by the fixation of a single carbon dioxide molecule, while 8 to 10 photons actually go into this process, it is apparent that photosynthesis can be about 30 to 38% efficient. But such efficiencies can seldom be measured except under ideal laboratory conditions. If a dense algal culture is used so that all of the light is absorbed and used in photosynthesis, efficiencies of 25 to 40% can be detected. (This is the basis for the calculation of an 8 to 10 quantum requirement.) Algal cultures on a commercial scale may sometimes achieve levels of 25% efficiency, but this is very unusual. Efficiency of crop plants is typically on the order of 1 to 2%, and this is also true for most commercial algal cultures. Efficiencies in individual leaves may be measured at values as high as 15 to 19%. Clearly, if the maximum possible photosynthesis efficiencies are around 30%, while those obtained in the field are around 1 to 2%, we have a considerable opportunity for improvement. If efficiencies could be brought only a little closer to the maximum possible levels, world food production figures could be doubled, tripled, or more. With the population explosion and dire prophecies of impending famine, this is clearly a worthy research goal.

From the standpoint of ecology, we should not overlook the importance of photosynthesis in maintaining our atmosphere in a condition suitable for animal life. Without photosynthesis and in the presence of continued respiration, carbon dioxide would soon build up to toxic levels, and oxygen would be depleted. At the moment, ecologists are attempting to assess the importance of the ocean's phytoplankton in maintaining this balance. Say that, as some have estimated, 70% of the oxygen in the atmosphere is produced by photosynthesis in the ocean and that the ocean's photosynthesizing organisms are all destroyed by such pollutants as mercury, DDT, or oil. Obviously, we would soon be in trouble. Yet the actual situation is difficult to assess. Some workers have estimated that only about 30% of the earth's photosynthesis occurs in the ocean; most oceans are quite sterile, being limited in their phytoplankton production by low nutrient elements (see Chapter 38). It is also possible that virtually all of the oxygen produced by photosynthesizing organisms in the ocean is used by respiring organisms in the ocean. If ocean photosynthesis greatly exceeds ocean respiration, then carbon compounds should be building up in the ocean and being depleted on land—and there is no evidence that this is happening. Perhaps we land organisms don't need the oceans after all! Let us go right ahead and pollute them to death. Obviously, this would be a foolish course to follow. The fact is, we simply don't know all the interactions between land and marine organisms. Certainly, there are many such interactions, and we must discover and measure them. It would be extremely unwise to do any more tampering with our planet's ecosystems than has already been done.

Essay 6–1 Chemistry of the Photosynthetic Pigments

The chemistry of chlorophyll has been studied extensively. Its molecular structure was determined during the first two decades of this century and finally confirmed by synthesizing the molecule in the laboratory—but only relatively recently (1960). The molecule is quite complex as metabolic molecules go. It consists of four subunits called pyroles. Each of these is a five-membered ring consisting of four carbons and one nitrogen. The four pyrole rings are held in a secondary flat ring with a carbon atom between each of them. In the center of this molecule is an atom of magnesium that is undoubtedly important in the photosynthetic process. The four pyrole rings hooked together, as in Figure 6–0a, form a molecule called a porphyrin. Porphyrins are important in life functions. Hemoglobin includes a porphyrin (with iron instead of magnesium), as do the cytochrome enzymes of respiration (also containing iron) discussed in Chapter 7.

The porphyrin part of the chlorophyll molecule forms a flat disc with magnesium at the center. In the different kinds of chlorophyll, there are different groups of carbon atoms extending out from the edge of this disc, as indicated in the figure. In some cases (chlorophyll b), these groups are more water-soluble than in other cases, accounting for a variety in solubilities of the different chlorophylls. All of the chlorophylls also have a long side chain called a phytol. This side chain is related in some ways to the fatty acids discussed in End Paper 4–1, and it is closely related to the carotenes discussed here. On one end it has an alcohol (—OH) group, and this is esterified to a carboxyl side chain on the porphyrin of chlorophyll.

The carotenoids consist of long chains of 30 carbon atoms in each molecule. They can be structured on paper by combining six units of five atoms each. In these so-called isoprenoid units, three carbons are attached in a chain with two others attached to one of the carbons at the end of the chain. Probably carotenoids (and other so-called isoprenoid compounds) are built up in the plant from these five-carbon isoprenoid units. In the carotenoids there are double bonds between some of the carbons in the chain (as discussed for fatty acids). The structure of β-carotene is shown in Figure 6–0b.

There are two kinds of carotenoids: carotenes and xanthophylls. Carotenes contain no oxygen, but xanthophylls do. Both carotenoids, because of the long carbon chains, are highly lipid-soluble, although

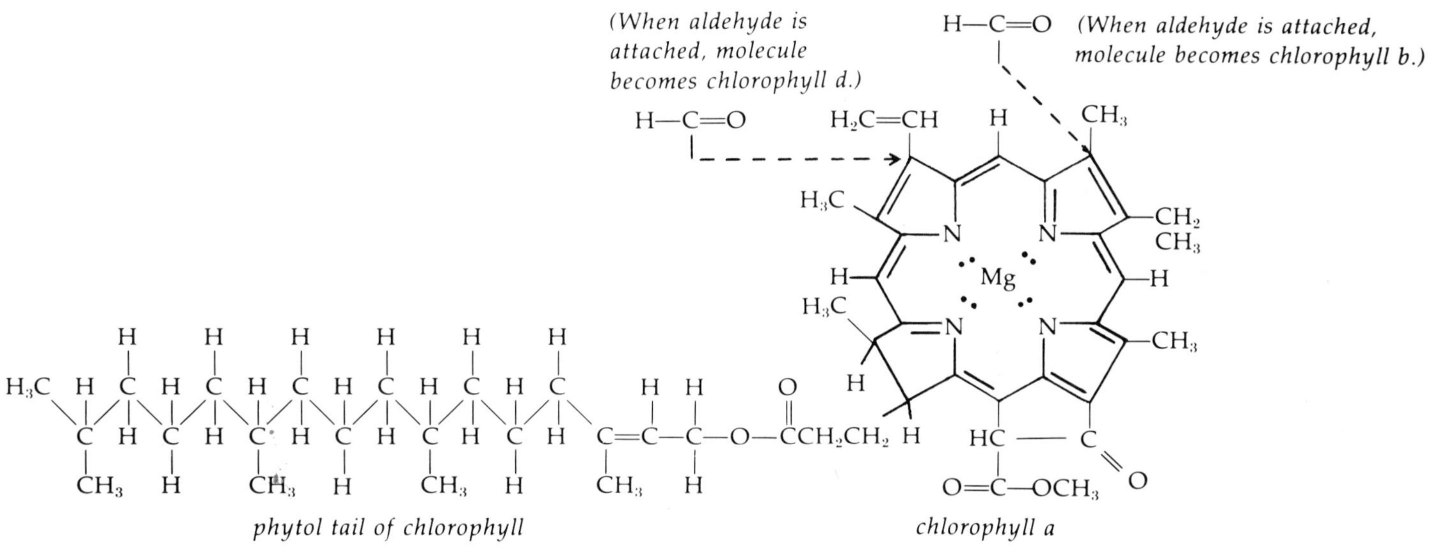

6–0a The molecular structure of chlorophylls a, b, and d. b and d are formed by substituting aldehyde groups for the groups shown.

the xanthophylls (because of their oxygen) are somewhat more water-soluble than the carotenes. These pigments are usually red, orange, or yellow, but they are widely distributed in both plants and animals and sometimes exhibit other colors. They are always present in chloroplasts. They account for the orange color of carrots, the red of tomatoes, the yellow of flowers; but they also account for the green of an avocado fruit, the pink of a shrimp, and the black of squid ink. Many of the fall colors of leaves in temperate zones are due to carotenoids. (Others are due to the water-soluble xanthophylls, usually present in the vacuole rather than in plastids.) The phytol side chain of chlorophyll is derived from a carotenoid, and vitamin A is produced by cleaving the carotene molecule.

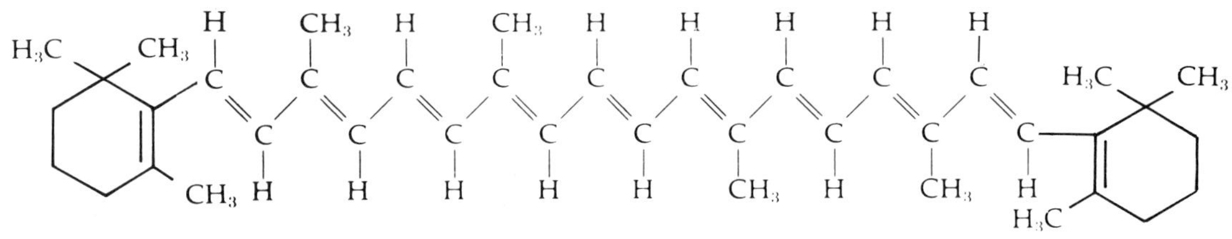

β-carotene

6–0b A representative of the carotenoids: β-carotene. It is abundant in chloroplast membranes and can split into two molecules of vitamin A.

End Paper 6–1 Photochemistry and the Nature of Light

We do not really understand the physical nature of light, but there are two theories or models that allow us to predict its behavior with a high degree of accuracy. The theories are not difficult to relate to each other in a mathematical sense, but in a physical sense they appear to be so different that they are mutually exclusive.

The first theory states that light is a wave phenomenon. Sound is clearly a wave phenomenon in which the energy is transmitted through a medium such as air by molecules hitting each other to produce waves, in a manner analogous to the way a wave is transmitted on the surface of a body of water. The physics of the transmission of waves through media has been studied in intricate detail, and the results can be nicely applied to light. The velocity (c) of light waves is extremely high, and it is essentially constant (300,000 km or 186,000 miles per second). Actually, although all electromagnetic radiation, including visible light, moves at this velocity in a vacuum, velocity is decreased slightly when light passes through a dense medium such as glass, making optical science possible.

The wavelength (λ) is the distance between the crests of energy. (In the waves on a body of water, it is the distance between the crests of two adjacent waves.) Light waves, then, are moving at velocity c with distances λ between them. If the crests could be counted as they went by, the number passing a given point in a given interval of time would be called the frequency (ν). Obviously, the distance between the waves and the velocity at which they travel will determine the frequency, and by a simple equation, $\nu = c/\lambda$. This equation says that dividing the distance traveled by a wave in a second by the distance between waves will give the number of waves that pass a point in a second.

By a simple algebraic transformation, it is apparent that the wavelength is equal to the velocity divided by the frequency: $\lambda = c/\nu$. This is not difficult to understand, and we can imagine how it might work perfectly well with the waves moving on the surface of a body of water.*

The wave model of electromagnetic radiation allows us to understand a number of important physical phenomena,

*Say that a water wave moves at 40 m/min. If we set up a vibrator that produces waves 1 m apart, then the frequency will be 40 waves/min or

$$\nu = \frac{c}{\lambda} \text{ or } 40 \text{ waves/min} = \frac{40 \text{ m/min}}{1 \text{ m}}.$$

On the other hand, if we set our vibrator to produce 80 waves/min, then the distance between waves will be 0.5 m or

$$\lambda = \frac{c}{\nu} \text{ or } 0.5 \text{ m} = \frac{40 \text{ m/min}}{80 \text{ waves/min}}.$$

particularly diffraction (see Essay 8–2). Furthermore, we have recognized the idea that the wavelength is an indication of the color. By applying the mathematics of wave motion (in much more detail than implied here), we can predict the behavior of light as it passes through lenses, minute holes, thin films, etc. But we are in trouble. There is apparently no medium (such as water or air) through which the light waves might travel. The ether was postulated to be this medium, but the Michelsen-Morley experiment performed in 1887 seems to disprove its existence conclusively.

The second theory or model of light suggests that light consists of a number of particles called photons or quanta. These particles travel through space at velocity c, and they have no mass (although the theory of relativity predicts that they will respond slightly to gravitational force). Around the turn of the century Max Planck, the great German physicist, showed that light energy comes in indivisible packets; that is, a photon or quantum (the terms are synonymous) of light energy cannot be subdivided. It is difficult to explain this finding in terms of the wave theory of light, but the quantum model can be easily applied to the facts of photochemistry, as we shall do here. It was also applied by Einstein to the so-called photoelectric effect, the principle utilized in photographic exposure meters and many other modern devices. Incidentally, the photon model of light will not explain the diffraction phenomena so nicely described by the wave model. Thus scientists have simply learned to live with this schizophrenic situation, using each of the theories where it works best.

Measurements showed that the energy of a photon is proportional to the frequency. This is stated by the simple equation $E = h\nu$, where h is Planck's constant (1.58×10^{-34} cal/sec or 6.624×10^{-27} erg/sec). The equation states that the higher the frequency, the more energy per photon. Since, as we just saw, frequency and wavelength are reciprocally related, we could write $E = ch/\lambda$. Written this way, we can see that the *shorter* the wavelength, the greater the energy of a single photon.

Note that we are not talking about total light energy produced by illumination. We are discussing energy per photon. The equation says that a photon of light in the blue part of the spectrum will have nearly twice as much energy as a photon of light in the red part of the spectrum. Nevertheless, the actual energy received by a surface illuminated with light is called the intensity, and it can vary over an almost infinite range. Red light can be much more intense than blue light and consequently transmit much more energy to a given surface. In the quantum model, this means that there are many more photons per unit area falling upon that surface per unit time. Nevertheless, we need to remember during the following discussion that photons of red light have only about half as much energy as photons of blue light.

The photoelectric effect was a demonstration that molecules and atoms absorb photons individually. How can this be? Typically, it occurs by moving an electron in an atom to a higher energy level, that is, by moving it farther out from the nucleus. Since the electron and the nucleus attract each other (one is positive, the other is negative), moving them apart requires an input of energy—the energy of a photon. As it turns out, electrons are apparently moving around the nucleus in orbits that are determined by a wave nature of electron motion. We won't try to explain this further, but the result is that *electrons can only occupy certain orbits* around the nucleus. If the photon contains exactly the right amount of energy to move the electron from one orbit to another, then it can be absorbed; if it does not contain exactly the right amount of energy, it cannot be absorbed. The science that studies all this is called quantum mechanics.

Of course, the situation is more complicated than we have implied. It is simple in such atoms as sodium heated to the gaseous state, where changes can only be brought about by moving electrons from one energy level to another. Thus only quanta at very discrete wavelengths can be absorbed. The yellow-orange absorption line of sodium is at the wavelength in the spectrum of 4893 A. This means that sodium atoms in the gaseous state can absorb quanta at 4893 A but not at 5900 A, for example. In molecules such as chlorophyll, however, the absorption bands are not nearly so discrete, as we saw in Figure 6–3. This is because there are several other changes in the chlorophyll molecule that can be brought about by quanta in addition to changing the energy levels of an electron. The molecule can vibrate slightly, rotate, etc. Each of these changes allows a broadening of the absorption bands.

What happens when a molecule absorbs a photon, causing a change in its energy level and causing the molecule to be in the so-called excited state? There are three things that can happen that are of interest to us:

(1) The molecule can fall back to the ground state, and the energy between the two states can be contributed to the overall kinetic energy of the system. In other words, the energy absorbed as a photon may be converted to heat, which is a very common result of photon absorption. It is going on when something warms in the sun light.

(2) As the molecule falls back to the ground state (e.g., as an electron falls to a lower orbit), another photon can be emitted; that is, as the molecule falls back to the ground state, it can give off light (or any radiant energy). The emitted photon is always less energetic (longer wavelength) than the original one absorbed. This process is known as fluorescence when the light is given off in an extremely short time (10^{-8} sec). If the light is given off after a longer interval of time (up to minutes), the phenomenon is called phosphorescence.

(3) As the molecule falls back to the ground state, the energy can be incorporated into a chemical bond. This is the photochemistry that we have been referring to, and of course, it is the fundamental reaction of photosynthesis.

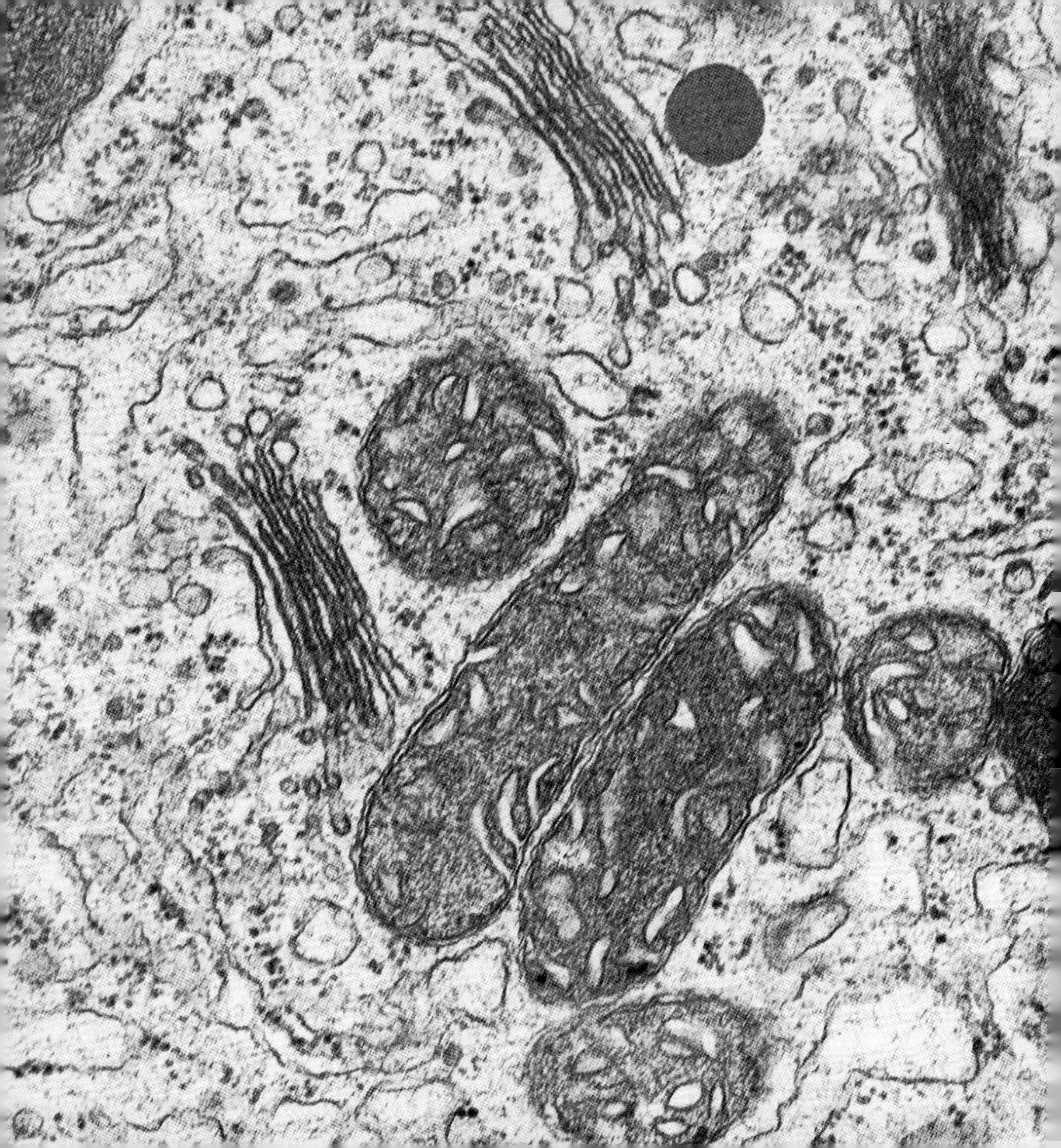

7

Respiration: Mitochondria

7–A *Four mitochondria can be seen in the center of this electron microscope photograph. Two have been sectioned so that the longitudinal structure of the mitochondria is visible, while the other two have been cross-sectioned. The double membrane that limits the mitochondrion is visible; so is the way in which the inner membrane projects toward the center of the organelle. The structures above and to the side of the mitochondria (composed of stacks of membranes) are dictyosomes. The loose sheets of membranes scattered throughout the photo are parts of the endoplasmic reticulum (ER), while the small black spherical bodies are ribosomes. The large, round structure near the top of the photo is a lipid droplet.*

It is night in our woodland acre. Sunlight and photosynthesis are no longer driving the ecosystem. All is quiet and serene. But the ecosystem is still functioning, as billions of trillions of molecules undergo changes each instant. Now, the driving force is the energy produced by respiration—largely the chemical bond energy of ATP. Actually, many components of our ecosystem were never driven directly by the energy captured in photosynthesis (NADPH and ATP). The animals, fungi, bacteria, and all nonphotosynthesizing organisms depended upon respiratory energy.

In cellular respiration, virtually everything is opposite to photosynthesis. Carbon dioxide is produced, and oxygen is consumed. Energy is released. Photosynthesis and respiration balance each other within the closed system of the planet earth. A summary equation for glucose provides a good example of cellular respiration:

$$C_6H_{12}O_6 + 6O_2 \xrightarrow{\text{enzymes}} 6CO_2 + 6H_2O + 710 \text{ kcal}$$

Chemically, respiration is a complex process. There are four intricate series of enzymatically controlled steps. At least 50 and probably more enzymes are involved. It took the first 60 years of this century to work all this out. Many brilliant minds were involved, and a number of important new techniques had to be developed. It was an exciting time for those involved, and several of the leaders were rewarded with Nobel prizes.

Two of the four phases of respiration (series of enzymatic steps) occur in the free cytoplasm of the cell. The other two phases take place in the mitochondria. The part of respiration that occurs in the mitochondria produces by far the most energy. It may be that these thermodynamically more efficient steps require a more rigid structure-function relationship so that substrate molecules don't have to diffuse randomly from one

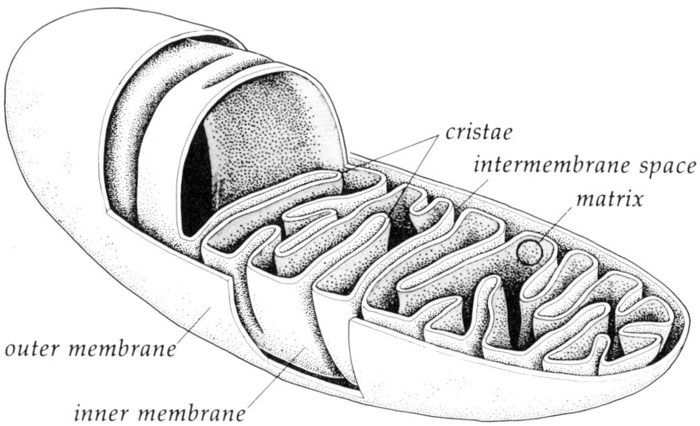

7–1 A three-dimensional, interpretive drawing of a mitochondrion, showing the inner membrane forming cristae.

enzyme to another. Rather, the enzymes may be arranged in close conjunction with each other so that the diffusion of substrates from one to the next is readily accomplished.

Mitochondria

The mitochondria in cells were first seen in about 1900 by the great German cytologist, Richard Altmann. In the light microscope, they appear as small spheres, rods, or filaments. They vary in size but are generally 0.5 to 1.0 μ in diameter and 1 to 2 μ in length. But their small size makes them difficult to observe in the light microscope, and nothing can be seen of their internal structure. By using thin-sectioning techniques and the electron microscope, however, mitochondria are observed to have a rather elaborate internal structure (Fig. 7–A). The mitochondria consist of both an outer and an inner membrane. The inner membrane is folded into the center cavity or matrix of the mitochondrion to form cristae (Fig. 7–1). The cristae may be in the form of shelves or in the form of tubular projections. The latter form appears to be more common in plants. The number of cristae is frequently an indication of the amount of respiration occurring. The greater the number of cristae, the greater the oxygen uptake.

The origin of mitochondria has long been a matter of conjecture. As in the case of plastids, more and more ultrastructural and biochemical data indicate that mitochondria arise from pre-existing mitochondria. For years it was postulated, particularly by French plant cytologists, that mitochondria and chloroplasts both developed from a common, undifferentiated particle. This concept has generally been discarded—although in many ways plastids and mitochondria are similar. In recent years mitochondria or mitochondrial-like particles have been seen in phase contrast motion pictures of living cells to apparently arise by budding of chloroplasts. Some biologists have proposed that mitochondria were originally free-living, bacterial-like organisms that became incorporated into the evolving protocells. The reasoning is similar to that involving chloroplasts, and the case is equally difficult to prove.

The appearance of cellular organelles will depend upon how they are prepared for electron microscopy. Mitochondria may be "positively" stained by using osmium or potassium permanganate. In such staining the membranes of the mitochondria appear dark against a light background. If phosphotungstic acid is used, "negative" staining results. In this case, the membranes appear light against a dark background. Negative staining has allowed us to see that the internal membranes are covered with stalked particles. These have been called elementary particles or oxysomes. They consist of a little head about 75 to 80 A in diameter on a stalk about 30 to 35 A thick and 45 to 50 A long. They are spaced at intervals of about 100 A. They occur in both plant and animal mitochondria, and they were first discovered in 1961 by Humberto Fernandez-Moran.

Mitochondria have been broken down and fractionated by David Green at the University of Wisconsin, allowing us to localize some of their functions to specific mitochondrial structures. Enzymes of the Krebs Cycle (one phase of respiration) occur in the matrix between the membranes, while the electron transfer system is located on the membranes and apparently, more specifically, on the oxysomes (stalked particles). This situation is similar to that encountered in chloroplasts in which the electron transfer system is localized on membranes, while the carbon dioxide fixation occurs in the stroma. There are several other enzymatic reactions of mitochondria besides those concerned with respiration and discussed here. Indeed, over half of the enzymatic activity of the cell is apparently localized in or on mitochondria. In our discussion we will be concerned primarily with the production of energy-rich compounds (ATP) in the process of cellular respiration. In this context we may think of mitochondria as the cellular powerhouses.

Studying Respiration

Of course, unraveling the 50 or more enzymatically controlled steps of cellular respiration was itself an extremely complex process. Before examining what was discovered, let us consider five of the principal approaches that were used in working out the metabolic steps of cellular respiration and that might be used in similar studies.

A search is carried out for the intermediate compounds. This illustrates an interesting and important point about how science must often work. One can randomly study all of the compounds of cells, which was done during the latter part of the last century and is continuing, perhaps at a decreased level of intensity, up to the present. But once we know about the different kinds of sugars, amino acids, etc., contained in cells, how do we relate these to metabolism? Often, this requires the creative act of devising hypothetical schemes that might utilize the intermediates. Frequently, the intermediates are known for species other than those being studied; so they must be looked for in the experimental material at hand. The point is: one must often suspect the answer before he can look for it.

One searches for the suitable enzymes. If the scheme is reasonable and supported by the presence of the suitable intermediates in the tissue, then one can begin to search for the enzymes that might convert one intermediate into another. This is done by adding a suspected substrate to a mixture of enzymes from the tissue (or perhaps a suspension of mitochondria) and then looking for the suspected product. If the product appears in the presence of the mixture of enzymes but not in its absence, then the particular enzyme of interest must be present in the mixture.

Inhibitors of specific enzymes are added. Once some of the enzymes had been found and characterized, it was discovered that certain of them were susceptible to specific inhibitors. The inhibitors could then be added to intact tissue to see if the respiratory process was inhibited and if the substrate of the inhibited enzyme tended to accumulate. Sometimes the inhibitors were specific because (as discussed in Chapter 5) they had structures similar to those of the precursor molecules. In such a case, these antimetabolites carry out their inhibition by competition with the substrate for an active site.

Suspected intermediates are added to intact tissue. If they are metabolized and if this occurs in the manner predicted, then their metabolism must be part of cellular respiration.

Isotopic tracers can be used. An example is the use of ^{14}C to discover the mechanism of CO_2 fixation in photosynthesis. Of course, this is a profitable approach, but interestingly, most of the steps of cellular respiration were worked out before isotopic tracers were available. Incidentally, it is possible to use not only radioactive isotopes but also nonradioactive isotopes having atomic weights different from those prevalent in nature. Such isotopes are detected with a mass spectrometer, an instrument that is more complex, expensive, and difficult to operate than those that detect radioactivity.

Using these and other methods, the four phases of respiration were worked out. Figure 7–2 summarizes these phases, while the figures that follow show the details of the four phases. It is interesting and enjoyable to follow the changes from step to step, gaining insight into the complexity of cellular respiration. But memorizing molecular structures, enzyme names, and other details is probably not profitable at this stage.

Pentose Phosphate Pathway (PPP)

Although ideas hinting at this phase of respiration were advanced as early as the 1930s, the actual work leading to discovery was not done until the 1950s. Hence this phase of respiration was the last to be worked out. As Figure 7–3 indicates, a carbon dioxide molecule is given off in the first step of the PPP, and

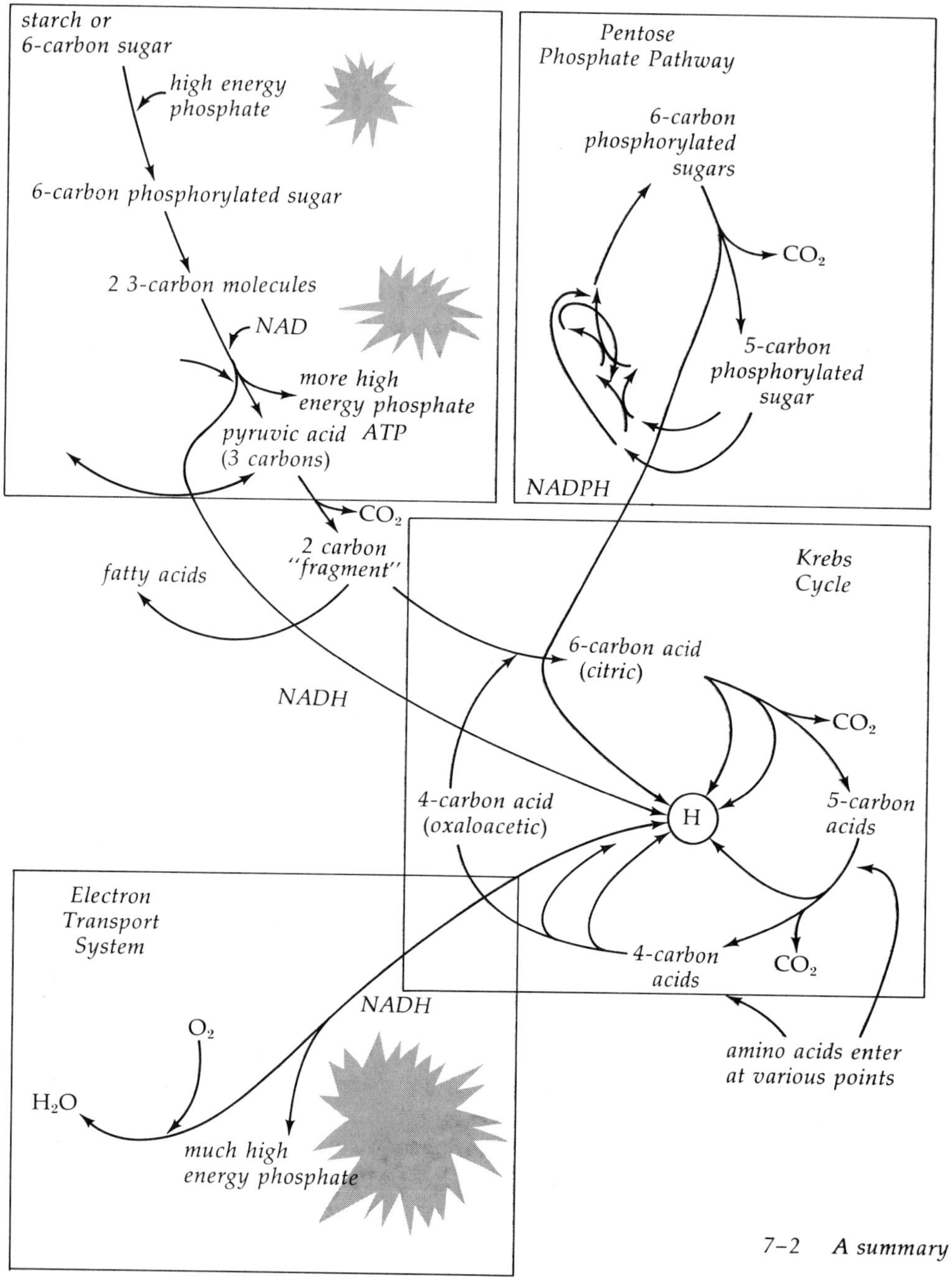

7–2 A summary of the four pathways of respiration.

NADP is reduced to NADPH as glucose (with phosphate attached) is oxidized. The product of this reaction is ribulose-diphosphate. All the other reactions (many of which are identical to those in the Calvin-Benson Cycle of photosynthesis) merely recycle the ribulose-diphosphate back to glucose. Some of the steps in the PPP are related to glycolysis, as seen by comparing Figure 7–3 with Figure 7–4. When the cycle has turned six times, six carbon dioxide molecules have been given off, and one glucose molecule has consequently

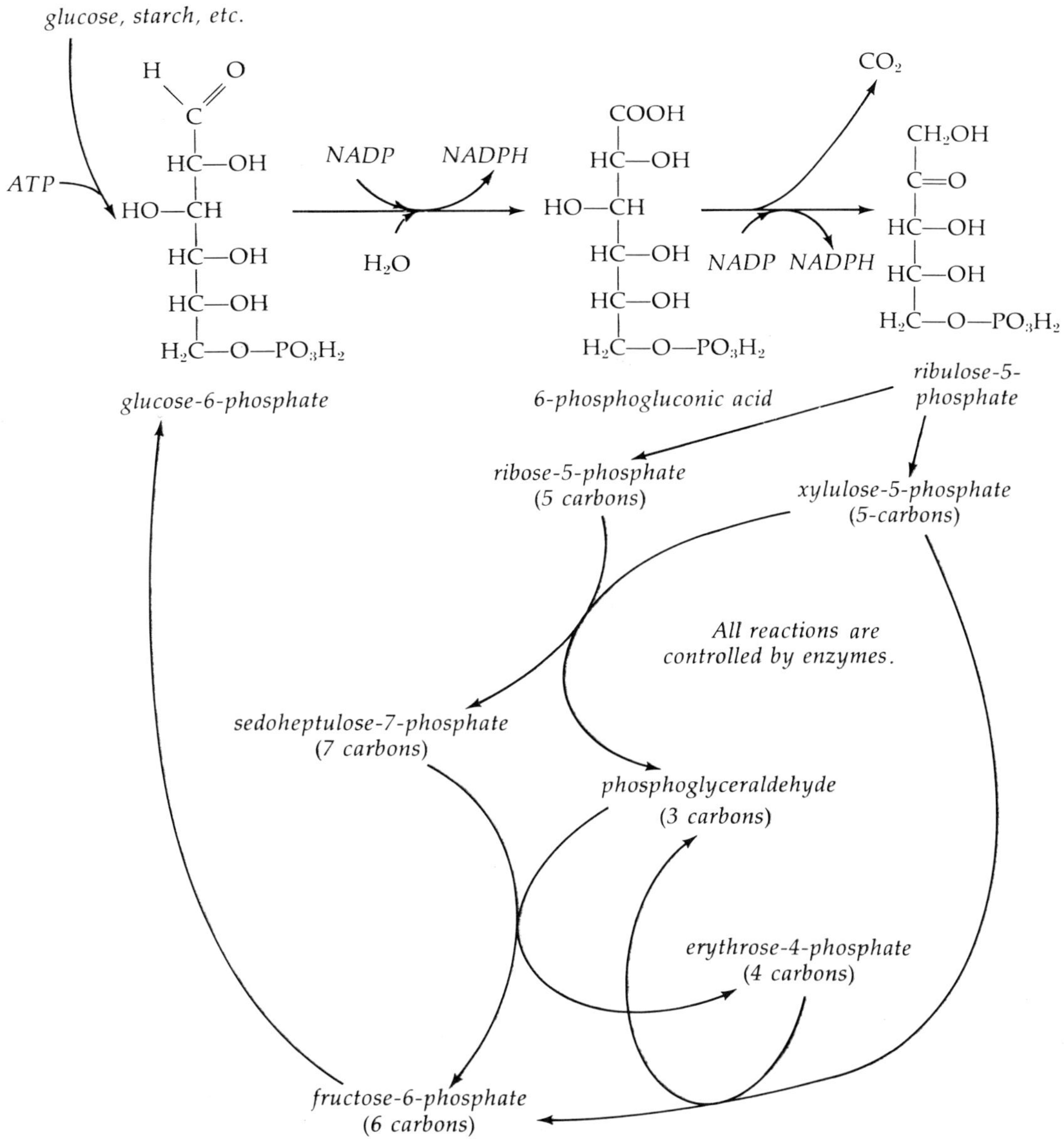

7–3 *The reactions of the Pentose Phosphate Pathway. Note that ATP is required to initiate the pathway by producing glucose-6-phosphate; that CO_2 is released at only one step; and that energy in the form of NADPH is only released at that same step. For every six molecules of glucose moved through the pathway, six CO_2 molecules are released, and five glucose molecules are produced.*

been used up. Six molecules of the energy-rich NADPH have been produced.

Glycolysis or Fermentation

This also starts with glucose (Fig. 7–4), which is first activated by adding a phosphate from ATP and producing ADP. It was the discovery of the phosphorylated sugars, which initiate glycolysis, that led to an understanding of this phase of respiration. The six-carbon phosphorylated sugars split, producing two phosphorylated three-carbon compounds. Although these do not have all the characteristics of carbohydrates, we may nevertheless think of them as members of this class.

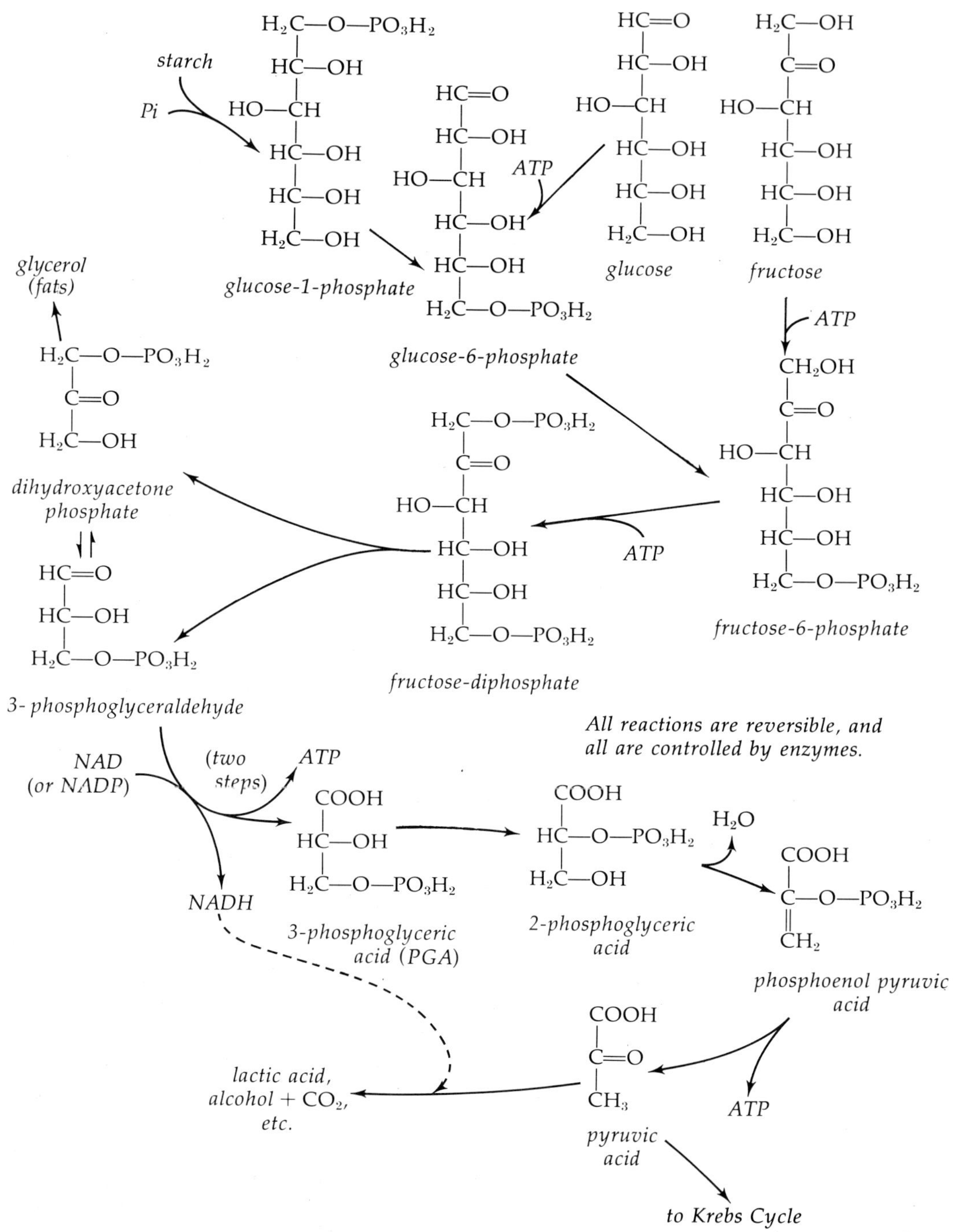

7–4 An outline of the steps of glycolysis. Note that some of the reactions, especially those involving PGA, phosphoenol pyruvic acid, etc., are similar to those of the Pentose Phosphate Pathway and the two carbon-fixation pathways of photosynthesis. In this diagram molecular structures are shown so that you can see the rearrangements of atoms that take place. Note, for example, the shifting of the phosphate group and the formation of the double bond between PGA and phosphoenol pyruvic acid. These rearrangements concentrate some of the energy of the molecule in the phosphate group so a molecule of ATP can be formed.

The three-carbon compounds are then changed through several enzymatic steps, some of which were already encountered in the discussion of the Calvin-Benson Cycle of photosynthesis. For example, in a manner opposite to that in photosynthesis, PGAldehyde is converted to PGA, producing (instead of using) the energy-rich NADH and ATP. We can summarize the products of glycolysis as follows:

(1) For each three-carbon molecule that goes through the glycolytic steps, a reduced coenzyme is

produced. This coenzyme is similar to the NADPH that we have already mentioned in relation to photosynthesis, but it has one phosphate molecule less. It is called NADH. For each glucose molecule entering glycolysis, two molecules of NADH are produced (one for each three-carbon molecule).

(2) Four molecules of ATP are produced (two from each three-carbon molecule), but two molecules were required for each glucose molecule to get the process going. Thus for each glucose molecule utilized, only two ATP molecules are produced.

(3) The final products are two molecules of pyruvic acid for each glucose molecule.

If oxygen is present, the NADH is oxidized by the system described on page 110 to produce NAD, water, and about three molecules of ATP. If no oxygen is present, the NADH becomes oxidized to NAD by reducing the pyruvic acid. In one case (e.g., yeast), the pyruvic acid is converted to alcohol (CH_3CH_2OH) and CO_2. When oxygen is limiting in the muscle cells of mammals, the pyruvic acid is converted by NADH to lactic acid. It is the lactic acid building up in the muscle that causes the stiffness following heavy exercise. Such exercise produces pyruvic acid faster than it can be used in the later oxygen-requiring reactions of respiration, and lactic acid consequently accumulates. In various bacteria and other microorganisms, the pyruvic acid is changed in still other ways. In any case, when oxygen is absent, the pyruvic acid undergoes some change—a process we call fermentation.

Most cells contain enzymes for both the PPP and glycolysis. What determines which of these respiratory processes will take place? Of course, the enzymes that are available will be important in making this determination. If there are many more enzyme molecules for the PPP than for glycolysis, the PPP would probably predominate. Apparently, the relative concentrations of NAD or NADP are also important in making this determination. NAD stimulates glycolysis and NADP stimulates the PPP. NADH functions best in the electron transport system (discussed later), while NADPH is often best used for synthetic processes (such as in photosynthesis). It will depend, then, upon what is happening in the cell. If synthetic processes are predominating, NADPH is converted to NADP, stimulating the PPP. If energy is being produced in the electron transport system, however, NADH is converted to NAD and glycolysis is favored. These are beautiful examples of feedback control—when NADPH is used in synthetic processes, increasing amounts are needed. But synthetic processes produce NADP, which in turn stimulates the PPP to produce more $NADPH^-$, thus filling the need (see summary in Fig. 7–5).

Krebs Cycle

This phase of respiration is initiated by an extremely important key reaction for metabolism. Pyruvic acid is acted upon and converted to a form in which it can enter the Krebs Cycle. One carbon is released as carbon dioxide, and the other two are combined with a sulfur-containing coenzyme called Coenzyme A to produce Acetyl Coenzyme A. In the process a molecule of NAD is reduced to $NADH^-$. (The two electrons combine with the NAD, but only one of the protons combines—the other being released as a hydrogen ion.) This reaction can be summarized as follows:

$$\text{Pyruvic acid} + \text{CoA} + \text{NAD} \rightarrow \text{NADH}^- + \text{H}^+ + \text{Acetyl·CoA} + \text{CO}_2$$

Numerous things can happen to the Acetyl CoA, which is why this is such an important reaction. It is used in the synthesis of fatty acids, as well as many other metabolic products. In respiration the Acetyl CoA combines with oxaloacetic acid, producing citric acid and regenerating the CoA (Fig. 7–6). Then the citric acid undergoes a series of at least seven reactions, regenerating the oxaloacetic acid. Twice CO_2 is produced (accounting for the two carbons of the Acetyl CoA that were originally transferred to oxaloacetic acid). Three times NAD is reduced to NADH. At one step another coenzyme, FAD, is reduced, producing $FADH_2$. At still another step, CoA is again involved. This time ADP is combined with inorganic phosphate (Pi) to produce ATP. At a number of steps in the Krebs Cycle, water molecules are added, replacing the hydrogens lost by the five reductions of NAD and FAD and the oxygen lost in CO_2. No oxygen is taken up in the steps shown in Figure 7–5, but the cycle will not operate unless oxygen is present. As we shall see in a moment, this is because the NADH and the $FADH_2$ must be acted upon by the electron transport system—the hydrogens eventually combine with oxygen to produce water.

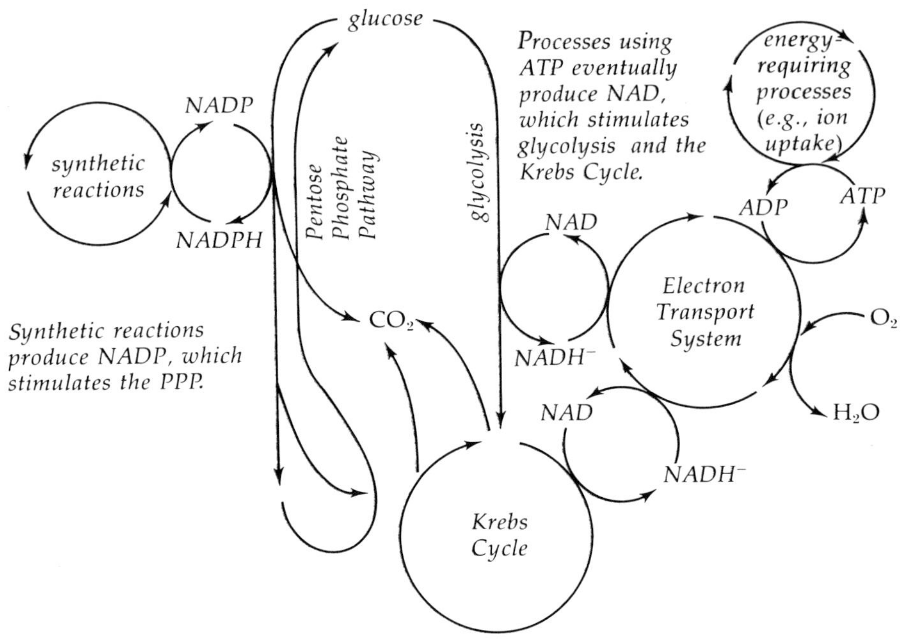

7–5 A schematic representation of some of the feedback reactions that control the relative rates of the reactions of respiration. If synthetic reactions are predominating, the Pentose Phosphate Pathway is promoted; ATP-requiring reactions promote glycolysis. ATP-requiring processes produce ADP, which stimulates the electron transport system and, in turn, glycolysis and the Krebs Cycle.

Electron Transport (Cytochrome) System

The electrons of NADH or $FADH_2$ (and to a slightly less efficient extent, NADPH) are transferred from one enzyme to another in the electron transport system (Fig. 7–7). Sometimes, the proton part of the hydrogen takes part in this transfer (the first steps particularly), but sometimes, only the electrons are moved—the proton is released as a hydrogen ion. Finally, after about six steps, the electrons and the protons (hydrogen ions) are combined with oxygen to produce water. This is the only point in all of respiration where oxygen enters in!

At three points along the electron transport system, ADP is combined with Pi to produce ATP. This is by far the most important source of ATP in the respiratory processes and also provides another example of feedback control. Operation of the cytochrome system is controlled by the amount of ADP (Fig. 7–5). When ADP is high, the electron transport system functions more rapidly, oxidizing more pyruvic acid and eventually more glucose. ADP becomes high when ATP is used in energy-requiring processes. Thus, as the cell works harder, more ADP is produced, speeding up respiration and making more ATP so the cell can work harder.

Interestingly, a number of the molecules in the electron transport system (the prosthetic groups of the cytochrome enzymes) are iron-containing porphyrins similar in structure to the chlorophyll molecule (although chlorophyll contains magnesium instead of iron). Cytochrome oxidase is also a porphyrin but contains copper.

Everything taking part in the electron transport system except NADH and NADPH (and also CoQ) appears to be tightly bound to the mitochondrial membranes. This is undoubtedly of considerable importance. An enzyme in the system is able to oxidize the one above it in the sequence because it has an oxidation-reduction potential that is more oxidizing than the enzyme it oxidizes. Thus cytochrome oxidase should be able to oxidize the first one in the system, avoiding all the intermediate steps. This doesn't happen possibly because the enzymes are localized so that they can only oxidize each other according to the proper sequence, as shown in Figure 7–7; that is, the terminal oxidase enzyme cannot oxidize the others because it cannot come in contact with them—also possibly because its active site does not have a configuration allowing it to react with any but its own substrate enzyme. If the structure-function relationship is not controlling at the level of the cellular or-

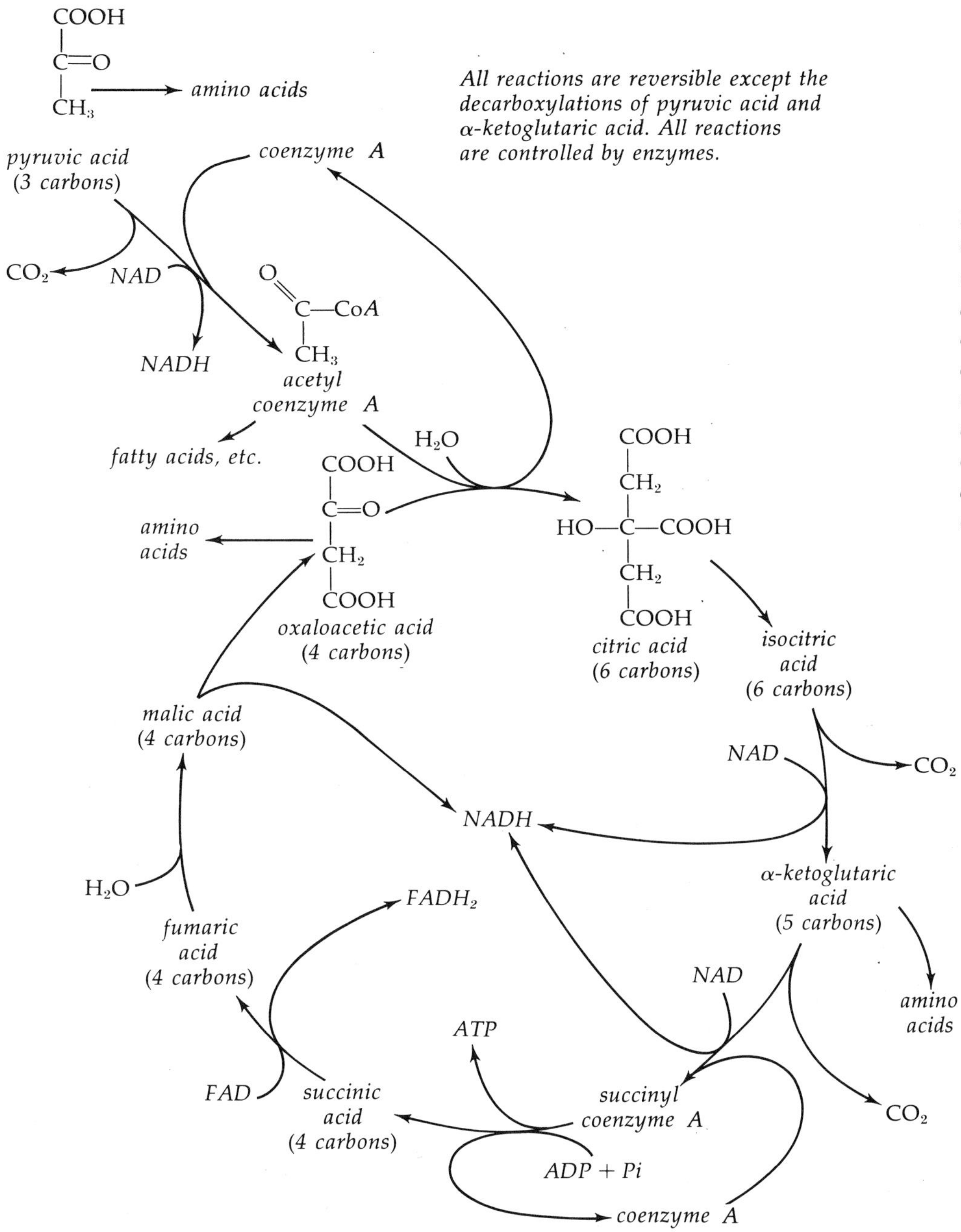

7–6 Reactions of the Krebs Cycle. Note important key compounds such as pyruvic acid, acetyl coenzyme A, α-ketoglutaric acid, and oxaloacetic acid. These compounds act as intermediates or precursors in the synthesis of fatty acids, amino acids, and numerous other compounds.

ganelles, it is certainly controlling at the molecular level of the active site.

ATP is produced in photosynthesis by a set of reactions similar to those of respiration—the transfer of electrons from one enzyme to another, possibly the same enzymes as those important in the cytochrome system. In photosynthesis, however, we are dealing with an overall reduction instead of an overall oxidation. This is possible because chlorophyll, by absorbing light, can act as an electron pump, pumping the electrons to the reduction side of the electron transport system.

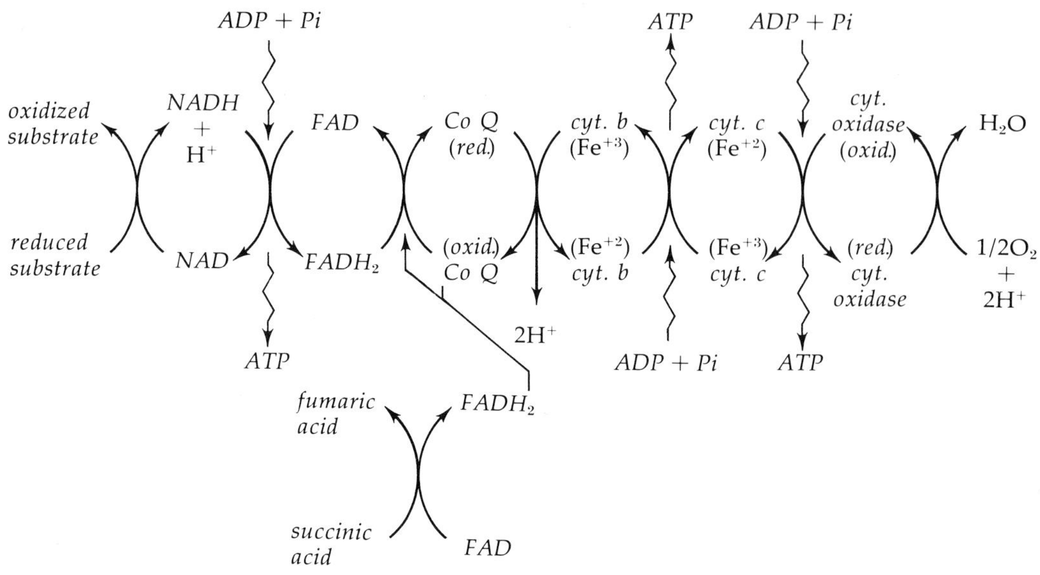

7–7 The Electron Transport System.

Energetics of Respiration

Where are ATP molecules produced in respiration? Two ATPs are produced in glycolysis when no oxygen is present. If the two NADH molecules produced for each glucose molecule are acted upon by the electron transport system, six more molecules of ATP will be produced in the presence of oxygen. Another six are produced from the two NADH molecules produced at the CoA step for each glucose (two pyruvic acids). Eighteen more are produced from the six NADHs released in the Krebs Cycle for each glucose (two pyruvic acids). Two are produced in the step from α-ketoglutaric acid to succinic acid, and four are produced from the $FADH_2$ produced in the Krebs Cycle. The total for a single molecule of glucose is 38 molecules of ATP, as summarized in Table 7–8.

Each molecular weight of ATP molecules (each mole) has about 8 kcal of energy available for use by the cell (depending somewhat upon such conditions in the cell as acidity). Hence, if 38 moles of ATP are produced per mole of glucose, then 304 kcal of energy per mole of glucose are saved in the respiration process for future use. If a mole of glucose is burned in a calorimeter, 710 kcal are produced; so the efficiency of respiration is $304/710 \times 100 = 43\%$. This efficiency is somewhat higher than that measured for photosynthesis, and it is high compared to many of the energy-transferring processes with which we are familiar. A piston steam engine, for example, is only about 10% efficient, although modern steam turbines may be 40% efficient. The 57% of the energy not contained in ATP is lost as heat.

Other Phases of Metabolism

That we have only scratched the surface of many metabolic steps closely related to respiration can be seen in the following summary:

(1) The hexose phosphates are metabolized in many ways. For example, they are the immediate precursors for several of the cell-wall components, for starch, and for sucrose.

(2) The three-carbon phosphates act as immediate precursors for such compounds as glycerol, various phospholipids, and some of the oils.

(3) Phenolic compounds and indole compounds, including the amino acid tryptophan and many of the growth hormones called auxins (see Chapter 30), are synthesized from phosphoenol pyruvic acid.

(4) Not only alcohol and lactic acid but also several amino acids (alanine directly) are synthesized from pyruvic acid.

(5) Acetyl CoA provides a two-carbon building block for several important compounds, including the fatty acids, waxes, isoprenoids (carotenoids, terpenes,

the phytol of chlorophyll), steroids (large complex organic alcohols), gibberellins (important growth hormones of plants), and several molecules containing a benzene ring (aromatic compounds).

(6) Oxaloacetic acid is also an important precursor for a number of compounds, including amino acids (aspartic directly) and alkaloids (including the nucleotides that constitute the links of DNA and RNA molecules).

(7) In an important reaction, ammonium is added to α-ketoglutaric acid to produce the amino acid glutamic acid. This molecule can then act as a source of amino groups in the synthesis of several other amino acids.

(8) Succinyl CoA provides four-carbon units that are used in the synthesis of such porphyrins as chlorophyll and the cytochromes.

7–8 *A summary of respiration energetics*

	ATPs/step	ATPs/glucose
From glycolysis	(2)	2
From NADH released in glycolysis	(3)	6
From NADH at CoA step	(3)	6
From the 3 NADHs released in the Krebs Cycle	(9)	18
In the step from α-ketoglutaric acid to succinic acid	(1)	2
From the $FADH_2$ in the Krebs Cycle	(2)	4
	Total:	38

Respiration in the Ecosystem

Again visualize our sleeping woodland ecosystem. Perhaps most of the cells are "resting"—their metabolic processes at a relatively low level of activity. But as the nocturnal owl swoops down to capture a mouse, nervous impulses are causing intense activity in numerous muscle cells in the owl's body, and the cells of his eyes, ears, and other senses are operating at full capacity. As the mouse glimpses the moonlight shadow of the gliding owl, many of his cells also begin to operate at full capacity. Countless molecules of ATP are entering into countless reactions—contracting muscles, intensifying hearing, preparing for fight or flight. The energy from the ATP is being used in these reactions, and molecules of ADP are being produced. As these quickly build up, the waiting NADH and NADPH molecules can rapidly begin to enter the electron transport system, and this in turn stimulates the other parts of respiration as more molecules of NAD and NADP become available. Eventually, glycogen (animal starch) is being broken down to provide more glucose molecules—all of these steps originally in response to the increasing ADP. If the process is most efficient in the owl, he may get his prey; if in the mouse, he may escape.

The same things are taking place in the plant cells, but they are less dramatic because there are no nervous systems or rapid-acting sense organs to trigger the response to danger. Nevertheless, the same principles apply. The dead remains of the mouse, excreted by the owl (who won this one) decay in the grass, releasing ions of nitrate, potassium, phosphate, etc. To absorb these, the grass root cells use up ATP molecules, producing ADP. This stimulates respiration in these cells, producing more ATP, resulting in absorption of more ions, and finally displaying a burst of plant growth.

Plants are responding in other ways during the night in our ecosystem. Biological clocks are functioning, measuring the hours of darkness. When sufficient hours have elapsed, metabolic systems come into play in some species (e.g., the cocklebur, Chapter 35) to synthesize molecules of a flowering hormone. Detection of suitable night lengths indicates a suitable season for flowering of these species. The reactions themselves—synthesis of flowering hormone in the leaf, transmission of it to the bud, and formation of flowers—all require the energy of ATP.

Consider once more the balance of things. In our sleeping ecosystem, oxygen is being used up and carbon dioxide is being produced. This respiration in both plant and animal cells provides the energy for all the intricate web of activities going on. But unless the oxygen is replenished and the carbon dioxide taken out of the atmosphere before it can build up to toxic levels, life will eventually cease. In the morning, when the sun comes up, photosynthesis will renew its tasks of energy capture, carbon fixing, and atmospheric purification. And respiration too will continue its life-supporting functions.

8

Cell Biosynthesis

The Nucleus—Control Center of the Cell

Virtually all living cells contain one or more conspicuous, roughly spherical structure—the nucleus. Most cells in the higher plants contain only one nucleus. The nucleus was one of the first cell parts to be seen with the microscope, since it can be readily stained with a large number of dyes. The nucleus, which contains one or more smaller bodies (the nucleoli) is the control center of the cell. It possesses the information necessary for the production of proteins, which are vital to the development and functioning of the cell.

Nuclei are surrounded by double membranes that contain numerous pores. The pores are circular in face view and provide a channel between the inside of the nucleus and the cytoplasm. Within the nuclear membrane or envelope is the chromatin, suspended in the more liquid nuclear sap.

A *A group of cells from the style of a cotton flower in which a single nucleus is conspicuous.*

B *The nucleus of a meristematic cell of a root tip. The chromatin is particularly obvious because the cell is preparing to divide.*

C *A freeze-etched preparation of a nucleus from an onion root tip, as seen with the electron microscope. This type of preparation reveals quite clearly the surface view of the nuclear pores.*

D *A nucleus from a developing fruit of* Capsella, *as seen in section with the electron microscope. The outer membrane of the nuclear envelope is continuous with the endoplasmic reticulum or ER found in the cytoplasm.*

The ER (endoplasmic reticulum) consists of sheets or tubes of membranes that permeate the cytoplasm. Some ER are found in all cells, but the amount varies greatly. There may be an enormous difference in the amount of ER that even two adjacent cells contain. Such differences reflect the relatively synthetic activity of the cells, with the more active cell having the greater amount of ER.

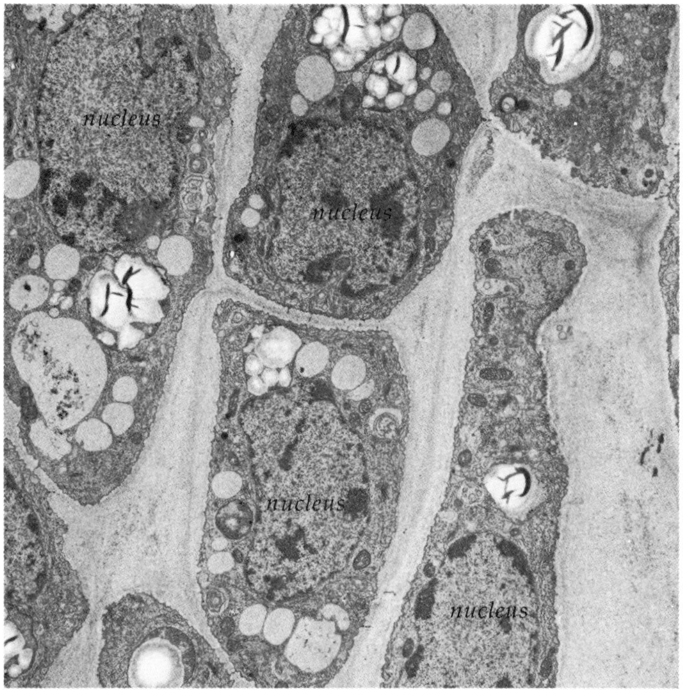
nucleus
nucleus
nucleus
nucleus
A

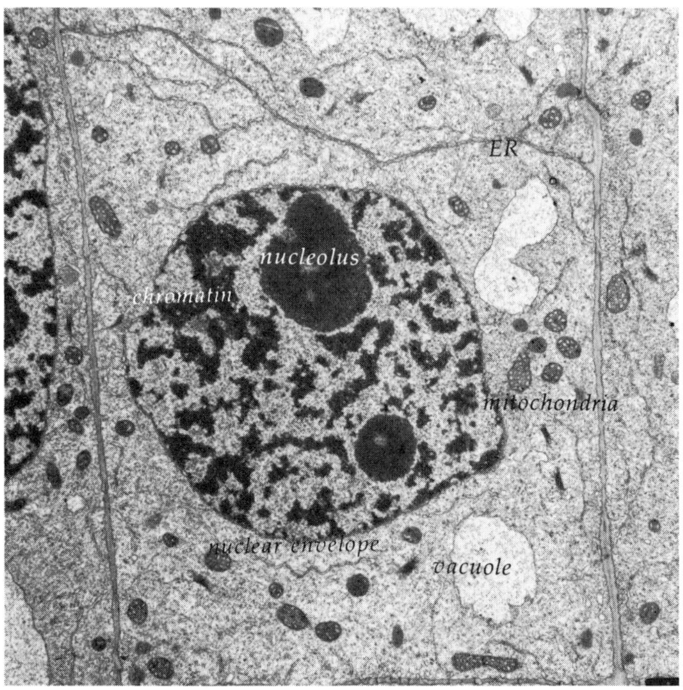
ER
nucleolus
chromatin
mitochondria
nuclear envelope
vacuole
B

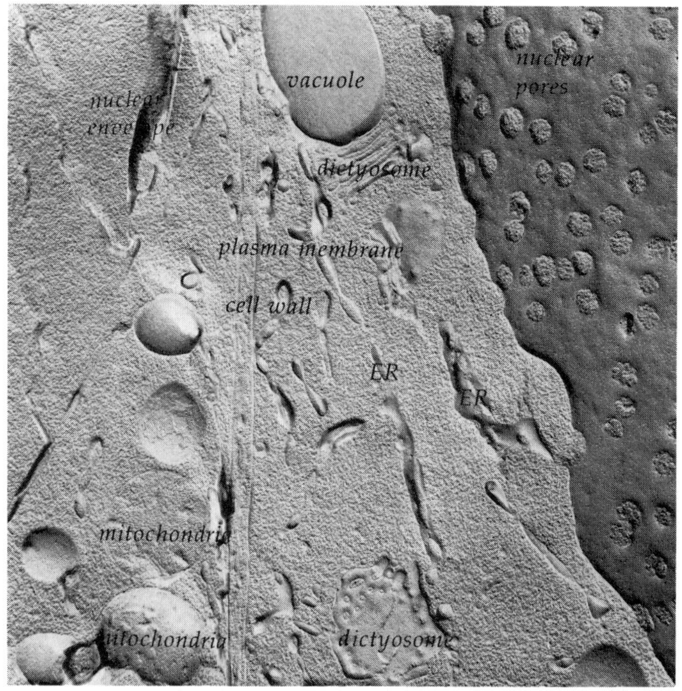
vacuole
nuclear pores
nuclear envelope
dictyosome
plasma membrane
cell wall
ER
ER
mitochondria
mitochondria
dictyosome
C

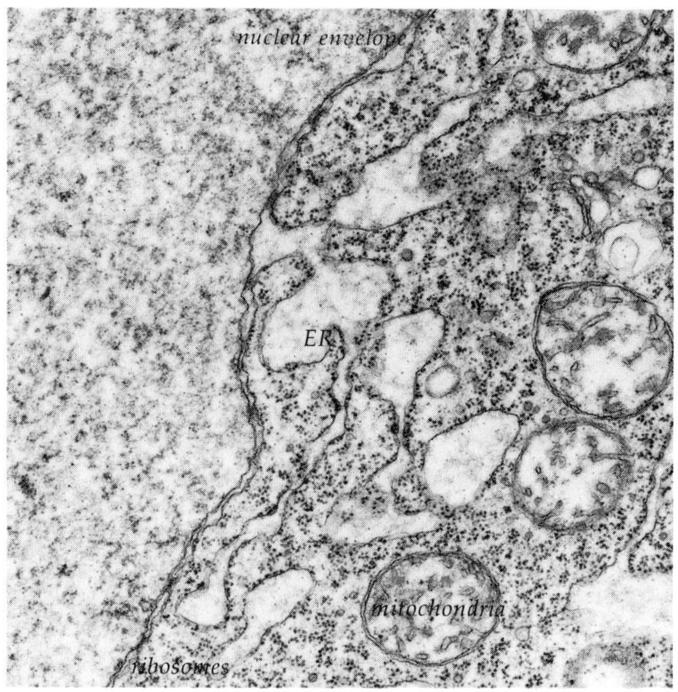
nuclear envelope
ER
mitochondria
ribosomes
D

E A section through two adjacent cells in the developing fruit of Capsella.

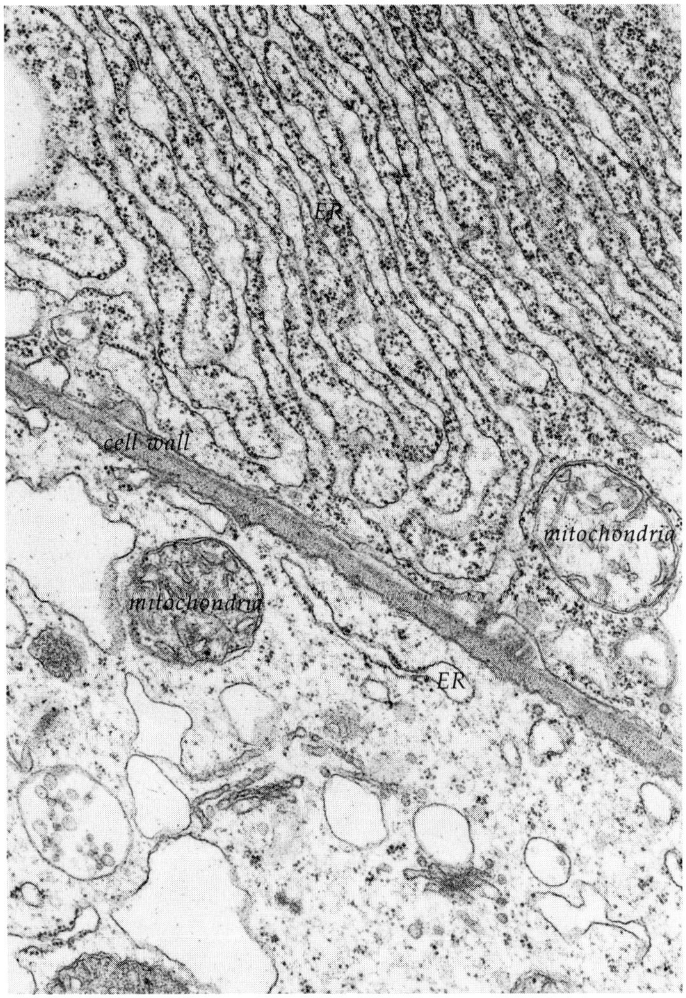

E

In the warm, tropical waters of the Caribbean grow the handsome green algae of the genus *Acetabularia*. Members of this genus have slender stalks (some 1 to 4 cm long) and flaring tops, which in some species are shaped like the bowl of a crystal goblet. The organism's single nucleus is located near the base of the stalk.

For a cell such as *Acetabularia* to live, it must do more than photosynthesize and respire; it must constantly produce proteins, lipids, and nucleic acids to replace parts, to grow, and to alter its metabolism as new demands are placed on it. Except when dormant (as in seeds or spores), a cell must continually synthesize new enzymes and other basic cell components to survive.

If we are to understand how cells and organisms function, it is vital to examine the problem of how they manufacture the compounds they need. The green alga *Acetabularia* provides a good introduction to this topic, since it is one of the favorite organisms of cell biologists concerned with the problem of cell development and the control of cell metabolism by the nucleus of the cell.

The Nucleus

The nucleus (Fig. 8–A) is a universal and prominent feature of all cells except bacteria and blue-green algae. A few living cells such as the sieve-tube elements of plants and the red blood cells of animals may lose their nuclei when they become mature, but all cells have nuclei at some time in their lives. Situated in the cytoplasm, it can occupy a major proportion of the cell volume in young, rapidly dividing cells (Fig. 8–B). In older cells it comprises a smaller part of the total volume of the cell and may become pressed between the vacuole and the wall. Although most cells of the vascular plant have a single nucleus, some cells have two or more. The latex tubes are one example of a multinucleate cell in a vascular plant. In many nonvascular plants, the cells have more than one nucleus. There are numerous cases in which there are large numbers of free nuclei in a mass of cytoplasm (e.g., the slime molds) or in which the number of nuclei are more limited (e.g., many algae).

The nuclear material in some cells such as bacteria and blue-green algae is not limited by a membrane

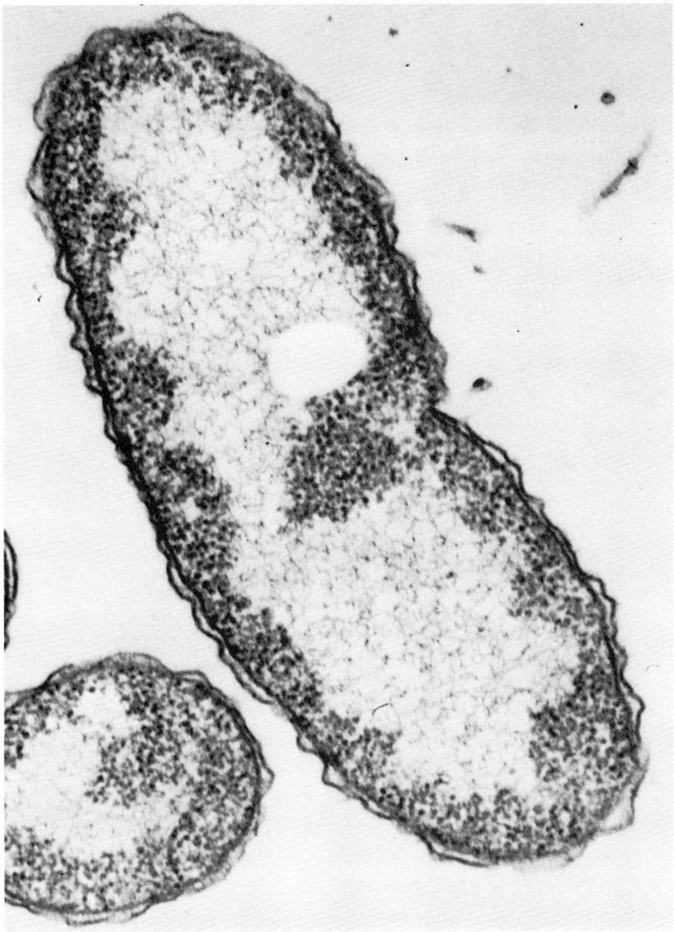

8–1 In the bacteria and the blue-green algae, the nuclear region is not membrane bound as it is in other cells. In this section of a bacterium the nuclear plasm is seen in the center of the cell. The region is fibrous in nature and contains the DNA of the cell.

(Fig. 8–1). In both examples DNA, lacking an associated protein (as in higher plants and animals), is clearly present and usually more or less centrally located in the cell. There is no membrane surrounding this material, thus limiting it from the cytoplasm. This condition is considered more primitive than that where a nuclear membrane is present. The significance of this lack of a limiting nuclear envelope is still not clear. Cells lacking membrane-bound nuclei are termed procaryotic, while those with true nuclei are called eucaryotic.

Within the nucleus, one or more bodies, the nucleoli (nucleolus, singular), are visible (Figure 8–B). These are roughly spherical in shape and may appear homogeneous or divided into two phases, one more dense than the other. In some cases the dense portion of the nucleolus can be resolved into a mass of closely packed, dense spheres, while the lighter material is fibrillar. The nucleoli are composed of protein and RNA, although lipids and small amounts of DNA may be present.

The nucleus is filled with liquid called the nuclear sap. Besides the nucleoli, nuclear ribosomelike particles and the chromatin are found within the nuclear sap. The particles in the nucleus are similar in size, shape, and composition to cytoplasmic ribosomes. Characteristic of the nuclear envelope is the presence of pores easily visible in the electron microscope (Figure 8–C). The pores in the nuclear membrane are usually large enough to permit the passage of ribosomes.

Control Center of the Cell

The major function of the nucleus is to control and direct the development of the cell. This control can be demonstrated in many ways, but one of the most dramatic is by experimentation using *Acetabularia*. As we saw, each alga is a single cell with a nucleus situated in the basal part of the stalk. If the cap or top is cut off, the stalk will regenerate a new cap. The following experiment can be performed using two species of *Acetabularia*, each having a differently shaped top. First, the top of one species is removed. Next, the base containing the nucleus is cut off and a base, complete with nucleus, of the second species is grafted to the cut stalk (Fig. 8–2). The stalk will regenerate a new top but not of the original type; it will be the type characteristic of the species whose grafted base contained the nucleus. Thus the information necessary for the formation of a regenerated top is present in the nucleus.

This is a simple version of the classical experiment performed by J. Haemmerling of Finland. His work with *Acetabularia*, which he demonstrated was a beautiful organism for this type of research, clearly showed that growth and development were under the control of the nucleus. Moreover, this direction was based on the hereditary composition of the nucleus—development was under the control of genes, the primary hereditary units of the cell. That growth and development

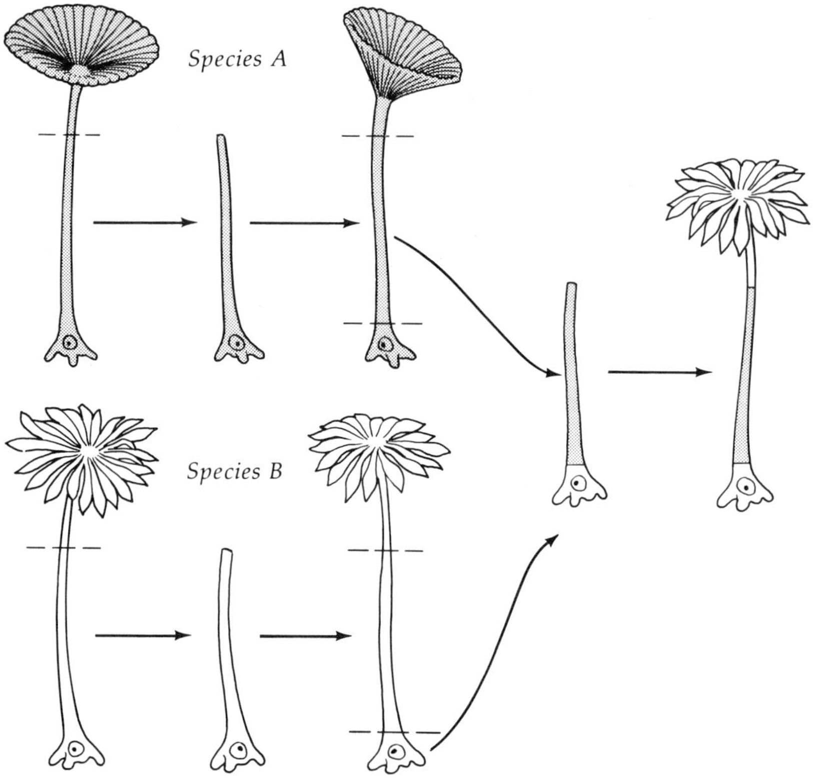

8–2 Outlined here is one of the many experiments performed with Acetabularia. *Two species are shown. When the cap is removed, each is capable of regenerating the same type of cap it had originally. But, if the basal part of the stalk containing the nucleus is exchanged, the stalk will produce a new cap that is the same type as the nucleus, not the type originally found on the stalk.*

were under genetic control was known for many years before Haemmerling's experiment, as was the fact that the genes were located on the chromosomes in the nucleus. What this experiment did was to confirm the location of control of development in the nucleus.

How can genes control development or any other cell process? This was a major question in understanding gene action that was beautifully answered by the research of George Beadle and Edward Tatum, then at Stanford University. They decided to use the pink bread mold *Neurospora,* which, being a fungus, is dependent on external nutrient supplies. First, they irradiated *Neurospora* to produce mutations; then they grew the plants in test tubes on a culture medium of known composition, eventually finding a number of mutant plants that required additional chemicals to live. Careful analysis revealed that these plants lacked enzymes necessary to utilize the original compounds in the medium. These plants required compounds that the missing enzymes prevented the plants from making themselves. Beadle and Tatum combined biochemical analyses with careful genetic analyses of the cells. Their conclusion, which they fully documented, was that the normal plants' genes controlled the production of the enzymes that were missing in the mutant. Moreover, they concluded that the relationship was one enzyme, one gene. Work by other investigators, using other organisms, extended the data to the general conclusion that genes regulate enzyme synthesis for all living things. Thus the statement is now more accurately one gene, one polypeptide chain (a part of an enzyme), but the original concept of Beadle and Tatum, for which they received the 1958 Nobel prize, is still valid.

Since, as we already know, all enzymes are protein, the question now becomes one of discovering how proteins are synthesized. Where, for example, does protein synthesis occur within the cell? There are several approaches to the answer. One is to grow cells in radioactive amino acids and then determine where in the cell they are incorporated by using autoradiographs (see End Paper 8–1). When this is done, the radioactive protein appears primarily in the cytoplasm—not the nucleus. True, some protein synthesis occurs in the nucleus as well as in the plastids and mitochondria, but the bulk takes place in the cytoplasm. Another approach is to grind up the cell and attempt to determine which

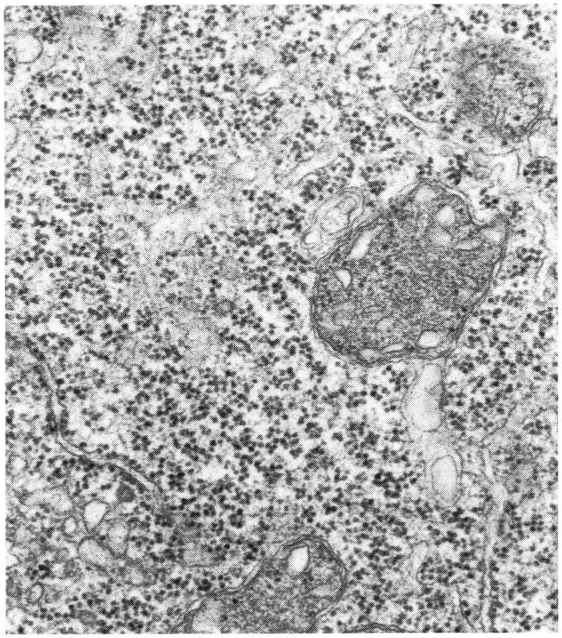

(*a*)

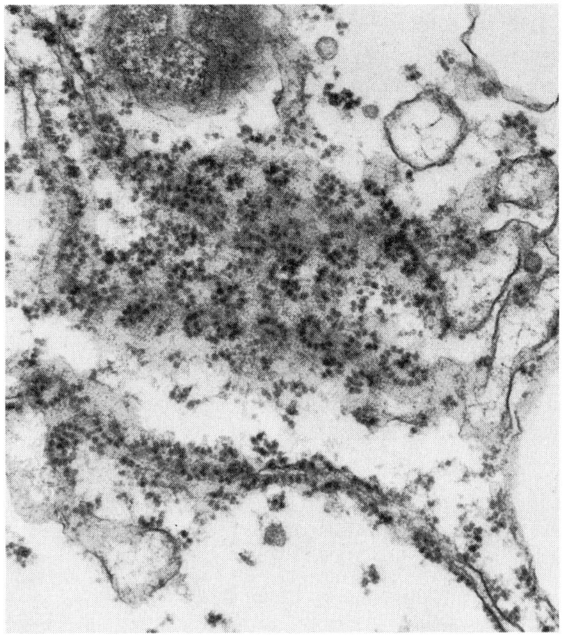

(*b*)

8–3 (a) The cytoplasm of this embryotic cell is filled with ribosomes that are "free." Some are found attached to a short segment of ER, as shown in the lower-left portion of the photo. Ribosomes can also be seen in the mitochondrion. These are smaller than the ones found in the cytoplasm and are similar in size to those found in bacteria (× 45,000). (b) A surface view of the ER that shows the ribosomes arranged in curved rows. Such complexes of ribosomes—either attached to the ER or free in the cytoplasm—are called polysomes (× 45,000).

of the cell parts is involved with protein synthesis. This proved far from easy but the answer was clear: the ribosomes.

Ribosomes

Having mentioned ribosomes in earlier chapters, let us now look at them in greater detail (see Figure 8–3). Ribosomes are small—the average size ranging from 150 (in bacteria and blue-green algae) to 200 A (in the cytoplasm of all higher plants and animals). They are composed of ribonucleic acid (RNA) and protein and account for approximately 85% of the total amount of cytoplasmic RNA. Ribosomes are found floating free in the cytoplasm or attached to the endoplasmic reticulum (ER). The unattached ribosomes are frequently found in clusters called polysomes. The individual ribosomes are composed of two pieces: one small (40 *s*, where *s* is a unit based on how fast a particle settles in a centrifuge tube) and one larger (60 *s*). Ribosomes, or at least precursors of ribosomes, are made in the nucleus and stored in the nucleolus until they presumably pass into the cytoplasm.

At the ribosome the amino acids are assembled into proteins. Amino acids are made in different parts of the cell: mitochondria, plastids, and cytoplasm. Energy is required for protein synthesis, since the first step is the combination of an amino acid with an ATP molecule resulting in an activated amino acid.

$$\text{Amino acid} + \text{ATP} \xrightleftharpoons[\text{Mg}^{++}]{\text{enzyme}} \text{amino acid–AMP} + \text{phosphate}$$

To make a specific protein molecule, more than activated amino acids and ribosomes are needed—specifically, information on the sequence of the amino acids in the protein. We have seen that the ultimate form of the protein molecule is dependent on the sequential arrangement of the amino acids. If genes are to determine protein synthesis, they must be involved in this process. But the genes are in the nucleus, and the site of protein synthesis is in the cytoplasm.

Nucleic Acids

How can the nucleus direct the cytoplasmic synthesis of protein? Looking at the nucleus and cytoplasm, we find one type of compound common to both: the nucleic

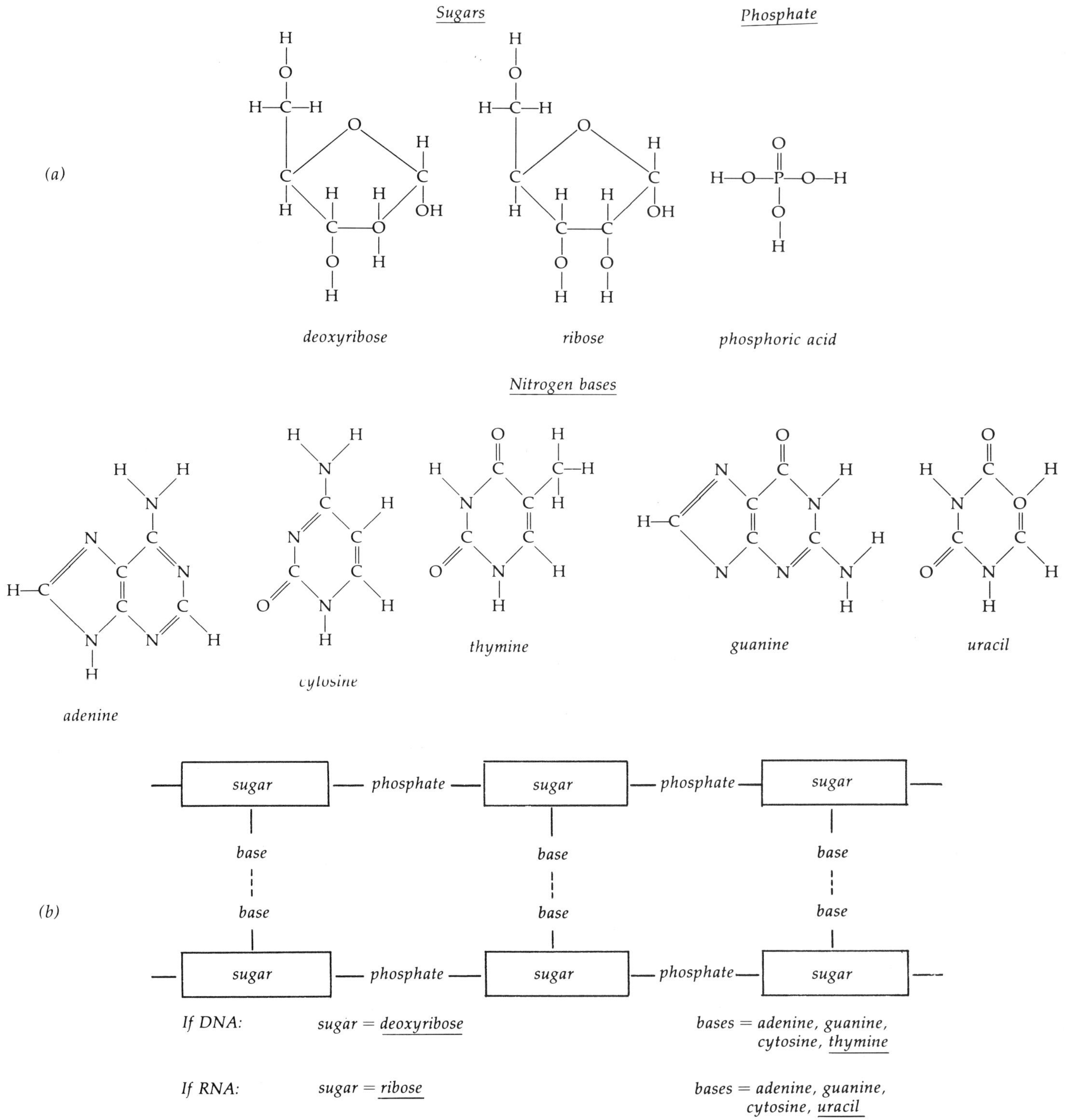

8–4 (a) Chemical formula for the parts of RNA and DNA. (b) A schematic diagram of a nucleic-acid molecule.

acids. There are two kinds of nucleic acids: deoxyribonucleic acid (DNA) and ribonucleic acid (RNA). Both DNA and RNA are polymers, consisting of repeating units of sugar, phosphate, and nitrogen bases (Fig. 8–4). The two nucleic acids have the same basic composition but differ in several important ways. One concerns sugar—deoxyribose is found only in DNA, while ribose is found in RNA. They are both 5-carbon sugars (p. 45), and the nucleic acids derive their specific names from them. Another difference is in the composition of the nitrogen bases. A total of five of these are present. Three are common to both DNA and RNA: adenine, guanine, and cytosine. In addition, DNA contains thymine, and RNA contains uracil. Thus each nucleic acid has four bases, three in common and one different. It was found early that the bulk of DNA is localized in the nucleus, while RNA is present both in the nucleus and cytoplasm. In the 1940s the Swedish biologist T. O. Casperssen and the Belgian Jean Brachet originated the idea that RNA was involved in protein synthesis and that there was a relationship between nuclear and cytoplasmic RNA. It was also suggested that DNA was involved with RNA synthesis.

Essay 8–1 The Double Helix

One of the truly great discoveries of the last 20 years was the determination of the structure of DNA. James D. Watson, a young postdoctoral fellow working in the laboratory of Francis Crick at Cambridge University, first conceived the basic structure of the molecule in 1953. Watson and Crick knew the basic facts necessary to construct a model of the molecule, but the essential key to the problem escaped them. This is one of the most difficult of all times for a scientist, and one that Watson describes well in his book The Double Helix:

"I went ahead spending most evenings at the films, vaguely dreaming that any moment the answer would suddenly hit me. Occasionally my wild pursuit of the celluloid backfired, the worst occasion being an evening set aside for 'Ecstasy.' Peter and I had both been too young to observe the original showings of Hedy Lamar's romps in the nude, and so on the long-awaited night we collected Elizabeth and went up to the Rex. However, the only swimming scene left intact by the English censor was an inverted reflection from a pool of water. Before the film was half over we joined the violent booing of the disgusted undergraduates as the dubbed voices uttered words of uncontrolled passion.

Even during good films I found it almost impossible to forget the bases.

. . . Generally it was late in the evening after I got back to my rooms that I tried to puzzle out the mystery of the bases. Their formulas were written out in J. N. Davidson's little book The Biochemistry of Nucleic Acids, *a copy of which I kept in Clare, so I could be sure that I had the correct structures when I drew tiny pictures of the bases on sheets of Cavendish notepaper. My aim was somehow to arrange the centrally located bases in such a way that the backbones on the outside were completely regular—that is, giving the sugar-phosphate groups of each nucleotide identical three-dimensional configurations. But each time I tried to come up with a solution I ran into the obstacle that the four bases each had a quite different shape. Moreover, there were many reasons to believe that the sequences of the bases of a given polynucleotide chain were very irregular. Thus, unless some very special trick existed, randomly twisting two polynucleotide chains around one another should result in a mess. In some places the bigger bases must touch each other, while in other regions, where the smaller bases would lie opposite to each other, there would exist a gap or else their backbone regions must buckle in.*

. . . My doodling of the bases on paper at first got nowhere, regardless of whether or not I had been to a film. Even the necessity to expunge 'Ecstasy' from my mind did not lead to passable hydrogen bonds, and I fell asleep hoping that an undergraduate party the next afternoon at Downing would be full of pretty girls.

. . . Not until the middle of next week, however, did a nontrivial idea emerge. It came while I was drawing the fused rings of adenine on paper. Suddenly I realized

the potentially profound implications of a DNA structure in which the adenine residue formed hydrogen bonds similar to those found in crystals of pure adenine. If DNA was like this, each adenine residue would form two hydrogen bonds to an adenine residue related to it by a 180-degree rotation. Most important, two symmetrical hydrogen bonds could also hold together pairs of guanine, cytosine, or thymine. I thus started wondering whether each DNA molecule consisted of two chains with identical base sequences held together by hydrogen bonds between pairs of identical bases. There was the complication, however, that such a structure could not have a regular backbone, since the purines (adenine and guanine) and the pyrimidines (thymine and cytosine) have different shapes. The resulting backbone would have to show minor in-and-out buckles depending on whether pairs of purines or pyrimidines were in the center.

Despite the messy backbone, my pulse began to race. If this was DNA, I should create a bombshell by announcing its discovery. The existence of two intertwined chains with identical base sequences could not be a chance matter. Instead it would strongly suggest that one chain in each molecule had at some earlier stage served as the template for the synthesis of the other chain. Under this scheme, gene replication starts with the separation of its two identical chains. Then two new daughter strands are made on the two parental templates, thereby forming two DNA molecules identical to the original molecule. Thus, the essential trick of gene replication could come from the requirement that each base in the newly synthesized chain always hydrogen-bonds to an identical base. That night, however, I could not see why the common tautomeric form of guanine would not hydrogen-bond to adenine. Likewise, several other pairing mistakes should also occur. But since there was no reason to rule out the participation of specific enzymes, I saw no need to be unduly disturbed. For example, there might exist an enzyme specific for adenine that caused adenine always to be inserted opposite an adenine residue on the template strands.

As the clock went past midnight I was becoming more and more pleased. There had been far too many days when Francis and I worried that the DNA structure might turn out to be superficially very dull, suggesting nothing about either its replication or its function in controlling cell biochemistry. But now, to my delight and amazement, the answer was turning out to be profoundly interesting. For over two hours I happily lay awake with pairs of adenine residues whirling in front of my closed eyes. Only for brief moments did the fear shoot through me that an idea this good could be wrong."

And, of course, it was wrong—at least as far as the bases that actually pair were concerned. But the concept was right, and this provided the key for the correct model of DNA that Watson worked out a few days later.

Is DNA the genetic material? The answer is yes, and the data are impressive. One of the earliest and most convincing pieces of evidence was the observation that a harmless form of pneumococcus bacteria could be transformed into a virulent form by DNA isolated from the virulent strain (Fig. 8–5). Next, there are the viruses. One type called bacteriophage (phage for short) infects bacterial cells. The phage consists of a strand of DNA surrounded by a protein coat. When the phage infects the bacteria, only the DNA enters the host, but this is enough to result in the production of a complete new phage by the bacteria (Fig. 8–6). When higher plant and animal cells were examined for DNA content, it was found that the amount of DNA per nucleus is constant, except before mitosis when it is precisely doubled, then returned to the original amount after division. Moreover, each of the gametes after meiosis has one half the DNA content of the original cell, but after the gametes fuse to form the zygote, the base amount is reestablished. In other words, the DNA content precisely follows the chromosome number. These and other data focused attention on DNA as the genetic material, but a gene is essentially a unit of information. The information in a gene is how to arrange

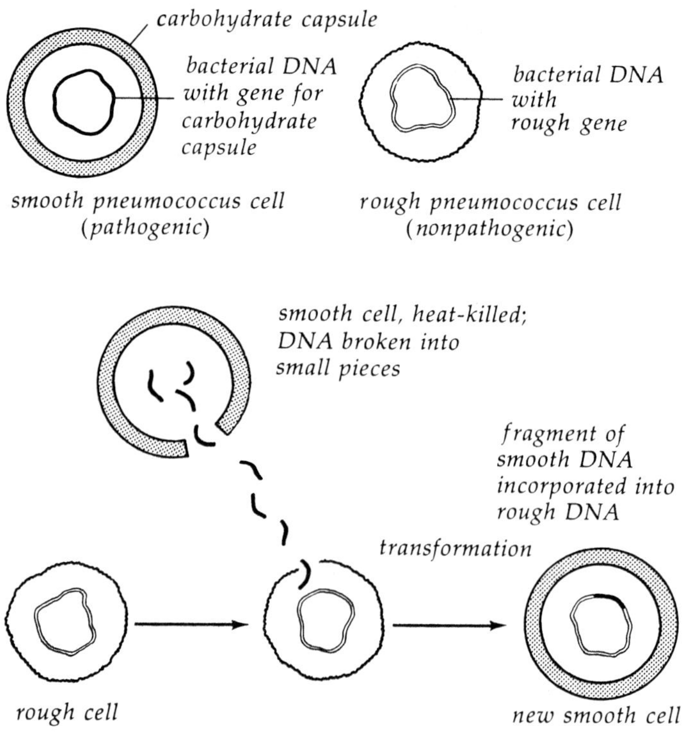

8–5 The transformation of rough, nonpathogenic pneumococcus into the smooth pathogenic type can be effected by DNA from the dead virulent strain entering and becoming part of the harmless variety's DNA.

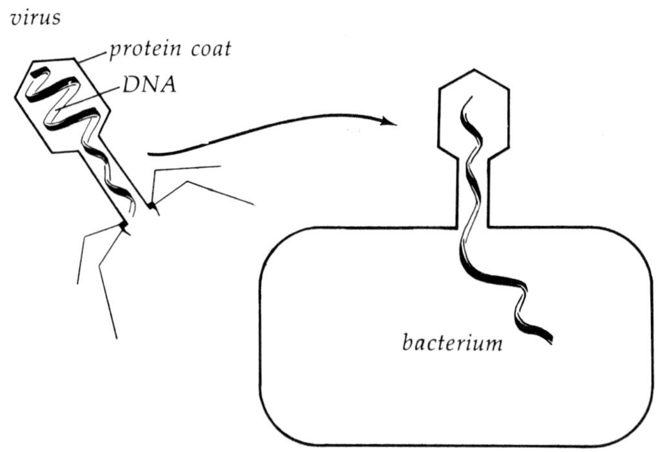

8–6 Bacteriophages are viruses that attack bacteria. Each phage particle consists of a coiled molecule of DNA surrounded by a protein coat. Experimental evidence, based on labeling with radioactive compounds and direct observation with the electron microscope, has shown that only viral DNA—not protein—enters the bacterium at the time of infection. Yet this is able to take over the synthetic machinery of the cell, forming new viruses.

the amino acids in a given protein; so to understand how DNA can contain such information, we must ask the next question.

What is the structure of a DNA molecule? The answer to this question won James D. Watson, Francis Crick, and Maurice H. Wilkins the Nobel prize. The data available to Watson and Crick included the general composition: deoxyribose, phosphate, and the bases adenine, guanine, cytosine, and thymine. They knew the molecular sizes of these components. In addition, the work of an American biochemist, Erwin Chargaff at Columbia University, demonstrated that the amounts of the bases were not random. For each molecule of adenine there was one of thymine, and for each molecule of guanine there was one of cytosine. Thus the bases appeared paired in some way, but there was no fixed quantitative relationship between the pairs. The most important piece of evidence they had available was an X-ray diffraction photo of a DNA molecule. The photograph they had was made by Wilkins by passing a beam of X-rays through a crystal of DNA and recording on film the way in which parts of the molecule diffracted the beam (see Essay 8–2). This procedure had been exploited by the American chemist Linus Pauling of the California Institute of Technology, who had received a Nobel prize for his work on chemical bonds and the determination of molecular structure of certain proteins. He too was approaching a solution to the structure of DNA at the same time Watson and Crick were. On the basis of these three lines of evidence, Watson and Crick proposed their famous model of DNA.

In this model the DNA molecule is a double stranded structure much like a ladder (Fig. 8–7). The sides of the ladder consist of the phosphate and sugar molecules. The rungs are the base pairs. If one side of

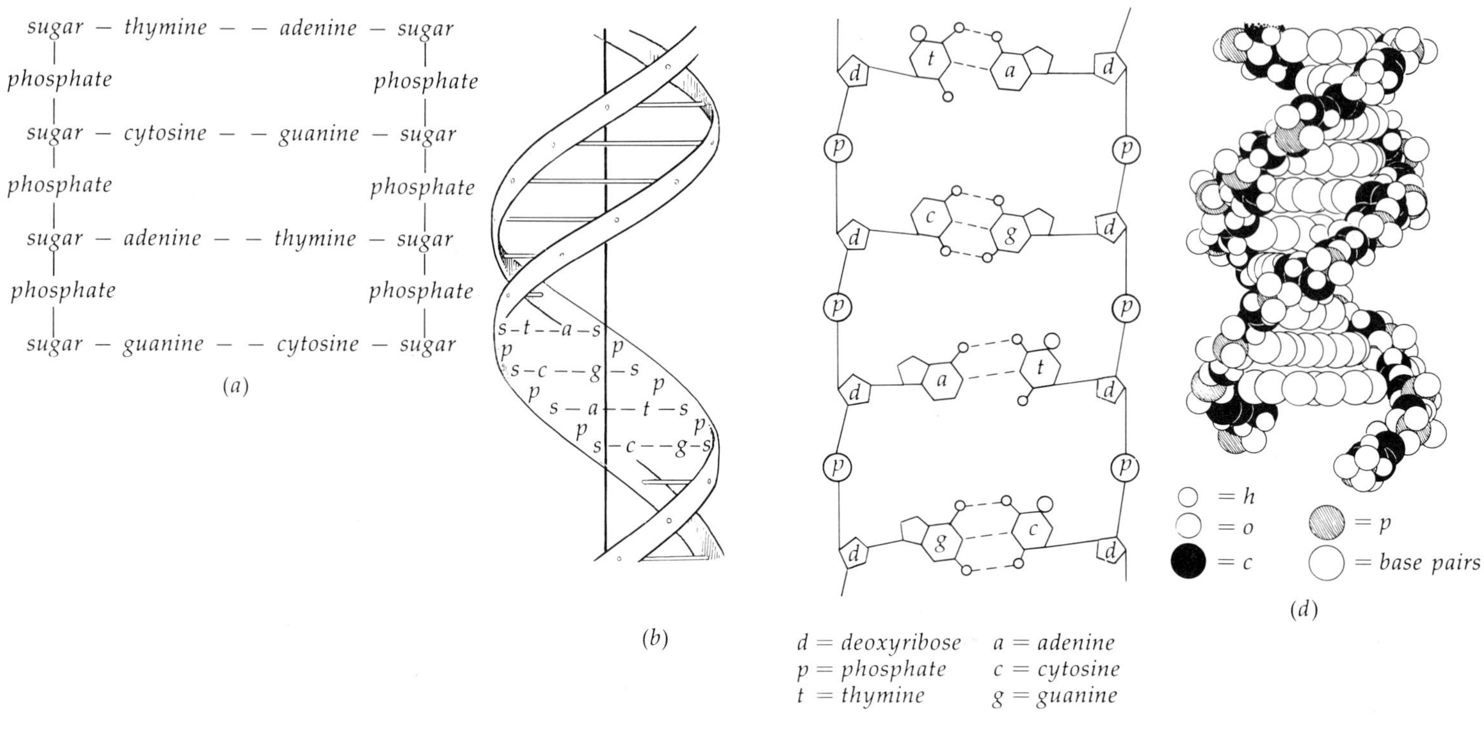

8–7 *Diagram of DNA (Watson-Crick model).*

a rung is an adenine, the other will be a thymine; if one side is a cytosine, the other is a guanine. But it makes no difference which base is on which side or the arrangement of the pairs along the ladder. The ladder is not straight but twisted in the form of a helix. This was not the first model of DNA proposed (nor was it the last), but it appears to be the right one. One of the major attractions of this model is that it offers an explanation of how the molecule can be precisely reproduced (Fig. 8–8). To get two DNA molecules from one, the original molecule splits down the middle and the base pairs separate. Next, nucleotides (base—sugar—phosphate) pair with the appropriate alternate that is part of the original DNA molecule. Adenine nucleotide pairs with thymine and guanine nucleotide with cytosine, or conversely. Finally, enzymes called DNA polymerases zipper together the nucleotides now arranged in appropriate position along the complementary strand of DNA. Arthur Kronberg, now at Stanford University, discovered the enzymes necessary for DNA synthesis and demonstrated that the following reaction occurred:

$$\text{4 deoxynucleotide triphosphates} \xrightarrow[\text{DNA template}]{\text{enzyme, } Mg^{++}} \text{DNA} + \text{4 pyrophosphate } (P_2O_7^{-4})$$

For this and other contributions, Kronberg shared the 1959 Nobel prize in medicine and physiology with Severo Ochoa, who in 1955 found an enzyme capable of synthesizing RNA.

The Genetic Code

How is information stored in the DNA molecule and how is it transmitted to the cytoplasm? As elegant as the Watson and Crick model of DNA is and as powerful as DNA polymerase may be, this question still remained unanswered. A single gene can be considered a portion of a long DNA molecule—one consisting of several thousand nucleotide pairs. Thinking this way, it is possible to see how the arrangement of the bases along the backbone of the molecule could correspond to the arrangement of amino acids in a protein. The problem, of course, is that the DNA is in the nucleus, and protein synthesis is taking place in the cytoplasm. Thus a messenger is needed, and it was found in the form of an RNA molecule called messenger RNA (m-RNA). A molecule of m-RNA is formed by a mechanism similar to that pictured for DNA duplication. First, the molecule splits down the middle; then nucleotides pair with those attached to the backbone (Fig. 8–9). Guanine and cytosine pair as before, but now uridine instead of thymine pairs with adenine. The nucleotides again contain the bases, phosphates, and sugars, but this time the sugar is ribose, not deoxyribose. Once finished, the new RNA molecule separates from the DNA strand and migrates to the cytoplasm. The synthesis of m-RNA represents the transcription of the genetic information. The concept of m-RNA was both introduced and proved by the 1965 Nobel prize winners, Jacques Monod and Francois Jacob, working at the Pasteur Institute in Paris with Lwoff. But now that we have the information (order of arrangement of nucleotides) transferred to the cytoplasm, we face another problem.

How is the information in the messenger RNA translated into the amino acid sequence in the protein? One immediate difficulty is that while there are some 20 amino acids, there are but 4 nitrogen bases. This means that there must be some type of code involved. Theoretical considerations show that the code must be based on more than single bases. The same is true for combinations of 2 bases (4^2) that would yield 16 information pieces, not 20. With combinations of 3 bases, there are 4^3 or 64 triplets—more than enough information for 20 amino acids (Table 8–10).

Intensive research by two groups, one led by Marshall Nirenburg at the National Institutes of Health in Maryland and a second, independent group led by H. Gobind Khorana at the University of Wisconsin, showed that the system is based on combinations of three bases. Nirenburg's group devised a method for synthesizing and testing triplets of nucleotides of known base composition and sequence. The Khorana group synthesized molecules of DNA or RNA in which known sequences of three nucleotides could be repeated many times. The data obtained from both groups, while not identical, agreed closely and thus forms the foundation for our current understanding of the genetic code. For their work, both Nirenburg and Khorana were awarded the Nobel prize in 1968.

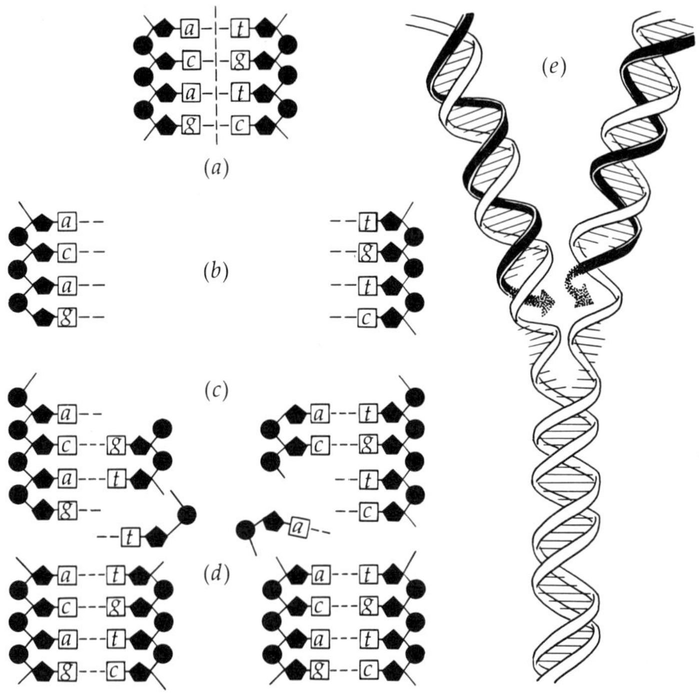

8–8 A diagrammatic representation of the postulated method of DNA duplication.

The code is based, as predicted, on triplets now called codons. Hence the sequence adenine-cytosine-cytosine is translated to the amino acid histidine, and uridine-guanine-adenine is the codon for methionine. But we saw earlier that $4^3 = 64$ triplets, which means that there is an excess of 44 triplets over the 20 needed for amino acids, if there were only one triplet per amino acid. Research, however, has shown that the code is degenerate in that more than one triplet can stand for the same amino acid. Thus UCU and UCC, where U stands for uracil and C for cytosine, both code for the amino acid serine. The entire 64 triplets and the amino acids they code for is shown in Table 8–11. This figure clearly shows that, in general, the first two bases are always the same for a particular amino acid, while the third base may vary.

Only three combinations of bases—UAA, UAG, and UGA—have not been found to code any amino acid. These are called nonsense codons, and their function was understood when it was realized that some m-RNAs carried more than one message. The sequence of codons necessary to make a single protein is called a cistron. Some mechanism must be available to start the translation of one cistron, end it at the right point, and start the next one. The nonsense codons appear to be involved in terminating the read-out sequence, while the amino acid methionine (that has a single codon) is involved in starting the sequence.

The site where the message is read is on the ribosomes in the cytoplasm. Actually, several ribosomes are involved with a single strand of m-RNA.

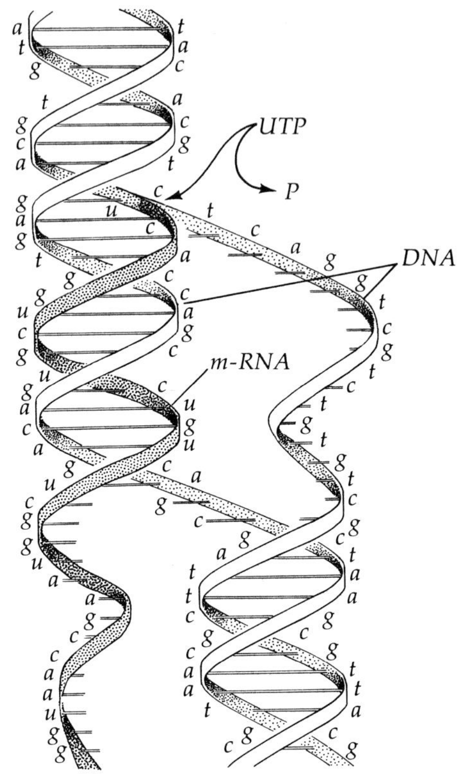

8–9 Synthesis of messenger RNA. The first step in the formation of m-RNA is the separation of a portion of the two strands of DNA. Then, the complementary bases are aligned opposite one of these strands [adenine (a) against uridine (u); guanine (g) against cytosine (c)], and the RNA is formed.

Protein Synthesis

How are proteins formed? The m-RNA is with the ribosomes, and the amino acids have been activated by combining with ATP. But one more compound is necessary to make the process go: transfer RNA (t-RNA). A t-RNA was isolated and its structure determined by Robert Holley, then at the California Institute of Technology, and for this he shared the 1968 Nobel prize. This RNA is relatively small and is amino acid specific; that is, there is a different t-RNA for each different amino acid. Moreover, each t-RNA has three unpaired bases exposed at one end that are the reciprocals of one of the code triplets on the m-RNA.

The sequence of events starts with the activated amino acids combining with their specific t-RNA (Fig. 8–12). At this point AMP is released.

Amino acid–AMP + t-RNA $\rightleftharpoons$
t-RNA–amino acid + AMP

Next, the t-RNAs with the attached amino acids migrate to the m-RNA ribosome complex (Fig. 8–13). Here, in a stepwise reaction, the ribosome moves along the m-RNA, and a t-RNA is matched to the corresponding codon on the m-RNA. The amino acid is attached to the growing peptide chain by an enzymatic reaction that depends on enzymes separate from the ribosome. After the amino acids are joined, the t-RNA is released, and the protein is synthesized step by step. When the last amino acids are attached, the protein floats free in the cytoplasm, assuming its final form.

The message is translated correctly because the ribosome starts at one end of the m-RNA and works, one codon at a time, along the molecule. This mechanism appears to prevent the message from being misread by having the triplets overlap. What happens when deletion of one or two bases from the DNA causes portions of the m-RNA to be missing? This results in a frame-shift. The ribosome starts down the m-RNA and, as long as the triplets are in the proper order, the ribosome proceeds. When it reaches a point where nonsense codons are present because the deletion of some bases has shifted the order of the triplets (Fig. 8–14), the ribosome stops translating and the peptide is discharged as an unfinished enzyme molecule. The same mechanism is used for m-RNAs with more than one cistron, but in this case one of the three stable

8–10 Nucleotide combinations for singlet, doublet, and triplet codes

			Singlets (4)				
		A	G	C	U		
			Doublets (16)				
AA	AC	GA	GC	CA	CC	UA	UC
AG	AU	GG	GU	CG	CU	UG	UU
			Triplets (64)				
AAA	ACA	GAA	GCA	CAA	CCA	UAA	UCA
AAG	ACG	GAG	GCG	CAG	CCG	UAG	UCG
AAC	ACC	GAC	GCC	CAC	CCC	UAC	UCC
AAU	ACU	GAU	GCU	CAU	CCU	UAU	UCU
AGA	AUA	GGA	GUA	CGA	CUA	UGA	UUA
AGG	AUG	GGG	GUG	CGG	CUG	UGG	UUG
AGC	AUC	GGC	GUC	CGC	CUC	UGC	UUC
AGU	AUU	GGU	GUU	CGU	CUU	UGU	UUU

8–11 The genetic code

Base sequences in RNA					
1st base	2nd base				3rd base
	U	C	A	G	
U	phenylalanine	serine	tyrosine	cysteine	U
	phenylalanine	serine	tyrosine	cysteine	C
	leucine?	serine	termination	cysteine?	A
	leucine, f-methionine	serine	termination?	tryptophan	G
C	leucine	proline	histidine	arginine	U
	leucine	proline	histidine	arginine	C
	leucine	proline	glutamine	arginine	A
	leucine	proline	glutamine	arginine	G
A	isoleucine	threonine	asparagine	serine	U
	isoleucine	threonine	asparagine	serine	C
	isoleucine	threonine	lysine	arginine	A
	methionine, f-methionine	threonine	lysine	arginine	G
G	valine	alanine	aspartic	glycine	U
	valine	alanine	aspartic	glycine	C
	valine	alanine	glutamic	glycine	A
	valine, f-methionine	alanine	glutamic	glycine	G

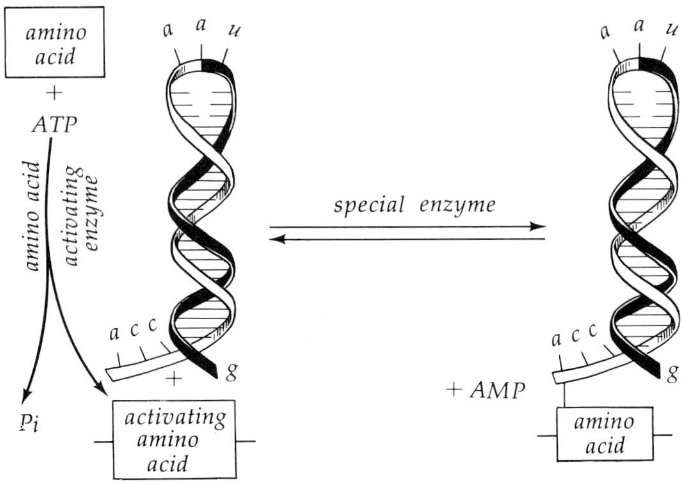

8–12 A diagrammatic summary of the possible sequence of events in the formation of proteins. Transfer RNA is a special helical form of RNA which transports amino acids to their proper site in the protein chain. There is at least one t-RNA for each of the 20 common amino acids. All, however, seem to carry the bases (acc), where the amino acids attach, and g at the opposite end. The attachment requires a specific enzyme and energy supplied by adenosine triphosphate. Unpaired bases in t-RNA (aau in the example) may provide the means by which t-RNA "recognizes" the place to deposit its amino-acid package.

nonsense codons is used to terminate translation, producing a whole, active protein molecule.

In all of the reactions involved in translating the message from DNA into protein, it is important to remember that enzymes are mediating the various steps. Ernest Du Paw, now at Stanford University, emphasizes this point in his book *Cell and Molecular Biology,* discussing the genetic code. He calls attention to the fact that while the organization of the triplets can be spoken of as a code (as we have), the sequences are not so much a code as a physiochemical mechanism whose operation depends on the specificity of several enzymatically governed reactions. The fact that the enzymes themselves are synthesized from a DNA–RNA code does not alter the fact that the code has meaning only when translated by other enzymes.

We are left with the paradox that, while we have broken the code for translating from a DNA language into a protein language, we don't understand either. Amino acid sequence has meaning because it confers enzymatic activity in a protein, but DNA language has meaning only when it is translated into protein language. We know only a few words (enzyme protein molecules) of the protein language and those only in part: the amino acid sequences of a few active sites. Molecular biology is an area of research in which the great advances of the last few years have led many to feel that all essentials are known. This is not true because many important questions remain not only unanswered but unasked.

How is protein synthesis regulated by the cell? We have seen that enzymes are synthesized in conjunction with m-RNA molecules coded from portions of a DNA molecule. What governs the production of specific m-RNA molecules by the DNA? Before we consider the answer, let us remember that enzymes frequently act in teams; we should inquire whether the synthesis of the team is coordinated or regulated by the same substance. An answer to these questions comes from the work of Jacob and Monod, mentioned in our discussion of m-RNA. Working with enzymes in bacteria that appear to arise in response to specific substrates, they found that the messages or genes needed to produce a team of enzymes were located next to one another along the length of the DNA molecule. Immediately adjacent to these genes was one that governed the operation of them all (Fig. 8–15). This gene they called the operator gene, and the combination of the enzyme (or structural) genes plus the operator gene is termed the operon. In many cases, there is only one structural gene per operon. We do not know how many structural genes there are per operon for higher organisms because the operon model has not been shown except for a single case in *Drosophila,* the common fruitfly.

In bacteria the action of the operator gene is governed by the presence of various substances in the cell that may be related to the action of the enzyme it regulates. Thus, in the enzyme Jacob and Monod originally studied, they found that the new substrate induces the synthesis of a new enzyme rather than releases the inhibition of the regulator gene, resulting in its production.

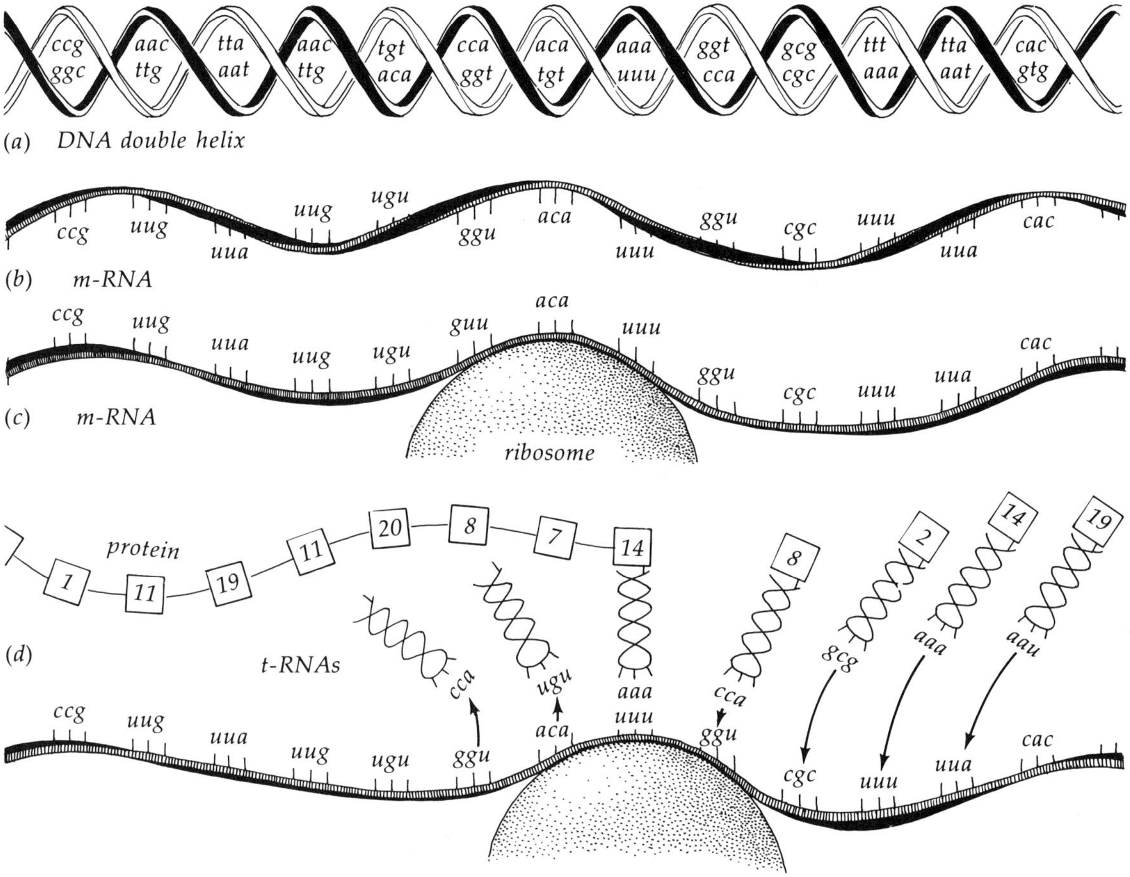

8–13 Synthesis of protein begins with the genetic code embodied in DNA. (a) The code is transcribed into messenger RNA. (b) In the diagram the message has been derived from the DNA strand shown in darker type. The m-RNA finds its way to a ribosome (c), the site of protein synthesis. Amino acids, indicated by numbered rectangles, are carried to proper sites on the m-RNA by molecules of t-RNA. Bases are actually equidistant, not grouped in triplets, and mechanism of recognition between t-RNA and m-RNA is hypothetical. Linkage of amino acid subunits creates a protein molecule.

What is the nature of the gene repressor? The answer appears to be dependent on what organisms we are examining. In the case of the bacteria that Jacob and Monod worked with, the answer is protein molecules that bind specifically to the operator; in the case of eucaryotic organisms, the answer is different. Earlier in this chapter, we made the point that bacteria and blue-green algae not only lack nuclei but the DNA present is not associated with proteins, as it is in eucaryotic cells that have membrane-bound nuclei. In eucaryotic organisms the research of V. G. Allfrey and A. E. Mirsky, both at Rockefeller University, and Ru-chih

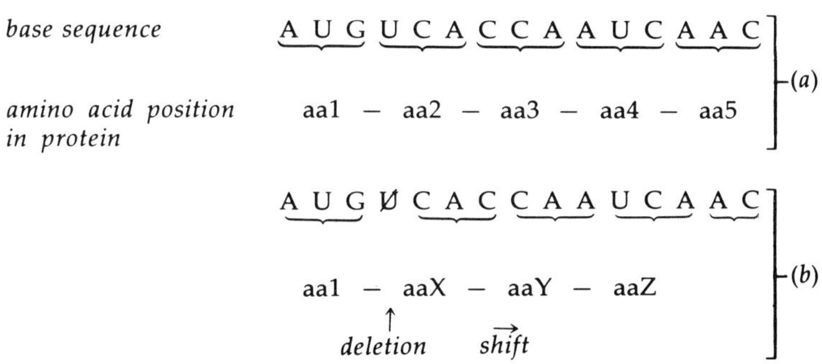

8–14 The effect of the deletion of a single base is shown in this diagram. The deletion of a single base in the sequence shown in (a) means that the reading is shifted one place to the right (b), resulting in the wrong amino acids being placed in the protein.

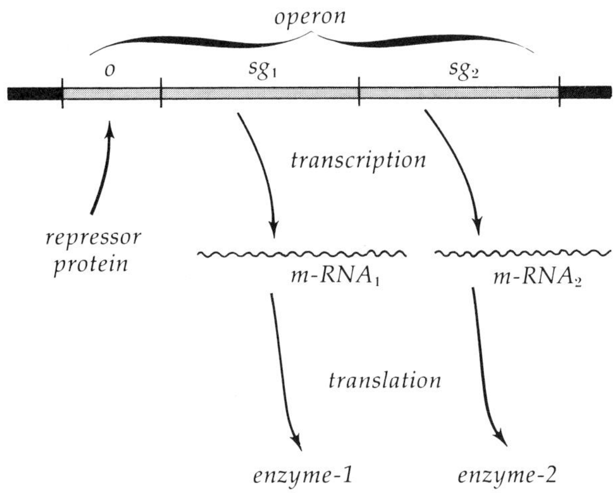

8–15 The operation of the operon, as postulated for bacteria by Jacob and Monod. The operator gene (o) is immediately adjacent to two structural genes (sg_1) and (sg_2). Only when the operator gene is open can the structural genes make the messenger RNA (m-RNA) necessary for enzyme synthesis. The opening and closing of the operator is regulated by repressors. The repressor in bacteria is a protein.

Huang and James Bonner, at the California Institute of Technology, indicate that histone proteins act as gene repressors. Histones are found in the chromosomes of all plants and animals, except bacteria and blue-green algae. Histones are rather simple proteins, and the amino acid sequence of the related histones of plants and animals are similar. One of their functions—they appear to have several in the nucleus—is to prevent DNA from being transcribed.

Research in Bonner's laboratory indicates that a low molecular-weight RNA, which they discovered, may be associated with activating gene transcription. This special RNA appears to act in conjunction with chromosomal RNA bound to chromosomal protein. This combination may determine which genes will be transcribed, perhaps by combining with the operator and preventing, by such a combination, a particular structural gene from being covered by a histone and repressed. Research in this area is developing rapidly, and many of the ideas and interpretations of the data are subject to change—as they always are. Bonner has made the interesting point that there appears to be a fundamental difference between procaryotic organisms and eucaryotic ones. In the former, genes appear to be turned on except when specifically repressed, while in the latter genes are turned off until specifically activated.

The understanding of the gene as a sequence of bases in a molecule of DNA has profound implications. One is that we can now understand the nature of mutations (see Chapter 13)—they are changes in the sequence of bases in the DNA molecule. Such changes can be induced by ultraviolet radiation, X-rays, or ionizing radiation from radioactive atoms such as those produced by atomic explosions. They may also be caused by certain chemicals or simply by heat. Only a minor change in the sequence is necessary to cause a change in the final protein molecule. Such a change might result in an enzyme being inactive or having a new specificity in a reaction. The understanding of gene action we have gained has enabled biologists to take a new look at a large number of both plant and animal diseases, particularly various forms of cancer. It also holds visions of providing man with a means of tailor-making species to his own specification. Some have extended these visions to man himself and have been either pleased or horrified by what they imagine could happen. A few have suggested that this line of research should cease until man's ability to cope with scientific discoveries reaches the level of sophistication of the discoveries themselves. Stopping or even slowing a field that has grown as this one has is impossible. The number of scientists is great, and as we have seen, they are found worldwide. Nevertheless, these discoveries pose many problems that may tax to the limit man's ability to deal with them.

Essay 8–2 X-ray Diffraction

A powerful tool for studying the structures of crystals is X-ray diffraction. It was originally used to verify the structure of such simple crystals as sodium chloride. Pauling used it to determine the more complex α-helix structure of protein backbones, and Watson and Crick used the X-ray diffraction data of Wilkins to test their models of the DNA double helix.

How does it work? Do the X-rays provide a photographic image that clearly illustrates the crystal's structure? Not really. Usually, one must first understand the structure of the crystal before understanding the X-ray diffraction photograph. A beam of monochromatic X-rays (one having only one wavelength) is projected upon a crystal. The atoms of the crystal scatter the incident radiation as if each were itself a source of radiation. Sometimes, the waves of energy sent out by different atoms are in phase with each other, causing a reinforcement and hence a dark spot on the photographic negative. Whether such reinforcement occurs depends upon the wavelength of the incident X-rays and the spacing of the atoms in the crystal. When the atoms lie in parallel planes, there will be a maximum reinforcement when the angle of incidence of the monochromatic X-rays bears a relationship to the distance between planes, as described by Bragg's Law (formulated by Sir William L. Bragg, Director of the Cavendish Laboratory where Watson and Crick worked out the structure of DNA). In crystals there are planes consisting of many atoms per unit area, and there are other planes consisting of fewer atoms per unit area. One can visualize this by looking down the rows of an orchard. Some rows have many trees per unit distance; others have fewer. Crystals are similar, except they are three-dimensional. In any case, the brightest spots on the X-ray diffraction image are produced by parallel planes with the most atoms per unit area of the plane.

With all these planes producing numerous spots of various intervals (Fig. 8–0), it should be clear why the diffraction image is not a picture of the crystal structure. To interpret it, the scientist must try to imagine possible crystal structures that fit the known information about the composition of the crystal. For example, Watson and Crick knew DNA must consist of four nucleotides and what their molecular structures are. They also knew how many atoms there were per unit volume in the DNA crystals prepared by Wilkins. Their challenge was to arrange these in various ways permitted by the laws of crystal formation. Once the double-helix crystal had been "invented," they could see where the planes of atoms were, and knowing the wavelengths and incident angles of Wilkins' X-rays, they could

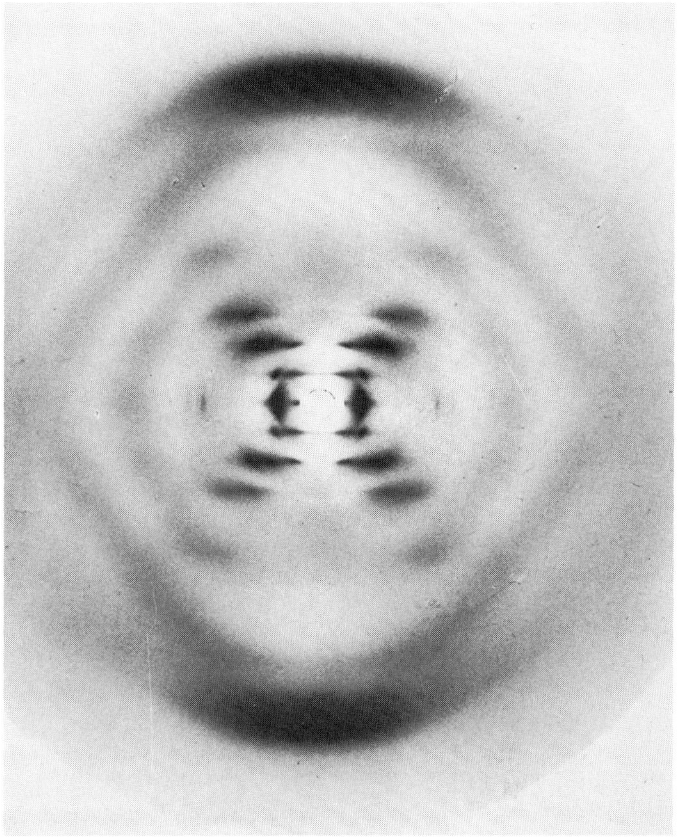

8–0 An X-ray diffraction pattern from a semi-crystalline fiber of sodium DNA at 92% relative humidity.

calculate the predicted positions of bright spots and compare their calculations with the actual photographic plates.

The whole procedure illustrates beautifully how science sometimes works. As Pauling emphasized in telling of his discovery of the α-helix, there is no coldly logical, mechanical process to deduce the crystal structure from the X-ray data. Rather, the scientist must first be a dreamer, an artist, an imaginative composer—he must be creative in the truest sense of the word. When his creation (the proposed crystal structure) has finally been born, then mechanistic science (Bragg's Law) can take over to confirm or reject the model.

Endoplasmic Reticulum

The ribosomes, as noted earlier, may be free in the cytoplasm or attached to the endoplasmic reticulum (ER). The ER is a membrane system present in all cells, although in highly variable amounts and form (Fig. 8–E). When sectioned and observed in the electron microscope, it appears as a series of parallel pairs of single (unit) membranes. These membranes form an elaborate branching network or reticulum—hence the name. Sometimes, the ER may be in the shape of narrow tubes, but the more common type consists of long, folded, saclike sheets of membranes.

There is a definite inside and outside to the ER. Usually, the ER has ribosomes attached to the side in contact with the cytoplasm. Ribosomes are only found on this side, never on the inner side. ER with ribosomes is called rough ER, while that lacking ribosomes is termed smooth. Rough ER is much more abundant than smooth, which may be completely absent from many cells.

The function of the ER appears to be storing and transporting various cell products, particularly protein. Large amounts of protein can be found accumulated within enlargements of the ER in a variety of cells (Fig. 8–E).

Combined biochemical and electron microscope studies, primarily of animal tissue, indicate that the protein is made, as you would expect, in conjunction with the ribosomes on the outer surface of the ER. As the polypeptide chain of the future protein is formed, it apparently passes through a hole in the center of the ribosome and continues through a corresponding pore in the ER. Through this tunnel, the new protein molecule is introduced into the cisternae of the ER.

Once within the ER, the fate of the protein molecule depends on its type and the developmental-metabolic condition of the cell. Some of the protein is stored; this is true in many seeds in which the cells are filled with stored proteins to be used when the seed germinates. Some of the protein is exported to other parts of the cell and outside the cell. Recent evidence, for example, indicates that many of the Krebs Cycle enzymes in the mitochondria are synthesized on the ER, transported to the mitochondria through the ER, and transferred into the mitochondria by some unknown mechanism, while the ER is closely appressed to the outer mitochondrial membrane. The enzymes in the ER may also be packaged in the form of microbodies (see Chapter 7). There are three other organelles in the cytoplasm that may be immediately related to the ER: lysosomes, spherosomes, and dictyosomes.

Lysosomes

The existence of lysosomes was first indicated in experiments with isolated mitochondria. As described in Chapter 7, mitochondria can be isolated from a cell and purified. Attempts to obtain completely pure and uniform mitochondria resulted in the discovery that particles of the same size and shape as mitochondria but containing different enzymes were present. These different enzymes are primarily types that mediate the breakdown of various cell constituents.

The lysosomes have a single external membrane and lack internal structure. The current concept of lysosome function is based on the belief that the enzymes present within them remain inactive until they are released (Fig. 8–16). The enzymes then become active, and the result is the breakdown of cell parts. The role of lysosomes in animal cells is better understood than in plants. The extent of involvement of lysosomes in plant cell development and function is still unknown.

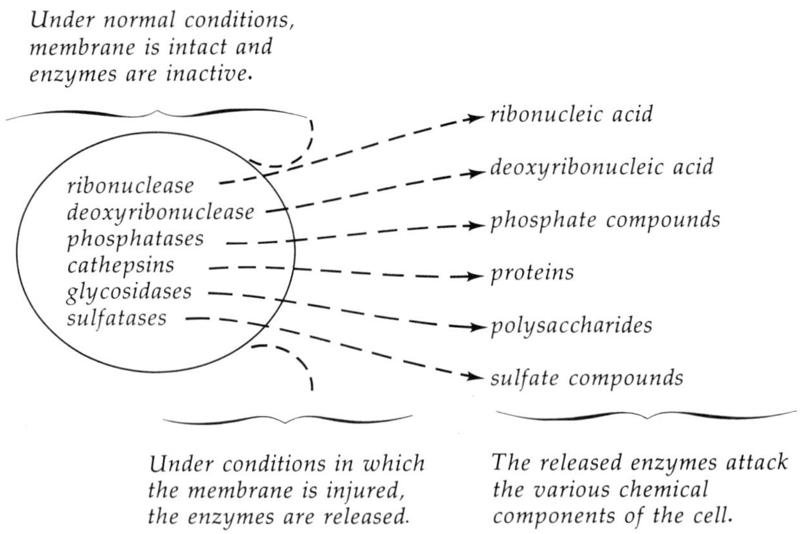

8–16 Diagram of the lysosome concept.

Spherosomes

When living plant cells are examined in the microscope, the cytoplasm appears to circulate at a rapid rate. The mitochondria, chloroplasts, and other cell parts can be observed to move about in the cell. With careful observation, small, completely spherical particles can be seen moving rapidly through the cell. These are the spherosomes.

Spherosomes are difficult to study because of their small size and their poor preservation in killed and fixed cells. A group of Swiss botanists in the laboratory of Professor Frey-Wyssling studied the origin and development of spherosomes. They believe that they are formed by the endoplasmic reticulum. Small, immature spherosomes are pinched off from the tubelike endoplasmic reticulum; these spherosomes enlarge, resulting in the single membrane-bound spherosomes found in plant cells. Frequently, enlargement continues, and a fat body develops. This development may indicate that the spherosomes are a special cell part involved in fat production. Many possible functions have been proposed for spherosomes, and their true function is by no means certain.

Essay 8–3 The Nobel Prize

The setting is magnificent: a beautiful city, handsome buildings, elegantly clothed people. The ceremony is impressive, with the Swedish King himself presenting the award: the Nobel prize. The winner of the prize receives a gold medal, an individually designed diploma bearing a citation, and approximately $80,000—less if he is awarded the prize jointly with others. The money is essentially immaterial because the Nobel laureate commands top salaries and has many other opportunities open to him to gain additional income. Few things are dearer to the hearts of university presidents and boards of trustees or regents than Nobel prize laureates!

The prize is based on the fortune of the Swedish chemist and engineer Alfred Nobel (1833–1896). He invented the modern form of dynamite and the devices necessary to detonate it. Other forms of explosives and oil exploitation added to his fortune, making it immense at the time of his death. Having never married, he left this fortune in trust to establish five prizes in the fields of physics, chemistry, physiology or medicine, literature, and peace. Nobel disliked biology in general and botany in particular. Only in recent years have biological discoveries of fundamental importance been recognized by Nobel prizes in chemistry and physiology.

The role of the Nobel prize in the advancement of science is an interesting one. No biologist would probably ever admit to openly pursuing a line of research simply to win the prize. Yet, in certain areas, it has the impact of making research extremely competitive—literally a race for data and conclusions against another scientist or group of scientists. Nowhere has this been more evident in recent years than in the field of nucleic acid research, and nowhere has this type of race been more candidly described than in The Double Helix *by James Watson (see Essay 8–1). He received the Nobel prize in 1962 for his work on the structure of the nucleic acid: DNA.*

The area of research covered in this chapter has produced many prizes: Beadle, Tatum, Lwoff, Watson, Crick, Wilkins, Monod, Jacob, Nirenburg, Khorana, Kronberg, Ochoa, and Holley. Has the awarding of Nobel prizes in this area helped future research? Perhaps they speeded up the progress and made some aspects of the work more like a race than is usual in most biological research. But the problems were of such importance and the background information had reached such a level that intensive research and rapid advances were clearly inevitable. This is strikingly true in the case of the structure of DNA: if Watson and Crick had not proposed the correct model in 1953, Pauling would have in 1954 or 1955. One can hardly imagine transfer RNA not being found even without Holley or messenger RNA not being discovered even if Monod had been killed leading his French underground unit in World War II. The pattern by which science advances has not changed in the 100 years since Darwin or the 300 years since Hooke.

The Nobel prize serves an important function in focusing public attention on current research and on brilliant, productive scientists. Those that have been chosen in biology are completely worthy of the honor. But for every Nobel prize winner there are many scientists who have contributed as much but are not so honored—Chargaff, Casperssen, and Brachet, to mention but a few in this area. As in so many situations, these awards are good as long as their limitations are recognized and their value not inflated. Unfortunately, this is difficult for most people, including scientists, to do.

Dictyosomes

The appearance of a dictyosome in the electron microscope is shown in Figure 8–17. A three-dimensional diagram of the typical dictyosome found in a plant cell is shown in Figure 8–18. The center of the dictyosome is composed of a series of discs with open, lacelike edges. Each disc is bounded by a single membrane. Around the edges of the discs are rows of small vesicles, which may be derived from the central discs. There are variations in this general pattern both in the size and number of the discs and the vesicles. One example of such variation in the structure of the dictyosome can be seen in the development of the cell wall. For example, in the root-cap cells, which surround the tip of the root, the young cells have dictyosomes similar to the one shown in Figure 8–17. As a cell becomes older, it develops a special, thick cell wall. During the growth of this wall the dictyosomes expand (and possibly multiply) and produce large vesicles, which appear

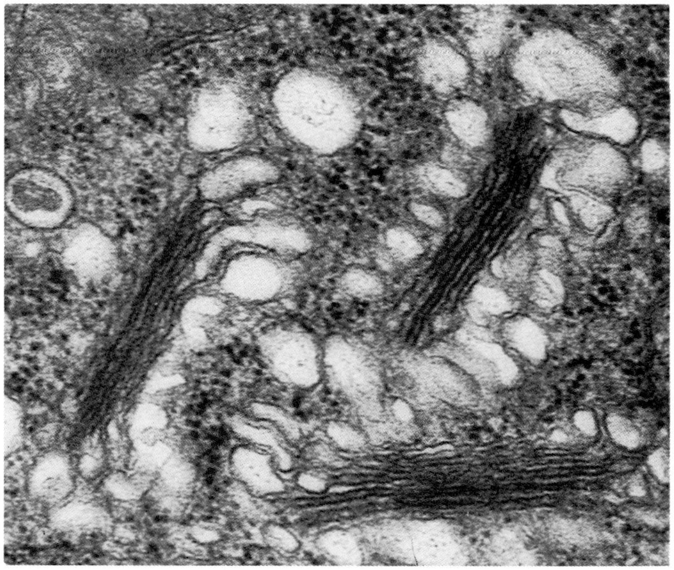

8–17 An electron microscope view of a group of dictyosomes in a plant cell.

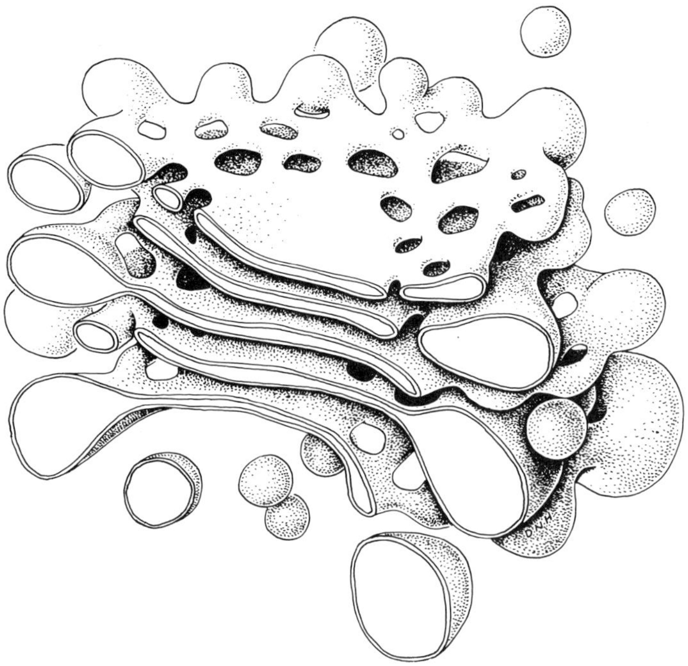

8–18 A three-dimensional drawing of a generalized plant-cell dictyosome.

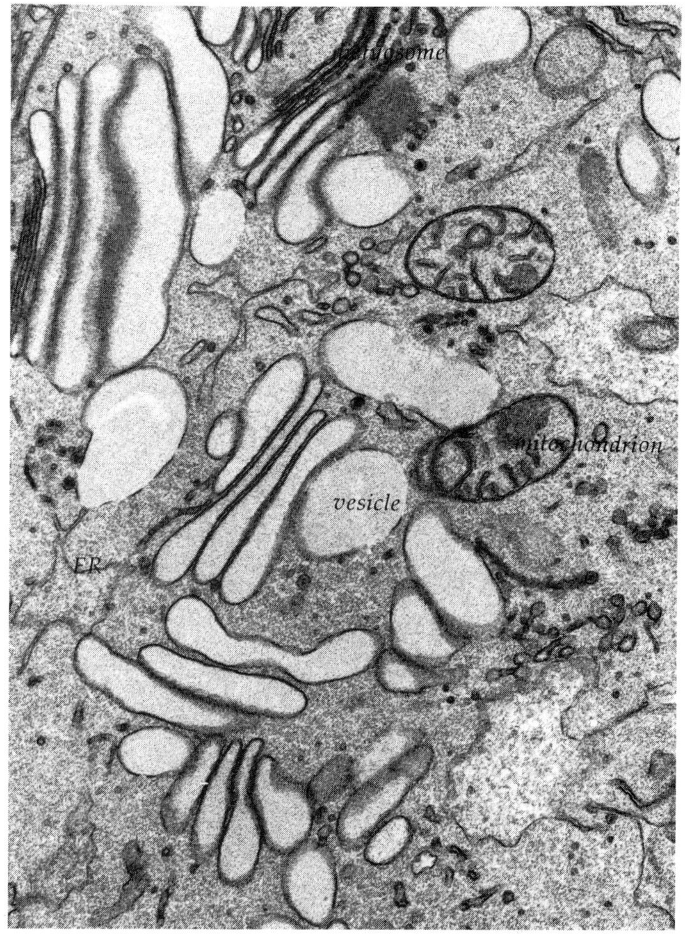

8–19 Dictyosomes in the root-cap cells of corn. The vesicles are produced at the end of the cisternae and then migrate to the edge of the cell, where they fuse with the plasma membrane of the cell and discharge their contents against the wall. The tissue is fixed in $KMnO_4$, *which removes the ribosomes and some other structures from the cells.*

to contain hemicellulose similar to that found in the wall. These vesicles apparently migrate to the surface of the cell and discharge their contents, which fuse with the other wall materials (Fig. 8–19).

Further data on the function of the dictyosomes were obtained from cells grown in compounds that contained radioactive atoms. These were compounds that were incorporated into the pectic substances and hemicellulose. Autoradiographs were made to follow the course of these labeled compounds through the cell. They first appeared in the dictyosomes and the dictyosome vesicles; later, when the vesicles fused with the plasma membrane, the radioactive compounds appeared in the wall. In other experiments radioactive lignin precursors were used, and again the dictyosomes were found to be involved in the deposition of this important wall material. There is no evidence of the dictyosomes being involved in cellulose synthesis; rather it is matrix materials such as lignin, pectic substances, and hemicellulose that are deposited through the mechanism of the dictyosome vesicles in higher plants. In algae the dictyosomes are the site of formation of a variety of wall materials, including platelike structures composed of calcium. The dictyosomes in animals are also involved in secretion of materials by the cell—in this case, protein and protein-carbohydrate combinations. The possible role of the dictyosome in secretion of protein by plant cells has not been fully explored.

End Paper 8–1 Autoradiography

The technique of autoradiography is extremely valuable to the cell biologist because it enables him to localize the sites and time of synthesis, as well as the movement, of many substances within the cell. Autoradiography is based on the ability of the nuclear physicist to produce radioactive isotopes of elements commonly found in cellular compounds. The most useful of these are the radioactive isotopes of carbon (^{14}C) and hydrogen (^{3}H, called tritium). The biochemist can synthesize these isotopes into a host of compounds that cells can use (Table 8–20).

Localization of these compounds within the cell is possible because the radiation from the isotope will expose photographic emulsion in the same way as light. The film in the camera has a layer of light-sensitive material composed of silver halide crystals. When exposed to light, some or all of the crystals (depending on the amount of light) are sensitized, and these sensitized crystals are reduced to free silver during the developing process while the unsensitized crystals are removed. This is why the exposed areas appear black in a photographic negative. For study on the cellular level, the tissue is sectioned and placed on microscope slides (Fig. 8–21a). It is then covered by a thin layer of photographic emulsion. This is done in the dark, and the coated slide is stored in the dark for hours or months. When the emulsion is developed, the site of the radioactive compound can be seen by the deposits of black, granular-appearing silver grains over the part of the cell in which the compound has been incorporated (Fig. 8–21b). The procedure has been refined to the point where it can be used on electron microscope preparations. The only difference is that at the higher magnification of the electron microscope, the grains developed by the usual methods appear as thin strands of silver. But special developing procedures result in tiny solid silver grains even at the greater magnification of the electron microscope.

8–20 Some important atomic isotopes used in biological research

Isotope	Percent natural abundance	Atomic mass	Lifetime ($t_{1/2}$ = time until half decays)
$_1H^{*}$	99.99	1.01	
$_1H^{2***}$	0.02	2.01	
$_1H^{3**}$		3.00	12.26 yr
$_6C^{10}$			19.0 sec
$_6C^{11}$			20.5 min
$_6C^{12*}$	98.89	12.00	
$_6C^{13}$	1.11	13.00	
$_6C^{14**}$			5770 yr
$_6C^{15}$			2.25 sec
$_6C^{16}$			0.74 sec
$_7N^{12}$			0.01 sec
$_7N^{13}$			10.0 min
$_7N^{14*}$	99.63	14.00	
$_7N^{15***}$	0.37	15.00	
$_7N^{16}$			7.35 sec
$_7N^{17}$			4.14 sec
$_8O^{14}$			73 sec
$_8O^{15}$			2.03 min
$_8O^{16*}$	99.76	15.99	
$_8O^{17}$	0.04	17.00	
$_8O^{18***}$	0.204	18.00	
$_8O^{19}$			29.4 sec
$_8O^{20}$			14 sec
$_{15}P^{28}$			0.28 sec
$_{15}P^{29}$			4.4 sec
$_{15}P^{30}$			2.6 min
$_{15}P^{31*}$	100	30.97	
$_{15}P^{32**}$			14.3 days
$_{15}P^{33}$			25 days

The table shows all of the known isotopes of hydrogen, carbon, nitrogen, oxygen, and phosphorus, illustrating the variety of isotopes. Only the ones marked with asterisks are of much importance.

Subscript preceding the chemical symbol indicates the number of protons in the nucleus. This number determines what each element shall be. Superscript to the right of the chemical symbol indicates the number of protons plus the number of neutrons in the atomic nucleus.

*This indicates in each case the most common, stable isotope.

**This indicates the radioactive isotope most commonly used in research. Half-times for these isotopes are of reasonable durations; others are too short. ^{3}H is called tritium.

***This indicates stable isotopes of different atomic weights from the ones most common in nature. They are used in research but must be detected with a mass spectrophotometer. ^{2}H is called deuterium.

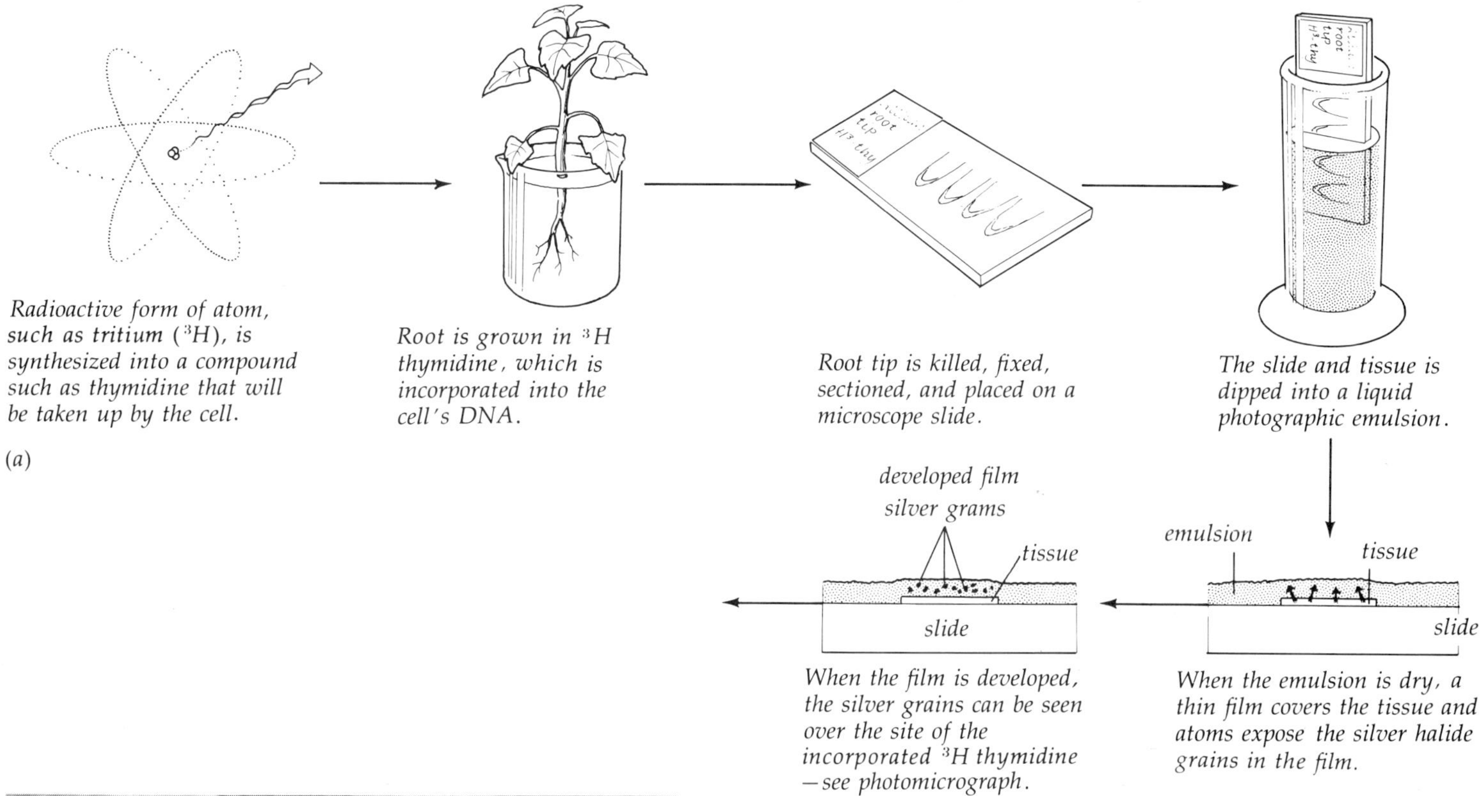

(a)

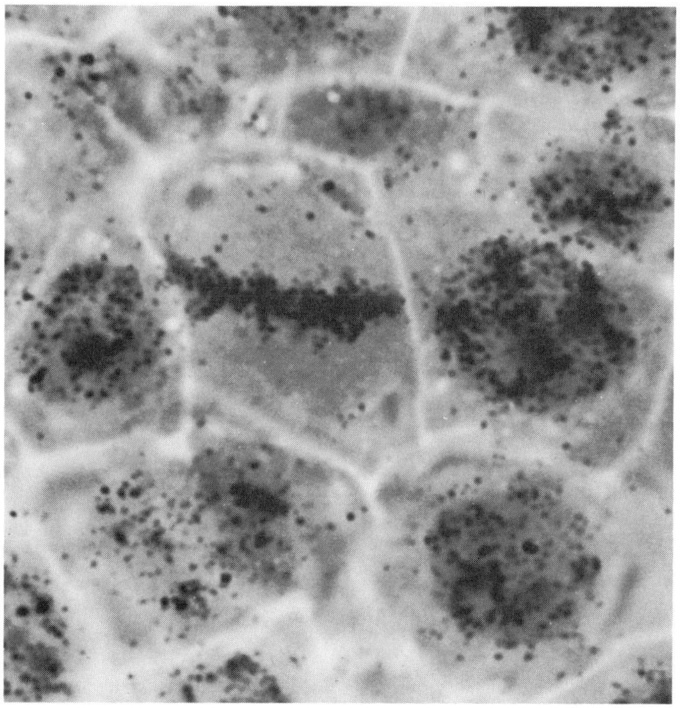

(b)

8–21 *(a) The technique of autoradiography as applied to tissue sections. The photomicrograph (b) illustrates the end result seen with a light microscope. In this case, the cells were allowed to incorporate ^{3}H-thymidine.*

9

Chromosomes, Mitosis, and Meiosis

From Little Acorns Grow

The discovery of mitosis in the late 1800s opened a new era in the study of the cell. Yet a hundred years later many aspects of this important phenomenon are still only partially understood. Research into the mechanism through which the nucleus divides involves the light and the electron microscopes. Photographs taken by both these methods are reproduced here to provide an introduction to this fascinating process.

Mitosis, as seen with the light microscope, is beautiful and graceful—quite literally a dance of the chromosomes. These photographs show a single nucleus from a developing seed as it underwent mitosis. A special optical system was used that gives a three-dimensional effect to the objects. Accompanying each photograph is a simplified diagram emphasizing the essential features of each stage.

A Prophase: chromosomes (ch) are visible; the nuclear envelope (ne) is still present; and the clear zone (cz) can be clearly seen, with the plastids (pl) and other cell organelles excluded from it.

B Transition to metaphase: the nuclear envelope is no longer present; spindle fibers (sf) can be seen; and the chromosomes (ch) are becoming aligned.

C Metaphase: the centromeres (ct) are aligned on the metaphase plate (mp); the poles (p) are well defined; the spindle fibers (sf) are numerous; and the chromatids (cd) are clearly unwinding.

D Early anaphase: the chromosomes are moving toward the poles (p), with the centromeres (ct) first; spindle fibers (sf) are evident; and the plastids (pl) and other cell organelles are clustered outside the spindle.

E Late anaphase: the chromosomes, each consisting of two chromatids (cd), have almost reached the poles (p) around which the plastids (pl) are clustered; spindle fibers (sf) are still evident, running from pole to pole.

F Early telophase: the chromosomes have reached the poles, and the movement of the chromosomes has stopped.

G Mid-telophase: the chromosomes are more relaxed in appearance; nuclei are being formed; and the early beginnings of the cell plate are evident (arrow).

H Late telophase: the forming cell plate is clearly visible (arrow), and the nuclei (n) are rapidly returning to the interphase condition.

Electron microscope studies have added a new dimension to our understanding of mitosis. This series of photographs is of dividing cells of the tobacco leaf, and unlike the light-microscope photographs, they are not all of the same cell. Because the cells must be killed and sectioned to be viewed with the electron microscope, the dynamic studies possible with the light microscope are impossible with the electron microscope. But the increased magnification of the electron microscope makes it possible to observe details that are invisible with the light microscope. In the case of mitosis, the electron microscope has revealed the presence of microtubules, which are important at various stages of the process and which are also involved in the movement of chromosomes.

I Early prophase: the chromosomes (ch) have begun to condense and the nuclear envelope (ne) is intact; the poles (p) are visible as collections of microtubules.

J Late prophase: the chromosomes (ch) are condensed, and although the nuclear envelope is present, microtubules can be detected (parallel to the long arrows) radiating from the poles (p).

K Transition to metaphase: the nuclear envelope (ne) is breaking into fragments, and spindle fibers (sf) can be seen extending from the poles (p) to the chromosomes (ch).

L Telophase: the nuclear envelope is developing in various regions around the chromosomes (arrows), and the cell plate (cp) is forming from the fusion of small vesicles.

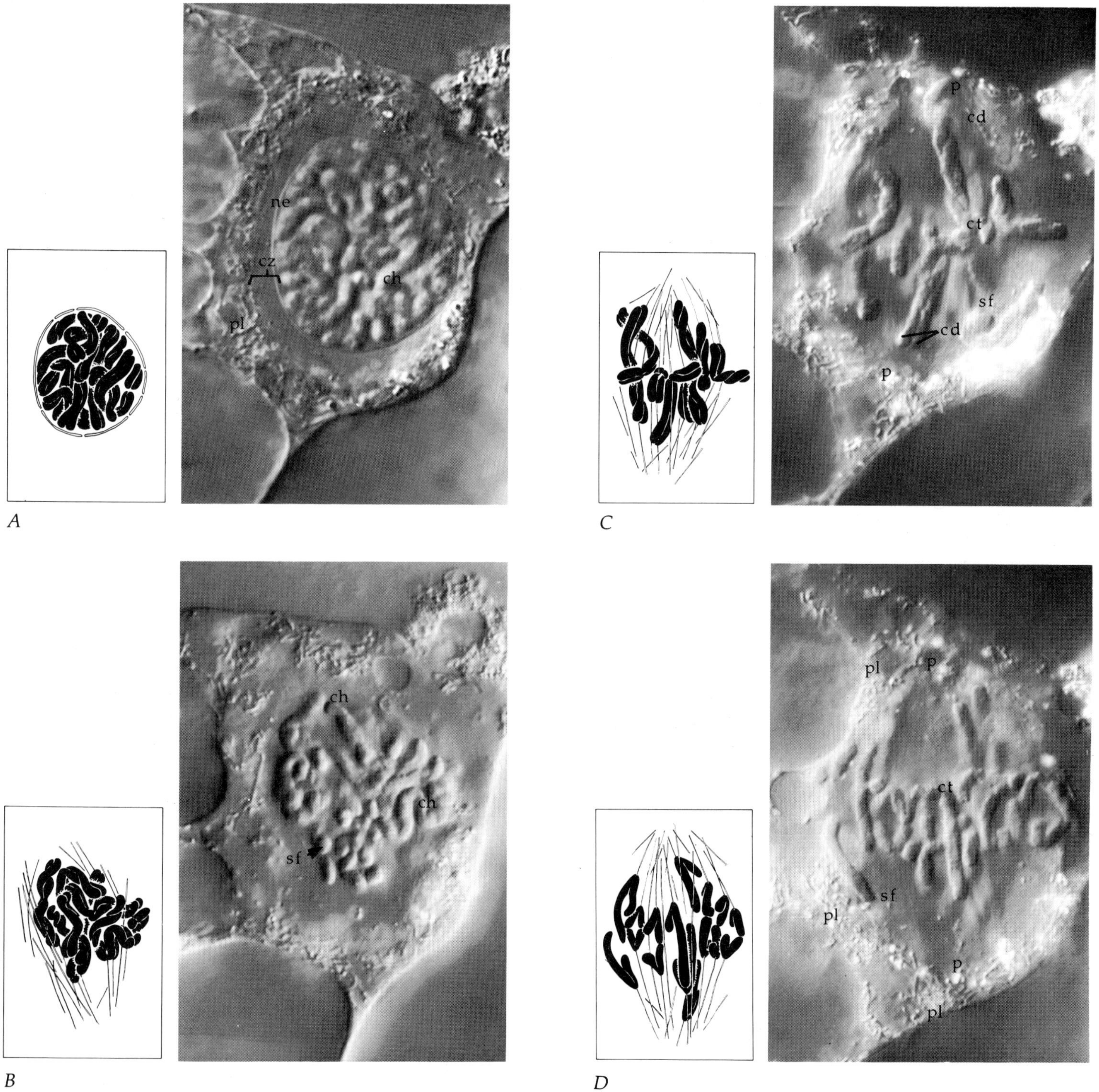
ne
cz
ch
pl
A
ch
ch
sf
B
p
cd
ct
sf
cd
p
C
pl
p
ct
sf
pl
p
pl
D

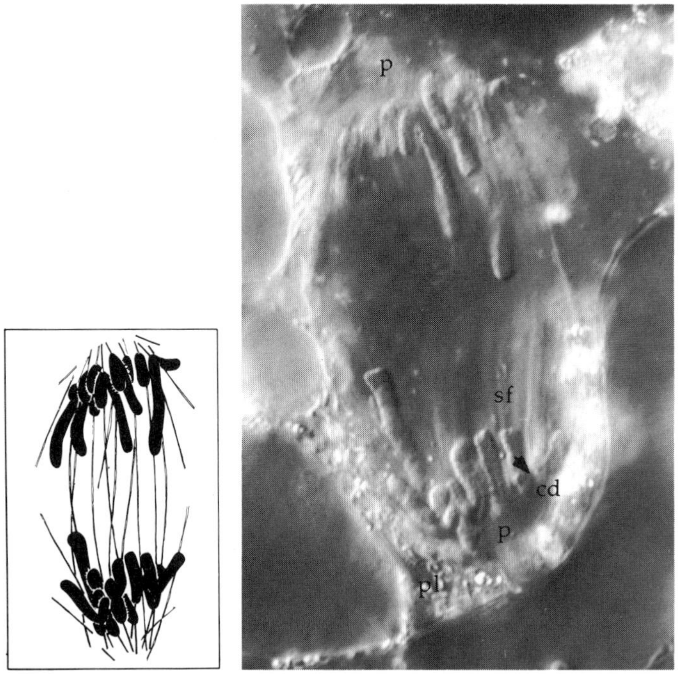

E

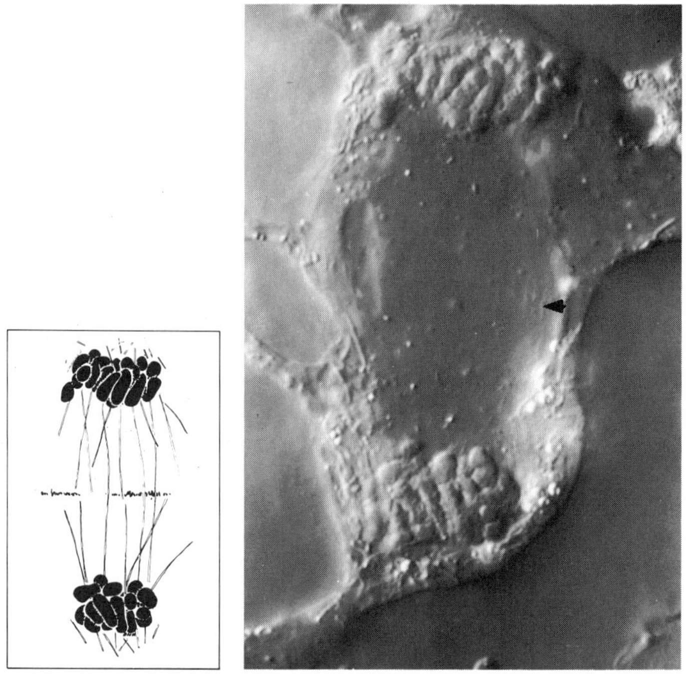

G

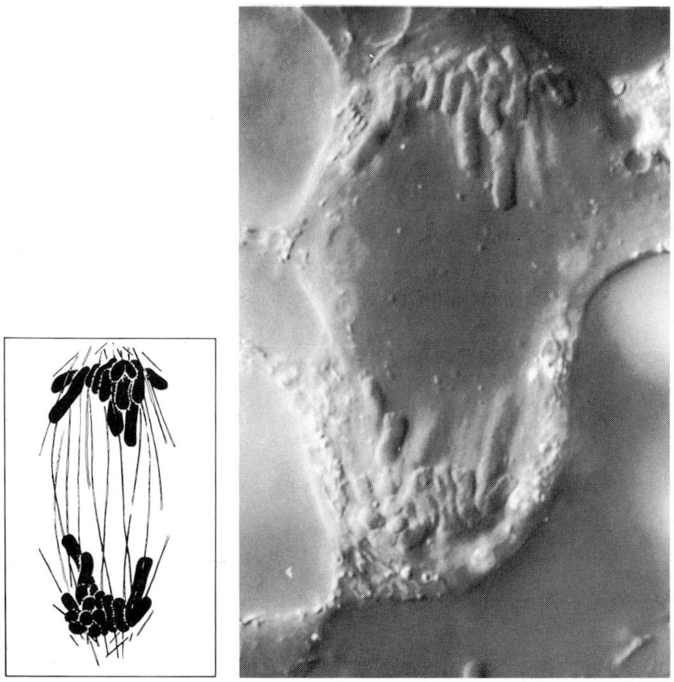

F

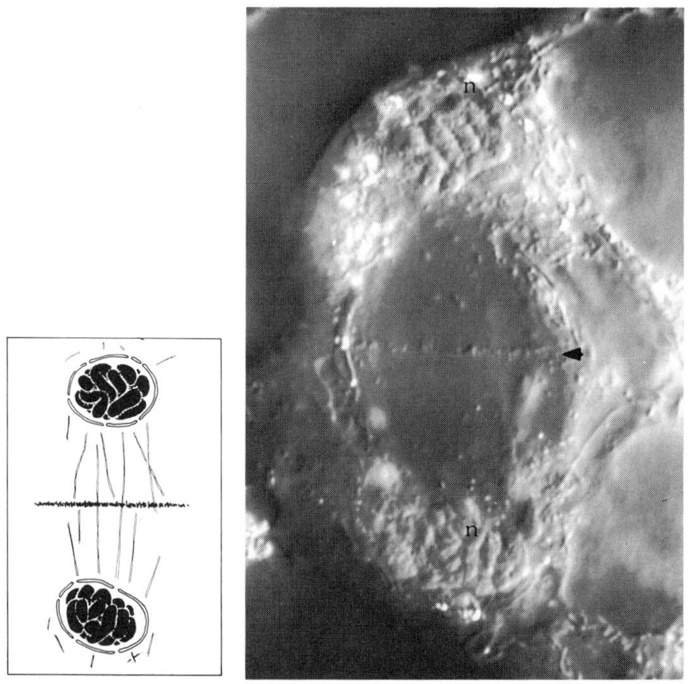

H

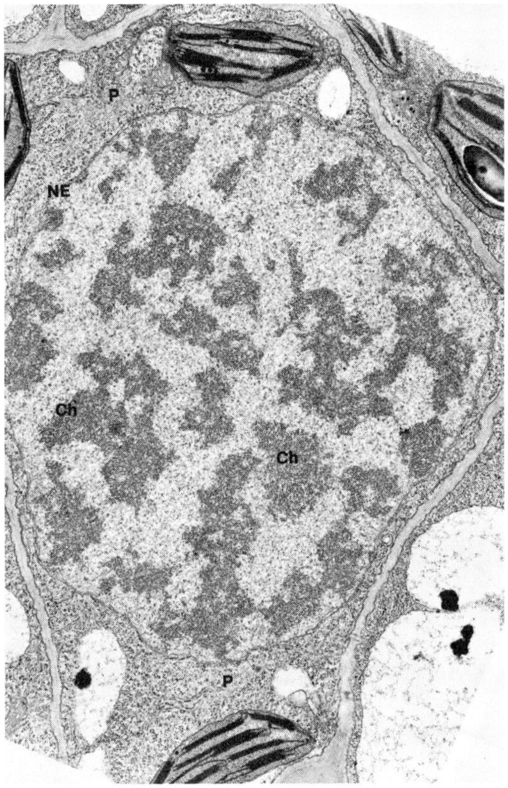

I

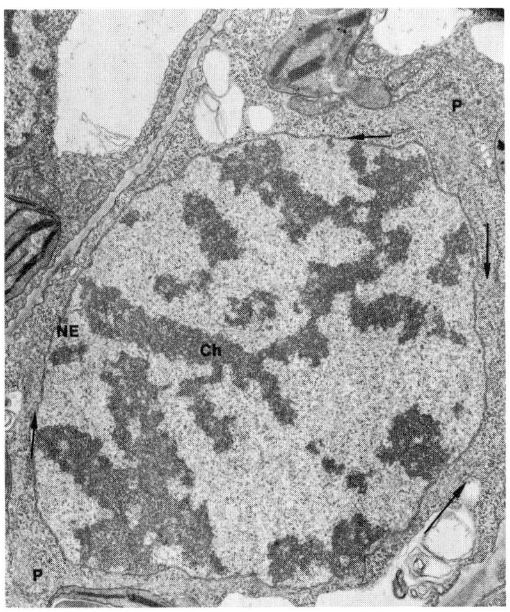

J

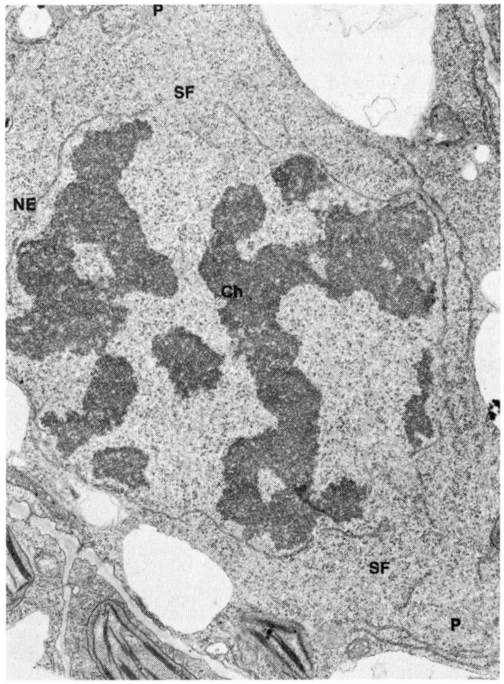

K

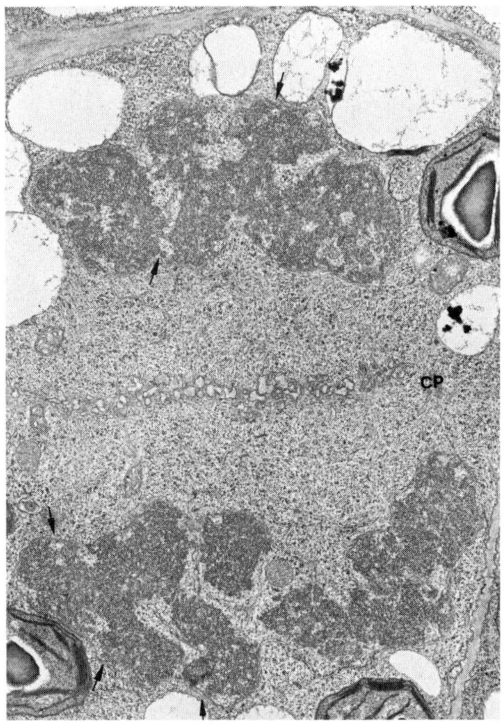

L

The giant sequoia redwood, General Grant, rises 267.4 ft from the forest floor. Over 3500 years ago it began life as a zygote 100 μ long. During these thousands of years, the division of the zygote—and the cells derived from the products of that division—have produced this massive individual composed of billions of cells. The cells of the meristems at the tips of the stems, roots, and along the trunk continue to divide and will do so until the tree dies.

In a stream meandering through the sequoia grove, a unicellular alga less than 50 μ long divides and produces daughter cells. Over the millions of years that this alga has existed, such divisions have maintained the continuity of the species.

From one end of the plant (and animal) kingdom to the other, all organisms must grow to survive, and the cells of which they are composed must be capable of reproduction. A key feature of this reproduction is the duplication of the chromosomes. The all-important information necessary for the production of proteins is part of the DNA that, in turn, is part of the chromosomes.

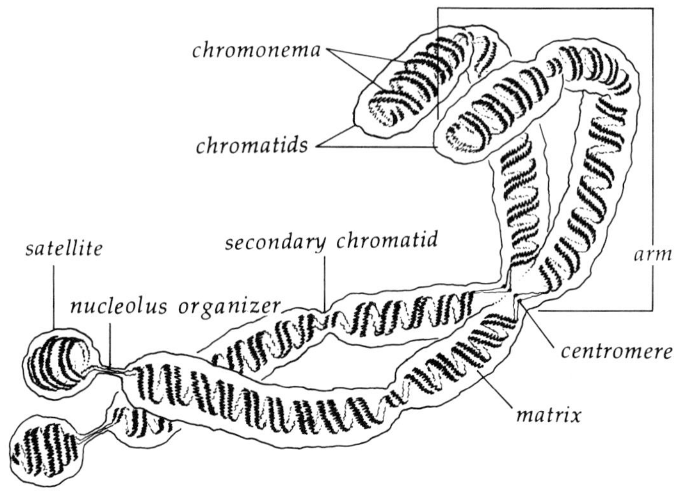

9–1 The parts of a metaphase chromosome as found in most plants. There are two chromatids joined at the centromere. At one end of the chromatids is a constriction that delimits a small segment of the chromatid, called a satellite. An additional, secondary chromatid may also be present. Within some chromatids can be seen still smaller structures, the chromonema. A matrix is present between and around the chromonema. The part of a chromatid between the centromere and the end of the chromatid is called an arm.

Chromosomes

The chromosomes of all eucaryotic cells are complex structures containing DNA and protein. There are several kinds of proteins, of which the most important are the histones. These contain large amounts of basic amino acids.

In the nondividing cell the chromosomes are long and threadlike, but as division begins, they coil and contract in size. Midway through division they reach their minimal length, and the details of their structure are clearly visible.

Each chromosome consists of two identical strands: the chromatids, joined at only one place—the centromere (Fig. 9–1). The centromere appears as a narrow band in the condensed chromosome, and the parts of the chromatids on either side of the centromere are called arms. The chromatids, seen with the light microscope, appear to be composed of coiled strands of chromatin, and a surrounding matrix.

Chromosome size, shape, and number vary greatly. The chromosomes of a given species are present in sets. If one set is present, the cell is called haploid. All gametophytes, and all gametes are haploid. If there are two sets, the cell is called diploid. All sporophytes, and all zygotes are diploid. In some cases, higher numbers of chromosomes are present, usually as even multiples (triploid, tetraploid, etc.). The number of chromosomes in a set, the *n* number, is characteristic of a species and varies widely among the species in the plant kingdom. It ranges from 2 to over 500 and has little if any bearing on the evolutionary status of the species.

The size of chromosomes in plants also varies tremendously. Some of the smallest are found in the fungi, where the length is measured in tenths of microns. At the other extreme, some plants such as the lily have chromosomes in the range of 10 to 20 μ

and *Trillium* as long as 30 μ. The size of chromosomes can also differ greatly within a given chromosome set.

Cell Reproduction—Mitosis

In the vast majority of organisms, cell reproduction or mitosis involves two synchronized processes (see the photo essay). The first, karyokinesis or nuclear division, is the division of the nucleus into two. The second, cytokinesis or cell division, involves the division of the cells so that each half receives one nucleus. In most cases, nuclear and cytoplasmic division are closely connected in time, but there are cases in which the nuclei undergo repeated divisions without the occurrence of cytoplasmic division. In the multinucleate cells that result, cell divisions may occur at a later time and are not connected with mitosis.

Nuclear and cytoplasmic division are two of the most complex processes in the biological world. Although the cell can hardly be considered static in its nondividing state, during division it passes through a dramatic continuum of change and movement.

To facilitate the study of karyokinesis, the process (Figs. A–L) has been divided into four stages: prophase, metaphase, anaphase, and telophase. The beginning of prophase is marked by changes in the appearance of the nucleus. The chromosomes become visible within the nucleus as the coils of the chromatids tighten and condense. Slowly, their structure and number become clear under the light microscope.

The end of prophase and the beginning of metaphase is characterized by the disappearance of the nuclear membrane and the nucleoli. The chromosomes move to a central position in the region formerly delimited by the nuclear membrane. Fibers now become visible that extend from the centromeres to the opposite ends of the region occupied by the nucleus, known as the poles; fibers also extend directly from pole to pole. This apparatus, composed of a mass of fibers, is called the spindle and the fibers, the spindle fibers.

After the centromeres of all the chromosomes are arranged on a plate, typically at the equator halfway between the poles, the centromeres divide. The chromatids, now completely separate, move to the poles. This period of movement, anaphase, ends when the new chromosomes reach the poles. At this point, telophase begins, and a new nuclear membrane appears. During telophase, the coiling of the chromosomes relaxes, and the nucleoli are reformed.

In most plant cells cytokinesis is coordinated with karyokinesis and begins while the chromosomes are moving to the poles. At this time, the fibers that extend from pole to pole become more dense in appearance in the center (equidistant from the poles). This region is known as the phragmoplast (Fig. 9–2). Small, round bodies, apparently produced by the dictyosomes or ER, begin to gather in this area and result in the formation of the cell plate. The cell plate, which is the first stage in the formation of a new wall, forms initially in the region of the spindle and then spreads outward toward the margin of the cell. Finally, the cell plate makes contact with the mature cell wall of the original cell, completing the process. Further changes occur in the cell plate, and it is then incorporated into the new cell wall.

This summary of mitosis indicates only the basic aspects of the process. For an elaboration of some of the many interesting details of mitosis, see End Paper 9–1.

Duration of Mitosis

The lengths of time involved in the stages of mitosis vary considerably under different environmental conditions and from organism to organism.

Prophase is usually the longest stage, lasting one to two or more hours. The duration of metaphase is much shorter, in the range of 5 to 15 minutes. Anaphase is usually the shortest stage, only 2 to 10 minutes. Finally, telophase may vary from 10 to 30 minutes. (These figures indicate *relative* lengths and should not be considered exact limits.)

Variations in Mitosis

The pattern of mitosis just described is characteristic of most plant cells. But differences in detail exist, in some algae, fungi, and vascular plants, and in most animals (particularly with regard to the spindle; see Fig. 9–3).

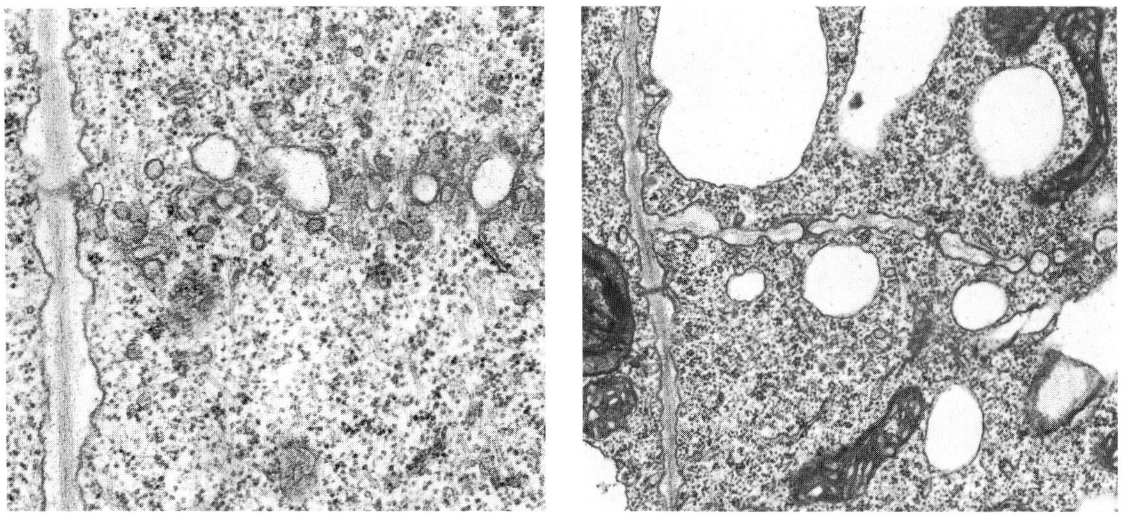

(*a*) (*b*)

9–2 This series of electron microscope photographs illustrates the early stages of formation of the cell plate in tobacco (Nicotiana). *As the chromosomes reach the poles, there is an accumulation of small vesicles at the equatorial plate. These fuse into larger vesicles* (a) *that will eventually form the cell plate and join with the existing cell wall* (b). *During these early stages, the microtubules are prominent and appear to play an important role in the formation of the cell plate.*

Adjacent to the nuclear membrane but in the cytoplasm of many of these cells is a pair of dense bodies called centrioles (Fig. 9–3b). Centrioles are particularly characteristic of animal cells but are also found in some plant cells. Each centriole consists of a pair of small cylinders with walls composed of 9 groups of 3 smaller tubules. The 2 small cylinders usually lie at right angles to each other. At the beginning of prophase, the 2 centrioles move apart, and the spindle begins to form. The spindle is usually extensive, with spindle fibers radiating out into the cytoplasm. In some plants, true centrioles are lacking, and the poles of the spindle consist of dense, granular bodies. When the chromosomes reach the poles and the nuclear membrane begins to reappear, the centrioles reproduce so that each new nucleus has 2 centrioles.

Centrioles appear to be self-reproducing bodies. Because of this, attempts were made to see if DNA were present in them. At first, the data indicated that they did in fact possess DNA, but more recent data have failed to reveal DNA in the centrioles studied.

Centrioles appear closely associated with the motility of the reproductive cells. These motile cells have flagella, and a centriole or a modification of the centriole is usually associated with the base of the flagella. In plants such as the ferns and cycads, only the sperms have flagella, and only the sperm cells have centrioles, which appear in the last divisions that produce the sperms.

Although we have discussed only the most general patterns of mitosis and cell division, many variations actually occur (Fig. 9–3a). Indeed, almost every conceivable modification can and does occur in some organism.

Meiosis

Besides mitosis, a second type of division is found in specialized cells of both plants and animals. In this type of nuclear division, called meiosis, the chromosomes are not divided so that each daughter nucleus has the same number of chromosomes as the original cell. Instead, they are reduced in number so that each daughter nucleus receives precisely one-half of the original number of chromosomes.

Meiosis is associated with the formation of the gametophyte generation that forms the reproductive cells or gametes. The egg and sperm in animals and a few plants are produced directly through meiosis. At fertilization or syngamy, the egg and sperm nuclei fuse, and the original chromosome number is reestablished. In most plants, meiosis is not directly linked to the formation of gametes. Instead, the nuclei and the cells resulting from meiosis divide several or many times by mitosis before sex cells are differentiated. The

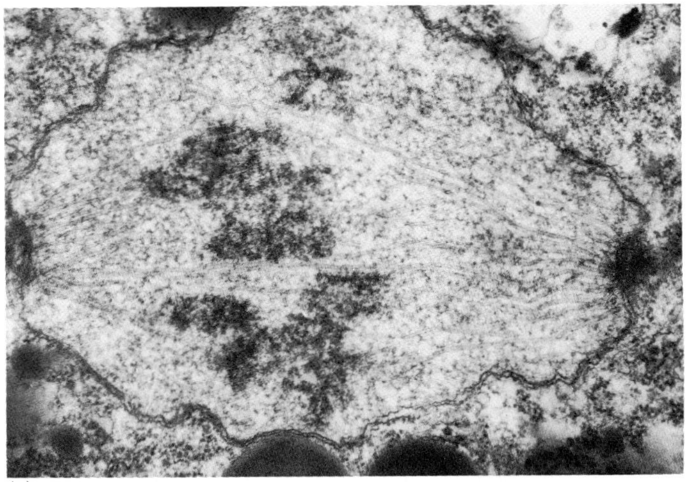

(a)

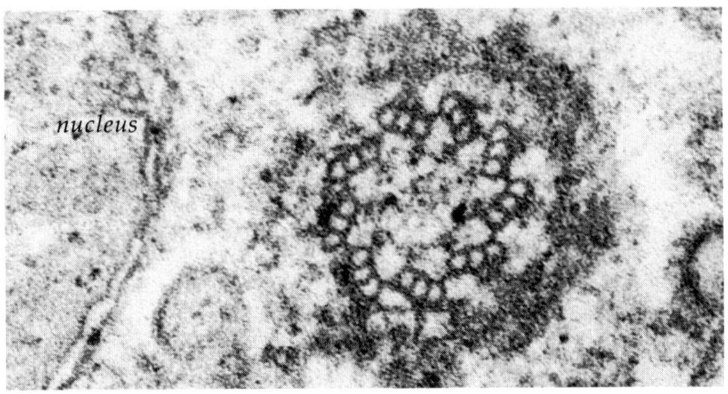

(b)

9–3 *(a) There are many variations on the common pattern of mitosis, particularly in the algae and fungi. One such variation is this electron microscope photograph of a dividing nucleus in the aquatic fungus* Catenaria anguillulae. *Prophase is not significantly different from that found in the more common type of fungus, except centrioles are present. These are short, cylindrical bodies found in the cytoplasm and are the poles of the spindle. At the end of prophase the nuclear envelope does not break down; instead it persists throughout the division. This photo shows metaphase, with the chromosomes aligned on the equatorial plate but the nuclear envelope still intact. By early telophase the chromosomes reach the poles, and the nuclear envelope pinches in two. Thus, the two daughter nuclei are formed without the nuclear envelope disappearing. (b) The structure of the centriole in* Catenaria anguillulae. *It consists of nine groups of three tubules each—a structure that is also found in flagella.*

green algae, for example, usually have the reduced (haploid) chromosome number during the major part of their life cycles and the full (diploid) chromosome number for only a short period. Thus meiosis must be considered as a reduction division and not as a division that always results in the formation of gametes.

The mechanics of meiosis are more involved than those of mitosis (Fig. 9–4). Two closely coordinated divisions constitute meiosis. In meiosis I the chromosome number is halved; in meiosis II the reduced number is reproduced. This means that every meiotic division results in 4 nuclei, each with one-half of the chromosome number present in the original nucleus.

The prophase of the first division of meiosis is long and elaborate. In the beginning stage of meiosis, the chromosomes are always present in the diploid number, that is, in sets of two called homologous pairs. Except during meiosis, the homologous chromosomes act independently of one another. At the beginning of meiosis, the chromosomes are at their maximum degree of uncoiling. The chromosomes appear to be longitudinally single rather than double (as mitotic chromosomes are), but their DNA content suggests that they are actually double at this time.

Next, the pairing of the homologous chromosomes, known as synapsis, takes place. Pairing begins at one or more points along the length of the chromosomes and then proceeds in a zipperlike fashion. The pairing is extremely precise and, when complete, gives the nucleus the appearance of being haploid.

The chromosomes have been decreasing in size, as a result of coiling, from the great length they displayed at the beginning of meiosis. The chromosomes are wound around one another, and in many organisms they can now be seen to consist of 2 chromatids each—the paired homologous chromosomes comprising a four-stranded unit. This condition is more clearly seen in the next stage as the forces of attraction between the homologous chromosomes lessen, and they move apart. They do not separate completely but are held together at one or more points by means of chiasmata. These are points where 2 nonsister chromatids break and then rejoin, resulting in the precise exchange of portions of chromatids between the paired homologous chromosomes (Fig. 9–4). Only 2 of the 4 chromatids are involved at any one point of contact, but if two or more exchanges occur, all 4 chromatids may be involved. The mechanism involved in these exchanges is not clear,

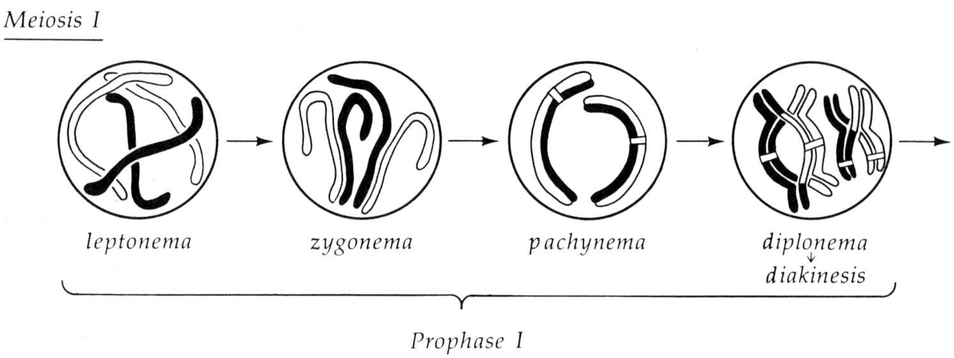

(*a*)

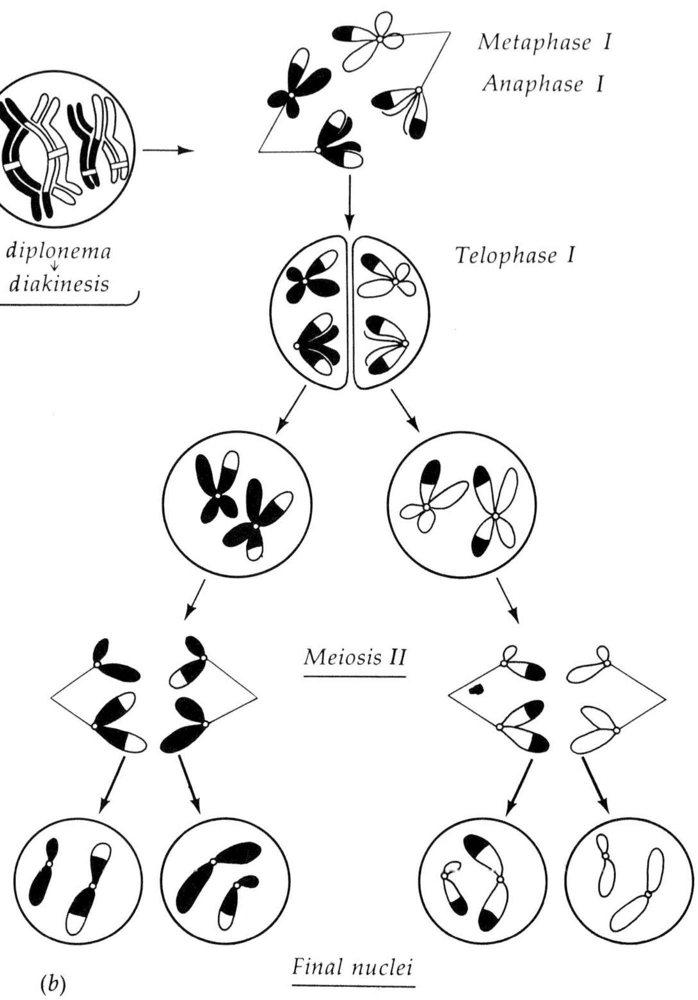

(*b*)

9–4 Stages of meiosis. The names are those applied to the various stages of division.

but the structure known as synaptonemal complex is probably involved (Fig. 9–5).

Coiling continues to shorten the chromosomes, and they reach their minimal size. The chiasmata tend to lose their original position and move toward the ends of the chromosomes. The chromosomes frequently migrate to the periphery of the nucleus near the nuclear membrane.

Metaphase I begins as the nuclear membrane breaks down and the spindle appears. The homologous chromosomes, which are still joined through the chiasmata, become oriented on the spindle, with the centromeres of each lying on either side of and equidistant from the equatorial plate.

Anaphase I begins with the movement of the chromosomes to the poles. The original centromere regions of the chromosomes do not separate; hence each of the homologous chromosomes moves to an opposite pole. This is the period of reduction in chromosome number, since each group of moving chromosomes is half the original number present in the nucleus. Each chromosome can be seen to consist of 2 distinctly separate chromatids, united only at their centromeres.

Telophase I is essentially the same as in mitosis: the nuclear envelope forms; the chromosomes uncoil; and a cell membrane is produced.

Meiosis II may begin almost as soon as meiosis I is complete, but in some cases, the two divisions may be separated by a time interval of hours to months. The prophase that begins the second meiotic division is uncomplicated compared with prophase I, and it closely resembles the prophase of mitosis. At metaphase, the centromere divides, and the chromatids finally separate to form 2 new chromosomes. These chromosomes are not identical, as in the case of mitosis, because of the exchange of segments of the chromatids between the homologous chromosomes during meiosis I. Anaphase and telophase are the same as in mitosis or meiosis I.

Meiosis and Genes

Meiosis is an elaborate mechanism for the separation of homologous chromosomes. These chromosomes

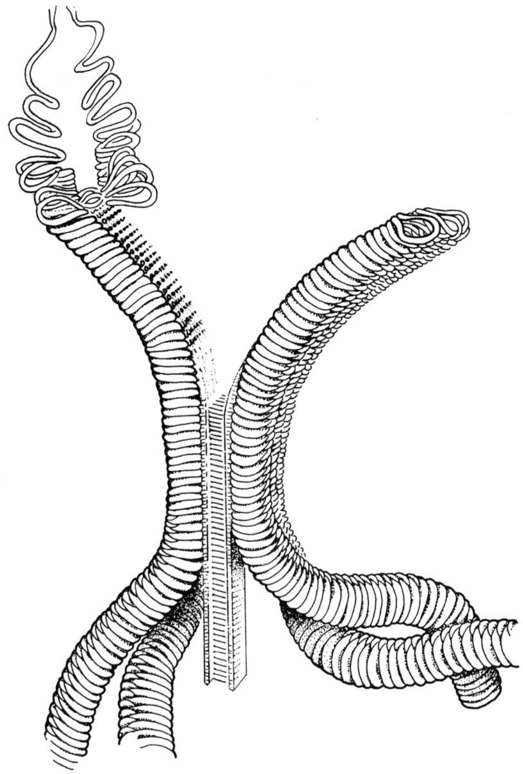

(a)

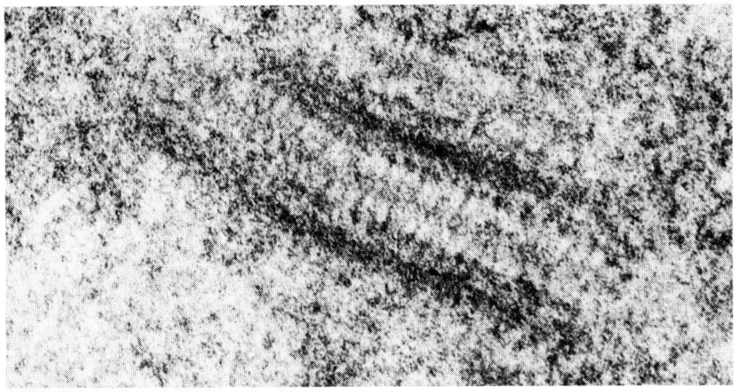

(b)

9–5 Associated with the pairing of the chromosomes (and probably the physical basis for it) is a structure known as the synaptonemal complex. It is a ribbonlike object consisting of two parallel dense rods, each 450 A in diameter, separated uniformly at a distance of 1000 to 1200 A, and connected to a median longitudinal element by transverse fibers. Consisting primarily of protein, the synaptonemal complex is probably formed by the spontaneous aggregation of specific protein molecules and may exist as an independent structure. The homologous chromosomes lie along opposite sides of the complex, each in association with one of the dense outer rods. The photograph shown here is an electron microscope view of a synaptonemal complex from a plant cell. The diagram relates the formation of the synaptonemal complex with the various stages of meiosis.

are united again when the gametes fuse. The question that immediately arises is: why is such an elaborate procedure necessary to secure such a simple result? The answer is that the result is far from simple and that the separation and recombination of chromosomes is highly significant to the organism. To understand this answer, it is necessary to examine more closely the nature of chromosomes and to discuss genes.

Genes are the units of hereditary information that determine the form and function of the cell and the organism. In diploid cells genes are present in pairs—one gene of each pair is on one homologous chromosome, while the second is on the other homologous chromosome. There are thousands of gene pairs in any cell, and they are part of the chromosomes. The genes on a chromosome are arranged in a fixed linear order.

As discussed in Chapter 8, genes act by controlling protein synthesis and consequently enzyme synthesis. Generally, there is 1 gene for each enzyme or polypeptide chain. Thus, if 10 enzymatic reactions are involved in the formation of a flower pigment, there must be 10 genes that govern its production. Genes, while usually highly stable, are capable of change. Such changes are called mutations and may result in relatively small or relatively great changes in gene action. The genes constituting a pair may be different. A gene pair, for example, may determine flower color. One gene may give red flowers, while another may give white. If the gene for red flowers is dominant, then the flowers will be red when the gene pair includes at least 1 "red" gene. Only plants with 2 "white" genes will have white flowers.

Even with this limited amount of information about the genes, several important features of meiosis can be readily understood. First, the importance of

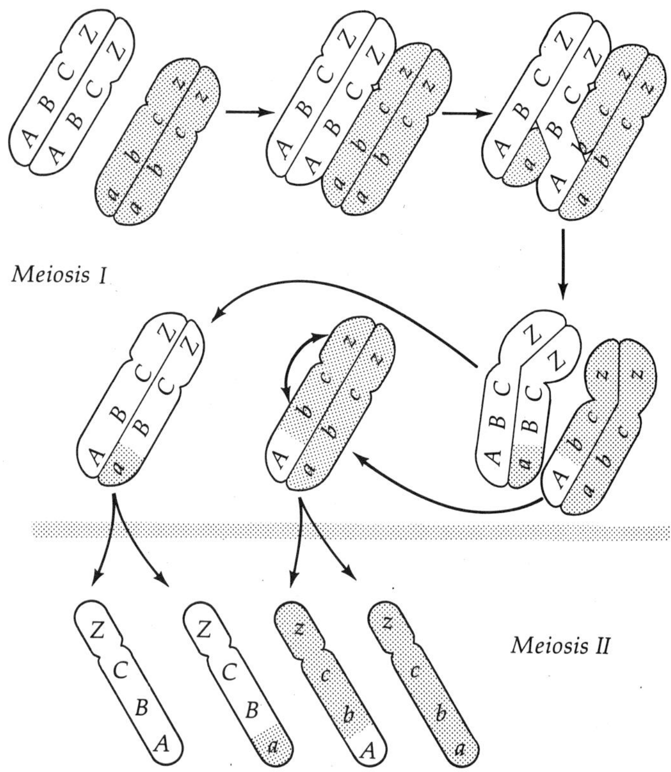

9–6 The importance of meiosis as seen by the distribution of genes. The letters on the chromosomes represent genes.

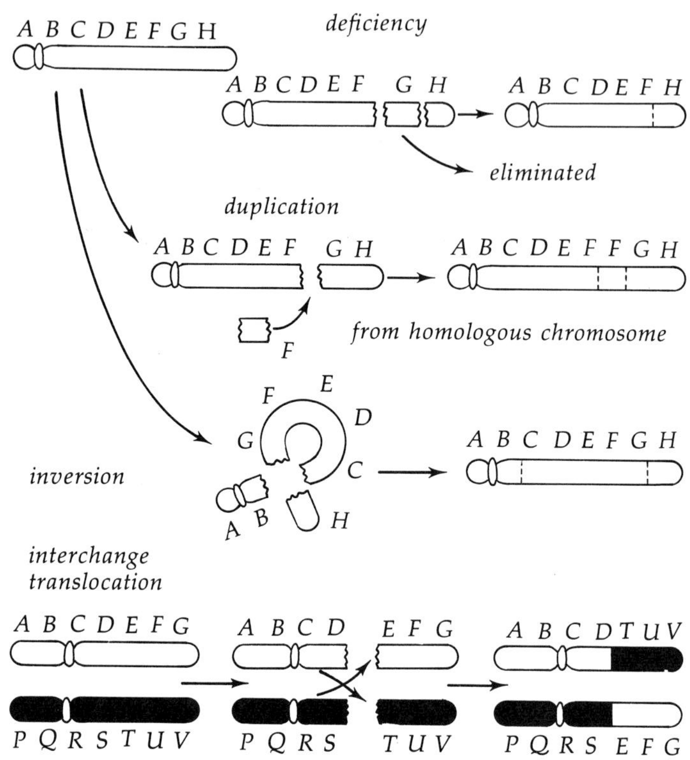

9–7 Diagrams showing how chromosome breakage and reunion can give rise to the four principal structural changes that chromosomes undergo: deficiency, duplication, inversion, and translocation.

the separation and recombination of the homologous chromosomes becomes clear. Through this mechanism, a recombination of genes can be achieved that is impossible to attain through mitosis. Such recombination results in the fullest possible expression of the genetic potential of an interbreeding population. Changes that occur by mutation can be expressed and distributed through the population, if they are advantageous. Second, the importance of chiasmata formation is evident. As a result of chiasmata formation, parts of chromatids are exchanged, and different genes are introduced into a chromosome, thus furthering the degree of genetic variety in recombination. Third, the second division of meiosis, through the separation of chromatids, increases further the number of possible recombinations. The diagram in Figure 9–6 illustrates these points.

Meiosis and gametic fusion produce individuals that are different from one another. These differences are the basis for natural selection and for the adaptation of the species to the environment. Consequently, meiosis is an extremely important process in the continued existence and development of the species.

Chromosome Aberrations

When the true meaning of chromosomes and chromosome pairing during meiosis became known, it was apparent that the genes were not always in the correct order. Aberrant chromosomes were found that lacked portions, had parts reversed, duplicated, or translocated. These four major forms of chromosome aberrations (Fig. 9–7) have various effects on the organism and play a role in evolution by rearranging the genetic material of the individual.

Deletions. A portion of the chromosome is missing and with it the genes found there. As can be readily imagined, most deletions are harmful and are frequently lethal.

Duplications. These are the opposite of deletions and, while less lethal, may cause an imbalance of gene activity that will reduce the survival potential of the organism.

Inversions. These occur when a chromosome breaks in two places, the middle segment becomes turned around, and the chromosome reforms. This results in a segment of the chromosome being reversed with respect to the undamaged chromosome. At pairing during prophase of meiosis I, various difficulties can arise that, with chiasmata formation, may lead to such complications as chromosomes with two centromeres moving in opposite directions or segments of chromosomes lacking centromeres altogether. Clearly, inversions can cause great harm to the organism.

Translocations. These are the exchange of parts between nonhomologous chromosomes. This can lead to genes being transferred to a completely different chromosome and to radically altered linkage relationships. Translocations can have a profound effect on the adaptive value of the chromosomes involved (see Chapter 13).

End Paper 9–1 The Details of Mitosis

In discussing mitosis in this chapter, we have stressed the essential unity of the process without becoming bogged down in the details. But many of the details that are known are fascinating. So we present some of them here for your interest and enjoyment.

Preparation for Mitosis

For mitosis to occur, a number of preliminary steps must take place in the nucleus and the cytoplasm. A variety of experimental evidence indicates that the control of division is largely a function of the cytoplasm. This control may take a number of forms, including the involvement of specific, growth-regulating substances (see Chapters 30 and 31).

Another factor important in the preparation of a cell for division is energy. Energy is necessary for mitosis. The nucleus is not the site of reactions that yield large amounts of energy. Furthermore, during mitosis itself evidence indicates that respiration is low. Consequently, before mitosis can occur, sufficiently large supplies of readily available energy must be accumulated so that the process can be carried through to completion. If such energy is not available, mitosis will not begin. Once mitosis has begun, it continues to completion and is almost entirely independent of environmental changes that affect the energy supply.

Duplication of the chromosomes is another important part of the preparation for mitosis. Mitosis itself is not the process of chromosome reduplication. It is the mechanism by which the separation of duplicated chromosomes is achieved and the full complement of chromosomes is distributed to the two new nuclei. The actual duplication of the chromosomes occurs before prophase.

The earliest indication that DNA synthesis takes place before mitosis starts came in the late 1940s from the work of Pollister and his students at Columbia University. He was instrumental in developing the visible light microspectrophotometer for the measurement of DNA.

In microspectrophotometry the aim is to measure the amount of a substance such as DNA by determining the amount of color produced when the substance is caused to react with a dye. A colored object absorbs light. Thus, if there is a way to measure the amount of light absorbed, the amount of the substance can be measured. In the case of a microspectrophotometer, light of a known color (i.e., wavelength) is passed through a microscope. The eyepiece of the microscope contains a sensitive photocell attached to an elec-

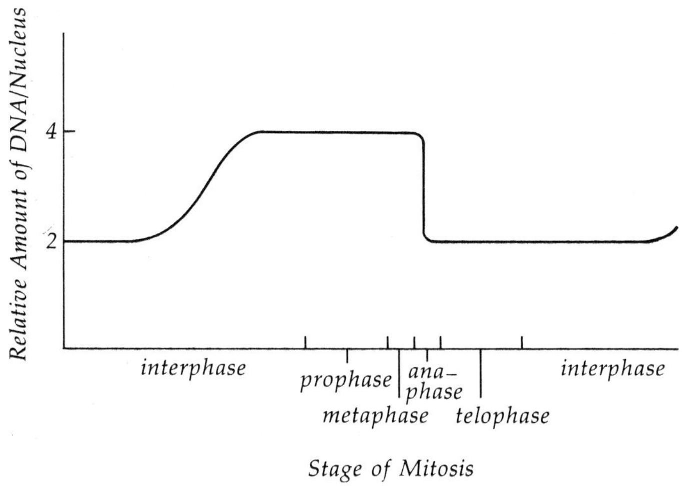

9–8 *Changes in the DNA content of a nucleus during mitosis.*

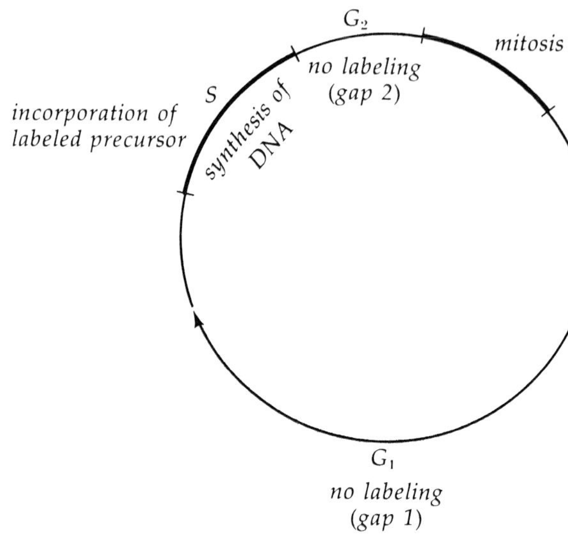

9–9 *The synthesis of DNA as determined from autoradiographic evidence.*

trical measuring device. One measurement is taken in a clear part of the cell, and a second is taken in the colored part. The difference in these two measurements represents the amount of light absorbed and, indirectly, the amount of the substance.

One of the men then working with Pollister was Dr. Hewson Swift, now at the University of Chicago, who applied the method to the measurement of DNA in plant nuclei. The type of data he acquired can be graphed (Fig. 9–8). From this graph it can be seen that DNA doubling occurs well before the beginning of prophase.

Another way to study DNA synthesis is through autoradiography, described in End Paper 8–1. In the case of DNA, the most widely used radioactive precursor is ^{3}H-thymidine. The reason for this choice is immediately apparent from the composition of the DNA molecule. The base thymine is found only in DNA, and most cells readily take up thymidine, which is thymine with an added sugar molecule. The autoradiographic data can be diagrammed, as shown in Figure 9–9. The G_1 (for gap) phase is a period in which no incorporation occurs following mitosis; the S (for synthesis) phase is when the labeled compound is incorporated into DNA as the new DNA is formed; and the G_2 phase is the second period of no incorporation or gap before mitosis begins. Thus the autoradiographic data and the microspectrophotometric data agree.

There is considerable variation in the duration of G_1, S, and G_2 in different species and in different cells of the same organisms. For example, the cells in the cortex of the roots of many plants enter a prolonged G_2 phase after the last mitosis; hence many nuclei have twice the DNA amount of the G_1 cells. Similar cells in the leaf remain at the G_1 amount following the final mitosis and never enter another DNA-synthesis phase. In some cells the G_1 phase is short and the G_2 phase long, while in others the reverse is true.

This cycle can also be repeated without mitosis occurring. This is true of many plant cells. The amount of DNA in the G_1 phase of a diploid cell is usually called $2c$. The letter c is used so as not to confuse the value with n, which represents the number of chromosomes. The term originated in microspectrophotometry, where the amount of light absorption can be accurately measured and reliable comparative data obtained but absolute quantitative data cannot. Thus it is impossible to say so many picograms of DNA are present, but it is possible to establish c values. In the G_1 phase there is $2c$ DNA present that goes to $4c$ DNA during S and remains there during G_2. Of course, the number of chromosomes remains the same during these changes in DNA amount. Since the DNA-synthesis cycle can occur several times without mitosis in many plant cells, $4c$, $8c$, $16c$, and even $32c$ values have been found in plants. In some cases, the chromosome number is known to be double or more the $2n$ number, but in others, there is no increase in number of chromosomes, only DNA content.

The proteins associated with the chromosomes, the histones, are also synthesized before prophase begins. Thus

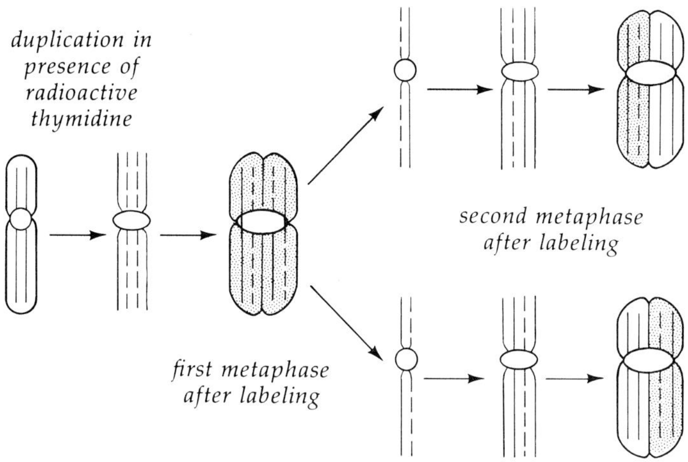

9–10 The mitotic duplication of chromosomes. The chromosomes labeled with tritiated thymidine are shown by cross-hatching.

the chemical components of the chromosomes all appear to be synthesized before mitosis occurs.

When chromosomes duplicate during interphase, the whole chromatid is reproduced along its length. The DNA present is conserved as postulated in the double-helical model of DNA; each original strand in the DNA remains intact but separates, forming two completely new strands. This was shown in the now classic experiment of Dr. J. Herbert Taylor, then at Columbia University. Taylor grew root tips of *Crepis capillaris* in ^{3}H-thymidine for 6 hours and then transferred them to unlabeled thymidine for various lengths of time. The roots were then killed; the chromosomes stained; and the cells squashed out on a slide. Autoradiographs were made that showed radioactivity in the chromatids by the presence of silver grains over them in the developed autoradiograph. If the nuclei had undergone only one division following the period of labeling, then both chromatids of each chromosome would be equally radioactive. If the roots had been allowed to grow for a longer period so that the nuclei had time to undergo two divisions, then only one chromatid of each chromosome would be radioactive. This is consistent with the concept of DNA replication based on separation of the original double helix of DNA and the synthesis of new complementary strands of DNA (Fig. 9–10).

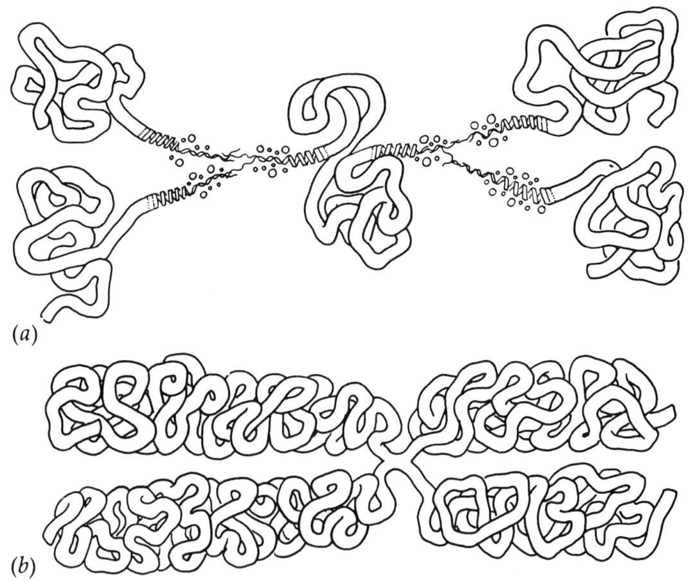

9–11 Chromosome ultrastructure in the folded-fiber model in which the interphase chromosome is visualized as a single chromatid consisting of a 230 A fiber composed of a single DNA helix wrapped with protein and held in place with a second protein. At replication (a) synthesis of a second fiber starts from either end and proceeds to the center. At metaphase (b) the sister fibers fold up as sister chromatids, possibly held together by unreplicated fiber segments in the centromere.

Ultrastructure of the Chromosomes

The question of chromosome ultrastructure is presently subject to heated controversy. Available data suggest that the fundamental elements of chromosome structure are long, DNA-protein fibers about 230 A in diameter.

How are these fibers arranged into a chromatid? At present, there are two competing theories: the unineme concept, where a chromatid is postulated as a single elaborately coiled strand of DNA protein, and the polyneme concept, where a chromatid is believed to consist of many strands of DNA protein. The unineme theory best fits the genetical, and some of the experimental, evidence, but the polyneme theory derives strong support from light-microscope observations and some experimental data.

The unineme concept appears to be receiving the most support from current research. In general, electron-microscope studies (such as those of Ernest Du Paw at Stanford and Jack Griffith at the California Institute of Technology) indicate that a single strand is the basis of chromatid organization (Fig. 9–11). But the dust has not yet settled, and

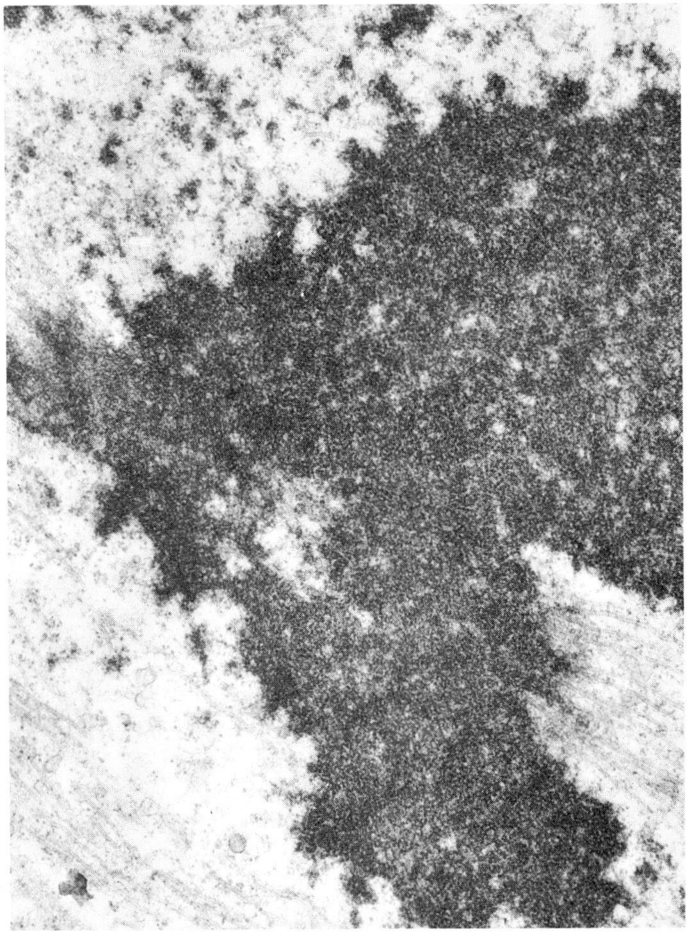

9–12 Spindle fibers attached to the centromeres of a metaphase chromosome. Each of the spindle fibers is composed of many microtubules.

although the unineme concept may be correct, it will not be completely accepted until it can be reconciled with all of the data.

Prophase

In the nucleus of the nondividing cell, the chromosomes are expanded to their maximum length. They may become so diffuse in appearance that they cannot be seen. With the beginning of prophase, chromosomes progressively shorten and thicken through a coiling mechanism that is not currently understood.

While the chromosomes are condensing in the nucleus, a clear area is formed around the nucleus. In this zone the mitochondria, plastids, and similarly large cell parts are excluded. This clear zone is ellipsoidal in shape, while the nucleus is spherical. The position of the poles and hence the plane of division are indicated by the location of the poles of the ellipse. With the electron microscope, microtubules can be seen in the clear zone. Early in prophase these are randomly arranged, but later they become oriented parallel to one another, eventually forming a sheath of tubules surrounding the nucleus.

Metaphase

With the breakdown of the nuclear envelope, the spindle now becomes clearly visible under the light microscope. At this magnification, fibers can be seen extending from the centromere of the chromosomes to the poles; these are the chromosomal spindle fibers. Additional fibers, the continuous spindle fibers, extend from pole to pole; these are not attached to the chromosomes. With the electron microscope, both types of spindle fibers can be resolved into collections of from 75 to 150 microtubules (Fig. 9–12). In the case of the chromosomal spindle fibers, the microtubules are attached to the chromosome at the centromere. These microtubules appear to originate at the centromere and form toward the poles. The continuous spindle fibers on the other hand are composed of microtubules found in the clear zone before the nuclear membrane breaks down. Thus the two types of spindle fibers, although similar in form, have different origins.

The microtubules of the chromosomal fibers are attached to two sites on opposite sides of the centromere. The chromosome is divided in this region, and a hole separates the two attachment sites. The chromatids (the future sister chromosomes) are joined by the region on either side of the centromere rather than at the centromere itself.

As the nuclear envelope breaks down, the chromosomes begin a rather intricate pattern of movements probably associated with the growth of the chromosomal fibers. The result of these movements is the alignment of the centromeres on a plane perpendicular to the axis of the spindle and midway between the poles. This is the metaphase or equatorial plate. If the chromosomes are large, only the centromeres are arranged on the plate; if the chromosomes are small, the entire chromosome complement may be arranged on the plate.

The composition of the spindle is known primarily from analyses of isolated spindles from sea urchin and starfish

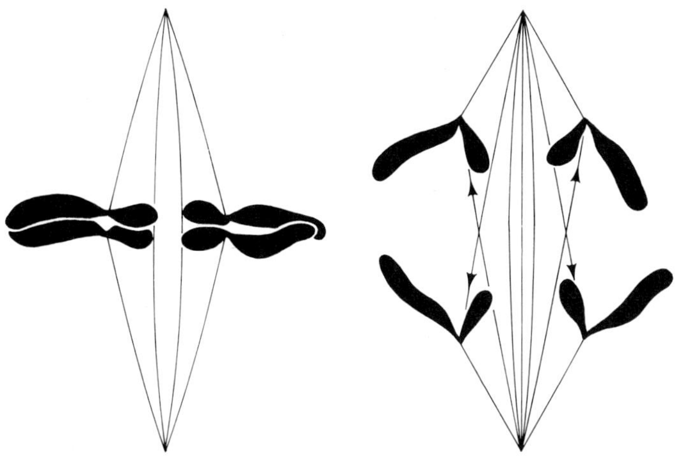

9–13 According to one current theory of chromosome movement, the chromosomes are pushed to the poles by the elongation of spindle fibers, not simply by their contraction. This diagram illustrates that concept. Note that the number of spindle fibers increases at the equatorial plate in anaphase.

eggs. These data and some supplementary evidence indicate that the spindle is composed mainly of protein. This protein appears to be a product of the cytoplasm rather than the nucleus.

Anaphase

After all the centromeres are aligned on the plate, the centromere region of each chromosome separates, and the two chromatids become independent of one another. At this point each chromatid becomes a chromosome. The chromosomes now begin moving toward the poles; one of each pair of new chromosomes moves to the opposite pole.

The movement of a chromosome depends on the presence of the spindle and the centromere. If the spindle is disrupted (as it is when the cell is treated with various drugs), no chromosome movement occurs. If the chromosome lacks a centromere, it will not migrate to the pole and will be lost to the nucleus. While the chromosomes are moving toward the poles, the poles move away from each other.

The mechanism of chromosome movement and the movement of the poles are not yet fully understood. At present, the data suggest that the chromosomes, rather than being pulled to the poles by the spindle fibers, are actually pushed. This is accomplished by the lengthening of the fibers that run from the poles to the centromere. If this is the case, then the number of fibers between the separating chromosomes should be twice the number of fibers between the chromosome and the poles (Fig. 9–13). This has proved true for a few species, where it has been possible to make the necessary counts. There is also evidence that interactions between the chromosome fibers and the continuous fibers are essential for chromosome movement.

The best impression of the movement of the chromosomes during anaphase can be gained through time-lapse microcinematography using a phase-contrast microscope. Outstanding films of this type have been made by Dr. Andre Bajer, now at the University of Oregon. He has used these films as the bases for detailed, kinetic analysis of the events associated with mitosis, particularly chromosome movement. Microcinematography has been combined with microbeams of ultraviolet light or slow neutrons to disrupt portions of the spindle. Analyses are then made to determine the effect of such treatment on chromosome movement.

Telophase

When the chromosomes reach the poles, they form a compact group, and the nuclear membrane forms. The chromosomes reverse their coiling, becoming longer and more diffuse in appearance. The nucleoli reappear, associated with special regions on specific chromosomes. These changes occur during this final stage of mitosis.

The outer membrane of the nuclear envelope and the endoplasmic reticulum (ER) are apparently continuous in the nondividing cell. During late anaphase and telophase, the nuclear membrane appears to be reformed from the ER. Thus the continuity between the nuclear membrane and the ER of the cytoplasm, which existed before division, is reestablished. But the details of nuclear-membrane formation are still unknown.

In anaphase and early telophase, with organisms favorable for cytological observation, each new chromosome appears to consist of 2 chromatids. During the period beginning with the end of one telophase and ending with the beginning of the following prophase, these chromatids are duplicated. The coiling and uncoiling of the chromosomes and the chromatids is an involved and complicated process for which the underlying mechanism is unknown.

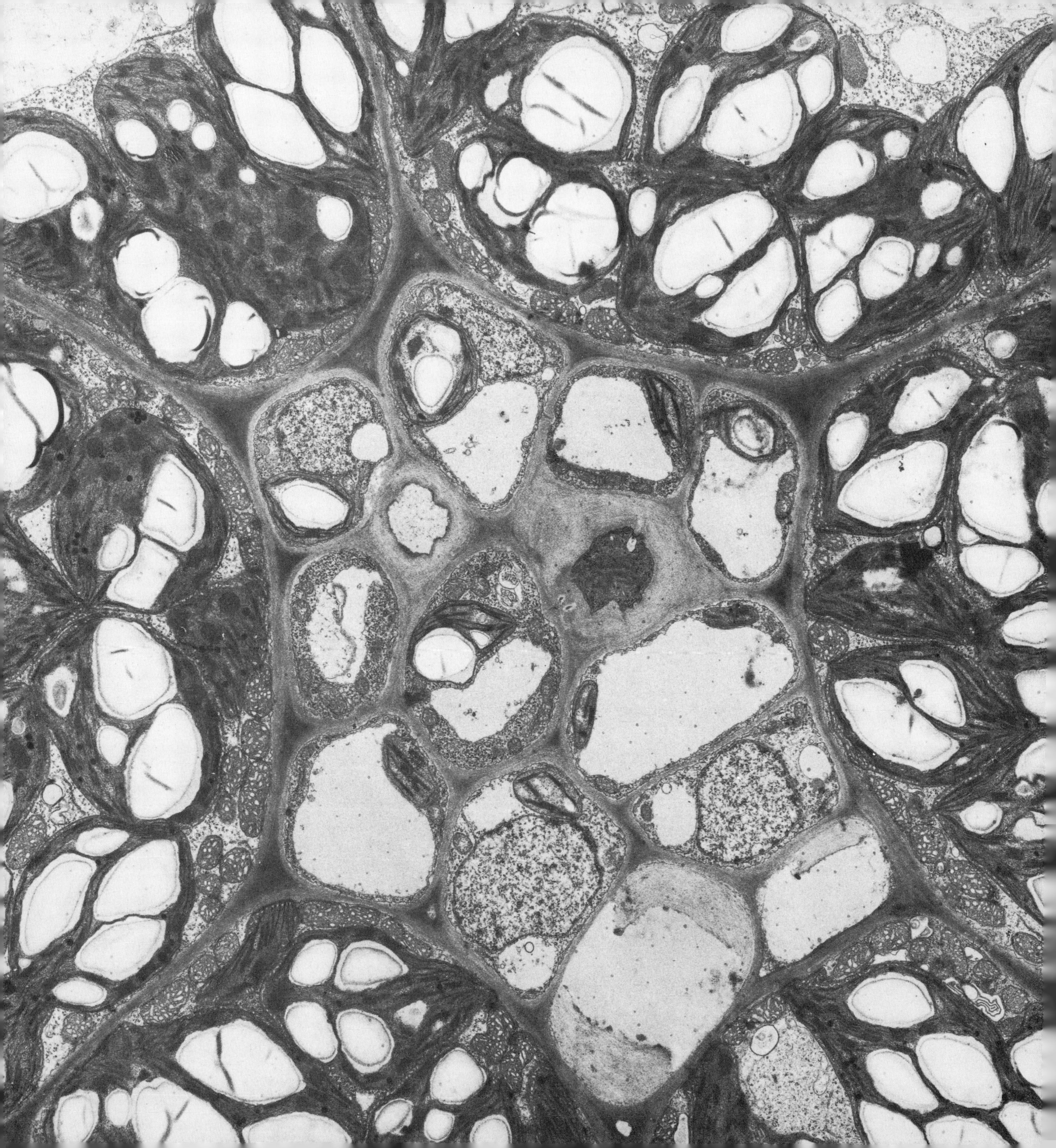

10

Cell Biosynthesis and Control

Cell research is remarkably like working a jigsaw puzzle. The major features are attacked first, and the remaining, more difficult parts are filled in later. Fortunately, the cell shows more organization than many jigsaw puzzles, but it is not easy to achieve a complete picture of the cell either for the scientist or the student. So far, you have learned about various cell parts and their functions, but we have not discussed the cell as a whole or integrated the ultrastructural and biochemical data with the information on chromosomes and mitosis. We will attempt to do this now.

We say attempt because the problem of expressing and integrating the data poses serious difficulties (Fig. 10–A). The first is that the processes carried out in the various organelles are interdependent and occurring at the same time. Thus one must think three dimensionally in space and in time. The second difficulty is that while the organelles are interdependent on one another, they are also independent in many of their reactions. A third problem is that when our knowledge of the cell is summarized (Fig. 10–1), it appears more complete than it is in reality. Finally, in discussions such as this we are describing the "average" plant cell, but the average cell no more exists than does the average man.

10–A A section through a small vascular bundle in a sugarcane leaf, as seen with the electron microscope, which illustrates the differences between cells and cellular interactions. The differences in structure between the surrounding, chloroplast-containing cells and the vascular cells in the center are clear even at relatively low magnification.

Cell Metabolism

Let us start by following a sunbeam into a green plant cell. The light strikes the chlorophyll in the lamella of the chloroplast and energy is trapped. This energy is used to fix CO_2 into carbohydrates in reactions taking place in the stroma of the chloroplast. The sugar produced may be converted to starch and stored in the chloroplast, or it may leave and be converted to starch by special starch storing plastids. It may also leave the chloroplast and be used by the mitochondria. Before we follow a sugar molecule out of the chloroplast, let us examine the chloroplast more fully.

Besides synthesizing sugar, chloroplasts also make amino acids and contain DNA and ribosomes. For a

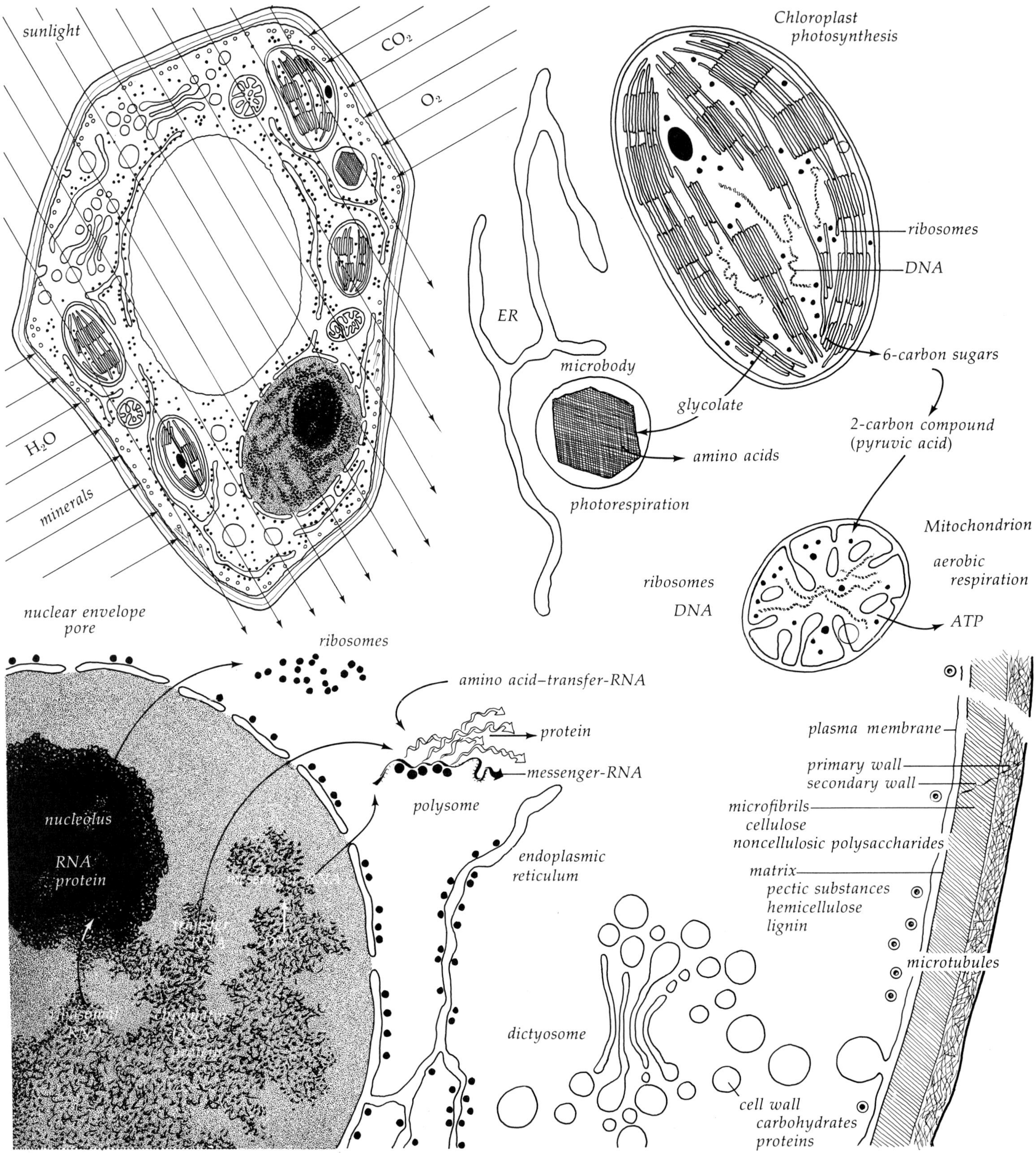
sunlight
CO_2
O_2
H_2O
minerals
Chloroplast
photosynthesis
ribosomes
DNA
6-carbon sugars
ER
microbody
glycolate
amino acids
photorespiration
2-carbon compound
(pyruvic acid)
Mitochondrion
aerobic
respiration
ribosomes
DNA
ATP
nuclear envelope
pore
ribosomes
amino acid–transfer-RNA
protein
messenger-RNA
polysome
nucleolus
RNA
protein
endoplasmic
reticulum
plasma membrane
primary wall
secondary wall
microfibrils
cellulose
noncellulosic polysaccharides
matrix
pectic substances
hemicellulose
lignin
microtubules
dictyosome
cell wall
carbohydrates
proteins

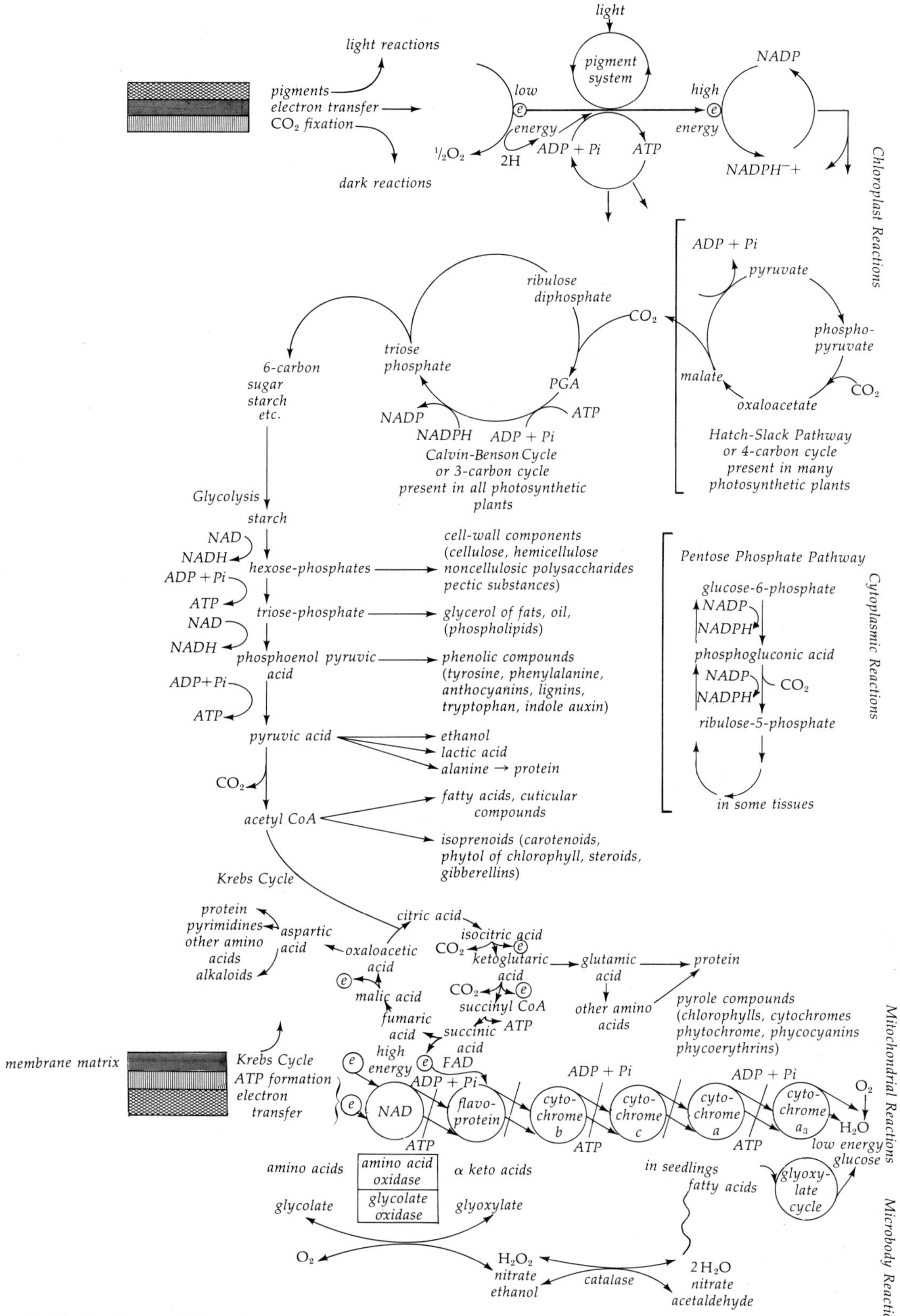

10–1 A summary of cell structure and function.

long time DNA was thought to be present only in the nucleus, but it is also present in both the chloroplasts and mitochondria. The ribosomes in the chloroplasts are smaller than in the cytoplasm but otherwise identical to them. The chloroplast has its own energy source in the ATP and NADPHs produced during the first stage of photosynthesis and appears capable of making its own protein from its own DNA, thus making it more independent of the nucleus than most cell parts. Yet it is still under at least partial control of the nucleus.

Excess sugar leaves the chloroplast and may enter a chain of reactions that end in a mitochondrion, where the sugar is degraded to CO_2 and water with the production of ATPs. Mitochondria, as well as chloroplasts, contain their own DNA and ribosomes. They too produce amino acids and can synthesize their own protein. Again, they are not completely free of nuclear control. A few of the ATPs produced by the mitochondria are used by the mitochondria (e.g., in protein synthesis), but most are available for other reactions in the cell.

Some of these ATPs are utilized by the amino acid-activating enzyme converting amino acids to the activated form. These are captured by their specific t-RNA molecules. The amino acid–t-RNA complex becomes attached to a strand of m-RNA on a ribosome. The amino acid becomes joined to the forming chain of amino acids trailing off into the cytoplasm. When the last of the message is read and the last amino acid added, the new protein floats away from the ribosome, and its final configuration is assumed.

If the new protein is an enzyme, it may begin to function immediately by breaking down or building up a compound. Its activity as an enzyme will be regulated by both the amount of the substrate and the end products present in the system. If the protein is a storage protein, it may be accumulated in the ER or special plastids. The new protein may also be part of new membranes that the cell is constantly synthesizing. Some of these membranes are involved in the formation of dictyosome vesicles.

The dictyosomes appear to be assembly areas rather than production sites; here cell wall components are packaged and sent to the plasma membrane. The membrane surrounding the dictyosome vesicles and the plasma membrane appear identical under the electron microscope. To package the transported material, the dictyosome must synthesize new membranes. To do this requires not only protein but also lipids. The fatty acids needed to synthesize lipids are made in the cytoplasm in association with the ER, but the assembly of the lipid molecules themselves takes place in the mitochondria. The dictyosomes form a close association with the ER, and the two may comprise a more or less continuous system.

The ribosomes are usually grouped together in small clusters or polysomes around a strand of m-RNA. Like the m-RNA, the ribosomes are produced by the nucleus. Ribosomes are half RNA and half protein. The RNA is presumably made in conjunction with chromosomal DNA, but the assembly of the ribosomes takes place in the nucleolus. Here the protein complement is added, and from here the new ribosome or immediate ribosome precursor moves to the cytoplasm.

Cell Duplication

The nucleolus consists of a dense portion and one or more much less dense regions. The dense portion consists of ribosomelike granules, other spherical particles, and long, fibrous material that is probably protein. The size of the nucleolus is related to the activity of the cell. When the cell is actively synthesizing protein, the nucleolus enlarges, but when the cell is less metabolically active, it becomes smaller. The nucleolus is formed by a special part of a particular chromosome.

The chromosomes are composed of DNA and protein—some of the protein being of a special class known as histone. Each chromosome consists of two chromatids, and each of these consists of strands of DNA coiled around each other. The DNA molecules are separated, at least locally, while m-RNA is being synthesized. This requires the bases, phosphate, deoxyribose, a special enzyme, and energy. Special regions of the chromosome or genes have the power to regulate the production of m-RNA by genes in its immediate proximity. The histones may operate by covering or uncovering large blocks of genes.

At some unknown stimulus the chromosomes cease RNA production and begin duplicating their DNA and protein content. This may occur at a wide variety of times in the life of the cell and does not necessarily take place immediately preceding mitosis. Changes in the cytoplasm related to mitosis do take place just before the beginning of division. Around the nucleus a region develops that is free of mitochondria, plastids, and other large cell organelles. This clear region, rich in protein that will be involved in the formation of spindle fibers, is wider at the places where the future poles will appear. Stores of ATPs are also built up so that once mitosis begins respiration need not continue.

Within the nucleus the chromosomes are thickening and shortening. The nuclear envelope breaks into fragments of ER, and the spindle fibers form. These are 200 A in diameter and run from the chromosomes to the poles or from pole to pole. Clusters of fibers attach to the centromere. During this period the nucleolus breaks down, and the component parts blend with the cytoplasm. When the centromeres divide, the chromosomes are moved to the poles by the spindle fibers. At the same time the poles are moving away from each other. In the cytoplasm, the mitochondria, plastids, and other organelles cluster around the poles and along the sides of the spindle. Roughly equal numbers are present at each pole.

As the chromosomes near the poles, additional fibers begin to appear in the central region formerly occupied by the chromosomes. Small vesicles produced by the dictyosomes accumulate in the same region. These begin to fuse, forming the new cell wall and plasma membrane. Strands of ER are trapped in the forming cell wall and mark the position of the plasmodesmata in the final wall. When the central area is divided by the cell wall, fibers continue to form at the edges, and more vesicles accumulate until the existing plasma membrane and wall is reached. The new and old plasma membrane and wall fuse, thus dividing the cell in two. Microtubules appear in the cytoplasm adjacent to the new wall. Microfibrils of cellulose are deposited, as are the pectic substances and hemicellulose—the latter two via dictyosome vesicles.

The chromosomes reach the poles, the spindle fibers break down, and a new nuclear envelope is formed from the ER. The chromosomes expand and become longer and thinner. In localized regions on particular chromosomes, nucleolar material is synthesized. This material accumulates, and a new nucleolus appears. In the expanded chromatin the DNA molecules unwind, and RNA synthesis begins.

Cytoplasmic streaming, which has slowed or stopped during mitosis, resumes, and respiration increases. Both the mitochondria and plastids synthe-

---

Essay 10–1 Turning Genes On: The Most Exciting Period of My Scientific Life

This essay was given its title by the author James Bonner, of the California Institute of Technology in Pasadena, California. Those who know Bonner and his work will be surprised that he can single out a particular period of his scientific life as the most exciting. Each will be sure that the time he knew or worked with Bonner was the most exciting in both their lives and by far the most important problem that either had worked on. This feeling of excitement and interest is characteristic of Bonner's laboratory at Cal Tech and of the man himself.

"During the 1950s I spent a great deal of energy and enthusiasm on trying to find out how hormones cause particular specific developmental pathways to start upon their course. I finally concluded that hormones must do their thing by causing particular genes not before expressed to become expressed—to be turned on. I therefore came to the conclusion that to study how hormones act one must study the material basis of the control of gene activity. In the winter of 1959, therefore, I and my colleague, Dr. Ru-chih C. Huang, now

Professor of Biology at the Johns Hopkins University, started a series of investigations to find out whether it would be possible to cause such isolated chromosomes, if they could be isolated, to synthesize messenger-RNA as they do in life. After many misadventures and false starts, we found that it is possible to recover the bulk of the DNA of plant cells as chromatin. In addition, we found that such chromatin, if provided with the four riboside triphosphates, could and would synthesize RNA, just as such chromosomes do in life.

We studied this matter with all seriousness. It appeared a marvelous opportunity to find out, by studying chromosomes exposed in public, how they work in private in the nucleus. We were dissatisfied with the low rate of synthesis of RNA by the natural RNA polymerase enzyme that is bound to chromosomes as we isolate them. In the summer of 1962 we therefore attempted an experiment of adding to isolated chromosomes of pea plants polymerase separately isolated and purified from the bacterium E. coli. *To our amazement and surprise, bacterial polymerase turned out to be able to transcribe the DNA of pea chromatin absolutely marvelously and to produce all sorts of RNA. How nice! We had discovered a means by which the DNA of chromosomes could be transcribed at a rate at least 1000-fold in excess of that permitted by the amount of RNA polymerase naturally bound to chromatin.*

Since chromatin is such a good template for transcription by RNA polymerase, we decided to try as a template protein-free DNA prepared from it. We prepared such DNA by dissociating the chromosomal proteins from DNA with cesium chloride. The high ionic strength medium of CsCl breaks the bonds between chromosomal proteins and DNA and permits the DNA to be pelleted by centrifugation separately from the chromosomal proteins. We then added RNA polymerase to chromosomal DNA. To our surprise and pleasure the chromosomal DNA proved to be a many-fold better template for transcription by RNA than did chromatin itself. Clearly, some fraction of the DNA of chromatin is not available for transcription into RNA. We had the feeling immediately that we were seeing in the test tube a reflection of the generally held notion that in any individual kind of specialized cell of the higher organism only a fraction of the DNA is transcribed, while the majority is held in some repressed state. This experiment then opened up the further question: what kind of chromosomal protein is responsible for conferring upon DNA the inability to be transcribed by RNA polymerase?

We settled this question also in the fall of 1962. Chromatin was dissolved in 4 M CsCl; the solution was spun in the centrifuge; and the chromosomal proteins were collected. From the proteins we extracted those characteristic chromosomal ones known as histones. These are characterized by the fact that they are soluble in acid because they contain many positively charged groups. Other chromosomal proteins are not soluble in acid. The histones we then recombined with chromosomal DNA. The experiment showed clearly that DNA combined with the histone fraction of chromatin is not able to be transcribed by RNA polymerase. DNA combined with the nonhistone proteins of chromatin is completely able to be transcribed by RNA polymerase—in fact, a major portion of the nonhistone proteins of chromosomes appears to be RNA polymerase molecules.

From the results of this experiment, published in the fall of 1962, we have the first clue to the study of the gene activity in higher organisms. Those genes that are repressed (turned off) are complexed with histones. Their DNA is not transcribable by RNA polymerase. Those genes that are active or transcribable by RNA polymerase (turned on) are not complexed with histones."

---

size more protein, RNA, and DNA and then probably divide. They do not all divide at the same time, but eventually their number approximates the number present in the original cell.

Control Mechanisms in the Cell

The cell is a dynamic entity, existing in an ever-changing state yet in equilibrium with its surroundings. To survive, the cell has a myriad of ways to control its growth and metabolism. Some are self-regulating (like many of the enzyme reactions), while others are regulated through gene action and m-RNA production. The cell is usually not growing in isolation and is regulated in many of its activities by other cells. Some regulation probably occurs through the interaction of special chemicals, called hormones, with the m-RNA production in the nucleus; other types may operate through altering membrane characteristics. The most basic controls are those that direct the development and differentiation of cells into different types of cells and that satisfy the need for integrated, harmonious action among different kinds of cells. There are also controls that adjust to such short-term changes in the environment as temperature, light, amount of water, and the demands of surrounding cells. Thus the cell has what may be thought of as coarse and fine controls.

What are they? Regulation of cell activity occurs on at least four levels. These are level 1—the replication of DNA; 2—the transcription of DNA to produce m-RNA molecules; 3—the translation of RNA to make the enzymes that the m-RNA codes; and 4—the self-regulation of enzymatic activity by interaction of enzyme molecules with substrates and inhibitors.

We know considerably more about some of these regulators than we do about others. In the case of DNA replication, we know little about the nature of the signals that start the process. This activity is related in some unknown way to the various plant growth hormones such as auxins and cytokinins discussed in later chapters.

At the level of transcription—the step of synthesizing m-RNA from DNA—we know little more than the fact that it is controlled by activator substances. These apparently keep certain genes free of histones, which would normally cover the genes and prevent their transcription. This is the same level of action as the low molecular-weight RNA and chromosomal RNA discussed in Chapter 9.

At the level of translation—the steps of making proteins in conjunction with m-RNA—we are equally uninformed. In most cases, it appears that m-RNA is made and rapidly used in protein synthesis. There are cases, however, in which m-RNA is sometimes stored for long periods before it is read. This is true of many dormant systems followed by periods of rapid growth. In the first stages of germination and growth of both seeds and pollen grains, protein synthesis is from m-RNA made before the cells became dormant. This allows the system to react more quickly to favorable environmental conditions and permits a rapid period of growth, which is then continued through the production of more m-RNA. The presence of this long-lived, inactive m-RNA poses various questions. Where, for example, is it stored in the cell? How can it remain unread? What activates it at the correct time?

Finally, at the level of regulation of enzyme activity, we have the greatest amount of information. We know that the level of activity of an enzyme depends upon the concentration of its substrate: the higher the concentration of substrate, the more all of the enzyme molecules present will participate in the relevant chemical reaction—to the point where all the enzyme molecules are busy all the time. If substrate production is not rapid enough, high enzyme activity will result in a decrease in concentration of the substrate, which in turn will result in a decrease in enzyme activity. This decrease in enzyme activity will cause an increase in the substrate concentration, and the process will be repeated as enzyme activity again increases.

We saw earlier that the vast number of enzymes operate in series or teams, and for these enzymes, end-product feedback control is important. In this type of control, the end product of the reaction is important in governing enzyme activity, and the concentration of the end product is determined by the activity of the next enzyme in line. For example, suppose there are 10 different enzymes involved in the synthesis of an amino acid such as histidine. In many such biosynthetic pathways, each of the enzymes is inhibited by high concentrations of the product of the last enzyme in the series. As the cell fills up with the product of this chain of reactions, all of the relevant enzymes become diminished in activity so that they no longer make the substance (in this case histidine) as rapidly. Should the substance be removed by metabolism, all of the enzymes in the chain speed up their action, restoring the supply.

This type of feedback control is based on allostery: the reaction of an enzyme molecule with special molecules that change the structure of the enzyme inhibiting or promoting its activity. At the same time the altered enzyme may be able to act on a new and different substrate. Thus allosteric reactions may not only inhibit but shift a metabolic or synthetic pathway.

This, then, is at least a partial picture of the general cell. Shortly, we shall examine the cell both as a free-living organism and as a part of multicellular organisms. Then we shall see how this basic pattern has been altered through evolution to adapt the organism to its particular environment. We will also look at cell growth and development and see how the pattern of cell division may be modified and how the other stages of cell development occur.

11

Darwin and Evolution

Saturday, June 30, 1860, was a warm day at Oxford University. The British Association for the Advancement of Science was holding its annual meeting. The meeting was nearing its end, but many had remained to hear a paper by a man with a peculiar accent, an American named Draper, entitled "The Intellectual Development of Europe Considered with Special Reference to the Views of Mr. Darwin." This typically Victorian title alluded to the publication a year earlier of Charles Darwin's book *The Origin of Species.* Public interest in Darwin's work was high—the first edition of 1250 copies sold out the day it appeared.

The long, narrow room in which Draper spoke was packed. In the center of the room sat a solid block of clergy. On either side were ladies in bright summer dresses and fluttering handkerchiefs. In the back of the room sat a group of rather rowdy undergraduates. The 700 people packed into the room were not particularly interested in Mr. Draper but in the men on the platform behind him. These included Bishop Samuel Wilberforce, Thomas Huxley, and Joseph Hooker. The audience knew that Bishop Wilberforce was against evolution: the idea that species could change and develop into new forms. They also knew that Huxley agreed with Darwin on the concept of evolution. The audience was therefore waiting impatiently for the debate between these two excellent speakers.

When Draper finished, several other speakers on the platform tried to comment on his lecture, but they were shouted down by the crowd who called for Wilberforce. The bishop arose (Fig. 11–1). He was a large, imposing man with a florid complexion, widely known as "Soapy Sam" because of his smooth speaking style. Jovial, popular, rather loud-voiced but an excellent orator, Wilberforce now belabored Huxley with wit and sarcasm. Understanding little science himself, he fully exploited the popular tendency to regard every novelty as an absurdity. Sweeping the audience with him, he finally turned to Huxley and with mock politeness begged to know "whether it was through his grandfather or his grandmother that he claimed his

11–A This is how Charles Darwin looked in 1856, at the age of 47, when he began the manuscript that would become The Origin of Species.

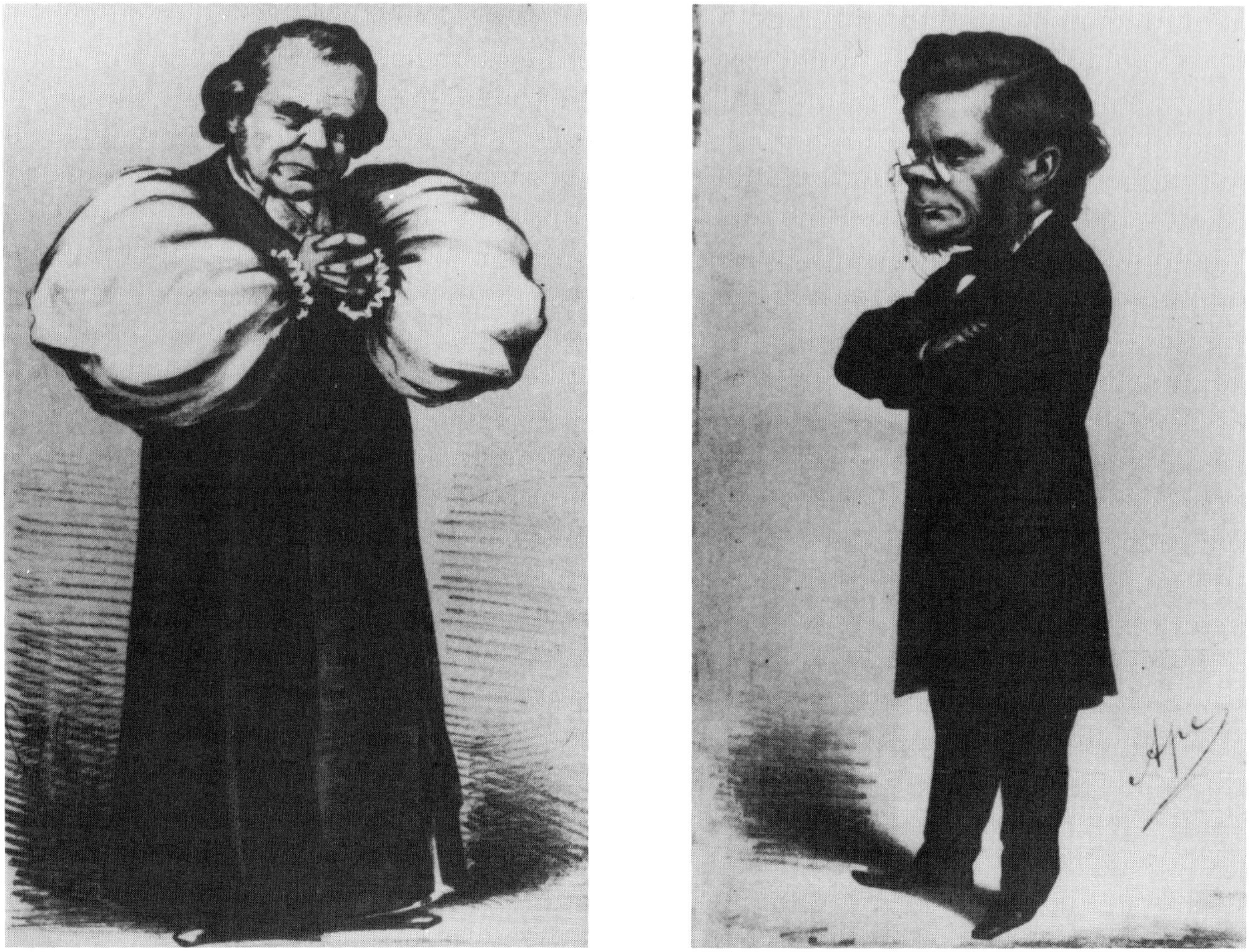

11–1 (a) Bishop Wilberforce *(b) Thomas Huxley*

descent from a monkey." In a roar of applause he sat down. During the applause Huxley, who had been worrying about how to present Darwin's case, turned to the man next to him, the famous botanist Joseph Hooker, and whispered, "The Lord hath delivered him into my hands."

The audience now called for Huxley, who quietly rose and went to the speaker's stand. He, too, was a tall man with black hair and black piercing eyes (Fig. 11–1). He was at the height of his powers, a brilliant man, with an excellent reputation as a scientist. Turning to Wilberforce, he said that he had come only in the interest of science and that he had heard nothing to prejudice Mr. Darwin's case. Darwin's theory, he said, was more than a hypothesis; it was the best explanation of the origin of species yet advanced. Touching on the Bishop's obvious ignorance of science, Huxley quietly and gravely said that he would not be ashamed to have a monkey for his ancestor, but he would be ashamed to be connected with a man who used great gifts to obscure the truth. The crowd roared its approval. Bishop Wilberforce knew he was crushed, and he left the platform. Women fainted. The whole affair was a grand success—except for the Bishop.

Darwin's Theory

What were Darwin's ideas? First, he presented an overwhelming mass of evidence that evolution had taken place and was the means whereby species originated. Over 20 years of carefully documented research went into the creation of *The Origin of Species.* Second, he presented a theory for the mechanism of evolution — the theory of natural selection. This theory is based on two facts: (1) all species overpopulate; and (2) variations exist among the individuals of a given species.

Darwin noted in his numerous observations that all plants and animals overproduce.* A single fern plant may produce millions of spores, each capable of forming a new plant. A single orchid may produce thousands or even millions of seeds. A single apple tree may produce hundreds of apples each containing several seeds, all able to grow into new trees. But the earth is not covered with ferns, orchids, or apple trees. Only a few percent of the spores or seeds germinate; only a small percent of these survive the early stages of growth; and only a small percent of these young plants reach sexual maturity and reproduce — in any stable population, only two for each set of parents.

But why do some species survive to reproduce and leave descendants while others do not? The answer to this, Darwin said, was in the observation that not all individuals of the species are the same. There is variation both between offspring of the same parents and between offspring of the total population. These differences may be marked or they may be minor, but they are observable in all living things. Some of the differences we see are a result of environment (see Chapters 37 and 39); for example, one plant is taller than another because it received more water or light. This type of variation is not inheritable and cannot be passed on to the next generation. Other differences can be inherited by the offspring, and these are the ones that are important to the theory of natural selection. Some of these variations make one individual better suited to a particular environment than another, perhaps allowing it and not the other to reproduce.

*Interestingly, this rather "obvious" point, as it applies to animal populations, is being questioned by several modern researchers. The *potential* for overproduction is there, but in many well-documented cases, as populations approach crowded conditions, various mechanisms (depending on the species) come into play to drastically reduce the birthrate. This does not occur in the plant examples that are cited.

The individuals best fitted to a given environment will be those that can reproduce and in reproducing continue the species. Thus there is selection among organisms. There is, said Darwin, a *struggle for existence,* and in this struggle it is *the fittest who survive* and reproduce.

The struggle need not be a tooth and claw combat; usually it is not. But the plant living on the desert is engaged in a struggle for water to survive; for example, if its roots can grow to a source of water, the plant will live long enough to produce seed. Those individuals best developed to obtain and retain water will reproduce, while the others, no matter how long they may survive as individuals, will eventually die and not contribute to the continuation of the species. A tree seedling growing on the forest floor is being put to the ultimate test by the amount of light reaching it through the canopy of leaves overhead. Once the seed has germinated and the root grown into the soil, the location of the seedling in the forest is fixed. If the light on the forest floor is marginal for survival, the seedling will live if some feature of its physiology or morphology allows it to utilize light more effectively than similar seedlings around it. Again, the important point is not merely survival but reproduction. This critical feature can be termed reproductive fitness.

Changes in the environment act as strong pressures to select for and against characters of an organism (Fig. 11–2). These do not include all changes in the environment but only those that reach some critical magnitude where they become limiting to the growth and reproductive capacity of the plant. Studies of ferns show that species that exist on islands are related to those found on the adjacent mainland, but they are not quite the same species. One consistent difference is in the size of the spores; the island species have larger and heavier spores than the corresponding mainland species. At first thought this seems all wrong. How could such heavy spores be blown to the island? The answer is that the light spored species colonized the island and evolved into the heavy spored species. But why should the character of heavy spores be selected? The answer is in the same winds that brought the light spores to the island. The wind carries the light spores from the island and drops them in the sea where they die, while the heavy spores, carried only a short distance, fall on the island and survive. The same thing is true for seed plants on islands where the island species have much heavier seeds than mainland species. Another example is illustrated in Essay 11–1.

11–2 One of the great strengths of the theory of natural selection is that it explains such puzzles as the Cactaceae and the Euphorbiaceae. The photograph shows a collection of what appear to be cacti, yet when they flower, it is clear that they belong to two widely different groups of plants. One, the familiar cacti (right), are found only in the western hemisphere, while the second, the euphorbs (left), are found in Africa and Asia. All are desert or semi desert plants, and their similarity in vegetative features had long puzzled botanists. But, according to the theory of natural selection, the differences arose because similar selection pressures resulted in the survival of populations with similar features that permitted them to survive and reproduce. This is a striking case of convergent evolution.

The same conditions explain the presence of flightless insects and birds on islands. In all cases the wind and surrounding water are strong selection factors in the island environment. These factors act to eliminate from the reproducing population individuals unsuited for this particular environment, and slowly new island species evolve from the reproducing survivors.

The change of a single character does not create a new species, but once isolated on the island, additional environmental factors will result in the selection of other characters that make the individuals more closely adapted to the island environment, ultimately resulting in organisms distinct from the original colonizers from the mainland. The concept of adaptation is an important one and means that the characters of an organism increase its reproductive potential, which is the ultimate critical factor for survival of the species.

Darwin's theory of natural selection is based on two facts: overpopulation and variation. From them he postulated that those individuals of a species best able to reproduce would be selected by environmental (physical, biological, or both) pressures. Eventually, the surviving, reproducing individuals would accumulate enough different features from the original species that they, as a population, would be a distinct species. Darwin distinguished between variations that could be transmitted to the offspring by the parents and those that could not. Only variations that could be inherited were meaningful in terms of the evolution of new species. The theory of natural selection focused attention on the question of inheritance with regard to variation. Which variations are heritable? How are variations transmitted? How do variations arise? Darwin had a grasp of heritable versus nonheritable variations, but neither he nor anyone else at the time had any idea how variations were transmitted or how they arose. In *The Origin of Species* Darwin presents an incorrect idea about how variations could be transmitted, but this has no bearing on his concept of evolution by natural selection. The theory is based on the observation that such variations occurred and that somehow they were heritable. For a fuller understanding of natural selection, the rules governing inheritance would have to be established, but they do not modify the concept that heritable variations appear and are selected for or against.

Darwin's theory of evolution is a triumph of synthetic scientific thought. He surveyed and organized a vast amount of material, and in *The Origin of Species* produced one of the most important scientific writings of all time.

Essay 11–1 Observations on the Evolution of Island Species

Dr. Sherwin Carlquist of the Rancho Santa Ana Botanic Garden, Claremont, California, is one of the foremost American biosystematists and botanical evolutionists. He has contributed to many fields but has a special interest in islands and the plants and animals that live there. His Island Life *is a popular account of the natural history and geology of islands, and his new* Hawaii: A Natural History *is a fascinating description of these popular islands. In the following account Dr. Carlquist describes the nature of his work on the loss of dispersibility in Hawaiian species.*

"Bidens pilosa, *a common weed throughout the United States and other regions, has an excellent mechanism for dispersal. The fruits are narrow and needle-like, and bear stiff hairs. Each fruit is tipped with two or three prong-like awns, and these awns are barbed like a fishhook. Thus, a fruit can penetrate fur or feathers and remain embedded there, whether it lodges at its basal end or tip end. These fruits catch easily in socks, and are not easy to remove, giving the plant the common name of 'beggar tick.'*

While investigating evolutionary phenomena in the Hawaiian flora and fauna, I was struck by the commonness of flightlessness in Hawaiian insects and even some birds. Flightlessness is common in island insects and birds—the dodo of Mauritius is an outstanding example. Apparently such flightlessness is related to a tendency for individuals to stay close to suitable food sources, and lack of predators which would make flight valuable for evading capture.

I was familiar with the odd shapes of fruits in Hawaiian species of Bidens, *and the fact that other plants in the Hawaiian flora have other alterations—chiefly gigantism—when compared with their mainland*

11–0 *The seeds of various species of* Bidens: (*a*) Bidens pilosa, (*b*) Bidens macrocarpa, *and* (*c*) Bidens waianaeensis.

relatives. In searching for an explanation, I wondered if perhaps there was a counterpart in plants to the flightlessness of island birds and insects.

I discovered that a few species of Bidens *in Hawaii do have fruits that look like those of the mainland species* Bidens pilosa. *These, however, grow on cliffs where seabirds nest, and so continue to be dispersed the same way their ancestors were. Other species, however, have evolved into forest areas. Here they cannot be dispersed by seabirds. Areas of suitable forest for each species are limited in the Hawaiian Islands. Thus a highly efficient dispersal mechanism is no longer of selective value, and production of mechanisms designed for transport in fur and feathers is not economical for the plant—and might even result in transport of fruits to unsuitable sites. The forest species of* Bidens *in Hawaii have lost the awns to varying degrees, and show gigantism of fruits. The larger size permits more food storage, and seedlings that germinate in shady forest regions would thus have a better chance of survival. There are not even any hairs or bristles on most Hawaiian* Bidens *fruits. Thus, they tend to travel only short distances from parent plants. This tendency keeps them within the habitat suited to the parental plants.* Bidens macrocarpa *has huge fruits with only useless vestiges of the awns.* Bidens waianaeensis *has fruits with an odd corkscrew-like shape. These curious changes would undoubtedly not take place on a continental area, where localities good for a weed are geographically scattered and where dispersing animals (or humans) are abundant. Thus, the Hawaiian species of* Bidens *have adapted to rain forest, and in doing so have lost contact with their dispersing agent, and have lost their original dispersing mechanism. Field studies showed me the ecological connection to these fruit shapes, and study of the entire Hawaiian flora showed many parallel examples in other groups of plants, confirming this interpretation."*

Development of the Theory

Darwin did not originate the idea of evolution; the idea had existed long before Darwin was born. There are glimpses of the idea of the mutability of species in a number of ancient writings, but it was not until the time of Linnaeus and the great surge of classification in the 1700s that suggestive relationships between organisms arose, leading inevitably to speculation about their origin. Linnaeus, while a staunch believer in the special creation of species in his early works, began to waver in this belief in his final books and papers. A French contemporary, Comte de Buffon (1707–1788), scattered ideas on evolution through his immense 44 volume work on natural history. Probably stimulated by Buffon, Erasmus Darwin (1731–1802), who was Charles Darwin's grandfather, and Jean Baptiste Lamarck (1744–1829) presented similar theories on evolution. Lamarck was more systematic and is better remembered today, although at the time Erasmus Darwin was well known in England. Both men presented data for the existence of evolution, and both formulated theories on the basis of the inheritance of acquired characteristics, which are characteristics developed during the lifetime of an individual that could be passed on to its offspring. Lamarck, for example, believed that over successive generations the neck of a giraffe grew longer because it stretched to eat high branches; this character was then transmitted to its offspring. In these theories there is a strong involvement of willing change on the part of the organism that is completely lacking in Darwin's and modern theories of evolution.

Great stimulation was given to the study of evolution by geologists. Features of the landscape had long been thought to have been the result of one or more great floods. With the work of the British geologists James Hutton (1726–1797), William (Strata) Smith (1769–1839), and later Charles Lyell (1797–1875), the concept was established that gradual changes in the world about us could eventually result in profound alterations of that world. These new concepts carried with them a new time scale, a vastly greater age of the earth than had been thought possible before. Without knowing it, the geologists were providing the biologists with time—millions of years of time—in which to allow evolution to occur (see Essay 11–2). Moreover, once it was established that the everlasting hills are, in fact, far from everlasting, doubt arises about the concept that species are unchanging and everlasting.

Thus, before Darwin ever appears on the scene, the stage is set. Actually, several investigators presented theories similar to Darwin's in the first half of the 1800s, but they were either unclear, incomplete, or ignored. Both scientifically and intellectually the climate was finally becoming favorable for a public and scientific debate on evolution, and Charles Darwin provided the focal point.

Would we have the theory of natural selection today if Darwin had never been born or had died before he published his great work? The answer is clearly yes! In 1854—16 years after Darwin first formulated his theory and while he was still gathering data for his book—he received a manuscript from a young biologist, Alfred Russel Wallace, in which he read his own theory. Darwin was thrown into a terrible dilemma: was all his work to come to nothing? He turned Wallace's paper over to Lyell and Hooker, men of great scientific stature who knew what Darwin was working on. They eventually persuaded Darwin to present the 1844 version of his theory, along with Wallace's paper, at the meeting of the London Linnaean Society on July 1, 1858. Darwin then threw himself into a fervor of work and finished *The Origin of Species.* How clearly Wallace understood the theory of natural selection is illustrated in End Paper 11–1.

Moreover, Wallace was not the only biologist to have the same idea as Darwin. The American botanist, Asa Gray, was developing the same concepts but was several steps behind Darwin. Gray and Darwin were regular correspondents. So, when it became evident to Darwin which turn Gray's thoughts were taking, he wrote his theory to Gray (in 1847), and Gray immediately accepted it.

Thus even without Darwin we would still have the theory of natural selection. This does not detract from Darwin's originality or achievements. Rather, it emphasizes the fact that scientific knowledge advances in a slow, frontlike pattern, and while great and brilliant men can act to hasten or strengthen this advance, the accumulation and understanding of data and ideas must be by the discipline as a whole. In the next chapter we shall see what happens when a genius gets a generation ahead of his time.

which is one of the two or more alternative states of a gene that occupy the same position on homologous chromosomes. Many genes have several alleles. Each gene of a series, called multiple alleles, has control over the expression of one of several forms of the same character. For example, snapdragon flower color was found to be controlled by a series of nine alleles, governing a range of colors from red through various paler shades to ivory color with red stripes. Each allele was apparently dominant to those producing colors lighter than itself. One of the most frequently cited examples of multiple alleles is the series controlling human blood types, described in standard biology texts.

Alleles are possible to understand in terms of DNA and protein as changes in the base sequences of the unit of DNA comprising a gene. Such a change would result in an alteration of the amino-acid sequence in the protein, affecting its action as an enzyme and changing the expression of the phenotype. This concept is also useful in the case of incomplete dominance. Say the gene for red flowers is a DNA molecule that codes for an enzyme that catalyzes the synthesis of a red anthocyanin pigment. Say the gene for white is simply a defective DNA molecule coding for a defective protein with *no* enzymatic activity. If both genes are the effective DNA molecule (for red), enough enzyme and hence enough pigment will be synthesized to produce red flowers; if only one gene is effective, only half as much pigment will be produced, resulting in pink flowers; no effective genes produce no pigment, resulting in white flowers.

Essay 12–3 The Importance of the Individual in the Advance of Science—A Contrary View

In the spirit of equal time for opposing views, we offer the following excerpt from A History of Genetics *by A. H. Sturtevant of the California Institute of Technology. Dr. Sturtevant is one of the founding fathers of modern genetics, and here is what he has to say about the role of Mendel in the development of genetics.*

"There is a widespread view that scientific discoveries are more or less inevitable, and that it makes little difference whether or not a particular individual makes a discovery at a given time: if the time is not ripe for it, it will not be understood and will have little or no effect on future events; if the time is ripe, then someone else will soon make the discovery anyhow.

The history of Mendelism is one of the often-cited examples here. According to this interpretation, Mendel's paper was not understood in 1866 because the time was not ripe; in 1900, when the time was ripe, the principles were discovered independently by three different people. To me, this account seems greatly oversimplified—though it must be admitted that the development of the subject would probably have been much the same, even to the dates, if Mendel's paper had never been written.

It is true that the paper was ahead of its time, but it was not difficult to understand, and it seems unlikely that it would have remained unappreciated for so long if it had appeared in a less obscure journal, or if Mendel himself had published the further cases that he reported in his letters to Nägeli. It must be remembered that Nägeli's failure to appreciate the paper in 1866 can be matched by Pearson's failure in 1904. Both were outstanding men, and both were actively studying heredity, but to both of them Mendel's results appeared as trivial cases involving a few superficial characters, obviously neither useful nor illuminating for any general theory of inheritance. It does not follow that no biologist was likely to have appreciated the paper if he had seen it before 1900—I have suggested . . . that Galton, for one, would very likely have done so.

As for the simultaneous discovery in 1900, I have pointed out . . . that it seems likely that the independent discovery was the finding of Mendel's paper, and that the actual working out of the principles without knowledge of Mendel's work was only accomplished once—by Correns—and even here it is not possible to be certain how clearly he understood the principles before he read Mendel.

In connection with the idea of the inevitability of scientific discoveries, it seems necessary to inquire into the meaning of the expression 'when the time is ripe.' The state of knowledge and opinion at a given time is obviously the result of individual intellectual efforts and can scarcely be thought of as inevitable. It does make a difference whether a discovery is made now, or next year, since the whole course of events in the intervening year is altered by the discovery; if it is not made now, there is a chance that the time may be overripe next year, since attention may have shifted to a quite different field.''

Gene interactions. Research with large numbers of genes revealed that a group of genes may be involved in producing a single character. These are genes working in teams. Each gene is a distinct unit that may have one or more alleles, but all members of the series must be present in the dominant condition to produce a given character. Flower color and form are frequently controlled by such combinations of genes. This situation is understandable now in terms of enzyme action and cell physiology. Twenty to thirty or more steps may be involved in the synthesis of a single pigment. Each of these steps requires a specific enzyme, and each of these enzymes is determined by a gene. Detailed work on the nutritional requirements of the pink bread mold (*Neurospora*) by Beadle and Tatum, discussed in Chapter 9, provides another example.

In another case, the team of genes acts on some quantitative feature of an organism such as corn-ear length, fat content of various seeds, skin color in man, or perhaps even human intelligence. If all genes are present in one form, the feature may be at its maximum. This could be, for example, the blackest human skin color; all genes in the other form may result in the whitest skin color; and intermediate mixtures produce intermediate skin colors. This is called quantitative inheritance.

It is also possible to have one gene modify the action of another. If, in a simple case, gene A produced an enzyme A′ that synthesized the substrate for enzyme B′ produced by gene B, any change in gene A that would result in the malfunctioning of its enzyme would directly affect the expression of gene B. An *albino* (animal or man) is a good example. In such a case, a homozygous recessive condition of one set of alleles inhibits the action of *all* of the other genes for skin color. This might be because a gene is defective that would otherwise allow synthesis of some precursor for all pigments. Subtler and more complex mechanisms in which one gene modifies another are known as epistatic genes or modifiers.

Some genes also control enzymes that are central to many characters of the organism. Thus a change in a gene producing an enzyme involved in lignin synthesis may cause a great number of different and unpredictable effects throughout the plant. This capacity of a single gene to affect many characteristics is known as pleiotropy.

Penetrance and expressivity. In the discussion of Mendel's experiments we assumed that any dominant gene is expressed, but we have just seen that the expression of some genes depends upon the presence of others. Many complications of modifier genes are now known, so two terms have been devised to describe the situation: penetrance is the percentage of individuals that, when carrying a given gene in a combination that might allow its expression, actually express that phenotype; expressivity denotes the manner in which the phenotype is expressed. Mendel's genes showed 100% penetrance and clear-cut expressivity; many other genes do not.

These concepts bring up an important point: *no gene acts completely by itself.* Its action is always influenced by the overall genetic makeup (genotype) of the individual in which it occurs. Even environment may influence penetrance and expressivity. An interesting example is the Himalayan rabbit. (The case of the Siamese cat is similar.) Normally, all these rabbits contain a dominant gene for black fur color, but this is only expressed at the tips of the feet, ears, nose, and tail. It turns out that expression of this gene is highly sensitive to temperature. If hair is removed from a patch on the back of the rabbit and the patch kept covered with an ice pack, hair is black when it grows out. For the black color to be expressed, temperature must be below about 33°C—as it normally is at the tips of the extremities.

12–5 A water buttercup (Ranunculus aquatilis) *illustrating the difference in leaf shape between floating and submerged leaves.*

A plant example of this type of interaction is the water buttercup (*Ranunculus aquatilis*). This plant grows in shallow pools, with some of the leaves submerged and some above the surface of the water (Fig. 12–5). The difference in the environment below and above water markedly affects penetrance and expressivity. The leaves below water are finely dissected, but those above water are entire.

We must remember that genes always act within some environment—the environment created by all other genes as well as the surrounding physical environment (see Essay 12–4). The total complex of genes in an organism is called the genome. Understanding the functioning genome—why genes act only in certain *places* (obviously genes for flower shape are expressed that way only in the flower) and only at certain *times*—is certainly one of the most complex, important, and challenging unsolved problems of biology. It is the problem of development: understanding how the genome in the zygote controls the formation of the adult with all its phenotypes. We shall return to the problem in later chapters.

Essay 12–4 On Making a Green Revolution

We have all read of the astonishing results through plant-breeding programs. The yields of essential crops here and abroad have been enormously increased, bringing both great benefits and great problems. But how are such results achieved? Here Wade Dewey, a successful plant breeder at Utah State University discusses some of the methods used and the results that have been obtained.

"Of the many contributions plant breeding has made to agriculture over the past 50 years, the two best known to the general public are high-yielding hybrid corn and the development of rust- and smut-resistant wheat varieties. These two achievements have involved crops that, although closely related botanically, represent distinctly different breeding groups. Corn is largely cross-pollinated; hence a corn variety is made up of individual plants that are highly heterozygous. Wheat is normally self-pollinated; thus most wheat varieties are composed of homozygous plants. More than anything else, a plant's method of reproduction determines its genetic makeup and its most appropriate breeding procedures.

Although superior corn and wheat varieties are developed via somewhat different routes (crossing, selecting, and testing), the underlying principles are basically the same. In any breeding program, the first, and possibly most important, step is the selection of parents. The number of possible genetic combinations within a species may be virtually limitless, but in any cross one is limited to the genes possessed by the two parents. Indiscriminate crossing wastes time, money, and effort and involves the same wishful thinking that characterizes the duck hunter who hazards a 'flock shot' rather than concentrating on a specific target. There are occasions when both the breeder and the hunter may hit a target at which they did not aim. Such a fortuitous find in a breeding program is referred to as a transgressive segregate, that is, a plant possessing characteristics not evidenced by either parent or a plant exceeding the parental range for a given character. The alert breeder keeps a sharp eye out for such unexpected superior segregates, but his main effort is concerned with planned crosses involving parents that together possess the characters sought in the new variety.

The actual crossing process is usually the least complicated step in a breeding program. It requires only a familiarity with the reproductive habits of the plant and a degree of dexterity in manipulating the floral structures. Wheat, for example, has perfect flowers (both sexes in the same flower). A cross is made by removing the three anthers with tweezers (emasculation) a few days before pollen is released and then fertilizing the flower with pollen from the desired paternal parent. Each flower is emasculated and pollinated individually, with each successful pollination resulting in a single seed. Corn, on the other hand, bears male and female flowers separately, making emasculation and crossing much easier. A single application of pollen to the female inflorescence (group of flowers—the ear) in corn may result in several hundred seeds. The relative ease with which the floral structures of corn can be manipulated has been a major factor in the success of hybrid corn; conversely the difficulty encountered in crossing wheat has been a major hindrance to hybrid wheat production.

The period of screening and selection during the segregating generations following the cross is actually the heart of a breeding program. During this period the breeder attempts to isolate superior plants by applying his knowledge of genetics, statistics, plant pathology, and agronomy in sifting through the innumerable genotypes that comprise the segregating populations. In self-pollinating species such as wheat, this segregating process goes on for six or seven generations following a cross, after which most plants are, for practical purposes, homozygous. By this time thousands of individual plants have been examined and classified. During these early years, selection is practiced largely for such obvious qualitative characters as disease resistance.

The final phase of a breeding program involves the evaluation of the breeding lines that have survived the preliminary screening during the segregating generations. This advanced testing is primarily concerned with the less obvious, quantitative characters such as yield and involves replicated trials and statistical analysis. In many cases this advanced testing phase takes nearly as long as the developmental phase.

If all goes well—and it seldom does—it commonly takes 10 to 12 years between the time a cross is made and the new variety emerges. Consequently, the breeder must anticipate the quality requirements, disease problems, cultural conditions, market demands, etc. that a variety will encounter some 10 years hence. But not every cross, no matter how carefully planned, results in a new variety. Truly superior new genetic combinations are relatively rare in crop plants that have been cultivated intensively for many years. It is not uncommon for a breeder to make and work with several thousand crosses during his professional career and to release less than a dozen varieties; yet the value of a single superior variety may far outweigh the entire investment made in the breeding program."

Mutation

How do the various alleles originate? It is fine to talk of alleles for tallness and alleles for shortness or multiple alleles of a gene, but how do they come about? According to our present theories, they originate through a change or mutation of an existing gene. The term mutation is a complex one with a distinctly odd history.

It was first used by De Vries, one of Mendel's rediscoverers. He was working with the evening primrose (*Oenothera*), when he found a number of characters appearing that were not present in either parental line. He believed that these new characters represented changes in genes and would be heritable, as were other genes. The concept that De Vries had was correct, but his data were wrong. He had made an unfortunate choice in selecting an organism with a highly anomalous type of chromosome behavior; virtually none of the changes he saw were true gene mutations. Not long after De Vries' work, Thomas Hunt Morgan, at Columbia University, found true mutations in his fruitflies.

The most common meaning associated with the term "mutation" is *a specific change in a gene.* Other meanings have to do with changes in chromosome structure that result in different phenotypic expressions. These chromosome aberrations, discussed in Chapter 9, are particularly important in certain phases of genetics. They have been studied intensively in many modern taxonomic investigations.

How do gene mutations occur? They occur through changes in the base composition of the DNA of the gene. It is now known that a mutation can involve as little as one change in a single nucleotide pair. Conditions that will facilitate such changes can produce mutations. The first discovery of such agents was by Hermann J. Müller, who in 1927 found that treating *Drosophila* with X-rays greatly increased the rate of appearance of mutations. (He received the 1946 Nobel prize for this discovery.) L. J. Stadler achieved the same result with barley the following year. Radiation from radioactive materials has the same effect. These are all forms of ionizing radiation, and their effects are proportional to the dose of radiation that the organism receives. The manner in which ionizing radiations produce mutations is not fully understood. The action may be directly on the DNA, altering or destroying a portion of the chain, or the radiation-induced mutation may be due to chemical changes produced by the ionizing radiation within the cell. Ultraviolet light, which is not an ionizing radiation, is also capable of inducing mutations. The wavelength of UV most effective as a mutagen is that which is absorbed maximally by DNA. In any case, the mutagenic agent must act on present or potential sex cells if the mutation is to be transmitted to future generations. Mutations in other cells (so-called somatic tissue) may cause recognizable phenotypic changes. Such a sport may be a branch with portions of the leaves lacking chlorophyll (variegated). To preserve this, vegetative propagation may be necessary.

Many chemicals induce mutations. Some are as simple as hydrogen peroxide (H_2O_2), while others are more complicated (the nitrogen mustard gases used during World War I). An effective mutagen in microorganisms is 5-bromouracil (BU), which is an analogue to thymine. (An analogue is a compound with a molecular structure similar to a natural compound, see Chapter 8.) When bacteria are deprived of thymine and grown in the presence of BU, the BU will be synthesized into DNA, replacing the thymine. At reproduction, errors occur in the synthesis of new DNA, as a result of the presence of the BU, causing mutations.

How do mutations occur in nature? There is apparently not enough natural radiation to account for the number of spontaneous mutations that occur. F. W. Stahl has proposed a theory based on the fact that the four bases (adenine, thymine, guanine, and cytosine) can temporarily exist in rare states that have an affinity for the wrong base and form mismated base pairs. Thus, in this condition, adenine may pair with cytosine instead of thymine. Subsequently, when the DNA molecule duplicates, it will give rise to one strand of normal DNA and one mutant strand in which an adenine-thymine base pair has been replaced by one of guanine-cytosine (Fig. 12–6). There is some experimental support for this theory, and it fits well with the general concept of the DNA structure of the gene.

The presence of certain genes can greatly influence the rate of mutations of the remaining genes. The natural rate of mutation is low, on the order of 1 to 100 per million gametes, and is highly variable not only with regard to organism but in terms of specific genes in a given species.

Ultimately, gene mutation is a physiochemical event that has the effect of increasing the number of different forms of the gene present in the population. Of course, since changes in DNA structure (hence mutations) occur at random, the vast majority of mutations might be expected to result in defective genes. Experience does show that almost all mutations are deleterious. Indeed, many lethal mutations are known. In the best known cases, the gene may be lethal when it occurs in the homozygous condition but not in the heterozygous condition. This might be expected if the gene controlled some enzyme essential to life. If one allele were normal, the enzyme would be produced; if both were ineffective alleles (homozygous), the enzyme would not be produced, and the zygote could not develop. Of course, if a mutation produced a gene lethal in the heterozygous condition, we would never know, since no zygote containing it could develop. It is conceivable that a certain percentage of mutations might have current or potential adaptive value in a population.* A value often mentioned in textbooks is about one "good" mutation out of every thousand.

---

*Even this has been seriously questioned. See Frank B. Salisbury, Natural selection and the complexity of the gene, *Nature* 224 (1969):342–343; Frank B. Salisbury, Doubts about the modern synthetic theory of evolution, *The American Biology Teacher* 33 (1971):335–338.

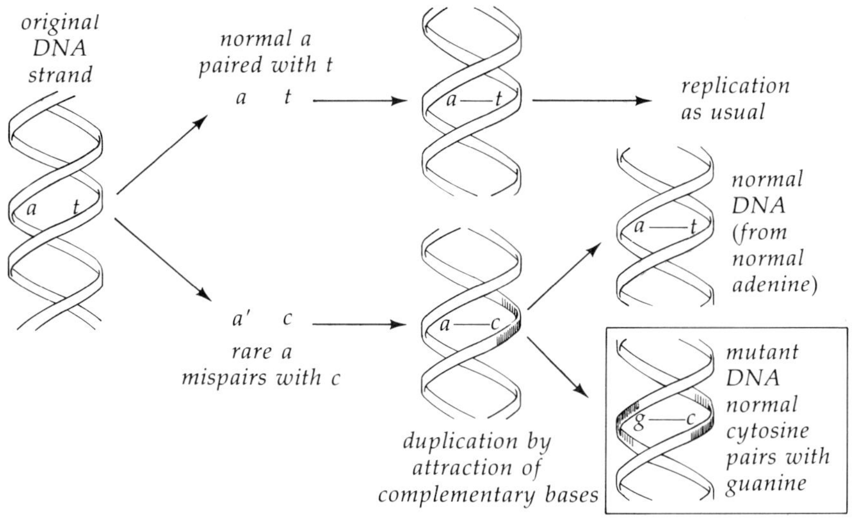

12–6 A diagrammatic sequence showing how a change of adenine from the normal to the rare (or tautomeric) state can produce substitution of a g–c base pair from one a–t pair, resulting in a mutation.

Among the many things modern man has to worry about is the possibility that his technology will increase deleterious mutations. This is a clear danger with radioactive fallout from nuclear explosions—a threat to all living things on earth. Frequent X-rays also pose the same danger to man, although with precautions (lead shielding) the gamete-producing organs could be protected. What about chemical mutagens in the environment? Is cancer related to mutations (X-rays are known to cause cancer)? We need answers to these and other questions.

Independent Assortment and Linked Genes

Mendel worked with seven gene pairs and found that at reproduction each separated and assorted independently of the others. Research in the early 1900s showed that this conclusion was not valid if more genes were considered. Groups of genes were found that acted as if they were linked together. The correlation of linkage groups with chromosomes was soon made, and the basis for Mendel's results soon found—each pea has seven pairs of chromosomes. That Mendel should have accidentally chosen seven genes on different chromosomes seems incredibly lucky, and various suggestions have been made to account for his "luck." One of the most reasonable, by Harriet Creighton of Wellesley College, proposes that Mendel made the choice of characters on the basis of a deep understanding of his experimental organism. The pea had been cultivated for generations by Austrian farmers, and it was probably long recognized that it didn't make any difference if the peas were tall or short, the seeds could be either green or yellow, the flowers white or colored, and so on. This type of knowledge would have been useful to Mendel in his selection of both organism and characters, allowing him to capitalize on clear-cut features of the garden pea. Others feel that Mendel simply ignored the data that did not agree with his conclusions.

After the discovery of linkage groups, the idea of independent assortment of genes proposed by Mendel was modified to a concept of independent assortment of chromosomes. Genes linked together as parts of the DNA molecule of the chromosome move together but are independent in terms of assortment of the other chromosomes and their genes. Mendel's experiment can be viewed by us as evidence for the independent assortment of chromosomes during reproduction, which was established by using marker gene pairs—one for each chromosome pair.

Soon after linkage was established, it was found that such linkage was not always complete; genes that appeared normally in one group would appear in another. The reason for this, as we saw in Chapter 9, is the behavior of the chromosomes during meiosis. Crossing-over occurs when there is an exchange between chromatids of different chromosomes during prophase of meiosis I. The fact that crossing-over takes place and that linkage groups are not fixed makes possible a higher degree of genetic recombination. Both the independent assortment of chromosomes and the rearrangement of genes on the chromosomes during meiosis greatly increase the amount of genetic recombination possible in a population.

End Paper 12–1 Hybrid Vigor and Plant Breeding

From the standpoint of profits to man, perhaps the most valuable discovery of genetics was that of hybrid vigor. Joseph Gottlieb Kölreuter, who did so much early work on plant reproduction and hybridization, discovered in 1763 that hybrids were often more vigorous than either parent. Darwin concluded in his book on plant reproduction (1876) that "cross-fertilization is generally beneficial and self-fertilization injurious." For example, when selections of corn were self-pollinated for several generations, they became progressively less vigorous, yielding less and less. When two of these purebred lines were crossed, the progeny exhibited a greatly increased vigor, and yields went up 25 to 40% or more. In 1910 George Shull coined the term heterosis to describe this situation.

Actually, the American Indians planted different kinds of kernels in the same hill, ensuring cross-fertilization. Such a primitive insight into heterosis could be quite an achievement, but most pre-Mendelian improvements of agricultural plants came about by *selecting* and thereby producing lines that were more and more homozygous (pure).

How does heterosis work? Why should hybrids be more vigorous than their pure-line parents? Geneticists still don't know the answers, but two theories are usually cited. Shull's term heterosis was derived from the heterozygosity or overdominance theory, which suggests that both dominant and recessive alleles contribute to the growth and vigor of an individual. This theory says that two dominant alleles are too much of a good thing, while two recessive alleles are not enough. The combination of a dominant and a recessive allele proves to be just right. The point in this case is that recessive alleles are not totally passive but contribute something in a positive way.

The second theory is often referred to as the dominance theory. It assumes that inbreeding eventually results in the combination of harmful recessive alleles in a homozygous condition. In combination with a dominant allele, a potentially harmful recessive allele cannot be expressed, but inbreeding greatly increases the likelihood that recessive alleles may become homozygous. Different recessives are homozygous in different pure lines; so crossing is likely to combine each of these with a dominant. According to this theory, there is nothing intrinsically bad about the *process* of inbreeding (as Darwin and many others thought); rather it is the *results* of the inbreeding that are harmful (the increase in deleterious, homozygous recessives). It is easy to show that when brothers and sisters or first cousins produce offspring, the chances of forming double-recessive homozygous-allele pairs are greatly increased. (For second cousins, this effect is rather minimal.)

An example is known that illustrates the dominance theory nicely. Each of several varieties of dwarf corn is unable to synthesize the growth hormone, gibberellin. The varieties differ in that each is unable to carry out a different step in the synthesis of gibberellin. Each variety is homozygous recessive for its step; that is, both alleles in the pair controlling that step are ineffective—unable to synthesize an active enzyme for that step. If any two different dwarfs are crossed, each supplies the missing allele (and hence its enzyme) for the other, gibberellin is synthesized, and the offspring are much taller and more vigorous than either parent.

According to the dominance theory, it should be possible to produce pure lines from which the recessives have been completely eliminated. Such pure lines should be as vigorous and productive as the hybrids. So far, this has not been achieved, but this could well be because of the very large numbers of genes involved and the fact that highly beneficial genes may be linked on the same chromosomes with deleterious ones. Even without linkage, it is difficult to combine large numbers of genes in such a way that all dominants are together. It was calculated that to provide an even chance for such a combination of 15 pairs of alleles, 744 million individual corn plants would have to be produced and examined. These would occupy about one and a half times the total corn acreage of South Dakota. Hence the theory has never really been tested.

Most geneticists believe that there is some truth in both the dominance and the overdominance theories of heterosis. Although the dominance theory is thought to be the most important explanation, most are willing to concede that recessive alleles may sometimes contribute to vigor.

Plant Breeding

The impact of controlled plant (and animal) breeding on man's food supply has been fantastic. Surely, the world would be very hungry now (and no doubt much less populated) if it had not been for the achievements of plant breeders. These achievements can be grouped into about five categories.

Yield. As just indicated, hybrid corn produces 25 to 40% more than pure lines. Assuming that about 30% of the more than 4 billion bushels of corn produced in the United States each year (end of the 1960s) is due to heterosis, this amounts to about 1.5 billion dollars. Probably, this would pay for all of the research on genetics from the time of Mendel until the present! The principle of heterosis is being applied to an increasing number of plants, including many vegetables, flowers, and now even wheat. It was easy to produce hybrid corn plants because rows of different pure lines could be planted next to each other and then the pollen-producing tassels removed from one. The corn produced on the plants without tassels will necessarily have to be hybrid corn. In recent years it has been discovered that male sterile lines can be developed. On such plants, no pollen is shed; so it is not even necessary to remove the tassels. Restorer genes must be introduced at the last cross; otherwise there will be no pollen to produce seed in the farmer's field. Plant breeders working with many crops (e.g., sugar beets) are looking for male sterile lines.

Wheat has been a particularly difficult problem because the wheat flower is normally self-pollinated like Mendel's peas. It is possible to produce hybrids manually, and that is what is done by plant breeders. But how can hybrid wheats be produced on a commercial scale? Again, the secret seems to be male sterile lines. If self-pollination does not occur, the stigmas (pollen-receiving part of the flower) come out of the tight wheat flower and become receptive to the pollen shed by nearby male-fertile plants.

Yields can be greatly increased by selection as well as by hybrid vigor. Wheat yields have been increased over the years in this manner, and recently a breakthrough was made in which wheat plants were produced with very short stems—the so-called dwarf or semi-dwarf varieties. The problem had been that really great increases in yield depended upon heavy fertilization with nitrogen-containing fertilizers, but this in turn led to rapid growth of the plant and long, spindly stems that would fall over in a moderate wind and be virtually impossible to harvest. Such falling over is called lodging. The dwarf and semi-dwarf varieties are not subject to lodging (or much less so) and thus can be heavily fertilized even under irrigated conditions. The result is a tremendous increase in yield—25 to 300% depending upon the previous baseline.

Semi-dwarfs were introduced into the Western world by Orville A. Vogel at Washington State University, and much development was done by Norman E. Borlaug at the International Maize and Wheat Improvement Center in Mexico City (supported by the Rockefeller Foundation), who received the 1970 Nobel peace prize for his work. Mexico has become an exporter of wheat rather than an importer. In the late 1960s a comparable achievement was made with rice at the International Rice Institute in the Philippines. Propagation of this rice throughout the Orient has, for the time being, averted what appeared to be an unavoidable famine of disastrous magnitude. This is the heart of the so-called "green revolution."

Resistance to pests. Plant breeding is surely the best way to deal with man's competitors for his agricultural crops: insects, plant diseases, and weeds. So far, little headway has been made by breeding in fighting weeds (we are almost completely dependent upon cultural practices and the application of chemicals), but plant breeders have been extremely successful in producing varieties resistant to plant diseases and to a lesser extent insects. In the 1930s the sugar beet industry in the western states was vanishing due to a virus disease called curlytop. This was spreading throughout all the fields, and it was quite apparent that growing sugar beets would be completely unprofitable within a short time. Plant breeders produced a curlytop-resistant variety of sugar beets and saved the day.

Smuts and rusts of wheat could be almost equally devastating if resistant varieties were not continually being produced. The problem is that the smut and rust organisms are "fighting back" by evolving strains that are virulent toward newly developed cereal varieties. Wheat breeders must continually work on varieties resistant to the new strains of smut or rust. Apparently, the smut or rust organisms are mutating and/or being selected, which accounts for their changing virulence.

We leave ourselves vulnerable to such things by wide use of single, high-yielding varieties of corn, wheat, and other crops. In the summer of 1970 a corn blight destroyed a significant portion of the crop in the United States. This was a blight that had not been a problem for many years, and efforts to breed resistance against it had diminished. The hybrid varieties being widely used were quite susceptible. Resistant varieties were readily available from the corn breeders' stocks so that the blight could be eliminated within one year—by feverishly producing resistant seed during the winter on the islands of Hawaii!

Quality. Breeding plants for high quality in some desirable feature is being increasingly emphasized. Perhaps the most important recent example is the breeding of high lysine corn. Lysine is an amino acid required in the diet of man but typically occurring only in low concentrations in most plant foods. Thus peoples of South America and other areas who are largely dependent upon plant foods rather than meat often suffer lysine deficiency. The symptoms are retarded mental and physical development. Toward the end of the 1960s commercial varieties of corn were produced with relatively high lysine—high enough to completely eliminate the symptoms of lysine deficiency.

In the United States breeders are involved in many problems that are more trivial than the life and death matter of lysine content. Apples are being bred that have an exceptionally deep red color, most appealing to the housewife customer. Wheat is being bred not only for yield and disease resistance but to produce flour of a quality considered desirable by the consumers. Forages are being bred for palatability and suitability to the animals that eat them. Sorghums, for example, must be bred with low hydrocyanic acid (otherwise they are poisonous to livestock), and alfalfas are being bred for low saponins (otherwise they cause bloating in cattle). Of course, a great deal of effort has gone into breeding flowers for their beauty.

Environmental adaptability. Plant breeders are producing varieties that can be grown outside of what had normally been accepted as their range. This requires breeding for frost resistance, drought resistance, etc. It could become increasingly important.

Mechanical harvesting. With increases in labor costs and stability in farm prices, it became essential to devise ways of mechanically cultivating and harvesting the various crops. Mechanical tomato harvesters were invented and constructed, but these didn't work well, since tomatoes normally grow on decumbent vines, the fruit ripening over several weeks. The answer was to breed upright, determinate varieties. The plants were smaller (must be planted more densely), but do not grow as vines, and fruit ripens all at once. Dwarf varieties of sorghum have been produced that can be harvested with a combine, allowing the very nutritious and high-producing sorghums to be used in animal feeding. Fruits have been developed for good shipping qualities. Some of us might argue with the latter's success, since the solid, easy-to-ship but tasteless tomatoes from Mexico provide a good example.

Crops were being improved by breeding long before the beginning of this century. Indeed, the real achievements were made by the ancients, who originally domesticated the species upon which we depend for food. But, until the rediscovery of Mendel's paper, plant breeding was strictly an art. Much of this remains, as is evident in watching a plant breeder at work. He must carry in his mind volumes of information about desirable features, and this he calls upon as he goes through his fields making his selections. Yet plant breeding has just about become a science, thanks to Mendel's laws and developing genetics. The crosses the breeder makes are not haphazard but according to the rules of his science. He studies chromosomes, pollen viabilities, and numerous other features of plant growth. He develops polyploids by treatment with colchicine and other compounds. He seeks germ plasm (genes) from all over the world to incorporate into his lines. Someday, he may even understand the genetic language, constructing genes as he needs them and engineering them into his lines directly. In any case, he is himself an interesting breed of scientist whose art involves the combining of humanitarian feelings with scientific insights.

End Paper 12–2 Sex Determination

The young science of genetics was intrigued with, among many other things, the question of how sex is controlled. As early as 1891, before the rediscovery of Mendel's paper, it had been found that certain insects produce two kinds of sperm in equal numbers, one of which has an extra chromosome not possessed by the other. It was suggested in 1902 that this was related to sex determination—eggs fertilized by one kind of sperm becoming males and eggs fertilized by the other kind becoming females. The extra chromosomes were then called sex chromosomes, while the rest were called autosomes. This led to active study of the chromosomes of many species of animals and eventually plants.

In most insects, in all mammals studied so far (including man), in many fish and amphibia, and in a few plants (those having flowers with the different sexes on separate individuals), the females have two identical sex chromosomes, called X; while the males have only one of these plus another chromosome, different in appearance, called Y. That is, females are characterized by XX sex chromosomes and males by XY. When the gametes are formed, all of the eggs produced by females have only X chromosomes. The males, on the other hand, produce sperm, half of which contain the X chromosome and the other half the Y. When an X-containing sperm combines with an egg, the zygote will be XX; a Y-containing sperm produces an XY zygote.

Are males determined by the presence of the Y chromosome and females by its absence, or are females determined by the presence of two X chromosomes and males by the fact that they have only one X chromosome? In *Drosophila* (the common fruitfly) it was found that femaleness is apparently determined by the proportion of X chromosomes to autosomes. Occasionally, at meiosis, the X chromosomes fail to separate, producing XX gametes and hence some XXY zygotes. These become females. But similar examples have now been found in other kinds of organisms, including man, and they prove to be males. Hence, in man, as contrasted to *Drosophila*, maleness is determined by the presence of the Y chromosome and femaleness by its absence. The type of sex determination found in *Drosophila* is found in some plants such as *Rumex acetosa*, while the type found in man is also found in *Rumex acetosella* and *Silene*.

Many of the lower plants exist in the haploid condition as gametophytes; that is, they have only one complement of chromosomes in each cell rather than two (see Chapter 1 and Sections IV and V). In a few of these cases, half of the gametophytes are female, producing eggs, and the other half male, producing sperm. A good example studied in some detail is the liverwort (*Sphaerocarpus*). The female gametophyte contains the X chromosome, while the male gametophyte contains the Y chromosome. The diploid sporophytes have an XY pair, but they exhibit neither maleness nor femaleness.

They produce spores that grow to become the gametophyte organisms. As might be expected, half of the spores contain the X chromosome and the other half the Y chromosome, although all are identical in appearance.

There are several other interesting mechanisms of sex determination. In butterflies, moths, birds, and some fishes, the males contain the XX, and the females the XY chromosomes. Grasshoppers and some other insects lack the Y chromosome entirely—females are XX and males XO, with the O implying no chromosome. In the honeybee fertilized eggs develop into female workers or into "queens," if the developing eggs are supplied with a special material called royal jelly. Males develop from unfertilized eggs and consequently are haploid. They are the drones.

Most of the flowering plants are bisexual, producing both male and female organs on the same individual, typically within a single flower. In some cases, it is possible to influence relative numbers of female or male organs (ovaries or anthers) by suitable manipulations. In a few other plants, sexes occur in separate flowers but on the same individual. These are more easily influenced. Age may determine sex in some of these species. For example, in squash only male flowers are formed at first, and finally only female flowers are formed. More male flowers are formed under long days—but the opposite is true in spinach. Warm nights produce more female flowers in cucumbers. In these plants female flowers are also increased by treatment with auxins but decreased by treatment with gibberellins (plant growth hormones, see Chapters 30 and 31).

A number of the plants that exhibit different sexes on different individuals have their sexes determined in a manner analogous to that of man and other mammals: females are XX, while males are XY. This is true of several dozen species, including *Elodea, Rumex, Humulus,* and *Silene* (*Melandrium*). In most of these cases, the X chromosome is relatively large and easily distinguished from its much smaller counterpart, the Y chromosome. But, in several species of plants (e.g., *Bryonia*), there are no visible chromosome differences between the sexes, even though sexes occur on different individuals. In these cases, it is still possible to imagine that sex chromosomes exist and differ in their genetic content in a manner similar to those cases having X and Y chromosomes but that all of the sex chromosomes are identical (or nearly so) in appearance.

In 1910 Thomas Hunt Morgan, who was to contribute an amazing amount of information about genetics based upon his studies with *Drosophila,* recognized that sex chromosomes could lead to some interesting modes of inheritance that might appear to contravene Mendel's laws. A recessive gene on a portion of the X chromosome not matched by the Y chromosome (which is usually smaller) might always be expressed in males but only be expressed in females when it occurred on both X chromosomes—that is, when it was homozygous. He found examples of this in *Drosophila* and introduced us to the concept of sex-linked inheritance.

Many examples are now known, the most frequently cited being hemophilia and red-green color blindness in man. Hemophilia is a condition in which the blood fails to clot so that even a minor cut can lead to death from bleeding. The slightest bruise may cause a break in the capillaries beneath the skin and extensive internal bleeding, particularly at the joints. The gene for hemophilia was carried by Queen Victoria and passed through several of the royal houses of Europe. Since the gene is fairly rare, it is not likely to occur in the homozygous condition in females; thus females with hemophilia are uncommon. When a female carries the gene, however, half of her sons may be expected to be hemophiliacs, and half of her daughters will carry the gene. The same situation is true for color blindness (the inability to distinguish between red and green), which is far more common in men than in women.

The plant *Silene dioica* (campion) provides a similar example. A gene occurs on the X chromosome that in its recessive form produces narrow leaves. A female plant might carry the allele but be broad-leaved, containing the dominant broad-leaved allele on the other of its two X chromosomes. Half of the male offspring from such a female plant might be expected to exhibit narrow leaves. There is a further complication: the allele for narrow leaves is lethal in pollen grains. Thus the males that show the narrow-leaf trait cannot pass it on. None of the pollen grains containing that allele will be viable. The trait is passed to half of the female offspring of females that carry it, but none of them will show it. Indeed, female plants never show the trait, since to do so, both recessive alleles must be present, and this is impossible because one would have to be received through a pollen grain.

There is a portion of the Y chromosome in man that has no counterpart on the X chromosome. Genes occurring on this segment are passed only from a father to all his sons. They cannot even be carried by females. The gene for "crocodile skin" in man is an example.

Examples of sex-influenced genes are known in which the expression of a gene is influenced by sex, although the gene does not occur on the sex chromosome. In man, for example, one kind of baldness is dominant in males but recessive in females. Heterozygous males exhibit baldness, but females must be homozygous to exhibit the trait. (In females the trait is seldom exhibited to the extent that it is in males, even when the gene is homozygous.)

13

Genetics, Evolution, and Systematics

Darwin's theory of natural selection for the origin of species focused attention on the problem of variation. Mendel and the geneticists of the early 1900s provided a basis for understanding how such variation might arise and how it is transmitted from one generation to the next. Under the circumstances, an early synthesis of evolutionary and genetic evidence would be expected, thus forming the basis for a new, deeper understanding of the mechanism of evolution. Instead, open hostility appeared between workers in the two fields, and it took over 20 years and other men to effect the final synthesis. The reasons for the differences in outlook of the two groups are instructive in understanding both the role of genes in evolution and the way science advances.

As we saw in Chapter 12, the Dutch botanist Hugo De Vries discovered new types of the evening primrose that differed markedly from their parents. These he found could arise in a single step. We saw also that he was wrong about the basis of the observation (they were not gene mutations but chromosomal alterations), but the observations themselves were valid. These observations suggested to De Vries that evolution did not occur through gradual changes, as postulated by Darwin, but by spontaneous origin of new types, which differed radically from their parents in several characteristics. In this context natural selection merely eliminates those types unfit to survive. The mutation theory, as it was called, also derived much support from the work of the Danish botanist W. Johannsen, one of the great geneticists of his time, who introduced the terms genotype, phenotype, and gene. In 1905 Johannsen tried to apply Darwin's theory to altering the genetic character of the garden bean. Selecting the largest and smallest beans from the seed lot of a single variety, he grew the offspring and followed the change in seed size during several generations. After an initial success in changing seed size in the first generation, he found that later selection for large and small size had no effect on the offspring. These data clearly supported De Vries'

13–A Darwin noted that many orchids produced both nectar and fragrancies that attracted insects, but he was puzzled by the absence of nectar in orchid genera like Ophrys. *The solution to this problem came only recently, when it was shown that these species are pollinated by pseudocopulation—male insects visit them because they mistake them for female insects. Obviously, in such a situation the presence of nectar would be of little advantage. It is probably just as well that Darwin did not discover the mechanism involved, since, as Julian Huxley has suggested, Victorian propriety might have been shocked by the facts!*

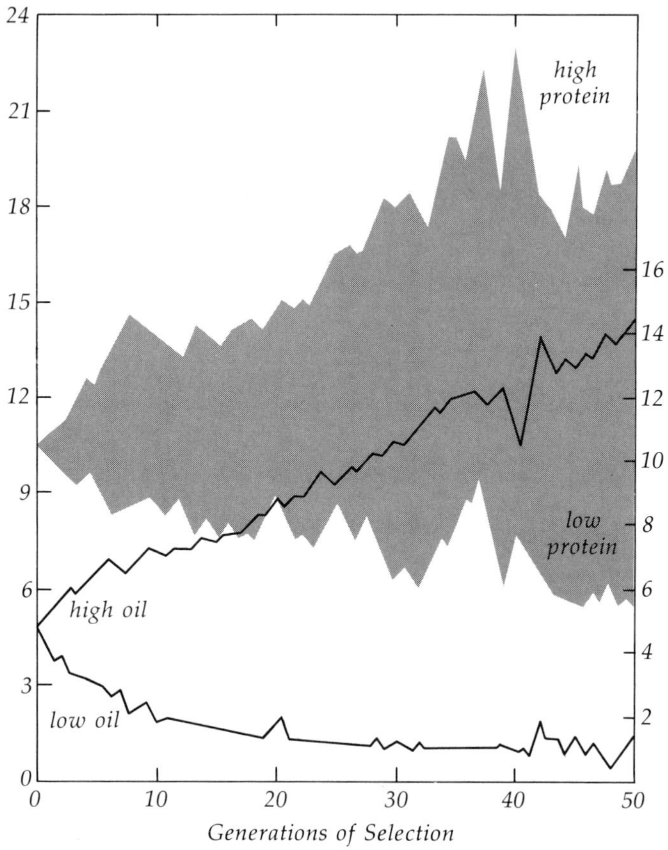

13–1 *The response to artificial selection (over 50 generations) for increased and decreased protein and oil content of corn kernels, starting with an open-pollinated variety of field corn.*

view and placed many of the early geneticists in opposition to the theory of natural selection.

But the mutation theory was based on data containing serious flaws. De Vries was not working with gene mutations, and Johannsen had selected the wrong organism. Beans, like Mendel's peas, are self-pollinating and are noted for their uniformity and constancy; differences due to genes have been eliminated from any variety by the breeders over the years. The variation in seed size that remains does not have a genetic cause.

Later, workers established the true nature of gene mutations and found that many mutations causing conspicuous changes make the organisms in which they appear less able to compete for survival. On the other hand, repetition of Johannsen's experiment with other organisms produced progressive changes over as many as 50 to 100 generations. One example of the type of experiment that did show progressive change with

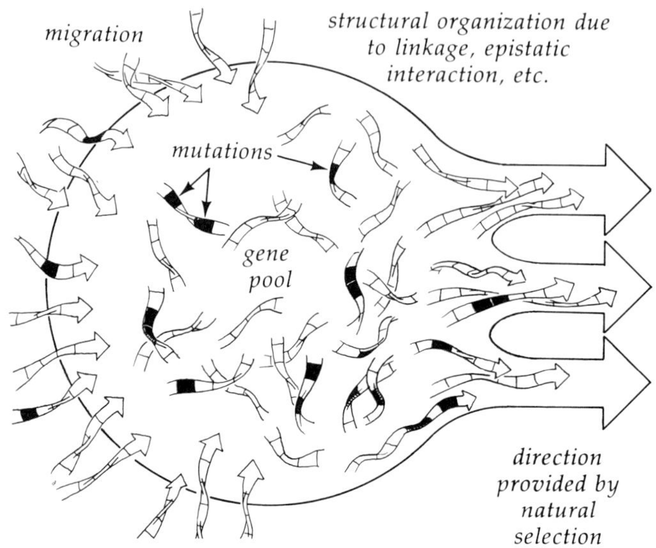

13–2 *A diagram illustrating how the four basic processes—(1) mutation; (2) genetic recombination, which results from intercrossing between individuals of the population as well as between them and the occasional new genotypes that enter by migration; (3) structural changes in the chromosomes, which affect linkage and epistatic interaction of genes; and (4) natural selection—interact to produce a progressive change in the population that keeps it adapted to the changing environment.*

many generations of selection was carried out at the University of Illinois and involved corn. Starting in 1893 field corn was grown for 50 generations, during which time the plants were selected for high and low protein content and high and low oil content of the kernels. The results are shown in Figure 13–1. The data show that the high protein line doubled the amount present in the original stock, while in the low protein line it was halved. The same was true for the oil. Corn is a cross-pollinating species and is highly heterozygous; consequently it gave the results expected. Johannsen had simply picked the wrong experimental organism.

Even while their results were incorrect, De Vries, Johannsen, and their associates were attempting to devise precise experiments to test their hypotheses, carefully interpret their results, and be as rigorous as the state of their science allowed. This tended to narrow their approach, particularly in the eyes of the Darwinian evolutionists. Many of these men, such as David Starr Jordan at Stanford University and H. F. Osborn at Harvard, were essentially naturalists with great interest in describing and classifying organisms and situations. They were the linear descendants of the naturalist side of Darwin and had great sensitivity to the organisms they were observing and to the overall picture of evolution. They also apparently had little interest in or appreciation for experimentation and the interpretation of the results obtained by experimentation. As G. Ledyard Stebbins (see Essay 13–1) of the University of California at Davis, one of the men who eventually synthesized the conflicting views, so aptly puts it: "The Mendelians were looking upward from the genes, the naturalists downward from the phenotypes, and neither had the least conception of the enormous complexity of biological processes which separates the two in higher animals and plants."

Gradually, during the first half of this century, a synthesis of data from both groups became possible as new evidence accumulated and additional minds analyzed it. That the inheritance of quantitatively (rather than qualitatively) varying character is due to the simultaneous action of many separate genes, each having a small effect by itself, was shown by E. M. East at Harvard and H. Nilsson-Ehle in Sweden. Equally important was the mathematical demonstration that evolutionary change does not depend solely on the introduction of new alleles. The important consideration is change in the frequency of all genes (and their alleles) in a population.

The Modern Synthetic Theory of Evolution

The theory that emerged from all of the controversy and the tremendous mass of data gathered from fruit flies, corn, mice, fungi, barley, guinea pigs, and many other plants and animals is that evolution is a result of a number of interacting processes: gene mutation, changes in chromosome structure and number, genetic recombination, natural selection, reproductive isolation, migration, hybridization, and chance.

These processes can be assembled into groups (Fig. 13–2). The first group consists of those processes

---

Essay 13–1 Experiments with the Origin of Species

Can man actually produce new species? There is no question that he can make new varieties by the score, particularly of horticultural or crop plants, but can he make a species that will survive in nature? Here, G. Ledyard Stebbins of the Department of Genetics, University of California, Davis, presents some of his ideas and experiments on this subject. Professor Stebbins is widely recognized as one of the most creative and productive plant geneticists and evolutionists. His book Variation and Evolution in Plants *(New York: Columbia University Press, 1950) is a major contribution in evolutionary thought. This is what he has to say concerning the making of species:*

"During the first part of this century, a major goal of evolutionists was to show how new species can originate by means of actual experiments. This would require the production of a population which is morphologically recognizable and distinctive, is reproductively isolated from all other populations, and which can become established and survive under natural conditions. There are two ways of solving this problem: (1) to re-constitute the existing species, and (2) to produce an entirely new species, that has never existed before, and to establish it in nature.

The first of these goals was achieved by Arne Müntzing while I was a graduate student in 1930 by the artificial synthesis of the allopolyploid Galeopsis tetrahit *from a hybrid between its diploid ancestors,* G. pubescens *and* G. speciosa. *Soon after I read his paper, I resolved to achieve the second objectives if I could. Partial attainment of this goal has been reached by many investigators, who have produced vigorous artificial allopolyploids, and have maintained them in the garden. The successful establishment of such populations in nature, however, has not yet been done.*

In my own research I have come near to success three different times. The first effort began soon after the discovery of colchicine as a method for doubling chromosome numbers. In 1941 I produced an artificial autotetraploid of a grass, Erharta erecta, *a South African native which has become naturalized in the Bay Region, particularly on the UC campus, Berkeley. In 1943 I established twin plots, one planted with seed of diploid* E. erecta *and one containing the artificial autopolyploid, on nine different sites in Strawberry Canyon and the surrounding hills. In two of these the species has survived and spread and in 1971 is far more abundant than it was 28 years ago. In one of these sites, a shady canyon, the plants are nearly all diploid but a few tetraploids can still be found. In the other, which is higher up on the slope, both diploids and tetraploids are abundant. Moreover, they differ from each other in their ecological preferences. The tetraploid thrives best in deep shade and well drained soil; the diploid tolerates partial sun and poorer drainage. So, we now have on this hillside two morphologically distinct populations having different ecological preferences which are completely isolated reproductively from each other. The tetraploid did not occur anywhere until I produced and established it. Can it be regarded as a newly made species or is it just another chromosomal race? I am not willing to answer this question.*

The second example is the segmental allopolyploid produced from the hybrid between woodland wild rye (Elymus glaucus) *and squirrel tail grass* (Sitanion jubatum). *The original diploid hybrid was almost completely sterile; the polyploid had up to 80 per cent fertility. Attempts to establish it in nature were only partly successful. One population lasted from 1952 to 1966 but is now extinct.*

The third example is a 'laboratory microspecies', which originated by hybridization without doubling the chromosome number. In 1941 I conceived the idea that if the sterility of an F_1 hybrid results from interaction between many different and opposite factors that come into it from its two parents, some of the later generation segregates from that hybrid should possess in a homozygous condition new combinations of factors which should make them fertile within themselves but forming partly sterile hybrids when crossed back to either of their original parents. A statistical estimate, explained in Chapter 6 of my book Processes of Organic Evolution, *is that if the parental species differ with respect to n pairs of complementary factors, the proportion of viable gametes produced by the F_1 hybrid would be $1/2^n$, a figure which indicates that if n is 6 or more pairs of factors, the hybrid will have less than 1 per cent fertility. If, however, these factors can recombine at random by independent assortment, $1/2^n$ of the progeny would be fully fertile. Among these fertile genotypes, $1 - 1/2^{(n-1)}$ will be new combinations partly isolated from both of their parents. So, if $n = 6$ or more, fertile derivatives will be very rare, but once they appear, they have a strong likelihood of forming partly sterile hybrids when back crossed to either parent and so becoming the initial stock of a new species.*

An opportunity for testing this hypothesis came in 1948 when I found that almost all hybrids between allopatric populations of Elymus glaucus *are partly or completely sterile. This taxonomic 'species' is actually an aggregate of scores or hundreds of morphologically recognizable, reproductively isolated microspecies or sibling species, two, three or even up to eight of which may occur in the same population and nevertheless remain recognizably distinct from each other. I found that with respect to seedlings and vegetative characters, some of these microspecies resembled populations of* Sitanion *more than they did other populations of* E. glaucus. *I suspected, therefore, that their number has been increased by natural backcrossing of the highly sterile F_1 hybrid to* Elymus, *followed by segregation of fertile, reproductively isolated genotypes containing new combinations of the sterility factors. This hypothesis was tested by placing 20 clonal divisions of a hybrid plant in the garden, interplanted with 20 clones of fertile* E. glaucus *so that the hybrid stigmas would be abundantly covered with* Elymus *pollen. From the hybrid clones 146,000 florets were harvested from which 4 back cross plants resulted. Three of these were as sterile as the F_1; the other was 10% fertile, and was morphologically unlike any plant previously seen in the garden, since it was intermediate between the F_1 and* E. glaucus. *In the second generation of selfing from this plant, a vigorous, fertile progeny was produced that had all of the morphological characteristics of* E. glaucus. *Nevertheless, when crossed back to its original* E. glaucus *parent, it produced a sterile hybrid. I had, in fact, produced a new microspecies of the* E. glaucus *complex by backcrossing the* Elymus-Sitanion *hybrid to its* Elymus *parent.*

The final test of establishing this new microspecies under natural conditions was never attempted. So perhaps, this also should be regarded as a near miss rather than the bull's eye. I haven't given up yet and may be able eventually to report complete success. But time is running out and there are lots of other interesting things to do."

---

that provide genetic variability: gene mutation, changes in chromosome structure and number, and genetic recombination. Without genetic variability, evolution cannot take place. Darwin understood this without understanding the basis of variability.

The second group consists of those processes that

guide populations of organisms into adaptive channels: natural selection and reproductive isolation. Natural selection, the process defined by Darwin, has already been discussed. Reproductive isolation is the separation of populations by barriers to gene exchange so that they differentiate into distinct species; it was also recognized by Darwin. We shall discuss reproductive isolation later, since it is a central idea in the current concept of evolution.

The remaining three processes—migration, hybridization and chance—are really accessory processes to the first five. They act to influence the workings of the principal groups. Migration, for example, of individuals from one population to another increases the amount of genetic variability available to a population. Hybridization between subpopulations of a species or closely related species has the same effect. Finally, the effects of chance are felt only in small populations where it may alter the direction of natural selections.

An analogy of the interrelationships of these processes in evolution is proposed by Stebbins in which he likens an evolutionary line of organisms to an automobile being driven on the highway. The gasoline is mutation that, as the sole source of new genetic variation, is essential but is not the immediate source of driving power. The motor is genetic recombination that, through the rearrangement of genes and chromosomes during sexual reproduction, provides the immediate source of variability upon which selection acts. The driver is natural selection directing genetic variability toward adaptation to the environment. The highway is reproductive isolation with its limits and direction indicators exerting control on the driver and permitting several automobiles to drive in the same direction at the same time.

The Gene Pool

To understand the role of the genes in evolution it is necessary to think not of the individual but the population. In this regard the concept of the gene pool of a population is important. The gene pool is the total of all the alleles in a population. Since many genes may have many different alleles, in a cross-breeding population (one in which the individuals are not self-pollinating or self-fertilizing) many allele pairs in an organism may be heterozygous. Hence individuals in such a population may be heterozygous for a large number of genes. Variation is introduced into the gene

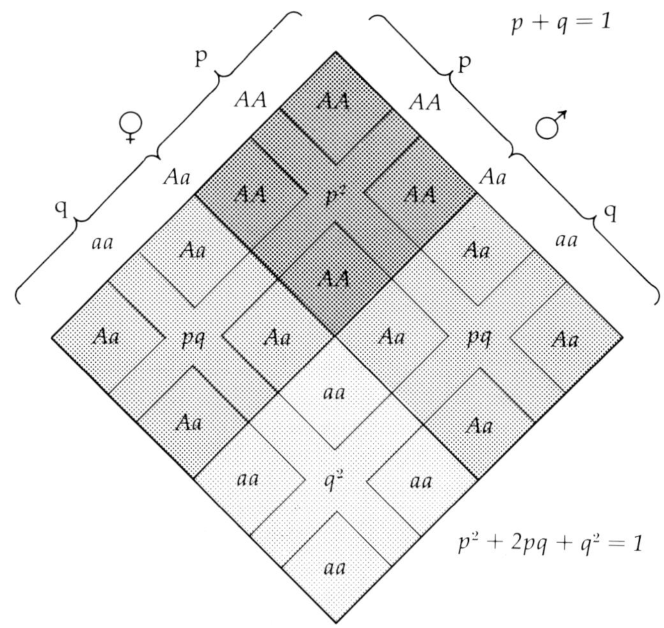

13–3 Possible combinations of gametes in a population consisting of AA, Aa, aa individuals, which illustrates the Hardy-Weinberg equilibrium principle.

pool by mutation and the recombination of genes on the chromosomes, both discussed in Chapter 12.

The effect of this high degree of heterozygosity is that there is always a large number of recessive genes and gene combinations that are only apparent when inbreeding or selection occurs. Recessive genes show only when they are in the homozygous condition and are not visible in the heterozygous condition.

Why aren't the recessive genes eventually eliminated from the gene pool by selection in favor of the dominant gene? The answer is that in a large, isolated, random-breeding population the proportion of genes will remain the same, generation after generation provided that there is no mutation and no selective force is applied. This was first appreciated, independently, in 1908 by the English mathematician G. H. Hardy and by G. Weinberg, a German physician. They presented an algebraic formula in which the frequencies of the alleles A and a are represented by the symbols p and q. Then the genotypes AA, Aa, and aa will occur in the frequencies p^2, $2pq$, and q^2 (Fig. 13–3). The basic importance of the Hardy-Weinberg binomial expansion is simply that it shows (under conditions described) that gene frequencies will remain the same (regardless of how high or how low) from generation to generation. But, for the Hardy-Weinberg law to apply and for the equilibrium in gene frequencies to exist, a number of conditions must be met.

The population must be large enough. In small populations any number of chance events can upset the equilibrium by removing individual genotypes from the population. This will not happen if the population is large enough. In applying the Hardy-Weinberg Law to actual situations, the population must be large enough so that sampling errors can be disregarded. Changes in small populations due to chance are called genetic drift.

Either there must be no mutation or the rate of mutation of a specific allele must be matched by a reverse mutation. Clearly, a change from A to a or conversely will alter the values of p and q.

There must be no selective mating. If there is some mechanism that ensures that only a part of the population will mate, the equilibrium will be altered.

There must be no migration of individuals either into or out of the population. This might also alter A and a.

Reproduction must be purely random. There must be no selection. Neither allele can have a differential effect on the reproductive capacity of individuals bearing it. If the three genotypes AA, Aa, and aa could somehow be identified in a real population, a comparison of their actual frequency and that predicted by the Hardy-Weinberg formula can be used to tell us about the forces acting to change the frequency of genes in a population. For example, in a population in which there was a great deal of inbreeding, the two homozygotes (AA and aa) would be more. On the other hand, if the heterozygote is higher than expected, it would mean it had features that made it better adapted than the homozygotes. As we saw, one theory of hybrid vigor is based on the superiority of heterozygotes (see End Paper 12–1).

Many examples of clearly superior heterozygotes are now known. Through the heterozygotes, the population maintains an extra supply of genetic diversity, allowing it to respond to changes in the environment. But a side effect is that deleterious recessive genes, even if they are lethal, are retained in the population. The lower the frequency of the gene in the population the less the possibility of its elimination by natural selection, since the chance of a cross resulting in a homozygous condition becomes correspondingly small.

Natural Selection and Populations

The Hardy-Weinberg Law made it clear that under constant conditions, there is an equilibrium in the genetic composition of a population. Evolution can occur only when this equilibrium is upset by a selection mechanism that results in the nonrandom reproduction of genotypes. This was seen by the British statistician R. A. Fisher and by Sewall Wright, a geneticist at the University of Chicago. Wright, a brilliant mathematical biologist, said that evolution is basically a statistical transformation of a population.

Some of the earliest controlled experiments on populations were done using the small, fast-growing fruitfly. These characteristics made it possible to grow many thousands of flies through many generations in relatively short periods of time.

Even easier to handle are the bacteria. These organisms have generation times in terms of hours, and being haploid rather than diploid, mutations are immediately apparent and not masked by the presence of a dominant allele. One experiment using bacteria was performed by Lederberg, whom we met in our discussion of the molecular side of genetics. Lederberg used the antibiotic streptomycin as a selection factor. Most colonies of bacteria grown on media containing streptomycin were killed, but a few survived. They had a genotype that made them resistant to the antibiotic. Lederberg, through an ingenious technique (Fig. 13–4), was able to show that the streptomycin selected for a gene mutation that was already present in the bacteria before they were ever exposed to the antibiotic. The streptomycin did not cause the gene mutation. Rather, the gene was there, and the streptomycin effectively altered its frequency in the population.

Exactly the same kind of effect has been shown in natural populations with regard to DDT and insects. An increased use of DDT has decreased its effectiveness against insects, since it acts as a selective factor in favor of resistant populations of insects.

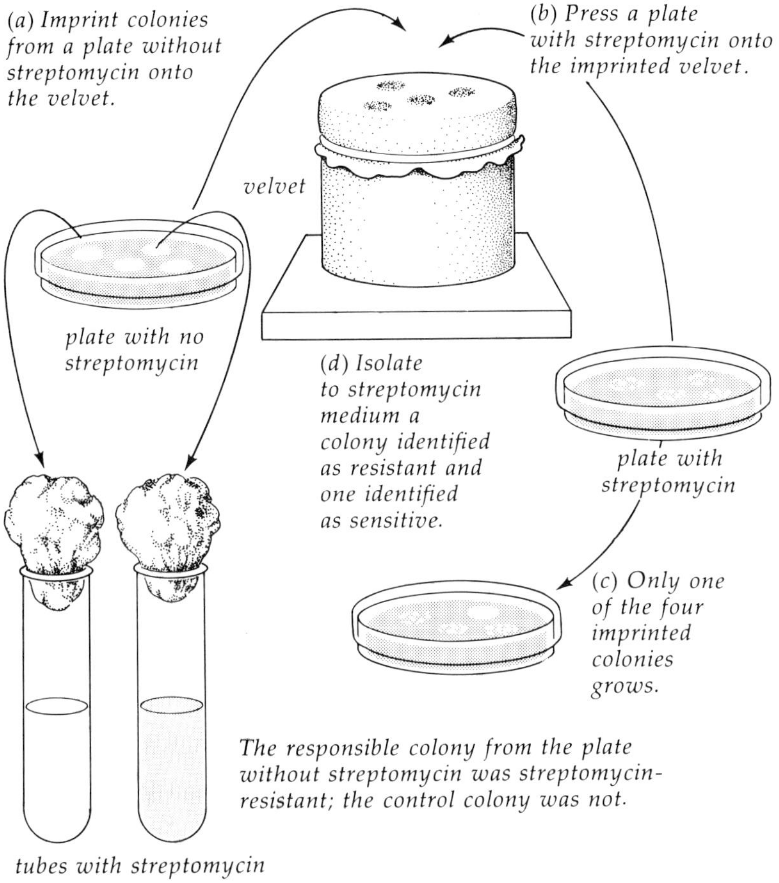

13–4 The fact that streptomycin resistance can occur quite independently of exposure of the bacteria to the drug was ingeniously illustrated by the duplicate-plating method developed by Joshua Lederberg. First, colonies of bacteria were grown in a petri dish on a medium lacking the drug. Second, some individuals of each colony were transferred to a new petri dish containing a medium in which streptomycin was present. The position of the colonies in both dishes was exactly the same because they were transferred through their adherence to a round piece of velvet, which exactly matched the dish. On the streptomycin plate, only the resistant colonies grew. Then, because the positions of the colonies on the + and − streptomycin plates were known to be the same, it was possible to show that the original colonies that had never been exposed to the antibiotic were, in fact, streptomycin resistant without ever having been in the presence of the drug. Thus the resistant colonies had grown from a single resistant bacterial cell, which had mutated to resistance before the streptomycin was applied.

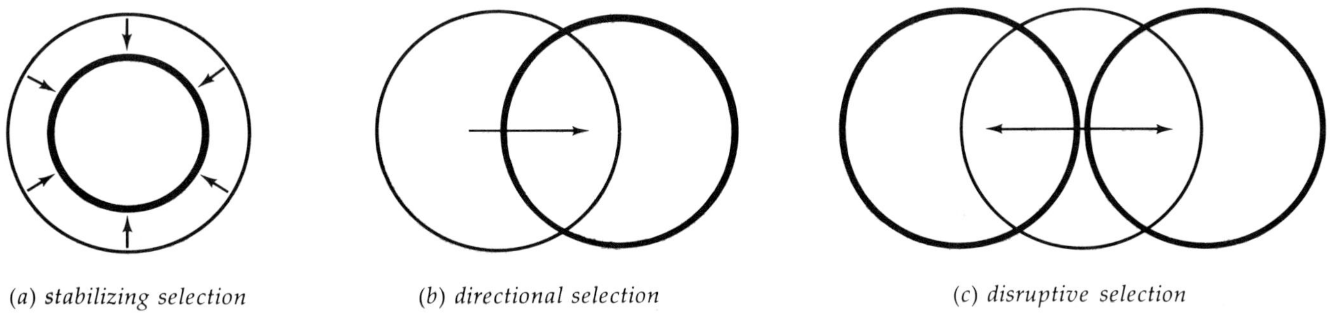

13–5 The three different types of natural selection.

As the result of treatment of a population of bacteria with streptomycin or houseflies with DDT, the susceptible individuals die, and the progeny result from the surviving resistant individuals. Clearly, under these conditions some genes will become more frequent in the population and others rarer.

Selection acts on the phenotype, which is determined by numbers and sets of interacting genes. Our discussion in Chapter 12 stressed the complex relationship of many genes in the final expression of a single character in the individual. The many alleles of these interrelated genes form a web that makes it difficult if not impossible, in most cases, to relate selection to a single gene or even a single set of genes.

Three types of selection have been recognized.

Stabilizing selection means the population is essentially in equilibrium with its environment (Fig. 13–5a). If this environment is relatively constant over a period of time, selection acts to eliminate those individuals that differ significantly from well-adapted genotypes. Selection is acting to minimize the disruptive influence of mutation and genetic segregation in a stable environment. A stable environment does not imply no variations in the environment but only relatively stable conditions over a fairly long period.

Directional selection represents the core of Darwin's theory of natural selection (Fig. 13–5b). It is selection that results in a regular change of the population in one direction with respect to certain adaptive characteristics. This type of selection takes place when the environment is changing progressively in a particular direction. The streptomycin changed the environment in a particular direction, and the resistant genotype was favored over the formerly well-adapted ancestral genotype.

Disruptive selection is essentially the reverse of stabilizing selection. Here the extreme genotypes are favored over the intermediate forms (Fig. 13–5c). The result of disruptive selection is the breaking up of a previously homogeneous population into several distinctive groups, each subject to different selection pressures in different parts of the environment.

Reproductive Isolation and Evolution

So far we have been examining how populations of organisms can be altered into various types of subpopulations as selection acts on the frequency of genes in the gene pool. But how does a subpopulation become a species? To understand the answer we must first discuss what we mean by species.

The definition of species is not easy, since it is a diffuse concept rather than a precise entity. Moreover, its meaning has changed over the years, and there are different problems in defining species of animals and plants. The original definition of species was morphological: *groups of organisms in which the individuals closely resembled one another but had clear-cut and significant differences in form from other related groups.* For the vast majority of plants and animals, this is still the principal means by which species are recognized.

In 1937 Theodosius Dobzhansky, a professor of zoology at Columbia University, suggested an objective definition of the species: ". . . *groups of actually or potentially interbreeding natural populations, which are reproductively isolated from other such groups.*" Ernst Mayr, a zoologist at the American Museum of Natural History in New York, elaborated on this concept in a book and several papers published between 1942 and 1963. According to him a species is defined by only one criterion: reproductive isolation. If organisms do or can interbreed, then they are members of a single species; if they do not or cannot, they are separate species. This is called the biological species concept. The old approach of defining species according to their morphological differences is referred to as the morphological species concept. It is assumed, of course, that differences accounting for morphological species reflect the reproductive isolation defining biological species. Dobzhansky, Mayr, and others were able to show that the biological species approach works quite successfully with animals. Mayr, an ornithologist, was able to apply the concept extremely well with birds—but then there are only about 8600 species of birds in the entire world.

This concept of species makes it possible to understand why some groups of organisms, which appear quite different, are termed races or subspecies, while other groups that look quite similar are designated as species. Races and subspecies are populations within a species, which for one reason or another remained isolated from one another so that their genotypes and the resulting phenotype are distinct. Yet they are still capable of complete interbreeding when the barriers are removed, and if this is done either naturally or experimentally, all difference will be lost with time. Man has made much use of the development of races through artificial separation of populations of domestic animals and plants. Dogs are a conspicuous example, but all other animals domesticated by man show the same division into races, as does man himself. The same is true of domestic plants, as is evident from the large numbers of different kinds of roses, chrysanthemums or food crop plants.

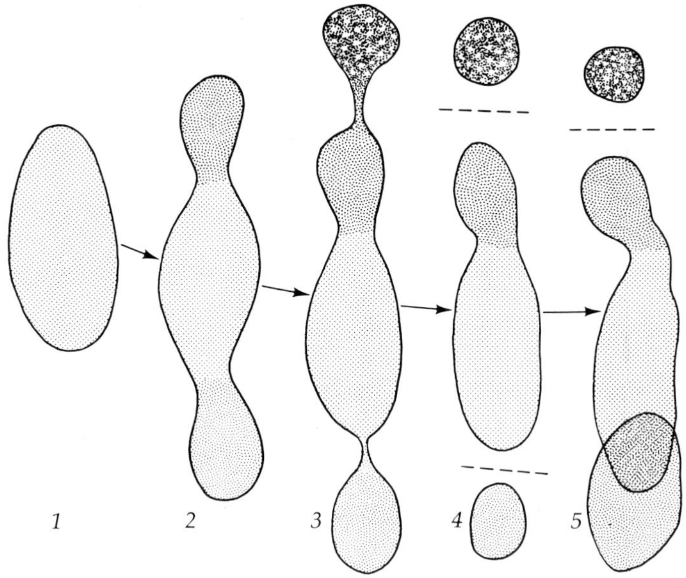

13–6 A diagram showing the sequence of events that leads to the production of different races, subspecies, and species, starting with a homogeneously similar group of populations.

The other extreme found in nature are some species of plants that are almost identical in appearance but because of one mechanism or another cannot mate. The pollen of one species may not germinate on the stigma of the other; so no seed can be set. In such sibling species the gene pools of the two species are prevented from merging, and the species remain distinct while growing in the same geographic region.

The barriers resulting in reproductive isolation may be either physical or biological and may be interrelated. How such barriers function in the formation of new races and species can be seen in Figure 13–6. In the first stage a single, relatively homogeneous population possessing a large gene pool is living in a reasonably stable environment. During the second stage there are changes in the environment such that significant differences exist within the range of the population. There is a migration of parts of the population to these new environments and the formation of races and subspecies. A further change in the environment results in more migration and the geographic isolation of some of the races or subspecies. By stage four some, but not all, of the isolated subspecies have developed (through either genetic or chromosomal alterations) changes that

13–7 A summary of the most important isolating mechanisms that separate species of organisms

Prezygotic mechanisms: Prevent fertilization and zygote formation.

Habitat: The populations live in the same regions but occupy different habitats.

Seasonal or temporal: The populations exist in the same regions but are sexually mature at different times.

Ethological (only in animals): The populations are isolated by different and incompatible behavior before mating.

Mechanical: Cross-pollination is prevented or restricted by differences in reproductive structures (genitalia in animals, flowers in plants).

Postzygotic mechanisms: Fertilization takes place, and hybrid zygotes are formed. But these are inviable or give rise to weak or sterile hybrids.

Hybrid inviability or weakness.

Developmental hybrid sterility: Hybrids are sterile because gonads develop abnormally, or meiosis breaks down before it is completed.

Segregational hybrid sterility: Hybrids are sterile because of abnormal segregation to the gametes of whole chromosomes, chromosome segments, or combinations of genes.

F_2 Breakdown: F_1 hybrids are normal, vigorous, and fertile. But F_2 contains many weak or sterile individuals.

control reproductive isolating mechanisms. Finally, by the fifth stage, environmental modifications permit formerly isolated populations to coexist again in the same region. Because of the reproductive isolating mechanism that separates them now, they can be recognized as valid species.

The various types of isolating mechanisms summarized by Stebbins are presented in Table 13–7, and a few examples are discussed here.

At step 5 (Fig. 13–6), we now have three separate species, the original parent species, one new species that shares a common environment with the parent species, and one new species that is geographically isolated and shares no common range with either species. Variations on this pattern are common in nature. The species occupying the common area are called sympatric species, while those that inhabit different areas are called allopatric species.

One type of mechanism is simple spatial separation. This may mean that the species are separated by

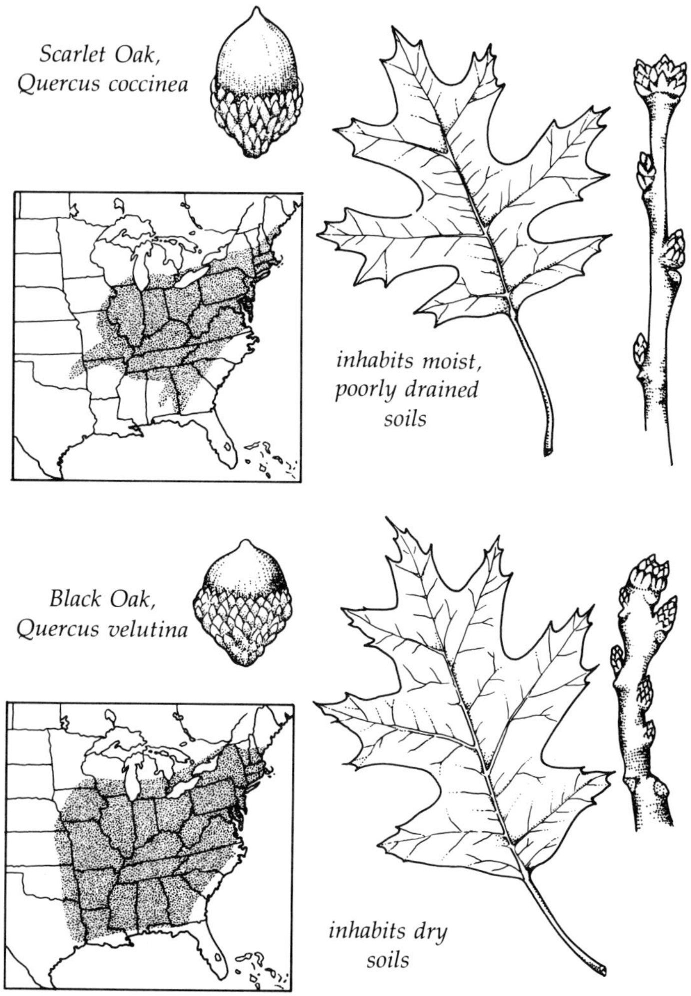

13–8 Two species of oaks isolated reproductively largely by differences in habitat requirements. Top: scarlet oak (Quercus coccinea) inhabits poorly drained soils. Bottom : black oak (Quercus velutina) lives in dry sites.

large masses of land and water, as are the different magnolia species in eastern North America and Asia. Another example of spatial separation, on a different order of magnitude, is the scarlet oak (*Quercus coccinea*) and the black oak (*Quercus velutina*). Both occur throughout the eastern United States (Fig. 13–8), but the scarlet oak is adapted to swamps or poorly drained bottom lands, while the black oak is adapted to drier, well-drained upland soils. Thus they are rarely close enough to cross-pollinate. In intermediate habitats hybrids may form, and the distinctions between species break down.

Isolation mechanisms may also involve such factors as the time that pollen is released. In one species of pine growing in central and southern California the shedding of the pollen begins in late winter, while the second species growing in the north sheds pollen in April. In some places the two species grow side by side but, because they retain their characteristic times of pollen shedding, do not cross-pollinate. Variation in the way pollination occurs is frequently involved in reproductive isolation. For example, in the case of a species called the scarlet monkeyflower (*Mimulus cardinalis*), hummingbirds are attracted to the bright red petals and collect the nectar. Both the anthers and the stigma are so situated that the forehead of the bird becomes covered with pollen, which is then transferred to other stigmas. In a closely related species, the Lewis monkeyflower (*Mimulus lewisii*), the petals are lavender and the stigma and anthers located within a tube composed of flower parts.

Bees are attracted to these flowers and must enter them to collect the nectar. In so doing the bees become covered with pollen that they carry to other plants. Yet, even if hummingbirds did collect nectar from *M. lewisii* —and they occasionally do—the position of the anthers and stigma prevents pollen from being rubbed off on their bodies. The same is true for the scarlet monkeyflower in relation to bees. Thus hybridization is avoided, even though hybrids can be produced if the two species are artificially crossed. In such hybrids the anthers and stigmas are so placed that neither bees nor birds can effect pollination, and although the hybrids are fertile, they cannot reproduce in nature because no pollination mechanism exists for flowers of this form.

Other species of flowering plants are prevented from producing hybrids by the fact that the pollen of one species either will not germinate or will die in the flower tissue before reaching the egg of the other species. This is an even more subtle isolating mechanism; yet it is found in a large number of plants.

One of the major reasons why species remain distinct and isolated is because they cannot produce fertile hybrids. When we discussed meiosis, we observed that during meiosis I the homologous chromosomes paired and that at anaphase of meiosis I the homologous chromosomes separated and moved to opposite poles. Thus an even and orderly reduction in the number of chromosomes took place. The pairing between the homologous chromosomes is exact. We also saw in Chapter 12 that the structure of chromosomes can be altered in a number of ways. With time the chromosomes of two species derived from a common parent species can change to where they can no longer pair at meiosis I. The unpaired chromosomes cannot line up at the metaphase plate, and they are distributed irregularly to the poles (Fig. 13–9). The resulting imbalance in chromosome complement prevents the cells from developing normally.

If the changes in the chromosomes are not as extensive, some pairing may occur along part of the length of the chromosomes. The degree of pairing can be used as a measure of the closeness of two species—the more pairing the more closely related the two species. Incomplete pairing leads to chromosome imbalance and cell death.

There are other bases for hybrid failure besides the inability to produce fertile offspring. For largely unknown reasons hybrids show what is called genic disharmony. Development is particularly affected by genic disharmony that shows up at the division of the zygote, at the early stages of embryo development, at germination or during seedling growth, and during the formation of eggs and pollen.

Obviously, the origin of reproductive isolating mechanisms is of great importance in understanding evolution. This origin is still only partially known. But we do know that each type of isolating mechanism is under the control of many genes and that these originated through mutation, genetic recombination, and natural selection of particular kinds of differences.

Polyploidy and Species Formation

In general, natural hybrids (crosses between members of populations with different adaptive gene complexes) are less well adapted to the environment than either parent. But when a new environment becomes established, the hybrid may be better adapted than either parent because its combined gene pool is derived from each parent. Yet, as we just saw, the hybrids are frequently sterile or partially sterile as a result of failure of chromosome pairing and separation during meiosis.

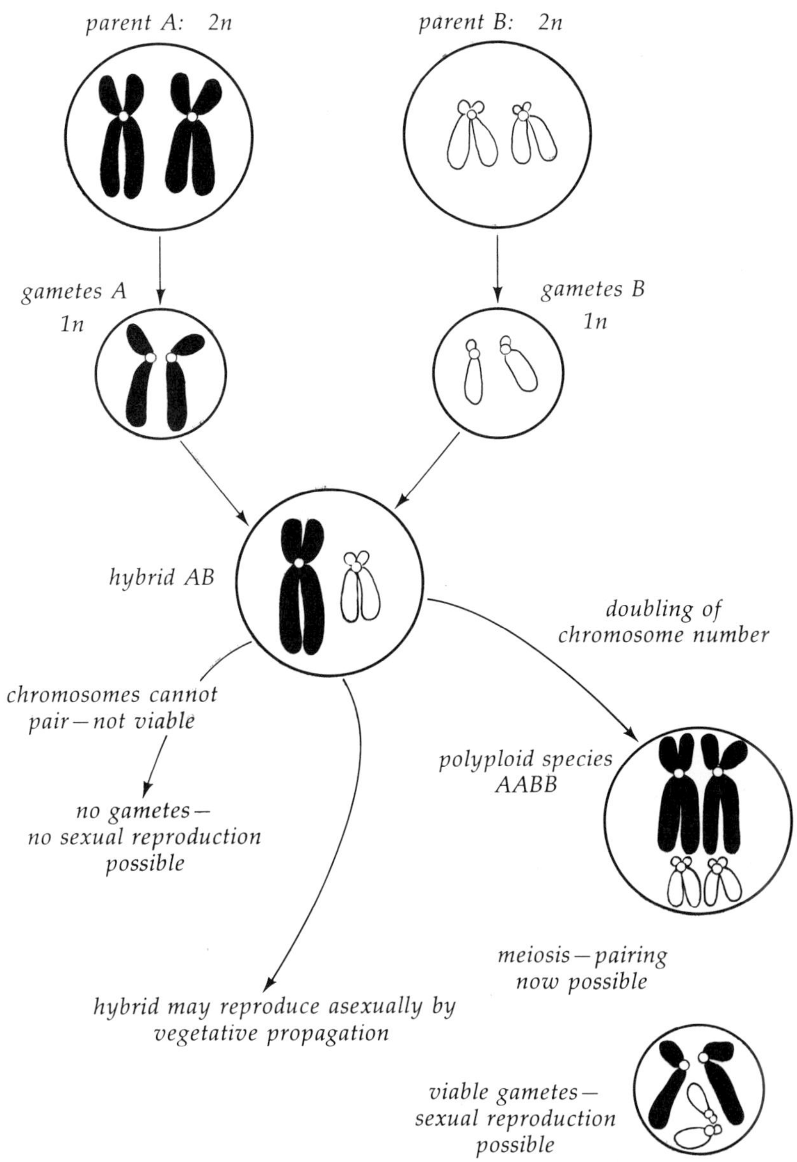

13–9 Chromosome behavior during the formation of a hybrid and a polyploid, which results from doubling the chromosome number of the hybrid. Note that the hybrid cannot produce viable gametes because the chromosomes lack homologous pairs but that the polyploid can because such chromosomes are now present.

A means of circumventing the whole problem of chromosome pairing in hybrids is to double the entire complement of chromosomes present. Automatically,

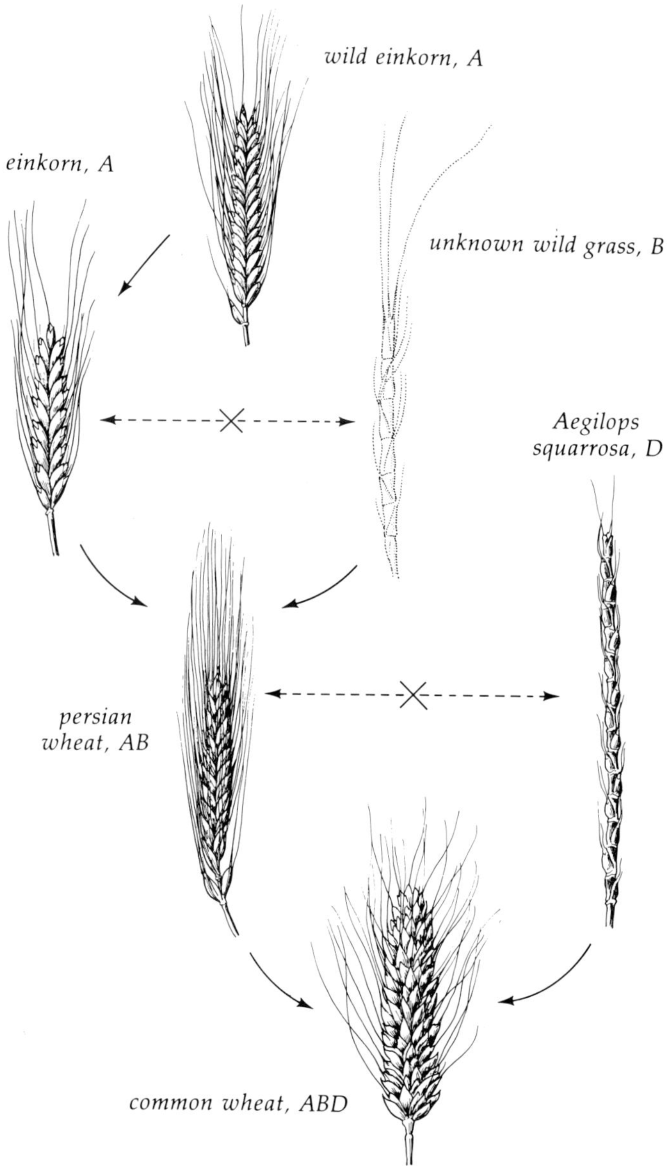

13–10 *The evolution of new species by polyploidy is clearly illustrated in the genetic history of wheat.*

all the chromosomes are present as pairs, and there is no difficulty during meiosis. Consider a case of two species, one with chromosome complement AA and the other with BB, that form a hybrid whose chromosome complement would be AB (Fig. 13–9). The chromosomes of complement A are so different from the chromosomes of complement B that meiosis in the hybrid is not normal, and the hybrid is sterile. Now, doubling occurs; the hybrid chromosome complement is AABB; meiosis is normal; and complete fertility is established. This is called polyploidy. The polyploid AABB is true-breeding and also isolated from the parental species because the offspring of a cross with either parent are sterile.

Wheat is a good example of the evolution of new species by polyploidy (Fig. 13–10). This important crop was one of the first brought into cultivation, with grains dating from 7000 B.C. coming from eastern Iraq. Several wild wheat species with 14 chromosomes (two sets of 7 designated AA) are found growing in that area. Also growing in the area is a wild grass with 14 chromosomes different from A, as judged by failure of the chromosomes of set A to pair with those of the grass; so these are designated B (BB). Several of the common cultivated wheats have a chromosome number of 28, and careful analyses of meiosis show that these wheats are AABB. Further combinations are known, since another common cultivated wheat has 42 chromosomes (6 sets of 7 chromosomes) that can be represented AABBDD—the last chromosome set coming from another wild grass. These wheats arose, as far as is known, from hybridization without man's direct intervention, but man did act as an agent of selection of the new species that were of greatest use to him.

Many new species have evolved through hybridization followed by chromosome doubling. But polyploid species appear to contribute little to progressive evolution, being merely variations on an established line of evolution. The large number of genes in the polyploid probably acts to dilute out the effects of new mutations and genetic recombinations.

Species and Systematics

Earlier in this chapter we presented two quite different definitions of species. The classical definition relies on the concept of morphological similarities, while the so-called "biological species" is based on the criterion of reproductive isolation.

The second definition is more precise than the first, but botanists have generally been unwilling to accept it. The reason is that strict application of the principle appears to go against strong intuitive feelings regarding the nature of many long recognized species. For example, *Platanus orientalis* is the sycamore of the Middle East, while *Platanus occidentalis* is the sycamore native to eastern North America (Fig. 13–11). Both trees are quite distinct from each other morphologically, and they have been isolated from each other probably for a million years or more (since the end of the Tertiary Period). Yet they readily produce fertile hybrids, the widely cultivated *Platanus acerifolia* (maple-leaf sycamore). Strict application of the biological species concept would say that these must be the same species, but botanists are not willing to go along with the idea. Of course, one might say that they had no potential to interbreed *in nature*—unless man is admitted to be such a part of nature that his interbreeding the two groups may be thought of as a natural act.

The problem is even more difficult when one encounters groups reproductively isolated in nature that are infertile experimentally but morphologically identical. The botanists would like to call these one species. But if we say, as we would have to in the sycamore example, that they have no natural potential to interbreed, then we would have to refer to them as individual species just because they are isolated geographically. *Larrea divaricata* (creosote bush) occurs in both North and South America, isolated by tropical Central America. It is interfertile and morphologically similar. Those who believe it should be two species anyway use the name *Larrea tridentata* for the North American form.

There are numerous species that are separated in their distribution into isolated groups, often with hundreds of miles between different groups. *Larix lyallii* (subalpine larch) growing in the Cascades of Washington and the Rockies of Idaho, Montana, and Alberta, Canada, is one example. Of course, it is not inconceivable that windblown pollen might cross the hundreds of miles of desert separating these different groups, but certainly not much genetic material is exchanged between them. They are essentially reproductively isolated. Does this mean that the ones in the Cascades should be thought of as a different species from the ones in the Rockies? Strict application of Dobzhansky's definition might say yes, but the botanical taxonomist is not enamored with the idea.

There are examples in which species appear to reproduce primarily by clones. Aspens in the Rocky Mountains occur as large clones that seem to be reproductively isolated from each other. Is each clone a species? Even more pertinent to the problem is the phenomenon of apomixis, in which plants reproduce by seeds but *without fertilization*. In many groups a

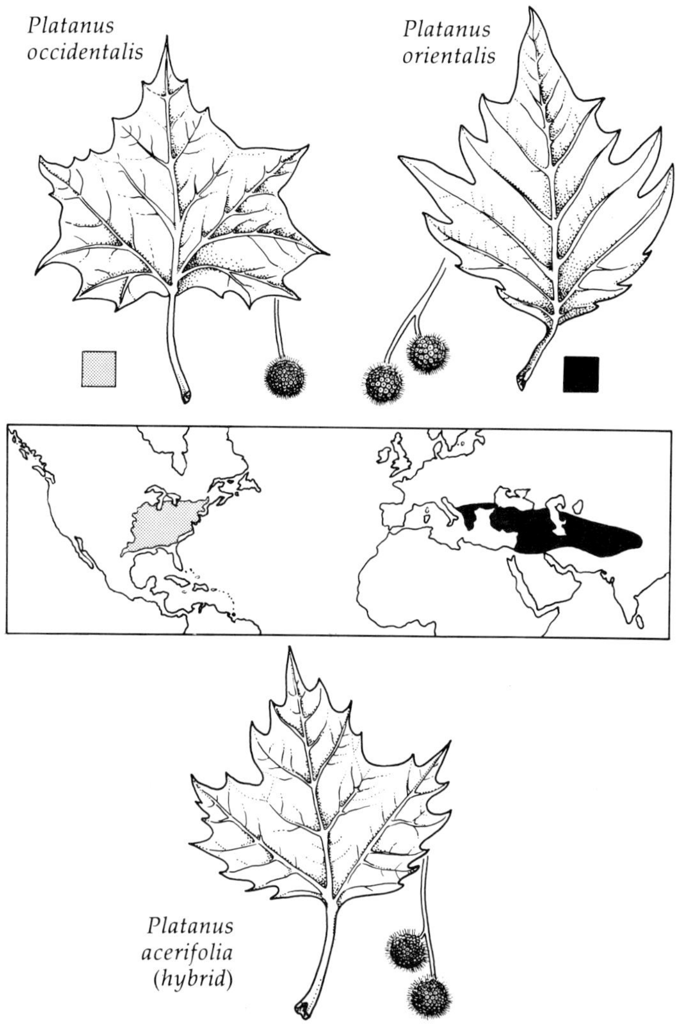

13–11 The native distribution of two species of sycamore (Platanus occidentalis and Platanus orientalis) and a widely cultivated fertile hybrid (Platanus acerifolia).

cell in the ovule may grow like a zygote, producing an embryo with the identical genetic makeup of the parent plant. There are several ways this can occur, but the end result is the same. In some species apomixis only occurs in the absence of fertilization; otherwise reproduction occurs normally. Still others exhibit obligate apomixis, the normal fertilization procedure being totally lacking. The phenomenon is rather widespread in the plant kingdom, including many grasses, composites, etc. Even the dandelion exhibits apomixis!

The point of all this is that plants reproducing by obligate apomixis are reproductively isolated from each other. Specimens of *Poa sandbergi* (a bluegrass) east of the Wellsville Mountains near Logan, Utah, appear slightly different from their counterparts over the mountain 8 or 10 miles to the west near Brigham City. And well they might, since they reproduce apomictically and have not exchanged genes for a long time; hence they are reproductively isolated. But are they separate species? Few taxonomists would be willing to admit this.

On the other hand, there are many groups such as oaks, birches, and willows in which many species freely interbreed in nature. If all species of birch (*Betula*) interbreed, then applying the biological species concept, there is really only one species of birch, and the rest are subspecies. The genus *Quercus* (the oak) might be divided into two or three species, consisting of the large number of present species that interbreed freely with each other at their borders. This also seems too radical for most taxonomists—it is lumping carried to an extreme. It might be, however, that even in such cases the biological species approach could have some merit. At least in these latter cases, we could apply the biological species concept if we wanted to, and by so doing we would be basing our species definition upon an objective statement rather than an indefinable, subjective intuition.

At the moment we do not know exactly what a species is. But this is for a good reason: species are not static. They change with time and in response to environmental alterations and selection pressures. Some may be in the process of splitting into more species; it is simply difficult to know what to do in such cases.

Emphasis in taxonomic studies has now shifted from a collection of type species (practiced by early taxonomists) to a study of *populations*. Application of the biological species concept certainly implies population interactions, which are studied using genetic and evolutionary theory. The attempt is not so much to study the species as an unchanging taxonomic unit (establishing type specimens, etc.) but to study the dynamic processes of speciation. This is one of the outstanding contributions of modern taxonomy.

Classification of Plants

The results of past evolution are the organisms that we see all around us. These plants and animals are present in a fantastic array of sizes, shapes, and colors. This diversity is such a part of our world that we frequently overlook it or take it for granted. Yet, as we saw in Chapter 1, even a small piece of ground can reveal a surprising number of different plants and animals. But this variety of life, a product of millions of years of evolution, poses problems in basic communication.

How do you communicate about hundreds and hundreds of plants, some of which no one has ever seen before? How do you arrange the known plants so that you can study them systematically? The quest for a system to use in classifying plants led to the work of Carolus Linnaeus (1707–1778), the Swedish biologist-naturalist (see Essay 13–2). At the heart of his scheme was the concept of species. To Linnaeus, a species was a group of plants with distinct characters that distinguish it from all other groups of plants. This is the classical definition discussed earlier. The value of the species concept is enhanced by the concept of the genus—a collection of closely related species. If you want to name a plant, you must use both the generic and specific name.

This is precisely what Linnaeus proposed: all plant and animal species should be designated by two Latin words. The first (the first letter of which is always capitalized) is the genus (generic) name, while the second (which is always lower case) is the species (specific) designation. Thus, if you want to talk about the white oak, the name you use is *Quercus alba*. If, instead, you intend to discuss the black oak, the name you use is *Quercus velutina*. Simply by seeing the names you know that the two species are closely related, since they both belong to the same genus, but have some independent, distinctive characters, since they belong to two species.

Why use Latin names? At first they were employed because Latin was the universal language of educated communication. Latin still held an important place in the 1700s when Linnaeus proposed his system of classification. Today, few biologists can read or write Latin; yet the formal or scientific name is still always latinized, and names of new species are latinized. Frequently the Latin is not good Latin, and many strange genus and species names are made by latinizing the last name of a famous botanist or a friend of the discoverer. Yet, odd as it may seem, Latin is used to reduce the confusion caused by common names. At first this seems contrary to common sense: how can *Zea mays* be less ambiguous than corn? If you are an American using the name corn in the United States, everything is fine. If, however, you were in England and talked about corn, your listeners would think you were discussing wheat, oats, and similar grains. In Holland corn is called Turkish wheat, and in parts of Africa it is mealies. If you were in Japan, you would not be understood at all. There are hundreds of similar examples where a common name either means different plants to different people or is meaningless.

There is a further complication with common names—many plants simply don't have them! Hundreds and thousands of fungi and algae were unknown before the microscope and therefore have no common names. Plants from distant lands frequently lack names that are suitable in another language. Less conspicuous plants everywhere and plants from unusual locations may never have been seen by man before. Finally, there are those groups of species that may be so similar that they share a common name but are actually distinct species (e.g., "cotton").

Essay 13–2 Linnaeus and the Origins of Systematics

Carl von Linné, as he was born in 1707 in southern Sweden, or Carolus Linnaeus, as he became in 1761 by act of the King of Sweden, is a man all too frequently credited with too little and too much.

Linnaeus, who has been called a phenomenon rather than a man, was a poor pastor's son who narrowly missed going to college. He eventually entered the University of Uppsala, Sweden, where he nearly starved before his interest and abilities in botanical research led one of the professors to take him into his home. In 1732, while still a student, he spent five months exploring Lapland for the Academy of Sciences of Sweden. He went to Holland to finish his education as a physician and while there finished the first edition of his well-known work, Systema Naturae.

After receiving his M.D. in 1735, he traveled to England and France on his way back to Sweden. Everywhere he went his charm and intelligence made him a host of enthusiastic friends. An example is his meeting with John Jacob Dillenius, a Swedish botanist who had long lived in England and was Professor of Botany at Oxford, as recorded in the diary of Peter Collinson:

"The learned Botanist, Dillenius, was at first haughty; conceiving Linnaeus' Genera, *which he had got half printed from Holland, to be written against him; but he afterwards detained him a month, without leaving Linnaeus an hour to himself the whole day long; and at last took leave of him with tears in his eyes, after having given him the choice of living with him till his death, as the salary of the Professorship was sufficient for them both."*

But return to Sweden he did—to the practice of medicine. In 1742, at the age of 35, he was made Professor of Botany at the University of Uppsala, a post he held until his death in 1778 although through many of the early years of his professorship he continued to practice medicine.

What did Linnaeus contribute to biology? It is almost easier to answer what he did not. He is usually cited as the founder of the system of binomial nomenclature (all organisms known by two Latin names—the genus and the species); yet others had done this before him. Frequently, Linnaeus is said to have been the first to systematically organize plants and animals, but in 1583 Andreas Caesalpinus of Florence placed 1520 plants into 15 classes on the bases of their seeds and fruits. Later, in 1700, John Ray, an Englishman, outlined a reasonable scheme to classify plants, and his French contemporary Joseph de Tournefort described some 8000 plants and placed them into 22 classes. Linnaeus' own system, while functional, was artificial and is no longer used.

What then did Linnaeus do? He acted as a focal point, as a giant magnet that attracted men and specimens. He constantly published descriptions of thousands of plants, animals, minerals, and diseases. Like a Norsk god he rained down names on the living world around him—a rain the humans eagerly absorbed. Linnaeus was a man of his times, and the time was one when the educated public had become fascinated with the sheer delight of naming. His prestige in England, a nation with close ties to Sweden at the time, was enormous. Men from all corners of the world—the American wilderness, the Near East, darkest Africa, and the steaming Orient—sent him plants and seeds. And, like a great captain, he contrived to send men out to collect: Forskäl to Arabia, Hasselquist to Syria, Kalm to North America, Thunberg to South Africa, Japan, and Ceylon, and Solander around the world with Captain Cook.

In his passion to collect and classify, Linnaeus laid the foundation for the work of the evolutionists of the next century. In his eager quest to understand the living world around him, he encouraged morphologist and physiologist. The excitement of Linnaeus, his co-workers, and his followers was at least that of the present-day molecular biologist. Without the order that Linnaeus introduced into the chaos of man's emerging knowledge of his world, little headway could have been made. But the genius of Linnaeus provided only the prelude to the explosion of scientific knowledge that was to follow.

But the system of using a genus and species name, the so-called binomial system of nomenclature, does more than provide a convenient way to talk about a plant; it makes available to you a host of information about the plant. The genus and species are the core of a much more elaborate system of ever-enlarging groups: a number of closely related genera comprise a family; related families belong to an order; orders make up classes; and classes in turn are members of a division (or in the case of animals, phylum). Thus the pineapple is *Ananas comosus* (the genus name comes from the South American Indian word for the plant: nana) and a member of the family Bromeliaceae. (Almost all family names end in *aceae,* although there have been many exceptions such as Palmae and Cruciferae that modern systems are eliminating.) The Bromeliaceae is a relatively large family and includes some 65 genera and over 1500 species. It contains species, like the pineapple, that grow directly in soil and many others that are epiphytes (Fig. 13–12). Epiphytes are usually not harmful to their host, although they may become so dense that they overshadow the plant they are on. This is true for Spanish moss, which is also a member of the Bromeliaceae. Many of the genera of the Bromeliaceae are short-stem epiphytes with their leaves forming a pitcherlike rosette usually partially filled with water. The Bromeliaceae are among the most conspicuous epiphytes of tropical rain forests, growing high above the ground in the branches of trees. In recent years they have become popular greenhouse plants because of their unusual and beautiful foliage and flowers.

The Bromeliaceae with five other families form the order *Farinales* (all orders end in *ales*). The Farinales and many other orders are placed together in the subclass *Monocotyledoneae.* The "monocots" in turn comprise one of the two great subclasses of the *Angiospermae* or flowering plants.

Let us look at one more example. Magnolias are trees or shrubs native to North and Central America, the Himalayas, and East Asia. One common magnolia is *Magnolia soulangeana.* In this case both names are derived from the names of famous botanists: the generic from that of Pierre Magnol (1638–1715), who was Professor of Medicine and director of the botanic garden at Montpellier, France; the species from that of M. Soulangean, who lived in France in the early 1700s. Another common magnolia is *Magnolia grandiflora;* here the species name is descriptive, meaning large-flowered.

The genus *Magnolia* and nine other related genera are placed together in the family Magnoliaceae. The beautiful tulip tree of the eastern forests is a member of another genus in this family (*Liriodendron*). The Magnoliaceae is in the order *Ranales.* This order contains 12 families including the water lily family, the buttercup family, the laurel family, and the barberry family. The order Ranales in turn is a member of the subclass Dicotyledoneae. The "dicots" and the "monocots" comprise the class Angiospermae.

Unfortunately, scientific names also have their drawbacks. Although it has been over 200 years since the publication of the *Species Plantarum,* taxonomy is

13–12 *The pineapple, the bromeliad, and the spanish moss shown in these photographs illustrate one type of evolution. The pineapple is a terrestrial plant and is found growing only in the soil. The bromeliad shown is an epiphyte, and although it still resembles the pineapple in many ways, it is adapted to growth in trees. Finally, the spanish moss is completely adapted to life far above the soil, and it no longer looks at all like the pineapple to which it actually is related.*

still a developing science. As taxonomists evaluate their specimens more and more carefully, they decide upon changes in the composition of genera and in the species designations. This is not a trivial matter. For example, of the 1400 species of the Front Range in Colorado listed in a handbook by William A. Weber, about 100 names changed between the 1953 and the 1961 editions. Many of these were common plants well known to local botanists. Frequently, the name change was not so much for a scientific reason (a reevaluation of relationships) but the result of the discovery (through library research) of a name that had been used earlier.

Numerous procedural questions have arisen since Linnaeus published the *Species Plantarum* in 1753 and initiated the binomial system and a host of scientific names still in use. These questions are now considered at International Botanical Congresses held every five years.

Problems in Organizing Species

How do we approach the vast number of plant species? That is the question that has bedeviled botanists as long as there have been botanists. Basically, there are two approaches: artificial systems and natural systems.

It is not too difficult to construct an artificial or arbitrary system for the plants within the boundaries of a given region. One simply groups together those plants having similar characteristics, and if the number of plants is limited enough, the groups are frequently quite distinct from each other. Theophrastus (372–287 B.C.), a student of the Greek philosopher Aristotle, was an accomplished botanist by anyone's standards.* In devising a system of classification for plants, he first divided them into woody and herbaceous groups. Throughout most of history, trees seemed distinct enough to merit a separate group in the plant kingdom. To the tree life form could be added herbs, shrubs, submerged emerged, or floating aquatics, epiphytes, etc. Obviously, it will be handy in any classification scheme to recognize these differences, although from an evolutionary standpoint, they may be quite artificial.

Using the angiosperms or flowering plants because of their great familiarity and abundance, we can see

*Theophrastus was familiar with the seed, stem, and leaf differences of monocots and dicots; the anatomy of stems, including annual rings of both xylem and phloem; the distinction between roots and rhizomes; the possible nature of sepals and petals as specialized leaves; the existence of flowers without petals; and the distinctions between hypogynous, perigynous, and epigynous flowers.

some of the problems involved in constructing a classification scheme. There are numerous features of angiosperms that might be used in an artificial or natural classification scheme. The difference will depend upon whether the features are applied arbitrarily and indiscriminately, regardless of other evidence indicating relationship, or whether the features are used only as indications of relationship. We might, for example, consider the vegetative characteristics of leaves (thickness, margins, surface characteristics, deciduous or evergreen behavior), stem anatomy, root morphology, anatomy, etc. All of these features are inevitably important in any classification scheme, but none of them nor any combination of the vegetative features is suitable for a truly comprehensive classification scheme. The characteristics of the flower seem to be much more important. Botanists have been primarily concerned with such features as the presence or absence of petals, numbers of petals, regularity or irregularity of the corolla, numbers of stamens, fusion of stamens to the corolla, numbers of carpels, and location of the carpels in relation to the other flower parts. These floral characteristics seem to provide the best basis for a classification scheme, either artificial or natural.

One of Linnaeus' most important contributions to botany was his recognition of the importance of flower structure in angiosperm classification. Linnaeus devised a totally artificial system of classification based upon 24 classes in the plant kingdom. One of these included plants without flowers, that is, all of the plants other than angiosperms. The angiosperm groups were divided first by the number of stamens and by their features, particularly their relationship to the carpels. Linnaeus was fully aware that the system was not a natural one; it threw together monocots and dicots into the same class, for example. Nevertheless, it provided an effective way to classify the plants known at that time. It dominated botany for at least a century and can still be found in floras written as late as 1961.*

This system had a profound influence on the development of botanical science because it greatly stimulated interest among both scientists and laymen. This interest, as it found its outlet in the collection and examination of the flora of the world, led to the accumulation of data upon which a more natural system would be based. Linnaeus published the fragment of a natural system (in *Classes Plantarum,* 1738), but he wisely recognized that data were insufficient for a complete natural system and stated that this would have to wait for the future. In his natural system, he enumerated 68 orders, each about the equivalent of a family in a modern classification scheme. When Linnaeus spoke of a natural system of classification before the time of the theory of evolution, he was referring to a system that would properly outline the plan followed by the Creator.

The modern concept of a natural system of classification of angiosperms rests (as we have seen for other groups) upon the theory of evolution, which says that all plants are related to each other by descent from a common ancestor. The goal of taxonomy is to define an organization that reflects natural evolutionary relationships; that is, to define a genetic system. The ideal approach would be to have fossils of all the types that ever existed and to organize them according to lines of descent. These would lead to the modern groups of plants, and the relationships between modern groups would be quite apparent from their lines of descent. Unfortunately, available angiosperm fossils represent only the minutest fraction of plants that have existed in the past. In our striving for a natural system, then, we must be content primarily with an examination of living species and an attempt to deduce from their modern characteristics the history of their descent.

It is always tempting to look for primitive living groups from which the more advanced species must have evolved. All taxonomists are aware that no living species could have descended from the individuals of any other contemporary species. Yet there is a tendency to overlook this and to create phylogenetic trees in which primitive groups form the basis (the trunk), and the most advanced groups are the uppermost twigs. The immediate implication for the student is that the advanced groups evolved from the primitive ones. The assumption in such cases is not that the groups are really primitive in the historical sense, since they are alive and growing today; rather they are primitive in the sense that they retain more of the features of the most historically primitive plants than do other modern groups. There is ample evidence to support the idea that modern features will coexist with primitive features in the same plant. Although ancient species *may*

*August Binz and Alfred Becherer. 1961. *Schul-und-Exkursionsflora für die Schweiz.* Basel: Benno Schwabe and Co.

have propagated themselves almost unchanged from the beginning (with new and more advanced species branching off), it is certainly *unlikely* that they would remain completely unchanged over vast intervals of time.

The real problem of any natural classification scheme purporting to arrange plants according to degrees of primitiveness or advancedness is determining those features that are primitive or advanced. As we shall see, it is not easy to decide, and in many cases, because evidence is totally lacking, it is impossible to decide. One then has the alternative of either being content with an incomplete system or deciding arbitrarily about the importance of certain characters. Virtually no one is content with an incomplete system; thus arbitrary decisions are inevitably made. As soon as an arbitrary decision enters into a classification scheme, the system becomes to that degree artificial. No system exists that is wholly natural. In a sense, we are in a position comparable to Linnaeus with his fragment, although we hope that our fragments (being based upon much evidence not available to him) are more inclusive than his. The majority of families and orders fall together fairly well and may be thought of as natural groups. A few are "catchall" families or orders, but these few are mostly in the older systems. Organization above the level of the orders proves to be much more difficult. Here is where the interesting controversies arise, and we shall examine them, at least briefly, in Chapter 29.

Evolution of the Higher Categories of Plants

Are the evolutionary processes the same for categories of plants above the level of the species, as they are for species and races? The answer is yes, and it is based on the three types of natural selection we discussed for species: stabilizing, directional, and disruptive. Disruptive selection followed by simultaneous directional selection in several different lines produces adaptive radiation. Conditions are favorable for adaptive radiation, for example, when a population acquires by evolution a new complex of adaptive characters that enable it to exploit an available environment more efficiently than can any of the other organisms already present. Adaptive radiation also occurs when organisms move into an essentially new environment (e.g., an island), with diverse ecological niches that are not filled.

Important in the evolution of the higher categories is extinction. Extinction of intermediate forms sharpens the differences between the existing groups. Because of this, the older lines of plants are usually much more easily defined, while the newer lines are more difficult to organize. They are still in the process of adaptive radiation, and the spectrum of characters present in the organisms makes it difficult to know how to evaluate them.

One of the remarkable features of evolution at all levels is the stability of many species over millions of years. Some blue-green algae appear to be the same as those that lived over a billion years ago; there are species of horsetails apparently identical to plants living in the great coal swamps of the Carboniferous some 130 million years ago; and the beautiful, broad-leaved gymnosperm, the ginkgo, has remained unchanged since the Cretaceous. In the case of these old species, it appears that the particular environments to which they are adapted have shrunk in extent and area through time, while the organisms have continuously remained in a stable equilibrium with their environments. This equilibrium is maintained by stabilizing selection.

14

Evolution of the Plant Kingdom

Plants of the Past

In Chapter 1 we took a walk through a woodland to introduce you to the plants that grow around us today. What would that woodland look like if we had taken our walk 25 million, 150 million, 325 million, 350 million, or 375 million years ago? Of course, we cannot do this, but the question fascinates men. Many of them have tried to reconstruct those past worlds, basing their attempts on the evidence (in the form of fossils) that organisms of the past have left. Fossils take many forms and vary greatly in what they tell the scientist. Unfortunately, in all cases, the fossil record is incomplete, and many pieces of data resist interpretation. Despite these handicaps, it is possible to present reasonable reconstructions of the past, like the ones we have reproduced here. While viewing these drawings, remember that the climates were different than now and that even the continents themselves did not have their present positions and shapes. Finally, try to put yourself in the pictures and imagine what these plants really looked like.

A All life originated in the sea, and it was many millions of years before either plants or animals were able to live on land. The first land plants and their immediate successors were far from imposing. They were no more than several feet high and lacked both true roots and leaves. Here we see some of these earlier land plants that were present in the lower Devonian period.

B Some 25 million years later, or by the middle of the Devonian, the plants are larger, more complex in structure, and more diversified. The ancestors to the present-day horsetails and club mosses are present, and some of the plants are now 6 to 7 feet tall. Early ferns are also present in this flora.

C The plant groups present in the Devonian continued to evolve and flourished during the Carboniferous period. This was a time of giant swamps, huge tree horsetails, and club mosses. Progymnosperms and ferns were also present, as were the first animal invaders of the land—the insects. Dragonflies with 30-inch wing spans and foot-long cockroaches lived among the rotting remains of the lush flora.

D The Mesozoic era, particularly the Jurassic period, was the age of gymnosperms and dinosaurs. This view of the Jurassic landscape of Utah, where the great dinosaur fossil beds are found, shows several types of dinosaurs and a variety of gymnosperms and ferns. Despite the grassland appearance of the plain in the background, these plants are not angiosperms because flowering plants have not yet evolved.

E The Mesozoic ended in a great period of extinction. All of the dinosaurs, the Bennettitales, many of the cycads and Ginkgos, and some of the conifers all died out. But, during this same period, the angiosperms evolved. These flowering plants included not only tree species, but short-lived annuals such as grasses. The landscape takes on a much more modern appearance. In this scene of the Tertiary period of the Cenozoic era in Middle Europe, we see a meadow that looks almost like one we could walk in today. Of course, we would not find palms at this latitude today and the animals (including a mastodon—an antlerless type of antelope) are surely not what we would expect to see.

A

B

C

D

E

Clouds hanging low on the dim horizon are torn by repeated bolts of lightning and claps of thunder. The continuous rain falls on a still warm earth. An irritating atmosphere devoid of oxygen and full of nauseous odors prevails. A scene from Dante's *Inferno?* No, merely a reconstruction of the earth at the time that life was beginning to evolve.

The Origin of Life

How did life on earth originate? For this discussion, we shall assume that life has evolved by means of physical and chemical reactions that are governed by the normal laws of physics and chemistry. You can look at this type of discussion as an intellectual exercise or as an indication of the way things actually happened. In either case, we should give serious consideration to this basic question because it is one of the most challenging areas in biology.

We must start with the origin of the earth, and to do that, we must consider the origin of the sun. The sun was apparently formed by the condensation of particles of ice, dust, and gases present in space. The internal temperature of the sun rose as heat and energy resulted from this condensation, eventually reaching the stage where it was high enough to drive thermonuclear reactions. It is from these thermonuclear reactions that we derive all of the energy on earth.

The earth and other planets were probably formed from the same materials and at the same time as the sun. The same method of formation was involved: local condensations in the same mass of ice, dust, and gases. The most common elements in this material (also the most common throughout the solar system) were hydrogen, oxygen, carbon, nitrogen, sulfur, and phosphorus. In the atmosphere of the young earth there were gases present that are no longer there, while some currently present were missing. The gases in this early atmosphere are believed to have been hydrogen (H_2), helium (He), methane (carbon and hydrogen, CH_4), ammonia (nitrogen and hydrogen, NH_3), and water (H_2O); missing were nitrogen (N_2) and oxygen (O_2).

As the earth cooled, the water that was formerly present in gaseous form in the atmosphere began to condense as water. Water fell on the warm earth, and while much of it probably evaporated, the remainder started to erode the surface of the earth. This water drained into the depressions and the shallow places in the crust, eventually forming the seas and oceans. The eroded minerals from the earth and much methane and ammonia from the atmosphere were dissolved in this water. In the atmosphere and (to a lesser extent) in this warm sea, high energy radiations from the sun (mostly ultraviolet light), cosmic radiations, and high energy discharges from lightning provided the energy for a variety of chemical reactions that yielded various simple organic compounds. These probably included simple sugars, fatty acids, glycerine, amino acids, and nitrogen bases. This stage can be duplicated in the laboratory by such means as passing electrical discharges through a mixture of H_2, CH_4, NH_3, and H_2O (Fig. 14–1). After waiting a few days, the gases and liquids can be analyzed and these simple compounds found. They are "simple" only by comparison with the advanced, complicated compounds found in our present living systems, and they are certainly more elaborate than the basic materials once present in the sea.

The evolutionary process continued under much the same conditions in the oceans and seas, gradually building up more complicated compounds: sugar molecules joined together to form polysaccharides; fatty acids and glycerine condensed to form fats and lipids; amino acids combined to form proteins that, in turn, gave rise to enzymes; nitrogen bases combined with sugars and phosphates to produce nucleotides that, in turn, combined to form nucleic acids. Two of the most significant points at this stage in the evolution of life are the formation of enzymes and the formation of nucleic acids. We saw how enzymes, through their ability to promote certain reactions, become exceedingly important. The nucleic acids are self-reproducing compounds with regular structures. Hence, when the nucleic acids combine with the proteins to form nuclear proteins, the bases of a cell have evolved.

With the arrival of the nuclear proteins, we have units that might be thought of as having a number of functions associated with life. These include reproduction, mutation (in the sense of permanent change in the pattern of reproduction), and nutrition. The latter is important because it means that we are now dealing with a molecule that can utilize the pieces of other molecules and the energy they contain to build up its own structure. Finally, there is the property of aggregation. By this we mean that there is the possibility of this complex molecule, made up of nucleic acids and proteins, assimilating, as a permanent part of the structure associated with it, other molecules in the form of shells or various types of external arrangements.

If we follow this line of reasoning, including the aggregation principle, we then have the nuclear proteins plus an organic shell. This shell may take two forms: either an immediate coating, as in the modern viruses, or a looser aggregation with a limiting membrane around the outside, as in cells. Undoubtedly, in the beginning neither modern viruses nor modern cells were formed, but something similar to these, which we might call protoviruses and early cells, did exist. The protoviruses and early cells can now be conceived of as having the properties of growth and development through their ability to synthesize from the molecules surrounding the various parts of their individual units. To do this, however, we must always think of these protoviruses and cells as having internal control over the processes going on within them. We also believe that at this point the process of fermentation developed, whereby energy could now be obtained in reasonable amounts from the utilization of simple sugars. If we think of fermentation as occurring at this time, a new complexity is added to our description of the origin of life. This is the evolution of CO_2 in the atmosphere. We now have a gas, CO_2, that was not present in the original atmosphere of the earth and that adds various potentialities to the future development of the cells.

The protoviruses and the early cells at first probably utilized the materials present in the seas around them, but as these decreased in amount, it is evident that other ways of getting food for growth and energy had to be found. One of the easiest ways is through parasitism, where one cell literally lives off another. Thus parasites probably arose early in the development of cells. Still another way is to actually engulf living cells, to digest the cells, and to utilize the various components of these cells for the structure of the engulfing cell. Such behavior is animalism, and such a cell would be a forerunner of the modern animal. Another approach is not to engulf living cells or living protoviruses but to utilize dead cells, which is called saprophytism. Finally, there is the possibility of using a mechanism, akin to photosynthesis, involving the oxidation of inorganic compounds instead of light for

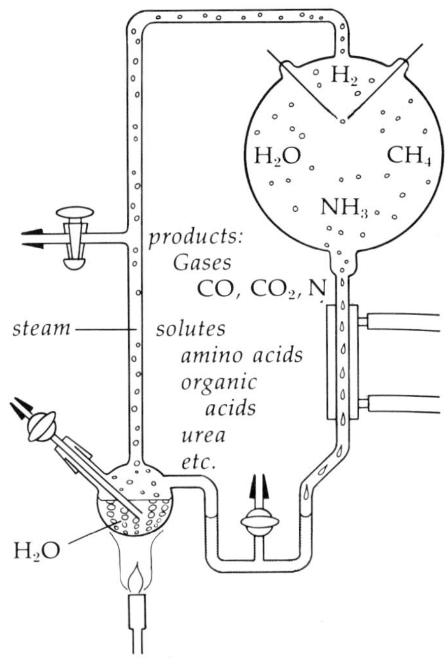

14–1 A diagrammatic representation of the apparatus designed by S. L. Miller to study the synthesis of organic molecules under conditions believed present on the primitive earth. Steam was passed through a mixture of gaseous methane (CH_4), ammonia (NH_3), and hydrogen (H_2) and then exposed to a high-energy electric spark before being recondensed as water. After a week of operating the apparatus, the water was found to contain amino acids.

the liberation of energy. This maneuver is called chemosynthesis. Such a procedure is still carried on by some cells, particularly in bacteria, where a variety of inorganic materials are used for sources of energy.

The presence of CO_2 in the atmosphere presents yet another possibility: under the proper conditions it can be combined with water and, using light energy, produce sugar. This, of course, is photosynthesis. To think of photosynthesis occurring at this point, we must first imagine some way in which chlorophyll could evolve. Chlorophyll and the various pigments are relatively complex molecules but consist of repeating units; thus it is possible they developed through evolutionary processes similar to those producing the complex molecules we have just discussed. Having the chlorophyll present together with the CO_2 in water, it might be possible to get photosynthesis and the first plant cells.

The result of photosynthesis is, of course, sugar on the one hand and oxygen on the other. The oxygen is released into the atmosphere; so we now have another gas appearing in the atmosphere that was not present before. This gas is entirely different from CO_2 and is highly reactive, thus making a great change in the condition of the atmosphere and the physical properties of the inorganic world. We can see this if we think of oxygen combining with one of the original components of the system, methane, resulting in the formation of CO_2. Oxygen combined with the ammonia present in the atmosphere yields gaseous nitrogen (and water), and an oxygen molecule combined with another oxygen atom gives ozone, which in the upper atmosphere filters out ultraviolet radiation. Finally, the availability of oxygen to the various types of organisms might allow the development of aerobic respiration through processes similar to those already discussed. Thus we can think of the time at which oxygen became readily available for utilization in all of these many processes as the oxygen revolution. It was a fundamental step and resulted not only in the formation of our present atmosphere but also in the development of the present biochemical systems that exist within living plant and animal cells.

When we look for evidence to support these ideas, we find that the atmospheres of Uranus and Neptune—planets with strong gravitational fields and low temperatures—consist of H_2, He, and methane. Jupiter and Saturn also contain gaseous ammonia. Looking at the fossil record on earth, we find that in the oldest rocks there are remains of cells that are remarkably similar to present-day blue-green algae and bacteria. Also interesting is the fact that the blue-green algae are remarkably tolerant of ultraviolet radiation. And, by duplicating the conditions believed to be present on the early earth, we can produce many biological compounds in the laboratory.

Finally, we would like to emphasize that for the purposes of discussion we have presented a very simplified version of the way in which early evolution may have occurred. Most of this discussion is highly speculative; we know little for certain beyond the formation of the amino acids. For example, from our discussion of photosynthesis in Chapter 6 it is clear that much more is needed for photosynthesis than simply chlorophyll and CO_2; a whole system of enzymes plus an elaborate structure is involved in the process as it occurs today. But the fact that these are fiendishly complex systems should not stop us from seeking ways in which they might have arisen.

Will we ever make life in a flask in the laboratory? Probably not, but the question raises the thought of life originating on other planets in other solar systems. It has been estimated that there are a billion planets capable of supporting life just in our own galaxy. Whether life can originate on another planet can only be answered by further exploration of space.

The Earliest Organisms

When did life appear on earth? The earth is calculated to be some 4.5 billion years old. The evidence for the earliest life comes not from morphological fossils but from chemical fossils. These are the stable organic compounds that are produced by living organisms and can be preserved for millions of years. For example, chlorophyll is degraded within a few hundred years to the stable compound porphyrin that can be easily identified in a rock sample. A variety of such chemical fossils exist, indicating that some form of life existed more than 3.1 billion years ago. The oldest rocks thus far discovered on earth are approximately 3.5 billion years old.

Until some 15 years ago, it was generally believed that there were no fossils in the oldest rocks. Reports of microfossils in ancient rocks had been made but not generally believed. Then a short paper, published in 1954 by S. A. Tyler and Elso Barghoorn of Harvard University, presented convincing evidence that such fossils did exist (Fig. 14–2). The work of Barghoorn and his associates established interest and credibility in the field, and the data have since rapidly accumulated.

The evidence is good that unicellular organisms were present on the earth at least 3 billion years ago. The first cells were similar to bacteria and blue-green algae. The available evidence indicates that they lacked organized nuclei and hence were procaryotic cells. Algal photosynthesis, in which oxygen is released (in bacterial photosynthesis it is not), is strongly indicated about 2.8 billion years ago.

The next step in evolution is from cells without membrane-bound nuclei (procaryotic) to cells with true nuclei (eucaryotic). The retention of evidence of a nucleus is an extremely difficult thing to imagine in a fossil, but there are fossils that strongly suggest that the living cells had nuclei. The available evidence indicates that the nucleate cell had become established as early as 900 million (0.9 billion) years ago. The establishment of such cells presented the potential for mitosis, meiosis, genetic exchange through sexuality, and an alternation of diploid and haploid generations.

The various developments took place during what is known as the Precambrian. The history of the earth is divided into various lengths of time and given names to help facilitate discussion. The largest divisions of time are called eras, and these are divided into periods, which may then be divided into epochs. We are living in the Quaternary period of the Cenozoic era (meaning recent life). The era before that was the Mesozoic (middle life), and the one before that was the Paleozoic (old life). The earliest period of the Paleozoic is the Cambrian; time earlier than this is called the Precambrian. The eras vary greatly in duration, the longest being the Precambrian.

By the end of the Precambrian, there were well-established lines of bacteria, blue-green algae, and eucaryotic algae. Single-celled and filamentous organisms were present, and the modern cell had evolved (Fig. 14–2).

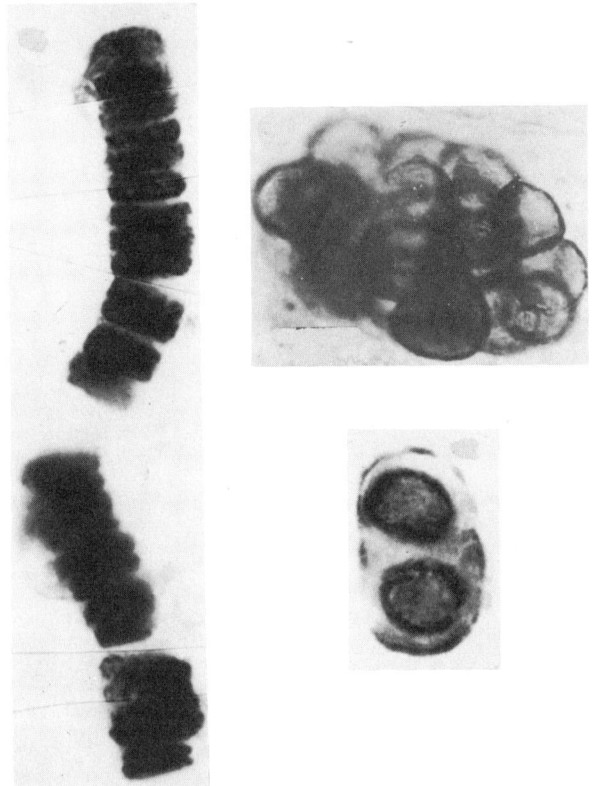

14–2 Some of the truly dramatic discoveries of modern paleobotany were the Precambrian microfossils. Here we show some of the fossils of blue-green algae and eucaryotic algae that have been recovered from the black chert of the Bitter Springs Formation in central Australia. These are from the Late Precambrian and are believed to be some 900 million years old. Recognizable microfossils are also known from the Early Precambrian—around 3100 million years ago.

After the Precambrian

For more than 2.5 billion years, life had been represented by unicellular organisms, but with the beginning of the Paleozoic era, life began to get more and more complex. The pace of development and the diversity of the organisms (both plant and animal) increase as we proceed through geologic time to the present. We shall return to evidence from the fossil record and consider the evolution of the various plant groups many times and in considerable detail in later chapters. Now, however, we would like to set the stage with a brief overview of the course of plant evolution as seen from fossil records (Fig. 14–3).

The beginning of the Paleozoic era (the Cambrian through the Silurian periods) is a time of life in the oceans and the seas. The algae diversified and increased in the complexity of the organization of their structures. Marine invertebrates evolved and represented the largest and most numerous animals present. Indeed, all of the major invertebrate phyla known today were present in the Cambrian. The climate then was warm and mild, a condition that continued through much of the Paleozoic era. The mapping of sedimentary rocks shows that broad, shallow seas covered great

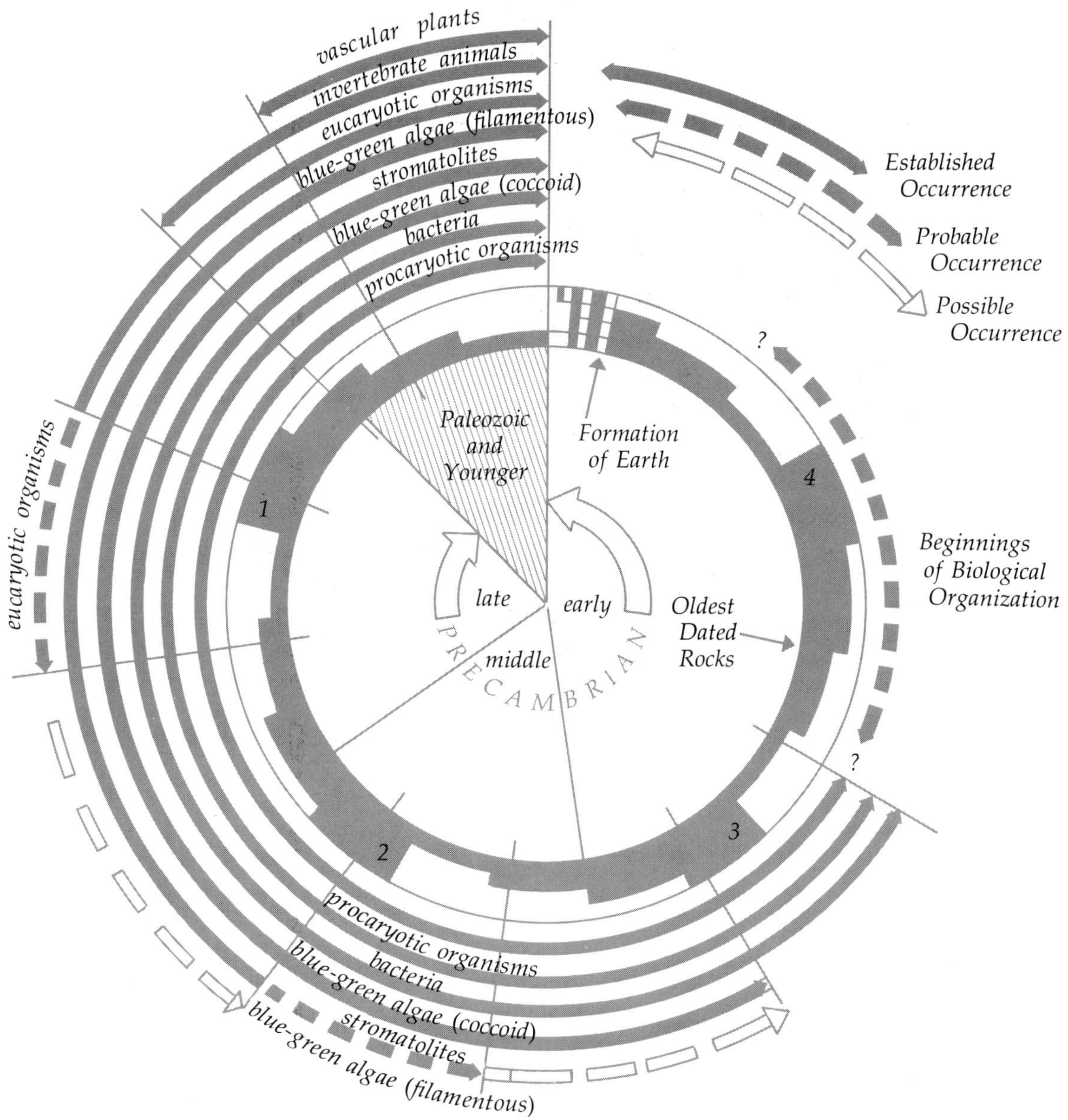

14–3 A geologic clock showing the known or inferred geologic distribution of various types of primitive organisms, based on morphological and chemical fossils. The numbers on the clock are billions of years from the present, which is located at 12 o'clock.

areas of what is now North America. These seas persisted in changing patterns, throughout most of the era.

At the end of the Silurian and the beginning of the Devonian period of the Paleozoic, plants became adapted to life on land. The earliest land plants were simple vascular plants that resemble but may not be related to the present-day Psilopsida. Their fossils are found with remains of horsetails and club mosses. Some of the horsetails and club mosses of the Paleozoic, particularly during the age of the great coal swamps of the Carboniferous period, were treelike plants. Ferns and early gymnosperms also appeared in considerable numbers during the middle Paleozoic. Fish were extremely numerous in the shallow seas. A few reptiles adapted to the land during the Carboniferous, but insects were the first animals to become numerous on land during this period.

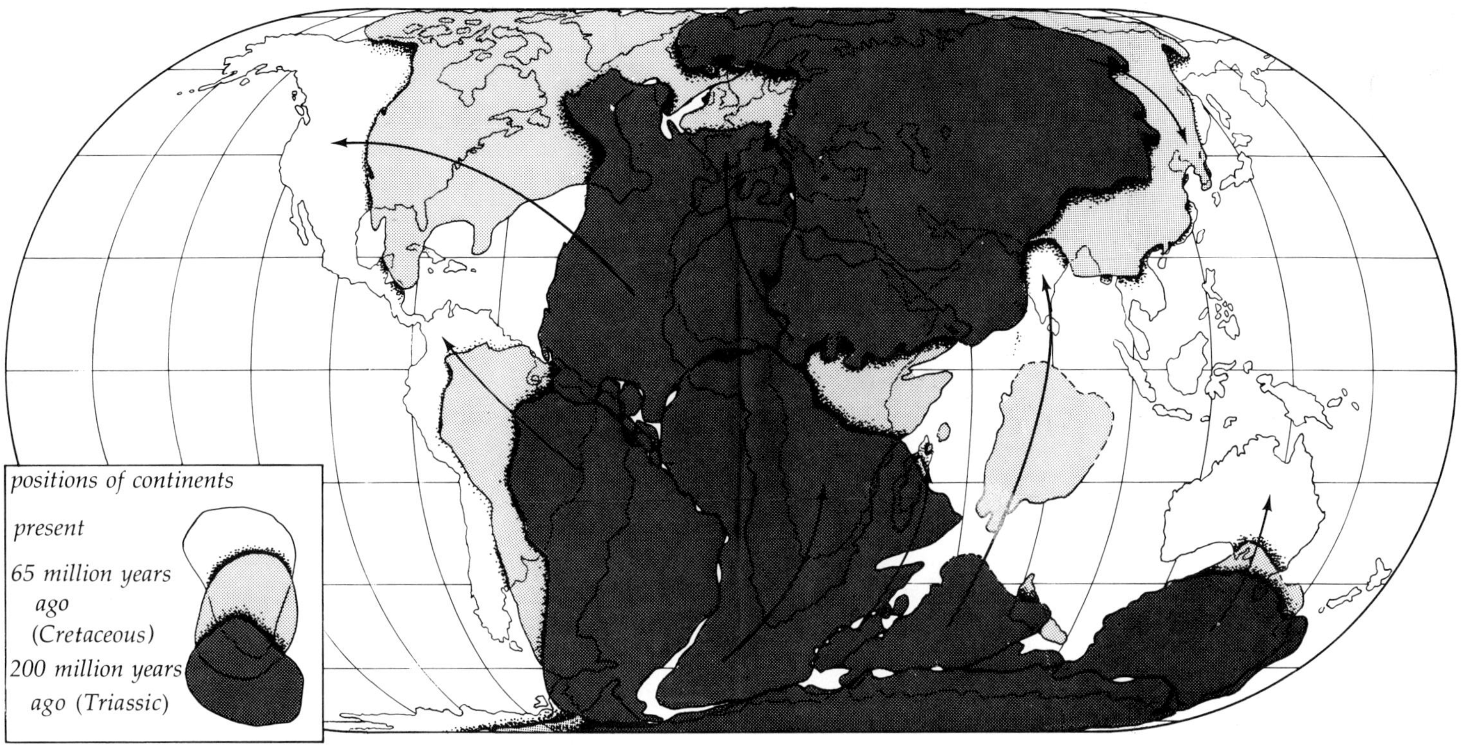

14–4 Evidence first from fossils and then from physical measurement indicate that the continents are far from stationary and are, in fact, drifting. This map summarizes the path of the various continents as they have drifted with time.

The end of the Paleozoic era is marked by a change from a warm, mild climate to a cold, dry one and a definite reduction in the extent of the seas. The coal forests were reduced and with them the great tree horsetails and club mosses. The reptiles now began to diversify, and the first mammal-like reptiles appeared.

The Mesozoic era is familiar to many, since it is the time of the great dinosaurs. The gymnosperms were now the prominent plant species and showed a great diversification of types. The climate in what is now the temperate zone was fairly mild. In the final period (Cretaceous) of the Mesozoic era, the flowering plants became more numerous, in terms of both species and individuals, and they spread into all types of environments. This increase in flowering plants took place during the period when the dinosaurs were becoming extinct and the early mammals were appearing.

The Cenozoic era began with a mild climate, and the subtropical rain forest extended over most of the earth. As the era went on, the climate became more temperate, and the plants that had been growing together began to be distributed on the basis of their ability to withstand cooler conditions. Giant mammals made their appearance in the fossil record of the Cenozoic only to become extinct shortly thereafter. During the mid-Cenozoic the great development of the herbaceous angiosperms, as contrasted with the older woody species, occurred. This period established many floras that have persisted virtually unchanged to the present day. At the end of the Cenozoic, man finally appeared on the scene. By then, flowering plants had become the dominant form of plant life.

Continental Drift

Figure 14–4 shows the distribution of water and land over North America for various periods. Such

PRECAMBRIAN	PALEOZOIC					
4600	570	500	430	395	345	280 — 225
PRECAMBRIAN 4030	CAMBRIAN 70	ORDOVICIAN 70	SILURIAN 35	DEVONIAN 50	CARBONIFEROUS 65	PERMIAN 55
Earth's crust forms and seas and the continental land masses appear. Evidence of granite intrusions; volcanic activity, mountain building; glaciation.	*Two great troughs in the East and West of N. America fill with sediments that will later build the Appalachians, the Rockies and other mountain ranges.*	*About 70% of N. America covered by shallow seas. Some volcanic and the eastern land masses, including mountains in New England, begins to rise.*	*Much of Eastern N. America inundated by a salty, inland sea. Volcanoes are active in New Brunswick and Maine.*	*Eastern N. America, from Canada to North Carolina, rises from the sea.*	*Large areas of Eastern and Midwestern N. America become a great swamp that is repeatedly submerged by shallow seas. Great swamp forests later become coal.*	*Violent geologic climatic disturbances mark this period. Great wind-blown deserts cover much of the continent.*
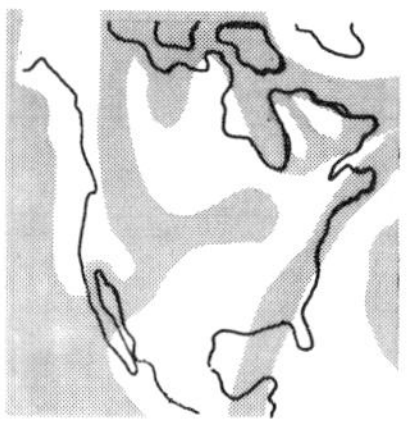	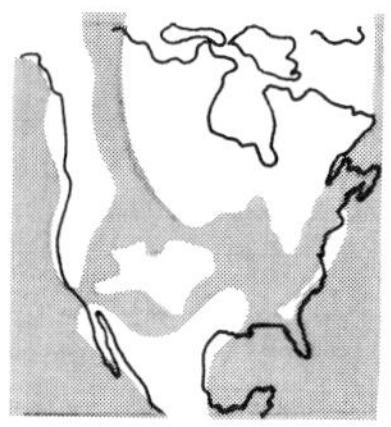	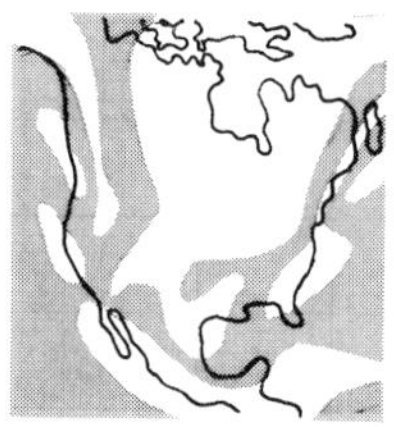	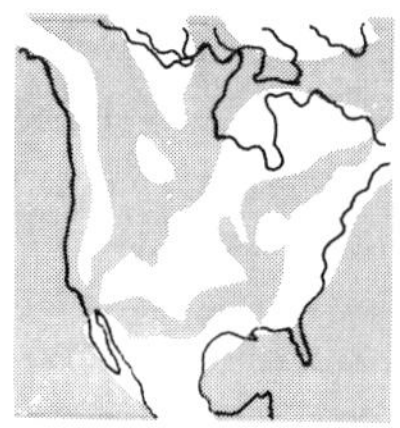	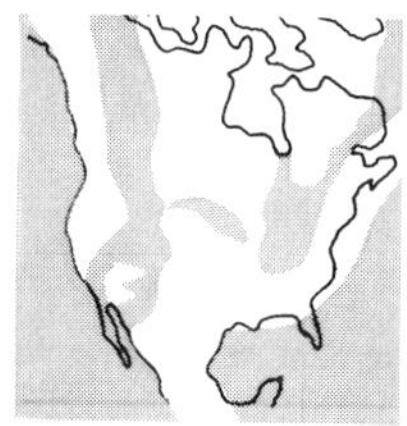	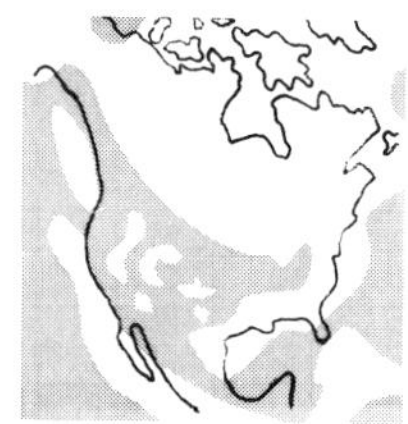	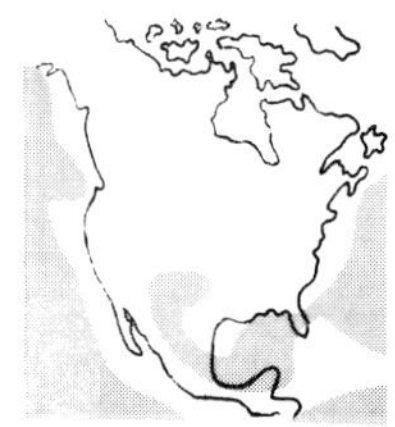
Earliest evidence of life in the form of microfossils and chemical remains indicate origin of life at 3 billion years ago. The organisms that have been found are algal-like and fungal-like plants and soft-bodied animals.	*Extensive fossil record begins in this period. Plant life was all aquatic and primarily algal. The animals were all invertebrates that, in countless varieties, filled the seas as the dominant life form. Trilobites were extremely common and numerous.*	*Plants remained aquatic and relatively little is known about them. Invertebrate marine life continued to dominate the living world, but the first vertebrate appeared. This was a jawless fish and was the forerunner of the animals that would dominate whatever sphere they entered.*	*For most of the period the plants are aquatic algae but near the end of the first land plants appear. Animal life in the sea similar to last period but masses of coral are present forming huge reefs. First land animals appeared—these were similar to the present-day scorpions.*	*Land plants evolved rapidly and tree club mosses and horsetails form forests in later part of period. Fishes become so numerous and advanced that this is frequently called the Age of Fishes. Early amphibians developing by end of period.*	*Giant club mosses, horsetails, progymnosperms, and ferns are present particularly in the huge swamp regions. Sharks abundant in the sea. Winded insects, often of huge size present on land. Primitive reptiles evolved from ranks of the amphibians to become the first vertebrates able to live full time on land.*	*Great age of extinctia. The tree club mosses and horsetails all disappear and many of the ferns. The gymnosperms evolve. Fish continue to evolve in the sea and vertebrates take over the land. Reptiles become dominant on the land.*

14–5 *A history of the evolution of plants.*

BEGAN MILLIONS OF YEARS AGO	190	136	65	2.5	0
ERA	MESOZOIC			CENOZOIC	
PERIOD DURATION IN MILLIONS OF YEARS	*TRIASSIC* 35	*JURASSIC* 54	*CRETACEOUS* 71	*TERTIARY* 625	*QUATERNARY* 2.5
MAJOR GEOLOGIC EVENTS	*The Appalachian Mountains, built in the Permian and once as high as the modern Alps, begin to erode.*	*The Sierra Nevada mountain range from S. California to Alaska are built by upthrust.*	*The Rocky Mountains, from Central America to Alaska, rise out of a sediment-filled trough. The sea for the last time inundates much of the continent.*	*The Columbia Plateau and the Cascade Range rise, and the Rocky Mountains reach their present height. There is also extensive volcanic activity in the Northwest.*	*Continent much as at present but covered periodically by glaciers of four ice ages.*
EXTENT OF SEAS OVER NORTH AMERICA	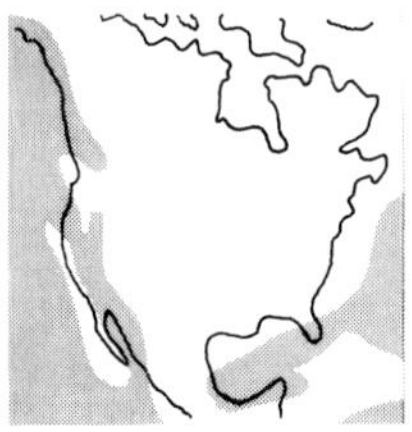	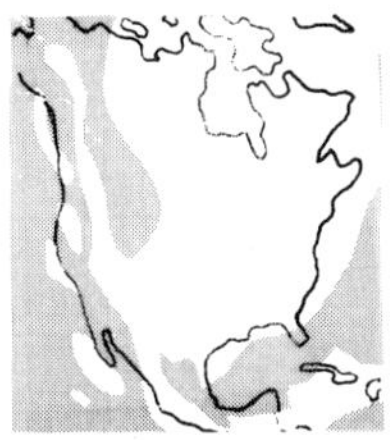	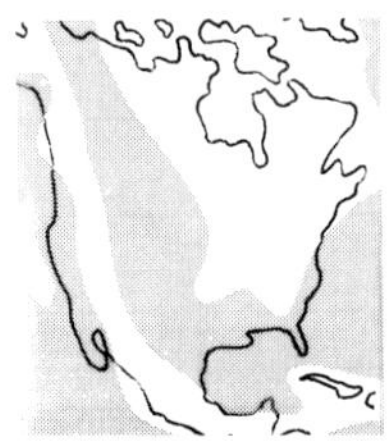	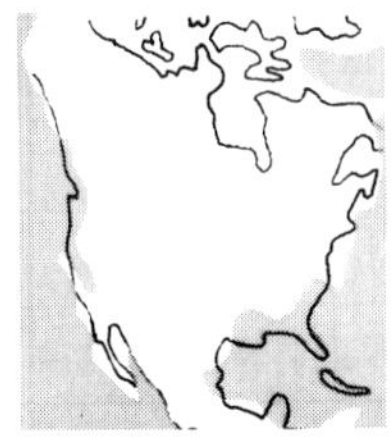	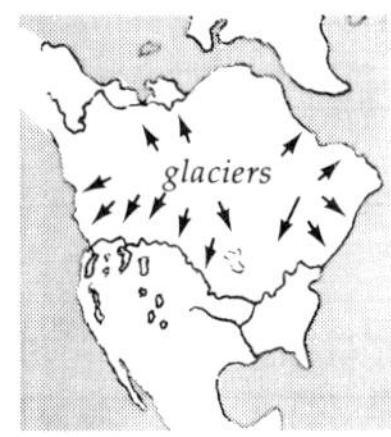
MAJOR BIOLOGICAL EVENTS	*Gymnosperms dominant in the flora. Many cycads and conifers present. Reptiles dominant with early crocodiles present as well as small dinosaurs. The first warm-blooded mammals appear that range in size from a rat to a small dog.*	*The gymnosperms continue to dominate the flora. Dinosaurs reach their peak and dominate the land. Primitive birds appear.*	*Angiosperms evolve and the first of the flowering trees appear. The cycads begin to die out and only a handful of species survive period. Dinosaurs become highly specialized and die out near end of period. Birds become highly evolved but mammals are still small and inconspicuous.*	*Angiosperms continue to evolve and herbaceous species are present in increasing numbers. Mammals take over dominance of the land with horses, rhinoceros, elephants, zebras, and rodents all appearing early in the period. Many mammals evolved to become extinct shortly after.*	*Ice sheets cause migrations of both plants and animals. Great development of animals and an increase in fern species. Most tertiary animals become extinct. Many giant mammals evolve only to become extinct. Man appears and evolves into the dominant organism.*

maps are based on the distribution of sedimentary rocks (rocks formed from sediments accumulating in bodies of water) and are of great value in understanding the oceans and land. But, when they are depicted, as we have done, over a map of the present North American continent, they give the misleading impression that the continent has always been in the same relative position and that only the relative distribution of land and water has changed. This is not the case because of the phenomenon called continental drift.

The theory of continental drift (in its most extreme form) is based on the idea that all of the present continents once existed as a single land mass, Pangaea (Fig. 14–5). This later separated into two larger super continents, Laurasia (comprising North America, Europe, and Asia) to the north and Gondwana (South America, Africa, Antarctica, Australia, and the Indian subcontinent) to the south. This initial separation required some 20 million years of drift, and by the end of the Triassic period (some 180 million years ago) the Australia-Antarctica mass had begun to move from the South America – Africa group with the Indian subcontinent beginning a long migration northward. During this period the Atlantic Ocean and the Indian Ocean began to be formed. Gradually, during the Jurassic and Cretaceous periods of the Mesozoic and continuing through the Cenozoic era to the present, the continents have continued to move and are moving today. They cover vast distances relatively rapidly, an average of 10 km per million years. North America moves westward the length of your body during your lifetime!

The mechanism of continental drift is based on the concept of continents as parts of plates of rock that form a covering floating over a more molten interior. Movement, usually called spreading, begins with a rift developing under a continent and molten rock from the interior pushing up. The rift spreads as more material is forced upward. At the same time the rift is developing, a trench is formed into which the edge of the plate is pulled away from the rift (Fig. 14–6). The rifts develop into oceans. The Atlantic and the Indian oceans are rift oceans, while the Pacific, as the ancestral ocean, is getting smaller.

The current theory of continental drift (not the mechanism just described) was proposed in the 1920s by the German geologist Alfred Wegener. But the idea is much older, having first been suggested in 1858 by the French paleobotanist Snider. He based his observations on the extraordinary similarity between the fossils found in coal on either side of the Atlantic. This led him to construct the first jigsaw puzzle that fitted together the opposing shores of both the North and South Atlantic oceans. The great bulk of evidence for a continuous land mass, which was later dispersed, comes from the southern not the northern hemisphere.

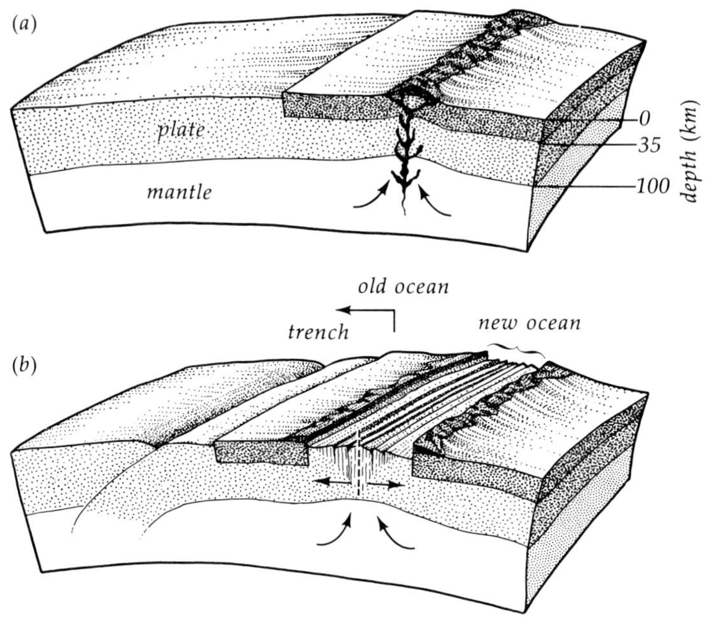

14–6 The exact method of continental drift is not yet known, but the most reasonable explanation is shown here. The process begins (a) when a spreading rift develops under a continent resting on a single plate; (b) molten rock (basalt) spills out from the mantle, thus pushing the two new continents apart. While this is happening, a trench develops into which oceanic crust of the new moving plate is pulled and consumed. The new ocean increases in size, while the old ocean decreases.

From the mid-1800s onward, fossils were found that formed a consistent flora in South America south of the Amazon and east of the Andes, in Africa south of the Sahara, all of India, Australia, and Tasmania, and, finally, in east Antarctica. This is known as the *Glossopteris* flora (after the most common genus found in it), which existed during the Permian. The important point is that not one species but a whole collection of species from various plant groups are present. *Glossopteris* is a seed plant of a type where it is impossible for the seeds to be carried over thousands of miles of ocean and still survive. Hence, either there were land bridges where there are now none, or a continuous land mass must have broken apart and the continents moved into their present positions. For many years a minority group of paleobotanists held to the idea of continental drift, often in the face of adverse comments from geologists. General acceptance of the theory has occurred only within the last five years and is based on the extensive exploration, mapping, and magnetic analysis of rocks of the ocean floor.

Major Groups of Plants

The end products of these millions of years of evolution are the plants that surround you. In our woodland walk we pointed out many of the major groups of plants that inhabit our world. Now let us again briefly examine them, looking for the fundamental similarities and differences among the major groups. Then we shall examine each in detail in Chapters 15 to 29.

Viruses

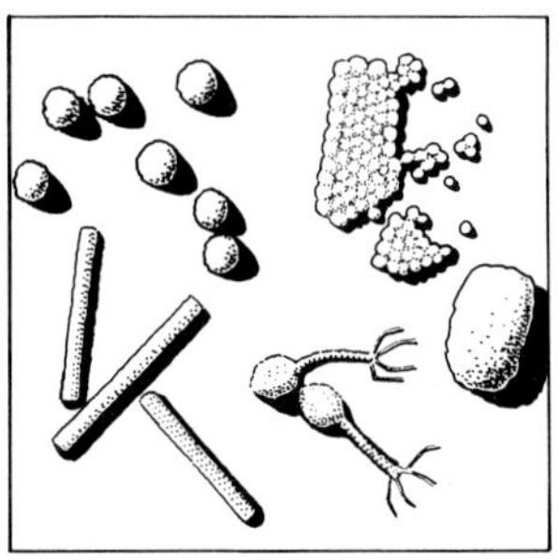

Viruses are essentially units of information that lack their own machinery to transcribe the message. As a result, viruses can exist only so long as there are living cells whose biochemical pathways they can utilize to produce more virus particles. Thus all viruses are both infective and parasites on organisms, ranging from bacteria to man and flowering plants. The critical component of all viruses is a strand of nucleic acid—either DNA or RNA but never both. The nucleic acid is the infective part of the virus and takes over the cell's own synthetic mechanisms. Viruses are small but tremendously important to man.

Bacteria—Blue-green Algae

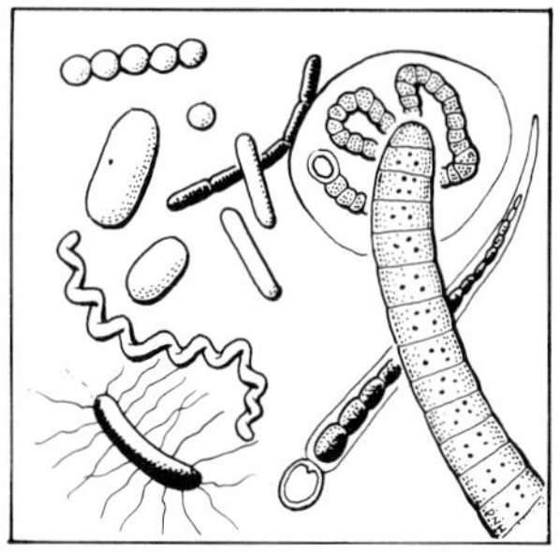

Bacteria and blue-green algae are true cellular organisms, but it is the organization of their cells that sets them apart from all other organisms. They lack membrane-bound nuclei, mitochondria, plastids, and similar organelles found in other cells. We have already termed them procaryotic organisms because of these characteristics. The fossil record shows they are ancient lines—lines that are old but still highly successful. Bacteria are both parasites and saprophytes, with a few species being photosynthetic. Blue-green algae are independent photosynthetic organisms. Some species of both the bacteria and the blue-green algae can convert atmospheric nitrogen into nitrogen-containing compounds that can be used by other plants. In terms both of nitrogen fixation and the recycling of elements through bacterial breakdown, the procaryotic organisms are of utmost importance to the ecosystem.

Algae

The various divisions of algae constitute a vast array of plants growing in all types of aquatic and some terrestrial environments. They range in size from minute one-celled species to some of the longest plants known. On the whole they are relatively simple plants in terms of their morphology, with many species consisting of single cells, filaments of cells, and colonies. The morphology of the marine algae is generally more complex than that of the freshwater species, but even the most elaborate, the giant kelps, are not as complicated as the true land plants.

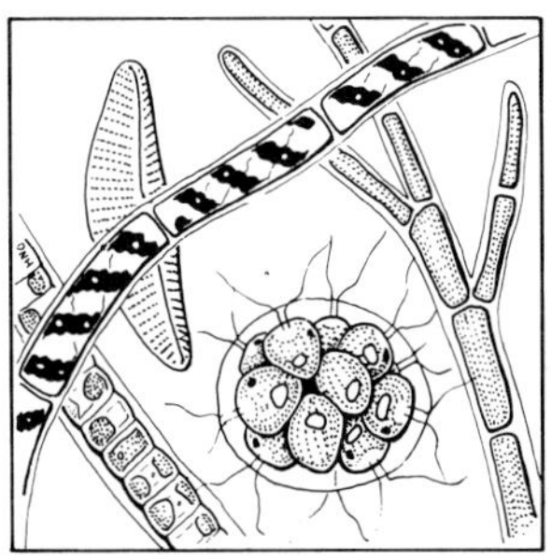

At one time all the algae were placed together in one large group. This was later decided to be artificial, and they were separated into various divisions. But there are several unifying features shared by all algal divisions. Typically, algae contain chlorophyll a (as do the blue-

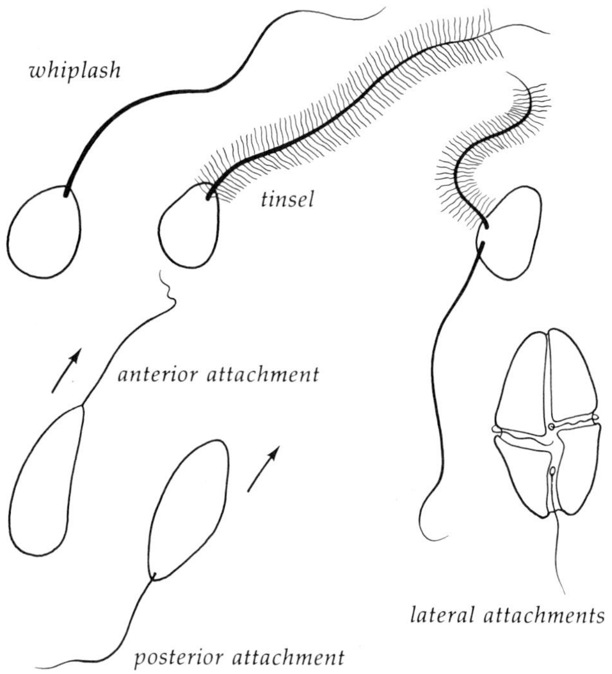

(a)

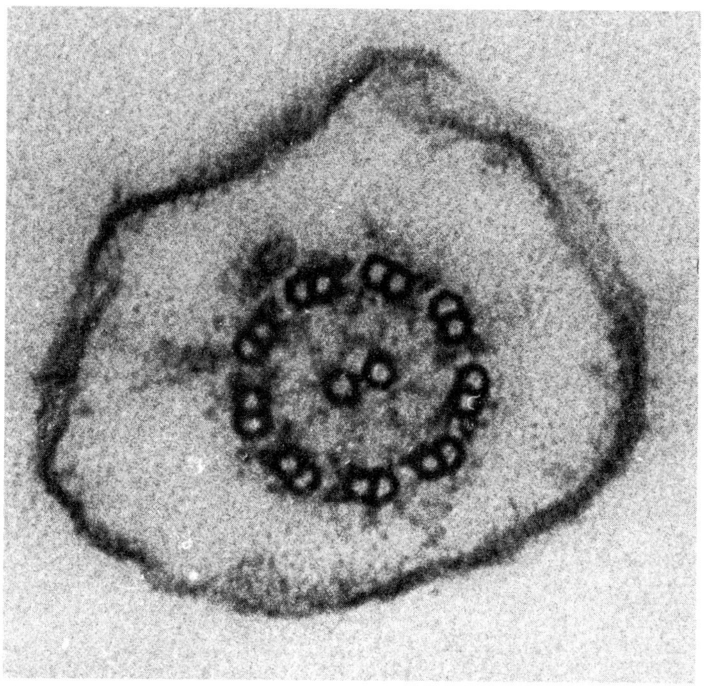

(b)

14–7 *The general appearance and the types of attachment of flagella are shown in (a). The ultrastructure of a flagellum is shown in (b). Flagella are usually 0.2 μ in diameter but vary greatly in length, from several microns to 150 μ. (Generally the shorter ones are called cilia and the longer ones flagella, but they have the same structure.) Each flagellum consists of 2 central fibrils surrounded by a ring of 9 double fibrils or microtubules. This 9 + 2 arrangement is remarkably constant throughout both the plant and animal kingdoms (although a few exceptions are known). The fibrils in the ring have arms on only one side. The whole flagellum is surrounded by a membrane that is an extension of the plasma membrane.*

Inside the cell the flagellum is attached to the basal body, which has the same structure as the centriole. The basal body is cylindrical, some 1200 to 1500 A in diameter and 3000 to 20,000 A in length. There is no outer membrane, and the ring of nine fibrils is the same as in the flagellum, except each consists of 3 (rather than 2) microtubules. The central tubules are reduced from 2 to 1 and are then absent altogether. The central tubules are missing, and at the end of the basal body closest to the cell surface, a terminal plate some 300 A thick is found. Thus the basal body is like an inverted tumbler open to the cytoplasm.

In the tinsel flagella the hairlike projections arise at right angles to the central cylinder. Each hair is about 80 A in diameter. The relation of the hair to the external membrane of the flagellum is not yet clearly understood.

green algae but not the photosynthetic bacteria) and produce oxygen as a by-product of photosynthesis (photosynthetic bacteria produce sulfur). The algae also have unicellular reproductive structures, the eggs and sperms not being surrounded by layers of sterile protective cells. Vascular tissue, composed of cells specialized for the conduction of water and foodstuffs within the plant, is also missing in the algae, except for the kelps.

The differences between the various divisions are marked and based on biochemical, morphological, and cytological criteria. The most striking of the biochemical differences is that of pigments and gives rise to the common names of many algae: green algae, brown algae, red algae, etc.

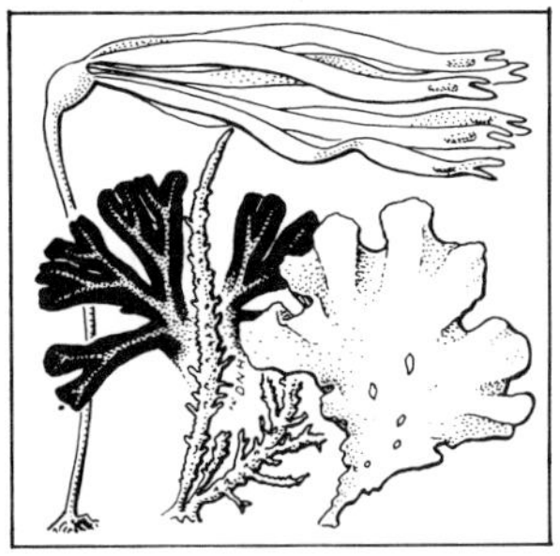

All algae have chlorophyll a, but many also have a second chlorophyll slightly different in structure, which is given a different initial (chlorophyll b, c, d, or e). Also present in addition to the green chlorophylls are the fat-soluble pigments, the carotenoids (carotenes and xanthophylls) that are yellow to orange in color, and the water-soluble pigments, the phycobiliproteins (phycocyanin and phycoerythrin) that are blue or red. Various modifications of these pigments are present in the different algal groups and provide major criteria for separating the algae.

Other biochemical differences are used to separate the algae. These include the form of the reserve food material and the composition of the cell wall.

One of the most important morphological features separating the algal divisions is the type, position, and number of the flagella present in the motile cells. Flagella occur in two basic shapes: a long, slender type appropriately called whiplash and a more elaborate type—in which hairlike extensions protrude at right angles to the central cylinder of the flagellum—called tinsel. The various arrangements of these two flagella types in the algae are shown in Figure 14–7. The intricate and beautiful ultrastructure of the flagellum can also be seen in this figure.

Finally, there are differences in the basic pattern of the life cycles in the various algae. In all sexually reproducing plants, there is an alternation between generations having the diploid chromosome number and those having the haploid chromosome number. The diploid or sporophyte generation produces haploid cells through meiosis. These may divide to form the haploid or gametophyte generation that, by mitosis, produces gametes. The fusion of two haploid gametes results in the formation of the diploid zygote, which is the first cell of the diploid, sporophyte generation.

The general pattern of the alternation of haploid-diploid generations is shown in Figure 14–8. This pattern is drastically altered in various algal groups. In some, the diploid generation is reduced to a single cell, the zygote. The opposite extreme is found in those algae in which meiosis gives rise directly to the gametes that fuse to form the zygote. These types of life cycles are also diagrammed in Figure 14–8.

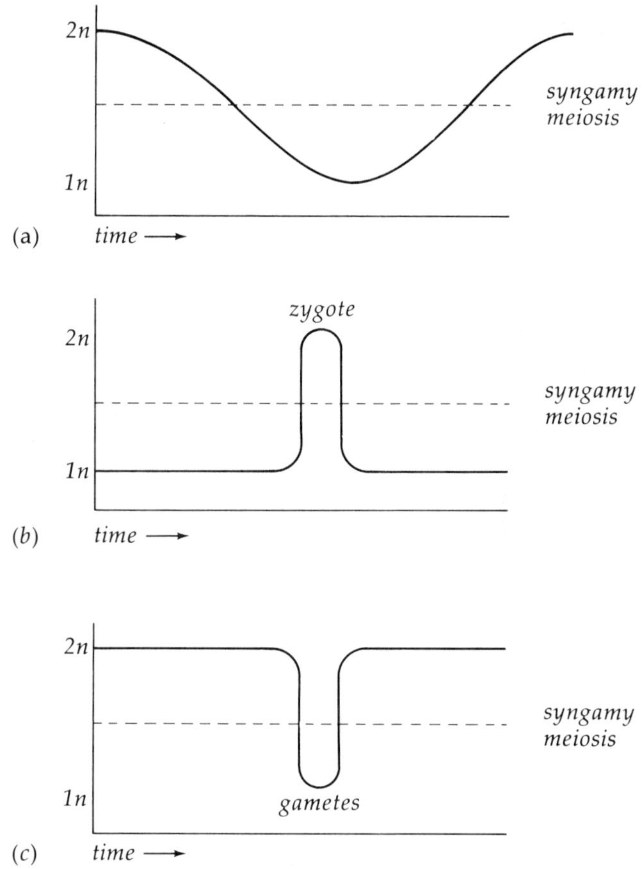

14–8 Generalized life histories found in the algae: (a) duration of haploid and diploid phases the same; (b) diploid phase confined to the zygote stage; (c) haploid phase confined to the gametes.

Following the current patterns in work with the algae (excluding the blue-green algae), we recognize these divisions:

Pyrrophyta	or	dinoflagellates and cryptomonds
Chrysophyta	or	golden algae and diatoms
Phaeophyta	or	brown algae

Rhodophyta or red algae
Euglenophyta or euglenids
Xanthophyta or yellow-green algae
Chlorophyta or green algae

The major features of each of these divisions in terms of the criteria just discussed (pigments, food storage form, composition of wall, type, number and position of flagella, structure of chloroplasts, type of nuclear division, type of life cycle, and general morphology) are given in Table 14–9. This table can be used as an introduction to the algae or as a summary (after you have read Chapters 17 and 18).

The algae constitute a truly fantastic group of plants that form the bases of the aquatic ecosystems on earth. Literally no life would be possible in the oceans, lakes, rivers, ponds, and other bodies of water without the microscopic algae that are the foundation of the food pyramid. They are the primary source of energy that filters up through a system of larger and larger animals, ultimately to man. For man to underestimate the importance of the algae in the ecosystem would be dangerous indeed.

Fungi

Until recently, mankind has been able to overlook the importance of the algae, but the impact of the fungi has been much more direct. A million dead in Ireland because of a famine caused by a fungal disease of potatoes is hardly something that escapes public notice. On the other hand, the equally important role of the fungi in recycling materials incorporated into the bodies of plants and animals has, until recently, been unknown to most people.

The fungi constitute another vast assemblage of plants with similarities that unify it and differences that divide it into a number of more or less clearly defined groups. No fungus carries on photosynthesis; all are either saprophytes or parasites. They all have relatively simple bodies in terms of morphology and cellular detail. Yet they differ in their structure and mode of reproduction to a marked degree, although the lines along which they would be divided are by no means clear to all mycologists (scientists who study fungi).

Generally, the fungi are divided into two divisions: Myxomycophyta or slime molds and Eumycophyta or true fungi. The slime molds lack cell walls, and many are capable of engulfing solid food particles. The true fungi, on the other hand, generally have a rigid cell wall, and although some are unicellular, most are filamentous. To obtain food, the fungi secrete enzymes that break down the material they are growing on and absorb the end products.

The Eumycophyta are divided into a number of classes on the basis of mode of reproduction, cellular structure, number, type, and position of the flagella (which are similar to those discussed for the algae), and composition of the wall. The most important classes are the Chytridiomycetes, the Oomycetes, the Zygomycetes, the Ascomycetes, the Basidiomycetes and the Deuteromycetes or Fungi Imperfecti. The characteristics of these classes are summarized in Table 14–10.

The Chytridiomycetes or chytrids are all aquatic fungi, the Oomycetes have both aquatic and terrestrial species, while the Zygomycetes, which includes the common black bread mold are all terrestrial. The Ascomycetes or sac fungi are also primarily terrestrial and include the yeasts, *Penicillium*, and various wood-rotting fungi. The Basidiomycetes include the common mushroom and wheat rust.

Land Plants

The groups above the algae and fungi are primarily land plants. The invasion of the land (discussed in Chapter 22) occurred during the early part of the Devonian, and it was not long, in geologic terms, before the land was covered with plants. The present-day land plants can be placed into two divisions: Bryophyta, a rather small group that includes the mosses, and the huge division Tracheophyta.

The Bryophyta or bryophytes, known to most people as the delicate beautiful mosses, are all relatively small. They are distributed worldwide, but, while found in both arctic and arid regions, they grow best in wet or tropical forests. There bryophytes grow profusely as epiphytes on trees and other plants. Many grow directly on the soil, with some developing over considerable areas in the permafrost regions of the arctic and subarctic. Bryophytes are commonly found as pioneer organisms on bare rock.

The moss plant that is frequently seen growing along the riverbank or between the bricks in the walk is a gametophyte or haploid plant. This is true for all the bryophytes. The gametophyte is the independent dominant generation. The sporophyte or diploid plant is smaller and is dependent upon the gametophyte. All bryophytes have motile sperms and a primitive embryo, which develops within the female sex organ (archegonium) in the gametophyte.

The bryophytes are set off from all the other green land plants because the gametophyte is the dominant generation. Because of this unusual feature (unusual for land plants, not for algae), the bryophytes have long been set apart from other land plants as possibly representing an alternative approach to life on land. Recent comparative studies have tended to place the bryophytes closer to the mainstream of land-plant evolution.

Vascular Plants

The feature that unites the Tracheophyta, a huge division including ferns, pines, flowering plants, and essentially all of the plants of which you are usually aware, is the presence of vascular tissue. The vascular tissue consists of the water-carrying xylem and the phloem that transports organic materials such as carbohydrates. Besides the xylem and phloem, the Tracheophyta contain many other specialized cells and tissues and have plant bodies that are much more elaborate than any of the algae, fungi, or bryophytes.

In all of the Tracheophyta the sporophyte or diploid generation is dominant and the conspicuous plant. The gametophyte or haploid generation becomes smaller and smaller as the species become more and more advanced. For example, in the ferns, the gametophyte, although small, can be seen easily and is a green independent plant; in the flowering plant, it is microscopic and wholly dependent on the sporophyte for food.

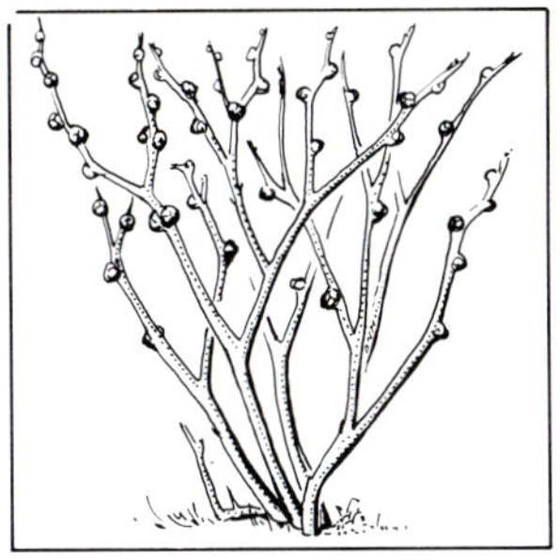

Psilopsids. The Psilopsida or psilopsids form a small subdivision of vascular plants that have long puzzled botanists. In the living psilopsids, which are mainly tropical epiphytes, the sporophytes are dominant and independent. They are true vascular plants with a xylem and a phloem. They lack both leaves and roots; the stem functions underground as a root and aboveground as the main photosynthetic organ. The gametophytes are rarely seen, since they are small, subterranean plants.

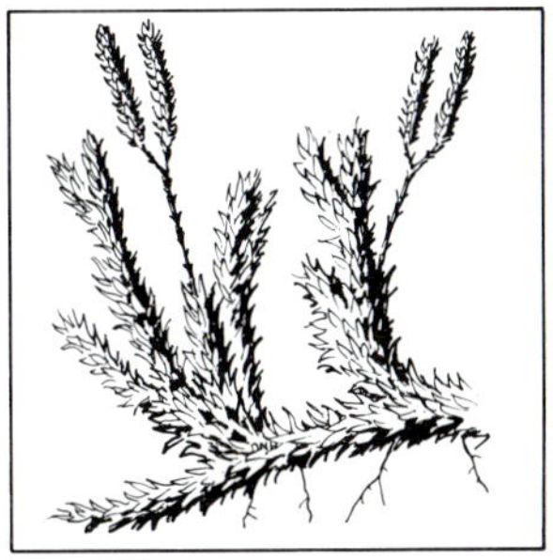

Lycopods. The Lycopsida is a subdivision that includes ancient plants whose fossil record extends back into the early Devonian. They developed a large number of species during the time of the Carboniferous, coal-swamp forest. Some species were small herbaceous forms, but many were tall trees. They flourished until the great swamps began to decrease in extent during the Permian. Now, there are only five genera of herbaceous lycopods in existence. These are worldwide in distribution and are commonly found on the floors of forests.

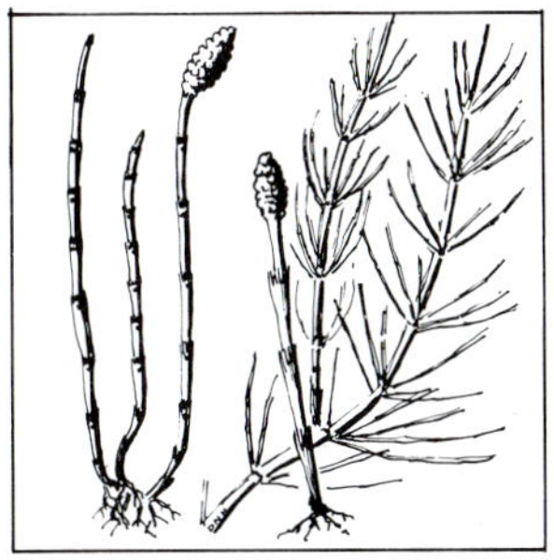

Sphenopsids. The cool forests of the Carboniferous contained many sphenopsids, of the subdivision Sphenopsida, as well as lycopods. They grew in a wide variety of shapes and sizes, many being good-sized trees. During the Permian they too declined in numbers and species; today there remains only a single genus, *Equisetum*, with some 25 to 30 species. The most familiar are the "horsetails" or "scouring rushes" that grow along the sandy riverbanks or railroad embankments.

14–9 A summary of the algae

Name	Cell characteristics	General morphology	Reproduction
Cyanophyta (blue-green algae) 150 genera 1500 species	Procaryotic cells—contain no nuclei or other cell organelles.	Single cells, filaments or simple colonies.	Asexual only by cell division—no form of sexual reproduction known.
Euglenophyta (euglena algae) 25 genera 450 species	Eucaryotic cells—gullet present as well as chloroplasts. Contain eyespot; colorless species present; uninucleate; thylakoids in series of two to five in chloroplasts; nuclear envelope intact during mitosis; microtubules found only inside nucleus.	Motile, single cells.	Asexual by cell division—sexual reproduction not definitely known.
Pyrrophyta (dinoflagellates) 125 genera 1000 species	Eucaryotic cells—may or may not have a rigid cell wall. Some have eyespot; thylakoids stacked in groups of three; many are colorless and some engulf food; uninucleate; chromosomes composed only of DNA (no protein); nuclear envelope intact during mitosis; chromosomes appear attached to nuclear envelope.	Mostly single cells—most motile.	Asexual by cell division—sexual reproduction rare. Haploid generation dominant.
Chrysophyta (golden algae, diatoms) 165 genera 5800 species	Eucaryotic cells—uninucleate. Some lack chlorophyll; thylakoids stacked in groups of three; spindle forms outside of nucleus, and after nuclear envelope breaks down, chromosomes become attached to completed spindle.	Mostly single cells—some in simple colonies.	Primarily asexual by cell division. Sexual reproduction common—gametes may be both motile, or one motile (male) and one nonmotile (female).
Xanthophyta (yellow-green algae) 75 genera 400 species	Eucaryotic cells—most uninucleate, some multinucleate.	Mostly single cells—some in colonies and filaments.	Asexual by cell division, zoospores and fragmentation. Sexual reproduction in some by two motile gametes, or one motile (male) and one nonmotile (female).
Phaeophyta (brown algae) 240 genera 1500 species	Eucaryotic cells—generally uninucleate although in specialized parts (i.e., holdfasts) they may be multinucleate. Some lack pigments and are saprophytes or parasites; thylakoids in chloroplasts stacked in groups of three; mitosis similar to higher plants except that nuclear envelope persists through division, except at poles.	Great diversity starting with microscopic filaments (no single cells or colonies known) continuing to plants 150–210 feet or more in length with well-differentiated holdfast, stipe, and blades.	Alternation of multicellular generations—with the diploid generation usually dominant. Sexual reproduction common; two motile gametes or one motile and one nonmotile.
Rhodophyta (red algae) 400 genera 4000 species	Eucaryotic cells—in simple forms they are usually uninucleate; in more advanced forms they are multinucleate. Thylakoids in chloroplasts are unstacked and are continuous from one end of the chloroplast to the other; mitosis similar to higher plants.	Great diversity from single cells to elaborate, although not massive, plants.	Alternation of multicellular generations—frequently with isomorphic bodies. Diploid generation may be essentially parasitic, remaining attached to haploid generation; female gamete usually stationary with male gamete being carried to it.
Chlorophyta (green algae) 450 genera 7000 species	Eucaryotic cells—many are uninucleate but some are multinucleate. Thylakoids in groups of two to six generally running length of chloroplasts but some species show granalike stacks; mitosis as in higher plants, although part of nuclear membrane persists throughout mitosis.	Single cells, filaments, colonies and multicellular bodies having a blade and holdfast.	Alternation of generations—haploid dominant or isomorphic generations. Both gametes motile, or male motile and female nonmotile.

Flagellation	Pigments	Cell wall	Distribution	Storage products
None	Chlorophyll a, β carotene, phycocyanin, phycoerythrin, xanthophylls.	Mucopeptide, pectin, protein.	Soil, fresh and salt water—both as bottom dwellers and plankton.	Glycogen Protein
One apical tinsel flagellum	Chlorophyll a & b, β carotene, xanthophylls.	Lack firm wall. Surrounded by flexible membrane composed of noncellulosic polysaccharides.	Most occur in freshwater—particularly polluted water.	Paramylon
Two flagella—one wrapped around cell; whiplash.	Chlorophyll a & c, β carotene, xanthophylls.	Wall, when present, consists of cellulosic plates.	Fresh, brackish, and salt water—also sand. Found living symbiotically with many animals.	Starch Fats Oils
1 or 2 flagella (depending on species or type of gamete); whiplash.	Chlorophyll a & c, β carotene, xanthophylls.	Cellulose and pectin and, in the diatoms, silicon in large amounts.	Found in both freshwater and marine environments, both as important part of the phytoplankton and on solid substrates.	Modified laminarin
Two flagella—one tinsel, one whiplash.	Chlorophyll a, β carotene, xanthophylls.	Cellulose and pectin (with some silicon).	Predominantly freshwater. Present both in plankton and on solid substrates; many are epiphytes on other algae and vascular plants.	Modified laminarin
Reproductive cells only. Usually two flagella on side of cell; one tinsel, one whiplash.	Chlorophyll a & c, β carotene, fucoxanthin, xanthophylls.	Cellulose, pectin, alginic acid, fucoidin.	Almost entirely marine—found particularly in the intertidal regions in northern waters.	Laminarin
No motile cells.	Chlorophyll a, β carotene, xanthophylls, phycoerythrin, phycocyanin.	Cellulose, pectin, agar carrageenan, mannan, some with calcium and magnesium carbonate.	Predominantly marine—widespread through all latitudes. Some are freshwater, a few terrestrial.	Modified starch
Two apical flagella—both whiplash.	Chlorophyll a & b, β carotene, xanthophylls.	Cellulose, pectin (occasionally with calcium carbonate). Some species lack cellulose in their walls.	Predominantly freshwater—but many common marine species. Also found in soil.	Starch

14–10 *A summary of the fungi*

Name	Cell characteristics	General morphology	Reproduction	Flagellation	Cell-wall composition	Distribution
Myxomycophyta 500 species	The true slime molds have a multinucleate body; the cellular slime molds are unicellular during feeding stage. Both types can ingest solid food particles.	Main vegetative stage of true slime molds is a thin sheetlike plasmodium lacking walls. Cellular slime molds have amoeba-like feeding stage.	Form spores in erect fruiting body. Syngamy and meiosis occur.	Two flagella of whiplash type.	All lack cell wall during feeding stage. Cellulose present in spore wall; no chitin present.	Found in moist soil, wood, dung, and decaying vegetation.
Chytridiomycetes	Uninucleate, unicellular species, as well as multicellular filamentous forms.	Most small, single cells; some are filamentous, multinucleate.	Motile gametes fuse to motile zygote; zoospores.	One posterior whiplash flagellum.	Predominately chitin.	Aquatic; many of the microscopic chytrids are parasites of diatoms.
Oomycetes	Multinucleate	Filamentous; hyphae form mycelium.	Both male and female gametes nonmotile; male gametes carried to egg by tube.	One whiplash and one tinsel flagellum, both anterior.	Cellulose present but no chitin; zoospores lack a wall.	Aquatic and terrestrial; many parasites of crop plants; cause late blight of potato and downy mildew of grape.
Zygomycetes	Multinucleate	Filamentous; hyphae form mycelium.	Fusion of hyphal tips of + and − strains.	No flagellated cells.	Predominately chitin.	All terrestrial; in soil, on dung, as saprophytes and parasites, particularly of insects.
Ascomycetes 30,000 species	Hyphae are subdivided by crosswalls with central pore; mitochondria, nuclei, etc. migrate from cell to cell.	Filamentous; hyphae form mycelium.	Hyphal fusion results in binucleate cells (dikaryotic) that form ascus; nuclei fuse, then undergo meiosis and produce ascospores; conspicuous fruiting bodies in many species.	No flagellated cells.	Chitin and insoluble glucans.	Many terrestrial, some aquatic. Saprophytes (pink bread mold, penicillium, wood-rotting forms), parasites (ergot) and yeasts.
Basidiomycetes 13,000 species	Hyphae contain crosswalls with pores in them; most cells are binucleate; clamp connections between cells.	Filamentous; hyphae form mycelium.	Fusion of nuclei in special enlarged cells (basidium); meiosis results in basidiospores. Fruiting body conspicuous in many species.	No flagellated cells.	Chitin present.	Terrestrial; in soil, wood, decaying vegetation, dung. Many are parasites: wheat rust, smuts, etc. Also includes the mushrooms.

Ferns. All the remaining vascular plants can be placed in a single subdivision: the Pteropsida. This grouping is largely based on the nature of the leaf and the way in which the vascular tissue leaves the stem and enters the leaf. The Pteropsida include the ferns, gymnosperms, and angiosperms.

When you look at most ferns, the part of the plant that you see is the leaf. The fern leaf is large and frequently dissected into leaflets, which may be further subdivided, giving the entire unit a beautiful lacelike appearance.

Spores develop in sporangia on the undersurface of the leaves or on specially modified leaves. The spores are light, wind-borne, and resistant. On germination they produce heart-shaped, green gametophytes. The antheridia and archegonia are formed on the lower surface of the gametophyte. Motile, multiflagellated sperm are produced, and the zygote is formed in the gametophyte. An embryo is formed in place and germinates to produce a new sporophyte generation.

The ferns are a successful group of land plants that, throughout their long history, have never formed a dominant part of the flora.

Pteridosperms or seed ferns. Many large plants with huge fernlike fronds were found as fossils of the Carboniferous and were believed to be true ferns. Additional fossil discoveries showed that these plants bore true seeds and pollen and therefore could not be ferns. They have been placed in a separate group: the pteridosperms or seed ferns. All of the pteridosperms are now extinct; none survived the Mesozoic.

Gymnosperms. These four groups of plants, the Cycadophyta, Ginkgophyta, Coniferophyta, and Gnetophyta, all have a common characteristic: the seeds are not contained within a special structure (the ovary found in flowering plants). The seeds are consequently termed naked, which can be misleading to the student, since the seeds of pines, which are gymnosperms, look similar to the seeds of many flowering plants. Later, in Chapters 27 and 28, the differences will become more meaningful.

The term gymnosperm is an ancient one, first used by Theophrastus around 300 B.C., when he wrote:

Some seeds again are enclosed in a pod,
Some in a husk,
Some in a vessel, and some are completely naked.

Gymnosperm in Greek means "naked seed." All gymnosperms are trees or shrubs, yet they vary greatly in form and the details of their reproduction. This variation is so great that many botanists no longer feel the gymnosperms are a natural group and have separated them into the following four groups.

Cycads. The modern cycads, in the class Cycadophyta, are little known plants, occurring in restricted tropical and subtropical locations. They resemble palms and tree ferns, most being small to moderate in height, although they may be trees some 60 ft high. Their leaves are tough and leathery, sometimes with sharp edges that can cut through clothing.

Ginkgo. Ginkgo biloba is the single surviving species of the Ginkgophyta that flourished from the late Paleozoic through mid-Mesozoic times. It is a beautiful tree, probably native to southeastern China. Now, this ancient tree is commonly found along city streets because it is practically smog resistant! It is highly prized for its fan-shaped leaves and general appearance. Despite the great difference in form between the cycads and ginkgo, the reproductive cycle is almost the same in its essential parts, including swimming sperms.

Gnetophyta. This small but highly unusual group of plants is composed of three divergent genera. *Ephedra* is a shrub of desert regions in both hemispheres; *Gnetum* is a tropical shrub; and *Welwitschia* is a strange desert plant of western South Africa. As different as these plants are, they still have features in common with the other four groups making up the gymnosperms.

Coniferophyta. Unlike the cycads, ginkgo, and the gnetophytes, the members of the Coniferophyta are familiar to everyone, since they include the pines, firs, spruces, hemlocks, larches, junipers, Douglas firs, yews, redwoods, and *Araucaria.* Most have needlelike or scale leaves, although a few species found in Australia and New Zealand have broad leaves. Most have cones —the junipers have fleshy scales and the yews a berry-like portion at the base of the seed.

The sporophyte (the tree or shrub) is a highly elaborate structure whose anatomy, physiology, and development we shall cover in later Chapters. The reproductive structures are cones: small, short-lived, delicate pollen cones, and large, long-lived, tough seed cones.

The conifers extend from the late Carboniferous to the present, reaching their greatest domination of the land flora in the mid-Mesozoic. Another related group of plants, the Cordaitales, were abundant in the early Carboniferous and overlapped with conifers in the late Carboniferous before they became extinct in the Mesozoic. The Cordaitales in turn overlapped the more primitive progymnosperms, which are known from the Upper Devonian.

Angiosperms—the flowering plants. The Cenozoic era in which we are now living has been called the age of the flowering plants, and the name is apt. The vast bulk of the plants surrounding you are members of the Angiospermae. These flowering plants are found everywhere—the grasses underfoot, the oaks by the side of the walk, the lilacs in bloom in the spring, the floating duckweed on the lake, cacti in the desert, the snow plants above tree line in the mountains and subarctic, and the huge trees of the tropical rain forest with the orchids and bromeliads growing on their branches. The distinguishing feature of this vast array of plants is the flower and the fruit that develops from the ovary of the flower.

The Classification of Plants

Are the labels put on the various plant groups the right labels or the only labels? It is possible to ask such questions because these labels are for man's understanding, but the groups are a result of evolution. Man strives to unravel the relationships between plants; he seeks to erect a phylogenetic scheme. Using all the evidence at his command (morphological, cytological, biochemical, physiological, or any other), many botanists have attempted to place all of the known species of plants into a framework of interrelated categories. The aim of such exercises (beyond the pure enjoyment it gives to many botanists) is to classify plants according to their evolutionary sequence, thus relating the species to their ancestors.

Clearly, this is no easy task, and one can readily predict that, knowing human nature and the fact that the data are far from complete, each man will arrive at a different scheme. Seven such schemes are presented in Table 14–11. They range in time from 1883 to 1970 and in numbers of divisions from 4 to 24.

The oldest system shown, that of Eichler (1883), is also the simplest. He saw the plant kingdom as four great groups: the Thallophyta with their relatively simple bodies and unprotected eggs and sperms; the Bryophyta as primitive land plants lacking vascular tissue but with jackets of sterile cells around the eggs and sperms; the Pteridophyta, true land vascular plants but lacking seeds; and the Spermatophyta as the most advanced land plants with seeds. There is a definite elegance to Eichler's scheme—a unity that seemed strong but was actually an illusion.

A tremendous mass of data, particularly fossil and microscopic observations, concerning plants past and present accumulated in the late 1800s and the first half of the 1900s. These data led many to question the four-division concept and resulted in Tippo's reevaluation of the scheme. He saw the fungi and algae as still related in a subkingdom but now raised Eichler's classes to divisions (a promotion equivalent to jumping from Colonel to General). The fungi, the Eumycophyta, were kept together in a single division, but the algae became seven separate divisions. In dealing with the land plants, Tippo grouped all of them in a subkingdom on the basis of a character not used by Eichler—the embryo. He divided this enormous group of plants into only two divisions: the nonvascular, the Bryophyta, and the vascular land plants, the Tracheophyta. Obviously, to Tippo the presence of xylem and phloem is of far greater importance than the other characters. The division Spermatophyta was abandoned, and the ferns, gymnosperms, and angiosperms placed in the same subdivision. Tippo's system was a major contribution, and it is still a major concept in phylogenetic thinking.

In 1957 Bold published a book on the morphology of plants in which he proposed a new scheme that, in some ways, is an extension of the concept Tippo applied to the algae. In Bold's approach, all of Eichler's and Tippo's classes and subdivisions become divisions. Bold feels that using a single common attribute such as xylem and phloem is open to serious doubt, and he abandoned the Tracheophyta as a separate division—even though this gave him a division (Ginkgophyta) with a single species. Also abandoned were the concepts of bryophytes and gymnosperms. The old Bryophyta is not part of Bold's system. He sees two separate divisions: one containing the liverworts (called the Hepatophyta) and one containing the mosses (called, by Bold, the Bryophyta). The Gymnospermae gave way to four separate divisions. What Bold is saying in his scheme is that we don't have enough facts to be certain of the relation of any of these, so let's keep them separate until we do.

On the whole, Scagel and his coauthors agree with Bold. They are obviously unwilling to go as far as he with the fungi and the bryophytes, but their general approach is basically the same. They also did not follow Bold in substituting Microphyllophyta for the old name Lycopsida (or Lycophyta), which Bold declared quite meaningless ("plant having the appearance of a wolf"). The name Microphyllophyta has not caught on; so we go on using the meaningless Lycopsida because it is an old familiar name. Bold also tried to correct the algal names. For example, Chlorophyta actually means green (chloro) plant (phyta) and not green algae. By changing it to chlorophycophyta, we have green algal (phyco) plant, which is much more accurate. He did this for all the algae but has fared no better than with Lycopsida. To most botanists the green algae are the Chlorophyta, and that is that!

Delevoryas is clearly more closely associated with Tippo's way of thinking than Bold or Scagel, but the two schemes (Delevoryas and Tippo) are not identical. Delevoryas has dropped the subkingdom Embryophyta, which was attacked by many on the grounds that some algae have embryos similar to land plants, although they are not dormant. He also dropped the Gymnospermae and substituted four classes for them.

The most radical departure from the older systems is Doyle's. He has rejected the concept of a unified class of water molds and has substituted a series of classes in its place. But far more radical is his treatment of the green land plants. He has placed them in a single division with the green algae. This results in over 75% of the known species in one division! The unifying features of this group are the types of chlorophyll (a and b), the type of flagella when present (all whiplash), and the type of starch. These are all truly basic features of the plants involved, but Bold (because he objects to the vascular tissue as a common feature being given too much importance) would clearly view them as much less compelling.

All of these schemes are basically conservative because they dealt with plants as one half of the living organisms; the animal kingdom represents the second half. A newer, and in some ways more radical, approach is to invoke multiple kingdoms. In such a system we can have the following distribution of living things:

KINGDOM MONERA:	procaryotic cells; the bacteria and blue-green algae.
KINGDOM PROTISTA:	unicellular or colonial unicellular eucaryotes—Protozoa, Euglenophyta, Pyrrophyta, Xanthophyta, Chrysophyta, Myxomycota.
KINGDOM FUNGI:	all of the true fungi.
KINGDOM PLANTAE:	multicellular and related unicellular plants; multicellular algae, bryophytes, and vascular plants.
KINGDOM ANIMALIA:	multicellular, eucaryotic animals.

While such schemes are excellent in emphasizing the similarity of the procaryotic organisms and the similarity of the protozoa with many algae, it is frequently at the expense of unity of the algal divisions and other groups. It is also unclear to many why the fungi warrant a whole kingdom to themselves.

What is the relevance of all of these schemes to the student? Why don't we just tell you which one is the right one, and you can memorize it! Clearly, there is no right one. The plants are there; they are where they are

14–11 Some examples of systems of classification of the plant kingdom

Eichler (1883)	Tippo (1942)	Bold (1957)	Scagel et al. (1965)	Delevoryas (1966)
	Subking.: Thallophyta			
Div.: Thallophyta				
Cl.: Fungi				
Schizomycetes	Div.: Schizomycophyta	Div.: Schizomycota	Div.: Schizomycophyta	Div.: Schizomycophyta
	Div.: Myxomycophyta	Div.: Myxomycota	Div.: Myxomycota	Div.: Myxomycophyta
Eumycetes	Div.: Eumycophyta		Div.: Eumycota	
	Cl.: Phycomycetes	Div.: Phycomycota	Cl.: Phycomycetes	Div.: Phycomycophyta
	Cl.: Ascomycetes	Div.: Ascomycota	Cl.: Ascomycetes	Div.: Ascomycophyta
	Cl.: Basidiomycetes	Div.: Basidiomycota	Cl.: Basidiomycetes	Div.: Basidiomycophyta
Cl.: Algae			Div.: Cyanophyta	Div.: Cyanophyta
Cyanophyceae	Div.: Cyanophyta	Div.: Cyanophycophyta	Div.: Euglenophyta	Div.: Euglenophyta
	Div.: Euglenophyta	Div.: Euglenophycophyta		
		Div.: Charophycophyta	Div.: Chrypophyta	Div.: Chrysophyta
Diatomeae	Div.: Chrypophyta	Div.: Chrysophycophyta	Div.: Xanthophyta	
			Div.: Pyrrophyta	Div.: Pyrrophyta
	Div.: Pyrrophyta	Div.: Pyrrophycophyta	Div.: Phaeophyta	Div.: Phaeophyta
Phaeophyceae	Div.: Phaeophyta	Div.: Phaeophycophyta	Div.: Rhodophyta	Div.: Rhodophyta
Rhodophyceae	Div.: Rhodophyta	Div.: Rhodophycophyta	Div.: Chlorophyta	Div.: Chlorophyta
Chlorophyceae	Div.: Chlorophyta	Div.: Chlorophycophyta		
	Subking.: Embryophyta			
Div.: Bryophyta	Div.: Bryophyta		Div.: Bryophyta	Div.: Bryophyta
Cl.: Hepaticae	Cl.: Hepaticae	Div.: Hepatophyta	Cl.: Hepaticae	
Cl.: Musci	Cl.: Musci	Div.: Bryophyta	Cl.: Musci	
			Cl.: Anthocerotae	
Div.: Pteridophyta	Div.: Tracheophyta			
	Subdiv.: Psilopsida	Div.: Psilophyta	Div.: Psilophyta	Div.: Tracheophyta
Cl.: Lycopodinae	Subdiv.: Lycopsida	Div.: Microphyllophyta	Div.: Lycophyta	Subdiv.: Psilopsida
Cl.: Equisetina	Subdiv.: Sphenopsida	Div.: Arthrophyta	Div.: Arthrophyta	Subdiv.: Lycopsida
	Subdiv.: Pteropsida			Subdiv.: Sphenopsida
Cl.: Filicinae	Cl.: Filicinae	Div.: Pterophyta	Div.: Pterophyta	Subdiv.: Pterophyta
Div.: Spermatophyta				Subdiv.: Spermopsida
Cl.: Gymnospermae	Cl.: Gymnospermae		Div.: Cycadophyta	
	Subcl.: Cycadophytae	Div.: Cycadophyta	Div.: Ginkgophyta	Cl.: Cycadophyta
	Subcl.: Coniferophytae	Div.: Ginkgophyta	Div.: Coniferophyta	Cl.: Ginkgophyta
		Div.: Coniferophyta	Div.: Gnetophyta	Cl.: Coniferophyta
		Div.: Gnetophyta		Cl.: Gnetophyta
Cl.: Angiospermae	Cl.: Angiospermae	Div.: Anthophyta	Div.: Anthophyta	Cl.: Angiospermophyta
			Cl.: Dicotyledoneae	Subcl.: Dicotyledoneae
			Cl.: Monocotyledoneae	Subcl.: Monocotyledoneae
Number of Divisions: 4	12	24	21	14

Harris (1969)	Doyle (1970)	This text (1972)	English names		Approx. number of living species
Div.: Schizophyta	Div.: Schizophyta	Div.: Schizophyta	bacteria		1600
	Div.: Myxomycota	Div.: Myxomycota	slime molds		500
Div.: Eumycophyta	Div.: Eumycota	Div.: Eumycota	true fungi		100,000
	Cl.: Chytridomycetes	Cl.: Chytridomycetes	water molds	fungi	
	Cl.: Oomycetes	Cl.: Oomycetes			
	Cl.: Zygomycetes	Cl.: Zygomycetes			
	Cl.: Ascomycetes	Cl.: Ascomycetes	sac fungi		
	Cl.: Basidiomycetes	Cl.: Basidiomycetes	club fungi		
Div.: Cyanophyta	Div.: Cyanophyta	Div.: Cyanophyta	blue-green algae		1500
	Div.: Euglenophyta	Div.: Euglenophyta	euglenids		450
Div.: Chrysophyta	Div.: Chrysophyta	Div.: Chrysophyta	golden algae, diatoms		425
	Div.: Xanthophyta	Div.: Xanthophyta	yellow-green algae	algae	5800
Div.: Pyrrophyta	Div.: Pyrrophyta	Div.: Pyrrophyta	dinoflagellates		1000
Div.: Phaeophyta	Div.: Phaeophyta	Div.: Phaeophyta	brown algae		1500
Div.: Rhodophyta	Div.: Rhodophyta	Div.: Rhodophyta	red algae		4000
Div.: Chlorophyta	Div.: Chlorophyta	Div.: Chlorophyta	green algae		7000
	Subdiv.: Chlorophycophytina (green algae)				
	Subdiv.: Embryophytina				
Div.: Bryophyta	Cl.: Bryopsida	Div.: Bryophyta			
		Cl.: Hepaticae	liverworts		
		Cl.: Musci	mosses	bryo-phytes	23,600
		Cl.: Anthocerotae	hornworts		
		Div.: Tracheophyta	vascular plants		
Div.: Tracheophyta		Subdiv.: Psilopsida	psilopods	ferns	3
Subdiv.: Psilopsida	Cl.: Psilopsida	Subdiv.: Lycopsida	lycopods	and	1200
Subdiv.: Lycopsida	Cl.: Lycopsida	Subdiv.: Sphenopsida	horsetails	allies	40
Subdiv.: Pteropsida		Subdiv.: Pteropsida	ferns		10,000
Cl.: Filicinae	Cl.: Pteropsida	Cl.: Filicinae			
	Cl.: Spermopsida		seed plants		
Cl.: Gymnospermae	Subcl.: Gymnospermidae				
		Cl.: Cycadophyta	cycads		100
		Cl.: Ginkgophyta	ginkgos	gymno-	1
		Cl.: Coniferophyta	conifers	sperms	520
		Cl.: Gnetophyta	gnetum		44
Cl.: Angiospermae	Subcl.: Angiospermidae	Cl.: Angiospermae	flowering plants	angio-	
Subcl.: Dicotyledoneae	Series: Dicotyledoneae	Subcl.: Dicotyledoneae	dicots	sperms	236,500
Subcl.: Monocotyledoneae	Series: Monocotyledoneae	Subcl.: Monocotyledoneae	monocots		48.500
10	11	13		Total:	461,283

through evolution; there must have been some common ancestry for present-day plants. Yet the record has been written on the winds of millions of years, and only the survivors and a few buried, incomplete, transformed remains are left to outline the great epic of plant evolution. Into the gaps man can throw only his imagination and ingenuity. You should not think of one author and his system as right or wrong. Rather, as you study the plant kingdom, attempt to understand the basis for the different schemes. In this way you will deepen your knowledge of plants at all levels and may even arrive at your own system—which will be just as right or wrong as the data on which it is based! In the meantime, this book will use the following system, which is not innovative, radical, or ingenious but which we hope will be functional.

THE PLANT KINGDOM

Division: Schizophyta	bacteria
Division: Cyanophyta	blue-green algae
Division: Myxomycota	slime molds
Division: Eumycota	fungi
Class: Chytridomycetes	water molds
Class: Oomycetes	
Class: Zygomycetes	
Class: Ascomycetes	sac fungi
Class: Basidiomycetes	club fungi
Division: Euglenophyta	euglena
Division: Chrysophyta	golden algae, diatoms
Division: Xanthophyta	yellow-green algae
Division: Pyrrophyta	dinoflagellates
Division: Phaeophyta	brown algae
Division: Rhodophyta	red algae
Division: Chlorophyta	green algae
Division: Bryophyta	
Class: Hepaticae	liverworts
Class: Anthocerotae	hornworts
Class: Musci	mosses
Division: Tracheophyta	
Subdivision: Psilopsida	psilopsids
Subdivision: Lycopsida	lycopods
Subdivision: Sphenopsida	horsetails
Subdivision: Pteropsida	
Class: Filicinae	ferns
Class: Cycadophyta	cycads
Class: Ginkgophyta	ginkgo
Class: Coniferophyta	conifers
Class: Gnetophyta	gnetum
Class: Angiospermae	flowering plants
Subclass: dicotyledoneae	
Subclass: monocotyledoneae	

End Paper 14–1 The Making of Fossils

Have you ever sat on the edge of a pond or stream on an autumn day and watched leaves drift down onto the water? Some float and are carried away on the gentle currents. Others slowly sink beneath the surface and fade from sight.

Have you ever wondered what happens to these? Most decay, serving as food for the bacteria and fungi present in the pond. A few will fall to the deeper parts of the pond, where the water is colder and less oxygen is present so that decay is delayed. They may become covered with a layer of mud, which further reduces decay. The layer of mud may become thicker and thicker, and more and more leaves may be embedded within it as years pass. The mud may then be slowly converted into stone, and the impressions of the leaves are now preserved in the rock. Thus the leaf you saw fall into the pond may become a fossil.

Fossils and Geology

Many fossils are compressions in which the plant material is pressed, and a mold of the buried part is made in the rock. In the case of these compressions or the impressions just described, little of the actual plant is usually left. But this is not always true. Many fossils preserve much of the original plant material, petrified wood, for example. In this case, the wood is infiltrated with mineral substances that crystallize and harden in the interior of the cells and in the intercellular spaces. It effectively embeds the plant parts, primarily the cell walls, in a hard matrix.

This is no molecule by molecule replacement of the substance of the wood by minerals, as is popularly believed, but an extremely good example of encasement. Microscopic detail can be preserved by this method, and the fossils can be treated in various ways so that the structure of the cells and tissues can be observed. One way is to polish the surface of such a fossil. The surface is then dipped into dilute acid that dissolves the rock but not the cell wall. The end result is that the walls project above the surface of the rock. Next, a thick solution of celluloid is poured over the surface, filling the spaces between the projecting walls. When the celluloid dries, it can be peeled from the surface of the rock. In this peel are embedded the pieces of wall in their original positions and thin enough to be observed with the microscope. Such preparations are aptly called peels and are used in the detailed study of fossil plants. Many fossils cannot be effectively etched with acid. These may have to be sawed and the sections polished thinly in order to be viewed with the microscope.

Not all fossils are so altered from the living plant as those found in impressions, compactions, or petrified parts. The silica walls of diatoms are unchanged by the death of the organism and are resistant to decay. When the cells die the walls fall to the bottoms of the lake or ocean and accumulate in ever-increasing thickness. The result is diatomaceous earth, which is extensively mined and consists of fossil diatoms. Fossil remains of other members of the golden-brown algae are also found in the ooze on the ocean floor at depths of 3000 to 10,000 ft. The walls of pollen grains are also extremely resistant to decay and accumulate in bogs and lakes.

There are still other means by which plant parts have been preserved or impressions of their remains made. In general, however, woody parts of plants, leaves, and fruits are both best and most often preserved. The nonwoody parts such as flowers and small plants lacking woody parts decay more readily and are less frequently found as fossils. Moreover, because of the way fossilization most commonly occurs, plants growing in lowlands near abundant water are far more likely to be represented in the fossil record than those growing in dry, highland locations. The fossil record is thus an incomplete one, although hundreds of thousands of fossils are known. Still, it is possible to confidently reconstruct many individual plants and even whole communities of plants that are now extinct.

To understand the fossil record, it is necessary to appreciate the way rock layers or strata are formed. Rain and wind are constantly wearing away the highlands, hills, and mountains and carrying the material to the lowlands, where it is deposited. There, layer after layer of various kinds of materials gradually builds up the level of the earth. As the accumulations become deeper, the pressure on the lower levels increases, and the sedimented material is consolidated into rock. Ultimately, upheavals occur, and the strata may be raised, bent, fractured, or upended. But the original relationship of the strata is nearly always maintained: the oldest layers are on the bottom, the youngest on the top. Consequently, fossils found in the deepest strata are generally the oldest, while those in the upper strata are the youngest.

Fossils are not found nor are they equally well preserved in all strata. For example, fossils from the Paleozoic Era are found in parts of the central and eastern United States where 250 million years of accumulated rock strata can be found in an almost regular succession. Strata from the middle of the Cretaceous to the present are well represented in the western states. This means that the fossil record is a series of discontinuous accumulations that must be placed in proper relationship to one another. This also means that the paleontologist must know where to search in order to find fossils. Obviously, the paleontologist must also be a first-rate geologist. Yet, no matter how diligently he searches, the fossil record will always remain incomplete and show a high degree of randomness in those organisms or parts of organisms that are preserved.

15

Viruses

TMV

The most famous and best understood of the plant viruses is the one that causes a disease of tobacco plant, known as tobacco mosaic virus or TMV. This was the first virus to be isolated and crystallized, as well as the first to be seen in the electron microscope. It was also the first virus to be taken apart, reconstituted, and still remain infective in the host. A vast amount of research has been carried out involving TMV, and this research has greatly increased our knowledge of viruses in general.

A Isolated TMV can be seen with the electron microscope after special treatment. First, solutions of the virus are sprayed on plastic-coated grids. Then, the virus particles are lightly covered or "shadowed" with a metal so that their structure is brought out under the electron microscope. TMV, treated in this way, appears as tiny rods. The spheres (880 A in diameter) seen in this photograph are polystyrene and are used to calibrate the size of the virus particles.

B TMV can also be sectioned and viewed with the electron microscope, as is done in this photo. Then, the virus appears to consist of rings or layers of material. The outer, less dense layer is composed of protein, while the inner, dark layer is composed of ribonucleic acid (RNA). Thus TMV is like all other viruses because it consists of a protein and a nucleic acid (either RNA or DNA, but never both).

C In cells heavily infected with TMV, the virus may accumulate in crystalline arrangements. All the viral particles are essentially the same length, and they become arranged in regular alternating rows within the cytoplasm of the cell.

A

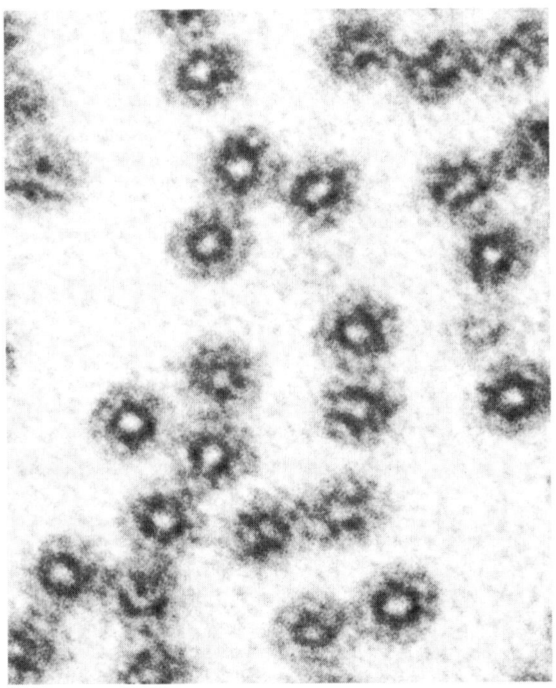

B

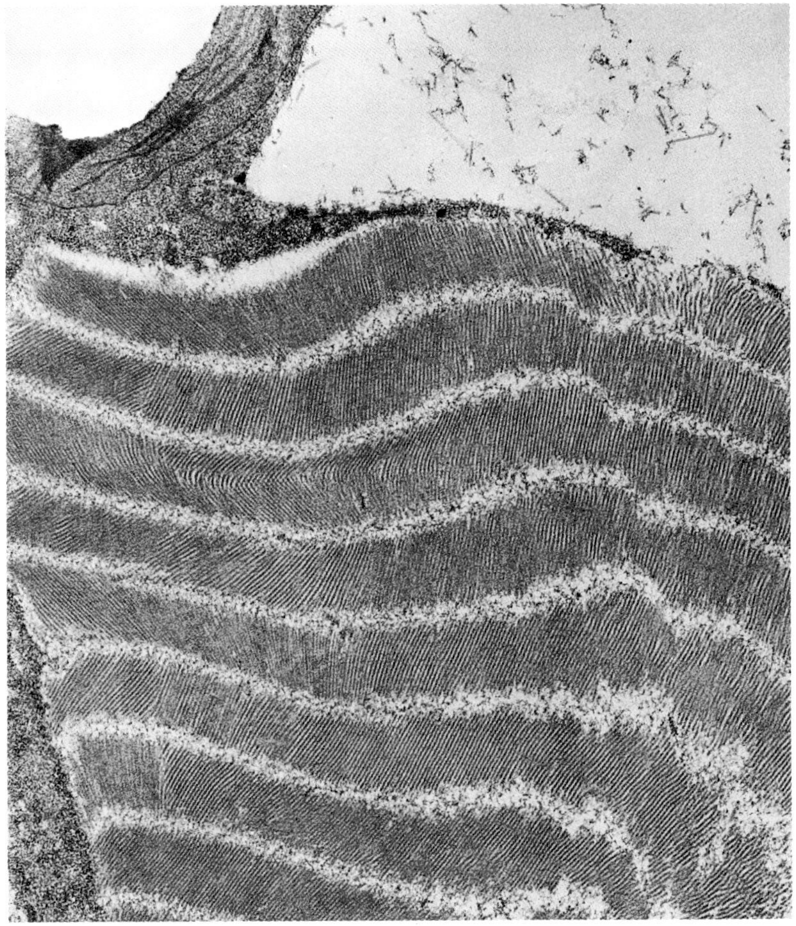

C

D How then does a virus move from one cell to the next or from one part of the plant to another? In the case of beet-yellow virus, particles in the plasmodesmata connect the cells, as shown in this photograph.

E Beet-yellow virus has also been found in the sieve tubes of the phloem and can obviously be carried through the plant by the same system that carries carbohydrates and other organic substances. This photograph actually shows virus particles in the openings or pores between two such cells in the leaf of a beet.

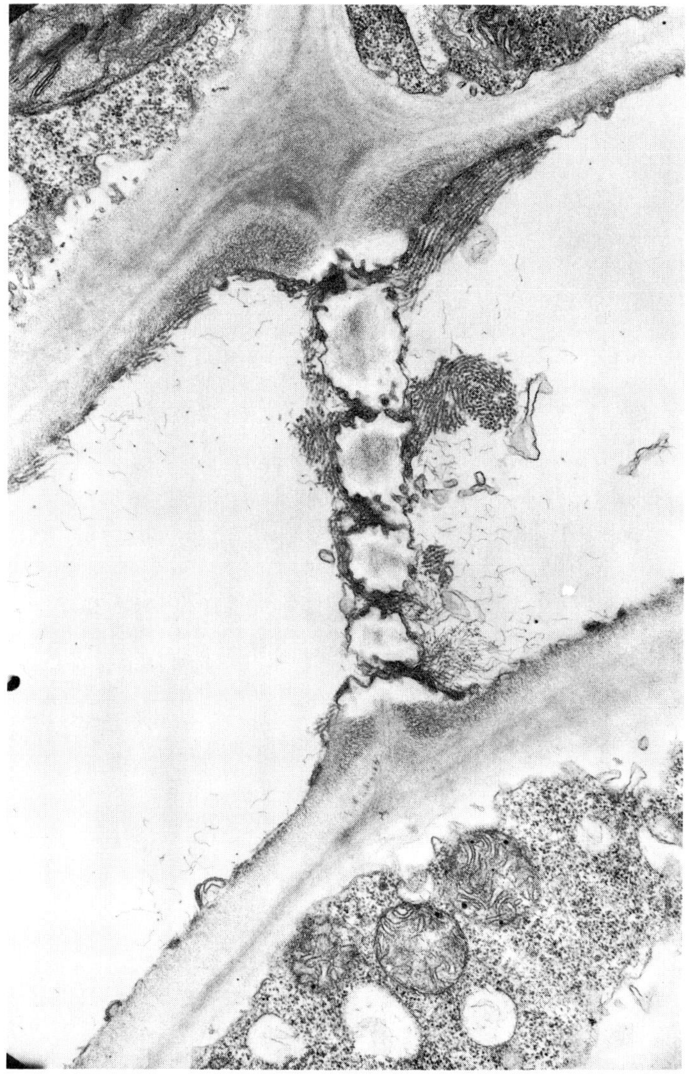

D

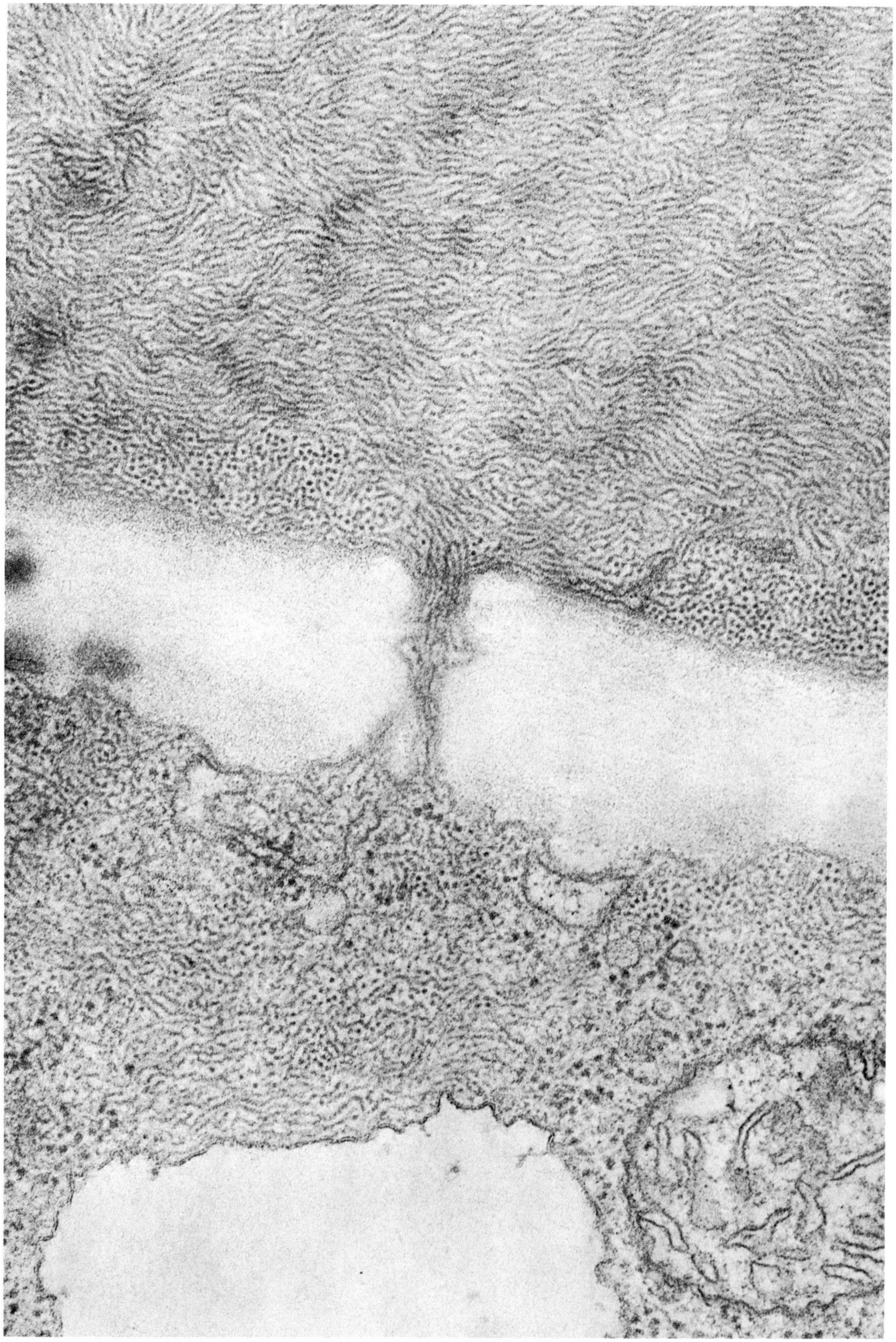

E

F Accumulations of virus particles can also be found in other parts of the cell. In this photograph concentrations of TMV are seen in the chloroplast of a tobacco cell.

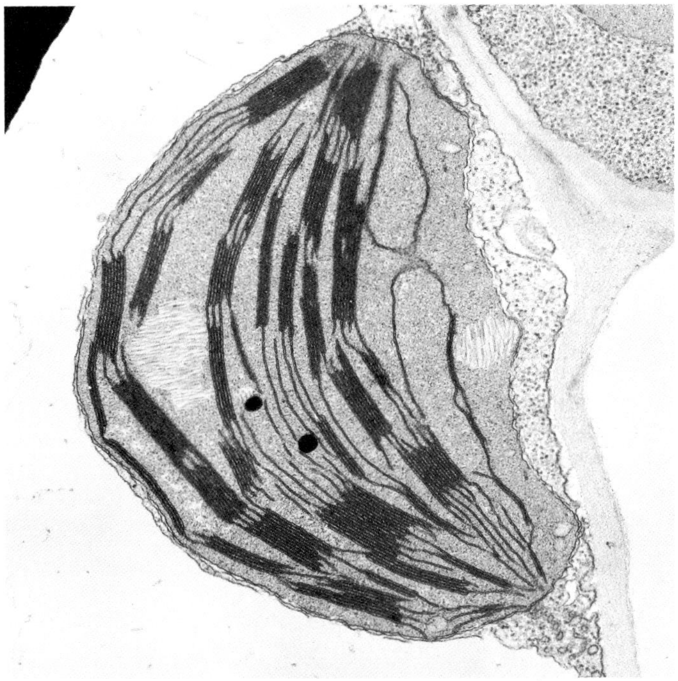

F

Forming the staple diet of hundreds of millions, rice is a way of life in Asia. The success of the rice crop means life or death. Most rice (*Oryza sativa*) is grown in flooded paddies under conditions where the roots are in an anaerobic environment that would kill most plants. In the paddies up to three crops per year can be obtained, and the fertility of the paddy can be retained year after year. The cultivation of rice requires special skills and is such that the more time and energy a farmer puts into his crop the greater the yield. Pound for pound, rice is the best energy source among all common cereal crops.

Man is not the sole consumer of rice. Many insects (among them leafhoppers and aphids) attack the plant. These insects feed and move on, frequently damaging the plant. The young rice plants may become stunted or dwarfed not because of the insect bite but because of the disease transmitted by the insect. A single female leafhopper may destroy many plants, since she is infective her entire life. In addition, each young leafhopper that the female produces is similarly infected—for up to seven generations. The insects are acting as passive transfer agents or vectors for the disease. This disease, rice stunt, is not a bacterial or a fungal disease but one caused by a virus.

What are viruses? Viruses are small, simple, and cannot reproduce outside cells. They are, in effect, units of information lacking their own machinery to transcribe the message. To do this, they must appropriate the machinery of a living cell. Viruses are packages of nucleic acid, either DNA or RNA (but never both), wrapped in a protein cover. The proteins protect the nucleic acid from the environment, are important in finding the proper host cell through a specific affinity for certain kinds of cells, and aid in the entrance of the nucleic acid into the host cell.

Because of their very nature all viruses are infective and parasitic. They cause a large number of diseases in animals and plants, attacking organisms ranging from bacteria to man and angiosperms. While simple in structure, they have a variety of shapes (spheres, cubes, bricks, irregular) and sizes (from 0.01 to 0.5 μ).

Viruses that infect plants contain RNA and are sometimes called the plant viruses. Those that live in bacteria and animals are usually, but not always, DNA viruses. In general, the RNA-containing viruses are simpler in organization than the DNA-containing viruses. The most complex ones are those that infect bacteria.

TMV

One of the most thoroughly understood viruses is the tobacco mosaic virus or TMV. This is a rod-shaped, RNA-containing virus that causes a common and damaging disease in the leaves of tobacco plants. TMV is famous because it was the first virus to be purified to the point where it could be crystallized. This was accomplished by Wendell Stanley, University of California, Berkeley, in 1935. The fine, needle-shaped crystals could be dissolved, and the virus could then infect healthy plants.

TMV consists of a protein coat composed of 2200 identical ovoid protein units termed capsomeres (Fig. 15–1). Each capsomere consists of 158 amino acids. The capsomeres are arranged in the form of a helix with their long axis on the radius. The RNA is also in the form of a helix and is fitted in a groove in the capsomeres near the center of the helix. The whole unit is some 3000 A long. It has been experimentally possible to strip the protein from the RNA and show that the RNA by itself is infective, although less than 1% as infective as the complete unit. The RNA and the protein can also be recombined to form fully infective particles. Moreover, under proper conditions the capsomeres will spontaneously aggregate around the RNA.

TMV and similar viruses have relatively simple mechanisms for entering the host cell. The proteins of the capsid interact in a highly specific way with the surface of the host cell. This interaction forms the basis of the identification by the virus of an acceptable host. But the capsid proteins do not react with every part of the surface of the host cell (even if it is the right host), only with certain receptor sites, and not all cells have such sites.

Many of the simple viruses seem to pass through the plasma membrane of the cell—protein coat and all. Some viruses are known to contain enzymes that aid in this penetration. Once inside the cell, the enzymes of the host quickly remove the protein coating, and the

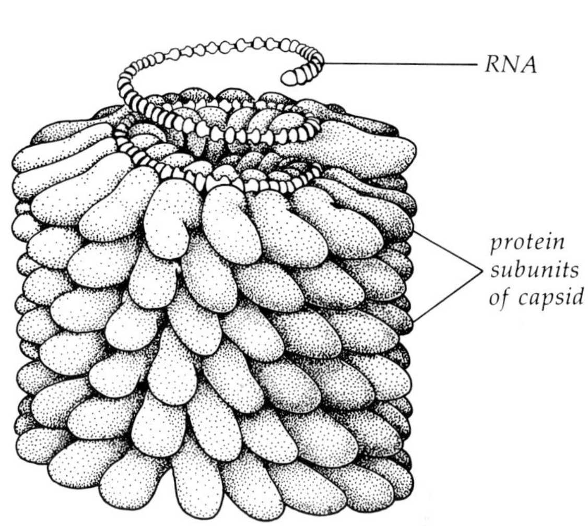

15–1 The tobacco mosaic virus (TMV) shows the arrangement of the RNA and the capsomeres composed of protein. The entire virus is much longer than shown here.

viral RNA is free to act. The viral RNA replicates and acts as a m-RNA that codes directly for viral proteins using host ribosomes. One of the unique features of RNA viruses is that they produce an enzyme that makes possible the replication of viral RNA. This enzyme, RNA replicase, is unknown in other organisms, none of which replicate RNA.

A more remarkable enzyme has been found that facilitates the synthesis of DNA from viral RNA. For many years this relationship was assumed to be impossible, but the work of Howard Temin and Satoshi Mizutani at the University of Wisconsin and David Baltimore at M.I.T., published in 1970, has shown the presence of the enzyme in at least seven RNA viruses. Clearly, the relationship of this enzyme to RNA replicase must be clarified. It is rather startling that in an area of research as intensively explored as DNA-RNA interactions, fundamental discoveries can still be made.

As well known as is the structure of TMV, there are still many questions unanswered concerning other aspects of its synthesis and its relationship to the host. For example, where in the host cell does TMV duplication occur? Ultrastructural studies have revealed complete TMV particles in the nucleus, cytoplasm, plastids (including chloroplasts), and even the vacuole, although this is considered an artifact. Does this mean that TMV is synthesized in all of these places? No, since complete particles could move from one part to another. There is evidence that viral RNA is made in the nucleolus, viral protein in the cytoplasm, and assembly occurs in the cytoplasm. This does not explain TMV in the plastids, but it does provide an explanation for the large accumulations of viral protein lacking viral RNA found in some infected cells. The suggestion has also been made that different types of viruses are made in different parts of the cell.

How can viruses move from one cell to the next? This is another important question that has not been answered in full. Plant viruses fall into two general patterns with regard to movement throughout the plant: those that utilize the phloem and thus move rapidly through the plant and those that move directly from one cell to the next. TMV is in the second class, which are relatively slowly spreading viruses. But how TMV moves from cell to cell is unknown. For several viral diseases particles actually have been seen within the plasmodesmata. The curious thing is that the normal internal structure of the plasmodesmata is missing, and it is filled with virus. Although a search was made, no TMV particles have been found in plasmodesmata.

Large masses of TMV and other viruses frequently occur in infected cells. These form crystalline structures easily seen in the light microscope. The various viruses form characteristic accumulations that can be used to identify the virus. It is curious that even large masses of TMV particles will not prevent a cell from dividing, with both daughter cells containing the virus.

Bacteriophages

The viruses that attack bacteria are called bacteriophages (phages for short) and vary considerably in size and form. One of the largest and most complex is *phage* T_4, which attacks *E. coli*. This virus consists of a hexagonal head some 700 A in diameter containing a double strand of DNA (Fig. 15–2). Most phages contain DNA; only a few have RNA in them. The DNA of T_4 is

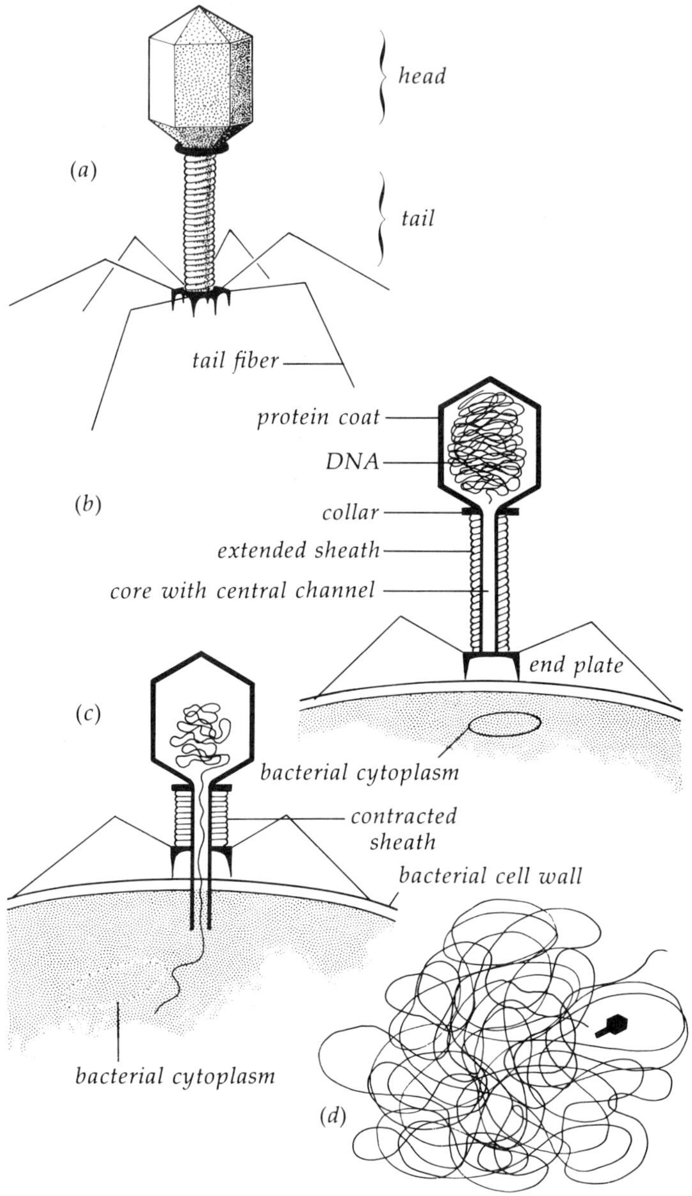

15–2 The morphology of a bacteriophage (T_4) and its mode of attack on a bacterium is diagrammed. The long tail fibers, shown extended in (a), apparently aid the virus in making contact with the bacteria. The end plate projections, shown in (b), help secure the virus, and the contraction of the sheath plunges the core of the tail through the wall of the bacteria. The DNA in the head is injected into the bacteria, as shown in (c). The DNA is highly coiled in the head; the amount of DNA in the head is represented by the line in (d) for a phage the size shown.

long, having a molecular weight of approximately 120 million, and is tightly coiled within the head. There is a protein inside the head associated with the DNA that is believed to help hold the DNA in its highly coiled condition. The head itself is composed of capsomeres of protein, each with a molecular weight of about 80,000. At the base of the head is a protein collar with a complex tail assembly attached. The tail consists of a sheath of 144 contractile protein subunits wrapped around a hollow protein core. The free end of the tail is attached to a flat, hexagonal base plate with 6 spikes protruding from it. Finally, extending from these are long, slender, twisted, rigid tail fibers.

The tail fibers are important because they attach the phage to the surface of the host cell (Fig. 15–2). Having attached itself to a suitable host by the tail fibers, the phage next digests a small hole through the wall of the bacterium by means of enzymes. The tail proteins now contract; the head collapses; and the DNA is shot into the bacterium. A "ghost" of the phage is left on the outside of the bacterium; a badly infected cell may look like a miniature porcupine with the empty ghosts adhering to it.

Only the DNA and a small amount of protein associated with it in the head of the virus particle enters the host—but this is enough. Almost immediately, the viral DNA begins to take over the direction of the host cell metabolism (Fig. 15–3). One of the first acts is the production of various enzymes and proteins necessary to carry out the first steps of viral reproduction. Among these is the production of DNAases that break down the bacterial DNA but do not attack the viral DNA. Not only does the viral DNA direct the degradation of the host DNA but it causes the production of enzymes that completely prevent the synthesis of new bacterial DNA. Instead, copies of the viral DNA, some 100, are produced.

Other enzymes coded by the viral DNA are involved in producing the proteins of the covering. An important facet of viral reproduction in the bacterium is timing, and many control factors are built into the mechanism. Thus, through elaborate and surprisingly well-understood mechanisms, the viral DNA directs the formation of new virus particles. One of the last enzymes to be produced is one that dissolves the host wall, releasing the new virus particles. The entire process, from infection to lysis with the production of several hundred new virus particles, takes *less than half an hour.*

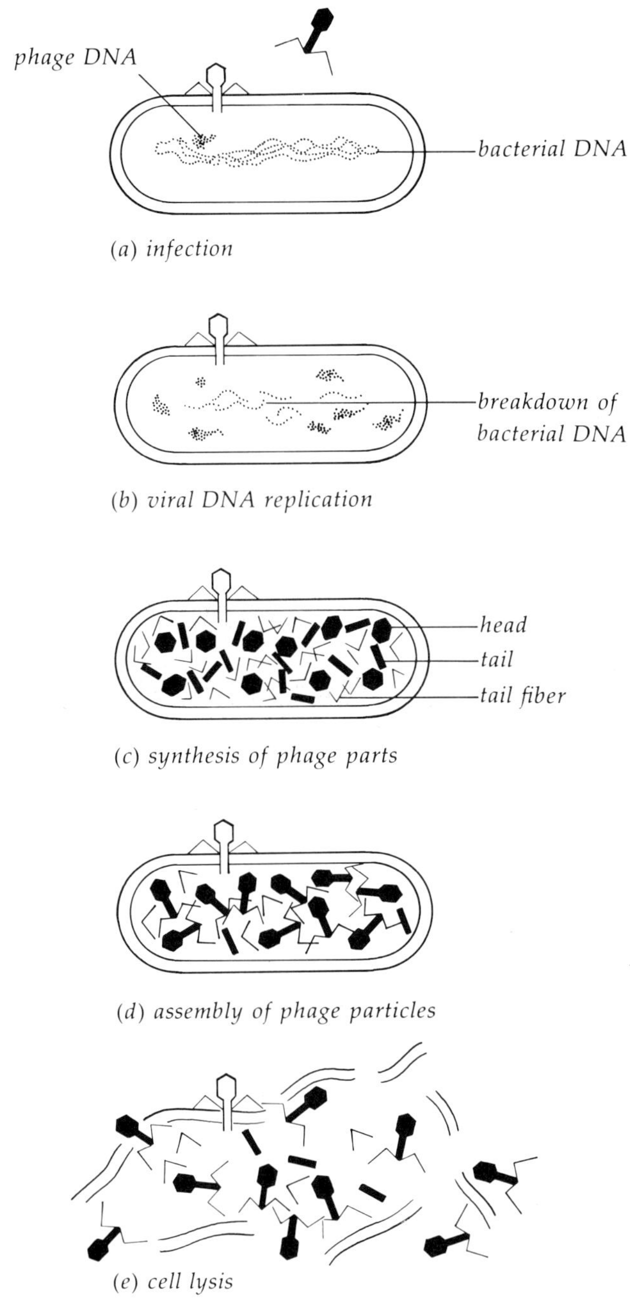

15–3 The developmental cycle of a virulent phage. The phage particles and bacterial cells are not drawn to scale.

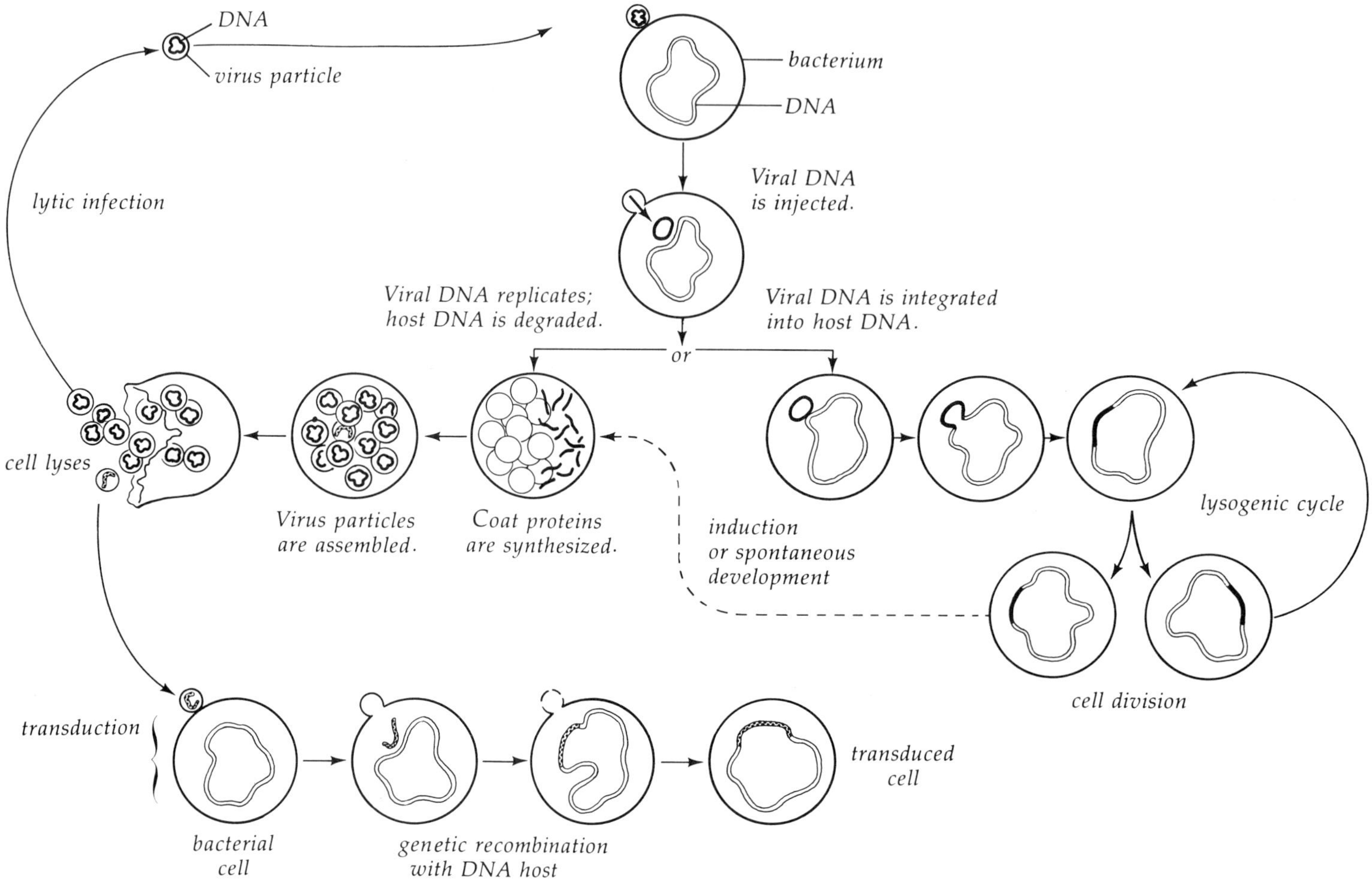

15–4 In the case of temperate phages, two developmental cycles are possible. The one shown on the left-hand side is the same as a virulent phage and results in the death of the bacterium and the formation of new viruses. The one on the right shows how the viral DNA is incorporated into the bacterial DNA and is replicated every time the bacteria divides. But under some conditions—even after many generations of bacteria—the virus may become virulent and destroy the cell. When the new virus particles are being assembled in the bacteria, a mistake may occur, and a piece of bacterial DNA may be coated with viral protein. This is capable of infecting a new bacterium, while the old bacterial DNA fragment becomes a part of the new host bacterium's DNA. This is transduction, shown at the bottom of the diagram.

Not all bacteriophages are as virulent as T_4. Some are known as temperate phages and are said to have a lysogenic phage-bacterium relationship (Fig. 15–4). In these viruses the DNA injected into the host does not take over the cell; instead the viral DNA becomes part of the bacterial DNA, forming one continuous loop of DNA. For many thousands of generations both the viral and bacterial DNA are synthesized together, with an immunity protein being formed that prevents both the independent replication of the viral DNA and the infection of the bacterium by other viruses. This integrated condition will persist until an agent such as X-rays, ultraviolet light, or some chemical frees the viral DNA. The liberated viral DNA then acts in the usual virulent manner, producing new virus particles and killing the cell.

There is the possibility of error in how the viral DNA is released. The breaks that liberate the viral

DNA may not occur in the right place, and a part of the viral DNA may be left behind with an equal size piece of bacterial DNA as part of the new viral DNA strand. The new viral DNA is reproduced, packaged, liberated, and infects more bacteria (Fig. 15–4). It may lack the information necessary to replicate and kill the bacteria, but the bacterial DNA, carried along with the viral DNA, may recombine with the DNA of the new host cell. As incredible as it seems, the virus may act as a carrier of genetic material involved in an indirect form of sexual recombination. This phenomenon, called transduction, has been widely used in bacterial genetics to investigate gene action.

Origin of Viruses

Where did the viruses originate? It is intriguing to think of viruses as the continuation of a line of early life—procellular life. As romantic as this sounds, most biologists concerned with the problem believe this is not the case. Instead, viruses are generally held to have some ancestral connection with procaryotic cells such as bacteria. In fact, the theory has been advanced that the viruses are a highly stripped down version of bacterial chromosomes or, in the case of RNA viruses, packages of bacterial m-RNA.

End Paper 15–1 Viruses and Plant Diseases

Viruses, particularly the phages, are the favorite experimental subjects of molecular biologists, but many of them are major pests to farmers. There are literally thousands of viral diseases of plants, many of them of the most virulent type. They take a toll measured in the millions of dollars each year in crops as diverse as barley, rice, potatoes, peaches, apples, and sugar beets. In 1959, for example, a virus named Barley Yellow Dwarf caused damage to 50% of the oat, 30% of the wheat, and 20% of the barley crop in South Dakota alone, and five other nearby states were affected.

Viruses are usually named on the basis of the symptoms they cause in the host plant. While this may seem indirect, so little is known about most viruses and so difficult are they to deal with as entities that this is the only way of approaching them. In general, plant-pathogenic viruses can be placed into two major groups: the *yellows* and the *mosaics.* The symptoms of the yellows are probably caused by hormonal imbalance and vascular disturbance, which result in yellowing, leaf curling, dwarfing, excessive branching, shortening of internodes, and the formation of galls. The symptoms of the mosaic viruses are due to chloroplast destruction and the death of tissues, which result in mottling and spotting.

The major agents in the spread of viruses are insects, especially the leafhoppers that transmit particularly effectively the yellows group and the aphids that transmit the mosaics. Other means are also possible: some viruses are soil borne; others are transmitted through the seed; still more are passed on by fungi, nematodes, and parasitic vascular plants such as dodder. TMV can be transmitted from an infected plant to a healthy plant simply by touching the leaves. It can even be transmitted through cigarettes made of tobacco from diseased leaves!

In the case of aphids, the insect inserts a long stylet directly into the phloem and drinks the carbohydrate-rich sap that also contains the virus. The virus is absorbed from the intestinal tract and carried to the salivary glands where it is injected into a healthy plant the next time the aphid feeds. The yellow viruses persist in the vector, while the mosaic viruses generally do not. A curious phenomenon found in some viruses is that they must have an incubation time, usually of several days, within the insect before they are able to infect a healthy plant. What occurs during this period is not known.

Virus diseases are particularly hard to control and virtually impossible to cure. Elimination of the insect vector is effective but usually not practical. Destruction of the diseased plants is also employed, often at great expense and labor. One of the problems is that one virus can attack many different species of plants and be carried by a large number of different insects. Viruses can also be transmitted from one generation to another through the seed—a situation that is rare in bacterial diseases—so that planting only from seed does not remove the virus.

It is possible to excise and culture the shoot tips of rapidly growing individuals. Under these conditions the virus may not be present in the meristem, and virus-free plants may be obtained. But this type of approach can be used only in limited areas of horticulture because of the expense. One of these is orchid growing where viruses are a constant problem, and vegetative propagation is important. In a few instances plants can be grown at temperatures high enough to kill the virus but not the plant. The greatest hope in the control of virus diseases in crops is through the breeding of resistant varieties.

16

Bacteria

Bacteria—Unseen Friends and Foes

Bacteria have caused untold misery to the human race, as well as providing man with many benefits. Yet for centuries their existence was completely unsuspected. The reason, of course, is that they cannot be seen by the human eye. Only after the invention of the light microscope were bacteria seen. But, even after the light microscope was perfected in the late 1800s, the structure of bacteria remained largely a mystery, and it was many more years before their real importance was understood. The reason, again, was their small size; even at a × 1000 magnification, they appeared to have little internal structure. Only the advent of the electron microscope and procedures permitting thin-sectioning allowed bacteriologists to resolve the ultrastructure of these friends and foes.

These photographs present a "rogues' gallery" of bacteria.

A Escherichia coli, *an inhabitant of the human digestive tract and the most widely and intensively studied of all bacteria.*

B Diplococcus pneumoniae, *the cause of pneumonia in man.*

C Bacteria can carry on photosynthesis, and here we show several of the many species that can: Rhodopseudomonas vannielii *(upper left), a nonsulfur bacteria;* Thiopedia sp. *(upper right), a purple sulfur bacteria;* Thiocapsa sp. *(lower left), another purple sulfur bacteria; and* Thiospirillum jenense *(lower right), a purple sulfur bacteria.*

D Sulfur bacteria utilize sulfur compounds as a source of energy and may deposit sulfur either inside or outside their cells. In the case of the colorless sulfur bacteria shown here, the sulfur is held within the cells, and the filaments of bacteria forming the rosette are shiny.

When bacteria were sectioned and examined with the electron microscope, they turned out to be structurally fairly simple cells.

E The bacterium shown in this photo is Escherichia coli. *It consists of a wall and a plasma membrane that surrounds a region lacking membranes. The center of the cell is occupied with the nucleoplasm, which contains the DNA. Ribosomes fill the region around the nucleoplasm.*

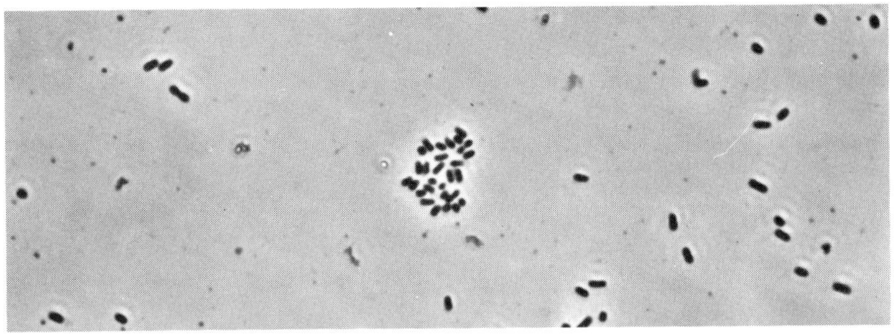

A

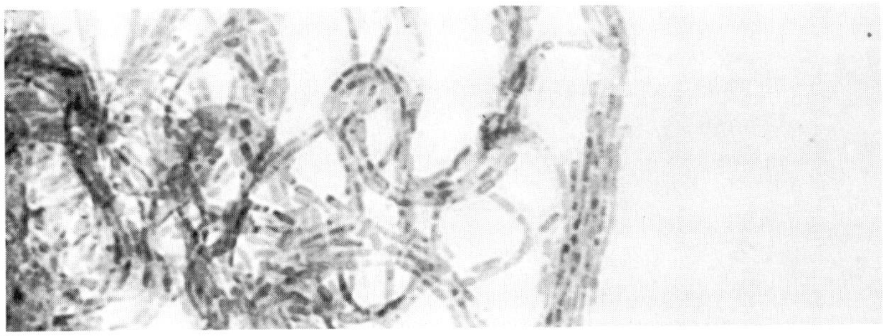

B

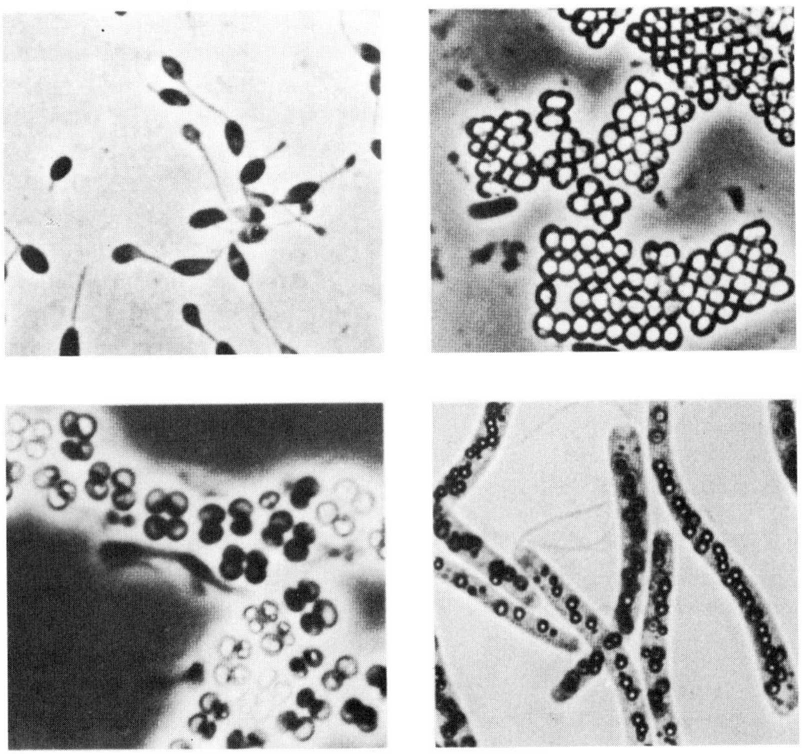

C

D

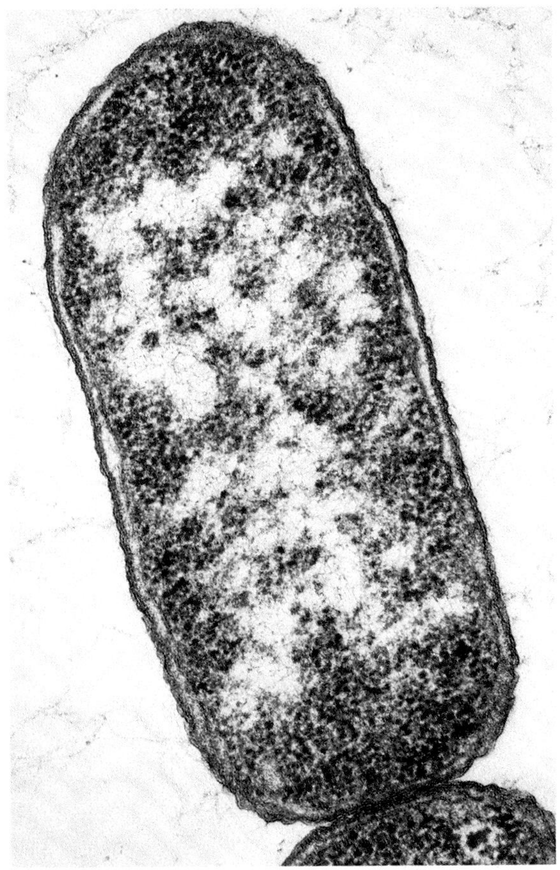

E

F Not all bacteria are quite as simple in organization. This bacterium is Rhodospirillum rubrium, *which contains photosynthetic membranes in the form of membranous sacs.*

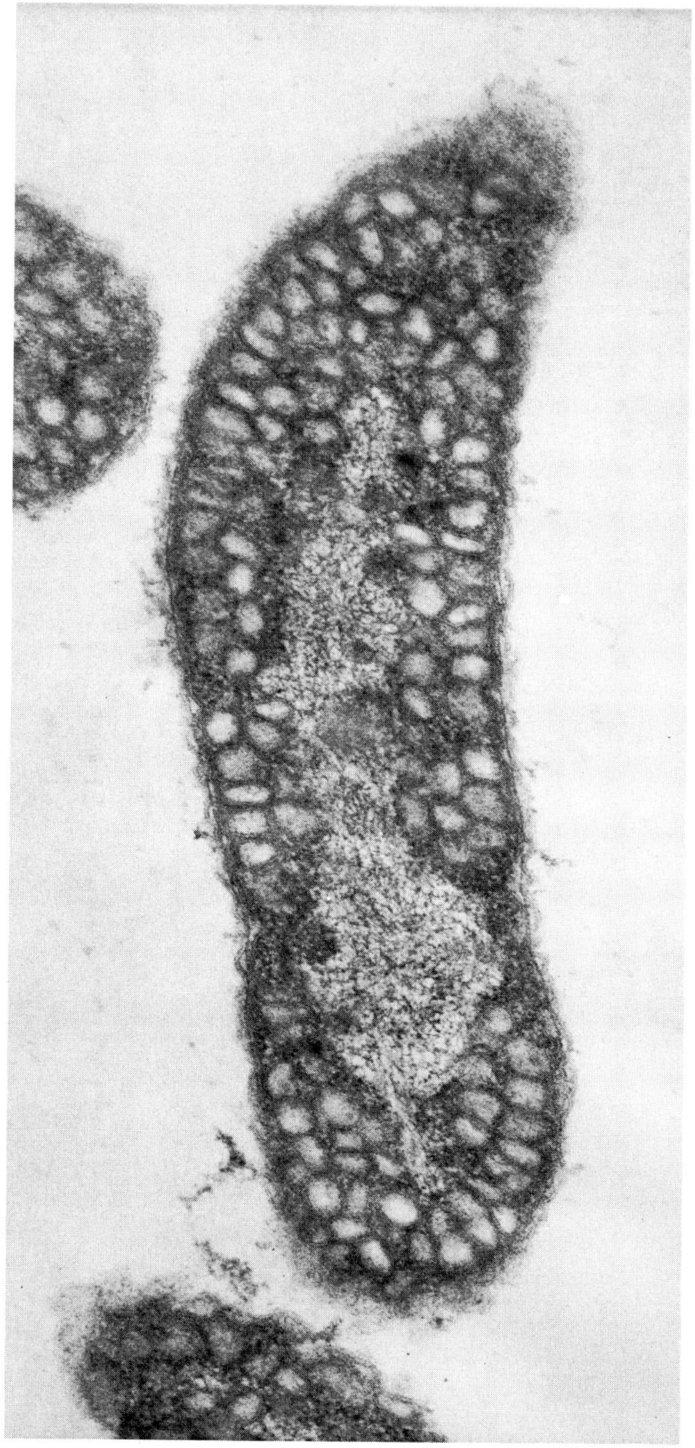

F

G The thickness of the wall and the structure of the plasma membrane can be seen in this freeze-etched preparation of E. coli. *This type of preparation achieves a three-dimensional effect so that the various surfaces of the cell can be seen. The cell wall is quite thick, and the plasma membrane has a curious pattern of what appears to be open holes. Chemical studies have shown that the composition of the bacterial wall is quite different from that found in most plants and that it contains considerable quantities of amino acids.*

H Bacteria may also have flagella that enable them to move at rapid rates for their size. These flagella may be located over the entire length of the bacterium, as in Salmonella typhosa *(right), or in a tuft at one end of the cell, as in* Ectothiorhodospira mobilis *(left). In the case of* Salmonella typhosa *the term flagella is used for the longer attachments, while the shorter ones are called pili.*

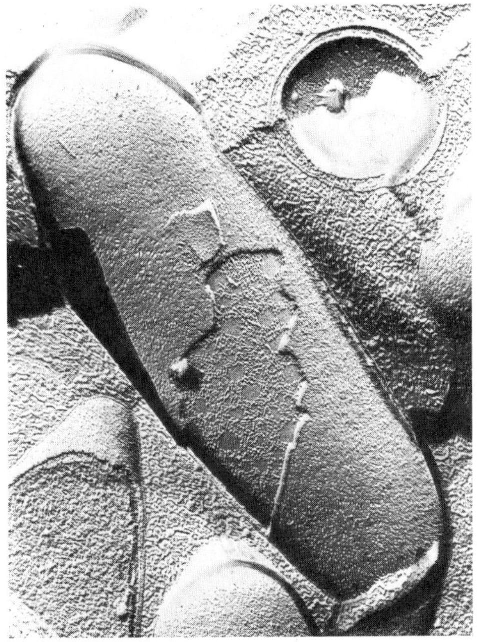

G

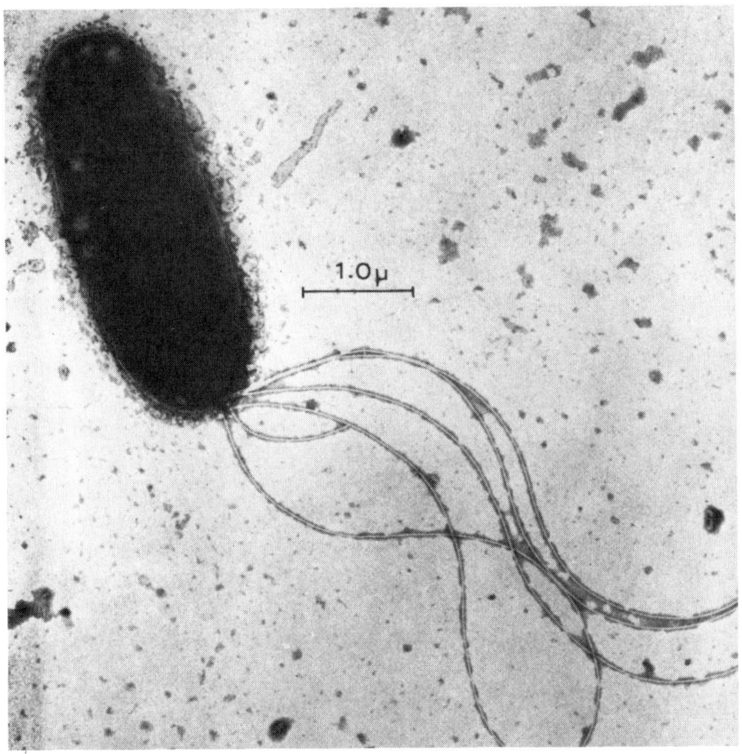

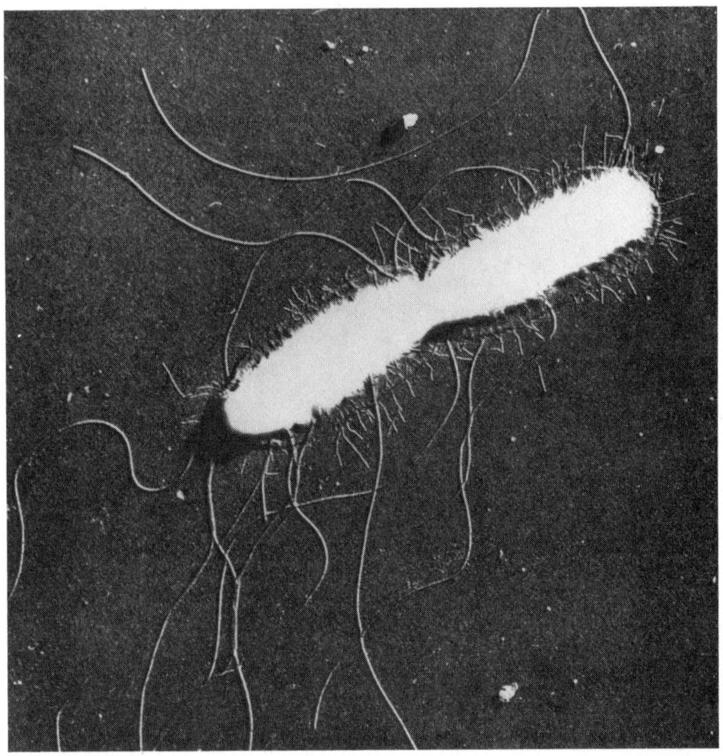

H

Death lay over the towns like a thick, black mantle. Corpses of rich and poor alike littered the streets. At least one in every four persons was dead, and in many regions entire towns had been wiped out. This was the Black Death that between 1348 and 1350 reduced the population of Europe from 80 to 60 million. The economic and cultural impact on Western Civilization was immense. Men turned either to carousing and ribaldry or to piety and devotion.

Strange and unusual treatments were tried, but none proved effective in combatting the dread plague. The most important reason for the ineffectuality of these attempts was the complete lack of understanding of its cause. The Black Death was not the act of an angry God or a revengeful Devil but was caused by a microscopic organism, a bacterium named *Bacillus pestis*. Transmitted to man by fleas from black rats, ground squirrels and similar rodents, the disease is relatively easy to control by using proper sanitary measures. But such measures were almost as scarce at the time as was the knowledge of the causal agent.

Throughout the ages, other epidemics (though none as serious as the Black Death) have been caused by bacteria. Asiatic cholera, endemic in India, is caused by the bacterium *Vibrio cholerae*. This dread disease is spread through contaminated water but can also be kept under control through proper sanitation. Again and again, until recent times, cholera has swept the world in great waves, killing countless thousands. Typhoid, the scourge of armies, is caused by another waterborne bacterium, *Salmonella typhosa*. Vaccination against this disease has proven an effective means of control. Other species of *Salmonella* cause paratyphoid and food poisoning. Since *Salmonella* species are found in almost all foods derived from animals (particularly eggs, after the shell is broken, and meat), care should be taken that such foods be adequately heated before they are eaten.

Currently, the bacterial diseases of syphilis (*Treponema pallidum*) and gonorrhea (*Neisseria gonorrhoeae*) are at epidemic levels in many American cities. These venereal diseases are also the scourge of armies, and the U.S. Army has put forth considerable effort to educate its personnel about the dangers of these diseases. Clearly, despite such attempts the diseases spread, even in the face of antibiotics, which at first offered almost complete control. Penicillin was hailed as the conqueror of these organisms, and for many years this belief seemed justified. Then, slowly, penicillin-resistant varieties began to appear, presumably because of mutations in the genetic material. Selection pressure in the form of drug treatment has resulted in the increase of these strains. Obviously, the workings of evolution may pose the ultimate threat to the sexual revolution.

Bacteria and Plants

Bacteria also cause many plant diseases that have resulted in crop losses measured in the hundreds of millions of dollars. One of the first plant diseases to be recognized as being caused by bacteria was fire-blight of apples and pears (Fig. 16–1). The disease is dramatic and devastating, blackening and blighting whole orchards in a single night. The bacterium, *Erwinia amylovora*, is spread by insects (particularly honey bees), birds, squirrels, and splashing rain. The transmission of a single bacterium may be enough to cause the death of the plant. The bacteria gain entrance to the tree through various natural openings such as stomata or any scratches or cuts in the bark. Once inside, they multiply rapidly and are carried to every part of the tree, blighting the blossoms, killing the twigs, and causing large brown rings on older leaves and elliptical cankers on the branches and stems. It is in the cankers that the bacteria overwinter, and when the cankers begin exuding a sticky fluid filled with bacteria, the bacteria then spread to other trees.

In the case of fire blight and similar diseases, the bacteria never enter the live cells of the host. They live instead in the intercellular spaces secreting enzymes that break down the middle lamella and result in maceration of the tissue and death of the cells. The bacteria then utilize the material in the dead cells. Bacteria causing other diseases can penetrate still living cells and hasten their death. Some bacteria can affect plants by invading the vessels. There they secrete enzymes that break down pectic substances and cause the formation of a slimy mass of material clogging the vessels and blocking the translocation of water and nutrients.

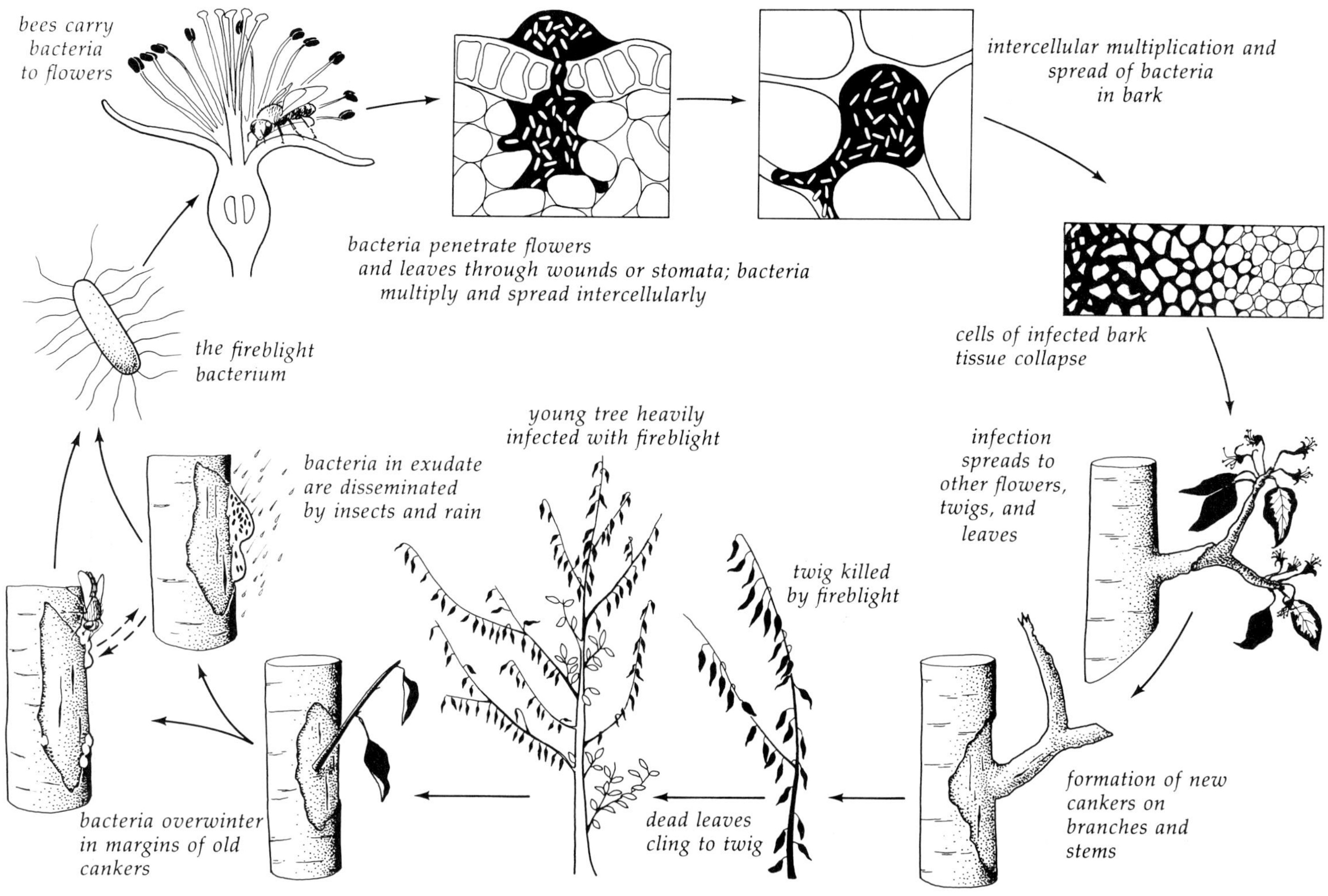

16–1 *The disease cycle of pear and apple fireblight, caused by* Erwinia amylovora.

There is also a second group of bacterial diseases that, instead of killing the cells, causes cell proliferation; these are the crown gall and related bacteria (Fig. 16–2). Crown gall is one of the most intensively studied of the bacterial diseases because of the unique way in which it develops and the apparent similarity of crown gall to cancer. When crown gall-type bacteria such as *Agrobacterium tumefaciens* enter the host through a fresh wound, they remain in the intercellular spaces and produce the plant-growth hormone IAA. Intensive cell division is initiated in the host shortly after infection. During this phase of rapid cell division, the host cells are somehow conditioned and made receptive to a stimulus (other than IAA) produced by the bacteria or by the host cells, responding to infection by the bacteria. This stimulus, known as the tumor-inducing principle, transforms normal plant cells into tumor cells. The remarkable result of this transformation is that the tumor cells subsequently grow and divide independently of the bacteria, and the growth of the tumor cells can no longer be controlled by the host plant. Despite years of research, the nature of the tumor-inducing principle and the various interactions that occur in the tumor cells are not known.

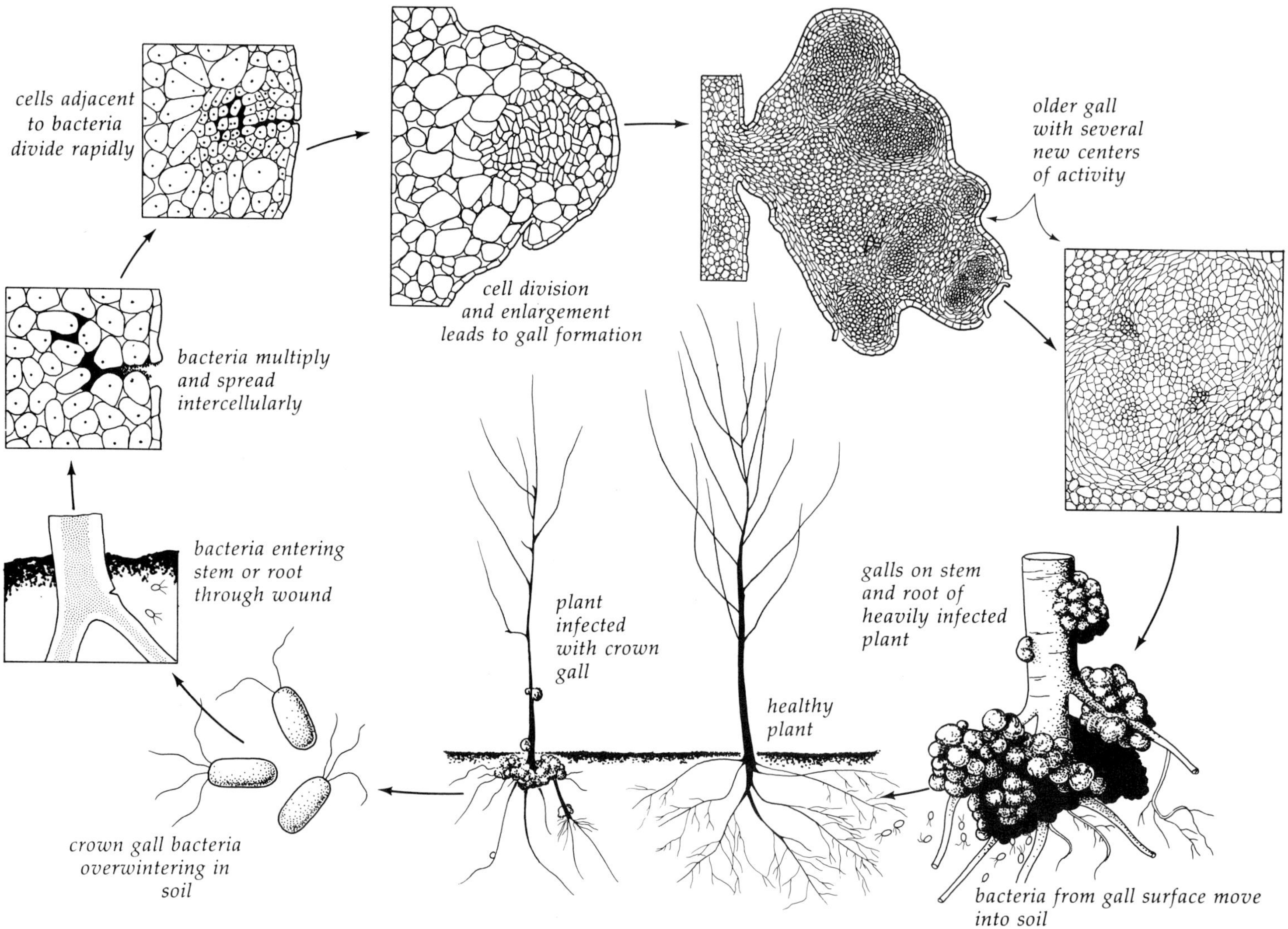

16–2 *The disease cycle of crown gall, caused by* Agrobacterium tumefaciens.

The bacteria that cause diseases in plants are facultative saprophytes, meaning they are organisms that are usually parasitic but can also live as saprophytes. Obligate parasites are organisms that grow and multiply only on living organisms. There are also facultative parasites, which are normally saprophytes but under certain conditions can become parasitic.

Bacteria and the Ecosystem

Bacteria are everywhere. They are found not only as disease-causing organisms in plants and animals but also in soil, ponds, lakes, streams, hot springs, gasoline storage tanks, food, snow, sewage, stomachs and intestines of animals, concrete, and stone.

In every gram of soil, some 100,000,000 living bacteria exist. A section of a piece of soil less than 25 mm in length is diagrammed in Figure 16–3. Thus a tiny piece of soil is a complex environment for a multitude of bacteria.

Large numbers of bacteria are also found in all types of water—from birdbaths to the open ocean. They live on the bottom, floating in the water, or as epiphytes, on all manner of plants and animals. The bacteria found in the lower depths of the ocean live at a temperature of 4 to 10° C and under pressures ranging as high as 500 to 1000 atm. Bacteria can also live at high temperatures. These are the thermophilic bacteria that grow at temperatures of 55 to 70° C; the extreme is a filamentous bacterium called *Flexibacterium* that flourishes in hot springs at temperatures up to 90° C.

On the whole, bacteria do not tolerate acid pHs; hence pickling in dilute acid prevents the growth of most bacteria. Some bacteria, however, not only tolerate high amounts of acid but actually produce sulfuric acid.

The primary basis for the universal distribution of bacteria is their versatility in metabolism. They can live in aerobic and anaerobic conditions and can utilize just about everything as an energy source. But, unfortunately for man, not quite everything. One of the most important roles bacteria have in the ecosystem is the breakdown of organic matter so that elements bound within it can be liberated and recycled. In nature there are few compounds that bacteria cannot attack, but various man-made plastics have so far resisted bacterial attack, thus posing major disposal problems.

Bacteria are also common in the digestive tracts of most animals and are necessary to the existence of many. These are the grazing animals such as cattle and buffalo, which have multiple stomachs containing enormous numbers of both bacteria and protozoa (Fig. 16–4). The bacteria and protozoa grow on the cellulose of the grass eaten by the animal, which has no mechanism for directly digesting cellulose. Later, the bacteria themselves are digested.

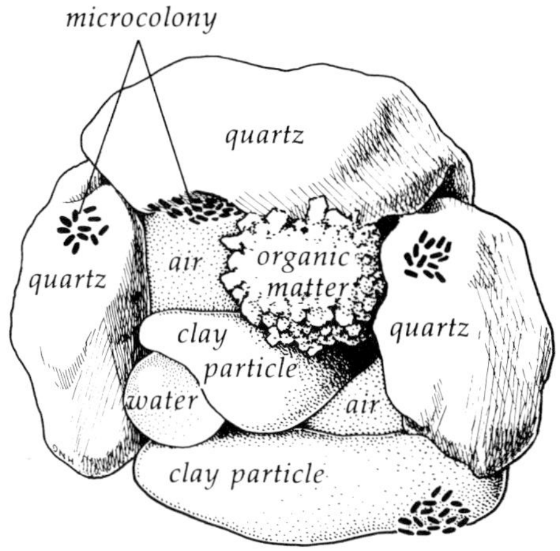

16–3 A section through a soil crumb showing the distribution of bacterial colonies. These are microcolonies of bacteria, not individual bacteria, and the data illustrate the complex and diverse environment represented by soil. The drawing shows how a soil particle is an aggregate composed of mineral and organic components, and how soil microorganisms are localized within it. Most soil bacteria are not found free in the soil solution but as microcolonies attached to soil particles.

The Nitrogen Cycle

We live in a sea of nitrogen—our atmosphere consists of 79% nitrogen. Yet this nitrogen is not immediately available to most plants and animals. Only some bacteria and blue-green algae are capable of utilizing atmospheric nitrogen; thus they assume a critical position in the ecosystem. But there is another side to this picture. Nitrogen fixation is essential but so is the breakdown of nitrogen-containing compounds. Such recycling is necessary to avoid making all of the nitrogen unavailable by sequestering it in an unusable form. Bacteria are necessary for this reutilization of nitrogen, as well as for its initial fixation. The reactions involved form an elaborate network, usually called the nitrogen cycle (Fig. 16–5), that is a good example of the way bacteria form an essential part of the ecosystem.

Let us begin a discussion of this important cycle by first considering the protoplasm of dead plants and animals. One group of bacteria begins by breaking

(a)

(b)

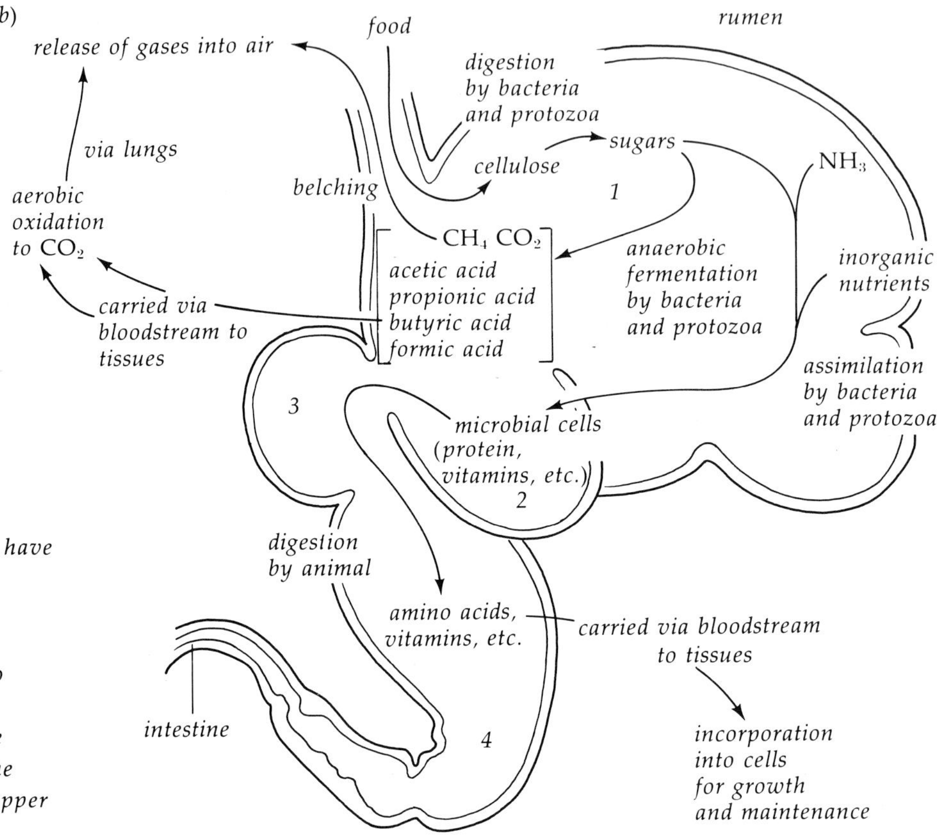

16–4 (a) Buffalo, cattle, and similar ruminants have four-chambered stomachs. In chambers 1 and 2, partially chewed plant material is stored, and digestion begins. The coarse plant parts are regurgitated, chewed as cud, then swallowed into chamber 3, where excess water is removed. The resulting pulp then passes into chamber 4, where conventional digestion largely takes place. (b) The biochemical steps involved in the process. (The upper part represents chambers 1 and 2.)

down the proteins into amino acids and other organic residues. In so doing the bacteria gain energy for their own growth. Other bacteria convert the amino acids into ammonia, which can be absorbed by some plants and incorporated into amino acids. Aquatic plants that grow with their roots in anaerobic conditions (e.g., rice) are particularly efficient in using ammonia. When oxygen is present, as in most soils, different bacteria change the ammonia into nitrites. Plants cannot use nitrites, but other bacteria are present in the soil that can convert nitrites into nitrates. Nitrates are readily absorbed by plants, which reduce them to ammonia, in which form they enter into new amino acids. When the plant is eaten by an animal, the protein is broken down into amino acids or polypeptide groups that are utilized by the animal. In each of these steps, from protein to nitrate, energy is released to the bacteria.

Also in the cycle are denitrifying bacteria. These bacteria liberate nitrogen from nitrites and nitrates. Their action returns nitrogen to the atmosphere. Denitrifying bacteria were formerly thought to be decidedly unhelpful, as far as man was concerned, because they deplete the soil of nitrogen compounds necessary for the growth of crops. In recent years, they have come to be considered in a different light, due to an increasing pollution of lakes and streams by nitrogen fertilizers. According to C. C. Delwiche of the University of California, Davis, the problem now is whether the denitrifying bacteria can cope with the increase.

The final linkage in our cycle is occupied by the nitrogen-fixing bacteria and blue-green algae. These organisms can take atmospheric nitrogen and directly incorporate it into their cells. While many nitrogen-fixing bacteria and blue-green algae live free in the soil, bacteria of the genus *Rhizobium* infect the roots of legume plants such as clover, alfalfa, and beans. The roots respond to these infections by producing lump-like growths or nodules in which the bacteria grow

(Fig. 16–6). Utilizing the carbohydrate produced by the host as its source of energy, the bacteria incorporate free nitrogen. Some of this nitrogen is made available to the host, but the bulk becomes incorporated into the soil, when the plant and its bacteria die.

Under natural conditions, grasslands and prairies have few plants that form nodules; instead they are characterized by nitrogen-fixing bacteria and blue-green algae that live free in the soil. In contrast, tropical soils show few free-living forms but many plants that form nodules. The legumes are definitely a tropical family, comprising not only the familiar herbaceous plants such as clover and beans but also many trees and shrubs. Over 90% of the legumes form nodules, and most grow in the tropics. Why should the nitrogen-fixing bacteria of the tropics grow in nodules, while they are free living in the temperate grasslands? The answer is probably connected with the amount of rainfall. Soluble materials are quickly washed from tropical soils by the large amount of rain falling on them, while the smaller amount of rain on the grasslands results in the accumulation of such substances. Under the ecological conditions present in the tropics, one can postulate that there were selection pressures on both the bacteria and legume that resulted in the evolution of the nodule condition. With such pressures lacking in the grasslands, the free-living forms remained.

Not only legumes have nitrogen-fixing bacteria associated with them. The nonlegume plant, alder, a northern tree species, provides a good example of the role of nitrogen-fixing bacteria in soil building. A study was made in Alaska of ground that had been laid bare by a receding glacier. The original organic nitrogen content of this soil was extremely low, but a pioneer vegetation of alder became established, probably as a result of the presence of nitrogen-fixing bacteria in its roots. As the alder cover developed and matured, the organic nitrogen level of the soil increased and rose to a peak. The alder was replaced by the climax forest of spruce, which could now grow on the enriched soil. These experiments also demonstrate the role of nitrogen-fixing bacteria as control agents affecting the whole nitrogen economy of the community of which they are a part.

This is true too for agricultural areas, where nitrogen is one of the essential minerals necessary for good crop yield. Man learned early to utilize legumes and their associated bacteria to retain soil fertility. In fields

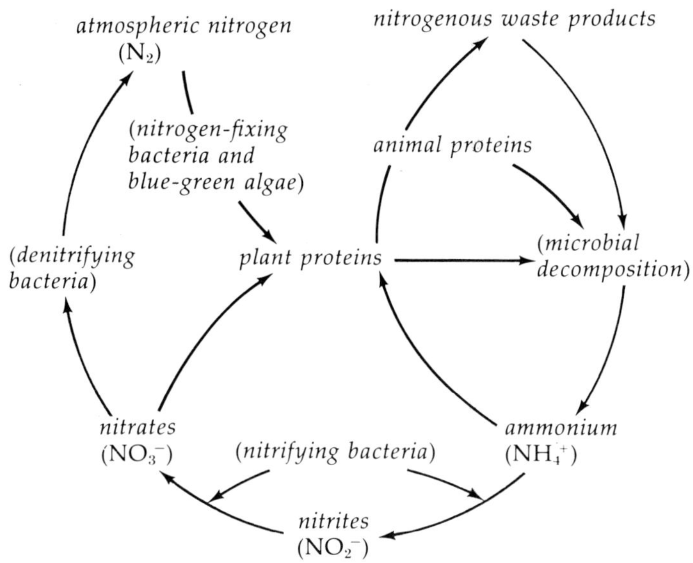

16–5 *The nitrogen cycle.*

under cultivation, the fertility of the soil is constantly being depleted as the crop is removed. This breaks the cycle of mineral nutrients, and the replenishment of these elements becomes necessary if farming is to continue. Fertilizers, either inorganic or organic, are one way in which this can be accomplished. Another method is to grow different plants in the field from year to year, in other words, the ancient practice of crop rotation. By including legumes and their bacteria in the schedule, the nitrogen level in the soil can be increased substantially. But, although crop rotation replaces the nitrogen in the soil, other elements have also been depleted. Hence this can and does lead to conditions where yield is reduced by the lack of other essential elements.

Crop rotation is less important now than formerly in parts of the world where intensive agriculture is carried on and land costs are high. In these areas chemical fertilizers are employed. Large amounts of nitrogen are added that have been fixed by a variety of industrial methods. Such fertilization is necessary to maintain the high levels of agricultural productivity needed to sustain increased populations, but it contains inherent problems. One of these is related to the excess nitrogen

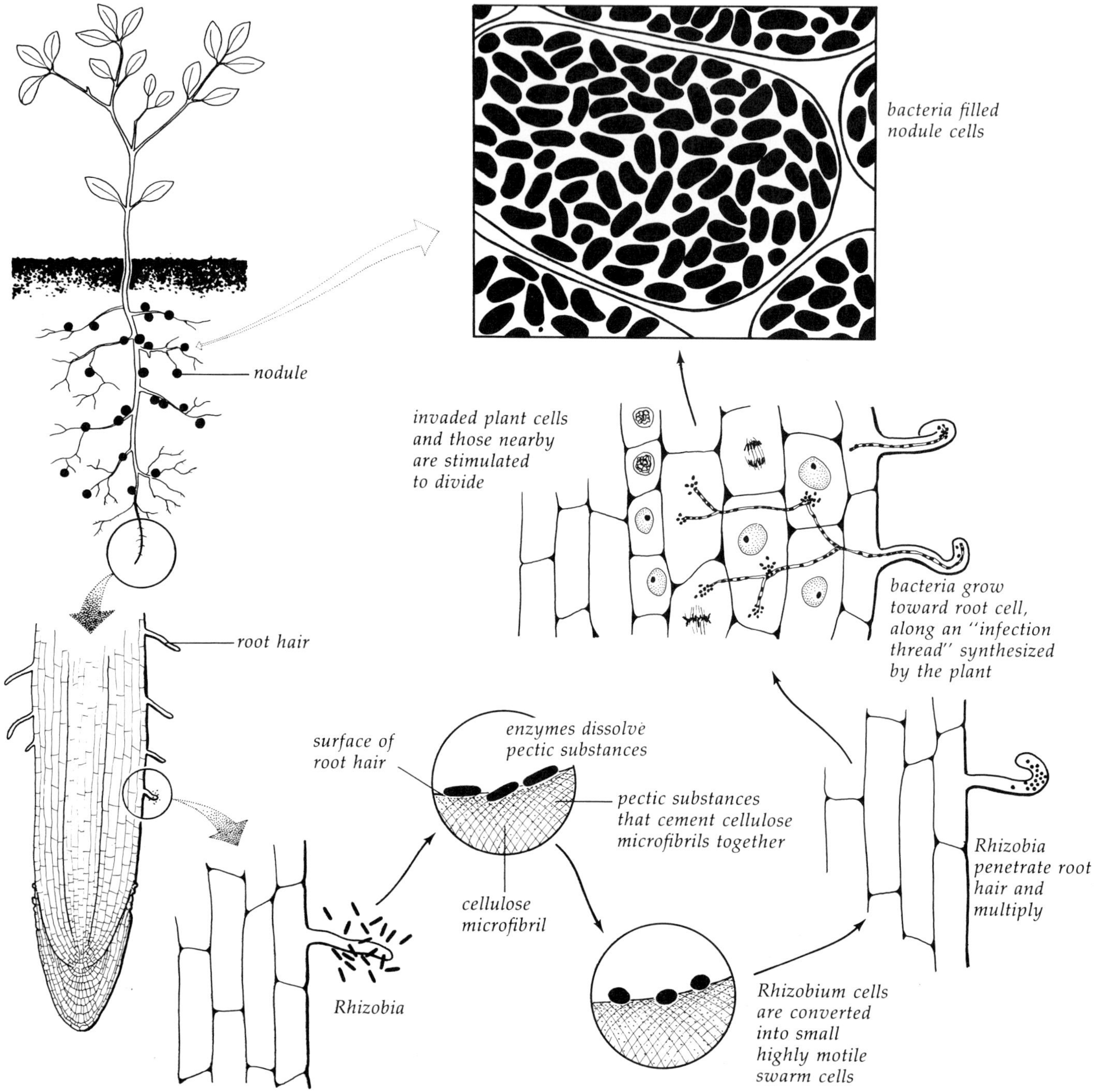

16–6 *The development of root nodules in clover, and the relationship of the bacteria to the root cells.*

compounds that are not absorbed by the plants. These are washed from the fields into streams and lakes, where they can upset the biological equilibria that are established in such bodies of water.

Before we look at networks, based on such elements as phosphorus, sulfur, and iron, that are similar to the nitrogen cycle, let us examine the structure and reproduction of bacteria. A good bacterium with which to start is one that is found in the intestines of many mammals, including man, and that is one of the most famous and probably the best understood—*Escherichia coli* or *E. coli.*

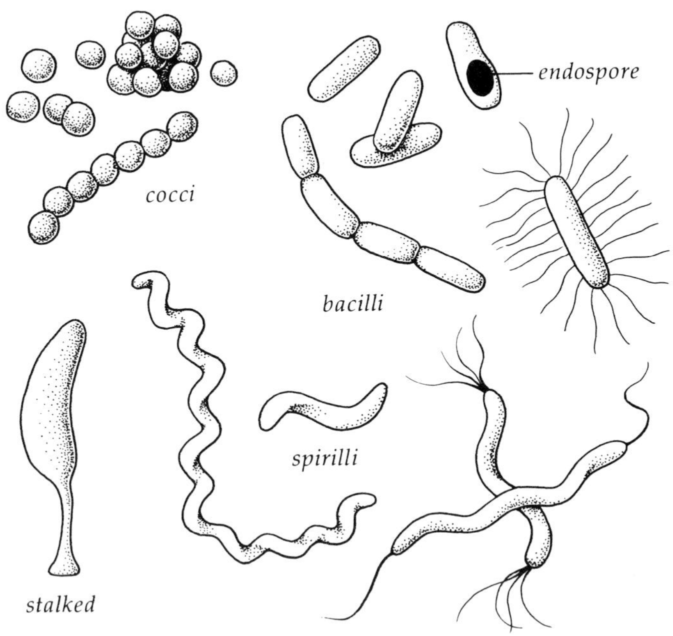

16–7 Various shapes of bacterial cells.

Bacterial Structure

Why is E. coli known the world over by biologists? The reason is that it is an almost ideal organism for experimental work in molecular genetics and bacterial biochemistry. In addition, *E. coli* is one host for the bacteriophage viruses. These viruses in turn form one of the primary subjects for virus and molecular biological research. As a result of these considerations, thousands and thousands of pounds of *E. coli* are grown annually, and researchers characterize themselves as *E. coli* physiologists or *E. coli* geneticists, devoting their entire careers to it.

Bacteria have a wide range of shapes and sizes (Fig. 16–7). Traditionally, bacteria have been grouped into three categories: rods or bacilli (*E. coli,* which is small and relatively simple in shape, is an example), spheres or cocci, and spirals or spirilli. But this is too limited a view of bacterial form. Bacterial shapes include filaments similar to fungi as well as multicellular chains and stalked forms.

Why do bacteria have various shapes? The immediate answer is that there is adaptive ecological significance to the shapes, but what can this be? First, it must be recognized that all nutrients entering the bacterium must do so by diffusion through the cell wall. The area of the wall will thus determine the efficiency of the uptake. In other words, rod-shaped bacteria will be more efficient than sphere-shaped bacteria in absorbing nutrients. Interestingly, most bacteria found in lakes and oceans, where concentrations of organic materials are low, are either filamentous or rod shaped, while the sphere-shaped bacteria are more common in the nutrient-rich sediments. Nutrition is not the only factor involved in bacteria shape; resistance to drying is another. Spherical cells are usually more resistant to drying than elongated ones, possibly because of more uniform pressures on the cell wall during drying. This idea is supported by the fact that spherical bacteria are the most common type isolated from the air. But what about the spiral bacteria and the stalked bacteria? The spiral shape has been related to the fact that these bacteria are usually motile, and moving in a helical pattern, as they do because of their shape, aids them in penetrating the debris and viscous media in which they live. Stalked bacteria, on the other hand, are epiphytes, and the stalk anchors the bacterium to the substrate. Both stalked and filamentous bacteria are common epiphytes on marine plants and animals. In both forms only one end of the bacterium or filament is attached to the substrate, thus allowing the organism as great an area for the diffusion of nutrients as possible.

Bacteria range in size from 0.5 to 5 μ, with most falling in the range of 1 to 5 μ. Consequently, they can be seen only with the aid of a microscope. The first bacteria were seen in 1683 by Antony van Leeuwenhoek. Leeuwenhoek, like his English contemporary Hooke, made his own microscopes. These had but a single lens and, although better than the compound microscopes then, did not offer any possibility of future development. Not until truly fine compound microscopes were available in the latter half of the 1800s could the more detailed structure of the bacteria be seen. It was during the 1870s that the great German botanist Ferdinand Cohn systematically studied large numbers of bacteria and produced the basis for our present system of classification. (Cohn also played an important role in the history of bacteriology because of his relationship with Koch—one of the most impressive of all microbiologists. See Essay 16–1.)

Essay 16–1 Robert Koch—Genius Recognized

In our discussion of Mendel in Chapter 12, we saw what happens to an unrecognized scientific genius, but what about the case where such genius is recognized? An excellent example is the German bacteriologist Robert Koch (1843–1910). The experiences of Mendel and Koch offer some striking similarities and some equally striking differences.

Koch was born in 1843, a member of a family with modest means. He studied medicine at Göttingen University, where he came under the influence of Jacob Henle, a pioneer in pathology. Receiving his degree, he practiced medicine in small German towns for several years. Then, in 1872, while continuing to carry on his medical practice, he began to study bacteria using a microscope his wife, Emmy, had given him on his 28th birthday. His approach was fresh and different, resulting in an array of experimental procedures and concepts that form the basis for much of modern bacteriology.

The first disease that he attacked was anthrax, a disease of both animals and man. During this investigation, he developed methods of inoculating test animals, the first incubator, and a hanging drop method that allowed him to obtain pure cultures of bacteria. With his crude, homemade equipment, he was able to isolate anthrax bacteria, culture them, and then use the cultured bacteria to reinfect healthy animals.

During the period that he investigated anthrax, he worked alone, but after establishing the cause of anthrax, he wrote to Ferdinand Cohn, professor of botany at nearby University of Breslau. Cohn had long worked with bacteria, immediately recognized the importance of the Koch research, and invited him to Breslau to present his data. There, his findings were acclaimed, and Cohn led a campaign to have Koch given a laboratory and research money. Koch became a member of the sanitary commission in Berlin and later professor at the School of Medicine in Berlin.

During Koch's long career, important discovery followed important discovery, many based on the development of new and simple procedures. These included the isolation of bacteria from one another by streaking mixtures of bacteria on a solid nutrient medium, where individual cells could form colonies. He first thought of this after seeing bacteria growing on an old slice of cooked potato someone had left on the laboratory bench. He also developed methods of solidifying liquid media with gelatin and established the method of obtaining pure cultures that is essentially the one used today.

Koch established the fact that bacteria are responsible for tuberculosis and Asiatic cholera. He investigated many diseases and, in his later years, traveled widely to assist in the control of bacterial diseases. Pasteur, the other giant in bacteriology at the time, and Koch had several fairly violent disagreements. They were contrasting personalities (Pasteur, a confident showman as well as a truly great investigator, and Koch, quiet and retiring) and citizens of nations (France and Germany) in frequent conflict at a time when nationalism was a strong force in both countries.

One of the outstanding contributions of Koch was his strong adherence to the concept that for a disease to be ascribed to a specific bacterium the following conditions had to be met: (1) the bacteria must be present in every case of the disease at the time the disease is active; (2) they must be isolated from the diseased organism and grown in pure culture in the laboratory; (3) the isolated bacteria when inoculated into a healthy subject will cause the disease; and (4) the same organism must be found in the inoculated organism and reisolated. He stated these steps in 1884, and they are known as "Koch's postulates." These ideas did not originate with Koch but with Henle, one of Koch's professors. Henle did not have sufficient means to establish them, while Koch did. Moreover, Koch insisted on their rigorous application. Koch's postulates should be applied not only to bacterial diseases but to any other activity ascribed to the presence of bacteria or other microorganisms.

Mendel and Koch are similar in that, working alone without elaborate equipment, they made major discoveries that they communicated to recognized professors in their fields. Mendel had bad luck with Nägeli, who did not realize the importance of the work Mendel had done. Koch, in contrast, had excellent luck with Cohn, who immediately recognized the importance of Koch's discoveries and techniques and worked unselfishly for him. The major difference therefore between Koch and Mendel is that Mendel was clearly ahead of his time but Koch was not. Pasteur, Cohn, Linde, Lister, and others had set the stage that Koch, with Pasteur, was to occupy. The time was ripe, the man was there, and a helpful, important friend materialized at the right moment to make sure he was recognized.

The typical bacterium, as seen in the light microscope, is surrounded by a rigid wall that may, in turn, be surrounded by a slime layer of varying thickness. The wall defines the gross morphology of the bacterium, and its composition is unlike the walls of most other organisms. The major components of the wall are peptidoglycans. These are compounds consisting of amino acids and sugar derivatives (Fig. 16–8). The sugars form a polysaccharide-based network connected by polypeptide (amino acid) cross linkages. Penicillin acts as an antibiotic, partly because it blocks the formation of the cross linkages. Also present in the cell walls of some, but not all, bacteria are lipopolysaccharides—compounds consisting of lipid molecules attached to polysaccharides. These occur as an outer layer on top of the mucopeptides.

The composition of the cell wall is the basis for one of the most famous and important of bacterial stains—Gram's stain. This stain, named after a Danish physician who published it in 1884, is achieved by staining bacteria with a dye called gentian violet. On the basis of this stain, bacteria can be divided into two large groups: those that stain purple, the Gram-positive, and those that do not stain, the Gram-negative. After many years of research, it is now generally conceded that the Gram-negative bacteria have the lipopolysaccharide layer in their cell walls, while the Gram-positive bacteria lack it. Although the Gram stain can be used to separate the bacteria into two groups based on the composition of their walls, it does not separate the pathogenic and nonpathogenic bacteria. The bacteria that cause diphtheria, tetanus, tuberculosis, pneumonia, and the pathogenic staphylococci bacteria are Gram-positive; Gram-negative bacteria include those that cause cholera, typhoid fever, gonorrhea, syphilis, and dysentery. What the Gram stain does do is to help characterize various physiological and biochemical features of the bacteria, including their reaction to antibiotics. Thus the Gram-positive bacteria are relatively resistant to penicillin and sulfa drugs, while the Gram-negative bacteria are susceptible to these agents. This resistance appears based on the lipopolysaccharide of the wall that protects the wall linkages. The Gram-negative bacteria are susceptible to streptomycin, which acts at the level of protein synthesis in the cell.

Projecting through the wall of the bacterium may be one or more flagella in a variety of patterns depending on the species (Fig. 16–9). The flagella of bacteria are not equivalent to those of higher organisms. They are only 100 to 200 A in diameter and consist of a single fibril in cross section composed of a single protein, flagellin. This protein belongs to the same group of proteins as the contractile proteins of animals. The flagella may be 3 to 12 μ long. Also present are pili, which are slightly different in form and composition, as well as being shorter and narrower in diameter. The

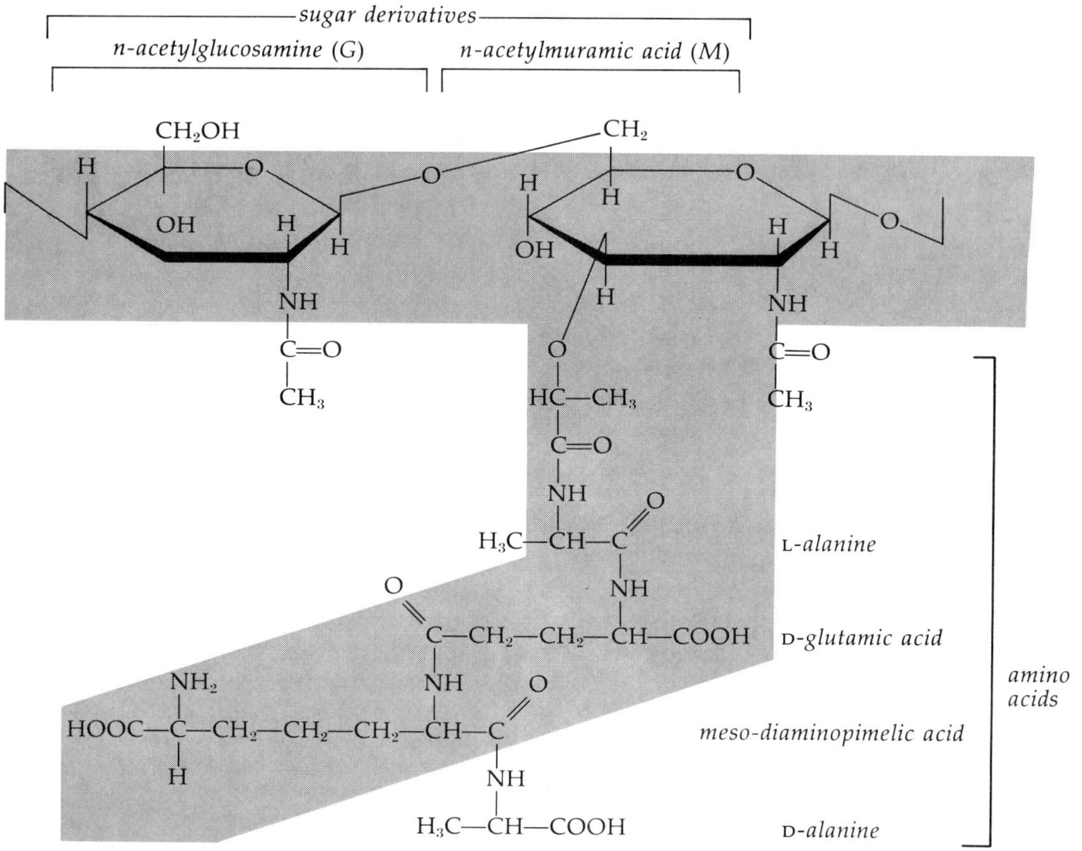

peptidoglycan—the basic repeating unit

16–8 The chemical structure of the bacterial cell-wall wall. Only the basic organization is shown. There are several modifications common in the basic peptidoglycan unit of different bacteria, and additional compounds may be present within the framework shown.

flagella allow the bacterium considerable powers of movement. Some bacteria appear to move by an unexplained gliding motion.

The interior of the bacterial cell proved difficult to resolve with the light microscope. With proper staining, a nuclear region could be distinguished from the cytoplasm, but that was all. The question of whether a true nucleus was present and true mitosis occurred was never successfully resolved until bacteria were examined in sections with the electron microscope. Even when viewed with the electron microscope, the bacterial cell is far from elaborate. In the section of *E. coli* seen in Figure 16–G, it is possible to distinguish the cell wall and the plasma membrane. The center of the cell is occupied by an irregular mass of fibrous material containing the DNA. This is called the nucleoplasm, and it lacks a limiting membrane. The only regular membranous structure found in bacteria is an array of concentric membranes associated with the plasma membrane known as the mesosome (Fig. 16–10). The cytoplasm is devoid of all membrane-bound organelles. Ribosomes are present as are accumulations of polyphosphates, which are a reserve supply of phosphorus. Even in much more complex bacteria, little more is found than assorted vacuoles, some filled with gas that permits the bacterium to float, and occasional stacks of membranes (Fig. 16–H).

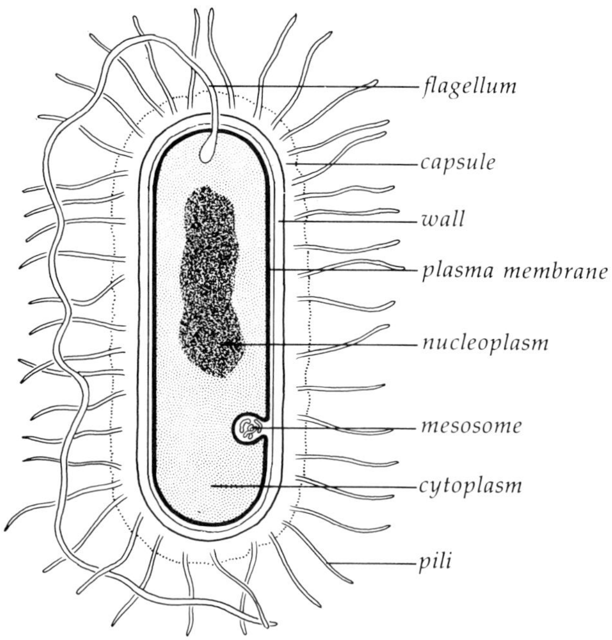

16–9 A diagrammatic representation of the structure of a generalized bacterium.

Cell Division in Bacteria

Bacteria reproduce by a simple means of cell division. The cells first grow in length by the synthesis of new wall material in the middle (Fig. 16–11). Then the cell wall and the plasma membrane in the middle of the cell grow toward the center of the cell, eventually dividing the cell in two. The wall then splits in the middle, and the two cells separate. Chromosomes are not present in bacteria in the same form as they are in higher organisms. Instead, DNA is present in the form of a continuous loop, at least in *E. coli*, approximately 1 mm long. There is evidence that bacterial DNA is attached to the cell membrane by the mesosome. The mesosome is also associated with cell division, and it has been suggested that each strand of daughter DNA synthesized before division is attached to separate mesosomes. The growth of the wall in the center of the bacterium would then separate the mesosomes and the attached DNA. The details of this mechanism are not completely known.

One of the remarkable features of bacterial cell division is the rapidity with which it can occur. Under favorable conditions, a bacterial cell may divide as

16–10 The relationship of the mesosome to the formation of the cell wall during bacterial division is clearly shown in this electron microsçope photograph.

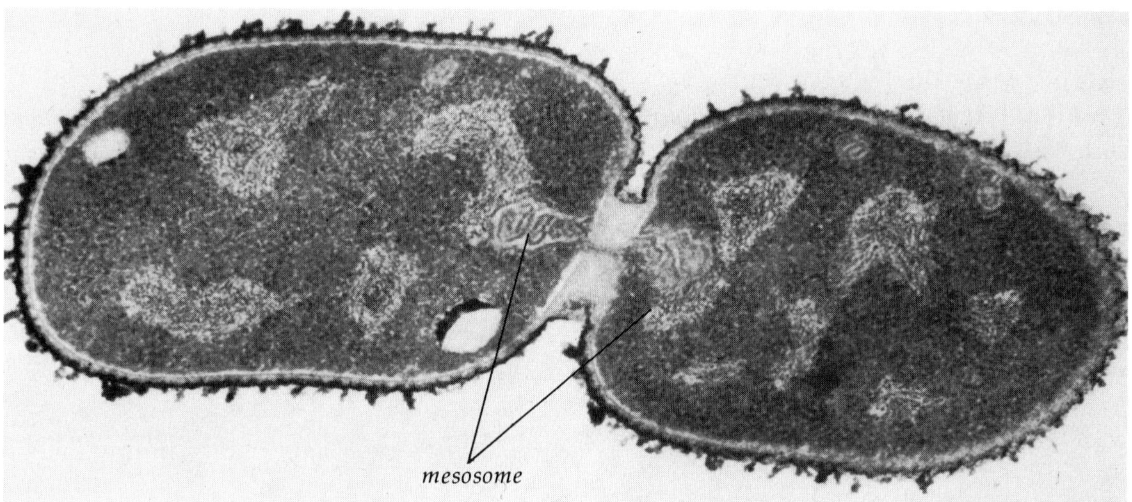

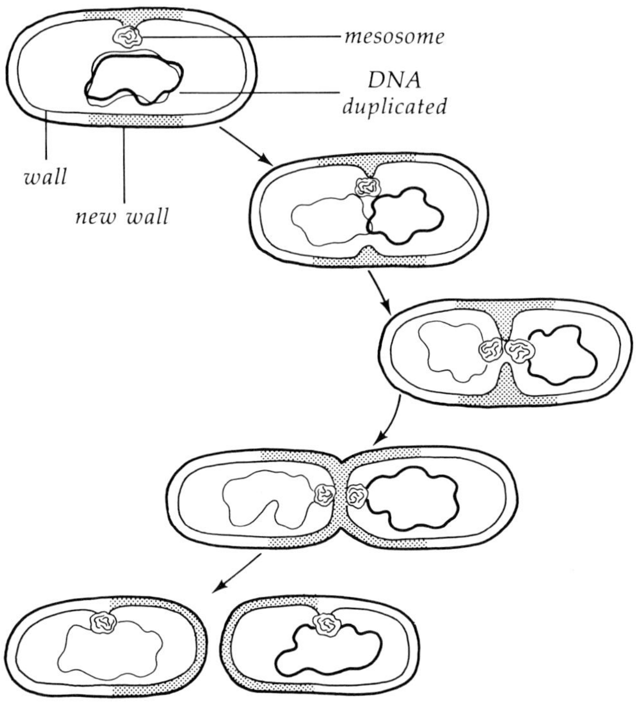

16–11 Cell division involves the duplication of the DNA and the synthesis of a new cell wall. The mesosomes are believed important in the separation of the DNA strands and in the formation of the new cross wall.

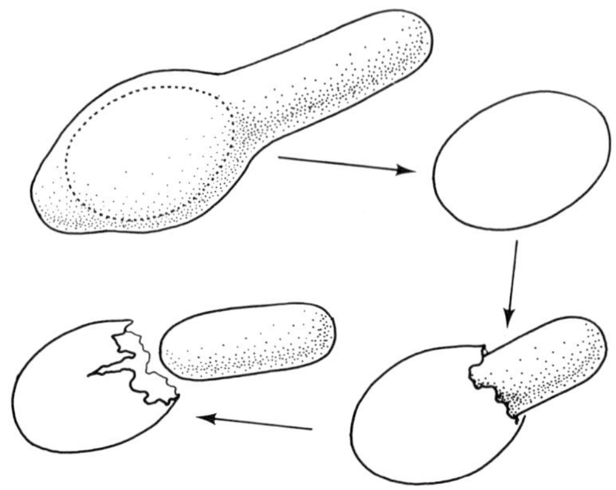

16–12 Spores in bacteria. The light oval regions in the bacteria indicate the presence of spores [× 2000]. These form within the original cell wall and are released when it breaks down. The spore wall then breaks down when the spore germinates, thus releasing a new cell.

often as once every 20 minutes. This means that, starting with a single bacterium, there would be over 4000 cells in 4 hours and some 4×10^{21} in 24 hours. This is why warm milk can sour in a matter of hours and why cooling, which slows the growth rate of the bacteria, helps preserve it.

Spore Formation

A feature of bacteria that increases their ability to withstand adverse environmental conditions is spore formation (Fig. 16–12). These are resting cells that are not directly comparable to the spores of other plants. In bacteria, spore formation occurs when a portion of the cytoplasm, frequently most of the living material in the cell, becomes concentrated into a relatively small mass that is then surrounded by a thick cell wall laid down within the original wall. The spores are highly resistant to all types of drastic extremes of temperature and moisture. Under the proper environmental conditions, the spore germinates. The spore swells; the old wall splits; and a single vegetative cell emerges. The new cell has its own cell wall and is capable of rapid growth.

Sexual Reproduction in Bacteria

Until 1947 all bacteria were thought to lack any type of sexual reproduction. But, through genetic analysis of mutant strains (first in *E. coli*) by Joshua Lederberg and Edward L. Tatum, sexual reproduction for bacteria was confirmed. By using ultraviolet radiation, they were able to produce two strains of *E. coli* each having triple mutants. The mutants involved nutrition characters: each strain required the addition of three specific amino acids to the culture media, if it were to grow, but the two strains required completely different sets of amino acids. Lederberg and Tatum were able to show that when the two strains were grown together, a small number of individuals appeared that required no supplemental amino acids. The only explanation for this result is that DNA was transferred from individuals of one strain to the other. Thus the original, unmutated or wild type of *E. coli* was reconstituted through the recombination of the genetic material.

Essay 16–2 Whiter than White—Enzyme Detergents

A major preoccupation of the average American housewife, according to television commercials (a dubious authority), is whiter, brighter laundry—or at least whiter and brighter than her neighbor's! In this frenetic quest for cleanliness the detergent manufacturers continually seek new products, new heroes in the struggle against grime. For a while they thought they had found one in the unlikely form of a bacterium called Bacillus subtilis. *This bacterium, like many others, secretes into its environment a number of enzymes that attack the surrounding material producing small, soluble compounds that the bacterium can absorb.* B. subtilis *is particularly prolific in secreting protein digesting enzymes active at alkaline pHs found in normal laundry conditions. The enzymes are effective in breaking down blood, meat, and other protein-based stains that are hard to handle with conventional cleaners.*

While enzyme detergents were first introduced into the United States in 1967–1968, they have been around for a lot longer—the first enzyme cleaner was produced in 1913 in Germany. Not until 1953, however, was B. subtilis *used as the source of the enzyme; earlier the pancreases of oxen and pigs were used. Methods were developed first in Switzerland and later in Denmark whereby* B. subtilis *could be cultured in huge vats and the enzymes extracted. The exact methods used as well as the exact composition of the enzyme concentrates are trade secrets and will not be revealed by the makers. It is known that the commercial extracts contain only 10% proteolytic enzymes and that the commercial extracts constitute only about 10% of the average box of detergent, meaning that the final product contains some 1% proteolytic enzymes.*

The enzymes were first introduced as presoaks and later into the detergents themselves. The success of these products was enormous—$75 million were sold in 1969—but then the market began to fall, for two reasons: health and effectiveness.

Not long after the enzyme products became widespread in Europe, respiratory troubles related to inhaled enzyme dust began to appear among factory workers in England. Later, similar problems appeared in American factories and were overcome by protective clothing and masks. English tests also indicated that women with skin problems found them exaggerated when they used the enzyme products, but American tests found no effect on healthy women. The health problem continued to worry many people; so the manufacturers developed a granular form of the enzyme. In this form, dust, believed to be the major source of irritation, is eliminated. A clear-cut answer about whether enzyme preparations constitute a health hazard in ordinary laundry use is still not available. Opinions on both sides of the question are diverse and sometimes heated.

The effectiveness of the preparations has also been questioned by various consumer groups. In general, they seem as effective as nonenzyme products, acting perhaps a little faster, but they are no miracle cleaner, as the early advertisements would have them.

The result of these concerns has been a trend away from enzyme detergents. Most of the major manufacturers have removed them from their mainline detergents and have confined them to presoaks. By 1971 sales were down to $25 million and still falling.

The concern about enzyme detergents may also reflect an increased interest in the environment by the average housewife—a concern that overrides even the threat of clothes that may not be whiter than white.

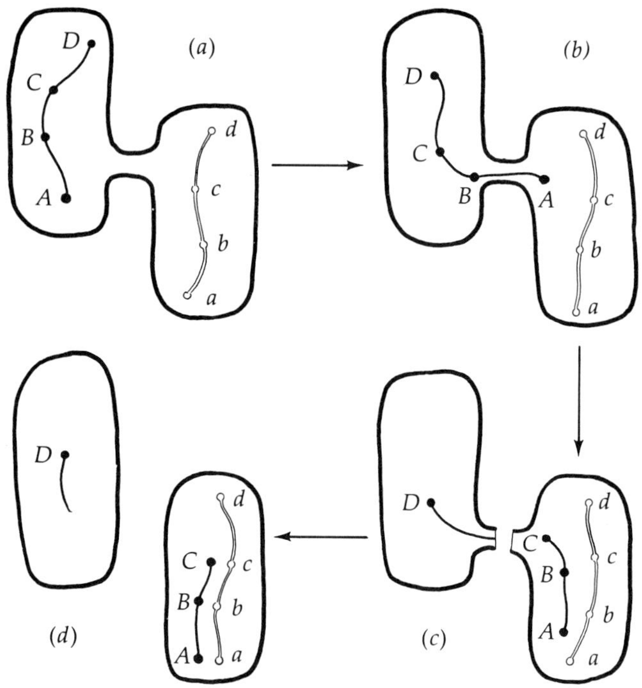

16–13 Conjugation type of sexuality in E. coli. *Only a portion of the DNA molecule is transferred, and the donor cell, with an incomplete DNA molecule, will die.*

Sexual recombination in *E. coli* is based on the transfer of a portion of a DNA molecule from one cell to another through conjugation. During conjugation two cells come together, and a bridge is formed between them (Fig. 16–13). Through this bridge passes a part of the DNA, called an episome, from one cell (the donor) to the second cell (the recipient). There is no exchange of DNA; so the donor cell always dies. When the recipient cell reproduces, the daughter cells have parts of the DNA from the donor cell substituted for parts of the original DNA of the recipient cell. This is how the daughter cells possess genetic information from two different parent cells.

Metabolism

When it comes to their metabolism, bacteria show a remarkable versatility in the compounds from which they can derive energy. Glucose forms the staple in their diet, but they can and will take on almost anything. Many bacteria secrete enzymes that act to break down proteins, fats, and other large molecules into molecules the organism can utilize. In their metabolism they do not necessarily carry the oxidative reactions to CO_2 and H_2O, discussed in Chapter 7. The acetic acid bacteria, for example, start with alcohol produced by the yeasts and convert it to acetic acid, releasing a significant amount of energy in the process. This, of course, is fine for the bacteria but not so good for the winemaker, who wishes to produce a good vintage not vinegar. Milk provides the sugar for another set of bacteria that converts it to lactic acid, again with the release of energy. The presence of the lactic acid gives the milk a sour taste that makes it undrinkable for most people. Yet it is not harmful and is necessary for the production of butter, buttermilk, and yogurt.

Some bacteria, being chemosynthetic, have more unusual ways of obtaining energy. These species obtain energy through a number of oxidation-reduction reactions in which electrons at high energy levels are moved to low energy levels. The energy liberated is used to synthesize the organic matter in the cell directly from CO_2. The bacteria use such starting compounds as ammonia, ferrous iron, sodium nitrate, and hydrogen sulfide.

Photosynthesis can be carried on by some bacteria, but it is quite different from that found in other plants. The bacteria have chlorophylls related to other plant groups, but they have their absorption maxima in the near-infrared portion of the spectrum. This means that bacteria can carry on photosynthesis in what would appear to us as darkness. Bacteria also photosynthesize in the absence of oxygen and release no oxygen. In photosynthetic bacteria the pigments, which, as in other plants, include carotenoid pigments, are located in membranes. These are not organized into chloroplasts but are formed into simple vacuolelike structures (Fig. 16–C).

An example of the type of photosynthetic reaction found in bacteria is that of the purple sulfur bacteria:

$$CO_2 + H_2S \rightarrow \text{glucose} + \text{sulfur}$$

This reaction is basically the same as that for photosynthesis involving water and resulting in the release of oxygen. Other bacteria make use of other hydrogen donors (e.g., organic compounds and molecular hydrogen) in place of H_2S.

Various sulfur bacteria are commonly present at the bottom of relatively still bodies of water such as lakes, ponds, some fjords, and even as large a body as the Black Sea. Leaves and other organic debris fall into such bodies and collect in a layer at the bottom. This layer is usually anaerobic due to the slow rate of oxygen diffusion to the bottom and the fact that aerobic bacteria use what oxygen is available. Anaerobic bacteria growing in this region break down the organic material producing sulfide from those amino acids and other compounds that contain sulfur. The sulfide is usually in the form of H_2S, which smells like rotten eggs. If ferrous irons are present in the mud of the lake bottom, the H_2S will react with them, and highly insoluble FeS is produced. If ferrous irons are not present or too much H_2S is produced, the H_2S can be oxidized by bacteria.

In anaerobic regions of the lake, where light can penetrate, the H_2S is used by photosynthetic sulfur bacteria as an electron donor, and sulfur is produced (see the equation just given). The light reaching this depth is considerably different from the light striking the surface of the water. The algae growing in the upper aerobic layers of the water absorb much of the light in the blue and red regions; hence, if the deeper-lying bacteria are to survive, they have to use the light that is not absorbed by the algae—mainly that in the far-red and near-infrared regions. Absorption spectra from a green alga, a blue-green alga, and a purple bacterium are shown in Figure 16–14; the complementary nature of the pigments involved are clearly shown. The level at which the photosynthetic bacteria live is determined by the clearness of the water. The relation of light, temperature, H_2S, oxygen, and depth to the growth of photosynthetic bacteria in two lakes is shown in Figure 16–15. Many of the conditions present in a lake or pond can be duplicated in the laboratory with a Winogradsky column (Fig. 16–16).

Under normal conditions an equilibrium is established, and H_2S never reaches the surface in amounts great enough to be a problem. One way in which this equilibrium can be upset is by the addition of sewage to the lake. This is rich in organic material and increases the number of sulfide bacteria, thus increasing the amount of H_2S. At the same time, the sewage increases the turbidity of the water thus decreasing the number of photosynthetic sulfur bacteria. The result of this combination of factors is to convert the pond or lake into a repugnant cesspool.

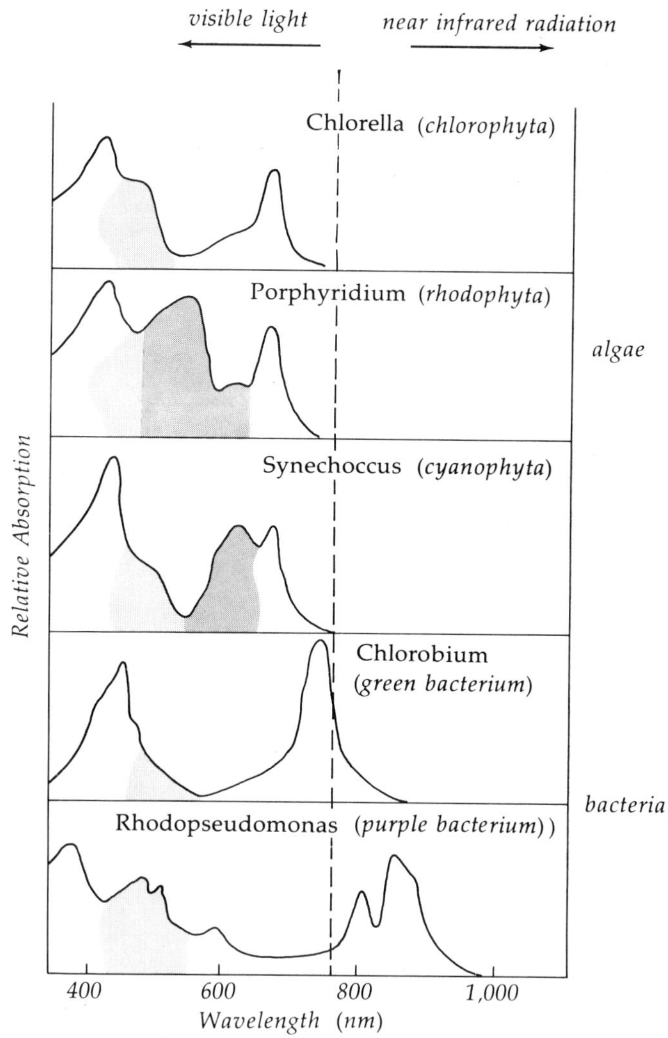

16–14 Absorption spectra of green, red, and blue-green algae compared with those of 2 photosynthetic bacteria. The green and purple sulfur bacteria have their main absorption peaks at wavelengths at which the algae do not absorb; thus the bacteria are able to capture light transmitted by the algae. The photosynthetic bacteria are frequently found in nature underneath the algae, where the main light available would be at the higher wavelengths.

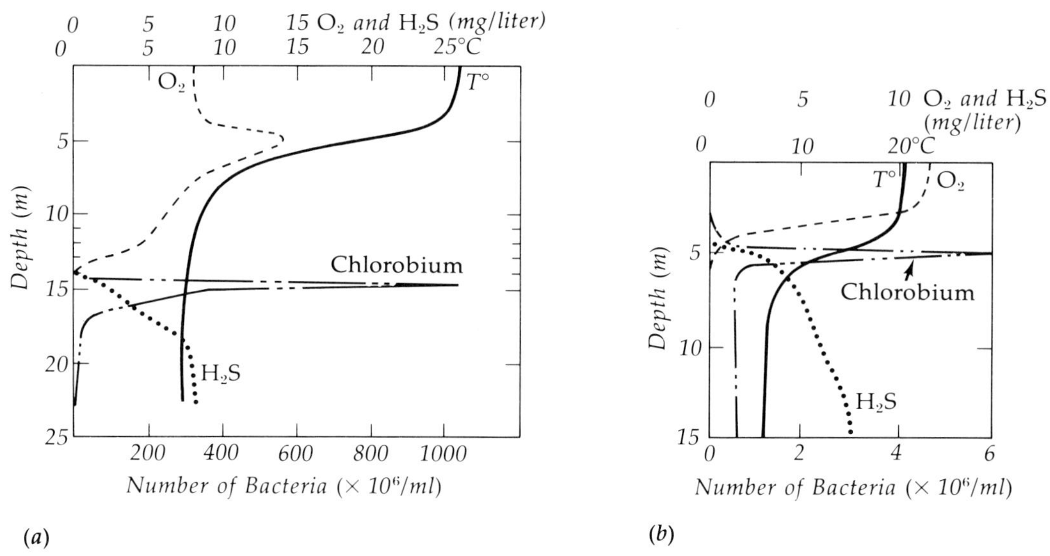

16–15 (a) Two lakes showing the distribution of photosynthetic bacteria as a function of depth and (b) other factors present in the lakes.

Essay 16–3 Bacteria and Yogurt

One of the oldest and most nutritious of all foods is yogurt. For centuries peoples of the Near East, Turkey, and the Balkans have made and enjoyed yogurt. More recently, millions in Europe and the United States have developed a taste for this natural food. Yogurt is essentially soured milk, but it is more than that. Two bacteria are normally found in yogurt: the common Lactobacillus casei, *which converts the lactose in the milk to lactic acid, thus making the environment more acid and unfavorable for the growth of pathogenic bacteria, and* Streptococcus thermophilus, *which adds a characteristic creamy flavor to the finished product.*

Yogurt is easily made with a minimum of equipment, and we now present one standard recipe, recognizing that there are almost as many ways to make yogurt as there are people who eat it!

Bring one quart of milk (to which may be added ½ cup of cream or 2 T powdered milk) to a boil in a pyrex, porcelain, or ceramic container (enameled iron pots are usable only if they are not chipped or cracked, since the presence of metals will inhibit the growth of the bacteria). Pour milk into a nonmetal container (a glass or earthen bowl with a cover is best). Dissolve 1 T to ½ cup of old yogurt (commercial yogurt may be used) in a little warm milk to act as the starter. Before adding the starter, test the temperature of the milk with your wrist to make sure it is no more than lukewarm. A thermometer is, of course, more accurate and should register 110 to 125°F when the starter is added. Mix in the starter, cover the bowl with a lid, and wrap a heavy towel over it. The bacteria will generate enough heat so that no more is needed; if you wish to speed up the process, you may place the milk in a warm place (i.e., in the oven with only the oven light as a source of heat or in a box warmed with a light bulb). Within 3 to 12 hours, depending on the amount of heat present, the mixture will become as thick as heavy cream or custard. Now, put it in a refrigerator to set for several hours before eating. And, in the immortal words of Julia Child, "Bon Appétit!"

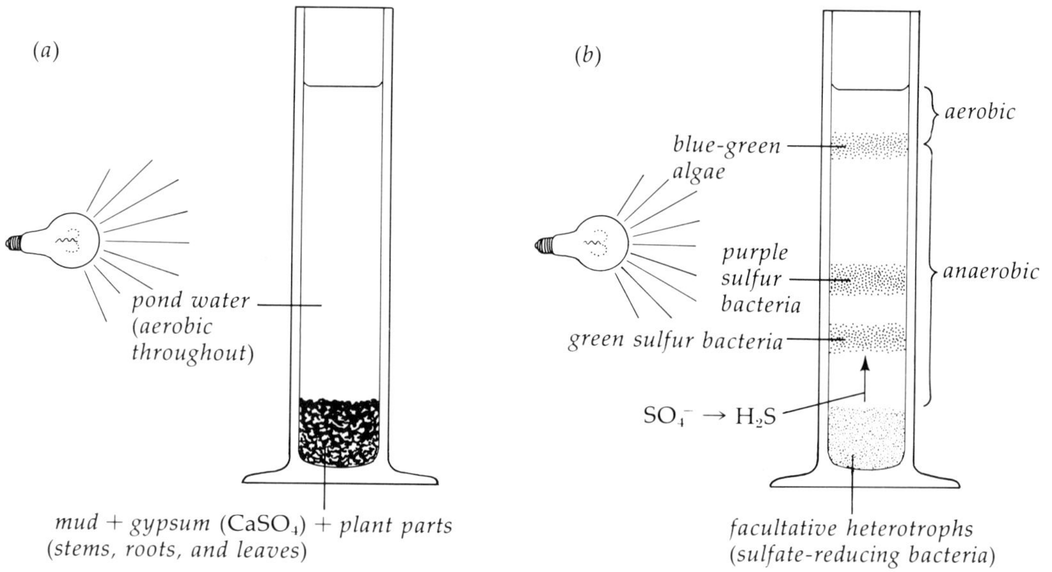

16–16 A Winogradsky column at the time it is initially set up (a) and after bacterial and algal colonies have been established in it (b).

These same sulfur bacteria can cause the corrosion of stone and concrete. One example of such corrosion is the damage being done to the impressive early Malayan temple of Angkor Wat in Cambodia. These magnificent ruins are overgrown by jungle and are slowly decomposing. The stone absorbs sulfide produced by bacteria in the soil. Other bacteria oxidize the sulfide to sulfur and sulfuric acid; the acid then corrodes the stone. Bacteria are also responsible for the corrosion of underground pipes by much the same mechanism.

In other circumstances the fact that bacteria can convert sulfur to sulfuric acid can be advantageous. One method of treating various plant diseases caused by fungi is to dust the plants with sulfur. Bacteria convert this to sulfuric acid on the surface of the plant, creating an environment that is too acid for the fungus but not strong enough to damage such plants as grapes.

The ability of bacteria to metabolize almost any type of material is being forcibly demonstrated by the current concern over mercury in our food. Mercury, used in large amounts in a variety of industrial processes, was dumped into rivers as waste. This was thought a safe procedure because mercury, the only common metal to be a liquid at room temperature, is most unreactive. Being a heavy metal, it sank to the bottom of the rivers, where it was covered by mud and muck—out of sight and out of mind. But, then, a disturbing thing began to happen: mercury began showing up in both freshwater and ocean fish (in general, the larger the fish the greater the concentration of mercury). The alarm was caused by the fact that mercury poisoning in humans is indeed terrible. It affects the central nervous system, and paralysis, blindness, and death may result. Where did the mercury come from? It came from bacteria that convert the inert, insoluble metallic mercury to soluble methyl mercury. This compound enters the food chain; the bacteria and algae containing it are eaten by protozoa, crustaceans, and insects that in turn are eaten by small fish; eventually the fish are caught and eaten by humans. At present, except where fish forms the major part of the diet as in Japan, the threat of mercury poisoning is not immediate. (One of the few suspected cases of mercury poisoning known in the United States was in a woman who, as part of a reducing program, ate 10 oz. of swordfish daily for 19 months.) But the long-range threat is real and cannot be ignored.

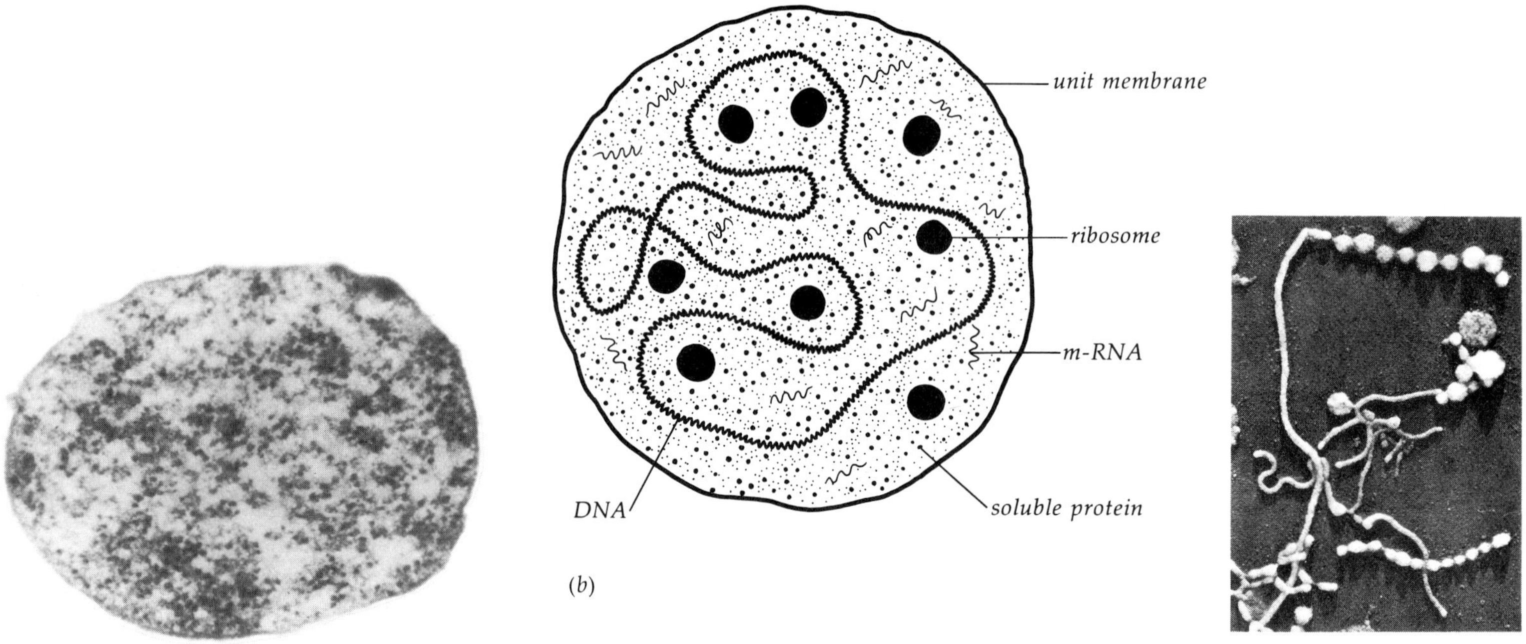

16–17 Mycoplasms are among the least complicated of all living things. Electron microscope studies indicate that they consist of little more than a plasma membrane, ribosomes, enzymes, and nucleic acids (a and b). Mycoplasms may have a wide variety of shapes, which appear to change quite readily. In this electron microscope photograph of metal-shadowed mycoplasm (c), the diversity of the shapes is clearly shown.

Mycoplasms

Bacteria are small, 1 to 2 μ in diameter, but there are still smaller organisms known as mycoplasms. (They are also known under the name of pleuropneumonia-like organisms or PPLOs, as a result of the type of disease some of them cause in animals.) These organisms range between 0.1 to 0.25 μ in diameter and contain, despite their small size, DNA, RNA, and a wide range of proteins and lipids. Their internal structures are almost nonexistent (Fig. 16–17): they have a plasma membrane (no cell wall), a few ribosomes, and a granular matrix. At times they may contain what appear to be smaller versions of themselves.

Until recently all mycoplasms were believed to be parasites, causing diseases in cattle, chickens, man, and vascular plants. But in 1970 a saprophytic mycoplasm was discovered in a burning coal refuse pile at the Friar Tuck mine in southwestern Indiana. This is a strange habitat indeed. The mycoplasm found there was both thermophilic (with an optimum growth temperature of 60° C) and acidophilic (with an optimum pH for growth of between 1 to 2). The organism was aptly named *Thermoplasma acidophila.*

Although little is known about mycoplasms, it is quite probable that many more will be discovered in future years. We can state this with some confidence, since microbiologists now possess the methods for their culture and observation. How important they will be except as disease organisms remains to be seen.

End Paper 16–1 Bacteria and Food

Bacteria do pose an immediate threat to us when the wrong kind occur in large numbers in our food. Unfortunately, the foods that we like are also the preferred foods of a great many bacteria. Let us look at some examples and some of the ways man has of preventing bacterial decay of food.

Consider an egg. This is a remarkably designed package of foodstuffs in a case, the shell, that allows gas exchange but is waterproof and impervious to bacteria, if handled correctly. Once the egg is opened, the contents provide an almost ideal culture medium for the growth of bacteria, and if the outside of the egg is not clean or the hands of the cook have not been washed well or the bowl is slightly dirty, bacteria can be immediately introduced into it. No harm is done if the egg or the food it is in is cooked; high heat will kill virtually all the bacteria. But, if the egg is used in a sauce, dressing, or custard, where it is either left raw or only gently heated, the bacteria will multiply rapidly, particularly if the mixture is left at room temperature or warmer. Starting with only 10 staphylococci bacteria, 600,000 can develop overnight, and the food eaten the next day could cause a severe case of food poisoning. By simply refrigerating the food, the bacterial growth would have been reduced to the point where the food could be safely eaten.

Simple bacteria growth in food does not in itself cause food poisoning or gastroenteritis. In our egg example we were careful to select a staphylococci bacterium that causes food poisoning. Many bacteria cause food spoilage, yet few cause food poisoning. Americans as a whole tolerate less food spoilage than most peoples. The other extreme are the Eskimos who relish a delicacy known as "titmuck." This is fish buried to allow bacterial fermentation to occur and is a semiliquid, foul-smelling substance even the Eskimos' dogs refuse to eat.

Food spoilage can be easily detected by smell, appearance, and taste—but poisoned food cannot. There are two distinct types of food poisoning. One is caused by bacteria that produce toxins. The most famous of these is the toxin produced by *Clostridium botulinum* that causes botulism. This toxin is so potent that 1×10^{-10} grams can kill a mouse. The bacteria in this case can form spores resistant to high temperatures and can grow in anaerobic conditions of essentially neutral pH. Another toxin-producing group of bacteria are the staphylococci, but they are less fatal.

The second type of food poisoning is caused by the presence of the bacteria themselves. This group includes *Salmonella* food poisoning that has been increasing steadily and receiving much publicity in recent years—so much so that Senator Magnuson of Washington introduced a bill in the Senate to have the name of the bacteria changed because he felt it was adversely affecting the sale of salmon!

Salmonella bacteria are rod-shaped and are natural inhabitants of the colon (bowel) of man and animals. They are passed out with feces and are introduced into food from soiled hands or contact with animal excreta. This is the only source, and it is incredible that in a highly industrialized, affluent society this type of food poisoning should exist at all, let alone be increasing at the rate it is.

Food preservation has been called nothing more than a race between man and the lower forms of life to see who gets the food supply first. Man's efforts in the race have been largely devoted to slowing down the bacteria. One of the most effective means is to keep the food cold so that bacterial growth is drastically reduced. Drying achieves the same end, since the bacteria cannot grow without water. Most bacteria also grow slowly or not at all in acid media; so pickling in vinegar, which is dilute acetic acid, is also a good way to keep food. Strong salt and sugar solutions also inhibit bacterial growth. All of these are ancient methods known to man for centuries. (Even freeze-drying was first developed by the Incas and is still used in the same form by the natives of the high mountains of Bolivia.)

Modern man has added many refinements to these procedures. We have developed mass means of freezing, cooling, drying, and canning (which is essentially heat preservation) food and of storing and shipping it. Modern food technology also relies on the addition of chemicals that retard the growth of bacteria. This is hardly a new procedure but one that has been receiving more attention in recent years. Such additives have been called poisons; in truth many are. Sodium nitrate, for example, has been used since the Middle Ages and, when used in small amounts, causes changes in food, usually meat, that inhibit bacterial growth; yet used in high concentrations it can be fatal. Many such additives have relatively little meaning in terms of ultimate food preservation but are there to ensure a particular color, texture, or manufacturing character to the food. These are clearly less defensible but are often of great economic importance. The repulsion that many people feel toward food additives and chemicals in general in agriculture has led in recent years to the establishment of natural-food stores. In general, the food runs to the less perishable types and usually requires the development of different tastes.

In the end the race is always between man and microbe to see who gets the food first. Without a doubt, bacteria and insects (with fungi running a strong third) are man's strongest competitors for domination of the ecosystem, and it is by no means clear who will survive longest.

17

Algae of the Soil and Freshwater

The Living Waters

In virtually every body of water—large or small, still or running, hot or cold—life is present. Plants and animals, both large and small, live there in a web of life spun around the algae. Different kinds of algae are found in various bodies of water: some are present only in freshwater, some are found only in salt water, and others live in both. In the soil and in freshwater habitats, the most commonly encountered algae are members of the Cyanophyta, or blue-green algae, and the Chlorophyta, or green algae. In many ways, the members of these two groups of plants superficially resemble one another, but this is highly misleading. The blue-green algae, like the bacteria, are procaryotic cells and lack membrane-bound organelles. On the other hand, the green algae are eucaryotic cells. Both the similarities and differences between these two important groups of plants will become apparent as we look at them with the light and electron microscope.

A *This is a common blue-green alga,* Oscillatoria, *as seen with both the light and electron microscopes. It is a filamentous alga in which the cells are arranged in files. The entire filament is capable of undulating motion—hence its name. The light microscope (left) shows the arrangement of the cells but little of their internal structure. The electron microscope (right) shows the ultrastructure of the cells. Thylakoid membranes run parallel to the end walls; the lighter areas contain the DNA; and the crystalline-like shapes are phosphate deposits. The cells are dividing. The wall is beginning to grow from the existing wall and proceed inward, until it reaches a similar wall growing from the opposite side, thus dividing the cell.*

B *Many blue-green algae grow as single cells or as loosely organized colonies of cells. Here is one of the most common of the blue-green algae,* Gloeocapsa. *The cells are surrounded by a thick wall and are resistant to drying.*

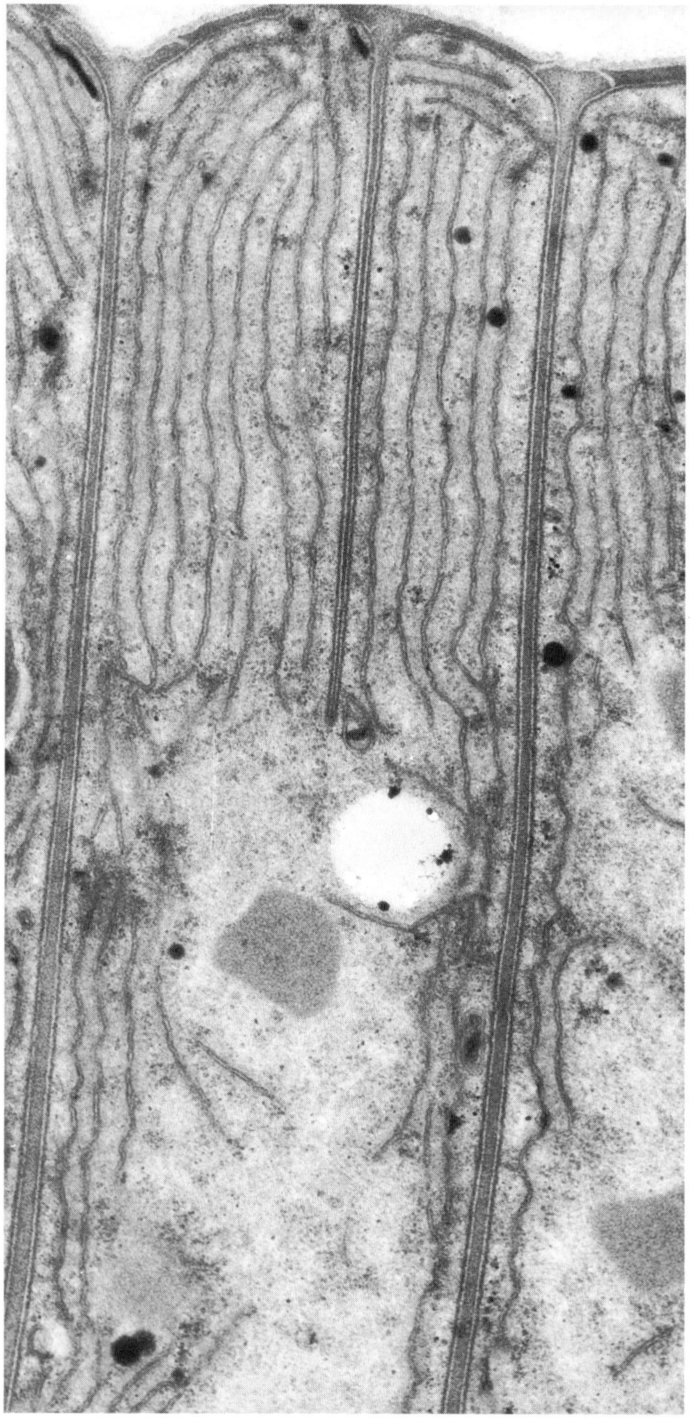

A

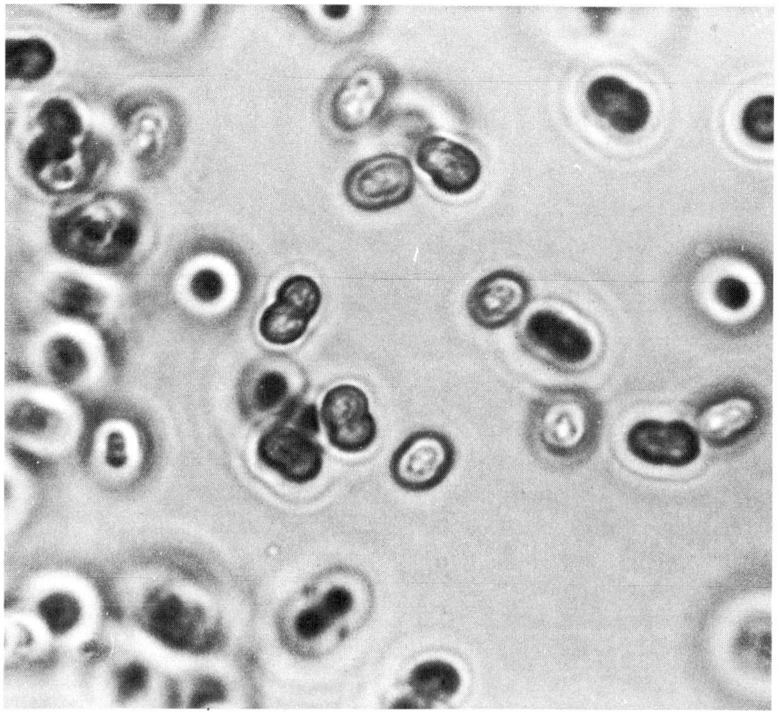

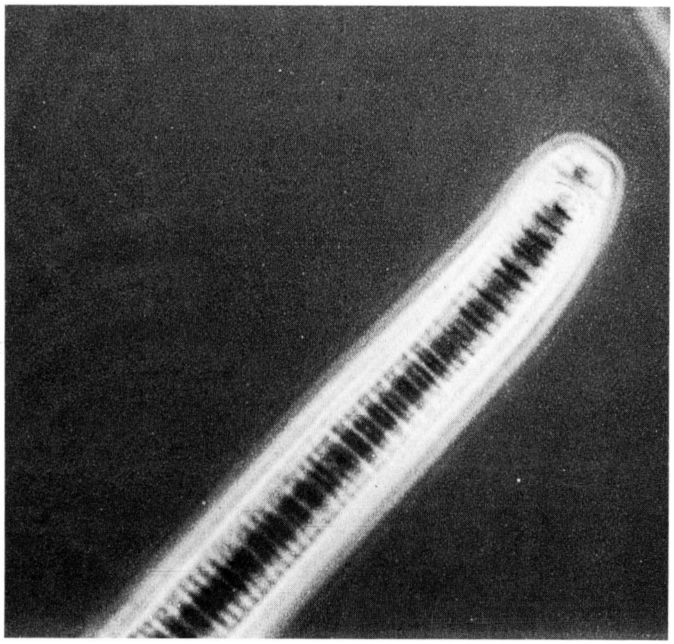

B

C *Another common blue-green alga is* Anabaena, *seen here with the light microscope. The larger of the cells in the filaments are called heterocysts and have special functions and structure.*

D Anabaena, *seen in section with the electron microscope, reveals cells with a handsome internal structure. The thylakoids wind throughout the cytoplasm, sometimes encircling the crystalline phosphate deposits. Many ribosomes are visible, while the lighter regions contain the DNA.*

E *The green algae also contain species that grow as single cells, colonies, and filaments, but the internal organization of the cells is vastly different and more complex. Here are two species of green algae as seen with the electron microscope;* Chlamydomonas, *a single-celled motile species, and* Oedogonium, *a common filamentous species. In both species the various cell organelles—nucleus, chloroplasts, mitochondria, and dictyosomes—are all present. Within the chloroplast is the pyrenoid, a structure that is involved in starch storage.*

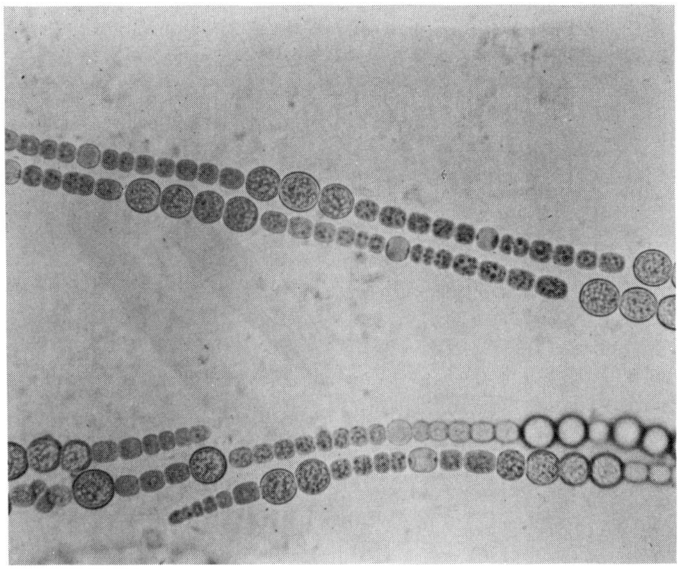

C

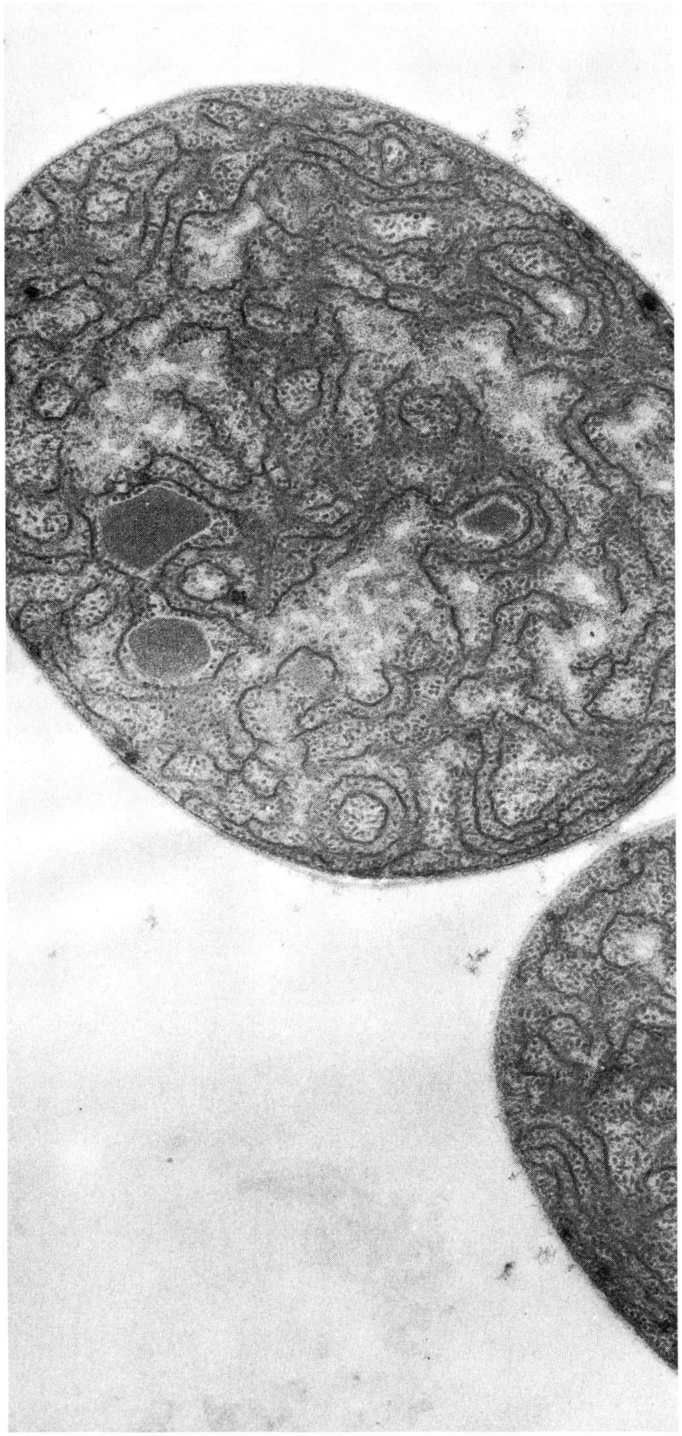

D

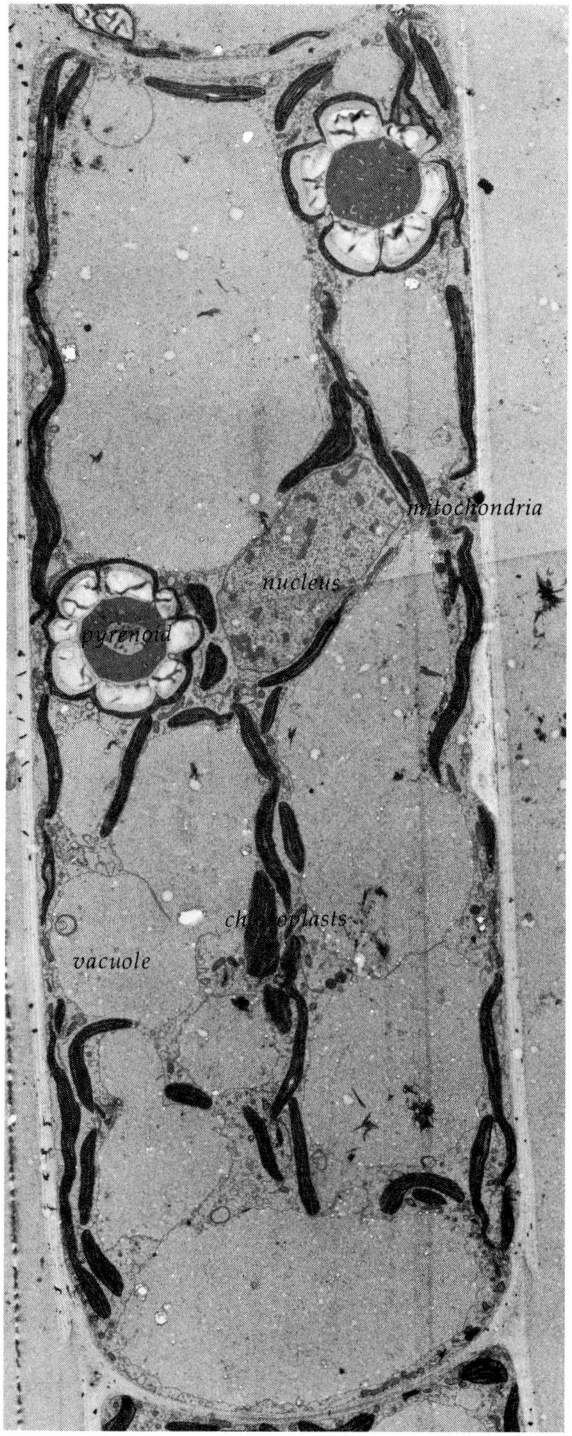

E *Oedogonium*

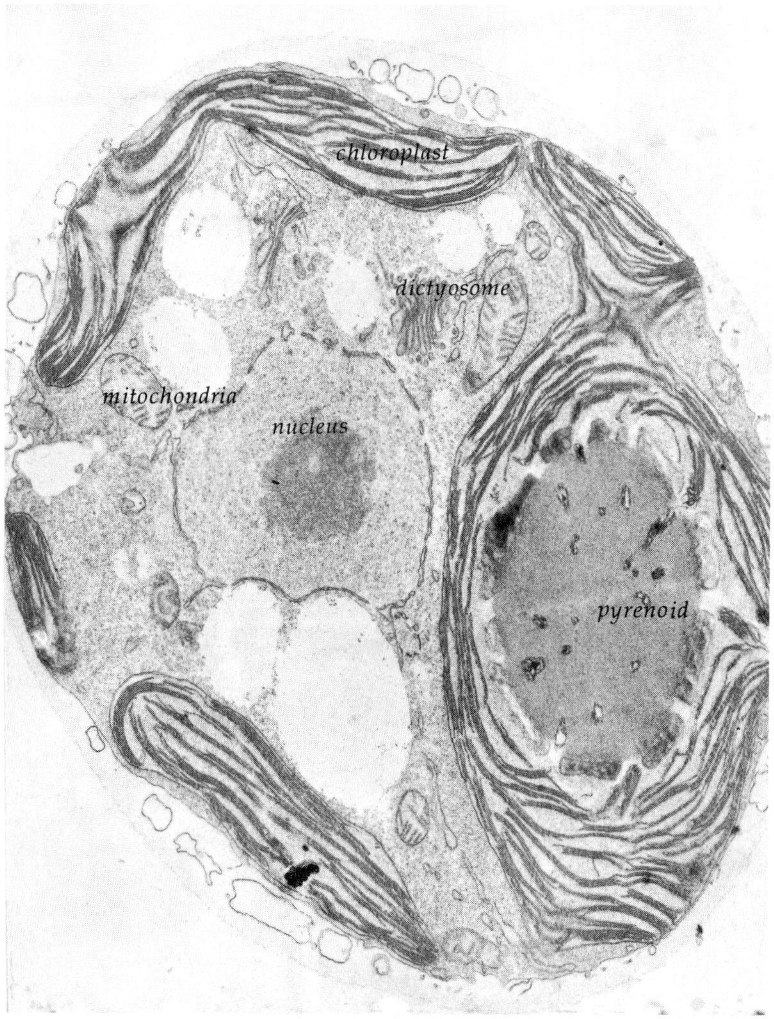

Chlamydomonas

F These light microscope photographs give some idea of the diversity of form of the green algae normally found in freshwater habitats. These include the filamentous Zygnema *with its stellate chloroplasts, the beautifully symmetrical desmid* Micrasterias, *the impressively spherical colonial* Volvox, *each cell complete with two flagella and an eyespot that appears as a tiny black dot, and the aptly named waternet* Hydrodictyon. *These are only a miniscule sampling of the actual variation of form that occurs in this large division.*

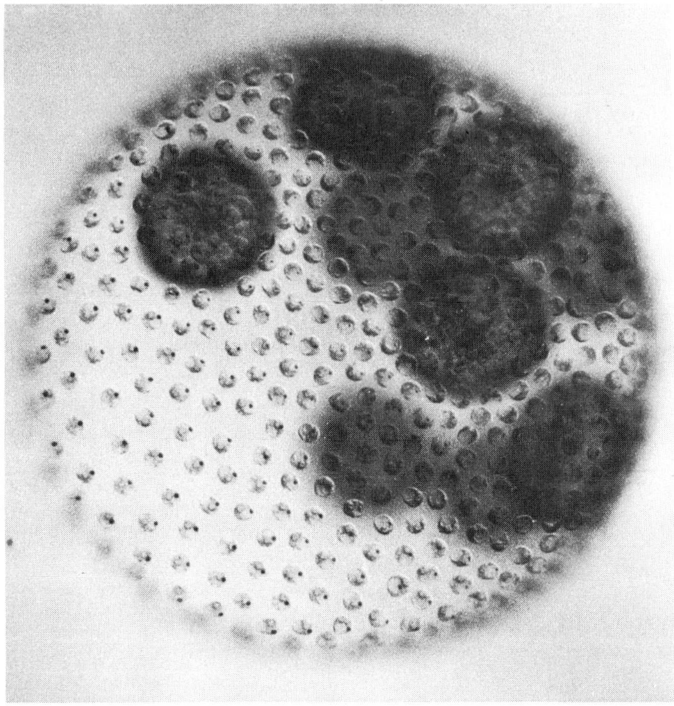

Volvox

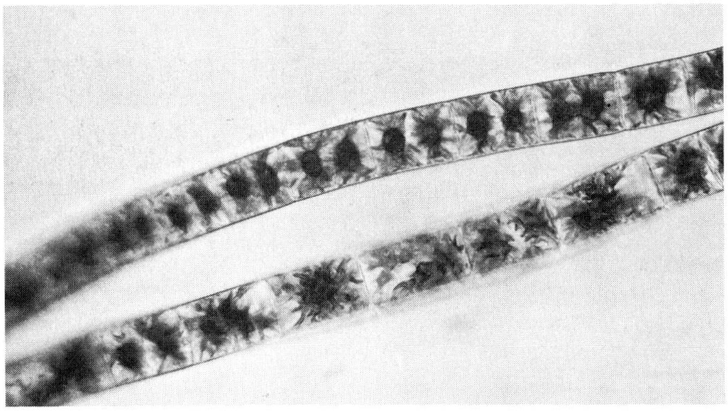

Zygnema

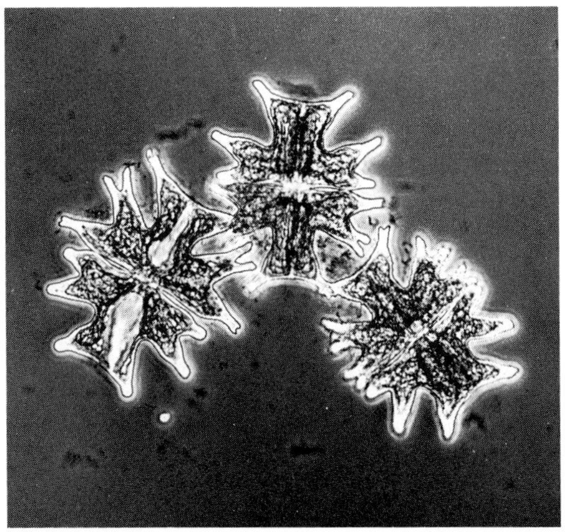

Micrasterias

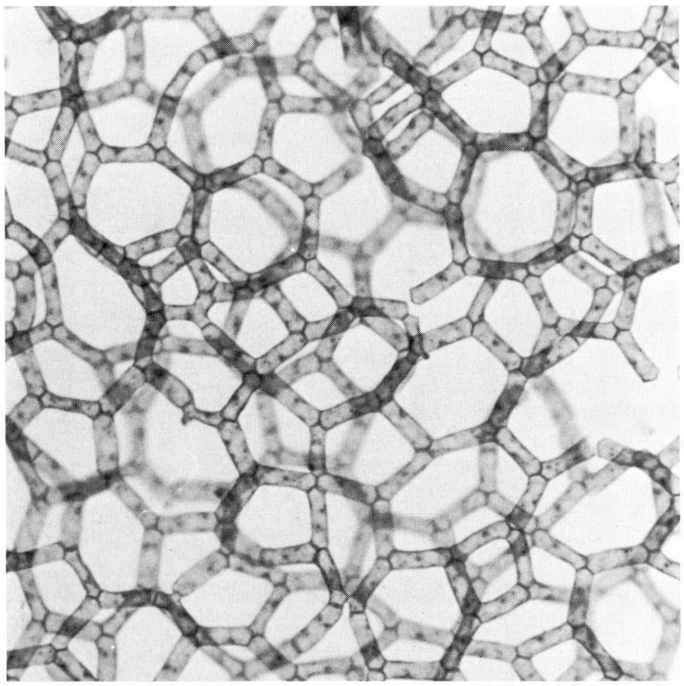

Hydrodictyon

The lake was dying. Its usually clear water was gray-green. Great mats of dying vegetation containing dead fish were washed up on shore. The whole area reeked with the odor of putrefaction. The once transparent waters were cloudy with floating, minute globules of material. The lake was the water supply for the nearby city, where people were now rationing what meager amounts of clear water they could secure. Blackmarket water was selling for $5 a pint. Those who could leave the city did, but there were few areas within 50 miles where clear water was available. It had been a hot, dry, sunny summer, and the people had enjoyed the good weather by using the large lake in great numbers.

Besides being an active tourist area the lake was the hub of agricultural land supported by irrigation from the lake's waters. The fruit processing plants, the distillery, the winery, and the local factory all bypassed the antiquated sewage plant and dumped wastes into the creeks that flowed into the lake. Animals drinking water from the lake were taken ill and dying in large numbers.

But all these conditions aided the growth of blue-green algae in fantastic numbers. The gray-green globules marring the lake waters were the blue-green algae in "bloom condition." These algae are able to grow in conditions that limit other organisms; some of the blue-green algae can fix atmospheric nitrogen. As long as other nutrients are present and the light and temperature are near optimum, the blue-green algae will flourish.

The lake and city described here are hypothetical; yet, if conditions remain unchanged, they could be true for many areas of the country within 20 years. Such conditions exist not simply because of man (blooms occur without man's intervention but on a vastly different scale) or because of a willful, destructive urge on the part of the algae (they are simply using the environment available to them) but because man has altered the environment in such a way that normally minor constituents of the ecosystem now become pests.

During a bloom the algae in the lake are intrusively evident, but they are present in the lake throughout the year. They also grow almost everywhere: in and on the soil, on trees, in ponds, puddles, birdbaths, ditches, streams, lakes, oceans, and snow, on houses and tombstones, on the backs of turtles, and in the hair of tree sloths. Their range in size, form, and color is almost overwhelming. Microscopic, single-celled species are present in the same group with beautiful and complex forms 1 m long. In some species during a season of growth, microscopic forms alternate with forms having bodies 40 m and more in length—some of the longest plants known. In color they appear brilliant-green, dusty-red, iridescent, olive-green, ruby-red, black, grass-green, brown, and just about every shade in between. Attempting to understand such diversity is not easy. But we shall approach this confusing jumble through a unified concept—the relation of algae to their environment. Such an ecological approach allows the form, function, and reproduction of the algae to be understood in terms of the environment in which they are found and their interactions with it.

Soil and Terrestrial Algae

One such environment is moist earth, perhaps near a creek shaded by trees or rocks. This humble habitat is the home of dozens of diverse kinds of algae. Especially common are a number of blue-green algae or Cyanophyta (see Fig. 17–B), including the microscopic *Gloeocapsa* consisting of masses of cells enclosed in a gelatinous sheath. In slightly wetter environments grow the filaments of *Oscillatoria*, composed of stacks of cells surrounded by a gelatinous matrix. These are but two of the many species of blue-green algae found in these locations (see Fig. 17–A, B).

The blue-green algae are procaryotic organisms (Fig. 17–A) like the bacteria. They lack organized nuclei, and they have no nucleoli, mitochondria, chloroplasts, or similar membrane-enclosed organelles. Cell walls in the blue-green algae are similar in composition and structure to bacteria and quite unlike those of other algae and higher plants. They contain mucopeptides and pectic substances but little, if any, cellulose.

A major feature of the blue-green algae is the presence of the phycobiliproteins, the red and blue pigments that together with chlorophyll give the algae their characteristic color. All blue-green algae are not blue-green in color; they may range from black to rusty-red, depending on the type and the amount of the particular phycobiliproteins present.

Growing near streams on the bark of trees and bushes (tinting them distinctly green) is an alga that superficially resembles the blue-green alga *Gloeocapsa.* This is *Desmococcus* (formerly known as *Pleurococcus* or *Protococcus*), a member of an entirely different division—the green algae or Chlorophyta (Fig. 17–A, B). An electron microscope section of a green algal cell reveals just how different are the organisms. Where *Gloeocapsa* lacks a nucleus, chloroplasts, and similar organelles, *Desmococcus* and other Chlorophyta have nuclei, nucleoli, chloroplasts, mitochondria, ER, and the other standard cell parts (Fig. 17–A). The chloroplasts of most of the Chlorophyta contain the same chlorophyll pigments, chlorophyll a and b, and carotenoids (usually in the same proportion) as vascular plants. Starch is the food storage compound in *Desmococcus,* as it is in most of the Chlorophyta. The starch is stored in the chloroplast and may be found surrounding a specialized structure within the chloroplast called a pyrenoid (Fig. 17–A). The cells of *Desmococcus* lack a vacuole, and the thick cell wall is not surrounded by a gelatinous sheath.

How do soil and tree bark, as an environment for plant growth, fulfill the survival requirements (water, light, oxygen, carbon dioxide, various minerals, and a reasonable temperature) of a photosynthetic plant? The factor that limits growth in these environments is water—not light—since both the soil and the bark are subject to drying out at almost any time. The amount of light and the availability of water are related, because sunlight will cause the evaporation of the water; the side of the tree or the portion of soil that receives the most light will be the driest. Thus, in the Northern Hemisphere, *Desmococcus* is found growing on the north sides of trees. Oxygen, CO_2, and minerals are all in reasonable supply as long as there is enough water. Temperature fluctuations are great between day and night and summer and winter, thus limiting cell growth. But survival in these environments is primarily dependent upon the ability of organisms to withstand drying out without marked changes in structure. Both *Desmococcus* and *Gloeocapsa* can survive loss of water without forming a special resistant cell or spore. How this is accomplished at the cellular level is not clearly understood. In *Gloeocapsa* the thick gelatinous sheath is considered important; so is the fact that both *Gloeocapsa* and *Desmococcus* lack central vacuoles and are densely cytoplasmic cells. Also, both are spherical cells, and, as in the bacteria, this form appears important in distributing the forces on the wall as the cell dries. How well the blue-green algae can withstand drying is shown by those species found in the Antarctic. There, temperatures drop as low as −88° C, and the cells are exposed to severely desiccating conditions as a result of low atmospheric humidity. That blue-green algae can survive long periods of desiccation is shown by the ability of some herbarium specimens to grow after 87 years of storage. But not all blue-green algae can withstand drying of the vegetative cells. This is true of many aquatic species, which develop spores that carry them through the dry periods.

The blue-green algae, along with the bacteria, found in the soil occupy a unique position in the ecosystem because of their ability to fix atmospheric nitrogen. Repeatedly, blue-green algae are found as pioneer species growing on sterile or underdeveloped soils. They grow in the ground uncovered by receding Antarctic glaciers and on sand dunes. In such locations they can grow because they utilize atmospheric nitrogen and inorganic phosphates in the substrate. The presence of the algae helps to change the texture of the substrate: when cells die, the material added to the soil through their decomposition adds to the nutrient content of the developing soil so that other plants can become established.

Pond Algae

The type of pond that we saw in our acre of woodland is an environment where a great diversity of algae grow in profusion. Many such ponds are temporary (existing only during and right after rains) and shallow (sometimes less than 1 m deep). In such ponds are found some of the most beautiful of the colonial green algae (see Fig. 17–B). These organisms are composed of various numbers of individual motile cells held together in a common gelatinous sheath. Some are hollow spheres (*Volvox*); some are solid spheres (*Pandorina*); others are flat plates of cells (*Gonium*). Growing in the same ponds are small, single-celled motile organisms such as *Chlamydomonas* (see Fig. 17–E), which are probably the primitive stock from which the more elaborate colonial forms evolved. *Chlamydomonas* (like the individual cells in the colonial forms) has 2 flagella per cell and may have a red pigment spot in the chloroplast (Fig. 17–E). This pigment spot is a specialized region of the chloroplast present in the cell and seems responsible for cell light sensitivity—the cells swim toward light of moderate intensity.

Also found in the pond are numerous filamentous algae (Fig. 17–F). These are composed of cylindrical cells arranged end to end, forming a strand. The chloroplasts present in such algae are highly variable. In the green alga *Spirogyra* the chloroplast is ribbonlike and arranged in a spiral. In others the chloroplasts may be platelike (*Mougeotia*), netlike (*Oedogonium*), or star-shaped (*Zygnema*). These filamentous algae may be found floating free or may be attached to the bottom by a basal cell modified to form a holdfast.

Floating in the water of the pond are many green and blue-green algae. These and other free-floating algae make up the phytoplankton (phyto = plant, plankton = drifting). Blue-green algae found in the phytoplankton include *Microcystis* and *Anabaena* (see Fig. 17–D); the latter is one of the nitrogen-fixing blue-green algae. Among the green algae present in the plankton are the desmids (Fig. 17–F). These beautiful, single-celled organisms are composed of two mirror-image halves.

Diatoms, members of the division Chrysophyta, compose a major part of the phytoplankton. Consisting of a single cell, each is surrounded by rigid, overlapping, highly ornamented walls that fit together like two halves of a box (see Fig. 17–11).These walls contain large amounts of silica (SiO_2)—as much as 95% of the wall—in addition to cellulose. Another member of the phytoplankton, one that is capable of swimming, is *Euglena*, a member of a small division Euglenophyta. These strange little organisms may be numerous and can color the water bright green. (Some of these don't contain chlorophyll and are therefore classed as protozoa by zoologists.)

Ponds and Algal Survival and Reproduction

Why are the algae growing in ponds and pools quite different from those in the soil? The answer quite clearly is because the environments are so different. In a pond water is plentiful (at least for certain periods of time), and the situation is truly aquatic. Oxygen and CO_2 are dissolved in the water, and both are usually present in amounts adequate for survival and growth, although the oxygen level is variable and dependent on a number of factors, including temperature. In ponds the water temperature is similar to the average air temperature and can be quite warm; it is also usually uniform with respect to depth. Light is also in good supply, although water with its impurities both reflects and absorbs a large amount of light. Enough light reaches the bottom of most ponds to support a population of bottom-dwelling species. Finally, ponds tend to be rich in minerals, and some contain large amounts of complex organic compounds such as vitamin B_{12}.

The algae and the pond must be considered together to understand how organisms are adapted to and interact with the environment. In early spring a dried-up pond is beginning to fill because of the rains, and a frozen pond thaws. In either case, the plants begin to grow from spores or other resting stages, the vegetative stages generally having been killed when the pond dried up or froze. In the case of the motile forms such as *Chlamydomonas, Volvox, Euglena*, the desmids, the diatoms, and the single-celled blue-green algae, reproduction by cell division is enough to populate the pond and to distribute the organisms throughout. The filamentous algae, particularly the attached forms, can also develop rapidly from the resistant stage by cell division. In addition, some have a means of dispersal known as the zoospore.

The zoospore is an asexual motile cell that aids in the dispersal and multiplication of the organism. One type of zoospore can be illustrated by the green alga *Oedogonium*. Here, the protoplasm of an individual cell contracts and pulls away from the encasing cell wall. A circular ring of flagella forms, and the cell wall breaks, releasing the motile zoospore. This swims to the bottom of the pond where it attaches itself to a solid object and undergoes cell division, producing a new filament. Obviously, if a large number of cells in only a few filaments did this, the organism could quite effectively spread in a pond. In the alga *Ulothrix* another type of zoospore formation is found. Here, individual cells divide mitotically, producing from 4 to 8 zoospores each equipped with 4 flagella. These zoospores are released from the confines of the original cell wall and may swim from a few hours to a few days. Eventually, they attach themselves to a stone or another plant and germinate. In some algae each cell can produce many zoospores but always in even multiples such as 4, 8, 16, etc. Obviously, in all these cases not all the zoospores survive—many are eaten by microscopic animals, while others fail to find the proper conditions for germination—but enough do survive to effectively populate a pond.

Zoospores are formed at times other than the period of early establishment in a pond. The condition under which zoospores are formed is instructive in terms of survival of the species. Cultures of a filamentous algae (e.g., *Oedogonium*), placed in the dark or in greatly reduced light, will rapidly produce large numbers of zoospores. The advantage of this to the plant in nature is clear: should the part of the pond it is growing in come in the shadow of overhanging vegetation for long periods, the motile zoospores, which are attracted to light, can establish new filaments in better lighted locations. Alternately, if similar algal cultures are growing on a medium containing minimal nutrients and if this medium is replaced with one rich in nutrients, large numbers of zoospores will be formed. In a natural environment this would give the algae the potential to establish new plants late in the season, if added nutrients become available (e.g., through rains washing minerals into the pond). Clearly, zoospore production is a versatile and valuable mechanism in the survival and dissemination of the species.

Filamentous algae can also be spread by fragmentation of the filament. Such fragmentation is particularly important in the blue-green algae, where special cells or weak regions develop to help the breaking-up of the filaments. Many green algae (e.g., *Spirogyra* and *Zygnema*) lack zoospores, and for them, as for the blue-green algae, fragmentation is an important method of dissemination.

Fragmentation and zoospore formation are effective ways for the organisms we have been discussing to survive in the pond environment. They do not, however, offer the organisms a means of adapting to long-range changes in the environment. Only the variation through mutation and the recombination possible in sexual reproduction can do this. It is therefore hardly surprising to find that most of these plants, the blue-green algae being an exception, have well-developed means of sexual reproduction.

Chlamydomonas. This alga offers a good example of sexual reproduction in the green algae (Fig. 17–1). The gametes are produced by the division of the vegetative cell. Each cell may produce 4, 8, 16, or 32 biflagellate gametes that are similar to but smaller than the original vegetative cell. Two morphologically similar gametes fuse to form a zygote with 4 flagella. Soon the flagella are withdrawn, and a thick, often strikingly sculptured cellulosic wall is formed around the zygote. This thick-walled zygote is resistant to extremes of temperature and moisture, and it is this stage that may survive the dry or cold periods. When water is again

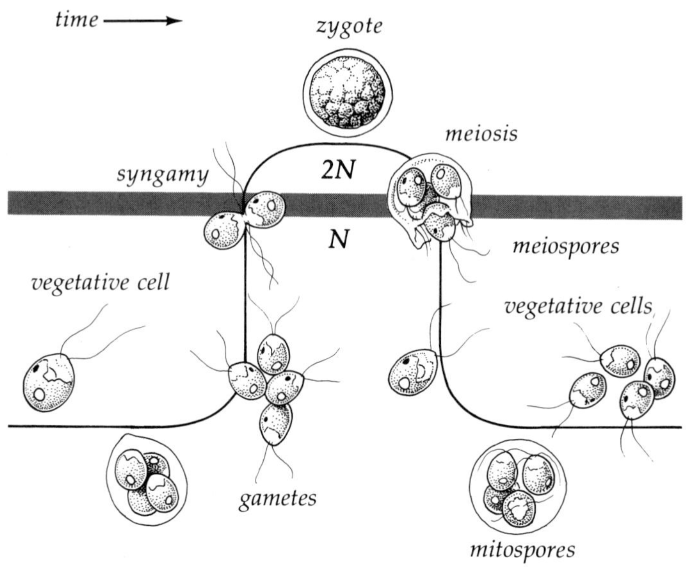

17–1 *Life history of* Chlamydomonas.

available and the temperature is warm enough, the zygote germinates. The first division of the zygote nucleus on germination is a meiotic one resulting in 4 small haploid cells, each with 2 flagella. These rapidly grow to the normal size of vegetative cells, and the life history is complete.

This relatively simple life history has some interesting features. The gametes that fuse are identical in appearance and are thus termed isogametes (Fig. 17–2). Yet they are not completely identical, since two produced by the same cell will not fuse. Experiments have shown that there are two strains in some species of *Chlamydomonas* (usually indicated as + and −), and only when gametes from these two strains are present does fusion take place. The strains must differ in some chemical rather than morphological way, but the result is that the organism is assured maximum variation upon recombination.

What makes *Chlamydomonas* produce gametes rather than vegetative cells? In one species, the answer is a depletion of nitrogen in the environment—a condition that may signal the end of the growing period. In another species the amount of photosynthesis, usually associated with day length or shading of the pond rather than nitrogen, is the determining factor. In any case, a specific condition frequently appears to trigger the sexual phase.

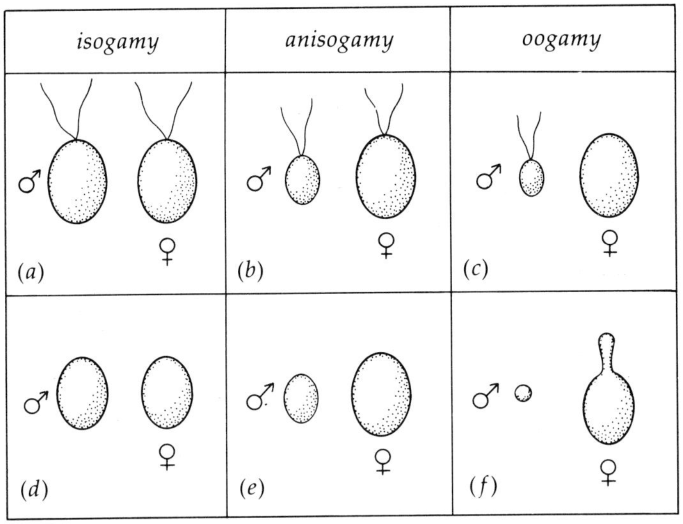

17–2 *Types of gametes. The gametes are either identical in appearance (isogamous, a and d) or dissimilar (anisogamous, b and e). Both gametes, only one, or neither may be motile. If one gamete is large and nonmotile and the other markedly smaller (motile or not), sexual reproduction is oogamous (c and f), with the larger gamete being regarded as the female (egg) and the smaller as the male (sperm).*

Ulothrix. A life cycle much like *Chlamydomonas* is found in this filamentous alga, which we considered earlier. Here, the vegetative cells of the filament each begin a number of mitotic divisions resulting in 32 or 64 biflagellated gametes. The gametes are released by the breakdown of the wall and will fuse with morphologically similar gametes produced by other filaments. The diploid zygote has 4 flagella for a short period, but these are lost when the thick walls form. Again, as in *Chlamydomonas,* the zygote is a resistant, dormant stage that carries the organism over periods unfavorable for growth. When the zygote germinates, it immediately undergoes meiosis, and 4 zoospores are formed. These swim until favorable anchorage is found; then they will each give rise to a new filament.

Oedogonium. In both *Chlamydomonas* and *Ulothrix* the isogametes are motile, but not all algae follow this pattern. In *Oedogonium,* an extremely common filamentous algae, the egg is large and immobile, while the sperms are small and motile (Fig. 17–3). One filament produces eggs similar to the vegetative cells except that the protoplast is pulled away from the wall, the cell itself is swollen and barrel-shaped, and a pore develops in the wall. The sperms are produced on separate filaments through the repeated mitotic division of specialized cells. Each sperm has a ring of short flagella at one end and, when released through the breakdown of the surrounding wall, swims to the egg-containing filament. The sperm is actually attracted to the egg by a substance produced by the egg. When a sperm reaches the pore, a fingerlike projection of the egg extends through the opening, becomes attached to the sperm, and pulls it through the opening where the fusion occurs. The zygote, which is now diploid, forms a thick wall and is once again the resistant, resting stage. Upon germination, the zygote produces, through meiotic division, 4 multiflagellate haploid zoospores, each capable of forming a new filament.

This type of reproduction, where one gamete is different from the other, is said to be anisogamic (Fig. 17–2). The most pronounced case of anisogamy is oogamy, where the female gamete is a large, stationary egg and the male gamete is a small, motile sperm. While common in the algae, anisogamy becomes the major means of reproduction in the remaining plant groups.

Even more elaborate arrangements for sexual reproduction are found in some species of *Oedogonium* (Fig. 17–3). In these the sperms are not formed directly in the original filament; instead a special zoospore is formed that swims to the cell that eventually divides to form the egg. The zoospore is attracted to the pre-egg cell, where it becomes attached and develops a filament only a few cells long, which is called the dwarf male and produces the sperm. The presence of the dwarf male triggers the division of the pre-egg cell and the formation of the egg by one of the new cells. The egg in turn secretes a gelatinous material that traps the sperms released by the dwarf male. The egg also secretes a substance that attracts the sperms to the egg. The interactions between the parts of this elaborate system must be controlled by various chemicals produced by the different cells at precisely the right times—a truly impressive mechanism that evolved to assure fertilization.

Spirogyra. In another filamentous alga, *Spirogyra,* a different mechanism is employed to unite the gametes during sexual reproduction. In some species, at the time sexual reproduction occurs, two filaments of

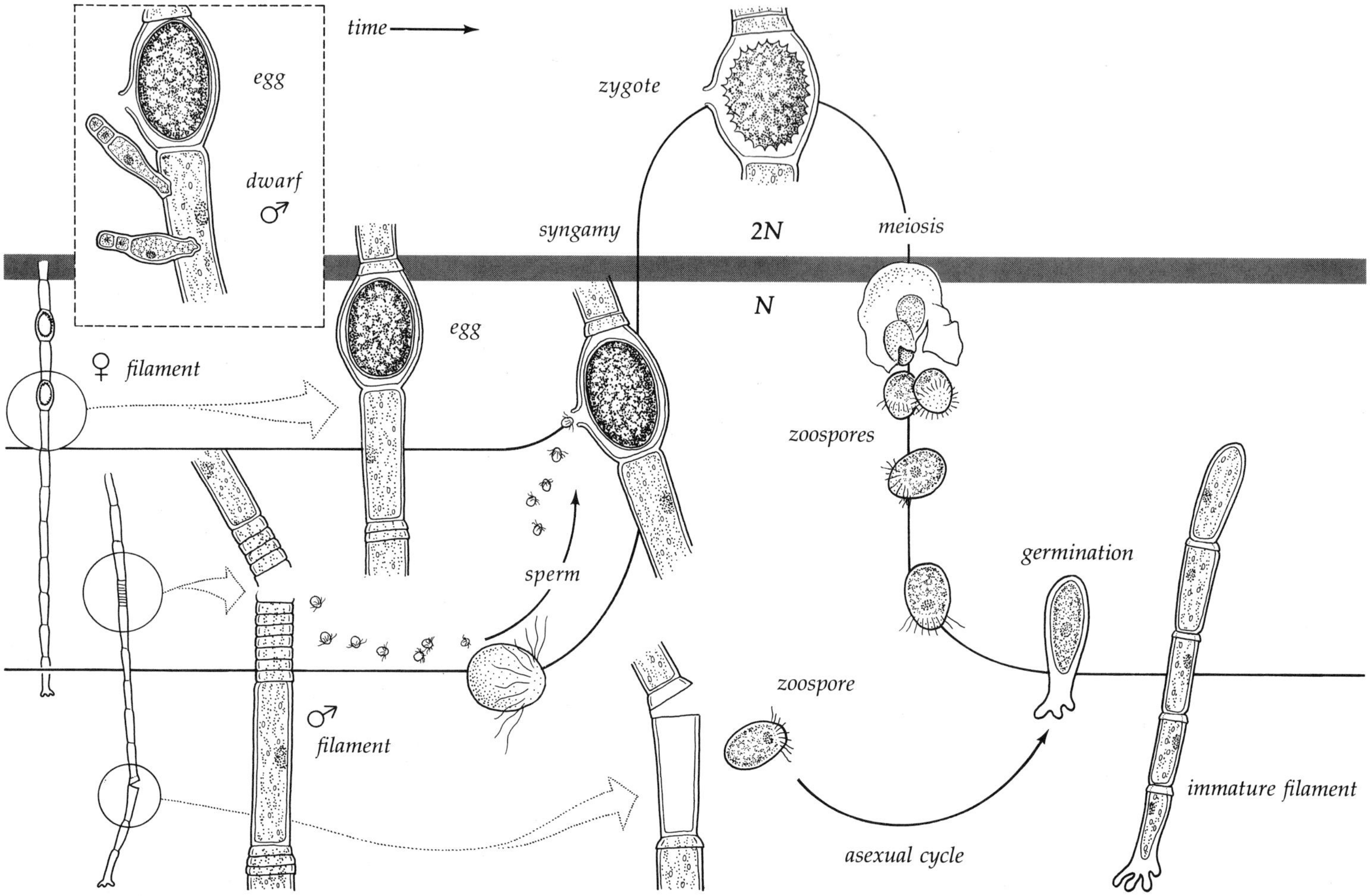

17–3 *Reproduction in* Oedogonium.

the proper strains become lined up against each other, cell by cell (Fig. 17–4). Short extensions grow out from the walls of the adjacent cells, and small tubes are formed. Then, the protoplast of one cell flows through the tube and into the connecting cell. The 2 protoplasts fuse, and the zygote is formed. The zygote once again is the dormant stage and, when it germinates, divides meiotically, producing 4 haploid nuclei. These do not form zoospores. Instead, all but one degenerate; that one divides repeatedly to form a new filament. The degeneration of 3 nuclei nicely reestablishes the original uninucleate condition of the filament. *Spirogyra* is a case of morphologically similar gametes, except one moves while the other is stationary. The process of transfer of gametes through connecting tubes is called conjugation.

Each of the organisms just discussed is well suited to the pond environment. All have a mechanism for establishing themselves (reproducing asexually and sexually) and of resisting the unfavorable periods of the environment. But conditions change in a pond as the season progresses. The optimal combination of light, temperature, oxygen, carbon dioxide, and minerals or other nutrients necessary for the growth of one species passes, and another species takes its place. This succession of algal species is increasingly important to man because it is part of the ever-growing problem of pollution.

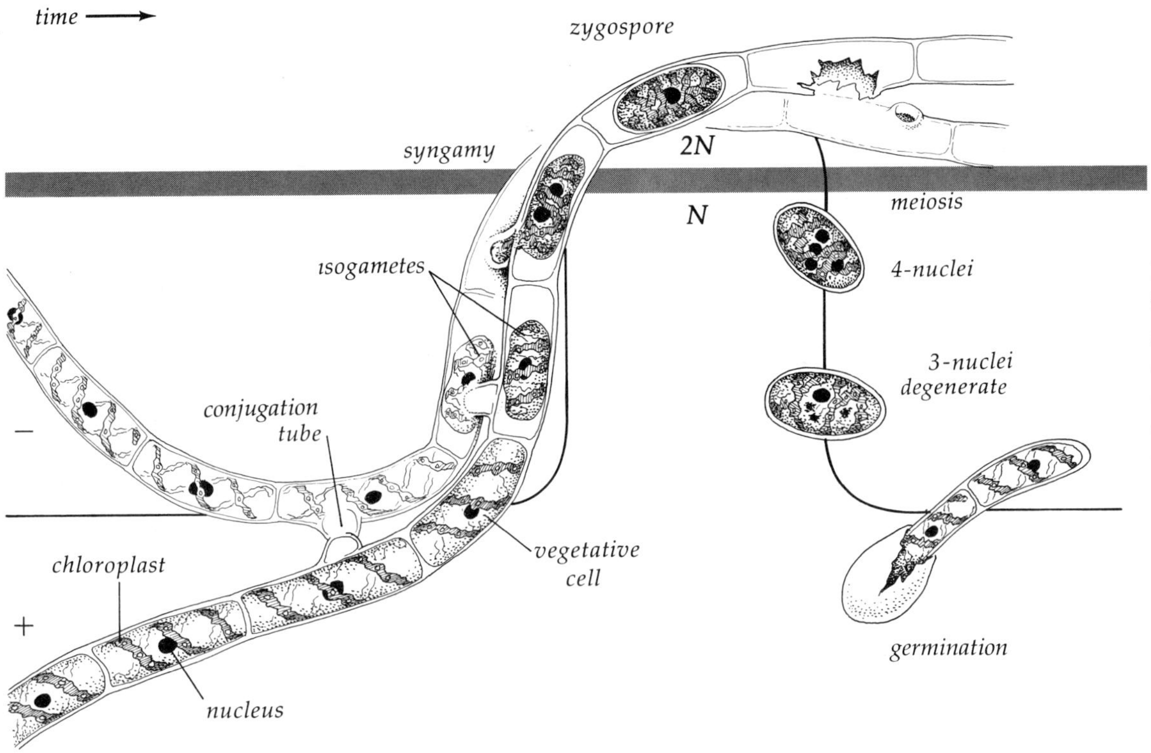

17–4 *The life history of* Spirogyra. *In this alga the gametes are only slightly different from the vegetative cells and are identical to each other—except that one moves, while the other does not.*

Lake Algae

Every lake is different—as every fisherman knows. Each differs in size, depth, clarity, mineral content, amount of shading, amount of wind-caused wave action, temperature, water flow in and out, and other physical and chemical factors that influence algal growth. In addition, each lake represents two quite different environments for algal growth. The shallow water around the edges most closely resembles the ponds. Here, light reaches the bottom, and various types of attached filamentous algae and bottom-dwelling diatoms and blue-green algae can survive.

The center of a deep lake presents a completely different environment. Only floating and swimming algae can live in the center of the lake. As we saw in our discussion of the bacteria found in lakes, the bottom is dark and occupied by bacteria rather than algae. In the upper layers of the lake light is present, and here we find the phytoplankton—diatoms, dinoflagellates, motile chrysophytes, green and blue-green algae. To what depth the algae grow depends not only on the amount of light striking the surface but the materials dissolved or suspended in the water and the number of algae present.

The planktonic algae live in a closely balanced equilibrium between the amount of food they can make through photosynthesis and the amount they use for energy through respiration. Measurements have shown that for lake algae the average ratio of photosynthesis to respiration is 1.05. This means that there is a slight excess of photosynthate available for growth and reproduction of the system. Thus relatively small changes in the amount of photosynthesis will influence the growth and number of the algae.

What factors influence the amount of photosynthesis taking place in a lake? One important factor is the depth to which light will penetrate the water. This depends on the amount of material dissolved in the water and the number of algae. If the algae are few in number, light can penetrate to a greater depth, and photosynthesis can occur as deep as 15 m. If the algae are numerous, they absorb the light near the surface and become concentrated in the upper few meters of water, effectively shading the water below them and inhibiting algal growth beyond this level.

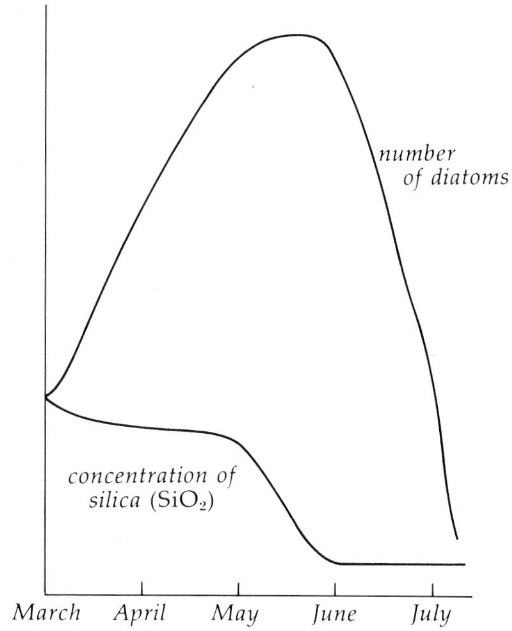

17–5 *This graph illustrates that the amount of silica available in a lake can be the limiting factor in the growth of diatoms. Other elements were measured, but none showed the type of correlation shown here.*

What determines the number of algae? Probably, the most important factor is the amount of nutrients available. These will vary from species to species and group to group, but one of the most clear-cut is the silica requirement of diatoms. Up to 90% of the wall of diatoms is composed of silica, and most species show a complete dependence on silica for division. The graph in Figure 17–5 shows the change in the number of two species of diatoms and the amount of silicon over a five-month period. As the number of diatoms increased in late winter and early spring, the amount of silicon decreased. By late spring, so much of the available silicon was incorporated into the cell walls of existing diatoms that further divisions were not possible, and the number of diatoms decreased as they died or were eaten by protozoa and invertebrates.

Nitrogen and particularly phosphorus are important nutrients that can limit algal growth, if they are present in too small quantities. While some of the blue-green algae can fix nitrogen, they still require a source of phosphorus.

Lakes that are low in nutrients and hence low in organic material are oligotropic. With time almost any lake will gradually increase in nutrients and organic material; this enrichment process is called eutrophication. It is a natural process that takes place whether or not man is present.

To understand these changes, let us follow the changes in a lake through the year (Fig. 17–6). Two physical factors are important: water is most dense at 4° C, and wind, through wave action, will cause a mixing and aeration of the water.

In winter the lake will be covered with ice at 0° C. The water immediately below will be slightly warmer (some 2° C), while the rest of the lake will be a uniform 4° C. Some light penetrates the ice, and some planktonic algae survive and photosynthesize immediately under it. The water at the bottom will be richer in nutrients than the upper layers because of the accumulation of the remains of most of last year's plankton organisms (both plant and animal). Bacteria living in the deep layers of the lake decompose these organic residues, and nitrogen, phosphorus, sulfur, and other dissolved nutrients accumulate.

As the ice melts, the surface water warms and becomes more dense, sinking below the colder, less dense layer immediately below. This results in periods when the lake is uniform in temperature from top to bottom and in which some mixing occurs. Further heating of the upper water and the stirring action of the wind completes what is termed spring overturn. The result of this overturn is that nutrients from the bottom layers become distributed throughout the lake. With the more or less stable temperatures of summer, the lake becomes stratified. Three zones are usually found. The upper zone is composed of warm, less dense water that is often several meters thick. The second zone is frequently not as deep and is characterized by rapid temperature change. The third and deepest zone is composed of the coldest water and is relatively stable.

The algae are present in the upper zone, where light is available, and as summer progresses, they will use up the available nutrients. Those materials in the deepest zone, even if abundant, will be unavailable because of the stratification. Another necessary component that may become limiting is CO_2. The large amount of photosynthesis occurring in this zone may deplete it of CO_2 and limit growth, if there is little wind. The action of wind in causing waves results in introducing CO_2 into this upper zone. Throughout the spring and summer, there is a succession of algae in this zone. Diatoms are present earliest and are succeeded by dinoflagellates and motile chrysophytes.

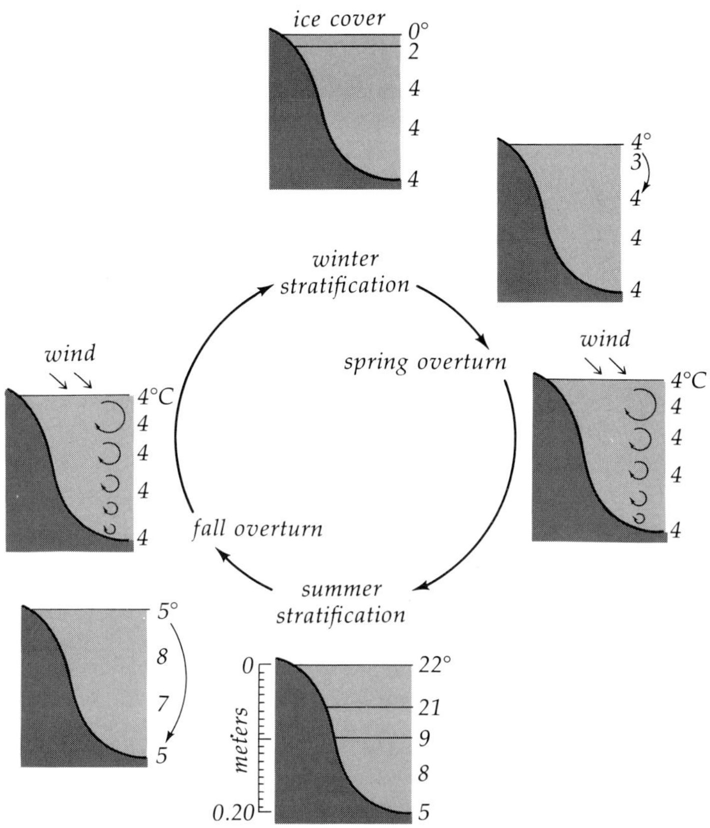

17–6 *The seasonal temperature cycle of a temperate lake.*

Finally, when nitrogen is becoming limiting for other species, the nitrogen-fixing species of blue-green algae may appear. All summer long the algae are being eaten by protozoa, insects, and fish. The excreta from these animals, as well as material from the algae, and remains of all the organisms collect on the bottom, where they are attacked by bacteria and, in shallower water, fungi.

As summer comes to an end, the days become shorter and the temperature colder. Both factors tend to decrease the amount of photosynthesis, and since the nutrients in the upper zone are largely depleted, the algae and the animal plankton decrease in numbers. The cooling of the surface water results in its becoming more dense and sinking to the bottom. This and wind action results in the fall overturn, and if the temperature is not too low, there may be a second, shorter period of algal growth. This period is limited by decreasing light (not only shorter days but more cloud cover) and decreasing temperature. Finally, ice forms over the surface, and we have completed our cycle.

With regard to nutrients, the lake is not at exactly the same place it was the previous year. There has been an increase in nutrient level as a result of nitrogen fixation, addition of phosphates from leaching of surrounding soil and other sources, and the slow accumulation of material from many places.

We have discussed the natural course of eutrophication or enrichment, but this phenomenon is also frequently mentioned in connection with water pollution. It is easy to see why man, deliberately or not, dumps all sorts of material into any body of water that he is near. Sewage is one of these, and even fully processed it adds calcium, sulfur, nitrogen, and organic carbon to the lake. Most important, it adds phosphates, primarily from laundry detergents. Another rich source of phosphates is from fertilizers applied to the surrounding farmlands. These are blown or washed into the lake either by rainwater or drainage from irrigated fields. Industrial wastes also play an important role in water pollution, even when they are not noxious.

This significant increase in amount of nutrients, particularly of phosphates, results in greatly increased algal growth. This itself is not bad, if the increase is not too great, because the additional algae eventually provide food for more fish. Most ponds and lakes can stand some enrichment, and the addition of fertilizers to lakes to increase yields of fish can be highly productive. When the level becomes too high, other factors come into play. Particularly in summer, when stratification is marked and high temperatures plus long days produce conditions favorable to algal growth, higher than usual phosphate levels result in massive blooms of algae (usually blue-green algae). Such concentrations of algae may result in a number of undesirable events.

Some blue-green algae such as *Microcystis* can produce toxic compounds that can cause the death of birds and mammals that drink the water in which the algae are growing. Usually the problem is not that drastic, but masses of the algae may clog the water filtration systems and impart a disagreeable taste to the water. The decomposing algae may give off extremely offensive odors. The bloom and the decay of the bloom also depletes the water of oxygen, which causes the death of both the animal plankton and the fish as well. Finally, as winter comes, the dead masses of algae sink to the bottom where bacteria decompose them. This depletes the oxygen under the ice and may result in massive deaths of fish.

Clearly, eutrophication is a major problem with regard to water pollution. The only permanent cure is careful control over the quality and quantity of the materials entering our lakes, but we have been reluctant to impose the necessary controls, particularly over industrial users and municipal sewage treatment plants. The last few years have seen an upsurge of interest and concern, but much more progress must be made—soon.

Essay 17–1 The Relationship of Bacteria and Blue-green Algae

So many points of similarity exist between bacteria and blue-green algae that there seems little doubt that they are related. Both groups are composed of procaryotic cells, and by this striking feature, are set apart from the remainder of the plants and animals. The cell walls are similar in composition and structure; probably because of this, members of both groups are susceptible to penicillin. Also, both are attacked by similar viruses and show tremendous range in the habitats they can utilize, particularly in terms of temperature. Both groups contain species that can fix nitrogen—the only organisms that can. The organization of their genetic material is similar; so are their ribosomes.

Yet the similarities between the bacteria and blue-green algae should not be pushed too far. Blue-green algae are not simply photosynthetic bacteria. With only a few known exceptions, all the blue-green algae are obligate, photosynthetic autotrophs. This means that they will grow only in the light; so no matter how much or how many sugars and other compounds you provide them with, they are unable to grow in the dark. There are a few unusual species that live within the colorless cells of other organisms, not in a parasitic relationship but a mutualistic one. In these cases the blue-green algae divide at the same time as the host. The photosynthetic pigments are also different between the bacteria and the blue-green algae. Blue-green algae contain chlorophyll a—the same chlorophyll found in all other photosynthetic organisms except bacteria.

Both the blue-green algae and the bacteria are extremely old groups of organisms; the oldest known fossils are bacteria and blue-green algae. They are both remarkably insensitive to ultraviolet radiation, which is believed to have been intense in the early earth environment. Their simple cell organization is what would be expected for primitive cells. But their relatively simple appearance does not mean they are unsuccessful, nor when we call them primitive, does it mean that they have failed to compete or are becoming extinct. On the contrary, the procaryotic organisms are estimated to comprise more than one half the bulk of all living organisms!

Algae and Rice Paddies

Before we write off all man's alterations of the environment as harmful or all aquatic blue-green algae as bad, let us return to the rice paddy discussed briefly in Chapter 15. Even more than a cleared and plowed field, a rice paddy is the ultimate in manipulation of the environment by man. Terraces and dikes must be built, water flow carefully controlled, and constant attention be given to the fields. Perhaps, the most spectacular feature of rice paddies is their productivity. When new paddies are established, the fertility of the soil appears to decline for several years. Then, a reversal occurs, and fertility increases, until a constant level is reached that will be retained for many years. These various changes take place without the addition of fertilizers, and although productivity can be improved by the addition of fertilizer, the paddies will continue to produce at a steady rate for hundreds of years.

Why? Because of blue-green algae—the same type of algae that polluted our hypothetical city water supply. As noted, many blue-green algae can fix atmospheric nitrogen, and some of these grow in the rice paddies. The blue-green algae in the rice paddy are performing the same function as the nodule-bacteria of

17–7 A summary of the freshwater habitats of algae

Habitat	Soil and trees	Temporary ponds	Lakes	Streams
Light	Medium to low	Relatively high bottom usually light	High on surface and edges low a few feet down	Medium to high
Water	Low	Plentiful for short periods but possibly lacking during long periods	Plentiful	Plentiful but variable flooding and drying up
O_2, CO_2	Plentiful	Usually good	Usually good on surface in spring and fall	Excellent
Nutrients (N, P, organic material)	Poor to fair	Usually excellent	Poor to good to excellent	Good
Representative algae	*Gloeocapsa* *Oscillatoria* *Desmococcus*	*Volvox* *Spirogyra* *Oedogonium* *Euglena* Desmids	Desmids Diatoms Dinoflagellates Motile chrysophytes *Anabaena* (a blue-green)	*Cladophora* *Stigeoclonium* Diatoms (attached species)

legumes in the case of ordinary fields. Rice can be grown year after year, several crops a year, with little or no decline in productivity because the nitrogen capacity of the paddy is maintained by the blue-green algae; hence no rotation of crops is necessary. Clearly, the lives of millions depend on these minute, microscopic plants.

Algae of Rivers and Streams

There is one last, major, freshwater, aquatic environment where algae grow that we should consider (Table 17–7). This is the environment provided by the numerous creeks, streams, and rivers flowing to the oceans and lakes. In running water of this nature there is normally more than enough dissolved oxygen and carbon dioxide. Nutrients may be highly variable, and the depth of the water may fluctuate greatly.

Under these conditions, it is hardly surprising that we find fewer algae and that many have evolved ways of attaching themselves to a suitable substrate. Many of the filamentous forms have well-developed, holdfast cells. One of the most common algae found in streams is the green alga *Cladophora.* This branched alga frequently looks like tufts of grass or long streams of green growing on submerged rocks in streams and creeks. The plant is composed of large multinucleate cells with calcium-rich walls.

Other filamentous algae secrete large amounts of gelatinous material that anchor them to branches, roots, and other structures in the stream. Diatoms, desmids, and other unicellular plants use similar methods to attach themselves to suitable substrates. Floating, phytoplankton diatoms, desmids, and unicellular green algae may also be present depending on the rate of flow. Some planktonic forms are simply carried down the river with the current.

Streams are subject to pollution as well as lakes. The running water provides a convenient means of washing your wastes into someone else's backyard. For years men have dumped all manner of materials into streams and rivers. Some, such as many industrial wastes, are toxic and simply kill all plants and animals downstream. Others, such as sewage, change the biological conditions in the immediate area of discharge but are gradually utilized as they pass downstream by a variety of bacteria, plants, and animals. Eventually, if the amount of discharge is not too great or does not contain toxic materials, the stream will return to its normal conditions.

Algae and Hot Springs

Blue-green algae, like bacteria, can grow in hot springs. Some species can grow at 75° C and will not grow below 30° C, thus making them thermophilic organisms.

How are the bacteria and blue-green algae of hot springs able to survive and grow at temperatures far above the point at which proteins are usually denatured? Thermophiles are found in three groups of organisms: bacteria, blue-green algae, and fungi. The bacteria and blue-green algae tolerate the highest temperatures and can colonize hot springs. The bacteria are chemotrophic and can live in water at 90° C (in the hot springs at Yellowstone Park, for example, where the altitude causes water to boil at 92° C). Species of blue-green algae live in water at 75° C and grow well at these temperatures. Thomas Brock of Indiana University, who has intensively studied the hot spring thermophilic organisms, suggests that we approach this puzzle by considering the proposition that life originated in conditions similar to the present hot springs. The thermophilic organisms would then be ancient relics of the past, and the question then becomes how did organisms adjust to life at lower temperatures. The data supporting this theory are by no means conclusive, but it is an interesting viewpoint. Brock also suggests that the limiting feature of life at high temperature is not impairment of enzyme activity but membrane stability.

Algae of Strange and Unusual Places

Before turning our attention to the algae that live in oceans (see Chapter 18), let us consider a few of the more unusual habitats in which algae are found. One of these is snow. Snow algae occur in both arctic and antarctic regions, as well as in the high mountains of most countries. These algae cause the phenomenon of red, yellow, or green snow. Red snow is generally caused by a species of the green alga *Chlamydomonas*, which contains a red pigment. It is usually seen in the resting stage in the snow. The red pigment is apparently a light shield that protects the cells from the high light intensity found in this habitat. Algae also grow as epiphytes on just about everything: other plants, animals, or epiphytes on epiphytes. One species lives within the air spaces of the leaves of some jungle plants, while others grow on the hair or scales of animals such as the three-toed sloth or ai. Some blue-green algae grow in special chambers in the leaves of a water fern, and others are found living in close association with other plants ranging from fungi, bryophytes, and lower vascular plants to gymnosperms and angiosperms.

End Paper 17–1 Cyanophyta, Euglenophyta, Chrysophyta, Xanthophyta, and Chlorophyta

In this chapter we have stressed the relationship of the freshwater algae to their environment. We have also emphasized the basic features and methods of their reproduction. But the algae are also a truly fascinating group of plants in terms of their cell organization, biochemistry, evolution, and development, some of which we shall discuss in this End Paper.

Cyanophyta

The ultrastructure of a blue-green algal cell is shown in Figure 17–D. The cell, while lacking cell organelles, is far from simple. The most conspicuous features are the parallel rows of membranes and masses of ribosomes and other particles. The membranes are photosynthetic thylakoids similar to those found within chloroplasts of eucaryotic green cells. Each thylakoid is essentially a flattened sac and is composed of 2 membranes. The thylakoids contain chlorophyll a and carotenoids. The phycobiliproteins are located in granules on the surface of the thylakoid.

In algae that carry out nitrogen fixation, this process also occurs in the thylakoids. Photosynthesis and nitrogen fixation are closely related in the blue-green algae. Respiration probably also takes place in the thylakoids, but the whole question of the nature and localization of respiration in the blue-green algae is not settled.

The cells contain a nucleoplasm, similar to bacterial cells, within which are located strands of DNA. Chromosomes of the type found in eucaryotic cells are missing; DNA without associated protein is present instead. Ribosomes are frequently present in large numbers and are smaller in size than the cytoplasmic ribosomes of other plants. The ribosomes of the bacteria and blue-green algae are similar in size.

Also found in the cell are various storage materials. These include glycogen particles (which are storage carbohydrates —"animal starch"), lipid or fat reserves, masses of protein, and granules of polyphosphate.

Many blue-green algae contain gas vacuoles (Fig. 17–8). These are found in floating species and are important in maintaining the buoyancy of the cell. The gas vacuoles are usually cylindrical in shape and can be rapidly formed by the cell. In some species the gas vacuole also appears to function as a light-shield protecting the photosynthetic pigments from light damage.

Blue-green algae display a number of forms, all of which lack flagella. Many are single-celled organisms, but many form multicellular colonies. Some of these colonies are in the form of squares or cubes of cells, while others form loosely organized globular colonies. Others form simple filaments, as in *Oscillatoria,* which derives its name from its slow, oscillating motion, easily observed under a microscope. In most of these filamentous blue-green algae, the cells are arranged on top of one another looking much like a stack of coins but surrounded by a gelatinous sheath instead of a paper wrapper. Some of the filamentous species show various types of branching. How such multicellular species may have evolved through changes in planes of cell division is postulated in Figure 17–9 .

In the filamentous species there are mechanisms for the fragmentation of the file of cells. This may result from the death of an individual cell creating a weak point in the file. Asexual spores similar to those found in bacteria also develop in some blue-green algae. There is no known form of sexual reproduction in the blue-green algae. The type of DNA recombination found in the bacteria has been reported in the blue-green algae, but the occurrence of this type of reproduction has not yet been firmly established.

17–8 Gas vacuoles in a blue-green alga, as seen with the electron microscope. The vacuoles can be seen both in cross section and a longitudinal view. The preparation is freeze-etched and reveals the structure of the membranes of the vacuoles (× 200,000).

Euglenophyta

The members of this small division (some 450 species) are unusual unicellular algae (Fig. 17–10). Some species contain chloroplasts, have either rigid exteriors or may be flexible, and possess flagella. Other species lack chloroplasts, have a flexible exterior, move in a creeping or gliding motion, and ingest food through a gullet. Some species are truly autotrophic, while others are heterotrophic. Still others may be either one, depending on the environment.

Nuclear division in the Euglenophyta is also unusual and departs markedly from the general type given in Chapter 9. During mitosis the nuclear membrane never breaks down, and the nucleolus divides, persisting throughout the division (Fig. 17–10). Microtubules are formed only within the nuclear membrane; the chromosomes lack centromeres; and it is unclear whether there are chromosome spindle fibers. This suggests that chromosome movement is autonomous, with the microtubules providing guidelines to the poles. The chromosomes remain condensed during interphase. All in

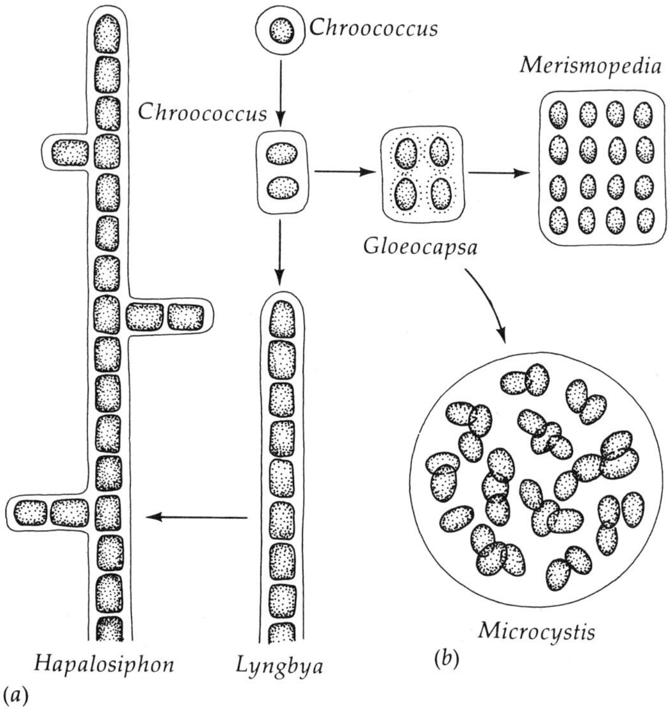

17–9 The organization of colonies of blue-green algae is directly related to the planes in which the cells divide. This diagram indicates a possible evolutionary line in the Cyanophyta, resulting from regular and irregular planes of cell division in filamentous (a) and nonfilamentous (b) species.

all, this is an unusual and, quite possibly, a primitive type of mitosis.

The chloroplasts contain thylakoids frequently grouped in stacks of 3 to 12. The pigments present are primarily chlorophyll a and b, plus carotenoids.

The cells reproduce while actively motile, and under unfavorable conditions, cells can encyst by forming a thickened cell wall. Sexual reproduction is not known.

Euglena is one of the most intensively studied of algae. The biochemical development of the chloroplasts and the chloroplast pigments have been carefully followed. The chloroplasts are known to divide, and they can be destroyed in a number of ways (irradiation and antibiotics) without injuring the cell.

Chrysophyta

This is a much larger division of algae than the Euglenophyta. It consists of 5800 species divided into two classes: Chrysophyceae or golden algae, with 300 species, and Bacillariophyceae or diatoms, with over 5500 species.

The golden algae are found mainly in freshwater and are frequently naked protoplasts equipped with flagella. In contrast, the diatoms, as we have already seen, have rigid walls containing silica (SiO_2)—as much as 95% of the wall—in addition to cellulose. The walls have various markings and sculpturings that make them beautiful under the microscope and are used to test the optical efficiency of microscopes. Diatoms in mass are anything but beautiful; they appear as a yellowish-brown scum or slime. They contain chlorophyll a and c, but the green pigments are masked by dominating carotenoids. The thylakoids within the chloroplasts are unusually arranged in stacks of three.

Diatoms are uninucleate and divide through mitosis, followed by a cytoplasmic division. Mitosis in diatoms is similar to the standard type discussed in Chapter 9, except that the spindle forms completely outside the nuclear envelope and when the envelope dissolves at the end of prophase the spindle appears to sink into the chromosomal mass. For a period of time, the protoplasm and new nuclei remain within the original cell walls, but then the upper and lower portions of the walls separate, each containing a protoplast and nucleus. Each new cell mass makes a new wall-half to complete the cells, each half making up the inner part of the box (Fig. 17–11). One half of the new cells are smaller than the original cell, and one half of their progeny are smaller still. Obviously, such loss of size can go on only so long. Size is restored during sexual reproduction. Meiosis occurs, and gametes are formed lacking walls. The fusion of 2 haploid gametes results in a zygote that increases in size to the maximum cell dimension of the species and then forms a new cell wall. Most diatoms have an absolute requirement for silicon and cannot divide unless it is present. Some species, under laboratory culture conditions, can be induced to divide in silicon-free water and will form cells lacking walls. After many generations, when these cells are given silicon they will form walls of the proper shape and maximum size for the species.

The walls of diatoms (Fig. 17–12) are extremely resistant to decay. While cellulose and lignin can be degraded by bacteria, silica is inorganic and is therefore changed to a less soluble form. As a result, the walls of diatoms accumulate on the ocean floor. Raised above sea level by geologic activity, such deposits form diatomaceous earth that is commercially mined and used in polishes, air and water filters, window cleaners, and in many other forms. The number of diatoms involved in these deposits can be imagined when more than 6,000,000 shells can be found in 1 mm^3 and the deposits may be 200 m or more deep.

Xanthophyta

Until fairly recently the Xanthophyta had been considered part of the Chrysophyta, but because of differences in pigments and other characteristics, many algologists have

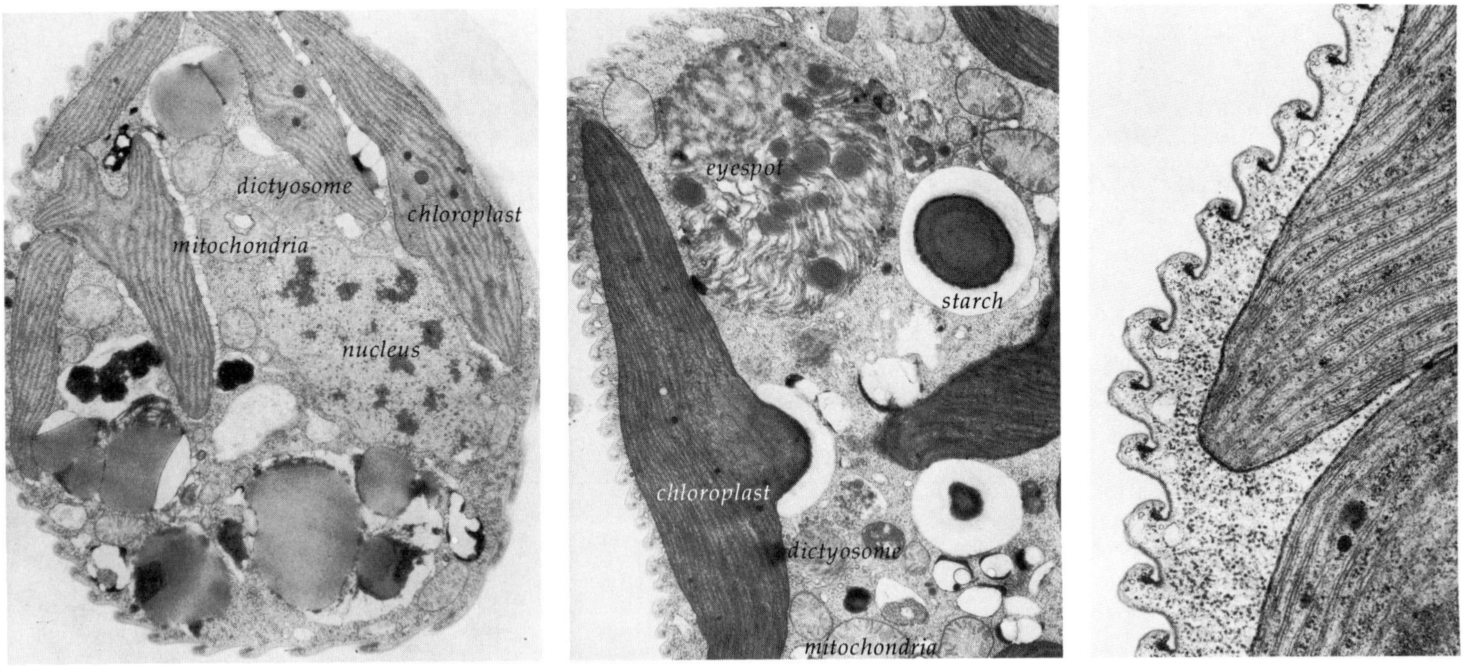

(a) (b) (c)

17–10 *(a) Many features of the Euglenophyta can be seen in this low-power view of a species in the genus* Colacium. *There is a single nucleus surrounded by chloroplasts and mitochondria. A large dictyosome is also present near the nucleus. The opposite end of the cell is filled with food storage material and a few mitochondria. The flagellum and associated structures are not visible in this section.*

(b) The swirling mass of membranes and discs in the upper left-hand corner of this photo is the eyespot, which appears red under the light microscope. The white ring below and to the right of the eyespot is starch, surrounding a small portion of the chloroplast. The relation of the starch to the chloroplast can better be seen in the plastid running down the left-hand side. The starch is present in the cytoplasm in the form of a dome over a special region of the chloroplast.

(c) This higher power view of the edge of the cell shows the nature of the surface and the structure of the chloroplast. No external wall is present, and the beautiful, wavelike pattern of the cell surface is associated with an area of dense material (in each trough) and a couple of microtubules (in each crest). Within the chloroplast, the thykaloids are stacked in groups 2 to 5, with 3 being the most common. The chloroplasts also contain large numbers of ribosomes.

placed them in a division by themselves. There are only 400 species in the division altogether, and most live in freshwater. The most commonly occurring forms are bulbous or threadlike and are found in ponds as plankton. Others are epiphytes, while a few are motile, having 2 flagella. The motile species are often small to tiny algae and hence overlooked. The Xanthophyta are a small group of plants, which we shall not discuss any further.

The cell walls of many green algae are composed of cellulose and pectic substances. Usually, the cellulose forms rather compact layers near the cytoplasm, while the pectin is present in thicker, outer layers. In some Chlorophyta cellulose is missing; instead strands of xylan, a polysaccharide composed of the pentose xylose, are found. In other green algae a mannose polysaccharide, called mannan, is present in place of cellulose. Curiously, mannan is not in the form of microfibrils (as are cellulose and xylan) but in granules. This makes a most unusual wall, one whose structure needs further study.

The division Chlorophyta is usually divided into two classes: the Chlorophyceae and the Charophyceae. The Charophyceae or stoneworts are a relatively small group of plants; the majority of the green algae are in the Chlorophyceae. This huge class can be divided into three lines that demonstrate how relatively simple changes in a basic process such as cell division can lead to widely diverse forms. These three lines of green algae are shown in Figure 17–13. Drawings of repre-

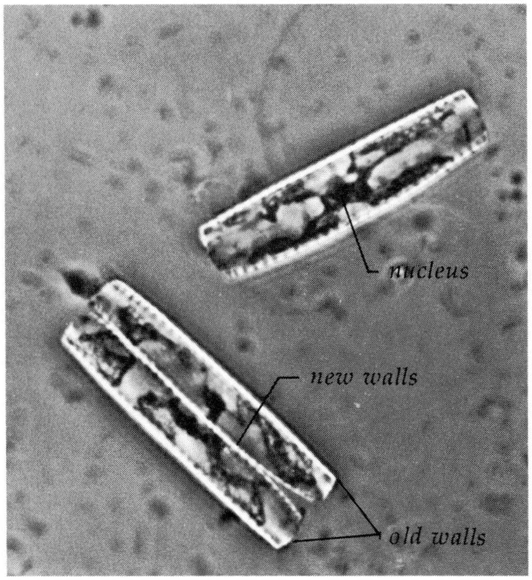

(a)

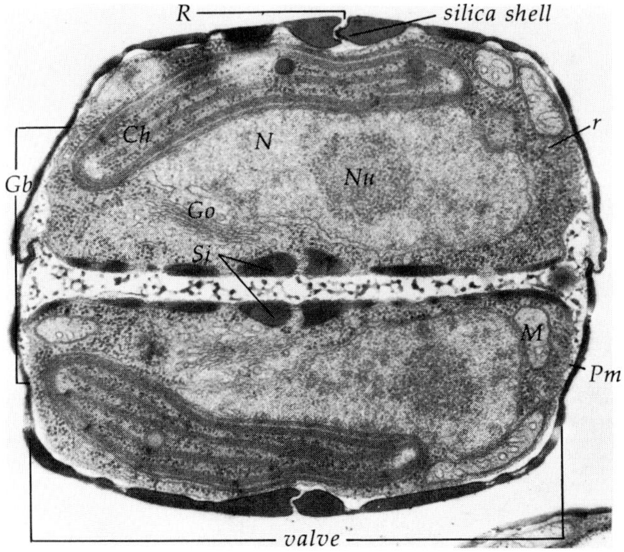

(b)

(c)

R=raphe
Gb=girdle bands
N=nucleus
Nu=nucleolus
Ch=chloroplast
Go=golgo
M=mitochondria
Pm=plasma membrane
r=ribosomes
Si=silica shell

(d)

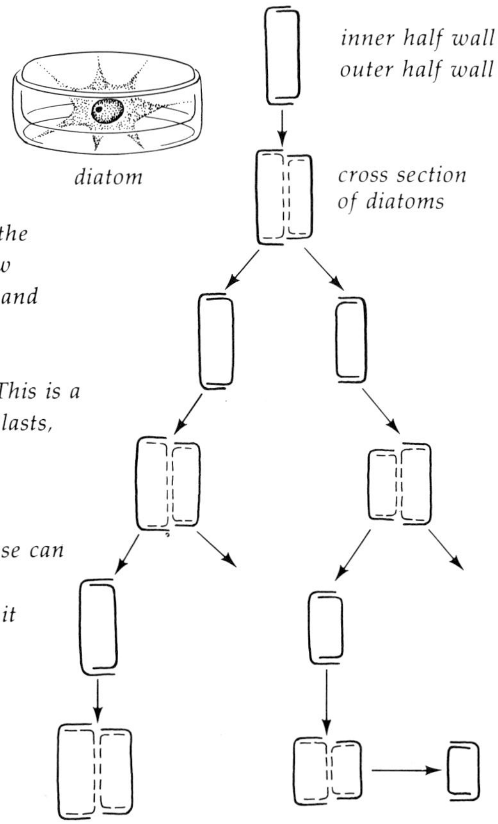

17–11 *(a) Diatom cells are surrounded by two overlapping walls. The nucleus is situated in the middle of the cell (surrounded by chloroplasts), which can be seen in this light microscope view of one of the largest of the diatoms. When the cell divides, a new wall is formed in the center, and the cells separate.*

(b) The structure of the dividing diatom can be better seen with the electron microscope. This is a section through the diatom Navicula pelliculosa. *The two nuclei are visible as are the chloroplasts, mitochondria, dictyosomes, and the ER. The new cell walls are in place, but the daughter cells have not yet separated (× 44,800).*

(c) Walls of diatoms are highly sculptured and are perforated with rows of tiny holes. These can be seen when the walls are cleaned of the cytoplasm and viewed with either the light or the electron microscope. Here, we see the same diatom shown in (b), but this time we are viewing it with the electron microscope while it is still whole. The larger photo is (× 14,000), while the inset depicts a pore with its perforations at (× 80,000).

(d) The cell-wall formation during cell division in a diatom. Since the inner wall is the only one that is synthesized each time, there is a gradual reduction in most of the cells as division continues.

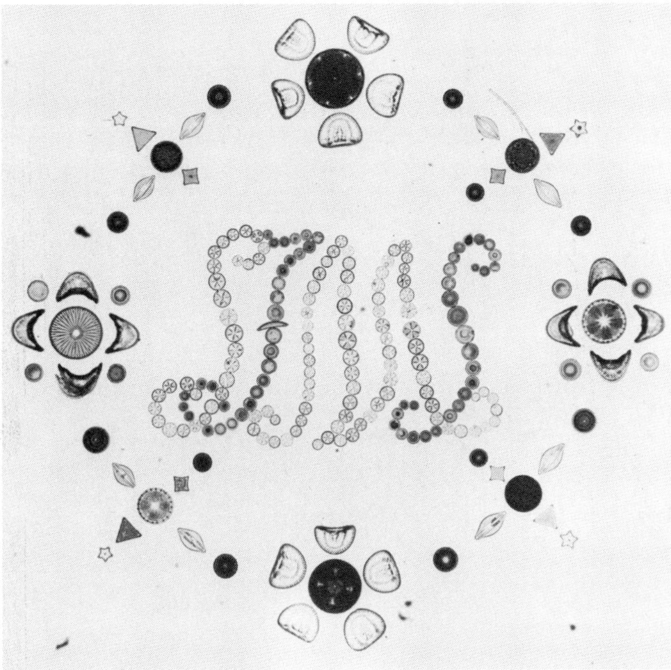

17–12 The beauty of the walls of the diatom resulted in an unusual art form, which was practiced at the turn of the century. Diatoms were collected, cleaned, and the walls carefully arranged in elaborate patterns on microscope slides. The one shown here is a commemorative arrangement made in 1870 for the San Francisco Microscopal Society and incorporates their initials as part of the design.

sentative species have been combined with diagrams showing the basic patterns of cell arrangements and number of nuclei per cell.

The starting point of all three lines is a single-celled species such as *Chlamydomonas*. In the first line (the Volvocine), division of the cells results in colonial forms in which each cell is a separate individual (although they may be joined by thin strands of protoplasm). Depending on the planes in which the divisions occur, the result may be flat colonies, cube-shaped colonies, or spherical colonies. Each species has a fixed number of cells per colony.

The second (or Tetrasporine) line is based on the daughter cells remaining together to form an organism in which they are interrelated and interdependent. Again, starting from *Chlamydomonas*, the first change is the loss of movement in most of the cells; motility is retained only in the zoospores and gametes of some of the species. When the cell divisions are restricted to a single plane, a filament results. Occasional divisions in a plane perpendicular to the file of cells can result in branching, while more extensive divisions result in a plant with a thicker, multicellular trunk. This is the type present in *Fritschiella* that is postulated as being in the line from which vascular plants evolved. Multistrands of cells growing side by side can originate from a modified branching pattern at the base of the plant and can result in flattened sheets of cells as in *Percursaria* and *Monostroma*. Divisions in an extra plane can give rise to *Ulva*, a marine green, sheetlike alga 2 cells thick.

Finally, in the third (or Siphonous) line, cell plate formation and nuclear division are not coordinated, resulting in plants that may have multinucleate cells (as in *Codium*) or plants that are one huge, multinucleate cell (as in *Valonia*, *Derbesia*, and *Bryopsis*).

The chart in Figure 17–13 by no means exhausts the diversity of form in the Chlorophyta. They represent a truly magnificent group of plants that are highly evolved, successful, and of great importance to the ecosystem.

Chlorophyta

The green algae or Chlorophyta constitute a group of highly diverse algae. With over 7000 species they are the largest of the algal divisions and show a tremendous range in size and shape. The freshwater green algae are single-celled, colonial, filamentous, or (rarely) leafy. The marine green algae are frequently larger than the freshwater ones.

The chloroplasts of the green algae are highly diverse in external form. Internally, the thylakoids are arranged in closely appressed groups of 2 to 6, and in some species well-developed grana, similar to those in higher plants, are formed.

During mitosis the nuclear envelope remains intact, except at the poles where holes form through which the microtubules of the spindle pass. In some species no centromeres are apparent, and the spindle fibers become attached to the length of the chromosomes. As mitosis progresses, the spindle and the nuclear envelope elongate, and when the chromosomes reach the poles, the nuclear membrane pinches shut behind them. Some, *Fritschiella* for example, form a cell plate at mitosis.

The chromosomes consist of DNA and histones as in higher plants. Polyploidy is common in the algae but not as well studied or understood as in vascular plants. In our discussion of polyploidy in Chapter 12, we used examples from vascular plants. This, unfortunately, may make it appear that it occurs only in these higher plants, but this is not the case. Polyploidy is common and important in many groups of plants, including the green algae.

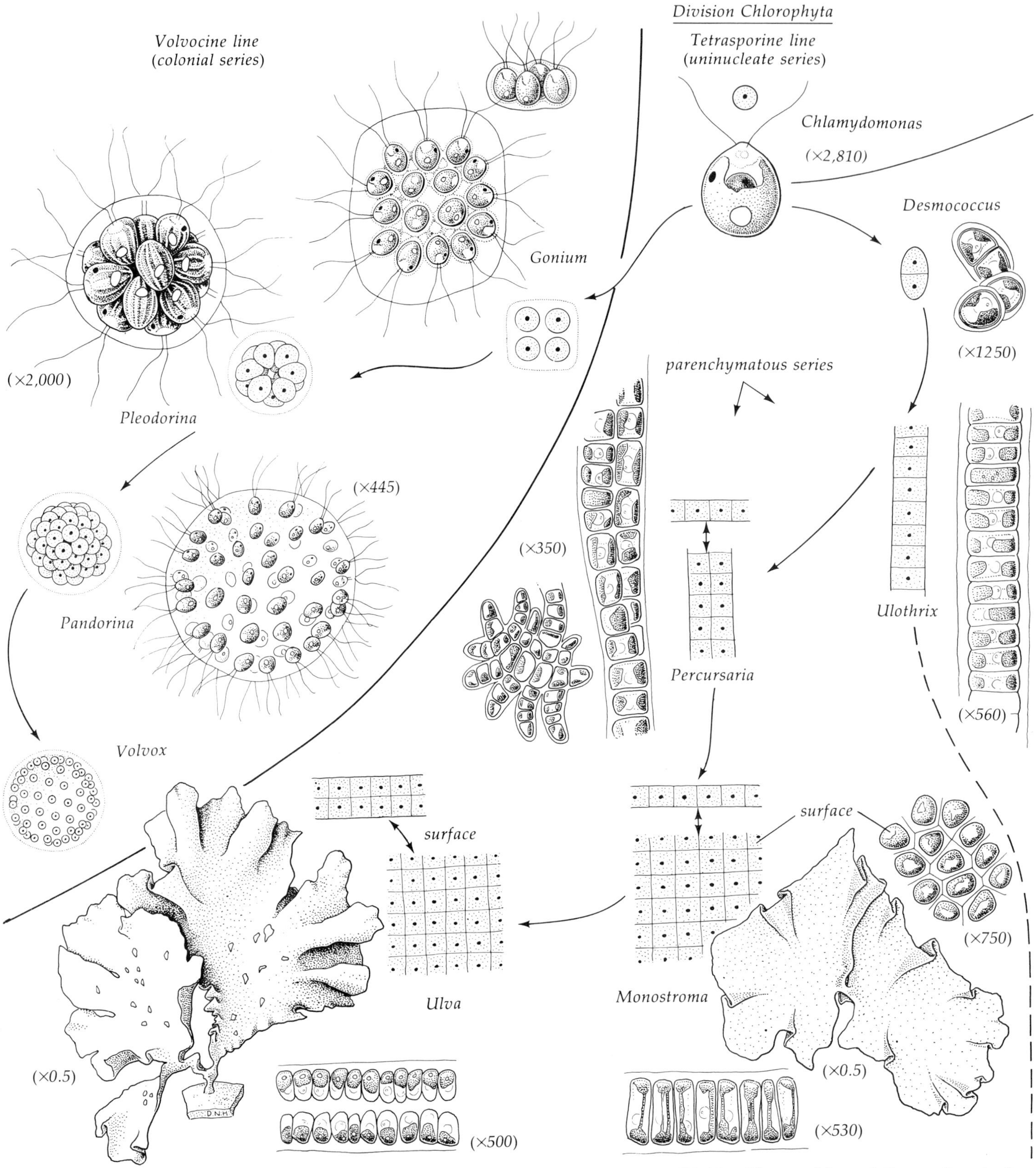

17–13 The possible evolutionary lines in the green algae.

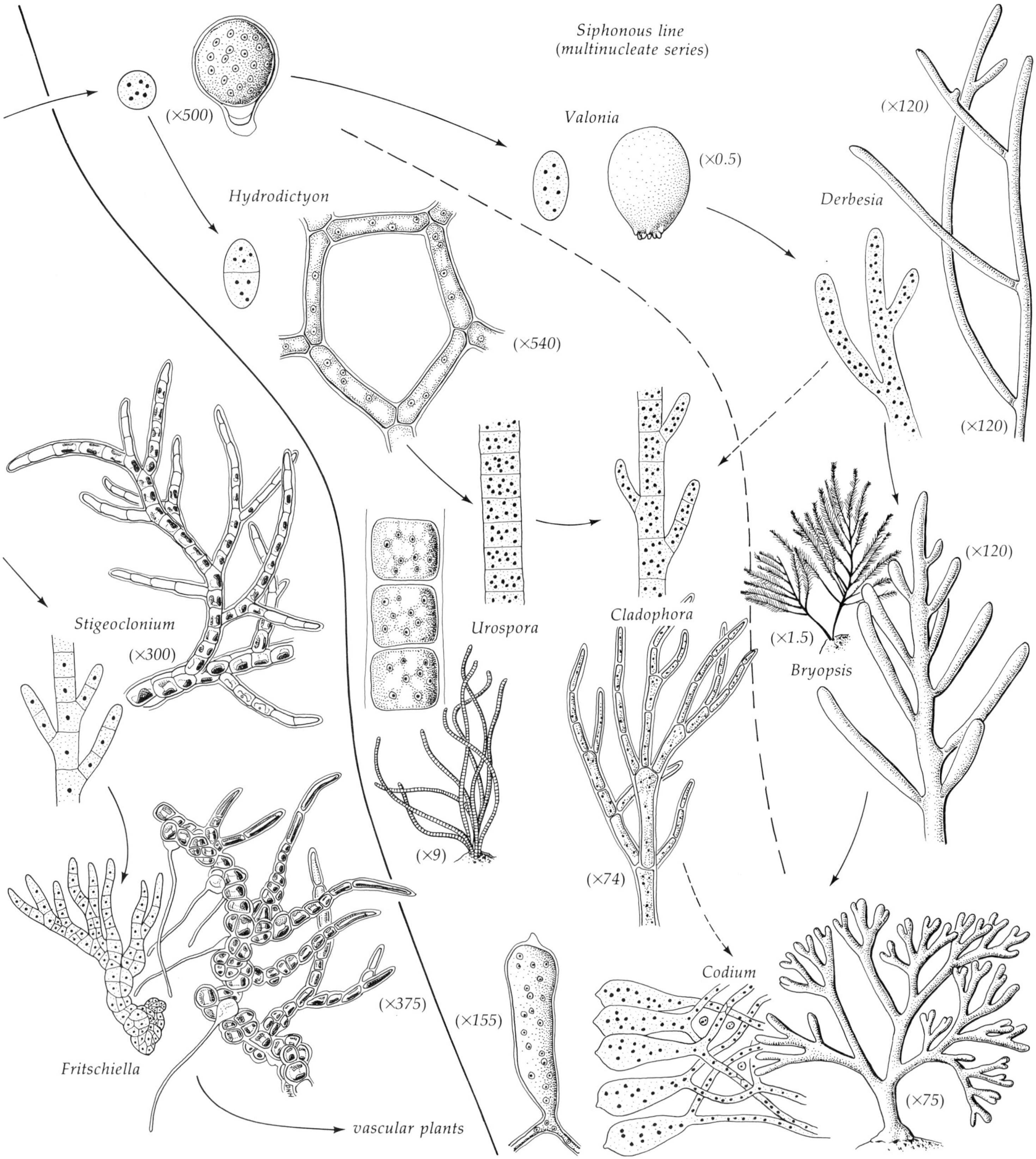

Siphonous line
(multinucleate series)
(×500)
Valonia
(×0.5)
(×120)
Hydrodictyon
Derbesia
(×540)
(×120)
Stigeoclonium
(×300)
Urospora
Cladophora
(×120)
(×1.5)
Bryopsis
(×9)
(×74)
Codium
(×375)
(×155)
Fritschiella
(×75)
vascular plants

18

Algae of the Oceans

Life between the Tides

If we were to visit a rocky shore in northern California, these are some of the common algae we would see. Similar plants are found along much of the Pacific Coast of North America, and members of the same families are found on rocky shores of the Atlantic Coast. As we walk toward the water's edge at low tide, Enteromorpha, Fucus, Ulva, *and* Porphyra *are among the first plants we would find on the beach. These plants can survive long periods out of the water and are remarkably resistant to drying. Living in even drier locations, literally in the spray of the waves, are marine blue-green algae and lichens.*

Enteromorpha *is a member of the green algae and is a light green color when seen on the beach.* Fucus, *in contrast, is a brown alga and is a dark olive drab.* Porphyra *is a representative of the red algae and is a dark red almost black color when growing on the rocks.* Ulva *is a member of the green algae and is a bright green.*

A Enteromorpha, *growing on a rock cliff near the ocean's edge.*

B Fucus, *photographed against a pebble beach, although it usually grows attached to rocks.*

A

B

Farther out, where the waves may still be breaking if the tide is not really low, we would see the large-bladed brown alga Laminaria. *This alga also has a well-developed holdfast and a stemlike stipe. Growing as an epiphyte on the blades of the* Laminaria *is a beautiful red alga,* Microcladia, *that has a lacelike body. Our photograph shows the* Microcladia *against a mass of sea foam. Such foam results from the breakup of marine diatoms in the churning waves and not from pollution, as is commonly thought. Also growing here is* Pelvetia, *a genus with a worldwide distribution that is closely related to* Fucus.

D Microcladia.

C Laminaria.

E Pelvetia.

On the tops of rocks and in other exposed locations bearing the full force of the crashing waves grows one of the most handsome of all West Coast algae, Postelsia palmaeformis, *or the sea palm. Stands of these plants look surprisingly like small forests of olive-green trees.* Postelsia *is a brown alga with a well-developed holdfast, a trunklike stipe, and a crown of graceful blades. Also growing in this area are algae with large flat blades that have an iridescent color and are appropriately named* Iridaea. *(A young plant is shown in the photograph here.)* Iridaea *is a red alga, although it usually looks green rather than red.*

G Postelsia palmaeformis.

H Iridaea.

F A stand of Postelsia palmaeformis.

Oceans are awe-inspiring. As the waves roll in and crash upon the shore, they appear to be embodied with a power and movement of their own. For centuries man has used the oceans as a source of food and a means of transport and commerce. But the plants of the oceans were largely ignored, except by coastal peoples in some parts of the world who ate them and by writers for Sunday supplements who told tales of death-trap "sargasso seas." As man has become increasingly interested in his environment and aware of the intricate interrelationships within the biotic world, the relationship of an ocean's plant life to its animal life has received more and more attention. The development of the aqualung by Jacques Cousteau and his masterful presentation of the marine world through the media have created an awareness of the importance of oceans by almost everyone, including those who live far from the oceans' shores.

18–1 Low and high tides on a Pacific beach. These two photos were taken on the same day at the same place and show the change in water level that is characteristic of the intertidal region.

Oceans as Environments

In the broadest terms, oceans consist of two basically different environments for algae: the vast open water and the shallower water by the ocean's edge. Only floating algae, marine phytoplankton, can survive in the open ocean—for the same reason that only phytoplankton is present in the center of lakes. Light penetrates only the upper layers of the water; hence no photosynthetic algae can exist in the dark depths of the ocean.

But, at the edge of the ocean, the water becomes shallow enough for algae attached to the bottom to grow. This environment is more complex in terms of algal survival because of the tides. As the tides move in and out, the amount of water covering the algae changes, thus affecting the amount and quality of light reaching the plants (Fig. 18–1). Moreover, the plants growing farthest up the beach will be uncovered and subject to drying during part of the day. Yet there are vast numbers of algae that grow in these intertidal regions and others that can survive in the subtidal regions, if the water is not too deep.

To understand these intertidal algae and their adaptation to this demanding environment, it is essential to understand tides. In most parts of the world there are two tides a day—twice the water rises and twice it recedes. This movement is directly related to the gravitational forces exerted on the earth by the moon and the sun (Fig. 18–2). In addition, the centrifugal force associated with the rotation of the earth–moon and earth–sun about their common center of gravity comes into the picture.

The combination of the gravitational pull of the moon and the centrifugal force on the ocean causes

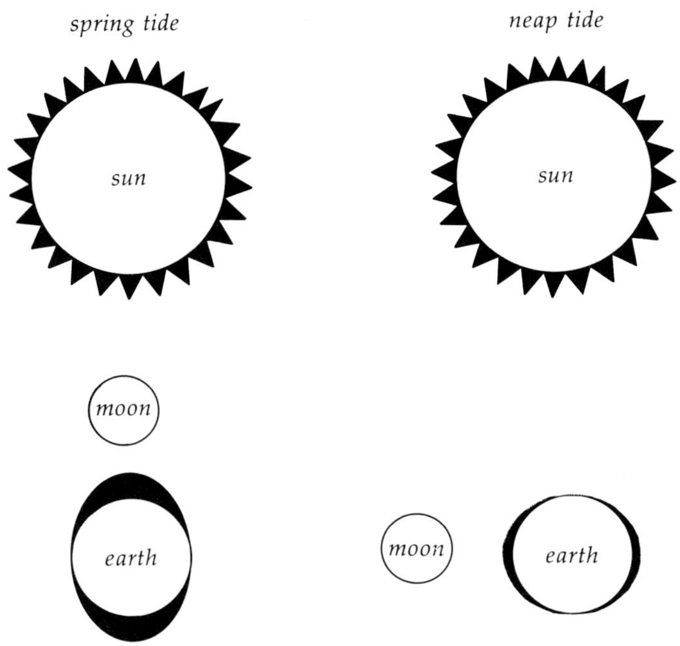

18–2 Position of the sun, moon, and earth at spring tide and neap tide—relative effect on tidal height being shown by the black band around earth.

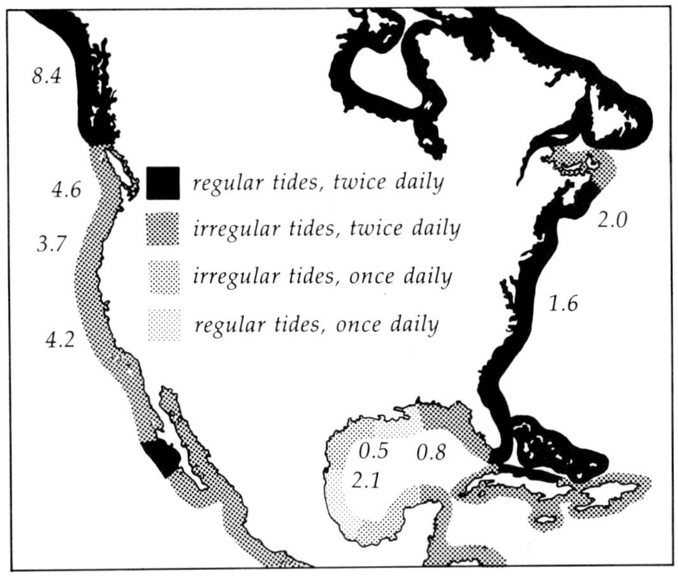

18–3 The occurrence of tides along the coast of North America. There is a wide range both in the regularity of the tides and the extent of water movement. The numbers represent (in meters) the maximum differences between low and high water. Note that the farther north from the equator, the greater the difference in the tides.

water to accumulate simultaneously on the side of the earth nearest the moon and on the side directly opposite. Then, as the earth rotates, this bulge of water moves against the shore, and high tides result. Since there are two such bulges of water, there are two high and two low tides a day. But the water does not rise an equal amount each day because the sun also influences the amount of water accumulated. Since the effect of the sun is not synchronized with the moon, the combined effect is more complicated than that which would be produced by either alone. When the sun, moon, and earth are approximately in line (at new moon and full moon), the forces are combined to produce extreme high and low tides. These are called spring tides and occur throughout the year. When the sun and moon form a right angle with the earth, their effects tend to cancel one another, and high and low water are not so extreme. These are called neap tides.

Other factors also influence the tidal flow of water. One of these is the shape of the ocean floor and the nature of the shore. Another is the latitude. The nearer the equator the smaller the tides, but the closer to the poles the greater the change between high and low tide. The result of all these interacting factors is a great variation in tidal action in different parts of the world. The differences found in the United States are shown in Figure 18–3. The East Coast has two regular tides daily; the West Coast also has two tides, but they are unequal (one is higher than the other). The average high tide in the West is also greater than in the East. Parts of the coast of the Gulf of Mexico have only a single tide, and the rise is relatively small. In addition, the monthly and seasonal changes we just mentioned are superimposed on the average daily pattern.

The nutritional aspects of oceans are also of great importance in algal survival. These are similar to the

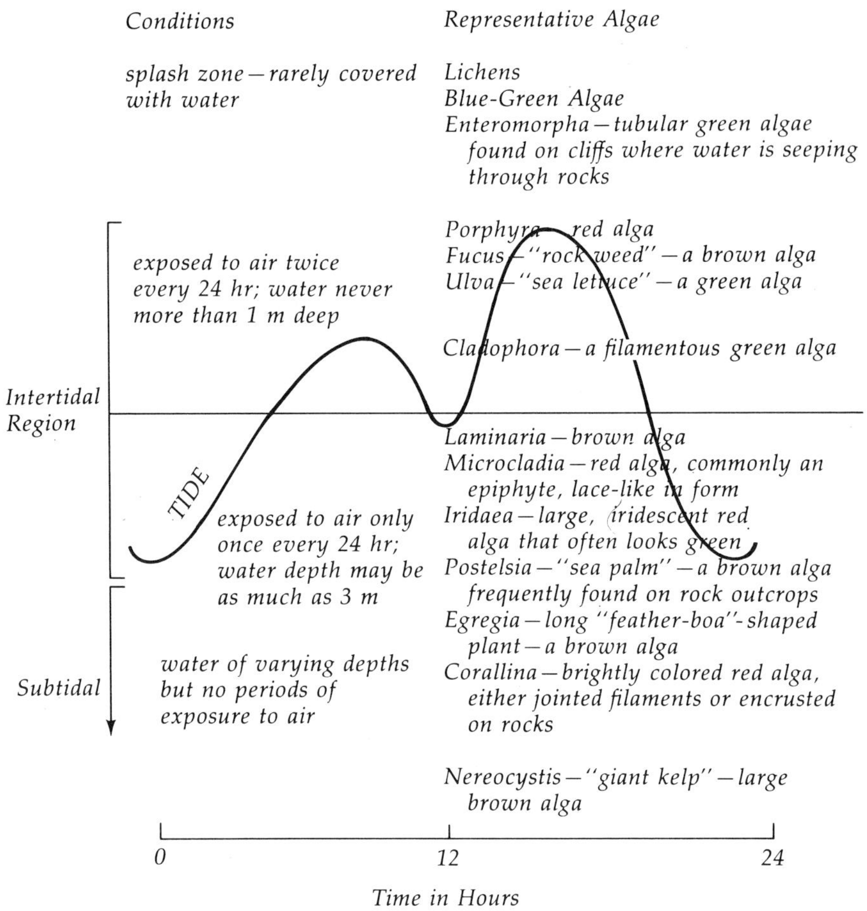

18–4 A summary of some of the conditions and plants found in the intertidal region. The line labeled TIDE represents the distance the water moves up and down the beach in a 24-hour period. This is for a coast, such as the Pacific Northwest, having two unequal tides a day. The plants that are listed are typical of a similar region. While the specific details of both the tides and the plants are different for other coasts, the general pattern is true for most intertidal regions. (The greatest differences would occur in tropical waters.)

ones discussed in relation to lakes. The remains of plants and animals, as well as all sorts of debris produced by them, accumulate in the deep, cold water at the bottom of the ocean. These layers of water are rich in the minerals necessary for algal growth. In temperate ocean regions there are seasonal changes that result in stratification in the upper layers of water (again similar to those found in lakes) and in which mixing of the water by wind and wave action is important in maintaining fertility. Such mixing, however, does not affect the deep, cold, nutrient-rich layers at the bottom. These move slowly in response to massive convection currents set up by the ice masses at the poles and the rotation of the earth. In places enormous upwellings occur as these currents flow against continents or are influenced by the ice at the South Pole or by the equator. Where these occur, the water is exceptionally rich in mineral nutrients necessary for algal growth, and it is in these regions where we find luxuriant algal floras.

Between High and Low Tides —Intertidal Environments

As we saw earlier, the edge of the ocean can be divided between the subtidal regions, where water is always present, and the intertidal region, where, because of the tides, water is present for only part of each day. The intertidal region can, in turn, be divided into zones. This division is more or less arbitrary because of the large number of variables involved. We will define only two: a high and a low intertidal zone (Fig. 18–4).

The high intertidal zone is characterized by substantial periods (usually two) in each 24 hours during which the ocean bottom is not covered with water. When the tide is in and the bottom covered, the water is usually not more than a meter or so deep. The low intertidal zone is characterized by deeper water, several meters deep, and fewer and shorter periods during which the bottom is exposed. The depth of the water is important because water absorbs light. Approximately 25% of the light striking the surface of the water is either absorbed or reflected, while an additional 25% is absorbed in the first meter. The red wavelengths are absorbed first; hence the deeper the light travels the bluer it becomes.

Another environmental feature of both the high and low intertidal zones is wave action. This is precisely the part of the shore on which the waves break. In many locations they pound with greater or lesser force day and night, year round; yet algae not only survive but thrive there.

Finally, it must be emphasized that even these broad, loosely defined zones cannot be considered uniform environments. There are differences in substrate; some are rocky, others sandy, and in some, outcroppings of rock or various kinds of reefs are present. The various sides of any projection offer different environments, since one is against the incoming waves and the opposite is shielded and protected. The top of the projection can be in another zone from the bottom and offer a different set of conditions for algal survival.

Most of the algae we will discuss are found growing on rocks. This is because the large algae cannot remain fixed in place if their holdfasts are unable to secure a stable foundation. In many cases, this is only a stone, but it helps anchor the algae. Small algae such as diatoms and dinoflagellates can live among the grains of sand, but sand presents a difficult problem for the larger algae, although some have adapted means to grow in more sandy locations. In these places the algae still require rock on which to anchor, but they are able to withstand the abrasive action of the sand against their holdfast and to survive where other, less well-adapted species would be washed away. On the other hand, these species can live only in these places because they cannot compete for light and the other vital necessities with species better adapted to rock surfaces.

This brings us to the last important factor in the intertidal zone: the influence of the plants that live there on one another and the impact of the animal population. Many of the marine algae are large and may form dense stands. This causes not only shading but changes in the light that penetrates beyond them. Thus, for plants to survive below such an algal canopy, they must be adapted to this particular biotically conditioned environment as well as to the other physical features of the habitat. The larger algae also act as substrates for smaller epiphytic species. For the epiphytes, the larger algae become a vital part of the environment.

Many of the marine invertebrates live by browsing on intertidal algae. They can greatly affect the number and distribution of various species of algae. Studies, for example, have shown that in some locations the sea urchin population ate about one third of the algal cover from the rock face each year. Limpets (snails) are also great algal eaters and play an important role by limiting the establishment of many algal species.

Marine Intertidal Algae

The principal algae found in the intertidal and subtidal regions are Chlorophyta, Rhodophyta, and Phaeophyta. The major difference between the freshwater and marine green algae is their difference in life history and the size and complexity of the vegetative body or thallus. On the whole, the marine Chlorophyta are considerably larger and more complex than their freshwater cousins. In terms of pigments, food-storage products, flagellation, plastid organization, nuclear division, and other features, they are similar (although one line of marine green algae differs significantly in a number of categories).

The Phaeophyta or brown algae are almost entirely marine, and most are macroscopic in size—there are no single-celled species. They contain chlorophyll a and c, β carotene, and an accessory carotenoid pigment, fucoxanthin, that masks the green color of the chlorophylls. This pigment is actually involved in the absorption of the light energy used in photosynthesis. The fucoxanthin absorbs light in wavelengths in which chlorophyll is not particularly effective and then transfers the energy to chlorophyll. In this way, fucoxanthin increases the range of colors of light the plant can use for photosynthesis. In view of the fact that the water is absorbing many of the wavelengths from the light available to the alga, the presence of an accessory pigment can make the difference between survival and death.

The walls of many brown algae consist of a thin inner layer of cellulose surrounded by a thick wall of mucilaginous materials containing phycocolloid, a complex, sulfur-containing polysaccharide. Phycocolloids help prevent the cell from drying, since they can absorb and retain 10 to 20 times their weight in water. They also protect the cells from freezing and act as shock absorbers and bumpers when the plant is bounced on the rocks by the surf.

Most of the brown algae are good-sized plants, and some such as the kelps are enormous in length (40 to 60 m). The thallus of many brown algae is differentiated into a rootlike holdfast, a stemlike stipe, and a leaflike blade. A similar type of thallus (but not as well differentiated) is also found in some marine green and red algae.

Red algae or Rhodophyta contain both marine and freshwater species, although there are few of the latter. The red algae are red because of accessory phycobilin pigments that are present in addition to the usual chlorophyll and carotenes. In general, the green chlorophyll is masked by the red of the phycobilins, giving the Rhodophyta a color that ranges from black-red to rosy-red. The phycobilins absorb the shorter, blue-green wavelengths of light that penetrate deepest in water and transfer the energy to chlorophyll for photosynthesis.

The walls of the red algae consist of an inner cellulose layer and an outer pectic layer, which contain phycocolloids.

Most species of Rhodophyta are filamentous. These may be simple filaments or intertwined into elaborate plant bodies. Many of these larger species are handsome plants, frequently collected and pressed solely for their attractive color and shape. Some are quite good-sized, but none are as large as the bigger brown algae.

One of the most distinctive features of the red algae is the complete lack of any flagellated cells. In contrast to all other groups of algae (except the blue-green) that have various types of motile reproductive cells, the red algae have none. Additional information on the cytology, morphology, and development of the Phaeophyta and Rhodophyta is presented in End Paper 18–1.

Now, let us examine some representative species of these divisions in the various intertidal zones in which they are found and see how the features just discussed represent adaptations to the environment. We shall also see how their methods of reproduction are adapted to the intertidal habitat.

Algae of the Higher Intertidal Zone

The plants growing at the highest point up the beach are the blue-green algae and lichens. These hardy organisms can live where there is little more water than that provided by ocean spray. In a slightly lower region, a slender, tubular plant named *Enteromorpha,* a green alga (Fig. 18–A), may be found. This plant is remarkably tolerant to drying and can survive just in the moisture thrown against the rocks by the spray. *Enteromorpha* is also found growing where springs of fresh water flow out of cracks in the rock and run down to the beach. Except for the blue-green algae, lichens, and a few hardy greens, conditions this far up the beach away from water are too dry for algal growth.

Ulva. Increasingly large numbers of algae are found farther and farther out. One of the most conspicuous is a bright green, flat, tissue-paper-thin alga called *Ulva* or sea lettuce. As its color loudly proclaims, it is a green alga. The leaflike blade of this plant is only two cell layers thick. At one end of the blade is a small holdfast that anchors the plant to the rocks. Although the blade appears fragile, it survives the pounding of the surf. The blade splits, and the edges may appear worn but that is the extent of the wave damage, until the plant is past its prime.

Are the individuals of *Ulva* growing in this zone members of the haploid or diploid generation? If *Ulva* were to follow the pattern of the freshwater green algae, the answer would be haploid. But *Ulva* does not follow this pattern. In *Ulva* the generations are isomorphic or identical in form, and a mature individual can be either haploid or diploid (Fig. 18–5). When the zygote germinates, it forms (under repeated mitotic divisions) a mature diploid plant. Then, under the proper conditions, some peripheral cells of this plant will undergo meiosis, producing motile haploid cells. They swim, settle, and germinate to develop into mature haploid plants. Again, under the proper conditions, the cells of the haploid plant divide mitotically and form motile gametes. All the gametes are motile, the males being slightly smaller in size and yellower in color and the females being larger and greener; only those from different plants will fuse. The thin-walled zygote soon germinates to form a new diploid plant. Thus *Ulva* has an alternation of isomorphic generations. (*Enteromorpha,* the tubular green alga we saw far up the beach, also has isomorphic generations.)

Another green alga that may be found growing in this zone is *Cladophora.* Species of this alga also grow in freshwater streams, while a different, more robust, species inhabits a marine environment. The holdfast,

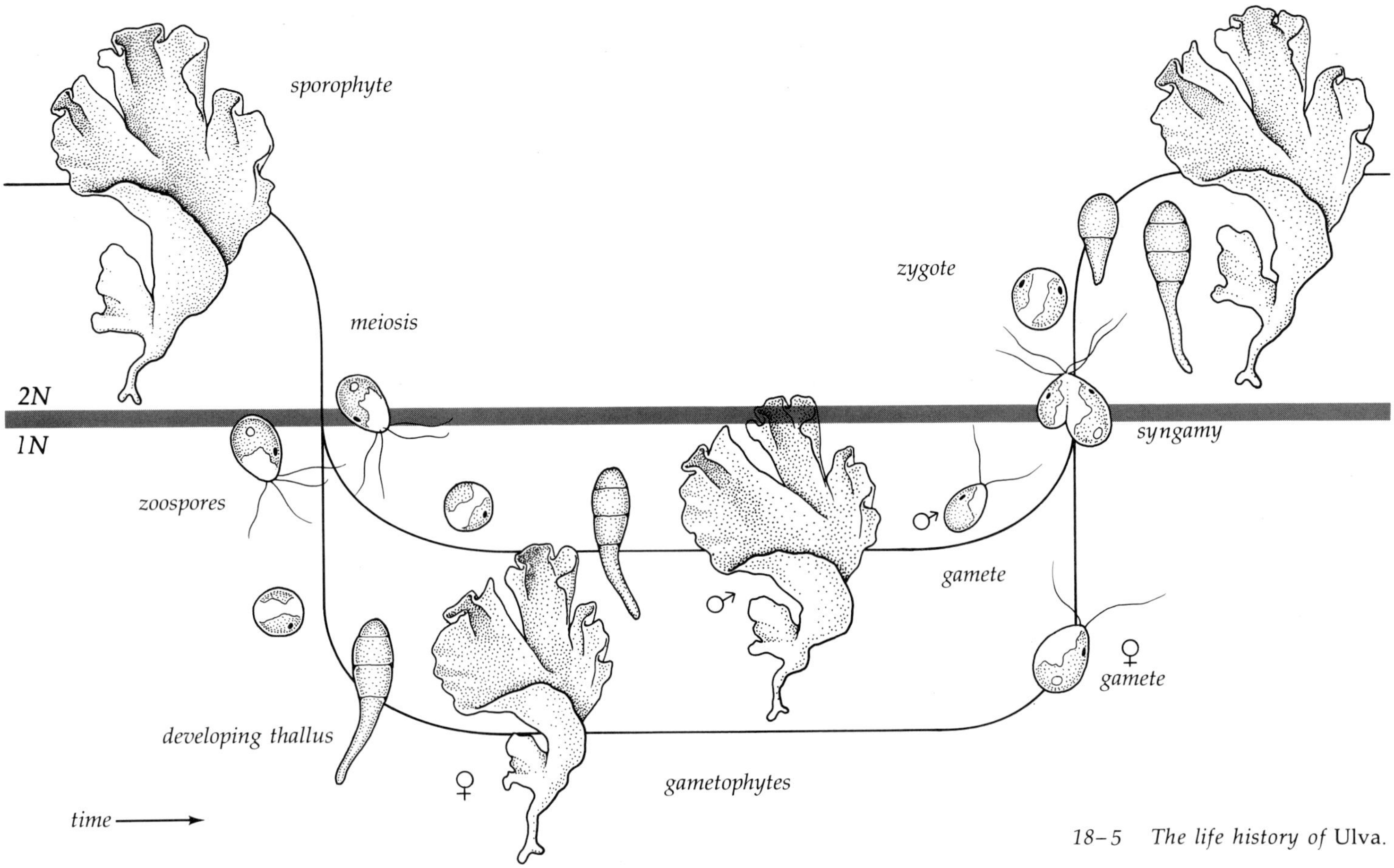

18–5 *The life history of* Ulva.

which kept it from being swept downstream, now prevents it from being washed out to sea.

Fucus. Growing on the rocks in large numbers is a somewhat flattened, rather coarse-bodied, equally-branched alga with a distinct, dark olive-drab color. This is *Fucus* (Fig. 18–B), commonly known as rockweed, and a member of the brown algae. In terms of basic cell requirements, *Fucus* is well adapted to life in the intertidal zone. The presence of the phycocolloids provides protection against a serious water problem, while the accessory pigment aids in light absorption when the plant is submerged.

In the case of *Fucus,* the plant normally seen is the diploid generation. The haploid generation is greatly reduced in size (down to single cells). *Fucus* represents the extreme in reduction in size of the gametophyte generation. The blades of *Fucus,* which are thick and leathery to the touch, contain small conceptacles in which the eggs and sperms are produced through meiosis from cells in the diploid body (Fig. 18–6). The eggs and sperms are released into the surrounding water, and the biflagellate sperms swim to the large nonmotile egg. One sperm fuses with the egg, and the egg, which previously lacked one, now forms a wall. The zygote sinks to the bottom, where the adhesive nature of the outer wall allows it to stick to rocks and other suitable substrates. If it fails to adhere, the zygote may be carried by currents far up to the beach or far out to sea.

The spherical, uniform zygote now becomes differentiated into a rhizoid end, which will form the holdfast, and an apical end, which will produce the stipe and blade. This differentiation is caused by a

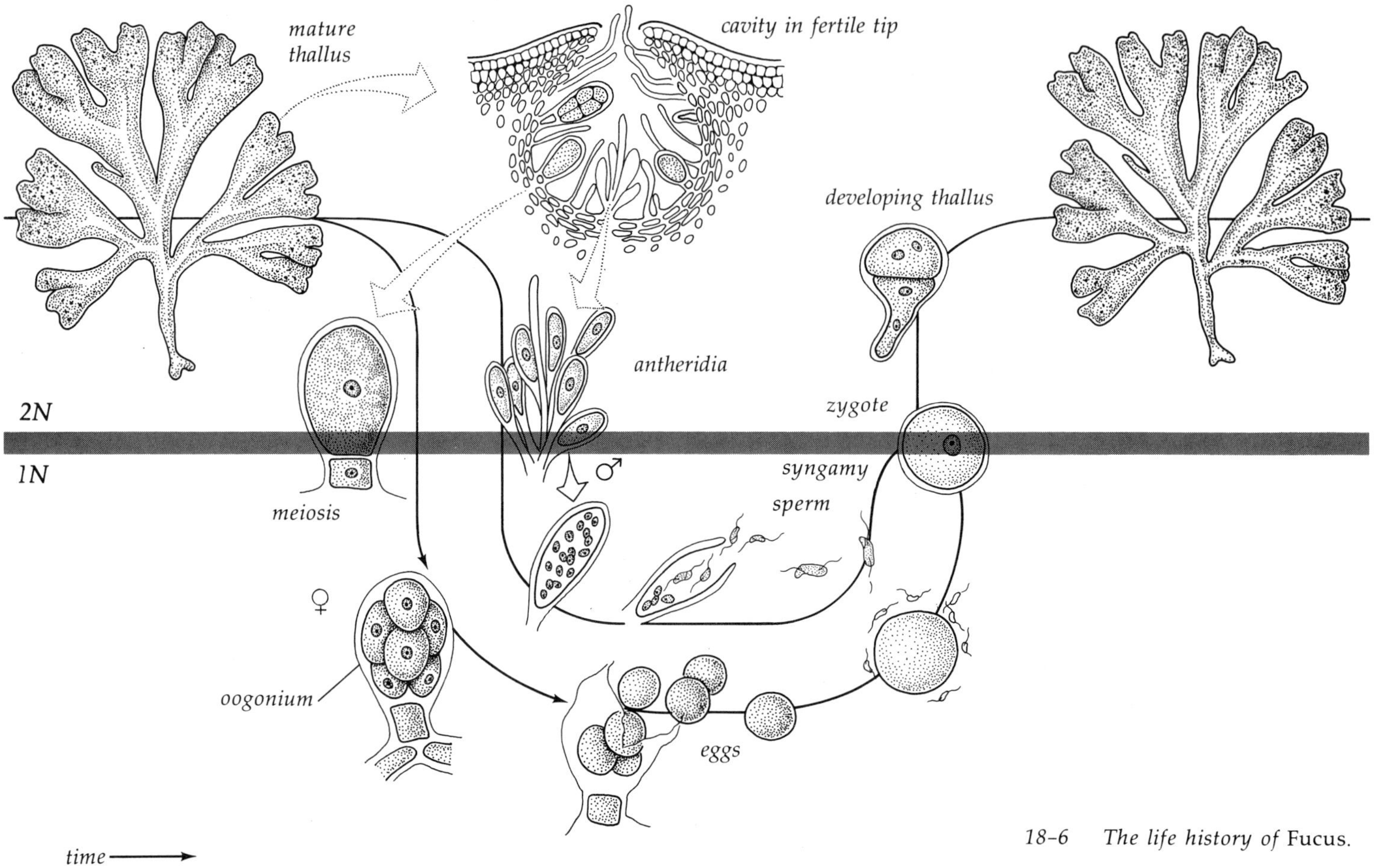

18-6 *The life history of* Fucus.

number of environmental factors, the most important being light—since the rhizoid end always forms away from light. Divisions of the zygote result in the formation of a small, pear-shaped plant that, through divisions of an apical cell and its derivatives, will give rise to the mature plant.

Many factors present in the intertidal environment will influence the establishment of the new plant. If it is too high in the zone, it may be stunted or die for lack of water; if too low, it may not receive enough light and die when it has exhausted its reserve of food. Also, browsing animals such as limpets and sea urchins destroy many of the young plants.

There are many species of *Fucus*, showing different patterns of distribution both within the intertidal zone and geographically. Some species of *Fucus* occur in quite deep water. A relative of *Fucus*, *Pelvetia*, is found in the upper parts of the intertidal zone and is also distributed worldwide. Some species of *Fucus* and *Pelvetia* are apparently so well adapted to the conditions of the higher tide zone that they are incapable of surviving prolonged periods of immersion in the sea. Other species found farther down the intertidal zone are able to survive, indeed thrive, on almost continuous submergence.

Wave action greatly influences both the distribution of the various species of *Fucus* within a given location and the number of plants. Plants on sheltered shores are more numerous and belong to species with longer blades compared with those on the exposed shore.

Finally, it should be pointed out that some brown algae such as the common *Dictyota* have isomorphic generations, the haploid and diploid plants being identical in size and shape.

Porphyra. Red algae are also found in the high intertidal zone, sometimes far up the shore. One of these is *Porphyra*. The species of this genera look much like a dark red version of *Ulva*, and the resemblance is carried to the structural level, since some of the species are two cells thick (others are one cell thick). *Porphyra* can become so desiccated that it feels crisp to the touch; yet, when returned to water, it shows no ill effects from drying. The mechanism underlying this ability to withstand such marked desiccation is not completely understood, but it probably involves the presence of the phycocolloids.

Porphyra and *Ulva* show the same general response to desiccation—a pattern quite different from *Fucus*. The bulky thallus of *Fucus* is capable of retaining more water than the thin leafy thallus of *Ulva* or *Porphyra*, both of which dry out rapidly when exposed to air. Plants of *Ulva* and *Porphyra* are capable of active photosynthesis for only a short period during exposure to air; yet, because of their relatively large surface area, they are as efficient photosynthetically as the bulky *Fucus*, which retains larger amounts of water but presents a smaller surface area.

The complete life history of *Porphyra* is unknown, but two stages have been discovered. One is the familiar leafy plant just discussed. This produces spores that may germinate on mollusc shells, developing into small filamentous plants that were once thought to be a completely different species. These small plants are perennials and always grow in deeper water than the leafy stage. The filamentous plants produce spores that, in turn, develop into the leafy stage, which is an annual. This means that the filamentous stage acts as a perennial source of spores that each year gives rise to the dense intertidal populations of *Porphyra*. Where syngamy and meiosis occur in the life history of *Porphyra* is still unknown, although there is some evidence that meiosis takes place in the microscopic filamentous plants.

The manner in which the red algae reproduce is more complex than the other algal divisions, but it is similar to the browns and greens in having isomorphic and heteromorphic generations. The details of the life history of a red alga will not be treated here, but some of the unusual features will be pointed out. One is that the red algae have no motile cells. Thus the gametes and spores must be passively carried on the currents. Another is that the cells functioning as eggs frequently have a long curved projection that acts as the receptive site for the male gamete, which is carried to it by water movement. Still another feature of many red algae is that they have three generations, usually two diploid and one haploid.

Essay 18–1 Norimaki—Sushi

Algae are collected for food in many parts of the world but are cultivated in only a few. One such country is Japan, where algae are eaten in a large number of ways: in soup, as a vegetable, in meat dishes, and even as a candy. One of the most important food alga in Japan is the red alga Porphyra, *and it is used, among other things, in making the popular sushi or rice roll. Acres of* Porphyra *are grown on nets suspended on bamboo poles. The nets are arranged so that the alga is exposed daily to the air, as it is in the intertidal zones. When the blades are the proper length, they are harvested and carefully dried. They are then packaged and can be obtained in markets in many parts of the United States.*

Norimaki-sushi—rice rolled in algae (nori)—is a common dish in Japan and is commonly found in Japanese lunch boxes. It is an extremely tasty food and not difficult to make. Here are the directions.

Preparing the rice:

2 cups short grain rice
2 tsp sugar
2 cups water
4 tbs white vinegar (rice vinegar is better)
1/4 tsp monosodium glutamate (MSG)
1 1/2 tbs sugar
1 tbs salt

Soak rice, water, and the 2 teaspoons sugar for one hour. Bring to boil in covered pan, then reduce heat to low, and cook 12 minutes. Remove from heat and let stand covered for 10 minutes. Combine vinegar, sugar, MSG, and salt and add to rice, tossing lightly to cool. Spread in shallow pan for further cooling. This amount of prepared rice will make 8 to 10 rolls, eight slices per roll.

In addition to the rice, you need:

1 package (5/7 oz.) dried seaweed. (Maru Ume brand is the best flavored.)
2 cans (3 3/5 oz.) Japanese broiled eels—they're really good but you can substitute shrimp paste, smoked oysters, anchovies, or other similar foods.
1 tablespoon Japanese-preserved red ginger, chopped very fine
1 bamboo sushi mat or a 9 1/2 × 9 1/2 inch square of heavy aluminum foil

Add 2 tablespoons sake to the rice before you start (you can drink the rest later with the sushi). Rinse the bamboo mat with cold water, pat partially dry with a paper towel, spread on counter top with strands of mat parallel to you. Toast one sheet of seaweed lightly over a gas flame or electric burner and place on the mat. Pat a thin, 1/2 inch, layer of vinegared rice evenly over the seaweed, covering all except a 1/2 inch strip at edge of seaweed farthest from you. Dip your hands in water as you spread the rice to keep it from sticking to your fingers. On top of the rice, sprinkle some of the bits of red ginger. Place the eel from one paper wrapping (there are four in a can) in a horizontal row across the center of the rice adding some of the liquid from the can. You can use an inch strip of shrimp paste in place of the eel. Thin strips of cooked vegetables (such as carrots or green beans) or canned pimiento may be placed with the eel for color.

Holding the center ingredients down with your fingers, lift the mat with your thumbs and roll over until the edge of the seaweed that was nearest to you meets the other end of the rice layer, the empty strip of seaweed forming the final seal as it overlaps the roll. Pull the mat away as you roll, using pressure to tighten the roll. Press in any loose ingredients at ends of rolls.

Wrap roll in plastic film for at least 1 hr before slicing. Cut into 1 in. slices with clean, sharp knife moistened with water before each cut is made. Excellent as a cold hors d'oeuvre served with sake.

Life in Deeper Water—Algae of the Lower Intertidal Zone

In the lower tide zone the rocks exposed at low tide may be several meters under water at high tide. The plants growing here are rarely subjected to prolonged periods out of water, and they may be under relative low light conditions for substantial periods. The algae that grow here have adapted to these conditions in a number of ways. Many have evolved long stipes and blades that essentially overcome the problem of water depth by carrying the photosynthetic cells close to the surface. On the blades of the large algae are epiphytes, getting a free ride in the sun. Below the larger algae, which are mainly browns, are the small reds with their blue-green, light-absorbing pigments. This is a zone filled with plants of all sizes and shapes.

18–7 On the rocks, at almost any level in the intertidal zone, one can find masses of material that look like thick mats of dried paint or tar. These are encrusted algae. Such forms are found among in the green, brown, and red algae, although they are most numerous (in terms of species) in the red algae. The encrusted algae all have thickened walls that form a matrix in which the cells are embedded. In some cases, the cells are arranged in plates; in others, as filaments; and in still others, as multinucleate masses. The encrusting material helps protect the cells from drying and the force of the waves.

Laminaria. The species of the genus *Laminaria* (Fig. 18–C) are found throughout the world in temperate and northern seas. They are Phaeophyta and have a variety of forms (ranging from one to several meters in length), although most can be described as having a long, flat blade, a sturdy cylindrical stipe, and a well-developed holdfast. The blade may split into strips as a result of wave action.

The plant just described is the diploid generation, and it is dominant in the life history of the species (Fig. 18–8). Meiosis occurs in large patches of cells on the blades, giving rise to biflagellate, haploid zoospores. These germinate to form two types of microscopic, gametophytic plants. One type, after producing a few cells through mitosis, forms a single egg per plant. The second type divides to form a smaller plant but with more cells, some of which produce biflagellate motile sperms. These swim to the egg, and one unites with it to form a thin-walled zygote. The zygote remains in place on the female gametophyte, soon dividing and developing into a mature diploid plant.

The sporophyte plants may live for several years. The blades grow continually through divisions of cells at the base, near the stipe. The ends of the blades are worn away by wave action and eaten by animals.

Other brown algae of the lower intertidal zone. A common Phaeophyta of the lower intertidal zone, particularly on the western coast of North America, is *Egregia*. This genus is related to *Laminaria* but has a quite different shape. The stipe is flattened, and numerous small blades are found attached at right angles to it. Some blades are hollow and act as floats, which keeps much of the plant near the surface of the water. A well-developed holdfast is present. The plants can reach 6 to 8 m in length and form dense stands.

One of the most attractive of the algae growing along the northern Pacific coast of North America is the brown alga *Postelsia palmaeformis* (Fig. 18– F,G) or the sea palm. This plant consists of a hollow stipe several centimeters in diameter and 0.5 m long, rising from a well-developed holdfast and crowned by a cluster of long, narrow blades. *Postelsia* grows on the tops of rocks, where they are closer to the surface but where they receive the full force of the breaking waves (Fig. 18– 9). The thick, mucilaginous walls apparently assist the plants in absorbing the force of the waves.

Beyond the point where the tides expose them to drying, grow the greatest of all brown algae—the kelps. This common name is derived from an old English word for all seaweeds and indicates various plants in different parts of the world. In England and Scotland it is used for the largest of the *Laminaria* species and several related genera found in the deep water. Along the Pacific coast of North America the term kelp—usually "giant" kelp—refers to *Macrocystis* and *Nereocystis*, which can reach lengths of 40 m or more. These impressive plants form large beds along the Pacific coast. *Nereocystis* has a well-developed holdfast and a long, hollow, cylindrical stipe ending in a large, spherical, hollow float containing an amazingly high amount of carbon monoxide! From this float, stream a number of long, narrow blades. All in all, it is a tremendously impressive plant and in many ways a world of its own. A single individual may have dozens of

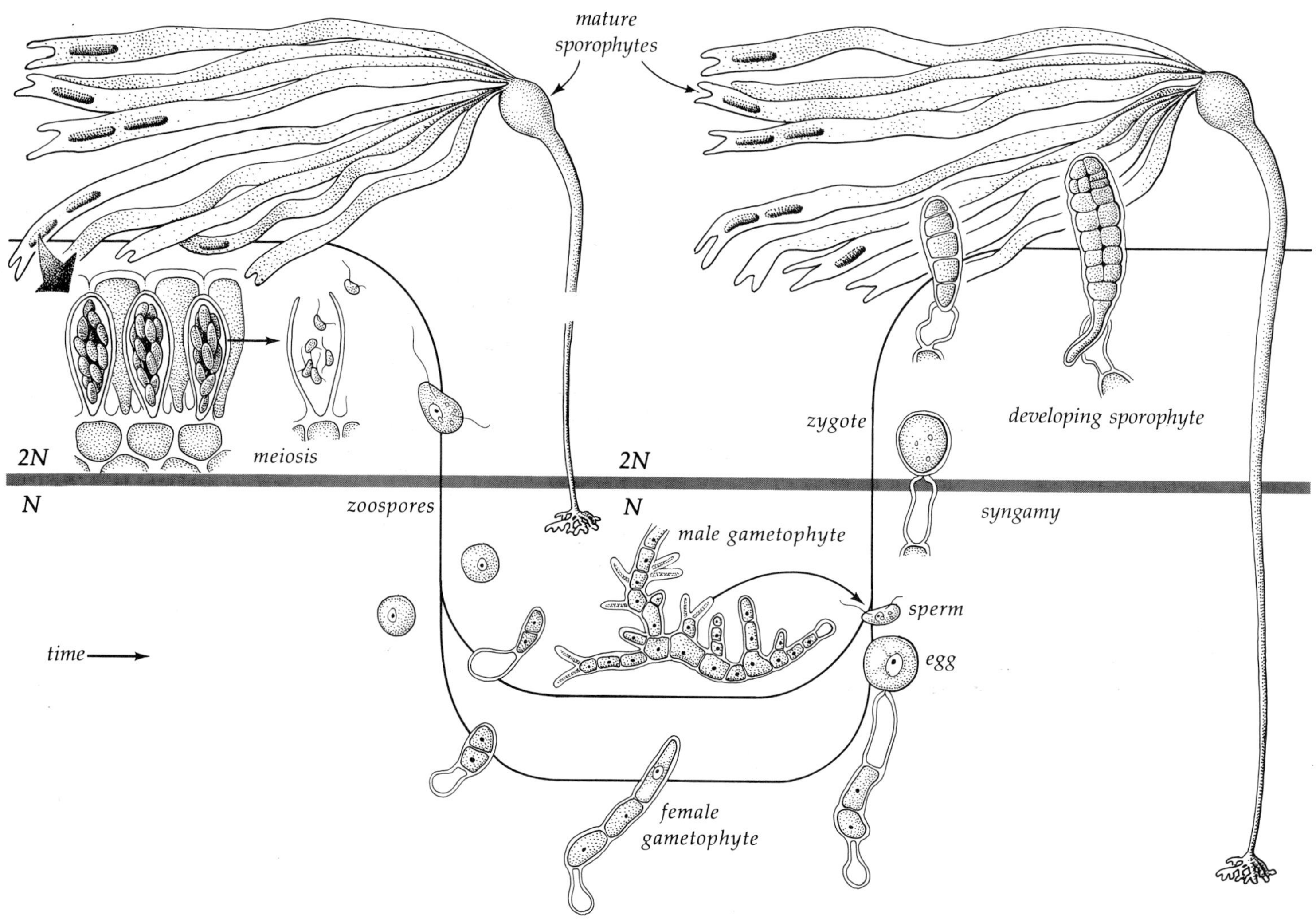

18–8 *The life history of* Nereocystis, *typical of the family Laminariales, which includes* Laminaria.

other algae (attached mainly to its stipe), as well as dozens of snails, small crabs, and other invertebrates living on it.

All of these large brown algae have a plant structure far removed from the simple cell aggregates of the freshwater algae. The development of a large body gives the plant certain advantages; yet it also brings problems. In *Fucus, Laminaria, Egregia, Postelsia,* and *Nereocystis* the evolution of holdfast, stipe, and blades has profound functional significance and has allowed the organisms to exploit a promising but difficult environment. On the other hand, the presence of holdfast, stipe, and blades means that there is a division of labor among the cells comprising the individual. The holdfast cells in *Nereocystis,* for example, are so far below the surface that they are probably less effective in carrying on photosynthesis. But, for the blades to be held in the best position for photosynthesis and not be washed ashore, the holdfast is necessary, and its cells must be kept alive. Moreover, the holdfast serves an important second function: it is a means of carrying the plant over winter. While many brown algae are annuals, as is

18–9 Waves breaking against a stand of Postelsia palmaeformis.

Nereocystis, many lose their stipes and blades, but the holdfast survives as in many *Laminaria* species. The holdfast has stored food reserves that allow it to overwinter and to grow in the spring. Under these conditions it is not surprising to find that in such algae there are special cells involved in the conduction of photosynthate from the blade down the stipe to the cells lower in the stipe and the holdfast.

Red algae of the lower intertidal zone. Many red algae are found growing in the lower intertidal zone, and they show a great range in size and shape. These red algae are a pinker red than those growing in the higher zone, which are often a dark, almost black, red. Some, like certain species of *Gigartina,* have large, tough blades that look and feel like a bath towel. Others have plain, smooth blades, much like *Laminaria,* but which have a greenish or red color that is iridescent when in water, as do species of *Iridaea* (Fig. 18–H). These two groups of red algae are fair-sized plants 20 to 60 cm long and grow attached to rocks.

Other red algae form remarkably beautiful, delicate, lacelike blades. Many of these are only 2 to 10 cm long, and some, like *Microcladia* (Fig. 18–D), are epiphytes on other algae, including other Rhodophyta. Still other Rhodophyta look like red or pink paint that has been poured on the rocks. These are the encrusting coralline red algae. Some coralline algae look like small, red platelets strung together and are often mistaken, when dead and sun-bleached, for minute animal skeletons. These are the articulated coralline algae. All of the coralline algae have large amounts of calcium carbonate in their cell walls. Both types of coralline algae are abundant, and the encrusting species are frequently found in regions of severe wave action. In some areas there are such large growths of these algae that whole beaches may be formed by small fragments of the plants. Some of the oldest known fossil red algae are corallines.

The red algae often grow in the shade of the brown algae. They appear able to survive in these conditions, thus exploiting an environment unavailable to other algae. The fact that their red pigment can readily absorb the blue to green light reaching their level is important in this ability to survive.

In tropical waters the brown algae are less common and smaller in size, while the red algae are more numerous. The red coralline algae are important in coral reefs. Corals are remarkable, stationary animals with delicate bodies encased in walls made of calcium carbonate. The thick calcium-impregnated walls of the encrusting algae are believed to help hold coral together. Without the cementing and protecting action of the algae, the coral in reefs exposed to surf would rapidly erode. Corals are also dependent on dinoflagellates. For coral to grow into reef-producing masses, food must be plentiful, the temperature must exceed 21° C throughout the year, and calcium carbonate, light, and oxygen must be abundant. As long as sufficient light is available, the algae photosynthesize using carbon dioxide, which is the by-product of the coral's respiration. In turn, the oxygen, which is the by-product of photosynthesis, is necessary for the coral's respiration. In addition, the coral receives carbohydrates made by the plant. All in all, a remarkable partnership!

Seasons and Intertidal Life

Marine algae, like their freshwater cousins, show periodic changes in number and predominance of species. Some of the species (e.g., many of the *Fucus* species) are perennial throughout the year. Some perennial species (e.g., some of the *Ulva* species) are reduced in winter to only a fraction of the size they are in the spring and summer, thus appearing to be almost annuals. For others (e.g., the *Laminaria* species), the blades, sometimes part of the stipe too, die in the winter but are produced again by the reactivated meristem in late winter and spring. Other algae are present only as germlings—young haploid or diploid plants that are frequently filamentous and that will later develop into mature plants. In general, the fall and early winter months are the time when the fewest algae can be seen. This depends, however, on latitude and the types of algae present in the flora. The rocks are never bare, and some species thrive in the winter months.

As the days lengthen in late winter and spring, the germlings develop into the mature plant. In the case of *Porphyra*, the filamentous form in the deeper water begins producing spores that are carried up the beach, where they germinate and start the annual leafy generation. The holdfasts and stipes that have overwintered now begin to produce new blades. Still later, thousands and thousands of spores are produced, and the various species fill every location available for growth.

Throughout the spring and summer the algae flourish, many continually growing new blades that are eaten away by animals or worn away by the surf. Some species begin their reproductive cycles during the summer, forming the gametophyte plants that in autumn produce zygotes that germinate into small, filamentous diploid plants clinging tenaciously to the rocks during the winter storms.

What influences the algae to reproduce? For some it appears that changes in nutrient level are important, but this is not true for all. Changes in temperature are critical for many, including *Ulva* and *Nereocystis*. Another important factor is the amount of light. In late summer and autumn the days are shortening, and less light is available to the organisms, a possibly significant factor in stimulating some of them to produce sexual rather than asexual cells. Light has effects on plant growth other than those simply relating to photosynthesis. Many plants can, in effect, measure the length of the day and the length of the night, and certain of their growth responses are related to changes in these two periods. Such responses are called photoperiodism. In flowering plants the induction of the flower is controlled, in many species, by the length of the days and nights (see Chapter 36). Recent work with marine algal species (*Porphyra* being one) indicates that some of these are also photoperiodically sensitive plants and form reproductive structures in response to the length of the day. The mechanism for such response in marine algae (or in flowering plants) is not understood. Definite conditions do control the reproductive cycle in these algae, and it would appear that this is a response to day length either as a function of total photosynthesis or as some more subtle mechanism through photoperiodism.

Algae of the Open Oceans

The open oceans appear, at first glance, to be a region bare of plant life; yet closer inspection shows how wrong this impression is. In our discussion of lakes, ponds, and rivers, we have already been introduced to the phytoplankton. Phytoplankton are also present in the oceans and are of tremendous importance to the animal life in these waters. The phytoplankton of the ocean are extremely complex in terms of the number of species present. The algal divisions present in the freshwater plankton are also present in the marine plankton.

The most conspicuous groups of the marine planktonalgae are the diatoms and the dinoflagellates. The number of diatoms, both in terms of numbers of species present and number of individuals per species, is absolutely fantastic—72 million diatoms have been calculated to be contained in 1 ft^3 of water. The diatoms that are present are similar in general morphology to their freshwater forms, although they frequently contain more spines and projections (Fig. 18–10). The spines and projections of various sorts aid in the flotation of the individuals in the ocean water.

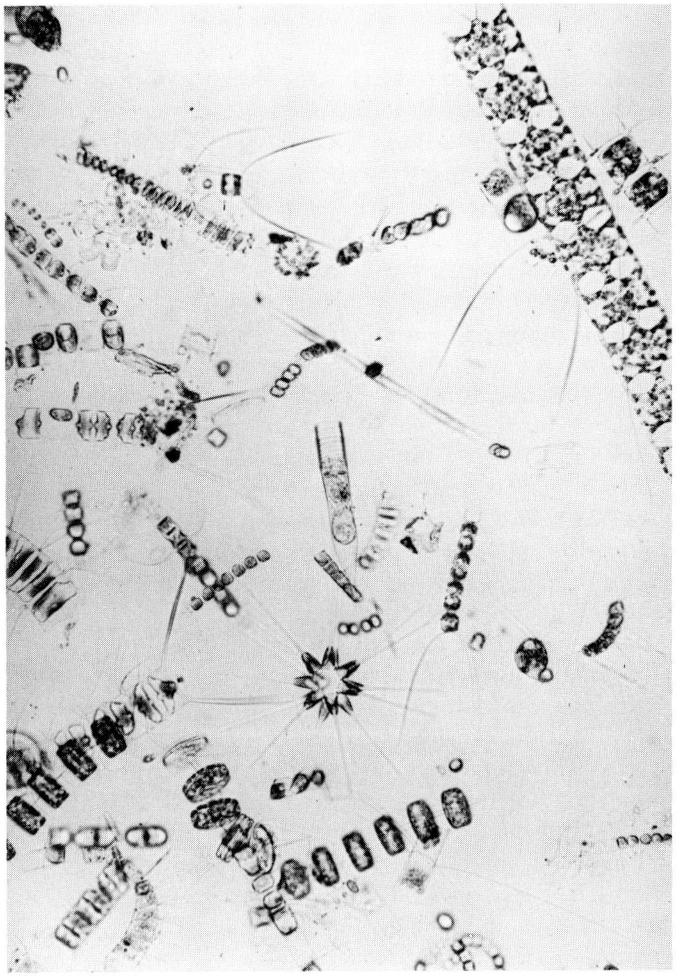

18–10 *Typical marine phytoplankton.*

The dinoflagellates belong to a completely different division of algae, called the *Pyrrophyta.* These are primarily single cells that either lack a wall, although they have a type of firm outer covering, or, more commonly, possess a cellulosic wall. Those that have a wall are called the armored forms; in these the cellulose layer is composed of a number of discrete, articulated, sculptured plates cemented tightly together in a definite pattern. The term "armored" is good because the organism looks like it is encased in medieval plate armor. The plates are frequently strikingly ornamented with pores, horns, spines, papillae, or winglike projections. They are beautiful organisms; some photosynthetic forms are a greenish-brown or golden-brown color. Their color comes from the presence of chlorophyll a and c, as well as a number of carotenoid pigments. Sometimes, the organisms are bright red. Most of them possess 2 flagella and are highly motile. Not all of the dinoflagellates contain photosynthetic pigments; some are saprophytes or parasites, able to ingest food. One of the more unusual of this variety of dinoflagellates is a form called *Noctiluca,* which means night light. This organism is common in the cooler seas, particularly in late summer and autumn. As an object passes at night through water containing this organism, the water will phosphoresce with a silver light due to the millions of *Noctiluca* present. The flagellum has been modified into an apparatus to seize its prey, which consists primarily of diatoms.

The marine plankton also contain members of the blue-green algae. They are not as numerous in numbers of species as in the freshwater habitats, but they are still capable of considerable growth. Under the proper conditions, they can cause extensive blooms in various parts of the ocean. The blue-green alga, *Oscillatoria,* has long been known to cause extensive blooms in parts of the Indian Ocean, where they can result in a mass death of fish. One species of blue-green algae produces a reddish bloom, and from this phenomenon the Red Sea got its name.

The phytoplankton of the ocean show seasonal variations, as do the freshwater forms. In both cases, the environmental factors causing these variations are similar: light, nutrients, and temperature. In the winter the amount of plankton is minimal. This is because in temperate climates the days are short and the amount of light is reduced, partially because the sun is low and most of its rays are reflected from the surface instead of penetrating into the water. There are adequate mineral nutrients (primarily nitrates and phosphates), but light limits algal growth. With the coming of spring the days lengthen, and the sun is ever higher in the sky, making more and more light available to the plants. Since the nutrient level is high and light is now plentiful, the algae photosynthesize and grow at remarkable rates. The number of planktonic algae sky rockets—diatoms increase in numbers as much as 60,000 times. By midsummer this growth rate falls off, since the nutrients are exhausted in the upper layer of water. Stratification of water occurs in oceans as it does in lakes—the upper layer being warm and the lower layers being cold. The nutrients made available through the death and decay of organisms sink to the lower levels. Thus they are out of reach of planktonic algae. The number of algae also decreases simply as a result of the enormous numbers eaten by the small animals, the zooplankton, present in the same layer of ocean. In autumn there is a brief burst of growth when mixing of the upper and lower layers of oceans occurs. This mixing occurs partially because the weaker autumn sunlight cannot maintain the higher temperature of the upper layer. As the layers mix, the nutrients accumulated in the lower layers now become available to the algae of the upper layer. But they come almost too late, since light is now becoming the limiting factor. Thus there is a brief fall growing period and then the decline of winter.

The interaction of light and mineral nutrients in determining the amount of plankton also determines the basic fertility of different regions of the oceans. The deep, cold layers of the ocean are the richest in mineral nutrients needed by the algae. Where these layers reach the surface, the phytoplankton are most abundant. As we saw earlier, upwellings occur in the Arctic and Antarctic oceans, along much of the western coast of South America, parts of the western coast of North America, the extreme northeastern coast of North America, the western coast of Africa, and the western coast of Australia. In all of these regions, phytoplankton are present in enormous amounts. But large expanses of open ocean are relatively sterile.

Phytoplankton and Marine Life

Why is the amount of phytoplankton important? Because it is the basis for all life in the open ocean. Whales have been one of the great lures for men, drawing them to the vast expanse of the Antarctic Ocean, the feeding grounds of the whales. The largest of all whales are the blue whales, and these live on creatures called krill. Krill resemble small lobsters without claws and are up to 6 to 7 cm long. The whales strain the krill out of the water by means of modified teeth. The krill, in turn, feed directly off the plankton. Thus a pyramid of life is established, with the plankton at the bottom, the krill in the middle, and the whale at the top. For each 100 kg of whale, 1000 kg of krill must be eaten. To produce this much krill, roughly 10,000 kg of phytoplankton must be utilized. This is a remarkably simple, direct, and efficient food chain and accounts for the spectacular size and numbers of this type of whale.

Other food chains or pyramids are more elaborate. Killer whales live off the seals that eat fish and squid, which in turn consume krill that ingest phytoplankton. As each level of organism requires energy for its own functions, food value is lost at each step of the pyramid (roughly nine-tenths at each step). Hence only 1 kg of killer whale is obtained at the expense of 10,000 kg of phytoplankton.

In other oceans and at different times, the food chains involve other organisms feeding off the phytoplankton. These, in turn, end in pyramids terminating in such large fish as tuna and sharks, or birds and mammals. In all instances, the phytoplankton form the base—they are the primary producers.

The problem of mercury in tuna and swordfish is a classic example of the food pyramid. As we saw in Chapter 16, bacteria convert the inert metallic mercury into methyl mercury that enters the food pyramid through phytoplankton and protozoa. These are eaten by krill or small fish, which are eaten by tuna and swordfish. The amount of mercury in swordfish is higher than in tuna and the explanation is simple: swordfish are bigger than tuna and must therefore eat more to survive, thus accumulating more mercury.

A similar pattern is true for DDT. This potent insecticide again enters the food pyramid at the lower levels and is concentrated in larger and larger animals.

Levels in fish-eating birds such as pelicans and various hawks have now reached the point where extinction is threatened because the DDT interferes with shell formation in the eggs. But DDT also has a direct effect on phytoplankton, reducing the numbers of some species as much as 50%. Mercury also kills many species of phytoplankton.

We need the products of the sea, we need them now, and we shall need them even more in the future. We cannot meet this need without the marine phytoplankton, and it is time that this simple but crucial point be kept in mind in all our uses of the ocean.

End Paper 18–1 Phaeophyta, Rhodophyta, Pyrrophyta, and Algal Evolution

Phaeophyta

The Phaeophyta contains some 1500 species all of which, except for three genera, are marine. The cells of the brown algae are generally uninucleate with relatively large nuclei. The chromosomes contain DNA and histones and are similar to those found in higher plants. The number of chromosomes range from $2n = 24$ to $2n = 64$. One species (*Sacchorhiza polyschides*) is known to have XY chromosomal sex determination (see Chapter 12) with the large X chromosome in the female gametophyte and a small Y in the male gametophyte. The nuclear membrane persists during mitosis, except at the poles where spindle microtubules originate near paired centrioles.

The cytoplasm contains numerous small vacuoles. The chloroplasts are usually disc-shaped, but plates, stars, and spiral, ribbon-shaped chloroplasts are found. Within the chloroplasts the thylakoids are loosely grouped in threes (Fig. 18–11). DNA in the chloroplast appears to be arranged

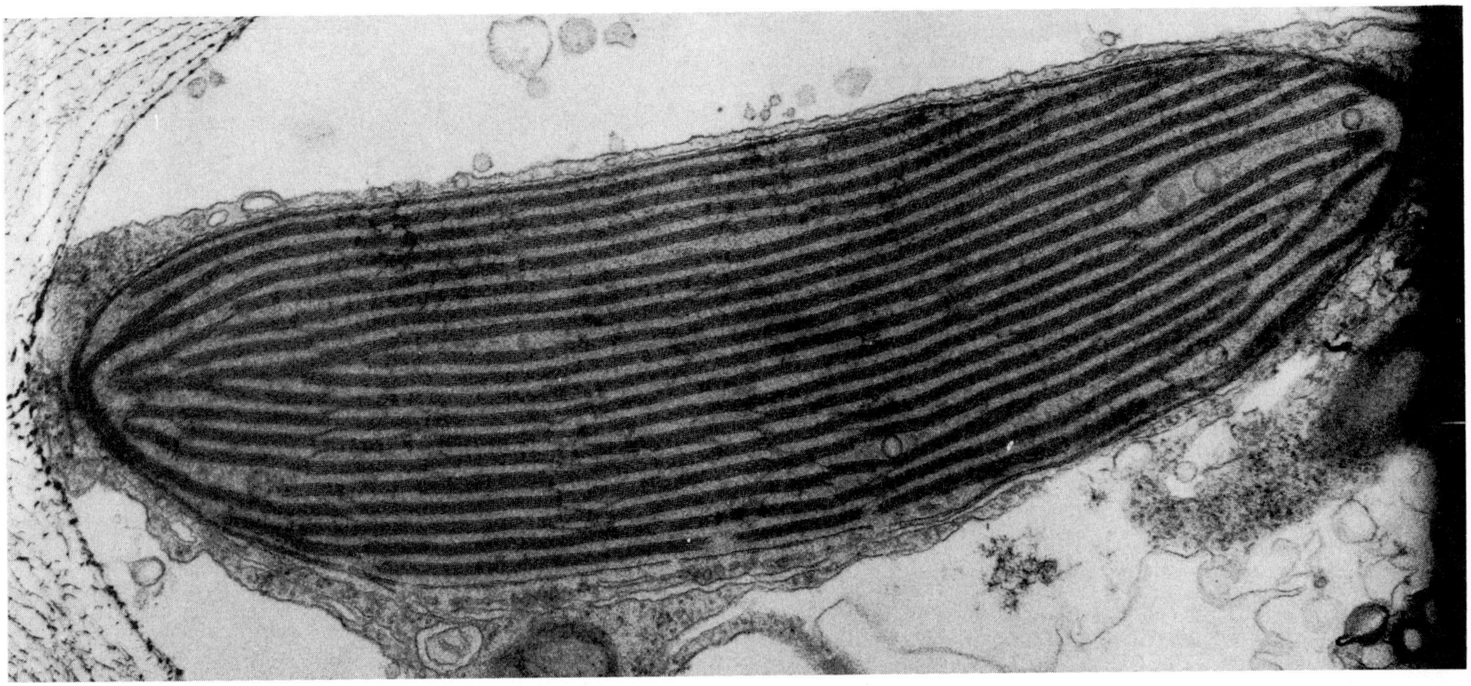

18–11 *A chloroplast from a brown alga, as seen with the electron microscope. The thylakoids are grouped in stacks of 5 and, in general, run the length of the plastid.*

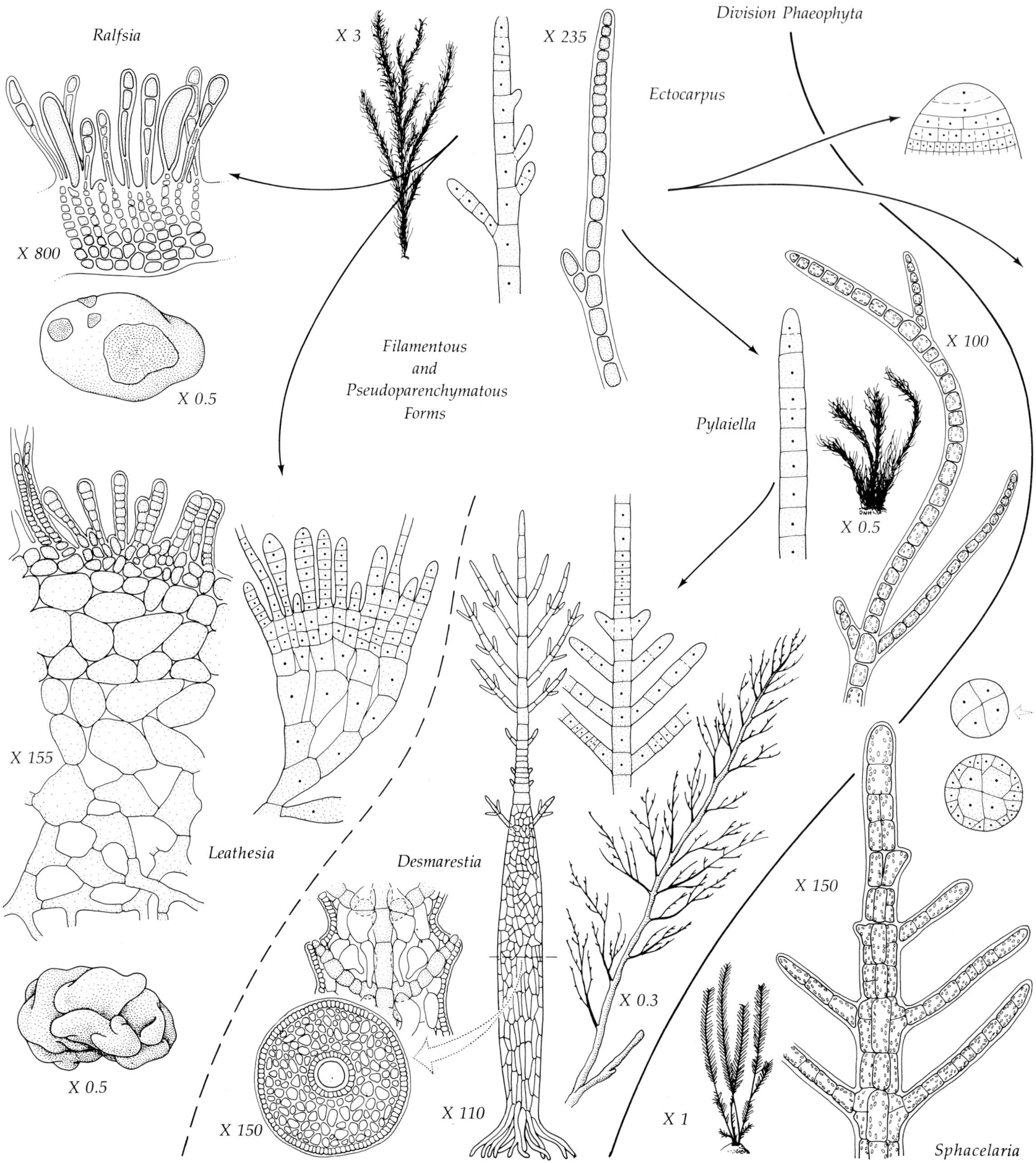

18–12 *The diversity of form in the brown algae is shown in this chart. The drawings illustrate the form of the entire plant, while the accompanying microscopic views show the cellular organization.*

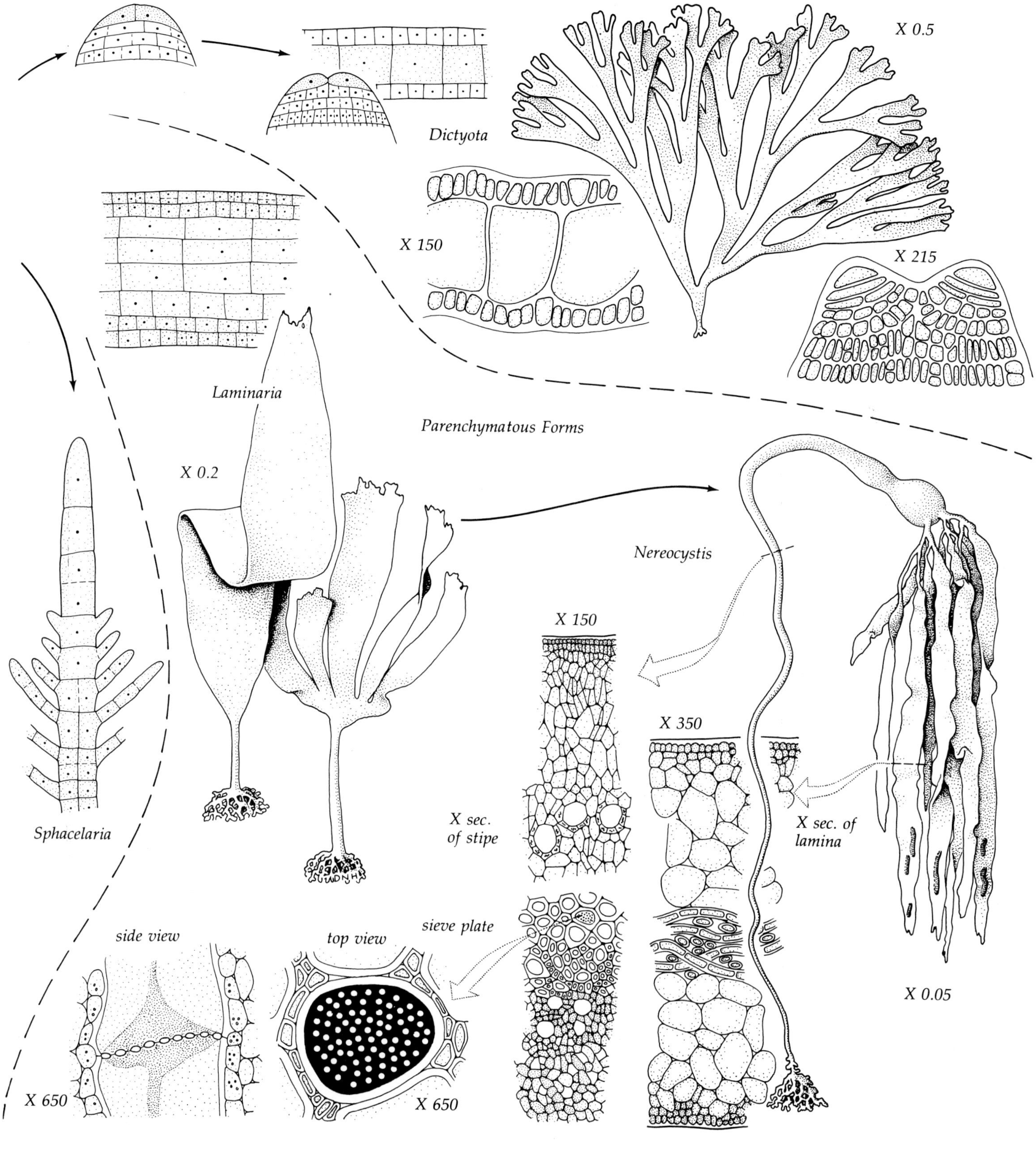
X 0.5
Dictyota
X 150
X 215
Laminaria
Parenchymatous Forms
X 0.2
Nereocystis
X 150
X 350
X sec. of stipe
X sec. of lamina
Sphacelaria
X 0.05
side view
top view
sieve plate
X 650
X 650

in a ring around the organelle. Laminarin, the food-storage product, is deposited outside the chloroplast and is different from starch.

No brown alga is motile, although motile cells are produced. These are generally laterally biflagellate, with the anterior flagellum being a tinsel type and the posterior a whiplash type.

The range in form of the brown algae, although great (see chart in Fig. 18–12), is based on relatively simple differences in patterns of development. The simplest brown algae are composed of cells in filaments: simple filaments, as in *Ectocarpus;* branched, intertwining filaments, as in *Desmarestia;* or multibranched filaments, as in *Leathesia.* In these filamentous forms, growth is based on divisions occurring in cells down the file of cells (i.e., intercalary) or at the base. Divisions are at right angles to the long axis of the file, except where branch filaments are formed. In other species cell divisions occur in other planes as well; hence flattened or cylindrical forms of algae are produced. Thus, while *Sphacelaria* is a filament at the tip, it has a fleshy or parenchymatous structure back from the tip. In *Sphacelaria* there is a single dividing apical cell at the end of the thallus. This is also true of *Dictyota,* except here divisions in other planes result in a flattened thallus. Equal division of the apical cell in *Dictyota* results in equal branching in the thallus. Additional planes of divisions result in thicker thalli such as in *Laminaria* and *Macrocystis.* These algae also have meristems at the intersection of the blade and the stipe, which produce the long blades. External form in algae can be deceiving in terms of cellular organization; for example, the encrusted alga *Ralfsia,* which looks like spilled paint on a rock, is actually a filamentous form.

The brown algae contain the largest and vegetatively most complex marine algae: *Nereocystis* and *Macrocystis.* These are huge plants, and they exhibit a high degree of complexity in the tissues that comprise them. Figure 18–12 shows some of the cellular features of the stipe and blade in *Nereocystis.* The cells of the outermost layer are small, compact, and rich in stored food. Below these are layers of larger cells through which run mucilage ducts surrounded by secretory cells. These layers of compact cells give way in turn to a region of loosely packed cells arranged in rough filaments. Beneath these are the thick-walled cells that include cells morphologically similar to those found in vascular plants involved in the movement of organic compounds through the plant—the sieve tubes. Morphologically, chemically, and physiologically the sieve tubes of *Nereocystis* are similar to those of vascular land plants. Since there is no indication that the brown algae gave rise to the land plants, we have here a remarkable example of convergent evolution. The blade is somewhat similar to the stipe in general arrangement of cells: small, compact cells on the exterior, larger cells below, and filamentous layers in the center. Also shown is a fertile portion of the blade.

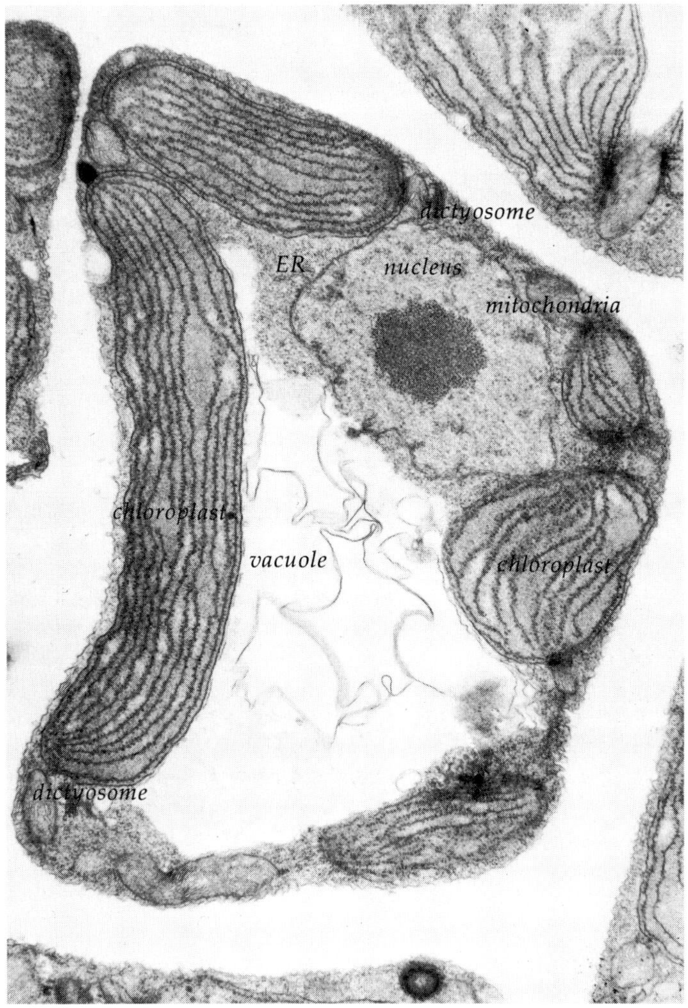

18–13 The ultrastructure of a cell of a red alga. The single thylakoid arrangement within the chloroplasts is typical of the red algae. The tiny dots on the surface of the thylakoids are granules of the phycobiliprotein accessory pigments. The nucleus, mitochondria, dictyosomes, ER, and vacuole can all be seen.

The cells in which the nuclei undergo meiosis elongate, and many spores per cell are produced by mitotic divisions of the haploid nuclei. Interspersed between the fertile spore-forming cells are sterile elongate cells. A cavity develops in the center of the stipe as the plant grows. The hollow stipe

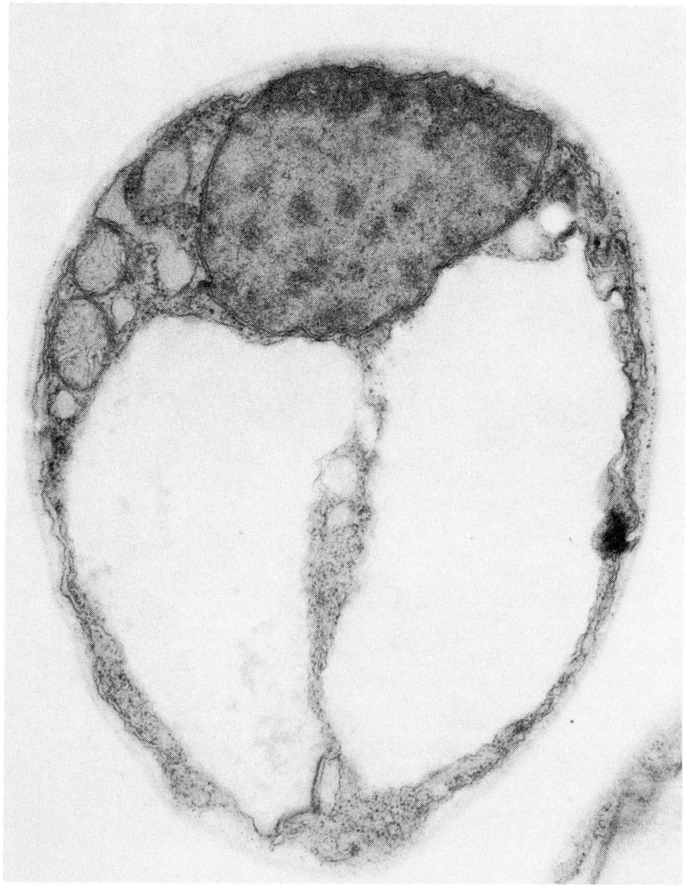

18–14 The ultrastructure of the male gamete in a red alga. The cell lacks any trace of flagellum apparatus and is greatly reduced in structure.

and the hollow float at its top help keep the plant more or less erect in the water. The gas in the cavity is rich in carbon monoxide, a deadly poison to living cells. The gas appears to be a by-product of the plant's metabolism, and a layer of dead cells lining the cavity are believed to protect the living cells from the gas.

Rhodophyta

There are nearly 4000 species of red algae in almost 600 genera. They are, to say the least, most unusual plants.

The red algae are the only algal division, except for the blue-greens, that lack motile cells of any sort. This may mean that they once had them and then lost the capacity to produce them or that they never evolved them in the first place. As yet, no true centrioles have been observed in the Rhodophyta, although centriole-like structures have recently been seen.

The cells in the simpler red algae are uninucleate, but in the more advanced forms, they are generally multinucleate.

In the simpler red algae there may be only one chloroplast, with or without pyrenoid-like structures. The more complex red algae generally have numerous small disc-shaped chloroplasts. The thylakoids are separate from each other—a condition that is considered primitive (Figs. 18–13 and 18–14). The phycobiliprotein accessory pigments are localized in 350 A granules on the outside surfaces of the thylakoid membrane. This is the same arrangement of similar pigments found in the blue-green algae.

The fact that the red algae and the blue-green algae both lack motile cells and have the same type of accessory pigments arranged in the same way on the thylakoids and the same arrangement of the thylakoids within the chloroplasts of the red algae or the cells of the blue-green algae has led some botanists to suggest that the two groups are related. The major difficulty is that the Rhodophyta are true eucaryotic cells, while the Cyanophyta are procaryotic.

Without doubt, some of the most beautiful plants in the world are found in the Rhodophyta. Many have exquisite red colors in a wide range of hues and delicate, lacelike forms. Not all red algae are handsome. Some are single cells or simple filaments, while others are flat sheets of cells. The Rhodophyta are divided into two subclasses: the Bangiophycidae and the Florideophycidae (Fig. 18–15). The Bangiophycidae contain the structurally simpler red algae: the unicellular genera such as *Porphyridium*, the filamentous genera such as *Erythrotrichia* and *Goniotrichum*, some rather simple, flattened genera such as *Bangia* and *Erythrocladia*, and the leaflike *Porphyra*. In contrast, the Florideophycidae are all multicellular forms, with the morphologically simplest being branched filaments of cells as in *Antithamnion*. The more complex forms are also basically filamentous, but the branches are fused to varying degrees and intertwine in a number of ways, resulting in a wide variety of forms. Growth of each branch in the filaments is primarily by division of a single apical cell. There may be either a main axis of a single row of cells or a main axis of a number of parallel or almost parallel filaments. With only this relatively simple basis of organization, the red algae have evolved a remarkably diverse morphology.

Pyrrophyta

There are some 1000 species of Pyrrophyta in 125 genera often grouped into two classes: the Dinophyceae (dinoflagel-

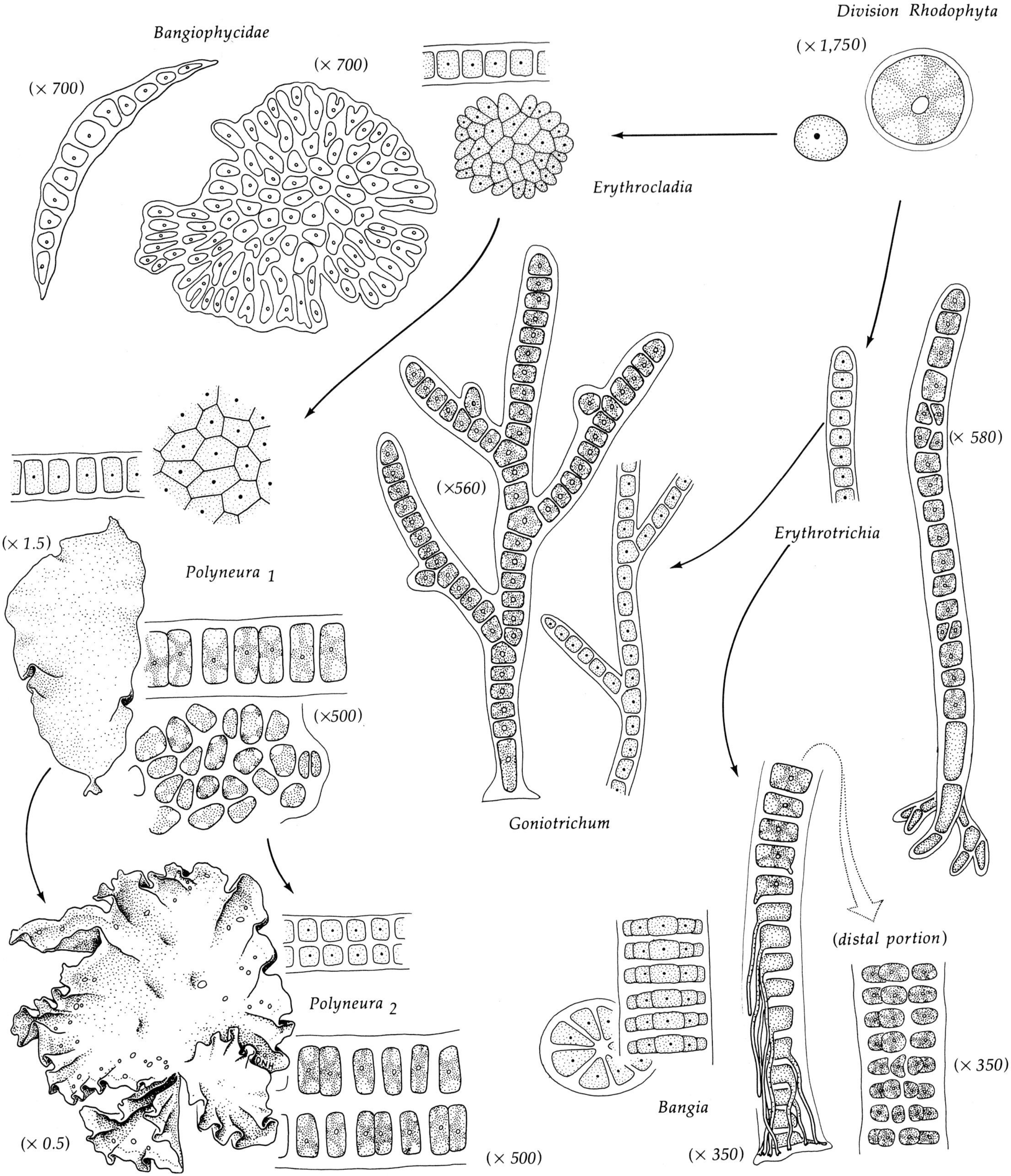

18–15 *The diversity of form of the red algae is shown in this chart. The basic unit of organization of many of the red algae is the filament; yet from this simple structural unit, remarkably complex plants can be formed.*

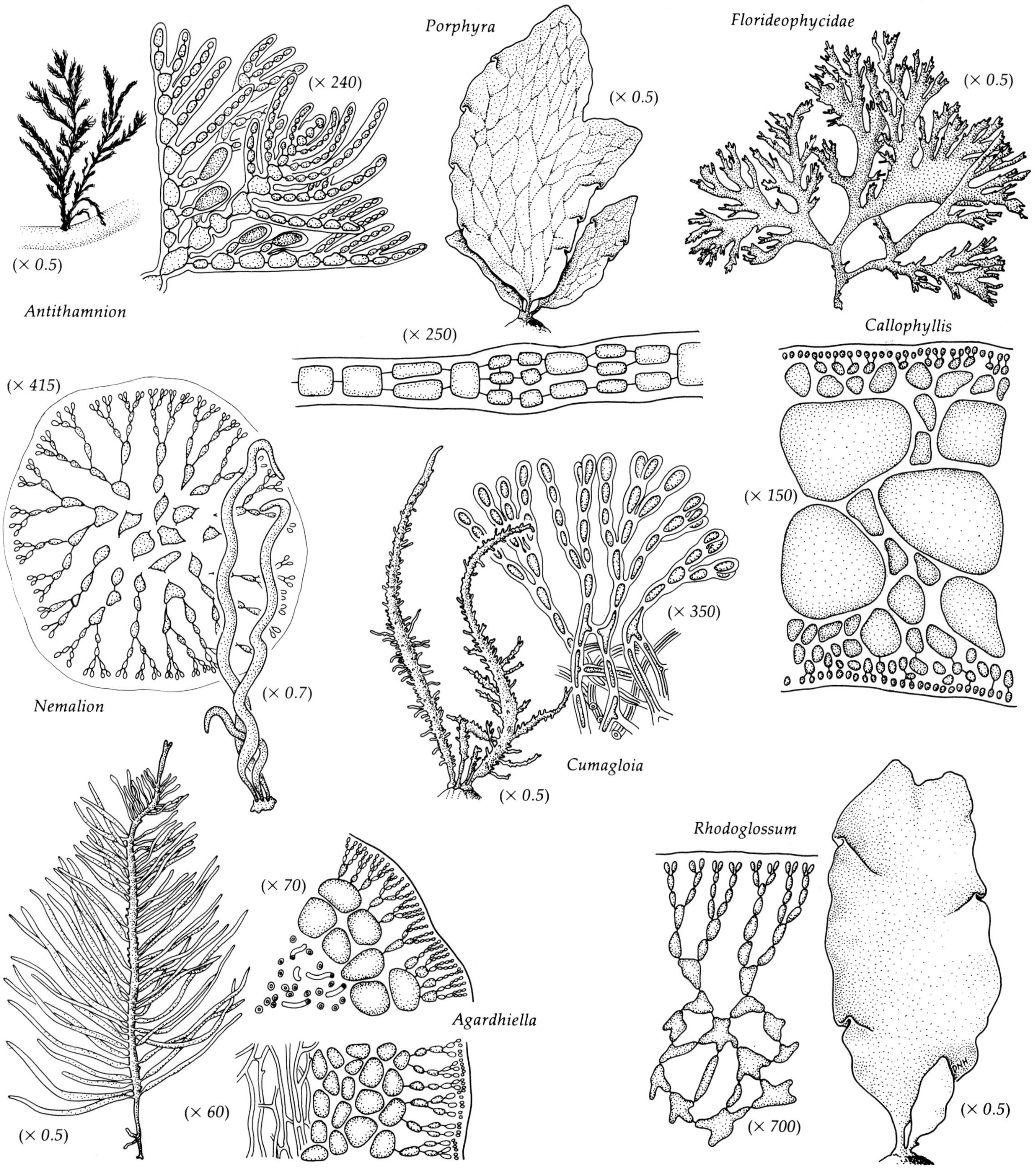
Antithamnion
(× 0.5)
(× 240)
Porphyra
(× 0.5)
(× 250)
Florideophycidae
(× 0.5)
Callophyllis
(× 150)
(× 415)
Nemalion
(× 0.7)
Cumagloia
(× 350)
(× 0.5)
(× 0.5)
(× 70)
Agardhiella
(× 60)
Rhodoglossum
(× 700)
(× 0.5)

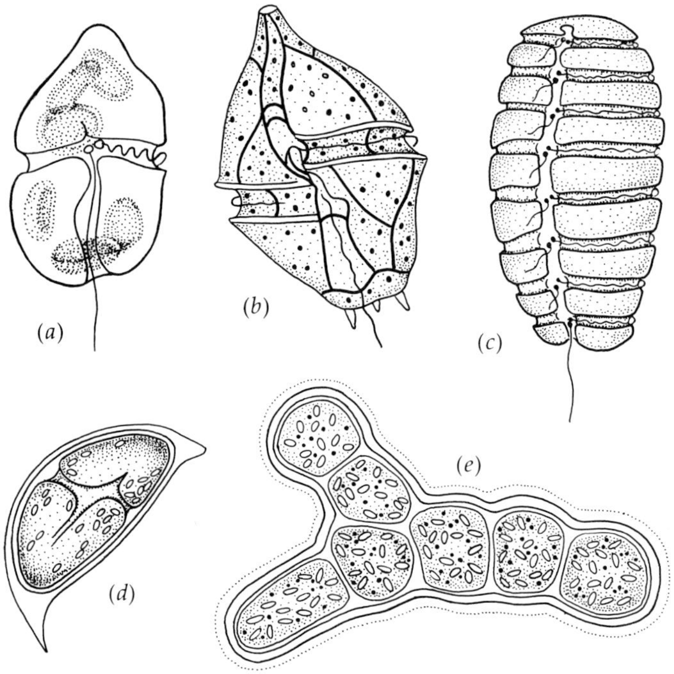

18–16 The morphological diversity in Pyrrophyta. (a) and (b) are motile single-celled species, while (c) is a motile colonial species. Single-celled and colonial nonmotile species are both shown in (d) and (e).

lates, Fig. 18–16) and the Cryptophyceae. Recent work indicates that the latter may be so distinct as to be recognized as a division of its own.

Most genera are unicellular motile flagellates. The cells are biflagellate, and in dinoflagellates one flagellum is ribbonlike, encircling the cell in a transverse groove, while the second trails posteriorly in a longitudinal groove.

The cells are uninucleate. The chromosomes in the dinoflagellates are most unusual for eucaryotic cells—they consist only of DNA but no protein. Thus they resemble the genetic material of procaryotic cells and appear to have a similar ultrastructure. No centromeres are present, and there is the same direct division of the DNA found in procaryotic cells. A well-defined nuclear envelope is present, but a most unusual mitosis is found. Early in division, the spherical nucleus is invaded by deep cytoplasmic invaginations that become essentially tunnels of cytoplasm through the nucleus. The nuclear envelope remains intact, and the tunnels, which transverse the nucleus parallel to the division axis, contain 8 to 20 microtubules. The nucleus now elongates and becomes dumbbell-shaped. The chromosomes, which have duplicated earlier, are attached to the nuclear envelope and separate, not necessarily in unison, as the nucleus elongates. At all times the nuclear envelope remains intact. Eventually, the nucleus pinches in two, and separate daughter nuclei are formed. At no time are the fibers inside the nucleus or attached to the chromosomes. A most remarkable and, probably, primitive division mechanism.

Chlorophyll a and c, as well as a number of carotenoid pigments, are present in the chloroplasts in which the thylakoids are in groups of three in the Dinophyceae and two in the Cryptophyceae. The Cryptophyceae also contain phycobiliprotein pigments similar but not identical to those found in the Cyanophyta and Rhodophyta. Here, however, they are not arranged in spherical particles on the *outside* of the thylakoids, as in the blue-green and red algae, but uniformly *inside* the thylakoids.

19

The Fungi: Water Molds, Bread Molds, and Mildews

The Fate of Frogs

The frog that ate the insect that ate the algae is dead. Its body floats in the pond over which it once ruled. But even in death there is life because the dead frog rapidly becomes the source of food for the many aquatic fungi that also live in the pond. Thus the frog is returned to the ecosystem, and the cycle of life continues. In all parts of our environment, the fungi are important agents in the breakdown and reutilization of organic material. Each is adapted to its particular niche, and those that live in the pond have developed particularly effective means of invading new hosts by using swimming zoospores. This photo essay will introduce various aquatic fungi to illustrate some of the unique features of the zoospores found in these organisms.

A *A dead frog covered with a lush growth of fungal hyphae, in this particular case,* Saprolegnia. *The hyphae grow into the body of the frog, breaking down the dead cells and using the resulting products for further growth and development.*

B *Eventually, cytoplasm and nuclei will accumulate in the ends of some of the hyphae, and zoospores will be produced. Such sporangia can produce large numbers of spores that are rapidly discharged when mature. This series of photographs of the fungus* Dictyuchus *was taken at 3 min. intervals and gives a good impression of the rate of discharge.*

A

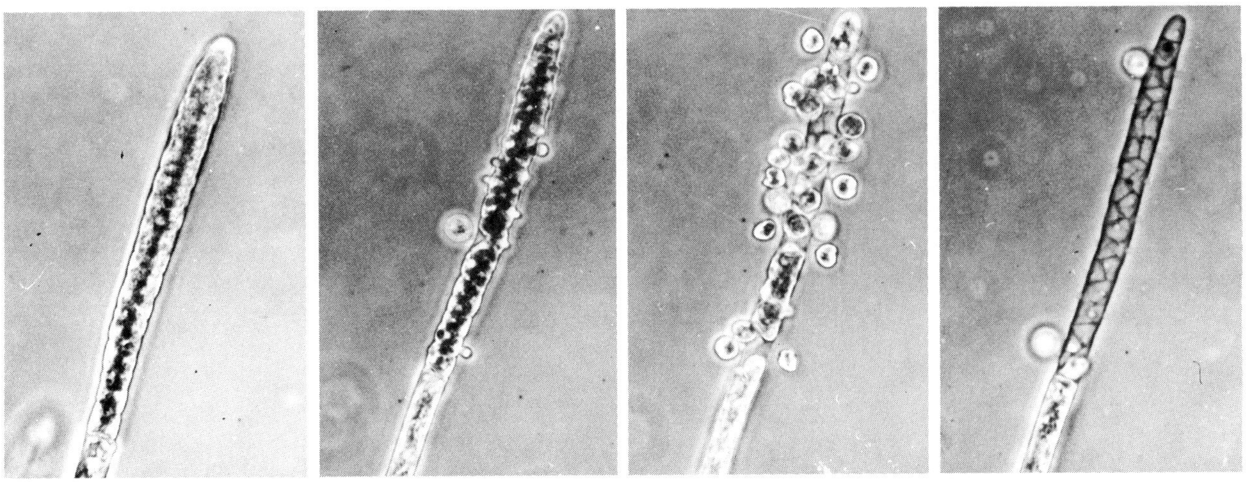

B

C The zoospores that are released are small, actively swimming cells. This beautiful photograph of a living zoospore of Allomyces *shows the single flagellum, the single nucleus (the larger, round light area containing a dark nucleolus near the point of attachment of the flagellum), and the surrounding nuclear cap.*

D An electron microscope photograph of the same type of zoospore (Allomyces) *shows the compact nature of its structure. Mitochondria surround the tip of the nucleus and the nuclear cap, which consists of tightly packed ribosomes. The dark spheres at the bottom end of the cell contain stored food.*

E An electron microscope photograph at higher magnification (of a different section) shows the relationship of the base of the flagellum to the nucleus. A large mitochondrion circles the base of the flagellum and in section appears to be two mitochondria.

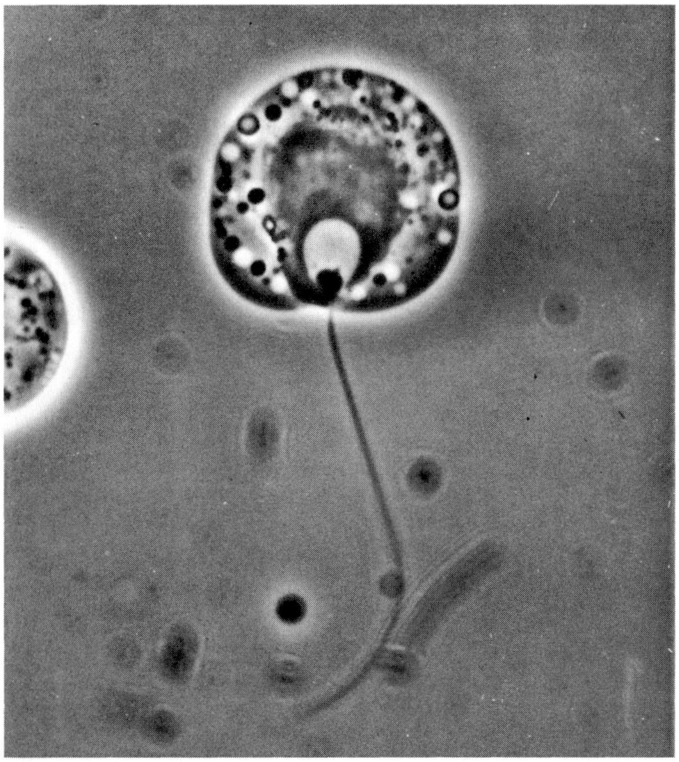

C

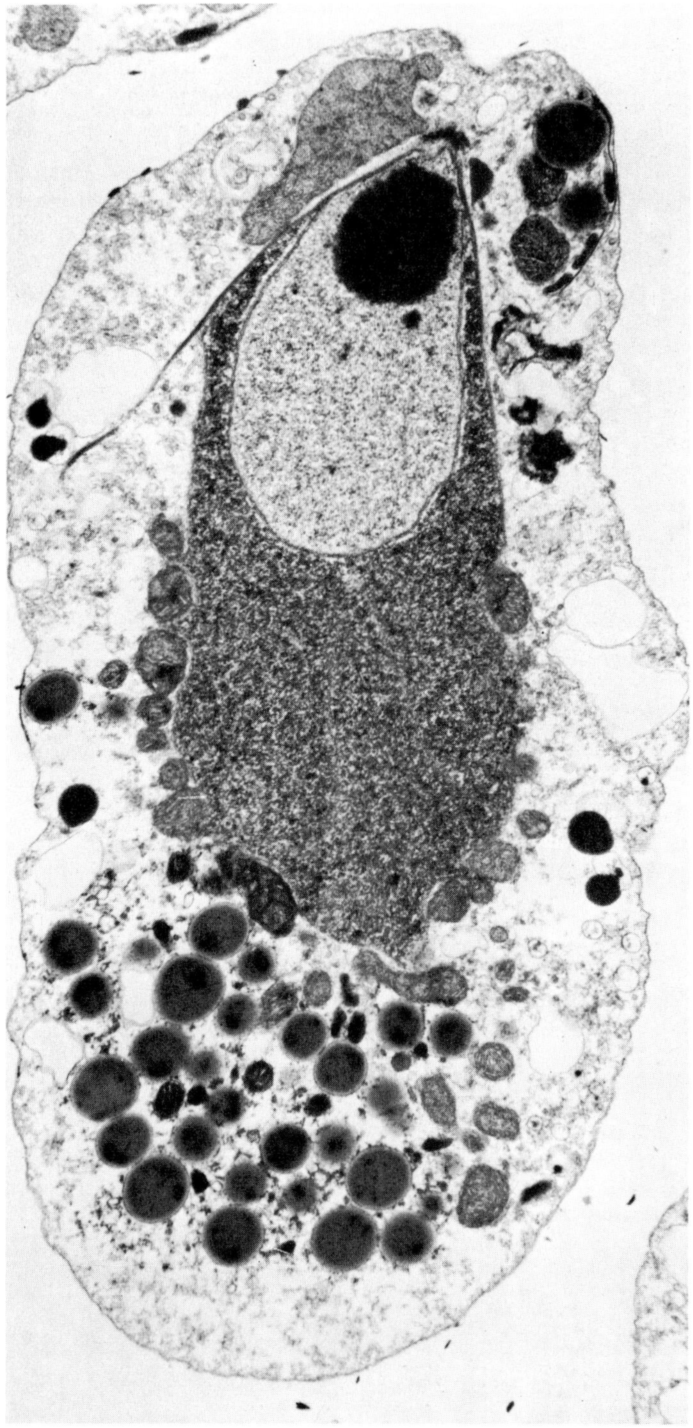

D

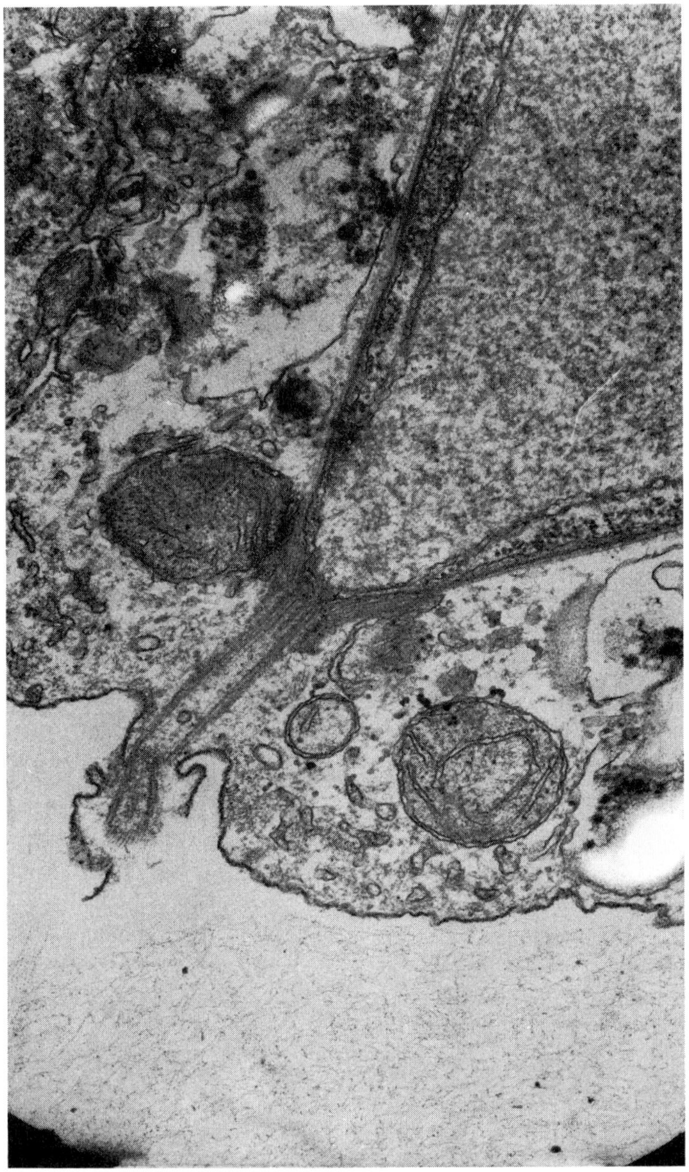

E

A leaf drops softly to the forest floor. Its role as a photosynthetic organ is over, but it is still an important part of the ecology of the forest. In the leaf are the many elements and compounds that were accumulated or synthesized by it during its period on the tree. If these materials remained in the leaf, they would be removed from the ecosystem and be unavailable to other organisms. But they are not. The leaf, or for that matter any dead organism, is a natural seed bed for the germination of fungal spores.

Fungi as Saprophytes

The spores of fungi surround us at all times. They are present in the air, in the soil, or on almost any surface that you can imagine. These minute units are frequently, but not always, single celled. They may be surrounded by thick walls, which help them survive desiccation, or they may be surrounded by fairly thin walls, which enable them to germinate readily when they reach a proper substrate. A leaf falling to the forest floor will quickly be covered with a variety of spores, some of which will germinate within minutes. When the spores germinate, they form tubular structures known as hyphae. These consist of a cell wall and cytoplasm containing such cell organelles as mitochondria and nuclei. Most of the cytoplasm is localized near the tips of the hyphae, and the more basal regions may be filled with large vacuoles (Fig. 19–1). Plastids are missing because the fungi are not photosynthetic and must therefore receive their energy from other sources. As they grow, the hyphae branch, forming a network that radiates outward from the spore. Under ideal conditions the hyphae will form an almost perfect circular colony as it grows over a flat surface (Fig. 19–2). In nature, of course, such growth is rarely seen, since the hyphae spread over the irregularities of the surface and between the units making up the leaf. The mass of hyphae formed by this growth is a mycelium (Fig. 19–2).

As the hyphae grow over the leaf, they secrete enzymes that break down the polysaccharides and other organic materials remaining in the leaf cells and

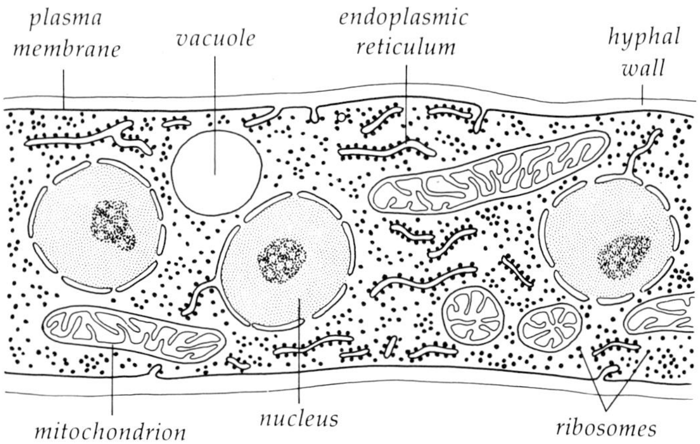

19–1 The ultrastructure of most fungal hyphae is simple: the most conspicuous parts are the nuclei, mitochondria, ER, and ribosomes.

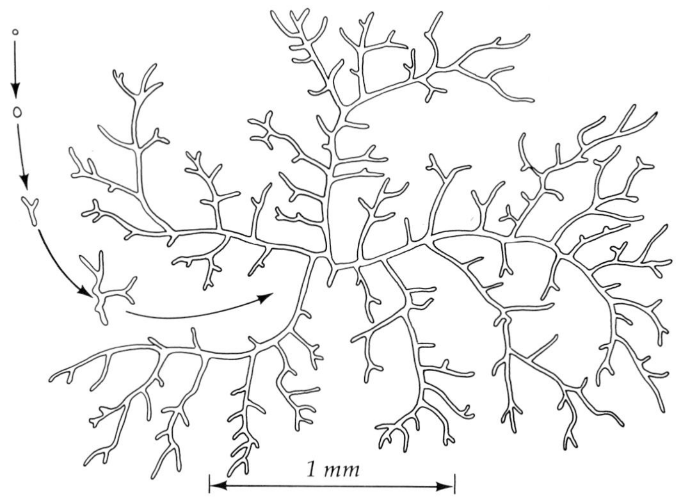

19–2 The growth pattern of mycelium, starting with a single spore.

absorb the more soluble compounds resulting from this breakdown. The fungi must live off the energy stored in the leaf because of its photosynthetic capacity. In utilizing the leaf's energy-containing compounds, the fungi return CO_2 to the atmosphere through the respiration taking place in their mitochondria. The first fungi to appear on the leaf are those whose spores germinate rapidly and have biological systems that are able to rapidly attack the sugars and starches present in the leaf cells. Later, additional spores will germinate, producing mycelia that will be able to digest the cellulose and lignin of the cell walls of the leaf cells. Some of the mycelia that attack the cell walls of the leaves may not, in fact, come from the germination of spores on the leaf but from mycelia that are always present in the upper surfaces of the forest soil.

The upper layer of soil under a forest or a grassland contains masses of fungal hyphae. These mycelia exist through the decomposition of the leaf material and other discarded, dead plant parts. Usually, in the breakdown of these materials, the fungi operate in close relationship with insects. When a tree falls in a storm or when branches are broken off, contributing to the litter on the forest floor, both are rapidly invaded by wood-boring insects. With these insects come the mycelia and spores of the wood-rotting fungi.

As the fungi break down the compounds present in the cells of the leaves and stems of the higher plants, some of the material is utilized by the fungi, and some is made soluble and utilized by other organisms, including other higher plants. In turn, the fungi die, the mycelia are decomposed by bacteria, and the substances that they absorbed and synthesized are again returned to the general ecosystem.

The type of nutrition that we have been discussing is based on the utilization of dead material. Plants that employ this type of nutrition are called saprophytes. From the overall perspective of an ecosystem, saprophytes are generally well-mannered and important members. They break down dead organisms, allowing the materials present to be recycled and reutilized. Unfortunately for man, a fungal saprophyte grows equally well on a log lying on the moist forest floor or on a supporting beam of a floor over a moist basement. Likewise, it utilizes the oak in a forest and the oak that is part of a sailing ship. In addition, many saprophytes can also use such man-made substrates as the insulation on an electrical wire.

Over the centuries, man has learned how to cope with wood-rotting fungi in building his structures. He has learned that adequate ventilation produces conditions that are too dry for fungal development. But there are times when good ventilation is difficult to achieve. One such situation is in the hulls of wooden sailing ships. England, dependent as she was on sea power and a navy composed primarily of oak ships, had a terrible problem with fungi that caused dry rot of the wood. In times of peace, when the ships were lined up side by side along the wharfs, the Navy Yard was known as "rotten row." Samuel Pepys, a noted diarist and Lord of the Admiralty, recounted how, in 1668, he saw mushrooms growing so thickly in the bottoms of Man of Wars that they had to be shoveled away. More ships of the British Navy were sunk by dry rot during the Revolutionary War than by the Americans. In one incident the HMS King George was heeled over for minor repairs below the water line when the whole bottom dropped out. A hundred men, including the Admiral, lost their lives in the accident. As the ships rotted, new ones were built to replace them, using the oak forests of England. Log by log, and tree by tree, the great forests dwindled until they were no more—gone into rotting ships! By the 1860s the situation was becoming critical, but the introduction of iron hulls saved the British Navy—and England's few remaining oak trees.

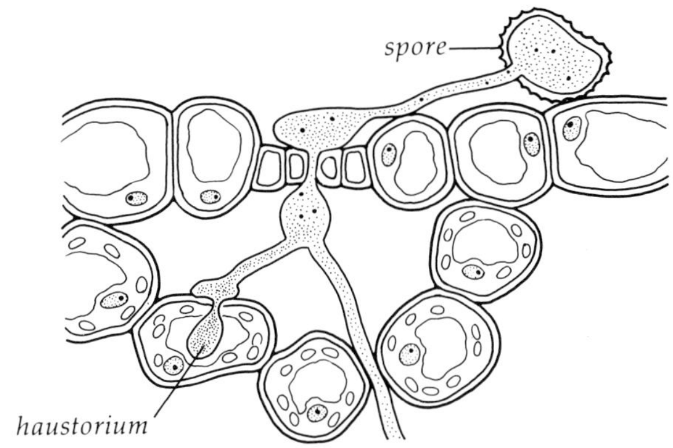

19–3 The germination of a spore on the surface of a leaf, and the penetration of the leaf (through the stomate) by the hypha. The hypha has developed a haustorium into one of the leaf cells.

Fungi as Parasites

Not all fungi are saprophytes. Some go one step further and attack not dead but living plants. In these cases, a spore falling on a leaf still attached to a plant begins to germinate and send out the usual hyphae (Fig. 19–3). What happens next is highly variable, depending on the species of fungus involved. In some cases, the hyphae remain on the surface of the leaf or stem, parasitizing only the outer layer of cells of the plant (Fig. 19–4). In others, as shown in Figure 19–3, the hyphae enter through the stomatal openings in the leaves and gain entrance into the plant. In still other cases, wounds that break the surface layer of the leaf are necessary to allow the fungal hyphae to penetrate deep within the organism. In addition, the hyphae, as they grow on or through the host plant, send out extensions into the living host cells (Fig. 19–4). These extensions or haustoria (singular: haustorium) absorb carbohydrates and other materials from the host cells. Eventually, the demands of the fungus overwhelm the host cell and cause its death. Depending on the amount of the infection and the growth rate of the fungus (along with the demands made on it by the fungus), the host plant may remain alive with minor damage for years or be killed within a few days.

Only a small percentage of the thousands of fungi is parasitic, but they have caused man a tremendous amount of difficulty. One of the most widely quoted examples (of great importance to both English and American history) is the potato disease called the late blight. The potato, which originated in South America, was introduced into Europe in the sixteenth century. Eventually, it found its way to Ireland, where it soon became almost the only crop in much of the country. The reasons for this were many. As a crop, it needs relatively little work and produces good yields (some 3 to 10 tons per acre). Moreover, potatoes are an excellent food and can be easily prepared in many ways. Soon, poor Irish farmers were eating 8 to 14 pounds of potatoes a day and little else. With a reliable food supply and consequent free time, the birthrate soared.

Then, in 1845 a disease appeared in the potato crops (first in Europe, then in England, and finally in Ireland). This disease, called the late blight of potatoes and caused by *Phytophthora infestans,* doomed both the field crop and the stored potatoes in Ireland. During seven days the following summer, the entire potato crop was destroyed, and one of the longest, most disas-

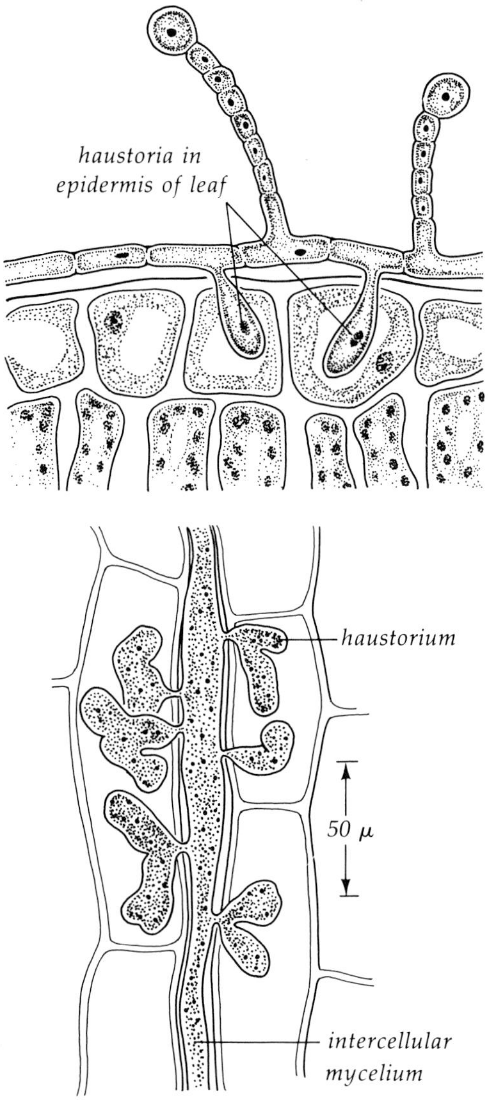

19–4 Haustoria take various forms in different fungi.

trous famines in modern times began. Between 1845 and 1851 almost a million people died in Ireland, and almost a million and a half migrated to the United States and Canada.

The social impact of the potato famine was great. The large number of Irish immigrants changed the complexion of American society. Another effect was the opportunity it gave John Peale, then Prime Minister of Great Britain, to repeal the corn laws, which were restrictions placed on the import of cheap grain into England. Although England was in the midst of the Industrial Revolution, it was still largely an agricultural nation. The landowners and farmers wanted produce kept high and cheap imported food kept out. The industrialists wanted cheap food for cheap labor. The repeal of the corn laws furthered the industrialization of England and the economic development of the United States. The westward migration to the Great Plains was stimulated by the price increase of grains. The demand for grain to export led, in turn, to the development of the railroad system, much of which was built by the Irish immigrants brought to America by the potato famine.

Divisions of Fungi

Thus far we have treated the fungi as if they composed one great group or division of plants. In fact, it was long considered valid to lump the algae and all of the fungi together into a division called the Thallophytes. This was based on the organization of their vegetative structures. Since they lacked specialized transport cells and did not have elaborate reproductive structures, they were considered to be more similar than the groups above them. Later it became clear that this was not a valid basis for distinction, and the fungi and algae were separated.

The fungi were divided into two groups: the true fungi or Eumycophyta and the slime molds or Myxomycophyta. Most of the fungi belong to the large group Eumycophyta, which was then broken into four classes based on reproductive structures: the Phycomycetes, the Ascomycetes, the Basidiomycetes, and the Deuteromycetes. The latter are also called the Fungi imperfecti and are not a natural group at all. They are simply a collection of fungi whose sexual phase cannot be identified and therefore cannot be placed in one of the other classes. This wholly artificial class is only a convenience to those working with the organisms.

In recent years it has also become apparent that the Phycomycetes is also an unnatural association of organisms but for quite different reasons. As research has continued in the Phycomycetes, it has begun to appear to many mycologists that the various groups making up this class are so distinct from one another that they should all be given equal status. Hence the trend has been to discard the class Phycomycetes and to replace it by six separate classes. Of these, three are the more important, both biologically and to man, and will be the only ones treated here. They are the Chytridiomycetes, the Oomycetes, and the Zygomycetes.

Both the Ascomycetes and the Basidiomycetes are well-defined classes and will probably not be broken up as were the Phycomycetes. In terms of numbers, the Ascomycetes are by far the greatest, followed by the Basidiomycetes, and then the various classes that formerly made up the Phycomycetes.

Now, let's look at the major classes of fungi, comparing the characteristics of each of these groups and the importance of their positions in the ecosystem.

Chytridiomycetes

The Chytridiomycetes are water molds characterized by motile cells that possess a single, posterior, whiplash flagellum. This class is also characterized by cell walls composed predominantly of chitin. Three major cell-wall compounds are found in fungal walls: cellulose, chitin, and glucans. Cellulose, discussed earlier, is found in walls throughout the plant kingdom.

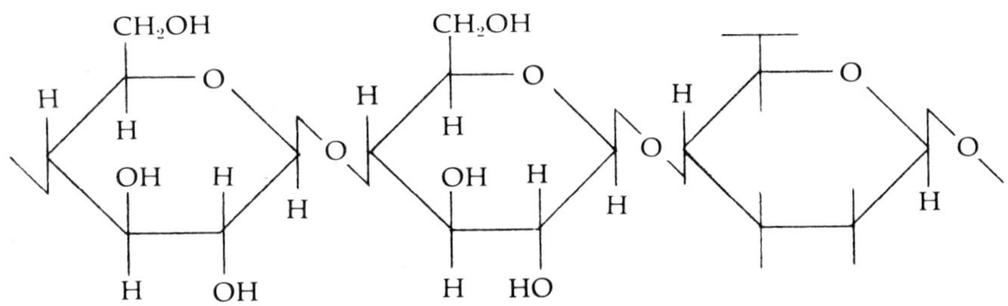

Cellulose

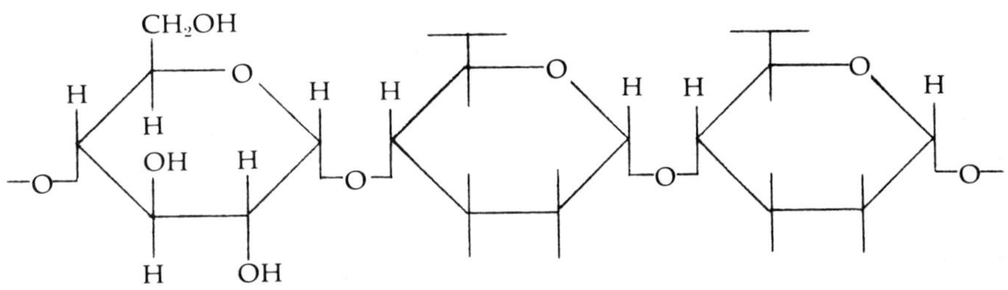

Starch (amylose unbranched, amylopectin branched)

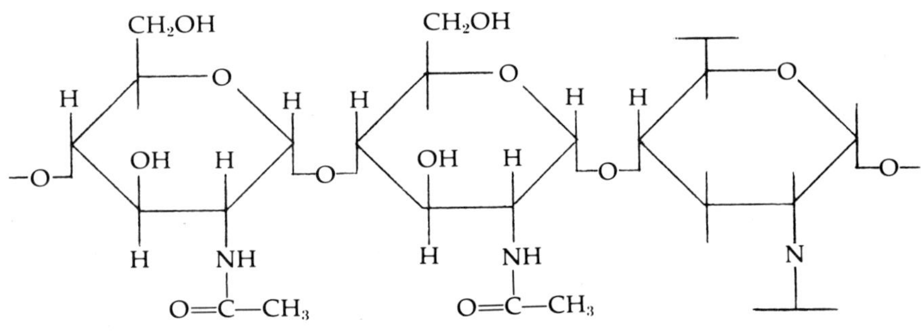

Chitin

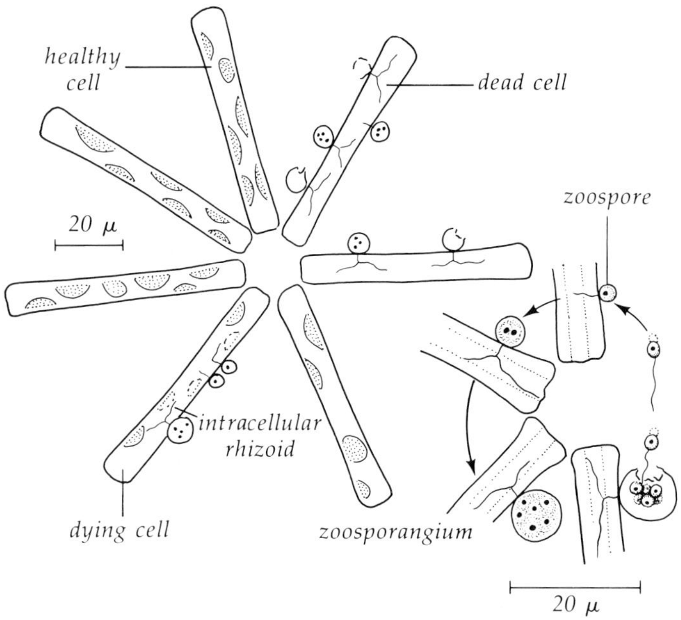

19–5 Rhizophydium planktonicum, *a chytrid that is a parasite on diatoms.*

Chitin is much like starch, but the basic unit is a glucose amine, which contains nitrogen. Glucan is also a polysaccharide and, like cellulose, has glucose as the basic repeating unit. The way in which the glucose molecules are joined together is different than in cellulose, resulting in quite a distinct compound.

Rhizophydium—the plankton eater. Most, but not all, of the Chytridiomycetes (chytrids, for short) are extremely small, single cells, although some are filamentous. Many chytrids are parasites of filamentous and unicellular algae, as well as other fungi—including other chytrids. One example of a tiny parasitic chytrid is *Rhizophydium planktonicum*, which uses a planktonic diatom (*Asterionella formosa*) as its host (Fig. 19–5).

The naked motile zoospores, swimming by means of their posterior whiplash flagella, attach themselves to the diatom cell. The settled zoospore forms a wall and grows a fine haustorium into the host cell. Several zoospores may attach themselves to the same diatom, which they rapidly kill by draining it of its metabolites. The attached spherical zoospore enlarges, and a number of new, naked zoospores are produced. The new zoospores are released when a lateral portion of the original wall dissolves. This chytrid is common in many lakes and is quite specific about its host—it grows on only one species of diatom.

Allomyces—a filamentous water mold. Not all chytrids are as small as *Rhizophydium* and not all are parasites. *Allomyces*, for example, is a filamentous saprophytic chytrid common in tropical climates (Fig. 19–6). It grows on the surface of decaying plant matter. The mycelium has hyphae that penetrate the substrate and absorb nutrients. Other hyphae extend away from the substrate, and reproductive structures are produced on these.

The cytoplasm in the hyphae of *Allomyces* is multinucleate and normally not divided into cells by cross walls. When reproduction is about to occur, cross walls form and gamete-producing structures or gametangia are formed. Male and female gametangia are always borne on the same filament. The male gametangia are orange colored, while the female gametangia are colorless. The gametes are formed through cleavage of the multinucleate cytoplasm into uninucleate male or female gametes. The mature gametes escape the gametangia through one or more pores that develop in the wall of the gametangia. The male gamete or sperm is small, orange, and swims rapidly through the water. The female gamete is larger than the male, colorless, and swims less actively than the male. The female gamete attracts the male by secreting a hormone called sirenin. (For a discussion of research on sex hormones in fungi, see End Paper 19–1.) The hormone is removed from the solution by the male gametes and inactivated. How the male gametes sense the hormone and react to a concentration gradient to reach the female gamete is unknown.

When the male gamete finds the female gamete, they fuse and a biflagellate zygote results. Each gamete has the haploid number of chromosomes (n); the zygote is therefore diploid ($2n$). For a while, the zygote swims by means of its 2 posterior flagella, but it eventually settles down on a suitable substrate, withdraws its flagella, and germinates into a new mycelium. This mycelium is similar in appearance to the plant that produced the gametes, except that its nuclei are diploid and the former were haploid. *Allomyces* therefore shows an alternation of generations that are isomorphic (identical in form).

The diploid generation also has reproductive structures, but these produce spores and are called sporangia.

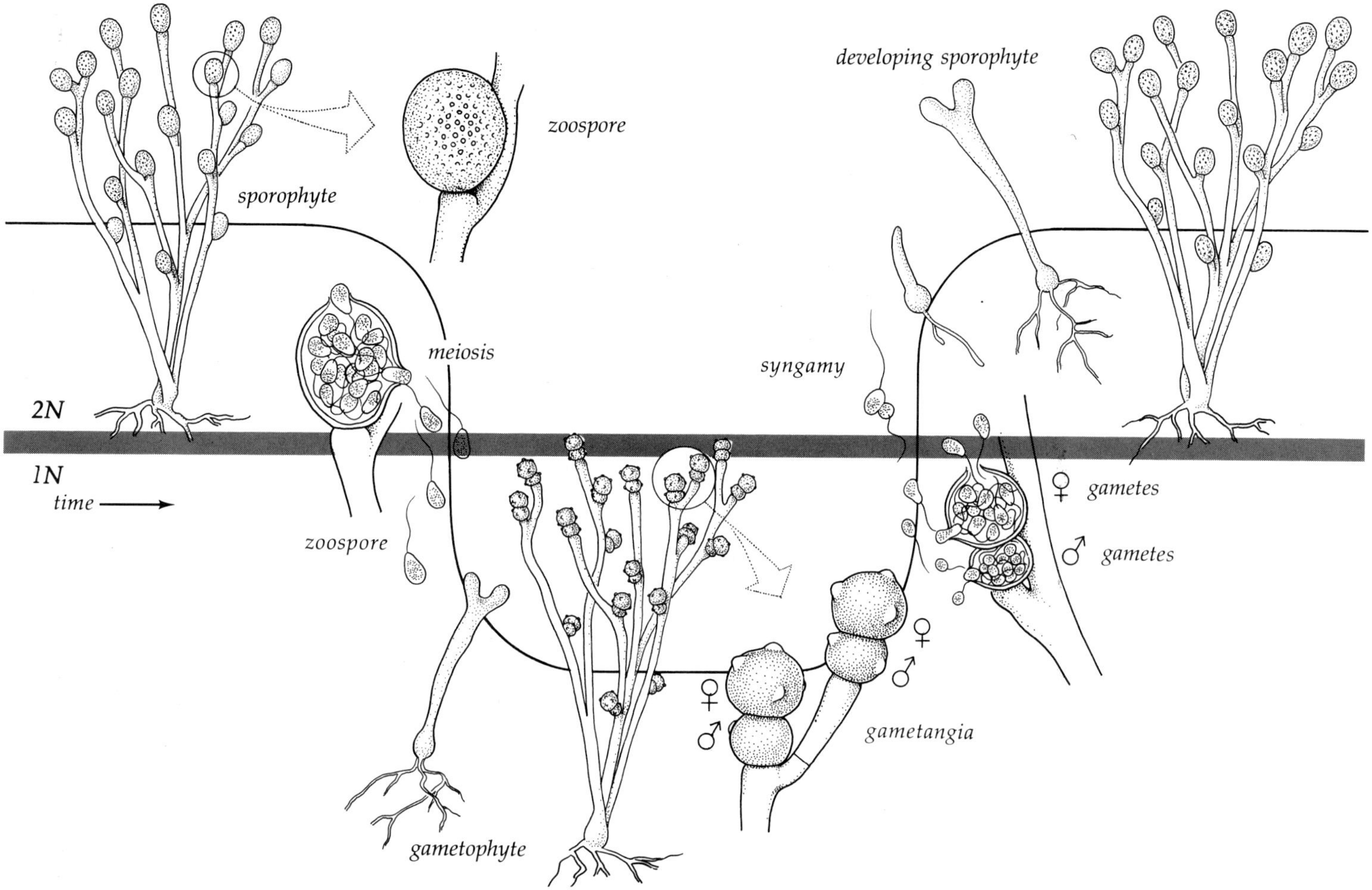

19–6 *The life history of* Allomyces.

Actually, two types of spores are produced: one is formed following a mitotic division and the nuclei are therefore diploid, while the second is formed following meiosis and its nuclei are haploid. Both spores are motile, possessing the single posterior whiplash flagellum characteristic of the Chytridiomycetes. When the diploid zoospores germinate, they give rise to diploid plants similar in all respects to their parent plant. The haploid zoospores produced by meiosis form a plant similar in appearance to the parent but containing haploid nuclei and capable of differentiating male and female gametangia and producing male and female gametes.

None of the motile cells (the gametes, zygote, or the two types of spores) are resistant to drying out; yet cultures of the plant could be established using dry soil. The spore that establishes the culture under these conditions is always a haploid zoospore, but such spores cannot survive the conditions present in the dry soil. In this case, it is not the spores but the sporangia, which produce the haploid zoospores, that are resistant to drying. Each resistant sporangium usually contains 12 diploid nuclei. When conditions are favorable, each of the nuclei undergoes meiosis to form 4 haploid zoospores. They emerge from the ruptured sporangium and swim around until they make contact with a suitable substrate, eventually producing haploid plants. Only when these plants form gametes and zygotes will diploid plants be found growing in the culture.

This sequence of events illustrates how well *Allomyces* is adapted to an aquatic environment in which periodic drying takes place. The resistant sporangium means that the plants can survive the dry periods. The fact that this is the diploid sporangium that will produce haploid zoospores means the plant can complete its sexual cycle rapidly (even if water is available for only a short period), thus securing for the species the evolutionary advantage of sexual reproduction. The fact that the diploid plants with their resistant sporangia and diploid zoospores are produced next means the plant is prepared for another dry period or can exploit the aquatic environment as long as water is available.

Essay 19–1 The Best Laid Plans of Mice and Men . . .

What makes a man go into science? What makes him study to be a botanist rather than, say, a chemist? One man's answer is presented here. He is Edward Cantino, now a professor of botany at Michigan State University, and well known for his work on the development and biochemistry of the water molds.

"Like many of my colleagues in science, I have experienced the unique kind of enjoyment that comes from planning a scientific experiment, setting it up, watching it unfold in the laboratory, and then finding that a bull's-eye was scored—or, at least, that results were more nearly right than wrong! But there is also another kind of gratification, more profound and deep-felt, that comes from being exposed to the personality of a very special individual and sharing with him his fascination and excitement with science. Those of us who can gratefully acknowledge having had this kind of satisfaction know first-hand how much it can contribute to—if not, in fact, create—the early glow, the first absorbing eagerness, that actually launches a scientific career. It can be one of the very richest of experiences, never to be forgotten.

In grammar school, I had always wanted to be a chemist. At St. Mary's High in the San Francisco Bay Area, I traded hard-to-get postage stamps for vials of chemicals to do 'experiments' and looked to the day when I would enroll at the University of California and study to be a chemist. In due time it came about, and with it four zestful years of long hours at the lab bench, watching Hildebrand's masterful performances in the chem auditorium, pitching pennies in the quant lab with Latimer, getting better-than-average yields in the synthesis of cinnamaldehyde for Steward. If my future was in science, undeniably it would be right there, amidst the cacophony of exploding test tubes and the balmy exhaust of sulfide generators in those incomparable, 'temporary,' pre–World War I laboratories!

But, then, one day I made a reckless move: I signed up for a course on an entirely unfamiliar subject—botany. It was being taught in an edifice called LSB, situated far from civilization in an uncharted eucalyptus forest at the distant end of the campus where few chem majors had set foot. After the first weeks in the course, a new teacher (he seemed hardly older than we students listening to him) took the podium to present a few lectures on something called 'alternation of generations' in lower plants. His name was Ralph Emerson, and he was to refashion all my plans! He introduced me to a whole new world of minikin creatures known as 'water molds,' and to the breed of biologist—'mycologists' by name—who had become addicted to them. He showed me how unusually multifarious these primitive fungi were; how they manufactured tiny cells with tails, with which they got around; and how to find the natural quarters where they went to rest and nest. He told me how such organisms could be hunted down and waylaid with dead flies, or bits of snakeskin, or fruits confined in little wire cages submerged in ponds and streams; how they could be brought into the mycological arena, herded into culture chambers, and domesticated; and how, after being rehabilitated, they had revealed their deepest secrets to the eager eyes of those aforementioned mycological peeping Toms whose happy lot it was to watch what they were doing—secrets about their development and vitality; how they reached full bloom; their unconstrained cohabitations and seemingly incestuous affairs; and many other tantalizing things about their curious and wondrous ways. I had been hooked by the inimitable personality of the water mold!

And so, through the force of his enthusiasm, his drive, his insight into the biological world, Ralph Emerson became my mentor—and I, his first graduate student. It led to a close and incomparable association that embodied my most unforgettable years in science, exciting years that indelibly marked the path I would follow in biological research for the next quarter century. (To this brief chronicle, one postscript seems appropriate: after leaving Berkeley in 1948 for my first job at the University of Pennsylvania, I isolated from a pond on the campus an undescribed water mold and subsequently christened it Blastocladiella emersonii; *this specific appellation, it seemed to me, would help ensure that the irresistible mycologist who had met the immovable chemist head on, and thereby promptly shattered a promising career, would not soon be forgotten!)*

If anything good has come from my 23 years of studying the antics of B. emersonii, *it has been in no small measure because I had the good fortune to have been inspired and excited by a major professor's love of experimental mycology and his consuming devotion to it. This does not happen nearly as often as it should to a young man or woman who is just beginning a career in science, but it is something well worth hoping for. And when, later on, the pupil-turned-professor has a graduate student of his own with whom, in turn, he shares his work, . . . well, that is partly what being a scientist is all about."*

Oomycetes

The *Oomycetes* are also morphologically a highly diverse class of fungi (water molds and mildews), containing both saprophytic and parasitic species as well as aquatic and terrestrial forms. One of the major characteristics of the class is the presence of zoospores lacking cell walls and possessing 1 whiplash and 1 tinsel flagellum. Both flagella are placed anterior to the cell. Another significant feature is the presence of cellulose and the absence of chitin in the wall.

Saprolegnia—the fly fungus. This is an example of a saprophytic, aquatic member of the Oomycetes (Fig. 19–7). This is a widely distributed fungus and can be easily captured by dropping a few dead houseflies in a dish of freshly collected pond water (almost any pond or stagnant ditch will do). Within a day or two each fly will be covered with a soft, white, fuzzy growth that is

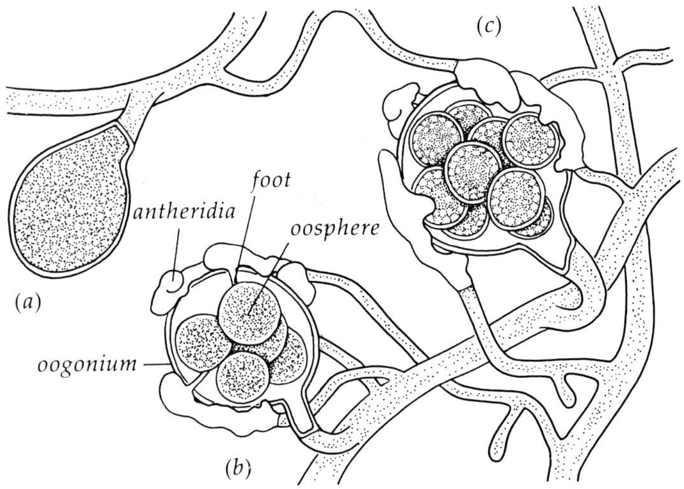

19–7 Sexual reproduction in Saprolegnia. *A young female gametangium is shown in (a), where the spherical eggs will develop. In (b) the male gametangia have grown to the female gametangium, and fertilization tubes have grown to the eggs. The spherical zygotes are shown, before their release, in (c).*

the mycelium of *Saprolegnia.* When viewed with a microscope the hyphae can be seen to be multinucleate and not divided by cross walls. Some hyphae penetrate the surface of the fly, while others project into the water. The ends of some of these latter filaments are cut off by cross walls. The cytoplasm in these portions becomes cleaved into uninucleate single cells that develop two flagella—typical of Oomycetes. These cells are motile zoospores, called primary zoospores because, after a period of motility, they withdraw their flagella and form a cell wall. In this state they are believed more resistant to unfavorable environmental conditions. Later, under the proper conditions, this cell will give rise to a single biflagellated zoospore, termed the secondary zoospore, which has 2 flagella (again, 1 whiplash and 1 tinsel) laterally inserted in the cell rather than at the ends. The secondary zoospore swims to a new substrate—another dead fly, other insect, or other animal—on which it can grow into a mycelium.

As long as there are adequate nutrients available, the mycelium continues to grow vegetatively and produce primary zoospores. When the source of food begins to decrease, the hyphae begin a series of changes that result in sexual reproduction. The female gametangia are almost spherical in shape and are partitioned off from the remainder of the hyphae by a cross wall. Within the female gametangia are differentiated a number of uninucleate eggs; each is spherical. The hyphal branches that will differentiate into the male gametangia grow toward the female gametangia and eventually make contact with it. Each hyphal tip becomes isolated from the rest of the plant by the formation of a cross wall. Male gametes are formed within the male gametangia. Next, fertilization tubes grow from the male gametangia into the female. The male gametes pass through these into the female gametangia, come in contact with the large spherical egg, and syngamy takes place. A thick wall develops around the zygote, and the dormant zygote becomes a resistant stage in the life cycle.

When the zygote germinates, it first swells and its wall becomes thinner. A short,clublike hypha is formed. Eventually, the tip is sealed off by the formation of a cross wall, and a small sporangium is formed. Zoospores are formed and released, starting the whole cycle anew.

Until recently it was not possible to say whether the nuclei of the hyphae in *Saprolegnia* were haploid or diploid. Clearly, the zygote nucleus is diploid, but did meiosis occur when the zygote divided or when the gametes were formed? If meiosis occurred when the zygote divided, all the nuclei would be haploid as in many freshwater algae; if meiosis took place at gamete formation, the hyphae would all be diploid. A major problem facing mycologists is the small size of fungal nuclei, usually 2 to 3 μ. Observations on the divisions giving rise to the egg and sperm nuclei had been contradictory: some investigators claimed they were mitotic, while others believed they were meiotic. Confirmatory evidence for the occurrence of meiosis during gametogenesis came from measurements of DNA content. These showed that the amount of DNA in the nuclei of the vegetative hyphae was twice that present in the gametes; consequently meiosis must be taking place during the formation of the gametes.

Downy mildew—the deadly Oomycetes. Not all Oomycetes are relatively harmless saprophytes. Some are effective and deadly parasites. These include *Phytophthora infestans,* the late blight of potato, and *Plasmo-*

para viticola, the downy mildew of grapes that has caused tremendous losses in the vineyards. *Peronospora parasitica* causes downy mildew on a wide range of hosts and is a pest commonly found on plants in home gardens. *Albugo candida* is the white rust on cabbage and related plants.

Plants become infected with downy mildew when airborne spores land on the young leaves and germinate in a drop of dew or rain. Upon germination the spore may give rise to a biflagellate zoospore, which forms a wall and then in turn germinates to form a sporangium. Alternately, the spore may directly form a hypha. The hypha grows along the surface of the leaf and into it through the stoma. Once within the host, the hyphae form a myçelium of unbranched hyphae that are multinucleate and lack cross walls. Large, branched haustoria are put forth into the host cells, which absorb nutrients from the cells for use by the fungus. In the more virulent mildews, the host cells are rapidly killed by this process. In the milder forms the host cells do not die, and the plant survives.

After the period of vegetative growth, the fungus develops straight hyphae that emerge from the stomata and grow outward. The ends of these hyphae branch several times, and the tip of each branch enlarges to form a spore. These are called conidia spores and the structure that produced them conidiophores (Fig. 19–8). The masses of conidiophores on the diseased plant give it a mildewed appearance.

The conidia spores are a most effective means of asexual reproduction and can rapidly carry a disease through a crop of plants. Sexual reproduction is also found in the powdery mildews and is similar to that found in *Saprolegnia*. The male and female gametangia are produced within the larger intercellular spaces of the host. The thick-walled zygote is the resting, resistant stage and is released only after the decay of the host plant. When the zygote germinates, it may either divide and give rise directly to zoospores or first produce a short filament, in the manner of *Saprolegnia,* and then form zoospores.

The downy mildew of grapes was almost as destructive of the vines in France as the late blight was of potatoes in Ireland. The disease was unknown in Europe, until it was introduced from the United States. The irony of the situation was that the vines were being imported to France to combat another pest introduced from America. This was a root aphid that caused the death of the European vines but to which the native

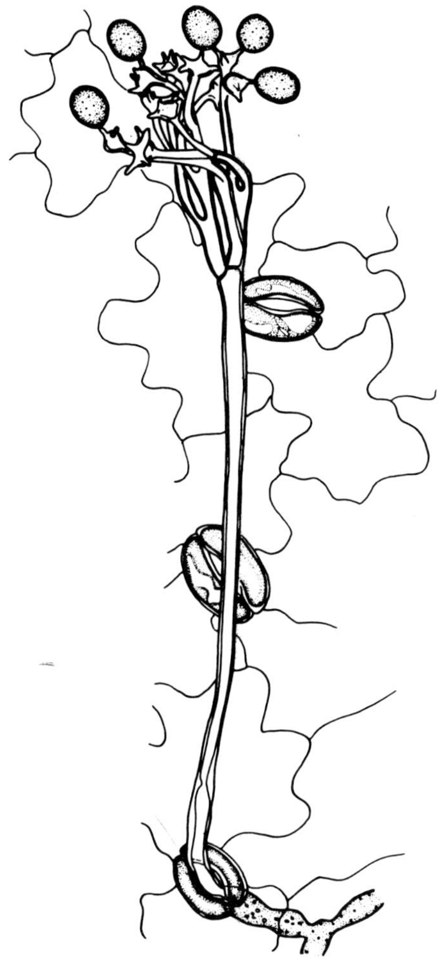

19–8 The conidiophore of the downy mildew as it emerges through a stomate of the host and produces conidia.

American grapes were much less susceptible. Root stocks of native American vines were grafted to the stems of the wine grapes. While this solved the aphid problem, it introduced the downy mildew. The disease became progressively worse during the 1870s, and in wet seasons it almost wiped out the harvest.

Then, one day in 1882 Professor Pierre Millardet of the University of Bordeaux was passing a vineyard, when he noticed that the vines along the road looked healthier than the vines in the field. They also looked a curious blue-green. The color, it turned out, was the result of the owner's spraying the plants with a mixture of copper sulfate and lime. He did this to discourage passersby from eating his grapes and in so doing provided Millardet with a clue to the control of the mildew. He called the mixture of copper hydroxide and lime that he developed the "Bordeaux mixture." This proved to be the first effective fungicide developed by man. Copper hydroxide is virtually insoluble in water but is slightly soluble in water containing germinating spores. This is due to the production of acids by the germinating spore so that enough copper is present in the drop containing the spore to kill the emerging hyphae. Thus the spore causes its own death, and the host remains unaffected. The Bordeaux mixture also proved effective against the late blight of potatoes and many similar plant diseases.

Zygomycetes

A common bread mold, a sharp-shooting fungus, and an insect parasite are all members of the class Zygomycetes. The member of this class most known to the general public is the black-bread mold *Rhizopus stolonifer.* It, like all the other members of the zygomycetes, is a terrestrial fungus completely lacking flagellated cells. The hyphae have cross walls, and chitin is a predominant component of the cell wall.

Rhizopus—the black-bread mold. *Rhizopus* can be found on almost any piece of moldy bread or soft fruit (Fig. 19–9). There it is a saprophyte—the hyphae secretes enzymes that help break down the substrate so that it can be absorbed by the fungus. But *Rhizopus* can also be found causing soft rot of sweet potatoes and strawberries. Here, the secreted enzymes kill the cells and partly dissolve the wall in advance of the growing hyphae. No haustoria are formed; the hyphae simply grow throughout the rotten tissue. Strictly speaking, *Rhizopus* in this case is still a saprophyte, but it is the cause of death of the tissue it is utilizing.

The structure of *Rhizopus* is more complex than it first appears. One part of the hypha grows rapidly over the surface of the substrate. At places hyphae extend into the substrate. Opposite these, hyphae differentiate sporangia that give rise to numerous dark-colored spores. These spores are light enough to float in the air for long periods and great distances.

Sexual reproduction occurs when the hyphae of two different strains (called plus and minus) grow together. Short lateral branches develop and come together; then their ends are cut off by cross walls to form multinucleate gametangia. These fuse by the dissolution of the wall, where the gametangia from the plus and minus strains meet. The multinucleate protoplasts fuse, and one nucleus from the plus strain and one from the minus strain pair and fuse, giving rise to a $2n$ zygote. This is repeated many times so that the structure that results is a multinucleate zygote. It develops a thick, warty wall and becomes the dormant stage of the organism capable of withstanding adverse environmental conditions. When the multinucleate zygote germinates, the nuclei first undergo meiosis; then a single sporangiophore is formed and haploid spores are produced.

Gametangia are formed only when plus and minus strains are present, and the developing gametangia quite obviously grow toward one another. These observations at first lead to the belief that hormones were involved in the process. Since 1924 the problem has been investigated using a related fungus, *Mucor mucedo.* Eventually, it was discovered that a group of compounds called trisporic acids A, B, and C were all involved. Trisporic acid C was established as the hormone responsible for the induction of the gametangia, but the entire picture is still not clear. Some earlier produced substance that is strain-specific must be formed, but its nature is unknown.

Pilobolus—the fungal cannon. In *Rhizopus* the spores formed in the sporangium are released when the walls of the sporangium break down and the wind carries them away. The large number of spores produced and the wide range of substrates that *Rhizopus* can use make this method of dispersal highly effective. In another zygomycete a different, far more elaborate method of spore discharge has evolved that is superbly adapted to its particular environment. This is the fungus *Pilobolus.*

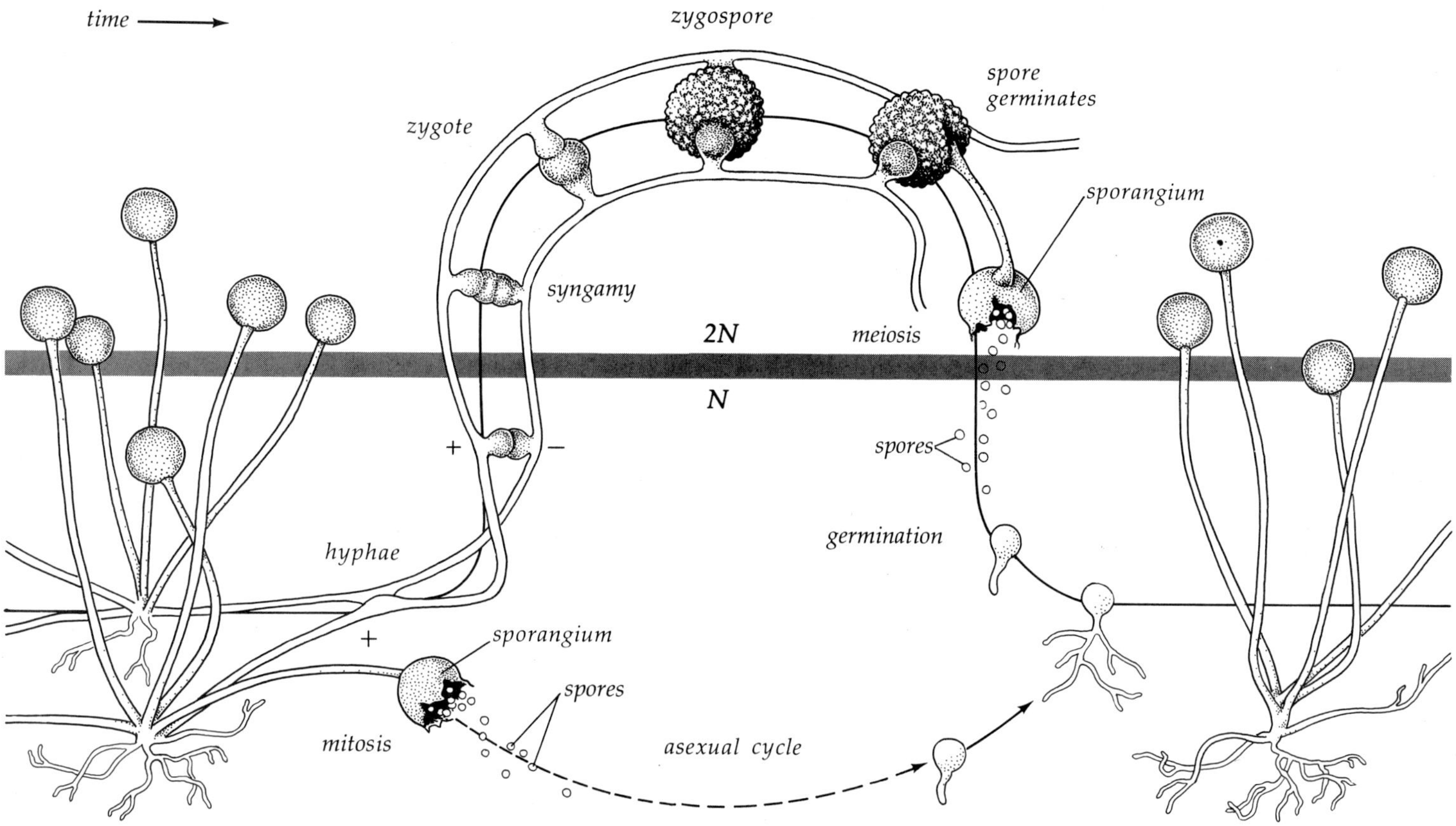

19–9 *The life history of* Rhizopus.

The natural habitat of this fungus is animal dung, particularly of grazing animals. Dung provides a good source of nutrient, and many fungi are found on all types. Many of the fungi growing on dung have windborne spores that simply land and germinate. This is a fairly haphazard approach to the problem of dispersal, and *Pilobolus* has evolved a much more sophisticated solution. In *Pilobolus* the spores for dispersal remain in the thick-walled sporangium, which has a slimy outer wall and a sticky bottom (Fig. 19–10). The sporangium adheres to the blades of grass and other plants and is eaten by the animal. When the dung is deposited, the sporangial wall breaks down, and the spore germinates, forming a mycelium. The problem now is how to get the sporangium from the dung, where it is formed, to the leaf where it can be eaten. The problem is complicated by the fact that the dung usually drops under vegetation, some distance from the leaves.

The mechanism that evolved in response to these conditions is a most remarkable fungal cannon that shoots the entire sporangium toward the light (Fig. 19–10). The sporangiophore of *Pilobolus* is some 0.5 to 1 cm long and consists of a basal bulb partially embedded in the dung, with a stalk that ends in a transparent bulb directly below the black sporangium. The transparent bulb acts as a lense to converge light on a

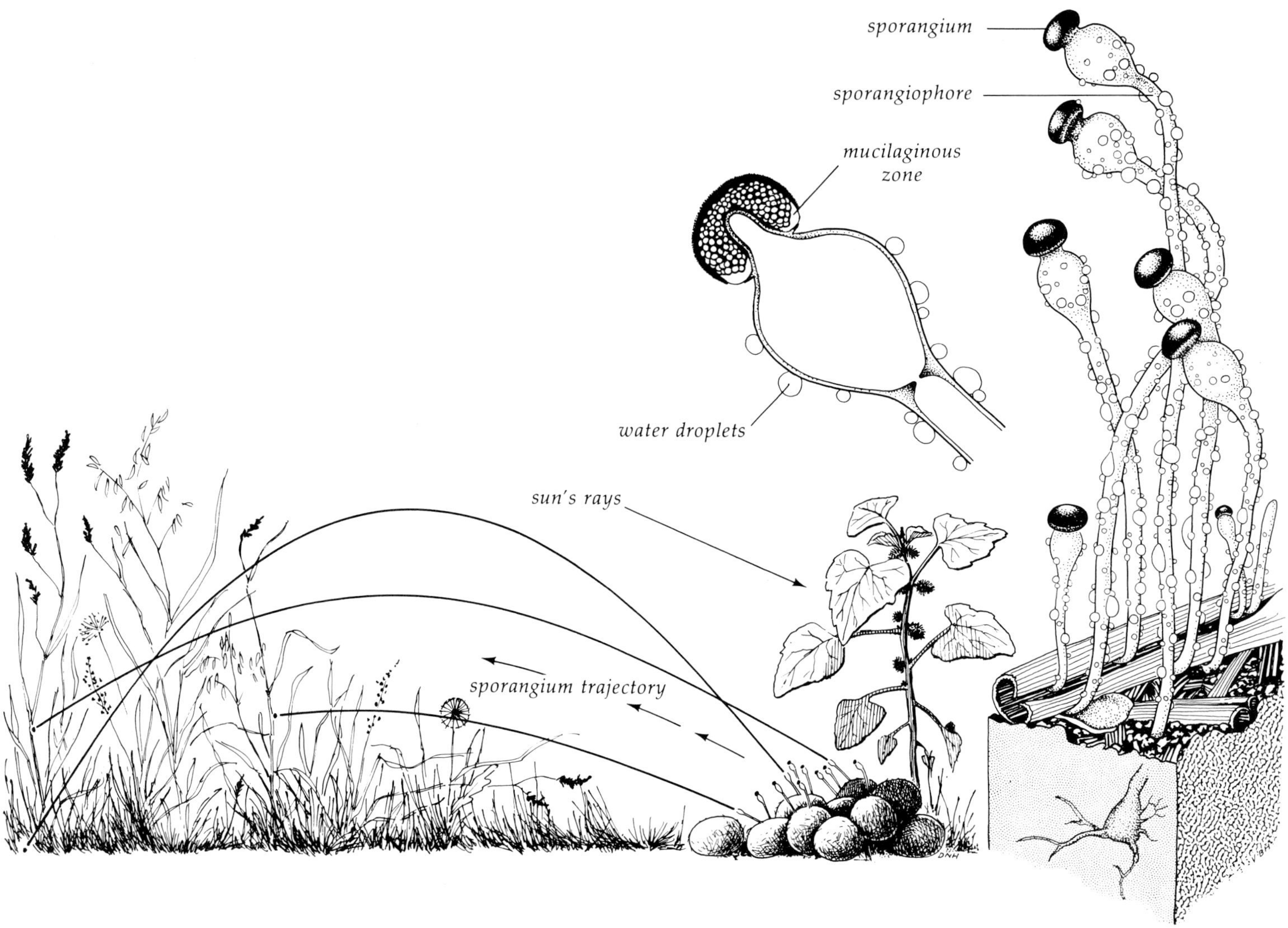

19–10 Pilobolus *is a common dung fungus that has evolved an elaborate mechanism to shoot sporangia toward the light.*

photoreceptive region at the base of the swelling (Fig. 19–10). The stalk thus bends toward the maximum source of light. The entire sporangiophore is a single, highly turgid cell. Pressure builds up within the sporangiophore during the day, and usually, by mid-morning it bursts along a line just below the sporangium. The whole sporangium may be shot a distance of up to several yards (Fig. 19–10). The discharge is violent: the sporangia reach speeds of close to 30 mph, and the vacuolar sap trails after the sporangium as the stalk collapses. The sporangium, as noted earlier, has a sticky ring of mucilage on its lower surface so that it readily adheres to the leaves. When the sporangium hits an obstacle such as a leaf, it flips over and becomes attached by the ring of mucilage on the underside so that the thick, dark, waxy top of the sporangium is up. The thick, waxy sporangium wall protects the thin-walled spores from desiccation, while its dark color prevents light damage to the colorless spores. Far from the spores being harmed by passage through the ani-

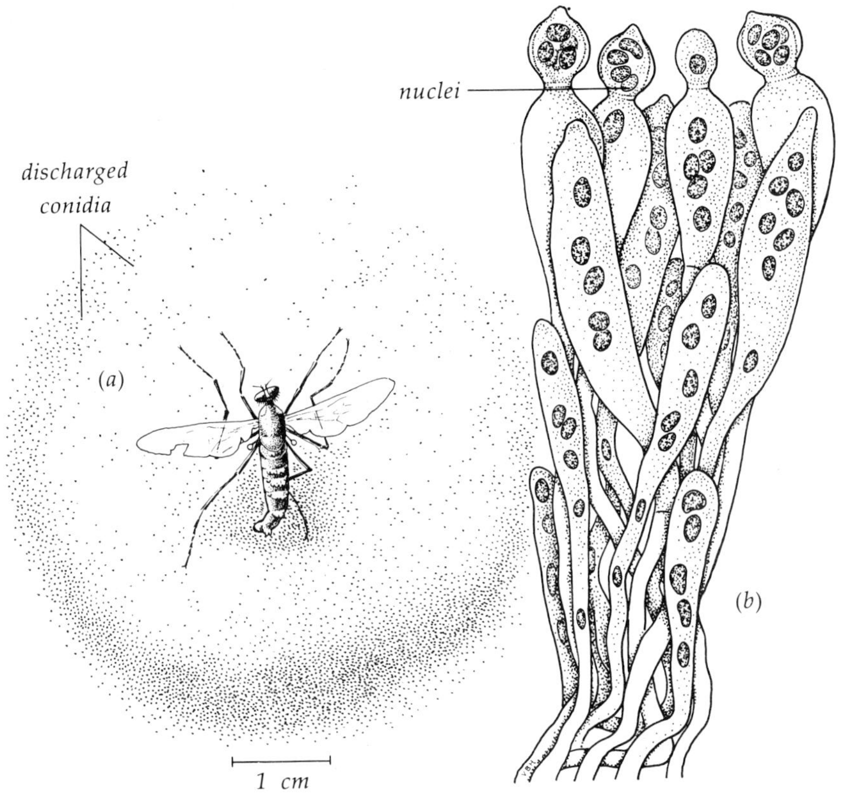

19–11 Entomophthora muscae *is a parasite of flies. The dead fly in (a) is surrounded by a halo of discharged spores. The hyphae grow through the skin of the fly and (b) shoot the multinucleate conidia away from the fly.*

mal, they are changed so that they will germinate shortly after the dung is excreted. All in all, *Pilobolus* shows remarkable adaptation to an unusual environment.

The collection and culture of *Pilobolus* is remarkably simple—if you have access to fresh cow or horse dung and don't mind collecting it! The fresh dung is placed in a glass or plastic dish, lined with moist paper toweling. The dish should be covered with a loose-fitting lid and kept in a cool room out of direct sunlight but where it receives a day-night cycle. In one to three days at most, *Pilobolus* will appear followed by a succession of other fungi.

Entomophthora—the insect killer. A number of the Zygomycetes are insect parasites. One of the most common is *Entomophthora muscae,* which lives on the ordinary housefly. Spores of this fungus stick to the epidermis of the fly, and hyphae grow into its body. There, the initial hyphae bud off spherical fragments that, in turn, grow into hyphae and may later fragment into single-celled pieces (Fig. 19–11). Gradually, the fly becomes filled with the fungus and finally dies, sometimes attached to window glass. Soon after the insect is dead, the single-celled pieces germinate and develop hyphae that push through the thinner parts of the fly's body. Each hypha forms a conidiophore bearing a single conidium at its end. In a manner somewhat similar to *Pilobolus,* the conidia are shot a distance of 2 to 3 cm. So many conidia may be discharged that they form a white ring around the dead fly (Fig. 19–11).

End Paper 19–1 Sex Hormones in Fungi

Reproduction in the water molds involves the production of sex hormones—chemicals that attract swimming gametes or interact with the hyphae to initiate the formation of sex organs and gametes. Two species that have been intensively studied are *Achlya* and *Allomyces.*

Achlya is an aquatic Oomycete with a method of sexual reproduction similar to that found in *Saprolegnia.* In both genera it is usual for the male and female gametangia to be produced by the same mycelium. Two species of *Achlya,* however, were discovered in which male gametangia were produced on one plant and female gametangia on separate plants (Fig. 19–12). It was also observed that unless the two plants grew in the same water, neither would produce reproductive structures.

The observations were the basis for a series of ingenious experiments, performed in the early 1940s by John Raper of Harvard University, in which he demonstrated the presence of a number of interacting sex hormones. Here is Professor Raper's recount of his experiments.

"The heavy production of sexual organ primordia on both ♂ and ♀ plants, while the two mates were still several millimeters apart, on agar media clearly established the presence and activity of diffusible sexual hormones. Even a careful study of the sequence of events in the interaction left unanswered the basic questions of who starts it all? How many hormones are there? Or, as a relevant limerick puts it,

> There was a young fairy named Bloom,
> Who went to a lesbian's room,
> They argued all night
> About who had the right
> To do what, when, with which, and to whom.

Insomnia, too, has its uses, and a sleepless night permitted the sort of concentration requisite to a solution: a slow, unidirectional flow of water through a series of small aquaria in which ♂ and ♀ plants could be placed in different orders. A few days later, the apparatus was ready: four tiny aquaria, each of ca. 2 ml capacity, made of microscope slides fastened together with instrument shellac, with capillary siphons connecting the vessels, and fed water by a constant leveling device at a rate of 1 ml/hr. A horizontal microscope completed the set-up. When the order of plants in relation to the direction of water flow was ♂-♀-♂-♀, reactions occurred only in the last two vessels with the production of antheridial and oogonial initials, respectively; when the order was reversed, ♀-♂-♀-♂, the previous ♂ and ♀ reactions of vessels 3 and 4 were shifted to vessels 2 and 3, and the antheridial initials of the ♂ plant in vessel 4 showed a strong tropic orientation toward the tip of the siphon from the reacting ♀ of vessel 3. Thus, in *Achlya,* as perhaps in certain other and even higher forms of life, the ♀ is the initiator in dalliance. [*Ed. note:* The chemical produced by the ♀ was named hormone A, and at the California Institute of Technology shortly before World War II, Raper laboriously concentrated the hormone but could not purify or analyze it at the time.]

During my first year at the University of Chicago after the war years as a radiobiologist, work was continued on the hormonal action of hormone A and the physiology of antheridial induction in *Achlya*. For this work, there was available a pitifully small supply of hormone A of high purity and standardized activity, and this vial of standard was dear to my heart—it being used only in critical experiments and then only in 0.01 ml portions.

Imagine my horror upon returning from a lecture to find my assistant, a fair-haired, first-year graduate student, on her hands and knees in the middle of the laboratory sucking up this precious liquid with a tiny pipette. She had dropped the bottle, which had broken, and had intuitively gone about the rational business of recovering what she could of the hormonal solution with the equipment at hand. In a mixture of shock at the obvious carelessness on the one hand and my admiration of her initiative in making the best of a totally unnecessary and bad situation on the other, I could only urge the completion of the task and enjoin her not to cry over spilt hormone. There was, of course, no possibility of precise comparison of the activity of the recovered hormone with the original, and it may well be that the quantitative aspects of the work with *Achlya* underwent a slight discontinuity as a result of the accident.

Forgiveness, however, was apparently not too difficult. Perhaps my failure to erupt into the violent display of temper that had been suggested in earlier and far less serious situations convinced her that I might be human after all. In any event, a couple of years later we were married, but over the years I've come to appreciate the monicker of "Spilly" bestowed on her by her family at a very early age.

Hormone A has now been isolated; its structure determined, its synthesis has been achieved by at least two different synthetic routes; and its chemical status has been legitimatized by the name *antheridiol* (Fig. 19–13). So that should be that. But one incident suggests that there may be more than one hormone A—perhaps many of them.

Two or three years ago, after Alma Barksdale and Trevor McMorris at the New York Botanical Garden had worked out the procedures for isolating the hormone and were working in collaboration with Arsenault and Biemann on its structure, Alma asked me if I would contribute to this effort a stock of hormone A that I had kept, sealed in an ampoule, since its extraction and enrichment from ca. 1600 liters of culture medium at Cal Tech in 1943. The material was a gram or so of yellowish, amorphous goo, but it contained very high activity, something of the order of 10^9 to 10^{10} physiological units—considerably more, as I recall, than they had had to work out the procedures for isolation. As I was only somewhat less anxious than they to see hormone A, the first identified plant sexual hormone ('twas not to be—Machlis and Co. won the race with the structure of *serenin* of *Allomyces* by a few months), the material was sent along posthaste.

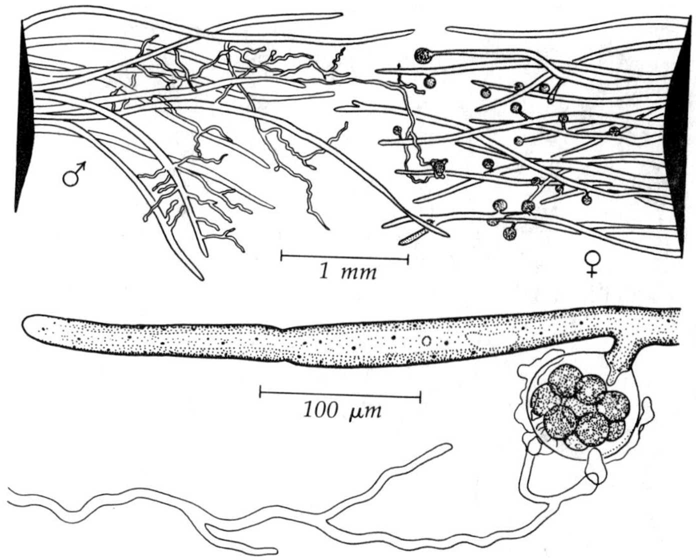

19–12 Achlya ambisexualis. *The male and female plants, which started on separate food sources, can be seen growing toward one another.*

Upon its arrival at the New York Botanical Garden, the material was put into solution, divided into two portions, and assayed. Little or none of its activity had been lost during its twenty-odd years'storage, but when the two batches were put through the procedures of isolation, virtually all of the activity was lost, less than 5% of the initial activity remained. Aside from everyone's annoyance at the loss of the material, the more important question remained: why should one sample of hormone differ so critically from another?

That different strains were the sources of the hormones involved may provide an answer. The sample that I provided came from a strain originally collected in North Carolina about 1934. This strain was lost in 1944 and was replaced in my work by one collected in the Fox River, Illinois, in 1946. This latter was the strain used by Barksdale *et al.* to isolate and eventually to identify hormone A or antheridiol. Both strains were assigned to the species *A. bisexualis,* but as Barksdale's work has shown, taxonomic determinations in this genus are somewhat problematic at best. The two samples of hormones must have differed in some chemical detail that did not affect their essential business of inducing sexual activity in male *Achlya*."

The work that Raper refers to as Machlis and Co. has to do with research carried out on *Allomyces* by Leonard Machlis of the University of California at Berkeley. Raper is a mycologist by training but Machlis is not. We asked Machlis how he started his work on *Allomyces.*

"The beginning goes back to 1946 when I joined the faculty of the Department of Botany at the University of California at Berkeley. Traditionally, most scientists continue work in the same general area as that of their doctoral research. Mine was on ion accumulation by barley roots. But World War II intervened between finishing my Ph.D. and returning to research in plant physiology. During the war years, I was first part of a group trying to produce rubber from a native California plant, then a civilian instructor on the use of the then new ahead-thrown rockets for hunting submarines, and finally with a civilian group developing guided missiles. When the war was over and after a sojourn of less than an academic year at the University of Illinois, I ended up in Berkeley in 1946. I suppose the sharp break from research during the war years made it rather easy to go into research on the nutrition and metabolism of *Allomyces* rather than resume studies on ion accumulation. In any event, influenced by the enthusiasm of my colleague, Ralph Emerson, for any work on his pet, *Allomyces,* that is what happened. This work began in 1946.

About 1956 I turned to the study of the sperm attractant. It is hard for me to remember exactly what led to this, but I do recall that I had become bored with what I had been studying. Also, I had on many occasions scraped male and female gametangia from the surface of nutrient agar plates, put them in water, and watched the gametes discharge. I noticed that the male gametes clustered around the undischarged female gametangia. Then when the females came out, there were the males, sitting ducks to be engulfed by the females and thereby form a zygote. I did a few simple experiments that showed beyond question that the female gametangia released a substance that attracted the male gametes to them.

In the meantime, I recognized that sperm attractants had been postulated or shown to occur in a rather large number of algae and fungi, but no one had succeeded in isolating and chemically characterizing even one of them. I decided this needed doing, and thus the research was under way. About this time I took a sabbatical leave, working in the laboratory of Professor Nils Fries at the University of Uppsala. My intent was to produce the sperm attractant and to learn something of its chemistry. I managed to accomplish two different things. The sperm attractant was given a name, sirenin—actually suggested by Dr. Vegis one day at tea time in the lab. And I recognized that I wasn't enough of a chemist to do much about the chemistry of sirenin."

When Machlis became interested in the nature of the sex hormone, he was faced with the problem created by the plants' being hermaphroditic (producing both male and female gametes on the same plant). If he were to use normal plants, the male gametes would remove the hormone from solution almost as fast as it was produced, thus making isolation almost impossible. He circumvented this problem by selecting an interspecific hybrid, developed from strains Emerson had in culture, in which 96 to 99% of the gametangia produced were female. Masses of these female plants were grown in 5 gallon carboys on a schedule in which 4 carboys of plants were harvested a day, 7 days a week. This was continued for 2 years. At the end of this time, Machlis had accumulated a total of 2 g of sirenin! Working from this supply, Henry Rapoport, a professor of chemistry at the University of California, Berkeley, was able to chemically characterize (Fig. 19–13) and later synthesize the compound—10 years after Machlis had started the research.

Machlis also used hybrids that produced predominantly male gametes for the bioassay of sirenin. In much biological research, particularly with hormones, the nature of the bio-

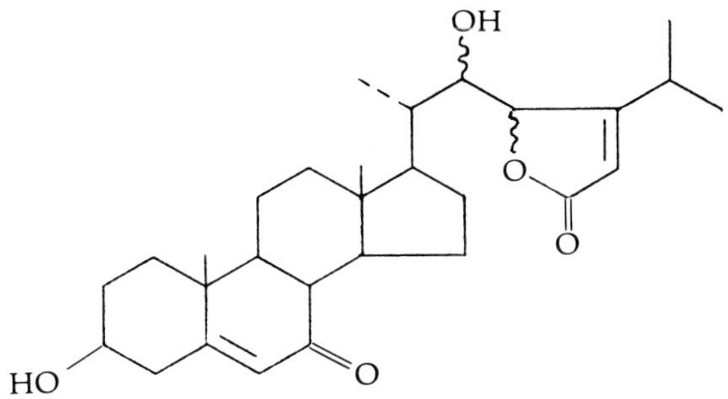

structure of antheridiol

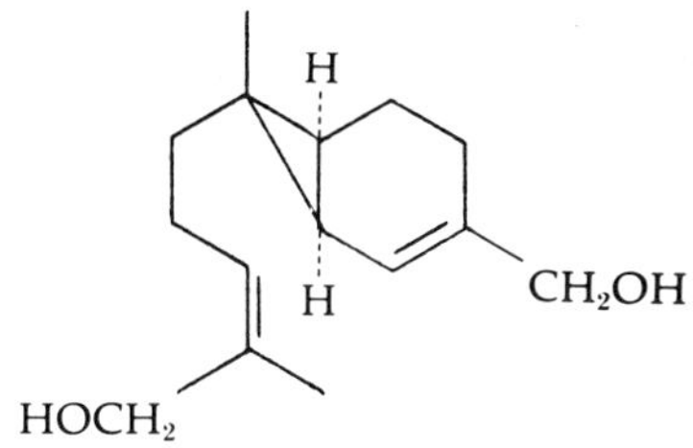

structure of sirenin

19–13 The chemical structure of antheridiol and sirenin.

assay is critical. Only when a means is found to measure how much of the compound is present can research on the problem be meaningful. In this case the bioassay consisted of constructing a cell whose bottom was a membrane. Below the cell is placed the sirenin solution, and in it, separated by the membrane from the sirenin, is placed a suspension of the sperm. This simple but effective assay is central to the problem of isolating, identifying, and studying sirenin. Obviously, the problem of sperm attraction in *Allomyces* is not solved when the hormone is isolated and characterized. Many questions such as the nature of the interaction of the sperm with the hormone remain unanswered, and many problems have arisen that were not even known until the research reached this point.

20

The Fungi: Yeasts, Mushrooms, Rusts, Imperfect Fungi, and Slime Molds

Mushrooms and Toadstools

On moonlit nights the fairies dance in a ring in the fields and woodlands, and where they dance, mushrooms spring up. When the dance is finished, they rest on the mushrooms. But the toads also come out and sit on the mushrooms. Those choosen by the fairies are edible by man, while those used by the toads (thus called toadstools) are poisonous to man. At least that is how one old folktale accounts for the appearance of mushrooms and toadstools. They are, of course, the fruiting bodies of the higher fungi: the ascomycetes and the basidiomycetes. There is no physical difference between what are commonly called mushrooms and toadstools, just as there is no easy way (aside from knowing which is which) to tell the poisonous from the edible species.

As an introduction to these groups of fungi, we now present a few of the many dozens of mushrooms and toadstools that you may commonly see. Those labeled A–B are ascomycetes, while those labeled C–I are all basidiomycetes.

A Sarcoscypha coccinea, *or scarlet cup, since the inside is a deep red.*

B Aleuria, *one of the sac fungi.*

C Morchella conica, *a true morels or sponge fungi.*

D Agaricus, *one of the common edible mushrooms.*

E Clavaria aurea, *a yellow coral mushroom.*

F Tremella, *one of the gelatinous mushrooms.*

G Polydopus, *a species of the numerous wood-rotting shelf fungi.*

H Calbovista subsculpta, *one of the puffballs.*

I Crucibulum levis, *bird's nest fungus.*

Under normal conditions the slime molds or myxomycetes are not exactly conspicuous features of the environment. The vegetative stage is found mainly in the litter of the forest floor or in rotten logs. They are hardly visible in such locations, but at times the thinly spread vegetative form will pull together in a slimy mass. One such mass is shown in the photograph labeled J. The fruiting bodies are formed from such masses (although usually not that large). At this point the slime mold becomes the architect of some of the most handsome structures found anywhere in the plant kingdom. All are small, but they exhibit a remarkable diversity of form, as shown in photographs K–N.

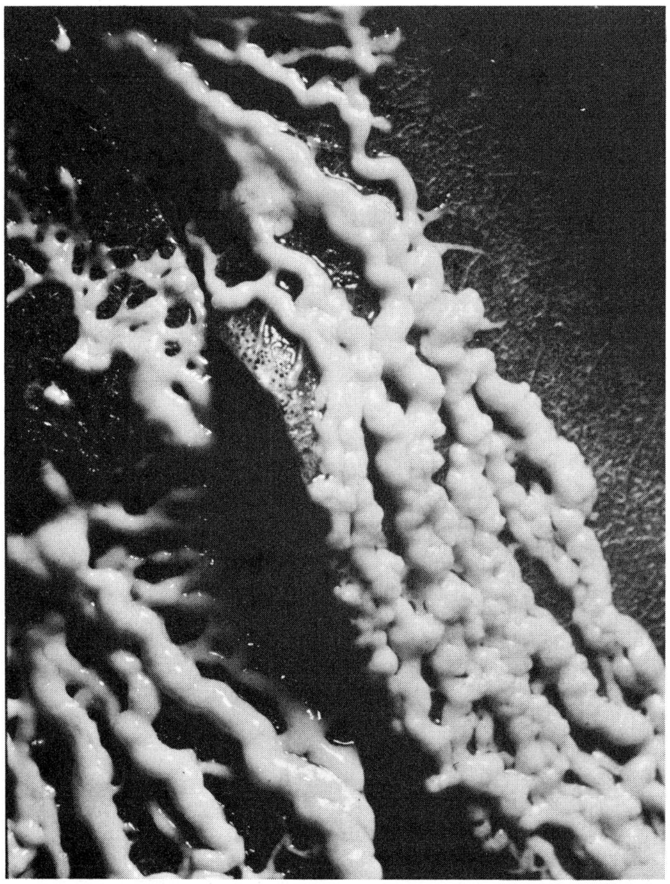

J The vegetative body of Leocarpus fragilis.

K Fruiting bodies of Trichia montana.

L Stemonitis fusca.

M Leocarpus fragilis.

N Comatricha typhoides.

The men trudged down the walks that led to their work places on the pyramid. They were tired but happy. The work had gone well, and soon they would receive payment for the day. They were not slaves but free men called by the Pharaoh to help in the construction. In general, they enjoyed the work as a welcome change from the monotonous labor in the fields and as a chance to gossip with men from other villages. Besides, the pay was good—four loaves of bread and two jugs of beer.

The beer was important, not only for the slight alcohol content, but as a part of their diet. By the time of the Pharaohs, beer had been made for over 3000 years. The beer that the Egyptian worker drank was different from the beer we drink today. Underbaked bread was made of coarsely crushed, germinated barley; the bread was then cut into small pieces, soaked in water in a large jug, and left to ferment (using yeast) for a day. For the Pharaoh and the nobles, the bread was strained out and the brew sweetened with dates, but the laborers drank the whole brew, thus receiving a far more nutritious beverage.

Ascomycetes

The yeasts responsible for fermentation are members of the Ascomycetes, a great class of fungi comprising some 30,000 species. The Ascomycetes are composed of highly diverse forms, containing such fungi as the yeasts, *Penicillium,* the deadly ergot, and various wood-rotting fungi. As diverse in form as the various genera and species may be, the Ascomycetes are a long-recognized natural class. None of the Ascomycetes have motile cells, and all have walls containing chitin and insoluble glucans. In most cases (the yeasts are an exception), the hyphae are subdivided by cross walls that have a central pore. This allows for the movement of cytoplasm from one cell to the next; not only cell organelles such as mitochondria but also nuclei have been seen to migrate from cell to cell. The cells that are formed may be either uninucleate or multinucleate. The sexual spores are produced on special structures called asci (ascus, singular).

Sexual reproduction in Ascomycetes. In most Ascomycetes (Fig. 20–1), sexual reproduction is based on the germination of different kinds of spores that appear identical but actually represent the different genetic strains, usually indicated by a + or − sign. Under the proper conditions a branch of one hypha enlarges, and many nuclei will accumulate in the swollen tip, termed an ascogonium. The tip of the second hypha enlarges only slightly, but again many nuclei are concentrated in it. The smaller tip, which is the antheridium, fuses with the larger, and the nuclei of the + and − strains mingle together. From this enlarged, swollen hypha many new hyphae form. As these hyphae grow, new cells are constantly being formed. The cells closest to the fused hyphal mass have several nuclei. As the tip of the hyphae grows farther and farther away, the cells become increasingly binucleate—always in pairs of + and − nuclei. Each of these contain 2 haploid nuclei so that as far as these organisms are concerned, the cells function as if they were actually diploid. This condition, where a cell contains 2 paired nuclei with one from each part, is called dikaryotic.* Eventually, the 2 nuclei of the subterminal cell or ascus (plural: asci) fuse, forming a single diploid nucleus, the zygote nucleus. Almost immediately, this nucleus undergoes meiosis, and each of the resulting 4 nuclei divide mitotically, forming 8 haploid nuclei. At the same time the ascus is elongating; eventually the 8 nuclei will line up in a neat row. A wall is formed around each nucleus so that 8 haploid ascospores are formed. The hyphae producing the ascospores are surrounded by many hyphae that do not produce spores and that are usually monokaryotic. This entire hyphal mass comprises a fruiting body or ascocarp.

---

*Cells containing 2 paired nuclei are called dikaryotic; those containing a single nucleus are monokaryotic. These terms refer only to the number of nuclei present, but it is possible to have two conditions with regard to the genetic makeup of the nuclei. All the nuclei may have the same genetic constitution in which case they are homokaryotic. If the nuclei present in the cells are derived from hyphal fusion and therefore are different genetically, they are heterokaryotic. Obviously, monokaryotic cells are always homokaryotic, but dikaryotic or multinucleate cells can be either homokaryotic or heterokaryotic. However, dikaryons are usually heterokaryotic. This terminology may seem quite confusing, but it is easy to understand if you remember that one set of terms refers to the *number of nuclei* and the other refers to *genetic makeup.*

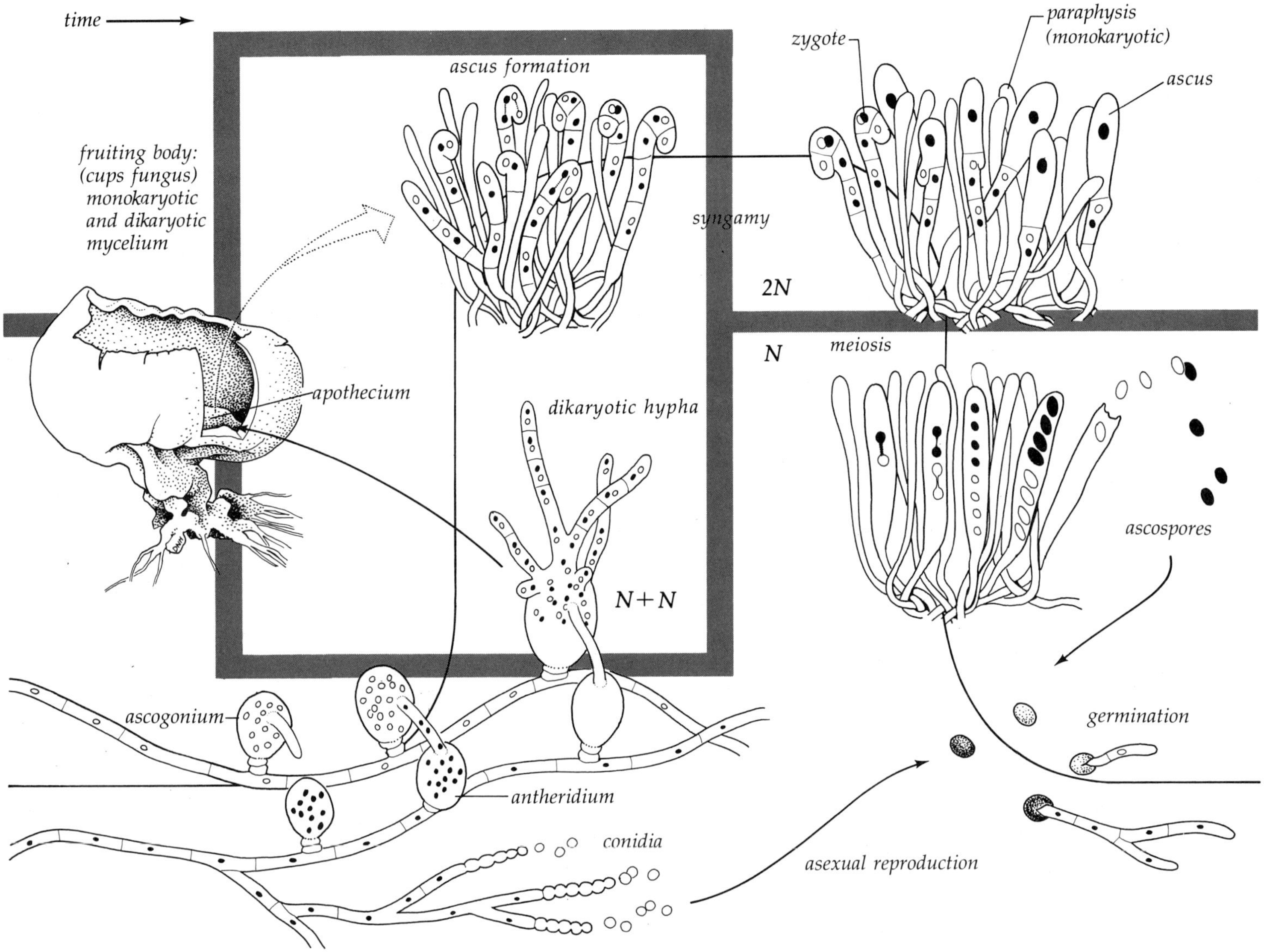

20–1 *The life history of an Ascomycete.*

Sexual reproduction is not common in *Penicillium,* but it is common in such Ascomycetes as *Neurospora* or the pink bread mold made famous by the geneticists.

Spore discharge in Ascomycetes. Each ascus is a single cell containing the ascospores, a thin layer of cytoplasm, and a large central vacuole in which the spores are actually suspended near the apex of the cell. As the ascus matures the hydrostatic pressure builds up. In some species such as *Pyromena* a lid at the tip of the ascus hinges backward; the ascus wall contracts; and the spores are shot into the air some 2 to 3 cm. In *Neurospora* a terminal spore is present in the ascus. The individual spores are extremely light, and once in the air are carried away by air currents. In the flask-shaped Ascomycetes such as *Sordaria fimicola* (Fig. 20–2), the ascocarp assumes the form of a bottle with a narrow neck. The sides of the bottle are composed of

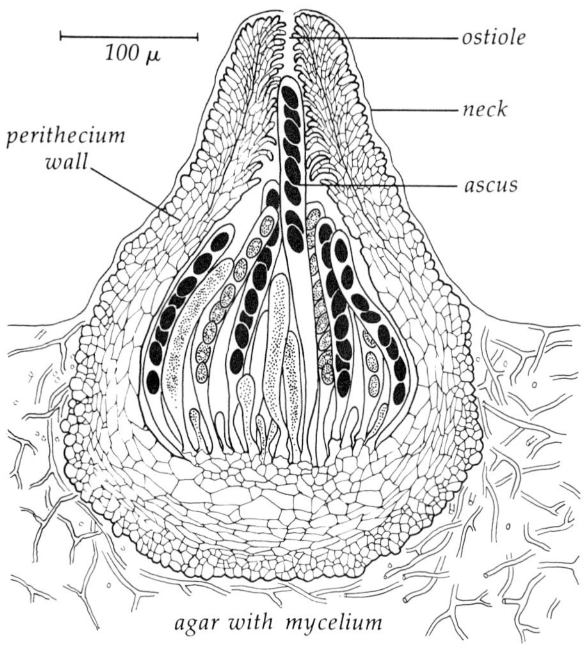

20–2 *The ascocarp of* Sordaria fimicola.

sterile (nonspore-forming) hyphae with the asci enclosed in the center. The neck of the bottle is large enough for only a single hypha to emerge at a time. Not all of the asci mature at the same rate. The most mature elongates first, fills the neck, and when the tip is clear of surrounding cells, discharges the spores up to 60 cm into the air. The empty ascus rapidly shrinks in size, and another takes its place. In the case of *Sordaria*, which is a dung fungus, the neck of the bottle is phototropic and bends toward the light, thus aiming the spores toward the light. *Sordaria* and *Pilobolus* present an interesting contrast: two widely separate groups faced with the same environmental conditions evolved different spore dispersal mechanisms but with the same end result.

Some of the most famous Ascomycetes, at least to gourmets, have no spore dispersal mechanisms. These are the truffles. The edible truffle, *Tuber aestirum*, is but one of a group of subterranean fungi that form underground ascocarps. The fruiting body of the truffle is irregular in shape, firm in consistency, and has a hard, black, lumpy exterior. It emits a strong odor that can be detected by dogs and pigs, which in France are used to hunt truffles. The spores are enclosed in a solid mass of mycelium and are never liberated. The smell is believed to attract rodents that dig up and eat the ascocarps; the spores then pass through the animal and are deposited with the feces.

Yeasts. Many claims have been made for yeasts—they are one of the most important of all domesticated organisms and they were the first to be domesticated. *Saccharomyces cerevisiae* (baker's or brewer's yeast) is grown around the world in complex, modern bakeries and breweries and in earthen pots containing dough or mash. The Egyptians and other ancient or primitive peoples used yeasts without understanding that they were dealing with a microorganism. Antony Leeuwenhoek was the first to see yeasts in 1680, but he did not recognize that they were alive. In the 1830s various biologists discovered that yeasts were

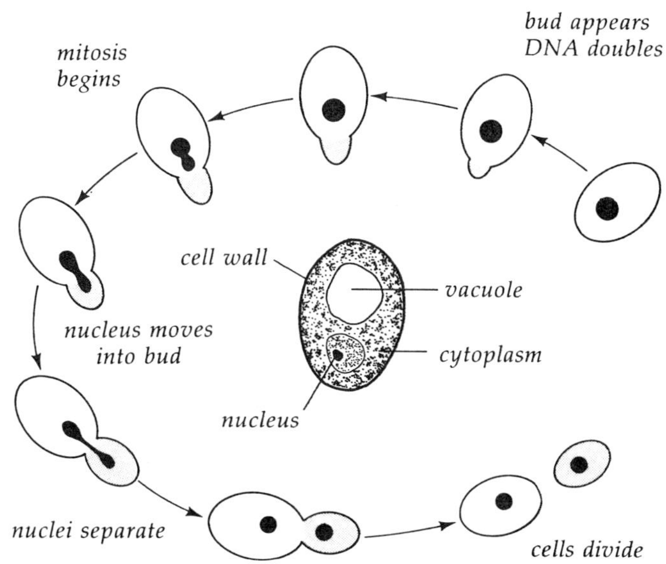

20–3 *A diagram of a yeast cell showing how it reproduces asexually by developing a small bud.*

living organisms, but it was Louis Pasteur who first established in 1857 their importance in the fermentation process.

Yeasts are single-celled organisms that are extremely simple in their organization (Fig. 20–3). But under the electron microscope, they can be seen to be eucaryotic cells. A yeast reproduces asexually by producing a small protuberance or bud on the side of the mother cell. The nucleus divides as the bud is forming, and one daughter nucleus moves into the bud (Fig. 20–3).

Penicillia. *Penicillium* are famous because they are the source of the first effective antibiotic—penicillin.

Other species of *Penicillium* are also important to man: *Penicillium roquefortii* and *Penicillium camemberti,* which are responsible for much of the flavor, texture and color of Roquefort and Camembert cheese. The veins of blue that characterize Roquefort or blue cheese are due to the spores produced by the fungus.

Ergot. Another Ascomycete, *Claviceps purpura,* utilizes a number of mechanisms to infect the flowers of the rye plant. Eventually, the fungus produces a dark purple structure that replaces the grain. This is called a sclerotium and is a mass of hyphae capable of surviving the winter (Fig. 20–4). The sclerotium, besides being surrounded by a thick wall, is also protected from being eaten by animals by the presence of several (actually close to 20) extremely potent and smelly alkaloids. When the ergots are harvested together with the rye grains, milled into flour, and eaten, the results are terrible: The victim burns with a fever; his body convulses violently; his reason is wiped away; he experiences violent hallucinations; and his fingers, toes, arms and legs blacken and drop off as a result of dry gangrene.

This is the Holy Fire or St. Anthony's Fire that swept through medieval Europe and into modern times—the last outbreak (a small one) occurred in France in 1951. Over the years thousands upon thousands of men, women, and children lost their lives because of ergot poisoning. The disease was particularly bad in southern France, where bread made of rye flour was the main staple of the peasants' diet. A Catholic nursing order, with hundreds of hospitals dedicated to St. Anthony, cared for centuries for sufferers of the disease, thus giving it one of its common names.

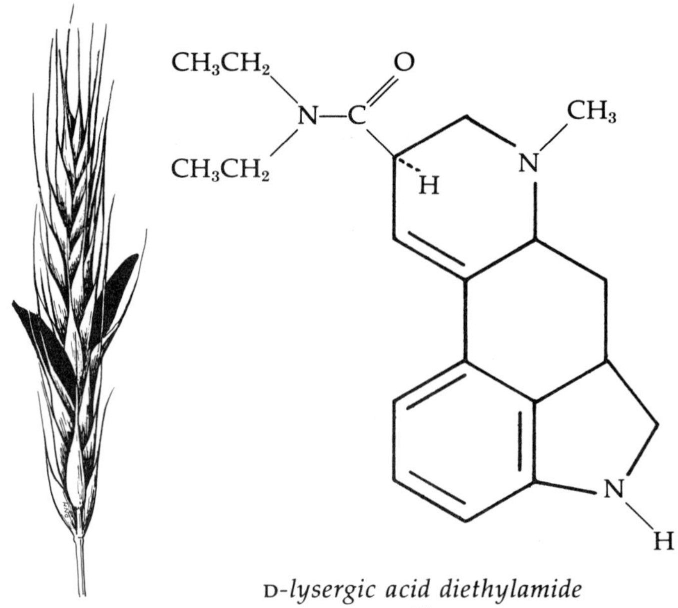

(a) D-*lysergic acid diethylamide* (b)

20–4 *(a) Sclerotium of* Claviceps purpura *on rye. (b) The chemical structure of LSD.*

The ergot also has the somewhat dubious distinction of being the source of lysergic acid from which lysergic acid diethylamide (LSD) was synthesized. Albert Hofmann, a chemist at the Sandoz Chemical Works in Basel, Switzerland, synthesized the hallucinogenic drug in 1938, but its powers were not discovered until 1943 (Fig. 20–4). Some people consider the effects of LSD to be so bad that it seems we are again being visited by a plague of Holy Fire—in a more sophisticated form. But the ergot story is not all bad. The same alkaloids that cause trouble in large amounts can be helpful in small doses. They are used to induce childbirth, control bleeding after delivery, treat high blood pressure and varicose veins, and ease severe migraine headaches.

Basidiomycetes

The Basidiomycetes contain many of the fungi of which we are commonly aware. There are some 13,000 species of Basidiomycetes in a vast array of forms. They

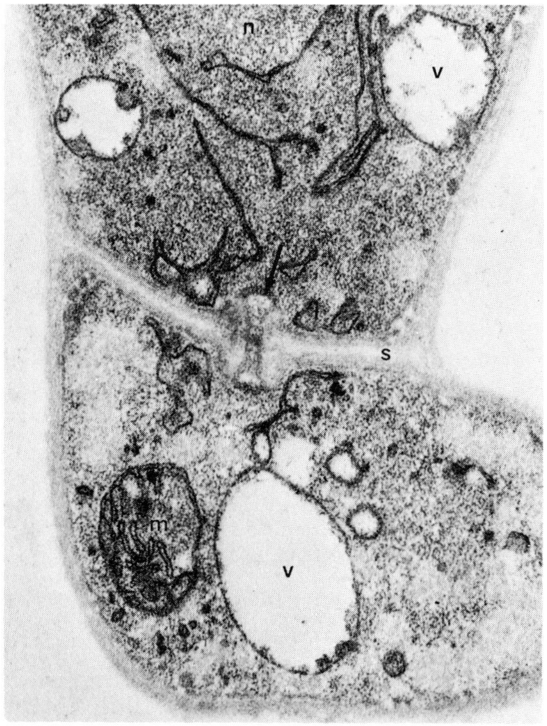

20–5 Cell structure in the Basidiomycetes. An electron microscope photograph of two cells showing a portion of a nucleus, several small vacuoles, a mitochondrion, and segments of the ER. The cells are separated by a wall in which a pore is visible.

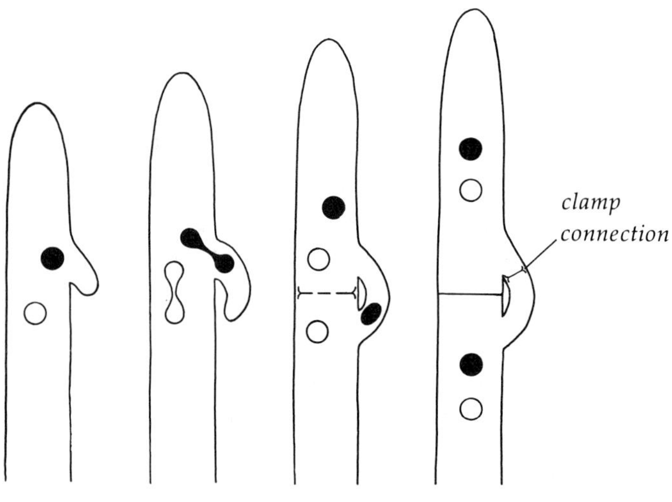

20–6 A diagram of cell division and the formation of clamp connections.

all have mycelia, consisting of branched hyphae, that contain cross walls with pores in them (Fig. 20–5). The hyphae fuse with one another, like the Ascomycetes, forming a three-dimensional network. Most of the cells are binucleate and dikaryotic with unusual structures called clamp connections present at the cross walls of many, but not all, species. The clamp connection is a mechanism that ensures that during cell division of a binucleate, dikaryotic cell both daughters receive one of each type of nucleus. How the clamp connection functions can be seen in Figure 20–6.

Mushrooms. Examining the life history of the common mushroom is a good introduction to this class. Again, we can start with the germination of the spores in the soil. Uninucleate cells are found first in the hyphae. But soon 2 hyphae fuse, and binucleate cells are formed (Fig. 20–7). Most of the mycelia of Basidiomycetes are composed of hyphae whose cells contain 2 nuclei, usually close together but still unfused.

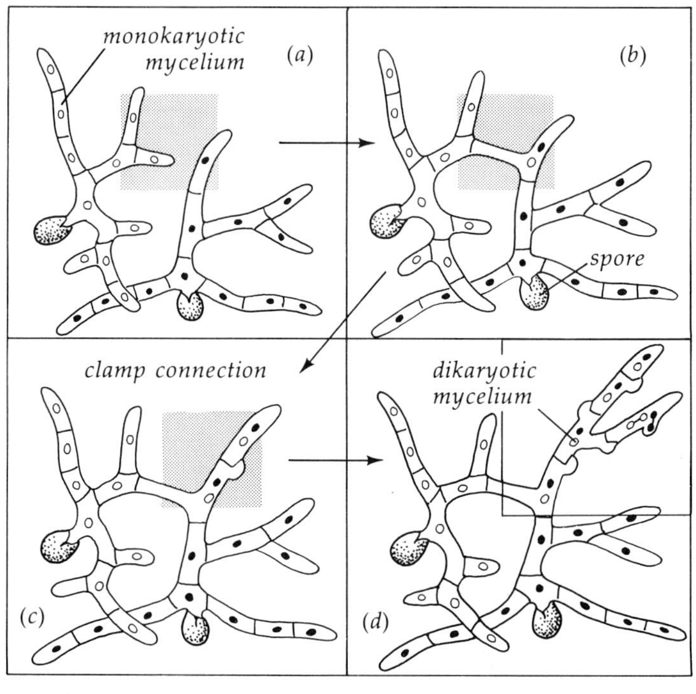

20–7 The fusion of two mycelia, composed of uninucleate cells, resulting in the formation of a binucleate mycelium.

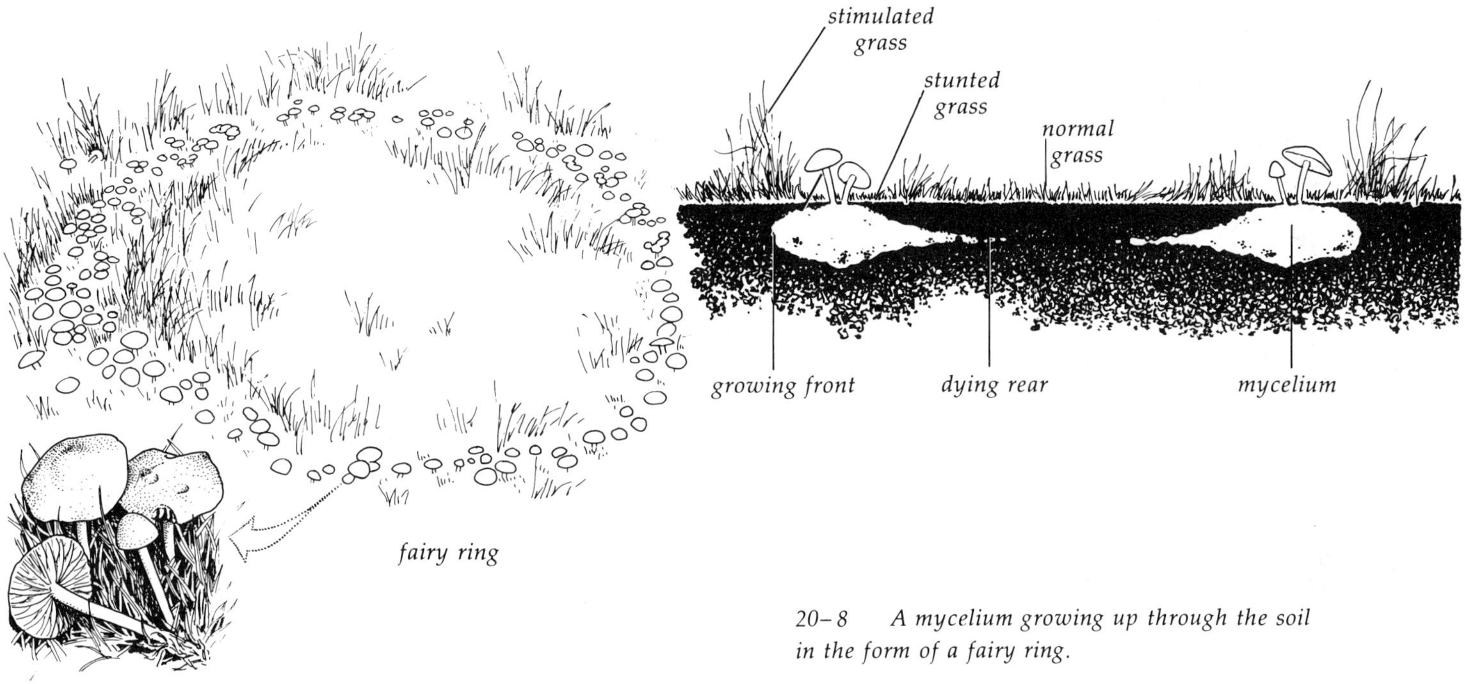

20–8 A mycelium growing up through the soil in the form of a fairy ring.

Any soil containing decaying plant material will probably be rich in mushroom mycelia: lawns, fields, or forest floors. The hyphae spread outward from the place where the spore germinates. A good example of the way a mycelium develops is one growing in a lawn (Fig. 20–8). Sometimes, it is possible to know where the mycelium is and to follow its growth without seeing it. What is seen is a ring of slightly taller and darker grass inside of which is a ring of stunted, lighter green grass. The grass both inside and outside these two concentric rings is normal. Each year the ring will get larger and each year a garland of mushrooms will form over the inner ring. This is the "fairy ring" that can be found the world over.

The explanation for the growth of a fairy ring is based on the physiology and morphology of the fungal mycelium. The center of the ring is the place where the spores germinated. From this point the hyphae creep outward in all directions, forming a solid circle of mycelium. Soon, the soil at the center becomes depleted of nutrients. While the hyphae at the edges continue to grow and the circumference of the mycelium increases, the hyphae at the center die; hence a ring is formed. The taller, stimulated grass is actually found slightly ahead of the advancing edge of the mycelium. The hyphae, as is characteristic of most fungi, are secreting enzymes that are breaking down the dead plant material in the soil to produce nutrients that cause the grass to grow taller and greener. The grass is stunted over the mycelium because it cannot compete with the fungus in extracting nutrients from the soil.

Sexual reproduction in Basidiomycetes. Early in the formation of the mycelium, hyphal fusion occurs, and dikaryotic cells are produced. This is really the start of sexual reproduction that will not be completed until the fruiting body is formed and the nuclei fuse. The mushrooms that spring up, sometimes overnight, are composed of hyphae and represent the fruiting body or basidiocarps (Fig. 20–9). Under the proper environmental conditions, the mycelium forms into a mass that becomes differentiated into a swollen basal part, a stemlike stalk, and a cap containing gills. All of this is formed below or at ground level and may be surrounded by a thin sheet of mycelium known as a veil. The mushroom becomes obvious when the stalk elongates and carries the cap above the ground, tearing the wall to pieces. While it is being carried upward, the cap

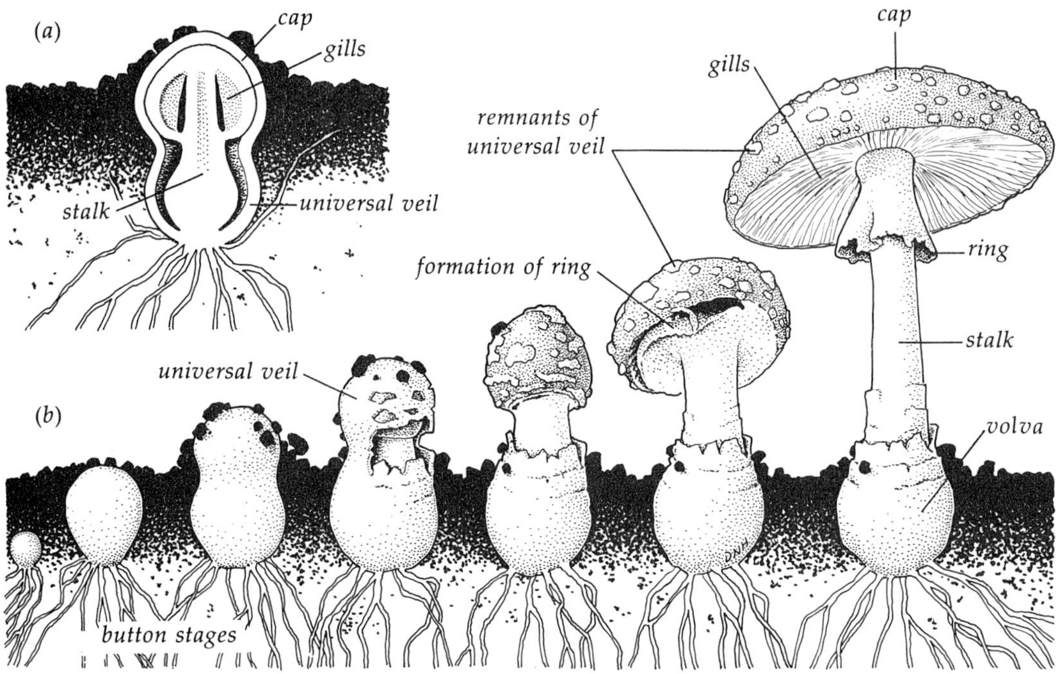

20–9 *(a) The development and structure of the fruiting body in Basidiomycetes. (b) The stages of development are shown for a species with a universal veil; however, this structure is not present in all species.*

expands outward in the way an umbrella opens. The gills line the lower surface of the cap; it is here that the spores, called the basidiospores, are formed.

Along the surface of the gills are many hyphal ends, the terminal cells of which enlarge, forming the basidium. The two nuclei present in the basidium fuse, forming true diploid nuclei (see Fig. 20–10). These nuclei represent the zygote stage and immediately undergo meiosis. The cells themselves continue to enlarge, and the 4 nuclei that are produced by the meiotic divisions in each cell migrate to the outer edges of the cells. Walls are then formed around the nuclei; each mature spore is located on the end of a short stalk or sterigmata. The spores are discharged a distance some 10 to 20 times their diameter. This gets them free of the hyphae and into the space between the gills, where they drop away from the cap. Once below the cap, air currents can carry the spores for miles.

When the spores are discharged the cap, stalk, and foot will begin to degenerate. In some species they turn into inky liquid, but in most cases they degenerate after spore discharge. Many insects also use mushrooms as seed beds, with the larva feeding on the caps. The underground vegetative mycelium remains alive and may produce a new crop of fruiting bodies annually for many years—even hundreds of years, under favorable, undisturbed conditions. Such enormous and ancient fairy rings are found among the stone pillars of Stonehenge.

The fruiting bodies of the Basidiomycetes assume a fantastic assortment of shapes in addition to the common mushroom or toadstool type (Fig. 20–11). They may resemble pieces of coral or be shaped like balls (aptly named puffballs because the spores are discharged in a puff when the ball is stepped on). Such puffballs may vary in size from less than one inch to a foot or more in diameter. Some Basidiomycetes have fruiting bodies that are relatively structureless and appear as jellied masses.

Some of the most conspicuous of the fruiting bodies are those produced by the shelf or bracket fungi (Fig. 20–11). These Basidiomycetes are wood-rotting forms and frequently live on the wood in the center of trees. This wood is composed of dead cells, and the fungus is actually a saprophyte, even though it resides within a living plant. The mycelium forms a tough, rigid fruiting body on the outside of the bark of the tree, which resembles a shelf or bracket. The upperside of this

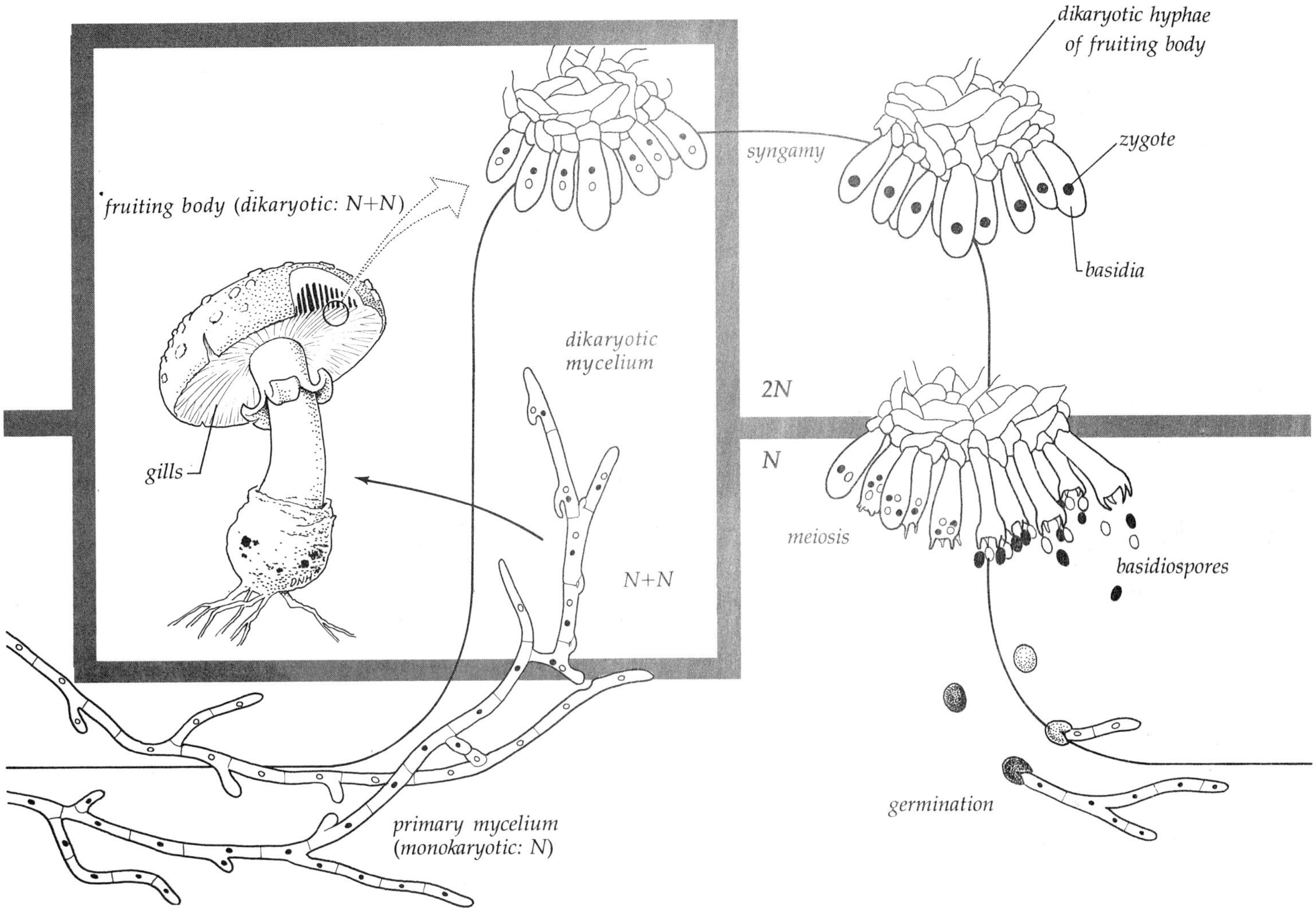

20–10 *Reproduction in the mushrooms.*

structure is solid, but the underside is permeated with numerous holes. These connect to cylindrical chambers that are lined with hyphae-producing basidiospores. The spores, shot from the basidia, fall through the opening at the bottom of the chamber and are dispersed by the wind. Unlike most fruiting bodies that persist only briefly, those of the shelf fungi remain intact for many years. Each year new growth occurs at the edges of the old structure, and spores are produced in this portion.

Not all Basidiomycetes rely on airborne spores. Some such as the stinkhorn, *Phallus impudicus,* utilize insects. These insects are flies attracted to the stinkhorn's fruiting body by the characteristic carrion smell and presence of slime in which the ripe spores are embedded (Fig. 20–12). The flies eat the slime and basidiospores, which they then unintentionally scatter far and wide.

Basidiomycetes also have representatives among the dung fungi, and they too have unusual spore dispersal mechanisms. One such fungus is *Sphaerobolus,* which can be found on old dung after *Pilobolus* and the Ascomycetes have grown. The basidiospores are formed into a spherical mass 1 mm in diameter and sit in a cup of vegetative hyphae consisting of two layers, the inner layer being highly turgid. This layer flips inside-out,

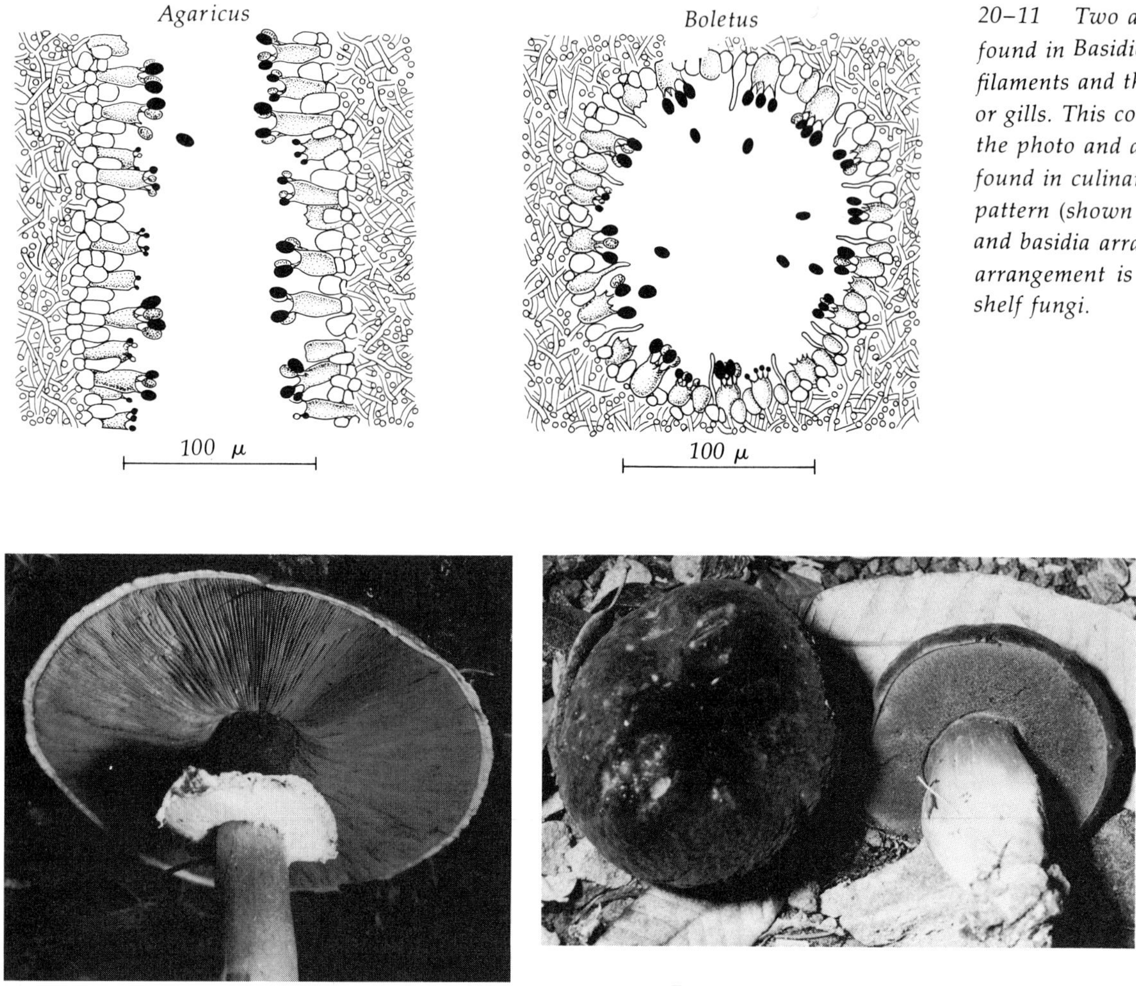

20–11 Two arrangements of the basidia are found in Basidiomycetes. In one the hyphal filaments and the basidia are arranged in rows or gills. This commonly seen pattern (shown in the photo and diagram on the left) is the one found in culinary mushrooms. The other pattern (shown on the right) has the filaments and basidia arranged in a cylinder. This arrangement is found in both mushrooms and shelf fungi.

shooting the spore mass several meters upward (Fig. 20–13). *Sphaerobolus* is also commonly found growing on rotten logs.

Besides using spores and the slow growth of the vegetative mycelia to carry the plant to new sources of food, some Basidiomycetes have developed special hyphal strands called rhizomorphs. These look surprisingly like roots of vascular plants and can grow rapidly over considerable distances to establish the fungus in a new host. One of the most common Basidiomycetes having well-developed rhizomorphs is *Armillaria mellea,* a tree parasite (Fig. 20–14).

No discussion of the Basidiomycetes would be complete without mentioning poisonous mushrooms. It has been estimated that of the 4000 species of mushrooms known some 30 to 40 may be poisonous. These species can be placed in two general groups: those in which the toxins cause vomiting by the victim and are usually not fatal, and those in which the toxins enter the bloodstream without complications and are fatal. The second group contains *Amanita phalloides* or the Angel of Death mushroom. This is a dangerous organism and should never be treated lightly—indeed, all poisonous mushrooms are serious business. Many wild mushrooms make good eating, but they should only be collected by a regional expert. There are no simple rules in collecting mushrooms, and there is no way to test for the toxins when you are preparing the mushrooms. Naturally, all mushrooms sold in grocery stores are cultivated species and therefore perfectly safe to eat.

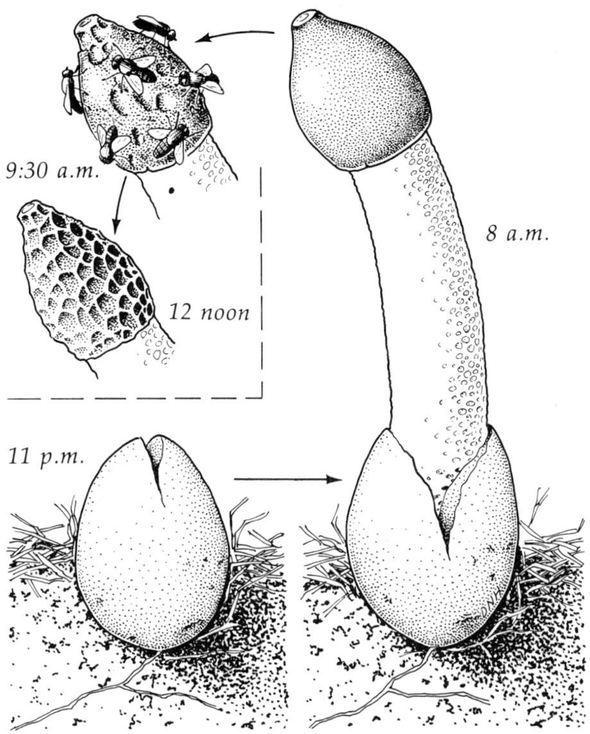

20–12 *The spores in some fungi, such as the* Phallus impudicus, *are dispersed by means other than the wind. In this particular fungus the fruiting body elongates during the night, and by morning the cap bearing a spore-slime mixture is well aboveground. By 9:30 the slime gives out a strong smell that attracts flies. By noon the flies will have removed all the spore-slime. These spores both adhere to the flies and pass unaffected through their alimentary tract.*

Rusts. Not all Basidiomycetes are saprophytes growing in soil and rotten wood. Some are the worst parasites of cultivated plants with which man must contend: the various rusts such as the white pine blister rust, the smuts, and the many other diseases of vascular plants. The life history of the rusts is discussed in End Paper 20–1. In these species the fruiting bodies are usually not conspicuous structures but appear as stripes or blotches of color on the stems and leaves of the infected plant.

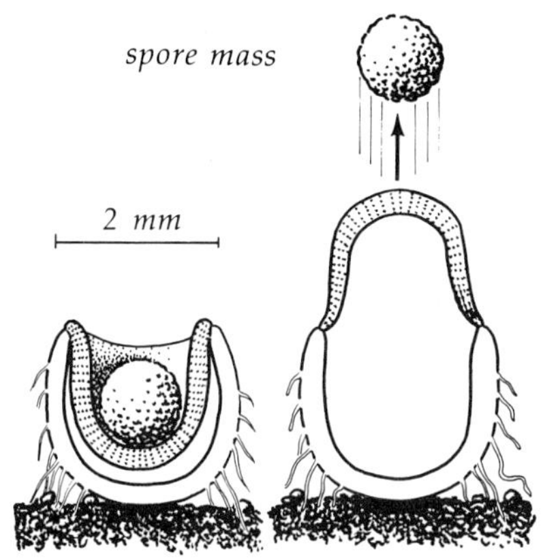

20–13 *Spore discharge in* Sphaerobolus.

Besides the rusts, the other large group of parasitic Basidiomycetes is the smuts. Their life cycles are like the rusts, except that there are no alternate hosts. The common name is derived from the masses of black, powdery teliospores that are produced, frequently on the fruit of the host. *Ustilago maydis,* a smut on corn, can be extremely destructive if the proper environmental conditions are present.

Deuteromycetes—Fungi Imperfecti

This division of fungi contains some 10,000 species; yet it has been called "a dustbin group." The label is accurate. It is an artificial collection of fungi based on the fact that these fungi reproduce only asexually, and assignment to other divisions requires knowledge of the mode of sexual reproduction. The name Fungi Imperfecti is derived from the concept that the sexual phase is the perfect stage; hence fungi lacking this stage are imperfect. If a sexual stage of a member of the Fungi Imperfecti is observed, it is placed in the appropriate division. Many mycologists believe that most of the Fungi Imperfecti are really Ascomycetes, if their vegetative characteristics are the criteria.

The Fungi Imperfecti constitute a diverse assortment of fungi including *Trichophyton interdigitale,* the cause of athlete's foot. Several other fungi that para-

(a)

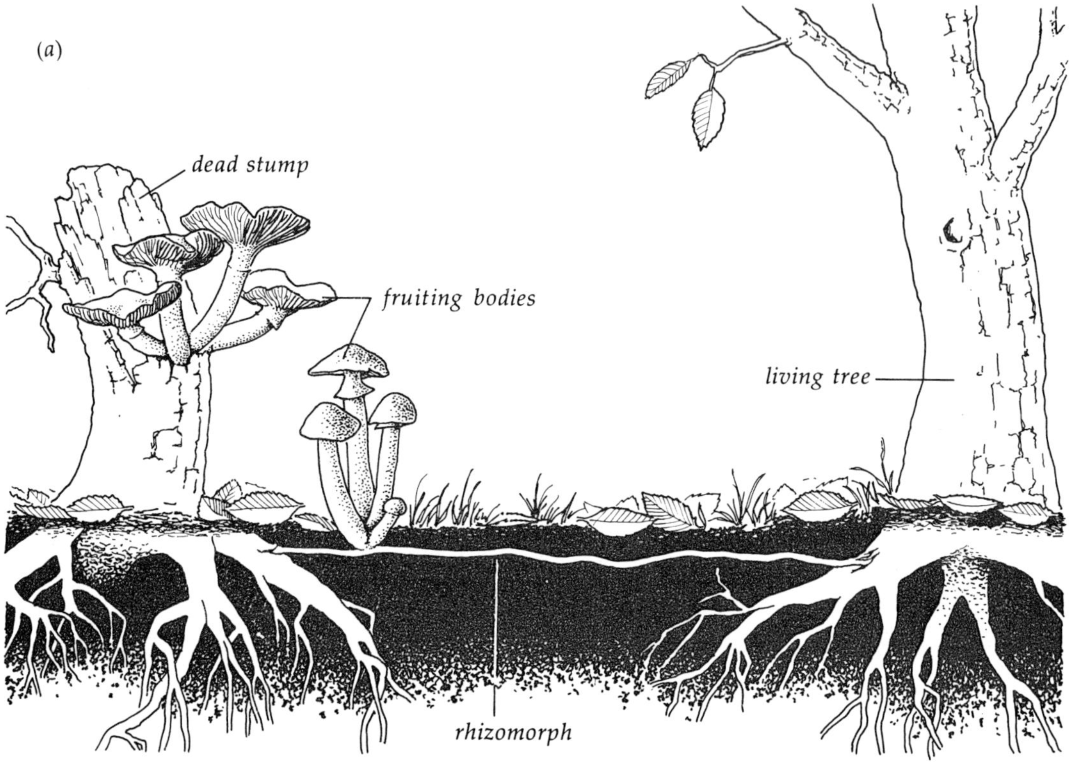

20–14 Armillaria mellea *spreads from one tree to another through the development of a rootlike hyphal mass, the rhizomorph (a). The remarkable resemblance of rhizomorphs to roots can be seen when they are sectioned and examined under the microscope (b).*

(b)

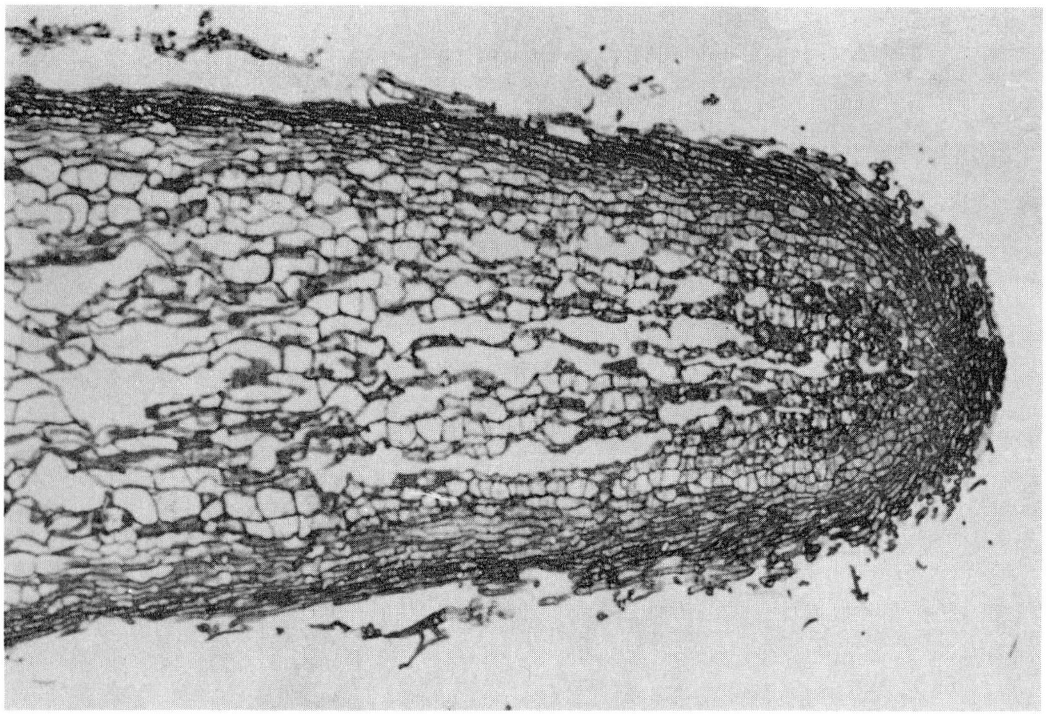

sitize man are members of this class. Cat and dog ringworm are also Fungi Imperfecti, as is head and skin ringworm of man. On the brighter side, one of the Fungi Imperfecti traps nematodes or eelworms. Nematodes are small worm-shaped animals only a few millimeters long that are major pests of orchard trees and many crops, since they infect the roots and cause great harm. *Dactylella bembicodes* (Fig. 20–15) captures nematodes with a trap consisting of a three-celled ring that is attached to the mycelium by a short cell acting as a stalk. The nematode enters the ring triggering the cells, which rapidly increase in diameter. The harder the nematode struggles, the harder it is held—frequently managing to be trapped at both ends. Hyphae then grow into the nematode utilizing the body as a food source. *Dactylella* can grow in the soil without nematodes to prey upon, but if nematodes appear or if the fungus is placed in water in which nematodes have been growing, the fungus will rapidly form traps. Thus the prey itself induces the formation of the trap. Hope is held that *Dactylella* would prove an effective biological control for nematodes, but thus far it has not.

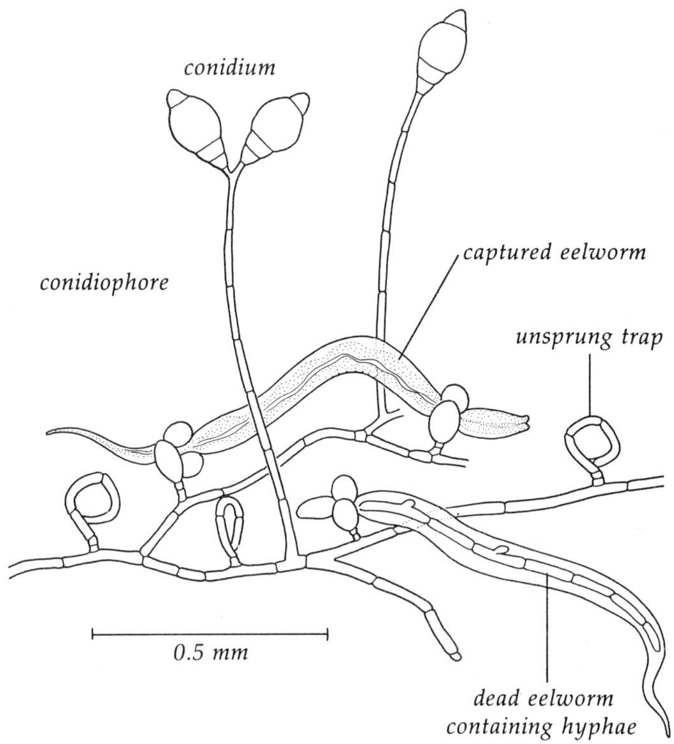

20–15 Dactylella *is a fungus that actually traps and then grows into nematodes. Three cells of the fungus form a ring, which then expands and traps the little animals.*

Myxomycophyta

The Myxomycophyta or the slime molds lack cell walls, except in the fruiting bodies, and ingest solid food particles. They exist as multinucleate masses or as uninucleate naked cells in moist soil, wood, decaying vegetation, or dung. The fruiting bodies and spores that they produce are similar to those of the fungi (Fig. 20–16c). Thus they sometimes resemble animals and sometimes plants. One of the founders of mycology, the great German botanist Heinrich de Barry (1831–1888), considered slime molds animals, and there are modern scientists who continue to agree with him and have even placed them in the protozoa. Yet most investigators who have worked intensively with them believe they are plants and place them in a separate division. Some, such as George Martin at the University of Iowa and Constantine Alexopoulos at the University of Texas, both renowned for their work with the slime molds, go even further and place all of the fungi in one division, the Mycophyta, making Myxomycotina a subdivision and Myxomycetes a class. We will treat them as a division with three classes (discussing only the first two): the Myxomycetes, the Acrasiomycetes, and the Labyrinthulomycetes. The first are the true slime molds, having multinucleate assimilative bodies, while the other two are composed of uninucleate cells.

True slime molds. A prime example of the Myxomycetes is the common slime mold *Physarum polycephalum* (Fig. 20–16). This fungus can frequently be observed beneath the bark of rotting tree trunks. It is yellow and has a netlike form. If some of the plasmodium, as this structure is called, is kept moist, it can be taken back to the laboratory, grown on moist towels, and fed oatmeal. The plasmodium will engulf pieces of the oatmeal in the same manner as it does bacteria, wood particles, and other organic matter in nature. This assimilation of food is identical to phagocytosis used

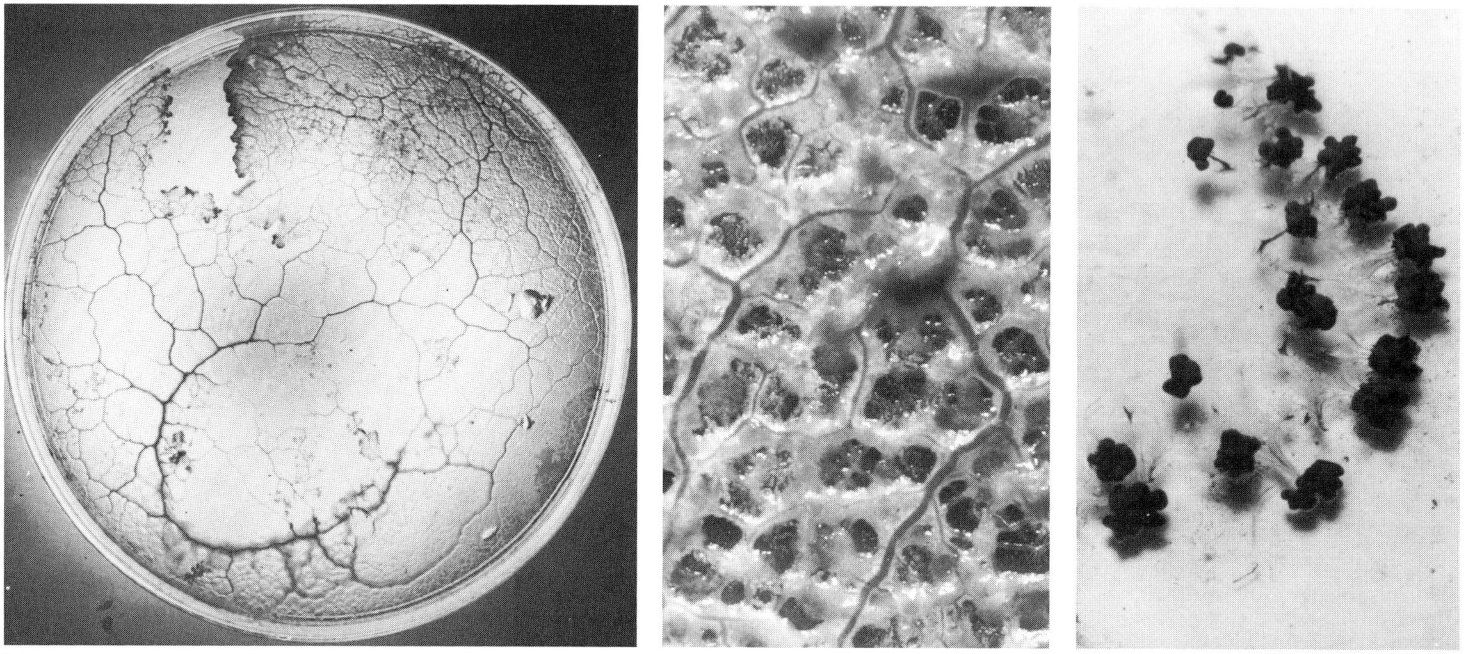

(*a*) (*b*) (*c*)

20–16 *(a) The plasmodium of* Physarum *as it crawls over the surface of agar. (b) A higher magnification of a plasmodium showing the detailed structure of the moving mass. (c) The fruiting bodies of* Physarum.

by amoeba. Liquids are also taken into the plasmodium by pinocytosis (see Chapter 4).

The plasmodium moves slowly over the substrate in response to gradients of moisture and dissolved nutrients. Typically, the migrating plasmodia are fan-shaped and are composed of a network of veinlike strands. The strands are more apparent toward the base of the plasmodium, while the advancing face is more uniform and lacks the veinlike organization. Each tubular strand consists of a semisolid, outer layer of protoplasm and a fluid inner portion. Fibers of contractile protein, identical to that found in animal muscle (actomyosin), are responsible for the streaming motion within the strands. The fluid protoplasm undergoes rapid rhythmic streaming; it flows in one direction for a few seconds, slows to a stop, then reverses and flows in the opposite direction. In the streaming protoplasm can be seen organelles, pigment granules, bacterial cells, and other particles ingested by the plasmodium.

When the conditions for vegetative growth become unfavorable (e.g., drying out and low temperatures), the plasmodium may produce a resistant structure, the sclerotium. This has a hard outer cover and may live for several years. Similar environmental conditions, together with decreasing food supply and other unknown factors, may stimulate the plasmodium to form spores instead of a sclerotium.

When sporulation is about to take place, the plasmodium creeps to an exposed location and forms several masses of protoplasm that become the sporangia. Each sporangium consists of a stalk and a spherical or rounded tip in which the spores form (Fig. 20–17). Meiosis occurs in the spores. The spores are surrounded by a rigid wall.

Upon germination, the spores of *Physarum polycephalum* form myxamoebae, which are similar to true amoebae and move only by amoeboid motion. In other species swarm cells are formed. These cells may either swim with whiplash flagella or move by amoeboid motion. Both myxamoebae and swarm cells can feed, divide, and form resistant cells. They also function as gametes. In the case of *Physarum* two myxamoebae

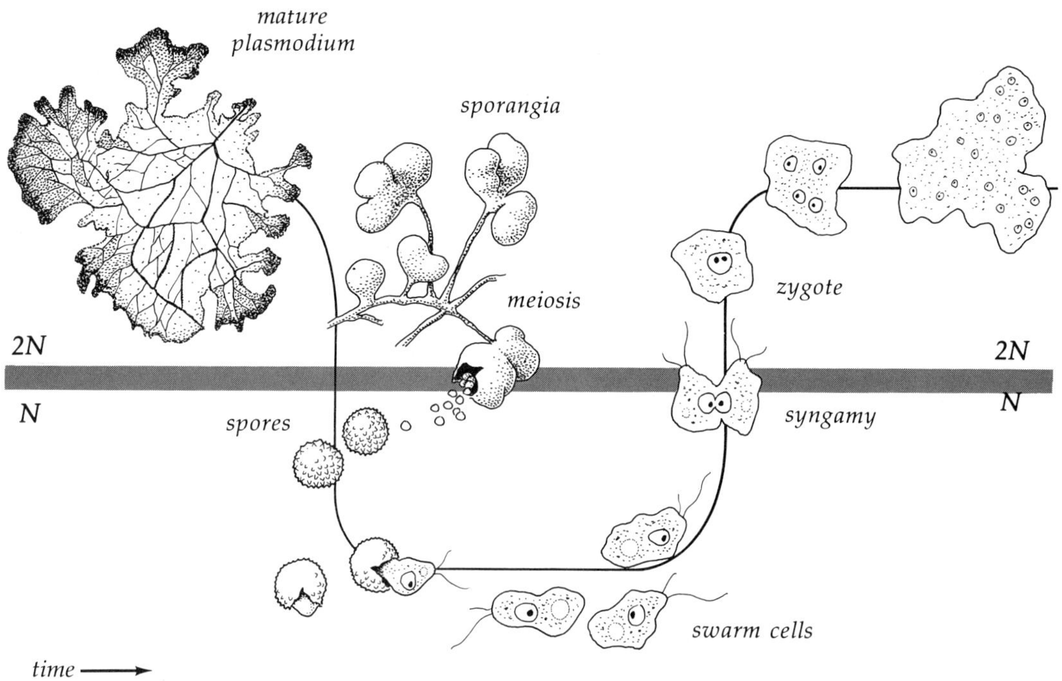

20–17 The life history of *Physarum polycephalum.*

fuse to form a zygote. The zygote nucleus divides mitotically in a matter of hours, and the daughter nuclei continue to divide (usually synchronously) forming a plasmodium. A plasmodium may form from a single zygote, or several zygotes or small plasmodia may coalesce, which occurs most commonly.

The sporangia of the true slime molds are small, beautiful objects in a range of forms (see the Photo Essay). Many consist of a filamentous stalk supporting tiny spherical, cylindrical, or feathery heads. In others the stalks are missing, and the sporangia consist of groups of globular or cylindrical bodies. Frequently, the sporangia are beautifully colored.

Cellular Slime Molds. Not all slime molds have true plasmodia. The ones that do not, which belong to the Acrasiomycetes and the Labyrinthulomycetes, are called the cellular slime molds. The best understood of these is *Dictyostelium discoideum,* which has proven to be an excellent organism for the study of various developmental processes (Fig. 20–18).

The vegetative stage of *Dictyostelium* is not a plasmodium but the free-living myxamoebae. These independent cells feed and divide, until the population density reaches a point where changes occur in the amoebae. Some of these begin to secrete a substance called acrasin. This hormone-like substance is actually cyclic adenosine monophosphate (cyclic AMP; Fig. 20–19), formed from ATP. The action of acrasin is to attract nearby amoebae to congregate where acrasin production is centered. As the amoebae come together, more of them produce acrasin so that amoebae are attracted over an ever-widening area. As the cells come together, they form a conical aggregation called a pseudoplasmodium. Unlike the true plasmodia, the cells in the pseudoplasmodia retain their individuality, and feeding does not occur.

In *Dictyostelium* the pseudoplasmodium, which is known as a slug, migrates. After a period of migration, marked by a characteristic trail of slime, the slug stops and rises upward. A sporocarp consisting of a stalk,

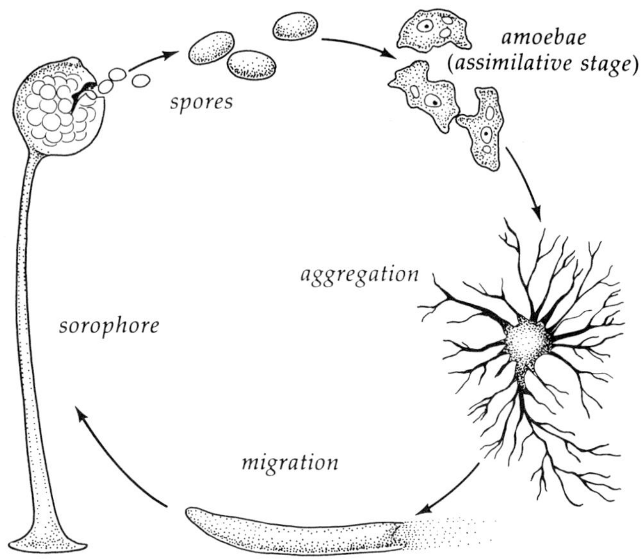

20–18 *The life history of* Dictyostelium discoideum.

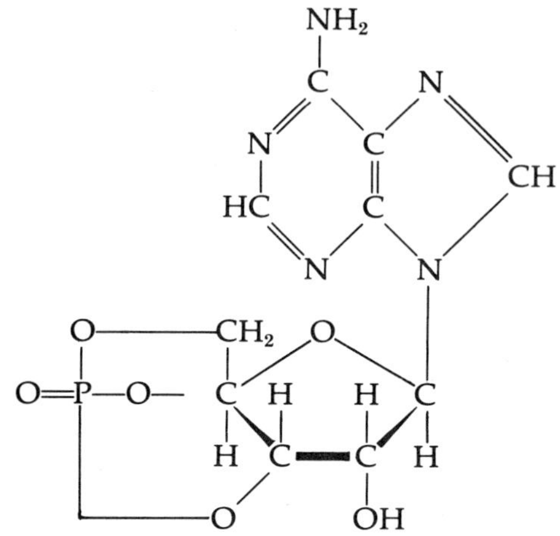

20–19 *The chemical structure of acrasin—cyclic AMP (adenosine monophosphate).*

basal disc, and a bean in which the spores are produced is formed by a remarkable series of interactions of the individual cells. The cells that formed the tip of the slug form a stalk and secrete a cellulose sheath that is strong enough to support the rest of the slug. This literally pulls itself up the stalk by adding new cells to the tip of the stalk. The cells that remain after the stalk is complete form a globular mass at the apex of the stalk. These form the spores, each of which is surrounded by a thin cellulose wall. Experiments have shown that although each cell is independent in the slug and all are morphologically similar, the position of a cell in the slug determines its position in the fruiting body and that the cells in the slug have some means of interaction so that they retain their correct position in the differentiated slug.

Upon germination the spores form amoebae. Where and if sexual reproduction takes place in the cellular slime molds is not known. The relation of the cellular to the plasmodial slime molds is also unknown.

End Paper 20–1 The Biology of Rusts

The classic example of a rust is *Puccinia graminis*, the black stem wheat rust (Fig. 20-20). This fungus has been called man's greatest competitor for wheat—a claim that is undoubtedly true. The black stem rust was probably present on the wheat plants man first brought under cultivation over 7500 years ago in the Fertile Crescent, a 1500-mile range of hills running from southern Palestine, north through Lebanon and Syria, across the south of Turkey, and down through northern Iraq to the Persian Gulf. The common barberry, a pretty bush with dark green foliage and bright red berries, evolved in this same region. The barberry is important to the fungus because it forms the alternate host on which sexual reproduction occurs. The basidiospores of the fungus germinate on the leaf of the barberry. These spores are haploid, and the hyphae that form are monokaryotic. These hyphae grow throughout the leaf and eventually produce two kinds of structures: spermagonia at the upper surface and aecia primordia near the lower surface of the leaf (Fig. 20–20).

The spermogonia are flask-shaped masses of hyphae. Some of these hyphae are short and produce small uninucleate cells called spermatia. The spermatia are suspended in a sugar solution in a spermagonia. Some of the hyphae are much longer and protrude beyond the opening of the spermagonia. These are the receptive hyphae. Insects, in quest of the sugary contents of the spermogonia, inadvertently carry spermatia from one spermagonium to the receptive hypha of another. There, if the mating types are right, the spermagonia will fuse with the receptive hypha. There are two mating types in wheat rust labeled simply + strain and − strain. Viable matings are made only between the two types. When, for example, a + strain spermatium comes in contact with a − strain receptive hypha, the two fuse, and the − strain nucleus of the spermatium migrates into the + strain cell of the receptive hypha. Then the − strain nucleus begins a series of rapid mitotic divisions. The new daughter nuclei migrate to the cells of the mycelium of which the receptive hypha was a part. This migration is possible through the pores in the cross walls of the cells forming the hyphae. The end result of this migration is that the mycelium is now dikaryotic, each cell containing two nuclei of different mating types.

The dikaryotization of the mycelium has the effect of stimulating the development of the aecia primordia. These masses of hyphae, formed earlier near the lower edge of the barberry leaf, now complete their development. They enlarge, grow out through the tissue of the leaf, and open to reveal rows of hyphae, each producing chains of binucleate, dikaryotic aeciospores (Fig. 20–20).

The aeciospores are airborne and rapidly spread to nearby wheat plants. In one case, a single barberry plant was the center for an infection of wheat 10 miles across! The aeciospores are thin-walled and germinate rapidly on the surface of the wheat epidermis. The dikaryotic hyphae that grow from the spores enter the plant through the stomata and rapidly proliferate within the host tissue. They soon grow to the point where they rupture the epidermis and form masses of dikaryotic spores, called urediospores (Fig. 20–20), at the end of short branches. The urediospores are capable of directly infecting healthy wheat plants, and it is through the urediospores that vast areas of wheat are infected. The color of the urediospores is orange to red, and from this the name rust is derived. Each mass of hyphae-producing urediospores is capable of making several thousand a day for several weeks. So many spores have been formed that in exceptionally favorable years for the production of spores, fogs and rains were blood red. The urediospores are killed by freezing during the winter on the northern prairies but can survive in Mexico and parts of the Southwest. The surviving spores are capable of reintroducing the disease in the spring; the prevailing winds carry the new urediospores farther and farther north.

This fungus has still another way to overwinter even in the North. This is through the formation in late summer of another type of spore—teliospores (Fig. 20–20). Changes in the physiology of the wheat plant itself trigger the formation of these spores. Each teliospore consists of two dikaryotic, thick-walled cells. The nuclei of each cell fuse so that the cells are now truly diploid cells. The thick wall enables the teliospores to survive the winter. The black spore walls are also the source of the common name for this specific rust, namely, the black stem wheat rust.

In spring, while the teliospores are still in place, they germinate, forming a short basidium. The nucleus now divides meiotically, and 4 uninucleate haploid cells are formed. Each of these cells produces a single, uninucleate, haploid basidiospore. It is these light, thin-walled, wind-transported spores that infect the barberry plant and continue the cycle. Half of the basidiospores are + strain and half are − strain, thus ensuring cross-mating in the formation of the dikaryotic stage.

This, then, is the remarkable life history of the black stem wheat rust. No less than five spore types and two hosts are involved! But just how important is wheat rust today? In an average year, if all the infected acres were put together, they would form a field as big as the state of Texas; the harvested wheat lost is equal to a pile as large as Mt. Vesuvius, 4000 feet high! It is no metaphor that man is in a life and death struggle with the rusts, and constant labor is required to keep ahead.

How can wheat rust be fought? The most obvious measure is to eradicate the barberry. The common barberry is native to neither Europe nor the Americas but was brought to these areas from Asia Minor because of its pleasant appearance, the medicinal powers attributed to tea made from its bark, and its berries, which could be made into jelly. In addition, there are two species of North American barberry that act as the alternate host, but the imported Japanese barberry will not. The connection between the common barberry and wheat rust was made as early as 1660, when the first barberry eradication law was put into effect in France. The first such law passed in the United States was in 1726 in the colony of Connecticut. Slowly, the barberry was eliminated, but although the picture improved, the rust persisted. The reason was the ability of the urediospores, as well as the teliospores, to overwinter.

The fight then continued through the wheat: could a wheat resistant to rust be found? The answer was yes. First,

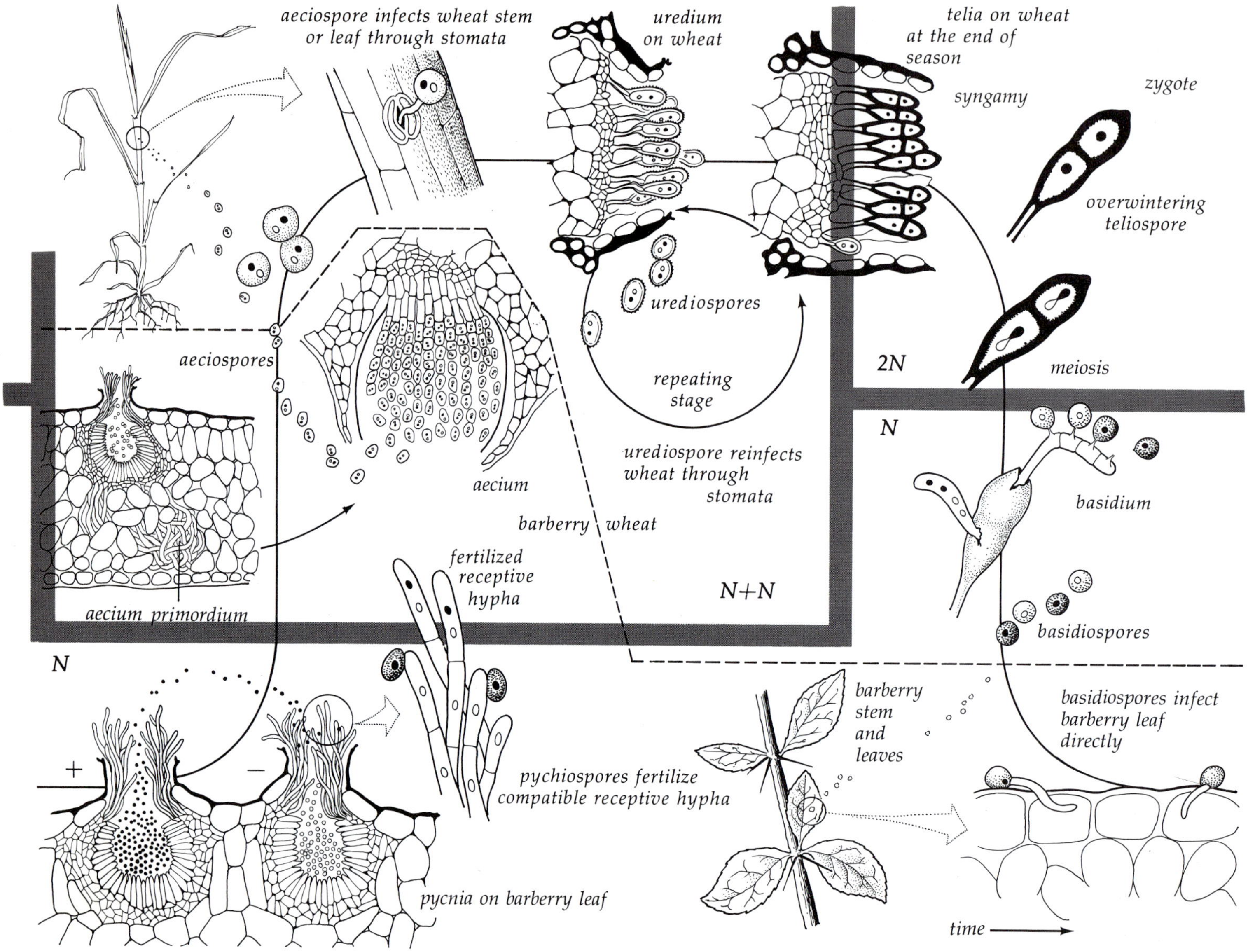

20–20 *The life history of black stem wheat rust,* Puccinia graminis.

there was Marquis, a wheat produced from Italian and Russian parents that remained resistant until 1916. Next, there was Ceres, which lasted until 1935 when rust wiped it out as a reliable strain. New wheats were bred again and again, but after a period of usefulness, they also became subject to the rust. In 1953, 35% of the bread wheat in the United States went down with rust. Are the wheat strains changing? No, the fungus is.

Wheat rust is not a genetically homogeneous population; it is composed of many different races. New races constantly arise through recombination during sexual reproduction and through mutation. It is only a matter of time before a population buildup of a race occurs that can grow on the new variety of wheat. Then, the struggle to breed a new resistant wheat is renewed. The removal of the barberry and the elimination of the sexual phase of the fungus should stop the rust from evolving new strains. Unfortunately, mutations in the fungus provide variation even in the absence of sexual reproduction.

The wheat rusts are by no means the only Basidiomycete rusts. The white pine blister rust (*Cronartium ribicola*) is another with alternate hosts: white pine and gooseberry or currant. In this case, eradication of the alternate host has saved the pines. Other rusts lack alternate hosts and have a simpler, shorter life cycle.

21

Fungal Ecology

Lichens

The lichens represent an extreme in biological togetherness that is hard to match. These plants are actually composed of two separate organisms—a fungus and an alga. Yet this association is so close that the lichens are given genus and species names, as if they were single organisms. Lichens are found in every kind of environment, but particularly under conditions where little else could survive. They have several distinct body plans, which are illustrated in these photographs. The most commonly seen lichens are the crustose types, which grow closely appressed to rocks as shown in A and B. Others may be leaflike in appearance and called foliose (C). Still others are shrubby or hairlike and called fruticose (D and E). For uncounted centuries man has used lichens to dye his clothes, and in the far North, these unusual plants provide food for his animals.

A *Various crustose lichens.*

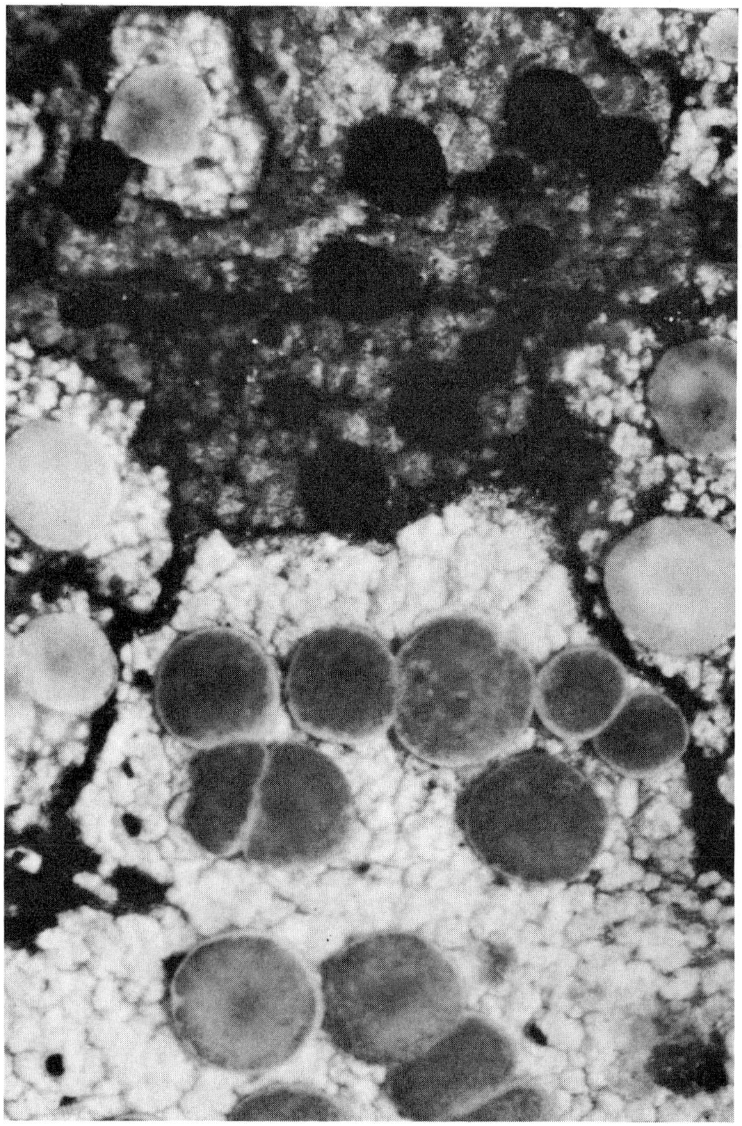

B Catellaria, Buellia, *and* Leoanora.

C Pilophorus.

D Phuscia stellaris.

E Usnea florida.

Stretching across the top of the world, across Europe, Asia, and North America, and continuing for thousands of miles is the tundra. The plants in this harsh environment are on the whole stunted in their growth and sparsely represented. Yet one plant does well under these conditions and is the mainstay for the animals and men of the tundra. This is the reindeer moss, which is not a moss at all but an extraordinary plant called a lichen. There are some 20,000 species of lichens, and they can be found on all continents from the equator to the Arctic and the Antarctic.

21–1 A lichen section reveals interwoven fungal hyphae that form a dense mat at the surface and a looser arrangement near the substrate. Between the dense upper layer and the lower layer are the algal cells.

Lichens

A lichen is a composite structure consisting of a fungus and an alga growing together. The fungus forms a network in which are embedded unicellular algal cells (Fig. 21–1). The great bulk of the fungi found in lichens are Ascomycetes, although a few are Basidiomycetes or Fungi Imperfecti. The algae are either green or blue-green algae. The algal species found in lichens are also found living free in nature, but this is not true of the fungi. These fungi can be cultured free of the algae in the laboratory, but they are never found by themselves in nature. The fungal hyphae are closely appressed to the algal cells and may actually penetrate them. Clearly, the fungus is obtaining carbohydrates and other organic compounds from the alga, but what is the alga receiving in return? The data on this point are not clear but do point to an exchange of minerals and perhaps certain organic compounds absorbed from the fungus. For many years lichens were used as an example of mutualistic symbiosis, the phenomenon described in Chapter 40 where two organisms living together both benefit from the arrangement. More recently, the claim has been made that the fungus is simply a parasite on the alga. The relationship is probably more than this, but its true nature is still unknown. Regardless of the nature of the interaction of the alga and the fungus, the combination is highly successful.

There are a number of body plans in the lichens. Some form a thin crust closely appressed to the substrate, looking almost like a layer of paint on a rock. (Figs. 21–A, B). These lichens are often seen growing on rocks, stone buildings or fences, and on tombstones. Other lichens have a distinctly leafy appearance, more upright in form and less securely fastened to the substrate. Some lichens form shrublike bodies with cylindrical or flattened branches that may be erect or hanging. From a distance, one of the hanging varieties looks like the Spanish moss found in the southern parts of the United States. But the former is a lichen, the latter is a flowering plant, and neither is a moss. There are also lichens that have the feel and texture of a mass of gelatin when wet.

Within the body of the lichen the fungus usually forms a relatively compact layer at the top and bottom, while the algal cells are held within a looser network of hyphae in the center. If the lichen is thick, the algal cells are always located close enough to the surface to receive adequate light.

Physiology and biochemistry of lichens. Lichens can survive prolonged drought. Tests have shown that some species of lichens can survive over 60 weeks without any water; yet no special resistant structures are formed. The water content of lichens closely parallels that in the surrounding air so that fogs and heavy dews can be meaningful in terms of lichen survival.

Many lichens have crystalline deposits on their surfaces. These are unusual organic acids generally known as lichen acids. Most of these compounds are colorless, but some have colors running from yellows through oranges and browns to red. Used as dyes, usually on wool, they are prized for the warm, earthy quality of the hues. The colors of genuine Harris tweeds are from lichen dyes. The lichen acids have also been studied because many of them are antibacterial agents. The hope was that they would be useful in medicine as antibiotics, but this has not been proven.

For many years it was believed that the lichen acids were found only in lichens and never in the fungus or the alga itself. Continued research has indicated that the substances are found in small amounts in the fungal species without the algal component being present. Yet, when the two components are together, the fungus produces vastly larger amounts. The basis for this reaction is unknown.

The rate of both respiration and photosynthesis is low in lichens. Lichens require a high intensity of light to survive.

Reproduction in lichens. Reproduction appears to be a difficult operation in lichens because two different organisms must reproduce at the same time. The most common method of reproduction is by fragmentation. Lichens fragment easily when dry, and small pieces can be carried over long distances. Many lichens also form soredia, which are clumps of algal cells surrounded by a few hyphal strands (Fig. 21–2). These are light and can be easily transported. In many lichens the fungi form spores, and in some only spores are formed with no means of vegetative reproduction. The fruiting of the fungus was long believed to be a vestige of the fungus's prelichen history and not involved in lichen reproduction. Only recently has the following series of events been established, and not all the stages are completely proven.

In the case of an Ascomycetes lichen, an ascospore is produced and released. When it lands in a suitable environment, it germinates and forms a mass of hyphae. If the hypha finds a suitable alga, it surrounds it, forming a new lichen. If it fails to find the right alga, the hypha dies. Growth is slow as the new lichen is established; it may take several years for it to be visible to the naked eye.

What determines if the alga is the right one, that is, suitable for lichen formation? What factors are necessary if the fungus and alga are to grow together successfully? These questions relate to the aggravating experience met by many investigators working with lichens: they could culture the fungus and alga separately, but they could not get them to grow together again! Finally, it was found that the lichen could be resynthesized only under conditions in which neither the fungus nor the alga could support themselves independently.

The work of Vernon Ahmadjian indicates that the algal cells incorporated into the lichen during formation are the ones not killed by the fungus. He observed that when the fungus made contact with the algal cells, it killed many but not all of them. Those that survived were able to live and reproduce in association with the fungus.

The growth of most lichens is slow, from 0.5 to 5 mm per year, but their life span is usually great. Some lichens are known to be hundreds of years old (calculated on their current rate of growth) and have been used in attempts to date archaeological remains. The age of the stone heads on Easter Island have been estimated at 400 years by using lichens. Many lichens have shorter lives, but life spans of 30 to 45 years are common.

Distribution and role of lichens in the ecosystem. There appear to be few environments too severe for lichens. Of the hundreds of plants living in Antarctica, all but a few mosses and flowering plants are lichens. In the Atlantic Ocean, 300 miles west of Scotland, a granite rock projects above the surface, barren except for lichens.

Forests are rich in lichens, provided the shade is not too great. Careful studies have shown that there is a progression of different species of lichen along branches of trees. In addition, there are patterns of lichen distribution from the base to the top of the tree. Finally, various species of trees have different lichen floras. This type of complexity of lichen distribution can also be found in grasslands, tundra, and even intertidal locations.

The distribution of lichens is determined by their general requirement for high light intensities, their ability to survive dry conditions, and the fact that some of them contain nitrogen-fixing, blue-green algae. Thus they are frequently found on bare rock surfaces, where they will grow until shaded out. Lichens are

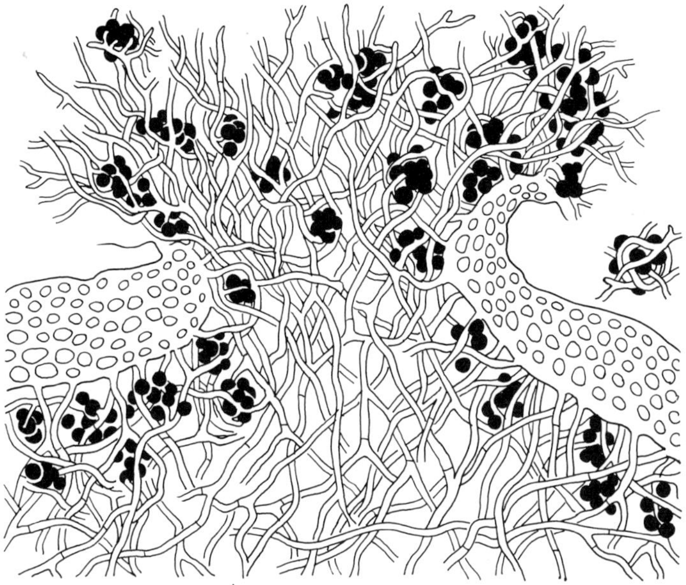

21–2 A cross section though a mass of soredia. Each is a group of algal cells held together by short fungal hyphae.

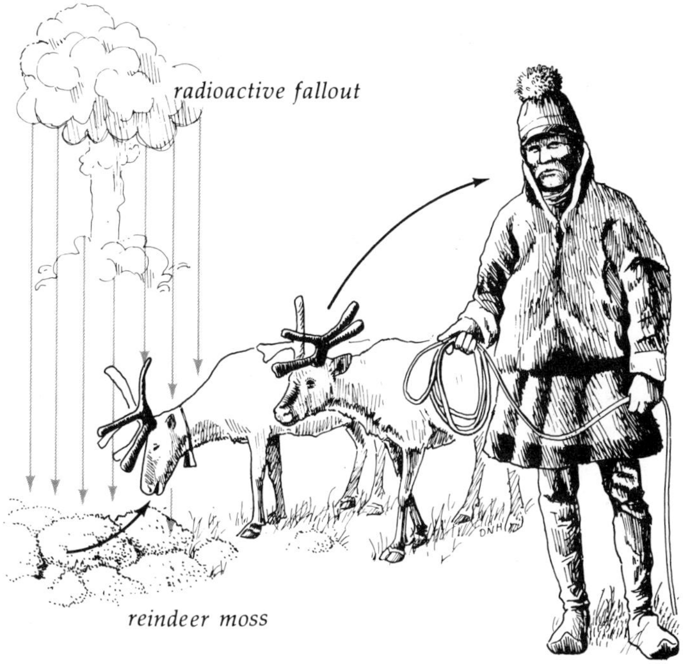

21–3 The path of radioactive isotopes—from fallout to a Laplander.

frequently said to be pioneers in plant succession, but how much they prepare the way for other plants is not clear. They grow so slowly and the amount of rock that they break down is so little that their importance has been questioned.

In the tundra, lichens form a dominant part of the flora and provide 60 to 90% of the diet of reindeer and other grazing animals. The Laplander eating the reindeer is thus dependent on the lichen (Fig. 21–3). In 1965 tests were made on Laplanders for radioactive materials in their bodies. Because of their remoteness from the sites of atmospheric bomb tests, which were then occurring, it might have been expected that their bodies would contain few radioactive elements (e.g., strontium, cesium, polonium, and iron). Yet, while the fallout in the area was low, the Laplanders had 55 times the amount of such materials as the Finns living farther south. The reason was the concentration of these elements by the lichens growing in mineral-deficient soils. The reindeer further concentrated the materials as it grazed and digested the lichens. The Laplander carried on the process by eating the reindeer.

Lichens also absorb and concentrate industrial waste material from the air. When the concentration becomes too great, the lichen dies. Lichens are therefore good indicators of air pollution, but they are also becoming scarce in many of the more industrial parts of the country. Able to withstand the most extreme temperature and moisture conditions, they are becoming victims of industrial pollution.

Mycorrhiza

Fungi interact not only with algae to form mutualistic symbiotic relationships but also with other plants. The most important of these is the association of the hyphae of various fungi (particularly Basidiomycetes) and the roots of vascular plants, mainly trees. Such a combination is known as mycorrhiza, which literally, means "fungus roots." The hyphae form a netlike arrangement around the roots causing them to have a thickened appearance (Fig. 21–4). The fungus is often one of those that forms the familiar mushroom fruiting bodies of the forest, although as a mycorrhiza it is a

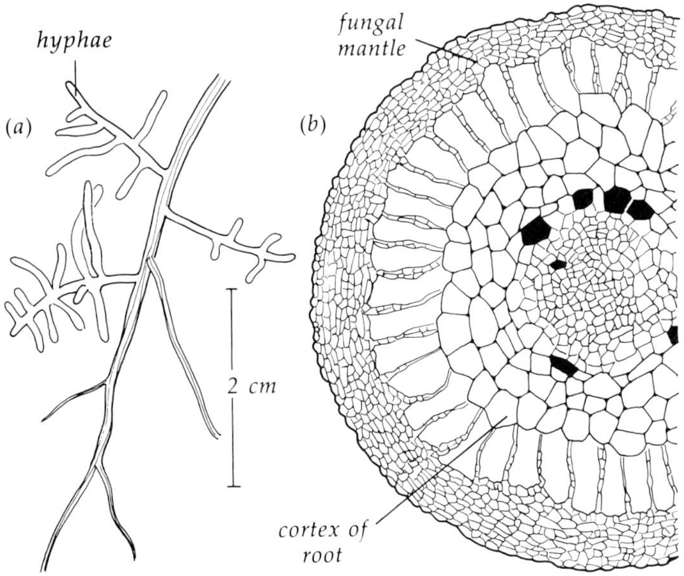

21–4 Mycorrhizal roots appear thicker than normal roots (a), and in section can be seen to have an elaborate mantle of fungal hyphae (b).

parasite living off carbohydrates it takes from the tree; yet studies show that the mycorrhiza aid in the absorption of salts and other materials from the soil by the tree. In this type of mycorrhiza the hyphae, although intercellular in the cortex of the root, project for a considerable distance external to the root. A second type of mycorrhiza also exists (Fig. 21–5). Here, the hyphae grow within the root itself and are not involved in mineral absorption. This type of mycorrhiza is found in germinating orchid seeds, corn, strawberries, and many tropical roots. In nature the tiny, poorly developed orchid embryos cannot grow until infected by the fungus. In the laboratory the effect of the fungus can be replaced with the plant growth hormone, indoleacetic acid (IAA). The conclusion is that the fungus is supplying the embryo with the growth hormone that it is unable to produce in adequate amounts during its early development.

Mycorrhiza are more important to the ecosystem than just aiding the establishment of orchid seedlings. They play a direct role in the growth of many trees. For example, when pines, which are not native to either country, were introduced into Puerto Rico and Australia, they grew poorly until soil containing mycorrhizal fungi was introduced. After the addition of the right kind of fungi to the soil, mycorrhiza developed, and the trees grew rapidly. There is little doubt that mycorrhiza are important to the growth of conifers around the world.

Recent observations by F. W. Went and N. Stark of the Desert Research Institute, Nevada, suggest that mycorrhiza are essential for tree growth in tropical rain forests with poor soils. Went and Stark observed that in the Amazon rain forests there were masses of mycorrhizal roots. Most of these were confined to a layer of litter (1 to 2 cm deep) that had accumulated on the forest floor. Below this thin layer were poor, yellow, water-logged clays. Such thin layers of litter and humus are typical of many tropical forests; they result not from the want of leaf and similar litter falling on the forest floor but from the rapid breakdown of this material and the leaching away by the rain of any soluble products.

In the humus layer Went and Stark found masses of mycorrhiza encircling and penetrating fallen fruits and branches, growing into termite galleries and growing under the bark of dead branches. The hyphae appeared to be involved in the breakdown of the woody material and other plant litter. The close association of the mycorrhiza with the degenerating material suggests that there is an immediate transfer of substances from the dead material to the root. Such a closed pathway would result in the maximum efficiency of recycled chemicals and reduce the amount lost by leaching. It is interesting to note that in the tropics nitrogen-fixing bacteria occur in nodules, while in temperate zones they are free in the soil. Again, the problem of leaching is circumvented in tropical plants by having the source of nitrogen built into the roots.

Forest Fungi

We began Chapter 19 with a discussion of the fate of a leaf as it fell to the forest floor. The point was made that the upper layers of the soil in a forest are quite literally permeated with living organisms. Bacteria, blue-green algae, various terrestrial algae, and, above all, the fungi permeate the soil and the debris lying on it.

One way of illustrating the role of fungi in the breakdown of forest debris is to look, even superficially,

(a)

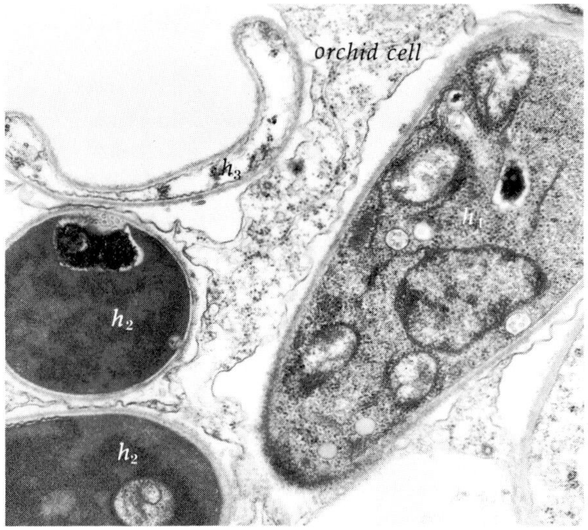

(b)

21–5 (a) Orchid seeds (0.5 mm in diameter) on a bed of leaves and debris. The thin threads of material running between them (see the arrows) are the fungal hyphae. (b) An electron microscope photograph of an orchid cell from a seed similar to the ones shown in (a). Several hyphae, at different stages of development, can be seen within the cell. The one on the right (h_1) is in a vigorous stage of growth, while the two dark hyphae on the left (h_2) are past the growth phase and filled with a protein storage material. Finally, the collapsed hypha on the top (h_3) is dead. The various changes in the hyphae are apparently regulated by the orchid. The fungus supplies the orchid with growth substances (which it later makes for itself) that are necessary for its early development.

at the amount of such material under the trees in various forests in different climates. In northern coniferous forests, where both the nature of the debris and the temperature act against fungal growth, the amount of the debris is large, although the amount of material falling from the trees is less than in the more southern deciduous forests. In a deciduous forest much more material (leaves, twigs, branches, etc.) reaches the forest floor, but because there is more rain and the fallen material provides a better substrate, the fungal flora growing there is luxurious, and the breakdown of the debris is rapid. This results in a lesser amount of visible material on the forest floor. In the tropical rain forest the situation, as we have seen, is even more exaggerated. The amount of debris reaching the floor is huge, but the high rainfall and temperatures result in almost instant utilization of the fallen material by the rich fungal flora. There is a clear relationship between the fungi, the soil, and the fertility of the forests. The faster dead plant and animal matter is returned to the soil, the sooner it can be reused by living organisms. One of the problems with reforestation, following intense forest fires, is that the heat has reduced the microflora. If, before some growth can be established, rains leach too many minerals from the soil, it may take decades to return to its former fertility.

Forest fungi may be roughly grouped into those forming mycorrhiza, those growing on trees, logs, leaves, and litter, those growing in open spaces, and those growing in the shade. There are no clear dividing lines between these rather loose habitats, and individual species are widely distributed through many or all of them. Basidiomycetes, Ascomycetes, Fungi Imperfecti, and Myxomycetes, as well as bacteria, are all prominent in the forest fungal flora.

The range of forest fungi with regard to host species specificity is great. Some fungi are found in many different forests; some are found only in coniferous or deciduous forests; and some are restricted to certain species. In the case of trees, some fungi attack only the dry, nonfunctional heartwood in the center of the trunk, while others grow only in the still functional, moist sapwood just below the bark (although usually not before the tree is dead). A heartwood fungus may enter a tree through a broken branch, a rotten root, or an insect burrow. Strictly speaking, the fungus is a saprophyte, since it digests dead cells of the wood. A hollow tree will obviously blow over more easily in a storm, and when it does fall, the fungi that helped cause its destruction usually continue to grow in the log.

Once the log is on the forest floor a series of intricate changes occur. For one thing, the sapwood begins to dry, and the bark loosens as the living cells immediately beneath it degenerate. Enough water is retained in the wood so spores sifting through cracks and openings in the bark can germinate. Fungal hyphae, bacteria, and insects all aid in the decay of the wood. There is usually a succession of various species that utilize different parts of the cell wall. The fungi that break down the more soluble parts of the cell wall are usually followed by those that digest cellulose and finally by lignin-attacking species.

Usually, a number of species is present at all times, and there is active competition between the various organisms. Many factors determine which fungi will be present in a given local habitat. One of these is the ability of the fungus to invade a new substrate either by the germination of spores or by the penetration of hyphae. The specific substrate may offer more resistance to the hyphae of one species than that of another, depending on the chemical composition of the walls; chemicals in the wood may adversely affect the germination of the spores of one species over another. As microorganisms grow in the decaying wood, they change the environment by secreting toxic metabolites that may, again, have more effect on one species of fungus than on another. According to David Park of Queens University, Belfast, the fungi most widely distributed and most successful in the forest are characterized by

(a) *rapid germination of spores and a high rate of hyphal growth,* both favoring rapid colonization,
(b) *good enzyme production,* which favors rapid substrate utilization,
(c) *production of substances toxic to other organisms,* which may reduce competition from other organisms for available substrates, and
(d) *tolerance of antibiotic substances produced by other organisms,* which allows growth even in the presence of competitors producing toxin or to succeed toxin-producing organisms.

Most frequently, fungi do not attack young, healthy, vigorous trees in an undisturbed environment. But, as the trees age, they become susceptible to fungal invasion; this is also true if there are detrimental changes in the environment. Aging is important not only in the trunks of trees but in their leaves as well. Young, healthy leaves secrete substances that inhibit the germination of fungal spores or, when they do germinate, they do not penetrate the leaves. As the leaves mature the number of fungi on the leaves, particularly species that are small and yeastlike in form, increase and the potency of the host's resistance mechanism declines. Thus, even before the leaf we discussed in Chapter 19 reaches the forest floor, its decay has begun. Again, changes in the environment, particularly the amount of moisture in the atmosphere, can drastically alter the picture (note the case of potato blight).

Grassland Fungi

The species of fungi found in grasslands and prairies are different from those seen in forests. Members of the Ascomycetes, Basidiomycetes, Zygomycetes, and Fungi Imperfecti are all found commonly among the grassland fungi.

The dead leaves, stems, and roots of grasses provide the nutrient source for these fungi. In Chapter 19 we discussed the growth of fungal mycelia through turf and the relationship between the mycelia and the turf. Many grassland fungi have fruiting bodies with longer stalks than forest species. This projects the cap above the tops of the grass, thus facilitating spore dispersal. The actual growth response that results in the longer stalk is probably regulated by relative humidity that is higher in the grass layer, thus allowing stalk elongation to continue for a longer period. Also common in grasslands are puffball fungi in which the spores are liberated by animals stepping on the thin surface of the mature fruiting body.

A special environment found in grasslands is created by dung. In Chapters 19 and 20 we saw fungi from widely diverse groups living on dung, but it is also true that although dung appears to be an almost ideal substrate for a great many fungi, only a relative few actually grow on it. The reasons for this illustrate the complex relationships between fungi and their environment.

The animal (e.g., a cow or horse) eats leaves on which the spores of the fungi are adhering (Fig. 21–6). As the ingested material moves through the animal, it is essentially sterilized by the digestive enzymes and high temperatures. The only surviving spores are those of a few resistant fungal species. Some of these fungi are so resistant to the conditions within the animal that they begin to germinate while still within the animal. Thus, when the dung is deposited, they have already grown somewhat and are subjected to little antagonism from other species whose spores land on the dung after it is on the ground. Moreover, the initial colonizers already in the dung have the further advantages of rapid growth, high enzyme synthesis, and efficient utilization of substrate.

As the earliest fungi utilize the dung, it changes to a substrate for later fungi. In addition, the moisture content is changing, and as the dung becomes drier its physical character changes. These alterations in the environment result in a succession of species growing on it. As we saw earlier, the first to appear within one or two days is the Zygomycete *Pilobolus* (see Chapter 19). This is followed by various Ascomycetes such as *Sordaria* (see Chapter 20), and these are then followed in old, dry dung by Basidiomycetes such as *Sphaerobolus* (see Chapter 20).

Each of these fungi from widely divergent groups shows common features, such as phototropic responses of the reproductive parts, that aid in the dispersal of

21–6 The pathway of spores of dung fungi.

the spores to the surrounding vegetation where they can be eaten. They also have large spores, frequently dark in color and coated with mucilage, that are resistant to the digestive enzymes of the herbivores and may, as we have seen, germinate in the animal. Thus the dung fungi have evolved a truly remarkable collection of unique features that enables them to successfully colonize and utilize a select environment.

Desert and Sand Dune Fungi

Deserts and sand dunes would appear to be unlikely locations for fungi, but they grow in both places. Wooden sticks have been shown to decompose within two years of being in desert soils as a result of fungal decay. Fungal hyphae also surround the roots of many desert plants, both annual and perennial, but are not mycorrhiza. In many desert soils the hyphae appear to hold the sand particles together and stabilize the soil.

This is also true for sand dunes. On dunes that have stopped moving the first colonizers are frequently blue-green algae, followed by a variety of fungi especially adapted to the harsh environment. These fungi utilize various debris (primarily animal), as well as dead algal colonizers, for their substrate.

Fungi are frequently found as early colonizers on bare parent rock and play an important role in the formation of soil. They soon follow the initial algal or lichen invaders and may actually precede them, remaining in the dormant spore stage until suitable substrate appears. As higher plants and animals arrive, the fungi are important in building up the organic matter in the developing soil and in adding stability through the presence of extensive mycelia growth.

Aquatic Fungi

Fungi, particularly members of the Chytridiomycetes and the Oomycetes, are found in all types of aquatic environments. Plant and animal debris falls into streams, lakes, pools, and ditches, where it acts as substrate for the water molds.

One study of a small portion of a stream in England discovered twelve species of aquatic fungi. These showed marked seasonal variation—an increase in numbers in the autumn, a maximum in the winter, a decline in the spring, and a low during the summer. The increase in the autumn was correlated with an increase in leaves, twigs, fruit, etc., which provide the substrate, and with a decrease in temperature and an increase in water due to fall rains. The summer decrease was correlated with the absence of substrate, low water, and higher temperatures.

As we saw in our discussion of the algae, ponds and lakes contain a number of different ecological areas. In the case of fungi, these are the open water of the mass of the lake, the edges (particularly with regard to submerged structures, e.g., branches which may be attached to the bottom), and the bottom.

In the open water are found not only the planktonic algae and animals but various floating objects. One representative of the latter category is pollen. In many ponds and lakes the amount of pollen, particularly from surrounding pine trees, may be so great it can support numerous chytrids (Fig. 21–7). These tiny motile fungi attack the grains shortly after they hit the water and complete their life cycle in two days. When the chytrid is through with the pollen grain, bacteria and protozoa invade.

Many aquatic fungi, particularly the chytrids, attack both plant and animal plankton. In some cases, the attacks can reach epidemic proportions, essentially wiping out a particular type of planktonic organism. This seems particularly true for certain of the zooplankton. In other cases, the fungi infect the organisms only after physical changes in the environment of the lake apparently weaken the host. Studies indicate that this is true for various algal species found in the plankton.

The same general pattern of a weakened or damaged individual being colonized by a fungus is true for fish. A common fungus found on fish is *Saprolegnia*. Once the fungus is established on the fish, the latter may die, but healthy fish are apparently not attacked.

Submerged branches, twigs, fruits, leaves, and other matter are readily attacked by fungi. The species of fungus involved depends on a number of factors. One is the depth the object is submerged, since oxygen decreases with depth. Another is the amount of wave action, since the abrasion of the substrate against the bottom can mechanically damage the fungus if the water is too rough. In general, shallow water with gentle wave action, which increases the amount of available oxygen, favors the growth and reproduction of fungi, particularly the Oomycetes. Ascomycetes and Fungi Imperfecti, which are also found in these same locations, appear less influenced by wave action.

The water at the bottom of a lake contains much less dissolved oxygen than in the upper levels. This fact is believed to be important in the distribution of fungi on the bottoms of lakes. The few species found here are mainly chytrids and Fungi Imperfecti with few, if any, Oomycetes. The number of fungi is usually highest in winter, when the oxygen content is greatest and the water temperature lowest.

There are also fungi that are found in salt water lakes, marshes, and in the ocean. Relatively little is known concerning marine fungi. Most of the studies made thus far have been conducted on submerged matter in shallow water. Oomycetes, Chytridiomycetes, Ascomycetes, and Fungi Imperfecti are all found in the ocean.

Aquatic fungi are extremely sensitive to water pollution. In streams and rivers heavily polluted by industrial wastes, there are few fungi. Where sewage is present, the number of species is usually markedly reduced, and there is a distribution of various fungi in relation to the cleanliness of the water. The lack of saprophytic fungi in such polluted waters must decidedly affect the entire nutrient cycle.

Thermophilic Fungi

In Chapters 16 and 17 we discussed the thermophilic bacteria and blue-green algae. There are also thermophilic fungi that can grow at or above 50° C—one even grows at 60° C. This places them well below the maximum temperature for bacteria and blue-green algae but above the bulk of other fungi and higher plants. These fungi are found in compost heaps, damp haystacks, and similar piles of moist plant matter. The first organisms to grow in such places are bacteria and fungi with ordinary temperature ranges. As the temperature in the pile increases as a result of respiration, the pioneer organisms can no longer survive, and the first of the thermophilic fungi and bacteria appear. As the temperature continues to climb, only the most hearty of the thermophiles can survive (e.g., the thermophilic bacteria that are still alive at 90° C). If the temperature continues to increase as a function of physical-chemical interactions, spontaneous combustion may occur.

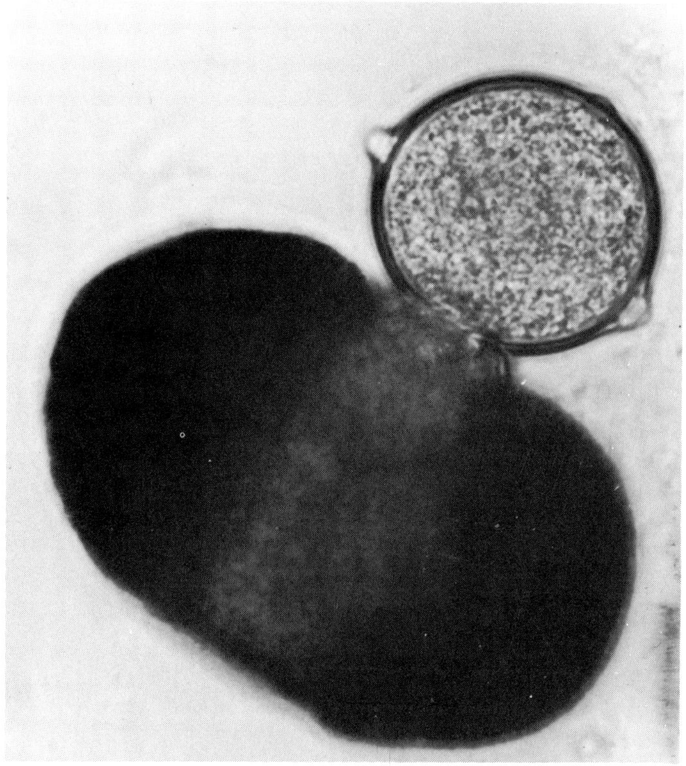

21–7 The mature thallus of the chytrid Rhizophydium sphaerothica, *as seen on a pollen grain of* Pinus.

Parasitic Fungi and the Ecosystem

The parasitic fungi also have an important place in the ecosystem. Under natural conditions where the parasite and host have evolved together, a fine balance exists between the two. Normally, only the old individuals or those weakened by adverse environmental conditions are killed by the parasite. Clearly, evolutionary pressures would act to eliminate the overactive pathogen. When man, in his quest for food, moves plants from their original habitat into one that favors not only the crop but the parasite as well, epidemics may occur. This was the case for the late blight of potato. In Peru, the native home of both the fungus and the potato, the disease is a minor one and not a serious threat to the crop. In Ireland the wetter, warmer climate turned a minor disease into a major disaster. The mildew of grape and the white pine blister rust are both examples of fungi to which the native hosts had evolved resistance.

One of the major threats implicit in the present deterioration of the natural environment is that we are losing valuable stocks of disease-resistant native plants. Plant breeders have long relied on such strains to build resistant genes into the genetic pools of domesticated crop plants. As more and more land is converted into crops, fewer of the native varieties and species remain. At the same time, we become more and more vulnerable to epidemics of parasitic fungi as more land is planted in fewer species of plants, and atmospheric pollution continually threatens the health of the plants. The recent corn blight is one example, which, although bad enough, is minor in comparison to an epidemic affecting the new wheat and rice varieties.

The fungi are of immense importance to man and the ecosystem. They are both helpful and harmful to man, but like so many features of our current environment, their detrimental effects are often due to man's manipulation of the environment. As we have stressed before, parasitic fungi are more opportunists than villains. If, before it is too late, we can learn how to better manage both the helpful and harmful aspects of the fungi, we shall be a long way in achieving a better life for man.

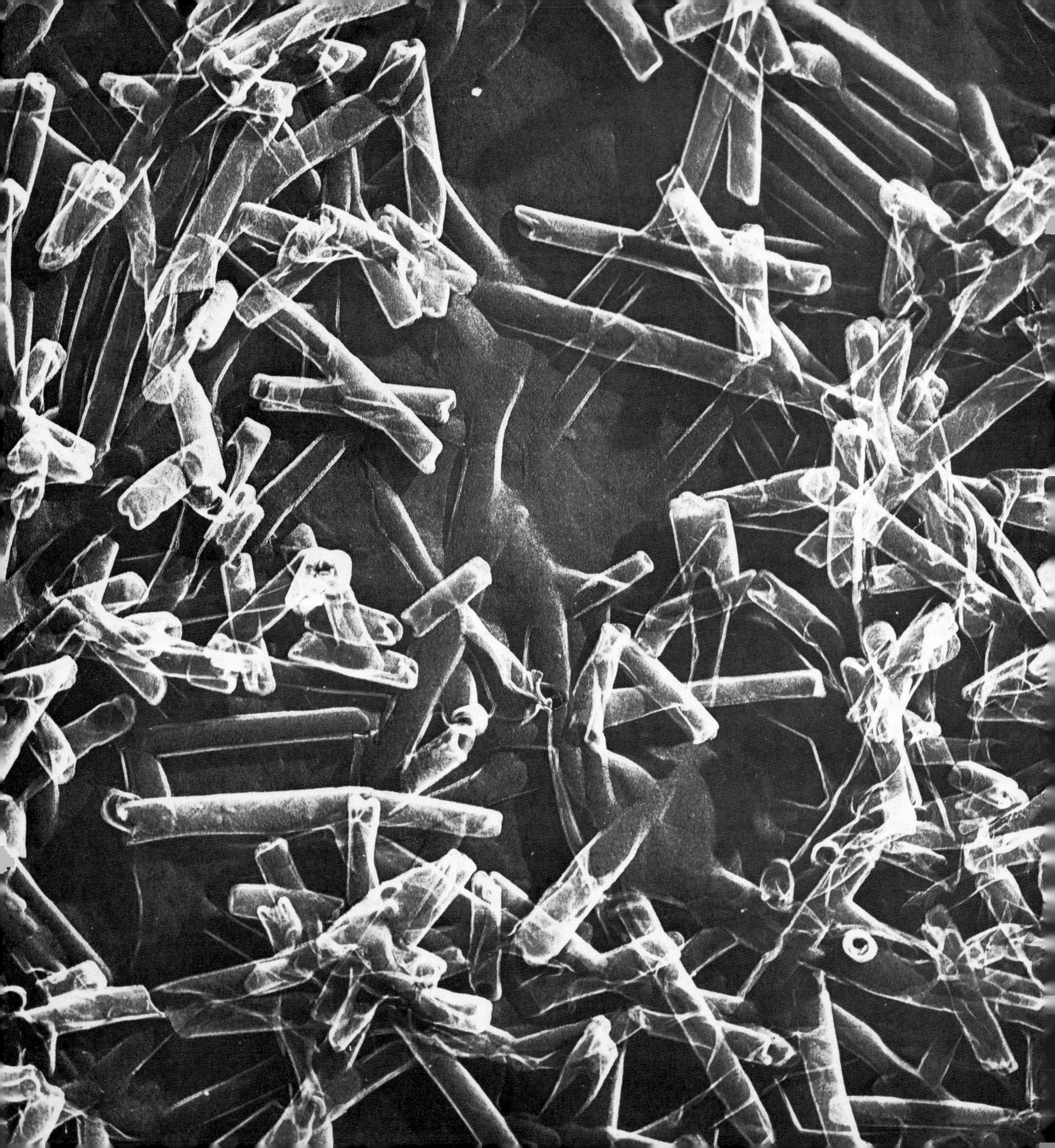

22

The Invasion of the Land

The oceans teemed with life. Invertebrates, often of great size, were present in vast numbers. Blue-green, green, red, and brown algae all crowded the lush, living seas. The sun shone brightly on the land, but there were no plants to absorb it. The land was barren. The time is the Silurian some 400 million years ago—3 billion or more years after life began on earth. During all this time plants and animals evolved in the oceans, but they did not invade the land.

Then, although the fossil record is scanty, near the end of the Silurian or during the early Devonian, small simple plants appeared on the land. Perhaps the first was a little plant known as *Cooksonia.* This plant had a slender, equally branching, leafless axis under 2 mm in diameter. Fossils of *Cooksonia* have been found in Wales, western Czechoslovakia, Siberia, and Canada. Other similar, relatively simple plants evolved during the next several million years, and for the next 5 million years they held dominion over the land. They were the vanguard of a great army of land plants that would literally cover the earth. From them and similar plants evolved the ferns, horsetails, club mosses, gymnosperms, and flowering plants that surround us.

Characteristic Features of Land Plants

What was it that made the land a more difficult environment for plants than the oceans? The answer is water or, more precisely, the distribution of water. In an aquatic environment, a plant is surrounded by water. This water contains all the dissolved minerals the plant needs, as well as CO_2 and O_2 (Fig. 22–1). The surrounding water also supports the plant, since it is a relatively dense medium in which the plant is buoyant. The water surrounding the plant provides all the water the cells need; hence the need to transport water is nonexistent. Virtually all of the cells carry on photosynthesis, and as a result there is no movement of

22–A *When plants finally became adapted to life on land, they evolved a number of important characters that enabled them to survive the conditions of their new environment. One of these was a special outer layer, the cuticle, that covered the external surface of the plant, helping to prevent loss of water. The outermost layer of the cuticle exudes a waxy secretion that accumulates in bazaar and beautiful forms when viewed with the electron microscope.*

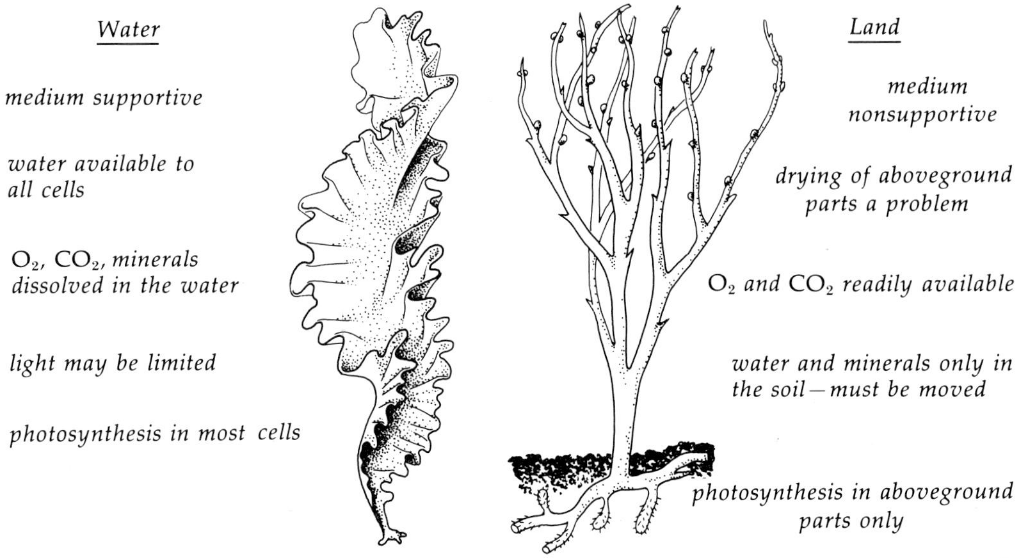

22–1 The difference between water and land as environments for plant growth.

carbohydrate from one part of the plant to another, except for a few large marine algae. It is advantageous for many algae to remain in place; thus a holdfast has developed.

On land, water is still present, but it is in the ground, since any surface water will rapidly evaporate into the air (Fig. 22–1). This is equally true of water in cells; an algal cell placed on land soon dries out. For a plant that evolved in water, this loss of water through evaporation is critical. It is hardly surprising that a common feature of all land plants is a special, relatively waterproof, outer-wall layer called a cuticle (Fig. 22–2). An accompanying feature of the cuticle is the pore or stomate. These are openings in the epidermis, the outermost layer of cells of a land plant. The outer walls of epidermal cells contain the cuticle. The stomatal openings are necessary for the adequate exchange of gases (CO_2 and O_2) between the outside and inside of the plant. In most land plants there are two specialized cells, the guard cells, that regulate the size of the opening.

Water loss can also be reduced by increasing the volume to surface ratio of the plant. In a flat structure the ratio of surface to volume is high, and water loss is great. In a cube or sphere the ratio is much less and decreases as the size of the object increases. But, as the size of an object increases, structural problems result from stresses that build internally. In a plant such stresses are frequently countered with changes in cell-wall structure or composition. Land plants contain an additional wall compound, lignin, that better enables them to resist these stresses.

Lignin is found in special support cells in land plants. Such cells are necessary because the plant, to successfully adapt to the land environment, must evolve structures that allow it to absorb water (underground) and light (aboveground). All land plants have evolved structures to absorb water from the soil. These may be simple rhizoids, which are elongations of cells in contact with the soil, or the more elaborate roots. Roots have several functions, only one of which (anchorage) they share with holdfasts. In addition, roots are necessary for the absorption of water and minerals from the soil and their transport to the aboveground parts of the plant.

Obviously, photosynthesis cannot occur underground, where there is no light. This environmental condition acted as a selective pressure that resulted in the evolution of aboveground, photosynthetic organs. In the earliest land plants only a vertical axis, the stem, is found. It is the major site of photosynthesis. In later land plants portions of the stem are modified into more specialized photosynthetic units, the leaves. The leaves and the stems that support them form a continuous system, the shoot (Fig. 22–3). One of the principal functions of the stem is to transport water to the leaves

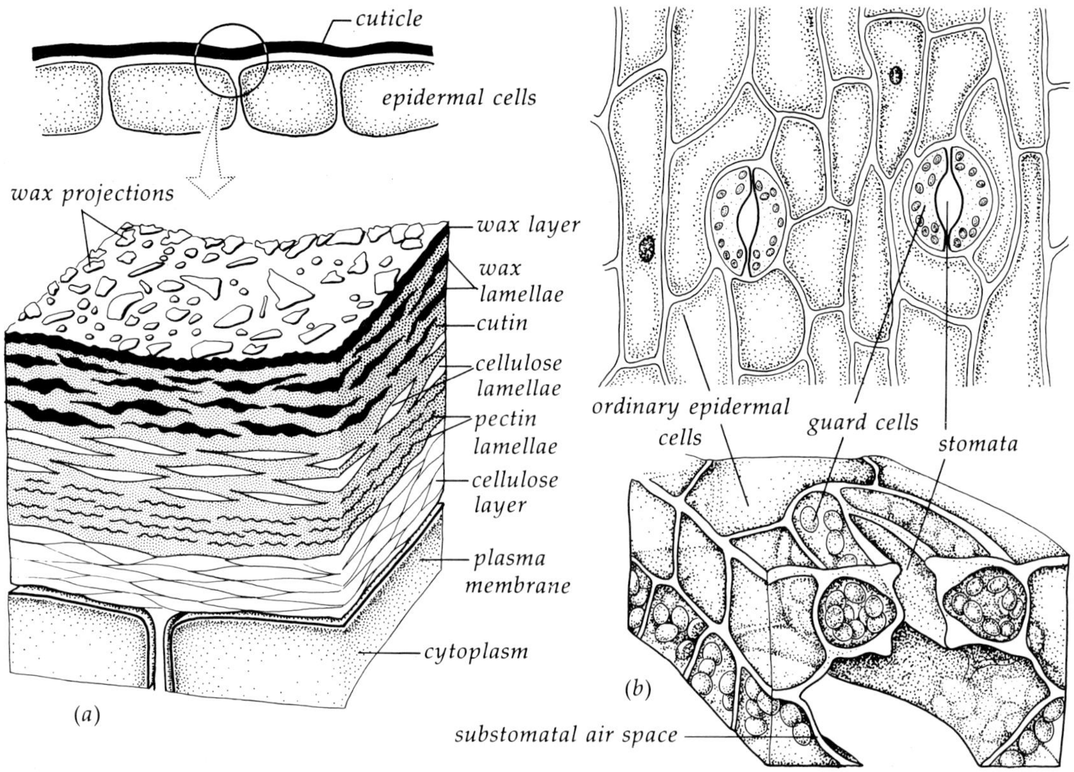

22–2 Two important features of plants, which evolved as they adapted to the land environment, are the cuticle and the stomates, with their surrounding guard cells. Diagram (a), summarizing the composition and construction of the cuticle, shows that it is a relatively elaborate structure. The stomates (b) are actually openings in the epidermis, and their size is regulated by the two guard cells.

The products of photosynthesis must also be transported from the leaves to the stem and the root. In response to these problems, land plants have evolved systems of specialized cells for transport and support. In vascular plants we find the xylem, which is involved in water movement, and the phloem, which is involved in the translocation of organic material. The cells of the xylem have thick walls containing lignin. The water-conductive cells of the xylem are devoid of any cell contents—they are literally pipes of water. The cells of the phloem are also highly specialized for transport, but in this case they are alive at the time they function. The xylem also functions as a major support system in many land plants. In addition, land plants have other cells that are modified as support elements.

Land plants have developed highly sensitive methods of detecting gravity and mechanisms for altering growth patterns with respect to gravity. This, too, is a necessary evolutionary step if plants are to develop on land.

Modifications in reproductive as well as vegetative structures have occurred in response to the limitations of the land environment. Clearly, the motile, swimming zoospore so widely found in the algae and aquatic fungi has limited value for the land plant. In contrast, spores with walls that prevent drying of the cell contents are of great survival value on land. All terrestrial fungi and green land plants have developed some type of resistant reproductive structure. These may be simple spores or the more elaborate seeds, but all have one kind or the other. Finally, in terms of sexual reproduction, a universal feature of green land plants is the enclosure of the cells that will form the gametes and the spores with jackets of sterile cells. The obvious function of the sterile cells is to protect the reproductive cells from desiccation.

One of the curious aspects of adaptation to land is the continued existence, for millions of years, of the motile, swimming sperm. Only the most highly evolved green land plants possess a mechanism for the delivery of the male gamete that does not require it to swim. This is the pollen and pollen tube found in the gymnosperms and angiosperms. In all other green land plants, at the critical phase of sexual reproduction when the

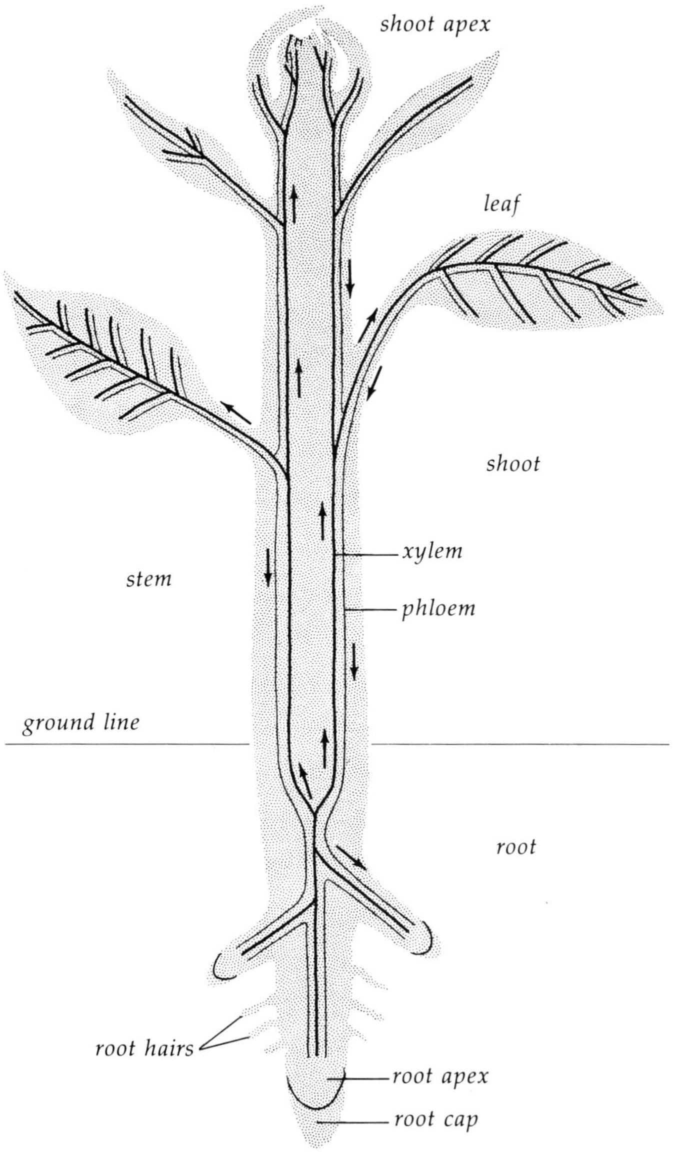

22–3 The general body plan of a vascular plant. The arrows indicate the direction of movement of materials in the xylem and the phloem.

sperm must reach the egg, water is needed for the sperm to swim! Obviously, this poses problems for the plants involved, but it must not be concluded that they are unsuccessful land plants because they possess motile sperm. The gymnosperms and angiosperms are simply better adapted and therefore more successful.

In all land plants fertilization occurs while the egg is still held within the female sex organ on the parent plant. The early development of the embryo also takes place there, where it is protected against the external environment (particularly against drying out). In the oldest land plants the embryo continues its development into a mature sporophyte in the gametophyte. In more recent land plants the embryo is provided with a food supply, protected with a tough outer covering, and released from the parent plant (now a sporophyte itself) in the form of a seed—one of the most profound and ingenious innovations that plants have evolved.

The common set of features evolved by green land plants in response to the land environment are:

1. *The cuticle*
2. *The pores or stomates surrounded by guard cells*
3. *Rhizoids or roots*
4. *Lignin*
5. *Vascular tissues—the xylem and the phloem*
6. *Resistant spores and seeds*
7. *The gametes enclosed in a jacket of sterile cells*
8. *The embryos formed and held in parent tissue*
9. *The highly refined, geotropic response system*

Obviously, not *every* land plant has all of these, but a majority do.

Origin of Land Plants

What were the ancestors of the land plant? The answer, which comes primarily from comparative studies of present-day plants (not from the fossil record) is quite clear: the green algae. This conclusion is based largely on the composition of the photosynthetic pigments. The green algae and all of the green land plants have the same kinds of chlorophylls—namely, chlorophylls a and b. This finding is considered particularly important in light of the fundamental nature of photosynthesis. In addition, the green algae and the green land plants all have true starch as their food reserve, which they store in chloroplasts or amyloplasts. Furthermore, when flagellated cells are present, they all have whiplash flagella. No group of algae have so many features in common with the green land plants as do the Chlorophyta.

The green alga that was most probably the ancestor

of the land plants was a filamentous species with a highly differentiated body. One example of this type is *Fritschiella* (Fig. 22–4). This alga lives on mud flats and has a body composed of a group of cells in the center from which arise branched, erect, photosynthetic filaments—as well as rhizoidal filaments that penetrate the mud. This general structure appears to be the one from which true land plants might have evolved.

When the land invasion occurred, was it accomplished by a single species, which then radiated out into many? Or, was it repeated many times by many species, each giving rise to a line of radiating species? These are almost impossible questions to answer because the fossil evidence is too meager to warrant a firm conclusion. In the view of Harlan Banks of Cornell University, a paleobotanist who has long worked in this area, the evidence points to a single invasion of green plants with the first plant being something related to *Cooksonia.* This concept is further strengthened by the universal similarity of the types of photosynthetic pigments, flagella types, and starch found in all green plants. Seemingly, there is no environmental reason why all of these features would be the same if the original plants were different. Many other botanists feel that it is improbable that so important an event should have occurred only once. They argue that the lack of a clearly continuous fossil record for the various major groups of land plants is enough to assign them different origins. The only possible resolution of these questions will come from more data (more fossils and more comparative studies), but it is far from certain that a definite answer will ever be available.

In addition to the green plant invasion, there were several other successful moves from aquatic environments to land. One of these was made by the fungi. Their adaptation to land followed that of the green plants, since they require them as a food source. The presence of spores and methods of reproduction not requiring free water made the terrestrial fungi well adapted to life on land. A variety of animals also followed the green plants onto land. The most famous of these are the vertebrates. The amphibians first made it during the late Devonian, while the reptiles followed in the Carboniferous. Insects and spiders adapted to land in the Devonian and became abundant in the Carboniferous period. Land snails also developed during the same period. Thus the land gradually became populated.

22–4 Fritschiella.

23
The Bryophytes

Mosses and Liverworts

Along the sides of paths and sprouting up between the stones themselves are tiny green plants that we all recognize as mosses. Mosses are only one representative of a larger group of plants known as the bryophytes. In addition to the mosses, the bryophytes contain the liverworts and hornworts—plants known to few people. But let us begin this pictorial introduction to the bryophytes with the mosses.

A Each moss plant consists of a small stem, surrounded by rings of leaves, which ends in a base that has rhizoids extending from it. These are not true stems or leaves because there is no vascular tissue present, as this cross section of the stem shows. The center of the stem is occupied with small-diameter, elongate cells, instead of vascular tissue. This particular plant is the Polytrichum juniperinum.

B Individual moss plants rarely grow singly; they are usually found in groups or mats. Concentrated growths of mosses and other bryophytes frequently occur where water is plentiful, as in the rain forest or on faces of rocks over which water is seeping. The plants seen in such mats are gametophytes. In the small group of mosses shown here, the long stalks extending from the smaller gametophyte plants are the sporophytes. Spores are produced in the capsules at the end of the stalks, and these will then germinate, forming the new gametophyte generation.

C The spores germinate to form a filamentous plant called the protonema. Small, budlike groups of cells will differentiate on the protonema, and these will then grow into mature gametophytes. These various stages are shown in these photographs (the arrow points to the bud).

A

B

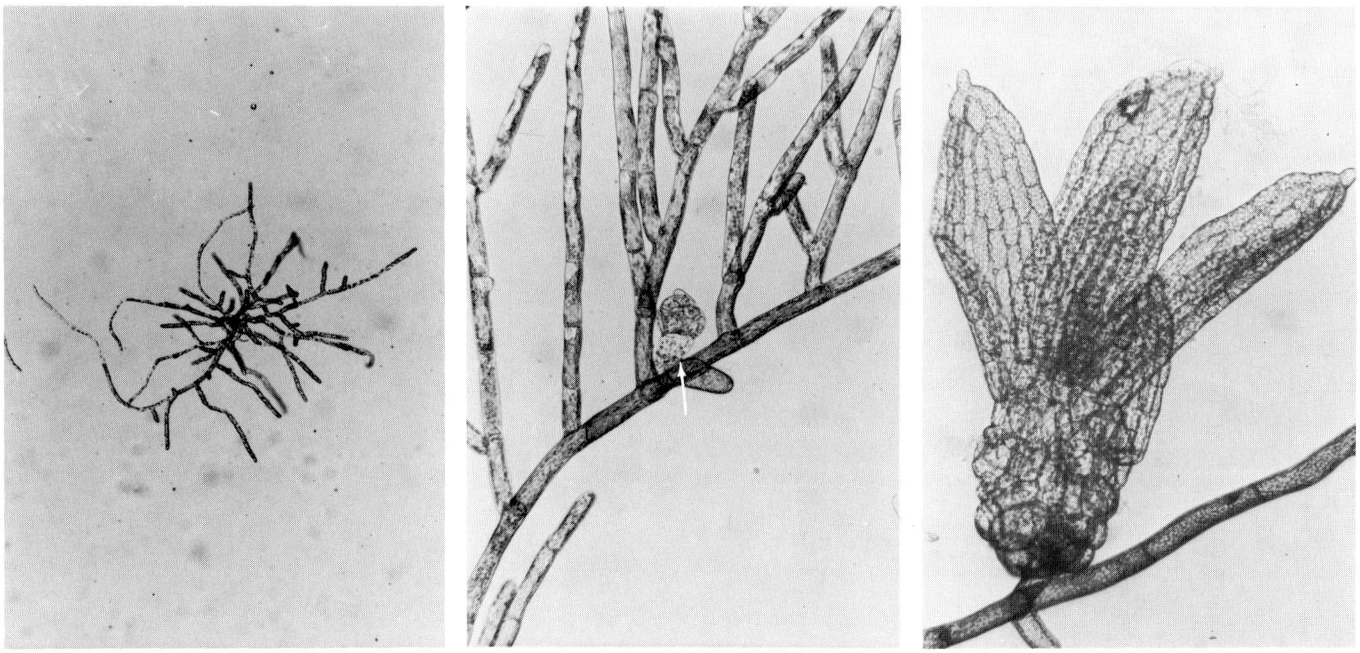

C

D Reproductive structures develop at the top of the leafy shoot of the gametophyte—the archegonia in which the egg is differentiated and the antheridia in which the sperm are produced. In these photographs we see the archegonia, at the tip of a mature gametophyte, at low magnification, and at higher magnification the egg and the neck cells are visible. Also visible at higher magnification are the antheridia with the sperm inside. When released, these sperm will swim down the neck of the archegonium, where one will unite with an egg to form the zygote. The zygote divides in place, and the sporophyte develops on top of the gametophyte.

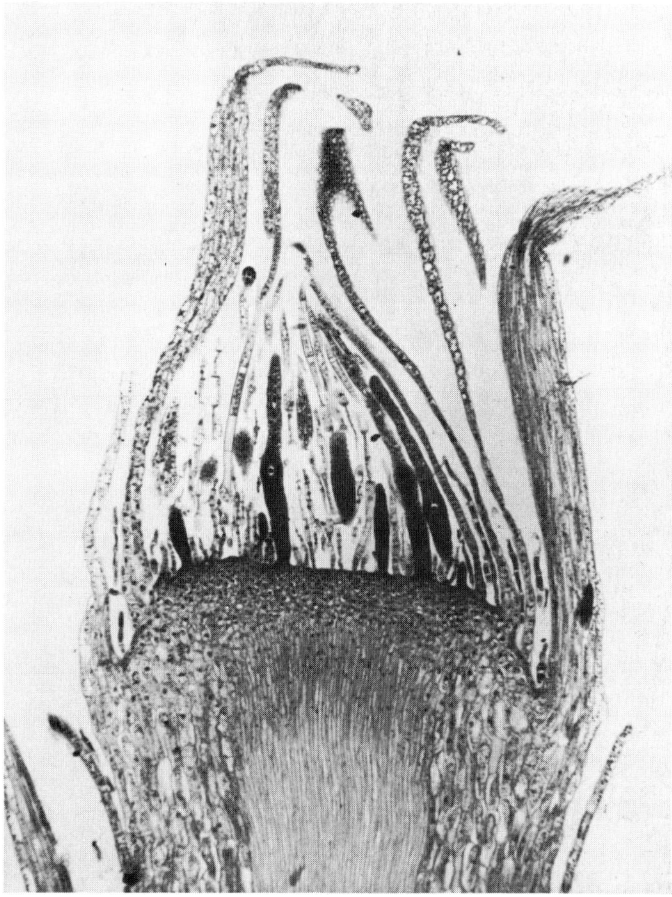

D

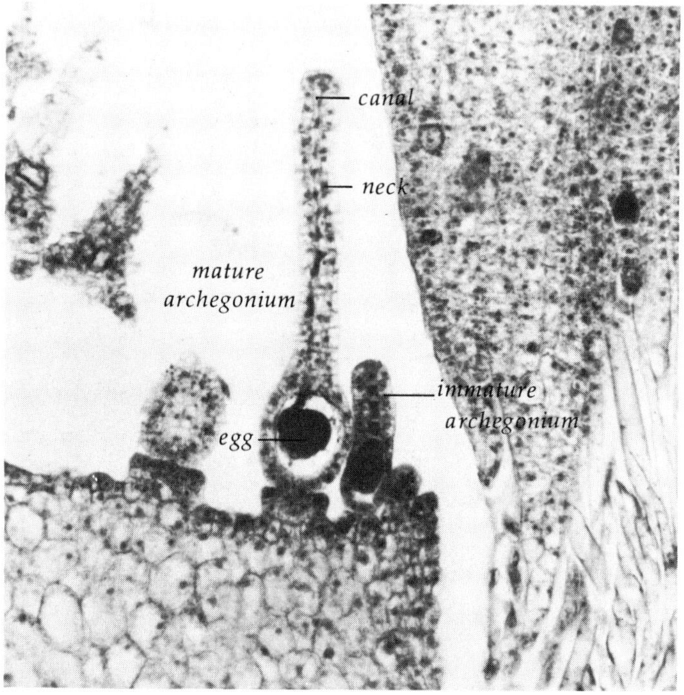

archegonia

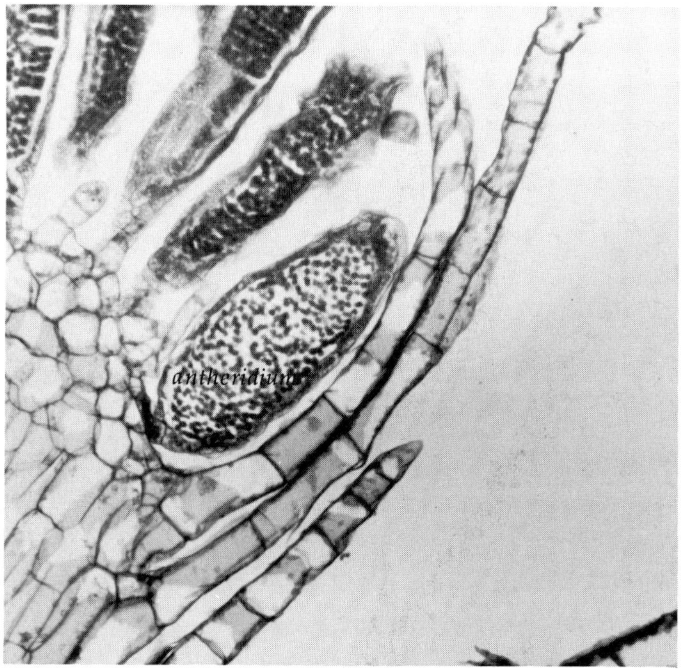

antheridia

Liverworts are relatively small plants that either resemble mosses or have a more prostrate body form from which the common name is derived. While they rarely make up a conspicuous part of our environment, they are found in all parts of the world. Here we present a few photographs of various species of liverworts to give you some idea of their morphology.

E *Two common prostrate liverworts are* Marchantia *(left) and* Conocephalum *(right). The spreading growth of such plants can be seen, as can the regular branching that occurs at the growing tips.*

F *This characteristic of regular branching can better be seen in this photograph of* Carsinia, *which is growing on moist vermiculite in a culture dish.*

F

E

G Not all liverworts are prostrate; they may also be leafy and look like a mat of moss. The basic structure of a leafy liverwort is best seen in Prionolobus *(left), where the delicate, stemlike axis supports leaves only one cell thick.* Frullania *(right) is more complex, each leaf having a saclike lobe that serves as a water reservoir. The branch to the left contains the antheridia.*

H Some liverworts live floating on the surface of lakes and ponds. Only one species is truly aquatic, meaning that it can reproduce in floating condition, but several such as the species of Ricciocarpus *shown here are amphibious, growing on both land and water.*

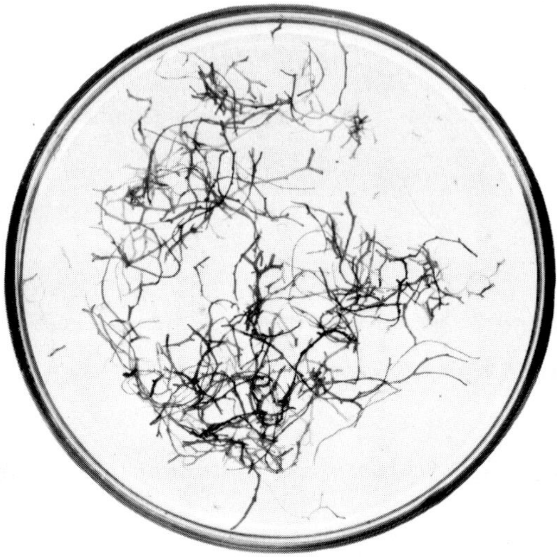

H

G

One day in May, 1950, two Danes, cutting peat in a bog, uncovered the body of a man with a noose around his neck. Thinking he had been recently murdered, they called the police. The police suspected it was not a recent death and called an archaeologist, who established that the body was 2000 years old. Yet his features are completely preserved, and he appears asleep rather than dead (Fig. 23–1).

The remarkable preservation of this Iron Age man is due to his burial place—a bog. A major factor in the environment of a bog is the high acid pH. This results because the original acid pH of glacially produced lakes is maintained and enhanced by a moss, *Sphagnum*, that grows around the margins of the lake. *Sphagnum* tends to increase the acidity through the release of hydrogen ions (protons) and the absorption of other cations. The plants also apparently produce antiseptic substances. (*Sphagnum* was used in surgical dressings through World War I.) The result is a remarkable decrease in the rate of decay of organic matter in bogs. This also means that there is a constant buildup of the remains of the *Sphagnum* over the years. Such accumulations are called peat, which can be cut into chunks, dried, and used as fuel. Many individual sphagnum bogs have been in existence since the close of the Pleistocene glaciation. Besides humans, thousands of years of accumulated pollen grains are preserved in the peat. A detailed analysis of the accumulated pollen can reveal the changes in the surrounding plant community over the years.

23–1 *Tollund man.*

The Bryophytes

Sphagnum is only one of the many mosses we see around us. Everyone is familiar with the mosses growing along the banks of streams, between the bricks of a walk, or on the sides of walls, houses, and flower pots. Worldwide in distribution, their tiny leafy shoots, beautiful green color, and softness to the touch have excited man's interest for centuries. Less familiar are the liverworts. These may be leafy or flatter in shape, sometimes forming mats of dark green over the soil, particularly in forests and under benches in greenhouses. Hornworts are rarer than liverworts, and few people are aware of them, although they occur worldwide. In fact, the degree to which people are aware of all bryophytes directly parallels their numbers—there are 14,500 species of mosses, 9000 of liverworts, and 100 of hornworts.

Most bryophytes are either terrestrial or epiphytic; only one is truly aquatic. They are generally quite luxuriant in wet, humid regions but are also found widely in northern and arctic areas. Many mosses are able to withstand periodic desiccation and can quickly return to their functional condition with the addition of water. Mosses and liverworts are capable of

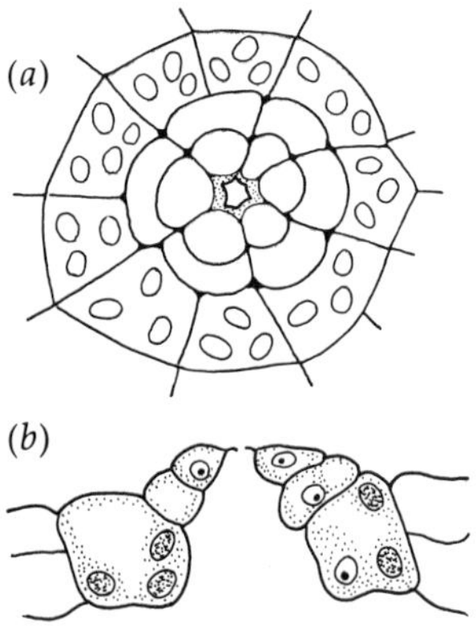

23–2 *Surface (a) and section view (b) of a pore and the surrounding cells in* Ricciocarpus natans.

surviving the physiological drought that occurs when they are frozen during the winter. Various mosses are found on acid soil and on heavy metal deposits. Several species are part of the dung flora, particularly of caribou. Bryophytes even grow on the faces of cliffs—over forty species were found on the famous bird cliffs off northern Norway.

The rain forests of the temperate and tropical latitudes are where bryophytes flourish. In the rain forests of the Olympic peninsula, the trunk and branches of every tree are covered with bryophytes and lichens—plants that are often found together. The protonemal mats of the mosses and the prostrate growth of the bryophytes trap water and minerals flowing down the stems. Bryophytes are also commonly found on well-rotted logs and form part of the sequence of plants involved in the breakdown of logs.

Bryophytes, like lichens, are frequently found as pioneer species on bare rock surfaces. Their ability to withstand drying is important in such locations. How important they actually are as pioneers of growth for other plant species is open to question, as it is with the lichens. They grow slowly under these conditions; little organic matter has been observed to accumulate; and they seem mainly to succeed themselves or lichens.

When we see a mat of mosses, liverworts, or hornworts, we are looking at a gametophyte generation. All of the cells are haploid, a condition familiar enough in green algae but decidedly uncommon among land plants. In the bryophytes, the gametophyte or haploid (n) plant is the dominant independent generation; the sporophyte or diploid ($2n$) plant is smaller and generally partially dependent on the gametophyte. In vascular plants the reverse is generally true: the sporophyte ($2n$) generation is dominant and the gametophyte (n) generation dependent.

Bryophytes are simply constructed compared with vascular plants, but even the simplest is far more complex than the type of green alga such as *Fritschiella* thought to be the progenitor of land plants. A cuticle is frequently present and so are pores (Fig. 23–2) or stomates. All bryophytes lack the true stems, leaves, and roots that exist in vascular plants, but the functional equivalents are present. Unicellular or multicellular rhizoids take the place of roots, and in many bryophytes there is a main axis bearing small lateral projections involved in photosynthesis. But none of these structures approach the complex organization found in vascular plants.

The Liverworts—Hepaticae

The liverworts contain both leafy species and those that grow more prostrate against the soil. The leafy liverworts far outnumber the prostrate species, but the latter are more frequently seen in the laboratory. Some of the most common of the liverworts belong to the genera *Ricciocarpus, Riccia,* and *Marchantia* (Fig. 23–4).

The plant body of the prostrate liverworts is generally fan-shaped, showing regular, equal (dichotomous) lobing and branching. The growth is from the front through the division of apical cells, while decay of the posterior portion of the body is usual. A midrib is commonly seen (Fig. 23–3).

The vegetative body is a dozen or more cells thick (Fig. 23–4). The uppermost layer of cells forms an epidermis, the outer walls of which contain cutin. These epidermal cells lack chloroplasts. Pores, which are openings in the epidermis, permit gas exchange by opening into chambers loosely filled with cells rich in chloroplasts. The extent and form of the chambers vary greatly: in some species of *Riccia* they are simple cylindrical pores; in *Marchantia* they are true chambers; and in *Ricciocarpus* they can become quite extensive. The nature of the pore can be either simple, as in *Riccia* and *Ricciocarpus,* or lined with files of small cells forming

23–3 Mosses, liverworts, and some ferns cover the trunks and branches of the trees in a rain forest.

tiny chimneys. Most of the photosynthesis in the liverworts is carried on by the cells contained in the air chambers and the cells immediately below. The cells composing the lower part of the plant are closely packed and generally lack chloroplasts; storage of food reserves is their function. The extent of this tissue can be considerable (*Marchantia*), intermediate (*Riccia*), or quite small (*Ricciocarpus*). Frequently, cells below this storage tissue contain fungal hyphae, but the relationship between the fungus and the bryophyte is unknown.

From the bottom-most layer of the plant extend unicellular rhizoids. The rhizoids are frequently two types: vertical and horizontal. The vertical rhizoids act like roots, absorbing water and minerals from the substrate as well as anchoring the plant. In the aquatic *Ricciocarpus* the scales hang below the plant and act as water and mineral absorbents; rhizoids are also present. In all species young scales overreach the growing point at the apex of the plant, protecting the meristematic cells at this point.

Liverworts can reproduce asexually by fragmentation or by a more elaborate method involving the production of gemmae. On the upper surface of the thallus of *Marchantia*, small, cup-shaped, multicellular structures called gemmae cups are formed (Fig. 23–5). Small discs of green tissue, the gemmae, grow from the bottom of these cups. When mature, the gemmae break off from the thallus and are carried away from the parent plant, frequently with the help of rain.

When sexual reproduction occurs, two multicellular structures are differentiated in the thallus: the male gametangium called the antheridium, which will produce the sperms, and the female gametangium or archegonium, which will produce the egg. The significant feature of both the antheridium and the archegonium is the jacket of sterile cells around the

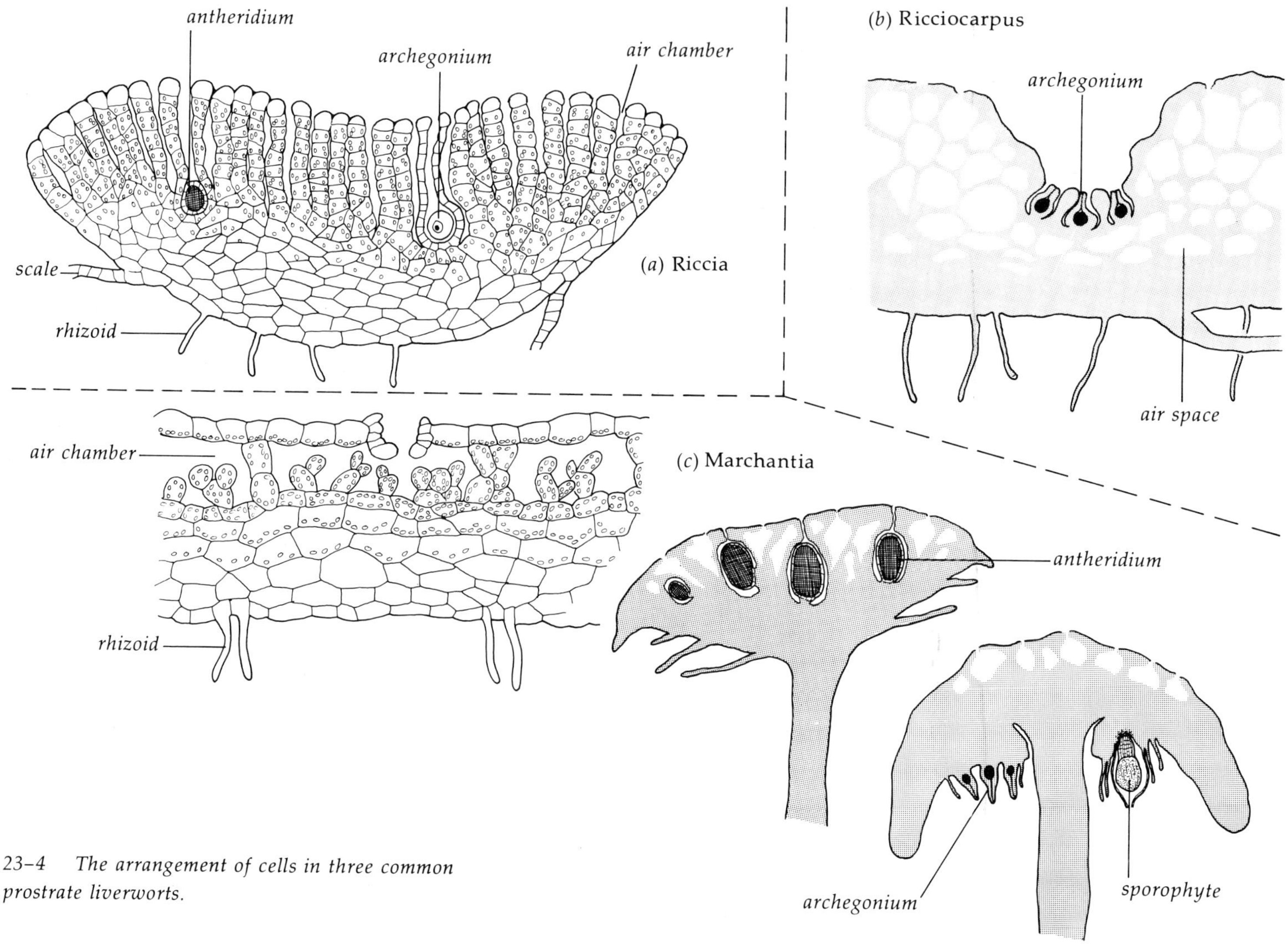

23–4 *The arrangement of cells in three common prostrate liverworts.*

developing gametes. This protection of the gametes by a layer of cells is believed to be a major factor in any plant's successful adaptation to life on land.

In *Ricciocarpus* and *Riccia* the archegonia and antheridia are produced directly on the upper portion of the vegetative plant. *Marchantia* is more complex: specialized erect stalks support discs on which are formed the antheridia and archegonia. The archegonia are formed on stalks with radiating finger-shaped projections (Fig. 23–6). Two types of stalks are never formed by the same individual; so there are male and female *Marchantia* plants.

The antheridia are composed of relatively small, densely cytoplasmic cells that form the biflagellated sperms. These cells occupy the central core of the antheridium and are surrounded by a single layer of nonreproductive or sterile cells. Since the liverwort plant is a gametophyte, sperms are formed by mitosis (not by meiosis as in animals). Several hundred sperms may be formed in each antheridium.

The archegonia are shaped quite differently from the antheridia, although both are composed of sterile and fertile cells. In the archegonium the sterile cells, again one layer deep, are in the form of a flask or bottle.

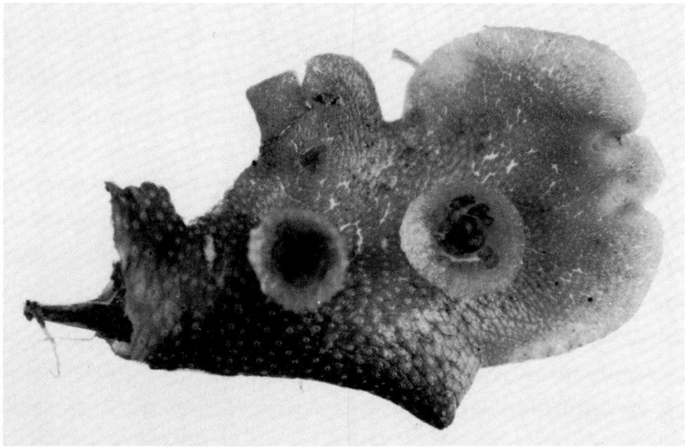

23–5 One means of asexual reproduction in some of the liverworts is through the production of tiny plants called gemmae. These are formed in the bottom of structures known as gemmae cups. This photograph shows two gemmae cups, with the gemmae visible in the bottoms.

23–6 In the liverwort Marchantia, *the archegonia and antheridia are borne on special branches or stalks. The antheridial stalks are characterized by fingerlike projections, while the archegonial stalks have disc-shaped heads.*

A column of cells differentiates in the center of the bottle. The bottom cell, occupying the base of the bottle, is the egg, while the cells occupying the neck are called the neck-canal cells. When the egg is mature, these cells degenerate, and the sterile cells at the top of the neck separate so that a clear canal is formed through which the sperms swim to the egg.

The mature sperms may be shot forcibly from the antheridium, or they may be released in a gelatinous mass. In either case, they are only freed when water is available for the sperms to swim in. The sperms are actively attracted to the opening in the neck of the archegonium and swim preferentially to it. They then swim down the neck to the egg, and although several sperms may reach the egg, only one will enter it. The fusion of the egg and sperm nucleus results in the formation of the zygote. In *Marchantia* the archegonia are usually fertilized before the elevation of the special stalk above the level of the main part of the plant. This means that the necks of the archegonia are immersed in a film of water close to the ground, which enhances their chances of fertilization.

When the zygote is formed, it divides almost immediately to produce an embryo that consists of a foot (an absorptive organ whose function is to transport nutrients from the maternal gametophyte to the developing sporophyte), a stalk or seta, and a capsule, which is the sporangium (Fig. 23–7). In *Ricciocarpus* and some other liverworts, the sporophyte is reduced to a spherical mass of cells that functions as a sporangium.

The cells occupying the center of the sporangium differentiate into spore mother cells or sporocytes. These cells enlarge, and their walls thicken. Meiosis occurs, and four haploid cells are formed, usually in a tetrahedral arrangement. These are the spores (Fig. 23–8). An elaborate wall, complete with characteristic wall-sculpturing, forms around each spore. The first step in the production of this wall is the deposition around each cell of a special wall composed of callose. This wall is not uniformly deposited; it has a pattern that acts as a mold against which the mature spore wall

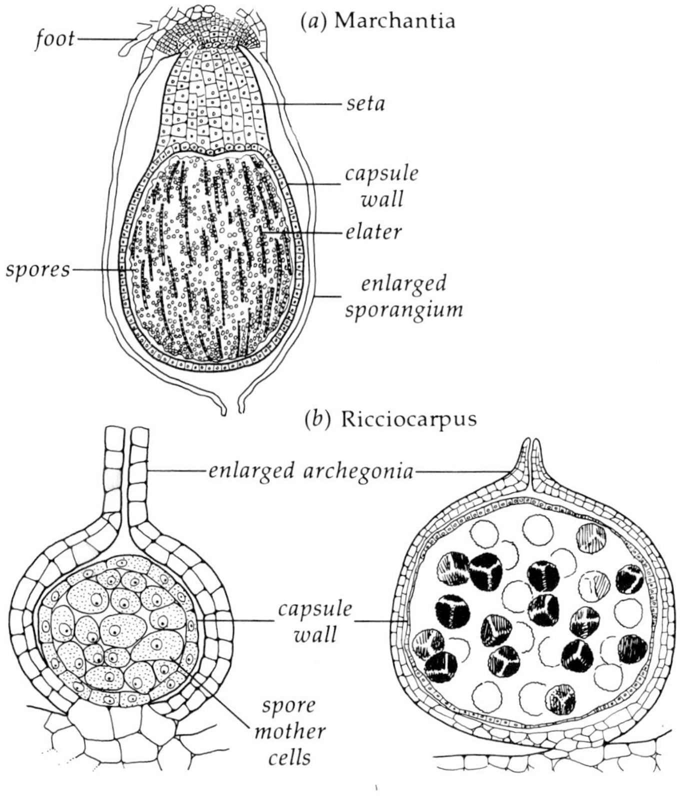

23–7 *The form of the sporophyte as seen in (a)* Marchantia *and (b)* Ricciocarpus. *The* Marchantia *sporophyte is immature insofar as the spore mother cells have not yet divided to form the spores, but the other parts of the sporophyte are mature. Two stages of development of the sporophyte of* Ricciocarpus *are shown, illustrating the reduced condition of the mature sporophyte.*

will be deposited. The callose also isolates the spore from the surrounding environment, thus further determining its development. The wall that is next laid down consists of two layers: the exine, which is deposited first between the callose wall and the plasma membrane, and the intine, which is deposited later. The exine is composed primarily of a special wall material found only in spores and pollen grains called sporopollenin. This is a remarkable substance, whose molecular structure is still highly controversial, and is one of the plant products most resistant to degradation. The striking permanence of sporopollenin has provided the paleobotanist with one of his most important tools: the analysis of fossil spores and pollen. The intine, which forms later than the exine, is composed of cellulose and noncellulosic polysaccharides.

Present among the spores within the capsule of many liverworts are specialized cells called elaters (Fig. 23–9). These cells are narrow, elongate dead cells at maturity, with spirally thickened cell walls. These walls coil and uncoil as a result of changes in humidity, and this movement aids in the discharge of the spores from the capsule. The mechanism basic to the movement of the elaters is essentially the same as that used in ferns for spore discharge and will be discussed later. With regard to spore discharge, elaters function in three different ways in the liverworts. They may simply curl and uncurl after the capsule opens, thus fluffing out the spore mass and aiding in its dispersal. They may remain with one end attached to the capsule wall after it opens, waving back and forth at the free end. This releases the spores gradually and disperses them most effectively. They may also act as tiny catapults, hurling the spore out by a violent unbending.

When they germinate, the spores form a short filament of cells and then develop into a new prostrate vegetative plant. All spores have a variety of food reserves that are utilized during spore germination. In the liverworts these are primarily starch and lipids.

Liverworts are not only relatively flat, prostrate plants; many are leafy in form and may be more or less erect. Most of the leafy liverworts are dorsiventrally flattened (in contrast to mosses, which are radially symmetrical) with two rows of leaflike, lateral, photosynthetic organs.

The Mosses — Musci

The small leafy shoot of the moss is the gametophyte (n) generation (Fig. 23–A). This shoot generally consists of three ranks of leaves helically arranged on the stem. They are only one cell thick except for the midrib, which is thicker but lacks true vascular tissue. The stem consists of an epidermis, a cortex, and a central column of elongate cells. A layer of starch-filled cells may delimit the cortex from the central region. In the most complex mosses, cells of the central column are differentiated into support cells, food-conducting cells,

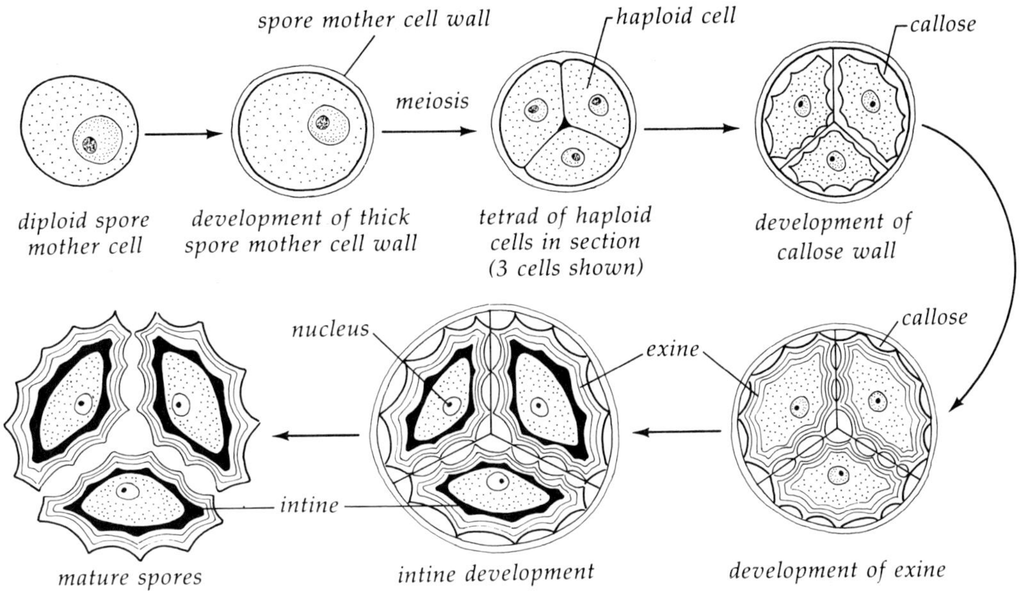

23–8 A diagram of spore formation in liverworts.

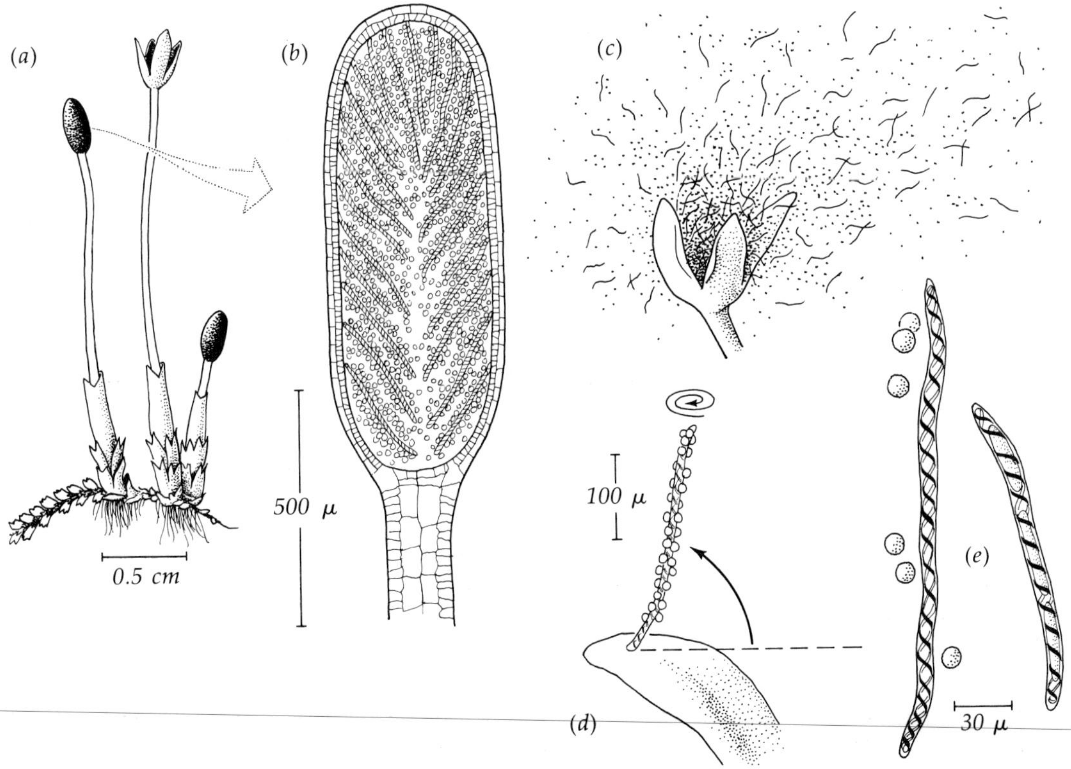

23–9 Spore discharge in Cephalozia bicuspidata, *a leafy liverwort, illustrating the role of elaters. Before discharge the stalk of the sporophyte elongates so that the capsule is carried into the air (a). Within the capsule (b) there are rows of elaters surrounded by spores. The capsule breaks open along four lines, releasing the spores (c). The elaters swing upright and turn on their axes (d). The elaters also increase and decrease in length (e), further increasing their effectiveness in dispersing the spores.*

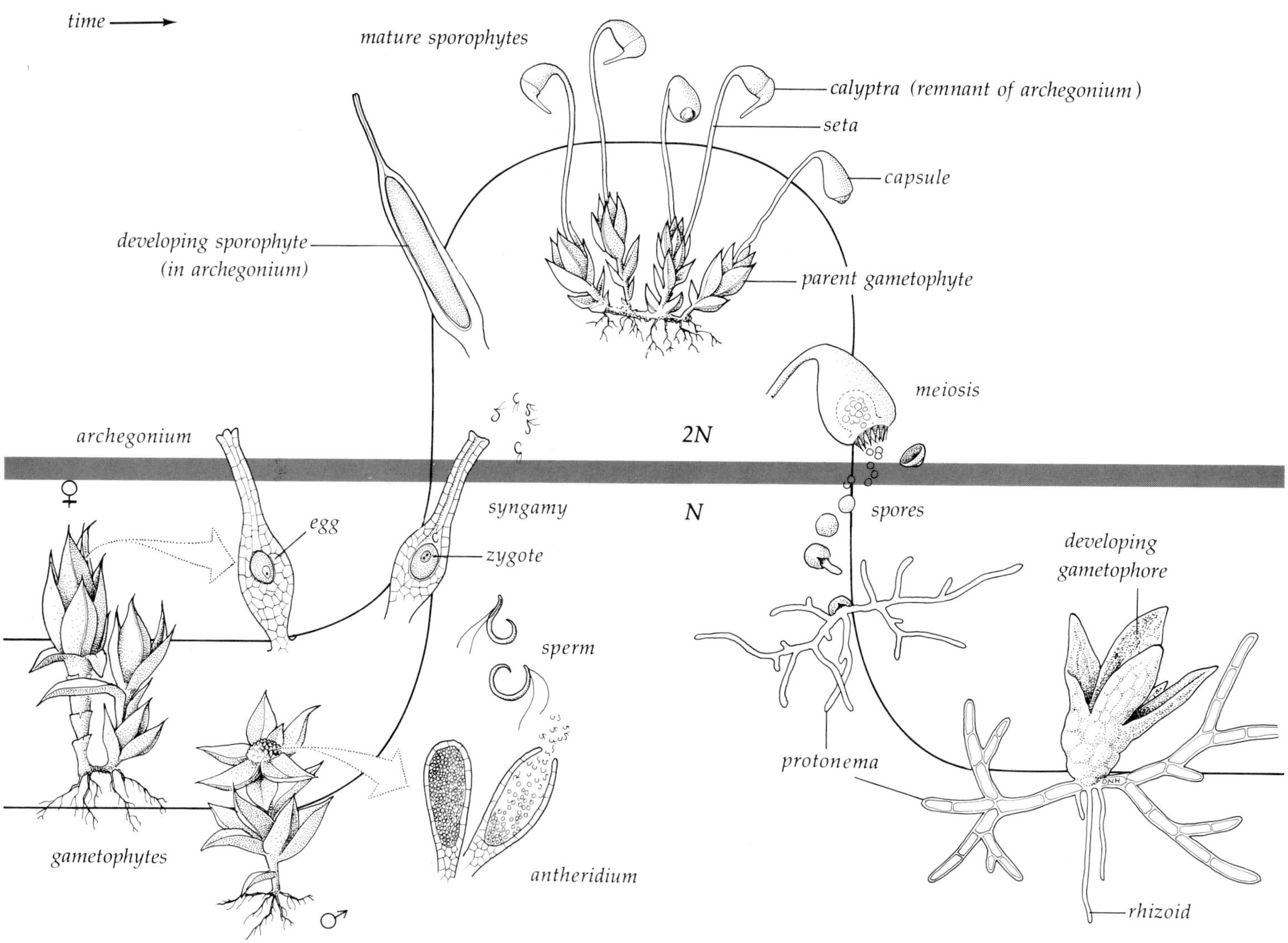

23–10 *The life history of a moss.*

and water-transporting cells, but none are the true vascular cells characteristic of the xylem and phloem. In some mosses water is transported in the central column, but in most, water is conducted by surface tension up the exterior of the plant and absorbed through the leaf cells. This type of water movement is possible because of the small size of the plant. Lignin is found only in the gametophyte cell walls of the giant New Zealand mosses that range from 35 to 55 cm high—ten times as high as most mosses. Multicellular rhizoids are attached to the bottom of the leafy shoot (Fig. 23–10).

The shoots develop from algal-like filaments of cells, known as protonemata (Fig. 23–10), which are produced by the germination of the spore. The protonema may be long lived and form green mats in favorable locations. In some mosses the protonema is quite elaborate and may have multicellular rhizoids, horizontal filaments, and vertical aerial filaments. The adult gametophyte is formed as an outgrowth of a cell that initiates in the protonema. This cell enlarges, and three planes of division are formed, resulting in a radially symmetrical shoot.

Specific conditions are required for the initiation of shoots by the protonema. Only certain protonemal cells can initiate shoots. Under normal conditions shoots are initiated only after a minimal surface area of the protonema is attained. In low light intensities sugar added to the protonema will result in shoot production. In the dark even the addition of sucrose will not initiate shoot formation, but the addition of kinetin (a cell division plant hormone; see Chapter 30) to the sucrose will produce shoots. Red light appears to be the spectral region necessary for the initiation of shoots, and it is this specific response that can be replaced by kinetin. The implication of these experiments is that light is necessary both for photosynthesis and for a morphogenetic process probably involving kinetin.

One protonema gives rise to more than one shoot, and the initiation of shoots does not inhibit further protonemal growth. This is the origin of moss mats, which may grow rapidly and persist over long periods of time. The protonema of a single spore has been known to cover an area 40 cm in diameter in several months' time.

When the shoot is mature, antheridia and archegonia are produced (Fig. 23–D). The factors initiating their formation are not understood, but they may be temperature-dependent in some species and photoperiodically controlled in others. Antheridia and archegonia are formed either at the tip of the main shoot or on special lateral branches. They may also be borne on the same plant or on separate plants. The forms of the antheridia and archegonia are similar to those found in the liverworts, although the necks on the archegonia are longer. The jacket cells of the antheridia of many mosses contain chromoplasts; hence at maturity they have a characteristic orange color. Interspersed among the antheridia and archegonia are filaments of sterile cells called paraphyses. They may be involved in protecting the antheridia and archegonia from drying, but this function is still unproven.

The sperms are biflagellated, highly modified cells. At maturity the chromosomes are tightly condensed in a ribbon-shaped nucleus that determines the shape of the sperm. Besides a single, long mitochondrion and the plasma membrane, the anterior flagella, and the cytoplasmic structures associated with them, little else is present. The egg is also a highly modified cell, rich in food reserves. Water is necessary for fertilization, and the actual process is essentially the same as in the liverworts. A sperm attractant is present, but its nature is unknown, although some data indicate that it may be as simple a compound as sucrose. More than one sperm may enter the egg, but only one will fuse with the egg nucleus to form the zygote. Once this occurs, any remaining sperms degenerate.

The zygote divides in place, surrounded by the cells of the archegonium (Fig. 23–D). As the embryo develops, one of the first parts to form is the foot. Special wall layers, formed in the cells at the base of the foot, are believed to facilitate the movement of materials into the sporophyte.

While the embryo is developing, the cells of the jacket of the archegonium are dividing. Eventually, the upper part of the archegonium is broken away from the bottom and carried upward on top of the growing sporophyte. This specialized structure is known as the calyptra. It forms a tight cover over the tip of the developing sporophyte, thus preventing the sporophyte capsule from developing prematurely.

The sporophyte of the mosses is similar to that of the liverworts, consisting of a foot, seta, and capsule. The seta is usually fairly long and slender, sometimes reaching several centimeters in length. The capsule is one to several millimeters long and is involved in the production and discharge of the spores. The tip of the sporophyte is covered by a lid or operculum. When the spores are mature, the operculum falls away, revealing a ring of teethlike units, the peristome (Fig. 23–11). The individual teeth are generally attached by their bases to the edge of the capsule. The tips move back and forth in response to changes in moisture content of the air and to wetting and drying. The peristome teeth close in humid weather or when they are wet, thus protecting the spores and preventing their release at a time when they could not be widely dispersed. During dry, windy periods the teeth open, and the spores are released and carried long distances by the wind.

The spores of mosses, like those of liverworts, have sculptured walls composed of intine and exine layers and containing sporopollenin. Germination may occur soon after the spores are liberated, or a period of dormancy may exist for some time. Spores are able to remain viable for long periods if germination conditions are unfavorable—in one case spores were still capable of germination after 55 years of storage. Most moss spores germinate only in the light. The spores on germination give rise to the protonema, and the cycle is started again.

23–11 The peristome teeth at the tip of the capsule in a moss.

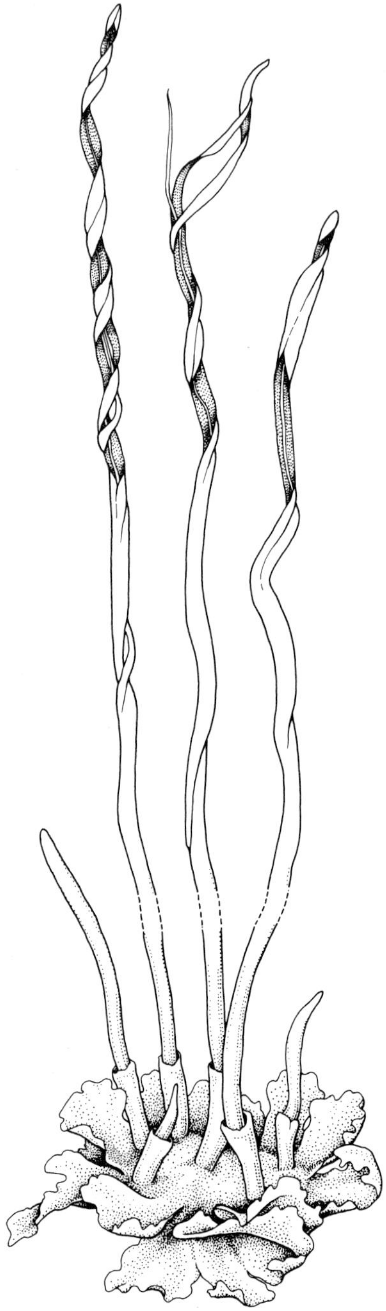

23–12 A hornwort, Phaeoceros laevis. *The structure at the base is the gametophyte, while the long twisting projections are the mature sporangia.*

The Hornworts—Anthocerotae

The hornworts superficially resemble the prostrate liverworts, but they exhibit marked differences in detail (Fig. 23-12). They are flat, inconspicuous plants found in unplowed fields or roadside cuts in many parts of the country. The gametophytes grow close against the ground and have a simple internal construction. The cells of some species contain a single large chloroplast complete with pyrenoid, an unusual feature in any plant but the alga. The antheridia are produced in small cup-shaped depressions of the vegetative body, while the archegonia are embedded in and completely surrounded by the cells of the thallus. One of the unusual features of the hornwort gametophyte is the presence of cells that greatly resemble the guard cells of vascular plants. They are thin-walled and do not open and close, but they are the only such cells known in gametophytes. The cavities beneath the guard cells are filled with slime and contain the blue-green alga *Nostoc*, which is a nitrogen-fixing alga.

The differences between the hornworts and the other bryophytes are more apparent in the sporophyte than in the gametophyte. After fertilization, requiring the usual presence of free water because the sperms are flagellated, the zygote divides in place. Subsequent divisions result in a cylindrical sporophyte of considerable length (as much as 2 to 5 cm). The sporophyte consists of a foot, a meristematic region immediately above the foot, and a long sporangium. The outer layers of the sporangium include guard cells and cells rich in chloroplasts. The center is occupied by a cylinder of sterile cells, the columella, that is surrounded by a ring of spore-producing tissue. As the sporangium matures, it splits in two along preformed lines of weakness. The twisting of the upper mature part of the capsule and the presence of elaters among the spores aid in spore release. Under favorable conditions the sporangium is constantly being synthesized at the base and maturing at the tip.

End Paper 23–1 The Evolution of Bryophytes

All of the bryophytes (the liverworts, hornworts, and mosses) have features that set them apart from the other green land plants. In each, the gametophyte is the independent, long-lived generation, while the sporophyte is the dependent, short-lived generation. Thus there is a marked alteration of generations, with the haploid dominating the diploid. This condition, while common in the green algae, is found in no other group of land plants. In most, the diploid generation is the increasingly important part of the life cycle. Within the bryophytes, however, there are marked variations in the dependence of the sporophyte on the gametophyte. In the liverworts, the sporophyte is largely dependent on the gametophyte. In some species of mosses, on the other hand, the gametophyte does little more than provide water and minerals; the sporophyte synthesizes all of its own organic material.

This observation has led some botanists to conclude that the independent, dominant sporophyte of the vascular plants arose from modifications of the dependent sporophytes of the bryophytes. The fossil record, although highly fragmentary, is against this general concept. There are also features of the anatomy of the sporophyte that are puzzling. One is the presence of guard cells. These are important for water movement and water loss in vascular plants but are nonfunctional in the sporophytes of bryophytes. Another feature is the presence of cells with secondary-wall thickenings in the sporophytes of some hornworts. These cells are not elaters and have been interpreted by the late Johannes Proskauer of the University of California, Berkeley, as the remnants of vascular tissue. Using such data, some botanists feel that the bryophytes have arisen through the modification of some of the first vascular plants. This means that the bryophytes evolved under environmental conditions that selected for the characters predominant in the gametophyte, while the vascular plants evolved under conditions favoring selection of a strong sporophyte. How this may have happened will be more understandable after we examine the lower vascular plants.

The concept of alternation of a haploid and a diploid generation runs through the plant kingdom. We discussed it in relation to the fungi and the algae, where it is much more obvious. The green algae from which the bryophytes are believed to have evolved are organisms with either strong haploid stages or equal haploid and diploid stages. We have just seen the condition in the bryophytes, and you can see why many botanists believe they or their vascular ancestor evolved from an isomorphic green alga.

Why should alternation of generations occur, and what is the basis for the differences in the form of the generations? One answer is that the extra genetic resources of the diploid condition offer the possibility of a more elaborate plant body. It is based on the observation that all elaborate plants, the vascular plants and the complex brown algae, are diploid organisms. Yet this answer is not good enough because experiments have demonstrated that the form of sporophytes in the bryophyte is not dependent on chromosome number.

An isolated piece of the seta of a sporophyte may be cultured and stimulated to divide. A new sporophyte is not produced but instead a diploid gametophyte. This in turn gives rise to diploid gametes that form tetraploid zygotes and sporophytes. All these structures with doubled chromosome numbers resemble the corresponding organisms with the normal chromosome numbers. Clearly, simply the number of chromosomes is not the answer to this problem in the bryophytes (or, for that matter, in the ferns and other vascular plants where similar experiments can be performed).

This does not rule out the possibility that doubling the chromosome number is important in evolution, since it doubles the genetic material available for mutation and increases the range of variation that can be selected for. What must be kept in mind is that the bryophytes are successful, worldwide organisms that have survived for millions of years and are occupying a variety of environmental niches.

24

The Anatomy and Development of Vascular Plants

The Architecture of Life

Ultimately, all aspects of plant anatomy and morphology are reduced to the characteristics of the cells making up the plant. In the end, of course, the total plant is more than merely the sum of the individual cells, but it is impossible to understand the whole plant without first understanding the cells. In this introduction to plant anatomy, we want to take you first on a microscopic and ultrastructural tour of some of the major cell types and then show you how these are put together to form the principal organs of the vascular plant.

A The meristematic cells of the root and shoot apex are responsible for the primary growth of the plant and are very important to its development. Pictured here is a meristematic cell from an onion root tip, as seen with the electron microscope. Such meristematic cells are relatively simple in structure and contain few cell organelles. The thin cell walls are also characteristic of such cells.

B The basic cell type is the parenchyma, shown here in a low-power electron microscope view. The large vacuoles are emphasized because of the accumulation of dark appearing tannins within them. The walls are thin, and the plastids contain starch—both common features in parenchyma cells.

C There are many variations on the basic cell pattern represented by the parenchyma cell. Some of them, as seen with the light microscope, are shown here: the collenchyma with localized wall thickenings, and two types of sclerenchyma—the multibranched sclereids and the fibers with their thick walls in concentric layers.

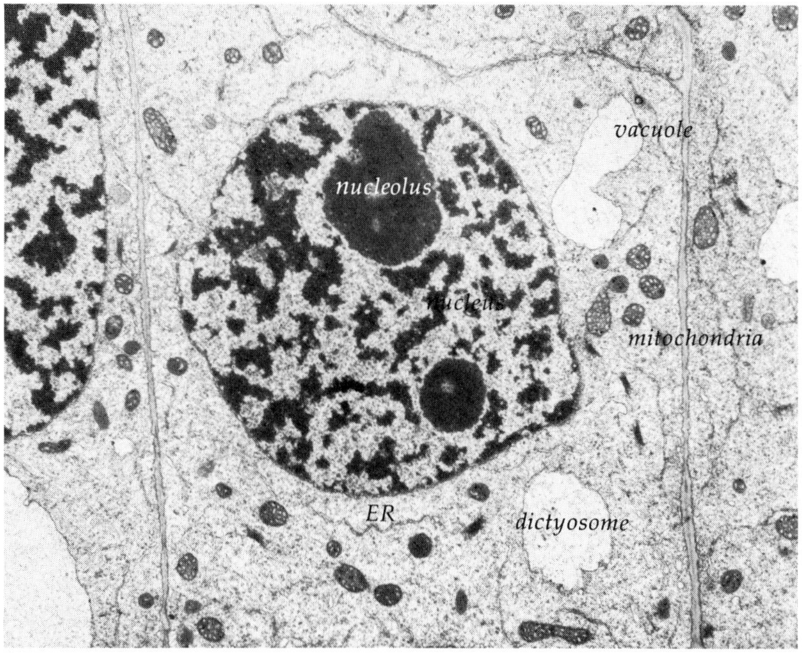
vacuole
nucleolus
nucleus
mitochondria
ER
dictyosome

A

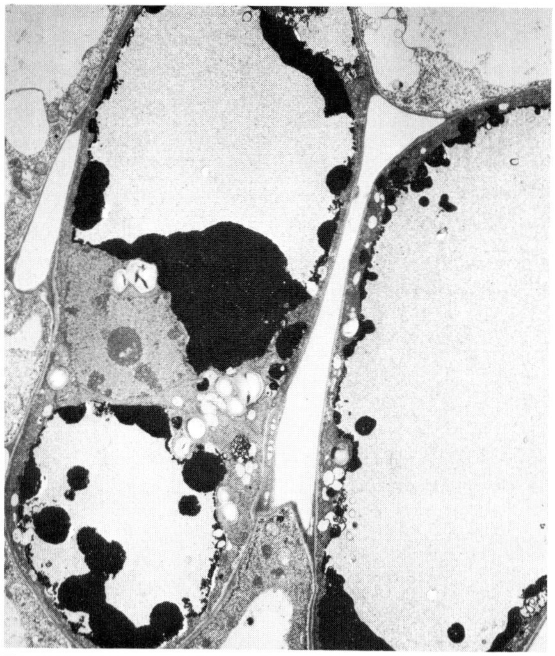

B

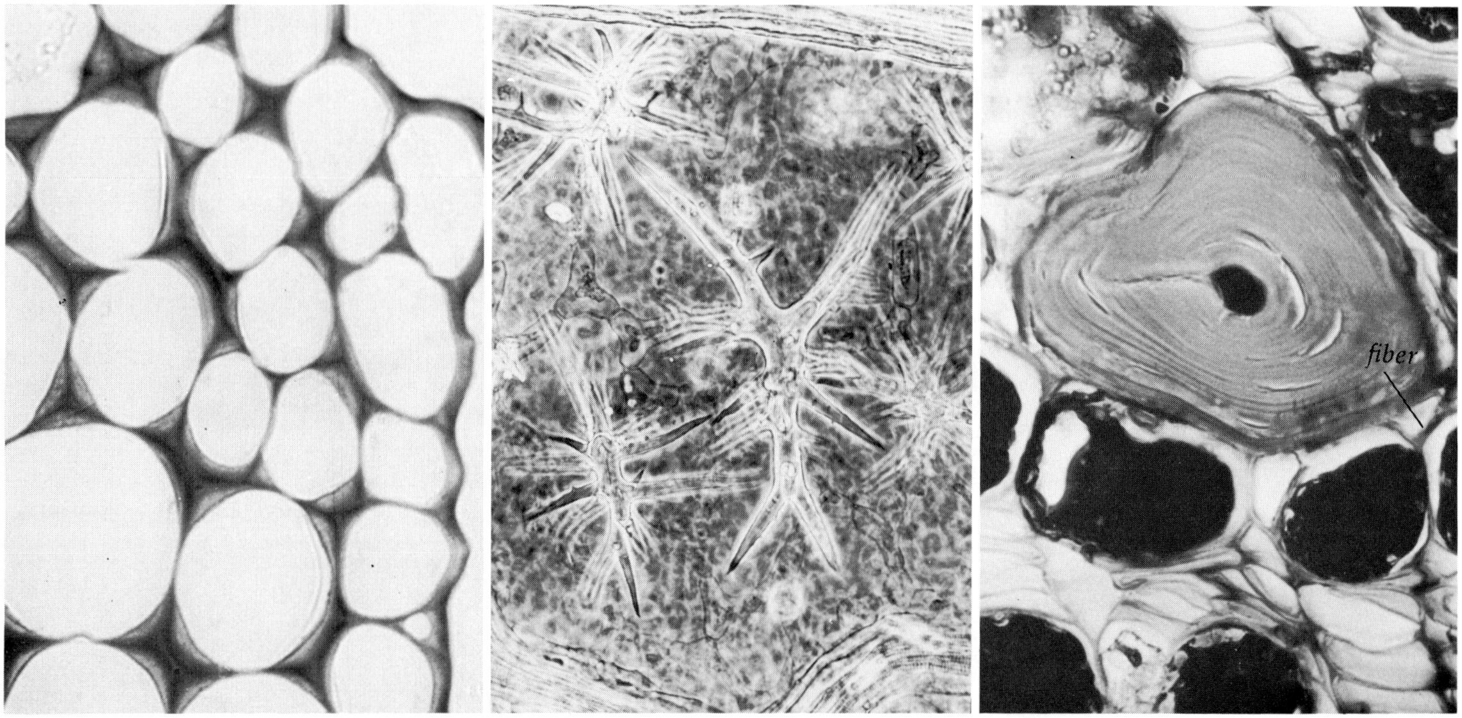
fiber

C

D Epidermal cells are modified parenchyma cells, and from these, the guard cells differentiate. They are elaborate cells. The specially thickened wall is important in the operation of the cells as is the presence of chloroplasts.

E Hair cells are also part of the epidermis and occur in a wide variety of shapes and sizes. Some are multicellular, as is the secretory hair shown in this electron microscope photograph. Others are single-celled, as the cotton epidermal hair that becomes the cotton fiber.

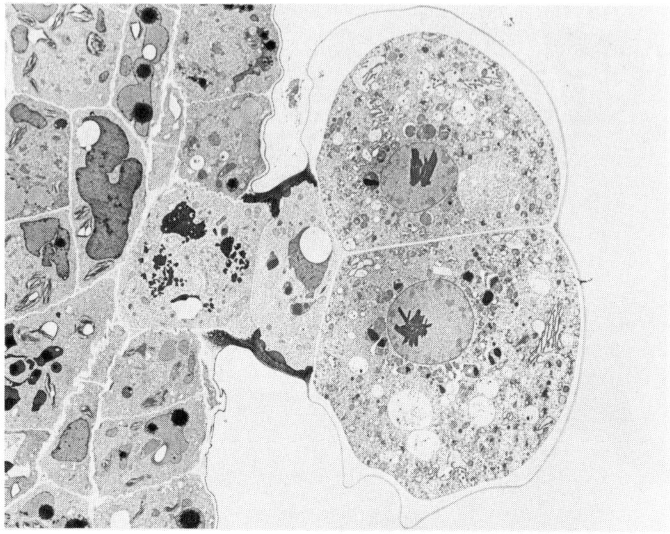

E

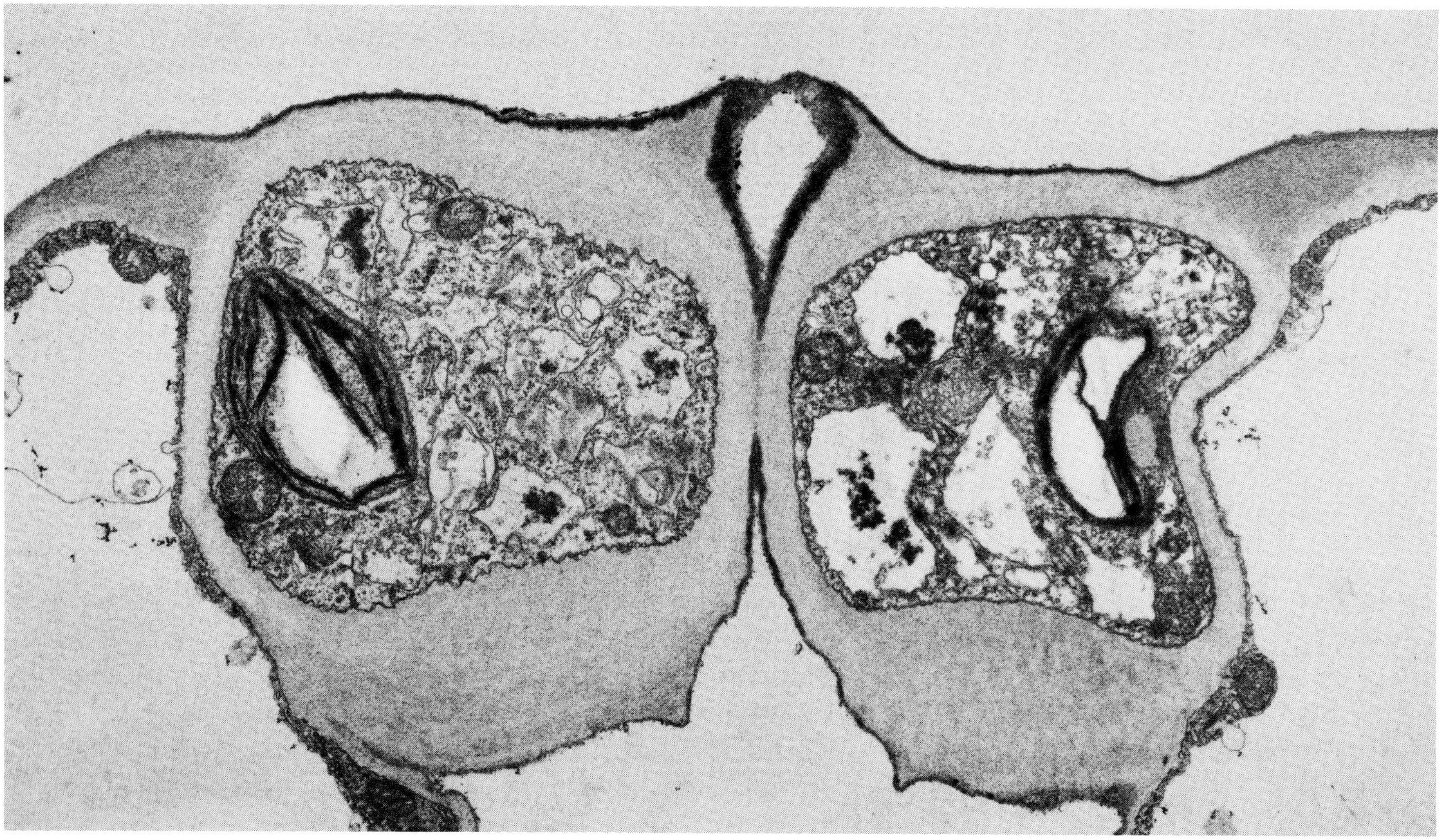

D

F The vascular tissues, both the phloem and the xylem, show marked differences from the cells we have seen thus far, particularly the tracheid vessels and the sieve cells. This difference can be seen in this electron microscope view of an immature and a mature tracheid. The general form of the tracheids and vessels can be seen in the piece of wood in which the cells have been separated from each other. While the cell is still immature, it contains a nucleus, a vacuole, and cytoplasm. But, with maturation (after wall thickening is complete), all of the contents of the cell degenerate and the cell becomes part of the pipeline.

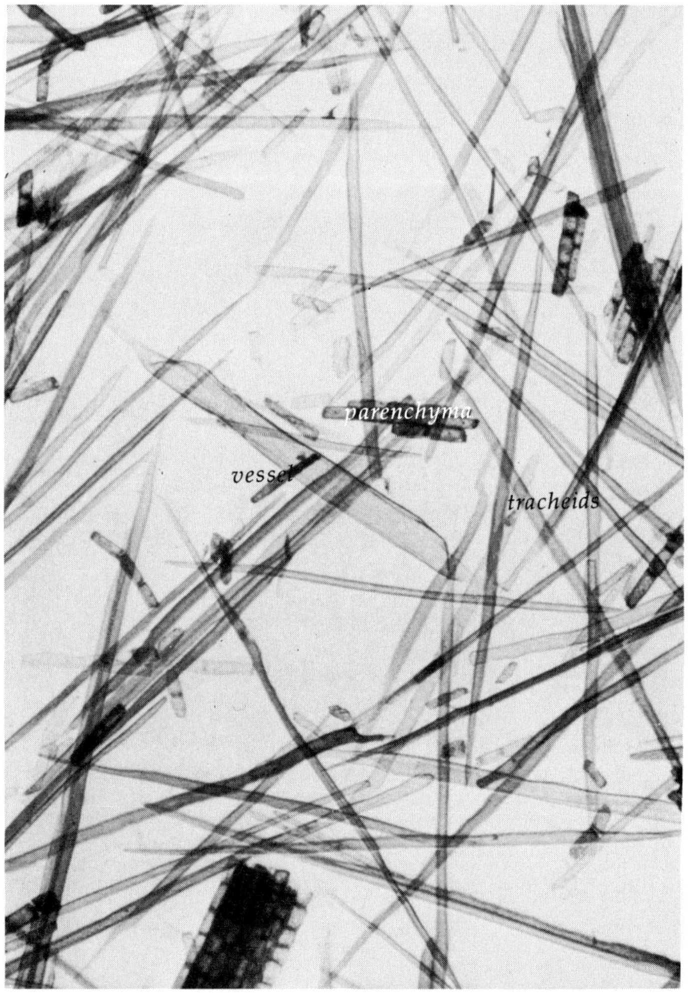

F

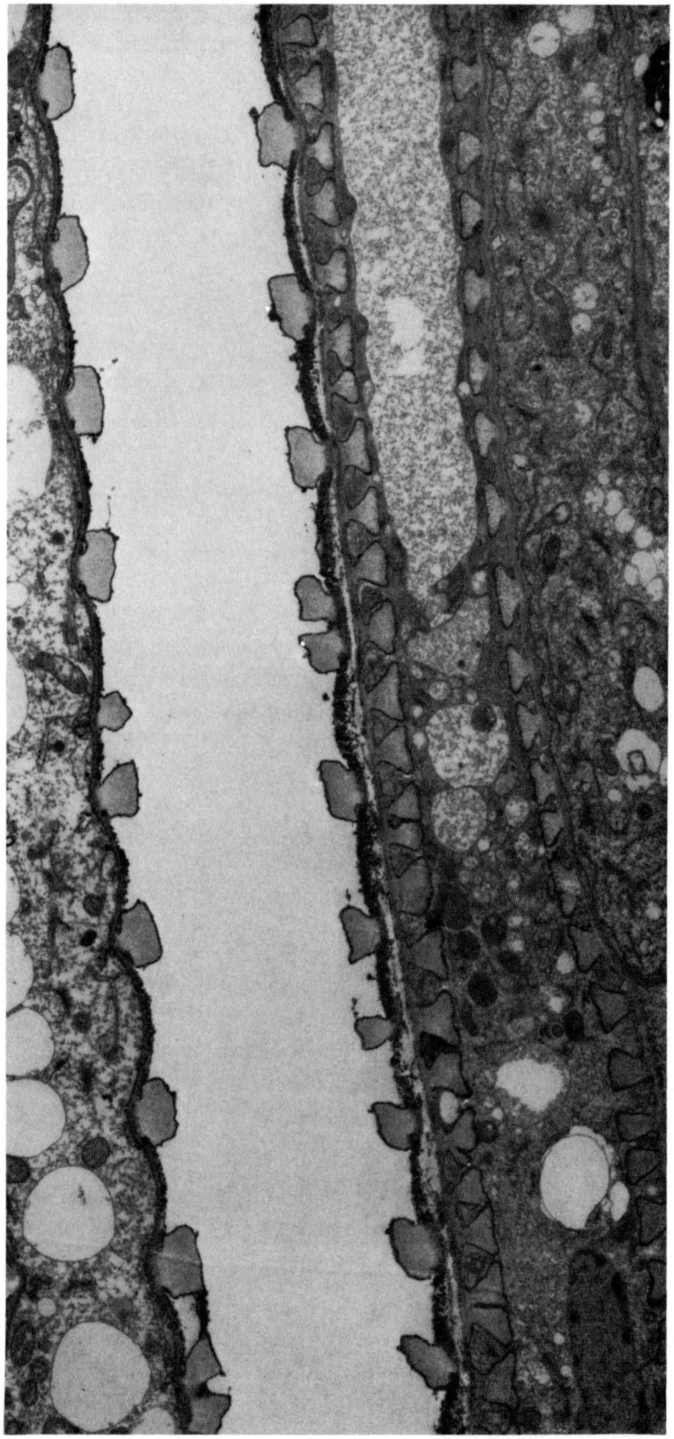

G Sieve tubes in the phloem are the conducting system for sugars and other organic compounds in the plant, and as such they are highly modified cells. When they become mature, they contain a modified cytoplasm but have lost their nuclei and vacuoles. The end wall of a sieve tube becomes highly modified into the aptly named sieve plate. The various features of sieve tubes are shown in these pictures of mature sieve tubes, as viewed with an electron microscope and a special type of phase-contrast light microscope, and in a cross section of stained preparations, as viewed with a regular light microscope. Companion cells are also shown in these photos. These are small cells that are developmentally linked to the sieve tubes. They are densely cytoplasmic and always retain their nuclei.

H Having seen some of the basic types of cells found in vascular plants, let us now see how they are put together into organs. First, let us consider the root. In this series of light microscope photographs we see cross and longitudinal sections of the root tip of an onion, together with a section through the mature portion of a buttercup (Ranunculus) *root. The various tissues have been labeled and will be discussed in this chapter. The root is surrounded by a single-layer epidermis, except at the tip, where the root cap covers the apical meristem and extends a short distance up the root. Inside the epidermis is the cortex, which is composed of parenchyma cells, and inside that is the stele, which is composed of vascular tissue. In many roots the center is occupied by xylem, while the phloem is found between arms of the xylem that extend outward. The endodermis marks the boundary between the cortex and the stele.*

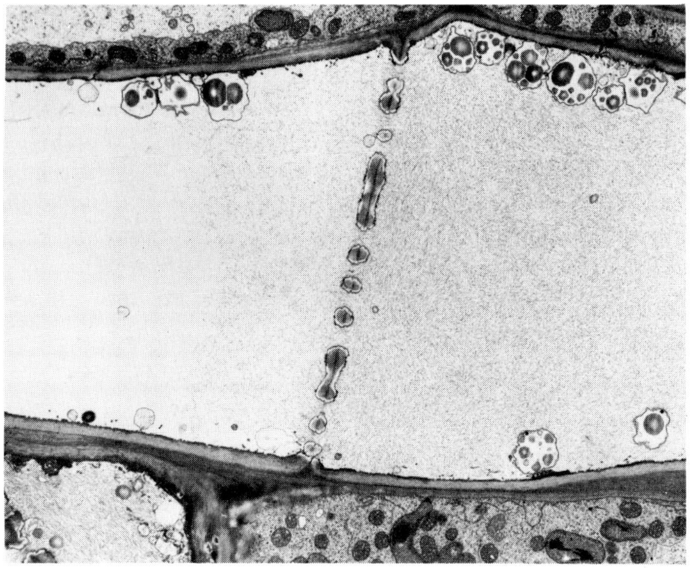

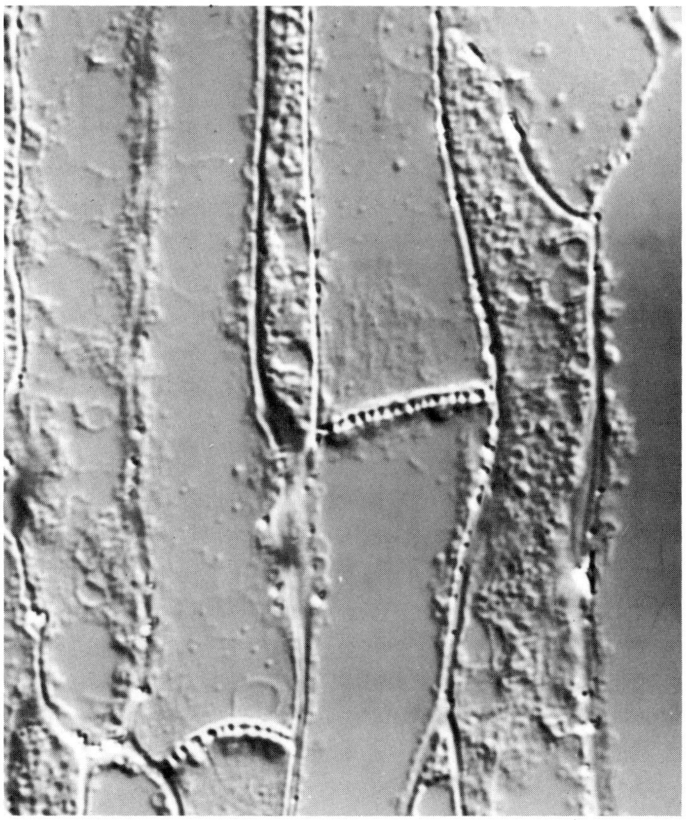

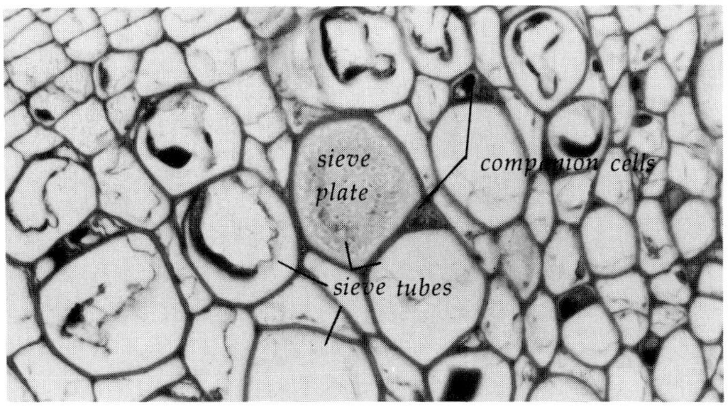

G

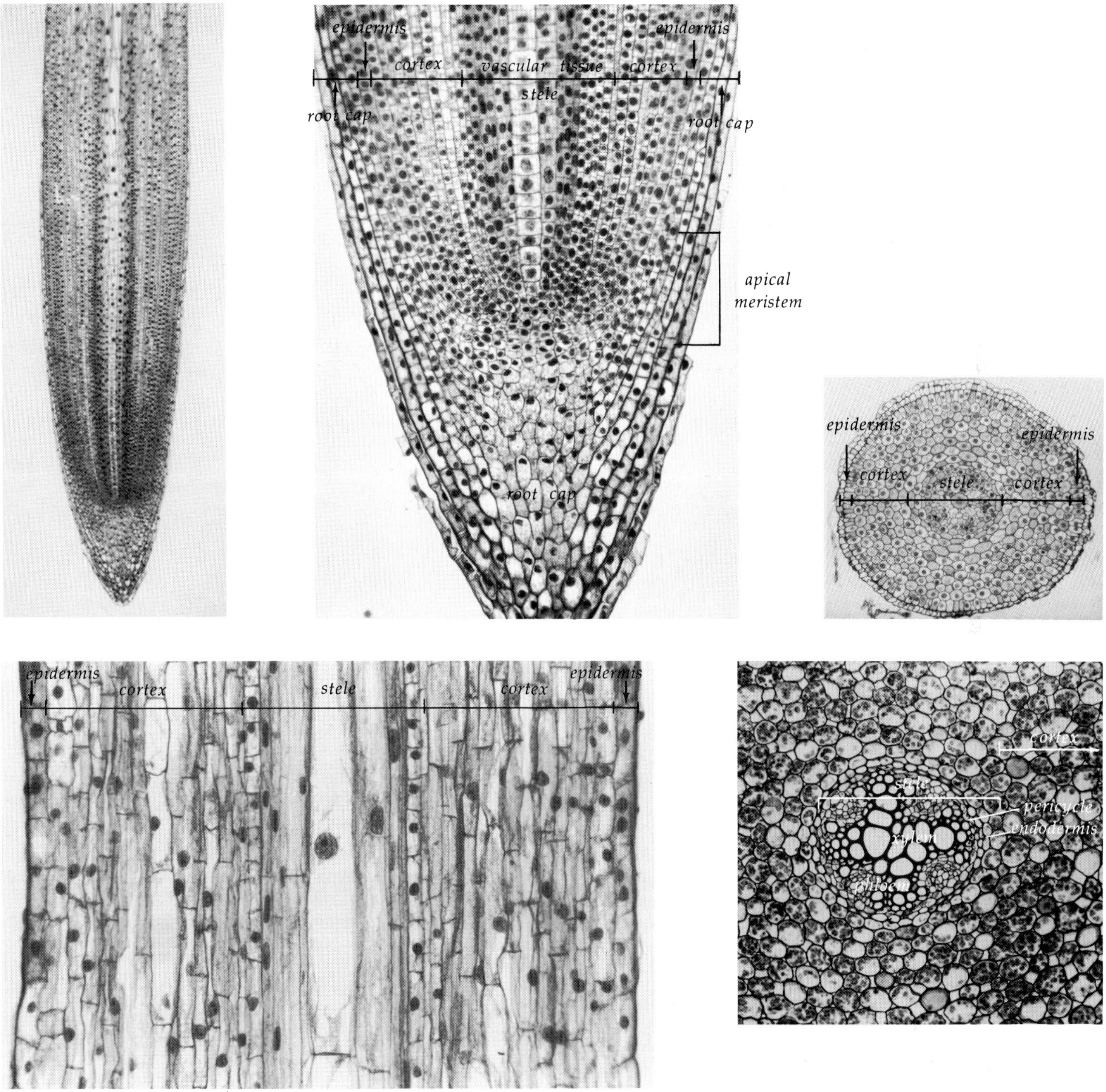
epidermis
cortex
vascular tissue
cortex
epidermis
stele
root cap
root cap
apical meristem
root cap
epidermis
epidermis
cortex
stele
cortex
epidermis
cortex
stele
cortex
epidermis
cortex
stele
pericycle
endodermis
xylem
phloem

H

I The apical meristem at the shoot tip, like that at the root tip, produces cells that form the stem and the leaves. As the leaves develop they form a protective cover for the meristem itself. The center of the stem is occupied by the pith, which is composed of parenchyma cells. In dicotyledonous plants the vascular tissue forms a ring or series of bundles between the pith and the cortex—the xylem is toward the inside and the phloem toward the outside. As the stem ages a new tissue appears in those stems that are able to grow in diameter. This is the vascular cambium and is a meristematic tissue in which the cells divide and give rise to additional xylem and phloem. In monocotyledonous plants the vascular tissue is arranged in the form of bundles scattered throughout the ground tissue. Vascular cambium is not present in such stems. Again, these photographs will serve as an introduction to some of the structures of the stem that will be discussed in this chapter.

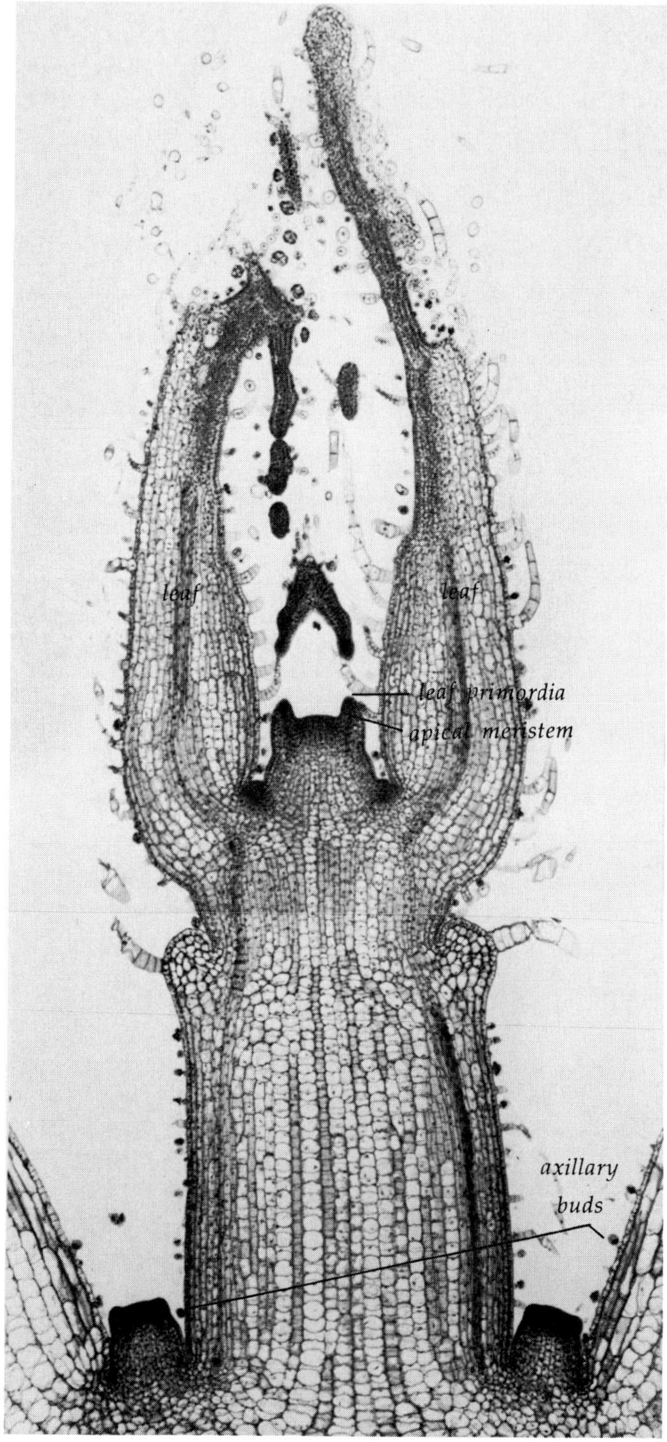

I

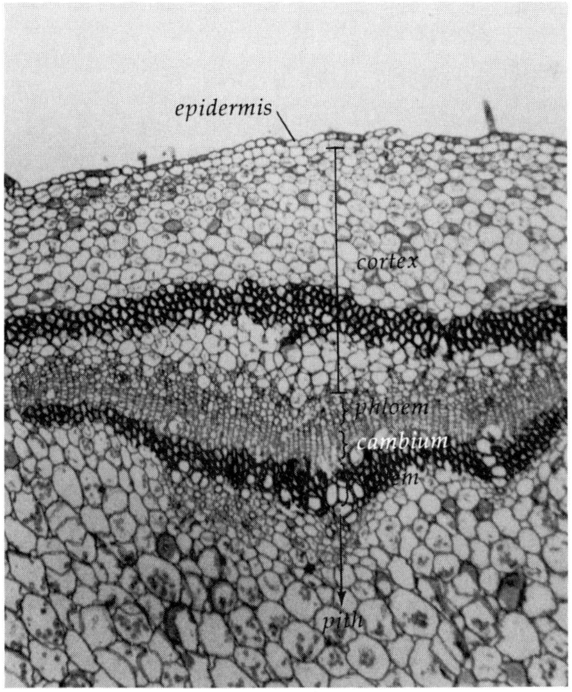
epidermis
cortex
phloem
cambium
pith

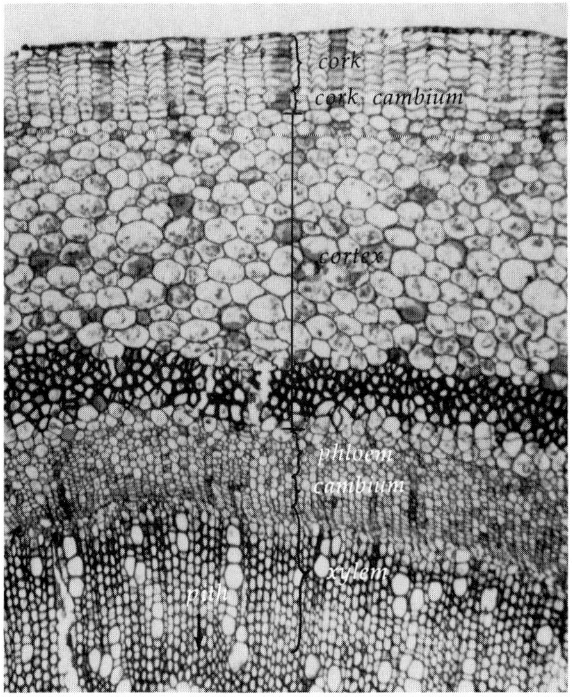
cork
cork cambium
cortex
phloem
cambium
xylem
pith

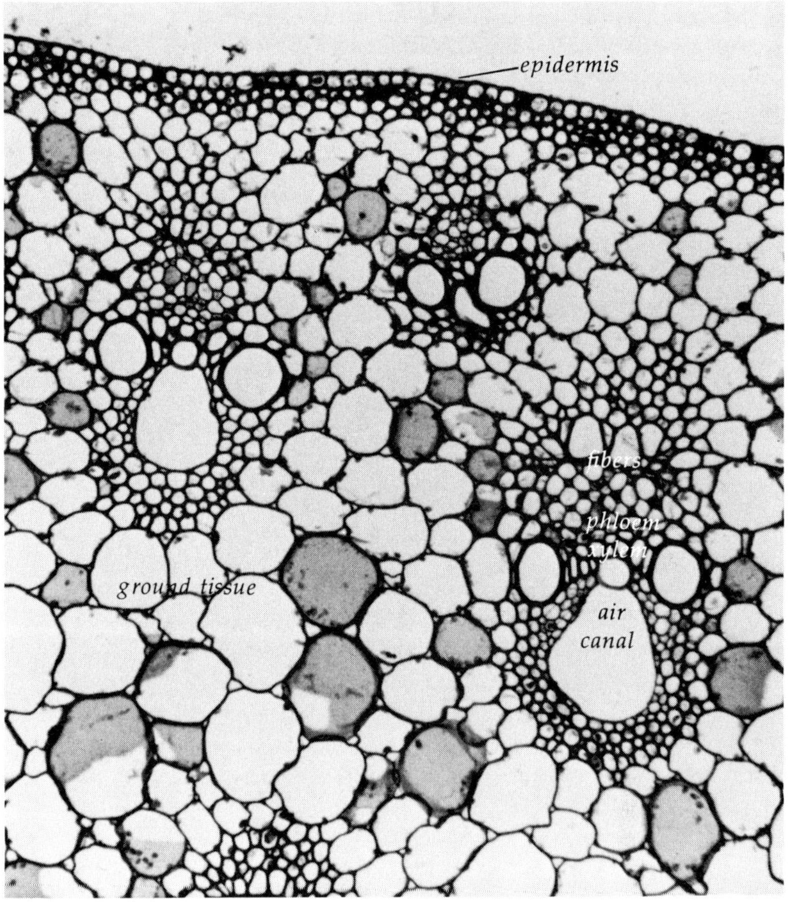
epidermis
fibers
phloem
xylem
ground tissue
air
canal

J The leaf is surrounded by an epidermis that contains numerous air holes or stomata, which, in turn, are surrounded by two guard cells. The number and arrangement of the stomata can be seen in this photograph of a segment of epidermis stripped from the leaf (left). Inside the epidermis the parenchyma cells show a number of arrangements. Frequently they are closely packed against the upper epidermis, while loosely packed against the lower epidermis (upper right). Strands of vascular tissue run throughout the leaf—the xylem to the top and the phloem to the bottom. The extent of the vascular tissue of veins can be seen in this section of a young corn leaf (upper left). The intimate relationship of the vascular tissue to the cells of the leaf can best be seen in sections cut parallel to the surface of the leaf.

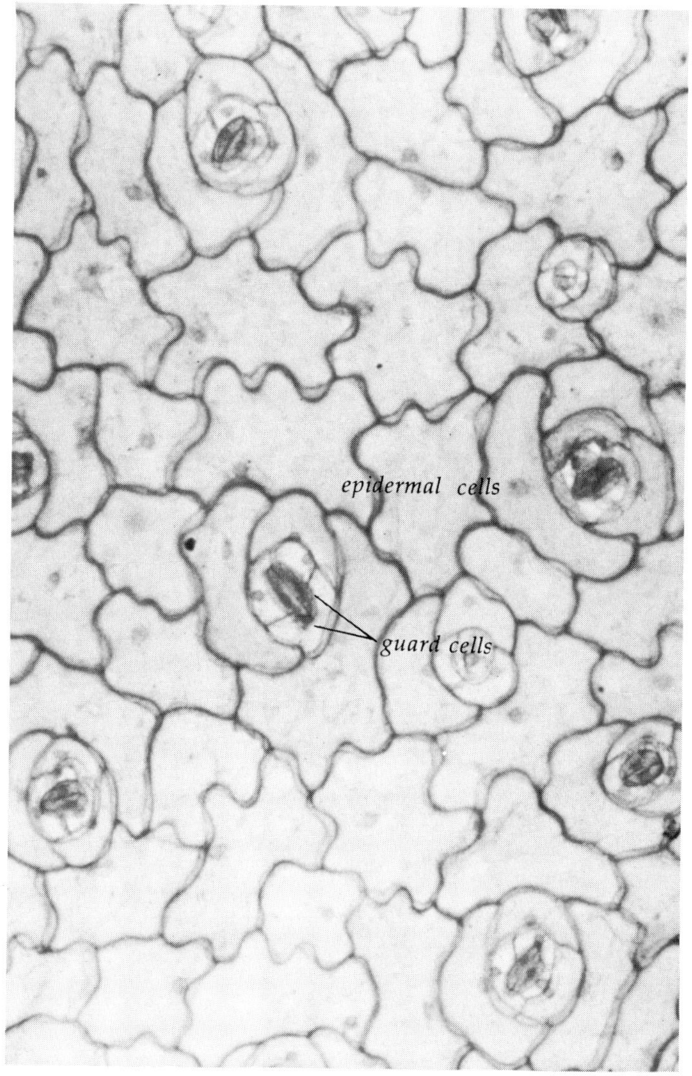

J

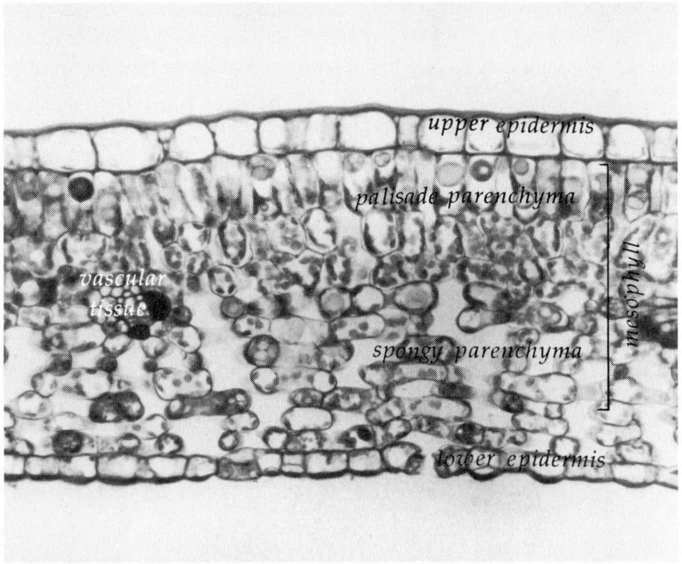
upper epidermis
palisade parenchyma
vascular tissue
mesophyll
spongy parenchyma
lower epidermis

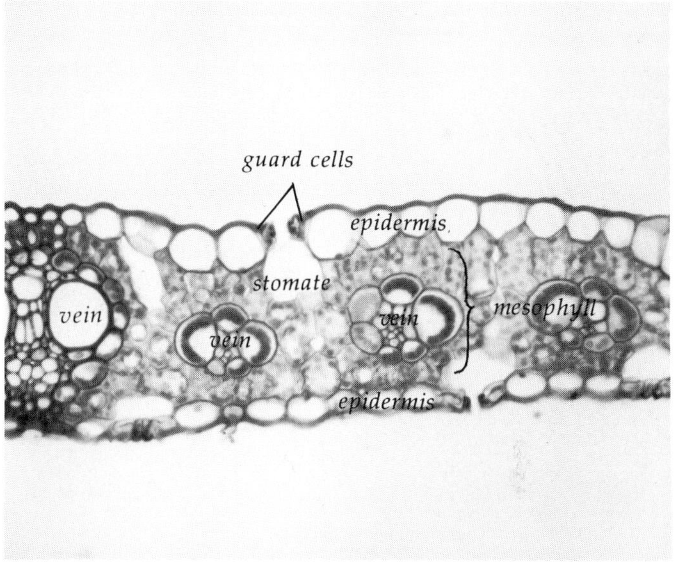
guard cells
epidermis
stomate
vein
vein
vein
mesophyll
epidermis

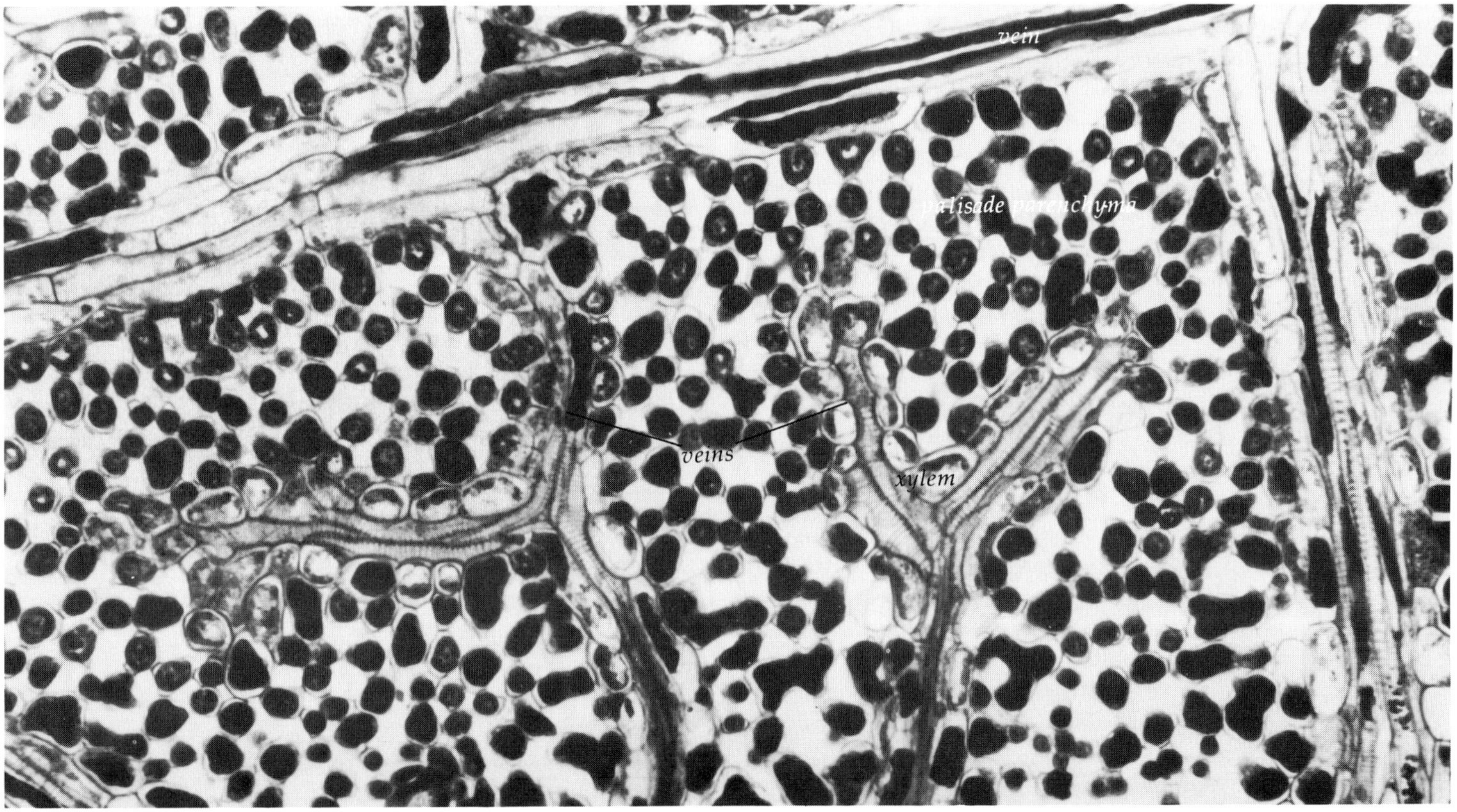
vein
palisade parenchyma
veins
xylem

All through the spring and summer, seedlings press upward through the soil and leaf litter on the floor of our acre of woodland. Some are from large seeds—the acorns that will form the great oaks or the winged seeds of the colorful maples; others are from tiny seeds that will become the short-lived spring flowers. From such seedlings arise the many complex species that compose so much of the woodland. The hallmark of these vascular plants is their remarkable structural adaptability to the environment, thus enabling them to conquer virtually all of the earth's land surfaces. The marvel of it all is that each one of this immense variety of complex plants (for that matter, any sexually reproducing, multicellular organism) begins as a single cell.

The Vascular Plant Life Cycle

The beginning is the union of the sperm and the egg. The resulting zygote divides, grows, and becomes specialized, until an embryonic plant can be discerned. What happens next depends upon the species and the environment. In nonseed plants the embryo grows immediately into the sporophyte. The seed habit provides much more flexibility. Some seeds are quiescent until moisture is ample and temperatures are right. Others remain dormant for varying time intervals (often years), germinating only when environmental conditions are suitable for the growth and survival of the seedling. For example, some seeds must experience the low temperatures of winter, thus ensuring germination only in the spring; other seeds germinate only after an inch of rain has leached an inhibitor from the seed coat, thus ensuring wet soil to a considerable depth. Germination, a highly critical stage in the life cycle of a seed plant, depends strongly upon environmental change; and this dependence often ensures that the developing seedling will have optimum conditions—that is, the dependence is adaptive.

The embryo or seedling responds to gravity by extending its roots into the earth and its shoot system into the atmosphere (Fig. 24–1). Cells at the root and shoot tips actively divide, enlarge, and specialize to produce the vegetative plant. The roots branch, and the shoots produce leaves. On the stem above each leaf stalk

24–1 *This young corn seedling shows the general form of the main axis of the plant.*

(petiole) is a bud that can grow into a branch. Often, the stems become thicker in diameter through cell division, enlargement, and specialization, producing

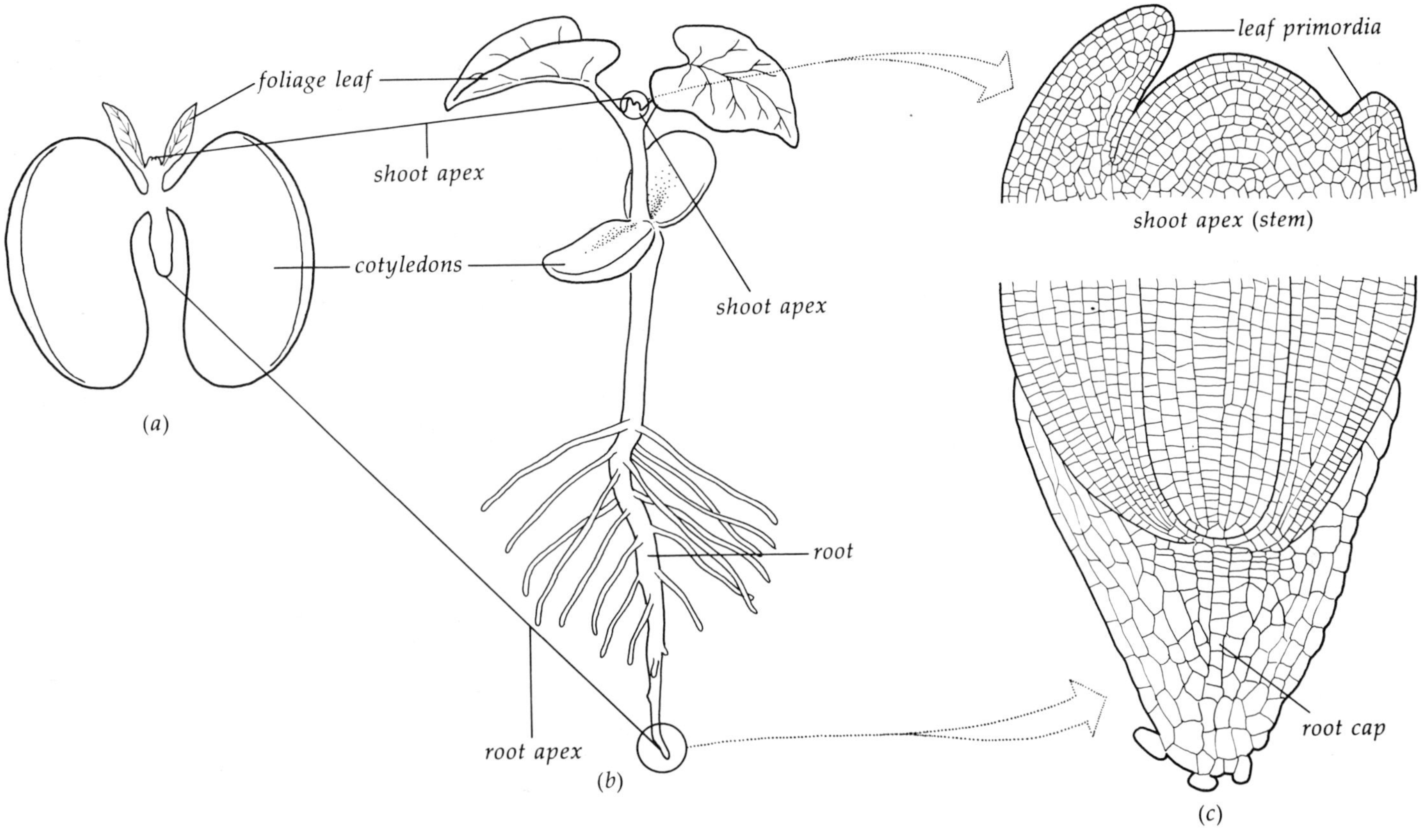

24–2 The position of the apical meristems in the (a) seed and (b) seedling. The general cellular organization of the meristems is indicated in (c).

wood and bark. The result of all this is a complex vegetative plant with an intricate system of branching roots and leaf-covered stems (Fig. 24–2). Again, what actually happens is a function of the species and its environment. Think of the stately pine, which has a single trunk and secondary branches, compared with the magnificent maple, which has a trunk that branches into numerous other trunks of more or less equal importance. Many plants of temperate regions anticipate the approach of winter (usually by detecting shorter days and longer nights): their buds and dividing cells become dormant, and their leaves often export many of their minerals, lose their chlorophyll, turn a brilliant autumn hue, and finally fall from the plant.

All of this is possible because the plant is a functioning, metabolic, open system, taking in carbon dioxide, water, and minerals and using the energy of sunlight to produce the myriad of molecules that make up its structure. This is the chemistry of life discussed in Chapters 5 through 10. But, to produce the final intricate structure, this chemistry must be exactly regulated and controlled in tune with the plant's life cycle.

At some point in its life—often during the first year (all annuals and many other plants) and sometimes only after 20 or 30 years (many trees)—the plant switches from the vegetative to the reproductive state, producing sporangia, cones, or flowers. To do so, the cells at the stem or branch tips must begin dividing, enlarging, and specializing in different ways according to individual sets of regulations contained in their genome. Again, this occurs in many species in response to such environmental changes as the low temperatures of winter, the lengthening days of spring, or the shorter days of late summer. As the reproductive structures mature, the sperm and egg again unite, and new embryos form, completing the life cycle.

Apical Meristems

Apical meristems are the key to understanding plant growth. These regions, near the tips of stems and roots, contain cells that continually divide, enlarge, elongate, and differentiate (Fig. 24–2). New cells are produced by division, and the elongation of the newly formed cells carries the meristematic cells downward in roots and upward in shoots. Differentiation converts the elongating cells to the mature tissues of the root and the shoot. Secondary or branch roots are continually formed in roots as new meristems arise relatively deep within the tissue. Leaf primordia are formed from surface cells on the flanks of the shoot meristem, and branch buds form in association with them.

Cells produced by divisions in the apical meristems become mature root, stem, leaf, or branch tissue, called primary tissue. The mode of growth that can produce leaves, stems, and roots for an indefinite time is said to be indeterminate. Think of what this means: *an organism growing in this way has the potential for immortality!* The apical meristems of the giant redwoods or the ancient bristlecone pines, for example, are literally thousands of years old. But common sense tells us most plants do not live that long and surely not forever. The answer to this paradox is that factors other than the potential of the apical meristem limit plant growth.

Take the growth of cultured root tips. During the mid-1930s Philip White discovered means to culture root tips. A short segment of root tip can be cut off a growing root and placed in a flask with a solution containing carbohydrates, minerals, and vitamins, which it normally receives from the shoot. Under these conditions the root tip grows well—so well that every few weeks it must be recultured by transferring a short segment of the root tip to a new flask. This has been done for years with some of the original roots, and measurements have been made of the amount the roots have grown. The accumulated growth of a single root since the 1930s is so great that if the root had remained intact, its length could encircle the earth. This root was from a plant that, under natural conditions, would have lived only a few months.

Another excellent example is the propagation of trees. Tree growth may be limited by reaching a size where the transport of materials becomes difficult or where its mass becomes so great that the cells can no longer carry the strain. In any case, trees have definite, genetically determined life spans. Yet some cultivated fruit trees do not produce seeds, or, more commonly, seeds are produced only by pollination from different varieties, resulting in hybrid offspring that yield unsuitable fruit. To circumvent this problem, growers remove a portion of a branch, then induce it to root, and grow it into a new tree. Sometimes, these cuttings are budded to growing root stocks. In either case, the fruit will be identical to the parent, being simply its extension. This highly successful procedure is exemplified by the world's navel orange trees, numbering in the millions and all coming from a single plant. Fruit trees in the Middle East have been propagated from biblical times to the present using these methods. Their apical meristems are as ancient as the redwoods.

Essay 24–1 Meristems and Idioblasts

How meristems are organized and how cells differentiate are two questions that have occupied many botanists during the past century. One, in particular, is Professor Adriance S. Foster of the University of California, Berkeley. Here, Dr. Foster relates some of his successful attempts in attacking these two problems.

"My early research interests were centered on comparative studies on the structure and development of the shoot apex in seed plants. It became increasingly clear to me that the shoot apices of gymnosperms represented a virtually unexplored field for intensive study, partly because of the lack of proper attention, in the nineteenth-century work, to the cytohistology

of apical meristems. This neglected problem seemed to me to require the major attention, in place of the overemphasis which had been given to the study of the geometrical arrangement of cells in shoot apices in general. I regard my study of the shoot apex of Ginkgo biloba, made in 1938, as perhaps my most useful investigation on apical meristems because I reached an interpretation of zonal structure (the presence of distinctive areas in the apex characterized cytologically and developmentally) that greatly influenced subsequent studies on the shoot apex in both gymnosperms and angiosperms. The study of such a 'living fossil' as Ginkgo then prompted me to make comparable types of cytohistological surveys on the shoot apices of various cycads, another surviving group of archaic seed plants. These studies, made on the apices of Cycas revoluta, Dioon and Microcycas calocoma, revealed types of apical organization which agreed in some respects with my discoveries with Ginkgo but which also emphasized the impossibility of interpreting 'all' shoot apices with reference either to (1) a rigid 'Histogen Concept' (Hanstein's Histogen Theory) or to (2) the Tunica-Corpus concept, which has been so widely applied in recent years to the description of cell arrangement and planes of cell division in angiosperm shoot apices.

A second major research interest, which has continued from graduate student days at Harvard to the present, is concerned with the ontogeny, anatomy, and phylogeny of foliar organs in the dicotyledons. In pursuing this enthusiasm, my studies have carried me into such diverse fields as the comparative ontogeny of bud scales and foliage leaves in Aesculus and Carya, the development and structure of foliar sclereids and other types of idioblasts in angiosperm leaves, and the study of venation patterns.

When I began my studies on sclereids in leaves, it soon became evident that conventional methods of microtoming and staining serial sections of various developmental stages could not give me the answers. I devised a method of clearing leaves, with the use of dilute NaOH (5–10%) followed by treatment with concentrated chloral hydrate; the cleared specimens were then stained in safranin. By means of this comparatively simple technique, I made the discovery that in some dicotyledons (Mouriria, Boronia, Boronella, and others) the branched sclereids are predominantly situated at the ultimate vein endings, in contrast to the more usual occurrence of idioblastic sclereids in a 'random' diffuse pattern in the mesophyll. 'Terminal sclereids' (i.e., those localized at vein endings) offer problems of considerable morphogenetic as well as taxonomic interest, and my studies have stimulated investigations by others on a variety of gymnosperm and angiosperm genera. In these more recent studies, as well as in my own work, cleared leaves enable one not only to determine accurately the distribution of sclereids in a given leaf type but make possible a realistic study of the total form and structure of the varied kinds of branched sclereids.

The technique of using clearing reagents followed by staining has also been adopted throughout my research, during the last 10 years, on leaf venation. Through a study of the literature, I decided that the best place to begin was provided by leaves with relatively 'simple' venation. I have concentrated on two highly remarkable angiosperms—Kingdonia uniflora and Circaeaster agrestis—both of which are 'unique' because of their open dichotomous leaf venation. My clearing technique has enabled me to survey in detail thousands of leaves of Circaeaster, using plants cultivated year after year, under controlled *conditions in a growth chamber. My results, I think, contribute to a better understanding of the morphological—and possibly the phylogenetic—significance of dichotomous venation within flowering plants, which as a group appear characterized by more complex types of pinnate and palmate reticulate venation.*

Progress in morphological and anatomical research is best accomplished by pursuing intensive studies—developmental and morphological—within the 'framework' of broad comparative studies. This philosophy is best illustrated by my own research on foliar sclereids and venation patterns, where the advances in my understanding have come about by combining intensive developmental studies with the results of broad morphological surveys. The adoption of simple techniques, such as the method of clearing leaves, has led to unexpected—and to me—fascinating discoveries."

Organization of Apical Meristems

Because of the importance of apical meristems to plant growth, an amazing amount of effort has gone into attempting to understand their nature. To the uninitiated, it might seem that an examination of sections under the microscope would immediately illustrate their organization. Unfortunately, such an examination of a shoot tip, for example, reveals what appears at first glance to be almost a jumble of disorganized cells. Even in roots, which initially appear more orderly, the ultimate interpretation is equally difficult. One problem with such analysis is that meristems must be killed and then sectioned for examination under the microscope. This means that the observer must attempt to reconstruct the structure of a dynamic system from a fixed point in time. We will see later what trouble this can cause. After years of patient observation and experimentation by numerous investigators, only a limited understanding of the organization of apical meristems has been gained.

Before discussing apical meristem organization, we should define a few necessary terms concerning planes of division (Fig. 24–3). Divisions occurring at right angles to the surface of the meristem are said to be anticlinal; those parallel to the surface are called periclinal. Divisions occurring in the plane of a radius (as in a cross section) are said to be radial; those at right angles to this plane are called tangential (parallel to a tangent of the circular cross section).

The apical meristems of roots and shoots may be organized in several different ways, depending upon the groups in which the plants are found. An organization that is simple to understand is that of the apical cell. In these systems (common in such plants as ferns and mosses), a single cell of the apex continually divides, and then the resulting cells undergo one or more divisions, ultimately producing the primary tissue of the plant (Fig. 24–4). This relatively simple arrangement is found only in the lower vascular plants and does not occur in the seed plants.

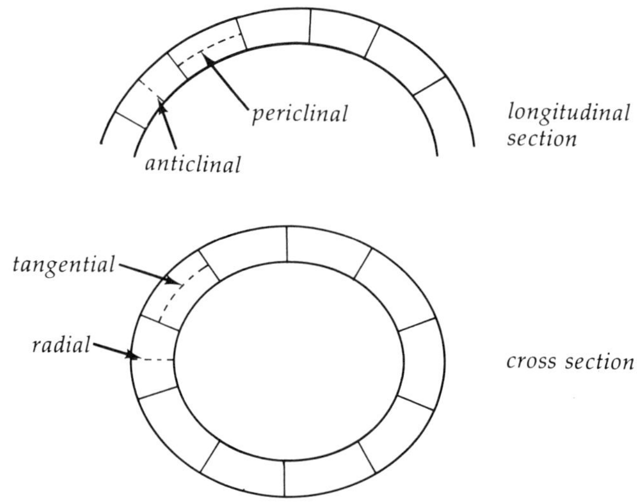

24–3 The planes of cell division found in plants and their names.

In the seed plants there are groups of cells that function as primary meristems: the protoderm, which gives rise to the epidermis, the ground meristem, which gives rise to the cortex and the pith, and the procambium, which forms the vascular cylinder that contains the xylem and phloem. The center of the root tip is occupied by a group of cells that are either not dividing or dividing extremely slowly. This region is called the quiescent center (Fig. 24–5). The significance of the quiescent center is not yet clear in terms of the growth

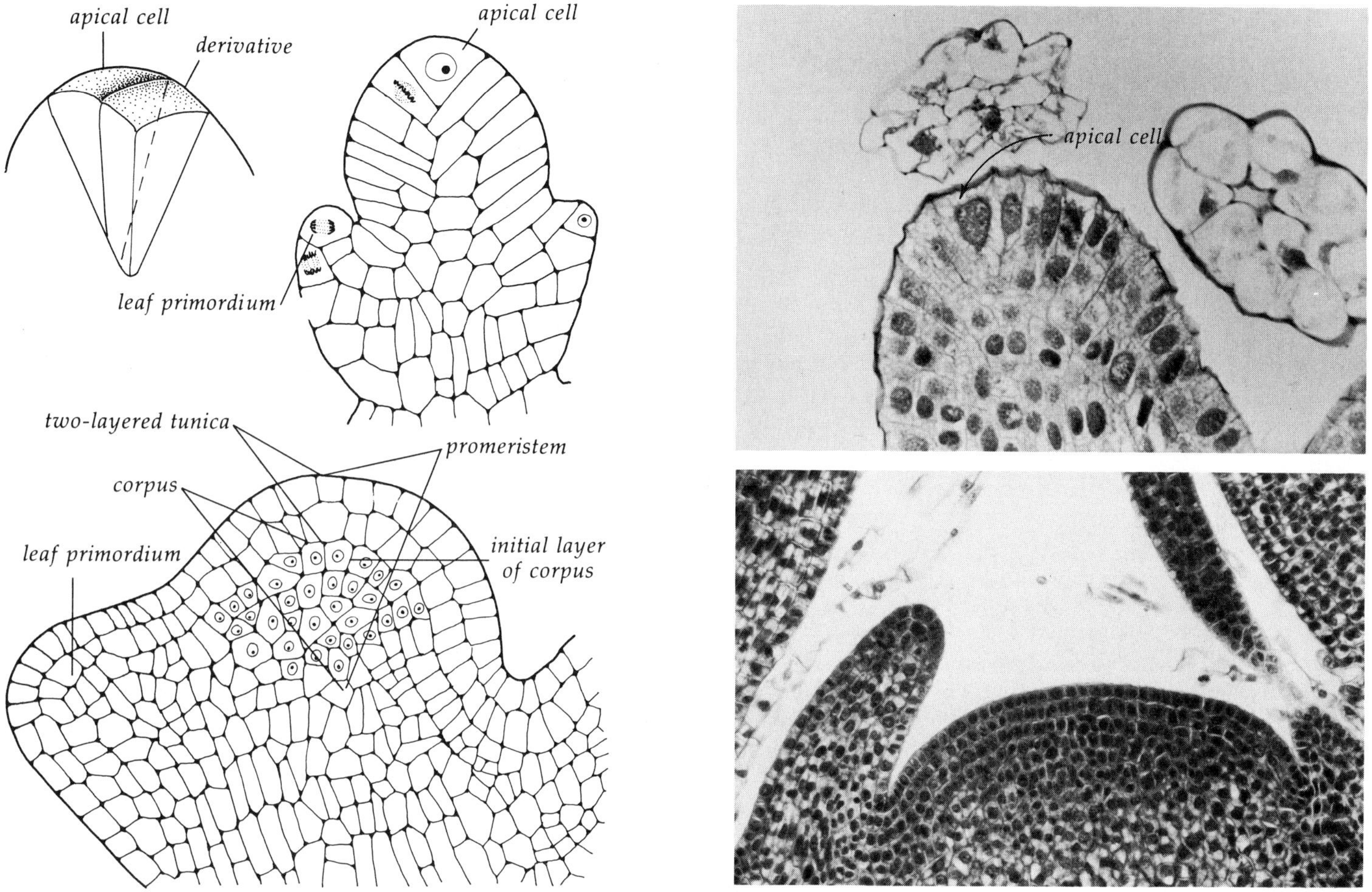

24–4 Diagrams and photographs illustrating the organization of the apical meristem; the apical cell type is at the top and the tunica-corpus type is below.

and physiology of the root. Experiments in which roots were subjected to large amounts of X-rays indicated that the cells of the quiescent center are much less subject to damage than the dividing and elongating cells. Moreover, the cells of the quiescent center are capable of dividing even though surrounding cells are damaged to the point where they can no longer divide. Thus the cells of the quiescent center are able to re-establish the growth of the root.

The existence of the quiescent center in roots has been firmly established, but its presence in shoot tips has given rise to heated controversy. A group of French botanists believed they had evidence that the cells at the center and tip of the shoot were quiescent (either nondividing or slowly dividing), and they termed this region the "meristem d'attente." They formulated a concept in which this group of cells was carried upward through the division of the cells below it. Thus, during most of the life of the plant, these cells did not divide, but when conditions were right for flowering, they were stimulated to divide and produced the flower. This concept was not widely accepted, and many researchers have collected data that indicate that these cells divide more rapidly than was first thought and are far from quiescent. But the meristem d'attente idea is interesting, and the data that support it cannot be ignored in future research on the shoot tip.

The theory of meristems that has been most widely applied to the shoots of angiosperms and some gymnosperms is the tunica-corpus theory. In this description

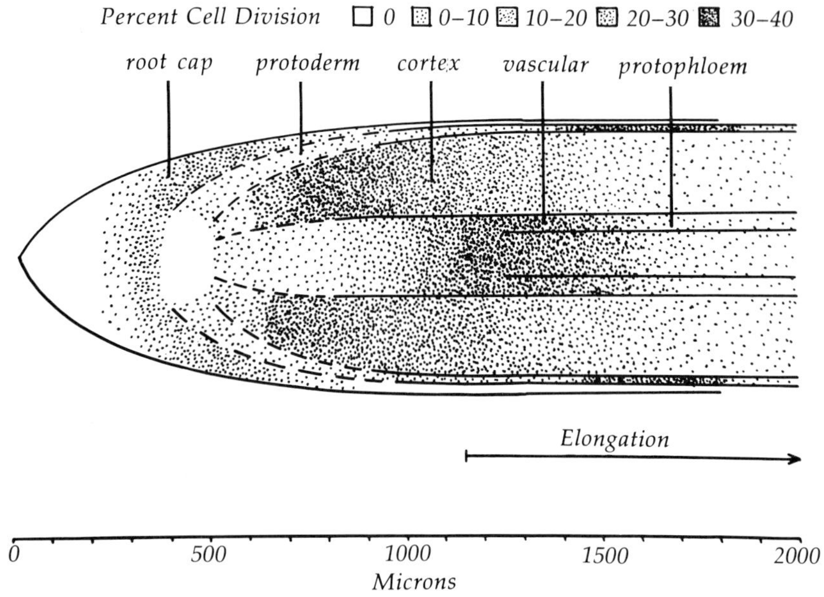

24–5 One of the ways to demonstrate the presence of the quiescent center is to plot the number of cell divisions taking place along the length of the root. When this is done (as it was for this onion root), we can see that the number of cell divisions in the quiescent center is essentially zero.

a central group of cells called the corpus is enclosed by one or more peripheral layers of cells called the tunica. The tunica cells divide only in an anticlinal plane, while the corpus cells divide anticlinally, periclinally, and obliquely. Each tunica layer has a number of so-called initials that may be primarily responsible for the other tunica cells. The corpus has one layer of initials. The theory does not imply any relationship between zonation in the meristem and the subsequent formation of specific tissues. For example, the outer layer of the tunica surely gives rise to the epidermis, but both tunica and corpus cells might account for the cortex.

Products of the Apical Meristems—The Primary Tissues

Regardless of the way that apical meristems are organized, they all produce the cells making up the primary plant body. These cells are of many different types, which differ markedly in form and function. Obviously, the cells are not arranged in a random fashion but are grouped together as tissues. Tissues may be simple, consisting of one cell type, or complex, containing a number of cell types.

The principal tissues of the vascular plant were defined in 1875 by the great German botanist Julius von Sachs. So astute were Sachs' observations and deductions that his system is still the most widely accepted. On the basis of topographic continuity, he defined three major tissue systems: the dermal, the vascular, and the fundamental or ground (Fig. 24–6).

The dermal system includes the epidermis. This is a layer of cells that surrounds the plant and acts as a protective covering. The vascular system contains two kinds of conducting tissues: the phloem, a complex tissue involved in the conduction of organic materials, and the xylem, also a complex tissue involved in the conduction of water in the plant. The ground tissue is found between the epidermis and the vascular tissue (where it is called the cortex) or sometimes in the center of the vascular tissue (where it is called the pith), and between the upper and lower epidermal layers in the leaf (where it is called the mesophyll—"middle leaf"). The main ground tissue is the parenchyma with two supporting tissues, the collenchyma and sclerenchyma, sometimes present.

Before we examine the organization of the primary plant, let us look at these cells and tissues both in terms of their physical appearance and their development. It should be noted that the concept of cell-tissue-organ that fits so well for most animal anatomy is much more subtle in plants where distinctions between the levels of organization are much less well defined.

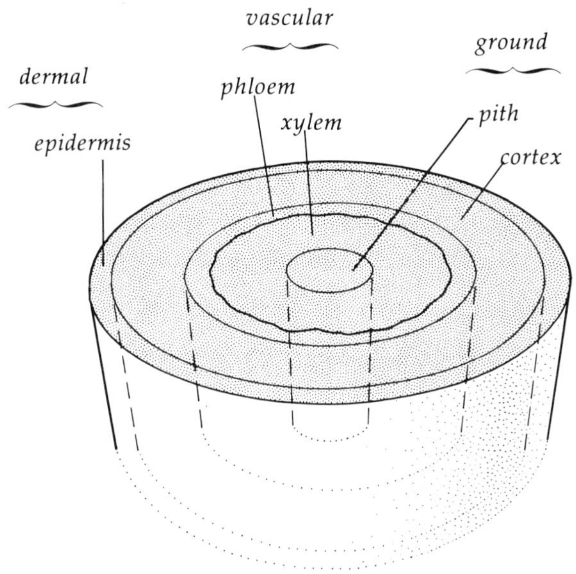

24–6 *The main tissue systems of the vascular plant.*

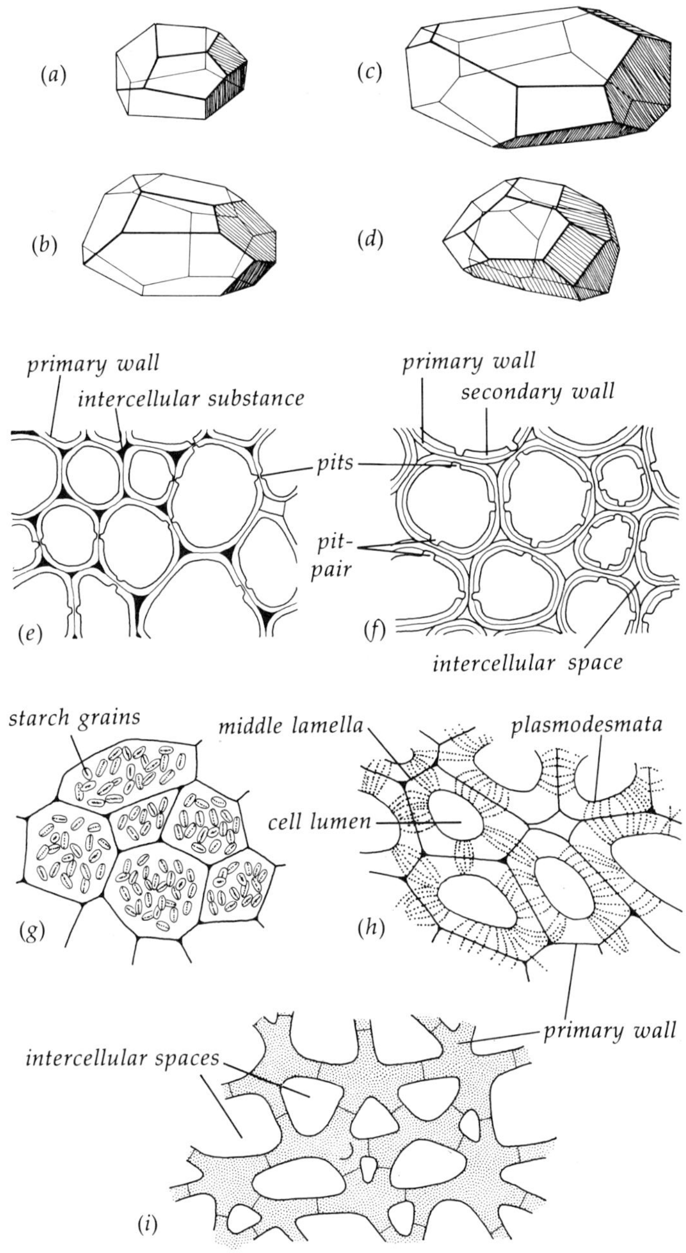

24–7 *Parenchyma cells frequently have 14 sides (b & c), but they may have as few as 10 (a) or as many as 17 (d). The walls of parenchyma cells may be thin primary walls (g), thick primary walls (h), or have a secondary wall (f). They may be filled with starch, like the ones present in potatoes (g), and they may have branches or arms, like the ones found in many leaves (i).*

Meristematic tissue. As we have just seen, the apical meristems of the plant are located at the tip of the root and shoot. The cells found here, frequently termed meristematic, are characterized by being densely cytoplasmic (containing few or no vacuoles), by having large nuclei in proportion to the volume of the cell, few poorly developed organelles, thin walls, and by actively dividing (Fig. 24–6). There are dividing cells that accurately fit this description, but so do the cells of the quiescent center that are not actively dividing. Many cells regularly dividing in the plant contain vacuoles (often quite large) but do not fit the general morphological concept of a meristematic cell.

Parenchyma. The most basic of all cell types or tissues found in vascular plants is the parenchyma (Fig. 24–7, 24–B). This is the "typical plant cell" so often drawn and discussed in botany and biology books. The general characteristics of parenchyma cells include, first, *a relatively simple cylindrical shape.* While many are roughly isodiametric (more or less equal in all dimensions), others are elongate. The average number of cell-wall surfaces or faces is 14, an "ideal" number for packed, three-dimensional figures. (Soap bubbles packed together may assume such a form, since maximum volume is accommodated by minimum surface.) Second, *the walls are usually thin and do not contain lignin.* In some cases, the walls may be lignified, but secondary walls are not formed. In other cases, the walls may be thick, but it is still a thick primary wall, not a secondary wall. Third, *the cell is alive at the time it is functionally mature.* The center of the cell is frequently occupied by a large vacuole crossed by strands of cytoplasm, while the bulk of the cytoplasm and nucleus is pressed against the wall. Most cell organelles are present, although chloroplasts may be lacking. Occasionally, the cells are binucleate or multinucleate.

Many parenchyma cells contain numerous chloroplasts. Such cells are called chlorenchyma. They occur not only in mesophyll of leaves but in stems and even in some roots, particularly epiphytic species. In leaves the chlorenchyma cells may be arranged within the mesophyll in tight, regular, upright rows, the palisade parenchyma, or in loose, irregular arrangements known as spongy parenchyma. Both types are usually present in the same leaf.

Parenchyma cells may also be filled with amyloplasts containing starch. When you eat a potato, you are eating starch-filled parenchyma cells. Starch-storing parenchyma cells are frequently roughly isodiametric in shape and may be so filled with starch that the vacuoles can hardly be seen. In still other parenchyma cells, the vacuoles may be filled with tannins, pigments, organic acids, or other substances.

Parenchyma cells are remarkably totipotent in their capacity to develop. They can develop into a number of other cell types even when they are mature. Moreover, they retain the capacity, present in the zygote, to develop into new plants. Single parenchyma cells have been isolated and have then grown into mature plants.

Collenchyma. This is essentially parenchyma with thickened walls (Fig. 24-C). These long cells with thickened primary walls function as support elements in the plant (Fig. 24–8). The walls are rich in pectic substances (45%), hemicellulose (35%), and cellulose (20%). No lignin is present in young cells but may occur in older cells. Collenchyma cells are alive at maturity and may contain chloroplasts.

Collenchyma cells are found particularly in young, rapidly growing organs. They are strong yet plastic,

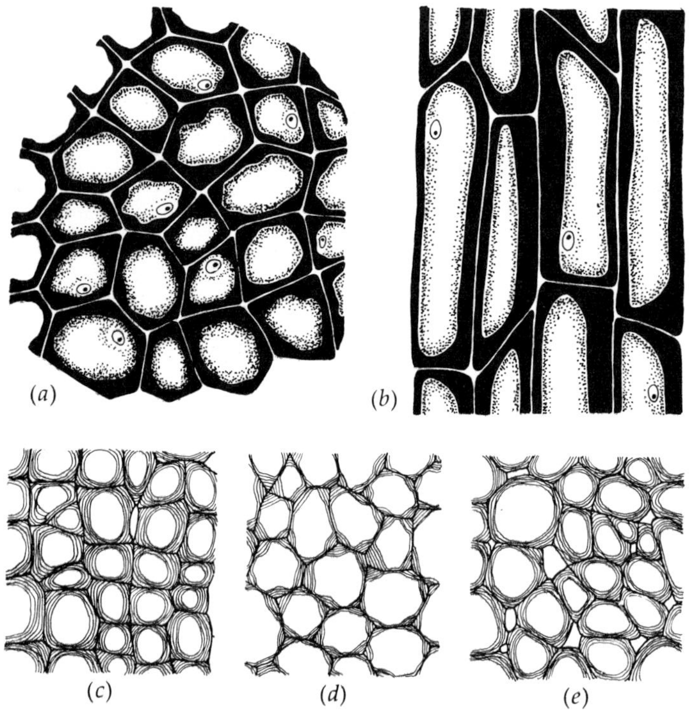

24–8 *Collenchyma cells as seen in cross (a) and longitudinal (b) section. The walls may be uniformly thickened (c), thickened only in the corners (d), or prominently thickened when adjacent to intercellular spaces (e).*

lending considerable strength to the plant. Experiments have shown that plants that are mechanically shaken during growth produce more collenchyma with thicker walls. Collenchyma is one of several cell types that have evolved because air is a much less dense medium than water; hence to grow erect requires an internal support system.

Sclerenchyma. It can be divided into sclereids and fibers. Sclereids are highly variable in size and shape (Fig. 24–9, 24–C). They range from isodiametric cells to H-shaped cells to cells with many long, odd-shaped arms. Fibers, in contrast, are long, narrow, uniformly thick-walled cells.

Sclereids usually occur as scattered cells in a wide range of locations, although they may occasionally

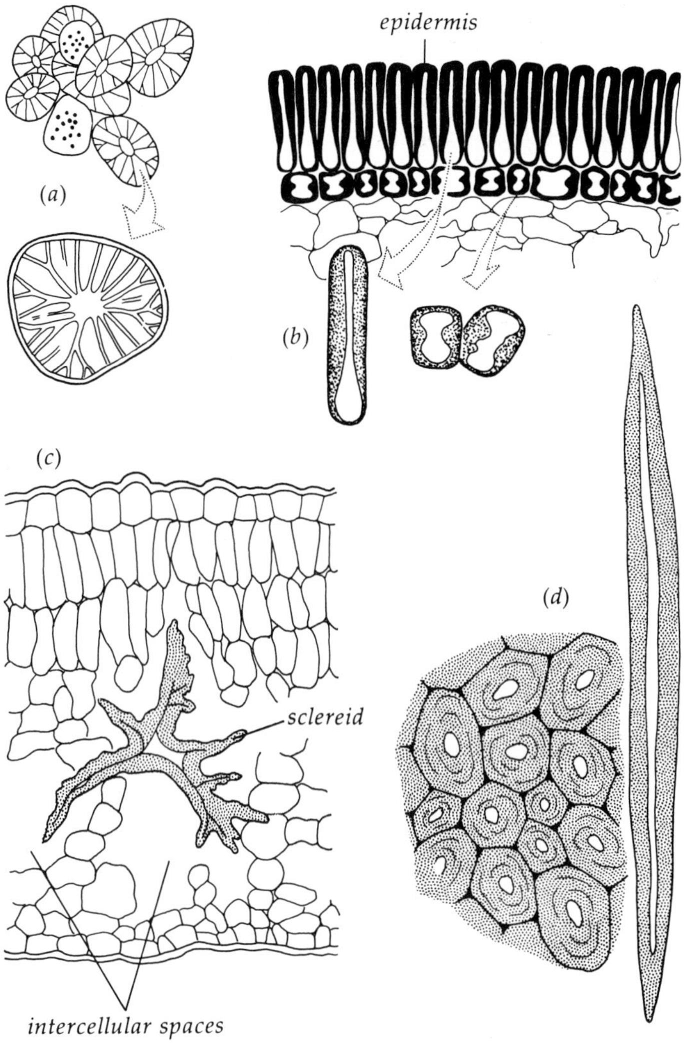

24–9 Sclerenchyma cells can be present as small, regularly shaped sclereids, like the stone cells (a) found in pears or like the cylindrical cells (b) occurring in the seed coats of many species. They may also be present in the form of the striking, many armed sclereids (c) encountered in leaves. Sclerenchyma cells also include fibers (d) that are thick walled in cross section but elongated in longitudinal section.

form layers (as in seeds). Sclereids develop from parenchyma cells. One of the first indications of the initiation of a sclereid is an enlargement of the nucleus. The cell then rapidly develops arms or projections of the wall, which grow into existing intercellular spaces or push between cells. The final form is dependent on the space available for the growth of the arms. When the growth of the arms ceases, the wall thickens. Finally, the nucleus and cytoplasm degenerate.

Specialized cells that differ markedly from the surrounding cells in size, content, or function are called idioblasts. There are many different kinds of idioblasts found in plants, and the sclereids are a good example. An intriguing problem is what determines where they are located. In some cases, the sclereid initials are related to the differentiation of the veins in leaves. In other cases, there appears to be a relationship to the surface of the plant. Usually, there is no evidence as to why sclereids form where they do.

Fibers are quite unlike sclereids in shape but are similar in their other features. They have thick, lignified secondary walls and are usually quite long. Some of the fibers we use for cloth are sclerenchyma fibers. Linen is woven from fibers obtained from the flax plant (*Linum usitatissimum*). In this case, the fibers are from 9 to 70 mm long and provide a durable, tough, yet pleasant-feeling material.

Tracheids and vessels. The xylem is the water-conducting system of the plant, and within the xylem the tracheids (Fig. 24–F) are the cells through which the water moves. In the most advanced land plants, tracheids are supplemented or replaced by vessels (Fig. 24–F).

Tracheids are relatively long, narrow cells with lignified secondary walls (Fig. 24–10). The oblique end walls are covered only by a thin membrane, and the cells are aligned end to end, facilitating the flow of water. The side walls contain numerous pits. Tracheids found in primary tissue have walls with characteristic bands of wall thickening. The nucleus and cytoplasm of a tracheid degenerate when the cell reaches functional maturity.

Vessel elements are shorter in length and larger in diameter than tracheids (Fig. 24–10). They too have lignified secondary walls, and at maturity, no longer contain cytoplasm or nuclei. One of the last acts in their maturation is the dissolution of major portions of the end walls so that when they are arranged in rows, water forms a continuous column through them. Vessels in

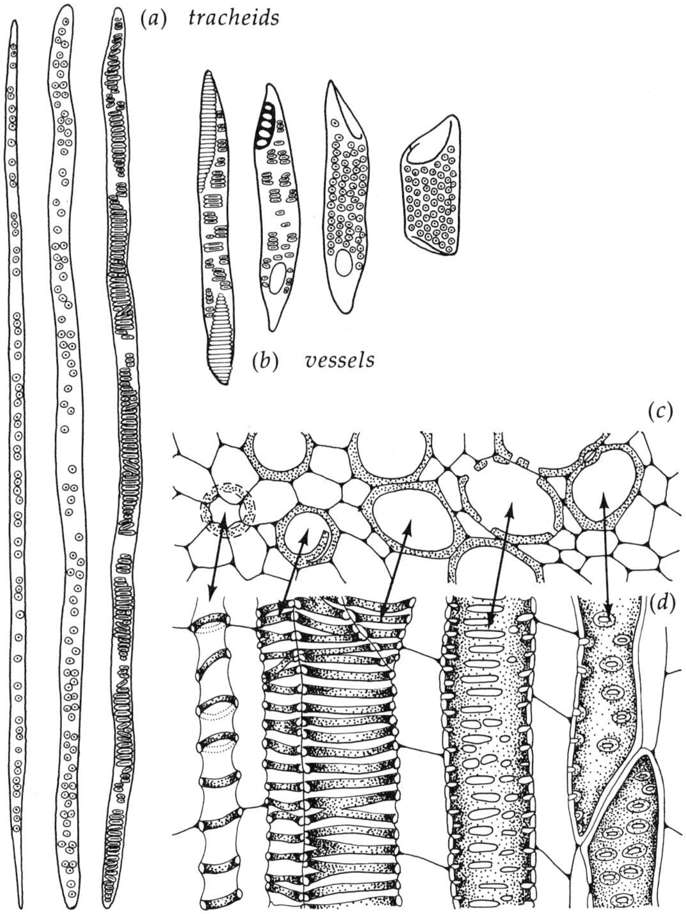

24–10 *Tracheids and vessel elements shown isolated from the tissue (a and b), in cross section (c), and in longitudinal section (d).*

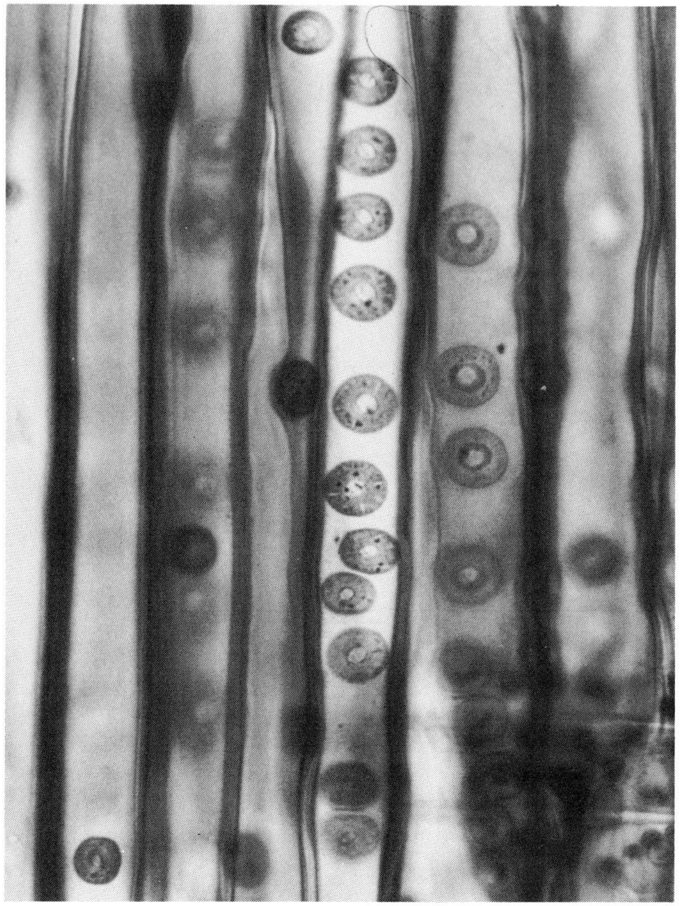

24–11 *Bordered pits in the tracheids of pine.*

primary tissue also have bands of secondary wall thickening. As in tracheids, the bands in the vessels formed earliest are rings; in later cells they are helical; still later they are reticulate; and finally the whole wall is thickened, except in the pits.

Pits (Fig. 24–11) are important structures in the walls of xylem cells. Two types are present: simple pits and bordered pits. In simple pits a thin portion of the walls of two cells is adjacent, but there is no actual hole in the walls. Instead, a pit membrane, comprising the primary walls and the middle lamella of the two cells, remains. In bordered pits there is a special enlargement of the wall around the depression, and the center of the pit membrane is thickened into a special structure called a torus. The face view of a bordered pit can be seen in Figure 24–11.

The differentiation of a vessel is shown in Figure 24–12. The pattern begins with the division of an already elongate initial cell. As the new cell elongates, the microtubules that lie parallel to the cell wall show little preferential distribution. Wall thickening is associated with the production of numerous dictyosome vesicles. At this stage the microtubules appear concentrated between the thickened areas of the wall.

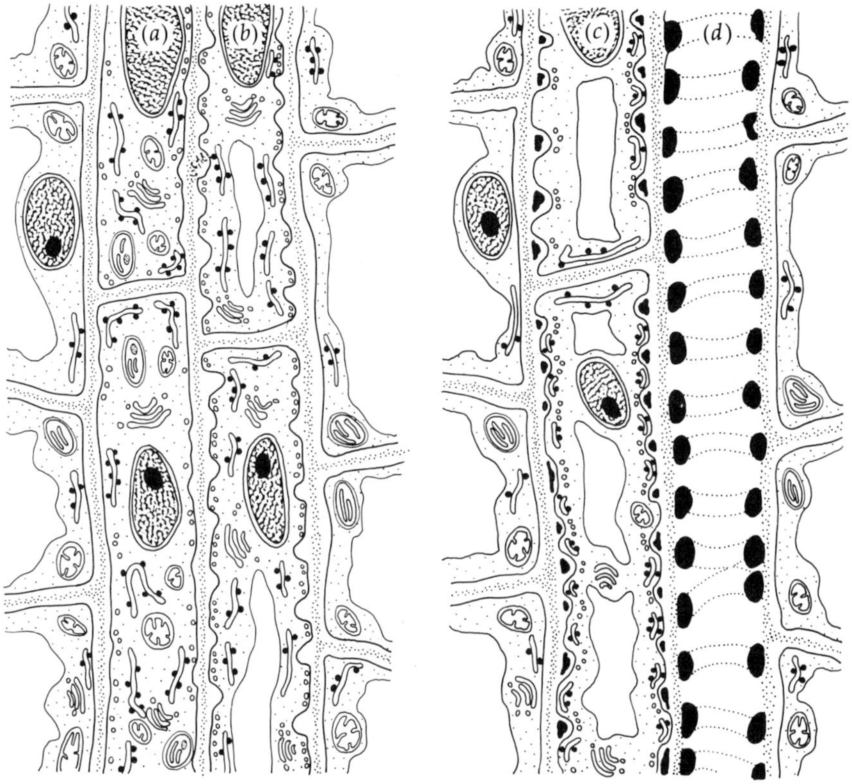

24–12 Differentiation of a vessel begins with the division of an already elongate procambial cell (a). This cell begins to elongate, and the microtubules, which lie parallel to the cell wall, show little preferential distribution. Wall thickening (b) is associated with the production of numerous dictyosome vesicles, and the microtubules appear concentrated between the thickened areas of the cell wall. Vacuolation continues, and ER is closely associated with the end walls. As vacuolation (c) nears completion, the microtubules concentrate over the wall thickenings, with segments of ER between them. Lignification begins in the thickened areas of the wall and gradually extends (d). The end walls swell and break down as the remaining cell parts disappear.

Vacuolation continues while wall thickening is occurring, and ER is closely associated with the end walls. As vacuole formation nears completion, the microtubules appear concentrated over the wall thickenings, while segments of ER are found between them. Lignification begins in the thickened areas of the wall and extends gradually. Finally, the end walls swell and break down as the remaining cell parts disappear.

End walls don't always disappear completely. Often, a perforation plate remains. This may have one large hole, a parallel series of openings, a network of openings, or a group of circular holes like in a sieve. A true vessel consists of a number of vessel elements (individual cells) lined up and connected by perforation plates or simple openings, where the end walls had been. Sometimes, the vessels are 1 to 5 m long, and vessels extending the height of an entire tree have been reported. Vessel elements may overlap somewhat (as do tracheids at the end walls), being connected by perforations. Water and solutions will pass through these perforations and through the perforation plates, but gas bubbles will not.

Vessels are clearly more efficient in carrying water than tracheids because of their larger diameter and lack of solid end walls. On the other hand, this larger diameter makes them much less effective as a supporting structure than the tracheids. In angiosperms, where vessels are a consistent feature of the xylem, xylem fibers have also developed. Both vessels and xylem fibers appear to have evolved from tracheids. The fibers are usually longer than tracheids, have thick, lignified secondary walls, and are dead when functionally mature.

Sieve cells and sieve-tube members. The phloem is the functional counterpart of the xylem, transporting solutions of organic materials rather than dilute mineral solutions, and the sieve cells and sieve tubes (Fig. 24–13, 24–14) are analogous to the tracheids and vessels.

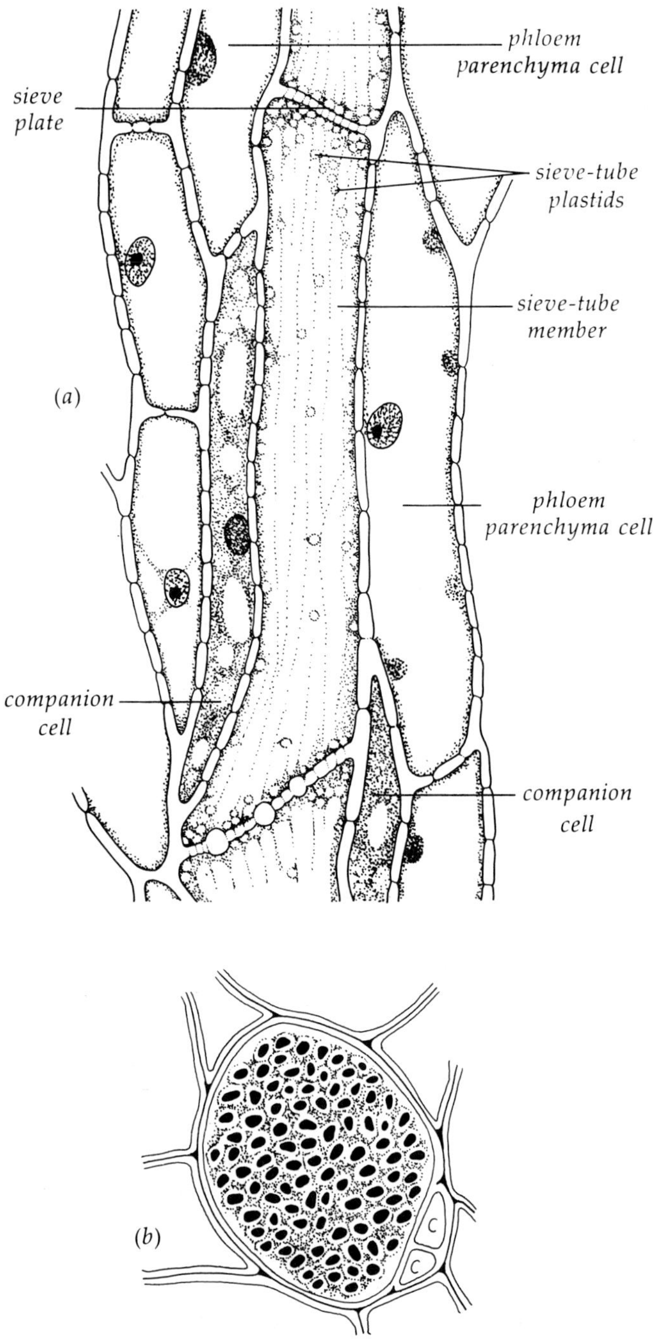

24–13 A longitudinal view of a mature sieve-tube member and companion cell (a). A face view of a sieve plate (b); the black areas are actually holes in the wall.

Sieve cells, usually longer and narrower than sieve-tube members or elements, are found in the more primitive land plants, while sieve-tube members are found only in the angiosperms. Sieve-tube members are aligned end to end to form sieve tubes.

The name sieve cells or sieve-tube member is derived from the appearance of the end walls of the cells, which actually look like sieves. Sieve-tube members are always associated with specialized cells called companion cells. These are much smaller cells and have dense cytoplasm. The two types of cells (sieve-tube members and companion cells) appear to be closely related physiologically; if the sieve tube ceases to function, the companion cell will die.

The differentiation of the sieve-tube member (Fig. 24–14) begins with an unequal division: the large cell will be the sieve-tube cell, while the smaller will be the companion cell. Vacuolation proceeds rapidly in the sieve-tube cell, and the cytoplasm becomes less dense. The plastids, which accumulate starch, are surrounded by ER. Segments of ER are found opposite the plasmodesmata joining the sieve tubes. Protein bodies, also called slime, appear in the cytoplasm associated with the ER. The cell continues to vacuolate during these changes, and the cytoplasm and vacuole become increasingly indistinguishable. Callose, a special cell-wall carbohydrate, appears at both ends of the plasmodesmata, and the plasmodesmata begin to enlarge. The nucleus now breaks down; sometimes the nucleolus will be extruded from the nucleus before it breaks down. As development continues, the plasmodesmata widen even further, and masses of callose form around the ends. The ER (associated with the wall), the plastids, and protein bodies disappear. Finally, the distinction between the vacuole and cytoplasm is completely lost, although a few organelles remain. The sieve pores in the sieve plate are wide, and the cell is filled with protein strands that run from one cell to another. Thus at maturity the sieve-tube cells still possess a living, although highly modified cytoplasm, but no nuclei.

Xylem and phloem. Both the xylem and phloem are complex tissue systems, consisting of many more cell types than tracheids/vessels and sieve cell/sieve tube members. The xylem contains, besides tracheids and vessels, various fibers and parenchyma cells. The phloem contains sieve cells or sieve-tube

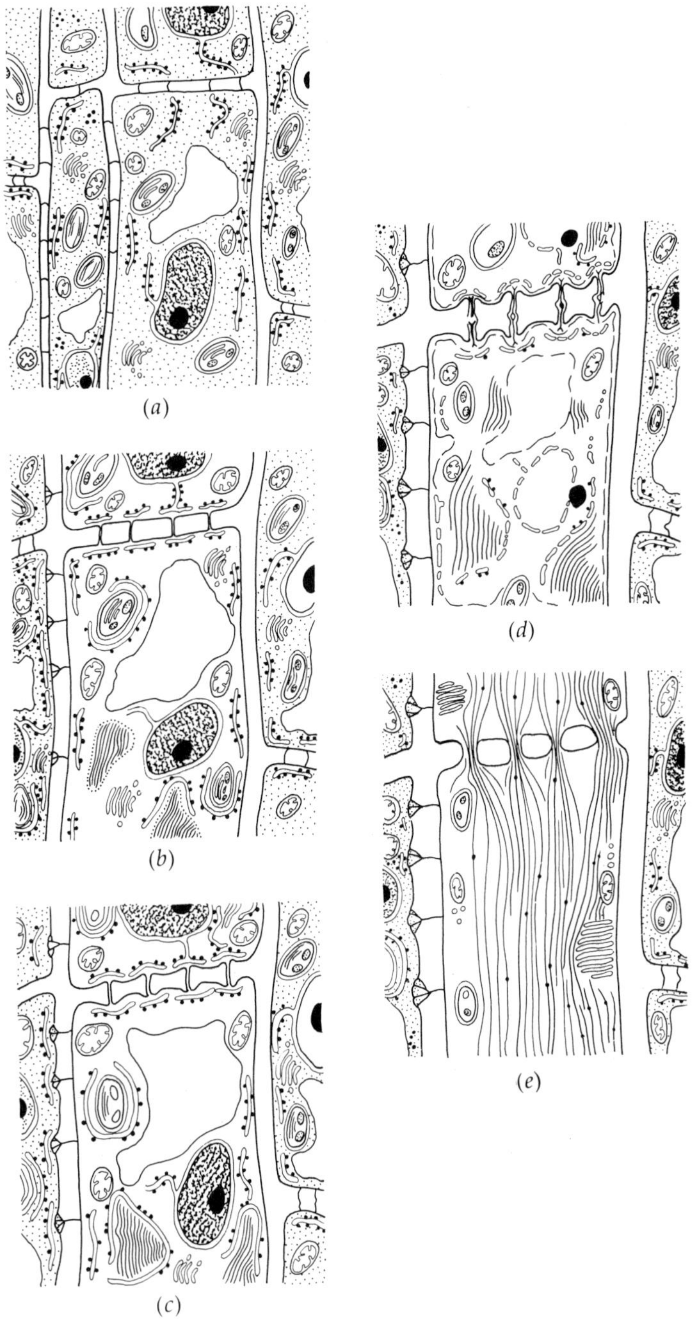

24–14 The differentiation of the sieve-tube element begins with an unequal division of a cambial cell (a). The larger cell will be the sieve-tube cell, while the smaller will form the companion cell. Vacuolation proceeds rapidly in the sieve-tube cell, and the cytoplasm becomes less dense (b). The plastids, which accumulate starch, are surrounded by ER. Segments of ER are found opposite the plasmodesmata joining the sieve tubes. Protein bodies (slime) also appear in the cytoplasm associated with the ER. The cell continues to vacuolate (c), and the cytoplasm and vacuole become indistinguishable. Callus appears at both ends of the plasmodesmata, and the plasmodesmata begin to enlarge. The nucleus now breaks down (d), with the nucleolus sometimes being extruded earlier. The plasmodesmata widen still further, and masses of callose form around the ends. The ER, both associated with the wall and the plastids or protein bodies, disappears. Finally, the distinction between the vacuole and cytoplasm is completely lost (e), although a few cell organelles remain. The sieve pores are wide, and the cell is filled with protein strands that run from one cell to another.

cells, companion cells, fibers, sclereids, and parenchyma cells. How these cells are assembled into the xylem and phloem found in roots, stems, and leaves will be discussed later.

Epidermis. This is the layer of cells covering the exterior of the vascular plant (Fig. 24–D). In general, epidermal cells have relatively thin walls, with the external wall being thicker than the others. The epidermis of pine needles is an example of extremely thick-walled epidermal cells. A characteristic feature of epidermal cells found on the aboveground parts of the plant is the cuticle. This layer of fatty substance, cutin, is secreted by the epidermal cells and covers the outer surface of the cell as a separate layer. In many plants the cuticle is covered with a deposit of wax.

The epidermal cells are alive when mature and contain nuclei capable of division. In flax and *Begonia* a single epidermal cell has been shown to give rise to a bud that can form a complete plant. Plastids are present, including chloroplasts in some plants. The vacuoles are usually large and may contain pigments or tannins.

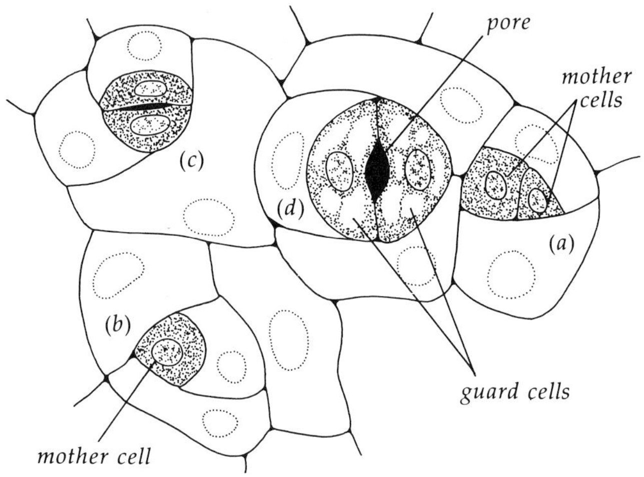

24–15 The development of guard cells: (a) an unequal division results in the formation of a small cell; (b) this cell enlarges and becomes the mother cell; (c) an equal division of this cell results in the formation of the guard cells (d).

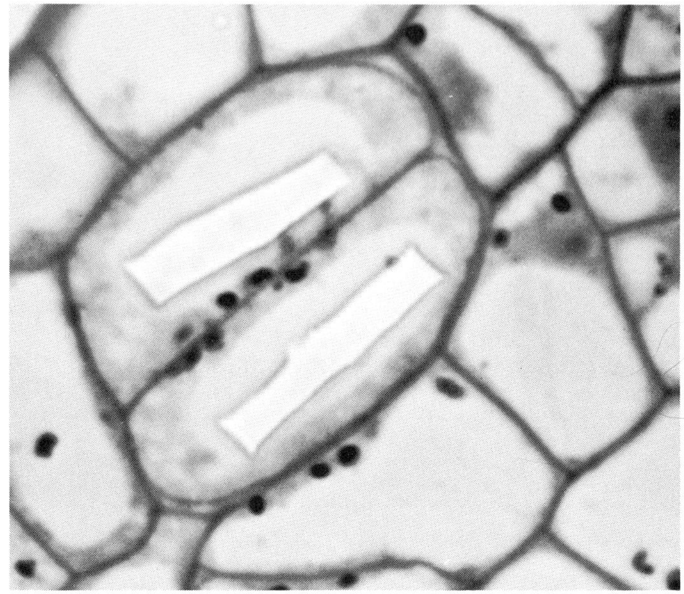

24–16 Some crystal-containing cells of a leaf. Both the cells and the crystals are similar in size. The two cells are organized as mirror images of each other.

A variety of specialized cells functioning as idioblasts are found in the epidermis. These include guard cells, root hairs, crystal cells, various types of hairs, and secretory cells.

The guard cells (Fig. 24–D) are a pair of specialized cells that surround the openings, or stomates, in the leaf. Changes in the shape of the guard cells regulate the size of the opening. The cuticle extends over the surface facing the stomatal pore. The wall of the guard cells is typically uneven in thickness. These uneven, thickened walls play a role in the opening and closing mechanism of the stomates.

The guard cells originate from a guard cell mother cell (Fig. 24–15). This cell is formed as a result of an unequal division of an epidermal cell; the smaller of the two daughter cells is the mother cell. When the mother cell divides, it does so in a precisely equal manner, giving rise to the guard cells. This picture may be complicated by a series of divisions that result in a number of subsidiary cells surrounding the guard cells.

Some crystal cells found in leaves have a pattern of development similar to those of guard cells. An unequal division produces a smaller cell that in turn divides equally to form two matching cells, each of which will contain a crystal (Fig. 24–16). In some grasses the mother cell may divide, forming two guard cells or a silica cell (cells filled with silica crystals) plus a cell with heavily suberized walls (sometimes called cork cells).

The hairs or trichomes forming part of the epidermis may be simple, single cells or elaborate multicellular structures frequently found on stems and leaves (Fig. 24–17, 24–E).

Root hairs are simple extensions of the root epidermal cells similar in principle to simple trichomes (Fig. 24–18). Sometimes, they reach 2 mm in length. They have a definite life span, being initiated some 100 to 300 μ behind the root tip. An unequal division of an epidermal cell results in a small cell, which will form the hair, and a large cell, which will be hairless. The cytoplasm of the hair initials is dense and becomes higher in RNA, protein, and various enzymes than the surrounding hairless cells. The DNA content of the hair cells increases up to eight times that of the adjacent

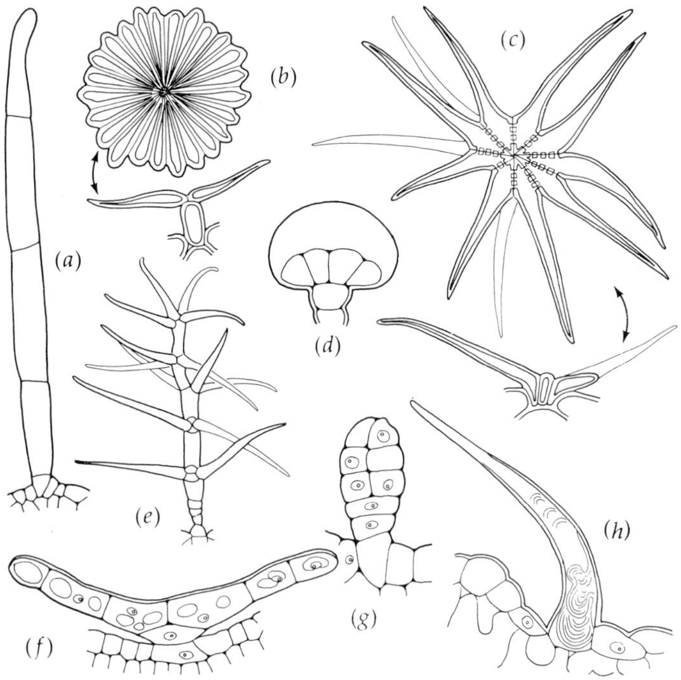

24–17 *Some examples of the variety of hairs found on stems and leaves.*

cells. The hair forms when a portion of the wall is modified and growth occurs only on one face of the cell. The hair extends by tip growth—only the tip of the hair actually forms new cell walls.

The presence of root hairs greatly increases the area of the absorbing surface of the plant in its contact with the soil. This can be illustrated by a study made of a single rye plant. The plant had 13,800,000 roots with a surface area of 2500 ft^2. On these were 14 billion root hairs having a total surface area of 4300 ft^2. The combined surface area of roots plus root hairs was 6800 ft^2—all packed into less than 2 ft^3 of soil!

Root hairs are present along about 1 cm of each growing root. They ordinarily live for days or, at most, weeks; only in a few species do they persist for a season. Old root hairs collapse, and the walls of the epidermal cells become suberized and lignified, if they do not slough off.

Some hairs found on stems, leaves, flowers, and fruit are simple hairs similar to those found in roots. A familiar example is the cotton fiber, which is actually an epidermal hair formed on the surface of the cotton seed. The walls of these cells are much thicker than in root hairs and are extremely high in cellulose. They become much longer, some 6 to 60 mm, depending on the variety of cotton.

Other hairs present on the aboveground parts of the plant are multicellular and have a wide variety of forms. Some are simple in shape; others are tufts of cells; many are in the form of a plate of cells often borne on a stalk; still others are glandular, secreting a variety of substances. Such secretory hairs release compounds such as nectar and various essential oils. One of the most unusual hairs is the stinging hair found in nettle (*Urtica*). The hair (Fig. 24–19) is like a fine capillary tube with calcium in the wall at the lower end and silica at the upper end. At the base a bulblike end is embedded in epidermal cells. At the upper end the tube bears a spherical tip that breaks off along a predetermined line when the hair comes in contact with an object such as human skin. The sharp edge that is left readily penetrates the skin, and the cell contents are forced into the wound. This material is complex in composition and highly irritating.

Secretory cells and tissues. We already noted that epidermal hairs could secrete a number of substances, and such cells are numerous in the vascular plants. The developmental stages start with the outgrowth of the wall of an epidermal cell. There is a clear polarization of the cytoplasm of this cell, with the nucleus and much of the cytoplasm occurring at the growing end of the cell. The cell now divides unequally, with the bulk of the cytoplasm in the upper daughter cell. This cell again divides unequally, and again the outermost cell contains more of the cytoplasm. The final division of this cell is precisely equal and perpendicular to the previous divisions. These two head cells enlarge and become filled with cytoplasm particularly rich in parallel strands of ER. The head cells are covered by a cuticle, and the secretion product, usually a volatile oil such as peppermint or balsam, accumulates between the wall and the cuticle. Presumably, the oil diffuses through the cuticle, since no pores are formed.

This is the type of hair that in *Cannabis sativa*, or marijuana, secretes a resin containing a series of complex alcohols, including tetrahydrocannabinols. These form the active principals in the drug marijuana, or

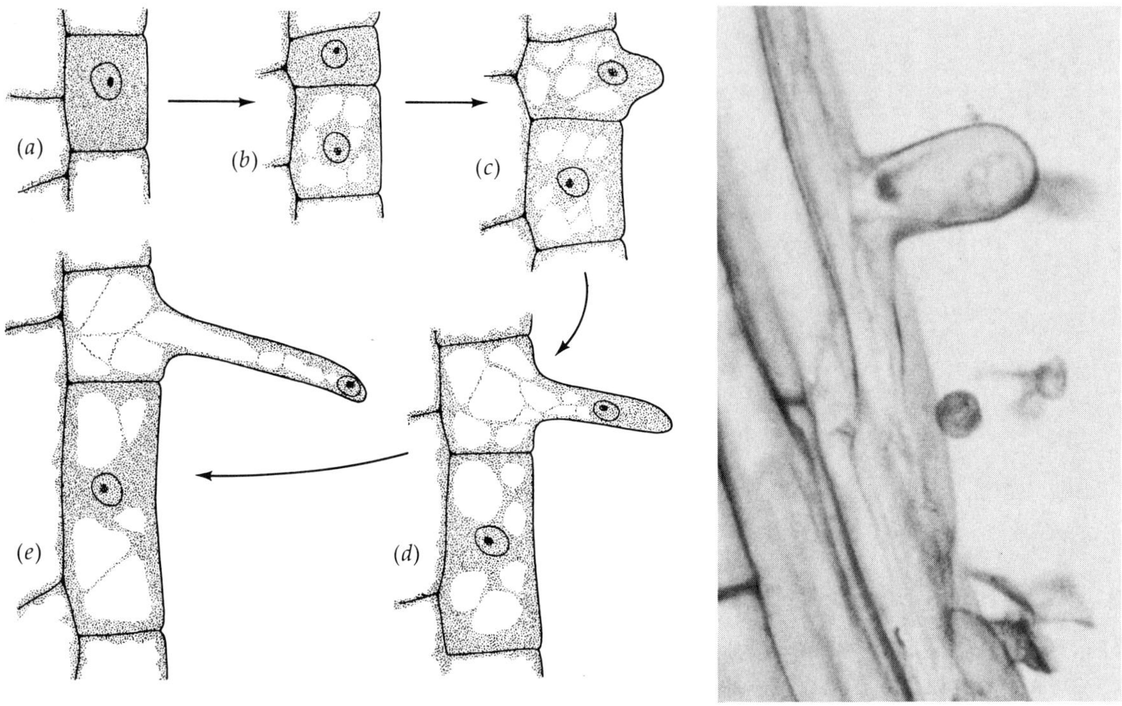

24–18 The stages in the development of a root hair. Right is a photo of an early developmental stage in a corn root.

"hashish," as it is known in the Old World. Such hairs are found only on the upper parts of the pistillate (female) plants.

Several types of secretory hairs are present on insectivorous plants, and a variety of substances is released. These substances include nectar to attract the insect, mucilage to trap it, and digestive enzymes to dissolve it. Dictyosomes are particularly active in the cells secreting mucilage, and the number of dictyosome vesicles in the cells can be correlated with the discharge of mucilage.

Nectaries are collections of cells that secrete the carbohydrate-rich solution known as nectar. These structures are usually part of the flower, but they may also occur on stems and parts of leaves. The object of the nectar is to attract insects, at which it is most effective. Unlike the case of mucilage, the dictyosomes of the active cells are apparently not involved in the secretion of nectar.

Not all secretory cells in the plant are on the surface or secrete substances external to the plant. Some are within the plant, and their secretions collect in internal cavities. This is true of the resin ducts found in conifers. In this case, a cavity is lined with living cells, the epithelium, that are rich in undifferentiated plastids and ER. These cells secrete resin into the cavity of the ducts, which can form complex systems in the plant. Myrrh and frankincense, important in the ancient Middle East as aromatic substances, are products of such internal secretory systems. Resin and products from resin (pitch, turpentine, rosin) were so important in the days of wooden ships that they were called "naval stores." Resin for naval stores was obtained by tapping pine trees—chipping through the bark and outer wood and collecting the material that drips out. The vast supply of naval stores in North America made the colonies critically important to England and to the British fleet.

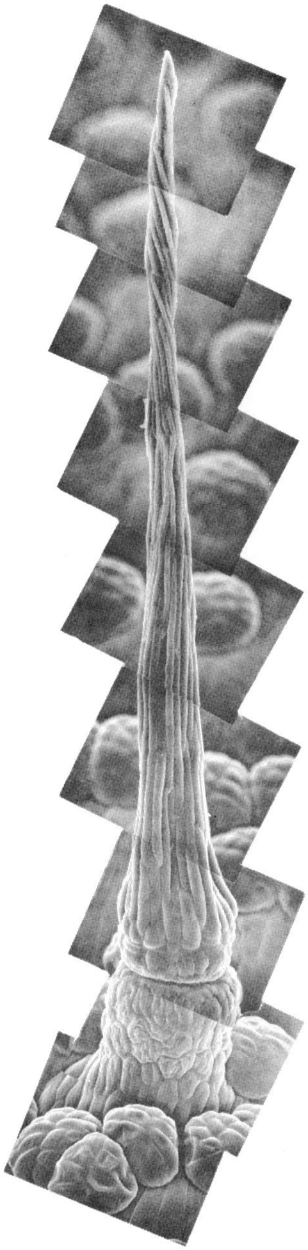

24–19 The trigger for the closing of leaves in the Venus flytrap is a modified epidermal hair. This image was made with the scanning electron microscope.

Another plant product that is obtained by tapping trees is rubber. The actual material drained from the tree is latex, which contains (besides rubber) carbohydrates, organic acids, alkaloids, terpenes, oils, resins, and various enzymes. Latex is found in many plants, not only *Hevea brasiliensis* (the rubber tree), and its composition varies greatly with the source. Latex is formed within the plant in laticifers, which are truly unique cells present in two main types: articulated and nonarticulated.

Nonarticulated laticifers originate from single cells in the embryo. As the embryo develops and the seedling grows, so do the laticifer cells. Eventually, a single laticifer cell may extend throughout the length of the plant from root to shoot. Such cells have rapid tip growth and are multinucleate. The greatest number of nuclei are present in the youngest parts of the cell. They divide more or less in synchrony in various parts of the cell in a wavelike pattern. The cell wall, except at the tip, is usually fairly thick. Each cell may be highly branched, but they generally do not fuse with other laticifer cells.

Articulated laticifers are formed from files of cells in which the end walls break down. This is the type found in the rubber tree and in the opium poppy, *Papaver somniferum*. Opium from which morphine and similar drugs are refined is formed in articulated laticifers.

Let us now consider how these cells and tissues, which are products of the apical meristems, are assembled into the parts of the vascular plants.

The Root

Most vascular plants have roots (Fig. 24–2). They are characteristic of seed plants, ferns, *Selaginella*, and *Lycopodium;* but *Psilotum*, the most primitive vascular plant, has rhizoids instead of true roots. A rhizoid is a hairlike structure that does not have the anatomical characteristics of a root, but both roots and rhizoids virtually always function as absorbing and anchoring organs. True roots may also store considerable quantities of food (e.g., carrots, sweet potatoes, turnips, beets —but not white potatoes, which are modified stems).

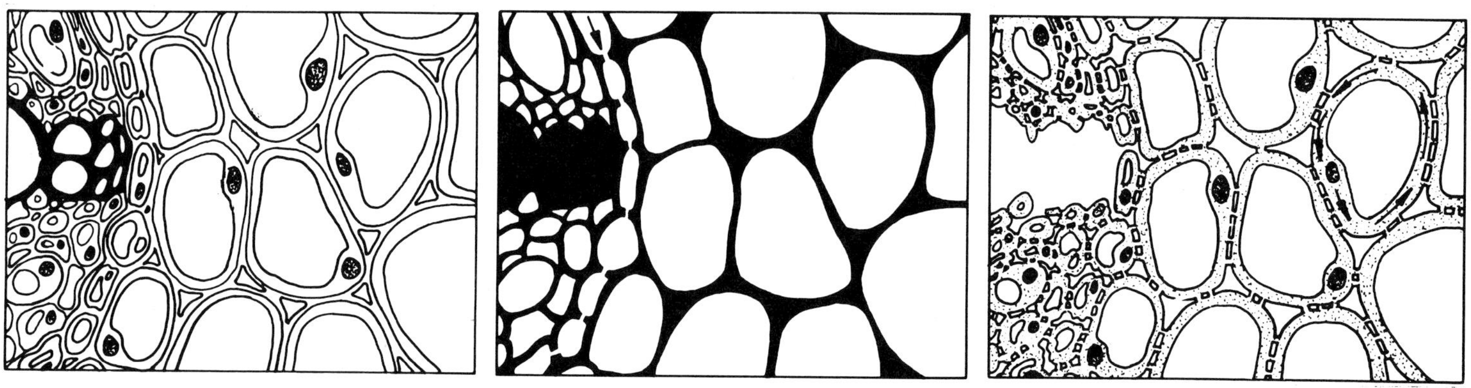

24–20 These three drawings illustrate the concept of the apoplast-symplast. The first drawing is of a portion of a section through a buttercup root, showing both the cell walls and the cytoplasm of the cells. The xylem and phloem are shown at the left and the cortex at the right. The pericycle and the endodermis (with its Casparian strips in the walls) separate the stele and the cortex. The second drawing shows only the cell walls and the intercellular spaces, including the water-filled xylem elements; this is the apoplast. Note that the continuity of the apoplast is broken by the Casparian strips in the walls of the endodermal cells. The third drawing shows only the cytoplasm, the nuclei of the cells, and the interconnected system that is formed when the plasmodesmata are included; this is the symplast.

The outside of the root is covered by an epidermis, usually one cell thick. Just inside the epidermis is a relatively thick layer of parenchyma cells, constituting the cortex. These cells usually contain starch and sometimes particles of protein or globules of fat. When the cortex is persistent, it sometimes contains sclerenchyma or collenchyma. The presence of conspicuous intercellular spaces is common among the parenchyma cells of roots.

Beside being important in food storage, cortex tissue is also the site of much lateral transport of water and other materials. The soil solution must move across the cortex into the vascular tissue, and dissolved food materials (e.g., sugars, proteins, fats) must move from the vascular tissues out into the cortex to be stored in these cells and for their nourishment. It seems likely that the soil solution moves through the highly porous cell walls of the cortex, but the food materials may move from one cell to another, possibly through the plasmodesmata. All of the cell walls of the plant, including dead elements such as those in the vascular tissues through which water may move rather freely, constitute a system referred to as the apoplast (Fig. 24–20). All of the cytoplasm of the plant, connected from cell to cell through the plasmodesmata, constitutes a phase of the plant called the symplast. As we shall see, solutions of sugars and other organic materials may typically move in the symplast, while solutions of inorganic solutes may move in the apoplast.

The movement of dissolved materials through the symplast might be accelerated by cytoplasmic streaming. The cytoplasm of many living cells can be seen to flow. The mechanism responsible for this streaming remains to be discovered, but obviously it would greatly speed up the rate at which dissolved materials could move across a tissue. Such substances could move from cell to cell by diffusion (a fairly slow process) and then rapidly within the cell by cytoplasmic streaming.

The innermost layer of the cortex is made up of a cylinder of cells called the endodermis. This layer of cells is in direct contact with the vascular cylinder inside—the stele. The endodermal cells are recognizable because their radial walls have a band of thickening

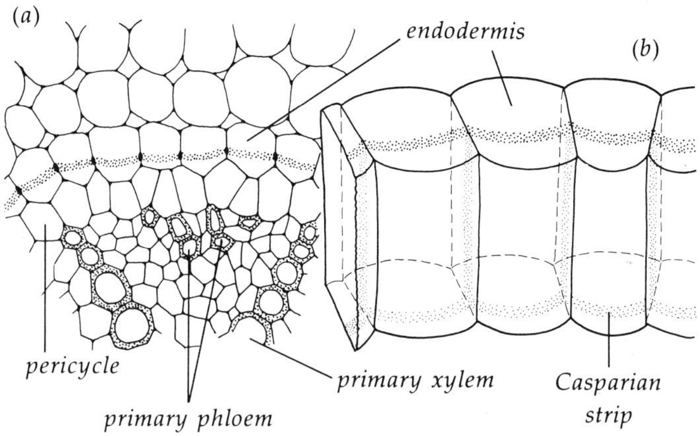

24–21 The structure of the endodermis. The relation of the endodermis to the cortex, the xylem, and the phloem can be seen in this drawing of the cross section of a root. The three-dimensional drawing of the endodermal cell shows the position of the Casparian strip in the wall.

called a Casparian strip (Fig. 24–21). The tangential walls are not so thickened. Substances moving from the cytoplasm of a cortex parenchyma cell to the cytoplasm of an endodermal cell (and perhaps on through into the stele) would pass through the endodermal cell itself. Substances moving through the cell walls in the cortex (the apoplast) into the stele would be expected to pass through the Casparian strips, but since these are impregnated with a relatively waterproof material called suberin, these substances must also pass through the cell itself. Thus the presence of the endodermal layer ensures that substances moving between the cortex and the stele (both in and out of the root) must move through the cytoplasm of living cells.

This being the case, the root can act like an osmometer with the endodermal layer serving the function of the membrane. If the water potential in the apoplast inside the endodermal layer is more negative than outside, the water will tend to move in as it does in an osmometer. In Chapter 4 we discussed the phenomenon of active accumulation. We learned that solutes may be taken into cells against a concentration gradient by the expenditure of metabolic energy obtained in respiration. Conceivably, solutes could be taken into cortex cells this way, and they would be taken in effectively because of the relatively high oxygen concentrations near the root surface. They then might move across the symplast (from cell to cell in the cytoplasm through plasmodesmata), crossing the endodermal layer into the stele. There they might tend to leak out of the symplast system because of lower oxygen concentrations. This would lead to a buildup in ion concentration in the apoplast system inside the stele compared with the apoplast system outside the endodermis. The result would be a more negative water potential inside, forming a perfect situation for so-called root pressures. If the top is removed from plants such as grapevines or tomato plants, liquid will exude from the cut surface, often under considerable pressure (5 or 6 bars), provided there is readily available water and the roots are well aerated.

Within the stele is where movement of materials occurs along the axis of the plant. Its outer layer, just inside the endodermis, is called the pericycle. This parenchymous tissue retains a meristematic potential. The meristems that grow into lateral roots (secondary roots) originate in the pericycle of virtually all plants. As the root meristem forms in the pericycle, it grows through the peripheral tissues of the mature root (the cortex and epidermis), finally reaching the soil (Fig. 24–22).

Inside the pericycle are the xylem and phloem. The xylem frequently forms a solid core in the center of the stele with ridgelike projections extending toward the pericycle, and the phloem alternates with the ridges of the xylem. The number of xylem ridges varies; in some species there are three ridges, in others four, and in others five or more.

The Stem

The meristem of the stem is enclosed in a bud made up of leaf primordia and young leaves (Fig. 24–1). Various types of hairs are usually associated with the bud. New meristems do not develop internally to form branches (although adventitious *roots* do develop this way on stems). Rather, the stem has an intricate system of leaves and buds. The point of attachment of the leaf is called the node, and the portion of the stem between the nodes is called the internode. The angle between the

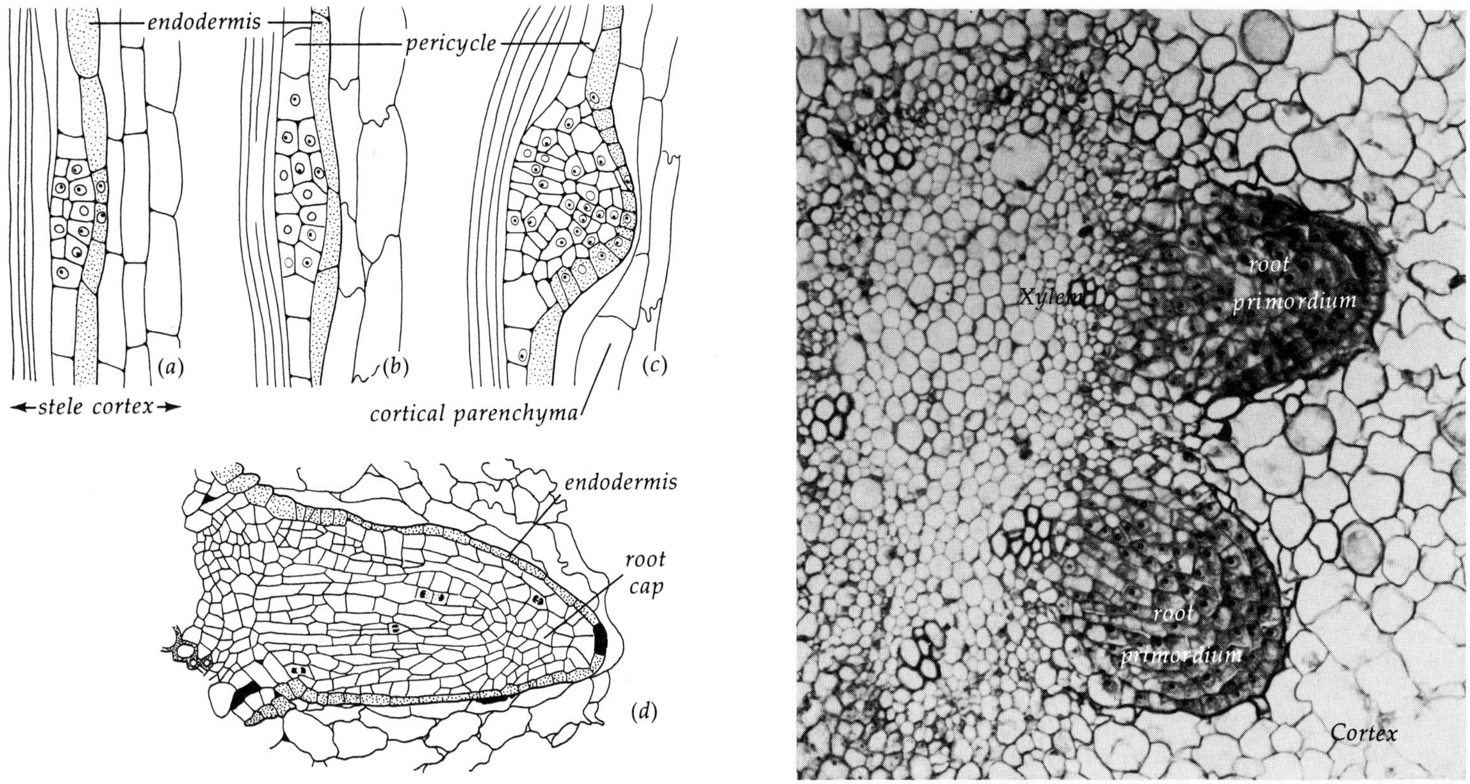

24–22 The origin of a secondary root starts with cell divisions in the pericycle that result in the establishment of a small mass of cells. These become the root primordia, which grows through the cortex to the outside.

leaf stem (the petiole) and the main stem is called the axil, and the bud that occurs in the axil is called an axillary bud.

The pattern of attachment of leaves to the stem is strongly dependent upon the species. Sometimes, only one leaf is attached at a node (alternate leaf arrangement); in other cases two (opposite arrangement) or more than two (whorled arrangement) may be attached. In some species the axillary buds remain dormant so that the plant consists only of a main stem with its leaves; in others the buds become active, producing highly branched plants. As contrasted with roots, stem apical meristems can form reproductive structures—flowers and fruits.

The stem, like the root, is composed of three tissue systems: the dermal, the ground, and the vascular. The dermal is represented by a single-celled epidermis, the cells of which are alive and capable of mitotic activity. Guard cells and stomates are present in the epidermis of stems, as they are in leaves, but are typically less numerous.

The vascular tissue is often arranged in bundles rather than as complete rings. This means that the parenchyma of the cortex is continuous with the pith. In many stems the cells of the outer layers of the cortex are rich in chloroplasts and are important elements in the photosynthetic capacity of the plant. Both the cortex and the pith may store large amounts of starch. Collenchyma and sclerenchyma are common in the ground tissue of stems, and laticifers and ducts of various kinds are also frequently found in this tissue.

The stems of most seed plants lack a well-defined endodermis. In a few species of seed plants and in the ferns, Casparian strips do occur, and in some the inner layer of the cortex is composed of cells rich in

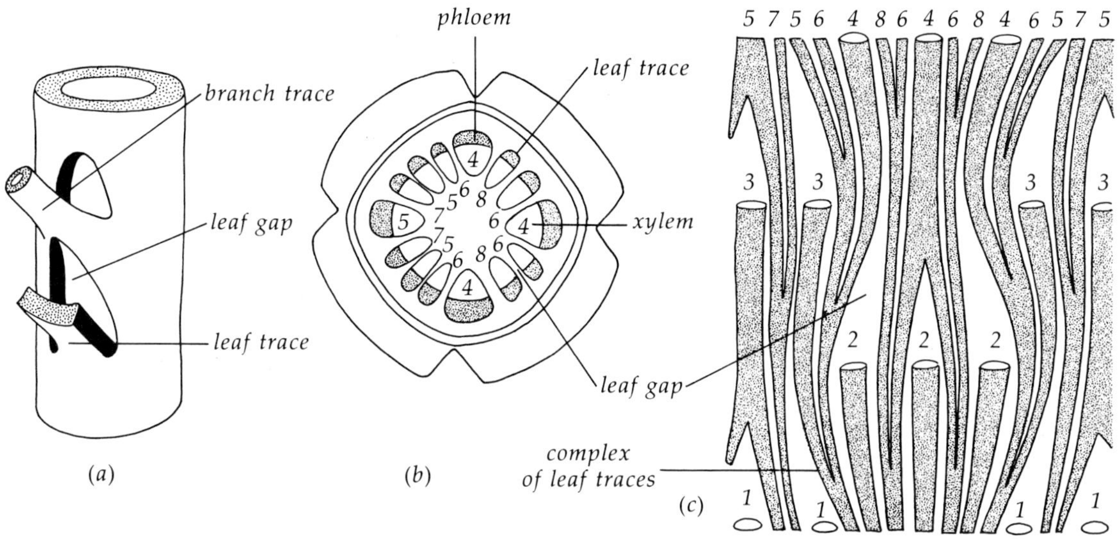

24–23 This three-dimensional diagram of the vascular cylinder in a stem (a) illustrates what happens when vascular tissue diverges to enter a leaf or branch. The presence of leaf gaps may break up the vascular cylinder into a number of discrete bundles (b). Such bundles are connected, as seen when the system is spread out and diagrammed in longitudinal section (c).

starch. In general, however, stems lack an endodermis, while roots have one.

The major components of the vascular tissue of the stem is the same as for the root: xylem and phloem. Usually, the arrangement of the two is different in stems than in roots. The phloem is toward the exterior of the stem, while the xylem is toward the center. A cap or surrounding layer of sclerenchyma may be associated with the vascular bundle, but there is no true pericycle in stems.

The organization of the vascular tissue in stems is highly variable—much more so than in roots. There are cases where phloem exists on both sides of the xylem (as in potato), or where the xylem surrounds the phloem (as in *Begonia*). There are also wide variations in the types of cells in both the xylem and phloem, depending on the group of plants and individual species. The vascular bundles may also be in a distinct cylindrical arrangement, or they may be scattered in the ground tissue. Many of these differences will be described when we discuss the various groups of vascular plants in Chapters 25 to 29.

The stem and the leaves are part of one continuous system, the shoot. This means that the vascular tissue of the stem is continuous through the nodes with that of the leaves. In the ferns, gymnosperms, and angiosperms, where this occurs, there is a gap at the node in the vascular ring of the stem, called the leaf gap, which is filled with parenchyma cells (Fig. 24–23a). The strand of vascular tissue leading to the leaf is termed the leaf trace. This seems simple enough, but the variations on this basic plan found in stems is staggering. One of the simpler variations is illustrated in Figure 24–23. In this diagram a segment of the vascular cylinder is shown as if it had been cut and rounded out flat. A little study, this diagram can give a good impression of the way vascular strands can diverge and come together in quite elaborate patterns.

The analysis of vascular patterns can sometimes be next to impossible. This was the case for the palms, which had defied analysis for over fifty years. To resolve the problem, an ingenious approach was employed by Martin Zimmerman of Harvard University. He selected a small species of palm and then sectioned and stained a long segment of the stem. This was the classical approach to the problem, but then Zimmerman photographed every fifth section with a motion picture camera. When the film was projected, it was as if you were rapidly moving up the stem, and you could see individual bundles of vascular tissue move out into the leaves. Even with this help, the analysis is still difficult.

The Leaf

Such primitive plants as *Lycopodium*, *Selaginella*, and *Psilotum* have a simple leaf called a microphyll. These organs may have only one vascular strand or none (as in *Psilotum*). A club moss microphyll is a good example. The leaves consist of a centrally located, unbranched vein or vascular bundle composed of tracheids and occasionally sieve cells. The blade consists of a mesophyll of uniform parenchyma cells containing numerous chloroplasts. The epidermis is one cell thick and supplied with stomates, sometimes only on the lower surface.

Such, more advanced vascular plants as ferns, gymnosperms, and angiosperms have megaphylls. These are the familiar leaves, although some like the needles or scales of gymnosperms do not fit the common picture of a leaf. The basic structure of most dicot leaves is a blade and a petiole (the leaf stem). Sometimes, there are several leaflets on a single petiole, producing a compound leaf.

The leaf has an upper and lower epidermis in which are located the guard cells and stomates. (In some species stomates are present only in the lower epidermis, while in others they are present in both.) Below the upper epidermis the palisade parenchyma is usually found. Frequently, one gains the impression from cross sections such as that in Figure 24–J that the cells are packed tightly together, but there is actually considerable intercellular space between them, as shown in a section cut parallel to the leaf surface in Figure 24–J. The spongy parenchyma below the palisade consists of cells more irregular in shape than those in the palisade and includes obvious large intercellular spaces. The palisade and spongy parenchyma form the mesophyll. The leaf veins are relatively elaborate vascular bundles, including phloem, xylem, and sclerenchyma. As the veins subdivide, the small divisions are called veinlets. Finally, each veinlet ends in a single tracheid.

Leaves originate near the tip of the shoot apical meristem. They form in a regular pattern, and the developing leaves overarch the apex. The question can be asked: what cells of the apex give rise to the new leaves? This question was answered for *Datura* (Jimsonweed) by S. Satina, A. F. Blakeslee, and A. G. Avery in Connecticut using plants treated with colchicine. This drug upsets the formation of the spindles during mitosis so that recognizable polyploid cells result. They observed that in one plant the polyploid cells occurred only in the outer layer of the tunica of the stem apex. In others the polyploid nuclei occurred only in the inner tunica layer. In still others these nuclei occurred in all of the corpus cells.

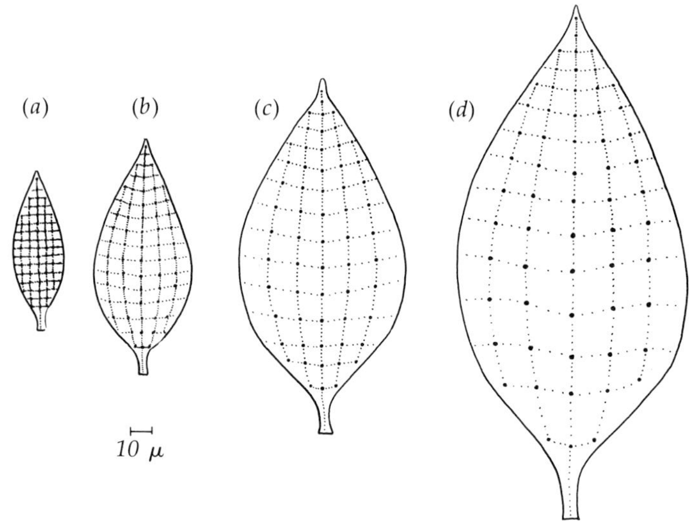

24–24 The four stages of development of a tobacco leaf, as seen in a surface view. When the leaf was one-quarter its final size (a), its entire surface was marked into 5-mm squares. Their unequal change in shape and size shows that growth is variable in different parts of the leaf; yet the fact that they have all increased in size indicates that growth has occurred throughout the leaf.

These observations were a strong substantiation of the tunica-corpus interpretation of stem apical meristems. Furthermore, one could follow the distribution of the polyploid cells during development of leaf primordia and leaves. All three layers contribute to the leaf: the outer tunica layer becomes the leaf epidermis; the second tunica layer contributes most to the bulk of the leaf tissue; and the corpus develops into the procambium and vascular tissue of the leaf.

How do the cells of the leaf primordium, certainly a simple organ, grow to become the complex organ that we recognize as a leaf? And how can we account for the vast variety of leaf shapes? Apparently, the growth of the leaf is dependent upon two processes: an elongation of the vascular elements and a growth of the tissue

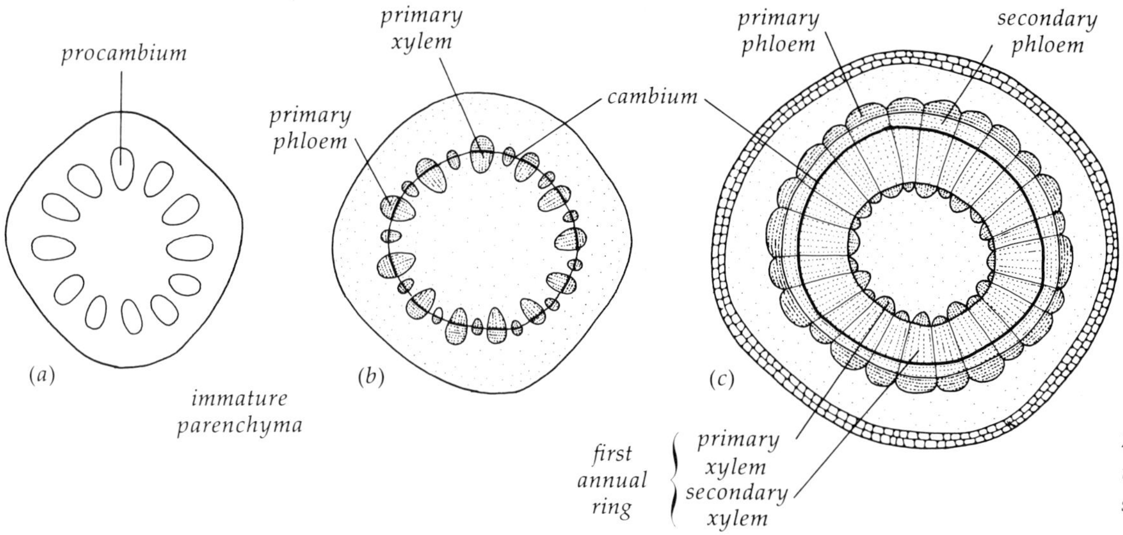

24–25 The development of the vascular cambium and the secondary vascular tissue.

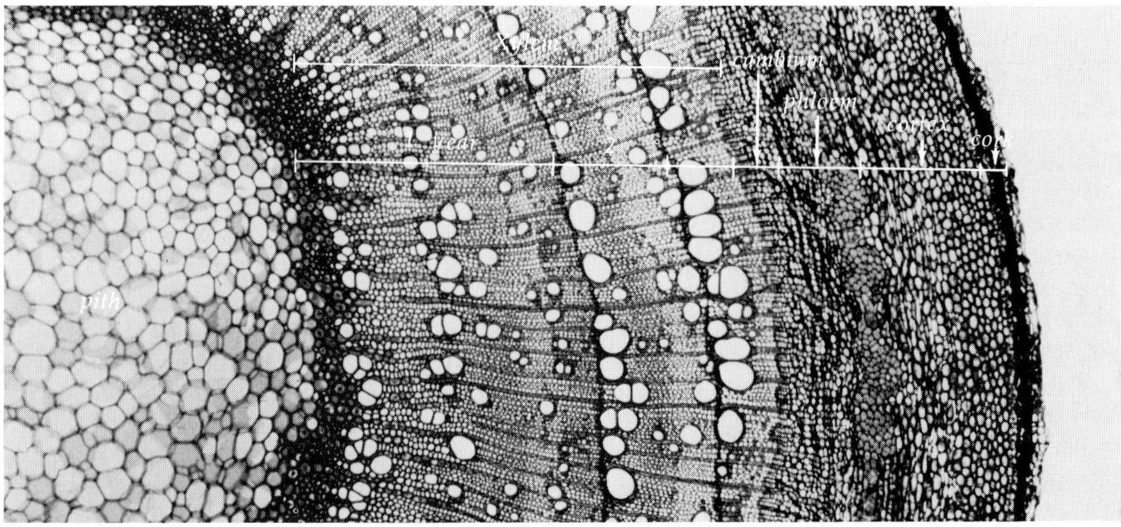

between, mostly in two dimensions. If the vascular elements grow more rapidly than the tissue between, an elongate leaf structure might result; if the tissue between grows fastest, a broad leaf might be produced. Obviously, numerous variations are possible, depending upon the relative rates of growth of the various vascular elements and other tissues. Of course, to produce the needles of conifers or the leaves of grasses, all of the tissues must grow primarily in one dimension, that is, by elongation. Typically, in broadleaf plants (especially dicots) leaf expansion is fairly uniform in all directions once the leaf primordium has developed into a young leaf. This can be shown by marking the young leaf with a crosshatch pattern of India ink and observing this pattern after the leaf has matured (Fig. 24–24).

The Vascular Cambium—Wood and Bark

Since leaves are usually determinate organs, their lifetimes are limited. Even in the tropics, leaves eventually senesce and die (sometimes it takes several years). This is also true of "evergreen" leaves in conifers.

The Vascular Cambium and Wood

As the tissues differentiate behind the apical meristems, those cells that will form the primary xylem and primary phloem are first seen as the procambium. In the course of development of the xylem and the phloem, a group of cells is left between them. This group of cells retains the ability to divide, and it will eventually develop into the vascular cambium. Cambia also differentiate in the parenchyma, thus separating the vascular bundles. Ultimately, the cambium in the bundles and between the bundles will form a continuous cylinder of cells around the stem (Fig. 24–25).

The vascular cambium is composed of two kinds of cells: the fusiform initials and the ray initials. Both are meristematic cells. The fusiform initials are long, prism-shaped cells, while the ray initials are short, cube-shaped cells. Both types of initials divide primarily in the tangential plane so that vertical rows of cells are formed. The fusiform initials are highly vacuolate cells, and the cell plate must develop through these vacuoles for the cell to divide.

The cells that are produced toward the center of the stem by the cambium differentiate into xylem, while those toward the surface develop into phloem. The products of the vascular cambium in some species may be quite simple—tracheids and parenchyma in conifers—or more complex, as in angiosperm trees (Fig. 24–26). The secondary xylem is the wood, while the secondary phloem forms a part of the bark.

Most vascular cambia in temperate zone plants are active only during part of the year. Growth begins in the spring, when water is available, and continues into the summer. The result is an annual ring characterized by relatively large cells produced as early or spring wood and small cells produced as late or summer wood. The boundary between the spring wood of one year and the summer wood of the preceding year is quite sharp. The result of this is a rather clearly defined annual ring that can be used to ascertain the age of the tree (Fig. 24–27).

There are some situations where this procedure is not completely reliable. Trees growing by a river, for example, may exhibit virtually no annual rings because they are always amply supplied with moisture. (Actually, temperature differences typically produce faint annual rings, even in this situation.) Sometimes, two "annual rings" are produced in a single year, if an unusually wet and cool period occurs in midsummer. Also, annual rings are not found in many tropical woods because growth conditions are too uniform throughout the year.

One of the many remarkable features of vascular cambia is that they are capable of reactivation of division for many years—3000 to 4000 years in the case of the redwoods.

The result of the activity of the cambium and the accumulation of xylem is that the diameter of the stem increases. As this occurs, there must be radial divisions of the initials so that the cambium remains intact. The older phloem becomes crushed by the expansion of the xylem; thus, while there is a massive accumulation of xylem in woody plants, there is little more than a year or two of phloem present.

As a woody stem continues to grow in diameter, the living parenchyma cells in the pith, the primary xylem, and the innermost rings of secondary xylem gradually die. The vessel elements and tracheids in these regions may also cease to be active in the transport of solutions. Such substances as tannins and various crystals may be deposited in these nonfunctional cells. Their color may gradually change to a darker, typically more reddish hue. Wood that has undergone these changes is referred to as heartwood in contrast to the active, functional sapwood. Of course, the amount of heartwood will gradually continue to increase year after year.

In the root the pericycle cells on the surface of the xylem ridges develop into cambium, as does a layer of undifferentiated cells inside the phloem strands and in contact with the concavity in the xylem cylinder (Fig. 24–28). Again, the cambial layer eventually forms a continuous cylinder, producing xylem to the inside and phloem to the outside, much as in the stem. Indeed, roots that have completed several years of secondary growth are difficult to distinguish on the basis of internal anatomy from stems of comparable ages.

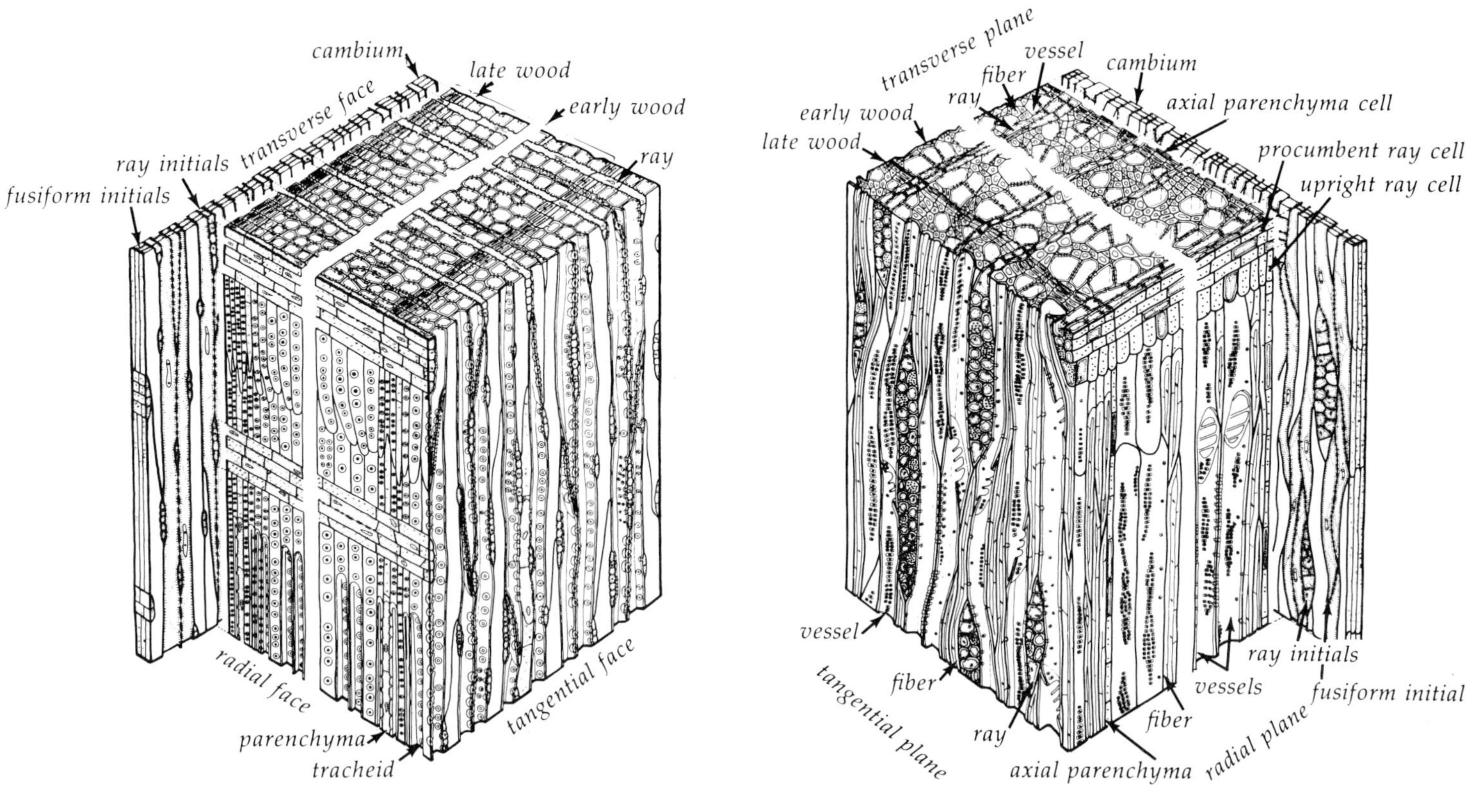

24–26 The cells produced by the vascular cambium may be relatively simple, as in the case of the majority of the gymnosperms illustrated in this block diagram (a) of the secondary xylem of Thuja occidentalis *(white cedar). On the other hand, they may be much more complex, which is true for the woods of many angiosperms, including* Liriodendron tulipifera *(tulip tree) shown in (b). The complex cellular composition of many angiosperm woods is responsible for the beautiful visual textures they exhibit, when they are made into furniture and other objects.*

While wood is essentially nothing more than secondary xylem, the bark is somewhat more complex, consisting not only of secondary phloem but also of a tissue called cork. As the stem grows in diameter due to the activity of the cambium layer, the phloem, cortex, and epidermal cells are pushed out, away from the center of the stem. In the process they tend to be crushed, split, and sloughed off. Only the phloem cells close to the cambium (few to several layers of cells) remain functional, since they are not subject to forces as great as those farther out. Clearly, these need some protection, unless they are produced rapidly and function only briefly.

Phloem cells are protected by the formation of a second cambium, the cork cambium or phellogen (Fig. 24–29). Rarely do epidermal cells become meristematic, but a layer of cortex cells does so more commonly. In a few rare cases, even primary phloem cells may become meristematic. In any case, these cells divide tangentially (mostly), producing a layer of cork cells

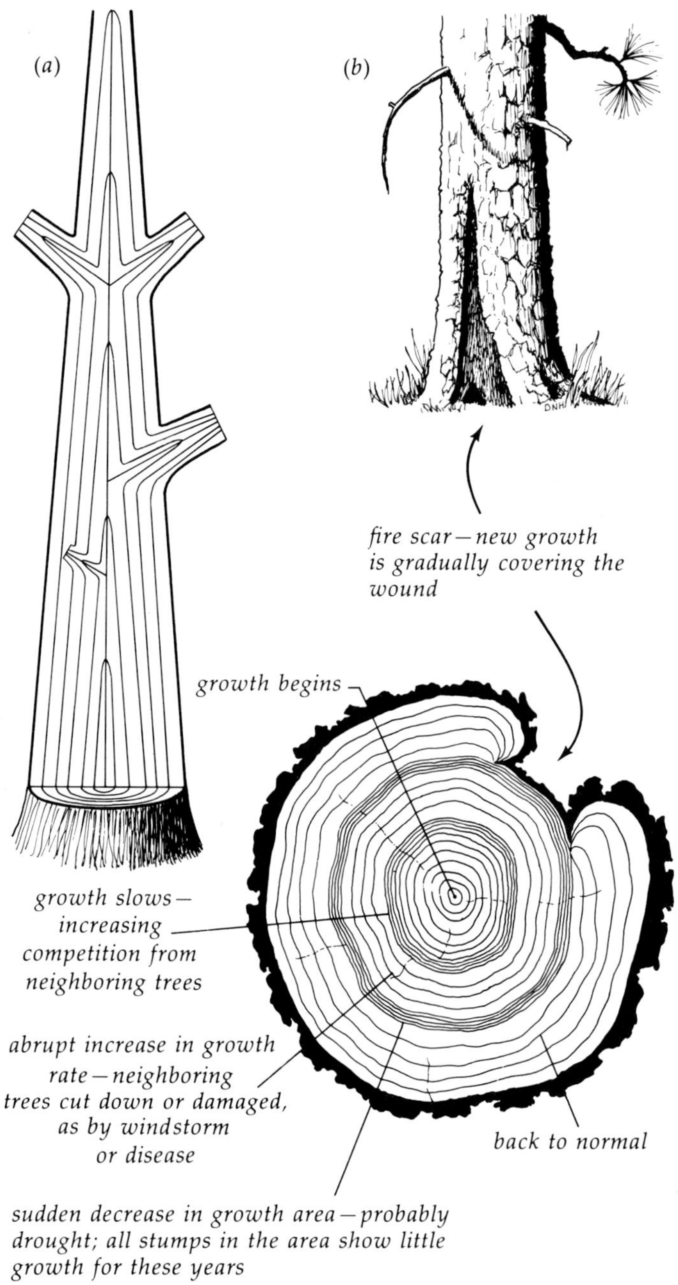

24–27 The number of annual rings varies with the distance aboveground from which the trunk is examined (a). This is a reflection of the way in which the tree grows. The width of the rings can also reveal a good deal of information concerning the past history of the tree (b).

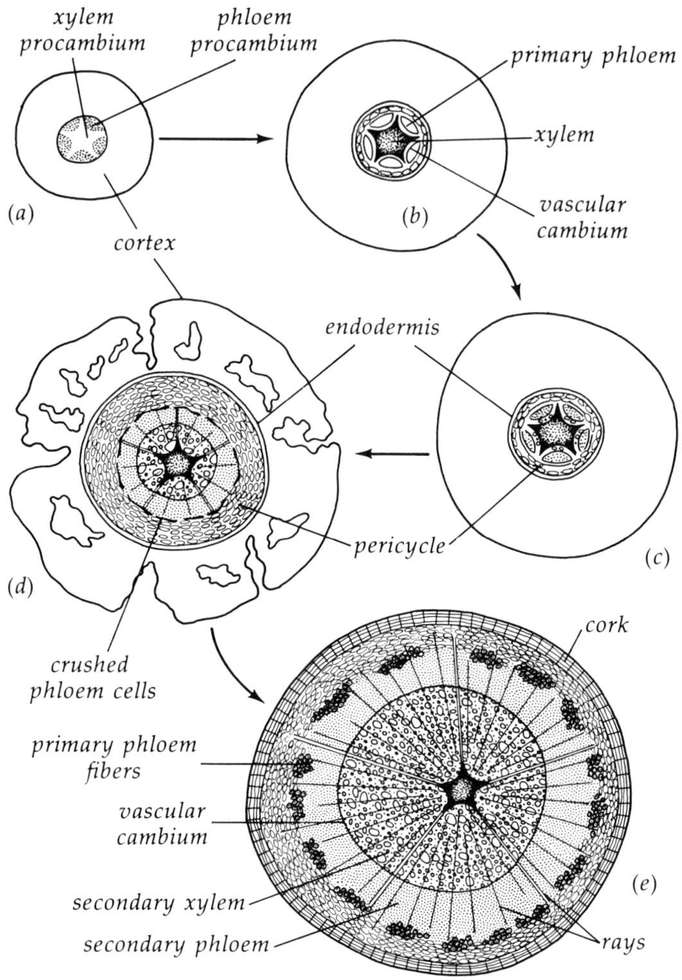

24–28 The development of the vascular cambium and secondary tissue in a root.

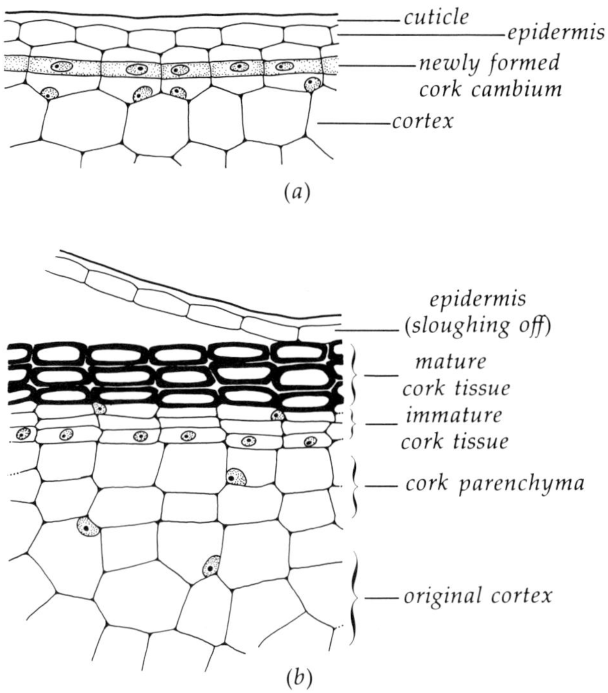

24–29 The development of the cork cambium and the cork. Top shows a young stem in which the cork cambium is beginning, while the bottom illustrates the layers of cork found in an older stem.

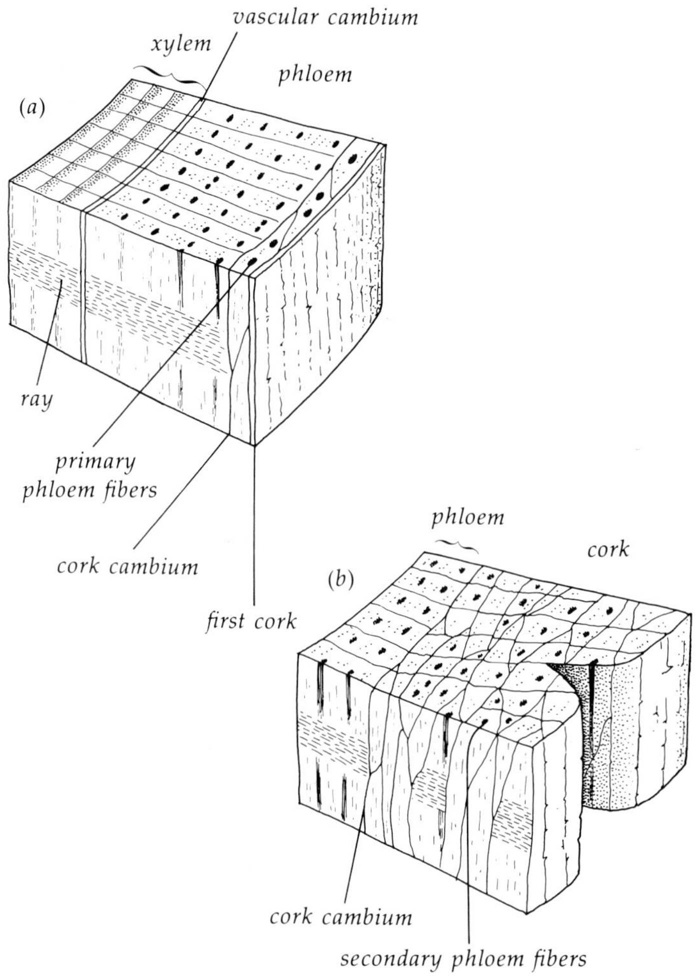

24–30 The relationship of the cork cambium and its products to the bark of trees. The patterns seen in the bark of different species is partly a result of the amount, character, and distribution of the phloem fibers and partly due to the distribution and activity of the cork cambium.

toward the outside. Usually, the cork cambium is not continuous around the stem; rather it is formed of overlapping arcs. Not only is cork tissue produced on the outside, but a tissue called phelloderm, consisting of parenchyma-like cells, is produced on the inside. Cork cambium, cork, and phelloderm are collectively called the periderm. The cork cells soon die and are impregnated with suberin. Any living cells outside of these (e.g., primary cortex cells, epidermis) can no longer receive nutrients from the inside and hence must also die. New cork cambia can continually be formed deeper into the cortex, until finally primary and secondary phloem cells are commonly producing the cork cambia. Everything outside of the vascular cambium (secondary phloem plus several layers of periderm) is referred to as bark (Fig. 24–30).

How are cells programmed to form the bark? Clearly, they are responding to their position in the stem, probably through some detection of environmental conditions unique to that position. Whether the cork cambium forms in response to the stresses caused by the growing xylem cylinder and whether the higher oxygen levels play a role will be examined in Chapter 31.

25
The Lower Vascular Plants

The Invaders

As you might have guessed, the ancient land plants were quite simple when contrasted with the elaborate vascular plants that we see around us today. Regardless of their primitive structure, they represent an enormous advance in plant evolution, and we are fortunate to have fossils of them from which schematic reconstructions can be made. Here, we present some of the fossil evidence and reconstructions of the major features of some of the most important of these early land plants.

A One of the best understood of the early land plants is Rhynia. *Excellent fossil remains of this plant reveal not only the general form of the plant but the cellular organization of the stem (top) and the spores (bottom).*

B Another important early land plant is Zosterophyllum. *The structure of the guard cells (upper left), sporangium (upper right), and vascular tissue (lower) can all be seen in the fossils.*

C Psilotum nudum. *There are no leaves or roots present. The sporangia (the hemispherical projections) are borne directly on the stem.*

Among the most primitive of the existing vascular plants is Psilotum. *Consisting of nothing more than a stem, it reproduces by spores, and the sperm must have free water in which to swim to reach the egg. In many ways,* Psilotum *resembles the fossils of some of the earliest land plants, yet there are no fossils that link* Psilotum *to those ancient plants.*

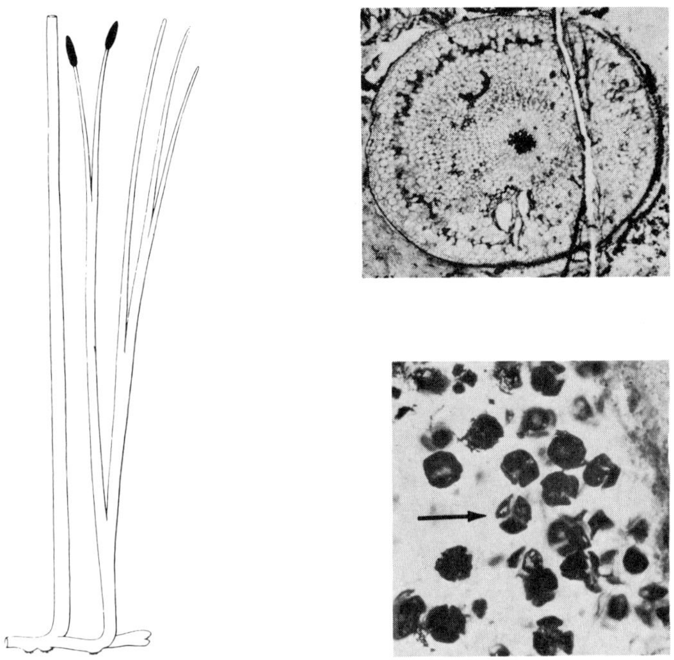

A

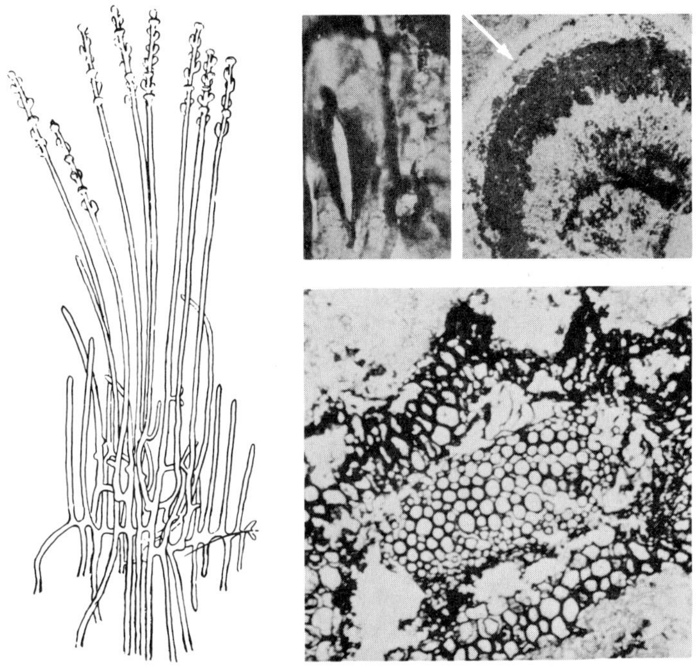

B

C

D A cross section of the stem of Psilotum nudum. *The vascular tissue can be clearly seen in the center of the stem. Photosynthesis is carried out by the cells lining the outside of the stem. The enlarged photograph shows some of these cells, as well as a pair of guard cells. The cuticle that is so characteristic of land plants can be clearly seen in the epidermal cells.*

E The gametophytes of Psilotum *are subterranean and colorless. They are able to survive because they are saprophytes. This photograph of a section of a gametophyte shows numerous fungal hyphae (arrows) within the cells of the gametophyte.*

F Two of the common lycopsida are Lycopodium *(left) and* Selaginella *(right).*

The lycopsida or club mosses are not conspicuous in our present ecosystem, although they once contributed to the great coal forests. Today, they are all relatively small, frequently handsome plants.

G Selaginella *is of interest because it produces two types of spores: large spores that will form gametophytes, which will form the archegonia, and small spores that will form gametophytes, which will develop the antheridia. In this cleared and stained cone of* Selaginella, *the megaspores can be seen on the left and the microspores on the right.*

Horsetails or sphenopsida are commonly seen in ditches and fields across the country. They are successful plants in our present ecosystem, although they lack the tree species that were present in the Paleozoic era.

H Equisetum sylvaticum, *as it is frequently seen growing in extensive stands, giving the appearance of a miniature forest.*

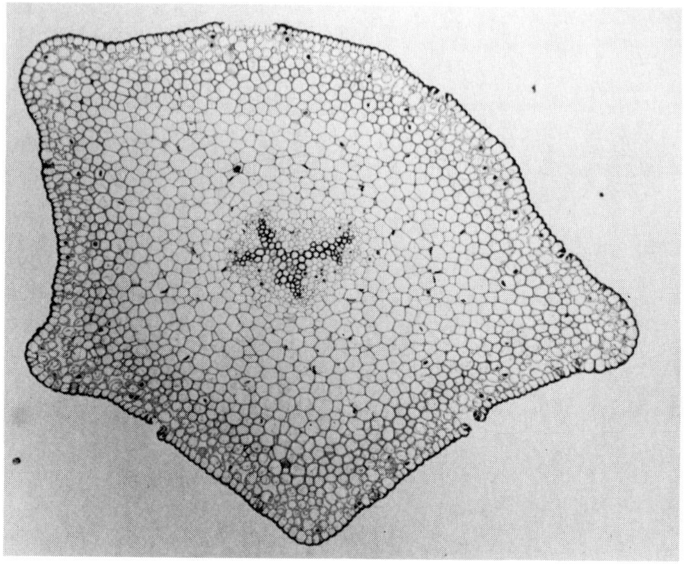

D

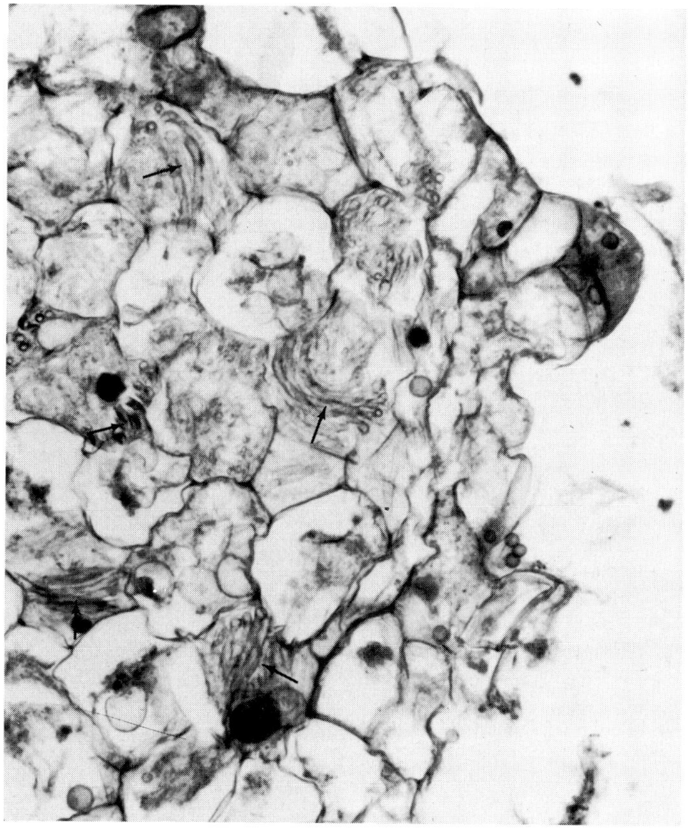

E

F

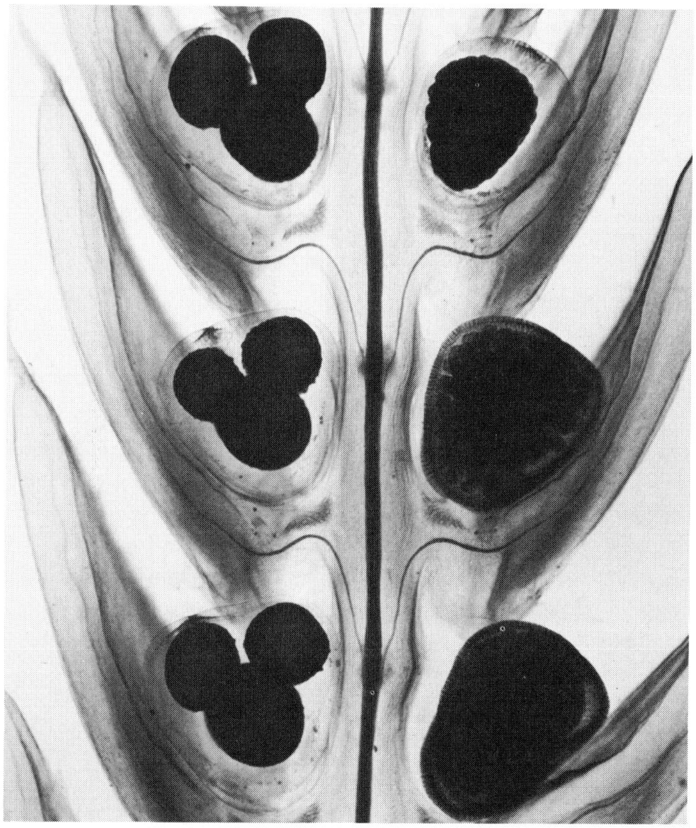

G

H

I Meiosis and spore production occur in sporangia grouped as small cones at the ends of reproductive branches.

J The large green plants are the sporophytes, as is the case for all vascular plants, while the gametophyte is small and inconspicuous. Here is one young gametophyte of Equisetum arvense. *It is green and grows on the surface of the soil, anchored by the long unicellular hairs that grow down from it.*

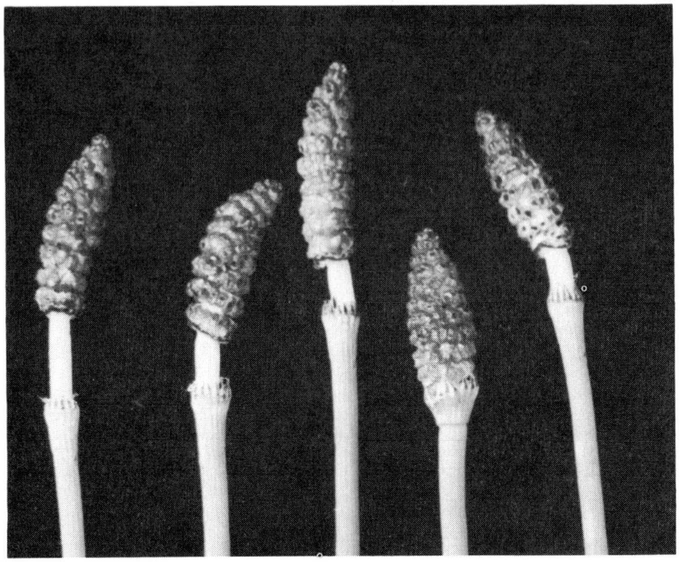

I

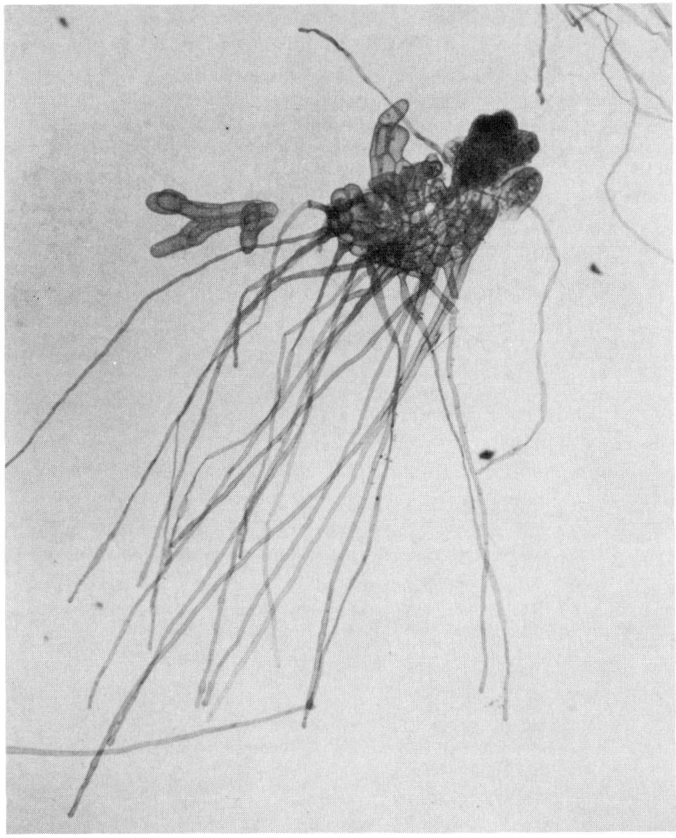

J

Some 400 million years ago, during the Lower Devonian, the shallow marshes on the edges of the ancient seas were filled with mats of slender, erect green plants (Fig. 25–1). Not one species but many occupied these favored locations, and in some cases, they left masses of beautifully preserved fossils. This is particularly true for the rocks from Scotland known as the Rhynie Chert.

The Earliest Land Plants

As we saw in Chapter 22, the earliest known fossils of land plants are of *Cooksonia,* a tiny, slender plant lacking roots and leaves and possessing terminal sporangia. So far as we can tell from our present fossil record, *Cooksonia* was the only land plant for several million years.

The Rhynie flora contain *Cooksonia* and other plants with a similar body plan (rootless, leafless stem with terminal sporangia), including a genus known as *Rhynia.* The fossils of *Cooksonia* are neither abundant nor well preserved, while those of *Rhynia* are both. It is therefore possible to accurately reconstruct *Rhynia* and to study it in detail (Fig. 25–A). For this reason Harlan Banks has proposed that all plants of this age resembling *Cooksonia, Rhynia,* and related genera be called *Rhynia*-type.

A second group of ancient plants was also present at the time of the *Rhynia*-type plants. These too were leafless and dichotomously branched, but their sporangia were borne along the sides of the stem (not terminal as in *Rhynia*). The stems grew in a characteristic pattern that produced "H-shaped" branches (Fig. 25–A); this pattern probably allowed the plant to spread outwardly from the center but still remain attached to the soil. This and similar plants have been placed in a genus named *Zosterophyllum* and represent a second line of early plants that Banks calls the *Zosterophyllum*-type.

Although both types are extremely ancient, the *Rhynia*-type is the older of two based on present fossil evidence, which is clearly incomplete. What, in turn, did they give rise to? As may be expected at a time when plants were being subjected to the new environmental pressures of land, which itself must have been changing, several new lines of plants evolved, only some of which proved enduring.

The *Rhynia*-type plants appear to have given rise to the *Psilophyton*-type plants. *Psilophyton* (Fig. 25–C) and *Rhynia* are the plants that formed the basis for the Psilopsida and that may or may not be represented in the present flora by two genera of primitive-appearing plants. *Psilophyton* is a famous fossil genus first described in 1859 by the Canadian botanist Sir J. W. Dawson. Within recent years reexamination of the fossil evidence on *Psilophyton* (both new finds and Dawson's original fossils) by Francis M. Hueber of the U.S. National Museum has clarified the structure of this important plant.

The *Psilophyton*-type plants are believed to form the ancestral stock for a number of lines of vascular plants. One of these lines may have led to the ferns (Fig. 25–2), and there is evidence that the present-day genera included in the Psilopsida would better be placed in the ferns (see Chapter 26). Another line appears to be the possible progenitor of the progymnosperms that in turn form the stock from which the seed ferns, gymnosperms, and ultimately the angiosperms may have developed.

The *Zosterophyllum*-type gave rise to the lycopsids. The lycopsids evolved two lines of species: one became trees while the second were small, herbaceous plants (Fig. 25–B). The sphenopsids show a similar pattern of evolution and origin. Starting as small plants in the Lower Devonian, they evolved a wide range of species, which included trees, in the Upper Devonian and Lower Carboniferous (Fig. 25–H). During these periods, which lasted 40 to 60 million years, the lycopsids and sphenopsids were part of the dominant plants of the land flora. In the Carboniferous their position of importance began to be assumed by the early gymnosperms, and by the end of the Permian they were no longer among the plants prominent in the flora. The treelike species became extinct, and the total number of species declined. But both the lycopsids and sphenopsids are still part of the present flora. Most people are familiar with at least one species of horsetail, as the sphenopsids are commonly known, and many know the handsome dwarf club mosses, resembling dwarf pine trees, that are members of the present-day lycopods. Almost no one, however, is

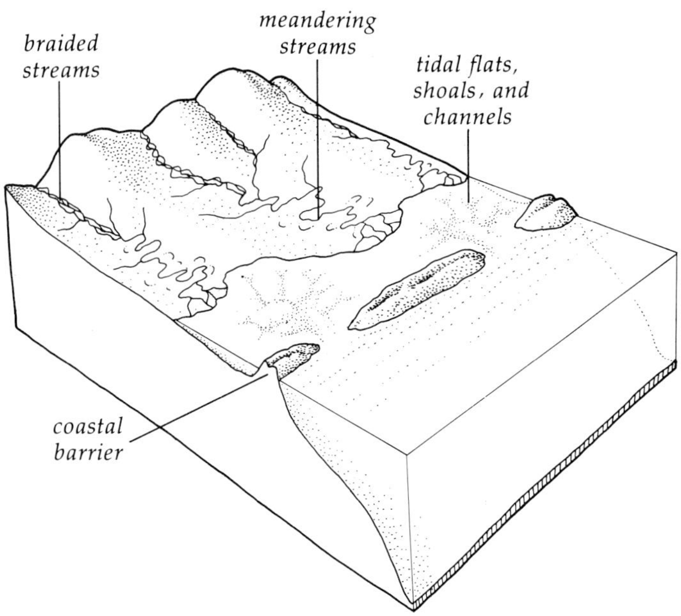

25–1 What did the region in which land plants evolved look like? This is a reconstruction of a region where the early land plants occurred. The earliest land plants such as Cooksonia *are believed to have evolved in tidal flats, shoals, and channels. The* Rhynie *flora, which is more advanced, apparently occupied the regions next to meandering streams.*

aware of two living species of psilopsids, sometimes called whiskferns, that are obscure species found in the tropics and subtropics.

All of these plants have features in common. All are true land plants with stems containing well-differentiated xylem and phloem. The sporophyte is the dominant, independent generation with the gametophyte only rarely seen and frequently subterranean. In the sporophyte the stem is the principal organ with both roots and leaves being poorly developed. The leaves are missing in some of these plants and scalelike in others. All have motile sperm and true embryos.

There are also differences between the various groups—differences in vegetative structure and methods of reproduction. To understand these ancient and interesting plants, let us examine each class separately.

Psilopsida

This group of plants is represented by two genera: *Psilotum* with 2 to 3 species and *Tmesipteris* with 8 to 10 species. These are extremely simple vascular plants. The psilopsid most commonly seen in botany classes is *Psilotum nudum,* since it grows well in greenhouses. This plant, which is native to the tropics and subtropics, consists of a stem that is erect and branched aboveground, while underground it is a rhizome (Fig. 25–C). Photosynthesis occurs in the green cells of the aboveground stem. There are pointed emergences on the stem, but these are not leaves (nor does the bulk of photosynthesis occur here). The stem consists of an epidermis with well-developed stomata and guard cells. The parenchyma cells of the cortex immediately below the epidermis are rich in chloroplasts and have large intercellular spaces that interconnect with the stomata. Below this layer the cortex consists of closely packed parenchyma. The center of the stem is occupied by a simple stele composed of xylem surrounded by phloem.

Meiosis occurs in the sporangia before the aerial spores are produced (Fig. 25–3). The sporangia themselves form on the side of the stem, not at the tip. All of the spores produced are of the same size. The gametophytes, which are formed on the germination of the spores, lack chlorophyll, are subterranean, and are always associated with an endophytic fungus. Antheridia and archegonia are produced on the same gametophyte; the sperms are multiflagellate.

Psilotum is surprisingly like *Rhynia,* although the sporangia of *Rhynia* were not lateral on the stem but terminal—all in all a remarkable similarity between plants separated by some 400 million years.

But there are problems. No fossil evidence linking the fossil psilopsids and the living species exists. There is a great gap between the end of the Devonian and the present. This may mean only that the plants had become so few in number that they were difficult to preserve as fossils. On the other hand, it may mean that the living psilopsids are unrelated to the fossil forms. They may, in fact, be primitive ferns, a possibility that is discussed in Chapter 26.

Lycopsida

Like the psilopsids, the lycopsids are ancient plants dating back to the Lower Devonian. Today, there are but five genera of surviving lycopsids, and all are herbaceous. But, in the Upper Devonian and Carboniferous, there were large numbers of lycopsids, many of them trees and some 90 to 100 ft tall (Fig. 25–6). The smaller herbaceous lycopods were also present at that time, and the fossils of some of these are similar to present-day species. The most common living lycopods belong to the genus *Lycopodium* with some 180 species, the genus *Selaginella* with over 700 species, and the genus *Isoetes* with over 50 species. While this makes a total of almost 1000 species, the Lycopsida do not occupy a conspicuous place in the current flora.

All of the lycopsids have a basic pattern of dichotomous branching (in which the two arms are more or less equal), and all have some type of true root (Fig. 25–4). They also have leaves that are called microphylls. This is a tricky term because it literally means "small-leaved," but although many of the leaves in the lycopods are small, many are not. Botanically, the word means a leaf in which the strand of xylem that extends into the leaf, the leaf trace, usually does not leave a gap where it leaves the xylem of the stem. These are considered primitive characteristics and are found in few vascular plants. The sporangia develop at the conjunction of the leaves and the stem. The sporangia may be grouped together into cones called strobili, or they may be spread out in the axils of leaves along the stem.

Essay 25–1 Chipping Away at Early Land Plants

In Chapter 9 we saw one answer to the question why a particular man enters a particular field at a particular time. Here, we present another answer to the same question but in quite a different field—paleobotany. Science is not only laboratories and libraries but also talking to the right people, including salesmen, at the right time. In an address as the Retiring President of the Botanical Society of America, Professor Harlan P. Banks, one of the truly great American paleobotanists, discusses the same question.

"As an undergraduate at Dartmouth, I was introduced to the joys of primitive land plants and their role in the phylogeny of vascular plants by the enthusiastic presentation of Rhynia *and* Horneophyton *by Carl L. Wilson.*

The attraction of these early land plants is the myriad of questions they raise: where are the fossils to be found—why—whence did they evolve—exactly when—in how many forms—how rapidly did speciation occur—what can we learn of their habitat—in what associations did they occur—are the fossils identical wherever found in the world—what can we learn of the paleogeography and paleoecology of Devonian time—how many whole plants can we reconstruct—how precisely can we pinpoint the time at which distinguishable subdivisions of vascular plants were evolved—what can we learn of the exact time of origin of major characteristics of vascular plants—what do Devonian plants contribute to phylogeny—what was happening in the animal world at the same time—to say nothing of the technical problems they pose for anyone who wants to learn their secrets.

Also, at Dartmouth William Patten was teaching a freshman course entitled Evolution. Patten devoted his research to a strenuous search for a missing link between invertebrate and vertebrate animals. His zealously pursued thesis was that a group of primitive armored fishes, the ostracoderms, filled the bill. But what caught my imagination were his exciting accounts of trips to Spitsbergen, Gaspé, and other localities whence he returned with an abundance of material. Among his collections from the Gaspé peninsula were beautiful fronds belonging to a supposed Devonian fern, Archaeopteris. *The sequel is obvious. My first collecting of Devonian plants was done on the Gaspé peninsula and included* Archaeopteris.

Subsequently, at Cornell I came under the influence of the master of teaching, research, and writing, the late Arthur J. Eames. Eames, at the time, was completing his Morphology of Vascular Plants: The Lower Groups *and was erecting the subdivision of vascular plants,* Psilopsida, *for the* Rhynia-*type of Devonian fossil and the living Psilotales. Equally stimulating was the enthusiasm of the late Loren C. Petry who reveled in the joys of a problem needing solution. Petry had collected on the Gaspé peninsula in the early thirties. Among his finds were calcitic nodules containing plants with preserved cell structure. I was lucky to get more of these and obviously began peeling the nodules in an attempt to reconstruct a plant. In the midst of this first project I got waylaid by a traveling salesman, a shoe salesman, from southeastern New York. As you may be aware, good salesmen are often extremely alert men always on the watch for something new. This particular one was attending a nut tree growers' meeting and stopped by to tell us about some fossil nuts being turned up by a retired engineer during road-building operations through Devonian strata. The nuts proved to be small masses of pyritized wood. But the engineer, Mr. William Deats, led me to a dense deposit of a lycopod that became* Colpodexylon deatsii *in my reconstituted thesis."*

The species of *Lycopodium* growing in the temperate zones are generally erect shrubby plants. Trailing or hanging forms are also common, particularly in the tropics where they grow as epiphytes. Probably, the most common *Lycopodium* is the one commonly known as ground pine, found in northern temperate forests. The aboveground stem of this handsome plant is highly branched and is covered with many small leaves of a beautiful green color. There is a horizontal, underground stem from which roots arise. In the stem a central strand of xylem is surrounded by phloem.

All the spores produced by the sporangia in the

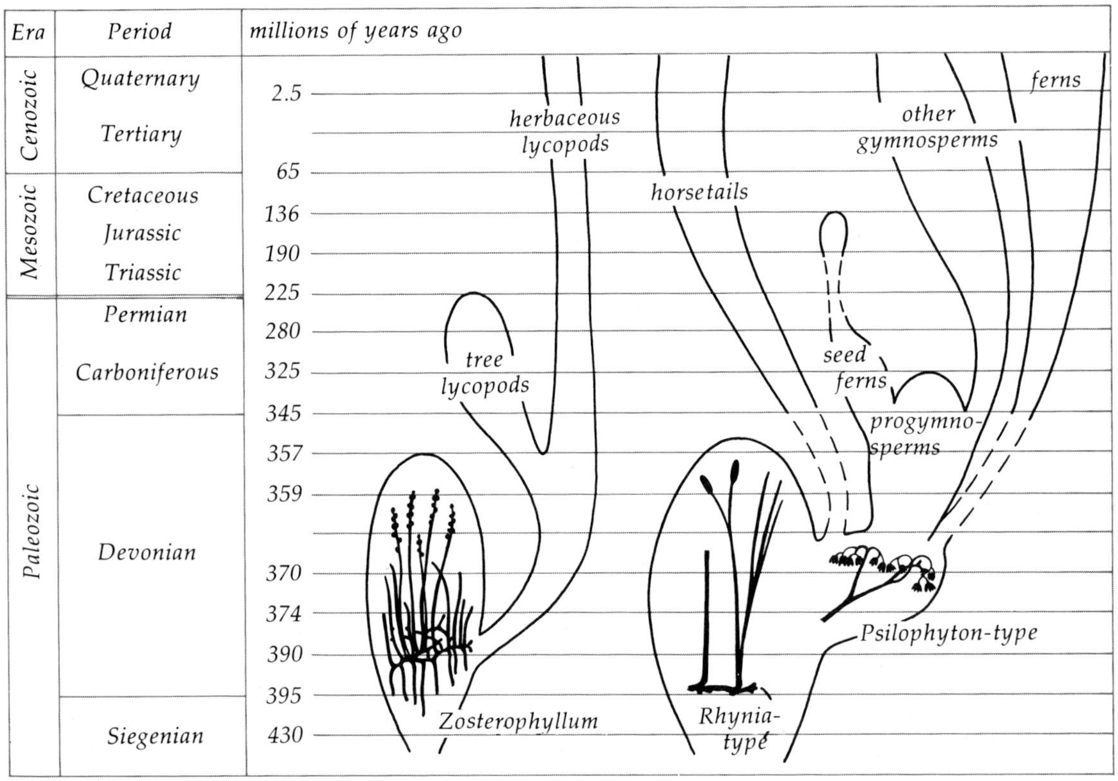

25–2 *The possible course of evolution from the two earliest types of vascular plants. In this diagram, which was prepared by Banks, the length of the Devonian period has been exaggerated by showing its subdivisions rather than presenting the period as a whole. Not shown on this chart are several groups of plants that evolved during the Devonian and became extinct by its end. One of these, the Nematophytales, are considered nonvascular land plants.*

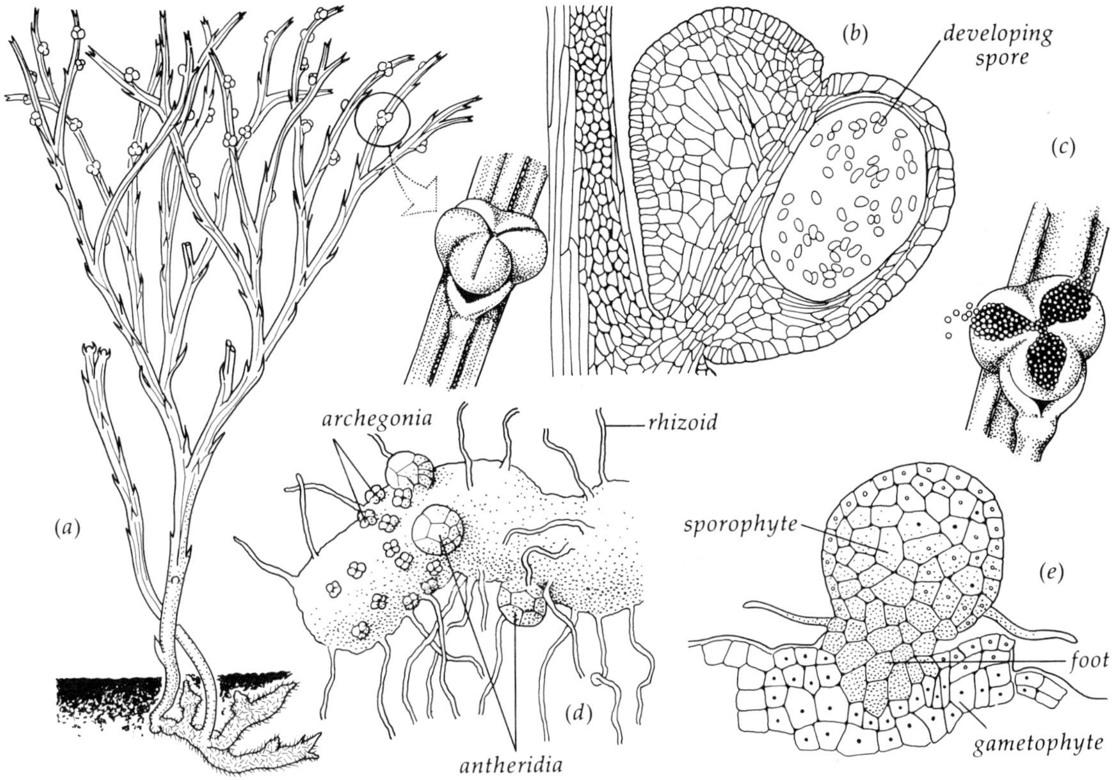

25–3 The details of the life history of Psilotum. *The haploid spores are produced following meiosis of the cells in the lateral sporangia (a & b). These are released when the sporangia open (c) and germinate to form the subterranean gametophyte (d). Antheridia and archegonia form on the gametophyte, and after fertilization the new sporophyte begins its development on the gametophyte (e).*

strobili are the same in size and shape. These form gametophytes that in some species grow at the surface of the soil and are green, while in other species they may be subterranean and lack chlorophyll. One of the fascinating observations about the gametophytes is that they will not develop unless they have an endophytic fungus associated with them. Once they are established, the gametophyte can survive for many years. Both antheridia and archegonia develop on the same gametophyte. Their form is much like that found in the bryophytes—the sperms are flagellated, and water is necessary for fertilization.

Selaginella, which resembles *Lycopodium* in general vegetative features, varies significantly in its pattern of reproduction from any land plant we have yet discussed. In the bryophytes, the psilopsids, and *Lycopodium,* the spores that are produced by the sporophyte are all similar in size and shape—they are homosporous. But in *Selaginella* two different types of spores are formed—the plant is heterosporous (Fig. 25–G). Two types of sporangia develop. One produces a few large spores called megaspores, which on germination give rise to the megagametophyte or female gametophyte. The other type of sporangia form many small spores or microspores, which germinate to produce the male gametophyte or microgametophyte.

The megaspore in *Selaginella* begins nuclear divisions while still present in the megasporangium. These are nuclear divisions without accompanying cell-wall formation; hence the megagametophyte that results is, at first, a multinucleate cell. Walls begin to develop at the edge of the megagametophyte nearest the area where the thick protective wall of the spore will rupture. Here, a much reduced archegonium is differentiated (Fig. 25–5). The microspores also begin their development within the microsporangium. Cell divisions result in the formation of a layer of sterile jacket cells immediately inside the spore coat and surrounding the cells that form the biflagellated sperm (Fig. 25–5). When the megagametophytes and the microgametophytes are mature, the spore coat that surrounds them splits, and the released sperms swim to arche-

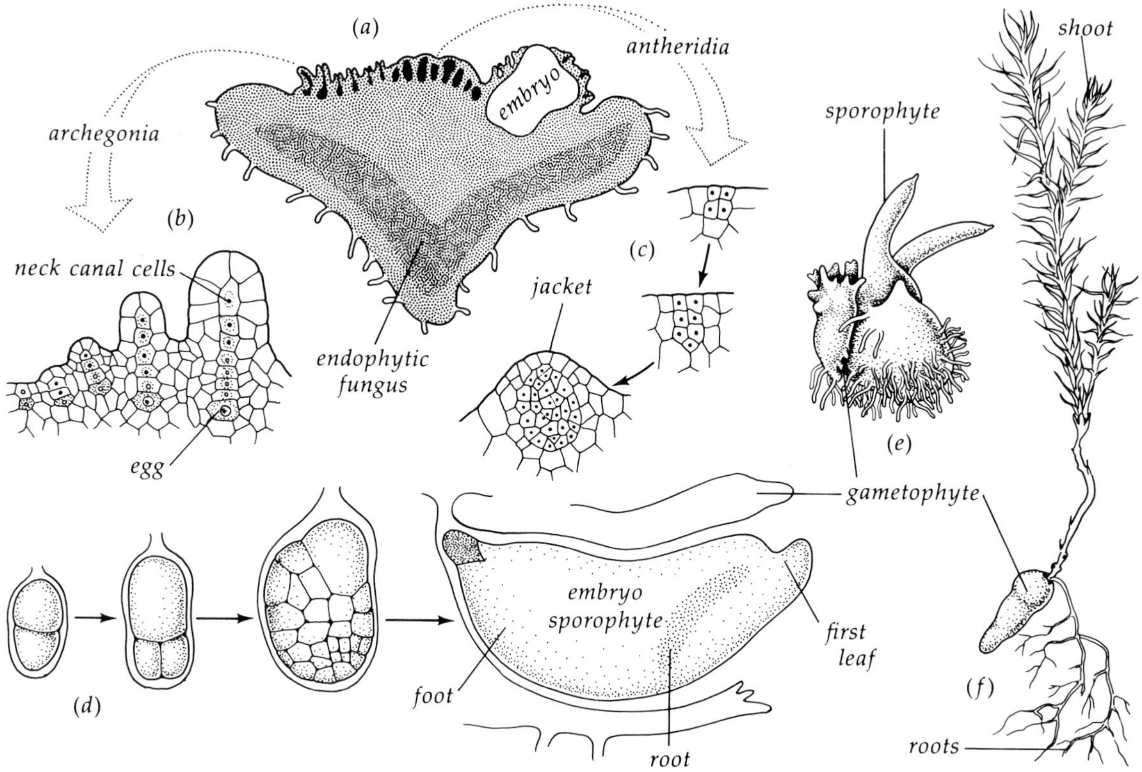

25–4 The details of the life history of Lycopodium. *A longitudinal section of the gametophyte (a) shows the position of the antheridia, archegonia, and one embryo. The stages in the development of an archegonium (b) an antheridium (c), and an embryo (d) are also shown. The further growth of the embryo into a mature sporophyte is shown in (e) and (f).*

gonia, down the neck canals, and fertilize the eggs. The details of this sequence of events vary from species to species. For example, in some species the megagametophytes and microgametophytes are still attached to the parent sporophyte when fertilization occurs, while in others they are shed earlier.

The occurrence of heterospores is of great significance in the evolution of plants. All of the gymnosperms and the angiosperms are heterosporous. This condition aids in promoting cross-fertilization and outbreeding. It is also the first step in the evolution of seeds. The existing evidence suggests that the condition of heterospory in the lycopods did not lead directly to the condition in the true seed plants. Instead, it appears to be another example of the independent origin of a major and useful feature in the life of the plant.

Why did such elaborate structures as the large lycopsids, which were among the first tree species, evolve? The obvious answer is that the tall stem gives the plant both an increased area for photosynthetic tissue and a better position to expose this tissue to sunlight in competition with smaller, shorter plants. In the evolution of these first trees (Fig. 25–6), we can clearly see the emergence of biological competition as an environmental factor in plant life on land. When the first plants adapted to the land environment, probably only the physical conditions of the environment determined the course of their evolution. The land was barren, and competition for available, habitable sites was nonexistent. Yet this condition could not have lasted long. Eventually, those plants in which genetic variation gave them greater height had an advantage, survived and reproduced.

Of course, this early evolution of the tree lycopsids occurred in the context of a particular physical environment and with a certain basic pool of genetic information. The environment in which they evolved was one in which water was abundant and the atmosphere water laden. The available evidence indicates that the climate was mild but not tropical. The tree lycopsids are essentially the same as the smaller species, although certainly different in detail. The vascular tissue was both primitive in form and small in amount. The growth appeared determinate; there was no secondary growth of vascular tissue. This is a primitive system for

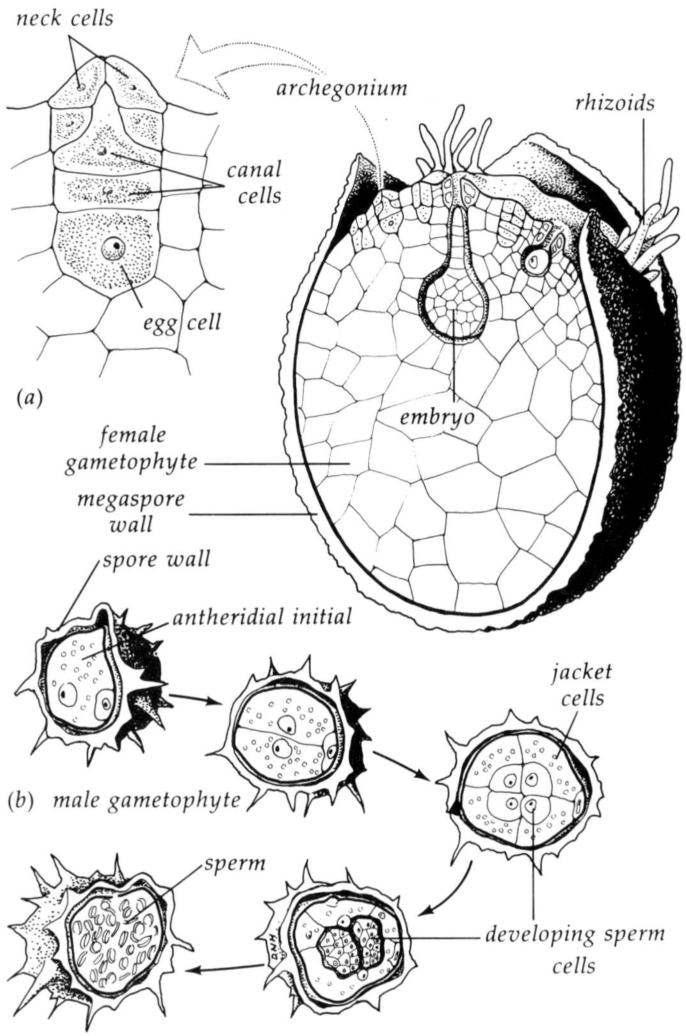

25–5 The development of the female and male gametophytes in Selaginella.

25–6 A diagrammatic reconstruction of a tree lycopod, Lepidodendron.

a tree, but under the conditions in which the lycopsids evolved, it was not only adequate but highly successful.

Unfortunately for the tree lycopsids, conditions did not remain the same. For 150 or more million years (certainly a respectable period of time), the tree lycopsids flourished, but then conditions changed. At the end of the Carboniferous and during the Permian, the great swamps dried up and the climate became drier as extensive glaciers formed. With the gradual disappearance of excess water, the tree lycopsids were in trouble. The amount of vascular tissue they contained apparently proved inadequate, and they became extinct.

Sphenopsida

The evolutionary history of the Sphenopsida closely parallels that of the Lycopsida. The earliest sphenopsids (horsetails) are found in the Devonian, and during the Carboniferous they were worldwide in distribution. It was also during the Carboniferous that they reached their greatest diversity in shape and size, ranging from small, herbaceous species (some similar to those existing today) to large arborescent species 50 to 70 ft high. During the Permian the larger species began to dis-

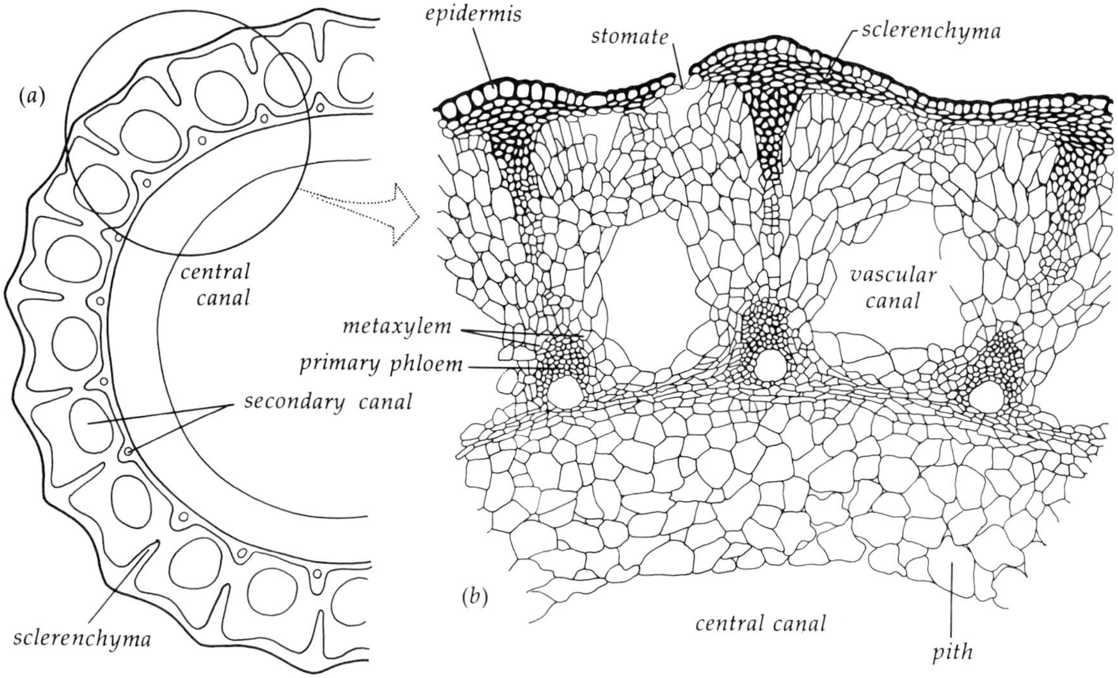

25–7 A diagram of half of a cross section of the stem of Equisetum *(a). There is a large central canal, as well as smaller secondary canals and bands of sclerenchyma. Part of the section has been drawn at higher magnification to show the details of the tissues (b). In this drawing, we can see the epidermis, stomate, primary phloem, metaxylem, and sclerenchyma.*

appear, and by the end of the Triassic only a few herbaceous species were left. Today, only one genus with 30 species survives, but this one genus, *Equisetum,* may not have undergone significant change from the Carboniferous, some 300 million years ago. Thus *Equisetum* may be one of the oldest living of vascular plant genera.

The appearance of the sphenopsids is characterized by the presence of whorls of small leaves or branches (Fig. 25–H). This gives the stem a particular jointed look. The leaves are small, pointed projections, and may lack chlorophyll. They are microphylls with a single unbranched midvein. The stem is the main photosynthetic unit of the plant. In some species what appear to be long needlelike leaves are whorls of thin branches.

The stem is marked by ridges of tissue composed of cells with thick walls rich in silica. The presence of these cells makes the plant useful for scouring pots but can cause injury to livestock that eat them. The center of the stem is hollow, being occupied by a large central canal and two other systems of canals. (The air canals do not extend through the nodes.) The general anatomy of the stem of *Equisetum* is shown in Figure 25–7.

The aboveground stems grow from underground rhizomes (which are stems not roots). Roots arise in whorls from the rhizome. At the tip of the stem there is a large apical cell covered by the newly formed leaves. Cell division also occurs in a ring of cells at the base of each internode. Horsetails break readily just above the node because of the thin walls in this meristematic region.

Like the lycopsids, the sporangia of the sphenopsids are grouped into cones (Fig. 25–8b). These are at the tips of stems or, in some species, at the ends of special stems lacking chlorophyll. The spores, formed after meiosis in the sporangia, are of the same size and shape; the sphenopsids are therefore homosporous. A characteristic of the spores of *Equisetum* is the possession of long, strap-shaped arms that are part of the outer wall. These elaters function in the same relative manner (they are hygroscopic) as the elaters found in liverworts and hornworts, but they are completely different in morphology and origin from the bryophyte elaters.

The spores germinate to form small gametophytes that are green and grow on the surface of the soil (Fig. 25–8c). The gametophytes look like small green pincushions, and both antheridia and archegonia form on the same plant. The sperms are multiflagellate, so water is necessary for fertilization. The zygote divides in place in the gametophyte, and an embryo develops that has a

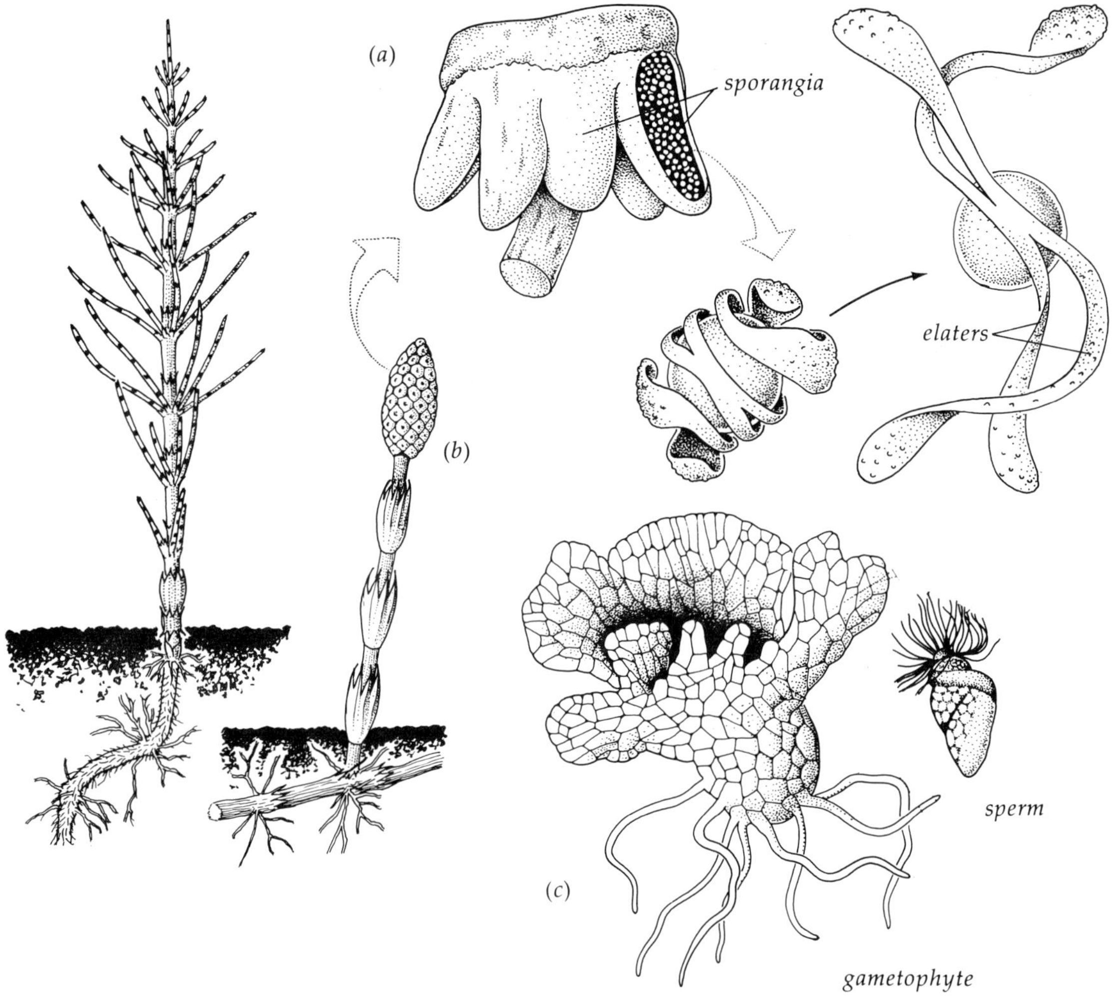

25–8 In Equisetum *the sporangia are clustered on scales (a) that are grouped into cones (b). The spores themselves have long extensions of their outer walls (the elaters) that aid in their dispersal. (c) The details of gametophyte and sperm development in* Equisetum.

nutritive foot as well as a shoot and root apex. A large gametophyte may support several sporophytes during their early development.

The present-day *Equisetums* are rarely more than 1 m tall. (The tallest is a tropical species that may reach 8 m, but it is supported by surrounding plants.) In the Carboniferous, however, there were tree size species such as *Calamites,* which were 10 to 14 m tall. These looked much like giant versions of present *Equisetum* species. Other extinct species of sphenopsids had wide, flat leaves that were attached in whorls to the stem.

26

The Ferns

Fronds, Fiddleheads, and Ferns

Because of the grace and beauty of the fronds, ferns have long captured the interest of man. Designs based on the fern leaf and the striking "fiddlehead," which it has as it unfolds, can be found in the artifacts man has produced through the centuries.

A Fern fronds and a fiddlehead.

A

If asked to describe a fern, most people would talk of large, lacy leaves growing directly from the soil, and with minor modifications, their description would be correct for many fern species. In these, the stem is underground, and all that is usually seen is the large compound leaf. There are, however, many variations on this general plan—three of which are illustrated here.

B Some of the most spectacular of the ferns are the tree ferns. These have upright, aboveground stems, which can reach significant heights, and possess leaves several meters long.

C The staghorn fern shown here illustrates another variation on the general body plan of ferns. These are epiphytes, which grow on the trunks of trees. They possess not only the regular fronds but modified leaves, which develop against the trunk of the tree. These grow over the stem and roots, not only protecting them but collecting humus in which the roots grow.

B

C

D As highly modified in their own way are the water ferns. Two species are illustrated here: Azolla *(left) and* Salvinia *(right). In such species the stems are greatly shortened, and the leaves are modified. In* Salvinia *roots are lacking—what appear to be roots are really extensions of the leaves. In* Azolla *the short, branching stems are densely covered with alternating bilobed leaves, and true roots are present.*

D

Under the canopy of the giant redwood trees few plants can survive, primarily because of the small amount of light that filters to the forest floor. One plant that does grow there is the sword fern (*Polystichum munitum*). This same fern is found in the conifer forests of the Pacific Northwest. Other ferns inhabit the great tropical forests—growing on the forest floor, on the trunks and branches of trees as epiphytes, and over the surface of trees as vines. In such forests there is a wealth of fern species; it has been estimated that 65 to 90% of the some 10,000 species of ferns live in the tropics. But ferns do not grow only in forests. They are also found on deserts, tundra, and in alpine locations.

The ferns are a large, successful group of plants but have never formed a dominant part of worldwide flora. The Carboniferous period was once called "the age of ferns" because of the large number of fernlike fossils that were found. Later investigations have shown that many of these fossils belonged to lycopsids, sphenopsids, and early gymnosperms (including the seed ferns). This is not to say that ferns were not abundantly present—they were—but they were not, as they are not now, large conspicuous plants that dominate the flora. The ferns of the Carboniferous grew under the large tree lycopsids, sphenopsids, and gymnosperms, as they now grow under the redwoods and other forest trees. And then as now, they successfully fill these particular environmental niches.

The lycopsids, sphenopsids, and ferns all evolved in the Devonian and flourished during the Upper Devonian and Carboniferous. All became drastically reduced in numbers and importance during the Permian; only a small number of the lycopsid and sphenopsid species have survived to the present. But the ferns show a different pattern. Some families of ferns persisted (and are now considered primitive ferns), while new fern species and families evolved during the Mesozoic and Cenozoic. There is evidence that there is continued evolution of new fern species today, particularly on tropical islands.

As a result of this long history and evolutionary adaptability, the ferns show a wide range of characters, some of which are similar to the lower vascular plants and others that are more advanced. Many ferns have underground stems or rhizomes and, in general, lack elaborately developed root systems. All have vascular systems with little or no secondary growth. All reproduce by spores and have independent gametophytes that produce sperms dependent on free water to reach the egg. It is in the fern leaf that some of the most advanced features of the plant are found.

Leaves

The leaves of ferns, gymnosperms, and angiosperms all have the same general features. But they may vary greatly in size and shape; compare, for example, a fern frond and a pine needle. These leaves are equipped with an elaborate supply of vascular tissue, and there is a marked gap in the vascular cylinder of the stem where a vascular strand diverges into a leaf (Fig. 26–1). This leaf gap is not found in the lower vascular plants.

Such leaves are megaphylls, as contrasted with the microphylls of the lower vascular plants. The available evidence from comparative and fossil studies, which are obviously incomplete, suggests that the two types of leaves may have different origins. The microphylls are believed to have developed from scalelike growths from the stem into which the vascular strand then developed. The megaphylls, on the other hand, are interpreted as arising from branch systems in which the tissue between the branches becomes flattened and extended, forming a leaf. The postulated development of the two types of leaves is shown in Figure 26–2. It must also be pointed out that it is possible to arrive at the form of the microphyll by the reduction in size and complexity of the megaphyllous leaf. The concept of the development of leaves through the development of a branch system is known as the "Telome Theory." This theory has been receiving considerable support, and the fossil record has yielded many data that are compatible with it.

The leaves of ferns vary enormously in size and shape. The fern leaf that most of us are familiar with is a large, compound leaf (Fig. 26–3). It consists of a stemlike stipe and a blade, which may be divided into pinnae that may, in turn, be divided in a number of ways. When such compound leaves are formed, frequently underground, the stipe and pinnae are tightly coiled. The leaf uncoils as the cells on the lower side of the leaf elongate more rapidly than those on the upper side. The coiled fern frond, called a "fiddlehead," is an interesting and frequently beautiful object as it uncoils in the spring (Fig. 26–A). In many ferns the leaves are annual, but several immature leaves, which will develop in future years, may be initiated and developing underground.

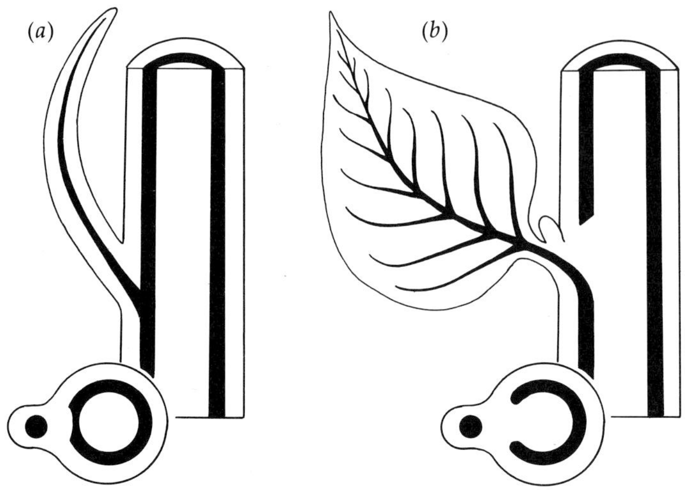

26–1 The relation of the vascular tissue as it leaves the stem and enters a microphyll (a) and a megaphyll (b). In the case of the megaphyll, there is a break in the cylinder of vascular tissue, which is occupied by parenchyma cells.

Not all ferns have elaborate, compound leaves. Some 10% of the species have simple blades that are not divided into pinnae. Studies of living ferns have established that the compound leaf is more primitive than the simple, bladed leaves.

Fern leaves also show a difference in their capacity to continue growth. In most ferns growth is determinate; the frond uncoils, expands, and growth ceases. In a few the meristem at the apex of the leaf (or in some other portion of the leaf) continues to grow for long periods. This results in a trailing or climbing leaf form. In some ferns this continued capacity for growth of the leaf tip has resulted in a means of reproduction. The end of such leaves forms a small plantlet when it comes in contact with the soil, and a new plant is formed.

The type of leaf found on a fern is correlated with the habitat in which the fern grows. Tropical areas with

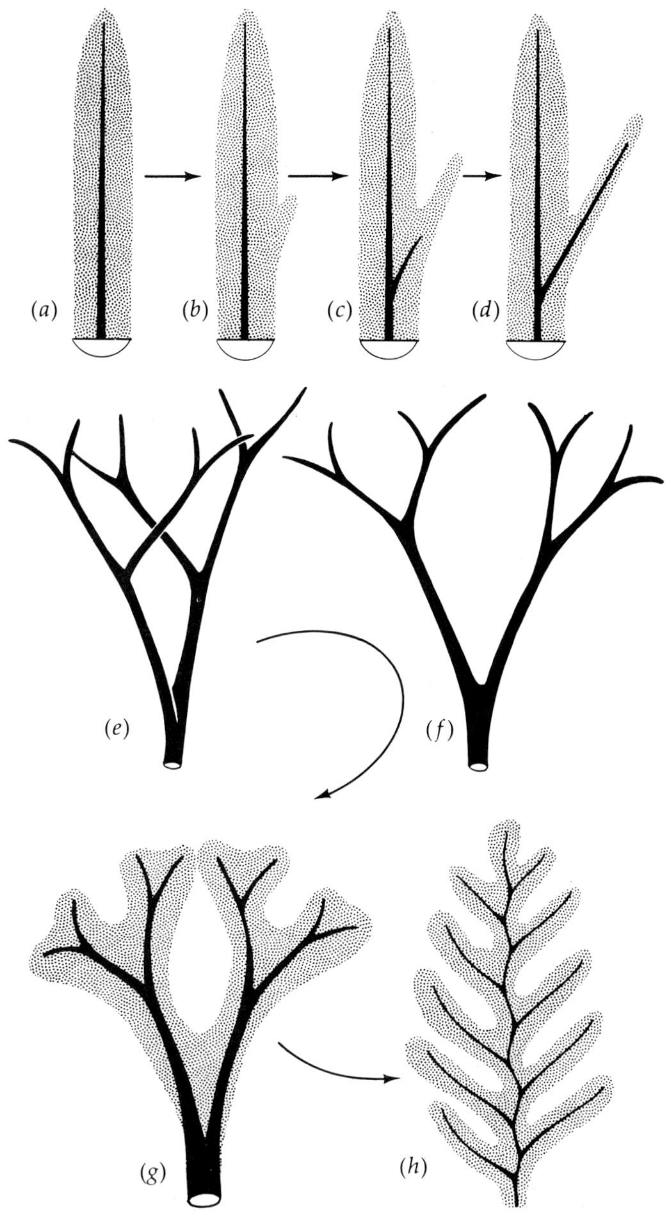

26–2 The possible evolutionary origin of microphylls and megaphylls. In the case of the microphyll (a–d), the formation is postulated through a projection of tissue from the stem into which the vascular tissue then develops. The possible origin of the megaphyll is shown in (e–h). Here, the starting point is a regularly branched axis that becomes flattened and extended by a webbing of parenchyma cells.

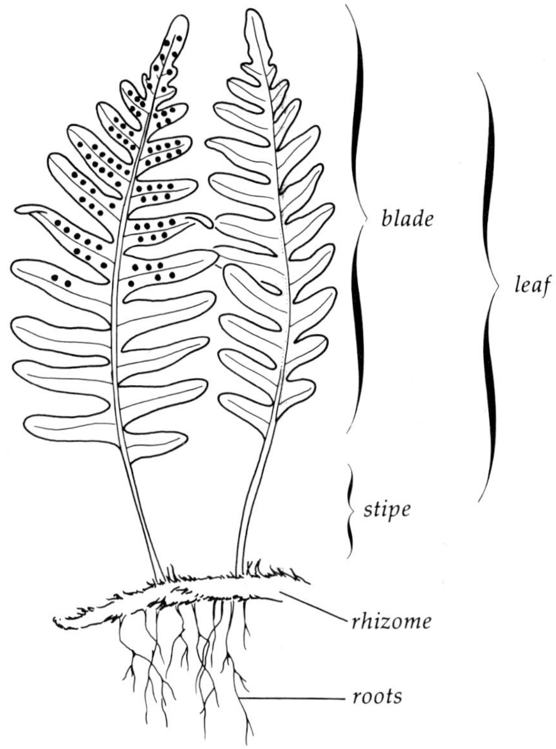

26–3 This large compound fern frond consists of a long stemlike stipe and a blade, which is frequently divided into pinnae that may also be divided in a number of ways.

adequate water and warm, year-round temperatures are ideal for the growth of ferns, and it has been estimated by Rolla Tryon of Harvard that 65% of all fern species grow here. Many of the species found in this habitat have relatively thin leaves with large amounts of surface area that persist for at least a year. The thickness of the leaf is closely correlated with the amount of moisture available. Filmy ferns exist in regions of high moisture, sometimes associated with cloud-forest conditions or waterfalls and streams. These ferns have extremely thin leaves, the blades of which may be only a single cell layer thick and so delicate that they must be kept constantly in high humidity to survive. There are over 600 species of filmy ferns, but only a few have ever been grown in greenhouses. The bulk of species having leaves showing indeterminate growth are found in tropical areas. These are ferns that climb or scramble over shrubs and trees.

Not all warm environments are wet; some are dry with only periodic moisture. These deserts or semideserts would appear to be poor environments for ferns, yet some 5% of the fern species are found there. They usually have small, leathery leaves that can generally survive for most of the year, becoming dormant during the dry periods and reviving when water is present. A few species have leaves that die as soon as water is no longer adequate and produce new leaves when it is. Such leaves tend to be thinner and broader than the persistent types.

Many tropical forests undergo pronounced dry periods, and in such forests are found many epiphytic ferns. Frequently, these species have simple leaves, although compound-leaved ferns are also present as epiphytes. In either case, the leaves are again generally small and leathery, becoming dormant in dry periods. The leaves of these species are usually simpler in form than leaves of the desert species. This can be related to the fact that such epiphytic environments are readily subject to drying, which may occur during periods of the day even during the wet season.

The alternate extremes to the tropical environments are the boreal and arctic environments. Here, water is available but variable during the warm period, which may be quite short. During the cold period the temperature may be well below 0°C for long periods. This is a rigorous environment for any type of plant life, but we find 5% of all fern species growing there. Most have compound leaves that die at the onset of the cold period. The leaves have a relatively large surface area and are fairly thin. They are thus adapted to the favorable conditions that exist during the warm periods and are discarded when the cold comes. A few ferns in this region are evergreen.

One of the most striking modifications of the fern leaf is found in the water ferns, particularly the floating water ferns. In one group, *Azolla*, the tiny branched stem is crowded with overlapping leaves about 1 mm long. Each leaf has two lobes; within the upper lobe is a cavity containing the blue-green alga *Anabaena azollae*, capable of fixing atmospheric nitrogen. Another floating water form is *Salvinia*. Leaves in whorls of three are arranged around a floating stem. The upper two leaves in each whorl are covered with unwettable hairs; the submerged leaves are finely divided and bear a striking resemblance to roots.

Stems

The stems of most ferns are either underground (rhizomes) or short erect structures near ground level. In a few species the erect stems may be barrel-like or quite tall, as in the tree ferns.

The internal anatomy of the stem is highly variable and has been widely studied as an important feature of fern evolution. In the most primitive ferns the stele is a solid mass of tracheids and resembles the condition in the lower vascular plants. This type of stele is called a protostele (Fig. 26–4). The xylem at the center of the protostele is surrounded by a ring of phloem that, in turn, is surrounded by the pericycle and an endodermis.

More advanced than the protostele is the siphonostele (Fig. 26–4). Here, the center of the stele is occupied by living parenchyma cells, the pith. The arrangement of tissues within the siphonosteles of ferns shows much variation. The phloem can, for example, be both internal as well as external to the xylem, and in some cases, a second endodermis may separate the internal phloem and the pith. The leaf gaps are usually conspicuous in the siphonostele, being marked by wide strips of parenchyma. Thus, unless there is a region where leaves are not present (as in the long internodes of some ferns), the siphonostele rarely appears as a continuous ring. The most highly evolved siphonosteles are those in which the primary vascular tissues appear as a ring or a scattered series of vascular bundles, called a dictyostele (Fig. 26–4). The strands of vascular tissue in a dictyostele consist of primary xylem enclosed in primary phloem and surrounded by an endodermis.

The tip of the stem is surrounded by the developing leaves. The stem apex is characterized by a large single apical cell, the derivatives of which give rise to the various tissues of the stem and the leaves (Fig. 26–5).

The rhizomes of ferns play an important role in the spread of the plant. Growing through the soil, they are less subject to drying out or being eaten by animals. By branching they can rapidly spread through an area and provide stiff competition to grasses and other plants. In New Zealand farmers and ranchers

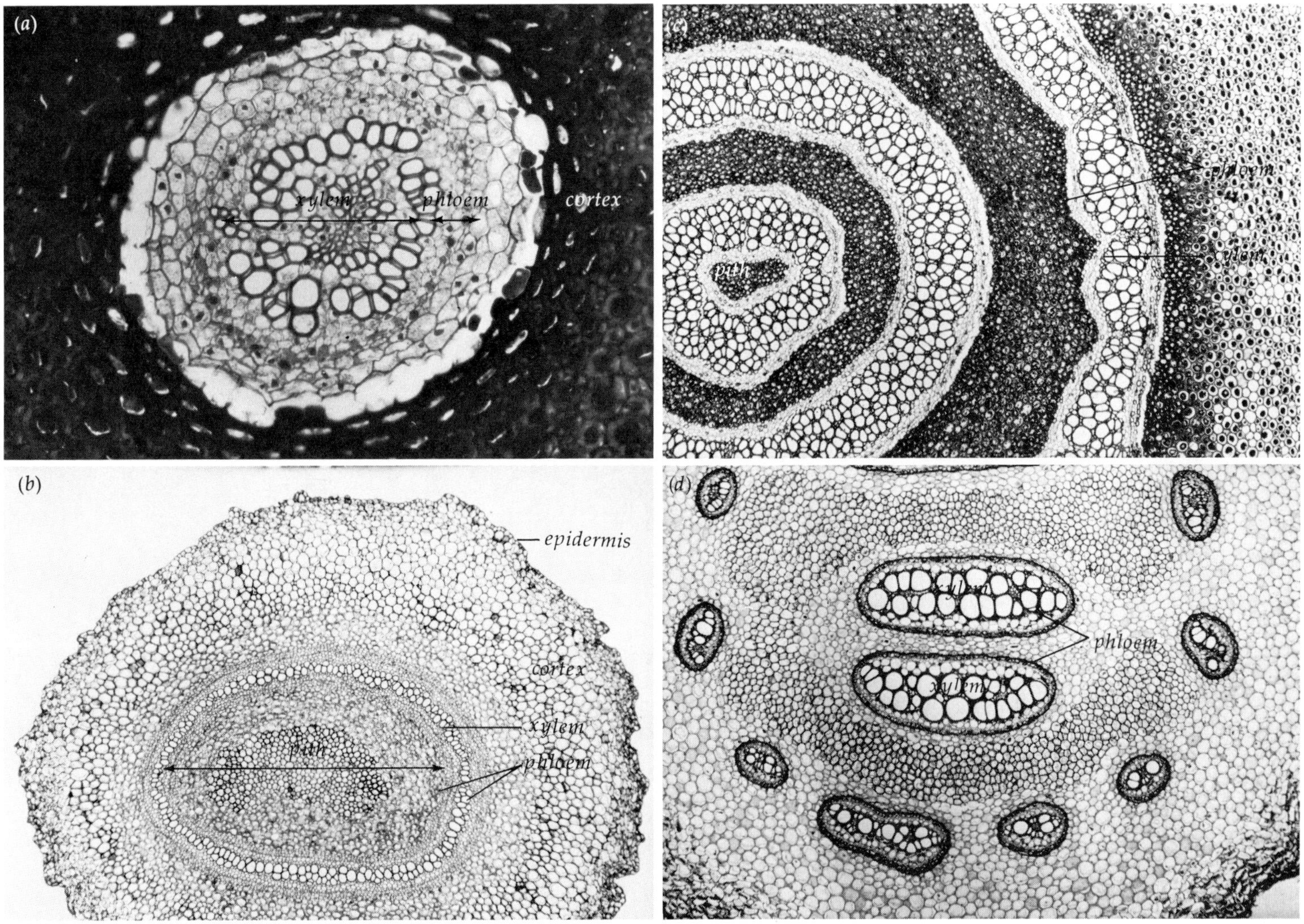

26–4 *The types of steles in ferns: (a) protostele in* Trichomanes reniforme, *(b) siphonostele in* Dicksonia, *(c) ringed siphonostele with internal phloem in* Matonia pentinata, *and (d) dictyostele in* Pteridium.

have spent generations trying to eradicate the bracken fern, since it becomes a pest on open, grassy hillsides where it crowds out the grass, ruining the pasture. Trying to dig out or grub up the rhizomes merely spreads the plants. Attempting to force the sheep to overgraze the ferns usually leads to sick sheep. The most successful attempts involve cutting the fronds back repeatedly so that the starch reserves in the stem will be depleted and the plant will starve. Some ferns have rhizomes so tough that they cannot be pulled out by hand, and wire cutters must be used to sever them.

We have already seen how the stem of floating aquatic ferns is much reduced in size. The opposite extreme is the stem of some of the tree ferns that may become 20 to 30 m tall. There is no secondary growth in the ferns by an organized cambium (except for one genus, *Botrychium*), even in the tree ferns whose trunks are covered by a mat of tangled, living, dead,

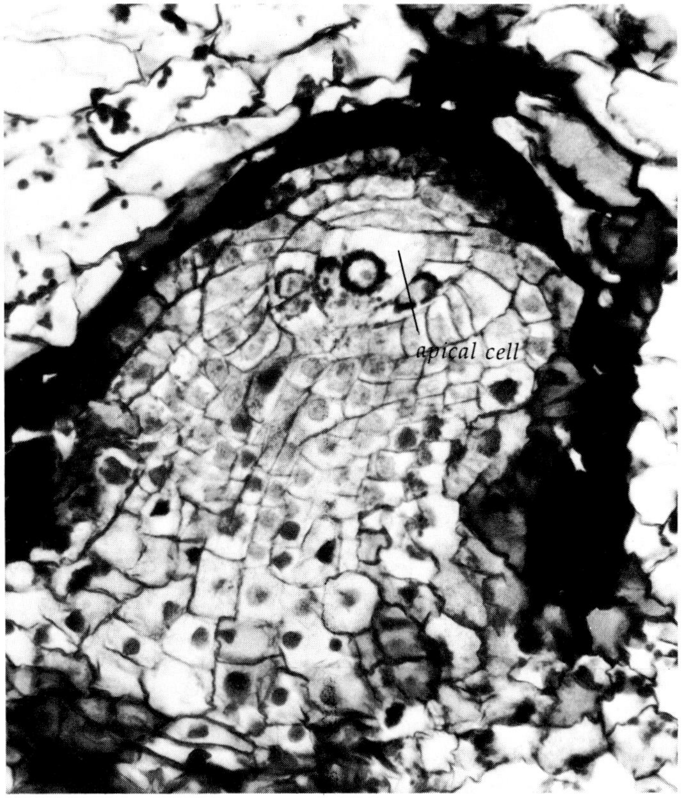

26–5 A section through a young root, showing the large apical cell that dominates the meristem.

and dying roots. The actual stem may be only a fairly small cylinder covered by roots.

Roots

In contrast with the stems, the roots of ferns are relatively uniform and simple (Fig. 26–6). The epidermis is thin-walled and may or may not have root hairs. The cortex usually contains thick-walled sclerenchyma cells, and there is always a well-defined endodermis. All fern roots have protosteles, regardless of the type of stele found in the stem. At the root apex is a single, pyramid-shaped apical cell protected by a root cap. This cell and its derivatives give rise to the cells that form the various tissues of the root.

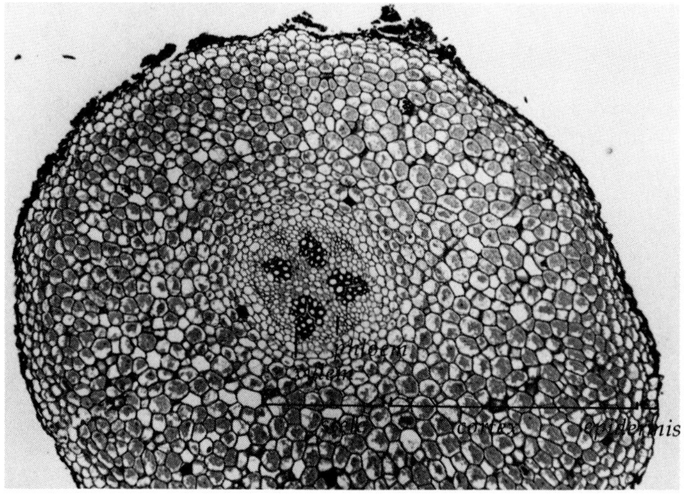

26–6 A cross section of a fern root.

Ferns have small, inconspicuous roots arising from the stem in the vicinity of the leaves. These are adventitious roots and originate from the pericycle of the stem, forcing their way through the tissues of the stem as they develop. A few ferns—some of the water ferns and the delicate, filmy ferns of the tropics—have lost their roots in the course of evolution and are now rootless.

Sporangia and Spores

The spores are produced in sporangia that are located on the leaves (Fig. 26–7). In some ferns these may be ordinary vegetative leaves, or they may be highly differentiated leaves that are essentially sporangial-bearing appendages. In other ferns an intermediate condition is present; part of a frond is modified into a sporangial-bearing region, while the remainder is vegetative.

The type of sporangium and the mode of its formation determine in which of two groups the ferns are placed—the eusporangiate ferns or the leptosporangiate ferns. The details of their development need not be dealt with here but the essential point is simple: the sporangia of the eusporangiate ferns are similar to those found in the lower vascular plants, while the sporangia of the leptosporangiate ferns

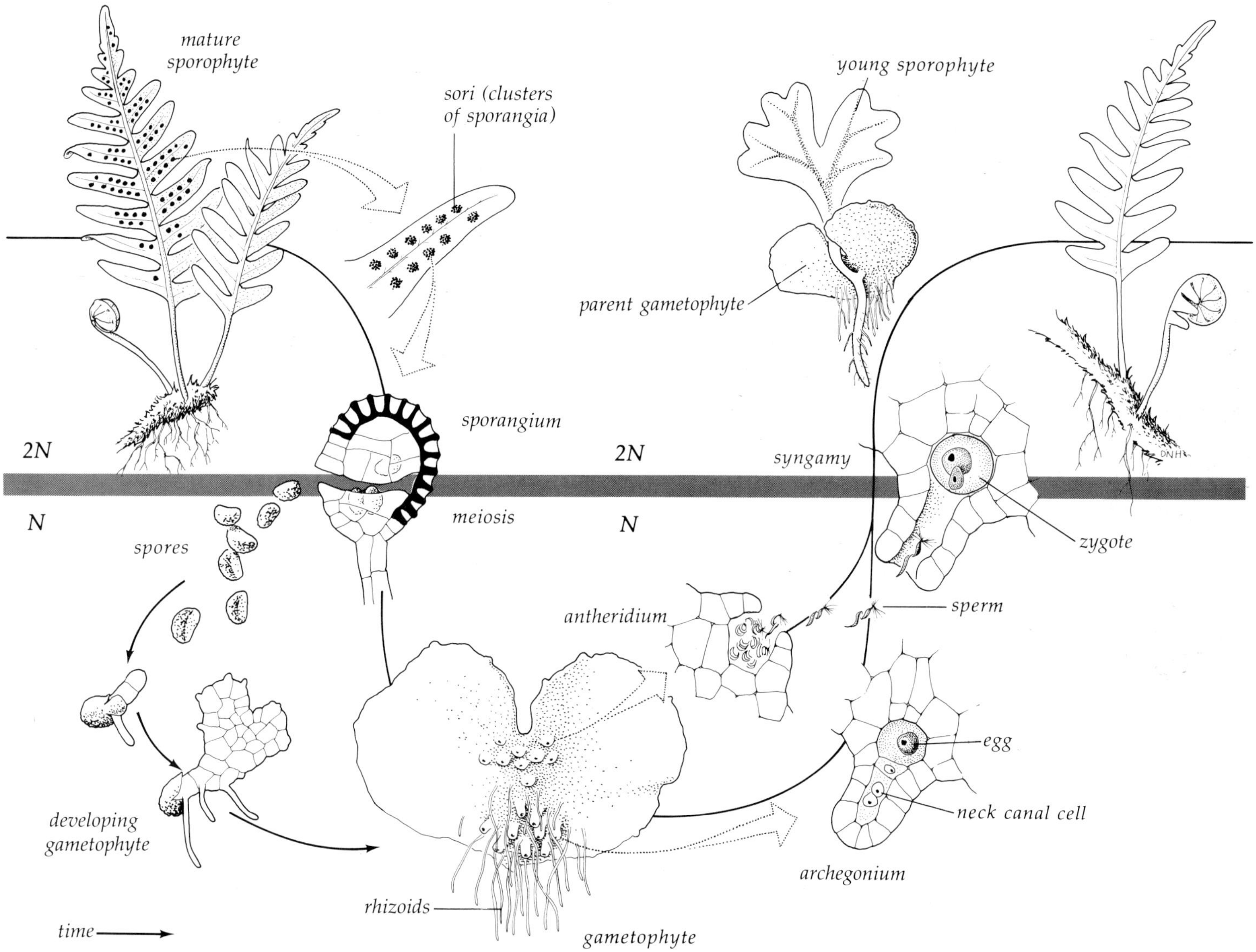

26–7 *The life history of a fern.*

represent a new and unique feature in the ferns. Consequently, on this (and other grounds) the eusporangiate ferns are considered by morphologists to be more primitive than the leptosporangiate ferns. Most of the common ferns are leptosporangiate, so we will concentrate on these.

The sporangia are often grouped together into sori (Fig. 26–8). Covering the sori in many ferns is a thin outgrowth of the leaf known as an indusium. The shape of the indusium is highly variable and important in fern classification. In a few ferns the edge of the leaf curls over to form a false indusium.

A sporangium in a fern consists of a flattened, box-shaped structure with a ribbed back held at the end of a stalk. The ribbed appearance comes from a file of cells called the annulus, and each cell of the annulus has a thickened wall around three sides. Thin-walled cells make up the rest of the sporangium wall

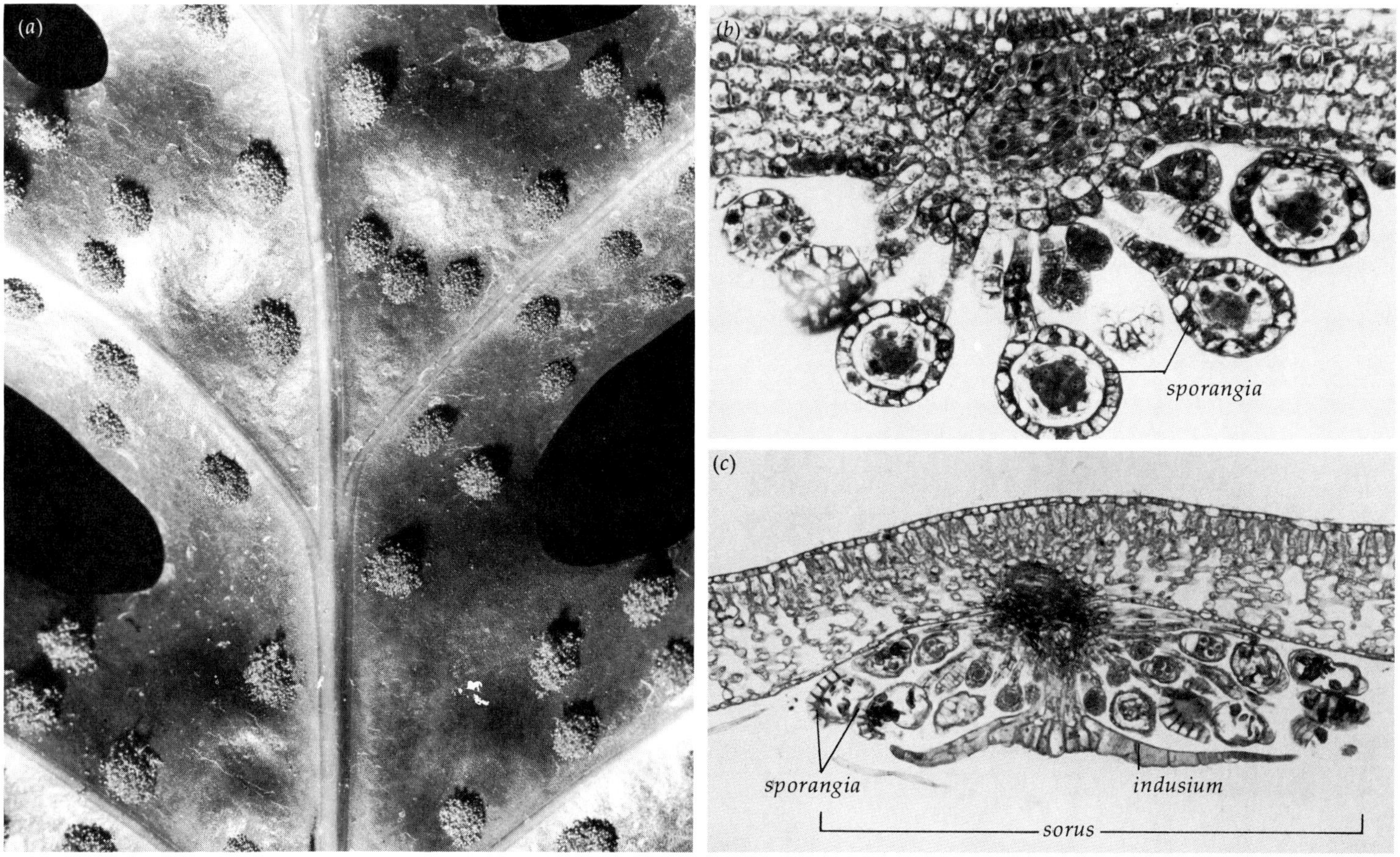

26–8 On the lower surface of fern leaves, round brown spots, known as sori, can frequently be seen. Upon closer examination these look like clusters of granular objects. When viewed with the microscope, they can be seen to be the sporangia in which the spores are formed. The mature sporangia function as dispersal mechanisms for the spores (as discussed in the text). The shape and distribution of the sori vary from species to species. In some, the sori are covered with an additional structure that resembles an umbrella and is called an indusium. Shown here are the sori on the surface of a leaf (a), a microscopic view of a sorus (b), and a similar view of a sorus covered with an indusium (c).

with two slightly enlarged lip cells in the front (Fig. 26–9). When the sporangium is mature, the cells of the annulus are full of water. The evaporation of this water causes the walls of the annulus cells to be pulled together, as a result of the cohesive forces of the water molecules and the adhesion of the water to the walls. The inward curving of the thickened annulus cell wall causes the sporangium to rupture along a line running through the lip cells. The spores are carried back into the top portion of the sporangium. As water continues to evaporate, a point is reached where the cohesive forces of the water decrease below the elastic force of the thick, radial walls. At this point, the remaining water in the annulus cavitates suddenly, releasing the tension so the walls return to their original position. This is a rapid, vigorous motion that throws the spores

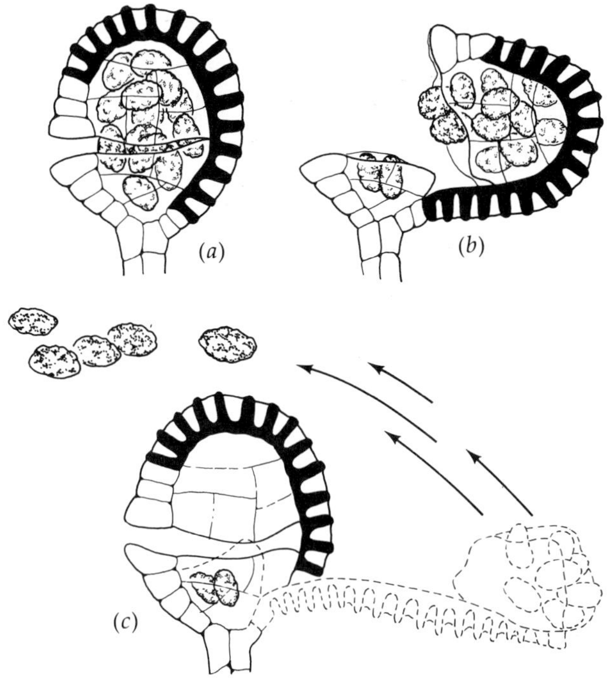

26–9 *The discharge of a sporangium in a fern.*

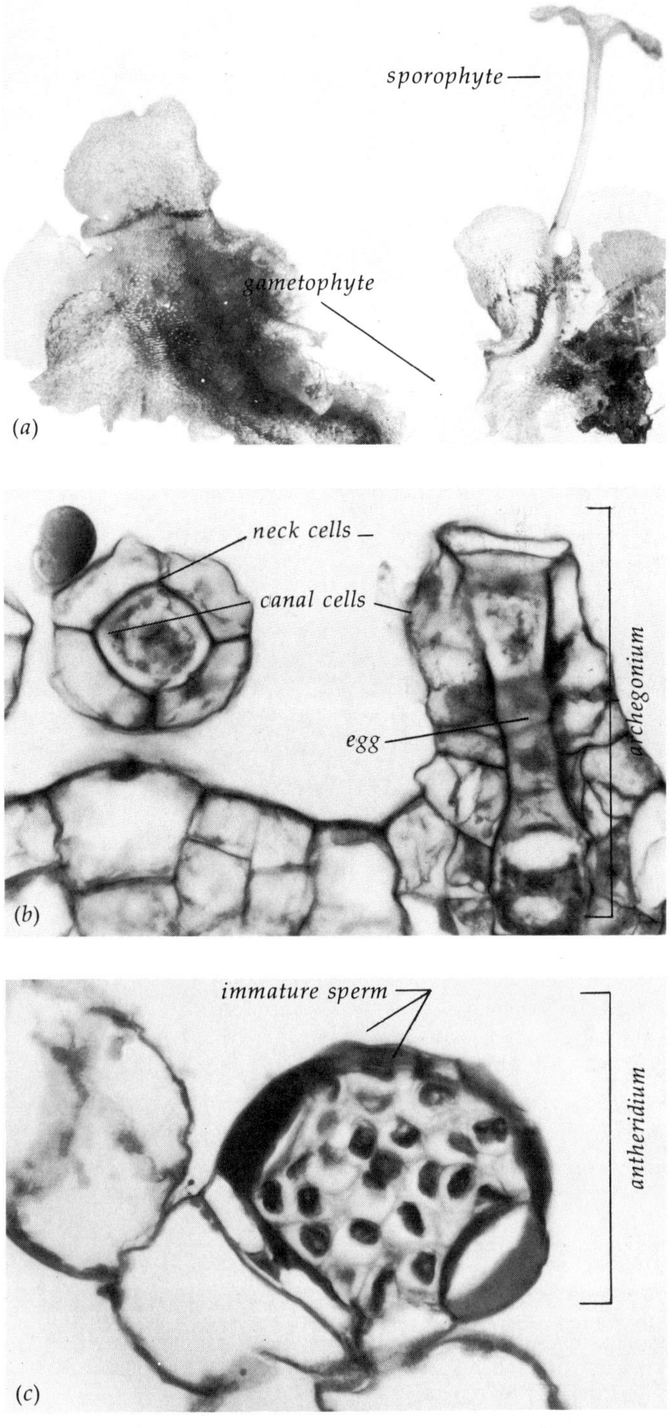

26–10 *The gametophytes and gametangia in ferns: (a) fern gametophytes, one with a young sporophyte growing from it, (b) archegonia showing the arrangement of cells in the neck and the egg at the end of the canal, which will be free of cells when the archegonium is mature, and (c) an antheridium containing immature sperms.*

into the air (Fig. 26–9). The sporangium with its annulus is a biological catapult. This mechanism is a variation on the general theme of wall-water interaction systems found in the fungi and the bryophytes. Such elaborate spore discharge methods are features only of the more recently evolved ferns.

The spores are light and can be carried by the wind for miles. An interesting adaptation to life on islands

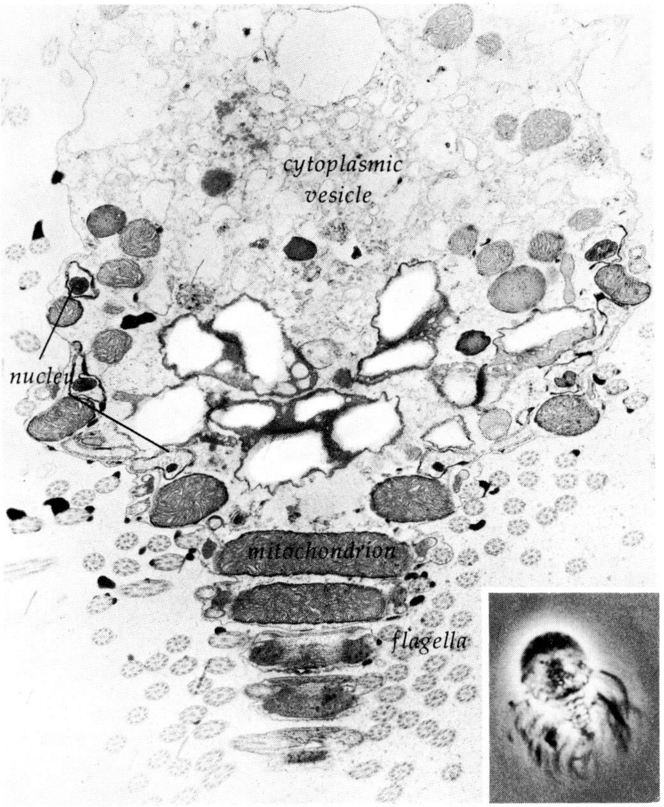

26–11 The structure of the sperm in ferns. The small photo is a light microscope view of a live sperm, while the large photo is an electron microscope view of a similar sperm. One long mitochondrion is coiled around the base of the sperm, where many flagella are attached. The nucleus is elongated and inserted between the coils of the mitochondrion. The entire cytoplasm of the sperm, except for the long coiled mitochondrion, the structures associated with the flagella, and the nucleus, are shed before the sperm enters the archegonium.

is found in the spore size of certain ferns. Those species found inhabiting the island have larger and heavier spores than species of the same genera growing on the mainland. Clearly, the heavier spores are less likely to be blown off the island into the ocean and therefore represent an advantage to those individuals. Most ferns are homosporic, producing but one type of spore; only a few are heterosporic.

Spore germination and gametophyte development. The spores germinate readily on damp soil. The reversible red-far-red system, known to involve phytochrome in higher vascular plants (see Chap. 35), controls spore germination in some ferns. Red light in these ferns induces germination, but in some ferns germination occurs in all wavelengths of light.

When the spores germinate, they divide in only one dimension to produce a filament of cells. This phase may persist for varying periods of time, but eventually the apical cell divides parallel, rather than perpendicular, to the direction of growth. This division begins the period of two-dimensional growth. Later still, three-dimensional growth results in the formation of the mature gametophyte.

Blue light is necessary for the transition from one-dimensional to two-dimensional growth to occur. As yet no chemical has been found to replace the requirement for blue light. There are known to be changes in RNA synthesis associated with the response to blue light. This has led to the suggestion that the light effect may be linked to gene repression. Other workers have felt that blue light affects the interaction between the rate of cell division and the rate of cell elongation. Under blue light the cell division is more rapid, and a change in the orientation of the cell plate results. There is a considerable body of evidence to support the cell division–cell elongation theory, but the exact mechanism that involves blue light is unknown.

The prothallus that develops eventually from the spore may have many forms: filamentous, tuberous, or platelike. The latter is commonly seen in laboratories and has a thickened midrib, an apical notch, two flattened lateral wings of tissue (giving the entire plant a heart shape), and unicellular rhizoids. Antheridia and archegonia are formed on the lower surface of the prothallus—the antheridia in the older parts of the prothallus and the archegonia in the younger portions near the apical notch (Fig. 26–10). The morphologies of the antheridia and archegonia are similar to those found in the lower vascular plants. The sperms are multiflagellate and motile (Fig. 26–11). They are attracted to the archegonium by a chemotactic substance, probably produced by its egg (Fig. 26–12).

Experimental evidence indicates that the substance is the malate ion. The neck of the archegonium is filled with a mucilaginous material from the breakdown of the ventral and neck-canal cells. This material contains the malic acid salts, and it is through

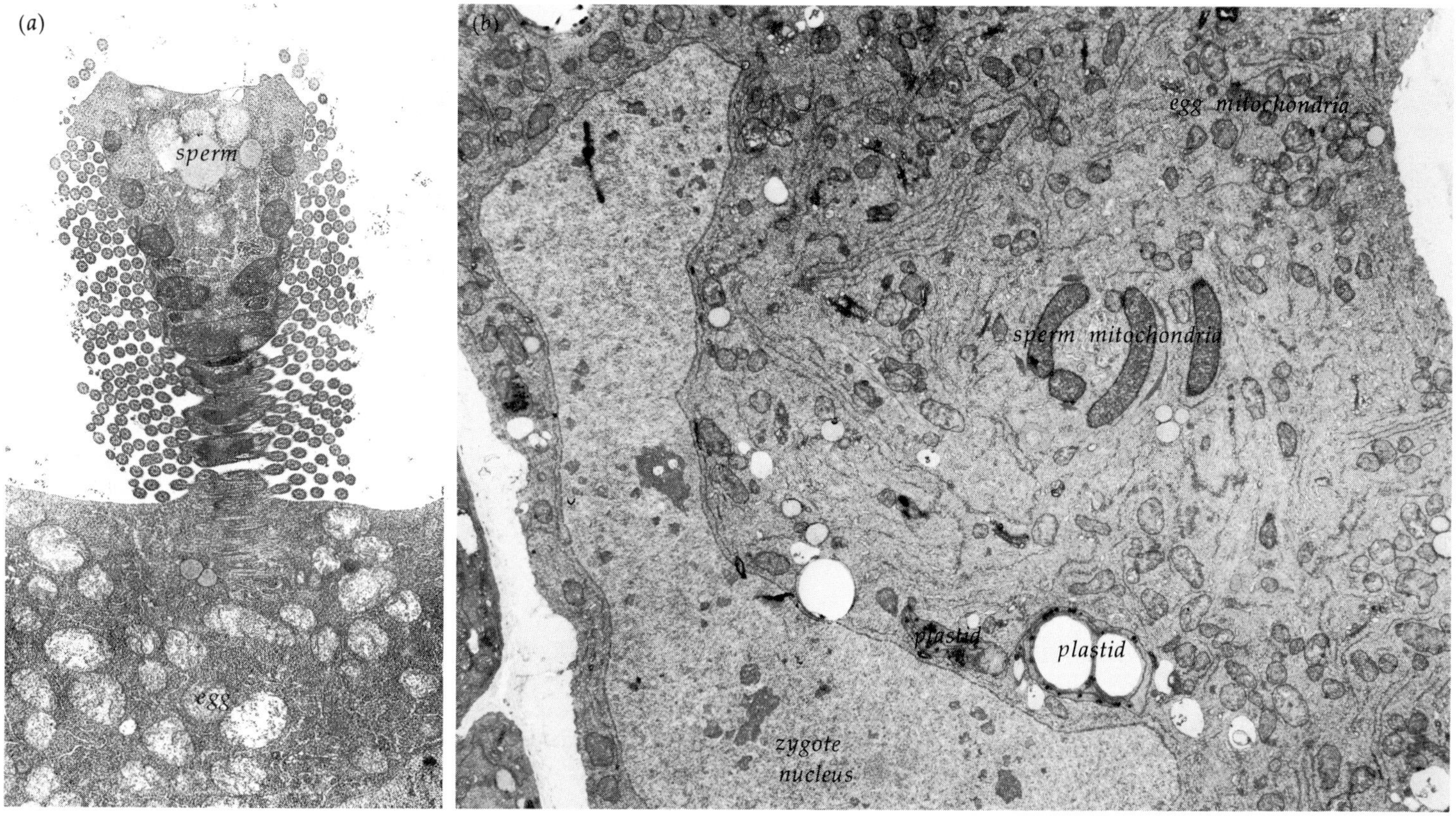

26–12 Fertilization in ferns: (a) a sperm penetrates the egg; (b) the zygote with the remains of the sperm cytoplasm is visible within. In contrast to the sperm, the egg cell is filled with various types of cell organelles.

this that the sperms swim. Several sperms may reach the egg, but only one will enter. The entering sperm penetrates the egg, and the remains of the many flagella and the mitochondrion present in the sperm can be seen in the egg cytoplasm. The nucleus and cytoplasm of the egg undergo a series of elaborate changes both before and after fertilization, the meaning of which is not clear.

The zygote divides within 24 to 48 hr. The first division is unequal and sets up a pattern of development that results in an embryo with a foot as well as root, shoot, and leaf primordia (Fig. 26–13). The foot is involved in the absorption of materials from the gametophyte during the early development of the sporophyte. Usually, only one sporophyte is formed on a gametophyte. The leaves of the young sporophyte are frequently different in morphology from those of the adult; they are generally less dissected than the mature leaves.

The ecology of the ferns has been said to be the ecology of the gametophyte, but this is only partially true. The statement stresses the fact that many fern gametophytes are delicate structures that require a fair supply of water and cannot readily survive desiccation. But there are many fern gametophytes that are tough plants capable of surviving through the winter. Moreover, the gametophytes of some species survive vegetatively, reproducing by gemmae, and

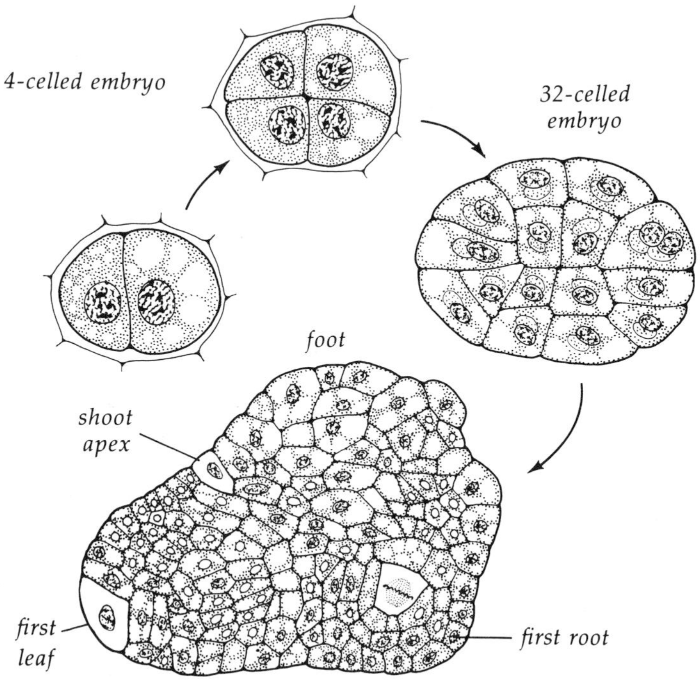

26–13 *The developmental stages in the growth of a fern embryo.*

there are ferns that lack sporophytes. The statement also minimizes the fact that the more ephemeral gametophytes are produced during the season most favorable for their growth and survival. Besides, ferns are perennial plants, and if the gametophytes are lost one year, the sporophyte survives to produce more spores the next year. Finally, there is the importance of vegetative reproduction in fern ecology. The underground stems are most effective in spreading the plants, and many ferns, although producing millions of spores, are distributed by the growth of the rhizomes. The leaves of some ferns, as noted earlier, are also capable of developing new plantlets.

End Paper 26–1 The Evolution of Ferns

Ferns are the only major plants of the present-day flora that clearly extend back to the Devonian. As we pointed out earlier, not all of the ferns currently growing around us had their origin in that distant time. Many are much younger—an estimated 90% evolved during the Cretaceous—but the remainder are quite ancient species.

What then is the origin of the ferns and how did the present species evolve? There are numerous possible ancestors, but none is certain. To answer this question, botanists have turned to both the fossil record and the comparative studies of living species.

Fossils have been found of what collectively can be called the preferns from Middle and Upper Devonian times. They show possible steps in the evolution of the frond, but whether such plants present ancestors to the modern ferns remains unclear. The general feeling among paleobotanists is that the early preferns evolved from a line of early psilopsids in the Devonian.

The similarity between some of the most primitive ferns and the present-day *Psilotums* is striking. David W. Bierhorst, at the University of Massachusetts, has made a number of intensive studies of several primitive ferns native to New Caledonia, an island in the South Pacific noted for the richness and uniqueness of its vegetation, including many rare and primitive species of plants. Bierhorst's careful work has revealed that these ferns have many characteristics similar to those found in living psilopsids. These include gametophytes that are not flat, green structures growing on the surface of the soil but cylindrical and subterranean, lacking chlorophyll but having fungi associated with them. This makes the gametophyte similar to that of *Psilotum*. An analysis of many features of these ferns and the living psilopsids has led Bierhorst to conclude that the living psilopsids belong in the Pteropsida (in the Filicales) and not in the Psilopsida. Although this concept has much data in its favor, particularly in stressing the possible relationship between the Pteropsida and the Psilopsida, enough questions and differences between the two remain that we have retained the conservative view and have not treated *Psilotum* as a fern.

Comparative morphological research, of which Bierhorst's is a good example, has played an important role in understanding the evolutionary relationships in the ferns. This method was emphasized by the celebrated English morphologist, F. O. Bower, who realized the importance of analyzing, as far as possible, the total morphological features of the ferns in attempting to analyze their phylogeny. He recognized as being taxonomically important such features as the external form and habit of the plant, the organization of the apical meristem, the nature of the vascular system of the shoot, the development and mature structure of the sporangium including the form and marking of the spores, and many other characters. After analyzing many species he drew up a list of what were primitive characters as opposed to more advanced characters. For example, in the largest order of ferns, the Filicales, Bower considered that slender, creeping rhizomes with protosteles are primitive, while larger, erect stems with dictyosteles are advanced.

In examining a group of ferns, the comparative morphologist will look at all of the features of the plant available to him and then attempt to rank the species, genera, or families on the basis of the state of the various characters. The primitive species will most closely resemble the ancestors of the ferns being considered, while the most advanced species will show the greatest divergence from the proposed progenitor. No one species in such a series would be expected to show all primitive or all advanced characters, and both primitive and advanced forms of different characters may exist in the species. Clearly, selection pressures acting on the gene pool will result in the differential evolution of certain characters in relation to the specific environment in which the population is growing. This may result, for example, in a fern with an advanced vegetative structure and a primitive spore discharge mechanism. One of the problems the morphologist faces is in determining whether a character is truly primitive or whether it appears primitive through evolutionary reduction.

In general, the ferns we have discussed in this chapter are advanced ferns of the order Filicales, sometimes called the true ferns. This is the largest order of living ferns, with 300 genera and about 9000 species. These have been grouped into from 9 to 33 families, depending on the viewpoint of the particular taxonomist involved. They constitute a complex group of plants, many highly evolved to special environments, making it difficult in numerous cases to determine natural affinities between them. The ferns that Bierhorst is examining are at one end of this large order, while the tree ferns and filmy ferns are found in other families of the same order. *Pteridium*, one of the most successful ferns in its ability to compete with flowering plants, is found in the family Pteridaceae in this order.

There are also two orders of water ferns, the Marsileales and the Salviniales. These are advanced ferns with a relatively short fossil record. *Salvinia* and *Azolla*, discussed earlier, are members of the Salviniales. These are floating ferns, while members of the Marsileales are water ferns with rhizomes that creep over the bottom of ponds or lakes. *Marsilea*, the commonest of the genera in both temperate and tropical climates, has leaves with long petioles so that the blades, which resemble a four-leaved clover, float on or near the surface of the water. *Marsilea*, as are all of the water ferns, is heterosporous and both forms microspores and megaspores.

The orders just discussed are all leptosporangiate ferns. There are two orders of living eusporangiate ferns, the Marattiales and Ophioglossales. The Marattiales have a long fossil record and were important in the flora of the Carboniferous. Intermediate between the eusporangiate and leptosporangiate ferns is the family Osmundaceae. These ferns are usually placed in the Filicales, although they have also been given an order of their own by some taxonomists. The Osmundaceae contain many common ferns including the cinnamon fern and the regal fern, but the total number of species is not large. Their remains are common among the fossils of the Mesozoic, but they appear to have declined since then. The Osmundaceae are not so much a missing link as an ancient side branch of evolution and may be viewed as living fossils.

The ferns are an ancient and interesting group of plants that, although intensively studied in the past, still contain many areas in which research is needed. On the whole, the ferns have survived well under man and appear to possess a gene pool capable of meeting further changes in the environment. Many fern species, however, have limited ranges and are being threatened by domestic animals and man. The extensive draining of marshes, ponds, and other sources of shallow water in the United States threatens many of the water ferns, particularly *Marsilea*. Furthermore, many islands of the Pacific, representing extremely old and isolated blocks of land, abound in species of plants not found elsewhere that are of great interest to biologists. One of the reasons that many of these ancient plants have survived on these islands is believed to be the lack of animals. On New Caledonia, for example, before man there were no mammals (except for a single species of bat), no marsupials, almost no birds, and few insects. Since man came, the rat, deer, and a few other animals have been introduced either intentionally or not. They have had relatively little effect on New Caledonia, but on other Pacific islands grazing and rooting animals such as pigs have eliminated many of the native plant species.

27

The Gymnosperms

Remembrance of Things Past

Cycads are ancient plants that have survived from the great age of the dinosaurs. With their long, palmlike leaves and their rough-textured stems, they are both graceful and ruggedly handsome. Although their leaves appear palmlike, their cones proclaim the fact that they are gymnosperms.

A This striking group of cycads illustrates both the general appearance of the group and the way in which they are used in landscape gardening.

The gymnosperms are dominated by the conifers—the evergreen, needleleaved trees with which we are all familiar. But these are only a part of the group in which there is wide variation in the form of the stem, leaf, and cone.

B Various species of pines are found in all parts of the country, covering many acres in the West and South. This is a limber pine growing in the Rocky Mountains.

A

B

C The characteristic reproductive structure in the conifers is the cone. Two types of cones are formed: one produces pollen, and the other produces the eggs and, later, the seeds. The pollen cones are usually small and persist for only a short time. They are initiated immediately behind the apical meristem, and the next years needles form in front of them. At first they are small, tight domes, but as they grow, they elongate and finally open to release the pollen (top). The amount of pollen released is enormous, as demonstrated when a branch bearing many cones is vigorously shaken (bottom).

C

D *In addition to the cycads and the conifers, the gymnosperms contain a number of groups of plants that bear reproductive similarities to the other members but that differ markedly in their morphology.*

E Ginkgo biloba, *like the dawn redwood, is a living fossil. The graceful branches and the beautiful leaves together with a remarkable resistance to smog have made* Ginkgo *a common city tree. Illustrated here is a leaf from a living tree, and a 17-million years old fossil of an almost identical leaf.*

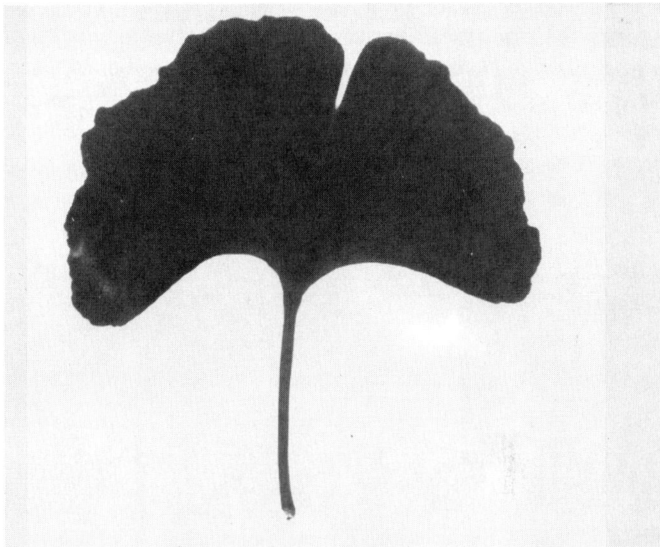

D

E

F Another of this collection of unusual gymnosperms is Ephedra. *There are some 35 species of this plant, and several of them are common to the western deserts. One such species is shown here growing in the desert, while a closer view shows the greatly reduced leaves and green stems with modified cones.*

F

G Finally, the most unusual of the group is Welwitschia, *a single species of which exists only in the Kaokoveld of South Africa. This plant grows two long, ribbon-shaped leaves that are continually produced by the apex. The leaves may split into segments and the ends become worn away. It has a long root that grows deep into the desert soil and, in the case of the greenhouse plant shown here, requires three pots to hold even a young plant.*

G

Paleobotany is one of the most interesting and most frustrating areas of botanical research. The finding of well-preserved specimens involves a high degree of luck, and then the data are, at best, fragmentary. The element of luck can be exaggerated since the professional paleobotanist is an expert in the detection of fossil beds and alert to the nature of the evidence he can expect to find. Yet the method of fossilization of plant material presents problems that may be all but impossible for him to overcome. When he does, the results may be far reaching. An example of this is the discovery of the progymnosperms by Charles Beck of the University of Michigan.

The Progymnosperms

The progymnosperm story must start with an understanding of the way fossils are classified. For example, the paleobotanist finds a number of essentially identical leaf fossils; he also finds a group of stem fossils at the same location. Do the stems and leaves belong to the same species? The investigator does not know, unless one is joined to the other in a fossil. In the absence of such definite proof, what is he to do in attempting to classify them? He does the only thing he can: he assigns a species name to the leaves and another to the stems. Grouping stem and leaf fossils with obvious similarities, he arrives at leaf genera and stem genera. Eventually, he may be able to collect enough data so that he can combine the leaf and stem species or, best of all, he can find a fossil in which the leaf is still attached to the stem. This is what happened to Beck (see End Paper 27–1).

He was able to show that the leaves of a genus called *Archaeopteris* were part of the same plant as a genus based on stem and root fossils, called *Callixylon*. The importance of this association is based on the fact that *Callixylon*, with a stem that showed secondary growth, had long been assumed to be related to the gymnosperms, while *Archaeopteris* was known to reproduce by spores. Thus a different type of plant was established by the fossil found by Beck, who has argued that these plants now called *Archaeopteris* are progymnosperms. The concept that the progymnosperms existed and were important in the late Devonian and

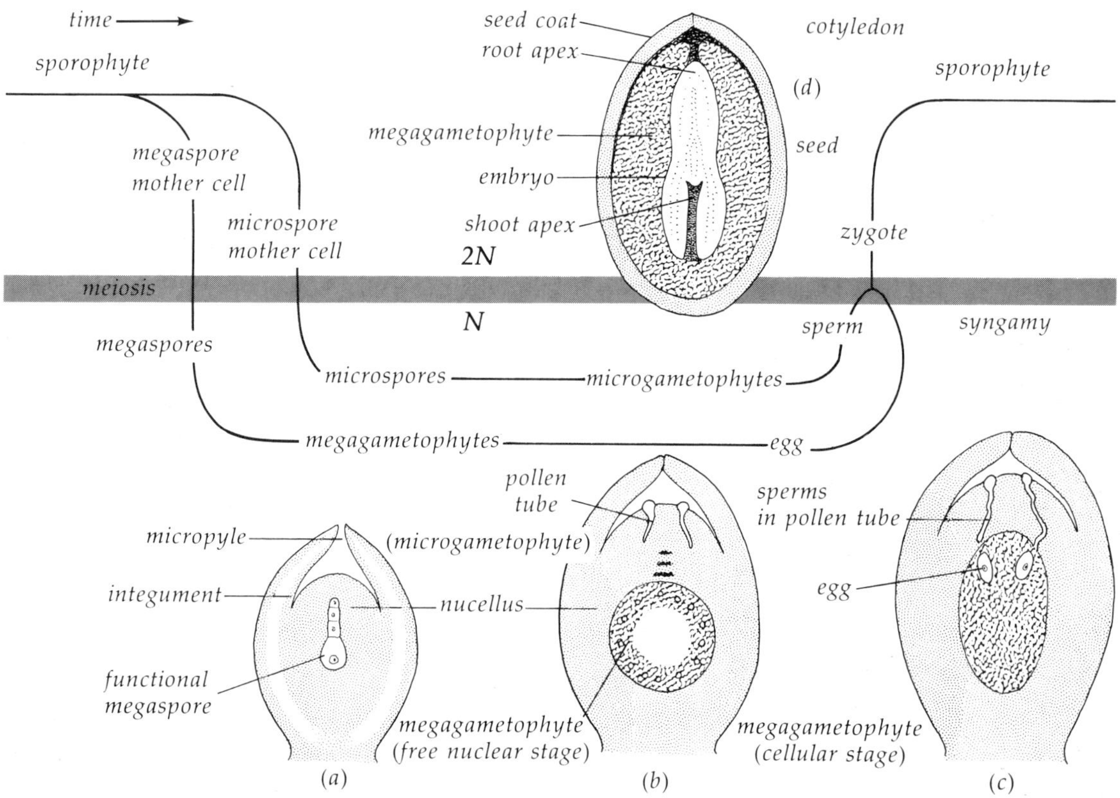

27–1 Stages in the development of the megagametophyte and the seed of the gymnosperms.

Carboniferous forests is now generally accepted and has added a new dimension to our understanding of the past.

The progymnosperms were plants that reproduced by spores, like the lower vascular plants and ferns, but produced secondary xylem, like the gymnosperms. The fact that the stems had secondary xylem or wood means that a well-defined vascular cambium was present. The presence of vascular cambium has profound implications in plant growth and development. Although elaborate, the growth of the tree lycopsids and sphenopsids was limited; it was finite, as if a pattern or plan unfolded as the plant grew and then it ended. But in the gymnosperms growth is continuous both in height and diameter. This means that the plants can compete successfully with the lycopsids, sphenopsids, and other large plants, even surviving conditions the others could not.

The progymnosperms form the stock from which many of the present gymnosperms appear to evolve. Whether all of the living and extinct gymnosperms have developed from them remains an open question and will be discussed later in regard to the various groups.

All living gymnosperms, as did all extinct gymnosperms (except the progymnosperms), produce seeds. The seed is a great advance over plants that do not have them. Combined with seed production is the modification of the reproductive cycle that eliminates dependence on an independent and vulnerable gametophytic generation.

Reproduction in the Gymnosperms

The gymnosperms, like the ferns and other vascular plants, produce spores through meiosis, but these are heterospores and are not released (Fig. 27–1). In the largest groups of the living gymnosperms, the cycads and the conifers, reproduction involves cones (Fig. 27–2). These are modified branches with a variety of sizes and shapes.

27–2 The pollen and seed cones of a cycad. The male cone on the left is mature, and the pollen has been released. But the seed cone or female cone on the right is still immature.

The male cones consist of a collection of microsporophylls on which the microsporangia develop. Within these, microspore mother cells or microsporocytes differentiate and undergo meiosis to form 4 haploid microspores. The nucleus of the microspore divides so that a small male gametophyte of only 2 cells is produced. This micro or male gametophyte is the pollen.

The female cones also consist of a group of megasporophylls on which develop the megasporangia. A megasporangium consists of a megaspore mother cell or megasporocyte surrounded by a mass of sporophyte tissue called the nucellus that, in turn, is enveloped by another layer of tissue, the integument. There is an opening in the integument called the micropyle. The megaspore mother cell undergoes meiosis to form 4 haploid megaspores; 3 of these degenerate, leaving a single functional megaspore. The nucleus of the megaspore divides mitotically to produce a multinucleate megagametophyte. Cell walls eventually form, and several archegonia differentiate, each consisting of an enlarged egg cell and 2 or 4 neck cells.

The pollen is carried by the wind to the female cone, where it filters down between the megasporophylls to the micropyle. There is frequently a drop of sticky fluid, the pollination drop, to which the pollen adheres. As this drop dries it pulls the pollen to the surface of the nucellus. The pollen germinates and produces a tube that grows slowly through the nucellus, drawing nutrients from the female tissue. During this development the cells within the pollen tube divide, and eventually 2 (occasionally more) sperms are formed. In the cycads and ginkgo these sperms are multiflagellated, while in the conifers they lack flagella and are nonmotile. In the case of the cycads and ginkgo, the pollen tube discharges into a cavity that forms immediately above the archegonia. The sperms swim to the eggs and fuse with them. In the case of the conifers the pollen tube grows directly to the egg and discharges the sperms into it. One of the sperm nuclei fuses with the egg nucleus, and the zygote is formed. Since several eggs can differentiate in each megagametophyte, more than one embryo is frequently formed. There is competition between the developing embryos, and usually only one survives in the mature seed.

The seed in the gymnosperms is composed of a resistant outer seed coat formed by the female sporophyte. This surrounds the remains of the female gametophyte, which may be extensive. The female gametophyte in turn encompasses the new sporophyte, the embryo.

The adaptive significance of this life cycle to life on land is tremendous. Consider just three major points.

The gametophytes now draw on the resources of the mature sporophytes to live. The gametophytes do not require preferred environmental conditions for growth. Wherever the sporophyte can survive, they can too.

27–3 *For many years the dawn redwood,* Metasequoia glyptostroides, *was known only from the fossil record. Then, it was found growing in China. Now we can compare the fossil leaves and cones with the living material—a dramatic illustration of the long continuity of the species.*

Water is no longer necessary for fertilization. Even where motile sperms are found, they swim in a special chamber in fluids supplied by the sporophyte, not in free rain or ground water.

The major reproductive agent is now the seed—a most remarkable structure. The seed contains not only a small edition of the sporophyte ready to grow but a built-in supply of food and a protective outer covering to minimize adverse environmental conditions.

The gymnosperms are completely adapted to life on land, but how successful have they been? Very! They have formed a significant part of the flora since the Carboniferous—over 345 million years ago. They have survived two periods of massive plant and animal extinctions, the Permian and the Cretaceous, and are still a major component of the vegetation around us. They have undoubtedly lost ground to the angiosperms, but they still cover millions and millions of acres of land. Let us now examine some of the individual groups.

The Cycads

When the dinosaurs ruled the earth in the Triassic and Cretaceous periods of the Mesozoic era, the cycads were worldwide in distribution, and there are good reasons to believe that the young leaves and seeds of the cycads provided food for some dinosaurs. The cycads and the dinosaurs reached their maximum development at the same time, and although the cycads did not become totally extinct, both decreased in importance in the Cenozoic era. The living cycads number only 9 genera and roughly 100 species, all confined to the subtropics and tropics.

The cycads present a striking appearance (Fig. 27–A). They have large, attractive, palmlike leaves that may be 1 to 2 m long and stems that are usually barrel-shaped but may, in some species, reach a height of 10 to 20 m. Cycads grow extremely slowly; one plant, some 2 m tall, was calculated to be at least 1000 years old! The stems of all cycads are normally unbranched. There are extensive areas of cortical tissue and pith in the stem, and secondary growth is not pronounced.

The reproductive cycle in the cycads closely follows the general pattern discussed earlier for gymnosperms. The female and male cones are borne on separate plants. In one genus, *Cycas*, the cone is loosely organized, but in all other cycads a more tightly organized cone is formed (Fig. 27–2). The size of the female cone in some species is tremendous, weighing 25 to 35 kg. The male cones are smaller but still respectable in size. Some of the details of cycad reproduction are shown in Figure 27–4. The sperms are multiflagellate and large—some can be seen with the naked eye. There is usually a period of 3 to 4 months between pollination and fertilization, and it may take a year for the seed to form.

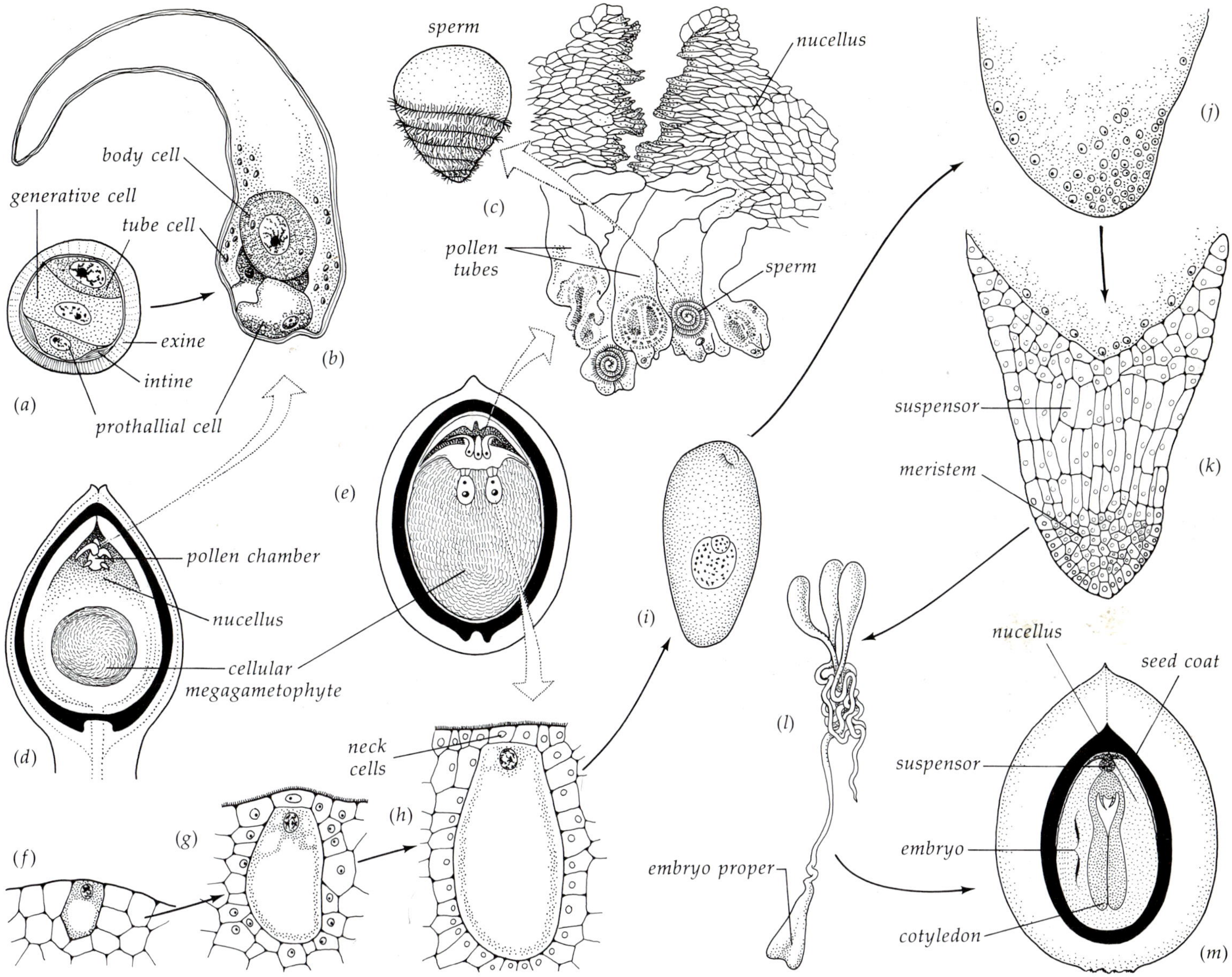

27–4 *The details of reproduction in the cycads. At the time it is released from the male cone, the microgametophyte consists of 3 cells (a). Within the female cone it develops into a tube, and the generative cell divides to form a stalk cell and a body cell (b). The body cell in turn divides to form 2 multiflagellated sperms (c). While this is occurring, the megaspore divides to form a free nuclear megagametophyte that, in turn, becomes converted to the cellular megagametophyte (d). The eggs are differentiated within the cellular megagametophyte through cell enlargement and division so that a large egg cell, surmounted by two small neck cells, finally results (e, f, g, & h). Syngamy occurs after the sperm swims to the egg, following the breakdown of the neck cells. The entire sperm apparently penetrates the egg. The remains of the flagella can be seen at one end of the cell (i), while the egg and sperm nuclei fuse near the center. Following the formation of the zygote, the embryo first develops through numerous free nuclear divisions (j), and only later do cell walls form (k). A layer of elongate cells differentiates at this time to form the suspensor (k). The enlargement of these cells forces the young embryo into the megagametophyte (l). Note in (l) that 3 eggs were fertilized and began to develop, but only one is successful. By the time the seed is mature, the suspensor has degenerated. The mature seed (m) consists of the embryo, surrounded by the remains of the megagametophyte, and a seed coat, formed from sporophyte tissue.*

27–5 The reconstruction of a pteriodosperm or seed fern. This species reaches heights of 3 to 4 m.

27–6 The reconstructions of extinct Bennettitales.

Early embryo development in the cycads is characterized by a long series of nuclear divisions without wall synthesis. Several hundred nuclei may be formed before cell walls are produced. Walls first form at what was the lower end of the zygote, while the upper end elongates, pushing the developing embryo into the nutritive megagametophyte tissue. Several embryos may begin to develop but only one survives.

The Pteriodosperms and Bennettitales

Before turning to the conifers, let us look briefly at two groups of extinct gymnosperms, the Pteriodosperms or seed ferns and the Bennettitales.

The seed ferns (Fig. 27–5) had large frondlike leaves that bore both seeds and pollen-producing organs. Before 1904 the many fernlike leaves found in Carboniferous rocks were believed to be true ferns. Then, two English paleobotanists, Daniel Oliver and Henry Scott, were able to prove that seeds were attached to the fernlike leaves, thus turning the "Age of the Ferns" into the "Age of the Seed Ferns." The seeds bore similarities to those of the modern cycads. The plants were not large —the tallest stood 5 m and others were considerably smaller. They were abundant during the Carboniferous and became extinct during the Cretaceous.

The Bennettitales are an extinct order of the cycads. They resembled the modern cycads in form and size. Like the cycads they evolved in the late Carboniferous or Permian, presumably from the seed ferns. Reaching the peak of their development in the Jurassic, they became extinct in the Cretaceous. Their general pattern of evolution closely parallels that of the dinosaurs.

The plants were either barrel-like or possessed relatively short, columnlike trunks (Fig. 27–6). The leaves were similar to those of the modern cycads. The striking feature of the Bennettitales is the presence of a flowerlike reproductive structure. But fossil evidence makes it clear that they were not true flowers; the seed, for example, is never enclosed in an ovary. The Bennettitales were once thought to be the ancestors of the angiosperms, primarily because of the form of the seed-bearing structure in some of the advanced members of the group. Additional evidence has shown that the resemblance is not that close, and the theory has generally been abandoned. There is no evidence why the Bennettitales became extinct or why some cycads have survived.

Essay 27–1 Two Lines or Three?

A frequent question of both botanists and students is how a group of plants interrelates in terms of evolution. Or who gave rise to whom? In the case of the gymnosperms we have a wide range of characters in a large number of groups, both living and extinct. Two ways in which these can be postulated as being related are shown in the accompanying diagram.

The two schemes are quite similar. Only one line is basically different—the one running to the seed ferns. In scheme 1 the seed ferns are indicated as originating from the progymnosperms, while in scheme 2 they are shown as coming from the preferns. At first this may seem an almost trivial point, but it is not. Scheme 1 says that the fernlike leaf evolved separately in the ferns and the seed ferns and that the concept of the gymnosperm as a natural group is valid. Scheme 2 says that the type of reproduction shown in the gymnosperms evolved independently several times and that it is better to speak of cycadophytes and coniferophytes in place of gymnosperms. At present the data are simply not adequate to say which one is right, but the discussions and arguments go on.

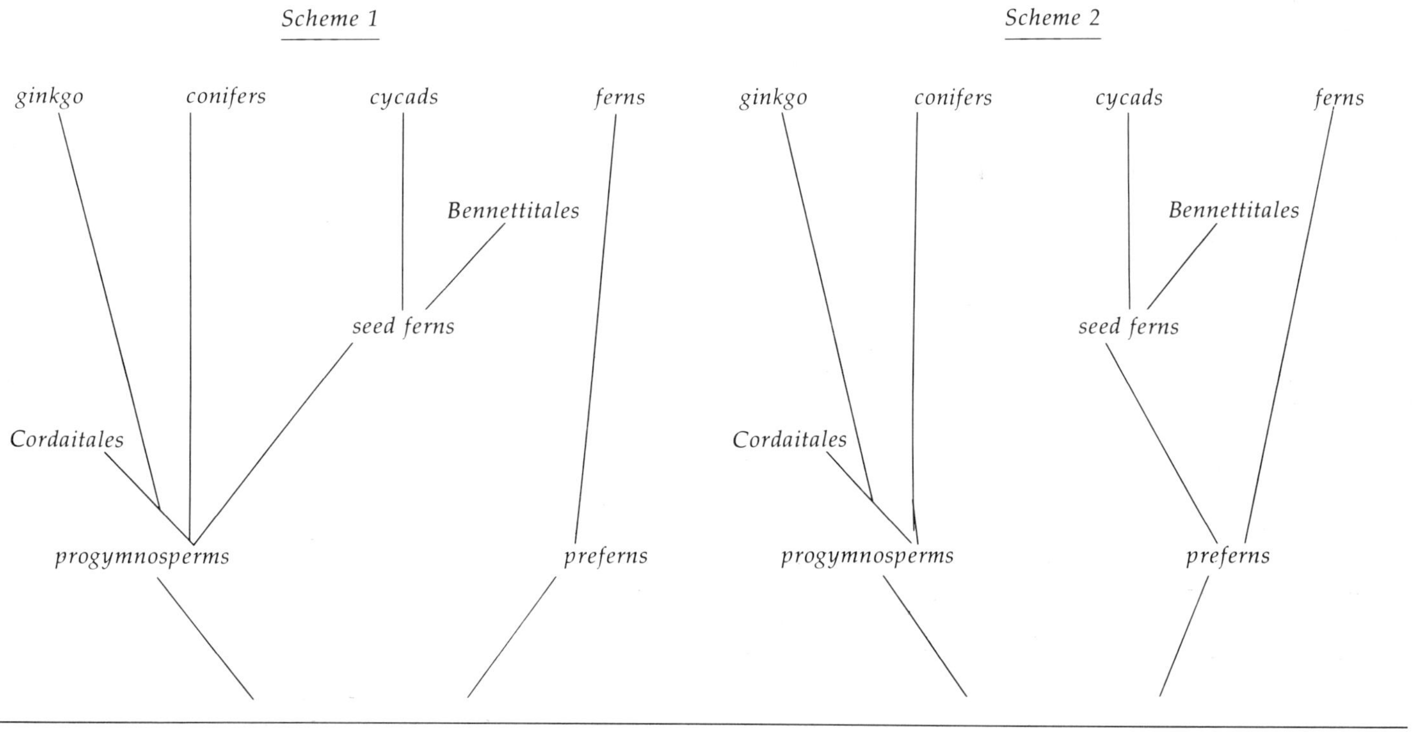

27–7 *The main morphological characteristics and distribution of conifers*

Family (no. of genera)	Range	General characteristics
Pinaceae (10)	Almost entirely Northern Hemisphere	Leaves needlelike; leaves and cone scales spirally arranged; bract and ovuliferous scale distinct; pollen grains mostly saccate.
Taxodiaceae (10)	China, Japan, Formosa, Tasmania, California, southern United States, and Mexico	Leaves and cone scales spirally arranged; bract and ovuliferous scale almost completely fused; pollen small with a papilla.
Cupressaceae (16)	Widespread in both hemispheres	Leaves and cone scales opposite; bract and ovuliferous scale strongly fused; pollen small with a small pore.
Araucariaceae (2)	Almost completely Southern Hemisphere	Leaves and cone scales spirally arranged; bract and ovuliferous scale completely fused; ovules solitary; pollen large and nonsaccate.
Podocarpaceae (7)	Mostly Southern Hemisphere, Central America, and West Indies	Leaves flat and broad with either a single vein or many veins; ovules terminal and single; pollen saccate.
Taxaceae (5)	Mostly Northern Hemisphere	Leaves flat and pointed; spiral; ovules terminal and solitary, with fleshy aril; pollen round and nonsaccate.
Cephalotaxaceae (1)	China, Japan, tropical Himalayas	Leaves flat and narrow; spiral; cone scales decussate; 2 ovules on ovuliferous scale; pollen circular and nonsaccate.

The Conifers

Vast forests of pine, spruce, fir, cedar, and other conifers stretch for miles across western North America, eastern and central China, parts of Australia and New Zealand, and in the great northern boreal forests that encircle the globe. Smaller stands of conifers are found in all parts of the world. Individual conifers (shrubs and trees) are widely planted as ornamentals, and ancient, individual trees were formerly honored with shrines in China. The conifers are a truly magnificent group of plants (Fig. 27–A). Their evolutionary history is longer than that of the cycads. But the cycads are essentially living fossils, while the conifers comprise some 50 genera and some 500 species. This is only about one-thirteenth the number of species represented by the ferns or one five-hundredth the number of angiosperms. Despite the relatively few species they are of great importance in many present ecosystems and have been important since the late Carboniferous period, 300 million years ago.

The families making up the conifers are listed in Table 27–7. The general range is given as well as their general characteristics. All conifers are shrubs or trees, and most have the single dominant axis so characteristic of the small pines, firs, Douglas firs, and spruces used as Christmas trees. In size they range from modest shrubs to the giant redwoods, 100 m tall and 4000 to 5000 years old. The leaves of all conifers are simple; most are needlelike or scalelike in form. In some, the leaves are much broader and are tough.

All the leaves, despite the small size of many, are megaphylls, as determined by their well-developed

27–8 The range in leaf size and shape in the conifers is greater than it first appears, as illustrated by this collective photograph of representative species.

vascular tissue (Fig. 27–8). The stems all possess a prominent vascular system. Secondary growth, which is found throughout the group, may frequently be extensive. As a rule, the xylem is anatomically uniform, consisting of tracheids and parenchyma cells. Cones are the most common reproductive structure (Fig. 27–D), but these may be highly modified from the typical form and may even be fleshy, as in juniper berries. In all cases, the female cones are larger and more long-lived than the male. In general, the two types of cones are produced on different trees, and all are wind pollinated. Instead of attempting to catalog all the various features of the many conifers, we will concentrate on one of the most important and familiar genera: *Pinus,* the pines.

Pinus. The pines belong to one of the ten genera of the family *Pinaceae.* In general, pines are trees with strong apical dominance. The trunk of a pine is predominantly secondary xylem, composed of regularly arranged tracheids with bordered pits and rays of living parenchyma cells (see Fig. 27–9). The rays run at right angles to the tracheids. In the xylem are resin ducts. These are lined with parenchyma and accumulate resins, compounds containing 20 to 30 carbon atoms per molecule. Resin production is a normal function of many gymnosperms, not only the pines, and is increased when the plant is wounded. The function of resin is believed to be that of a "scab," which prevents insect attack and infections by fungi; large quantities can be and have been produced. Amber is fossil resin, and large deposits are found along the Baltic. Insects trapped in the fresh resin are beautifully preserved in the amber for thousands of years.

The cork cambium, as well as the vascular cambium, is highly active, producing the thick, characteristically patterned barks of the pines. The thick

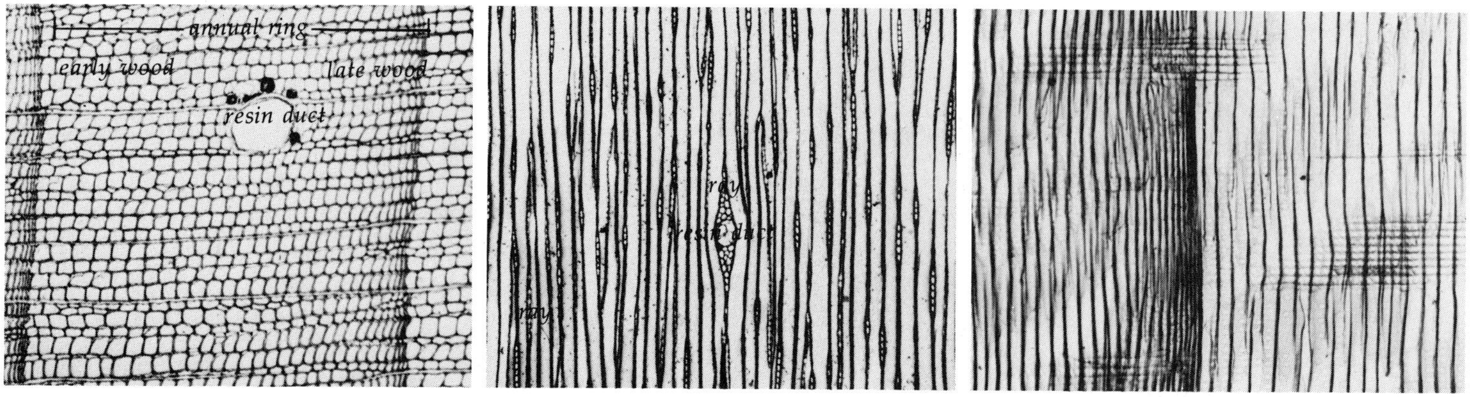

(a) (b) (c)

27–9 *Three views of a segment of secondary xylem in a pine: (a) transverse, (b) tangential, and (c) radial.*

bark is ecologically important in reducing fire damage in old trees, preventing insect, fungal, and bacterial infection, and reducing water loss. The outer cells of the bark have highly suberized walls and contain tannins and other phenolic compounds.

The leaves of the pines are all needlelike in shape. They are produced in groups of one, two, three, or five at the apex of short shoots that grow from the long shoots or branches. The cedar needle, shown in cross-section in Figure 27–10, is a remarkable leaf. One or two vascular bundles occupy the center of the needles. The xylem and phloem are surrounded by the transfusion tissue. This tissue is composed of special transfusion tracheids and parenchyma cells that form a complex, interconnected system in direct contact with the vascular tissue and the surrounding endodermis. The transfusion tissue is presumed to function in the transport of material between the vascular tissue and the mesophyll of the leaf. The mesophyll is composed of rather tightly packed parenchyma cells with incurving walls. Many of the mesophyll cells become filled with tannins. Resin ducts are present in the mesophyll. The epidermis is thick-walled and heavily cutinized and overlies one or more layers of thick-walled hypodermal cells. The guard cells and stomata are sunken. All these characteristics are those of leaves adapted to dry or xerophytic environmental conditions.

The needles of pines live from 2 to 14 years. They are shed by an abscission layer forming at the base of the short shoot that produced them. The needles are shed gradually over the year so that the trees never lose all their leaves at one time. But, as anyone who lives around pines knows, a large number of leaves are shed each year!

The root system in most pines is extensive but shallow. They are thus adapted for growth in rocky, shallow soils but are susceptible to being uprooted by heavy winds. The roots of most pines show a marked growth periodicity, growing in the early spring and summer and entering a dormant period in late summer that lasts through the winter. The root-hair zone is short in pines, and there is frequently an extensive mycorrhizal association found in the younger parts of the root.

The reproductive cycle of the pines closely follows the general pattern given for gymnosperms (Fig. 27–11). Pines are one of those gymnosperms in which both male and female cones are produced by the same tree. The male cones are formed in clusters at the base of terminal buds. The cones are initiated in late fall; meiosis occurs in early spring; and the pollen is shed in the spring. The pollen usually has two small wings or air bladders. Enormous amounts are produced, and in a favorable year the trees may have a distinct yellow color as a result of the released pollen falling on the leaves. After pollen release the small male cones rapidly dry and disintegrate.

The female cones are formed near the ends of the youngest branches and are larger than the male cones. At the time of pollination the female cones are still fairly small, and the scales of the cone are separated. The pollen falls between the scales and is held by a pollination droplet that draws the pollen through the micropyle. Further growth of the cone results in the coming together of the scales, sealing the pollen in-

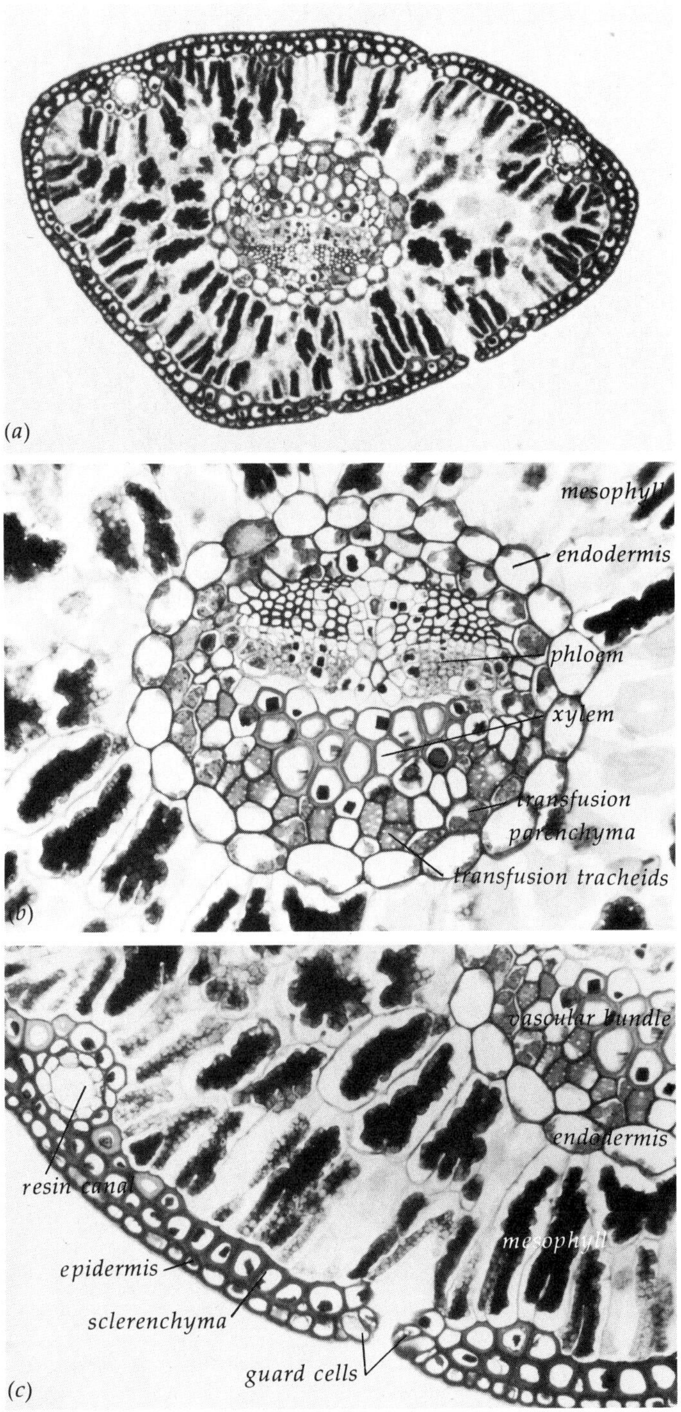

27–10 The needles of the conifers are complex, tightly organized structures, as seen in these photomicrographs of the needle of Cedrus deodora: *(a) a cross section, (b) a vascular bundle, and (c) mesophyll and epidermis.*

side. At the time of pollination the megagametophyte is far from ready for fertilization—it may be only at the megaspore mother cell stage! The pollen tube and the megagametophyte develop slowly over the year, and 12 to 14 months after pollination the pollen tube reaches the megagametophyte. Two nonmotile sperms are formed and discharged directly into the egg. One sperm nucleus fuses with the egg, the second degenerates in the egg cytoplasm.

The first two mitotic divisions result in 4 free nuclei at the end of the zygote away from the micropyle. Later, walls form, and 2 tiers of cells are established. More divisions and elongation result in long, narrow files of cells being formed. These files of cells may separate longitudinally, and each file forms a separate embryo. Polyembryogeny is thus established from a single zygote (as in identical human twins). Only one of the several embryos will survive. The developing embryos are pushed ever deeper into the old megagametophyte tissue by elongation of the suspensor cells (Fig. 27–12).

The mature embryo consists of a radicle, hypocotyl, and as many as 8 needlelike cotyledons. This is surrounded by the remains (often considerable) of the megagametophyte and a tough seed coat produced by the parent sporophyte (Fig. 27–13). The seeds also have a single wing and are usually shed in the spring, some two years after the female cone was initiated. A final period of growth of the female cone results in the separation of the scales and the release of the seed. In some cases, the cones are only opened after being subjected to heat. This results in the release of the seed on a bed of ash left from the fire that opened the cone.

When the seed germinates, the cotyledons first remain within the seed and absorb the food reserves still present in the megagametophyte (Fig. 27–14). Later, the remains of the seed are shed, and the cotyledons unfold and carry on photosynthesis. Pines are one of the plants in which the synthesis of chlorophyll can occur in the dark. But most pine seedlings require rather high light intensities for survival. They do well in clearings, old fields, and burnt-over areas but frequently fail to survive under the canopy of an established forest. Pines are one of the major pioneer species, even tolerating almost bare rock, where the seedlings germinate on ledges and in crevices.

Cordaitales. These are extinct, primitive conifers that evolved from the progymnosperms. They were

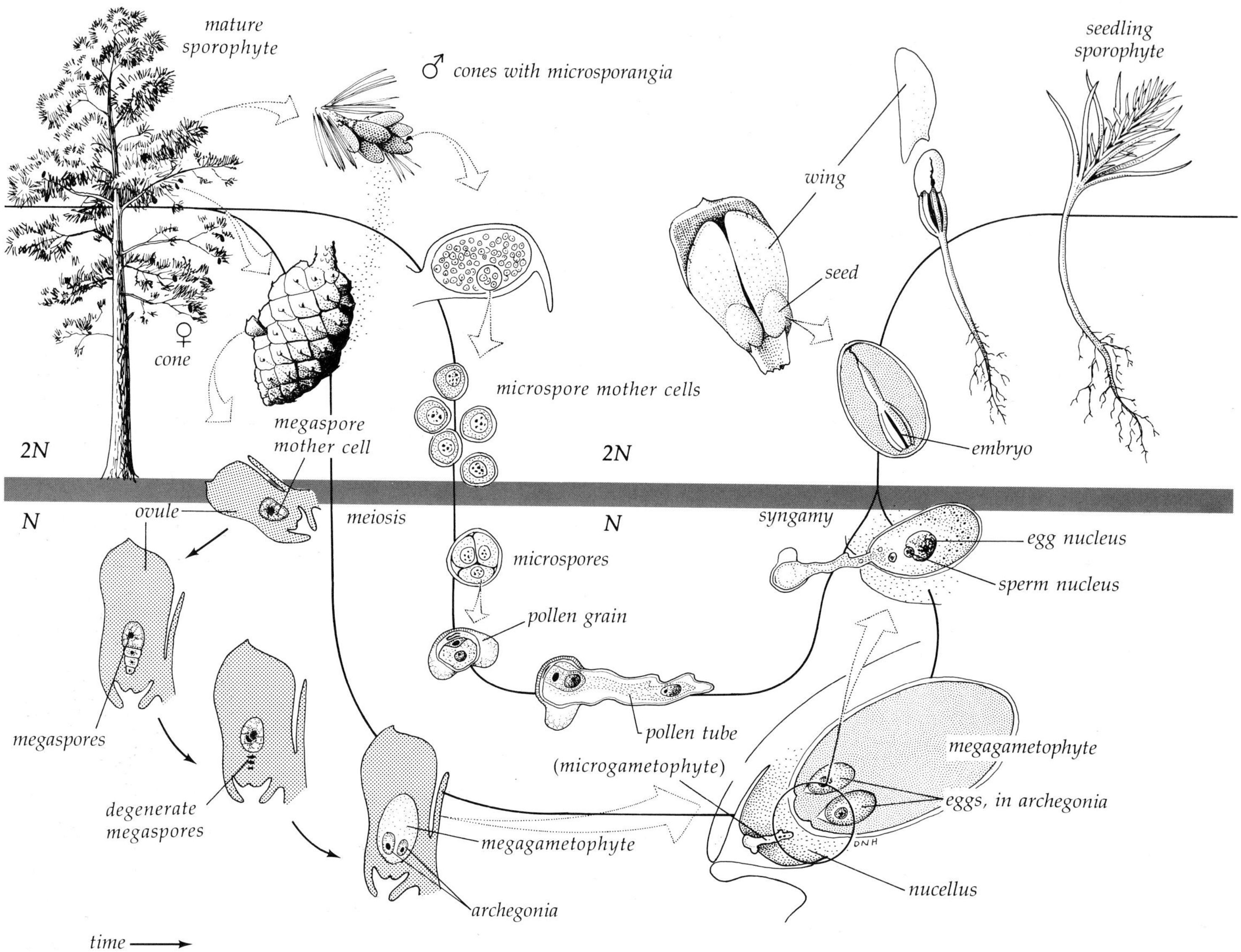

27–11 *The life history of a pine.*

tall, treelike plants standing some 30 m or more in height, thus making them the tallest trees in the Carboniferous forests (Fig. 27–15). Their leaves were thin and strap-shaped, sometimes over 1 m long and 15 cm wide. A well-developed cambium was present, but no Carboniferous specimen shows growth rings, suggesting that seasonal changes in climate were not pronounced. The reproductive structures were loosely organized and cannot properly be called cones. The Cordaitales became extinct during the Triassic. They are of interest primarily as the probable ancestor to both the modern conifers and ginkgo.

The Ginkgo

Ginkgo is a living fossil—a relic plant that may have survived because of man's interest in its majesty

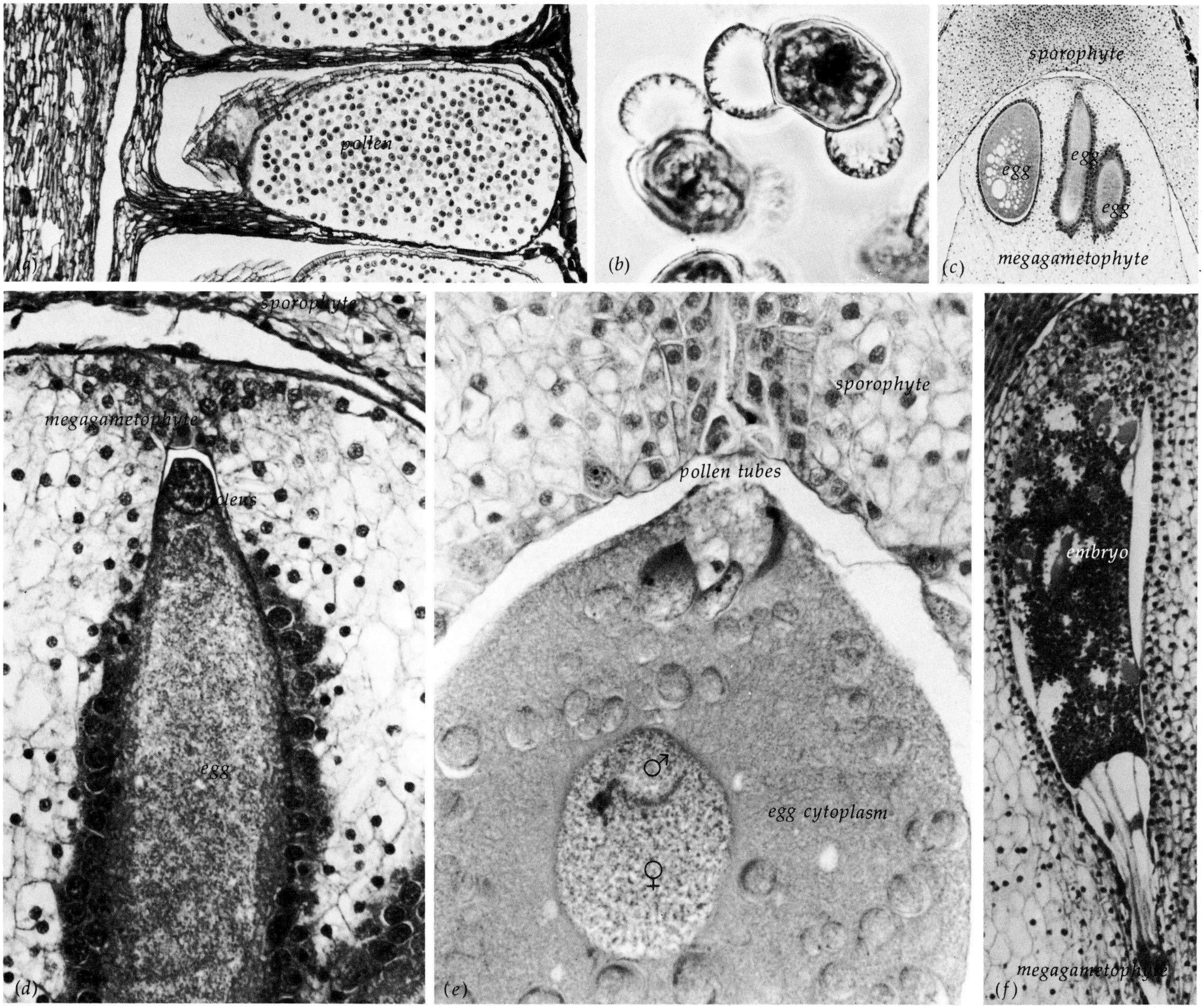

27–12 *Reproductive structures in pines. (a) Pollen is produced within sporangia that are collected into cones. (b) The pollen is winged and carried by the wind to the megasporangia. (c) Eggs develop within the megagametophyte, which is completely surrounded by the sporophyte. (d) An enlarged view of an egg. Several eggs are usually present in each megagametophyte. (e) The zygote: the remains of the pollen tubes can be seen in the sporophyte at the upper end of the megagametophyte, and within the zygote nucleus, the outline of the male nucleus is still visible. (f) An early stage in the development of the pine embryo. The remains of the egg cytoplasm is seen in the upper portion of the photo. The cells at the tip of the embryo are being forced downward into the megagametophyte through the elongation of the cells immediately behind them.*

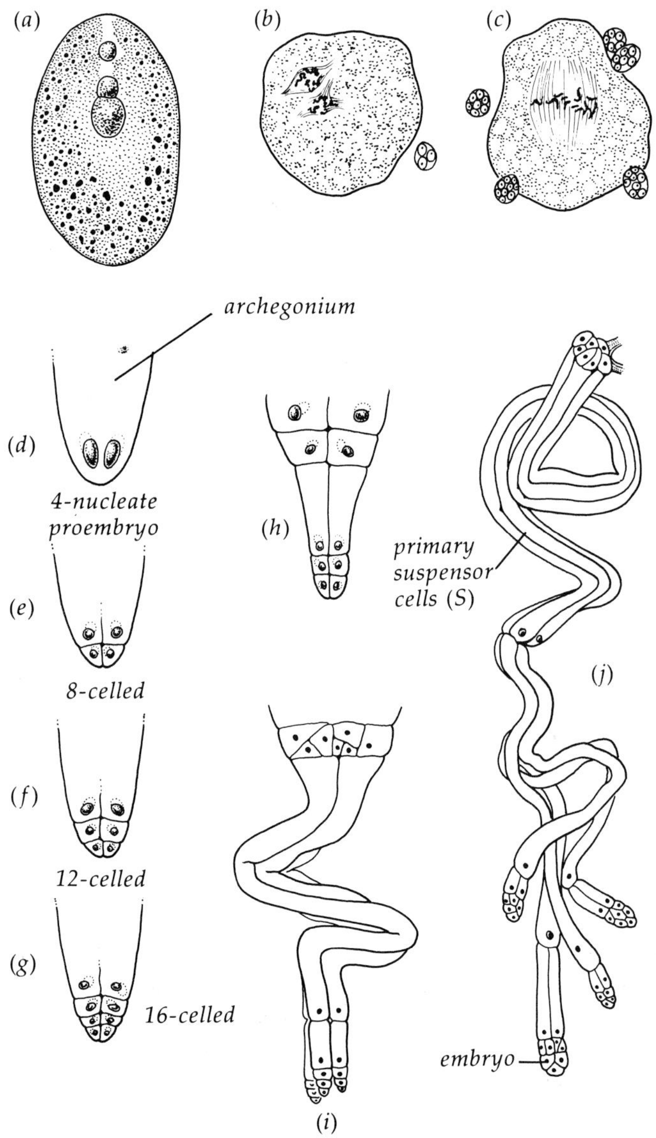

27–13 Fertilization and early embryology of pines. The fusion of the egg and sperm nuclei is shown as is the division of the zygote. The first divisions of the embryo nuclei are not immediately followed by wall formation. Eventually, 4 tiers of cells are formed. The cells of one of the tiers elongate and press the young embryo, which develops from the apical or bottom-most tier, into the surrounding megagametophyte tissue. Several embryos result from the separation of the files of cells in the early stages of development, but only a single embryo will reach maturity.

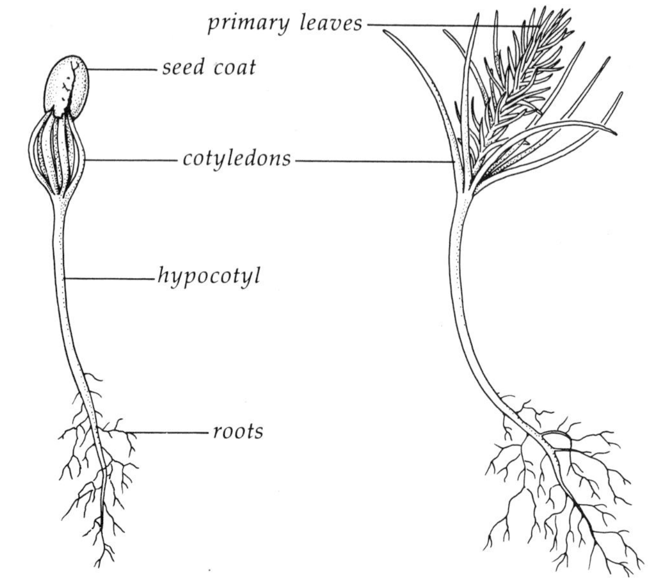

27–14 Stages in the development of the seedling of Pinus edulis. *The numerous cotyledons form a whorl, while the primary leaves are arranged spirally.*

and beauty. The interesting, fan-shaped leaves have been found as fossils in rocks from the Triassic and Jurassic periods of the Mesozoic era. During these periods there were a number of ginkgo species, world-wide in distribution. These gradually declined in number and range until, if any ginkgo trees do exist wild today, they are found only in the mountains of south-eastern China. The present living species, *Ginkgo biloba,* has been called the oldest living genus of seed plants.

Ginkgo biloba is a tall, stately tree that, when older, has several, gracefully curving branch stems uniting in a thick, basal trunk. The beautiful fan-shaped leaves are borne on distinct, short shoots and are shed each year. The male and female reproductive structures are always produced on separate trees. These structures differ markedly in external form from those of the cycads and conifers. In the female the ovules are found at the top of a stalk and are essentially naked. The reproductive cycle follows that given for gymnosperms in general. Large multiflagellated sperms are produced.

The seed becomes surrounded by a fleshy seed coat that, unfortunately, smells like rancid butter. Since *Ginkgo biloba* is widely grown in cities, because of its ability to survive urban environmental conditions, it is desirable to plant only male trees. The trees are quite large before they form reproductive structures, so other methods of sex determination in younger plants have been sought. One way is to examine the chromosomes, where the presence of X and Y chromosomes determines the sex of the tree.

The Gnetales

By any measure, this group of gymnosperms constitutes a strange collection of plants. It is composed of only three genera: *Ephedra, Welwitschia,* and *Gnetum,* each in its own family, its own order, and its own class (Figs. 27–F, G).

Ephedra is a genus with some 35 species, all of them either shrubs or small trees. Most are desert plants with green stems and reduced, scalelike leaves. *Gnetum,* in contrast, is a genus with some 30 species found in tropical forests. The species of *Gnetum* have broad leaves and are climbing vines or trees. *Welwitschia* inhabits a single desert area (the Kaokoveld) in South Africa. Over the surface of the desert, trail the long, strap-shaped leaves that are continually being produced by the top of the long underground stem. The only part of the stem visible is a woody, concave disc that produces both the leaves and reproductive structure.

The gnetales are of interest to botanists primarily because their xylem contains both vessels and tracheids, a prominent angiosperm characteristic. Some of their reproductive features also approach those of the angiosperms. For these reasons some morphologists have argued that the gnetales formed the ancestral stock from which the angiosperms evolved. The general consensus now is that this is unlikely and that the advanced features are simply a case of convergent evolution with the angiosperms. In any case, the three genera are so different from each other that some authors have suggested each should constitute its own division, as does ginkgo.

27–15 *A reconstruction of a Cordaitales.*

End Paper 27–1 The Discovery of the Progymnosperms

Fossils are where you find them, but the places where they are found are not always the places you might think. In this chapter we discussed the importance of the fossil found by Charles Beck of the University of Michigan in establishing the concept of the progymnosperms. Here, Dr. Beck describes the finding and identification of that critical fossil.

"The fossil specimens, which formed the basis of our discovery, were collected in 1957 on a dump heap just outside a large bluestone quarry in the western edge of the Catskill Mountains near Sidney, New York. I had visited this quarry the previous year and had found the foreman, Mr. Fernley Cole, to be not only very cooperative but also interested in the fossils he had often observed in the rock. He found it impossible to believe me, however, when I told him these were the remains of plants that had lived 370 million years ago. As support for my statements about the age of the rock and how the plant and animal fossils came to be embedded in it, I brought him a textbook of historical geology. You will no doubt recognize that the enlightenment of Mr. Cole was not my only motive. He was fascinated by

28

The Angiosperms

Consider for a moment a world without flowering plants. The first feature we would immediately miss would be the bright colors. We would be surrounded by a sea of greens—a wide range of greens—but still only green. Missing would be the reds, yellows, blues, oranges, violets, and all the other hues found in flowers and present in many fruits. But the color of angiosperms is not only in their flowers and fruits. In fact, the greatest display of color is in the autumn, when whole trees and forests are a blaze of red, yellow, or orange. Only ginkgo and two or three other gymnosperms have leaves that change their color in the fall.

A little later, we would notice that the seasons would be much less sharply marked to our senses. Spring would not be heralded by early flowers and sprouting seeds, nor would summer feel as full without the waving fields of grass and flowers. Autumn would be less the harbinger of winter, lacking the change of leaves, and winter, while just as cold, would surely seem less bleak without the bare-limbed trees.

A World of Angiosperms

Thus the flowering plants or angiosperms are an important part of our environment, appealing to our senses in many ways. They affect our sight as well as our smell and taste. This is no mere accident of evolution but an important facet of the life of angiosperms. When we come to the flowering plants, we come to plants that have truly evolved, not only to life on land but to life on a land filled with animals.

The colors and aromas of flowers, which so delight us, attract insects and other animals necessary for pollination. The colors and aromas of fruits entice us and a wide range of animals to eat them and spread the seed. The spines and hooks on seeds would be useless without fur. The intercalary meristems of grass, which provide us with a summer of both green lawn and lawn

28–A The contrast between the angiosperms and the gymnosperms is beautifully illustrated in this photo of a cherry tree in full flower in front of a tall stately Monterey pine.

mowing, allow the plant to survive grazing by cattle, deer, or buffalo. In addition, we find that these are the first vascular plants to trap and digest animals—only the fungi and the angiosperms do that!

Of a total of about 300,000 plants, some 275,000 species are angiosperms, grouped into several hundred families. This largest group of plants in the world comes in every conceivable size and shape: there are angiosperm trees that challenge the great conifers in size; there are tiny floating aquatic angiosperms that are little more than drifting leaves; there are great climbing vines that can kill large trees; there are the beautiful and delicate plants that bloom in the spring; there are shrubs of many sizes and shape; and there are the miles of grasses that cover our prairies. Not only do angiosperms assume almost every shapes and size, but they grow in almost all climates. The cacti and certain other angiosperms are adapted to live in all but the most arid deserts. The tropical rain forests are a fantastic collection of angiosperms, including literally thousands of species of orchids, many growing as epiphytes on flowering trees and vines. Other angiosperms grow near the tops of all but the highest mountains and to the edge of the great polar ice packs. Some have adapted to life in the intertidal zones, and many inhabit marshes and the edges of lakes. Temperate, arctic, tropical, wet or dry, these are all environments successfully invaded by angiosperms. Before we examine them in more detail and see why they are so successful, let us ask where they came from and when.

The History of the Angiosperms

The Cenozoic has rightly been called the "Age of the Mammals and the Angiosperms." The Mesozoic was the heyday of the dinosaurs and the gymnosperms, but during the last period of that era, the Cretaceous, a great dying out occurred around the world (Fig. 28–1). The dinosaurs, the bennettitales, many of the cycads and ginkgos, and some of the conifers all went down, but during this same period the mammals and the angiosperms arose.

Darwin called the origin of the angiosperms "an abominable mystery," and so it is, except that the mystery may not be so great as some would make it. The problem is the fossil record of the angiosperms. First, there is essentially no record of angiosperms before the Cretaceous, and second, when fossils of angiosperms do appear in the Cretaceous, they show a wide diversity of full-blown angiosperm features. When this type of situation exists, it is clear what will happen: botanists (as any scientist would) take the available evidence, apply their imaginations and scientific prejudices, formulate conflicting theories, and defend them mightily.

Two general schools of thought have been established, which may somewhat lightly be called "the early risers" and "the late risers." The early risers, led by Daniel I. Axelrod, believe that the history of the angiosperms extends farther back into the geologic past than the Cretaceous, probably into the Triassic and Permian periods. They maintain that the lack of fossils does not mean that the plants were not there but that they grew in such places as uplands, where conditions were not favorable for their preservation. This group argues that there is simply not enough time in the Cretaceous for so vast an evolutionary diversity to occur without a long pre-Cretaceous development.

The late risers believe that there is sufficient, indeed ample, time for such development within the Cretaceous. They point out that the Cretaceous extended 70 million years, while the Devonian, which saw the invasion of the land and the evolution of a complex land flora, lasted only 50 million. In addition, they point to the fact that the same type of claim was made about the appearance of land plants, namely that not enough time was available, until detailed analysis of the strata revealed a pattern of increasing complexity that now seems quite reasonable. The point, too, that evolution in the angiosperms is probably more rapid than in the gymnosperms is a valid one, based on the relative rates of reproduction. A rapidly reproducing pine takes at least two years to set seed, while a slow angiosperm may take several months. The late risers feel that although it may have been possible that angiosperms evolved in highlands where fossilization would be difficult, it is most unlikely that some angiosperm pollen would not be preserved. The pollen of angiosperms is remarkably distinctive and can be carried for miles by the wind. Fossil pollen has been found from the Mesozoic and as far back as the Carboniferous of the Paleozoic, but none of it is angiosperm pollen. Only when other kinds of angiosperm fossils are found is angiosperm pollen also present.

The one safe conclusion from all of this is that we do not know the time of origin of the angiosperms.

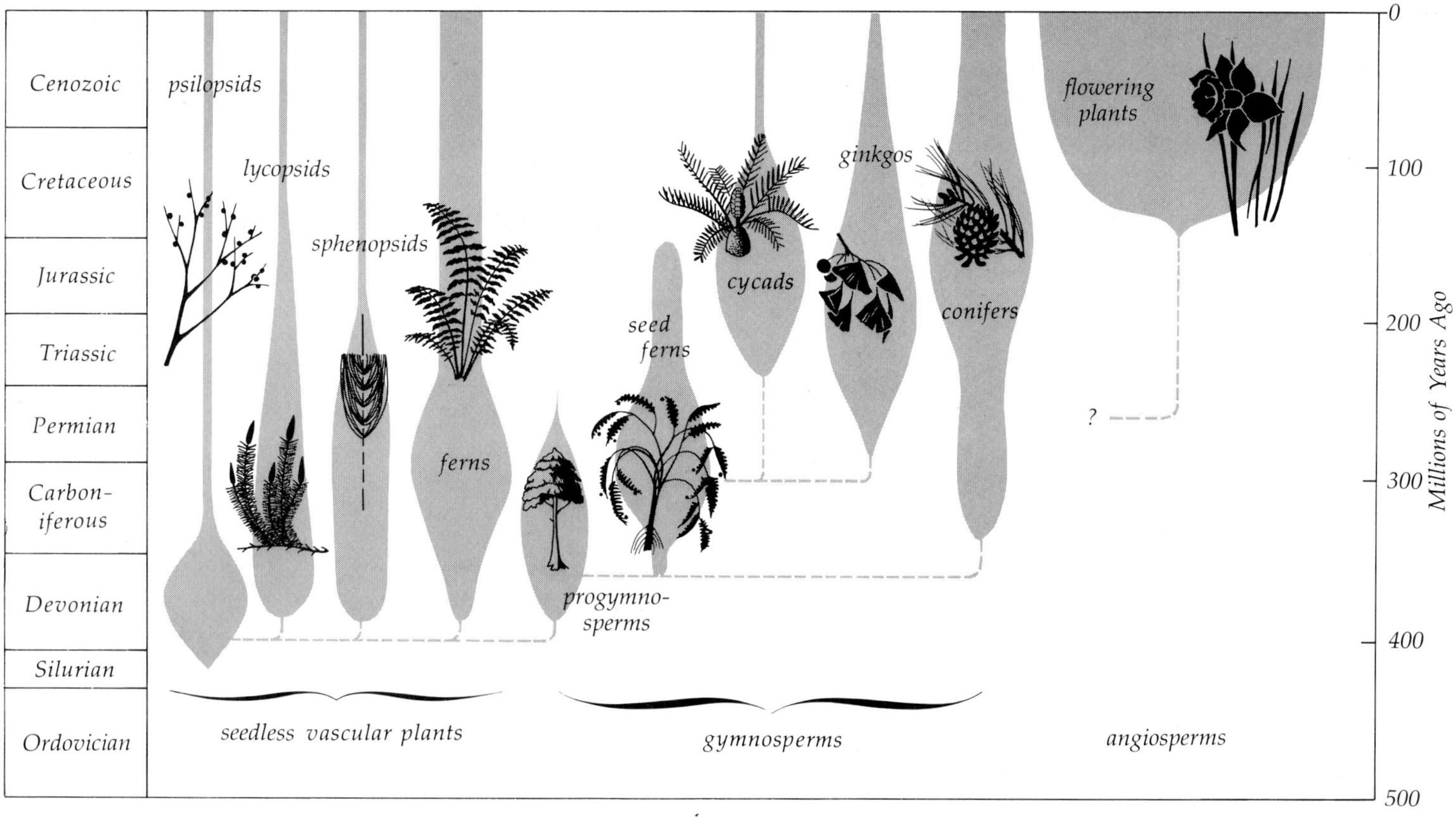

28–1 A summary of the major groups of vascular land plants and their possible relationships.

Current evidence more strongly supports the early Cretaceous than some older period. We also have no idea which plants were the ancestors of the angiosperms. A long list of plants could be made, including the bennettitales and the gnetales, but all have been ruled out on various grounds. At this point the angiosperms join the gnetales, ferns, seed ferns, and many other groups for which we simply do not know the ancestor.

The Success of the Angiosperms

When the angiosperms did appear on the face of the earth, why were they so successful? The answer appears to lie in their combination of features—some already present in other plants and some new—that provided the angiosperms with an almost bewildering capacity to adapt to most environmental conditions.

Let us first consider angiosperm trees. They displaced the conifers from the great forest areas that stretched from the Atlantic seaboard to the Mississippi. Only in the far north and south of this original eastern forest did conifers remain important. Although there are a few exceptions on both sides, a number of morphological and physiological factors may have contributed to this superiority. One may be that angiosperm trees have vessels. These make the xylem more effective in water transport. Second, they have broad leaves, which have a tremendous capacity for photosynthesis, and although such leaves transpire enormous amounts of water, the presence of vessels in the xylem results in the leaves having an ample supply of water—if the soil can provide it. Third, they have sieve tubes (and, of course, companion cells), a more advanced way to transport the excess photosynthate from the broad

*28–2 Characters distinguishing dicots from monocots**

Dicotyledoneae	Monocotyledoneae
1. Two cotyledons or seed leaves in the embryo	1. One cotyledon or seed leaf in the embryo
2. Principal veins of the leaves branching out from the midrib or from the base of it, not parallel, forming a distinct network	2. Principal veins of the leaves parallel to each other
3. Flower parts arranged in twos, fours, or usually fives	3. Flower parts arranged in threes
4. Root system characterized by a taproot, that is, a large primary root with branch roots growing from it	4. Root system fibrous—that is, without a principal or taproot, the evident roots being of about equal size
5. Stem with the vascular bundles in a single cylinder ("ring" as seen in cross section)	5. Stem with the vascular bundles scattered apparently irregularly through pith tissue
6. The cambium adding a new cylindroidal layer ("ring") of wood each year or each growing season	6. Stem and root without a cambium, not increasing in girth by the formation of annual cylindroidal layers of wood

*From Lyman Benson, *Plant Classification* (Lexington, Mass.: D. C. Heath & Co., 1957). Used by permission.

leaves to storage in the stem and root. Fourth, they have a mechanism to drop the broad leaves that are such an asset in summer and such a liability in winter. Fifth, they have a rate of reproduction two to four times as rapid as that of conifers; the seeds are often better supplied with food; and the seedlings can survive with less light. Finally, the taproot system of angiosperms allows them to tap deep sources of water and better to resist storm damage.

Conifers, on the other hand, are limited by a vascular tissue composed of tracheids and sieve cells. Their needle and scale leaves are superbly designed for conditions where water is scarce and temperatures extreme, but they lack the photosynthetic capacity of the broad leaves of the angiosperms. Make no mistake, conifers are highly successful plants, and in dry, cold, or rocky regions they are better able to survive than the angiosperms. The great western coniferous forests are proof of this. But in the middle zones and the tropics they lose ground to the angiosperm trees, although they still cover thousands of square miles of the earth's surface.

Angiosperm evolution did not stop at trees—that was only one line. Losing the massive secondary growth of trees, many angiosperm species evolved that live for a single season or have green, aboveground parts only during periods favorable for growth. The annuals replaced mass with speed—the speed of reproduction. Most take only weeks or months (sometimes even days) to complete their reproductive cycle. We have already mentioned the desert annuals that can go through a reproductive cycle on a single rainfall. Or consider the spring flowers in the eastern forests. These can grow and reproduce between the time the temperature is warm enough for growth and before the emergence of new leaves on the trees above makes it too dark for them to survive.

Such rates of reproduction are based on many factors: the formation of the flowers must be fast and pollination effective; the pollen tube must rapidly reach the megagametophyte; embryo development must not be delayed; and an efficient mechanism for seed dispersal must be present. All of these are true for angiosperm species, which many times have co-opted ani-

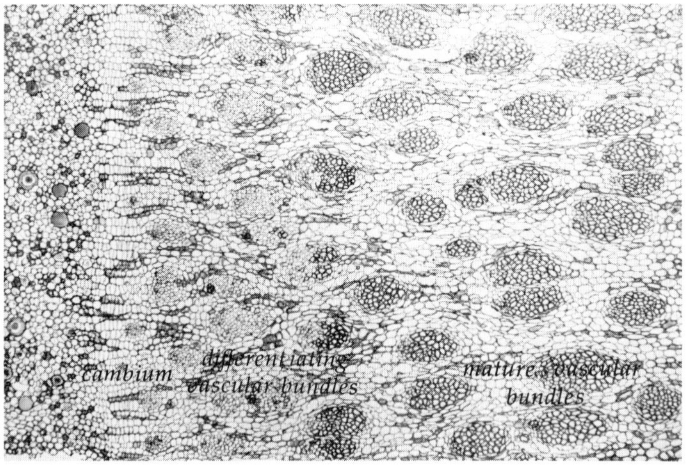

28–3 Secondary growth is not common in monocotyledonous plants, but it does occur in some. One example is Yucca, the striking desert plant. In this case, the cambial derivatives do not differentiate directly into xylem and phloem but instead form a tissue from which vascular bundles develop.

mals to achieve the necessary ends. Insects, birds, and bats supplement wind pollination, and the development of fruit aids in the dispersal of the seed not only by animals but by wind and water.

Perhaps, even more important than all of these considerations is the amazing adaptability of the angiosperm sporophyte. The leaf alone has been transformed in a myriad of ways. Some are reduced to scales barely visible to the eye; others are huge (10 to 12 ft long); still others are modified into thorns or traps that catch insects by snapping shut, drowning them, or engulfing them. Stems can be modified into photosynthetic organs, thorns, or tendrils or made into living water barrels or thin, tough, wirelike strands that can scarcely be cut. Roots, too, can be changed into storage organs, grasping structures to hold the plant to the bark of other plants, or sucker, parasitic growths that rob the host of its food.

The Sporophyte

Angiosperms can be divided into two broad groups or subclasses: the *Dicotyledoneae* and the *Monocotyledoneae.* The names refer to the number of cotyledons (storage seed leaves) present in the seed. But the differences between dicots and monocots go far beyond the fact that one has two such leaves, while the other has only one. These differences are summarized in Table 28–2. The Dicotyledoneae are the larger of the two with more than 200,000 species, compared with less than 50,000 species of Monocotyledoneae.

All angiosperm leaves are megaphylls. In dicots the veins form a network with the principal veins branching out from the midrib or the base. The leaves may be either undivided or dissected into lobes or compound leaves. The monocots have leaves that tend to be less divided and are frequently long and narrow. The veins in these leaves are parallel to each other.

The roots of dicots are characterized by a taproot. Usually, in this type of root there is a main, frequently thick, root with branch roots growing from it. The carrot is a rather exaggerated example of a dicot taproot. The monocot root is fibrous with the various roots being roughly the same size. The roots of grasses are typical monocot roots. The monocots also produce many underground structures that look as though they were roots but actually are not. The bulbs of such monocots as tulips and onions are short stems surrounded by fleshy storage leaves. Other monocots develop rhizomes (underground stems), but the potato, which is a rhizome, is a dicot.

The stems of dicots are frequently woody because of the presence of vascular cambium. The vascular bundles in dicots form a single cylinder. As in conifers, the vascular cambium in dicot trees produces a new ring each year. The wood in dicots is complex. Vessels are found in all but the most primitive dicots. Tracheids are also found in the xylem, as are fibers and parenchyma cells. There is a pith at the center of the stem, and in perennial dicots the cortex is usually lost and replaced by a periderm. In annual dicots a vascular and cork cambium may not develop at all (although in some they do).

The monocots lack secondary growth; both vascular and cork cambia are never formed. The vascular tissue is in the form of discrete bundles scattered throughout a fundamental tissue. Each vascular bundle is composed of several vessels of various diameters, sieve tubes with companion cells, and a surrounding sheath of sclerenchyma cells. Even in large tree monocots such as the palms vascular and cork cambia are not formed. Instead, a diffuse mass of cells, called a thickening meristem, divides to increase the number of cells but not in the

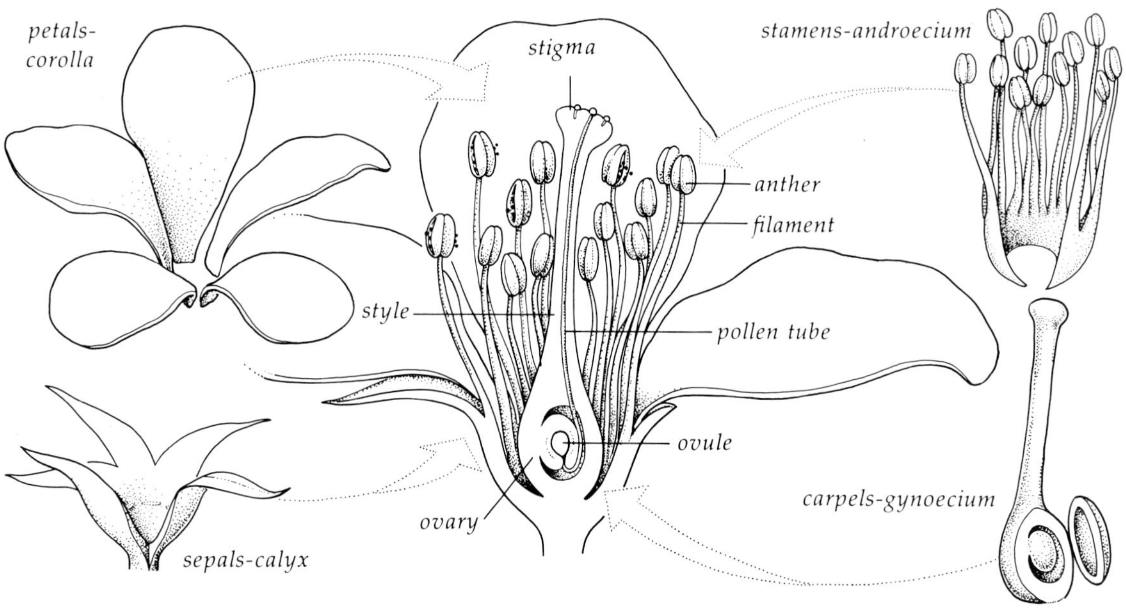

28–4 *The parts of a flower.*

precisely arranged manner of vascular cambia (Fig. 28–3).

The number of flower parts in dicots and monocots are different, but to understand the significance of this it is necessary to discuss the most unique part of the angiosperm—the flower.

The Flower

What do you see when you see a flower? This is a tricky question because it can be answered on many levels and in many senses that are not mutually exclusive. The answer may be God's finest creation or evolution's highest achievement (depending on your viewpoint). You might be seeing a modified stem system or a collection of modified leaves (depending on how you interpret the origin of the flower). You might be seeing a complete flower, a flower with missing parts, or a collection of individual flowers (depending on the particular flower). No matter how you view the flower it is a thing of interest, if not always beauty, and an ingenious mode of reproduction wonderfully adapted to the land environment.

An ideal flower consists of four rings of parts set on the end of a stem (Fig. 28–4). This flower stalk is known as the peduncle; the tip of this structure is the receptacle and forms the base of the flower. The outermost ring of parts is composed of sepals, known collectively as the calyx. The next ring of parts is composed of the petals, known collectively as the corolla. Together calyx and corolla form the perianth. The sepals and petals are not directly involved in reproduction. They are sterile parts but may be vital for reproduction, as in cases where the brightly colored petals attract insect pollinators. In general, petals are more brightly colored and prominent than sepals, which tend to be inconspicuous, green, and leaflike.

Inside the petals are the stamens. Each stamen consists of a usually slender stalk, the filament, bearing at its end a boxlike structure, the anther. Meiosis occurs within the anther, producing the microspores. These divide to form the two-cell microgametophyte or pollen. The stamens are known collectively as the androecium.

At the center of the flower are the carpels. Each carpel consists of a stigma at the top on which the pollen lands, a style in the center through which the pollen tube grows, and an ovary at the base containing the ovules with their megaspore and megagametophyte. The carpels are known collectively as the gynoecium. The carpels and parts of carpels may be free or joined together in a wide range of ways.

If a flower has all of these parts, it is called complete, but if it lacks any one it is called incomplete. In terms of the ultimate reproductive function of the flower, only stamens and carpels are necessary. As a result flowers

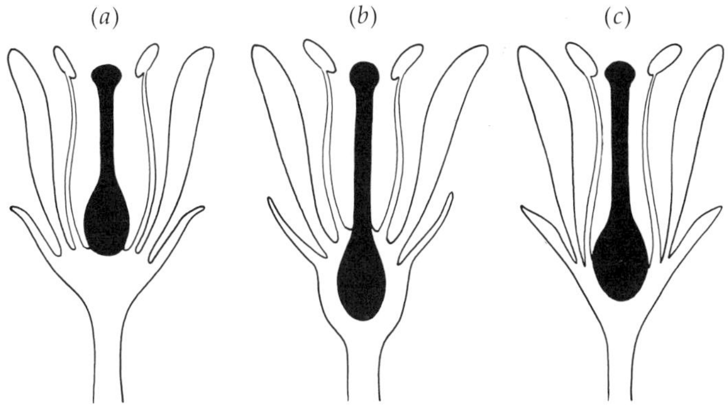

28–5 *Diagrams of (a) hypogynous, (b) epigynous, and (c) perigynous flowers.*

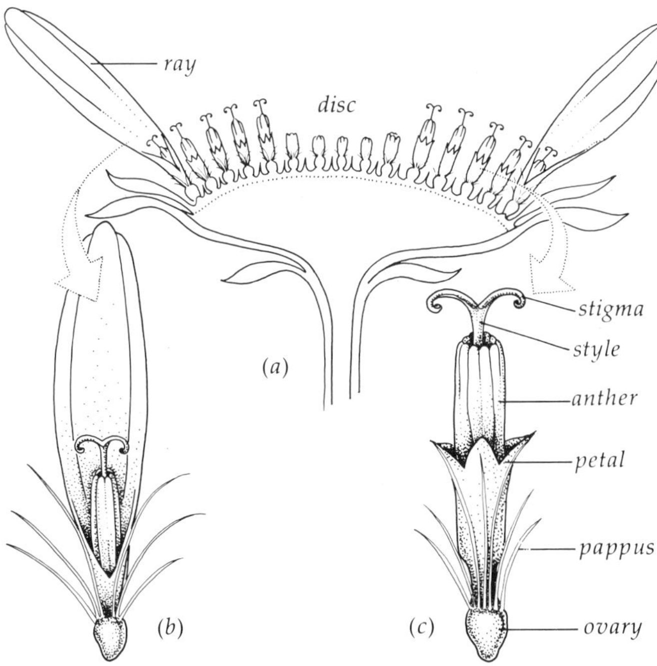

28–6 *Composite inflorescence of the type found in sunflowers. A cross section of an inflorescence is shown in (a), while a single ray flower is shown in (b) and a single disc flower in (c).*

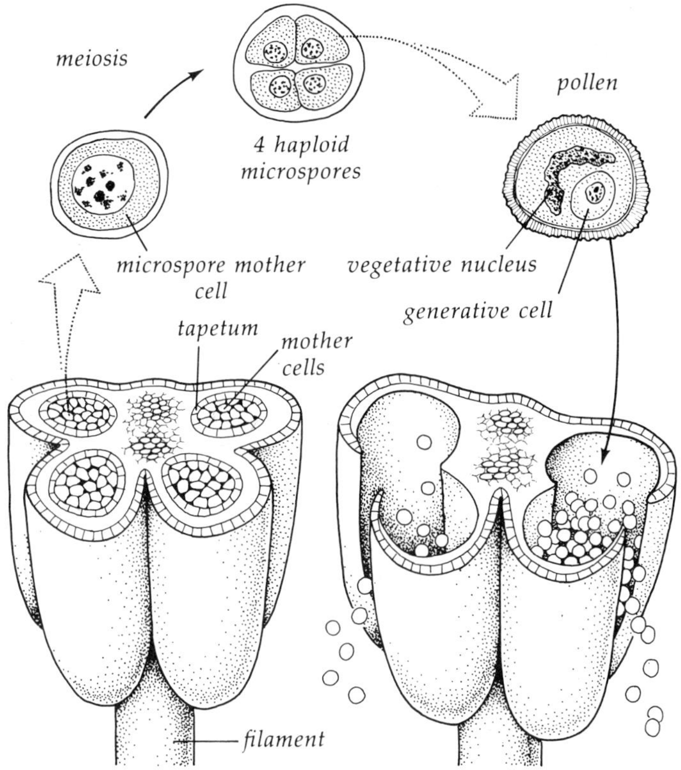

28–7 *The development of the pollen grain.*

that have both are called perfect, but if they lack either one they are called imperfect. This terminology leads to such interesting descriptions as "perfect incomplete flowers" (i.e., flowers having both stamens and carpels but lacking sepals or petals).

The presence or absence of floral parts is only the beginning of the variation in floral morphology. There is, for example, the way in which the parts are arranged in relation to the top of the flower stalk or receptacle. The various parts may have their bases on the same plane (Fig. 28–6) in which case the flower is hypogynous. In this case, the ovary is said to be superior because it is above the receptacle. In epigynous flowers the receptacle surrounds the ovary, and the ovary is called inferior. There are also instances where a sleeve-like extension of the stem surrounds the ovary without being attached to it, producing the perigynous flower.

The way in which the floral parts, especially the petals, are arranged around the axis is also subject to

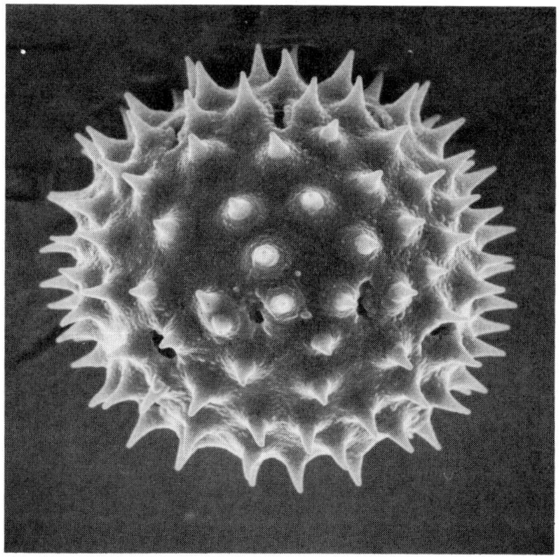

(a)

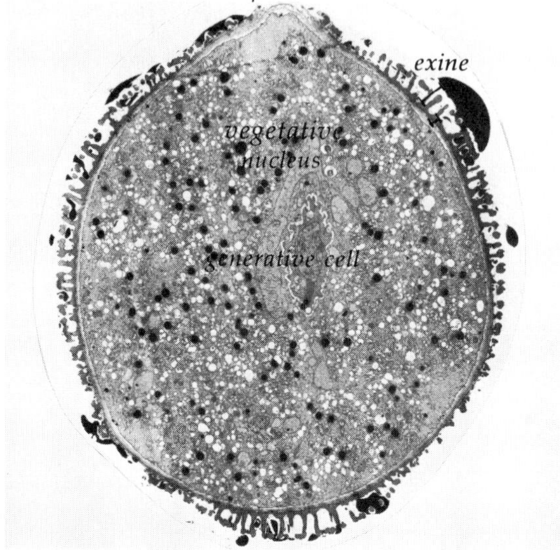

(b)

28–8 The wall of the pollen grain (a) is a beautiful object, particularly when viewed with the scanning electron microscope (which reveals its structure in three dimensions). The wall consists of two major layers: the intine, which is against the plasma membrane, and the exine (b), which is the outer wall layer. The intine is relatively simple in structure and has the usual cell-wall composition (cellulose, pectin, etc.). The exine is structurally much more complex. It consists of two major and several minor layers that form a wide variety of shapes. The exine is composed of sporopollenin. The wall has special regions from which the pollen tube will grow. These regions appear to be blocked out during wall formation by pieces of ER that are appressed to the plasma membranes at these points. The mechanics of the formation of the elaborate exine are not fully understood, but the presence of callose is believed to be important (although it does not determine the pattern). Within the pollen grain (b) is the generative cell that will form the sperm. Closely associated with the generative cell is the vegetative nucleus, which later controls the growth of the pollen tube.

variation. If all parts are equally spaced around the center, they are radially symmetrical, while irregular flowers exhibit dorsiventral symmetry, if the flower can be divided into left-right mirror images. These arrangements may also be called actinomorphy (radially symmetrical) as opposed to zygomorphy (bilaterally symmetrical).

Petals and sepals may be separate, or they may be fused at their bases (united calyx and corolla). The number of all the parts may vary, as do the size and shape of individual parts.

Flowers may occur singly, or they may be grouped together in groups called inflorescences. There are two major types of inflorescences. In one, flowers mature at the base of the inflorescence as new ones form and open at the top. These are called indeterminate. In the other all of the flowers are formed at about the same time, and the apical meristem is completely differentiated into flowers. This type is termed determinate.

There is a special type of determinate inflorescence that is confusing to many people. At first glance such flowers as the daisy and sunflower appear to be single flowers, but they are actually inflorescences or composites of flowers. Each consists of many small flowers arranged on a flat head at the end of the stem (Fig. 28–6). In many such inflorescences there are two kinds of individual flowers: the central disc flowers and the marginal ray flowers. The disc flowers usually have an inconspicuous corolla, while the ray flowers generally have a much enlarged corolla extending on only one

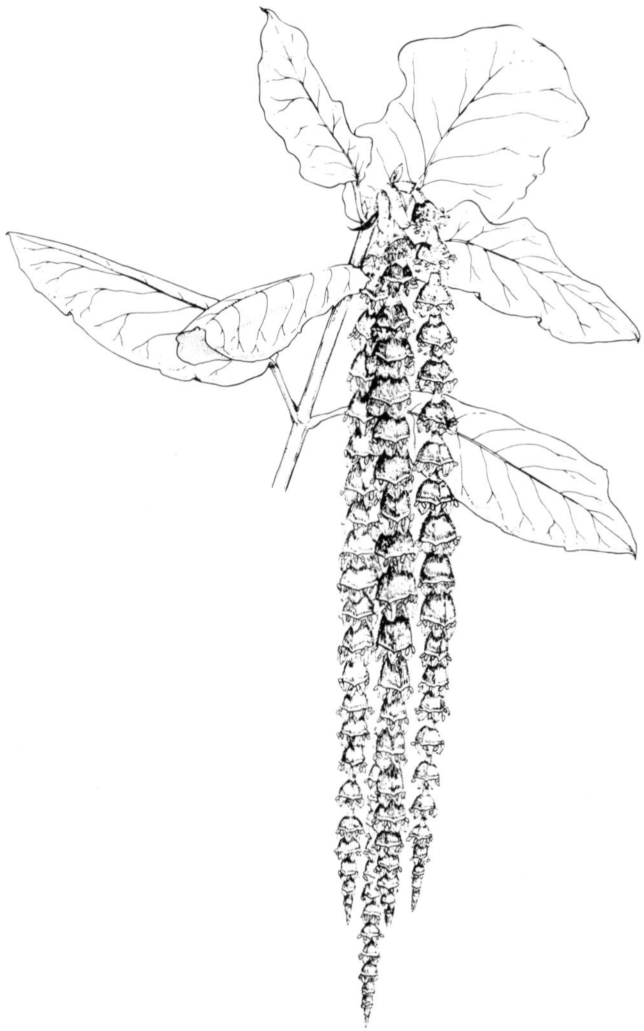

28–9 *Wind-pollinated flower.*

side. The next time you pull "petals" from a daisy, you will find that you are actually taking off individual flowers rather than simple petals!

One of the most complex and important types of inflorescence is found in the grasses. An elaborate but precise terminology has been developed to describe this particular form of inflorescence in flowers.

All of these possible variations result in an array of form that makes flowers the beautiful objects they are. When you see various flowers, take a few moments to examine them more closely. You will be rewarded with an understanding and heightened awareness of some of the most fascinating objects in nature.

The Stamen and Pollen

The anthers of the stamens are the places where pollen formation takes place (Fig. 28–7). Each anther is lobed into four parts. At the center of each lobe a column of cells, which are the microsporocytes or microspore mother cells, is differentiated. These are surrounded by a special tissue called the tapetum.

The microspore mother cells undergo meiosis, and 4 haploid spores are produced. An elaborate wall forms around the spores (Fig. 28–8), and the nucleus divides to form 2 nuclei. One of these enlarges, frequently becoming highly lobed or indented. This is the vegetative nucleus and functions as the nucleus of the pollen grain and pollen tube. The second nucleus becomes surrounded by a small amount of cytoplasm, a plasma membrane, and usually a thin wall. This is the generative nucleus of the generative cell (Fig. 28–8). The generative cell divides, usually in the pollen tube but in some species in the grain, to form the 2 sperms. Angiosperm sperms have been called naked nuclei, but this is clearly not the case—they are true cells. The pollen grain and pollen tube constitute the entire male gametophyte in the angiosperm.

Most pollen is released as single grains when the anther splits open. In some cases, the pollen may be held together in the original tetrads, or the entire mass of pollen may remain together, as in the orchids.

Pollination. The pollen must get from the anther to the stigma. If it does not succeed, reproduction fails. We have seen that the gymnosperms rely on wind to carry the pollen, but the great majority of flowers are pollinated by insects. In wind-pollinated angiosperms we find flowers in which the petals are reduced, and no nectar or scents are found. Instead, the anthers may be arranged in long filaments that sway in the wind, releasing the pollen (Fig. 28–9). The stigmas are similarly arranged and may be feathered or branched at the end. The pollen is usually produced in large amounts, and the individual grains are generally smooth and round.

In many angiosperm species the plants are pollinated by animals: insects (bees, moths, flies), birds,

(a)

(b)

28–10 *(a) The moth-pollinated flower* Lilium martagon *is shown being visited by a hawk moth that is taking nectar from a petal pouch. (b) The bee-pollinated flower* Salvia pratensis.

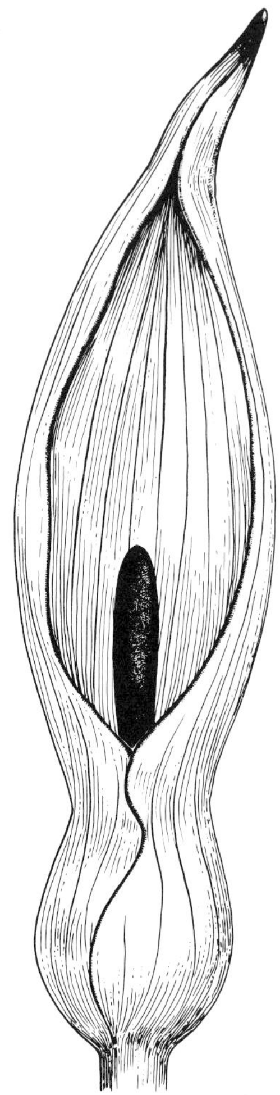

(a)

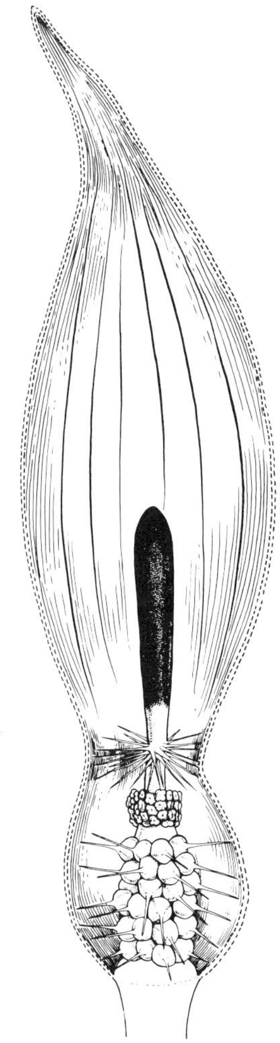

(b)

28–11 *An example of a fly-pollinated flower is* Arum maculatum. *In this case, what appears to be the flower (a) is actually an inflorescence composed of many flowers (b). These are arranged so that a temporary flytrap is formed in the base of the flower, which ensures the fly will be covered with pollen.*

and bats. Among these the common pollinators are insects. Such flowers produce nectar, which is about 25% glucose, on which the insects feed. The insects are attracted to the flowers by their color and scent. Thus many insect-pollinated flowers are brightly colored, the color depending on the particular insect involved, and sweet smelling. But not all smell good to us. Those that attract flies sometimes have the odor of and may look like rotten meat.

A wide variety of mechanisms has evolved to enhance the possibilities of pollination, based on the particular characteristic of each insect. For example, in many bee-pollinated flowers a landing stage is present, since bees can collect the nectar standing but not flying (Fig. 28–10). Moth-pollinated flowers are usually open and frequently are radially symmetrical, since these insects feed while hovering (Fig. 28–10). They also have different scents, may be open only at night (if pollinated by night-flying moths), and have the nectar deep in the flower, where only the long tongues of the moths can reach it. Fly-pollinated flowers, besides the bad odor, may have traps that confine the flies in the flower for a day (Fig. 28–11). The flies during this time crawl over the stigmas, depositing pollen collected on their bodies. The next day they are released by the opening of the anthers of the flower in which they were trapped. Crawling over the open anthers on their way out of the flower, they pick up new pollen on their bodies and are ready to start the cycle again.

Bird-pollinated flowers have bright, contrasting colors and lack scent. Bat-pollinated flowers, in contrast, usually have dull colored flowers that open at dusk and have mouselike smells. The flowers are arranged so that the bats can reach them, with the flowers sometimes suspended from long stalks (Fig. 28–13).

Perhaps, one of the most ingenious of all pollination mechanisms involves neither wind nor animals but water. The water plant *Vallisneria*, commonly called ribbon grass, has a female flower that floats on the surface of the water with the stigmas pointing outward (Fig. 28–13). Hundreds of small male flowers consisting of two stamens are produced underwater and released. These float on the surface of the water and act as tiny boats carrying the pollen to the stigmas.

Stigma, style, and pollen tube. The carpel consists of three parts: stigma, style, and ovary. Within the ovary are the ovules, and within each ovule is a megagametophyte.

28–12 The bat-pollinated flower of the sausage tree Kigelia.

The stigma is important in the reproductive cycle because it is here that the pollen lands or is deposited. It is on the stigma that the pollen germinates and that the tube begins its growth toward the ovule (Fig. 28–14). Many stigmatic surfaces are covered with hairs and secrete fluids that aid in germination of the pollen. Not infrequently, substances are also present that

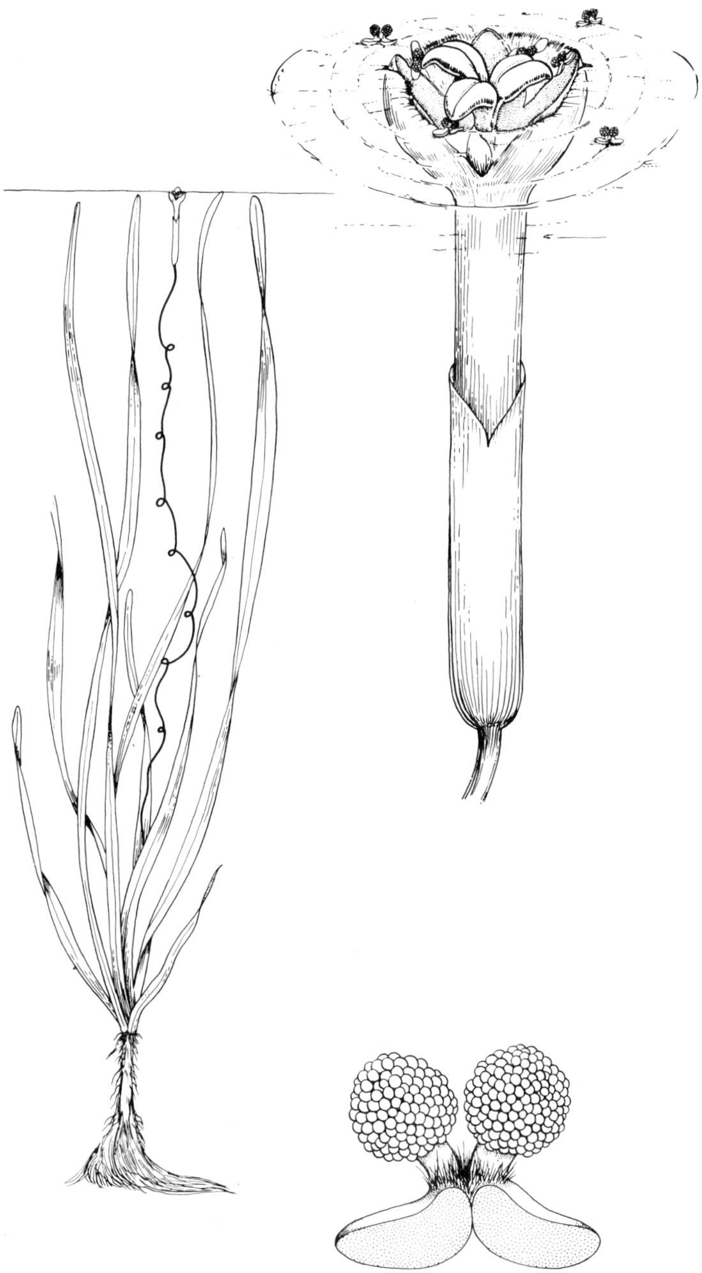

28–13 *Water-pollinated flowers of* Vallisneria americana. *The plant with the female flower floating at the surface but attached to the underwater parts of the plant is shown in (a). The male boat plants (b) are released and float to the female flowers (c).*

prevent the germination and growth of incompatible pollen, a mechanism that aids in outbreeding.

The style, which may vary from a few millimeters to many centimeters in length, is hollow in some species and solid in others. In the hollow styles the pollen tube grows down the surface of cells lining the canal. These have specially thickened walls, and, as the work of Walter Rosen at the State University of New York at Buffalo has shown, the pollen tube absorbs various compounds from these cells. In the solid styles the pollen tubes grow through the walls of special cells in the center of the style (Fig. 28–15). Lilies are good examples of plants with open styles, while cotton and petunia are plants with solid styles.

There is considerable evidence, much of it from Rosen's laboratory, that chemical substances are involved in directing the pollen tube during its growth. The nature of the substances is unknown, but they are believed to be organic molecules. Some plants (snapdragons, for example) have pollen tubes that grow toward increasing calcium concentrations, but this is not true for all pollen.

The work of Joseph Mascarhanus at the State University of New York at Albany has told us much about the way pollen tubes grow in terms of protein and nucleic acid metabolism. First, it is important to realize that most (but not all) pollen tubes grow exceptionally rapidly. In cotton, for example, the time from pollination to fertilization is some 12 hr, and the style is over 2 cm long. Many plants have even faster growing pollen. The pollen grain contains great quantities of ribosomes, both attached to the ER and in clumps in the cytoplasm. Mascarhanus has shown that although the vegetative nucleus makes m-RNA, it does not synthesize ribosomes. Moreover, there is enough m-RNA stored in the pollen so that the tube can grow a considerable distance before new m-RNA is necessary.

The growth of pollen tubes is strictly at the tip. Studies, again by Rosen, indicate that the wall near the tip is modified to absorb materials from the style. All pollen contains large amounts of stored compounds, not only carbohydrates and fats but also protein and RNA. Many kinds of pollen apparently need and obtain little nourishment from the style.

In most cases, as the pollen tube grows down the style, the generative cell divides to form 2 sperms. The chromosomes frequently become strung out down

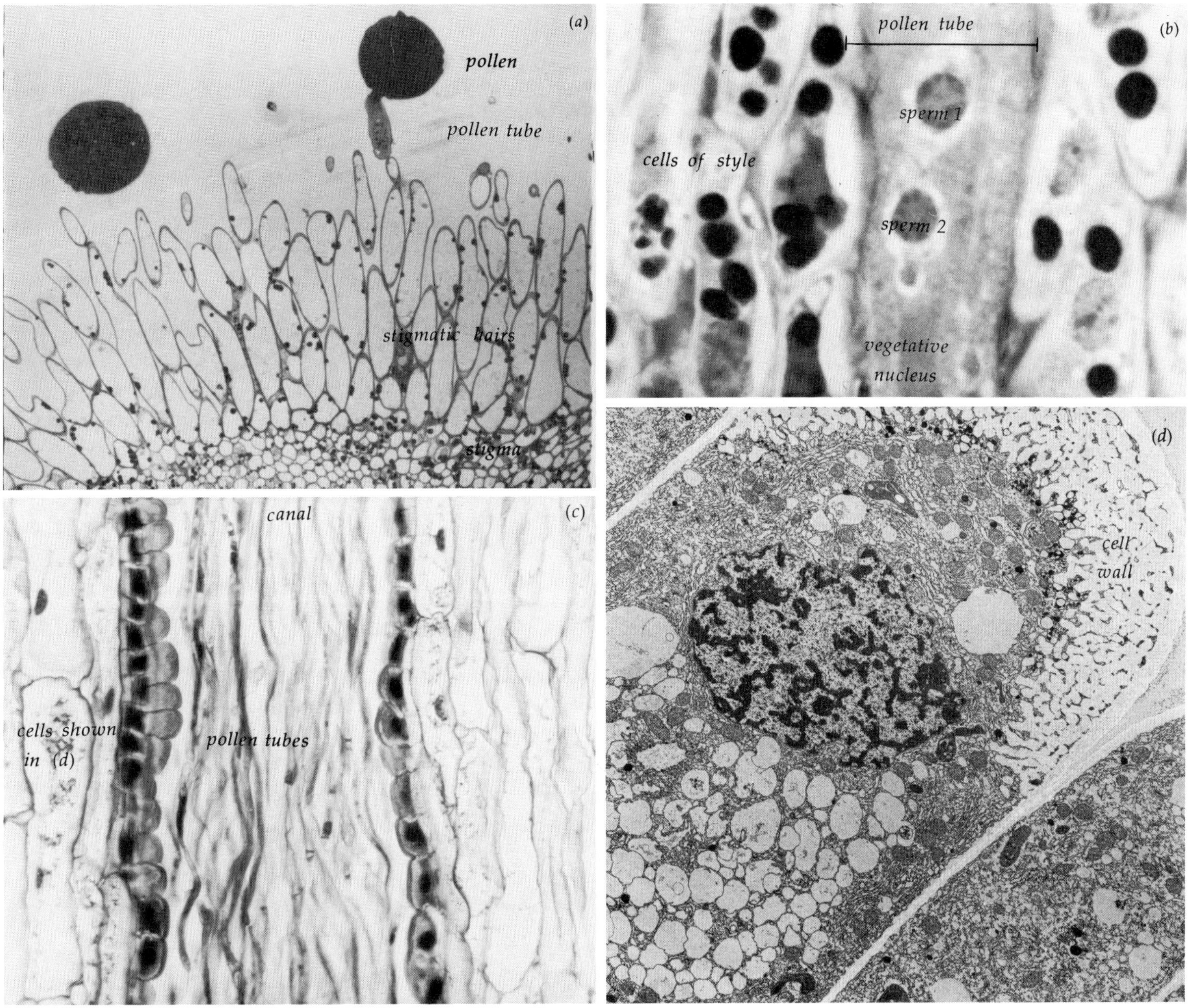

28–14 *Usually minutes after the pollen grain comes in contact with the surface of the stigma, it produces a tube that grows through the stigma and down the style (a). In the case of flowers with solid styles (b), the pollen tube grows through the walls of a special tissue called the transmitting tissue. The generative cell is carried down the tube as it grows and divides in the tube to form the two male gametes or sperms (b). These are true cells, although they are greatly reduced in structure. In flowers with hollow styles the pollen tube grows down a canal (c). This canal is lined with special cells that have an elaborate wall on the side toward the canal (d). Photos (a) and (b) are of cotton and, (c) and (d) are of the lily.*

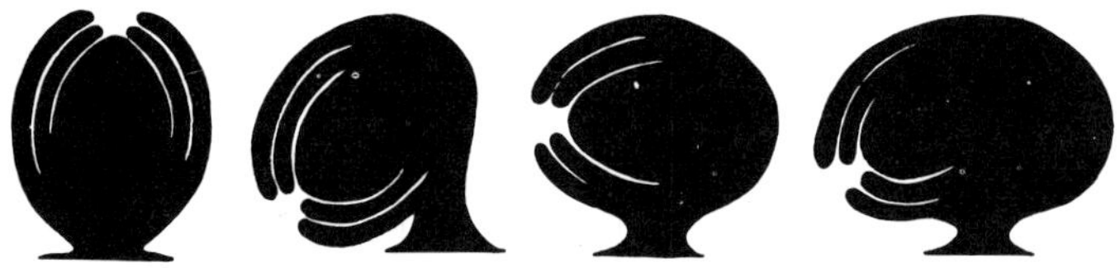

28–15 The types of ovules: (a) orthotropous, (b) anatropous, (c) amphitropous, and (d) campylotropous.

the tube, as mitosis occurs in the confined space. The two sperms formed are true cells, although greatly reduced in structure (Fig. 28–14). In most species, they consist of a nucleus, plasma membrane, some reduced mitochondria, a little ER, a dictyosome or two, some ribosomes, and a handful of small vesicles but no cell wall. Some microtubules may also be present. Plastids are usually absent, although they have been present in a few species studied.

Ovule and megagametophyte. Within the ovary, at the base of the carpel, are the ovules. The ovary is hollow, and the ovules are attached to the ovary by a short stalk, the funiculus. The way the ovules are attached varies considerably from species to species. The ovules can also have different shapes, depending on their attachment to the ovary. These are shown in Figure 28–15.

Each ovule consists of a nucellus, surrounded by two integuments (the outer and the inner). The relation of the development of these tissues to the development of the megagametophyte is shown in Figure 28–16. Note that there is a hole where the integuments do not come together. This is the micropyle, and through it the pollen tube grows on its way to the nucellus.

Megagametophyte development begins with the differentiation of a megasporocyte or megaspore mother cell within the nucellus. This cell enlarges and undergoes meiosis. Of the four megaspores so produced, three degenerate. The fourth begins to enlarge

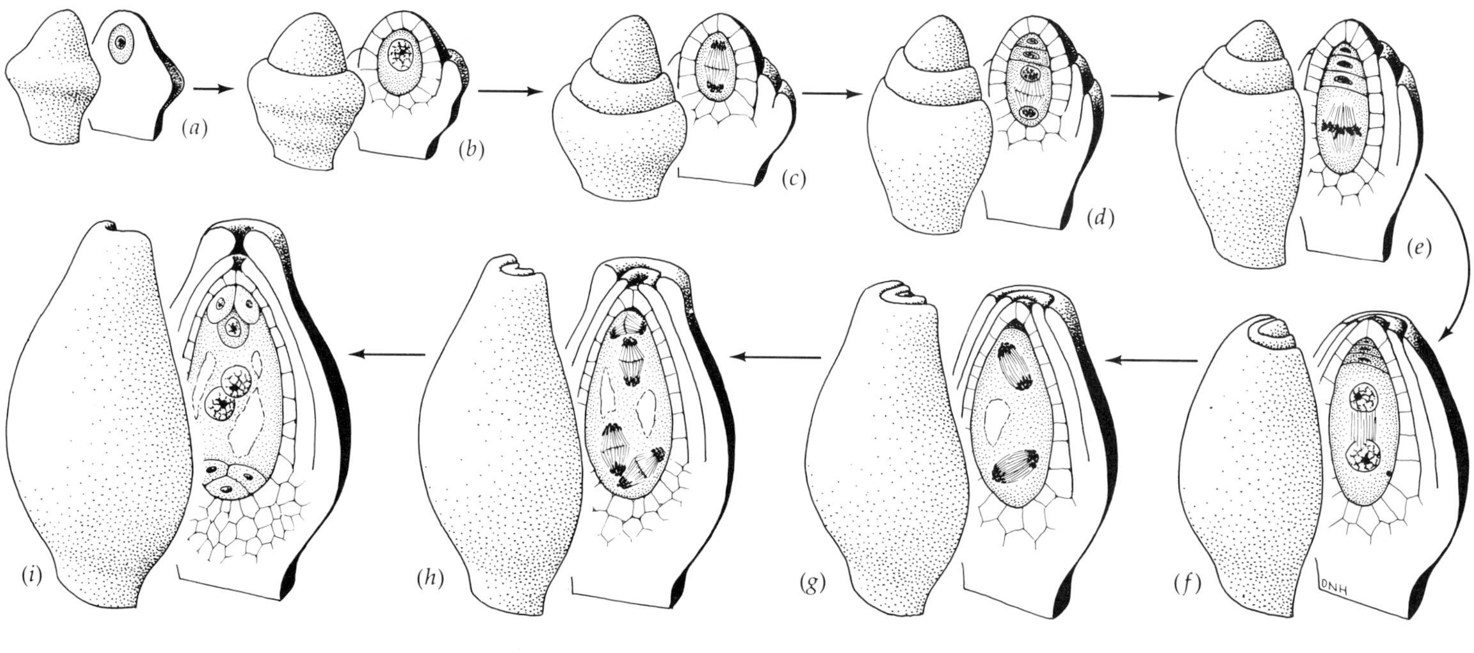

28–16 The development of the ovule: the growth of the integuments (a–i), the megaspore mother cell (a, b), meiosis (c–d), 3 megaspores degenerating with the nucleus of the remaining one dividing (e, f), the development of the female gametophyte (g–i). (At the micropyle end are the 2 synergids and the egg cell, and at the chalazal end are 3 antipodal cells. The remainder of the gametophyte is occupied by binucleate central cell.)

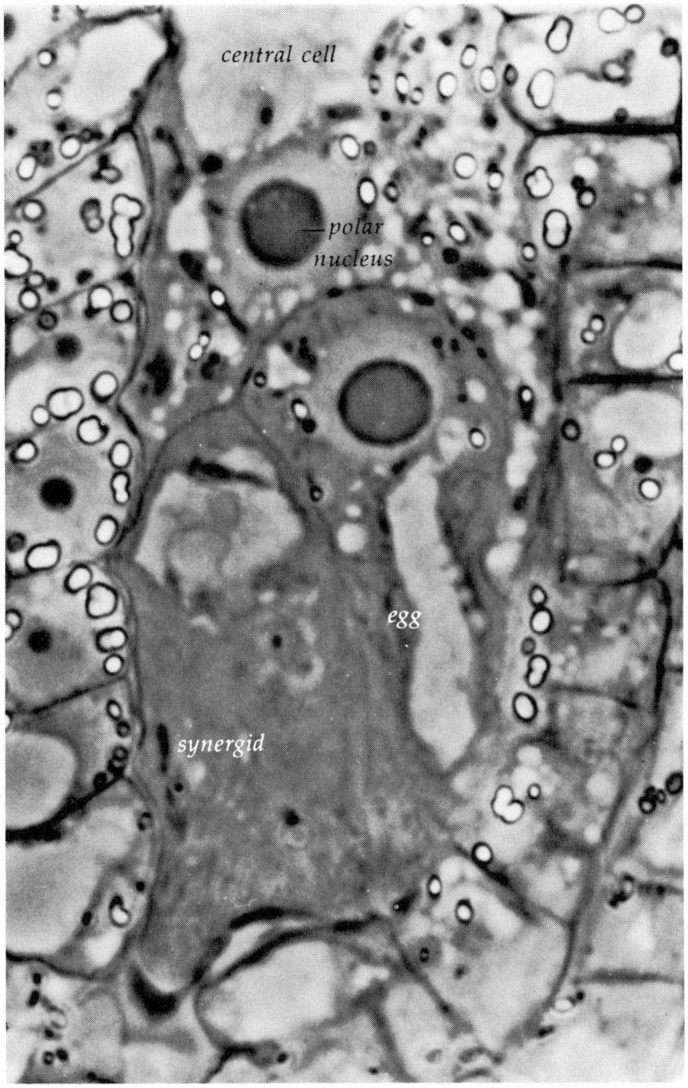

(a)

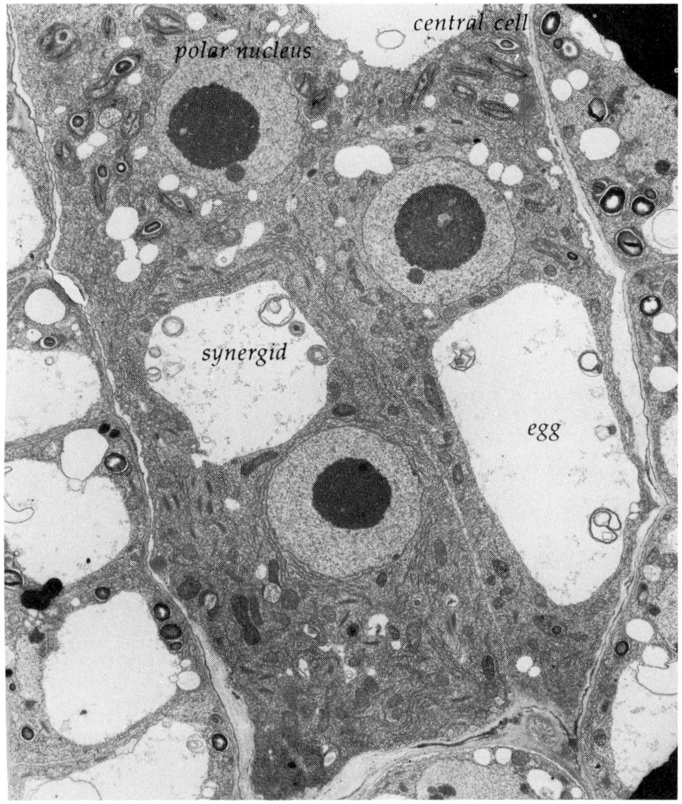

(b)

28–17 In angiosperms the egg is associated with 2 cells called synergids. These 3 cells, termed the egg apparatus, are shown here on both the light and electron microscope level for Capsella. *(a) The egg apparatus and polar nucleus as seen with the light microscope. (b) A similar arrangement of cells as seen with the electron microscope.*

and undergo a series of mitotic divisions. These divisions result in 8 haploid nuclei that are organized into 7 cells.

3 cells are located at the end of the gametophyte near the micropyle—the egg and two synergids (Fig. 28–17). In some species, the egg may be a cell with dense cytoplasm and a few, small vacuoles, while in others it may have a large vacuole and relatively little cytoplasm. The wall surrounding the egg is thin and in some species only partially covers the cell, since the end away from the micropyle lacks a wall.

The synergids are cells filled with cytoplasm rich in mitochondria, dictyosomes, ER, plastids containing starch, and other cell parts. They were once thought to have little function in the megagametophyte, but we now know that they play an important role in the entrance of the pollen tube into the megagametophyte.

At the other end of the megagametophyte or embryo sac, as the megagametophyte is commonly called, are 3 cells called the antipodals. Their function is still unknown.

The egg, 2 synergids, and 3 antipodals account for 6 of the 8 nuclei present in the embryo sac. The remaining 2 are present in the large central cell, which

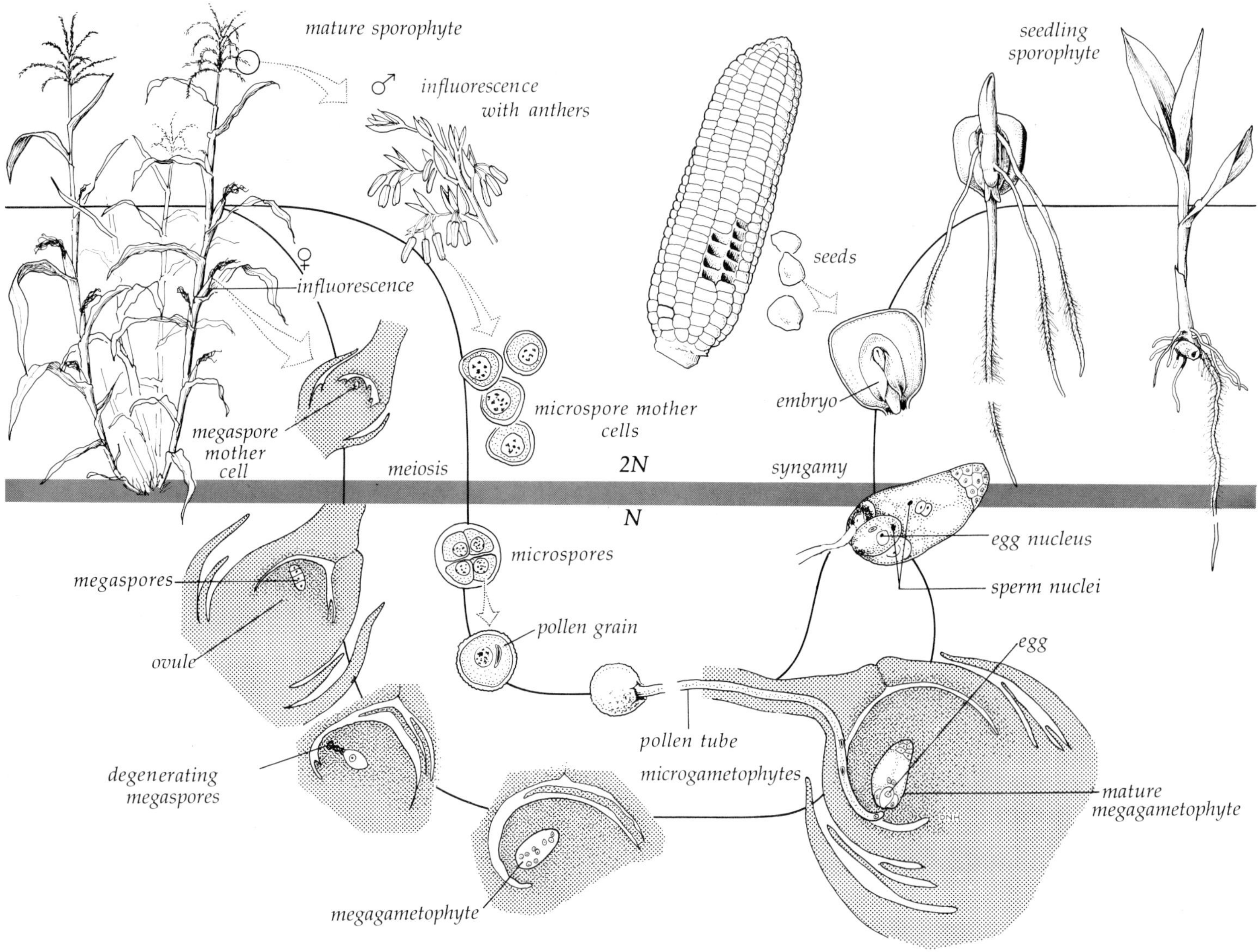

28–18 *The life cycle of an angiosperm.*

is binucleate. These 2 nuclei have a special name: polar nuclei.

There are many variations of embryo sac development in which the mature megagametophyte is different from the one just described. One is the type found in the lily and discussed in Figure 28–19. In others more than 8 nuclei are produced. Many polar nuclei may be present, or many antipodals may be formed. We will not go into the details of the numerous variations but simply point out that they do exist and that they do not affect reproduction.

Fertilization

The pollen tube grows down the style, enters the ovary growing along the wall, and grows up the funiculus of the ovule into the micropyle. It continues growing

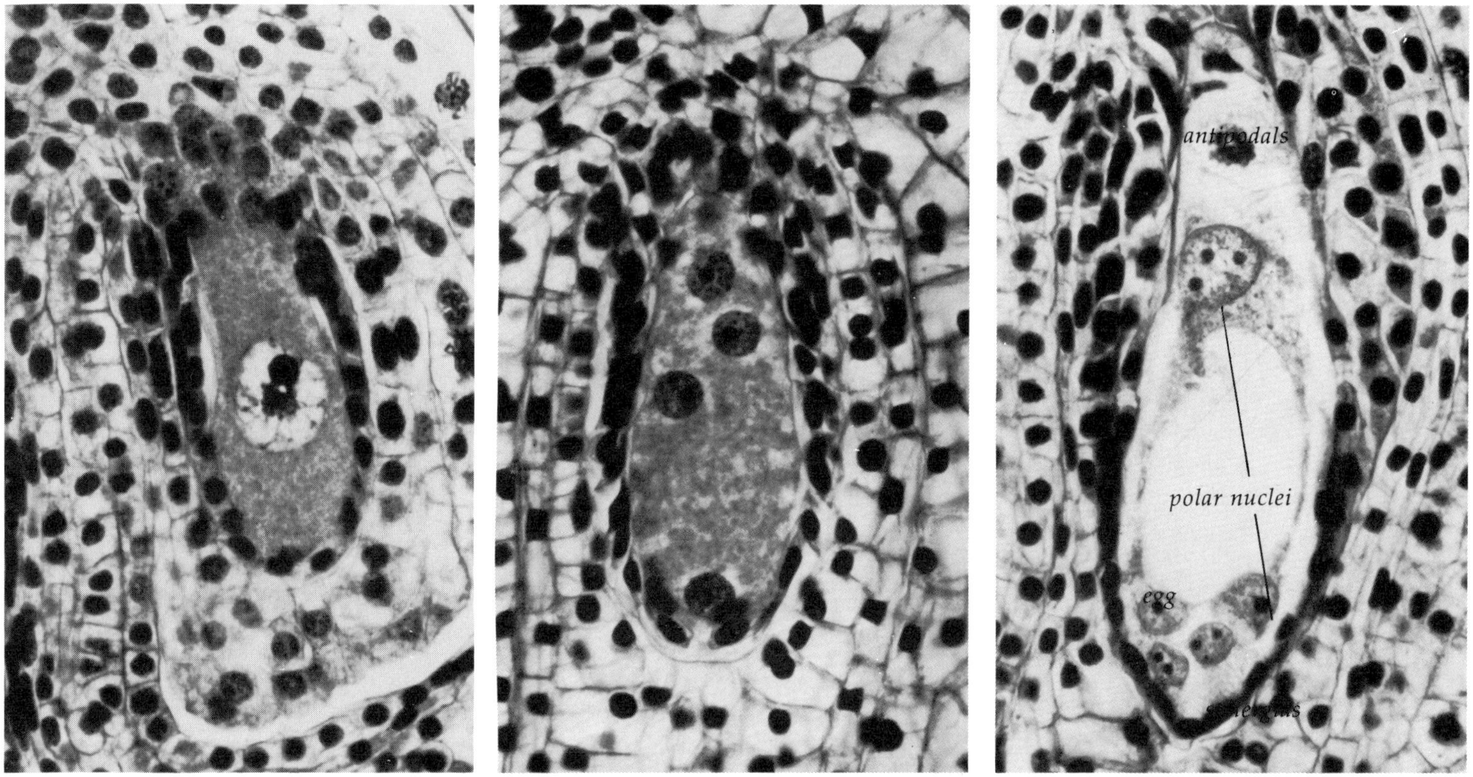

(a) (b) (c)

28–19 *The megagametophyte development described in the text is the one found in the vast majority of species. It is sometimes called the "normal type," although many variations are normal only for the species in which they occur. A modification commonly seen in elementary biology laboratories is that found in the lily. The difference starts with the division of the megaspore mother cell (a). Meiosis is normal. But walls do not form between the nuclei, and none degenerate (b). One migrates to the micropylar end of the embryo sac, while the other 3 migrate to the opposite end. All the nuclei now divide, but the 3 that are together do a curious thing—all the chromosomes line up together on the metaphase plate. This results in a division producing only 2 nuclei instead of 6. The division of the cell at the micropylar end is normal: the embryo sac now has a total of 4 nuclei. Each of these 4 nuclei divides again. Hence the usual 8 nuclei are produced, and these are organized into the regular 7 cells found in the normal type (c). Thus the end result is that the mature megagametophytes of the two types look identical. In the lily type, however, the nuclei of the antipodals and one of the polar nuclei are triploid, not haploid. This makes no difference in reproduction, since the egg is the standard haploid cell.*

through the nucellus and reaches the embryo sac. Here, the tip of the pollen tube grows into 1 of the 2 synergids and discharges (Fig. 28–20). The 2 sperms and the vegetative nucleus are discharged into the synergid. One of the sperms, or at least its nucleus, enters the egg, and after the egg and sperm nuclei fuse, a zygote is formed. The second sperm, or its nucleus, enters the central cell, and the polar nuclei and the sperm nucleus fuse to form the primary endosperm nucleus. This is double fertilization, a phenomenon that occurs only in the angiosperms.

Zygote and embryo. The zygote may resemble the egg if the egg is densely cytoplasmic to begin with. If the egg has a large vacuole, then the zygote usually

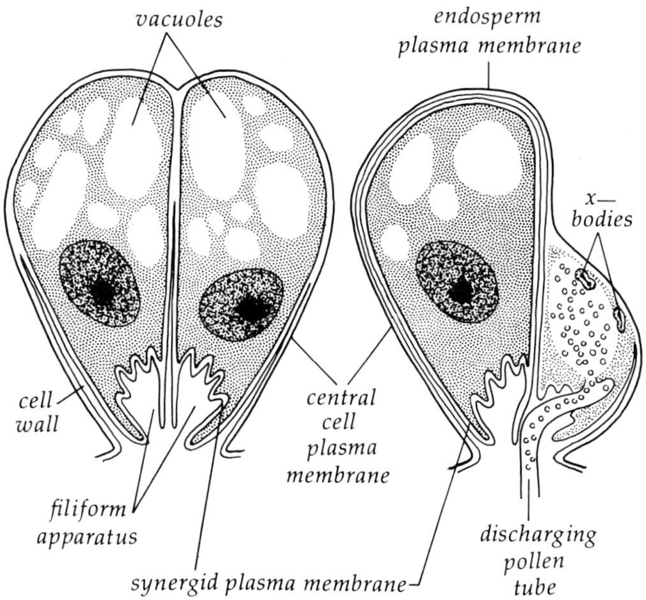

28–20 A diagrammatical summary of the relation of the synergids to the discharge of the pollen tube. The pollen tube enters one of the two synergids by growing through the filiform apparatus, which is a proliferation of the cell wall of the synergid. The pollen tube itself does not enter either the egg or the central cell; rather it discharges into the synergid it has penetrated.

shrinks in size and becomes highly cytoplasmic before division takes place.

The zygote may divide soon after formation, but in most cases there is a lag of several days before division occurs. The division of the zygote is always unequal (Fig. 28–21). The small terminal cell forms the part of the embryo that will continue to divide and develop into the mature embryo of the seed. The large basal cell divides a limited number of times and forms the suspensor. The suspensor apparently functions in the conduction of materials to the developing embryo.

Division of the cells of the embryo first gives rise to a globular mass of cells (Fig. 28–21). The first tissue to differentiate is the epidermis (called the protoderm in the embryo). Then, two centers of division are initiated that establish the beginnings of the cotyledons. This gives the embryo a heart shape. Continued cell division and some elongation result in an elongate embryo. Now, the shoot and root apical regions are clearly defined, and the beginnings of vascular differentiation can be seen. In the mature embryo 1 or 2 well-developed cotyledons are present; a shoot tip with several tiny leaves is formed; and a root tip complete with root cap is present. In most seeds the cotyledons are the main food storage organs.

Endosperm. The second sperm nucleus fuses with the two fusing polar nuclei, resulting (in the normal type embryo sac) in a triploid ($3n$) nucleus—the primary endosperm nucleus. In the lily this nucleus will not be triploid but pentaploid (one triploid polar nucleus + one haploid polar nucleus + one haploid sperm nucleus). In some species the endosperm may be of even higher ploidy levels; in a few it is only diploid. Regardless of the number of chromosomes, endosperm development is essentially the same.

Usually, the primary endosperm nucleus divides soon after formation, and the resulting nuclei continue to divide rapidly (Fig. 28–21). These are free nuclear divisions, and no cell walls are formed. The divisions are also roughly synchronous, and hundreds of nuclei are eventually formed. Cell-wall synthesis then takes place, commonly occurring more or less simultaneously throughout the endosperm. In some species, wall formation occurs at the time of nuclear division, and a cellular endosperm is produced.

The endosperm is clearly involved in the nutrition of the embryo, but the exact relation is unknown. If the endosperm fails to develop, as it does in many infertile crosses, the embryo fails to develop. In some of these cases, the embryo is known to be normal because free of the endosperm it can be cultured to maturity.

The endosperm is usually used up during the development of the embryo, but in some cases (the classic example being the castor bean), the endosperm is still a significant portion of the mature seed.

The Seed and Fruit

The angiosperm seed (Fig. 28–22) consists of the embryo (the new sporophyte generation), surrounded by the remains of the endosperm (a special structure that is neither gametophyte nor sporophyte), and a seed coat, produced by the parent sporophyte from the integuments. The bulk of most seeds is occupied by the cotyledons, the cells of which are filled with stored food. In seeds that contain large amounts of endosperm, the cotyledons are more leaflike in appearance, and the endosperm cells are filled with stored food including protein, fats, oil, and starch.

The range in seed size in angiosperms is great. The smallest are found in the orchids where close to 4 million dustlike seeds have been counted in a single fruit. The largest seed is that of the double coconut, *Coco-de-mer*, which weighs 20 kg.

While the seed is developing from the ovule, the ovary is developing into a *fruit*. Actually, the term fruit is often loosely used, and in many cases, the fruit involves more of the flower than just the ovary. Pollination followed by fertilization is usually necessary for fruit formation, but in some cases, the formation will occur without fertilization. Commercial bananas, for example, have no seeds because of this phenomenon, called parthenocarpy. Such growth regulators as IAA or gibberellins (see Chap. 30) can artificially induce parthenocarpy and can be used to obtain seedless tomatoes.

Essay 28–1 The Study of Fertilization in Angiosperms

The history of the study of fertilization in angiosperms is long and colorful. Theophrastus, a student of Aristotle, knew that the Arabs and Assyrians had a special ceremony in which a man climbed a male date tree, brought down the male inflorescence, and carried it to the female flowers, where it was rubbed against them to ensure a good supply of dates. For hundreds of years, despite this early insight, the role of pollen in seed production was unclear. The grains were too small to be studied with the naked eye.

The discovery of the microscope opened a new approach to the problem, but even then progress was slow. Not until 1824 did Giovanni Amici, who was actually a mathematician and astronomer, discover pollen tubes. The role of these tubes was far from clear and was made even less so by the work of Matthias Schleiden. He is the man who, with Schwann, formulated the cell theory. In that case he was right, but in the case of the pollen tubes he was not. Schleiden believed that the pollen tube grew down into the ovule and that the tip of the tube itself became the embryo. The ovule was merely an incubation chamber, not the contributor of the egg. Schleiden and Amici became involved in a violent argument—hardly uncommon in science—and finally, in 1849 the truly great German botanist, Wilhelm Hofmeister, settled the question in a beautiful piece of research that showed the true role of the pollen tube. Even so, double fertilization was not discovered until 1898 by Sergius Nawaschin, a Russian, and independently by Léon Guignard, a Frenchman.

For many years, despite numerous careful studies, the details of entrance of the pollen tube and the mechanics of fertilization were unclear—some still are. As noted earlier, the structure of sperms was unclear until recently. Now, with the aid of the electron microscope, it is clear that they are true cells, not merely nuclei. The synergid was believed to be destroyed by the pollen tube, and the pollen tube was said to discharge directly into the embryo sac. Again, through the use of the electron microscope, we know that the pollen tube does not destroy the synergid and that there is a series of interactions between pollen tube and synergid. We also know the mechanism of nuclear fusion involved in the formation of the zygote and the primary endosperm nucleus.

There remain a host of unanswered questions. How much of the sperm enters the egg? How does only one nucleus enter the egg and one the central cell? What mechanism directs the tip of the pollen tube into the synergid? What stops the growth of the pollen tube? What causes pollen tube discharge? And so on.

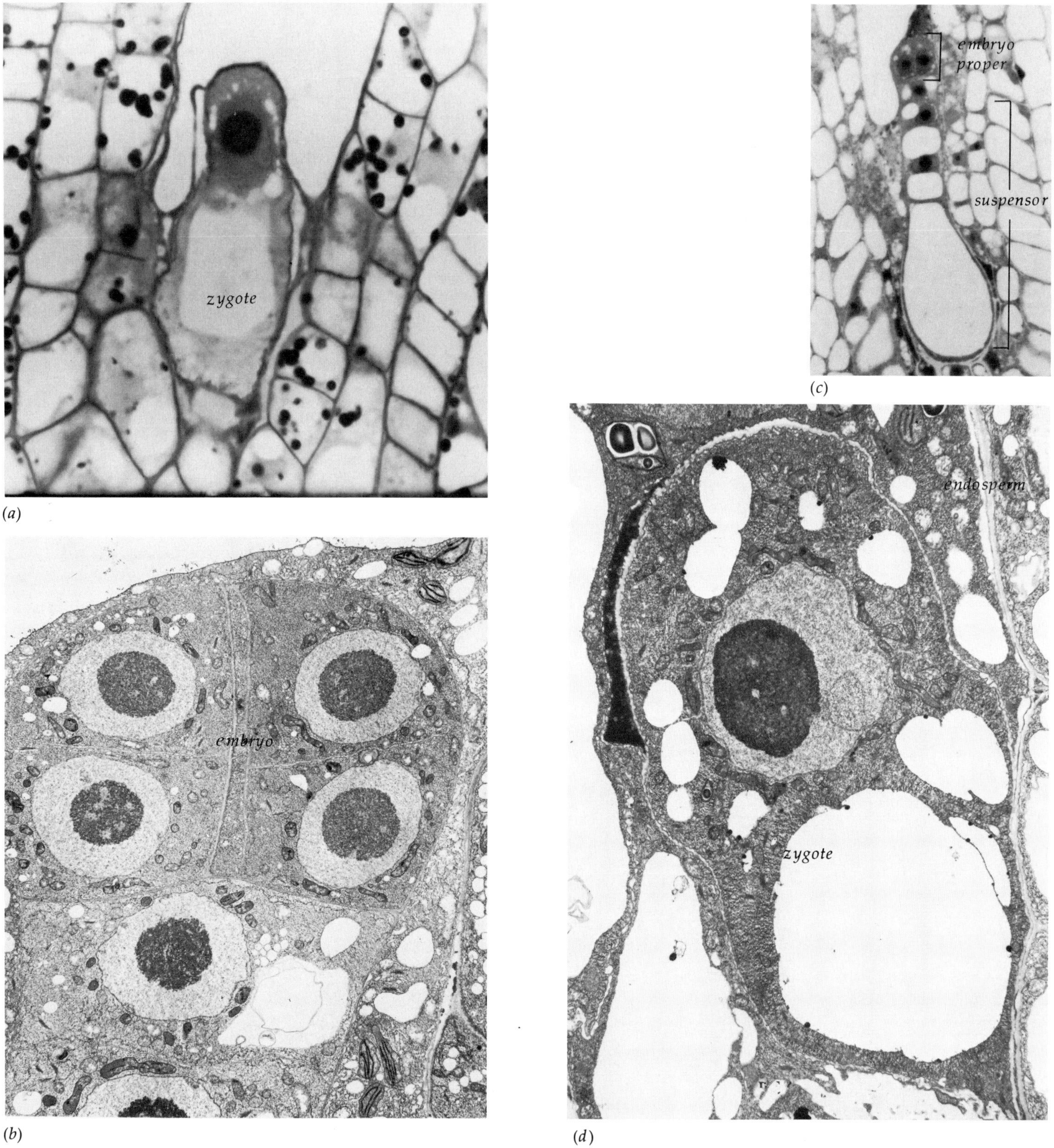
zygote
(a)
embryo
(b)
embryo proper
suspensor
(c)
endosperm
zygote
(d)

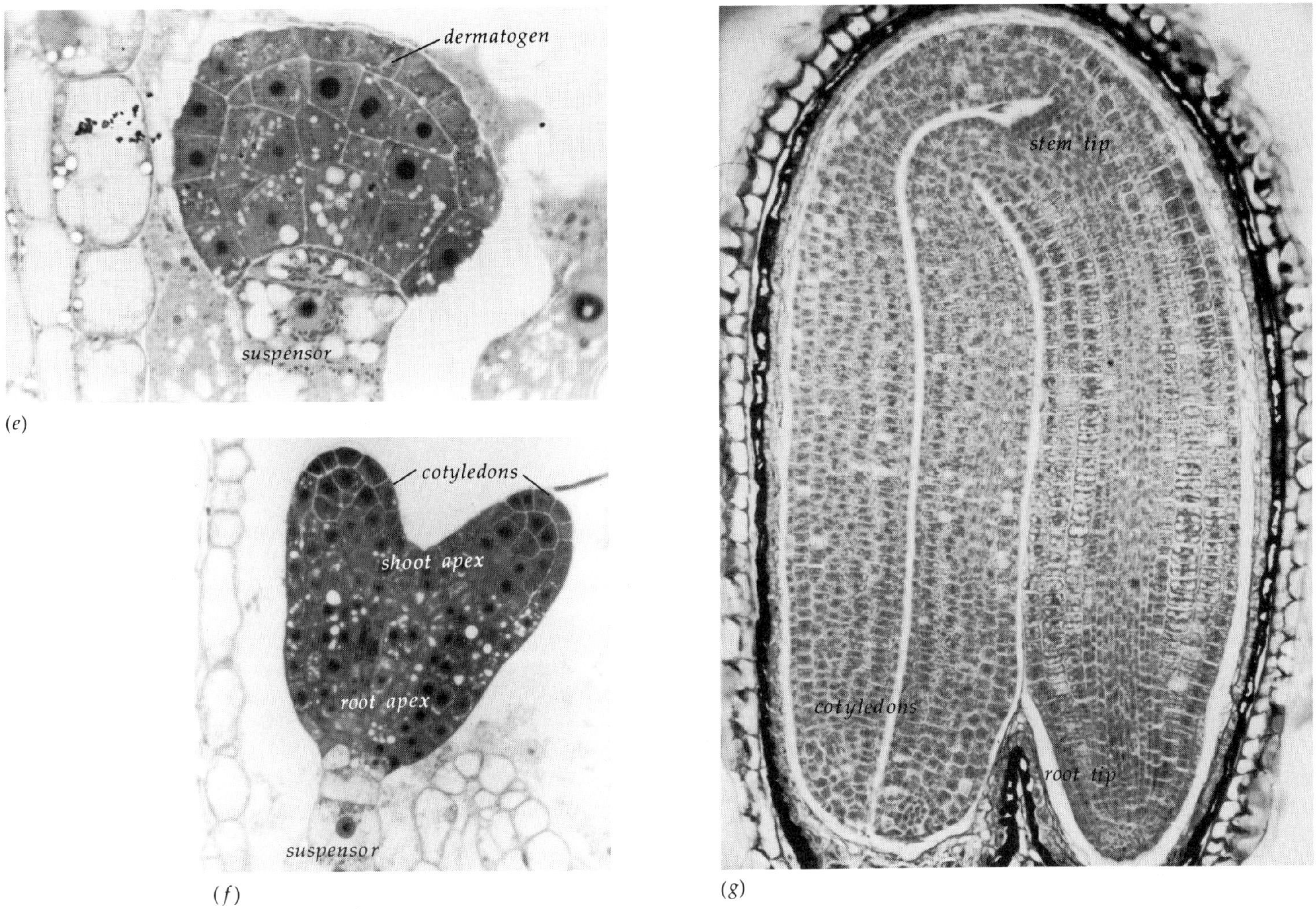

28–21 *The embryo development of* Capsella bursa-pastoris *or shepherd's purse is widely studied both in the laboratory and in the classroom. Noted for a regular pattern of development, it illustrates the general course of growth of angiosperm embryos. Here we present photographs from both the light and the electron microscope illustrating some of the major features of embryo development in* Capsella. *The egg was shown in Figure 28–17 and should be compared with the photo of the zygote shown here (a, b). In* Capsella *the zygote elongates before the first division. The first division is unequal (as it is in all angiosperms), thus resulting in a small terminal cell and a larger basal cell. The small terminal cell goes on to form the embryo proper, while the basal cell forms the suspensor, which functions only during the development of the embryo (c). The terminal cell undergoes a regular series of divisions that result in the formation of a roughly spherical mass of cells known as the globular embryo (d). In this globular embryo the future epidermis (the dermatogen) is differentiated (e). Further development of the embryo results in the formation of 2 cotyledons and the beginning of the axis (f); this is the heart stage. Continued cell divisions and elongation convert the embryo to a more elongate structure; this is called the torpedo or walkingstick stage. In* Capsella, *because of the shape of the ovule, the embryo becomes bent, and the cotyledons lie parallel to the axis (j).*

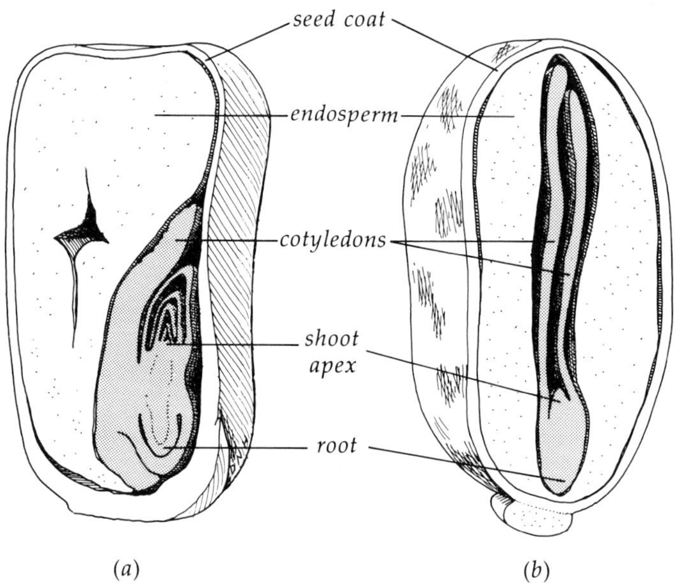

28–22 The structure of the seed of a monocotyledon (a—corn) and a dicotyledon (b—castor bean). Many dicot seeds, however, lack the large endosperm so characteristic of the castor bean.

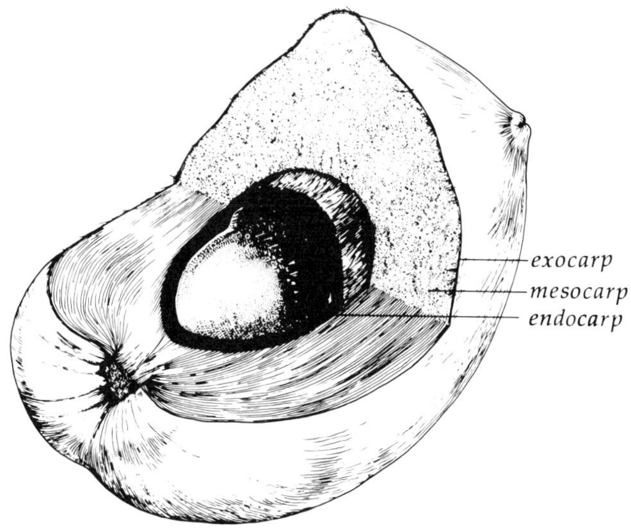

28–23 The coconut fruit Cocos nucifera *cut to show three layers of pericarp: the thin, hard exocarp, the fibrous mesocarp, and the stony endocarp.*

The ovary wall forms the pericarp of the fruit, which may be dry or fleshy, fibrous or stony. There are three distinct layers to the pericarp, which are nicely illustrated by the coconut (Fig. 28–23). The milk of the coconut is liquid endosperm, while the edible coconut meat is solid endosperm. Surrounding the coconut meat is the inner layer of the fruit wall, the endocarp. This is a stony layer in the case of coconut. The middle layer of the pericap, the mesocarp, is thick and fibrous. The exocarp is the outer layer, which is thin and hard.

Obviously, the layers of the pericarp can be modified in numerous ways in different fruits. The various types of fruit are classified artificially in Table 28–24 and illustrated in Figure 28–25.

Simple fruits are formed from a single ovary; aggregate fruits such as raspberries and strawberries are formed from a single flower with numerous but unfused carpels; and multiple fruits are formed from several flowers—the pineapple being a good example. The relation of the various parts of the flower to the mature fruit can best be shown graphically, as we have done in Figure 28–26. The development of the receptacle is important in many fruits, called accessory fruits. Apples, strawberries, and figs are good examples.

Seed Dispersal

The object of the fruit is the protection and dispersal of the seed. In this it succeeds magnificently. Many possible means of transport has been utilized by some angiosperm species. To see a beautiful example of evolution resulting in morphological structures superbly adapted to specific environmental conditions, let us examine a few dispersal mechanisms involving both fruit and seed.

Some are quite simple; orchid seeds are so small and light they can be carried many miles by the wind. For heavier seeds we see fruits in the form of wings,

28–24 *An artificial classification of fruits*

Fruits dry—pericarp dry when fruit is mature

Indehiscent fruits in which the pericarp remains closed at maturity.

Achene—a small single-seeded fruit with a relatively thin pericarp; except for its attachment by the funiculus, the seed lies free in the cavity of the ovary: buttercup (*Ranunculus*), crowfoot (*Potentilla*), water plantain (*Alisma*), buckwheat (*Fagopyrum*).

Caryopsis—achene-like except that it is derived from a compound ovary and the seed coat is firmly united to the pericarp: characteristic fruit of the Gramineae.

Cypsela—achene-like but derived from an inferior, compound ovary: characteristic fruit of the Compositae.

Nut—like an achene but derived from two or more carpels, and with a hard or stony pericarp: hazelnut (*Corylus*), basswood (*Tilia*), acorn (*Quercus*).

Samara—a winged achene: elm (*Ulmus*), ash (*Fraxinus*), tree of heaven (*Ailanthus*), hop tree (*Ptelea*).

Schizocarp—the product of a compound ovary that splits apart at maturity into a number of one-seeded portions termed mericarps: maple (*Acer*), many Umbelliferae, Labiatae, Malvaceae, Geraniaceae.

Dehiscent fruits in which the fruit splits, or opens in some manner to release the seeds.

Follicle—derived from a single carpel that splits at maturity down one side, usually along the ventral suture: columbine (*Aquilegia*), peony (*Paeonia*), larkspur (*Delphinium*), milkweed (*Asclepias*).

Legume—also from a single carpel, but dehiscing down both the dorsal and ventral sutures to form two valves: characteristic fruit of *Papilionaceae*.

Silique—the product of a superior compound ovary of two carpels, the pericarp separating as two halves leaving a persistent central portion with the seed or seeds attached to it: characteristic fruit of Cruciferae.

Capsule—from a compound ovary; various types of dehiscence are found (e.g., longitudinal, porous, or circumscissile); in general the dehiscence is from top downward, and the separated portions (valves or teeth) remain attached: characteristic of numerous families.

Fruits fleshy—pericarp partly or wholly fleshy or fibrous

Drupe—carpels one or more but usually single-seeded; mesocarp fleshy but endocarp hard and stony: cherry, peach, plum (*Prunus*), coconut (*Cocos*), olive (*Olea*).

Berry—one to several carpels, usually many-seeded; both mesocarp and endocarp are fleshy: one-seeded, nutmeg (*Myristica*), date (*Phoenix*); single carpel and several seeds: baneberry (*Actaea*), barberry (*Berberis*), *Mahonia;* others with more than one carpel: grape (*Vitis*), tomato (*Lycopersicon*) and potato (*Solanum*), *Asparagus.*

Pome—derived from a compound inferior ovary; much of the fleshy portion is the enlarged base of the perianth tube with only the central part composing the pericarp; both the exocarp and mesocarp are fleshy while the endocarp (the core) is stony or cartilaginous: characteristic fruit of the Pomoideae, apple (*Malus*), pear (*Pyrus*), quince (*Cydonia*), mountain ash (*Sorbus*).

such as those of the elm, maple, and ash (Fig. 28–27). In some cases, such as that of the common city tree, catalpa, the seeds themselves are winged. Plumed fruits and seeds are also effective in wind dispersal and are employed by many weeds such as milkweed and thistle. Willows and poplars, like cotton, have seeds covered with woolly hairs, which aid in their wind dispersal.

Another approach to seed dispersal is that of plant mechanisms involving water. Again, the coconut is a fine example of this type of dispersal. The thick, buoyant pericarp both protects the embryo and allows the seed to float to a new location over hundreds, even thousands, of miles of open ocean. Other fruits, less impressive in size, follow similar pathways on oceans, rivers, and lakes.

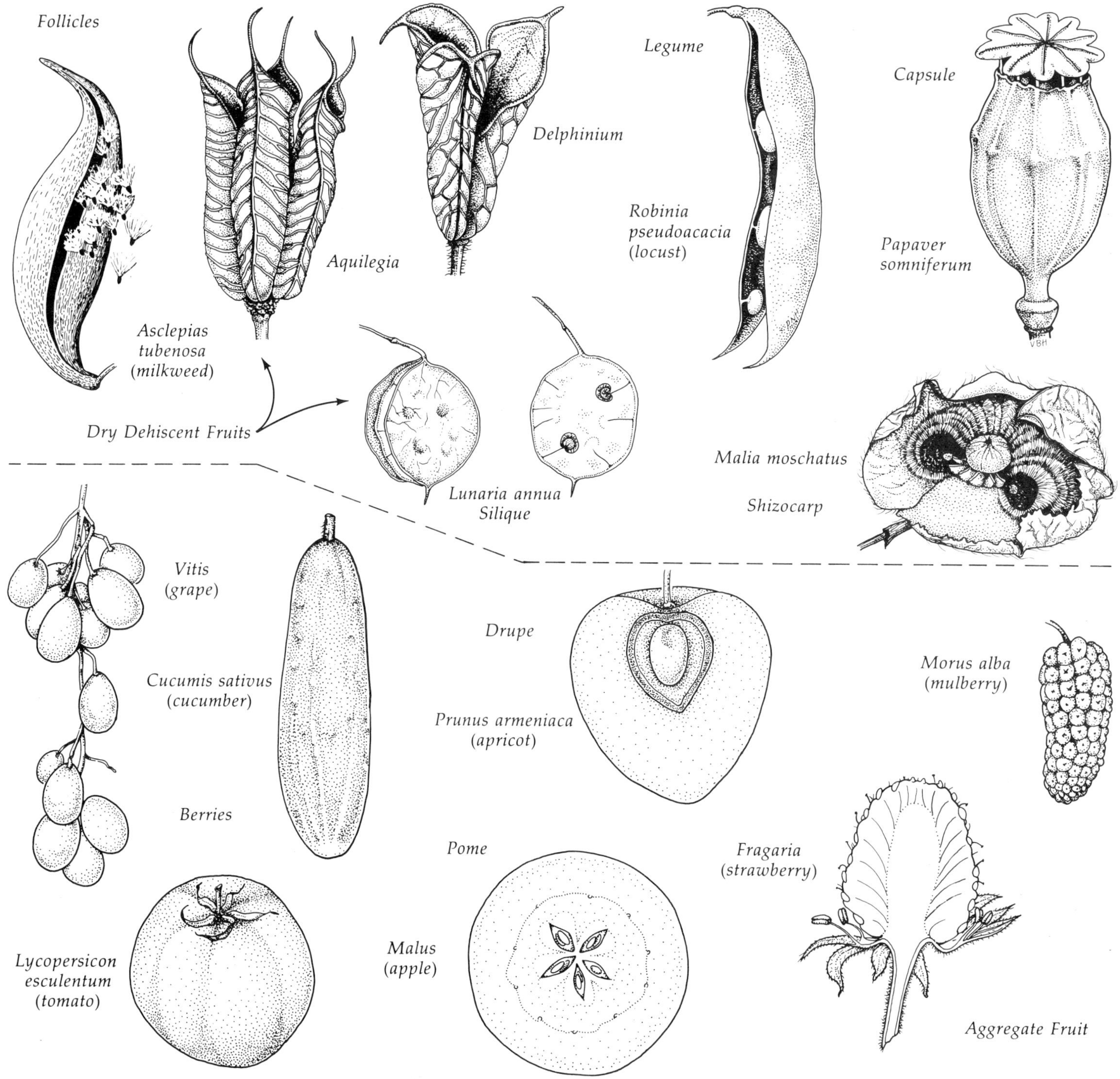

28–25 Some examples of various types of fruits.

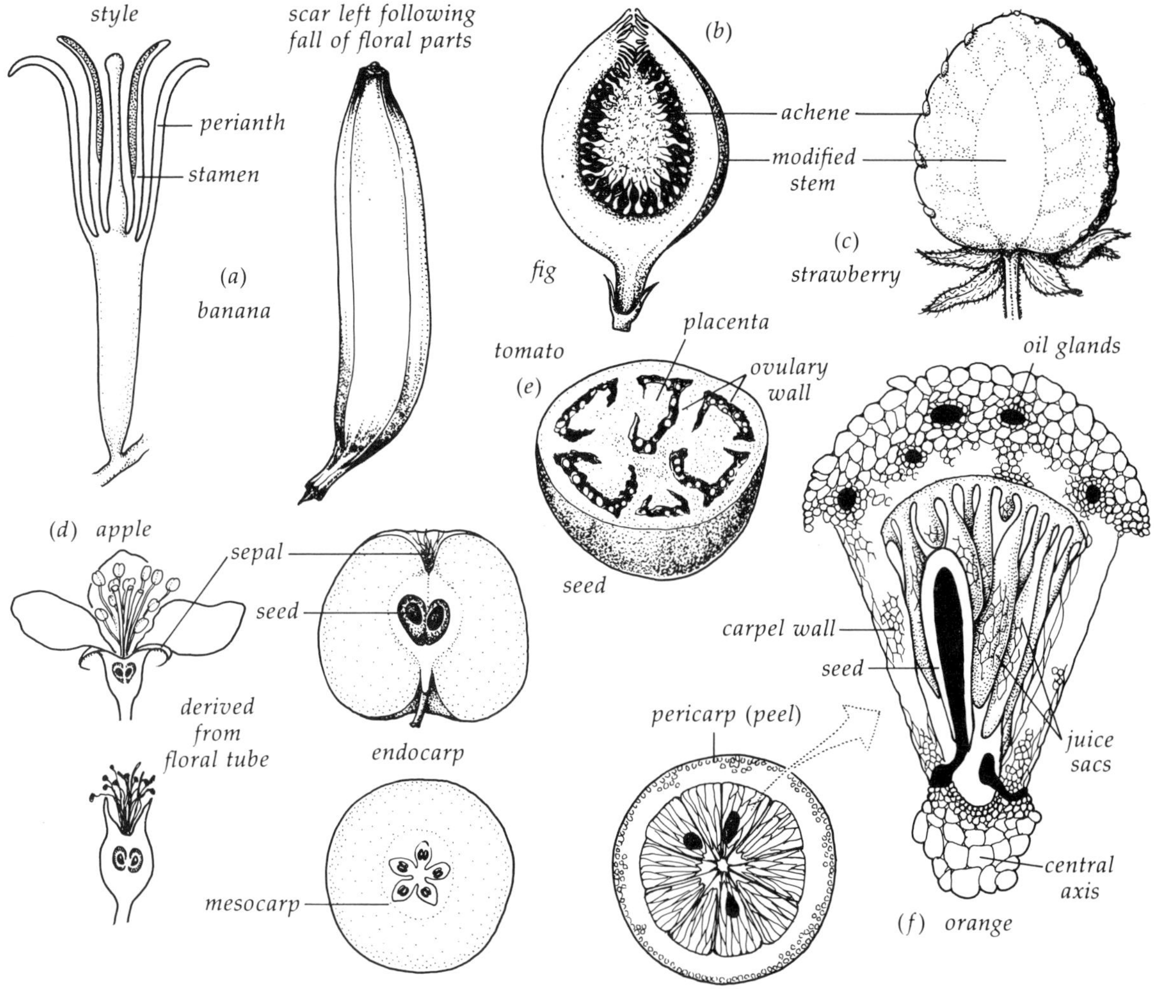

28–26 The relation of various parts of the flower to the development of the fruit.

One of the most impressive adaptations to life along the edges of oceans, also involving seed dispersal, is the mangrove. Mangroves live in mud flats flooded with salt or brackish (i.e., partially salty) waters of tropical and semitropical sea coasts. The plants have stilt roots that emerge from the trunk above high water; these support and aid in aeration of the underwater roots that grow in mud almost completely devoid of oxygen. The seeds of the mangrove germinate while still attached to the parent plants—a most unusual thing in angiosperms. A root is produced that may grow as much as 30 cm in length and be over 1 cm thick, before the seeds are dropped. When they are released, the germinated seeds fall like arrows with weighted ends and are planted in the mud. If they fail to become rooted immediately, they may be carried away by the tide. In this case, the root acts as a keel, thus increasing the chances it will root when carried in by the tide.

Some angiosperms (e.g., the saxifrages) have also evolved splash-cups—devices we saw develop independently in several fungi and the liverworts. Here, the rain falling in the cuplike fruit splashes the seeds away.

When we consider the role of animals in the dispersal of angiosperm seed, we reach the ultimate in plant adaptation to a land occupied by animals. Here, in an evolutionary sense, the plants are utilizing the striking motility of animals for their own survival.

Perhaps, the most obvious of these methods involves the eating of fleshy fruits, resulting in the seeds being passed through the alimentary tract and then

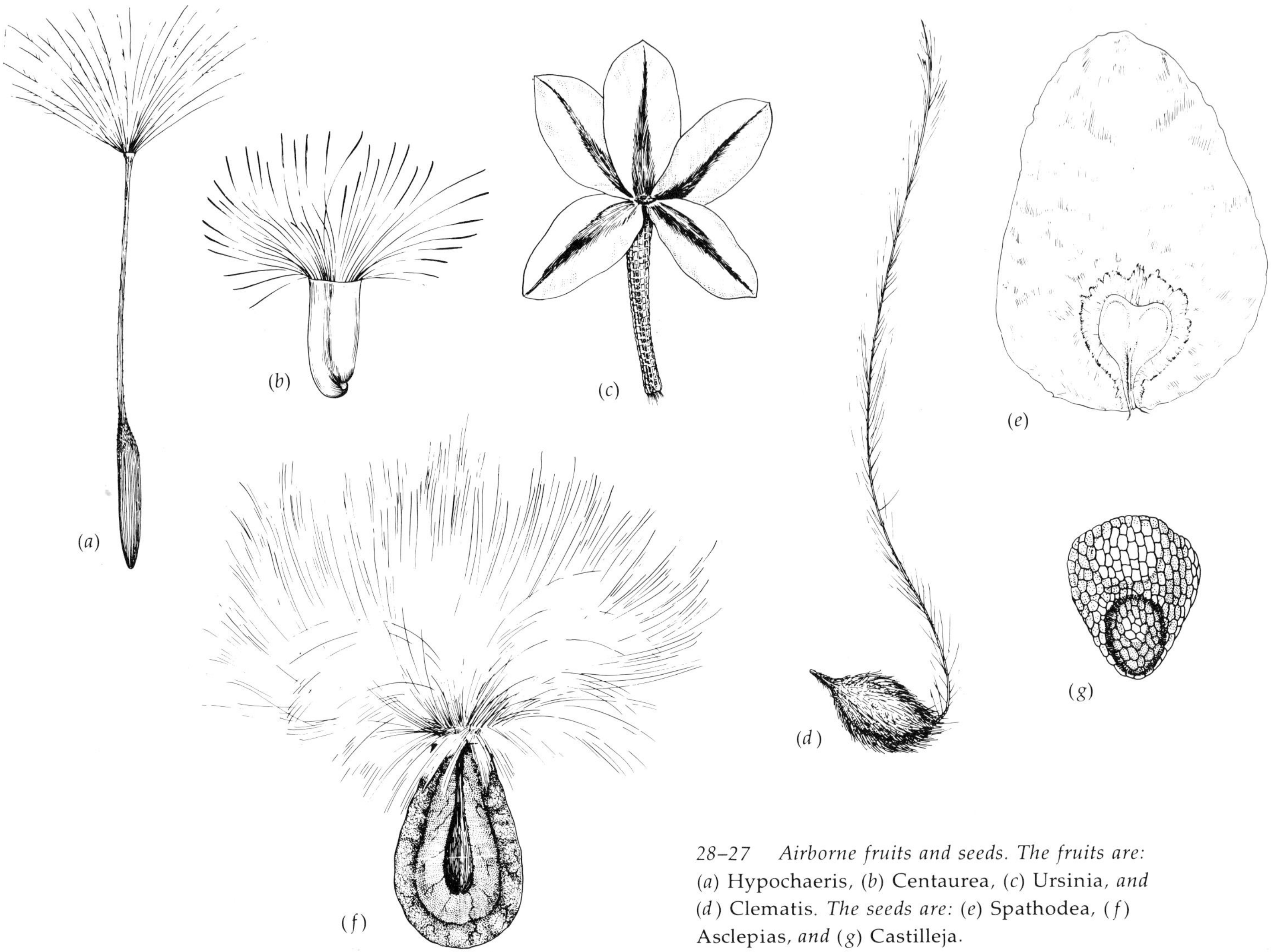

28–27 Airborne fruits and seeds. The fruits are: (*a*) Hypochaeris, (*b*) Centaurea, (*c*) Ursinia, *and* (*d*) Clematis. *The seeds are:* (*e*) Spathodea, (*f*) Asclepias, *and* (*g*) Castilleja.

dropped by the animal. Far from harming the seed, this passage through the animal scarifies the seed so more can germinate. Man has domesticated many of these plants because the fruit, quite obviously, is good to eat. Thus we find the strawberry, raspberry, currant, gooseberry, and cherry all in this group.

Even more common are those angiosperms in which the fruits or seeds have developed means of adhering to fur and feathers. These means include hooks, spines, prickles, hairs, and sticky external layers. All manner and sizes of these structures have evolved in many diverse families. Some representative examples are shown in Figure 28–28.

Nuts are another means of seed dispersal involving a plant-animal relationship. Here, it appears that part of the process is based on having some but not all of the seeds eaten. In the case of oaks, for example, squirrels collect and eat the acorns. They frequently bury (although from the plant's point of view it is planting) acorns that they never dig up. These germinate, and the new plant becomes established, away from the parent.

This type of interaction between plant and animal in seed dispersal is also characteristic of the most sophisticated of all such interactions—man and plant in agriculture. Here, some grasses evolved to the point

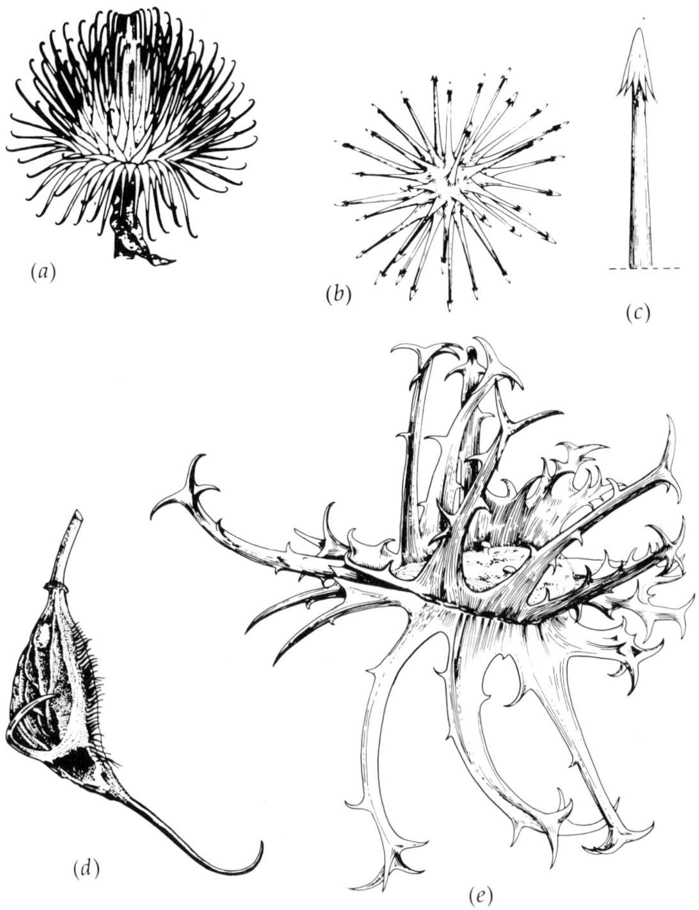

28–28 Animal-dispersed fruits often show spines and hooks of remarkable size and number. The spines of Arctium minus *have hooked tips (a).* Acaena *has spines with grapnellike tips (b & c). Two highly evolved forms with large spines are* Proboscidea *(d) and* Harpagophytum *(e).*

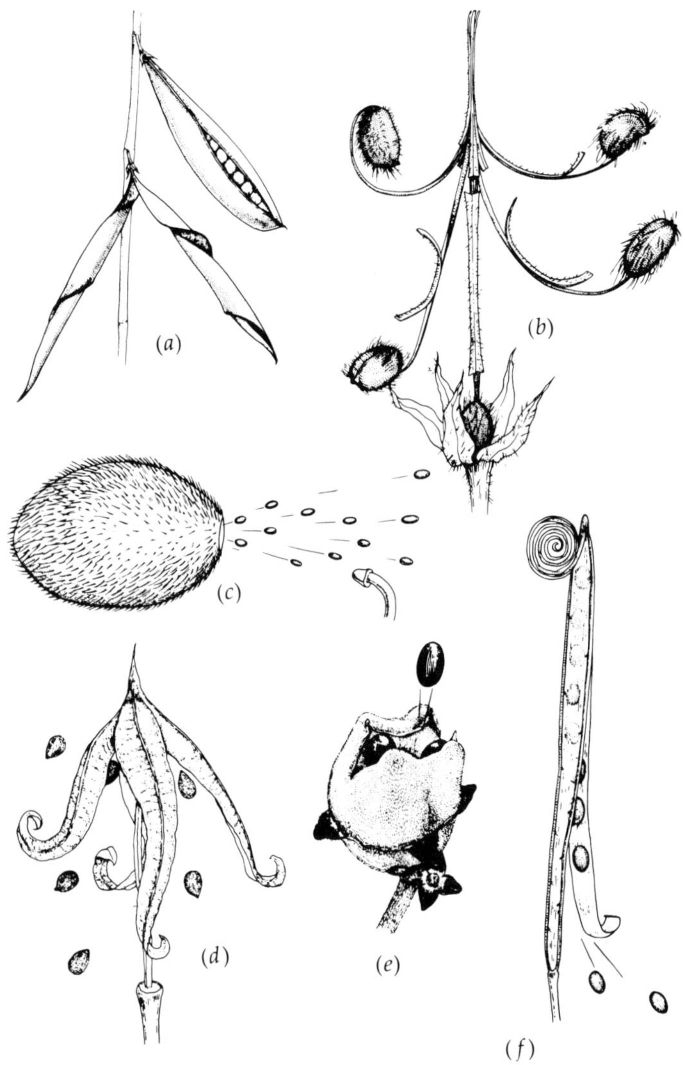

28–29 Some examples of mechanically dispersed seeds: (a) Vicia gigantea, *(b)* Geranium bicknellii; *(c)* Ecballium elaterium, *(d)* Impatiens aurella, *(e)* Hamamelis virginiana, *and (f)* Cardamine hirsuta.

where their seed could be used as food for man. Once man not only collected the native grass seed but began to plant it, he was on the level of the squirrel, although his act was more deliberate. Then, man began to both carry the seed and replace certain plants with other species he wanted. He also continually and more deliberately modified by selection the plants he had chosen, something that no other animal has consciously done. At the same time man himself has become increasingly dependent upon these same plants for his survival.

Finally, before we leave the topic of seed dispersal we must consider those plants that literally shoot their seeds away. Some of these are shown in Figure 28–29. Here, we have fruits that throw the seeds out by various twisting (Fig. 28–29a), coiling (Fig. 28–29f), slinging (Fig. 38–29b), and snapping (Fig. 28–29d) mechanisms. In addition, there are fruits in which internal pressure builds up, and the seeds are spit out through a small opening at one end (Fig. 28–29c). Finally, in witch hazel (Fig. 28–29e), a springlike mechanism tosses the seed a distance of 14 m or more.

29

The Angiosperms: Taxonomy and Major Groups

The Herbarium

For generations botanists have taken to the field to collect plants for study. They learned early that carefully dried and protected plants could be kept almost indefinitely. This meant that plants collected many years and miles apart could be compared and studied. Thus began the great collections of plants, accumulated over the centuries in many parts of the world. Such collections are called herbaria and may contain hundreds of thousands of specimens, although many are much smaller and more specialized. Herbaria are centers of taxonomic research, but they are used by a wide variety of botanists and agronomists.

Specimens are not only stored and studied in herbaria, they are also processed from their fresh condition to the final preserved form. These photographs will introduce you to some of the operations that take place in an herbarium.

A The fresh specimen is placed between sheets of newsprint. The folded newsprint containing the specimen is placed between blotter paper and corrugated cardboard (for ventilation) in a plant press. The straps are tightened to provide about 20 kg pressure so that the plants will dry flat. The pressing and drying of the specimens is often done in the field, if laboratory facilities are not close by.

B When thoroughly dry, the specimen is mounted on stiff white paper. To do this, the specimen is first placed on a plate on which an adhesive has been brushed, and then it is laid on the paper.

A

B

C A label is also attached to the lower, right-hand corner of the herbarium sheet to which the specimen has been glued. The label is as important as the specimen itself, since it not only provides the name of the plant but where it was collected, when, and by whom.

D When the glue is dry (usually this takes place under weights so the specimen will remain flat and be firmly attached to the paper), the specimens are filed in metal cabinets containing insecticides. Such cases can be seen behind Professor Arthur Holmgren, taxonomist at Utah State University, shown here utilizing the specimens in taxonomic studies.

C

D

There is something of the collector in all of us. So we can appreciate the impression of beauty created by an ingenious and logical organization of things. Furthermore, we gain a sense of satisfaction from knowing something about the world around us—whether our familiarity is with vintage cars, stamps, coins, rocks, early American architecture, the paintings of Van Gogh, the music of Bach, or the world of plants surrounding us. The angiosperm taxonomist has an opportunity to enjoy such things perhaps more than others. He can collect for a lifetime, achieving a marvelous assemblage of flowering plant specimens but knowing that still others await his search; he can organize with singleminded devotion, knowing that he may be discovering nature's story of evolution; and he can become intimately familiar with an extremely important phase of his environment. He is an indispensable cog in the botany machine. Without his work most botanical information would be meaningless, nonrepeatable, and incommunicable. All botanists simply must know the plants they talk about.

The angiosperms provide the best opportunity for the botanical taxonomist. Of the major plant groups, only the fungi approach the angiosperms in numbers of species, and there may be three times as many angiosperms as fungi. Actually, the number is so great that familiarity with such a large group might appear hopeless, or even, without careful organization, impossible. Yet the group can be organized. The angiosperm flower varies only in a limited number of important ways, although there are numerous other, more minor variations. By learning the principal variations and ways to combine them, even the beginning student can build a foundation that will allow him to recognize many flowering plants upon seeing them for the first time. Of course, his recognition will not be at the detailed level of the species, but he will be able to place them in familiar, broad categories. (It is like recognizing Gothic architecture as contrasted with other styles, even though one may not know the exact period and architect of a given structure.) On his own, the student can develop the ability to recognize the major families and genera and many species of the plants growing in a restricted region, such as the Wasatch Mountains of the Rocky Mountain chain or the Everglades of Florida. Plants of such regions usually number only in the hundreds rather than in the hundreds of thousands, and anyone with sufficient interest can master that many plants. The main thing is to develop a suitable philosophy of approach to an organization, as discussed in Essay 29–1 and End Paper 29–1.

The Techniques of Taxonomy

The goal of taxonomy is to relate living organisms into natural schemes representing their relationship to each other through evolutionary descent. The question is how to go about discovering natural relationships. The approach is to accumulate all of the data possible about plants. Virtually every morphological or physiological feature of a plant is gene controlled, and genes are units of inheritance in evolution. Hence, virtually every plant feature might conceivably provide clues to relationships.

The taxonomist today must be far more than a pigeon-holer. He must think analytically and dynamically in terms of evolutionary processes—and synthetically and comprehensively in terms of present relationships. This is no meager challenge. Biology has specialized to a high degree, but since the taxonomist can profit from the application of many modern specialties, he must become adept in some of them. We shall summarize the principal areas of modern biology that are currently contributing to taxonomic studies. All are combined in an approach called biosystematics.

Morphological studies. Traditionally, the taxonomist has primarily been concerned with the macroscopic morphological features of his specimens: vegetative and floral characteristics. He collects morphological information about his specimens in at least three different ways.

The principal approach utilizes the herbarium, where dried specimens are attached to herbarium sheets and filed for future reference (see the Photo Essay). The herbarium is an attempt to overcome the limitations of time and space; plants bloom in the field only for a brief period during each year and only at specific locations. To give an indication of distribution and variation within populations, a good herbarium

Essay 29–1 The Role of the Taxonomist in Biology

The work that scientists do is often poorly understood or misunderstood by the public. Here, an outstanding taxonomist, Dr. Lincoln Constance of the University of California at Berkeley, examines the role of plant taxonomists in studying the living world around us.

"While many modern biologists are concerned with the unity of fundamental structure, physiological processes, and mechanisms of heredity that run through all organisms, the taxonomist concerns himself primarily with the extraordinary diversity that organisms have achieved. By studying their form and function, he seeks to correlate these with their survival under complex conditions and their distribution, both past and present, over the variable surface of the earth.

The variety of organisms is so overwhelming that no one person can hope to know more than a tiny fraction of them. The attempt to bring all living things into an ordered arrangement must be a cooperative endeavor, which not only embraces a worldwide fraternity of systematists, but extends backward over several centuries. The taxonomist of each decade stands upon the shoulders of his predecessors, building upon the discoveries and observations of the past and adding his small contribution to the developing edifice of classification. Working chiefly with materials from the living world but occasionally assisted by evidence from the fossil record, his ultimate objective is to devise arrangements that harmonize with the presumed course of organic evolution. His evidence is largely circumstantial, and his conclusions are subject to verification and correction by his contemporaries and his successors.

His information comes from the world around him, from the library of millions of preserved specimens accumulated in herbaria, and the small but important fraction of exotic species that can be successfully nurtured in botanical gardens. From decade to decade his methods have undergone steady change, most strikingly to utilize an ever-expanding range of data—from morphology, anatomy, and paleobotany, to palynology, cytology, genetics, and comparative biochemistry.

Continued exploration, especially but not exclusively in the tropics, still yields previously unknown organisms. His efforts to account for these and fit them 'into the system' necessitates his familiarity with the work of all his predecessors and leads him to become something of an historian of science. Taxonomic publications do not suffer the fate of obsolescence which has overtaken so much biological writing, and all previous work must be continually re-examined and re-evaluated before he can hope to make a substantial contribution. But if he does his work well, he has the satisfaction of knowing that he has brought a better understanding to one small corner of the teeming biosphere."

collection should include numerous specimens of each species, each one labeled carefully as to time and place of collection and other details.

The field is not only the source of herbarium specimens, but it is also an important area for direct study. In spite of the limitations of time and space, field studies overcome the limitations of dried material and limited specimens. Mass collections can be made, and many things can be observed in the field that are not apparent from herbarium specimens. Field studies are used especially to investigate variations such as intergradations of phenotypes within species (see Essay 29–2). Isolating mechanisms may be examined, and taxonomic conclusions may be related to important aspects of ecology.

The third area of morphological activity for the taxonomist is microscopic morphology. This field is, of course, much more difficult than gross morphology, but it is a potentially fertile one, since there are so many things to study. These include trichomes, scales, glands, pollen grains (much work has been done here, particularly in identifying pollen found in bogs), embryos, epidermis, stem and root anatomy, and finally, structures inside the cell. Differences between species become less important at the cellular level; they are there but often subtle and difficult to detect.

Paleobotany. We have encountered many examples of the use of fossils in devising classification schemes for the large groups (e.g., divisions). Unfortunately, fossils are not of much help in studying angiosperm taxonomy, since, as we have seen, there are so few fossils compared to the number of living species.

Probably, the most important contribution of paleobotany is in providing evidence for ecological and geographical isolation of living taxa and their ancestors. Modern evolutionary theory strongly emphasizes the importance of isolation in speciation, as we have seen. Paleobotanical studies can provide evidence for such isolation of different groups during geological time.

Ecology. As in the case of paleobotany, ecology is important in helping us to understand isolation mechanisms. In Chapter 37 we shall discuss the principle of ecotypes. Studies have shown that a given species might have several different forms throughout its range of distribution, each morphologically or physiologically different from the others. When these forms were brought together in a common garden, they continued to exhibit their differences, indicating that the differences were genetically controlled rather than mere responses to varying environments. Obviously, the formation of ecotypes within species must be related to the principles of speciation studied in detail by taxonomists. Here is an important area of overlap between taxonomy and ecology.

Genetics. This is one of the most powerful modern taxonomic tools. The approach is to perform crosses to prove or disprove some theory of relationships. In 1897, for example, G. Heeger discovered what appeared to be a new species of shepherd's purse, genus *Capsella*. It had ellipsoidal capsules instead of the characteristic heart-shaped ones of the common *Capsella bursa-pastoris.* The new form was named *Capsella heegeri* in honor of its discoverer. It attracted considerable attention and was thought to be an example of speciation by mutation. The American geneticist George H. Shull crossed the new species with the traditional one. The cross was highly fertile, and all of the progeny had the heart-shaped capsules. In the F_2 generation, the forms segregated at about 15 heart-shaped capsules to 1 *heegeri* capsule. This could be nicely explained by a simple Mendelian inheritance involving two sets of alleles. Further genetic studies supported the idea. If the two forms differed by only two genes and were completely interfertile, there was no reason to call them separate species.

Application of genetics in taxonomic work is far more extensive than the simple shepherd's purse example might imply. For example, crosses may be made in all possible combinations and permutations of several species and subspecies within a genus. Perhaps, at the most basic level, genetics is used in taxonomy because it is an essential part of evolutionary theory, which itself is the basis of modern taxonomy. Evolution may be defined as a change in gene frequency within populations as a function of time, all of which is fundamentally genetics.

Cytology. Many modern taxonomists have become deeply involved in cytological studies, particularly of the chromosomes. These studies provide one good example of clear-cut species differences at the cellular level. Chromosome number is studied in preparations of dividing cells, usually from root tips or from pollen mother cells. Chromosomes from about 10% of the flowering plant species have now been counted. Chromosome numbers are not always constant within a species, although they are usually reliable.

Sometimes, chromosome morphology is valuable in taxonomic studies. Most commonly, chromosome lengths are examined, and relative distances between positions of the centromeres and ends of the chromosomes are tabulated. The number and type of chromosomal aberrations present at meiosis may also be important. Again, cytological studies have provided insights into evolutionary mechanisms.

Chemistry. If genes are similar (related), then the chemical products controlled by them should also be similar, offering yet another approach for the taxonomist. We have examined a few examples of this approach as applied particularly to the pigments that characterize the different divisions of algae. Virtually any plant constituent might be measured in such a taxonomic study, including, for example, cell-wall constituents, fatty acids, or amino acids. Paper and other forms of chromatography have made studies of this type possible. In many cases, it has been possible to "fingerprint" a species by chromatographic techniques, allowing the taxonomist to distinguish between it and another species. One problem is that concentrations of these constituents may be influenced by environmental factors. So, to use them in taxonomic studies, their stability and reliability must first be discovered.

Even more important, perhaps, is the examination of enzymes and genes. Such studies, involving sophisticated biochemical and molecular biological techniques, have been used on a relatively small scale but could become extremely valuable in the future.

Taximetrics. Several modern taxonomists have developed rather complex applications of mathematics and statistics to classification problems. An attempt is made to program a computer to make the comparisons that a "good taxonomist" might make. Huge amounts of data relating to plant characters can be manipulated, using a computer program, to arrive at mathematical evaluations of relationships.

The computer manipulations are quite objective, but establishing the computer program is less so. One must arbitrarily decide which characters are important in distinguishing between two or more species and then program the computer according to these decisions. Obviously, computers will distinguish between species in the same way as the programmer! The advantage of taximetrics lies not in helping to make basic decisions about the importance of characters but in handling large amounts of data. Furthermore, vast quantities of information relating to plants can be stored in a computer and retrieved almost instantly by punching a few keys.

Publication. As in any field of science, the taxonomist must publish his results. Often, his publications are concerned with accurate and brief, usually almost telegraphic, descriptions of the taxa with which he is working (e.g., species, genera, families, classes, etc.). If he is indicating relationships, he will probably do this with figures such as phylogenetic trees (e.g., Fig. 29-4, in the End Paper), or he may construct keys.

Even when these keys do not indicate relationships, they are frequently constructed to enable other taxonomists to identify plant (or animal) materials according to the schemes established by the taxonomist. Most modern keys are dichotomous: the user of the key always has a choice to make between two (and no more) possibilities. When he makes his choice, he is then confronted with another set of two possibilities, and he continues this procedure until the plant has been identified (Fig. 29–1).

The most common publication is the treatise or monograph of a species, genus, or perhaps a family. The author reports his studies of a group based on the examination of specimens at many herbaria, on field studies, and on correspondence with other taxonomists all over the world. This is perhaps the taxonomist's most profound work. If the summary is less detailed, it may be referred to as a synopsis. The most extensive project of a taxonomist is the publication of a flora, in which the plants of an entire region are described and classified. The region may be as small as Yellowstone National Park, in which the flora may be a hip-pocket paperback designed for tourists to take into the field, or as large as the Rocky Mountain Region, a flora consisting of several hardbound volumes. A flora of North America, coordinated by the American Institute of

Essay 29–2 Hybrid Oaks: An Exercise in Discovering and Solving Problems

Lyman Benson is the perfect combination of scientist and teacher. He is motivated both by a desire to understand relationships in the plant kingdom and by a desire to transmit this understanding to students—in the classroom and in several books. After receiving a Ph.D. from Stanford University, he spent six years at the University of Arizona, and ultimately became a world authority on the cacti. In 1944 he found his ideal niche among the students at Pomona College. How these studies have helped him reach his own research goals is indicated in this essay.

"Research in applied science is undertaken primarily to secure a solution to a problem of importance to human beings, but pure science is primarily for satisfaction of human intellectual curiosity. In applied science, the existence of a problem, such as exterminating an insect pest, is obvious, and the chief question is the method of solution. In pure science another factor overweighs all the rest. This is detection of the existence of a problem.

Detection of problems comes more easily for some persons than for others, but in no case does awareness come simply out of a clear sky. There must be a long period of observation, correlation, and cogitation of many related facts. After much thought, apparently out of the blue, the existence of a problem becomes evident, as does, often, a possible method for its solution. This leads to research. The course of research usually reveals a host of side problems, some of which may be more interesting than the original one. An illustration may be found in the history of a paper published in the American Journal of Botany in 1967. This was entitled Evolutionary sorting of characters in a hybrid swarm. I. Direction of slope.

I cannot remember a time I was not aware of various kinds of Californian oak trees. A deeper scientific interest developed after 1927 and a spring quarter course in classification of flowering plants. Interest might not have continued had not Jepson's Manual of the Flowering Plants of California *been taken along on a camping trip in the redwood belt the following summer and used to identify plants as a matter of interest and recreation. In the fall this was followed by the discovery that at*

Stanford University it was possible to receive course credit for independent work of this type, and ultimately this led to degrees in the biological sciences. By graduation in 1930, I had collected about 2,000 plant specimens and developed an awareness of the correlation of different species, including those of oaks, with different geographical and climatic belts.

In about 1935 I discovered a peculiar oak population occurring along the dry inner foothills of the western side of the San Joaquin Valley in Kern County, California. The first oaks in the foothills above the grassland at that point formed an oak woodland in miniature composed of trees with elfin versions of the trunks and forms of the blue oak common in the northern Californian foothills. However, the leaf and acorn characters were similar to those of the desert scrub oak, which occurs in the hills just above the Mojave and Colorado deserts in southern California.

At Pomona College in 1944, the oaks growing along the western edges of the deserts and around the southern end of the San Joaquin Valley were chosen as subjects for class research. The oak problems could be seen to be real because the professor did not know the answers. This lifted the studies out of the category of classroom exercises and into research, because the object was to determine the answer to a question rather than to find out what the professor thought about it.

Study began as a taxonomic problem concerned with the proper classification of the blue oak of northern California, Quercus Douglasii, *and the desert scrub oak, which has been named* Quercus turbinella *subspecies* californica. *The object in view was to clarify the status of these two taxa in view of a bewildering array of natural intergrading populations of oaks, including the one first observed in Kern County in the 1930s and others that had been encountered in an area 250 miles long stretching from the Coast Ranges in southern Monterey County to the desert edge in San Bernardino County. Many side problems came to light, and the observed factors relating to the overall question became so complex that classification still remains more or less unsolved. Students debated the problems back and forth, and their discussion continued to sharpen the questions as gradually more and more data accumulated, particularly through numerical assessment of population characters (techniques of Edgar Anderson). Characters of the members of populations were graded on a six-point scale, and these were translated into graphic representations.*

After a time, differences were noticed in the character combinations occurring on the north- and south-facing slopes of the same hill. On a series of washboardlike east-west ridges near Lebec, these differences were found to be repeated time after time. The records were compared with those of populations of the blue oak in the Sierra Nevada foothills near Yosemite, far away from areas of occurrence of the desert scrub oak, together with records of studies of the desert scrub oak along the desert edge, far away from the blue oak. As represented in bar diagrams, the characters of populations of the desert scrub oak gave a very short bar, those of the blue oak a very long one. Those from the hybrid swarm (as they were now interpreted) in the vicinity of Lebec, gave bars of intermediate length—but a considerably longer bar for the north-facing slopes than for the south-facing. This, then, indicated that in the hybrid swarms the north-facing slopes supported plants not of the blue oak, but approaching it, while the south-facing slopes supported plants not of the desert scrub oak, but approaching it.

In about 1963 around the campfire another question was raised. Were our methods too coarse? Did study of essentially north- and essentially south-facing slopes reflect the finer degrees of sorting that occurred according to direction of slope in a hybrid swarm? We selected ecologically similar slopes running as nearly as possible exactly toward directions at one-eighth intervals around the compass.

The results of this study formed our 1967 American Journal of Botany paper. They showed the closest approach to the blue oak to be on the northeast-facing slopes, reached by the sun early in the morning while it is at a low angle and while the ground is still cool, and the closest approach to the desert scrub oak to be on the southwest-facing slopes, in the direct rays of the sun in the afternoon when the ground is already preheated. This is important in the growing season, which occurs in the late spring as the winter rainy period is ending. Water on the southwest slope evaporates rapidly, leaving a dry soil. Moisture persists much longer on the northeast slope, where the growth of the oaks and of other plants is more lush; more leaf mold is formed, and better soil is built up by soil organisms. In winter this is also a colder slope, and, if any plants are eliminated by frost, this will occur here. Thus, because of a combination of ecological factors, the plants able to live on the northeast slope are different from those surviving on the southwest.

However, differences in the hybrid populations are detectable at one-eighth intervals around the compass! This indicates evolutionary sorting of characters to be rapid, because it is correlated with the, geologically speaking, ephemeral slopes of the same hills, and selection in the hybrid swarm must be very sensitive to the environment. This would indicate evolution (in this case) to occur much more rapidly than has been supposed, because the slopes in question are transitory insofar as geologic time is concerned. (Other studies showed that a few of the characters we studied were influenced directly by environment—but not enough to influence the conclusions.)

This research project, which started out to determine how certain taxa should be classified, was diverted into an ecological-taxonomic-evolutionary study of the effect of a certain group of environmental factors upon hybrid individuals and of their evolutionary response to the environment. The studies also answered the original question of the 1930s, indicating that the plants observed then in the elfin oak woodland were also of hybrid origin."

Biological Sciences, is presently being prepared with the cooperation of hundreds of taxonomists. These floras—small or large—provide the basic source of information for the student interested in the plants of a region.

Some Representative Orders of Angiosperms

Developing a satisfactory evolutionary system of classification for the angiosperms has proved to be a difficult and challenging task—one that is far from complete (see End Paper 29–1). The problem concerns what principles to apply in devising the system. Was the first angiosperm flower a simple structure, perhaps only a carpel and a stamen (possibly on separate plants)? One line of taxonomic thought held to this idea and therefore considered contemporary, wind-pollinated flowers of this type (e.g., willows, beefwoods, and cattails) to be most primitive. Or did the first flower have numerous sepals, petals, stamens, and separate carpels? Another system considers the order that includes such modern flowers (the Ranales, with its magnolias, buttercups, etc.) to be most primitive. Most systems agree that the monocots were differentiated from the dicots at an early stage of evolution, but then what happened? Was the next important evolutionary step a union of the petals to form a tubular corolla? Or was it the development of an inferior ovary? Flowers are known with all combinations: separate petals with superior or inferior ovary; united petals with superior or inferior ovary. So, if two groups were separated according to united or separate petals, then inferior (or superior) ovaries developed independently in both lines—or if inferior ovaries came first, then united petals developed independently in different lines. The same is true for several other features such as irregular corollas, separate sexes, flowers in heads, etc. There seems to be no way out of the dilemma posed by parallel or convergent evolution in the angiosperms.

End Paper 29–1 discusses the evolution of the two schemes based on the questions we have just raised. One, presented in its most complete form by Adolf Engler and K. Prantl in publications dating from 1897 to 1936, considers the simple, wind-pollinated flowers to be most primitive and the union of petals in one line to be an early step. The other scheme, formulated by Charles E. Bessey (between 1897 and 1915) in a manner that had a strong impact on taxonomists, considers the magnolia and buttercup flowers with their many petals, stamens, and carpels to represent the most primitive type (Fig. 29–3 in the End Paper). This system has been further elaborated, and it is now the basis of most current efforts. Arthur Cronquist has devised a recent (1968) scheme (Fig. 29–5 in the End Paper) based largely on Bessey's work, and Lyman Benson has also developed a version well suited to teaching (Fig. 29–2). By determining the position of the ovary, the status of the corolla, and whether the plant is a monocot or a dicot, the plant can be located within one of the six main areas on Benson's chart. Further study develops familiarity with these areas so that the plant can be located more exactly.

For the most part, classification into orders and families is less controversial than arrangement of these orders and families into an evolutionary scheme. Nevertheless, Cronquist lists 74 orders and 354 families, while Benson lists 69 orders with 295 families, plus 12 others with uncertain positions in his system. Obviously, a resumé, even of all the orders, could fill many pages of text and might become rather monotonous reading. But, to provide a sampling, 23 groups were chosen because of their importance in classification schemes and their familiarity to many students—and in a few cases—because of their nonfamiliarity (although these may be important and often large tropical groups). We shall emphasize the floral structures important in classification and throw in a few comments on such matters as distribution, economic importance, etc. The names are according to Benson, sometimes with alternative possibilities in parentheses. Check the position of each order on Benson's chart (Fig. 29–2).

Magnolia

Ranales (Magnoliales): Buttercup order. This evidently ancient order is composed of 26 families, more than in any other order. There are about 5500 species. Some of the familiar families are the magnolias, the laurels, the buttercups, the barberries, and the water lilies. There are herbs, shrubs, trees, and vines. Members of this order typically have separate (not fused) flower parts in fives (although some are in threes) and a radially symmetrical (regular) arrange-

29–1 An example of a dichotomous key, utilizing a few leaves from ornamental trees. Pick a leaf from the illustration. Begin with the choices listed "A." If the second choice applies, go to "B"; if its second choice applies, go to "K"; and so on. To see if you have keyed the leaf correctly, compare the numbers in the key with the letters on the photographs (see the "answer" key on the bottom of the facing page).

A. Leaves needlelike, in bundles of two	1. *Pine* Pinus *sp.*
A. Leaves with a flattened blade	
B. Leaf simple (one blade on the petiole).	
C. Leaf entire (smooth margin—no teeth or lobes)	
D. Leaf elliptic	2. *Magnolia* Magnolia virginiana
D. Leaf not elliptic	
E. Leaf triangular (deltoid), rounded at base,	3. *Lilac* Syringa vulgaris
E. Leaf not triangular, tip inversely angled	4. *Tulip tree* Liriodendron tulipifera
C. Leaves toothed or lobed	
F. Leaves toothed	
G. Leaf uneven at base (oblique)	5. *Linden* Tilia *sp.*
G. Leaf even at base	
H. Leaves heart-shaped, acuminate tip	
I. Leaf small (1 to 2 in. at base), angled at base (cuneate)	6. *Birch* Betula papyrifera *var.* subcordata
I. Leaf longer (3 to 4 in. at base), straight at base (truncate)	7. *Cottonwood* Populus *sp.*
H. Leaf elliptic, acute tip, strongly oblique (uneven) at base	8. *Elm* Ulmus *sp.*
F. Leaf lobed	
J. Leaf palmately cleft, tips of lobes acuminate	9. *Maple* Acer *sp.*
J. Leaf pinnately lobed, tips of lobes rounded	10. *Oak* Quercus *sp.*
B. Leaf compound (three or more leafllets on one petiole with bud at base)	
K. Leaf palmately lobed, large (12 to 15 in. across)	11. *Horse chestnut* Aesculus hippocastanum
K. Leaf pinnately lobed, smaller	
L. Leaf with 3 to 5 leaflets	12. *Box elder* Acer negundo
L. Leaf with 5 or more leaflets	
M. Tips of leaflets acuminate	
N. Leaf with 5 to 7 leaflets	13. *Ash* Fraxinus *sp.*
N. Leaf with over 7 leaflets	14. *Sumac* Rhus typhina
M. Tips of leaflets rounded	
O. Leaf singly compound, leaflets toothed	15. *Roseleaf mountain ash* Sorbus *sp.*
O. Some leaves doubly compound, leaflets entire	16. *Honey locust* Gleditsia triancanthos

Key to the photographs

a = 7 *b* = 9 *c* = 16 *d* = 2 *e* = 13 *f* = 15 *g* = 8 *h* = 12 *i* = 11
j = 10 *k* = 1 *l* = 4 *m* = 5 *n* = 6 *o* = 14 *p* = 3

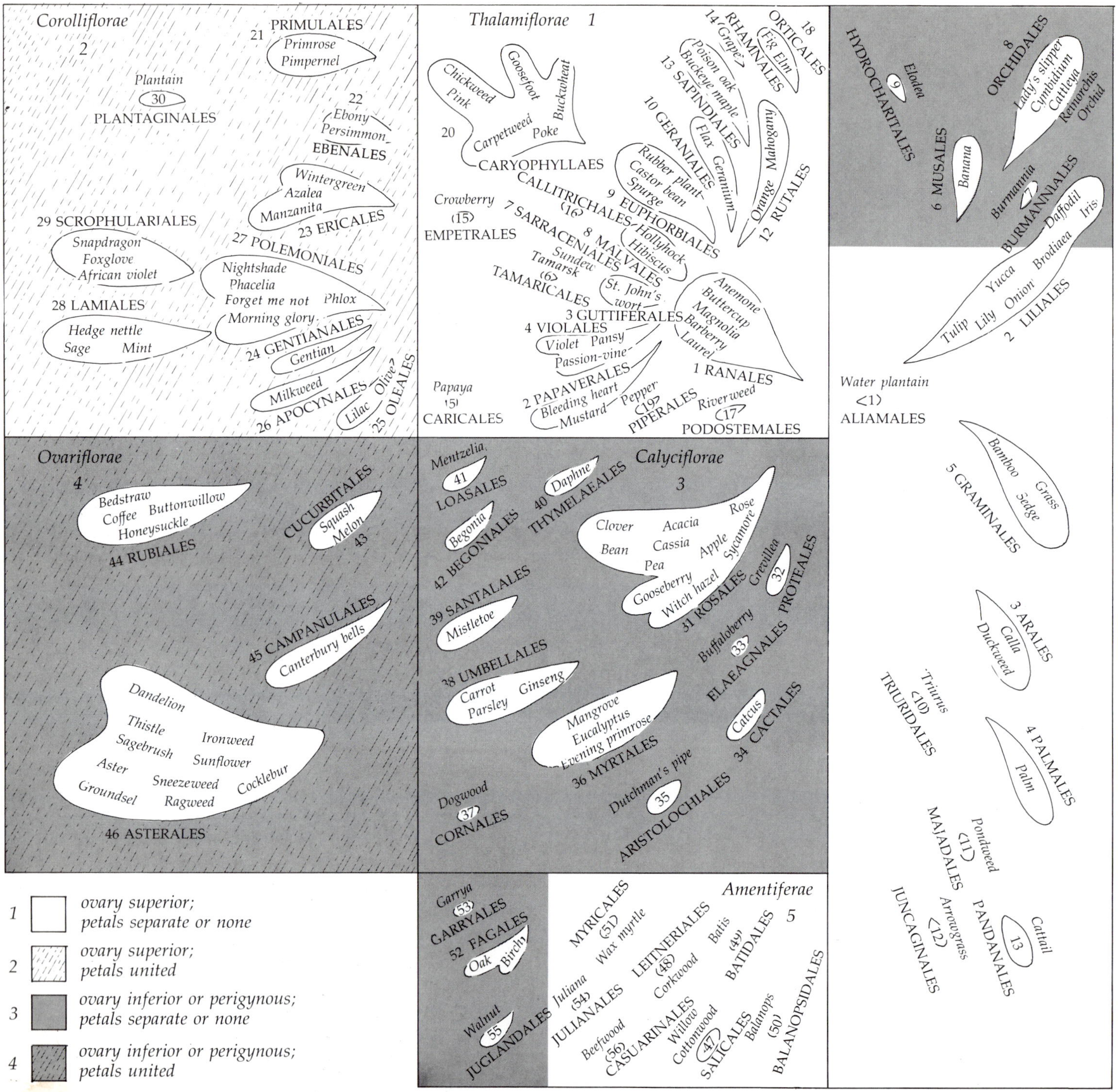

29–2 *An arrangement of the orders of flowering plants according to Lyman Benson. Sizes of the figures indicate the number of species in the orders, and their placement indicates approximate relationship.*

ment of flower parts; numerous petals, stamens, and carpels arranged spirally; and superior ovaries (hypogynous). But there is considerable variability in the Ranales concerning virtually all of these generalizations. The larkspurs and the monkshoods have highly irregular flowers, for example, and sometimes flower parts are fused (carpels in the water lilies, for example). These departures from the typical floral characteristics reflect much evolutionary development within the order.

Most modern systems of taxonomy assume that most or all other orders of dicots were derived from the Ranales, since this order exhibits the largest number of features considered to be primitive. The magnolia flower has essentially all of the general characteristics of the order (Benson's key words: free, regular, numerous, spiral, and hypogynous). One member of the Ranales, the recently discovered genus *Degeneria*, has carpels that are little more than folded sporophylls, not even fused at the edges. There is no special stigma for pollination, just a stigmatic crest running along the edges. A more primitive pistil could hardly be imagined. Flower parts in threes, exhibited by three families in the Ranales, is typical of monocots rather than dicots. The stamens show primitive characters, as do the pollen grains. About 100 species of angiosperms have primitive xylem tissue without vessels, and all of these species are in the Ranales. Endosperm is also abundant in the seed, which is another primitive character.

California poppy

Papaverales: Poppy order. This is another rather primitive order, having most of the same characters as the Ranales. Although carpels may be united, other floral parts are typically free. Most flowers are regular, although in the Cruciferae (the mustard family), stamens occur in two rows, the inner one having 4 stamens and the outer only 2. Flower parts are numerous in many species, although stamens may be as few as 6 or less. Carpels are numerous but may be as few as 2. If parts are numerous, they are arranged spirally. Ovaries are superior (hypogynous). Some members of this order such as the poppy (*Papaver*) have floral characteristics that differ from those of the Ranales only in that the ovaries and styles are united. Other members such as mustards differ from the Ranales in several important ways: the carpels are coalescent; the stamens are in two cycles of 2 and 4, and both stamens and carpels occur in cycles rather than a spiral.

Crown of thorns

Euphorbiales: Spurge order. These also have flowers resembling the Ranales in many ways, although the carpels are always coalescent (fused) and the stamens sometimes are. Flower parts are typically arranged in cycles rather than spirally and usually are limited in number (e.g., 5 or 10 stamens).

The order is an interesting one, including only two families, one of which is the Euphorbiaceae. This is one of the four or five largest flowering plant families, ranking after the composites, legumes, and orchids and about equal to the grasses. The genus *Euphorbia* is one of the largest in the world. Most members have milky sap, and some produce high-quality latex, as does the chief rubber tree of commerce, *Hevea brasiliensis.* The family is worldwide but mostly tropical in distribution. It includes the poinsettias grown at Christmas time, the castor bean, and eastern hemisphere species of *Euphorbia* that grow in deserts and superficially resemble the cacti of the western hemisphere. Tapioca is obtained from the roots of *Manihot,* after removing a poison. *Sebastiana* includes the Mexican jumping beans, with their rapid-moving moth larvae inside.

Pink geranium

Geraniales: Geranium order. The geraniums were considered by Bessey (and especially by Clements, see End Paper 29–1) to be an important center of evolution for the superior ovary line. The carpels are coalescent but separate at fruiting time in a highly characteristic way. Flowers are usually in parts of 5 or 10, and carpels may be in parts of 5 or 3. Parts are arranged in cycles. Most flowers are regular, but upon close examination the garden geranium (*Pelargonium*) is seen to be bilaterally symmetrical.

Caryophyllales: Pink order. This is the last of the orders with superior ovaries we shall consider that appears to be closely related to the Ranales. Carpels are fused; stamens are limited in number and arranged in cycles. In most families (including the four-o'clocks but not the pinks—carnations), petals are absent. This led Engler and Prantl to consider the Caryophyllales to be a transitional group between the primitive plants lacking petals and the more advanced groups. Under Bessey's system, the most primitive (Ranales) had petals; so this order had to be derived from ancestors with petals. This illustrates well the problems of finding even a beginning point in a supposed evolutionary series.

Heath

Ericales: Heath order. This order takes us into the area of Benson's chart defined by the superior ovary of the Ranales but having united petals · (a fused corolla tube). Stamens and carpels are limited in number and arranged in cycles. A few members of the group have slightly to completely inferior ovaries (illustrating that this character occurs "all over the map"), but this is atypical. Although the members of the family group together rather well on the basis of their flower shapes, vegetative characteristics, etc., there is considerable variation within the family. Some flowers lack petals, and a few do not have fused petals. Most of these variations are in the heath family, the Ericaceae. This family includes the colorful rhododendrons and azaleas, as well as a variety of other plants. There is wintergreen, pipsissewa, the parasitic Indian pipe, American laurels, various heaths, trailing arbutus, barberries, manzanitas, cranberries, huckleberries, and blueberries. A number of these plants grow near the timberline or in moist forests. Some of them grow above the timberline in the alpine zone. Several subfamilies have been proposed to account for all this variability.

Gentianales: Gentian order. Gentians are either herbs, shrubs, or trees. The stem usually has some phloem internal to the xylem. The sepals are coalescent, as are the petals, which typically form a distinctive and beautiful bell-shaped corolla. The stamens are attached to this corolla. The gentians most familiar to many of us grow in alpine meadows, often blooming late in the summer after all other alpine flowers have gone to seed. The deep indigo color of some alpine gentians is most striking, while others are a pale cream color.

The Gentianales is a rather controversial order. Everyone seems to recognize a gentian, but no really consistent character separates this order from its neighbors, particularly the phlox order. The corolla of the gentians tends to twist in the bud—but so do some of the phloxes—and some of the gentians don't! While leaves of most gentians are opposite and those of the phloxes alternate, there are exceptions. Yet, if the gentians are combined with the phloxes, then on equally good grounds, the olives and dogbanes might be included, and if that is done, then probably the mints and the snapdragons should also be included. The result would be a large order consisting of a jumbled hodgepodge of families. Thus most authors continue to live with the gentian order, although few are happy with it.

Scarlet gilia

The Polemoniales: Phlox order. Both sepals and petals are coalescent, forming calyx and corolla tubes. Stamens are attached to the petals, and the carpels are coalescent with superior ovaries. The corolla is usually regular. There are 5 stamens, which alternate with the petals. There are usually 2 carpels, but there may be 3, 4, or even 7. Parts are arranged in cycles. Most members of the order are herbs, and flowers are remarkably uniform within the order, as contrasted with some of the orders already discussed. The regular corolla and presence of 5 stamens alternate with the petals is highly characteristic. The order includes the morning glories, phloxes, borages, nightshades, and others. The Boraginaceae are highly characteristic in that the ovary is 4-chambered, each chamber forming a nutlet in the fruit. The Solanaceae or nightshade family is an especially important one, including such commercially valuable plants as potato, tomato, eggplant, red pepper, and several important drug plants. Nicotine is produced by tobacco, and belladonna and atropine from *Atropa*. Some such as petunia and *Datura* are cultivated for their showy flowers. Vascular bundles of the Solanaceae are unusual because strands of phloem occur on both sides of the xylem.

Horse mint

Lamiales: Mint order. The mints are similar in many ways to the phlox order, but the corolla has become markedly bilaterally symmetrical with a 2-lobed lip on top and a 3-lobed lip on the bottom. Fruits are closely similar to those of the borages, breaking into 4 nutlets. Stamens are reduced to 2 or 4; whereas all those of the Polemoniales have 5. Stems that are nearly square in cross section and opposite leaves provide two other distinguishing features of this order. The mints (Labiatae) are familiar because they are a source of oil. Many of the well-known herbs used in cooking, such as sage, thyme, savory, etc., are in the mint family.

Pentstemon

Scrophulariales: Snapdragon order. These are very similar to the mints, particularly in the 2-lipped, bilaterally symmetrical corolla. The erect upper lip of the corolla attracts insects, and the lower 3 petals form a landing platform. Often, the stigmas are sensitive to touch and clamp together when the insect enters the corolla tube. Pollen brought from other flowers is deposited on the stigmas, but when the insect leaves, they are closed so that the pollen it has picked up will not be deposited on the insects on their way out. The principal distinction from the mints is the fruit, which is nearly always a capsule with 1 or 2 seed chambers (instead of 4). Close relationship between the phloxes, mints, and snapdragons is quite apparent. Engler and Prantl included these in one order. Yet they are not really difficult to distinguish. The Scrophulariales are one of the largest orders and have spectacular flowers that are often cultivated as ornamentals. Examples include the snapdragon, monkey flower, Indian paintbrush, pentstemons, and elephant head.

Cactus flower

Cactales: Cactus order. This is an interesting group, characterized by extreme succulence of the stems, with the leaves reduced to spines. The numerous sepals and petals coalesce to each other at their bases, forming a unique and characteristic floral tube. Carpels also coalesce. There are numerous stamens. Flowers are epigynous (inferior ovaries). Adaptations are in the direction of survival under desert conditions.

Wild rose

Rosales: Rose order. This is the second largest order among the flowering plants, containing about 18,000 species—13,000 of them in the pea family (Leguminosae). It is a difficult order, since it includes an extreme range of variation. Many features of many representatives seem closely related to the Ranales. Often, for example, there are numerous stamens, but sometimes the number is 10 or less. Carpels are often numerous, but only one may be present. Most species are perigynous, and for this reason, Bessey thought the order was intermediate between the Ranales and the evolutionary line typified by inferior ovaries. Again, however, there is considerable variation, with many species being completely hypogynous (superior ovary). Important families include the succulent stonecrops (Crassulaceae), the roses (Rosaceae), sycamores (Platanaceae), saxifrages (Saxifragaceae), and peas (Leguminosae).

Fireweed
Sweetgum

Myrtales: Myrtle order. This large and interesting order appears to be closely related to the Rosales. Petals are rarely coalescent, and stamens, arranged in a spiral, are frequently numerous. Virtually all of the flowers are epigynous (inferior ovary) —usually extremely so. The main difference between these flowers and those of the Rosales is that the carpels are strongly coalescent in the Myrtales, including the styles and usually the stigmas. Sometimes, the ovaries are slightly coalescent in the Rosales, but typically they are not; the styles and stigmas never are. An accessory phloem internal to the xylem is common in this order, as are flower parts in groups of 4, although they may occur in groups of 3 to 15. Some of the 18 families are familiar, but many tropical ones such as the Brazil nuts or the mangroves are unfamiliar to us in the United States. The majority are trees or shrubs with a few interesting herbs.

The mangroves are distributed throughout the tropics, growing with their roots immersed in brackish water or even in sea water. The tree is supported by prop roots, which arise from the main trunk and from the large branches and which interlace with others from other trees. As we have seen, the seed germinates while the fruit is still on the tree and floats with the root pointing downward after it is released from the tree. The myrtle family includes about 2500 species, distributed throughout the tropics but particularly abundant in Australia and the American tropics. An example is the eucalyptus, widely planted and now growing wild in California but native to Australia.

The order includes the pomegranates and the water chestnuts, but perhaps the group most familiar to us is the evening-primrose family. This is one of the most attractive of all plant families, having 4 showy petals and 4 sepals with stamens in two series of 4 each. There are 4 carpels fused into a single pistil with 4 neatly divided stigmas. The ovary is extremely inferior, sometimes being 5 or 10 cm below the other floral parts. *Epilobium angustifolium* is the fireweed, a brilliant lance of red flowers growing widely on disturbed areas in the mountains of North America and Europe. It is one of the first plants to move on to a glacial moraine after the ice melts. Some of the *Oenotheras* have large yellow flowers 2 or 3 in. in diameter that close at dawn and open at dusk. As you might suspect, they are pollinated by moths.

Umbellales: Parsley order. This is another group with epigynous flowers, coalescent carpels, and a limited number of stamens and other flower parts (usually in 5s). The order is characterized by the arrangement of the flowers in umbels or compound umbels. It is further characterized by the extreme reduction of the sepals, which are typically minute or even absent. The order includes a number of plants grown for food or for flavoring, most of them in the Umbelliferae (Apiaceae) or parsley family. Many in the group produce aromatic oils that both smell and taste pleasant. Representatives include carrot, parsnip, parsley, and celery cultivated for food; plus anise (the source of licorice), caraway, dill, and fennel cultivated for flavoring. The poison hemlock was cultivated or at least collected in ancient Greece to be used in executions such as that of Socrates.

Campanulales: Bellflower order. The coalescent petals often form a rather lovely, bell-shaped corolla. The ovary is typically inferior, but there are a few exceptions. If the ovary were superior, it would be fairly easy to classify this order somewhere close to the Polemoniales (phloxes). Indeed, *Mertensia* in the borage family of the Polemoniales is referred to in much of the American West as the bluebell, and its corolla is similar to many of those in the bellflower group.

As a matter of fact, the bellflowers also resemble some of the Gentianales with their superior ovaries. It would therefore be easy to suggest (and has been) that they arose from the Polemoniales (or perhaps the

Gentianales) by the evolution of an inferior ovary. Yet they show other similarities to the Umbellales and the Rubiales, and their fused stamens are very similar to the Asterales. Obviously, decisions as to their evolutionary background will have to be arbitrary at this point. Incidentally, the problems are reminiscent of the honeysuckles in the Rubiales. Their corollas resemble, sometimes rather closely, members of the mints or Scrophulariaceae, but the ovaries are inferior.

Sunflower

Asterales: Aster or Sunflower order. The Asterales were considered by Bessey to be the most advanced in the epigynous line of evolution. Their characters depart from those of the Ranales in every significant way. All have flowers with clearly inferior ovaries, and the limited number of petals, anthers, and carpels are all coalescent. The stamens arise from the corolla tube. The most characteristic feature of this group is the arrangement of flowers into heads, each of which appears to the casual observer as a single flower. In a sunflower, for example, there are numerous disc flowers forming the center of the head. Each of these is a radially symmetrical, not too unusual, small flower; sepals are absent, replaced perhaps by hairlike appendages called pappus. Around the center of the disc flowers are ray flowers, each being strongly bilaterally symmetrical with its corolla extended into a long, straplike petal. These are arranged around the disc, not unlike the arrangement of separate petals in a single buttercup flower. The entire head has a number of bracts (special floral leaves) around the base (called an involucre), which are reminiscent of the sepals on a single flower. Either disc or ray flowers may be absent from the head, and either stamens or carpels may be absent from disc or ray flowers. Sexes may be in separate flowers or even on separate plants. There are several special features—the pappus, for example, which becomes the "parachute" on a dandelion seed.

The Asterales includes only one family, the Compositae (Asteraceae) or sunflower family. This is the largest of all plant families, including perhaps 20,000 species. Nearly all members of the group are herbs, but a few shrubs are known, and the dandelion tree grows in Mexico. The group is divided into 15 so-called tribes. Some of these have showy flowers such as the sunflower, marigold, everlasting, or aster tribes. Others such as the thistles are less showy, and some have flowers such as the ragweed, sneezeweed, sagebrush, or cocklebur that are not showy at all. Representative members occur worldwide, often in difficult habitats such as deserts, alpine areas, etc., but also in the more ordinary habitats.

Horsetail tree

Casuarinales: Beefwood order. This group is presented as a rather unfamiliar example of the Amentiferae, the group thought by Engler and Prantl to be most primitive. All members of the group have their flowers arranged in conelike catkins, and all are woody. The carpels of the Casuarinales are coalescent, and the ovaries are superior. There is only 1 stamen and 2 carpels; petals and sepals are absent. Beefwoods are trees that resemble pines from a distance. But the apparent "needles" are really branchlets about the same size as the pine needles of a longleaf pine. These branchlets are green and photosynthesize. Each one is grooved and has several obvious nodes surrounded by minute, scalelike leaves. The general appearance is strikingly similar to a miniature horsetail or scouring rush (*Equisetum*, p. 457). The flowers occur in catkins, which become woody and conelike, and the development of the megaspores in the ovule is unusual because several develop, each giving rise to a megagametophyte. These features were once thought to be primitive and perhaps to provide missing links between angiosperms and gymnosperms. Most modern authors consider these "primitive" features to be coincidental, regarding the flower as a result of long evolutionary reduction. The trees range from the Indo-Malayan region to Australia but are cultivated in tropical and subtropical lands, including California and the Southern states. In Florida they have escaped from cultivation and grow wild.

California live oak

Fagales: Beech order. This order provides an example of the Amentiferae with inferior ovaries. Carpels are coalescent, but stamens may sometimes be numerous. All species are trees or shrubs, and the plants are monoecious, having staminate flowers in catkins and pistillate flowers, solitary (as in oaks), or in clusters or catkins (as in beech). The order includes a number of familiar North American trees such as the birches, alders, oaks, chestnuts, and hazelnuts. The relationships are obscure.

Alismales: Water-plantain order. The basic floral structure of this monocot group is like that of the Ranales. Flower parts are free (not fused), radially symmetrical, numerous (stamens may be as few as 6 and carpels as few as 6 to 9), placed in a spiral, and with completely superior ovaries (hypogynous). The Ranales and the Alismales represent the closest approach of dicots and monocots to each other. Indeed, it is easy to find several more or less monocotyledonous characters among the Ranales: fibrous root systems, flowers arranged in 3s (in three Ranalian families), strongly differentiated sieve tubes and companion cells (resembling those of corn and other monocots), and even an occasional single cotyledon (due to abortion of the second cotyledon). Separation of the Ranales and the Alismales is based upon the other strongly monocotyledonous characters of the Alismales: parallel-veined leaves, vascular bundles irregularly scattered, no cambium, and a single cotyledon. Recent anatomical work has uncovered other differences such as the structure of the stomates.

The two families of the Alismales are widely distributed over much of the earth, and both are aquatic or marshland herbs. In spite of their wide distribution, they are not well known to nonbotanists, possibly because of their restriction to aquatic environments not often visited by the ordinary person. *Sagittaria,* the arrowhead, is perhaps the best-known representative in the United States.

Blue camas lily

Liliales: Lily order. The features of this order are more or less representative of all monocots, which can be readily characterized by contrasting them with the Liliales. Nearly any plant in the Liliales has hypogynous flowers, 3 sepals, 3 petals, 6 stamens in 2 cycles of 3 each, and 3 carpels coalescent and forming a compound pistil, usually with a single style. Actually, these characters best fit the Liliaceae family. Some families in the Liliales such as the Iridaceae (irises) have completely inferior ovaries (epigynous), a feature leading some authors to establish them as a separate order. A striking feature of most Liliales is the similarity between petals and sepals. Often, these are both colorful and identical, being separable only by their position on the receptacle. (Sepals are by definition always the lowest, enclosing the bud.) The similar sepals and petals are referred to collectively as a perianth. In a few cases, sepals are joined to petals at the base to form a perianth tube closely resembling the corolla tube of many dicots.

The Liliales include many showy members such as the lilies (over 200 genera and perhaps 3000 species, worldwide in distribution), spiderworts, bromeliads, amaryllis, yam, irises, and rushes (only in older systems).

Yellow ladyslipper

Orchidales: Orchid order. This order includes only one family, the Orchidaceae. It is the third or possibly the second largest plant family—the estimated number of species varies from 10,000 to 17,000. Most are in the tropics, and many probably remain to be discovered (although some named species are probably duplicates). The orchid flower is uniquely distinctive. It is strongly epigynous, has 3 sepals, and 3 petals which are usually quite different from the sepals. At least 1 of the 3 petals is highly differentiated from the others, producing the characteristic insect landing platform. The style arising from a 1- or 3-chambered ovary is attached to the stamen or stamens. Often, there is only 1 stamen, which may frequently include only half an anther. The orchids have another interesting feature, the pollinium, a waxy mass of pollen transported as a unit by a visiting insect to the next flower.

The fused stamen-style and the pollinium are almost—but not quite—unique to the orchids. Both features also occur in the milkweeds, a family of dicots with virtually no other distinguishing features in common with the orchids. This is mentioned as one of the most striking examples of parallel evolution. A single feature in common between two such distantly related families would be striking in itself, but having two such features is indeed impressive.

The orchids have a number of other unique characteristics. The perianth tube above the strongly inferior ovary is twisted about 180°; so the flower is "upside down." The seeds of orchids are extremely minute, with the embryo consisting of only a few cells. In many orchids no endosperm develops. The cotyledon and the plumule and radical develop only after the seeds have germinated. Each flower produces a large number of seeds (sometimes millions), which may make the pollinium with its large number of pollen grains a necessity.

Most orchids in the tropical rain forest grow as epiphytes high on the branches of trees. Often, they have aerial roots that absorb water from the frequent rain. Of course, as we have seen, there are many other species of epiphytes besides orchids. There are also a number of orchids growing in temperate and even arctic North America. These include the lady slippers or moccasin flowers, rattlesnake plantain, coralroot, and a number of others. Most are restricted to moist habitats in mountain meadows or on forest floors. A few (e.g., coralroot) are saprophytic, living upon decaying material. In spite of the generalization that most require ample moisture, *Habenaria unalascensis* grows not only in moist situations but also in the hot, dry chaparral belt in Southern California.

Chinese fan palm

Palmales: Palm order. The palm flower does not differ much from the lily flower. There are commonly 6 stamens (in two series) and 3 carpels. Carpels of some genera are coalescent, but many are not. One significant difference is that palm flowers are often unisexual, being either monoecious or dioecious. The perianth parts are small and scalelike. Often, the base of the inflorescence (group of flowers on a stem) is subtended by a spathe, that is, a large leaflike bract that at first encloses the entire unit (as in the Jack-in-the-pulpit, a monocot in an order not discussed here). In palms the spathe is usually yellow or yellowish brown. The endosperm is abundant and either oily or bony. The milk of a coconut is liquid endosperm, and the coconut meat is solid endosperm. Palms may be divided into fan palms with a simple palmate leaf or feather palms with pinnate leaves. But this division is nontechnical and can't be carried very far.

There is only one family in this order, and it is one of the largest among flowering plants. It is distributed throughout the tropics and subtropics. A few species occur in the United States: the native fan palms of southern California and Arizona and the palmettos of Florida and the other Southern states.

Besides the striking relationship to the lilies, evolutionary relationships of the palms are not clear. All of the palms are woody, however, in contrast with most other monocots (except the yucca—a lily, the bamboo—a grass, the dragon tree, and a few others).

Clump of bamboo

Graminales: Grass order. This order includes two families, the sedges and the grasses, both of them very large. Some botanists have suggested that the two should form separate orders, and this may well be valid. Probably, there are over 4000 species of grasses, mostly herbs but including woody, canelike plants such as the bamboos and giant reeds. Certainly, the grass family includes more individual plants than any other higher plant family. Grasses are the dominant plants of the prairies, but they occur virtually wherever plants of any kind can grow: in deciduous and coniferous forests, shrubby chaparral, the desert, the alpine and arctic tundras, wet places and dry places. The grasses are the most important plant family to man because they include such staple food plants as wheat, rice, oats, barley, rye, corn, sorghum, the various millets, sugarcane, and bamboo. The pasture and range grasses are far more numerous than the cereals and are also of inestimable value to man.

The grass plant usually has underground rhizomes with culms above; the culms are usually hollow but solid at the nodes. Sometimes (e.g., corn) pith fills these stems. Leaves are basal or alternate and usually elongate, growing at their bases by intercalary meristems. The herbaceous grasses are either annual or perennial.

The grass flower is quite unique, but its resemblance to the lily flower is still rather obvious. The ovary is superior, and there are typically 3 stamens and 3 coalescent carpels. (Sometimes, there are 1 to 6 stamens and 2 carpels.)

The grass flowers (each called a floret) are arranged in a spikelet (a number of flowers attached closely to a single axis), which is enclosed at the base by a pair of specialized, scalelike bracts (floral leaves) called glumes. Each floret is enclosed in a lower bract called the lemma and an upper bract called the palea. There are 2 small, thin, almost transparent scales, probably the reduced sepals and petals. Sometimes, a glume is missing, and sometimes there are extra lemmas or paleas, regarded as sterile because they do not enclose a flower. Actually, because of these three kinds of bracts, the flower is hardly visible at all, except at the time of anthesis, when the 3 anthers are extended out on long, threadlike filaments, releasing their pollen in the wind. This normally begins during the hours before daylight and completes by midmorning—as many hayfever sufferers well know! The 2 or 3 stigmas are long and feathery, ideally adapted to capturing the windblown pollen. Grasses are even further adapted in that they often produce seeds by apomixis in the absence of pollination. The fruit is a caryopsis, a special fruit found only in this family and characterized by fusion of the coat of a single seed and the ovary wall. It is, of course, the nutrition provided by these caryopses upon which man is so completely dependent.

End Paper 29–1 The Development of an Evolutionary System of Classification

Over 2000 years ago, Theophrastus observed and discussed such detailed aspects of floral structure as the relative position of the ovary. In the middle of the eighteenth century (1753), Linnaeus published his detailed and effective system of plant classification based upon floral structures. Even before Linnaeus, in 1703, John Ray published a plant classification system based primarily upon a division between woody and herbaceous plants (as Theophrastus and others had) and secondarily upon the distinction between monocots and dicots. Following Linnaeus, in 1789, Antoine Jussieu published an arrangement of plants based upon cotyledon and corolla characters, and also upon the distinction of hypogynous, perigynous, and epigynous flowers (noted by Theophrastus but neglected during the intervening centuries). Augustin Pyrame de Candolle elaborated this approach in successive publications, beginning in 1818 and continuing for over half a century. He included developing information about the vascular system and continued to emphasize the importance of the position of the ovary. This fairly elaborate system then provided the foundation for two almost diametrically opposed systems.

From 1897 to 1936 Adolf Engler and his associate K. Prantl at the University of Berlin developed an extensive classification system. They decided that the most primitive flowering plants among both the dicots and the monocots were those with the simplest flowers—the ones without petals and with very few stamens and pistils. They were small, simple, and wind-pollinated. They felt that the two subclasses (monocots and dicots) arose independently from an ancestral stock of unknown hypothetical and extinct gymnosperms. The living plants most nearly like this ancestor were the Amentiferae in the dicots and the Pandanales in the monocots.

The Amentiferae are trees or shrubs including willow, poplar, cottonwood, oak, beech, alder, birch, beefwood, wax-myrtle, walnut, and hickory, found mostly in the Northern Hemisphere. It seemed that the simple, unisexual flowers of the Amentiferae, occurring usually in catkins, resembled the strobili of gymnosperms. Revisions of Engler and Prantl's system by Richard von Wettstein suggested that the beefwoods resembled either *Ephedra* or *Equisetum*. Their small branchlets are green (about the size of pine needles) and jointed in a manner strongly reminiscent of miniature *Equisetum* (horsetail) plants. The beefwood flower could perhaps be derived from a strobilus similar to that of *Ephedra* (Mormon or Mexican tea; see Chapter 27).

Pandanales include the cattails, burweeds, and screw pines—all monocots having simple, wind-pollinated flowers somewhat comparable to the Amentiferae. Engler and Prantl did not summarize supposed lines of phylogenetic development from a single order as an evolutionary starting point; they thought of modern angiosperms as being composed of many fragmentary lines of evolution. The course of development was believed to have begun with flowers having no petals and to continue with the addition of sepals, then separate petals, and finally coalescent petals. The primary emphasis in their evolutionary sequence was upon petals. Secondary changes that occurred in the development of the flower had to do with the position of the ovary (hypogyny, perigyny, and epigyny), but plants with each of these features were scattered throughout the major groupings of Engler and Prantl.

The other direction of development, based upon de Candolle's system, was taken by George Bentham and Joseph Dalton Hooker. As in the case of Engler and Prantl, their classification scheme (devised between 1862 and 1893) was based upon an extensive study of a great section of the world's flora. Their system continued to emphasize the position of the ovary rather than the development of the corolla. Between about 1897 and 1915, Charles E. Bessey at the University of Nebraska published and championed a system of classification growing logically from that of Bentham and Hooker, de Candolle, and Jussieu. Bessey based his system upon an assumption quite contrary to that of Engler and Prantl. He felt that the most primitive angiosperm flower was an insect-pollinated one having numerous sepals, petals, stamens, and pistils. The living group having flowers most closely related to this ancient angiosperm ancestor is the order Ranales, including the buttercups, anemones, peony, magnolias, tulip trees, and others. The magnolia flower comes especially close to the imagined angiosperm ancestor. Evolution from this point on proceeded fundamentally in three directions: a monocot line, a line characterized by inferior ovaries (perigyny or epigyny), and a line characterized by superior ovaries (hypogyny).

Bessey's system was summarized in a figure thought to resemble a cactus and referred to familiarly as *Bessey's cactus* (Fig. 29–3). This was an extremely important contribution because it provided something students could readily grasp and apply while becoming familiar with the plants they encountered. In each of the three lines, evolution was said to have proceeded in parallel fashion by *reduction* in number of flower parts, *fusion* of petals in the corolla, development of an *irregular corolla*, and *separation of sexes* into separate flowers on a single plant (monoecious) or on separate plants (dioecious).

With these ideas in mind, the student could memorize Bessey's cactus without too much difficulty, examine a flower in the field, and mentally locate the flower on Bessey's chart. Determination of whether it was a monocot or a dicot and the position of the ovary immediately located the plant on one of the three lines. Further study of the corolla, number of parts, and sexuality then allowed the student to zero in on the plant's position. The chart was further elaborated in 1928, by Frederic Clements and his wife Edith, by running a series of lines horizontally across the three vertical lines of evolution, thus indicating the condition of the corolla and other features (Fig. 29–4). Once this chart was memorized, the job of field recognition was even simpler.

It is interesting to compare the systems of Engler and Prantl and Bessey in terms of the philosophy of classification systems. The problem in either case is one of parallelism. In Bessey's system, it is assumed that epigyny (inferior ovary) appeared only once among the dicots and that evolution then proceeded by reduction of petals, their fusion, and finally

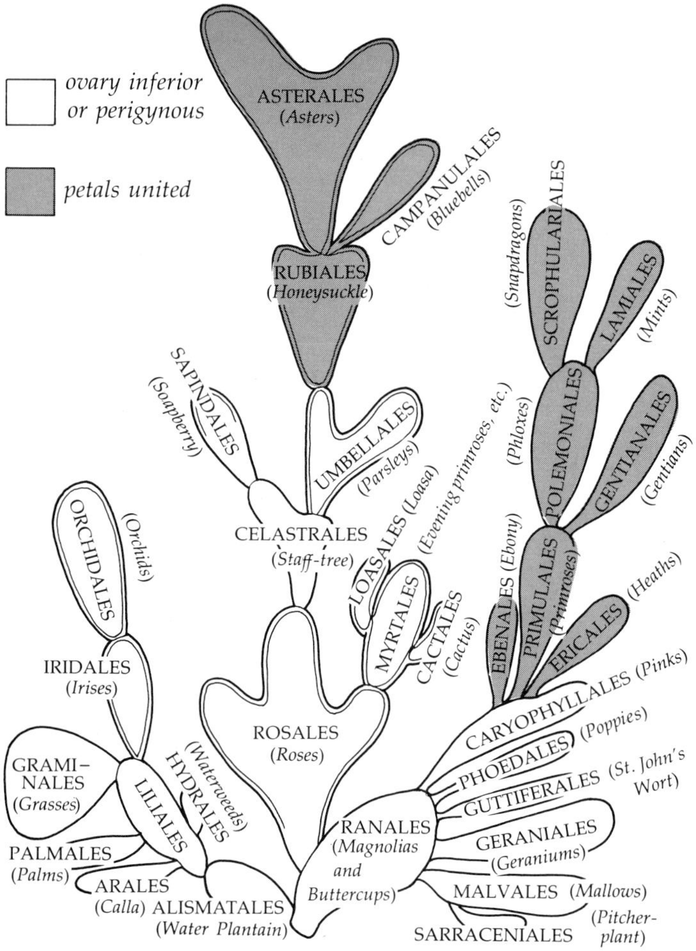

29–3 Bessey's cactus is a chart of the origin and relationships of the orders of flowering plants according to Charles E. Bessey. The left branch represents the monocots; the middle branch represents mostly flowers with inferior ovaries (except for the Ranales and some in the Rosales); and the right branches represent flowers with superior ovaries. Middle and right branches are both dicots. Proposed relationships are indicated by position in the chart, and the area of each figure is approximately proportional to the number of species occurring in the order. The chart in this form (showing perigynous or epigynous flowers and sympetalous flowers by shading) was adapted by Lyman Benson from Bessey's original chart.

formation of irregular corollas, such as in the honeysuckles. In the other line, epigyny never appeared (superior ovary remained), but petal evolution was paralleling that in the first line (i.e., reduction in number, fusion, and irregularity, such as in the snapdragons). The form of the corolla in certain honeysuckles and snapdragons is amazingly similar, but Bessey would have considered this as an example of parallel evolution. On the other hand, Engler and Prantl tended to distinguish lines by corolla form and accepted epigyny as an example of parallel evolution in their various lines.

Most evidence in recent years tends to support Bessey's system over that of Engler and Prantl, at least as far as the origin of the angiosperms goes. Based upon extensive anatomical evidence, it now seems quite unlikely that the Amentiferae could have evolved directly from the gymnosperms. A host of gymnosperm characteristics would have to be lost and many others strongly modified to account for this. On the other hand, Bessey suggested that the primitive Ranales arose from the extinct Bennettitales, but further evidence on the Bennettitales reproductive structures fails to support this approach. As we saw in Chapter 28, the origin of the angiosperms is simply unknown, but features of wood anatomy studied by I. W. Bailey and other evidences do support the primitive nature of the Ranales. Most workers are also convinced that primary arrangement according to ovary position, as in Bessey's system, is more logical and easy to handle in terms of the many observable features of plants (e.g., stem anatomy) than other systems. Yet as we saw in our summary of representative orders, even this feature is not always reliable. Inferior ovaries tend to appear here and there on everyone's map of the angiosperms.

It is important to recognize that regardless of the outcome of arguments about possible evolutionary lines, the work of Engler and Prantl in studying the world's flora was of superlative quality in the treatment of orders and families. It reflects both a broad and a deep knowledge of plants, and in many instances the orders are more nearly natural than Bessey's. Improvements of Engler and Prantl's system by Wettstein and Alfred B. Rendle at the British Museum of Natural History are even stronger in their treatments of orders and families.

The Engler-Prantl system was by far the most comprehensive and best-published organization of the plant kingdom up to its time. In response to this, plant specimens were arranged in nearly all the world's herbaria according to this system, and it was used in the vast majority of floras that have since been published. Until perhaps a decade ago, many botanists, particularly in Europe, defended the scientific merits of the system—although American botanists were mostly converts to Bessey's system. It has now been conceded that the basic assumptions of Engler and Prantl (Amentiferae as primitive) can no longer be defended, and virtually everyone follows some approach related to Bessey's

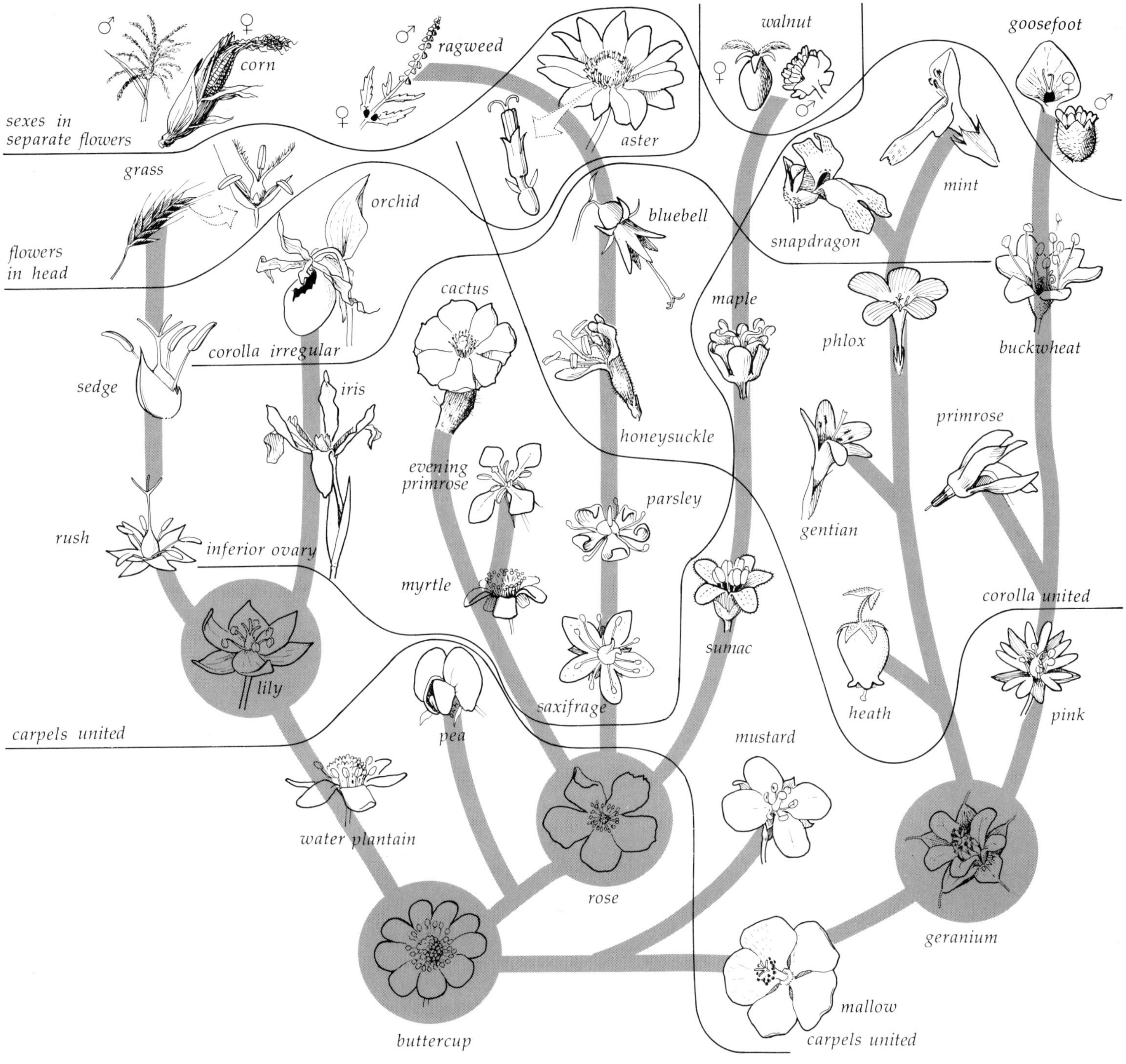

29–4 A chart showing the relationships among the flowering plants that is based upon Bessey's but modified by Frederic and Edith Clements. The lines represent levels of advance in each of the proposed evolutionary lines.

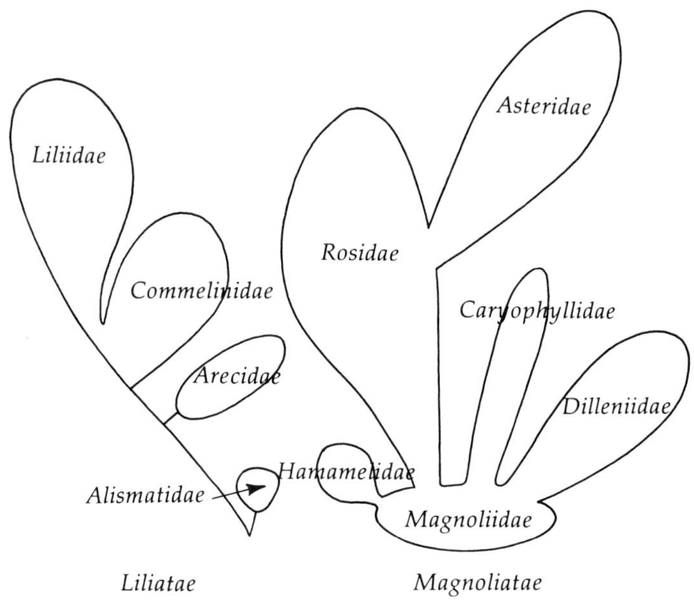

29–5 A chart illustrating a natural system of classification of flowering plants, as suggested by Arthur Cronquist. This chart is clearly related in concept to Bessey's cactus. But monocots and dicots are separate, and the groups show only the subclasses according to Cronquist. Hence it is less detailed than the other charts.

system. Although a few herbaria are changing to some newer system, many will probably remain arranged according to the Engler-Prantl system for a long time to come, considering the tremendous expense and effort required to rearrange them, the fact that the system is known by so many botanists, and that taxonomic systems are now recognized as being temporary at best.

It would be a mistake to imply that the major systems just discussed were the only ones ever proposed. For example, John Hutchinson of the Royal Botanic Gardens at Kew, England, constructed a system (1924–1948) resembling Bessey's except showing two lines of development from the Ranales, one woody and the other herbaceous. The considerable merits of his system were obscured because the division into trees and herbs splits many orders in a manner considered unnatural by most taxonomists—as does any approach based too rigidly upon a single feature. In 1968 Robert Thorne of the Rancho Santa Ana Botanic Garden in California presented a synopsis of a system he has been working on since the early 1950s, which has generated considerable interest among taxonomists.

Arthur Cronquist, the Senior Curator of the New York Botanical Garden, published a book in 1968 containing a somewhat original classification of the angiosperms (Fig. 29–5). His system was conceived independently but is similar to that of Armen Takhtajian, a Russian evolutionist, and Walter M. Zimmerman of Tübingen, Germany. (A joint paper was published by all three in 1966.) Cronquist considers the angiosperms to form a division called Magnoliophyta (Anthophyta in an earlier botany text). The monocots and dicots are considered classes within this division and given the special names of Magnoliopsida and Liliopsida. The dicots (Magnoliopsida) are in turn divided into six subclasses, which is an innovative approach. Cronquist imagines the first of these, the Magnoliidae, to be the basal complex from which the other angiosperms have been derived. Whereas Bessey had two principal lines of evolution in the dicots from the Ranales (comparable to the Magnoliidae), Cronquist has four. The largest of these is the Rosidae with its offshoot of the Asteridae (Fig. 29–5). Within each of the six subclasses, orders are listed, and these correspond fairly well with those of Engler and Prantl and others, although unique names have sometimes been suggested. In the monocots (the Liliopsida) there are four subclasses having their ultimate origin from the Magnoliidae through the Alismatidae, as suggested by Bessey (Fig. 29–3). Cronquist is fully aware that no living species (except polyploid species) evolved from any other living species, although the student could get this impression from the phylogenetic-tree approach used by all the authors discussed so far. Clearly, the overall scheme is strongly reminiscent of Bessey's, as Cronquist states.

Lyman Benson, a taxonomist and botany teacher at Pomona College in California, has attempted to organize the angiosperms to produce a system as valuable in teaching as that of Bessey. He is convinced that phylogenetic trees can only be misleading in the present state of our ignorance. Rather than connecting the various groups in a figure such as Bessey's cactus (which strongly influenced Benson as a young student), he outlines each group on a surface, implying that the distance between groups is proportional to the degree of relationship. The surface is then divided into six major areas, two of these being further subdivided (see Fig. 29–2). In spite of the fact that Benson's chart does not visually resemble Bessey's cactus, the relationship between the two soon becomes obvious. Five of the areas are devoted to the dicots, one of these being the Amentiferae, which is further subdivided into plants with superior or inferior ovaries. The remaining groups of the dicots are the four possible combinations of superior or inferior ovary with separate or united petals. The monocots are also divided into those having superior or inferior ovaries. As in the case of Bessey's cactus, the student can immediately place the flower he is examining into one of these main categories. For each order Benson lists the departure of floral characters from those of the

Ranales (dicots) or the Alismales (monocots). By learning these the student is then better able to recognize specimens.

Benson's chart has one particularly appealing feature: phylogenetic relationships do not imply descent of one modern group from another living group. A student of Benson's (Walter Bertsch) suggested that his chart might be thought of as representing the surface of a rising sea of time. The areas representing orders were analogous to the coral islands in the ocean that build upon each other, remaining at the surface, as the water level rises (perhaps due to the bottom sinking). The surface of our imaginary ocean of time is the present at any given moment, and it rises steadily as past time accumulates below. The bases of the coral islands tell the story of the evolution of their surfaces. If we could see deep into past time in Benson's ocean, we could discern the true relationships between orders. Our only real hope to do so lies in the fossils, but these are incomplete. The alternative, taken in this chapter, is to deduce from the nature of each of the islands (the orders) their most probable relationships to each other.

30

Development: Plant Hormones

We have covered much ground since our first visit to the woodland ecosystem. We have examined both the structure and function of the cells that compose the occupants of our acre. Diversity has been of prime interest to us: we have looked at how it passes from one generation to the next (genetics) and at the theory of evolution, which attempts to account for its origin. We have devoted 15 chapters to summaries of the diverse kinds of plants and their interrelationships. By now it should be apparent that this diversity in our woodland and in all other ecosystems is great enough to occupy a lifetime with scientific study. But there are other aspects of our woodland that deserve equally intense study.

The ecosystem is a dynamic entity not just because its cells are functioning in metabolism, which is the basis for all the other functions, but because of *development.* If we visit our acre of woodland in late winter or early spring, the deciduous trees and shrubs are all bare, and only the dead remains of last year's herbs cover much of the ground. A week or two later, buds are swelling on the trees, and seeds are germinating in the soil. As the year progresses, leaves and flowers will unfold on the shrubs and trees, and the herbs will grow and flower—some reaching heights of 1 or 2 m. Mosses, liverworts, ferns, and even lichens, fungi, bacteria, algae, and other organisms (including most animals) will also increase their size during the growing season. Of course, many will die, being consumed or decomposed by other members of the ecosystem.

Growth is certainly one of the most striking phenomena observable in any ecosystem. In Chapter 24 we described the anatomy of growth as it occurs in the vascular plants. The key to understanding plant growth is the concept of *meristems* and the manner in which they function by cell division, enlargement-elongation, and specialization. But how does this work? Why do cells divide, enlarge-elongate, and differentiate—and at the right times in the right places? How does a cell become an embryo, then a seedling, and finally a mature flowering plant? How do leaves and flowers

30–A Plant growth is a varied and marvelous thing. Among the angiosperms, there are numerous patterns of seedling development, and among the temperate deciduous species, the revelation of full foliage from a dormant bud can be observed every spring. The orientation of the plant is strongly dependent upon gravity and light, and some plants respond rapidly and in a striking way to water. This photograph is of young buds on a lodgepole pine (Pinus contorta) responding to gravity by growing in a vertical direction. Later in the season they will assume a horizontal position, growing plagiotropically.

grow out each year on perennial plants? These are the problems of development or morphogenesis (morpho = form; genesis = origin). To acquire an understanding may well require as much time, effort, and creativity as traveling to the stars, but its achievement could be as important to mankind.

In this and the next chapter we shall briefly survey what is known about plant development and what is being done to find out more. In the remaining chapters of this section, we shall then discuss some functions of the mature vascular plant: its exchanges with the environment, its movement of materials inside, and its responses to several factors of the environment.

The Study of Growth

One begins simply by measuring the size of a plant and plotting the results as a function of time (Fig. 30–1). The resulting graph may have features that suggest something about growth mechanisms. One might measure the *height* of a plant, its *weight*, or its *volume*. What is actually measured depends upon the point of view. For example, height of trees or meadow grass may be important to the forester or the livestock man; a farmer often counts his yield in tons per acre; and an apple grower may be concerned with the volume of his individual fruits.

When growth is measured in any of these ways, the resulting curve often has a typical and familiar shape: an S-shaped curve. Such growth curves may be thought of as having three principal phases. The first is a compound interest phase, during which size increases more rapidly as the organism gets larger—just as money in the bank (by adding interest) increases by a larger actual amount each year when the principal is larger. During such an increase, plotting the logarithm of size as a function of time yields a straight line (Fig. 30–1), which is called the log phase of growth. A maximum or grand phase of growth is then reached, sometimes extending over a considerable interval of time (weeks). Finally, growth begins to slow during the senescence phase.

The next step is to measure growth rates. The growth curve expresses total size as a function of time, but a *growth rate implies a change in size for a given interval of time.* For example, an organism may grow 3 cm/day, 1.5 kg/week, or 3.6 cm^3/day. Growth rates can be measured as the change in size between two measuring periods. But, if the time intervals between measurements are fairly long, such growth rates could be misleading, since these rates might change considerably during the interval between measurements. The growth rate can be obtained from the size curve by simply placing a ruler tangent to a point on the curve. The slope of the ruler indicates the extent of the change in size for any desired time interval. Thus a kind of instantaneous growth rate can be obtained for any point in time represented along the size curve, as illustrated in Figure 30–1.

When a rate curve is obtained from an S-shaped, total-size curve, a bell-shaped curve is often the result (Fig. 30–1). During the log phase of growth, the rate curve has the same shape as the size curve. This property of such curves is well known in calculus. The implication is that the larger an organism becomes, the more rapidly it is becoming even larger. If the maximum or grand phase of growth is represented by a straight line in the size curve, then the rate curve will also be represented by a straight line, but in this case, a horizontal line indicates that the rate is not changing as a function of time (Fig. 30–2). Such a horizontal portion of the rate curve is a common feature of plant growth. During the senescence phase, the rate the organism is increasing its size is decreasing, making the right-hand portion of the bell-shaped rate curve.

When nice curves using measured data (as in Figs. 30–1, 30–2) can be plotted, it is an overwhelming temptation to try to explain the curves with mathematical models. Essay 30–1 discusses some of the attempts made to understand growth by using mathematical models. Unfortunately, model building has had only limited success, although we can describe growth a bit better than we might have done a few years ago. We know that organisms are open systems, never achieving equilibrium but maintaining internal conditions suitable for metabolism and other life processes (that is, exhibiting homeostasis) and that organisms grow in size and increase in complexity from the zygote to the mature organism. But how do they do all this? We are learning more and more of the details, but we are still unable to synthesize a comprehensive picture.

The feature of increasing complexity is especially interesting. In Chapter 5 we discussed entropy and the Second Law of Thermodynamics. The conclusion was that randomness is always increasing in the universe, while organization is decreasing; that is, entropy (one measure of randomness) is always increasing. Yet multicellular life seems to be a way to "beat the system."

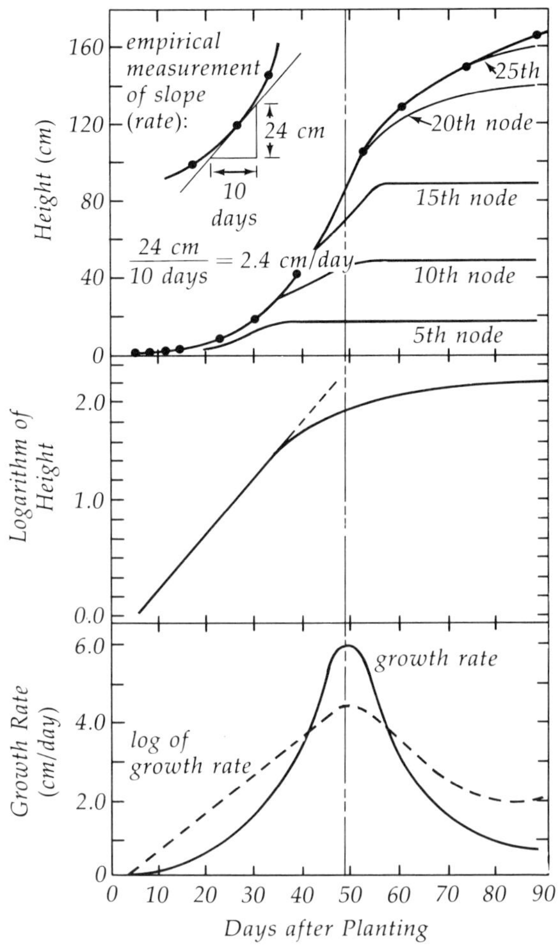

30–1 The growth curves for a corn plant expressed in different ways. Top curve shows size (height) as a function of time. Slope of this curve may be measured empirically by laying a ruler tangent to the point on the curve in question, as illustrated in the upper left. Measuring slope in this way results in the curve labeled growth rate at the bottom. Middle curve shows the data of the top curve plotted on a logarithmic scale, and dashed curve at the bottom shows logarithm of the growth rate. Note curves for heights of individual nodes on the corn plant (upper curve). Elongation is occurring below the tenth node, for example, until at least the time the twentieth node has become apparent. At that time elongation is occurring in more than the upper half of the stem (about the upper 50 cm).

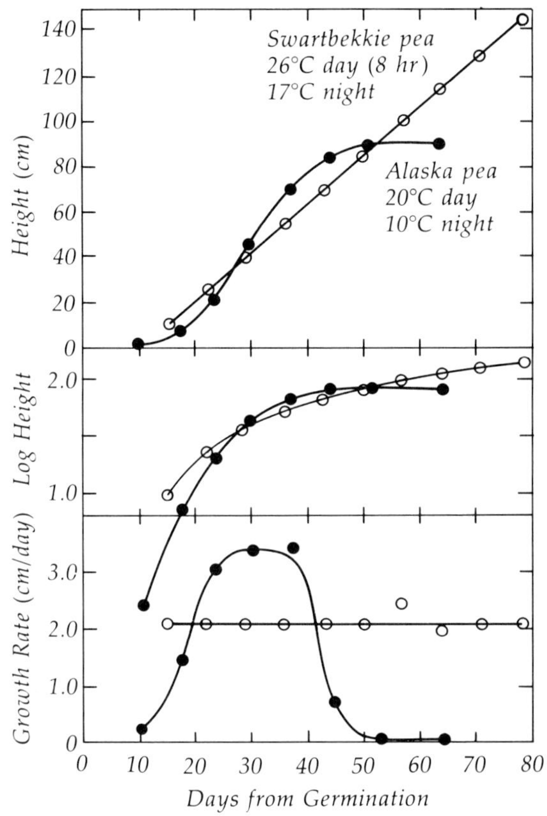

30–2 Two examples of plant growth curves that deviate from the fairly ideal curves of Figure 30–1. The curves are plotted in the same way, but the most important difference is that maximum growth rate occurred over a long interval of time, particularly in the Swartbekkie peas.

The zygote is much simpler than the adult. There are many kinds of cells in the adult, each apparently more complex than the zygote and each still containing the complex genetic message of the zygote itself. Or think of the evolutionary history of life on earth. Surely, the level of complexity is much higher now than it was 4.5 billion years ago. Life then appears to be programmed to decrease entropy, while all other spontaneous processes in the universe are increasing it.

Some biological philosophers consider this to be a serious, unsolved problem. Others feel that the problem is solved by recognizing that vast quantities of energy must be expended as the organism grows from the zygote to the adult or as life evolved from a simple form to its present complexity. Nevertheless, even if this is the answer (others are not so sure), life is indeed a unique manifestation in the way that it utilizes energy to decrease entropy.

How does it do it? What is the developmental program contained in the zygote that controls the unfolding of events until the organism grows old and dies? How does it function in cells and tissues and in response to environment? We need and are obtaining both vast quantities of objective data and deeply creative insights. As this process of accumulation of facts and examination of the data proceeds, we hope to approach more and more closely an understanding of what is happening. Perhaps some day we can formulate a testable model or set of equations that will explain the phenomenon of development.

Essay 30–1 Model Building and an Analysis of Growth

A valuable approach in science is to build models that simulate reality and then test the models by further comparison with available facts. The models may involve physical structures or abstruse mathematical concepts. Measurements of growth (Figs. 30–1, 30–2) have provided excellent data for mathematical model building.

The mathematician can account for the log phase of growth by a simple equation. He represents increment change in size by dG and an increment of time by dT. Hence dG/dT is a symbolic way of indicating growth rate. The simple, compound interest equation says that

$$\frac{dG}{dT} = kG$$

where G indicates size and k is some rate constant. The equation simply says that rate of growth is proportional to size.

But this relatively simple equation will not explain either the maximum phase or the senescence phase. To account for the senescence phase, a modification must be made. Perhaps growth rate is proportional, not only to size but also to the difference between size at any moment (G) and the final maximum size (G_f). Thus

$$\frac{dG}{dT} = kG(G_f - G)$$

Here, as G begins to approach G_f, the difference between the two begins to approach zero. When zero difference is reached, the growth rate equals zero.

This is the equation for an autocatalytic reaction. In such a reaction, rate is increased as the products of the reaction increase (products catalyze or speed up the reaction). Of course, the reaction will always be proportional to the amount of substrate present, since this is what must actually undergo the reaction. If G is the product and $(G_f - G) = G_s$ is the substrate, then one can see how such a reaction might be described by the equation. As the reaction proceeds, G is increasing and making the reaction go faster. But G_s is being used up, and eventually this will slow and finally stop the reaction completely.

If we were to apply the autocatalytic reaction model to an understanding of plant growth, we might assume that in plant growth (1) the final size (G_f) is determined by and (2) the rate of growth is dependent upon the availability of substrates. Neither of these assumptions is satisfying. Perhaps final size is somehow determined genetically, but experience tells us that it can be markedly influenced by temperature, available nutrients, etc. In addition, we know that plant growth does not stop because substances become limiting. It is easy to provide ample nutrients, but senescence sooner or later occurs anyway (unless a cutting is made, thus rejuvenating the meristem). Furthermore, the equation and the model do not account for the steady grand phase often observed for days or even weeks. The model, then, has told us only a little bit about plant growth.

Another problem is that the equation applies best to an autocatalytic reaction in a closed system. Yet organisms are clearly not closed systems. Light energy is continually entering, as are nutrients and water. Some of the energy is retained in the system as chemical bond energy, but much is lost to the environment as heat. Water is continually evaporating from the plant, and carbon dioxide is given off in respiration and oxygen in photosynthesis. The compound interest equations were formulated many decades ago, but Ludwig von Bertalanffy, a theoretical biologist, has taken the problems into account in more recent (1952) mathematical models based upon more advanced concepts in calculus. Von Bertalanffy's complex equations do a much better job of accounting for the grand phase and other features typical of living organisms. For example, homeostasis, the tendency in living things to maintain the status quo (e.g., levels of nutrients) can be predicted.

The Hormones of Cell Elongation: The Discovery of Auxin

We have attempted to segregate the steps of growth and morphogenesis into a number of components. At the cellular level these are division, enlargement-elongation, and specialization. Cell division is a relatively complex process, as is differentiation, but cell enlargement and then elongation seem to be somewhat simpler. Apparently, a cell enlarges because its wall becomes softer; it then takes up water by osmosis (internal water potentials are typically more negative than those outside); and the result is a swelling or growth of the cell. This is a good place to initiate our discussion.

Tropisms provide good examples of plant growth resulting largely from cell elongation. If a potted plant is laid on its side, the stem soon (sometimes within three hours or less) begins to grow upward, away from gravity. It is said to be negatively geotropic. Roots (especially the taproot), on the other hand, tend to grow toward the earth's center of gravity. They are said to be positively geotropic. Roots, branches, and rhizomes that grow nearly horizontally are said to be plagiotropic. If a plant is illuminated from one side, it grows toward the light, being positively phototropic. Roots growing away from the light are negatively phototropic. Not only does the stem respond to light, but leaves tend to orient themselves so that they either offer a flat surface to sunlight or are parallel to the sun's rays. Time-lapse photography shows plant stems waving in more or less random motions called nutations. Such motions in vines cause them to grow around any solid object that they contact.

A simple system used in studies on geo- and phototropism is the coleoptile of grass seedlings, particularly the oat (*Avena sativa* L.). The coleoptile is a leaflike sheath of tissue enclosing the first true leaves in grass seedlings (Fig. 30–3) and protecting them as the seedling penetrates the soil. It is formed as part of the embryo and consists of numerous small cells. The coleoptile grows largely by enlargement and elongation of these cells with only a few cell divisions. Growth of the coleoptile is positively phototropic and negatively geotropic, although it is inhibited by bright light so that its growth stops when it emerges from the soil surface. At that time the leaves inside penetrate through the coleoptile tip.

As early as 1880 Charles Darwin studied the phototropic responses of grass coleoptiles (Fig. 30–4). If he covered the tip with a small cap or removed the tip completely, the coleoptile did not bend toward the light—even though the actual bending (due to more cell elongation on the dark side than on the light side) occurred at some distance below the tip. Apparently, the tip itself was responding to the light coming from the side, sending a message to the cells below that resulted in increased elongation of cells on the dark side.

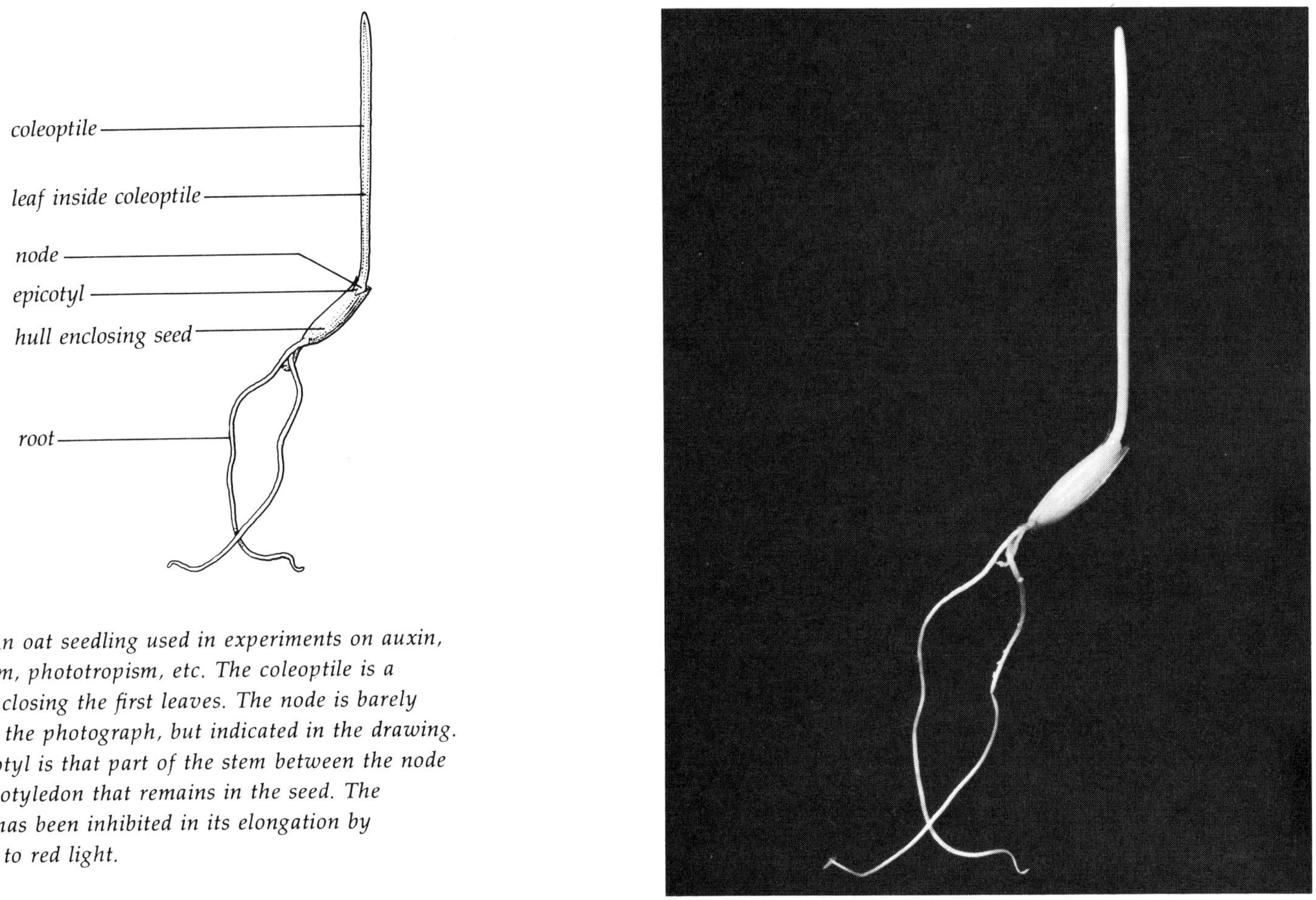

30–3 *An oat seedling used in experiments on auxin, geotropism, phototropism, etc. The coleoptile is a sheath enclosing the first leaves. The node is barely visible in the photograph, but indicated in the drawing. The epicotyl is that part of the stem between the node and the cotyledon that remains in the seed. The epicotyl has been inhibited in its elongation by exposure to red light.*

What is the message? In animals we would immediately think of a nervous impulse, an electrical stimulation of some kind. But plants do not have nerves (so far as we can detect), and although they may produce electrical currents, these have not been successfully related to the phototropic response. Another possibility is that the tip produces some kind of *growth hormone*. This could be a substance that inhibits growth and is concentrated on the light side of the coleoptile, or it could be a growth-promoting substance concentrated on the dark side. From the time of Darwin until 1928 (nearly 50 years), it was not even possible to distinguish between these two basic hypotheses of a nervous versus a hormonal impulse. A number of ingenious experiments were carried out by investigators located primarily in Europe. These experiments were simple but elegant in their conception.

P. Boysen-Jensen performed experiments from 1910 to 1913 that had a strong bearing upon the problem (Fig. 30–4). He showed that if a thin slice of mica (impervious to substances or electrical impulses) were inserted on the light side of the coleoptile just below the tip, bending still occurred toward the light; inserted on the dark side, bending did not occur. This seemed to rule in favor of the growth-promoting impulse. Boysen-Jensen also showed that phototropic bending would occur even when the coleoptile tip was cut off and then replaced on top of a thin layer of gelatin. A chemical substance might be expected to pass through such a layer, while an electrical stimulus might not.

Another worker who contributed to these experiments was A. Paal in Hungary. He showed that coleoptiles growing in the dark (or under a dim red light) could be caused to bend when a section was removed just below the tip. They would bend toward the side from which the section had been removed. If gelatin were put in place of the section, bending did not occur. A piece of mica placed on the side opposite to the gelatin

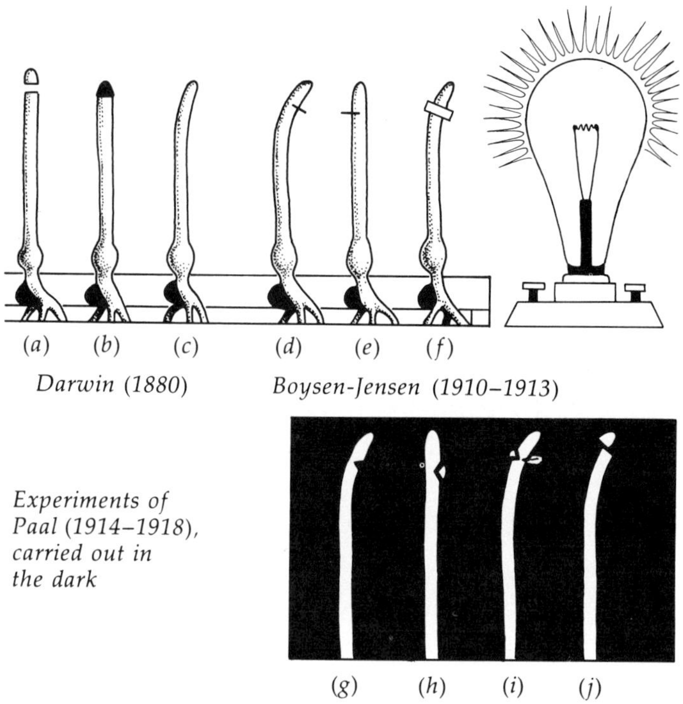

30–4 Some experiments preliminary to the discovery of auxin: When Darwin removed an Avena *coleoptile tip (a) or covered it (b), the coleoptile failed to bend toward the light, as an intact coleoptile will (c). Boysen-Jensen observed bending when a sheet of mica was inserted into the light side of the coleoptile (d), but not when it was inserted into the dark side (e). He could also observe bending when the tip was separated from the coleoptile below by a small layer of gelatin (f). Went was especially influenced in his experiments by those of Paal, who observed bending in the dark when a portion of coleoptile tissue was removed from one side (g), but not when it was replaced or when the hole was filled with gelatin (h). Removing the tip and then replacing it with gelatin on one side and mica on the other side (i) caused bending, as did replacing the tip on one side of the coleoptile (j).*

caused bending toward the mica. When the tip was removed and then replaced to one side of the coleoptile, bending occurred away from the side where the tip was replaced. These four experiments were all in agreement with the idea that a growth-promoting hormone is produced in the tips of coleoptiles, causing normal growth of the cells below, even when coleoptiles are in the dark. Unilateral light might in some way cause this substance to move toward the dark side, but light is not required for its production.

In the mid-1920s a young graduate student, Frits W. Went (see Essay 30–2), was working in his plant physiologist father's laboratory in Utrecht, Netherlands. There was much talk among the graduate students and professors in this laboratory about the hypothetical stem-growth substance. Some of the participants contended that photo- and geotropism could still be accounted for in terms of a nervous or electrical impulse, but Went was convinced, largely through the experiments of Boysen-Jensen and Paal, that a growth hormone must be involved. In the heat of one of their arguments, Went said: "Very well, I'll prove to you that the growth substance exists." And he did.

Essay 30–2 Why a Biologist? Some Reminiscences

Frits Went is introduced in this chapter but encountered in several others. His fame was assured when, in his early twenties, he demonstrated auxin, described in this chapter. Following this he spent five years in Java, a Dutch possession, where he received invaluable experience in tropical field botany. Almost 26 years were then spent at the California Institute of Technology, where hormone work continued and interests developed in desert physiological ecology. There he built the first phytotron. He moved in 1958 to St. Louis, Missouri, where he constructed a climatron, and in 1964 he moved to Reno, Nevada, where he continued his desert research. He has long been a perceptive observer of the workings of science.

"Years ago I tried to discover what made a biologist be a biologist. Soon I found out that there are about as many different motivations as there are biologists, but a few were more prevalent than others. Every human being is born with an enormous amount of intellectual curiosity. If this has not been blunted through unfortunate experiences in early or later youth, then an inspiring teacher may guide this urge for understanding toward biological problems. Although this curiosity could be directed toward any other intellectual endeavor, early acquaintance with plants or animals can direct an inquisitive mind—even without teacher-direction—toward the mysteries of life: the problems of growth, form, function, environment, and heredity. The orderly mind may be attracted toward taxonomy or biophysics, whereas the mind intrigued by complexity may select ecology, and the mind trying to understand interrelationships may become a physiologist. The mechanically inclined person might attack biological problems with delicate instruments, whereas the artist might try to solve problems of shape and color in nature.

There is an equal plethora of methodological approaches to the solution of biological problems. Since Francis Bacon in the seventeenth century pointed out the inevitability of cause and effect, the experimental or inductive approach has taken precedence over the old Aristotelian deductive approach, where through pure reasoning life was forced into a straitjacket of axioms or preconceived ideas. Our creative facilities may sometimes have been blunted by an overcritical attitude, but not a few bright minds have been misled by their or others' sophisticated hypotheses. Such complicated but artificial fabrications, like the phlogiston theory of the eighteenth century and the ether theory of the nineteenth century, have held back real understanding. It is difficult for us to say which present-day ideas may be discarded in the centuries to come, but the Second Law of Thermodynamics might be one of them.

An important aspect of the motivation to become a biologist may be humanitarian—the urge of the person to take a positive part in the well-being of his fellow man, as exemplified by physicians. Most disciplines in biology have an input to society, particularly in pathology, agriculture, and ecology. And let us not forget one of the most important motivations of all—teaching, to transmit the knowledge gathered by our culture to the generations to come, who will have to stand on the shoulders of the preceding generations. This attitude is quite contrary to that of a professor a hundred years ago, who admonished his successor never to tell the students everything he knew!

In my own case, ever since I chose botany over chemistry or engineering as my life's endeavor, I have been intrigued with the form and function of a plant and with its place and role in the environment. When walking in the country, I am forever wondering about why a particular plant grows in a particular spot; why it has its particular shape; why it does not grow 100 m farther on; why some plants grow in the desert or tropics, while others cannot; why a limited

number of plants have become weeds; why some plants are still very similar to others that lived 200 million years ago, whereas most are very recently evolved; why some trees can be tapped for sugar, whereas most cannot. And I want to look inside the plant to find how it grows and functions, why it branches the way it does, and how it responds to its environment in such a precise way. I know that some of these questions have been answered, but upon closer scrutiny I often find that the explanation is insufficient, based on wrong observations, or is blatantly anthropocentric. Thus nature is full of unsolved—or unsatisfactorily solved—problems. Intellectually, I am intrigued with all of them, but it is quite obvious that my time and energy are utterly insufficient to attack all of them in a scientific way. So I have to select. Often, I carry out a very simple experiment. If the result of such a preliminary experiment is equivocal, it means that the effect of the treatment is doubtful and would require very much work to come to a satisfactory conclusion. If, on the other hand, this simple experiment gives a positive answer, the chances that it will work in more sophisticated settings are very good. Thus my first experiment on auxin, the first on root formation, and the first on artificial rain in germination worked right off the bat, and I followed them up with other experiments, which produced a series of papers adding a positive contribution to the particular subject. And by the time that the law of diminishing returns set in, I had another subject of more immediate interest under investigation.

Early in my career I found out that an experiment should never be carried out to prove something but to test *an effect or to* find out about results. *After I had developed a quantitative test for auxin, I could use this to test the behavior of auxin in the plant. I thought then that I could prove that auxin moved in the plant by decrement, but when I carried out the tests, I found a strict polarity in the auxin transport. This polar auxin movement has been questioned by many scientists because it has no proper physical counterpart. They are willing, based purely on theoretical considerations, to question experimental facts. But in our day and age we have progressed from the Aristotelian to the Baconian viewpoint: theoretical considerations should be properly rooted in experiments.*

A similar case of hitching the horse behind the cart occurred in the explanation of the phototropic curvature. This was known to be due to differential growth of the lighted and shaded sides of a stem. The Blaauw theory of phototropism very plausibly explained it as a differential photochemical reaction of the two sides of the stem. But when the experiment was performed, it turned out that the dark side grew almost as much as the light side was retarded. The phototropic response turned out to be due to a deflection of the auxin stream from the apex toward the dark side of the stem. While the description of my experiments on the light-deflected auxin transport was in press, I was (pleasurably) surprised by a paper by N. Cholodny who had deduced this phototropic mechanism on theoretical grounds, which later was referred to as the Cholodny-Went theory of phototropism.

Whereas all my auxin experiments were based on more-or-less logical sequences of deductions leading to the crucial experiments, my ecological work was mostly a set of questions asked of nature, after observations of nature had posed the problems. These experiments became possible because of our construction of a phytotron, in which a number of the most important environmental factors such as temperature, light, humidity, wind, or rain could be controlled. With this new tool I could establish, for example, that the profuse flowering of the desert in certain years depended upon the precise germination response of the seeds of desert annual plants to temperature and rain and not upon a mystic 'survival of the fittest' or a 'struggle for existence.' Thus in the last 30 years phytotrons have contributed much to make ecology an experimental rather than a descriptive science—now that the extreme complexity of the organism in response to its total environment can be broken up into smaller and experimentally manageable subunits. But, although it is satisfying when laboratory experiments give us a better insight into the mechanism of life, for me the greatest thrill comes when these new insights make me understand what goes on in nature, when the laboratory knowledge is applicable in the field. Thus nature not only provides the inspiration; it is also the ultimate arbiter. The laboratory is only an interlude between perceiving and understanding."

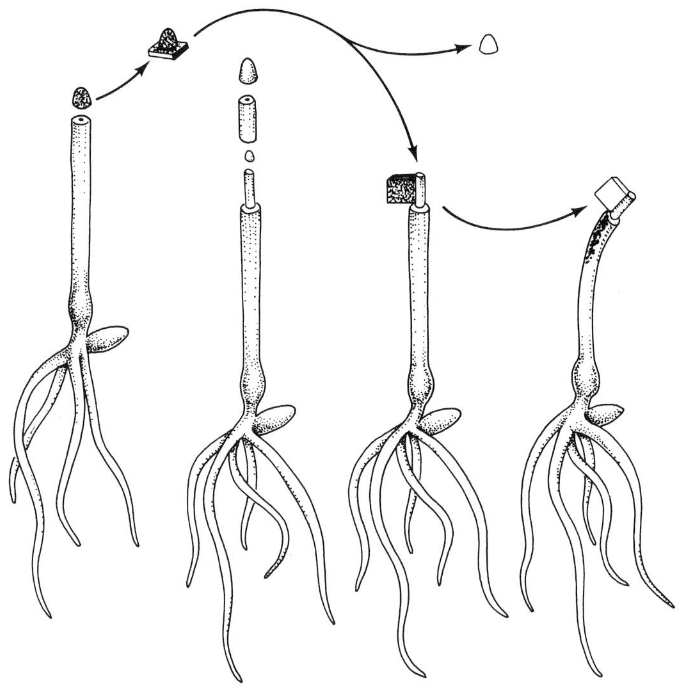

30–5 The experiment performed by Frits Went to demonstrate the presence of auxin in the tip of an oat seedling. The auxin is schematically indicated by heavy stippling. The tip is removed and placed on a block of gelatin. Auxin moves into the gelatin. Another seedling is prepared by removing the tip, waiting a short period of time, and then removing the tip again (a new "physiological tip" sometimes forms). The leaf inside the coleoptile is pulled out, and the gelatin block containing the auxin is placed against it. Auxin moves into the coleoptile on one side, causing it to bend.

His experiment (Fig. 30–5) combined elements of those of Boysen-Jensen and Paal. He removed the tip from a coleoptile, placed it on a small block of gelatin, and after an hour, discarded the tip and placed the small block of gelatin to the side of a coleoptile that had been decapitated and prepared in the manner illustrated. All of this was carried out in the dark or under a dim red light known to have no effect upon bending of the coleoptile. When the gelatin was placed to one side of the coleoptile stump, the cells below grew more than those on the other side, and curvature resulted. It is so unlikely that an electrical or nervous stimulus could exist in an isolated block of gelatin that the chemical explanation was clearly demonstrated.

Went spent a number of weeks working out the techniques of this experiment. Humidity is critical. If it is too low, the gelatin blocks dry out, and if it is too high, they dissolve in water absorbed from the air or become infested with bacteria. Went's experiment finally worked on April 17, 1926, at 3:00 A.M. He ran home to wake his father and tell him that the experiment had succeeded. Although interested, his father suggested that if the coleoptile would bend at 3:00 A.M., it would probably also bend the next day—and it did. Went called the substance he had isolated in his gelatin block auxin, which is now loosely defined as any substance active in Went's curvature test or in other related tests.

Went performed a number of other experiments within a few weeks following his initial discovery. Allowing the auxin to diffuse into blocks of gelatin, he showed that more auxin could indeed be measured on the dark side of a laterally illuminated coleoptile tip than on the light side (Fig. 30–6). He further showed that auxin moves only in a polar direction through coleoptile tissue (polar transport); that is, it moves only from the tip toward the base, even when the section is inverted so that gravity would presumably cause movement in the other direction.

One of the most valuable contributions of Went was the refinement of his curvature experiment into a quantitative test for auxin. The degree of curvature of the coleoptile after a given interval of time at controlled temperature and humidity indicated the amount of auxin in the small block of gelatin. A block of auxin upon which two tips had been placed caused twice as much bending, for example, as a block on which only one tip had been placed. The significance of Went's simple experiment and the ones that followed is that it was the first clear-cut demonstration of a plant-growth substance. This demonstration plus the ability to measure auxin *quantitatively* (the curvature bioassay) radically changed the entire course of plant physiological research.

Went's experiment led to the development of the concept of plant hormones. (Animal hormones were already known.) Such substances are now defined as organic molecules synthesized in one part of an organism and moved to another part, where they have a controlling or regulatory effect, causing the physiological response. Hormones are active in extremely small amounts. Nutrients such as sugar would fit the definition, except they are not active in small amounts. Mineral nutrients are excluded because they are not organic.

Compounds exist that influence plant growth and development in a manner similar or identical to that of hormones. These agents, which need not be natural plant components and are not necessarily transported, are called plant growth regulators. This broad term includes true plant hormones as well as synthetic growth regulators, which act much like hormones.

Went's assay for auxin opened the way to attempt its isolation and to study its chemistry. It was calculated that the amount of auxin in the coleoptile tip was so slight that 20,000 tons of tips would be required to obtain 1 g of pure auxin. Fortunately, other sources could be used, and in 1934 F. Kögl and his associates in the Netherlands isolated a compound from urine that caused decapitated coleoptiles to bend in Went's test. This proved to be indoleacetic acid (IAA). Several workers soon found IAA in plant tissues, and it has since been identified or at least implicated in a large number of species. Probably, Went's coleoptile auxin is also IAA, but other compounds that cause the bending response also occur in coleoptiles, although in smaller quantities.

Numerous substances have now been found that exhibit auxin activity. A few of these occur naturally in plants (some indole relatives of IAA); many others do not. Naphthalene acetic acid (NAA), indole butyric acid (IBA), and 2,4-dichlorophenoxyacetic acid (2,4-D) provide good examples (Fig. 30–7). These synthetic growth regulators are often used in laboratory experiments and in agriculture. Typically, they have different pharmacological effects from the natural auxins. For example, IAA is readily inactivated in the plant, but NAA and 2,4-D are not (or only much more slowly).

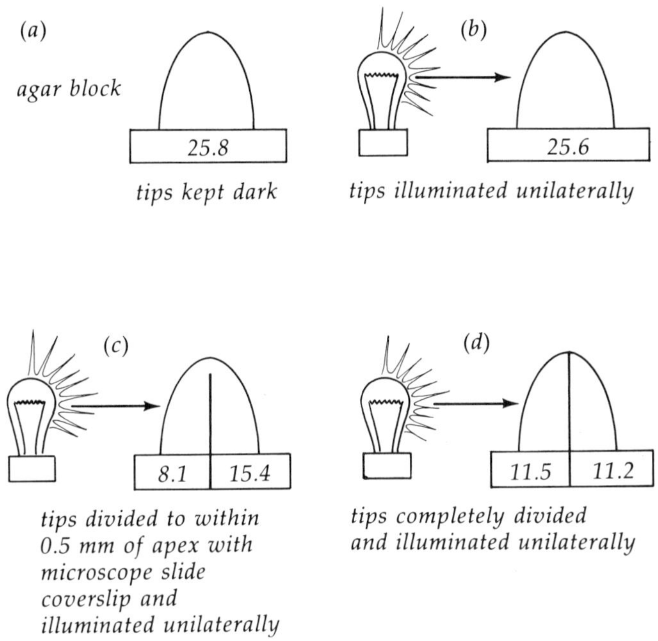

30–6 A duplication of Went's experiment shows that unilateral illumination of coleoptile tips leads to a transport of auxin from the illuminated to the shaded side and does not cause the destruction of auxin. Corn coleoptiles were used by W. R. Briggs. Numbers on the agar blocks represent the degrees of curvature caused by application of the blocks to decapitated oat coleoptile stumps. In the partially split tips, part of the auxin was transported laterally above the dividing barrier, but in the completely split tips this was not possible.

Auxin Action

As we mentioned, Went showed that phototropism could be explained by the distribution of auxin in response to light coming from one side. Recent experiments have confirmed that the auxin (or possibly a precursor) actually moves to the dark side, rather than being destroyed by light or synthesized more on the dark side.

But how is the auxin redistributed in response to light? One approach is to study the wavelengths of light that cause phototropism, that is, to measure an action spectrum for the process, as was done with photosynthesis. Coleoptiles bend only in response to blue light (Fig. 30–8). Apparently, some compound that absorbs blue light acts directly or indirectly on the distribution of auxin. Any compound absorbing only blue light appears yellow. Both riboflavin and carotene are yellow pigments of plants, and for several decades plant phys-

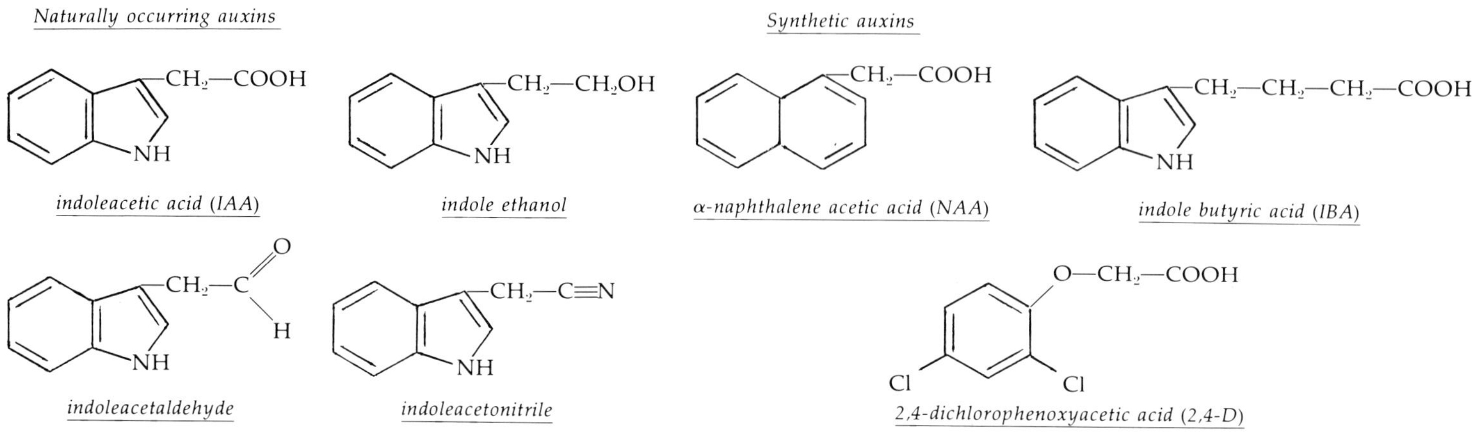

30–7 *Molecular structures of some naturally occurring and some artificially synthesized auxins.*

iologists have attempted to see if either of these can be responsible for phototropism. In recent experiments plants were produced (by feeding them certain inhibitors) that contain only about 20% as much carotenoids as untreated controls but comparable riboflavin. Both the treated plants and the controls were equally sensitive to light, indicating that the carotenoids may be of minor or no importance. Riboflavin may prove to be especially important in the phototropic response to ultraviolet light, as indicated by the action spectrum of Figure 30–8.

Herman Dolk, working in the laboratory where Went originally discovered auxin, performed the first experiment to show that geotropism can also be explained by auxin distribution. By letting auxin diffuse into gelatin blocks, he compared the amount of auxin on the lower side of a horizontally placed coleoptile with that on the upper side and found more on the lower side. Similar results have since been obtained with true stems.

Again, the response to gravity occurs in the tip, and again we remain at a complete loss for an explanation. Such compounds as indoleacetic acid, having molecular weights in the neighborhood of 100 to 300, must exist in true solution. In such a state their kinetic activity is too high for them to respond to gravity. There must be something in the cell large enough to respond to gravity. It would be called a statolith. One can show by calculation and by direct measurements that particles large enough to respond to gravity in a water medium are large enough to be visible with the light microscope; that is, they are larger than colloidal particles. In some way the statoliths must control the distribution of auxin in the stem.

As early as 1901 (before the discovery of auxin), it was suggested by B. Nemec, a Czechoslovakian plant physiologist, that the statoliths were amyloplasts containing starch grains. He showed that these particles do settle and that the time required was about right to account for the geotropic response. In 1966 K. V. Thimann (see Essay 31–2) and his students, then at Harvard, treated coleoptiles with another growth hormone (gibberellin), resulting in a complete loss of starch grains, so far as they could detect. The coleoptiles still responded to gravity, although somewhat more slowly. Apparently, after 60 years the starch statolith hypothesis had to be discarded, although a few workers are looking for compromises. No clear theory remains to fill its place.

An interesting special case is the plagiotropic (horizontal) growth of many tree branches. Often, branches of pines, spruces, firs, etc., grow at a specific angle in relation to the vertical. If the branch is bent away from this angle (or the trunk is moved from the vertical), a special kind of xylem tissue called reaction wood is formed, tending to force the branch (or trunk) back to its normal angle (Fig. 30–9). In gymnosperms compression wood forms on the bottom of a branch bent down,

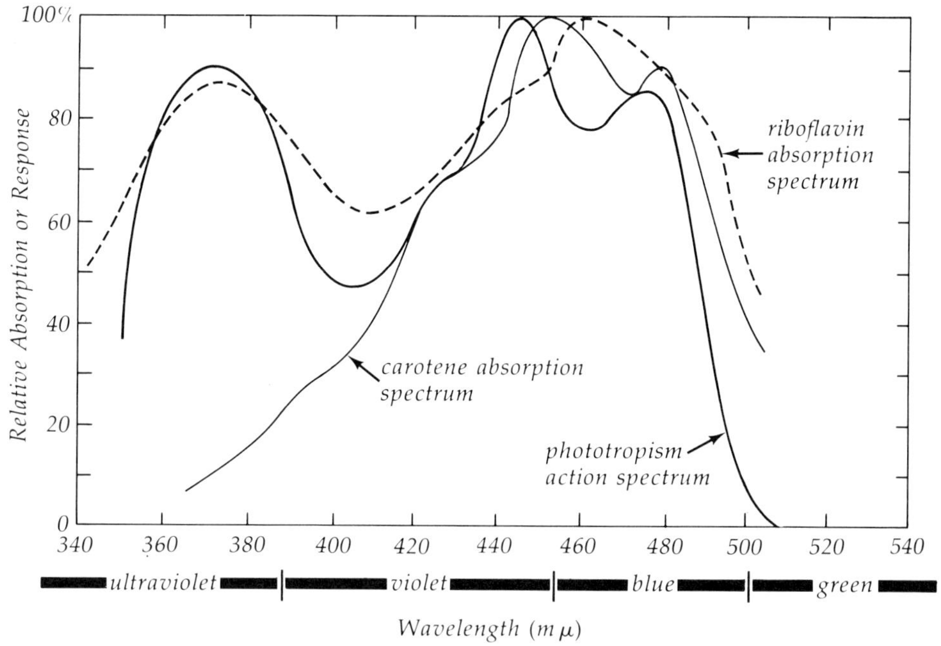

30–8 *The action spectrum for phototropism compared with the absorption spectra of riboflavin and carotene.*

pushing it back up, or on the top of a branch bent up, pushing it down. In angiosperms tension wood forms on the top of a branch bent down, *pulling* it up, or on the bottom of a branch bent up, pulling it back down.

Through light and electron microscope studies we are learning how the reaction wood produces compression or tension. It may have to do with the manner in which water molecules are adsorbed by the materials in the cell wall. It appears that auxin gradients can cause the formation of the reaction wood (compression wood in response to high auxin levels; tension wood in response to low auxin or auxin antagonists). Of course, the real problem is to determine how the cells detect their angle in relation to gravity. What kind of statolith arrangement could account for branches responding differently from the leader?

Another question concerns the responses of roots, which are opposite to those of stems. Does auxin in roots move toward the light and away from gravity? Probably not. It seems more likely that auxin *inhibits* the elongation of root cells (Fig. 30–10) but moves the same way in response to light or gravity. One possibility is that root tissue already contains an inhibiting amount of auxin so that *decreasing* its concentration promotes cell growth. On the other hand, root growth might be controlled by substances other than auxin. For example, certain inhibitors occur in some roots at concentrations that increase with distance behind the tip. Just at the point where cell elongation ceases, the concentration reaches the level that is inhibitory to root-cell elongation, when these inhibitors are applied externally. Some of the inhibitors are compounds, called unsaturated lactones, such as coumarin or scopoletin. It is conceivable that they may act upon the tissue response to auxin. Incidentally, it has recently been found that auxins sometimes have a polar movement in roots that is *opposite* to that in stems. They move preferentially toward instead of away from the tip. This observation may well make invalid all of the suggestions just offerred.

Auxin has been implicated in many responses besides phototropism and geotropism. Once indoleacetic acid had been identified as an auxin, it was possible to apply this material to plants in an almost random fashion, observing the results. Auxin was found to induce cell divisions and influence the growth of axillary buds, developing fruits, and the abscission of leaves. Most of these responses are more complex than the relatively simple cell elongations we are discussing here; so their detailed discussions will be postponed until the next chapter. But it is important to realize that auxin, in a manner typical of most of the known plant-growth hormones, acts in various ways depending upon species, concentration, responding organs, techniques of application, etc.

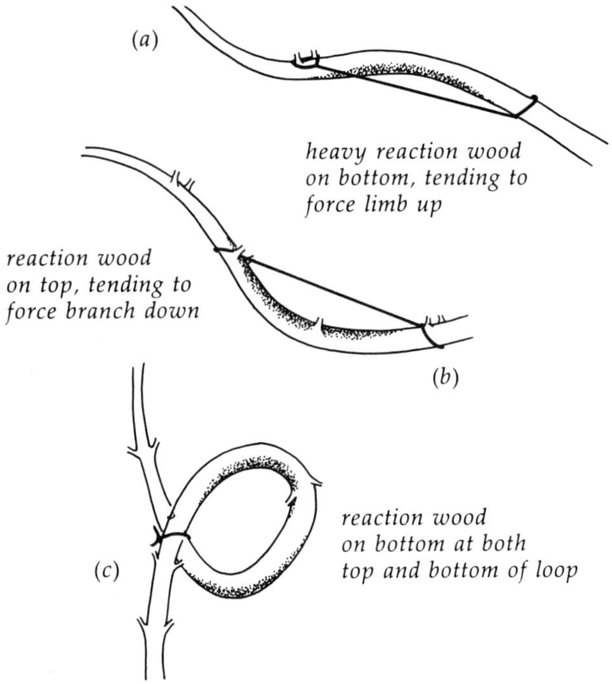

30–9 A summary of the experiments on the formation of reaction wood (indicated by stippling).

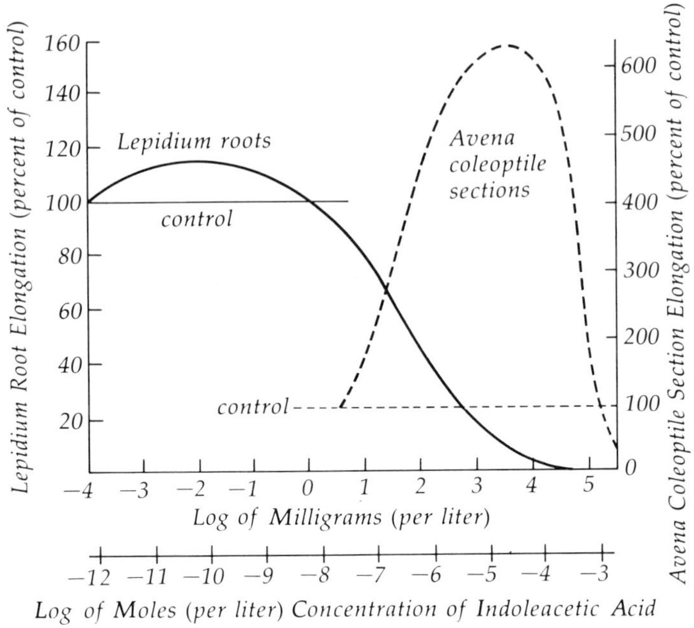

30–10 The growth of roots or coleoptile sections as a function of the auxin concentration of the solution with which the seedlings or sections are treated. Note that inhibition of root growth occurs at concentrations that maximally promote stem growth.

Ethylene

As early as 1878 it was noticed that seedlings grown in the dark sometimes exhibit curvature and twisting of the stems. Six years later Hans Molisch suggested that such effects might be caused by contaminants in illuminating gas and smoke—an early example of pollution! In 1901 D. Meljubow showed that the early twistings in horizontal growth of stems could be caused by traces of illuminating gas in the laboratory air, and he identified ethylene ($H_2C{=}CH_2$) and acetylene ($HC{\equiv}CH$) as the active fractions. Ethylene is far more effective than acetylene.

Following these discoveries, the growth of seedlings in the dark as affected by ethylene was investigated. Young pea seedlings grew in a horizontal plane (plagiotropically) when exposed to ethylene, and if they were reoriented to an upright position, they again curved until the stems had reached a horizontal plane. If ethylene was removed at any time, the stems reverted to the usual negative geotropism, curving upward. In the presence of strong light, the response to ethylene was greatly diminished or abolished. In 1924 it was discovered that ethylene hastened the ripening of green lemons, and it was soon found that many fruits responded in a similar way. Furthermore, the fruits themselves gave off the gas so that, for example, ripening apples in the same room with green bananas greatly hastened the ripening of the bananas. In 1935, during the period of high interest in auxin, it was suggested that ethylene may also be a naturally occurring growth regulator, but the suggestion was not actively pursued.

During the 1960s many plant physiologists finally became interested in ethylene, concluding that it is an important plant-growth hormone. The fact that it is a gas rather than a soluble organic molecule leads to a number of interesting features not common to the other hormones. A hormone to be effective as a regulator must be produced in response to some environmental stimulus or perhaps in response to the developmental program as it unfolds. It must then be moved from the control center to the site of action. Finally, it must be eliminated or inactivated so that its concentration will not build up to inhibitory levels. (Most hormones are effective only in a narrow range of concentrations—neither too low nor too high.) Auxin is continually destroyed or otherwise inactivated, for example, keeping concentrations within the right levels. With ethylene such destruction might not be essential, since the gas tends to diffuse out of the tissue and escape into the atmosphere. Apparently, ethylene is produced in growing meristems (as is auxin), and its first responsibility may be to keep these cells in a meristematic condition.

Is it possible that auxin acts only to induce the formation of ethylene, which then causes the observed response? In many cases this could be true, as has been suggested for epinasty (a twisting of stems and petioles), inhibition of the opening of the hook formed at the tip of many seedling stems, swelling of roots and stems, root-hair formation, and a host of other responses. On the other hand, ethylene inhibits a number of auxin effects such as the promotion of cell division. Ethylene seems to cause cell expansion but no division, and often the expansion is isodiametric (in all directions) rather than elongation. When coleoptile sections are treated with auxin (IAA), both ethylene production and elongation are promoted. When the IAA is removed, growth stops almost immediately (a most remarkable observation in itself), but ethylene production continues for an hour or two. Production of ethylene in response to auxin is inhibited by cyclohexamide (an inhibitor of protein synthesis), implying that auxin induces the formation of some protein enzyme in turn responsible for ethylene formation. Clearly, although auxin and ethylene are closely related, all of their effects cannot be mutually explained. In any case, plant physiologists are now willing to think of ethylene as a naturally occurring plant-growth regulator of considerable importance in normal plant development. And, as in the case of auxin and the other hormones discussed in the following sections, its effects are numerous and complex.

Gibberellins

At approximately the same time that Went was performing his experiments in Utrecht, a Japanese plant pathologist, E. Kurosawa, was studying the foolish-seedling disease of rice. This disease is due to a fungal infection, and the symptoms are rapid elongation of the stem, causing the plant to become extremely tall so that it falls over and eventually dies. Kurosawa was able to produce the symptoms of the disease with an extract from the pathogenic fungus. During the 1930s the active principle was isolated from such an extract and called gibberellin. The extract was found to contain two or three active fractions. Considerable study of this material followed its discovery. But investigations were carried out largely to determine the pathological effects of the fungus—not to investigate the possibility that gibberellins might be naturally occurring growth hormones for higher plants. At that time plant physiologists did not become aware of and interested in the substances.

Virtually nothing was known of these substances in the Western World, although abstracts describing the Japanese work had appeared in Western journals. In the early 1950s a group of researchers in England and another in the United States finally took cognizance of the Japanese work and communicated with the Japanese scientists, obtaining samples of gibberellins. In a few years the substances were being produced commercially from fungal cultures by American pharmaceutical companies. Study became intensive throughout most of the Western World, with the first papers appearing in 1954. A peak in exploratory work was probably reached in 1959, but research on these chemicals is still extremely active.

The chemical structures of the gibberellins were worked out at the University of Tokyo and at the Butterwick Research Laboratories at the Imperial Chemical Industries in England. Four gibberellins were isolated from fungal material and named gibberellin 1 to gibberellin 4 (abbreviated GA_1 and GA_4). GA_3 was extracted in the largest quantities, and it alone was called gibberellic acid, although all of the presently known gibberellins are acids. By 1965 the number of known gibberellins had increased slowly to about 14. With the advent of thin layer electrophoresis, gas chromatography, and mass spectrometry (especially the combination of the last two), the number of known gibberellins increased rapidly. A review of technical papers published before 1970 gives the molecular structures of 29 known gibberellins. The structures of GA_3 and a few other representatives are shown in Figure 30–11. Some of these have now been totally synthesized in the laboratory, although the steps involved are so complex that

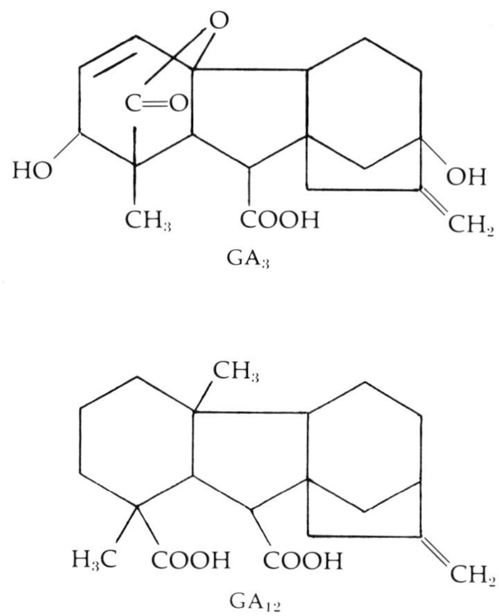

30–11 Molecular structures of some representative, naturally occurring gibberellins.

the GAs used in physiological research will probably be obtained only from natural sources for several years to come. Of the 29 known gibberellins, at least 19 are known to occur as natural products of plants. They have been found in angiosperms, gymnosperms, ferns, algae, and fungi, but not in bacteria. Certainly, they are of universal occurrence in the higher plants.

The most striking effect of the gibberellins is upon stem elongation, as in the foolish-seedling disease (Fig. 30–12). Some plants with stems that normally elongate at a nominal rate grow six times faster when treated with gibberellins. In certain genetic dwarf plants, not only is the dwarfism overcome, but their elongation is more rapid than in normal plants. Moreover, rosette plants, which usually do not elongate at all, often do elongate or bolt after gibberellin treatment. A few species have failed to respond, but in some cases, a species that failed to respond to one gibberellin subsequently responded to another. With 29 available gibberellins and 236,000 species of angiosperms, it is not surprising that all combinations have not been tried! The physiologist is overwhelmed with the possible modes of treatment, combinations of substances, and available plant materials, making it difficult to arrive at good solid conclusions.

Apparently, the effects of gibberellins are usually on cell elongation, although effects on cell division are also known. For example, when a rosette plant such as a beet with all its leaves radiating from a crown at the top of the root is treated with gibberellins and then sends up a bolting (flowering) shoot, cell divisions must occur to produce the cells that will elongate. Whereas auxins applied to intact plants seldom increase the growth of the plant, gibberellins applied to intact plants rapidly increase growth. This occurs over an unusually wide range of concentrations; in higher plant tissues, for example, gibberellins are found in concentrations of 0.001 to 500 μg/gram of fresh weight.

Certain plants that inherit the trait of being dwarfs grow as well as their normal counterparts when they are treated with gibberellins, suggesting that the normal growth of a plant is dependent upon some naturally

30–12 *The effects of a growth retardant (B–9) and a growth-promoter, gibberellic acid (GA_3), upon the growth of bean plants. The plant on the left was treated with B–9; the plant on the right with gibberellic acid; and the plant in the center is the untreated control.*

produced gibberellin. In such dwarfs study has now pinpointed defective genes (mutations) that normally control various steps in gibberellin biosynthesis.

Apparently gibberellins are produced in higher plants largely in the young leaves, in the roots, and in large quantities in developing seeds. They move throughout the plant in the phloem system, particularly when they are produced in the young leaves, but they can also move through the xylem system, especially when they are produced in the roots.

So-called bound gibberellins have been isolated. A few of these compounds have now been identified to be the usual gibberellins combined with other molecules such as an acetate or a glucopyronate radical. Apparently, bound gibberellins are important in the transport of gibberellins within the plant.

We have much to learn about the role of gibberellins as natural plant-growth hormones. For example, do the gibberellins produced by roots have important regulating effects in the shoots? We know that roots also produce cytokinins (see the next section) and that these are important shoot-growth regulators. Yet gibberellins are also produced by the young leaves, complicating the situation somewhat.

"Anti-gibberellin" compounds are also known. Some of these are synthetic compounds, including the growth retardants commonly called AMO, CCC, and phosfon (all having long chemical names). It is known that these three compounds prevent the synthesis of GA in the plant. Apparently, they do not interfere with GA action. Compounds are at least suspected that do interfere with GA action. These have been extracted from plants, but they have not as yet been identified.

As in the case of the auxins and ethylene, gibberellins have many effects on plant growth, besides the promotion of cell elongation. They influence fruit development and promote seed germination in dormant seeds and flowering in certain plants, particularly those normally requiring exposure to low temperatures and/or long days (see Chapter 36). In some cases, the gibberellins influence the same responses as auxin. Of course, stem elongation is the obvious example, but both auxins and gibberellins also cause fruits to develop without fertilization.

One aspect of gibberellin research has been rather disappointing: so far gibberellins seem to have only limited agricultural application. While stem elongation can be sharply increased, the dry weight of the plant seldom increases. The result is a tall, thin plant instead of a short, thick one. (One exception is the wide commercial application of gibberellins to grapes, causing increased yield and improved shape.) Perhaps, when we learn the proper techniques, we will discover some, more valuable agricultural applications of these compounds.

Obviously, we have not solved the problem of the control of cell enlargement and elongation, but at least we have made some headway by discovering compounds that promote the process. The situation with cell division is similar. We do not really understand the controls involved, but we have discovered a number of compounds that promote or inhibit division. In the remainder of this chapter, we shall discuss some of the experiments with substances controlling or influencing cell division. In the next chapter we shall consider still more details of hormonal action in differentiation, returning finally to the question of how hormones and growth regulators might function in the plant.

Organ Culture and Root-Growth Hormones

In the 1930s attempts were made to grow plant organs separate from the mother plant. A root (or other organ) might be isolated under sterile conditions and placed on a medium such as agar. Since the root normally has no photosynthesizing tissues, we would expect a requirement for an energy substrate. Sucrose could be added to the medium to meet this requirement. We would also expect that the major and minor mineral nutrient elements would be required, and they, too, were added.

On such a medium excised roots grew for a brief interval of time and then ceased their growth. Apparently, something besides sugars and minerals was required. In 1934 Philip R. White, using tomato roots, found that the cultures could be caused to grow indefinitely, when the medium was provided not only with sucrose and minerals but also an extract of dried brewer's yeast. It was later shown that the active substances in the yeast were the B vitamins, thiamine and pyridoxine. Extending the work to other species, it was found that pea roots required thiamine and nicotinic acid, while the common bindweed (*Convolvulus*) required only thiamine. Some roots will not grow even when provided with a number of B vitamins, amino acids, etc.; the suitable required factors remain to be discovered. Monocot roots are especially difficult to culture.

It seemed possible that in normal plants the B vitamins required by the roots are synthesized in the shoots and translocated to the root. In 1938, experiments by James Bonner demonstrated this to be the case. The vitamins are moved in the phloem tissue. If this is removed from around the stem (the stem is said to be *girdled*), the B vitamins accumulate above the girdle. Since cell divisions cease, it might be said that the B vitamins are required for cell division. Since they are made in one part of the plant and translocated to another, where they are effective in small quantities, they fit the definition of a hormone. Actually, biochemical research has shown that these molecules play an important role as coenzymes for certain of the enzymes of intermediary metabolism. For example, thiamine is a part of the coenzyme for the enzymes that evolve carbon dioxide in respiration.

Here, we are faced with an interesting test of our ability to handle terms. A *vitamin* is a compound required in the diet of an animal, the animal being unable to synthesize the substance. Whole green plants do synthesize all of these substances; consequently the term vitamin is meaningless for green plants. The root, on the other hand, is heterotrophic when cultured by itself, and in this sense, the term vitamin may begin to take on meaning for plants. In any case, many vitamins prove to be *coenzymes* in their mode of action. And, in the case of the root, these *vitamin-coenzymes* clearly act as *hormones*, confusing the situation even more. We may correctly speak of the root-growth hormones as cell-division factors, which probably act because of their role as coenzymes and which are equivalent to some of the vitamins required in animal nutrition.

The Wound Hormone

The root-growth hormones may be thought of as cell-division factors because cell division ceases in their absence, but this could be an indirect response. For example, some necessary precursor for cell division might not be synthesized in the absence of the B vitamins. In the case of the wound hormone, we apparently are dealing with a substance that is more directly implicated as an activator of cell division.

It has often been noticed that cutting plant tissue (such as potato tuber) causes cell division a few cells below the cut surface. Cells that have become relatively differentiated begin to divide again, producing a tissue similar to the cork cambium (see Chapter 24). If the cut surface of the tissue is washed immediately after cutting, the cork cambium layers fail to form. Something is washed off that would otherwise diffuse into the tissue and result in the formation of a lateral meristem. As early as 1913 and long before it had been isolated, Gottlieb Haberlandt in Austria referred to this substance as the wound hormone.

The formation of the wound meristem may be used as an assay for the presence of the wound hormone. Washings from the cut surface are applied back to the washed surface, producing the meristem. Such washings may then be fractionated, and individual fractions may be applied back to washed surfaces. The fraction that produces the meristem must contain the wound hormone. Through successive fractionation, the material may ultimately be isolated.

Such an approach was first taken with bean pods, and an active material was isolated and called traumatic acid. Probably, the substance is different in other tissues. For example, in potato tissues one wound hormone appears to be chlorogenic acid, although the situation is relatively complex. The substance is a phenol and readily oxidized. Its oxidation product(s) might be the substance(s) actually important as a wound hormone. In any case, we have a precedent for substances (in this case released from cut cells) that can move to intact cells, causing them to again become meristematic. Clearly, intact cells must be programmed to respond to wound hormones in this way.

The Cytokinins

Since the late 1930s plant physiologists have had another powerful tool for the study of cell division and differentiation—the culture of isolated tissues rather than organs. Again, a medium is used such as agar or gelatin, which has been supplemented with sugar and the essential mineral elements. The tissue may grow in a more or less disorganized way (much like a tumor), producing a mass of cells. One can add various substances to the medium to produce various effects.

In 1955 Carlos Miller (then a postdoctoral fellow) and others in the laboratory of Folke Skoog at the University of Wisconsin discovered that pith tissue from tobacco stems would grow on a medium containing sugar, minerals, vitamins, glycine, and auxin. To be more exact, the cells expanded, but they failed to divide. Sometimes, an individual cell would become as large as the head of a pin and could readily be seen with the naked eye. Microscopic examination indicated that nuclear divisions had taken place within the cells, but since no new cell walls were formed, each giant cell contained a number of nuclei. Miller found that the addition of extracts from aged herring-sperm DNA or from yeast caused the cells to divide, allowing the tissue to grow normally for an indefinite period of time. Using as their assay the occurrence of cell-wall formation, they were able to isolate the active material from the extracts. They named their material kinetin, and it proved to be 6N furfuryl adenine (Fig. 30–13).

Several synthetic compounds (mostly adenine derivatives) that have molecular structures similar to kinetin and that will also cause cell division in the

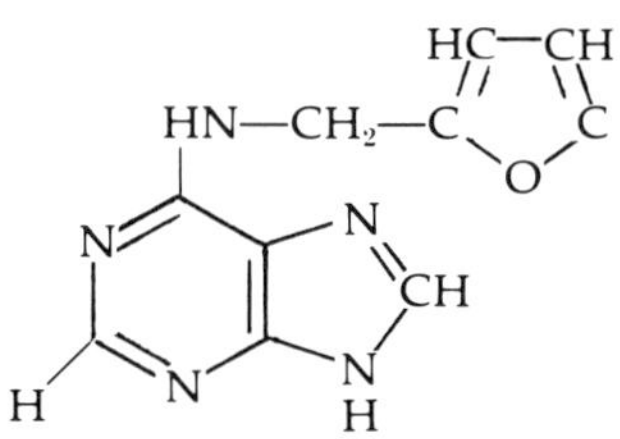
kinetin

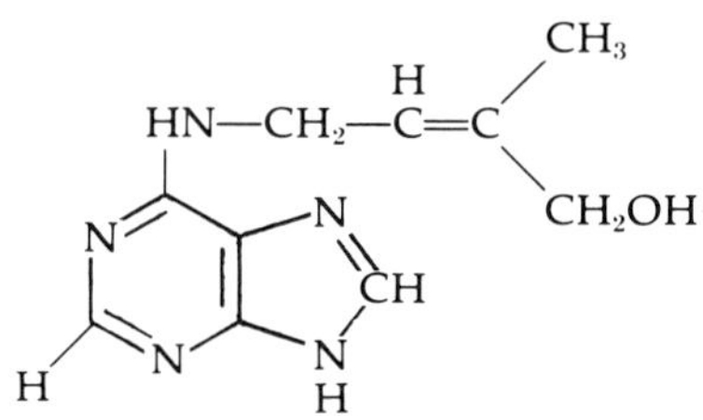
zeatin

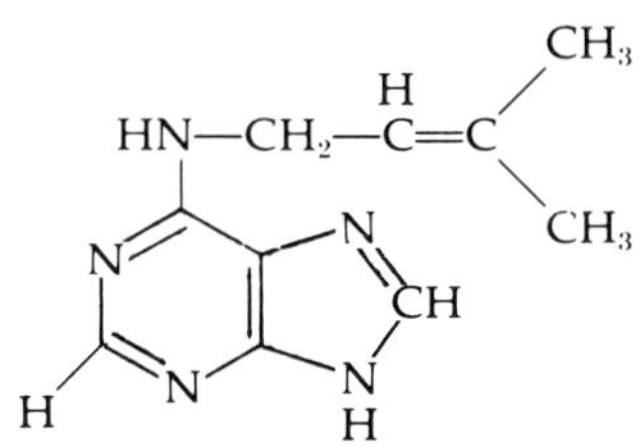
6-(γ,γ-dimethylallyl) amino purine (also called isopentenyl adenine)

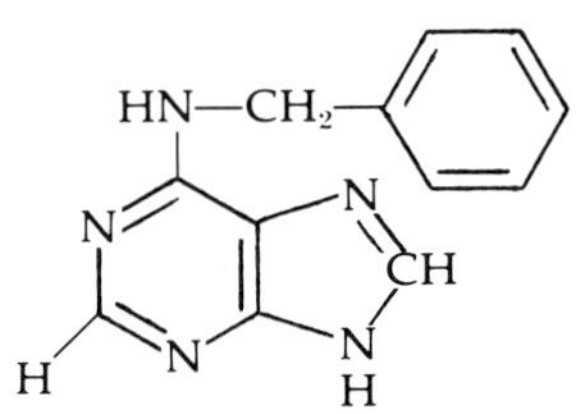
benzyladenine

30–13 Molecular structures of four cytokinins. Kinetin and benzyladenine are synthetic compounds; the other two are naturally occurring.

bioassay were soon discovered (Fig. 30–13). Kinetin itself has never been found in plants, but several compounds with similar activity have now been isolated from plant tissues. The first of these was isolated by Miller from corn and was later (1964) characterized by D. S. Latham and his coworkers. The substance was called zeatin. Originally, substances that caused cell division were called kinins, but these materials are now called cytokinins, since the term was used by animal physiologists in another context. In a review written by Skoog and published in 1970, 9 naturally occurring cytokinins are listed, and it is stated that at least 100 synthetic cytokinins have been produced. Compounds of at least one class accounting for many of the natural and synthetic cytokinins have an adenine nucleus similar to that of kinetin with some group substituted in the number 6 position on the molecule (Fig. 30–13).

Early work by Johannes van Overbeek (see Essay 30–3) and others showed that coconut milk contains a powerful stimulator(s) of plant growth. The milk of a coconut is the liquid endosperm and hence the food storage material for the developing coconut seedling. As such it might be expected to include potent growth-regulating substances. A group of researchers at Cornell University under the direction of Frederick C. Steward has studied the effects of coconut milk in detail. They found that carrot tissue could be maintained on a standard medium containing auxin but that the tissue only approximately doubled in size, after which growth ceased. When coconut milk was added to the medium, the carrot tissue grew rapidly and extensively for an indefinite time (Fig. 30–14).

The problem of discovering the active constituents of coconut milk has proved a difficult one. After failure in many initial attempts to isolate an active fraction from small quantities of coconut milk, the Cornell researchers obtained 660 gallons of milk from the coconuts blown down in a Florida hurricane. From this, they then extracted about 26 pounds of a dark heavy syrup, which they fractionated into a number of substances (many were identified as amino acids or other nutrients). From this extract they successfully obtained a few purified compounds that were effective in their bioassay. After further work with extracts and sources other than coconut milk, they concluded that at least three types of substances, acting jointly, are necessary for the growth-supporting activity of coconut milk and comparable substances.

A source of organic nitrogen. It could be either the amino acids obtained from the coconut milk, or some other source such as milk protein (casein) that has been hydrolyzed into its constituent amino acids.

A neutral fraction. It "stimulates" or "catalyzes" the third, active fraction. Such a response is called synergism (meaning that the two fractions applied together are more effective than the total activities of both fractions when each is applied alone). This neutral fraction consists of sugar alcohols such as d-sorbitol and mannitol and the alcohols called inositols.

The actual promoters of cell division. A number of interesting compounds have been isolated. These

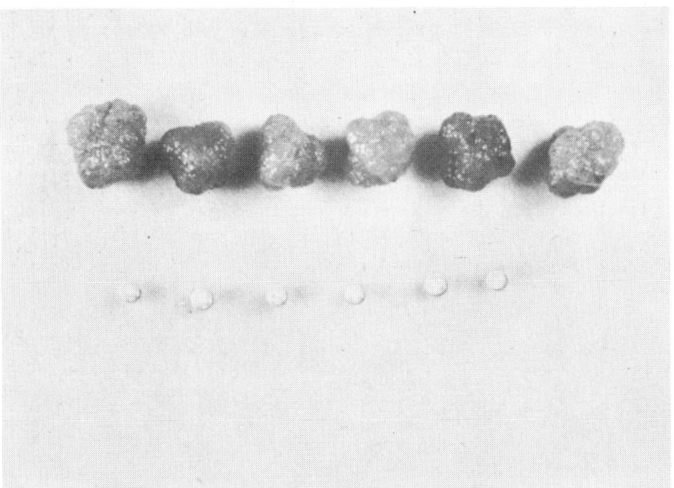

30–14 The growth of explanted pieces of carrot-root phloem. The upper row shows the large green explants grown on a basal medium supplemented with coconut milk. The lower row shows cylindrical orange explants grown on a basal culture medium without coconut milk. The explants have been removed from their growth flasks and placed on the petri dish to give some idea of size (diameter of petri dish 100 mm).

cytokinins all have activity in the carrot assay (provided the neutral fraction is present), but none of them alone or in combination yet matches the activity of the whole coconut milk. One of the active compounds proved to be 1,3-diphenylurea, a surprising discovery since urea and its derivatives were thought to be exclusively animal products. Another substance proved to be a colorless relative of the pigments found in flower petals. More substances have since been at least tentatively identified. In 1968 Latham reported that most of the cytokinin activity in coconut milk could be attributed to zeatin and one of its derivatives.

Essay 30–3 Coconuts and Cytokinins

Johannes van Overbeek is a plant physiologist who has strongly influenced the development of that science during the past several decades. Like Frits Went, he began at Utrecht in Holland, spent time in the tropics, and completed exciting research at the California Institute of Technology. He spent many years working on agricultural chemicals for industry, but he never lost his love of teaching and finally wrote a highly readable, little book on plant physiology addressed to high school students. This story is from his Caltech years.

"Most valuable in my scientific life were the research opportunities in the broad field of biology afforded me as a graduate student in Utrecht in the Netherlands. Even before I joined the plant hormone team of Went the elder, I had already discovered the secret of the tension in the catch muscle of clams (it is the hinge band) and the elusive symptoms of boron deficiency in Gramineae (white stripes). Once accepted as a graduate student in Holland, there were no endless regulations to cope with, no grade point ratios to worry about. There was just learning for the sake of scholarship. So I took many courses without credit ranging from surface chemistry to tropical agriculture. These learning experiences served me well in later years, as the following story will illustrate.

In 1940 I met the geneticist A. F. Blakeslee on a ferry between Anacortes and Friday Harbor in the San Juan Archipelago near Seattle, Washington. He was making wide crosses with Datura *and needed the hybrids for genetic analysis. Often no viable seed resulted, yet he felt that the embryos were viable, but that the endosperm was incompatible. So I suggested that he isolate the embryos and raise them on an artificial nutrient. He invited me to his laboratory in Cold Spring Harbor, Long Island, to develop the technique for him. I was an assistant professor at Caltech at that time and well familiar with media for isolated roots. But, whatever known medium I tried, the* Datura *embryos would not grow in it. Finally, I became rather desperate.*

One afternoon, after a good swim, all of a sudden I recalled my lessons in tropical agriculture and in particular that the water inside the coconut is liquid endosperm. It might contain the factor that was eluding me so far. At once I sent a laboratory assistant to the A&P store in Huntington to get some coconuts with the water inside. That same afternoon we removed the liquid aseptically and added some of it to the vials that contained the isolated embryos on agar slants. After the experiment was done, the embryos were placed in a shaded part of a greenhouse, and I left on a lecture tour to Columbia, Harvard, and several other universities in the East.

Upon my return I went straight to the greenhouse to look at my embryos. The results were astounding: the embryos had grown so much that they nearly filled the space in the small vials. My first thought was that someone had played a prank on me, and when told that Folke Skoog had visited Cold Spring Harbor during my absence, I felt pretty sure that he had played a trick on me! So my associate, Marie E. Conklin, and I repeated the experiment, and again we obtained fabulous growth upon addition of 10% coconut milk to the medium. This was probably the first time that cytokinin action was observed, although we did not know, of course, what the new factor was. When I returned to Caltech, A. J. Haagen-Smit, my graduate student, Ralph G. H. Siu, and I attempted to isolate the new growth factor, but it was always found in the dirtiest fractions. It was not until 1955 that Skoog's group finally identified the nature of the cytokinins."

Extracts from many other sources have also proven to be active. Any developing seed that has a liquid endosperm seems to have activity. These sources include immature bananas, walnuts, horse chestnuts, corn, and even certain plant tumors. Potato tissue contains an inhibitor of cell division that seems to antagonize the effects of the promoters. Steward suggests that normal plant growth might result from a balance between inhibitors and promoters of the type he has discovered. Could tumorous growth be an instance in which the promoters far overshadow the inhibitors?

Besides the promotion of cell division, cytokinins have a number of highly interesting effects on plant growth. Typically, the response will be a function of the tissue studied and the cytokinin applied. A number of bioassays have been developed, frequently depending upon the particular process under study. There are several bioassays, for example, based upon the ability of cytokinins to cause cell division. Miller had developed a very sensitive test using callus produced on soybean cotyledons.

Cytokinins have an important effect upon the senescence of leaves. If a leaf is removed from a plant, kept moist, and supplied with the usual nutrients, it will soon turn yellow and die—unless adventitious roots can form on the petiole. It was suggested in the 1930s that these roots produce a hormone essential for the maintenance of a healthy green leaf. It is now known that cytokinins prevent the yellowing of detached leaves; so one sensitive bioassay measures the chlorophyll content of leaf discs as influenced by a suspected cytokinin. Exudates from roots are highly active in this bioassay, showing that roots do produce cytokinins that, by being translocated to the leaves, act as hormones.

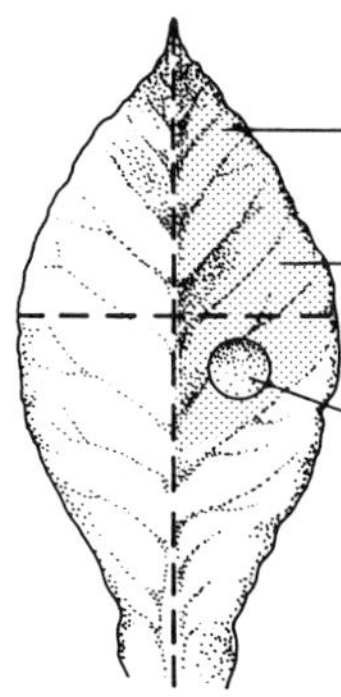

30–15 The effects of cytokinin upon translocation in a tobacco leaf.

Cytokinins have still another interesting effect in leaves. If a drop of a cytokinin solution is added to a yellowing leaf, not only does the spot become green but such substances as amino acids move into the spot from other parts of the leaf—even from other leaves (Fig. 30–15).

In the early work in Skoog's laboratory it was discovered that cytokinins might promote the formation of buds on tissue cultures. If auxins were provided at high levels, roots tended to form, but cytokinins overcame this effect, producing buds. (Promotion of bud activity can also be observed in other ways.)

A few years ago it was discovered that several t-RNA molecules contain cytokinins. As a matter of fact, the cytokinins are present only in those t-RNA molecules that recognize codons beginning with U (see Chapter 8). The possibility immediately became apparent that cytokinins act by becoming incorporated into t-RNA molecules. The case for this is indeed strong, but evidence is still ambiguous. Several possible modes of action of cytokinins have been proposed, but so far we are left with no really satisfying answers.

31

Development: Differentiation

Around and Around We Go—Symmetries and Phyllotaxes

Plants exhibit many kinds of symmetries. Spirals are seen not only in the arrangement of disc flowers in a sunflower and the scales on a pine cone but also in the order of arrangement of leaves on a stem. Often, branches exhibit considerable symmetry, and frequently the leaves on a plant are arranged so that sunlight is maximally absorbed, with only parts of certain leaves being shaded.

A *A mountain pine* (Pinus mugo) *cone. Note the similarity of arrangement of the scales to that of the disc flowers in the sunflower.*

B *A giant sunflower* (Helianthus sp.) *grown for seed. Flowers at the outer rim of the disc are past maturity, while those in the center are just beginning development.*

A

B

C *A coleus plant* (Coleus sp.) *with an opposite-leaf arrangement. Coleus is in the mint family and has square stems. The apical buds have been removed so that the axillary buds have become active.*

D *A purple or velvet plant* (Gynura aurantiaca) *with a spiral-leaf arrangement. Leaves are numbered beginning at the top. This arrangement is called "alternate" as contrasted with the "opposite" leaves of the coleus plant. There are many kinds of alternate arrangements.*

E *Looking down on the velvet plant. If a leaf is chosen (e.g., no. 8), then the one above it (e.g., no. 3) subtracted from the first gives the number 5. In counting from the first to the second (from no. 8 to no. 3), the spiral goes twice around the stem, so the leaf arrangement is then said to be $^2/_5$. Actually, in looking down on the top of the stem, the leaves are arranged so that no leaf is immediately above any other, but note that in Figure D leaf no. 2 is immediately above leaf no. 7 in its origin at the node. Leaf no. 4 is immediately above leaf no. 9, and leaf no. 3 is immediately above leaf no. 8.*

D

C

E

F *A fittonia* (Fittonia argyroneura) *plant. The leaves seem to have moved around so that no leaf is totally shaded.*

F

When visiting our woodland ecosystem, we are not immediately impressed with the fact that cells have been dividing, enlarging, and elongating. Such processes by themselves lead only to masses of cells, as in a tumor. The striking thing about the woodland is its variety of form. To produce the intricate patterns of stems, roots, leaves, flowers, fruits, or even fungal fruiting bodies, cells must be differentiating at appropriate times and in appropriate places—all according to the developmental program. Morphogenesis is taking place in a highly characteristic manner for each species.

Even at the cellular level, differentiation is much more complex than enlargement and elongation, if only because there is such a variety of end products. The intricate changes that occur in cell-wall structures provide a simple example of differentiation. In tracheids or vessels the entire cell contents break down and disappear after the secondary cell walls have been formed; in sieve tube elements the breakdown is partial. Leaf mesophyll cells with their numerous chloroplasts are strikingly different from guard cells, root-hair cells, or pollen mother cells, to name a few. Yet cellular differentiation is just the beginning of the problem. Cells differentiate to form tissues and organs, and all of this is coordinated to produce a complete plant, with roots, shoots, and reproductive systems. Even this is just a beginning, considering the multitude of different species and varieties. We are faced with an almost infinite array of end products to the process of differentiation—which logically follows from the essentially infinite possibilities for gene combinations.

The Problem of Morphogenesis

How do genes control the development of a mature organism from a zygote? In this chapter we shall examine the problem of differentiation or morphogenesis, first in a theoretical way, then by surveying some of the experimental approaches to it, and finally by returning to the question of the mode of action of growth regulators and some of their practical applications.

By the processes of cell division, enlargement-elongation, and differentiation, cells not only become highly

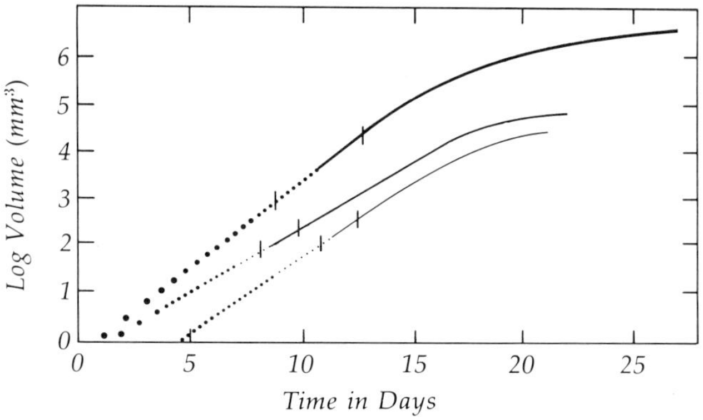

31–1 The relationship between cell division and cell enlargement in the growth of Cucurbita *fruits. Dotted parts of the curve indicate growth by cell division, and solid parts show growth by cell enlargement. Vertical bars indicate the time of change over from one form of growth to the other. Note that growth rate remains constant in spite of this change.*

specialized, but they do this in the right places at the right times, producing the complex, mature organism. Indeed, there is reason to believe that morphogenesis is more than growth and development of individual cells. Consider the tendency to restore the whole organism. When the vertical leader shoot from a fir tree is removed, for example, one (sometimes two) of the nearly horizontal axillary branches near the top grows vertically, assuming the role of leader and restoring the growth form of the tree. If branches are removed from one side of the tree, growth is typically more vigorous in that area, tending again to restore the original shape. Of course, this happens as *cells* respond to the wounding caused by removal of the leader or a branch. Nevertheless, the end result approaches a restoration of the original whole.

This holistic feature of morphogenesis is further illustrated by plant symmetries. Plants often exhibit highly complex symmetries of leaf and branch arrangement, positions of flowers in the inflorescence, or parts on an individual flower. Sometimes, the symmetries are obtained by a special growth process during which different parts of the plant grow at different rates in relation to each other, the ratio of the rates remaining constant. Frequently, the "symmetries" do not appear in the usual sense of the word but rather as a spiral arrangement of leaves on a stem or parts of a flower. Our ultimate understanding of morphogenesis must account for these phenomena.

Edmund W. Sinnott, a botanist who was active from about 1913 to 1965, was deeply concerned with the problems of morphogenesis. Near the end of his life he wrote a book, *Plant Morphogenesis,* that provides the most complete available summary of morphogenetic information relating to plants. Sinnott emphasized the holistic concept of organismal morphogenesis. In one study he observed the growth of squash fruits. Growth rate was measured continuously, and cells were examined (taking samples) to see whether the processes of cell division or cell enlargement were predominating at any moment. As Figure 31–1 indicates, cell divisions were frequent during the early part of the growth of the fruit. But these eventually ceased, and the later stages of growth were due only to cell enlargement. Yet the curve expressing growth of the fruit as a whole shows no sharp break between the periods of cell division and cell elongation. These subprocesses contribute only to the overall holistic growth of the fruit.

What is the developmental program? Science often attempts to understand nature by constructing models. Sometimes the models are real, and sometimes they are conceptual. With the rapid expansion of modern technology, we are provided with some new and exciting opportunities for model construction.

Computer technology may some day help us understand differentiation through analogical processes. For example, a computer may be programmed to carry out a series of computations or functions. In principle a factory could be run by such a computer. Not only would specific manipulations be performed in the proper order, converting raw materials into finished products, but quality control would be achieved by feedback systems that would monitor the quality of the product, making continual adjustments in the procedures to maintain this quality. The product quality in turn would be monitored by transducers, instruments that respond to such physical parameters as temperature, pressure, size, weight, or presence of specific chemicals, converting (translating) the intensity level of a given parameter into some stimulus (probably electrical) to which the control system (the computer) could respond. The response would be according to a preset program, changing the procedure so that the product quality, as detected by the transducers, would more nearly approach the desired levels. (A thermostat controlling house temperature is a good example of such a feedback system.) All this would be part of a master program.

A plant might be like this, except that the "factory" not only builds products but builds itself as well, duplicating itself at intervals. The master program must be contained in the genetic material (the genome) of the zygote. The machinery consists primarily of the enzymes, and the program must be a control over the genes that are actively making enzymes in certain cells and at certain stages of the life cycle. Many transducers can be inferred, and some are known. The pigment that absorbs light, causing a plant stem to bend toward the light source, is an excellent example. There must also be transducers that detect the levels of auxin, sugars, cytokinins, and a host of other materials or conditions, causing the cells to divide, grow, or differentiate in suitable ways controlled by their genetic programs.

While we can visualize the transducers and some of the steps going on in the plant "factory," we know nothing about the operation of the genetic program. It is logical to infer that such a program must exist, and this inference will probably lead to fruitful experimentation and the final revelation of the program itself. In the meantime, we can only sharpen the formulation of our questions about the program and seek to devise experimental access to it.

Is the cell totipotent? In 1958 F. C. Steward's work with carrot cultures and cytokinins led to a finding that is highly significant for the morphogenetic program. As was already known with other systems, rotation of carrot-tissue cultures in a liquid medium (continually aerated and never exposed to gravitational forces coming predominantly from only one side) produced a milky suspension of individual cells that had apparently broken off from the growing mass of carrot tissue. Given the proper environment, it appears that any one of the single cells in the suspension might develop much as a zygote, finally producing an entire plant (Fig. 31–2). W. H. Muir, Albert C. Hildebrandt, and Albert J. Riker showed in 1958 that single cells might be expected to produce entire plants; in 1959 Armin C. Braun at Rockefeller University in New York City elaborated on this, using tobacco crown galls. The observations have now been extended to several species. Incidentally, a cell will only become an embryo when it is in the presence of other cells or when it is grown on a medium that has previously supported other cells (a conditioned medium). The medium itself can originally be simple and completely defined.

This beautiful research allows us to conclude unequivocally that the individual cell is totipotent: it is capable of producing an entire plant. Clearly, each cell, including tumor cells, contains the entire genetic program. The trick is in getting the cells to dedifferentiate so the program can play out from the zygote stage. (Perhaps, isolating a cell in a suitable medium initiates the "zygote program.") Actually, we have long believed that all cells were totipotent, since we know that the chromosomes are duplicated at each mitotic cell division and since cells from virtually any part of the plant can be induced to become meristematic, producing shoot and finally root meristems capable of growing into a mature plant. Virtually any plant part can act as a cutting, if suitable species are employed.

Though the cell itself is totipotent, containing the entire developmental program, we must not overlook the importance of the environment in the expression of this totipotency. When embryos are excised from developing seeds of *Datura* (Jimson weed), their future development is strongly dependent upon the chemical environment. If the embryo is large enough, it will develop into an entire plant when it is placed on the standard medium. Smaller embryos also require amino acids and vitamins. Even smaller embryos require all of these things plus coconut milk. (It was this observation by Johannes van Overbeek and co-workers in 1942 that led to the discovery of the importance of coconut milk; see Essay 30–3.) Still smaller embryos fail to grow on any medium that has so far been artificially provided. Yet the medium provided by surrounding tissues inside the ovule is clearly sufficient; we merely lack complete understanding of the chemical and physical nature of this environment.

Incidentally, it is important to realize how totally heterotrophic a zygote (or a cell lacking chloroplasts in Steward's culture) really is. The higher plant is autotrophic, needing only water, carbon dioxide, oxygen, and minerals, but the zygote, which will become the higher plant, also requires a host of growth factors, many of which have yet to be defined.

The fact that expression of the genetic program always depends upon environment is especially well illustrated by the experiments with ferns discussed in Chapter 26. Sporophytes could be made to develop

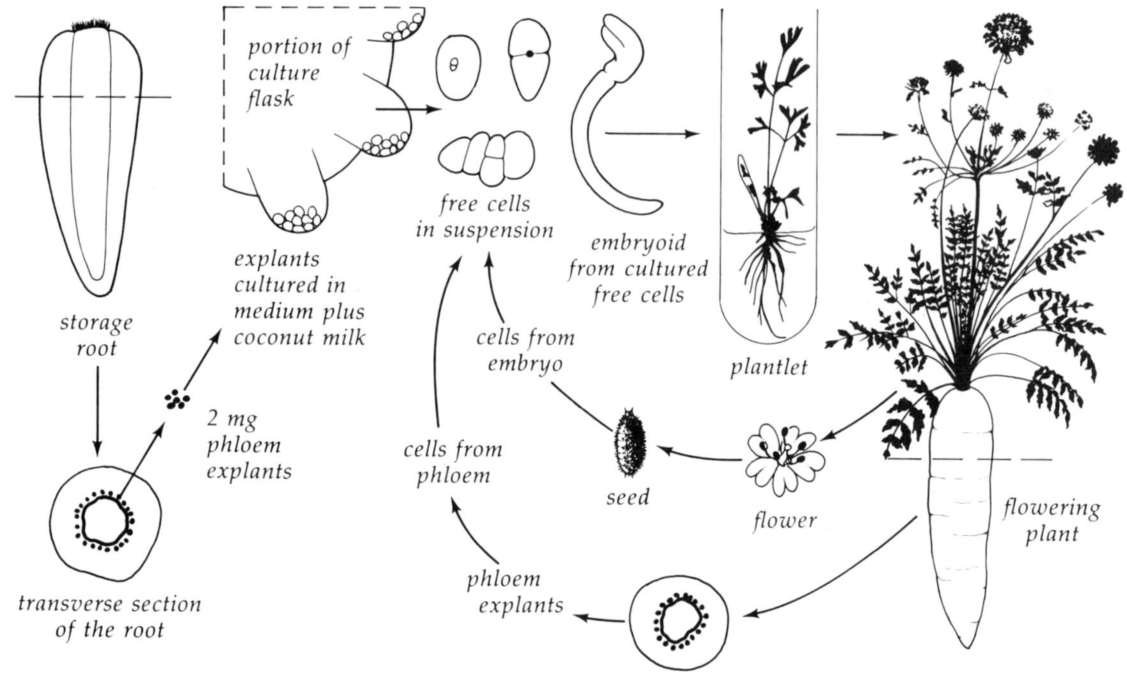

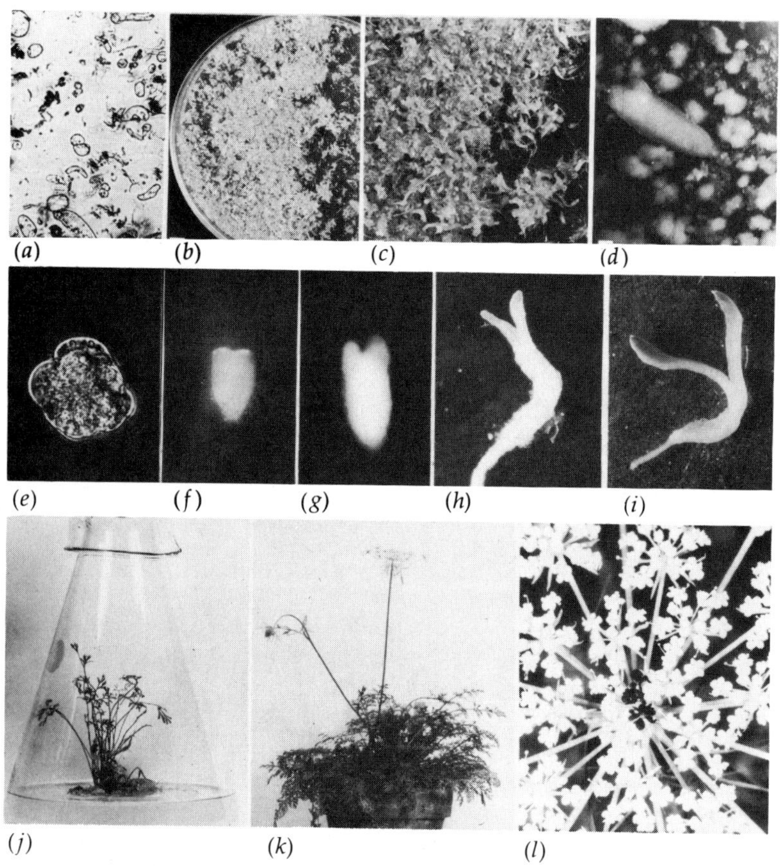

31–2 The "life cycle" of a carrot carried through F. C. Steward's experiments. Top—initially, explants are taken from a carrot storage root. These are cultured in the nipple flask with basal medium plus coconut milk. The resulting free cells in suspension will develop into embryos and small plants in a test tube. Finally, they may grow to a perfectly normal mature plant from which phloem explants could again be taken. Alternatively, free cell suspensions may be developed from flowers, seeds, or embryos. Bottom—photographic illustrations of the steps shown in the diagram. (a) Cultured carrot cells in liquid medium. (b) Embryoids developing in a petri dish on which a suspension of carrot cells was dispersed. (c) Higher magnification of the plate shown in b. (d) Globular masses of cells developing in a liquid medium from cells such as those shown in (a). (e) Globular stage of embryo development. (f) Heart-shaped stage. (g) Torpedo stage. (h), (i) Cotyledonary stages. (j) Developing plant grown from an embryoid on agar. (k) Such a plant after six months growth. (l) Flowers of a plant reared from cells as in (a).

31–3 Callus formation on stems in response to applied auxin. Right—the cocklebur stem on the left has had the tip removed and auxin (1% NAA) in lanolin placed on the stump. Callus formation is very local. The other two plants with leaves had the leaves dipped in a relatively concentrated NAA solution (1×10^{-3} M), and callus has subsequently formed on the stems below. The other two stems, showing heavy callus, were wounded and then treated with the auxin paste. Left—the tomato plant shows extreme callus formation.

from cells with *n* chromosomes (usually the gametophyte), and 2*n* gametophytes could also be produced. When the archegonium is cut so that the zygote is not tightly confined, it will often develop into a rather flat, thalluslike body similar (though not identical) to the gametophyte but markedly different from the normal sporophyte. Environment always influences expression of genes, sometimes in striking ways.

Approaches to Differentiation: Galls and Tumors

One situation is known in which it is possible to observe cellular division and growth with little differentiation. Several treatments, including the application of an auxin (e.g., in a lanolin paste) to the stem of a plant, result in an apparently uncontrolled cell division and enlargement, producing a mass of virtually undifferentiated, tumorlike tissue (Fig. 31–3). A certain hybrid between two species of tobacco spontaneously produces such a callus growth (hence referred to as genetic tumors). As we saw in Chapter 16, certain well-known plant tumors, the crown galls, are induced by the presence of a bacterium (*Agrobacterium tumefaciens*). The bacterium is essential for the formation of the initial tumors. But secondary tumors arise at a distance from the primary tumor, sometimes in the absence of the bacterium—a situation common to malignant animal cancers.

The study of calluses, galls, and tumors can provide insight into differentiation in at least two ways. First, we can learn something about the normal control of differentiation by discovering the factors that lead to loss of this control in the production of tumorlike growths. Second, these calluses and tumors can often be cultured on an artificial medium, where it is possible to study their nutritional requirements not only for maintenance of the undifferentiated mode of growth but also for the induction of differentiation. In either case, we may be studying factors that affect the morphogenetic program.

The induction of the crown-gall tumor has been studied rather intensively for many years now, although the problem is far from solved. The initial step is a wounding of the tissue. The wounding reaction, already described, follows this whether the infective bacterium is present or not. The bacterium infects the plant optimally only by interacting with the plant cells between 60 and 72 hr after wounding. At this period a burst of mitosis occurs in the wounded cells. Infection can occur before and after this period, but only at a reduced level. Hence susceptibility to the bacterium is a transient state.

It has been suggested that a tumor-inducing principle is formed during this interval of time. If the tem-

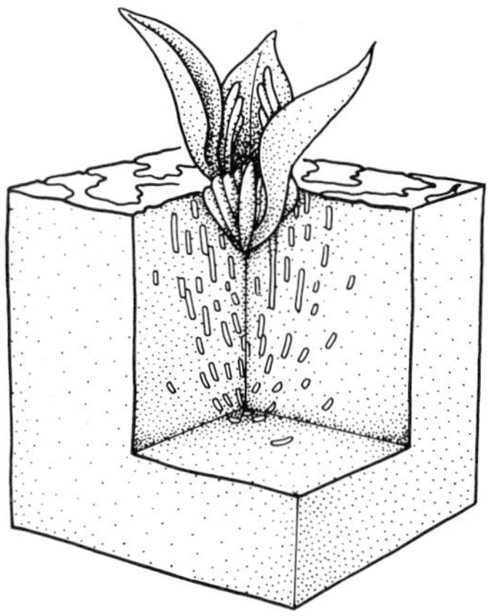

31–4 The effects of grafting an apex of lilac into a piece of lilac callus. Note that most of the developing vascular strands tend to be scattered. Their general orientation, however, seems to be strongly influenced by the presence of the bud, and in one case at least, a strand in the callus has become continuous with the bud. The approximate size of the callus is about 1 cm.

perature is held above 32° C, formation of this principle is prevented, although neither the bacterium nor the plant is damaged in any perceptible way. Following infection, cells begin to divide and grow in an uncontrolled way; typically they are polyploid (at least tetraploid and as much as 16-ploid—a situation not uncommon in animal tumors). Everything happens in the tumor at random. Mitotic figures are scattered throughout the cells, as are some xylem and phloem cells.

If the tumors are produced during the optimal 60 to 72 hr, they become relatively independent in their nutrient requirements, needing only the basic nutrients and sucrose. They are as autotrophic as possible without being capable of photosynthesis. Tumors formed before the optimal 60 to 72 hr (prevented during this time by higher temperatures) are small but will grow on a culture medium, provided it is supplied with various substances such as myoinositol, auxins, amino acids, nucleotides, and cytokinins.

Much of the work on crown galls has been carried out by Braun at the Rockefeller University. He suggested in 1958 that tumors form when various synthetic pathways in the cell, which are normally repressed, become derepressed during tumor induction. The cells appear to increase in their capacity to produce the substances that are the controlling agents in normal growth, thus resulting in increased cell division and enlargement.

Differentiation of Vascular Elements

In tumors, vascular elements do appear, randomly scattered throughout the tissue. In the developing root or shoot apex, vascular elements are distributed in highly predictable ways. Usually, phloem cells differentiate to the outside, and then xylem cells to the inside. Differentiation occurs in the cells that contact the ends of xylem and phloem cells already differentiated. In an axillary bud, however, differentiation of vascular elements occurs first in the bud (separated from the vascular elements of the stem), with a connection finally being made. Clearly, there is much more developmental control than in a tumor.

Callus tissue cultured on a defined medium provides a simple experimental system for the study of differentiation. If a bud is grafted onto a tissue culture, there is a tendency for vascular elements in the culture to become oriented in relation to the grafted bud, that is, to connect with this bud (Fig. 31–4). Somewhat the same effect could be obtained by implanting a block of agar containing auxin onto such a tissue culture. Finally, in 1963 Ralph H. Wetmore (see Essay 31–1) and John P. Rier of Harvard and Howard Universities reported that sugar (sucrose or glucose) concentration controls the formation of xylem or phloem, provided that auxin is present. Low sugar results in xylem formation; high causes sieve-tube formation; and intermediate levels (2.5 to 3.5%) produce both xylem and phloem with a cambium between. When auxin and sugar were applied to the callus surface, nodules of vascular tissue formed with phloem on the outside, xylem beneath, and much pith in the center. In the developing shoot or root apex, sugar arrives via the phloem tissue; thus it is reasonable that the cells differentiating at the end of phloem cells differentiate into phloem cells. The auxin would be coming from the meristematic cells at the very tip.

Convincing evidence has already been reported indicating that auxin controls the normal development of xylem and phloem in roots (John G. Torrey, 1957) and in stems (William P. Jacobs, 1956). Jacobs, for example, found a qualitative and a quantitative relationship between xylem formation in *Coleus* stems and the auxin produced by the leaves.

The experiments of Wetmore and Rier also demonstrated that the distance from the surface was important—phloem cells differentiate closest to the surface. Here, oxygen or carbon dioxide concentrations could be in control, but other factors might also be responsible. It is important and impressive to realize that xylem and phloem cells differ markedly from each other, although whether one or the other differentiates is apparently controlled by a relatively narrow range of environmental factors.

Essay 31–1 What Happens Where and Why—The Experimental Study of Development

There are as many ways to attack a question as there are people to ask it. This is certainly true for the study of cell development, which has been approached on a morphological, physiological, and biochemical level. Few investigators have been as successful as Professor Ralph Wetmore of Harvard University in applying joint attacks to this critical biological problem. Here he discusses some of his work on cell differentiation in shoot meristems and in tissue cultures.

"The driving urge that forced me to become a biologist early in my college years was my fascination with the mechanisms of evolution and the possible explanation of why today's plants exist, while whole groups of other plants have become extinct and are known only as fossils. On the biological staff at Harvard, Professor I. W. Bailey and I, as plant morphologists and anatomists, turned our research time and energy for some years in the 1930s to comparative anatomical studies of wood, or xylem, of angiosperms. Our purpose was to determine the value of xylem as an added but unexploited source of additional knowledge on the taxonomy of the angiosperms, thereby possibly better understanding what might have been critical changes over time in the evolutionary patterns of the vascular conducting systems.

We had assembled large documented collections of microscopic slides, some 35,000 of them, representing the secondary or cambium-produced xylem of over 300 families of angiosperms. The comparative studies from our laboratories had established beyond doubt that evolutionary trends could be recognized from correlations of extensive data, as well as the direction of these trends. But all of these data were assembled from woody plants—trees, shrubs, and woody vines—possessing a circumferentially increasing growth layer or cambium. Yet a cambium is not necessary for the successful and vigorous survival of vascular plants. In fact, the cambium can only arise from the procambium of the primary body, that is, that part of a plant derived from its shoot and root apical meristems.

By the late 1930s, it had become very clear that if we were to make continuous headway in our studies of anatomy in relation to taxonomy, we must learn something of developmental patterns in herbaceous plants, and why they persist in such numbers. Here began a testing of new techniques and new approaches. Ernest Ball, then working in our laboratories, was demonstrating that when he amputated a very minute terminal piece of the apical meristem of lupine (Lupinus albus) *or nasturtium* (Tropaeolum majus), *two woody herbs, each explant would grow on a sterile,* agar-semisolidified, nutrient, mineral-salt medium, containing also minor elements and sugar. He further demonstrated that these terminal segments were capable of rooting and then, and only then, characteristically developed into normal, whole, vegetative plants.

Could we use this sterile, nutrient culture technique for a progressive study of development of different herbaceous types in order to determine whether each could produce a cambium, and if not, why not? Could we establish a possible role for leaves or substances produced in and diffusing from leaves in the differentiation of primary vascular tissues of the leaves and of the stem below the leaf attachment? What would happen if surgical puncturing or removal of one or a few leaf primordia were carried out or if single leaf initials were isolated by longitudinal cuts carefully made so as to separate individual primordia from those laterally adjacent? Would the development of leaves and of these vascular tissues in the stem be modified? If chemical activation of the incidence of foliar and subjacent vascular tissue could be controlled, would quantitative modification of the endogenous activators help determine whether a cambium could be induced in the leaf bases or the stems? Would punctured leaf primordia develop at all? These and many other surgical experiments were tried in our laboratories and in those of other experimental morphologists, especially those of Professor C. W. Wardlaw in England and Professor Ball who had moved to North Carolina. Many increasingly informative investigations were simultaneously carried out in sterile culture, the medium being varied by the addition of exogenous hormonal substances, some natural, some analogous, and some substances proven to be inhibiting.

In our laboratory, it became apparent in the mid–1950s that if a sterile bud with a base cut wedge-shaped were grafted into an agar-filled, V-shaped cut in the top surface of a piece of callus from the same species and if the callus were planted in an adequate sterile medium, not only did the graft grow but vascular tissue developed in the callus, the graft bridge. Following these observations we found that if the V-shaped cut were filled with agar-solidified medium instead of a bud of the same or even a related species, the vascular tissue still developed. Its disposition in the callus, whether close to the cut or farther toward the periphery of the callus, was characteristically in a rectangular cross-sectional, circular, or triangular pattern, depending on the three-dimensional shape of the callus. Moreover, the vascular tissue consisted of xylem only, phloem only, or phloem surrounding xylem even with a cambium between the phloem and xylem, depending on the concentration of sugar in the V-shaped cut. Lower concentrations of sucrose favored xylem formation, higher stimulated phloem, and intermediate concentrations characteristically gave both, often with a cambial layer between. If, however, similar concentrations of indole-3-acetic acid and sugar were added to the medium and none on the top of the callus, vascular tissue tended to appear along the flanks and bottom of the callus in random pattern, but none appeared at the top.

Since 1960 much has been added, from our own and other laboratories, to our knowledge of factors involved in the differentiation of vascular tissue. Plants not believed to have vascular tissue, many fern prothalli, for example, can be biochemically stimulated to produce it by supplying deficient sugar in the medium. In other words, the hereditary potentiality is equally present in the prothallus and the sporophytic fern plant. Callus can be made friable by manipulating the medium and growing the culture on a shaker mechanism. These single cells or small clumps of cells can likewise be activated to produce vascular elements. They require, however, a cytokinin, that is, a hormone stimulating cell division. Other increases in knowledge along morphogenetic lines come fast at the present time. The important thing to note is that the plant is a growing and developing product of its own inherent enzymatic metabolism. If it lacks certain components needed for a certain process or it does not have it or them in adequate concentration or it possesses inhibitors to certain phases of those processes involved, the results cannot be achieved. We discover and understand these variables slowly, but each step is a step forward, even when the results appear negative. In the plant, anatomy is not distinct from physiology; structures present are used in physiological processes for which the heredity genome sets the stage. Evolutionary survival determines the success of the natural experiment. Structure and function must evolve together. Newer fields of endeavor—biochemistry, molecular biology, electron microscopy—all allow deeper probing and permit possible interpretations of evolutionary phenomena in organisms.

Yet, even now, we do not understand why herbs are herbs."

results:	*no growth*	*callus and roots, no shoots*	*callus only, no roots or shoots*	*shoots and callus, no roots*
treatments and concentrations:				
auxin (IAA)	*0*	*2 mg/liter*	*3 mg/liter*	*0.03 mg/liter*
kinetin	*0*	*0.02 mg/liter*	*0.2 mg/liter*	*1 mg/liter*

All flasks contain White's nutrient agar.

31–5 A demonstration of the effects of auxin (indoleacetic acid—IAA) and kinetin upon formation of roots and shoots in tobacco pith tissue cultures.

Meristem Development

As contrasted with the callus experiments, the meristem is a highly ordered and controlled system. It is normally capable of producing an entire plant. But developmental scientists ask where the control is exercised. Since phloem cells form in contact with phloem cells and xylem in contact with xylem, we immediately wonder whether differentiation occurs in response to the presence of already differentiated cells, or do the dividing cells of the meristem, completely undifferentiated, have the power to control the entire genetic program? Since vascular elements in axillary buds do not form in contact with stem vascular elements, we might initially be a bit suspicious about the idea that cells differentiate only because they are in contact with already differentiated cells.

The experimental approach is to culture excised meristems from stems and roots on media of known composition to see if it is possible for an excised meristem to produce an entire plant. Rather early (1946) Ernest Ball succeeded in this endeavor with the apical meristems of lupine, provided they included the three youngest leaf primordia. Apical meristems of club mosses, horsetails, and ferns would produce an entire plant in culture media even without leaf primordia (as shown by Wetmore in 1954). Apparently, angiosperm meristems are more heterotrophic (have fewer synthetic capabilities) than those of nonseed plants, but in neither case was vascular tissue necessary. Growth of root meristems on culture media was more difficult but was finally achieved. In this case, it was necessary to add a number of nutrients not required by the stem meristems with their leaf primordia.

We can conclude that the cells of the meristem are indeed totipotent and capable of producing an entire plant, although this is completely dependent upon the presence of suitable nutrients. The presence of nutrients seems somehow less mystical than the presence of already differentiated cells. Nevertheless, our conclusion is a compromise. Differentiation is always a matter of the genome being expressed within the context of a certain environment, in this case a nutrient environment.

Meristem Formation

In the intact plant, root and shoot meristems form in a variety of situations. As we have seen, branch roots normally appear in the pericycle, and axillary buds appear in the forming leaf axils. Adventitious root meristems also form on the stems of numerous species, and occasionally shoot meristems form on the roots of cer-

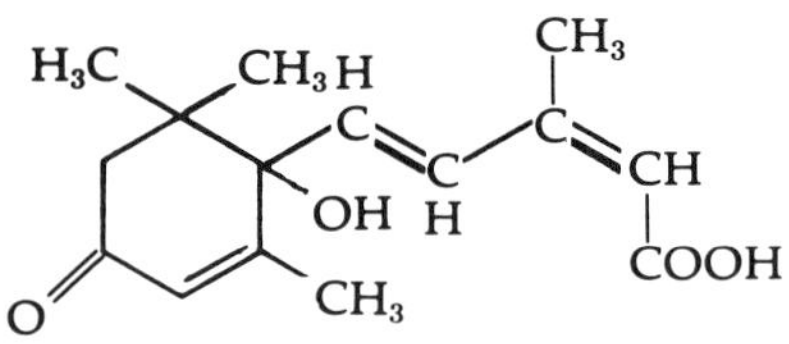

abscisic acid

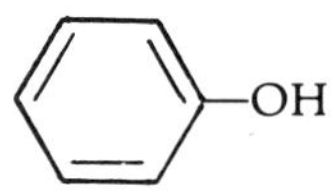

phenol

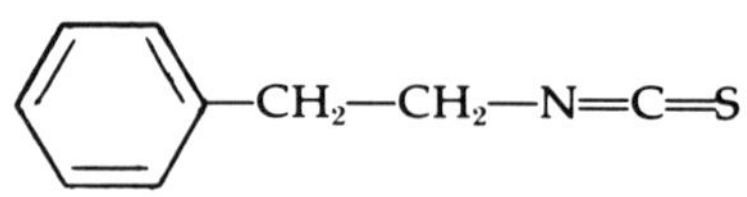

β-phenyl-isothiocyanate

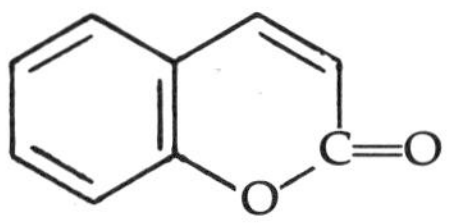

coumarin

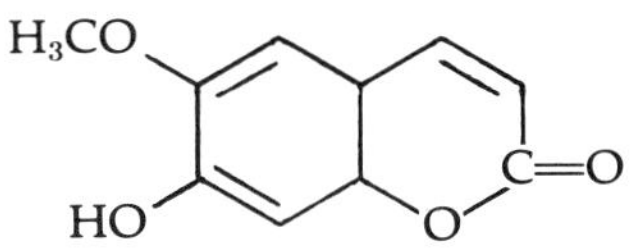

scopoletin

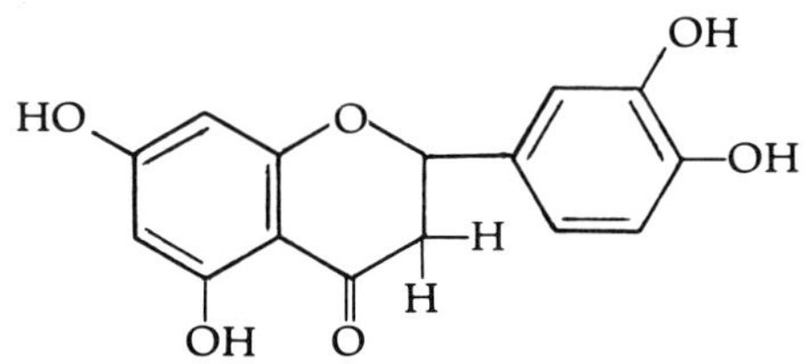

naringenin

31–6 Molecular structures of certain plant growth inhibitors.

tain species (e.g., the poplars, which are notorious for sending up suckers anywhere in the yard). Sometimes, adventitious buds also appear on mature stems. If we could control some of these processes, we would have achieved some understanding of them.

Some callus cultures (e.g., the genetic tumors described) occasionally form spontaneous buds or roots. In the late 1940s it was discovered that lowered oxygen levels resulted in increased bud formation in such cultures. It was then found that addition of adenine to the culture medium considerably promoted the formation of adventitious buds. Perhaps, lowered oxygen pressures might in some unknown manner result in increased adenine levels. But the real breakthrough, as we saw in Chapter 30, was the discovery of kinetin (an adenine derivative). While adenine increased the number of buds on such cultures only slightly, kinetin increased the number greatly (Fig. 31–5). Furthermore, when cytokinins were high, buds predominated; when auxins were high, adventitious roots predominated.

This observation matches results obtained in experiments using plant cuttings rather than tissue culture. Shortly after the discovery of auxins, it was found that many cuttings could be induced to form adventitious roots by treatment with auxins, sometimes in combination with such other substances as sucrose and a mixture of amino acids. Furthermore, it was found that treatment with cytokinins sometimes induced the formation of adventitious buds. An African violet leaf, for example, forms adventitious roots on the petiole rather readily, and often this is followed by the formation of adventitious buds. Adventitious bud formation can often be considerably stimulated by treatment with a cytokinin.

Apical Dominance: Bud Dormancy and Abscisic Acid

The presence of an active, growing bud at the tip of a stem often inhibits the active growth of the axillary buds. If the tip is removed, the axillary buds grow out —a pruning practice long used by horticulturists to produce a more bushy form.

Soon after the discovery of auxins, it was found that axillary bud growth in many species could be inhibited by application of auxin in a lanolin paste to the cut stump of the removed tip (Fig. 31–6). Apparently, the apical bud produces auxin, which in turn inhibits the growth of the axillary buds. But it must certainly be more complex. To begin with, auxin is usually a stimulator of growth, so how does it manage to inhibit in this case? Furthermore, auxin is completely ineffective in some species. In cocklebur it can be shown that the young leaves near the apical bud are inhibiting the growth of the axillary buds and that the extent of the inhibition is determined by light conditions and photoperiod. Removing the young leaves (even if the apical bud is left) allows the axillary buds to grow out, and applying auxin has no effect.

Essay 31–2 A Search for the Root-Forming Hormone

Kenneth V. Thimann, who was educated in England, came to the United States in 1930. He spent five years at the California Institute of Technology and then went to Harvard University, where he remained until 1965, when he became Provost of Crown College at the University of California, Santa Cruz. He has strongly influenced the development of plant physiology, especially in the field of plant hormones—as this account of his search for the root-forming hormone indicates.

"The search for the root-forming hormone was probably the most dramatic period of my scientific life. I had recently come to Caltech as a junior biochemist. Herman Dolk joined us soon afterward from Utrecht, where he had been working on the role of the 'growth substance' in geotropic curvature, and we agreed to go into partnership in an effort to try to isolate and identify this mysterious substance. I would be the chemist, and he would be the botanist; I would make the large-scale extractions and fractionations, and he would do the tests. But after about a year of this collaboration, there came that tragic weekend when Herman took his wife and daughter (and my mother, who was visiting that winter and spring) out on a ride to see the spring flowers in the desert, and he went off the road at Rosamond and was killed. Fortunately, the others survived, and Frans Dolk, who was a skillful technician, helped me to go on with the work, but it meant that I had to become plant physiologist as well as chemist. We worked along for a year doing tests at the end of each working day, and measuring them at night, until we had some very potent and partially purified extracts from a Rhizopus *culture medium, extracts that contained millions of 'units' of growth substance per cubic centimeter.*

Then Frits Went arrived as Dolk's replacement, late in 1933. Frits had been working in Java with Bouillenne on the postulated root-forming substance. Following up the 1925 experiments of van der Lek, which showed that roots on grape and willow cuttings tended to form directly below young leaves or growing buds, they had found that certain rice-polishing extracts applied to the upper end of Acalypha *cuttings would promote rooting below. So Frits and I decided to try for the isolation of the root-forming 'hormone' (a term introduced later), using a simple test with cuttings from pea seedlings. Imagine our great pleasure when the concentrated* Rhizopus *extracts turned out to be rich not only in growth substance but also in root-forming substance! Bill Cooper, a horticulturist at the Redlands Laboratory nearby, had been in the lab, and he tried out my preparations on a more horticultural subject, lemon cuttings. They produced magnificent roots. In Utrecht, Professor Kögl and his colleagues were also working on the growth substance, and they sent us one of their fractions ('Kögl V'), which was also highly active. In every purification step that I made, both kinds of biological activity went in parallel, and I soon began to suspect that they might both be due to the same substance. This would be flatly contrary to Sachs' ideas, which were in vogue at that time, of special and unique 'organ-forming' substances.*

Then came Kögl, Haagen-Smit, and Erxleben's discovery that the most active elongation-promoting fraction from human urine was the known compound indole-3-acetic acid. My preparations almost always contained nitrogen, and they gave deep red colors with strong acids—suggestive evidence of indoles. Also, I had found earlier that Rhizopus *only gave high yields of growth substance when certain specific peptones were in the medium, so an amino acid was under suspicion as the precursor. Tryptophane, of course, would fill the bill. Just at this time I got a small yield of crystals from one of the most active fractions; they gave the Salkowski test and melted at about the right temperature for indole acetic acid. They had tremendous activity both for elongation and for rooting.*

To clinch the matter, synthesis would be necessary. Joe Koepfli of the Chemistry Department had been interested in our project; he had worked with Perkins on alkaloids, which had given him a feeling for plant biochemistry. Accordingly we decided to work together on a synthesis. It went well, and late in October (1934) we obtained the pure substance (IAA). It was highly active in forming roots. Not only were both 'hormones' thus identified, but one substance clearly had two quite different biological activities—one depended on cell elongation and the other on cell division and organization. It was a surprise, but we also found that IAA had other activities, namely, bud inhibition and cambial activation. But those are another story."

Apparently, cytokinins are also involved. If pea stem segments are floated in an auxin solution, the axillary buds remain dormant even though the tip is removed. When cytokinins are added to the solution, the axillary buds begin to grow out. We are therefore confirmed (to an approximation) in our generalization that cytokinins promote bud growth, while auxins promote the formation of roots.

Bud dormancy may be due not only to apical dominance but to changes within the plant in response to the environment. In this case, all of the buds, including the terminal one, become dormant. (We shall discuss the factors leading to bud dormancy in Chapter 34, but part of the story is highly pertinent to our present discussion.) We have long suspected that bud dormancy might be caused by an inhibitor produced in the leaf, often in response to environmental conditions (e.g., a combination of short days and low temperatures). Thus it seemed reasonable to search for such an inhibitor by making extracts from the leaves thought to produce it.

In the early 1960s this and related approaches were being followed by three groups of investigators: Frederick T. Addicott and co-workers at the University of California in Davis, who were interested in finding a substance that might accelerate leaf fall, particularly in cotton plants; K. Rothwell and R. L. Wain at Wye College, London University, who were interested in a substance that accelerates flower drop in yellow lupines; and Phillip F. Wareing and co-workers at the University of Wales, Aberystwyth, who were interested in the substance that might cause bud dormancy. Each of these groups obtained an active fraction. It was subsequently discovered that all three groups were working with the same material, now called abscisic acid. Only minute quantities of the substance have been purified from plants (a total of only 29 mg by the middle of 1970). But the structure has been determined (Fig. 31–7), and synthetic abscisic acid can be produced in quantity in the laboratory. The synthetic material is not quite identical to the natural substance (it is a mixture of isomers rather than only the one found in plants), but it is as effective as the natural material when applied in the various test systems. Incidentally, light of moderate intensity alters the structure of even the natural isomer; this probably happens in the plant and could be important in the natural effects of abscisic acid.

Using various means of identification, abscisic acid has now been found to be a natural component of at least 35 species of angiosperms, 2 gymnosperms, and 1 fern. The substance is present in very small quantities (0.01 to 4 ppm), and it is active at about the same concentrations when applied to plants. So far, measured quantities are about the same in all tissues throughout the year, so correlations with bud dormancy have not become apparent.

(a) (b)

31–7 *Sun leaves (a) and shade leaves (b). Note the larger size, longer petioles, darker color, and more sparse distribution of the shade leaves.*

As we have seen with the auxins and the gibberellins, plant growth regulators are often active in several systems. Abscisic acid is no exception.

The applied chemical will induce buds of many plants (especially trees) to go into winter dormancy. The active apical bud becomes transformed into a resting bud with typical scales.

Abscisic acid promotes leaf abscission. It was known that when the blade is removed, leaving the petiole attached to the plant, growth of the abscission layer at the base of the petiole is strongly promoted. If auxin or a cytokinin is applied in a lanolin paste to the cut end of the petiole, formation of the abscission layer is inhibited. The picture becomes complex because relative auxin concentrations in the stem and the leaf blade have proven to be quite important. Furthermore, it was long known that ethylene caused premature abscission, and now it is known that abscisic acid applied to the blade produces the same effect. (This was the point of Addicott's original research leading to his isolation of abscisic acid.) In this case, ethylene is acting *opposite* to auxin and similar to abscisic acid.

If discs are cut from the leaves of various plants, they turn yellow and senesce much more rapidly in the presence of abscisic acid than in its absence. The substance apparently results in a more rapid protein breakdown and a decline in RNA. Abscisic acid sprayed onto intact leaves is much less effective in inducing senescence, probably because cytokinins (and gibberellins?) are being continually supplied to leaves from the roots, and these retard senescence.

Abscisic acid prevents flowering of certain species and promotes flowering of others. It tends to induce tuber formation in potatoes—possibly an indirect effect due to the inhibition of vegetative growth.

Abscisic acid inhibits germination of several nondormant seeds. This implies that normal dormancy in seeds as well as buds could be due to the presence of

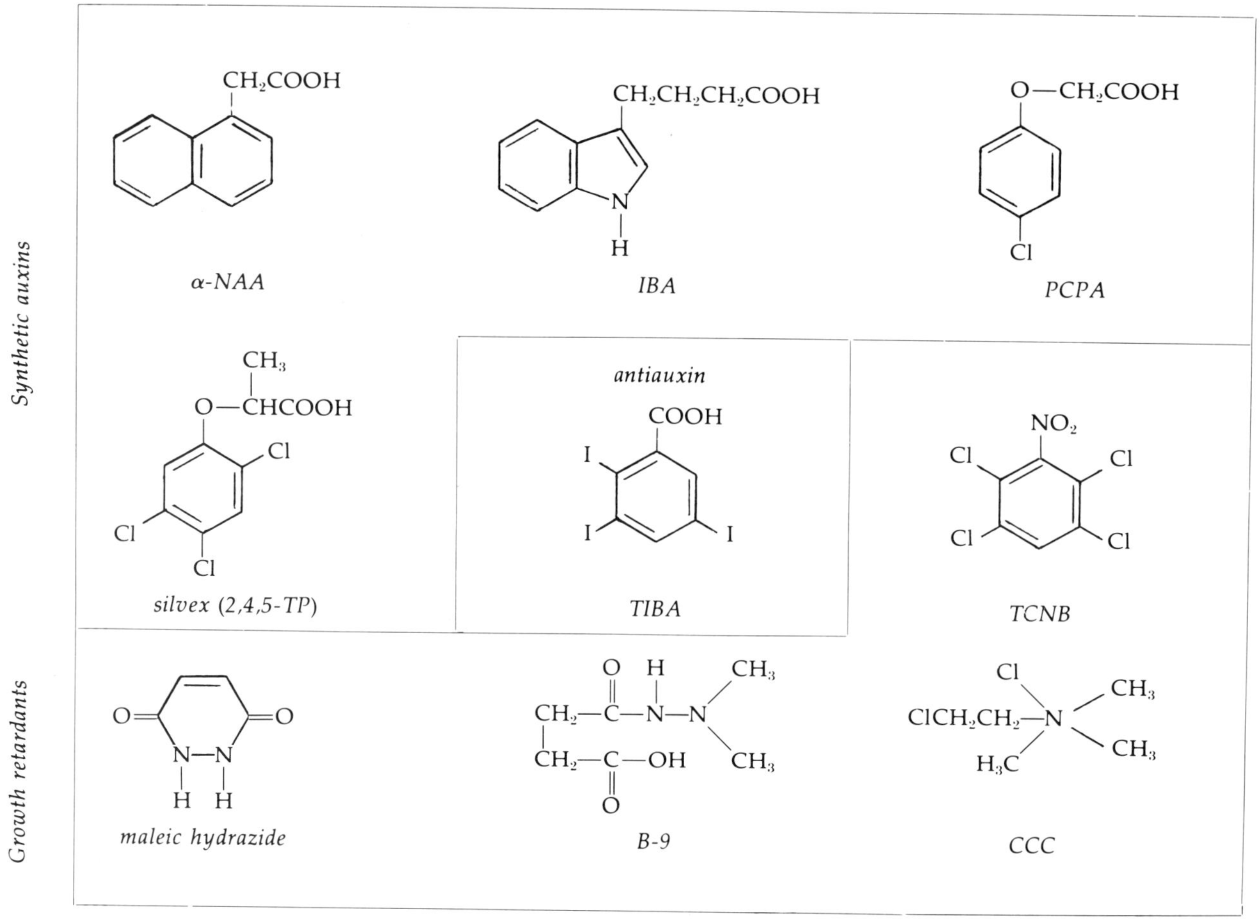

31–8 Molecular structures of a few representative growth regulators used in agriculture.

abscisic acid. Abscisic acid has been obtained from the coats of such dormant seeds as ash, peach, and rose.

Again, the picture is more complex. Other inhibitory substances have also been obtained from such tissues. David R. Walker and his co-workers, then at the University of North Carolina, obtained a growth-inhibiting material called naringenin from seed coats of dormant peach seeds. Several other inhibitors are also known to be natural components of plant tissues. Some of these are unsaturated lactones such as coumarin and scopoletin (Fig. 31–6). The effects of inhibitors such as abscisic acid are often overcome by promoters such as the gibberellins.

We are thus faced once again with the conclusion that plant growth is a response to a balance of several growth-regulating substances (several promoters and several inhibitors). Auxins, gibberellins, and cytokinins sometimes promote both cell enlargement-elongation and cell division, while abscisic acid and probably other inhibitors inhibit these same processes.

Leaf Formation and Development

We have discussed the formation and development of leaves from the anatomical standpoint. At present little is known about the physiology of this interesting morphogenetic process. But it is known that leaf development may be rather strongly influenced by environment. The most commonly observed phenomenon is the growth of sun and shade leaves (Fig. 31–7). Leaves grown in the sun are usually thicker (two or more layers of palisade tissue) and somewhat lighter in color than leaves grown in the shade. A spectacular example of the effect of environment upon leaf development can be observed in the aquatic buttercup *Ranunculus aquatilis.* Leaves submerged below the surface of the water are finely divided and differ in appearance from leaves that form above the surface.

The positioning of leaves on the stem poses a challenging problem of morphogenesis (see the Photo Essay). On some plants two leaves occur opposite each other at a node. More typically, leaves occur in a spiral arrangement around the stem with only one at each node. Leaf arrangement is called phyllotaxy. Sometimes, alternate leaves are on opposite sides of the stem so that to pass from a leaf to one directly below it involves one circuit of the stem and two leaves. This may be expressed by the fraction $^1/_2$. On other plants the spiral passes around the axis for each three leaves, represented by $^1/_3$. Another example is the $^2/_5$ type, in which two circuits involving five leaves are completed before one leaf occurs directly above another. The leaf arrangements that have been observed fall into the following series:

$$\frac{1}{2} \quad \frac{1}{3} \quad \frac{2}{5} \quad \frac{3}{8} \quad \frac{5}{13} \quad \frac{8}{21} \quad \frac{13}{34} \quad \frac{21}{55}$$

The fractions represent that fraction of the circumference of the axis traversed by the spiral in passing from one leaf to the next. The series is identical to a mathematical series known as the Fibonacci series, generated by summing the numerators and the denominators of two fractions to obtain the next fraction in the series.

The higher fractions approach the limit of 0.38197, or the ideal angle or golden mean. If a line is divided by that fraction, the small portion of the line bears the same relation to the larger portion as the larger portion bears to the original line. If successive leaves were formed at exactly this angular distance around the stem (0.38197 × 360° = 137°30′28″), no leaf would ever be exactly above any other leaf. It is not surprising that artists, mathematicians, and mystics have all been interested in leaf phyllotaxy!

How can we account for leaf phyllotaxy? Models have been formulated. Perhaps, a forming-leaf primordium inhibits the formation of the next one. If this inhibitory effect decreased gradually as the leaf primordium develops so that the youngest primordium had the greatest inhibitory effect, then the formation of a new primordium might tend to occur at about the Fibonacci angle from the most recently formed primordium. If the effect weakened rapidly rather than slowly, larger angles might be accounted for. In an alternative model, a leaf primordium forms where there is the most available space on the apex. No one has performed a definitive experiment to decide between these two models. It is possible that the best model has not yet been devised.

Leaf phyllotaxy provides us with a challenging mystery not easily accommodated with any presently available theory, including the broad idea that morphogenesis is controlled by a genetic program working through transducers and translaters of environmental stimuli. Leaf phyllotaxy seems to represent a highly intricate and sensitive system that is not greatly influenced by environment.

Flower Formation

One of the most outstanding events in development is the formation of flowers. Floral structures were discussed in Chapters 28 and 29, and the environmental factors controlling the formation of flowers will be discussed in Chapter 36. But the topic is mentioned here because of its relevance to development.

It is easy to demonstrate that flowers often form in response to specific environmental factors such as temperature or increasing (decreasing) day (night) length. As we have seen, leaves are the receptive organs perceiving day length, but the bud becomes transformed into a flower, implying that a flowering hormone is transported to the bud. Hence, as in so many of the other instances, development seems to be under the control of (or at least strongly influenced by) chemical factors: hormones and growth regulators.

In the case of the flowering hormone (called florigen), it has yet to be isolated, but current work suggests that this could be achieved fairly soon.

In the meantime, numerous growth regulators have been applied to plants, testing for their effects upon flowering. Auxins frequently inhibit flowering, although in a few cases (notably the pineapple) they exercise a promotive effect. Abscisic acid promotes flowering of black currants and a few other species but not the majority of species. Ethylene inhibits the flowering of cocklebur and promotes flowering of pineapple—consistent with the observation that ethylene effects are usually similar to those of auxin.

Probably, the most impressive effect of a growth regulator upon flowering is that of gibberellin. Many species that normally require either long days or low temperatures may be induced to flower in the absence of these stimuli by treatment with gibberellins. Many species that require *both* long days and low temperatures do not respond to gibberellin treatments, unless one of the environmental factors is provided. The vast majority of species that require short days do not respond. Nevertheless, many rosette plants (typically the growth form of those requiring long days or low temperatures) send up a flowering stalk (they will bolt) in response to gibberellin. Several gymnosperms can also be induced to form cones through gibberellin treatment.

Fruit Development

When auxins became available in the 1930s, it was found that they could be used to induce development of some fruits (e.g., tomatoes, cucumbers) without fertilization or the formation of seeds (parthenocarpic fruit development). Auxin applied in a lanolin paste to an ovary often resulted in development of the ovary into a fruit, even though pollen was withheld or the stigma removed. It was subsequently found that developing pollen tubes tended to produce relatively large quantities of auxin and that this was apparently the cause of normal fruit development in certain species. It was also shown that developing seeds produce relatively large quantities of auxin. Fruits of many species (e.g., grapes, apples, stone fruits) will not develop in response to auxin—so, as usual, the picture is complicated. In recent years it has been found that gibberellins cause the parthenocarpic development of these fruits and also of some that are promoted by auxins (e.g., tomatoes and figs). With grapes it has been possible to increase significantly the yield and quality of the crop by application of gibberellins in the field. Ethylene and cytokinins may also be involved, and it is well known that the ethylene produced by ripening fruits is important in the ripening process itself.

The Mode of Action of Growth Regulators

In our discussion of development, it has become apparent that much depends upon the presence or absence of growth regulators, both promoters and inhibitors. If we are to understand development, it is essential (among other things) to understand the mode of action of plant growth regulators.

How do plant growth regulators interact with the morphogenetic program to achieve their results? This question has been asked since their earliest discoveries. In the 1930s it was shown, for example, that auxins caused their effects on cell elongation by softening the walls of cells so that the cells can take up water osmotically and expand plastically and irreversibly (rather than elastically). But how do auxins soften cell walls? Only now are we beginning to reach levels so basic that we may be able to describe what actually happens in the interaction between the growth regulator molecule and its primary target. In any case, we can imagine that growth regulator molecules might be acting at several points.

Plant-growth regulators may be acting as coenzymes—substances also known to be active in small quantities. It is reasonable to believe that a given developmental step involves the action of some enzyme, and if this enzyme were promoted or inhibited in its activity by the presence of a given growth regulator, the action of the growth regulator would thereby be understood. Probably, the root growth hormones do act as coenzymes. These B vitamins are known to act as coenzymes in several metabolic reactions, and they are probably acting that way in their effects upon root growth. In spite of much effort to demonstrate a similar role for other plant growth regulators, no evidence for their coenzyme activity has yet come to light.

Plant-growth regulators could act upon enzyme activities by functioning as allosteric promoters or inhibitors. So far no evidence has appeared to support this possibility, but we should keep our minds open. Incidentally, there is current interest in a cell-wall protein called extensin. This appears to cross-link several cell-wall components, restricting cell expansion. Could auxin in some way break these cross-links by acting as a coenzyme or allosterically?

Auxins in particular and perhaps growth regulators in general might act through an effect upon membranes. Salt uptake is promoted, for example, in the presence of auxins, and this could be because the auxins are acting upon membrane machinery. Or they could be acting in more indirect ways. Some evidence presently supports the direct view. For example, uptake through membranes can be stimulated by auxin in very short times, less than 10 min and perhaps less than 1 min. It has been suggested that this does not give time for an indirect effect and that therefore the growth regulators must be acting upon the membranes themselves.

Growth regulators might activate or deactivate genes. Since morphogenesis is ultimately a function of the genetic material and the genetic program, it is reasonable to imagine that growth regulators might influence morphogenesis by turning some genes on and others off. In recent years investigators all over the world have been eagerly looking into this possibility and obtaining some positive results.

One of the first impressive examples concerned the germination of barley seeds. When the seed germinates, amylase enzymes are released by the aleurone (protein) layer of the endosperm, and these enzymes then digest the starch of the endosperm. The products of starch digestion provide an energy source for the rapidly growing embryo. What causes the release of amylase enzymes? Studies in 1960 indicated that gibberellins were released by the embryo as it became active and that these gibberellins in turn acted upon the cells of the aleurone layer, resulting in the release of amylases and other enzymes. It was possible to treat isolated aleurone tissue with gibberellin, causing release of amylases. Using such a system Joseph E. Varner and Ram Chandra showed that release of α-amylase was inhibited by actinomycin D, a compound that inhibits the transcription of RNA on DNA templates, and by other inhibitors of protein formation such as puromycin and chloramphenicol. Furthermore, radioactive phenylalanine (an amino acid) is incorporated into the α-amylase when aleurone layers are treated with gibberellins (but not otherwise). Hence it is clear that the α-amylase is not already present in the seed, merely being released in response to gibberellins (β-amylase is so released); rather its release apparently depends upon active genes producing m-RNA and on α-amylase being synthesized at the ribosomes in response to the m-RNA. Here, at least, is one gene that is turned on by a growth regulator.

Work with other growth regulators, including the auxins, has also provided evidence for a direct effect upon gene activation. Results with abscisic acid support the idea that this compound may act through a deactivation of genes. Virtually all of the responses to abscisic acid may be thought of as inhibitions of normally ongoing plant functions. Abscisic acid is a sort of plant tranquilizer, so to speak. Hence it is logical that this compound may act through a repression rather than an activation of the genetic material.

It also seems entirely possible that growth regulators may act in some manner that has not even been suspected. When we think of the many efforts to fathom auxin action during the 1930s and how little was actually known of plant metabolism at that time, we should remember to keep our minds open, realizing that virtually nothing is known at present about the developmental program and how it acts. It may well be that some growth regulators act as coenzymes (the root-growth hormones), that others influence allosteric enzymes (none yet discovered), that others act directly on membranes (the auxins and gibberellins?), and that some act directly upon the genes (auxins, gibberellins, cytokinins, and abscisic acid). A single compound could act in various ways, and perhaps the most basic mode of action remains to be discovered. (Hormones in animals seem to act at the membranes, releasing a secondary substance called cyclic-AMP, which then activates certain genes. Could a similar situation occur in plants?)

Application of Growth Regulator Knowledge

It is important to basic science to understand the mechanism of action of growth regulators. We obtain considerable satisfaction when our understanding is expanded—like solving a puzzle or learning how an engine runs. At the same time we might be able to use this understanding to solve some of the problems in the world today.

Numerous scientists in applied fields, particularly agriculture, have started to seek ways to apply the information already available about growth regulators. The descriptive facts about growth regulator activity have suggested certain applications. For example, in the greenhouse trade auxins (indolebutyric acid, particularly) are often applied to cuttings to stimulate root formation. These may even be applied (sometimes in combination with B vitamins) to stimulate root growth on newly transplanted plants. Since auxins influence abscission of both fruits and leaves, they have been used to prevent or to speed up fruit drop. (The response depends upon concentrations and season.) Much of the basic work on abscission has been supported by the cotton industry, since discovery of a growth regulator causing leaf abscission would greatly facilitate the use of cotton-picking machines. Unfortunately, abscisic acid does not work well when applied in the field. Gibberellins promote fruit growth, so they are applied on a commercial scale to grapes. They are also used in the malting process, since they stimulate the synthesis of amylases, an essential component in this process.

Essay 31–3 Bonsai

Ralph M. Johnson, a biochemist by training, is presently Dean of Science at Utah State University. Working with plants and applying some of the principles discussed in this chapter provides him with both a creative outlet and a welcome change from his professional routine.

"Bonsai means 'tree in a pot.' They are not artificial but real trees to be pruned, root-trimmed, fertilized and watched with loving care. Centuries ago the Japanese admired the appearance of trees growing in crevices high in the mountains. They were windblown and gnarled, never growing to their potential height. The Japanese began collecting them as special decorations for the court of the emperor. Aristocrats began paying high prices for good specimens. Trees were passed on from father to son and even given names. Soon, bonsai collecting became a common practice in Japan, where collectors met and compared techniques.

Bonsai may be only a few inches in height—3 feet tall is generally regarded as the maximum size—so the tree must constantly be trained, branches wired to desired shapes, buds pinched, and roots pruned and repotted. These practices limit growth by removing stem and root meristems and by restricting the absorbing roots, but a balance of nutrients (in small amounts) must be supplied to maintain a healthy metabolic condition. The object is to achieve a natural-looking tree in miniature. As the art developed—for it is an art form expressing the personality of the owner—five main styles were delineated: formal upright (top is erect), informal upright (top is bent slightly forward), slanting (bottom branch slants in opposite direction from trunk, top is bent), semicascade (tree is bent almost horizontally), and cascade (tree hangs down over edge of pot). Special containers and tools very plain in design were developed. The fancy pots, figurines and stands are commercial or only for display. Bonsai take many years to complete. Twenty-five years is considered minimum. What a feeling of serenity and peace one gets through working with growing trees, training and trimming them into works of art!

Trees suitable for bonsai must have flexible branches and trunks so they can be trained to the shape desired. But start with as old a tree as possible. Elegance of shape is an important consideration. Small-leaved deciduous trees or small-needle evergreens with alternate branching are most satisfactory. Good evergreens for bonsai are Cryptomeria, *black pine* (Pinus thunbergii), *Japanese white pine* (Pinus parviflora), *Sargent juniper, and Ezo spruce. Good deciduous trees are Japanese grey-bark elm* (Zelkoda serrata), *flowering quince, trident maples, and cotoneasters. The best source of trees is a nursery. Collecting specimens from the mountains is more romantic but difficult, and the trees are more likely to die.*

Spring is the best time to pot deciduous trees. Fall is best for evergreens. Deciduous trees should be repotted every year with considerable root-pruning. Evergreens should be repotted every three or four years. The pot should be unglazed, of very fine clay, with large drain holes. It should supplement the beauty of the plant by its simple form. It should be shallow but wide to confine root growth to a small space and to restrain the growth of foliage. The pot should be able to withstand wide changes of temperature, since the trees, with the exception of semitropical trees such as azaleas and pomegranates, are kept outdoors all year.

Wiring to train branches to the desired shape is done with copper wire or coated wire. Rewire each year, so wires will not mar the bark. Wiring damages buds, so wire after leaves reach full size.

31–0 Juniperus chinensis *'Kaizuka' (twisted juniper)—raft style.*

Bonsai are potted in sandy loam mixed with peat moss and are fertilized lightly once a week in spring and fall. They are watered by placing the pot in water to a depth just below the upper edge of the pot. They are kept in variable shade and sun, protected from strong winds. In winter the pot is buried in peat moss up over the base of the bonsai.

The ultimate goal is a healthy plant of exquisite shape. Both health and shape are dependent upon the possessor of the bonsai. Truly, such possession is a responsibility."

Synthetic compounds that mimic the action of growth regulators or tend to it have been thoroughly investigated. Many examples are now known: maleic hydrazide, phenoxyalkancarboxylic acids, substituted benzoic acids, a chemically complex group of growth retardants (CCC, AMO 1618, B-995, etc.), and substituted fluorine-9-carboxylic acids called morphactins. In addition to these there are the herbicides (Table 31–9), a host of compounds belonging to a number of chemical groups. These compounds are used to kill plants, and in some cases they selectively kill one kind of plant while not harming another kind. 2,4-dichlorophenoxyacetic acid (2,4-D) was the first of these compounds to be discovered. It is highly active in many auxin assays (so it is an auxin), but it kills many broad-leaf plants (dicots: dandelion, mustard, numerous familiar weeds) while not harming many narrow-leafed species (monocots: corn, wheat, Kentucky bluegrass, etc.). 2,4-D has been extremely valuable in agriculture, and its discovery instigated a widespread search for other materials that might have similar activities. Several derivatives of 2,4-D were discovered (e.g., 2,4,5-T, effective against woody plants and presently being banned because of a possible cancer-inducing action).

Many other compounds were found that bear no chemical relationship to 2,4-D but are nevertheless highly effective in one system or another. There is a derivative of picolinic acid, for example, called picloram or tordon that is extremely toxic to plants, even in minute quantities. There are also chlorinated organic acids such as trichloroacetic acid or dalapon (2,2-dichloropropionic acid) that are effective against many grasses but less so against dicots (e.g., sugar beets, beans, etc.). The triazines are effective against many weeds but have virtually no effect upon corn, so they are valuable herbicides with this crop. Of course, there are many other commercially valuable herbicides.

Some Final Thoughts

Just how much power can man handle? Consider what might be possible if we *really* understood and could control the developmental program. Perhaps, we could make plants to order, achieving maximum efficiency of production with the highest quality. Perhaps, we could grow food for 20 billion people or more. But what price would we pay? Surely, we would have to destroy many of the world's natural ecosystems to make room for the people and the food. What would be the ultimate consequences? How many herbicides can we scatter on the earth's surface before unpredictable reactions become catastrophic? 2,4-D seemed harmless. It is degraded in the soil, so its effects are temporary. But it is also easily blown around, and airplane application on a field of wheat may result in death of the trees in a nearby park or the beans in a nearby field. There have been so many such incidents that some agriculturists are now advocating a ban on 2,4-D. Other herbicides appeared much more dangerous from the start. Tordon remains in the soil for years, keeping it sterile of plant life, and may be washed into other areas where it is not wanted. Its use has been strongly restricted.

In Viet Nam we have a situation in which we can evaluate the effects of vast quantities of herbicides and defoliants sprayed into an ecosystem for long periods of time. Many scientists feel that irreversible damage has been done by these activities. Time will tell. Certainly, the present situation illustrates the potential dangers of improper application of scientific knowledge. We are babes in the wood as far as our present understanding of growth regulators and the developmental program goes. What could we do to an enemy—and to his environment—if we really understood these things? Can we destroy an enemy's environment while keeping our own intact? Can we compartmentalize our planet? Or are our ecosystems and biomes all tied together within the planetary ecosystem called earth?

Should we restrict all herbicides and growth regulators as "unnatural" intruders in the environment? Only if we want to sharply increase our food prices, be satisfied with much lower quality food, and return a sizable percentage of our population to the land (weeding and otherwise working in the fields). Perhaps, these things are all desirable. Perhaps, you would like to emulate your forefathers by working the land. But who will make the decisions? And who will research the facts upon which those decisions are made? What kind of personal action can you take to help the situation? These are questions that we all must face—and soon. Otherwise, we may finally be the victims of progressing events rather than their controllers.

Soil sterilants Kill all plant life (nonselective), temporarily.
- chlorates — sodium chlorate ($NaClO_3$)
- borates — sodium tetraborate ($Na_2B_4O_7$)
- arsenicals — arsenic trioxide (As_2O_3), sodium arsenite (Na_3AsO_3)

Contact herbicides Kill tissues they contact, nonselective.
- diquat (6,7-dihydrodipyridol [1,2-a:2′,1′-c] = pyrazindinium salt)
- paraquat (also a pyrazindinium salt)
- DNBP (4,6-dinitro-o-*sec*-butylphenol)

Auxin-type herbicides Mimic natural hormones, other actions as well, usually selective against dicots.
- chlorophenoxy herbicides
  - 2,4-D (2,4-dichlorophenoxyacetic acid)
  - 2,4,5-T (2,4,5-trichlorophenoxyacetic acid)
  - MCPA (2-methyl-4-chlorophenoxyacetic acid)
- benzoics
  - dicamba (2-methyl-3,6-dichlorobenzoic acid)
- picolinic acid derivatives
  - picloram (tordon) (4-amino-3,5-trichloropicolinic acid)

Chlorinated aliphatic acids Less hormonelike; often contact kill observed; selectivity against grasses.
- TCA (trichloroacetic acid)
- dalapon (2,2-dichloropropionic acid)

Aminotriazoles Inhibits chlorophyll synthesis (plants appear white); little selectivity.
- amitrole (3-amino-1,2,4-triazole)

Triazines Tolerated by corn, asparagus, etc., due to enzymatic detoxification.
- simazine (2-chloro-4,6-bis[ethylamino]-6-triazine)
- atrazine (2-chloro-4-ethylamino-6-isopropyl-amino-1,3,5-triazine)

Substituted ureas Applied before germination, selectivity due to placement on soil away from crop (triazine can be used the same way).
- monuron 3(4-chlorophenyl)1,1,dimethyl urea

Carbamates Also pre-emergent herbicides; highly volatile, must be soil mixed, absorbed selectively through a vapor phase.
- IPC (isopropyl N-phenylcarbamate)
- EPTC (ethyl N,N-dipropylthiocarbamate)
- chloro-IPC (isopropyl N-[3-chlorophenyl]-carbamate)

Acetanilides Pre-emergent, short-lived, mostly used against grasses.
- alachlor (2-chloro-2′,6′-diethyl-N-[methoxymethyl] acetanilide)

Substituted benzamide Highly selective against monocots (e.g., perennial quackgrass).
- kerb (N-[1,1-dimethylpropynyl]-3,5-dichlorobenzamide)

Substituted uracils Pre-emergent, low selectivity (use special placement), photosynthetic inhibitors.
- bromacil (5-bromo-3-*sec*-butyl-6-methyluracil)

Miscellaneous compounds
- trifluralin (N,N-di-n-propyl-2,6-dinitro-4-trifluoromethyl aniline) Pre-emergent, long-lived, selectivity against grasses (used in beans and cotton), a root pruner.
- diphenamid (N,N-dimethyl, 2,2-diphenyl acetamide) Less persistent and phytotoxic (used in tomatoes, peppers, etc.).
- dacthal (dimethyl-2,3,5,6-tetrachloroterephthalate) Also less phytotoxic, used in vegetable crops.
- endothal (7-oxabicyclo[2,2,1]heptane-2,3-dicarboxylic acid) Used against aquatic weeds.

For more information, see George E. Barrier (Editorial chairman). 1970. *Herbicide Handbook of the Weed Society of America* (2nd ed.). Geneva, New York: W. F. Humphrey Press.

32

Transfers between Plants and the Environment

We have taken many different looks at the structures and functions of our woodland acre. In earlier chapters we investigated cellular activities, and in the last two chapters we examined some of the developmental processes that account for the diversities of structure. Now, we shall consider some of the functions of mature plants not directly related to growth or cellular events—although it is difficult to divorce any plant activity from development and cellular function.

Plants (all organisms) are open systems. Minerals, water, oxygen, carbon dioxide, organic materials, and energy both enter and leave the plant. Minerals are absorbed by roots, and some leave the plant when leaves fall or in other ways. Carbon dioxide may enter the leaves, where it is utilized in the photosynthetic process, and leave the roots, where it is produced by respiration. The situation may be reversed for oxygen (leaving leaves, entering roots). The leaves may absorb light energy and release energy in the form of heat or in the conversion of liquid water to vapor. Roots may absorb a small amount of heat energy if the soil is warming, or they may lose energy if the soil is cooling. Organic compounds such as amino acids dissolved in the soil solution may enter the plants through the roots, while such other organic compounds as volatile oils may leave via the leaves. All of these exchanges are important in the functioning ecosystem.

We shall begin this chapter by discussing an aspect of the exchanges occurring between the roots and the surrounding soil, and then we shall discuss exchanges between leaves and environment. In the next chapter we shall consider movements going on *within* the plant.

32–A *Plants are continually exchanging matter and energy with their environment. For example, when a cocklebur* (Xanthium strumarium) *plant is subjected to a high heat load—in this case from two photoflood lamps—leaf temperature rises well above air temperature, and the plant begins to lose considerable water and energy by evaporation. This evaporation from plants is called transpiration.*

The Mineral Nutrition of Plants

We have already discussed one of the most important aspects of the mineral exchange between a

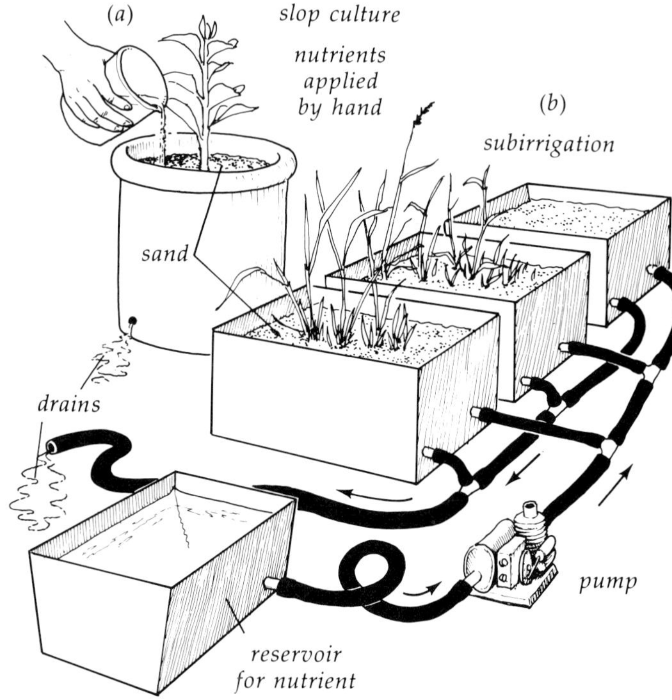

32–1 Methods of growing plants by hydroponics. The photograph shows a plant grown by solution culture. Before the photograph was taken, the glass jar had been wrapped with aluminum foil to exclude light. Note aeration device to the right of center. The drawings indicate methods of sand culture: either adding nutrient solutions by hand or automatically.

plant and its environment: the passage of ionic substances across membranes (see Chapter 4). This is the real exchange problem, and it is best considered at the cellular level. There is also an extensive lore of considerable theoretical importance and profound practical significance relating to the specific elements that may be required for plant growth. Research on the mineral nutrition of plants was initiated well over a century ago, using whole plants.

Early arguments were concerned with whether a plant was made of earth or water (two of the early "elements" of the Greek philosophers). An experiment of Jan Baptista von Helmont (1577–1644), published by his son in 1648, is often mentioned in relation to these arguments, although the experiment was conceived of earlier by others (e.g., Nicholas of Cusa, 1450). Von Helmont planted a 5 lb willow tree in 200 lb of dry soil and patiently watered it with rainwater for 5 years. The final weight of the tree was 169 lb, while the soil lost only 2 oz in weight. With inescapable logic it was concluded that the tree was made of water. The 2 oz lost from the soil were quite reasonably ascribed to experimental error. Of course, in the seventeenth century no one knew about the carbon dioxide in the air and certainly not that it might be contributing to the growth of a plant. In 1699 John Woodward in England performed an experiment that cast a little doubt on the conclusions of von Helmont. His mint plants grew for a while in rainwater but grew much better in water from the Thames River and best of all in a water extract of soil. Thus the soil seemed to have *some* importance. (Today mint plants might be instantly killed by the polluted water of the Thames!)

Woodward's experiment foreshadowed the approach that eventually clarified the situation. In 1860 Julius Sachs (one of the first modern plant physiologists

32–2 The composition of Knop's nutrient solution compared with a more modern solution

Knop's solution		Evan's* modification of Shive's solution		
salt	molarity	salt	molarity	ppm
$Ca(NO_3)_2 \cdot 4\ H_2O$	0.003	$Ca(NO_3)_2 \cdot 4\ H_2O$	0.005	
KNO_3	0.002	K_2SO_4	0.0025	
KH_2PO_4	0.0015	KH_2PO_4	0.0005	
$MgSO_4 \cdot 7\ H_2O$	0.0008	$MgSO_4 \cdot 7\ H_2O$	0.002	
$FePO_4$	trace	KCL		9.0 Cl
		Fe-versenate		0.5 Fe
		$MnSO_4$		0.25 Mn
		H_3BO_3		0.25 B
		$ZnSO_4$		0.25 Zn
		$CuSO_4$		0.02 Cu
		Na_2MoO_4		0.02 Mo

*From H. A. Evans and A. Nason. *Plant Physiology* 28:233–254, 1953.

and, in a sense, the father of the science), and in 1865 W. Knopf in Germany, approached the problem by growing plants with their roots immersed in mineral solutions. Chemistry had advanced enough so that the common elements had been discovered, and their chemical properties were well known. Relatively pure salts were available commercially. The process of photosynthesis was understood in its broad outlines; so the old earth-versus-water argument had been resolved. Agricultural chemists were aware that applying mineral fertilizers to soils could increase crop yields. (There was a minor controversy relating to the degree to which atmospheric nitrogen or nitrogen compounds such as ammonia might be used by plants. This conflict has still not been completely resolved.)

Sachs and Knopf dissolved small amounts of various salts in water and then attempted to grow plants with their roots immersed in these solutions (Fig. 32–1). The object was to discover what mineral elements had to be added to achieve suitable plant growth. It was ultimately found that 3 salts (6 mineral elements), listed in Table 32–2, were sufficient: calcium nitrate ($CaNO_3$), potassium phosphate (KH_2PO_4), and magnesium sulfate ($MgSO_4$). The so-called major nutrient elements were calcium, nitrogen, potassium, phosphorus, magnesium, and sulfur. (Formulas also include atoms of hydrogen and oxygen, but we usually think of the plant obtaining these from water and carbon dioxide.) It was also known at that early period that plants require iron but in quantities smaller than the other elements. Actually, a 3-salt solution does not provide as suitable a combination of elements as formulations using at least 4 salts (Table 32–2). In any case, it became quite clear that higher plants required certain mineral elements as well as the obviously required water and carbon dioxide, by then known to be important in photosynthesis.*

It also appeared that higher plants needed *nothing* more than the elements provided in the salts, water, and carbon dioxide; that is, they did not have a requirement for some kind of "vitamin" or other substance elaborated by living organisms. Of course, the experiments of Sachs and Knopf were not completely rigorous in excluding all microorganisms from the environment of the plant's roots or shoots. A skeptic could always say that such organisms (undoubtedly present) were necessary for the growth of the plants. Here we have the basis for an argument that has continued to the present, although it is seldom discussed among scientists who are familiar with the facts. The argument is perpetrated because of a popular cult known as *organic gardening*. Spokesmen for this group proclaim that plants grow

*Chemical symbols of the major mineral elements including hydrogen, oxygen, carbon, and the minor element iron (Fe) can be easily remembered by the mnemonic: C HOPKNS CaFe, Mg, where Mg stands for mighty good!

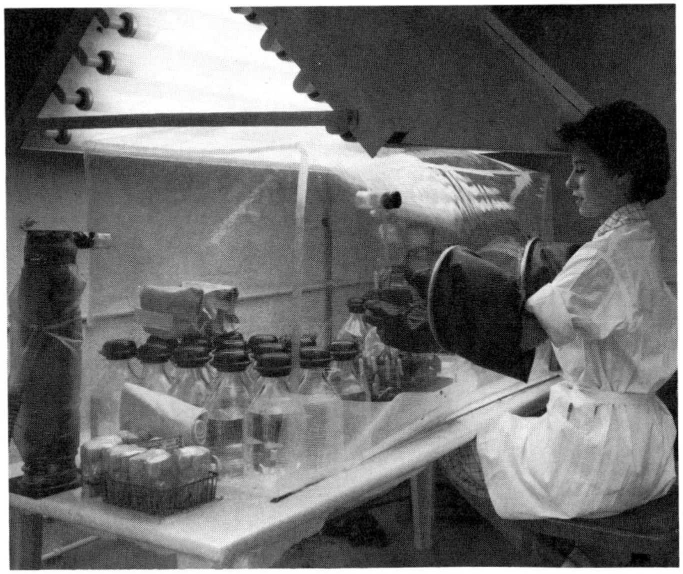

32–3 A plastic gnotobiotics tent used by R. R. Baker in studies of plants grown under germ-free conditions.

best in a soil containing large quantities of organic matter, that the organic matter per se is essential for this excellent growth, and that "artificial" mineral elements (as well as insecticides, etc.) are fundamentally bad for plants and for the people who eat the plants.

At least one careful scientific investigation having bearing on the problem has been carried out. Ralph R. Baker and his colleagues at Colorado State University grew plants in plastic tents that had been sterilized on the inside by such substances as peracetic acid (Fig. 32–3). Air passing into the tents was filtered to remove all bacteria or other microorganisms. Plants were illuminated through the clear plastic with a battery of fluorescent lamps. Under these conditions it was possible to grow plants in the complete absence of any other organisms and to compare their growth with plants grown in similar tents and in a traditional soil containing organic matter and microorganisms. It was clearly demonstrated that higher plants are indeed autotrophic, requiring only the basic mineral elements plus water and carbon dioxide; those in the sterile nutrient solutions grew as well as controls. (This work is being continued at the Lunar Receiving Laboratory of the Manned Space Center in Houston, Texas, where Charles Walkingshaw has grown—among other things—a mature, germ-free corn plant with tassels and ears!) Thus Sachs' and Knopf's experiment was vindicated.

What about the organic gardener? It should be perfectly obvious that soils containing large amounts of organic matter are excellent substrates for plant growth. The decaying organic matter releases the essential mineral elements (and others), and the humus is very valuable in retaining water in the soil. Furthermore, the decaying organic matter releases certain compounds called chelates that promote the absorption of iron and other elements by the plant roots. The adsorptive properties of the humus also contribute to a maintenance of optimum levels of available nutrients by the process of ion exchange. Cations (positive ions such as Mg^{2+}, Ca^{3+}, K^{+}) may be exchanged for hydrogen ions released by the plant root.

On the one hand, the organic gardener is clearly wrong in principle, as demonstrated in carefully controlled laboratory experiments. Higher plants are autotrophic and do not require magical substances provided by the humus. But, at the same time, he is clearly right in his practices. Improper use of chemical fertilizers has resulted in such problems as salt buildup in irrigated soils and the nitrogen enrichment of water in streams and lakes, leading, as we have seen, to eutrophication due to an overgrowth of algae. Furthermore, it is difficult to see how addition of organic matter to soils can ever result in anything but good for plants—if it is applied with a little wisdom. We are presently inundated with garbage and other organic wastes. Surely some means can be devised to replenish the earth by returning a significant portion of this material to the soil. The problem at the moment is that it is much more economical to use inorganic fertilizers than it is to process organic wastes.

At the beginning of the twentieth century, after commercial salts became available at levels of purity not previously possible, it became apparent that the six elements plus iron were not enough for a nutrient solution supporting plant growth. Apparently, the earlier salts contained impurities essential for plant growth. A number of careful studies were initiated. Criteria for essentiality were established: normal growth and reproduction of a plant cannot occur in the absence of an essential element; nor can this absence be overcome by substituting some other element; nor can an essential element promoting plant growth do so only by overcoming the toxic effect of some other element.

32–4 Elements essential for most higher plants and internal concentrations considered adequate

Element	Chemical symbol	Form available to plants*	Concentration in dry tissue (percent)	Relative no. of atoms compared to molybdenum
Molybdenum	Mo	$MoO_4^{=}$	0.00001	1
Copper	Cu	Cu^{+}, **Cu^{++}**	0.0006	100
Zinc	Zn	Zn^{++}	0.0020	300
Manganese	Mn	Mn^{++}	0.0050	1,000
Iron	Fe	Fe^{+++}, Fe^{++}	0.010	2,000
Boron	B	**$BO_3^{=}$**, $B_4O_7^{=}$	0.002	2,000
Chlorine	Cl	Cl^{-}	0.010	3,000
Sulfur	S	$SO_4^{=}$	0.1	30,000
Phosphorus	P	**$H_2PO_4^{-}$**, $HPO_4^{=}$	0.2	60,000
Magnesium	Mg	Mg^{++}	0.2	80,000
Calcium	Ca	Ca^{++}	0.5	125,000
Potassium	K	K^{+}	1.0	250,000
Nitrogen	N	**NO_3^{-}**, NH_4^{+}	1.5	1,000,000
Oxygen	O	O_2, H_2O	45	30,000,000
Carbon	C	CO_2	45	35,000,000
Hydrogen	H	H_2O	6	60,000,000

*If one of two forms is more common than the other, this is indicated by boldface type.

Chemical procedures were developed to an amazing level of sophistication to provide solutions deficient in one element or another.

As a result of these studies, it was found that at least seven other elements are required for plant growth: chlorine, iron, manganese, boron, zinc, copper, and molybdenum (Table 32–4). These are called minor elements—not because they are less essential but because they are required in smaller amounts. Certain plants also require sodium and others selenium (which, when present in significant concentrations, is toxic to cattle eating the plants). Still other elements may be essential for plant growth, including silicon, aluminum, cobalt, and possibly such rare earths as gallium and vanadium. Once it is possible to eliminate an element from a nutrient solution to the point where the plant will no longer grow properly, it is logical to conclude that the plant requires the element. On the other hand, if a plant does well in a solution *thought* to be deficient in an element, one can never be sure that the element is nonessential because one can never be absolutely certain that the solution does not contain the element (although modern spectrographic and atomic activation analysis allows a reasonable certainty of nonessentiality in a few cases). Incidentally, most animals require sodium, iodine (part of the hormone thyroxin), and cobalt (part of vitamin B_{12}). Cobalt is essential for the bacteria in the nitrogen-fixing nodules of legumes. Chromium and nickel also appear to be essential in trace amounts.

Another approach to the study of essentiality is to analyze the plant to see what elements it contains (Table 32–4). Of course, the fact that a plant contains an element certainly does not prove that the element is essential to the plant's growth. Nevertheless, once an element has been shown to be essential, analysis may give some indication of the optimum amounts required by the plant. Actually, we know that at least 60 of the known elements occur in plants (although only about 15 are known to be essential). It appears that virtually any element that can occur in solution may be taken up by plants.

Still another approach (possibly the best one of all) is to understand the biochemical role that a given essential element plays in the plant. For example, the fact that the chlorophyll molecule includes an atom of magnesium accounts for the essentiality of this element. The presence of iron in the cytochrome enzymes is

32–5 *Functions of mineral elements and symptoms of deficiencies*

Mineral	Functions of Mineral Elements	Symptoms of Mineral Deficiencies
Nitrogen	Part of amino acids, proteins, coenzymes, nucleic acids, chlorophyll, ATP.	Lack of growth or limited growth with chlorosis (yellowing) or loss of leaves in severe cases. Purplish coloration due to accumulation of anthocyanin pigments. Entire plant affected, older leaves most.
Phosphorus	Part of sugar phosphates, nucleotides, nucleic acids, phospholipids, coenzymes.	Dark green, stunted plants. Accumulation of anthocyanin pigments. Delayed maturity. Entire plant affected, older leaves most.
Potassium	Coenzyme. Necessary for protein synthesis. Essential for stomatal function.	Mottled chlorosis, necrosis (spots of dead tissue, especially at tips and margins, between veins). Older leaves most affected. Weak stalks, roots more susceptible to disease.
Sulfur	Part of proteins, coenzyme A, thiamine, and biotin.	Chlorosis of young leaves, usually no necrosis. Veins remain green, tissue between light green.
Magnesium	Part of chlorophyll. Activates numerous enzymes. Maintains structure of ribosomes.	Mottled or chlorotic leaves, may redden. Leaf tips turned upward. Older leaves most affected.
Calcium	May be essential in synthesis and stability of middle lamella. Maintains structure and permeability of membranes. Part of α-amylase.	Inhibition of root development and death of shoot and root tips. Young leaves and shoots most affected.
Iron	Essential part of cytochromes, ferredoxin. May be activator of certain enzymes.	Interveinal chlorosis of young leaves, stems short and slender. Buds remain alive.
Chlorine	Activates photosynthetic enzymes.	Wilted leaves, chlorosis, necrosis. Stunted, thickened roots or club-shaped roots near tips.
Manganese	Formation of amino acids. Activates many enzymes. Electron carrier, catalyst.	Chlorosis of young leaves, necrosis between veins; smallest veins remain green. Disorganization of lamellar membrane.
Boron	Carbohydrate translocation.	Death of stem and root apical meristems. Leaves twisted, pale at bases. Swollen, discolored root tips. Young tissues most affected.
Zinc	Chlorophyll formation, indoleacetic acid production, part of certain enzymes.	Reduction in leaf size and length of internodes. Distorted leaf margins. Chlorosis. Older leaves most affected.
Copper	Electron carrier, part of certain enzymes. Nitrate reduction.	Young leaves dark green, twisted, wilted, misshapen; tip remains alive.
Molybdenum	Electron carrier. Essential in nitrogen fixation.	Chlorosis or twisting and death of young leaves.

likewise positive proof of the essentiality of iron. Of course, knowing that a given element performs some function in the plant does not eliminate the possibility that it may perform additional functions. Table 32–5 shows functions for the essential elements. Surely, many other functions will be discovered, as we continue to learn about the machinery of life.

A philosophy diametrically opposed in practice to that of the organic gardener is the current emphasis in some agricultural circles on the use of hydroponics. In this procedure the methods developed during the middle of the last century by Sachs, Knopf, and others are applied on a commercial scale. Manufacturing companies are in business to supply the needed equipment. Tomato plants, for example, may be grown in greenhouses in large troughs filled with sand or gravel and watered with solutions of the major and minor elements. Solutions are circulated automatically, changed frequently, and carefully monitored. Temperatures, humidities, and other environmental factors are controlled and monitored in the greenhouse. It is a completely artificial agriculture—the exact opposite of organic gardening.

This is a challenging concept for the plant physiologist who would apply known principles. It is not without difficulty, however, and there is good reason to believe that a hydroponic procedure is less economical than some of the more traditional methods. Why, for example, should one bother to circulate costly solutions with expensive equipment when plants grow so well in soil? The use of a greenhouse may be economically feasible, since it allows the growth of plants year round, but it is difficult to think of a scientific justification for the use of hydroponics in the greenhouse. Perhaps there are some advantages in controlling certain diseases, etc., since the composition of the nutrient solution can be better known than the composition of any soil. Yet plants do grow extremely well in soils, particularly when they are fertilized judiciously and cared for by the best available horticultural and agronomic practices.

Essay 32—1 Nitrate, Soil, and Men

The role of man in the ecosystem has become increasingly important and increasingly subject to heated debate. Here, A. W. Roach from North Texas State University recounts his personal experience with the problem and offers his own interpretation of the causes behind our current ecological crisis.

"When I was a newly coined Ph.D., cruising up and down the eastern crosstimbers in a 1950 Ford, I liked to stop at the City Cafe in Aubrey, Texas for coffee when I was working stands in that area. I used to talk to old-timers, primarily to get as much past history as I could on the vegetation.

From them I learned of better days in Aubrey—30,000 cotton-bale years—yet, through the fly-specked window, I could see the rusting gin (yawning foundations where prosperous stores once stood), and I think it finally dawned on me why this was so.

It begins with the simple fact that man is dependent upon plants for his existence. He must eat every day, and the energy he requires is fixed in the leaves of green plants or is channeled through various food chains, always beginning with autotrophic plants. In turn, the plants upon which he depends are chained to a particular set of edaphic factors. The critical link in the chain is nitrate, an ion that is normally manufactured in the living bodies of microorganisms before it is usable by green plants.

Let us speculatively say that the amount of nitrate fixed per year per acre is enough to make the protein for one plant. That plant must die and return its protein to the soil for reuse through decay cycles for next year's two plants. Thus slowly the storehouse is built up, year after year. We call it the A_1 *layer of the soil, and the overriding duff of partially decayed organic matter, the* A_0 *layer. The latter breaks the hammering force of* *raindrops; it keeps the upper A_1 interstices open so that runoff is less than 1% and infiltration 99%. Sufficient A_1 to support a number of crops of high nitrate ion requiring cultivated plants may take thousands of years to build.*

Let us also call A_1 biotic potential. Of course, this varies according to macroclimate and overlaying vegetation—lowest in the tropics and highest under semi-arid grasslands. The potential may never be realized as carrying capacity (bushels of a crop per acre or pounds of meat per acre) because of subtractive factors, the most subtractive being man himself. This stems primarily, I think, from a belief that he is not subject to ecological law as are, say, other organisms. Why this is so is difficult to fathom. He has been a practicing agriculturist for some 200 centuries; yet it has only been a few decades ago that the first book on resource management appeared. Somehow man became directed toward a world of ideas and processes and divorced himself from the earth from which, like Antaeus, he drew his strength.

His population, except for localized situations, more-or-less responded to the dictates of his environment up to the nineteenth and twentieth centuries. Then two cultural revolutions, long-brewing, burst with full force upon mankind—the industrial and sanitary revolutions.

The Industrial Revolution supplied machines that needed labor to work them plus the impetus to import, transport, and stockpile raw materials. This was the crucial factor; famines ended and great cities were born. The simple expedient of boiling water and the improved diet caused life expectancies and population curves to skyrocket. The European burst from the bounds of his homelands to spread fingers of death, as one conservationist has put it, through the rich soils of half a dozen continents. With him he took the axe, fire, gun, and moleboard plow—the four fundamental tools of our modern riches. The frontier economy seemed to warrant a rapid spending of the richest soil—the more rapid the spending the more rapid the transition toward a more complex society. What if corn was exhausting the soils of the Atlantic seaboard in North America; there was more land to the west, wasn't there? Forests fell before the axe and fire; the A_0 layer disappeared beneath the plow. Runoff increased from 1 to 70%; infiltration dropped from 99 to 30%; and with the runoff went tons of A_1.

Great Britain had a head start on the Industrial Revolution. In the nineteenth century she accounted for 80% of the world's manufacturing. Small wonder her population became exponential. She drew on the eroding lands from half the world. What if coffee was ripping down the hillsides of the Sao Paulo in Brazil, or if the famous steaks at Simpson's carried the Fe, K, and nitrate from the blowing Pampas of Argentina? Trade balances were food, weren't they? Prosperity was industrialization, and business methods were equated with carrying capacity. Mankind then passed into its modern phase, and the semantic confusion of the word went with the process. Land is something worth so many dollars an acre or smells sweet after a rain or sticks to your boots. Biotic potential and the life-giving biophysical reality of the A_1 were lost in the confusion except to a few.

The history of our future is already written in immutable ecological law. Something has to give—somewhere! It has been said that civilization is built on 9 inches of topsoil. That cupboard is about bare. Realization has been postponed by a technical capacity to write promissory notes against the arable lands of all continents. The notes are now falling due."

Other Root Exchange Problems

In quantity the most important transfer between the root and its environment is the uptake of water. This occurs according to the principles discussed in Chapter 4. Water moves into the root in response to a gradient in water potential (more negative water potential inside), which is established in various ways. Usually, water moves into the root in a liquid form. When this is the case, the uptake of mineral elements will be at least somewhat dependent upon the rate of uptake of water by the plant, since the water brings dissolved minerals to the root surfaces. Occasionally, in dry soils, water apparently distills across a vapor barrier from soil particles into the roots. In such a situation, mineral uptake might well be independent of rate of water uptake.

Once the water and minerals are in the root, they may follow the path of least resistance through the cell walls, intercellular spaces, and xylem elements (the apoplast). Since the cell walls contain a predominance of negative charges, some of the positively charged ions (the cations) will be adsorbed on these negative sites. Furthermore, water and dissolved minerals, via their own independent absorption mechanisms, also pass across living cell membranes into the symplast phase of the plant. Once in the symplast, ions may move from cell to cell through plasmodesmata. By the process of cytoplasmic streaming, in which cytoplasm tends to circulate within the cell (according to mechanisms not yet understood), the movement of ions can be accelerated, since cytoplasmic streaming occurs much more rapidly than diffusion. In most plant cells ions in the cytoplasm (the symplast) may be transported across the tonoplast (vacuolar membrane) into the vacuole, but once materials have entered the vacuole of cells, they are lost to cellular metabolism, apparently remaining there indefinitely. Movement into the vacuole is a one-way street.

Movement of minerals into the plant as a whole proves to be less of a one-direction event than might be assumed. Certainly, many of the minerals taken up by plant roots are contained within the plant in vacuoles or incorporated into various molecules of the protoplasm. A considerable quantity, however, may leave the plant. Perhaps most of this quantity is shed with the leaves of deciduous plants—or as roots die and decay in the soil. But some water and minerals may diffuse out of living roots into the soil. Nitrogen-fixing plants are not only benefited themselves by the soluble nitrogen they obtain from the atmosphere (usually with the help of mutualistic bacteria in nodules), but much of this fixed nitrogen diffuses out of the roots into the surrounding soil, where it may be advantageously utilized by other plants. We shall shortly discuss how large quantities of water move from the soil into the roots, up through the plant, and finally into the atmosphere. This continual flow of water through the soil-plant-air system aids in the upward movement of dissolved mineral elements in the plant. Some of these elements may return through the phloem system to the roots and other parts of the plant. Some of the dissolved salts, however, are carried all the way to the leaf surfaces. There, they are sometimes excreted

32–6 Tamarisk leaves (Tamarix pentandra) *collected at Barstow in the Mojave Desert, California, showing heavy depositions of salt excreted on the surface.*

from the leaf, where they remain on the leaf surface until they are removed by rainwater or some other means. In a few cases, such as the tamarisk (*Tamarix pentandra*) of the southwestern deserts, crystals of sodium chloride are deposited on the leaves. These can be tasted or even seen with a hand lens (Fig. 32–6).

Gas Exchange between the Leaf and Its Environment

Since the root is normally confined in the soil, exchanges between it and the environment are necessarily at close range. A consideration of gas exchange between the root and its environment is mostly a study of the factors controlling oxygen and carbon dioxide levels in the soil. The leaf, on the other hand, existing in the open air, exchanges not only gases with this air but also heat by molecular collisions on its surface and

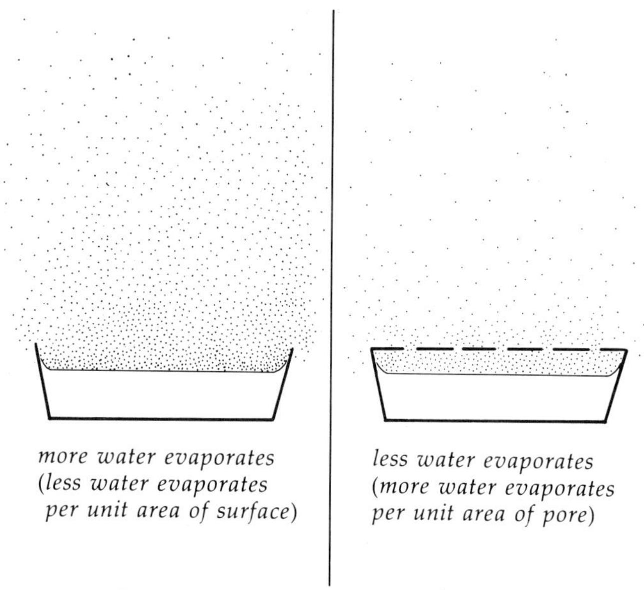

32–7 A schematic illustration of evaporation from an open surface and through small openings.

by radiation to and from the infinite reaches of outer space. A number of these exchanges are of primary interest: the carbon dioxide taken up in photosynthesis and released by respiration; the oxygen moving in an opposite direction; water vapor and volatile organic and other materials moving in and out (but mostly out); and energy being exchanged in various forms.

In this section we shall discuss the movement of gases into and out of the leaf. In each case, the movement results because of a gradient in chemical potential. Rate of movement is proportional to the partial pressure difference (inside and outside the leaf) that is responsible for the chemical potential gradient. If water vapor pressure is higher inside the leaf than outside, for example, water potential will be higher inside the leaf, and vapor will tend to move out at rates dependent directly on the vapor pressure difference and, inversely, on the distance involved. On the other hand, if carbon dioxide is being utilized inside the leaf, its partial pressure there will be less than that outside, and it will diffuse into the leaf if it can (i.e., open stomates, permeable cuticle). Volatile organic molecules such as terpenes or essential oils (e.g., flower odors or mint oil) evaporate in the leaf's intercellular space or on the leaf surface, building up partial pressures higher than those in the surrounding atmosphere. Again, diffusional movement occurs.

The basic driving force controlling these movements of gases is the same regardless of the gas being discussed. If chemical potentials (usually controlled by partial pressures) are high at one point and low at another, diffusion occurs toward the low point unless it is somehow obstructed. Since most studies have investigated the principles controlling the movement of water vapor, our discussion in this section will be exclusively concerned with that gas (Fig. 32–7).

Substances diffusing from the surface of the leaf follow simple principles—the plant exercising control only over the rate at which the substances are produced on the surface. On the other hand, substances evaporating from the surfaces of cells within the leaf into the intercellular spaces must pass through the stomates before reaching the outer atmosphere. The opening and closing of the stomates exercises considerable control over the passage of such substances out of or into the leaf; that is, although the chemical potential gradient may exist, movement can only occur if the stomates are open. This then is the basic mechanism with which we must deal: a leaf with intercellular spaces connecting to the outside world through minute pores (the stomates), and the physical-chemical laws governing evaporation of liquids and diffusion of gases.

The evaporation of water from plants is called transpiration. Because water is often limiting in the ecosystem, transpiration has been studied in detail since at least the late eighteenth century, when Stephen Hales considered the topic. It has long been known that large quantities of water are lost by transpiration. Amounts lost at any time depend upon both species and environment, but an often quoted example is that a corn plant, in growing to a final volume of 2 liters, loses 200 liters of water by transpiration; for 1 g of dry matter produced, about 500 g of water are transpired.

To obtain the basic data upon which theoretical discussions can be based, methods for measuring transpiration are required. Some of these are fairly complex and involve the measurement of humidity before and after a known volume of air is passed over the leaf surface. A simpler method is to use a potted plant, water it, seal the pot and the soil surface (perhaps with polyethylene plastic and aluminum foil), and then weigh the entire potted plant at intervals, assuming that virtually all of the loss in weight can be accounted for by transpiration. (There will be a small gain in weight due to photosynthesis and growth and a small loss in weight due to respiration; these are quite insignificant compared to the amount of water lost in transpiration.) All but one leaf may be removed from the plant, and the stem may also be wrapped with plastic.

Of course, this method would not work with a large plant such as a tree, so lysimeters are used. In these, plants are grown in several cubic yards of soil in a container placed on a weighing device of some kind. One of the best methods is to place the container on a large plastic "pillow" containing liquid. The liquid is connected to a standpipe, and the level of liquid in this pipe indicates the weight of the soil, water, and plants in the lysimeter.

Another method often used with large plants, both in the field and in the laboratory, is to remove a leaf from a plant and to weigh it immediately on a sensitive balance, noting the weight at intervals of a few minutes or less. It is assumed that transpiration continues at approximately the same rate for at least a short time after the leaf has been removed. Actually, experiments in which this technique is combined with the potted-plant technique suggest that removing the leaf may influence transpiration almost instantly. Nevertheless, at least an approximate value can be obtained in this manner, and sometimes no other technique is applicable. Unfortunately, in many of these methods we are faced with the problem of upsetting a process by attempting to measure it. For example, when leaves or whole plants are placed in transparent containers of plastic or glass and humidities are measured, environmental conditions are often much different inside than outside the container.

Diffusion through Small Pores

Quantitative measurements of transpiration indicate that much more water vapor passes through the stomates than one might imagine, considering only the area of these small pores. The amount of water typically lost from a plant by transpiration may be about half as much per unit leaf area as the amount of water lost from an equal area of free-water surface, although the stomatal area is only about 1% of the leaf area. This striking observation implies that gases pass by diffusion through small pores much more efficiently than might be expected. The physics of this phenomenon is well understood: a gas that diffuses through a small pore encounters a steep pressure gradient (or concentration gradient), resulting in rapid diffusion.

In most plants the stomates are open during at least part of the day, but they usually close at night, retarding the water loss, although some loss of water does occur by transpiration through the cuticle (typically 5 to 10% of stomatal transpiration). In certain species such as the succulents that grow in the desert, the mechanism has been even further perfected. Stomates are open at night when the air is cool, relative humidity is high, and transpiration is low. Carbon dioxide is absorbed and converted into organic acids in a process requiring little energy and progressing efficiently in the dark. The following day the stomates are closed, reducing transpiration during the hot, dry part of the 24-hour cycle. Photosynthetic processes then convert the carbon dioxide fixed in organic acids into carbohydrates and other products of photosynthesis. The overall process is somewhat less efficient than the usual photosynthesis—but it is a good compromise for keeping stomates closed during the day.

How do the stomates work? As indicated in Chapter 24, when the guard cells take up water and become more turgid, the stomates open. Much study has gone into an attempt to elucidate the mechanisms controlling this response. For the guard cells to take up water, either water potential in the cells must be lower than in surrounding cells, or there must be some active pumping mechanism capable of forcing water into the guard cells against a water-potential gradient. It can be shown that a water pump of this type in plants would require too much energy; thus there must be some osmotic mechanism. If pressure could be relieved in the guard cells, this would make the water potential more negative, and water would tend to move in. But opening of the stomates occurs as the guard cells increase in pressure, forcing the surrounding cells to yield; hence this mechanism seems unlikely. On the other hand, if solutes were to increase in the guard cells, this would also lower the water potential so that water would move in osmotically.

What could cause a buildup of solutes in the guard cells? In an attempt to find out, plant physiologists have amassed many facts relating to stomatal activities. Stomates open in the light (except for many succulents). They contain chloroplasts and may therefore be photosynthesizing in response to this light. Does photosynthesis produce sugars (the required solutes) rapidly enough to account for observed rates of opening? Apparently not, according to the best calculations. Furthermore, stomates can be caused to open *in the dark* by exposing them to CO_2-free air. Thus opening in the light may not be due to photosynthesis itself but may be an indirect response brought about by the lowering of carbon dioxide concentrations due to photosynthesis.

But how could lowered carbon dioxide concentrations result in an increase in solute concentration and consequent stomatal opening? A theory was proposed in 1923 to answer this question. It suggests that the lowered CO_2 concentration and the resulting lowered acidity in the cytoplasm influence amylase enzymes so that starch is broken down to sugar faster, with this sugar constituting the required solute. CO_2-free air in the dark also makes the cytoplasm less acid as carbonic acid diffuses out of the cytoplasm (as CO_2) into the air.

The theory was clear-cut and supported by much evidence, but new data began to cast doubts upon its validity. In the late 1960s interest developed in earlier findings that potassium ions accumulate in guard cells as stomates open. Indeed, potassium ions proved to be essential in this process (one of the few plant functions known to be dependent upon this major nutrient element). By 1968 it was suspected that enough potassium was taken up by guard cells to change the osmotic potential and thus account for stomatal opening. Recently, careful measurements of potassium with the new technique of electron-probe analysis, which allows measurements of several elements in individual guard cells, has confirmed that the osmotic effects can be accounted for by potassium ions (and some unknown anion, not containing either Cl, P, or S). No change from starch to sugar is necessary to account for stomatal opening. Hence the early theory can be abandoned, but the question of how light or carbon dioxide-free air can have their effects remains unanswered. At least we can now concentrate our attention on the mechanisms of potassium movement.

Incidentally, stomates close not only in response to darkness and high carbon dioxide levels but also in response to water deficits within the plant. It may be that when the plant begins to wilt, water also leaves the guard cells, causing the stomates to close. Retarding transpiration when soil water is limiting is clearly to the plant's advantage.

Over the years considerable effort has been expended to find chemicals that might control the opening and closing of stomates. Several such chemicals have been found. Phenylmercuric acetate is perhaps the best example. This compound, applied at low concentrations (e.g., 10^{-4}M), may cause stomates to close and to remain closed for two weeks to a month or more. As might be expected, transpiration is strongly retarded — a finding that could be very important to agriculture and to the retention of water by watersheds. Of course, photosynthesis is also retarded, limiting the value of the procedure for agriculture, but the effects on photosynthesis are often less than those on transpiration. This could indicate that more CO_2 than water vapor penetrates the cuticle, which is known to be true for many species. All of this is in the preliminary stages of research, but future applications could be very important.

What "Good" Is Transpiration?

Asking why a given process should exist for a plant's benefit implies that purpose is displayed in living things. This is the doctrine of *teleology*, and it is quite abhorrent to logical positivists and other scientific philosophers. Surely, one cannot develop a logical scientific argument if he allows himself to use teleology. Nevertheless, scientists have often made interesting observations and discoveries by investigating the "reasons" for a given process. In evolutionary terms we assume that features in organisms exist because of the process of natural selection. If they are detrimental to the organism, they will eventually be eliminated; if they are beneficial, they will be selected.

Why was transpiration selected? Some have suggested that transpiration serves no beneficial function; it is merely an inescapable (for most plants) by-product of photosynthesis and respiration. Plants must exchange carbon dioxide and oxygen with the atmosphere — this was selected in evolution — but as a consequence, they also lose water vapor. The stomatal mechanism at least retards the loss of water vapor during periods when the plant can't photosynthesize (at night) or when water is so limiting that the plant is in danger of death. Of course, the succulents, with their special carbon dioxide fixation at night, have found an even better solution to the problem. But even these must open their stomates to the desert atmosphere, although this occurs only under the most favorable conditions.

Others have suggested several ways in which transpiration might benefit the plant and consequently be more than a mere unavoidable compromise with the need for gas exchange.

Transpiration of fairly large quantities of water from the leaf surfaces results in a relatively rapid flow of water from the soil to the leaves. Rates as high as nearly 45 m/hr have been measured. It was suggested that this flow of water (the transpiration stream) carries with it the needed nutrient elements. Yet, as we shall see in the next chapter, this upward movement of water in the xylem can occur even in the absence of transpiration (e.g., in an atmosphere of 100% relative humidity—no vapor pressure gradient—and under a very low incoming heat load). Fluids are moving from the leaves to the roots and other parts of the plant in the phloem tissues, and some would naturally return in the xylem even in the absence of a transpiration stream.

Plant cells grow best when water potentials are somewhat below zero, that is, when plant cells are experiencing a relative water deficit. Completely turgid cells are said to grow more poorly than those that are continually losing a little water. This contention has been difficult to prove or to disprove. Evidence exists on both sides of the argument.

Leaves are cooled by evaporation of water in the transpiration process. For many years it was stated that the cooling effect must be minimal at best and that it could be of no practical consequence to the plant. This turns out not to be entirely true. Some plants, always of a type requiring an ample and continuous water supply, do cool their leaves significantly by transpiration. But to discuss the topic intelligently, we must investigate some of the principles of heat transfer between an object (specifically, the plant) and its environment.

Energy Exchange: Heat Transfer and Transpiration

The study of energy exchange between a plant and its environment illustrates the approach of biophysics—the application of physical principles (particularly mathematics) to biological problems. Often, the mathematics are not really difficult, provided one is willing to apply a little effort. *Heat transfer* is an excellent example and is explained in detail in End Paper 32–1. Here, we shall present only some of the conclusions taken from this study—not the arguments supporting them.

Under steady-state conditions, most of the energy exchange between a plant and its environment can be accounted for by radiation (absorption and emission by the leaf), convection (air in contact with the warm leaf being heated and then rising, or air warming a cooler leaf), and transpiration (evaporation of water requiring an energy input: about 600 cal to evaporate 1 g of water). If the leaf is absorbing more radiation than it is emitting (as a leaf in sunlight), the excess energy is dissipated from the leaf by convection and transpiration.

Loss of heat by convection is proportional to the difference in temperature between the leaf and the surrounding air—the greater the difference, the more heat is lost by convection. It is also inversely proportional to the boundary-layer resistance. The boundary layer is that layer of air in contact with the leaf that is influenced by the leaf. If the leaf is warmer than the air, the air close to the leaf will be warmer than the air farther out, and the layer of air that is warmer due to convection from the leaf is the boundary layer. If the boundary layer is thick, it will be relatively difficult for heat to penetrate it by convection; that is, its resistance is high. The boundary layer is thinner for small leaves and with increasing wind velocities; hence heat transfer by convection is more efficient under these conditions.

Transpiration is first proportional to the difference in vapor pressure of water inside and outside the leaf. Then, it is similar to convection, being inversely proportional to boundary-layer resistance. It tends to be more rapid for a thin boundary layer and hence for small leaves in a wind. But transpiration takes place mostly through the stomates; so it will be inversely proportional to leaf resistance, as well as boundary-layer resistance. If stomates are closed, leaf resistance is high and transpiration is greatly reduced. While only leaf shape and wind velocity influence boundary-layer resistance, numerous factors such as light intensity, CO_2 concentration, temperature, and water status in the plant may influence stomatal aperture and hence leaf resistance. Furthermore, vapor pressures are strongly influenced by temperature, which in turn is influenced by absorbed radiation. Many interactions must be considered in heat-transfer studies.

Ecology and Energy Exchange

Transpiration and heat transfer provide an excellent insight into some of the interactions between a plant and its environment, that is, into an important phase of plant ecology. Although transpiration has long been studied by ecologists, the principles of heat transfer were largely ignored. Of course, transpiration is so strongly influenced by heat transfer that studies ignoring these principles can hardly be accepted as valid without some further confirmation. Yet the interest that has developed in recent years allows us to appreciate the possibilities, some of which we shall now present.

Deserts. One extreme situation is the desert, where high temperatures are combined with high radiation, low humidity, and little available water. It is to the plant's advantage to conserve moisture and to maintain a relatively low leaf temperature. Many desert plants have small leaves (see Photo Essay 39), providing a thin boundary layer with consequent efficient convective heat transfer. But such a situation might also produce a high rate of evaporation unless leaf resistance is high. Some desert plants have sunken stomates (and other mechanisms), resulting in high leaf resistance and retarding transpiration; they may also be light gray in color, reflecting much of the sun's radiation. Desert succulents have the ability to close their stomates during the day and to fix CO_2 into organic acids at night.

If dissipation of heat must depend upon convection, then the leaf can only (at best) approach the air temperature, always being somewhat warmer than the air when radiant energy is high. Desert air temperatures are often extreme; so the ideal solution might be air conditioning by evaporative cooling—a technique applied widely by man living in desert cities. A few desert species with access to unlimited water (their roots reach the water table) achieve this. Typically, they have large leaves (e.g., the native palm of southern California, *Washington filifera*), and their high rate of transpiration results in leaf temperatures several degrees below air temperatures. This is possible because of the low ambient humidity, which provides an ample vapor pressure difference even when leaf temperature is relatively low, and because the large leaf has a thick boundary layer, resulting in a low rate of convective heat uptake by the leaf from the surrounding, warmer air. Leaf temperatures of large cocklebur leaves (not native to the desert), growing along a ditch bank in southern Arizona, were found to be as much as 9 to 11° C cooler than the surrounding air, which was 36° C under a clear, midsummer sky.

Alpine tundra. Alpine plants are faced with the opposite situation. Cool, fairly humid, gusty air combines with solar radiation intensities that may be extremely high. Water is seldom limiting, but evaporative cooling is no advantage in a situation where ambient temperatures are often well below the range considered optimum for metabolic processes. Several hundred measurements of leaf temperatures in the alpine flora indicated that when the sun is irradiating the leaves, their temperatures may often be as high as 30° C, which may be as much as 20° C above the temperature of the surrounding air. Alpine plants also typically have small or finely divided leaves, but the plants occur close to the ground (e.g., below 10 cm), where the wind velocities are greatly reduced (see Photo Essay 39), and a typical rosette plant form results in a thick boundary layer determined by the entire plant instead of by individual leaves. High leaf temperatures clearly indicate that transpiration does not provide much cooling power under these conditions. The reasons why are not always readily apparent, so integrated field and laboratory research would be in order.

The ecological problems of transpiration and heat transfer are somewhat less challenging in environments less extreme than those of the deserts and the alpine tundra. Nevertheless, careful study should provide some interesting insights. The extreme variety of leaf shapes even in moderate environments is certainly suggestive. Studies carried out during the first half of this century often provided data that seemed quite contradictory. Some reported that wind increased transpiration (as it always increases evaporation from a free surface); others said that wind decreased transpiration. When radiation loads are relatively low and leaf resistance is also low, transpiration will certainly be increased by wind. If leaf temperature is below air temperature, increasing wind velocity always tends to increase transpiration. Yet it is now clear that transpiration may indeed be decreased by wind when the radiation heat load is high, particularly if leaf resistance is also high. Under such conditions, the leaf temperature may be far above the air temperature, accounting for a high transpiration rate. The wind cools the leaf, and this cooling effect is more important in reducing transpiration than the wind is in increasing evaporation. It is now time to reinvestigate ecological problems of transpiration, taking into account the principles of heat transfer. (Incidentally, these principles were formulated clearly by H. T. Browne and F. Escombe as early as 1905, but they have been ignored until recently.) Some of the ecological factors influencing transpiration and heat transfer that may occur in various environments and some of the expected plant responses to these environments are summarized in Table 32–8.

32–8 *Theoretical energy budgets in several environments for a cocklebur leaf (5 cm long) with wind blowing across the leaf perpendicular to the midveins*[a]

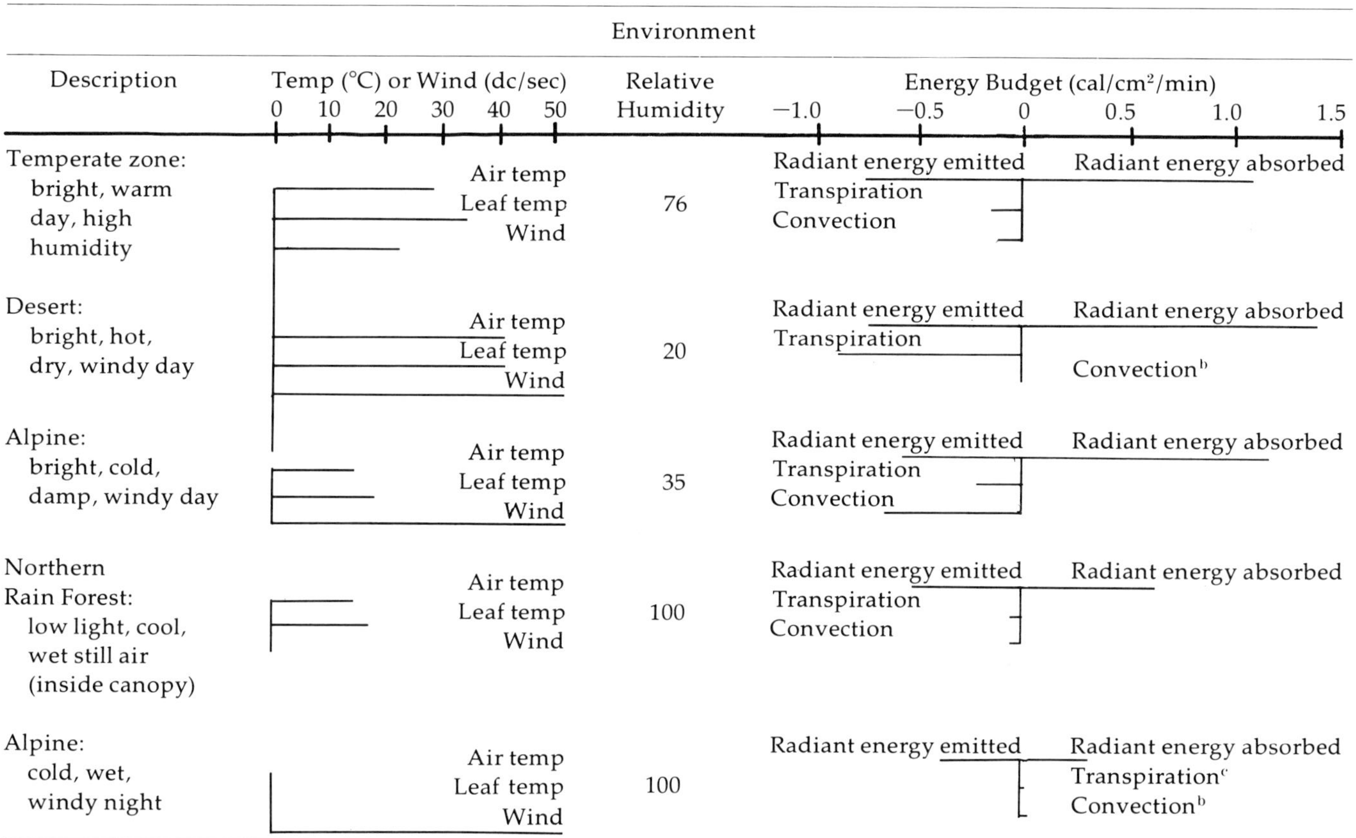

[a]This information obtained in part from the laboratory of David Gates at the Missouri Botanical Gardens in St. Louis. The data on radiant energy exchange between the atmosphere and the leaf were so obtained, but the convection and transpiration data were calculated from the energy budget equation by B. G. Drake.
[b]Leaf is *gaining* heat by convection because it is below air temperature.
[c]Frost is forming on the leaf.

End Paper 32–1 Energy Exchange: Some Mathematical Concepts

The principles of heat transfer between a plant and its environment become extremely complex when we attempt to apply them under all conditions, but by limiting our viewpoint, they can remain relatively simple. Consider a plant in the daytime under a relatively clear sky and in a steady breeze with constant air temperature and humidity. When environmental conditions are not constant, leaf temperature changes (heat is being stored or lost), complicating matters; so let us consider only steady-state conditions (constant leaf temperature). The leaf is absorbing heat in the form of radiation (visible light plus infrared radiation) and radiating some of this energy in the form of infrared radiation. The amount of energy absorbed is greater than the amount lost; we will call this excess the net absorbed radiant energy (Q). A small amount of the absorbed energy (1 to 3%) is used in photosynthesis (and perhaps other processes), but this is small enough to ignore in our simplification. Most of the energy absorbed by the leaf is dissipated by two physical processes: convection (C), in which air in immediate contact with the leaf is warmed by conduction, expands, and becomes lighter (so it rises), and transpiration (V). If the leaf is absorbing energy by any process, its value is positive; if losing energy, negative. Hence we can write the energy balance equation as

$$0 = Q + C + V.$$

Under our conditions, Q is positive, and C and V are negative. We could move Q to the other side, showing that the net absorbed radiation is equal to the sum of the energy lost by convection and by transpiration, or

$$-Q = C + V, \quad \text{or} \quad Q = -(C + V).$$

One of the main problems in making measurements that allow us to apply this equation concerns the energy absorbed by the leaf. In the visible part of the spectrum, blue and red are strongly absorbed by leaves, but green is largely transmitted or reflected (hence the green color of leaves). The near infrared part of the spectrum is reflected and transmitted even more than the green. Thus photographs taken with film sensitive to the near infrared show plants that appear brilliant white (Fig. 32–9). Nearly all of the far-infrared part of the spectrum is absorbed by plants—and by most objects. (They are "black" in that part of the spectrum, but no film is sensitive to far-infrared.) Absorption spectra of different plants are difficult to measure and differ in detail, but virtually all follow the general pattern described here (an example is Fig. 32–10).

The amount of radiant energy a leaf absorbs depends to a limited extent upon the species and physiological condition of the leaf (since the pattern is similar for most leaves) but much more so on the spectral distribution of the light falling on the leaf (Fig. 32–11). At night the atmosphere, clouds,

(*a*) (*b*)

32–9 *Plants showing high reflectivity in the infrared portion of the spectrum. Photograph (a) was taken with ordinary panchromatic film, which gives gray tones similar to the color intensities experienced by the human eye. Photograph (b) was taken on infrared sensitive film through a dark red filter that excludes most of the visible spectrum. Note the brilliant white appearance of the vegetation and compare in particular the clumps of grass on the right.*

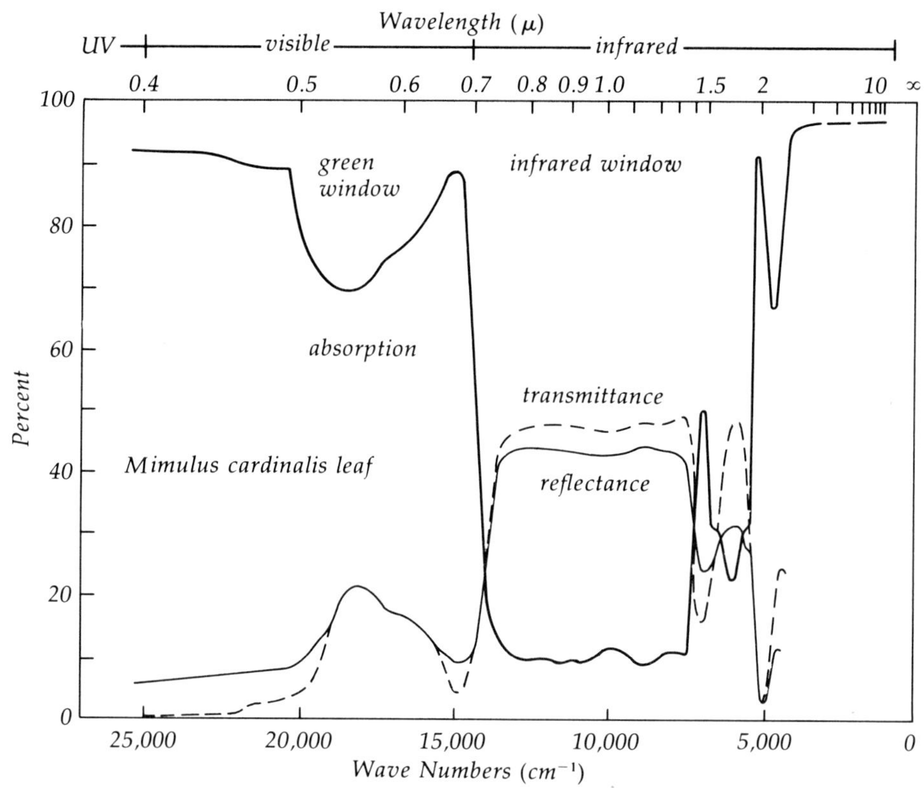

32–10 Absorption, transmission, and reflection spectra of a leaf. Note especially the "windows" in the green and in the near-infrared protions of the spectrum. The leaves are thin and light green. Note that emission spectra are plotted on a linear wavelength scale for· the sources and a linear wave-number scale for the leaf spectra.

and other objects in the environment are emitting far-infrared wavelengths, and these are almost completely absorbed by plants. Light from an incandescent bulb is rich in the near infrared that is not absorbed by leaves, while fluorescent light is strongly absorbed, since it is rich in blue and sometimes red. The peak of the solar spectrum (plotted as wavelengths) is in the green, much of which is not absorbed (relatively large amounts of near infrared are also unabsorbed). In general, about 50% of sunlight is absorbed by leaves, but this can vary over a wide range. Absorption of fluorescent light is somewhat higher, and absorption of incandescent light is considerably lower than absorption of fluorescent (due to the large quantities of unabsorbed near infrared). The energy absorbed by a leaf depends upon its absorption spectrum and the exact spectrum of the radiant energy falling on it—both rather difficult to determine and always somewhat variable.

The energy emitted by the leaf tends to be somewhat easier to calculate. This is done by applying *Stephan's Law*, in which the fourth power of the absolute temperature (T) is multiplied by the Stephan-Boltzmann constant ($\delta = 8.132 \times 10^{-11}$ cal/cm²/min/deg K⁴) and the emissivity (e—typically 0.95) to obtain the energy emitted (Q_r—in cal/cm²/min, when the above constant is used). Thus

$$Q_r = e\delta T^4.$$

Applying this formula, we see that a leaf at 0° C emits about 0.41 cal/cm²/min; at 20° C, 0.55 cal/cm²/min; and at 30° C, 0.62 cal/cm²/min. The incoming radiant energy may lie between 1.0 and 2.0 cal/cm²/min (sunlight plus far infrared from the surroundings), of which perhaps 75% is absorbed by the leaf (0.75 to 1.5 cal/cm²/min). So the amount of energy emitted by leaves as long-wavelength radiation is a sizable percentage of the amount absorbed from sunlight (perhaps 30 to 60%, depending upon conditions). The quantity (Q) in the heat balance equation refers to the difference, that is, to the excess of absorbed energy (Q_a) that is not radiated (perhaps 40 to 70% of that initially absorbed). Thus

$$Q = Q_a - Q_r \quad \text{or} \quad Q = Q_a - e\delta T^4.$$

The amount of energy lost from the leaf by convection is directly proportional to the difference in temperature between the leaf (T_l) and the surrounding air (T_a) and inversely

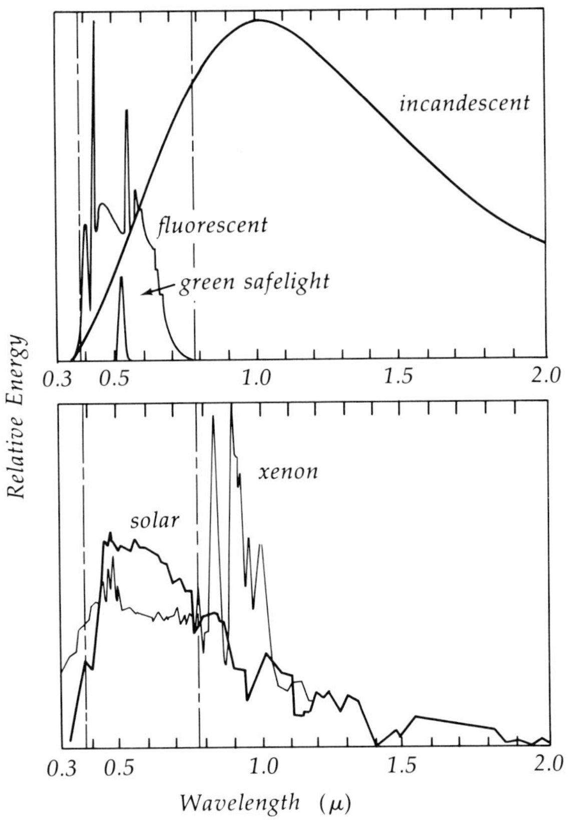

32–11 *Emission spectra for five sources. Note the incandescent lamp peak at about 1.0 μ, the mercury emission lines in fluorescent lamp spectra, the infrared peaks (0.85 to 1.05 μ) from the xenon lamp, and the solar peak in the midpart of the visible area (indicated by dashed vertical lines).*

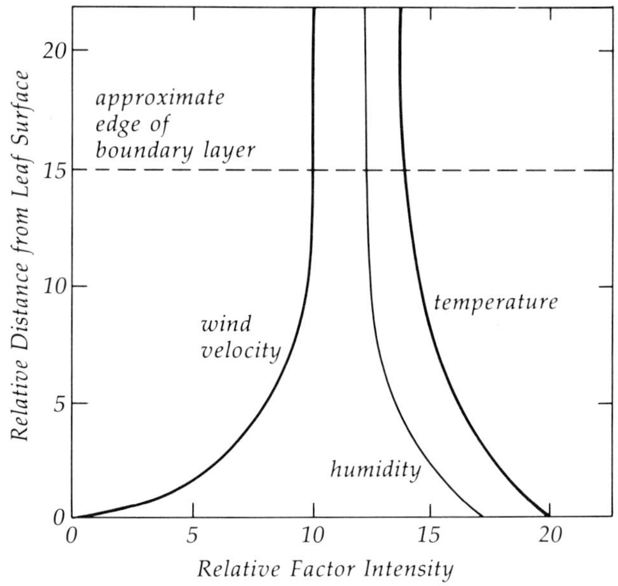

32–12 *Relative factor intensities as a function of distance from a surface, illustrating the principle of the boundary layer.*

proportional to the resistance (r_a) to convective heat transfer from the leaf to the air. Thus

$$C = \frac{T_a - T_l}{r_a}.$$

The resistance (r_a) is determined by the *boundary layer* conditions. The air in immediate contact with the leaf will be at almost the same temperature as the leaf; farther away, the air temperature may be quite different. The layer of air in which the temperature gradient is influenced by the leaf is called the boundary layer (Fig. 32–12). It is easy to imagine that the boundary layer gets thicker from the edge of the leaf toward the middle and that it gets thinner at all points with increasing wind velocity. Other factors being equal (leaf and air temperature), the thinner the boundary layer, the lower the resistance to convective heat transfer and the more rapid the transfer of heat by this mechanism. *The average boundary layer is thinner for small leaves and for increasing wind speeds;* so resistance to convection is less under these same conditions. The leaf having the thinnest boundary layer is a single needle of a conifer, although these needles occur in groups on the twig, and a group of needles collectively determines boundary-layer conditions.

Leaves occur in a great variety of shapes, and these shapes undoubtedly strongly influence boundary-layer conditions. Some of the largest leaves are those of the fan palms. As we saw in this chapter, their extremely thick boundary layers provide an advantage. Some of the smallest leaves are those of desert plants in dry soil. The thin boundary layer allows for most efficient convective heat transfer; so the leaf seldom gets much warmer than the surrounding air.

When 1 g of water at room temperature evaporates, about 600 cal are taken from the surroundings to convert the liquid to gas. Thus evaporation and transpiration are cooling processes. The laws governing transpiration are similar in many ways to those controlling convective heat transfer. The driving force is not the difference in temperature, as in convective heat transfer, but rather the difference in the vapor pressure of water inside the leaf and in the surrounding air beyond the boundary layer. (Of course, vapor pressure is strongly influenced by temperature.) Transpiration is proportional to this difference in vapor pressure (Fig. 32–13).

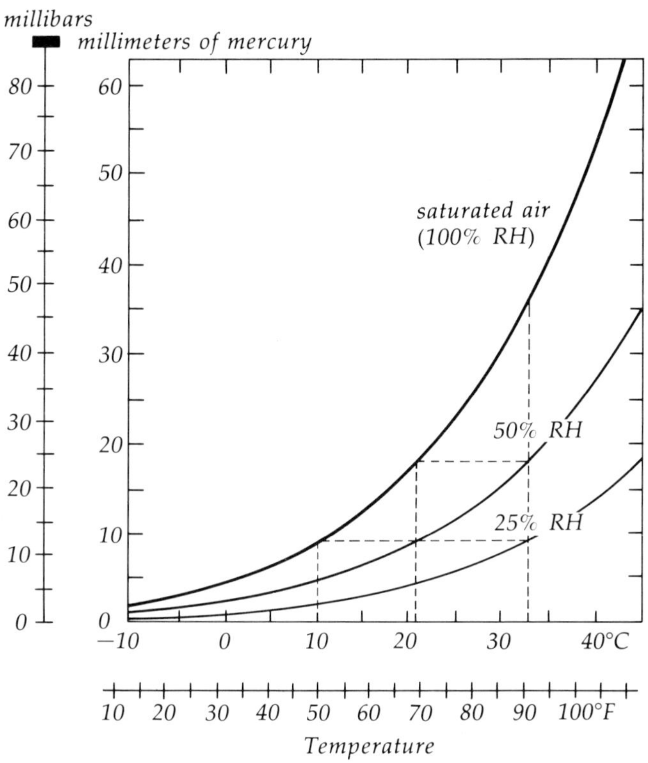

32–13 Relationship between moisture in the atmosphere (partial pressure of water vapor) and temperature for saturated air and for air at 50% and 25% relative humidity (RH). Dashed lines indicate, for example, that air at 10° C, 100% RH, has the same amount of moisture as air at 21° C, 50% RH, and air at 33° C, 25% RH.

In the simplest case, if leaf and air temperatures are the same, the driving force is proportional to the difference in relative humidities of the air in the leaf (nearly 100%) and the air outside (ambient air). If the leaf is warmer than the air, then the vapor pressure of the air inside is much higher than that of the ambient air, although the relative humidity in the ambient air may be quite high (due to the relationship between vapor pressure and temperature). This can be an important effect because high leaf temperatures (30 to 40° C) can increase transpiration much more than low relative humidities at low leaf and air temperatures.

As in convective heat transfer, transpiration is inversely proportional to the resistance to transfer of water from the leaf to the air. Effects on the boundary layer are closely similar in the case of transpiration and convection; that is, small leaves and increasing wind lower the boundary layer resistance to both convection and transpiration. In the case of transpiration, however, there is another important resistance: the leaf. This is rather complicated, but the most important part of leaf resistance is due to the stomates. When they are closed, resistance is extremely high; when open, relatively low. It is here that the physiological factors become important.

To summarize, transpiration (V) is proportional to the difference in vapor pressure between the air (P_a) and the inside of the leaf (P_1) and inversely proportional to the boundary layer resistance (r_a) plus the leaf resistance (r_1):

$$V = \frac{P_a - P_1}{r_a + r_1}$$

We could substitute the three equations for Q, C, and V in the heat balance equation, obtaining an expanded and more complete version. But doing so would not fully explain additional complications, which are often important. One of these, already mentioned, is that various environmental factors interact with each other in several complex ways. Net absorbed radiation (Q) determines the difference in temperature between the leaf and the surrounding air, and this in turn influences rates of convection, differences in vapor pressure, and consequently, transpiration. The second complication concerns the effects of such factors as light, temperature, carbon dioxide concentration, and water stress upon the stomates. This effect is not indicated in the equations, except as r_1 tends to vary in response to it.

33

Transport in Plants

We have seen that now much dynamic activity actually exists in our quiet ecosystem. The placid plants are not as inanimate as a casual observer might suppose. Energy and materials are continually entering and leaving the ecosystem and its members. Except for penetrating light rays, most of this exchange occurs at the surfaces. But now it is time to look below these surfaces because much is going on inside the plant. This activity is by no means confined to the molecular level. Fluids are flowing both up and down in roots, stems, and leaves. Water and minerals absorbed by the roots move to the shoots, and organic materials synthesized in the leaves move to the roots.

Transport Mechanisms

We have already examined the anatomy of the xylem and phloem elements through which this transport takes place. Now we shall examine the actual mechanics involved in this movement. Virtually everyone is familiar with the heart, which is a pumping device, and with the arteries, capillaries, and veins, which form a closed circulatory system. In some ways, the circulation of materials within plants is similar to that in animals, but in other ways it is significantly different. Rates are much faster in animals. The blood may move as much as 100 m/hr in the arteries close to the heart but much slower in capillaries. When the plant is actively transpiring, water may move upward nearly as fast—often over 40 m/hr. This is about as fast as the tip of a sweep second hand on a clock about 25 cm in diameter. Dissolved substances move much more slowly in the phloem of plants, on the order of 20 to 200 cm/hr. These slower rates compare with the tip of a minute hand on a clock about 30 cm in diameter. Some substances move even more slowly in the plant. The flowering hormone moves only about 0.2 to 0.4 cm/hr, about the speed of the tip of an hour hand on a watch.

The driving forces responsible for movement of fluids in plants and animals are essentially the same: *gradients in pressures.* In animals the gradient is created

33–A Transport in vascular plants is vital, since the cells of the various parts of the plant must be supplied with a wide range of materials not readily available to them. Few things demonstrate more clearly the importance of the vascular tissue and the transport that occurs within it than the cleared leaf shown here. The network of veins, consisting of strands of xylem and phloem, extends into every part of the leaf. Few cells are more than 2 or 3 cells away from a vein ending.

by the contraction and expansion of the heart, producing high pressure in the arteries near the heart and much lower pressure (even slightly negative pressures or tensions) in the veins emptying into the heart (Fig. 33–1). In plants the pressure gradients are produced by the transpiration of water from the leaves (creating extremely negative pressures in the xylem there) and by the osmotic uptake of water in the phloem cells of leaves (creating high positive pressures there). The magnitudes of the pressures are much greater in plants. Blood pressures seldom exceed $^1/_5$ of a bar, and negative pressures in the blood near the heart are only a minute fraction of a bar below zero. In the plant, due to the rigidity of cell walls, pressures and tensions may reach several bars, both above and below zero (Fig. 33–2). Understanding the circulation of fluids in animals is at least initially simple because of the pumping mechanism of the heart. But plants have no such moving parts, so discovering how their transport system works has been a more difficult task.

The Ascent of Sap

How does water get to the top of tall trees? Several explanations come instinctively to mind. One of these is capillarity. We all know from personal experience that water will rise a certain distance up a tube of small diameter. Capillarity occurs because of the curved miniscus formed by the water surface at the top of the water column in the tube. The edges of the miniscus reach upward because of the mutual attraction between the walls of the tube and the molecules of water. This attraction is strong enough to raise the column of water in the tube to a height that can be calculated by computing the diameter and the composition of the tube. In the cellulose tubes of tracheids and vessel elements constituting the xylem, water can rise by capillarity only about $^1/_3$ of a meter. Furthermore, the tubes in the xylem tissue are filled with water; hence there are no minisci. Capillarity, in the usual sense, simply cannot play a role.

Another explanation that one thinks of in relation to any problem of water movement involves pumps. There are two types of general pumps used to move water. An example of the first is the ancient hand pump in which a vacuum is created at the top so that atmospheric pressure forces the water up from below. The trouble with such a pump is that atmospheric pressure at sea level can lift a column of pure water only about 10.33 m, and some trees may be as tall as 130 m. The second type of pump is at the bottom of the pipe and forces the water up from the bottom. If the pump is powerful enough, water can be raised to any height. Such a pump is present in plants. It is based on the fact that the endodermal layer in roots plus a buildup of solutes in the fluid in the xylem forms an ideal osmometer (as we saw in our discussion of root anatomy). The end results are root pressures that have been measured as high as 5 or 6 bars—enough to force water to the top of a 60-m tree, provided there is no frictional resistance to the flow of water. But, of course, there is frictional resistance. The problem with root pressure being the universal mechanism for the movement of water is twofold: it is not strong enough, and it is found in only a few species of plants under certain environmental conditions (e.g., when transpiration is essentially nil and when the soil is fairly saturated with water).

The correct answer to our question appears to be the transpiration-tension-cohesion theory developed between 1894 and 1915 by several investigators, particularly Henry Dixon and John Jolly. The critical point in this theory is the inherent cohesion of water. While adhesion is the attraction of unlike molecules to each other (e.g., water to cellulose), cohesion is the attraction of like molecules to each other. The theory holds that water deficits are created in the leaves by transpiration and that the columns of water in the xylem are lifted intact in response to these deficits, the columns being held together by the cohesion of the water molecules. Ultimately, the cohesive tensions are transmitted to the roots and eventually the soil, thus drawing water from the soil through the plant to the leaves. The tops of the columns are held to the cellulose walls of the leaf cells by adhesion.

This theory offers a simple and unified explanation for the movement of water in plants of any size. Moreover, it is open to experimental questioning. Let us now examine some of these questions.

Are the driving forces of transpiration great enough to lift columns of water up a tree? Unless the atmosphere is nearly saturated with water, transpiration will tend to occur, and the driving forces that are created will exceed those required to reach the top of the tallest trees. Data depicting this possibility are summarized in Figure 33–2. They show that there is a steep gradient from high to low (negative) water potential, going from

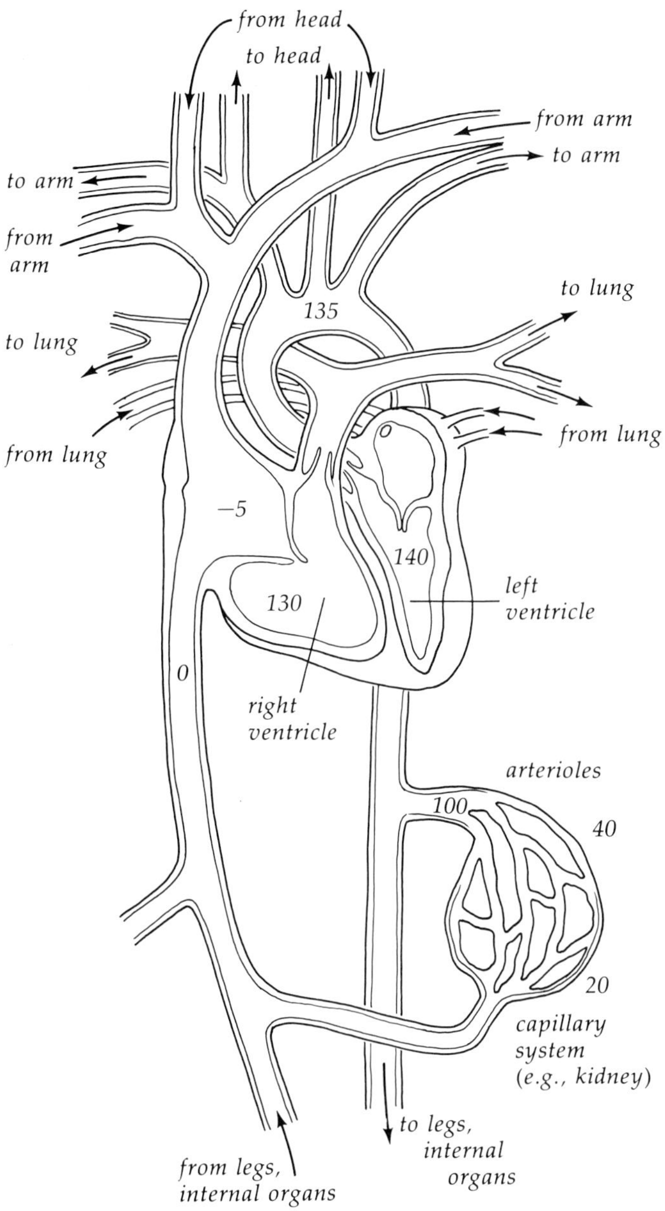

33–1 A schematic illustration of blood circulation in the human being. The figures refer to the pressure (expressed in mm of mercury—the units used in measuring blood pressure). Note the very high pressure in the left ventricle, the decrease in pressure throughout the system, finally the negative pressure where the blood reenters the heart.

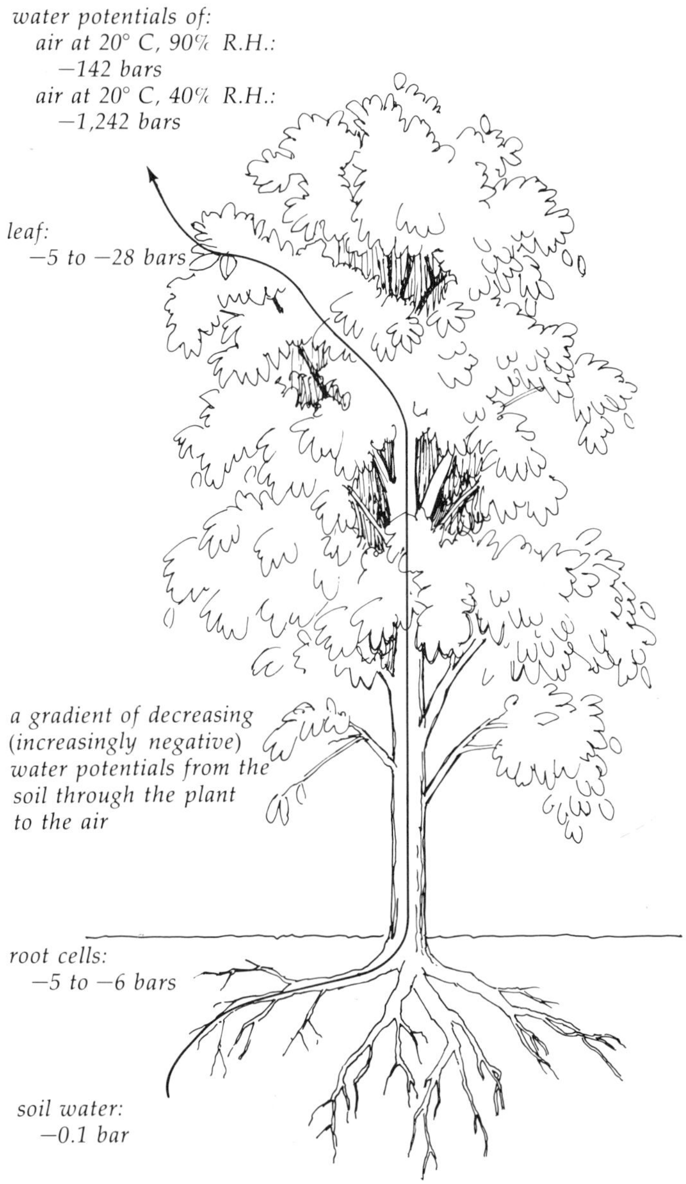

33–2 A schematic illustration of water potentials from the soil through a tree and into the atmosphere.

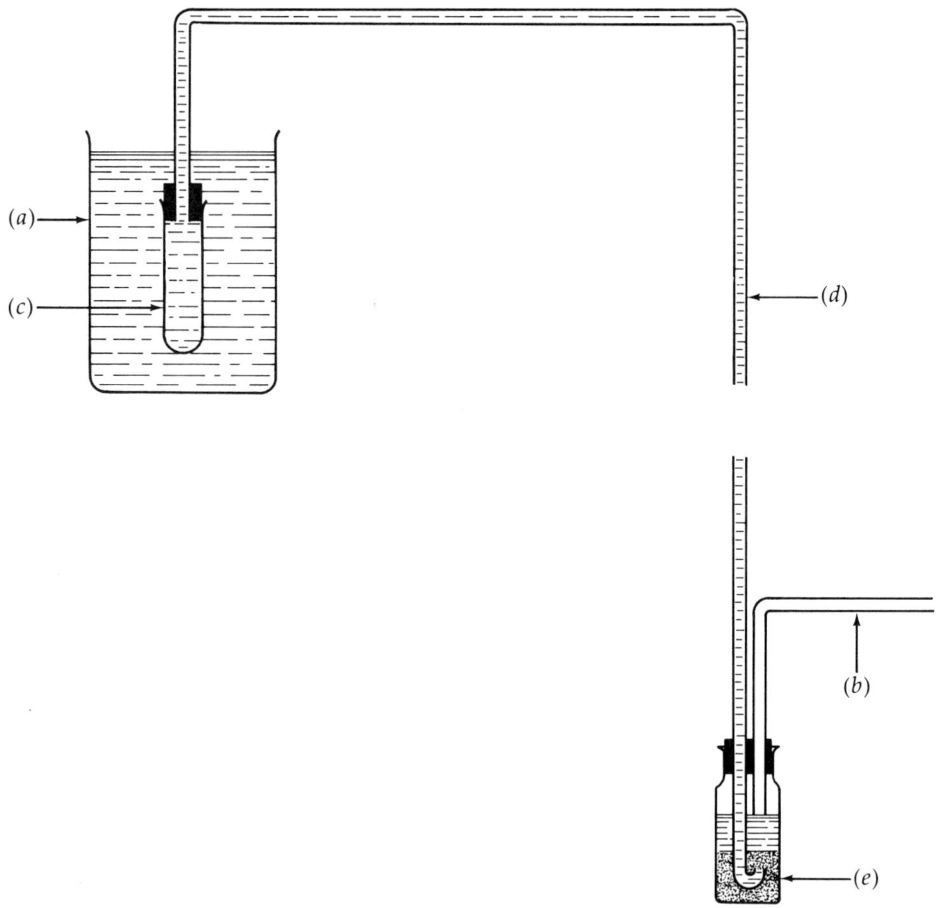

33–3 Askenasky's apparatus for demonstrating the cohesive forces of water. First, a beaker containing boiled water (a). This is allowed to filter through the apparatus for some time, removing all air bubbles from the tube (d). To initiate the experiment, the beaker of water is removed, and evaporation begins to occur at the surface of the porous clay cup (c). Evaporation will cause water to rise into tube (d), pulling with it some of the mercury (e) in the bottle. Air enters through the ventilation tube (b). If the mercury rises higher in the capillary tube (d) than can be supported by atmospheric pressure, then the cohesive property of water (and the adhesion between water and mercury) is demonstrated.

the soil water through the plant and into the atmosphere. When the relative humidity is 100%, water potential in the air is zero. As the humidity drops a few points, water potential drops tens, hundreds, and even thousands of bars below zero. The exact amount depends upon temperature. Water in the leaf is never below about −30 bars, so the big step in the water-potential gradient is usually between the leaf and the atmosphere.

Are the forces of adhesion strong enough to support the columns of water? The tops of the water columns are held to the walls of the leaf cells by the adhesion of water molecules to the hydrophilic surface of the cellulose. The adhesion of water to such surfaces is referred to as hydration or imbibition. The forces of hydration holding water molecules to hydrophilic surfaces can be measured by a number of methods, which show they are on the order of −1000 bars—far in excess of the 30 that are required (assuming 10 bars to raise water to the top of the tallest tree and another 10 to 20 to overcome the friction along the way).

Nature often demonstrates the forces of hydration. When plants grow, they expand, partially because they synthesize materials capable of hydration that then adsorb water and swell. Growing roots will often crack rocks, concrete, or asphalt. Sprouts of crabgrass or other plants are commonly seen coming up through asphalt. In one extreme situation a family of mushrooms actually exploded through a concrete driveway with the sound of a rifle shot!

Are the cohesive forces strong enough to keep the water columns from breaking under the from −20 to −30 bar tensions found in the xylem? In an ordinary pump pulling from above, if an attempt is made to raise water above the level supported by barometric pressure, the water cavitates. That is, when the column has been raised to the barometric height by pulling from above, the pressure drops to zero, and under these conditions water vacuum boils at any temperature above freezing. Such cavitation does not occur in the xylem, and for over 70 years plant physiologists have argued as to why it does not.

Dixon and others showed that if a glass tube were filled with boiled distilled water (containing no dissolved gases) and attached to a porous clay cup at the top, it is possible to demonstrate the cohesive properties of water. This experiment is shown in Figure 33–3, and it is sometimes possible in such an apparatus to raise the mercury at the bottom of the water about twice as high as the barometric height. This is far short of the heights required to raise water to the top of the tallest trees, but it at least demonstrates the principle.

The problem with this experiment is that plants do not contain boiled distilled water—they contain sap with many dissolved minerals and gases. Moreover, the experiment is a hard one to do and works only under ideal conditions. How then can it work in the plant? The answer is in the small size of the actual tubes of water in the xylem. In 1950 Lyman Briggs, who studied the cohesive properties of water by spinning capillary Z-tubes in a centrifuge (Fig. 33–4), showed that the smaller the tube diameter, the greater the force required to break the column. In a capillary of 0.5 mm, air-saturated tap water does not cavitate under tensions of −20 bars. In columns as small as the tracheids or vessel elements of xylem tissue (the apoplast), possible cohesive forces reach levels of −264 bars, certainly high enough to make the proposed mechanism a feasible one.

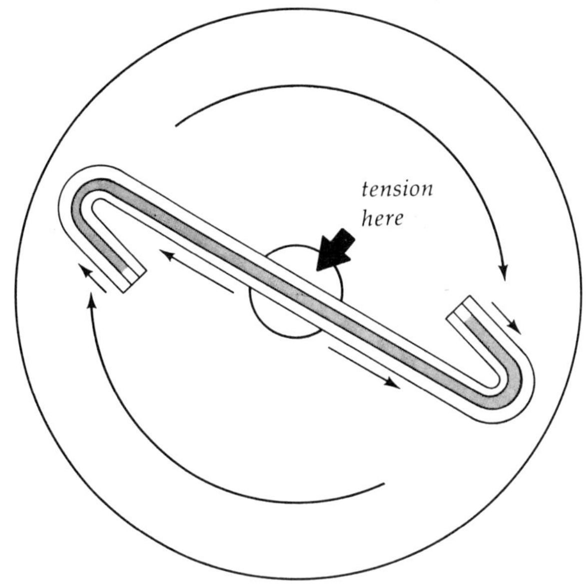

33–4 A method of measuring the cohesive properties of water utilizing a centrifuged Z-tube. Small arrows indicate the direction of centrifugal force and the principle of balancing due to the Z-tube.

But what if the column does break? The columns might, for example, break in some xylem elements when the tree is subjected to high winds. Why doesn't an expanding bubble of water vapor fill all of the xylem tubes in the trunk of the tree, cutting off the transpiration stream entirely? Apparently, this fails to happen because of the minute pores connecting the tracheids of vessel elements. These pores are simply too small to permit the passage of a bubble. The shape of the bubble is maintained by forces of surface tension that increase as the bubble gets smaller and the curvature of its surface increases. When the bubble is extremely small, the forces are correspondingly great, and it is most difficult to deform the bubble enough to pass through the minute openings between the xylem cells. Of course, liquid water passes freely through these openings. Thus, the ascent of sap in trees is possible because of the minute openings or pits that act as check valves between xylem elements when air bubbles are formed.

When trees in northern climates freeze in winter, dissolved air is forced out of the freezing sap, causing cavitation in the xylem tubes. Microscopic examination has demonstrated that the xylem elements contain an abundance of air bubbles after having been frozen. We do not know how such trees as the conifers can overcome this problem, but several suggestions are possible. Perhaps the columns of water form again during the spring rains, when the gradient is reversed and water may be absorbed by the leaves, moving into the xylem and perhaps the roots. Furthermore, as new xylem cells are formed by division of cambial cells, they are initially filled with water. These and other problems may be solved by using techniques and knowledge developed by hydraulic engineers in their study of water movement through the confined spaces of the soil.

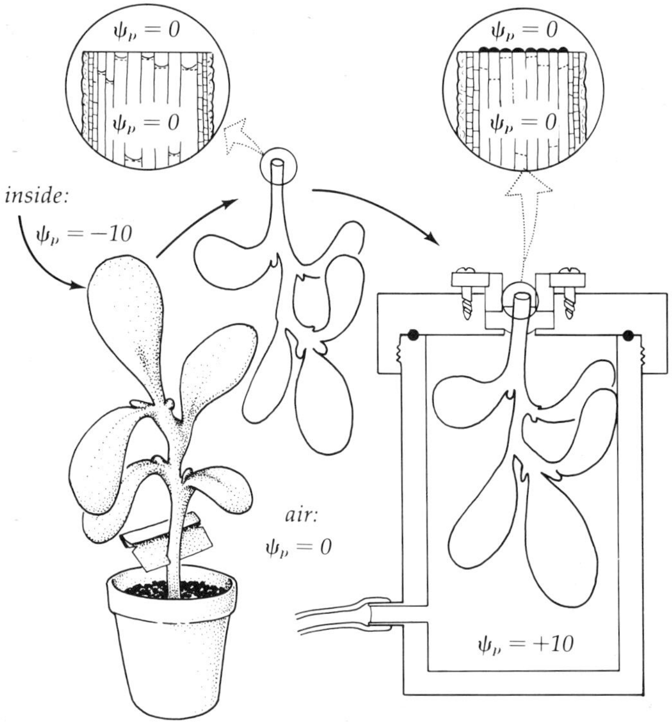

33–5 The method of measurement of tension within a plant stem utilizing a pressure bomb. When the plant part is cut away, the minisci of water at the cut surface retreat into the xylem tubes of the stem because the water in those tubes was under tension. The plant part is then placed in the pressure bomb and pressure applied until the minisci return to the cut surface, as observed with a microscope. At this point, the absolute value of the pressure being applied is equal to the absolute value of the tension that existed in the xylem tubes before they were cut.

Can tensions be measured in xylem tissues? This is the final—and one of the most critical—questions that can be asked about the cohesion-tension theory. If the theory is correct, then it should be possible to measure the tension in the sap when tall trees are transpiring rapidly. That the sap is under tension can be demonstrated or measured in a number of ways.

One of the easiest involves herbaceous rather than woody plants. If the stem of a rapidly transpiring herbaceous plant is submerged in a colored liquid and cut or punctured, subsequent dissection shows that the color moves rapidly up and down the stem from the point of the cut—exactly what one would expect if water in the xylem were under tension. In the case of trees it has been found that the diameter decreases slightly when evaporation rates are highest—as though this decrease were caused by increased tensions in response to the rapid transpiration.

Rather recently, two different methods have been developed to obtain quantitative measurements of sap tensions in the xylem. George G. Spomer, then at the University of Chicago, devised a small probe that inserts into the trunk of the tree and measures water potential. Spomer measured negative water potentials reflecting negative pressures (tensions) on the order of −3 to −6 bars.

Per F. Scholander of the Scripps Institute in California has another method. When a branch is cut from a tree, the xylem sap recedes some distance into the xylem elements and below the cut surface. This indicates (as in the experiment with the colored solution) that the sap was under tension. Scholander reasoned that the amount of tension could be measured by putting the branch in a pressure bomb and increasing the pressure until xylem sap could be seen with a microscope at the cut surface (Fig. 33–5). This pressure should then equal the absolute value of the tension before the branch was cut. Using a high-powered rifle, Scholander and his associates shot branches from a tall Douglas fir, making immediate measurements with a pressure bomb. The results were exactly as predicted: the measured tensions depended upon the height of the branch.

Translocation of Solutes

Having discussed the movement of water from the soil through the apoplast of the plant and finally into the atmosphere, we are now ready to consider the other major aspect of transport in plants: the movement of assimilates (products of assimilation such as sugar, amino acids, etc.) through the symplast phase of the plant. While tensions become common in the xylem, pressures are the rule in the phloem. This is a situation that must be reconciled.

The main organic constituent moved in solution throughout the plant is *sugar*, frequently sucrose. Of course, amino acids, various hormones, and even substances as complex as proteins and virus particles also move within the plant. But the basic observation is that movement in the plant takes place from point of origin (source) to point of utilization (sink), either up or down. This has been demonstrated in numerous ways. As early as 1675 Marcello Malphigi performed girdling experiments (repeated by Stephen Hales in 1727). These were among the first plant physiological experiments ever performed. In such experiments bark is removed from a branch or the trunk at various points. If a belt of bark is removed from the trunk (Fig. 33–6), the remaining bark below the girdle tends to shrivel and die, while that above the girdle swells and grows. If the girdle is cut on a branch between leaves and developing fruits, the fruits will not develop, but the bark swells on the leaf side of the girdle.

In recent years it has been possible to use labeled sugars (formed in the leaves by photosynthesis with radioactive carbon dioxide) to further demonstrate this movement from source (photosynthesizing leaves) to sink (developing fruits, growing roots, apical meristems, etc.). A source can also be storage tissue, where insoluble starch is converted to soluble sugar. Sinks can be any cells that store or use sugar in metabolic processes such as respiration. Such cells occur not only in the roots, seeds, etc., but also in cortex, pith rays, and so forth. Any cell that does not release stored food and that does not contain chloroplasts (e.g., an epidermal cell on a leaf) acts as a sink.

33–6 *The effects of girdling the trunk of a tree by removing bark from around the circumference. Note the swelled bark above the girdle compared to that below. The trunk has been cut to reveal the annual growth rings. Note that an entire year's growth was laid down above the girdle but not below.*

Mechanisms of Translocation

How does assimilate translocation occur in the plant? This problem has challenged plant physiologists since the days of Malphigi, and it has not been solved to everyone's satisfaction, although two or three mechanisms have been proposed that must operate at least part of the time in most plants. One approach has been to think in terms of diffusion, which would account for movement from source to sink. Yet this is a slow process, so additional mechanisms must be postulated to account for observed rates. The cytoplasmic streaming hypothesis is an example of such a mechanism. Rates of movement are accounted for by cytoplasmic streaming, but diffusion plays a role in moving materials from one cell to its adjacent cell. We have already considered the process in discussing movement of dissolved substances through the symplast from the outer surfaces of the root into the stele (see Chapter 24). As Figure 33–7 indicates, such a mechanism would permit movement in both directions in the same phloem cells but always from the source (point of highest concentration) to the sink for the particular substance being translocated. This attractive model surely represents what happens in any instance where substances can move from cell to cell and where cytoplasmic streaming occurs, such as in cortical cells, photosynthesizing cells in the leaf, and young sieve-tube elements. Unfortunately, no convincing evidence has ever been put forth that cytoplasm streams in mature sieve-tube elements, the site of most assimilate translocation. Furthermore, cytoplasmic streaming is too slow to account for observed rates of movement in the phloem.

The other approach suggests that assimilates move through the sieve-tube elements in a bulk flow of

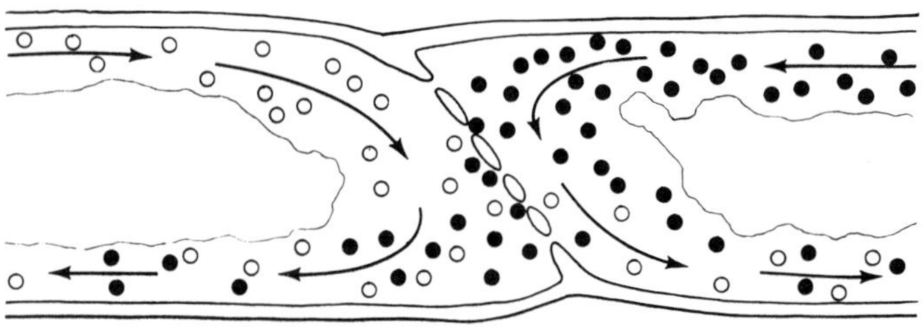

33–7 A schematic illustration of the cytoplasmic-streaming hypothesis of solute translocation. Note that diffusion of white or black particles across the sieve plate will occur in response only to their own concentration gradients. Movement of the two materials might occur simultaneously in opposite directions. Rate of movement is accelerated by streaming.

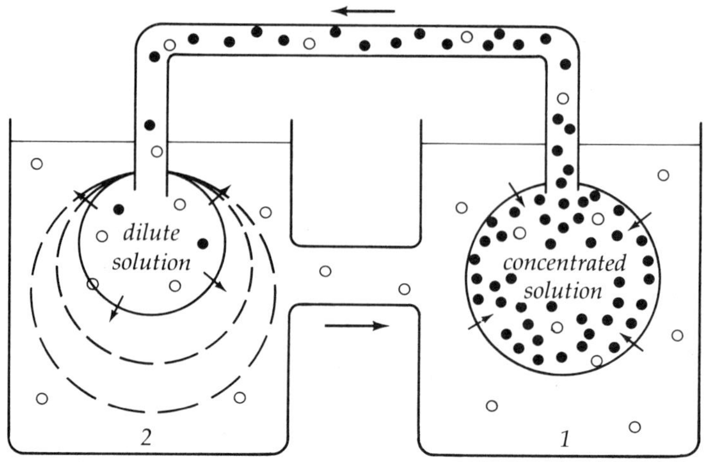

33–8 A model illustrating the pressure-flow theory of solute translocation as proposed by Münch. Note that concentration of black particles will control the rate and direction of flow, but white particles (which are much more dilute) will move along in the resulting stream. Dashed lines on the left imply that flow may occur due to expansion (growth) of the second osmometer (tissue), as well as movement out through the membrane.

material in response to pressure gradients. The mechanism that might create this mass flow is demonstrated by the model in Figure 33–8, which was suggested in 1930 by E. Münch, a German plant physiologist. Two osmometers are connected by a tube. Both are surrounded by a dilute solution. One osmometer contains a high concentration of solute, while the other contains a more dilute solution. Concentration of the solution surrounding the osmometers is low in comparison with the concentration inside. The osmometer containing the concentrated solution takes up water by osmosis, developing a pressure. This pressure is transmitted to the other osmometer (with the speed of sound), where it tends to increase the water potential until it is above that of the surrounding solution. Water then diffuses out of the second osmometer, relieving the pressure and resulting in a flow of liquid from the first osmometer to the second. Anything dissolved in this liquid is simply carried along in a bulk flow.

When Münch's pressure-flow hypothesis is applied to the plant (Fig. 33–9), the first osmometer is analogous to a producer cell in which osmotic potential is kept negative by photosynthesis, hydrolysis of starch, etc. A consumer cell in which sugars are continuously being removed by respiration and other processes represents the second osmometer. The phloem elements (the symplast as a whole) represent the connecting link between the two osmometers, and the surrounding solution occurs in the apoplast. We should note that the plant system may never reach equilibrium, because sugars are continually being produced and used up, while Münch's physical model reaches equilibrium when enough dissolved material is moved from the first osmometer to the second to make the concentrations equal in both.

In the model, the pressure could also be relieved by expansion of the second osmometer, indicated by the broken lines in Figure 33–8. In the plant this might occur when a fruit grows, for example. If a second substance is present at low concentration (besides the one at high concentration accounting for the osmotic potential in the first osmometer), it would be carried along with the first, regardless of the concentration gradient of the second material.

Münch's model works nicely in the laboratory. If all the conditions that make it work in the laboratory occur in the plant, then the mechanism must also work in the plant. This is the principle of the Uniformity of Nature, which states that when something happens under one set of circumstances, it will happen again under *identical* circumstances. The essential components of the model and consequently the plant are: an osmotic gradient resulting in a pressure gradient and the possibility of flow between the two (through sieve tubes—

the symplast). If these conditions are exactly met in the plant, then pressure flow must operate in the plant, which does not imply that there might not be modifications due to additional circumstances or that some other mechanisms might not also be operating.

The osmotic gradient was discussed briefly in relation to the ascent of sap (see Fig. 33–2). Osmotic potentials of leaves are typically much higher than those of roots or other sinks. Numerous measurements have been made to substantiate this observation. The presence of an osmotic gradient certainly implies a pressure gradient, but such a gradient is much more difficult to measure. Yet it is clear that liquids in the sieve-tube elements are under pressure, so we are probably justified in admitting that pressure gradients must occur.

Can substances flow through the sieve-tube elements at sufficient rates to account for the observed facts of assimilate translocation? Many studies have been made in which phloem tissue is carefully cut, and sap then exudes from it. Indeed, there are instances in which phloem exudate is commercially important. A palmyra palm in India may exude 11 liters of sap per day, and this sap may contain as much as 10% sucrose with 0.25% dissolved mineral solutes. Flow in such cases may be as rapid as 100 to 500 cm/hr through the sieve tubes. This rate of flow (faster than the normal rate in the intact plant) probably occurs because the pressure is relieved when the phloem elements are severed.

In recent years these studies have been greatly refined and extended by the use of "phloem-sucking" insects. Certain aphids insert their stylets directly and selectively into the sieve-tube elements of the phloem (Fig. 33–10). Sap is then forced into their bodies by the pressure in the sieve-tube elements; they exert no active sucking processes themselves. Exudate even collects on the surfaces of their bodies, where it is consumed by ants. An aphid is allowed to insert its stylet into the phloem, and then the insect is cut off, leaving the stylet in place. Sap continues to exude through the cut stylet for a considerable period of time (sometimes days). In one case in which rate of movement was about 100 cm/hr, it was calculated that the contents of 100 sieve-tube elements must be emptied every minute, implying that the entire contents of each sieve tube must move through the sieve plate in less than 1 sec. These rates are much more rapid than cytoplasmic streaming, as measured in cells where it is known to occur.

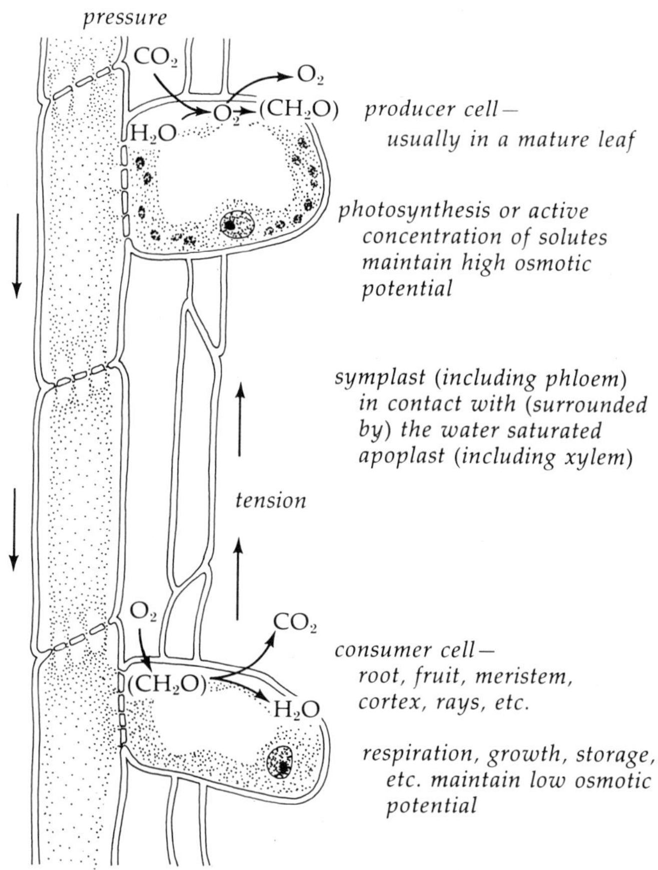

33–9 *The pressure-flow theory of Münch as it may be applied to the plant.*

Problems of Translocation Mechanisms

The remaining serious problem in understanding plant translocation is that rates are simply too fast to be easily accounted for by any present theory. At rates of 150 cm/hr, taking into account the fact that the sieve-plate openings make up only about 1/11 of the overall, cross-sectional area of the sieve tube, we must conclude that substances move through these openings at 1650 cm/hr. This estimate is based on several assumptions, but if it is even close, it is a phenomenal rate for such small openings. Yet in severed phloem elements the rate may be many times faster. Actually, such extreme

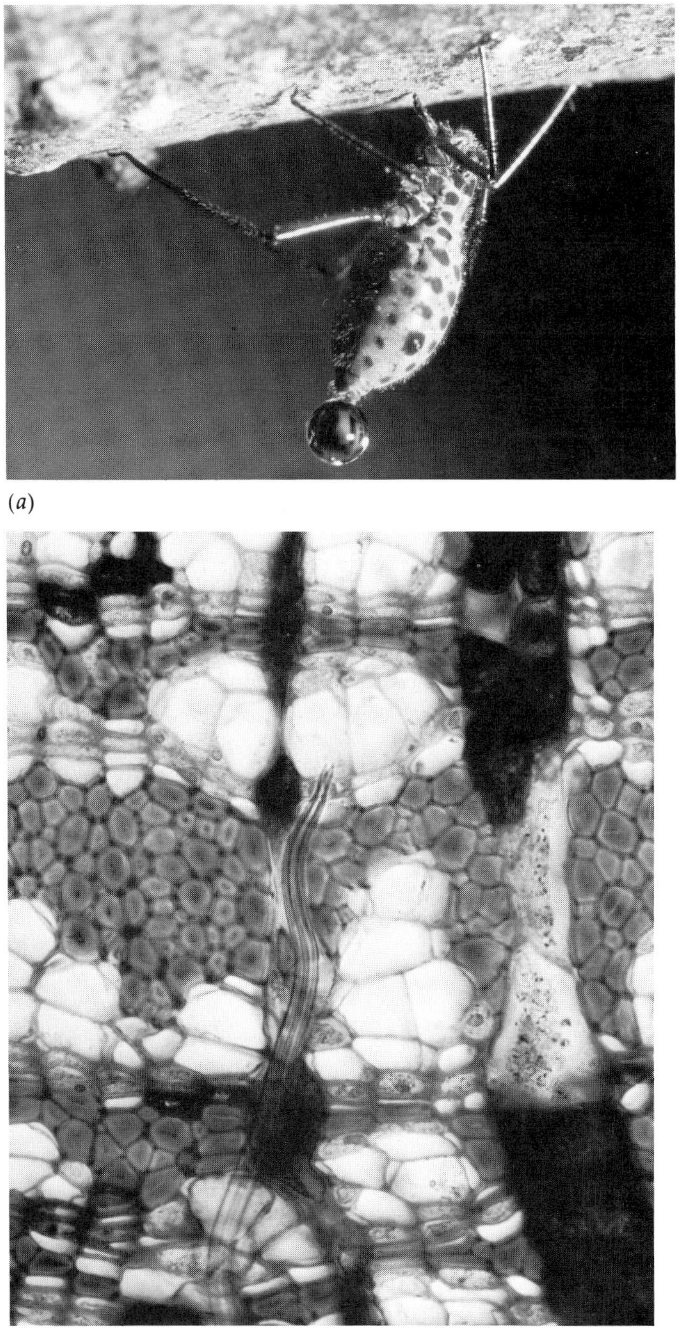

(a)

(b)

33–10 Study of translocation in the phloem by the use of aphids. (a) An aphid hanging upside down on a branch of a tree. Note the droplet of honeydew being exuded from the insect. (b) A cross section of the tree showing an aphid stylet that has penetrated to a sieve-tube element.

rates often result in plugging the sieve plates, probably by dislodging the slime bodies. Most controversy about proposed translocation mechanisms originates because the rates through the sieve plates seem impossibly high. Yet the rates themselves cannot be questioned. They are the demonstrable facts.

A number of studies have shown that the sieve-tube elements are functional only when they are alive and metabolizing. If they are killed or if respiration is inhibited in any way, movement through them ceases. Münch's model does not help us understand this requirement for metabolism. What does it all mean? Obviously, phloem cells are quite different from other plant cells. The nucleus is gone; the tonoplast is indistinct or missing; and there are usually companion cells present. Yet phloem cells are alive as contrasted with the dead xylem tubes.

What does the fact that phloem cells are different from other cells have to do with translocation? The problem that we face in studying the physiology of phloem cells is that the study procedures easily disturb cell functions. Preparation of sieve tubes for light and electron microscopy may well change these cells completely. Since liquids in the cells are under pressure, any kind of cutting or sectioning relieves this pressure and results in a movement of the substances inside.

Accumulation of substances such as sugars in the sieve-tube elements has been demonstrated in some systems. Apparently, sugars may be actively moved into the phloem cells near the source and then actively secreted at the sink. They may also be metabolized along the route. Here, at least, is one understandable aspect of the metabolic requirement. But does metabolism make the rapid rates possible in some manner not yet suspected? Very probably it does.

Importance of Bulk Flow

How important is a pressure-flow mechanism in the overall translocation of organic solutes? If natural and applied solutes usually move in the plant along with the sugar (the main osmotic component), we might feel that the mass flow mechanism is certainly an important one. If, on the other hand, substances move strictly according to *their own* concentration gradients, as might be accounted for by cytoplasmic streaming, we would be tempted to conclude that pressure flow is important only in certain conditions.

A great deal of work (extending over the last four decades) supports the conclusion that different substances nearly always move in the same direction in the phloem. Early work in the 1930s involving the movement of plant viruses was extensive and impressive. More recently, much experimentation has utilized the weed killer 2,4-D. Since this compound causes bending and other visible responses in the plant, it is easy to apply to one site and subsequently to find out where it goes in the plant. Since 1945 numerous researchers have shown that 2,4-D always follows the movement of sugars. If the leaf is darkened for a period of time and 2,4-D is applied, it will not move out until the leaf has either been allowed to photosynthesize or has been treated externally with sugar. Nor will 2,4-D move from one leaf into another that is exporting sugar.

This basic conclusion has been fortified by the use of many radioactive minerals and organic molecules, including various growth regulators, antibiotics, and sucrose. In some cases, it has been shown that such organic molecules will even move against their own concentration gradient as long as they are moving with the sugar. In other experiments, radioactive (tritiated) water and labeled solutes have moved together in the same direction. Often, the solutes in the water appear to move at different velocities, as though the sieve-tube elements were acting somewhat like a paper chromatogram. Some of the solutes may be retarded along the way, as dissolved materials are adsorbed and partitioned in chromatography. In some of these cases, mechanisms such as cytoplasmic streaming simply do not account for the observed results. Since diverse substances follow a mass-flow pattern, we are justified in concluding that pressure-flow mechanisms must be important in the plant.

Bidirectional Movement

In spite of the overwhelming number of experiments supporting mass flow, occasionally bidirectional movement in the phloem elements is reported. If it could be shown that one substance moves in one direction in a given phloem sieve tube while another substance is moving in the opposite direction, Münch's pressure-flow mechanism could not apply in that case. Actually, this has never been rigorously demonstrated, but there is considerable evidence for apparent bidirectional movement within a single vascular bundle.

Much of the older research utilized dyes, but in 1960 Orlin Biddulph and R. Cory at Washington State University approached the problem with considerable elegance by using radioactive tracers. An upper leaf on a bean plant was exposed to radioactive carbon dioxide in the light, producing radioactive sugar. Radioactive phosphorus was applied as phosphate to the leaf immediately below. After translocation had proceeded for some time, plants were harvested, frozen, and dried. The position of the strong-rayed phosphorus could then be determined on a radioautograph (frozen dried plant placed on a sheet of X-ray film—Fig. 33–11), shielding out the weak rays from the carbon with a layer of aluminum foil. After many months the phosphorus, with its short half-life, had decayed away, and the position of the carbon (long half-life) could then be determined. In a majority of trials, Biddulph and Cory found bidirectional movement of these two substances. In some cases, the bidirectional movement took place in separate bundles, and Münch's pressure-flow mechanism could always be used in explaining these cases. In other instances, bidirectional movement occurred in the same bundle, but individual phloem elements could not be resolved.

The bidirectional movement probably occurred in young phloem tissue, since cytoplasm is known to stream in immature phloem elements. Unidirectional movement was always most pronounced, but Biddulph and Cory concluded that both mechanisms must operate in the bean plant. There is no reason why they should not.

So far, we have concluded that both a pressure-flow mechanism and a cytoplasmic-streaming mechanism must work at least in some plants some of the time. Nevertheless, some scientists remain unconvinced. Recent studies with tritiated water indicate rates far too slow for mass flow. Yet, as we have seen, many earlier studies imply ample flow rates. Certainly, there is a rapid exchange between water molecules in the symplast and those in the apoplast. Possibly, only a few of the water molecules that leave the source in the leaf actually reach the sinks, and Biddulph suggests that the apparatus used was not sensitive enough to detect them. Sugar molecules would not exchange as readily.

Lateral movement from sieve tubes of sugars as well as water is not difficult to demonstrate with radioactive tracers, although it is much more restricted than longitudinal movement. Both lateral movement and its restricted nature are expected because of the symplast concept. If there are consuming cells in the neighbor-

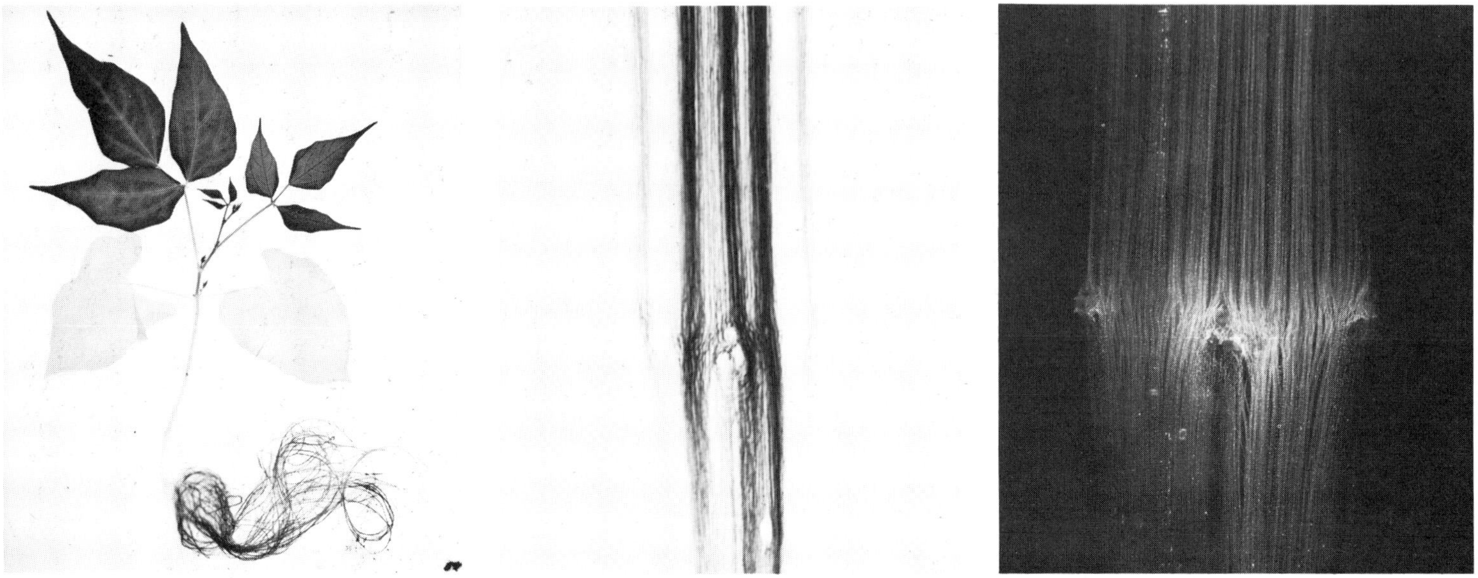

(a) (b) (c)

33–11 *Some examples of autoradiography. (a) The sulfate containing* ^{35}S *was applied to the roots 4 hr prior to harvest of the plant. Radioactive sulfur penetrated every part of the plant but tended to bypass the first set of mature leaves. This is typical for this element in bean plants. (b) Carbon dioxide containing* ^{14}C *was administered around the terminal leaflet of the first trifoliate leaf of a bean plant. The* ^{14}C *sucrose produced in photosynthesis travels toward the root through the phloem bundles leading directly from that leaflet (with little mixing with streams from other leaflets or other leaves). The autoradiograph has been enlarged considerably so that passage of the radioactive tracer in individual phloem tubes is apparent. (c) Here the bark was removed from the stem and immersed in aniline blue dye, which couples with callose (a material in the phloem cell). It is then illuminated with ultraviolet light so that the aniline blue fluoresces. This is photographed through a yellow-green filter. The phloem bundles through which the tracer is moving can be identified by comparing (b) and (c). The significance of this comparison is that translocation occurs not by a diffusive-type movement but by direct flow in discrete phloem bundles, as postulated by the Münch hypothesis of mass flow.*

hood of a sieve-tube element (such as phloem parenchyma, rays, cortex, or pith cells), the lateral movement through plasmodesmata would be a natural result of the phloem's function as a distribution system.

Some Additional Observation

There are several other complications in solute translocation. For example, phosphorus, potassium, sodium, chlorine, sulfur, and some nitrogen compounds are quite mobile within the plant, while calcium, magnesium, iron, and cobalt seem to be bound in plant tissue so that they are essentially immobile after they arrive in the leaf via the transpiration stream. Organic molecules also vary in their mobilities, as indicated in studies using radioactive weed killers and other growth regulators. Again, we are reminded of the analogy of the chromatogram.

In sugarcane, radioactive phosphorus moves out of the treated leaf through the phloem system into the root, back into the apoplast, up through the xylem, into the leaf, and around again a number of times. Other compounds are also capable of circulation within the plant, but calcium, magnesium, and cobalt usually do not circulate, although movements of calcium and mag-

33–12 *Response of the sensitive plant. The tip of one leaf is stimulated in such a manner (in this case, by using a flame) that the entire plant is not jarred. After 14 sec the petiole of that leaf has collapsed, and many of the leaflets have folded up. As the stimulus is transmitted back along the stem, other leaves collapse, and their leaflets fold up. Some of this activity can be observed even after $1\frac{1}{2}$ min.*

nesium have recently been detected. In 1628 William Harvey's discovery of circulation in animals was followed by a search for an equivalent circulation in plants. Since nothing comparable to the heart was found, it was concluded that such circulation did not exist in plants. Now we discover that if individual substances are considered, circulation of a sort does take place.

Mimosa, the sensitive plant, folds up its leaflets and its leaves upon contact. If one touches it with some object, the whole plant is likely to be disturbed, and all its leaves and leaflets fold up almost at once. If, however, one is extremely careful to contact only a portion of the plant (one way is by using a flame, as in Fig. 33–11), then the folding progresses from the point of contact back through the plant body. There is no nervous system in plants, but this response occurs because a chemical stimulus is translocated from the point of contact. The photographs in Figure 33–12 demonstrate this phenomenon and indicate that some activities of higher plants are easily noticeable.

Finally, how can we reconcile the ideas of high pressures in the symplast and high tensions in the apoplast? One might imagine that water potentials in the symplast would be much higher than in the apoplast, so water would move to the apoplast (xylem), causing the symplast (phloem) to wilt and pressure flow to cease. But that would be ignoring the effects of solutes on water potentials. Since the symplast usually does not wilt, it must contain solutes at concentrations exceeding those of the apoplast by enough to account for the pressure differences. Since the difference in measured osmotic potentials between the symplast and the apoplast seldom exceeds 30 bars, the pressure differences in nonwilted plants must be even less—but then that is a rather fantastic pressure difference compared with the fraction of 1 bar encountered in animals!

34

Plant Response to the Environment: Temperature and Dormancy

34–A Plants in temperate zones are able to "anticipate" the coming of winter. They respond to the shortening days and cooler nights of early autumn by initiating a complex series of metabolic and anatomical changes. Chlorophylls break down, and the carotenoid pigments become more evident. These fall colors are augmented by an increasing synthesis of anthocyanin pigments. Nitrogen and other minerals may be exported from the leaves into the stems and roots. A layer of cells (the abscission layer) begins to develop at the base of the petiole. When this development and the other metabolic changes are complete, the leaf falls from the branch. Living cells in the exposed buds and stems, which must survive the winter, undergo subtle changes (not yet fully understood) so that they will not be damaged by the impending frosts.

Let us look again at the plants in our acre of woodland. During the four-season cycle, the plants are subjected to a variety of environmental forces and conditions that, for the most part, are continually changing. Gravity is the most constant, but plants still respond to it—their stems grow away from the earth's center and their roots grow toward it. Light also influences the direction of growth of plant parts, but it influences many other things as well. Certainly, it is anything but constant. Intensities vary from the total darkness of the root environment on a moonless midnight to the sparkling brilliance of a clear day in June—or in snowy January. The quality varies also. Infrared rays come not only in sunlight but from water and carbon dioxide molecules in the atmosphere; their intensities depend on the temperatures of the gases. Ultraviolet sunlight intensities depend upon elevation, air pollution, time of day or year, and other factors. In the visible spectrum reds predominate just before sunrise and after sunset, but blue predominates in the open shade. Light from a blue sky is predominantly blue! Green and far-red outweigh red and blue after light has been filtered through the green (red- and blue-absorbing) leaves of the canopy. As we shall see in Chapters 35 and 36, plants respond to these subtle effects in numerous interesting ways.

Temperatures vary over a wide range, both during the daily cycle and throughout the year. On a clear night in midwinter, heat is radiated from the ecosystem into space, and temperatures may drop to tens of degrees below zero. Winter or summer, when the sun's rays strike plant, animal, or soil surfaces in the ecosystem, their energy is converted to heat, and temperatures rise (although the rise will be less if clouds intercept these rays). Cold air masses may move over the area from the far north, lowering temperatures even under bright skies. To an approximation, temperatures are predictable according to season, but we have all experienced the odd exceptions to this rule. Thus

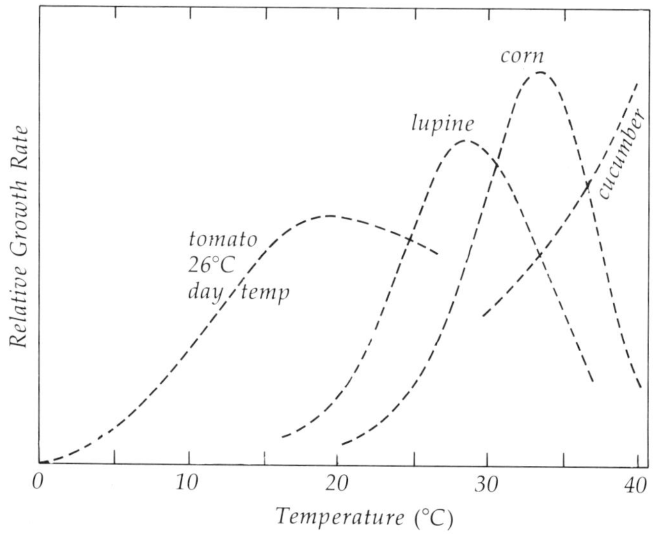

34–1 The effects of temperature on plant growth. Dotted lines show representative plant-growth curves as a function of temperature. Note that optimum temperatures are extremely dependent upon species.

plants must be adapted to survive the rigors of rapid and unpredictable temperature fluctuations. (In Chapter 32 we saw how some leaves may respond to their energy environments.) But plants can also utilize some of the more dependable temperature characteristics of the four seasons to adjust their life cycles. Weather is always cooling during fall and warming during spring, even when some days seem to go against the trend. A plant might average temperature conditions over a period of days or weeks and thereby detect the seasons.

Plants are also exposed to a great variety of moisture conditions throughout the year, and even such apparently minor factors as gas composition of the atmosphere vary according to location in our ecosystem or time of day. (Carbon dioxide may decrease significantly in still air and bright sunshine when photosynthesis is active.)

In this chapter we shall begin by emphasizing one aspect of the interaction between plants and environment: the response to temperature. In the following chapter we shall consider the response to light and the biological clock, a timing mechanism within organisms that synchronizes and prepares them for the natural cycles of changing environmental factors. Although our discussion will be built around the environmental factors of temperature and light, specific plant responses are clearly the principal topic. It will soon become apparent that a given plant response is seldom if ever influenced by a single environmental factor. In Chapter 36 we shall leave the environmental-factor approach and discuss a single plant response—the initiation of reproductive development—as it may be influenced by several environmental factors.

Temperature

In Chapter 5 we discussed the effects of temperature upon enzyme activity. Now we shall see that the responses of whole plants to varying temperatures are closely similar. There is often a minimum temperature below which the plant will not grow (e.g., 0 to 5°C); there is an optimum temperature at which the plant grows best (e.g., 28 to 35°C); and there is a maximum temperature above which the plant dies or at least will not grow (e.g., 45 to 52°C). These three temperatures, which vary from species to species, are often referred to as the cardinal temperatures (Fig. 34–1). Physiologists have long reasoned that the cardinal temperatures for plant growth are determined by the cardinal temperatures for critical enzyme systems in the plant, as controlled by the Q_{10} values for enzyme action and enzyme destruction (see Fig. 5–3). If life function is above all a matter of enzyme activity, then this is certainly a reasonable conclusion.

But it soon became apparent that such a simple explanation for plant response to temperature did not readily fit *all* of the observed facts. In the 1940s and early 1950s Frits Went carefully documented a phenomenon that was probably already at least suspected by most people who had had experience in growing plants under controlled temperatures (e.g., in a greenhouse). Went called the phenomenon thermoperiodism. He showed that growth of many species is best when day and night temperatures are alternated. For example, optimum growth temperatures for tomato plants (depending somewhat upon age and variety) may be 26°C during the day and 18°C during the night. Went studied responses of many species and varieties and compared them with average annual temperature patterns for several localities. A few plants (e.g., cocklebur) grow optimally under constant temperatures, but thermoperiodism is the rule for many plants. The African violet responds oppositely to most species, doing best when night temperatures are somewhat higher than day temperatures. Thus Went used to say that there were two classes of people in the

United States: those who can grow African violets and those who cannot. They correspond with those who leave their thermostats on a high setting at night, and those who lower the thermostat and open the windows!

Went studied a number of plants growing in the mountains and deserts of California. Some of these had unusually low maximum night temperatures. *Baeria charysostoma* is a small annual composite commonly seen during spring in mountain valleys, foothills, and occasionally the western Mojave Desert. Grown under short-day conditions, plants survive only for two months when the night temperature is 20°C. At lower temperatures they grow for at least 100 days. They die rapidly at night temperatures of 26°C—ideal for African violets! The concept of 26°C as a lethal temperature is surprising, to say the least. The *Baeria* plants flourish when the day temperature is well above 26°C, provided only that the night temperature is low enough. Other species acted similarly in Went's experiments.

What kind of mechanisms could account for plant death at this temperature? Findings such as Went's indicate that temperature may act upon the plant in some manner other than a mere control of enzymatic reaction rates, although temperature will clearly influence enzymatic reactions even if it does other things as well. Could there be direct effects of temperature on the DNA–RNA–protein system? Or are there as yet unknown temperature transducers? Exposure of certain species to some temperature (frequently a low temperature, sometimes a high temperature) may be necessary to trigger some morphogenetic response such as the breaking of dormancy or the initiation of flowers. We shall consider the effects of temperature (and other interacting factors as well) upon dormancy in this chapter and upon reproductive development in Chapter 36.

Bud Dormancy

Winter dormancy (often called rest), displayed by the buds of many species in temperate regions, is broken by prolonged exposure to low temperatures. Days to months may be required at temperatures below 10°C. In a study with fruit trees at Utah State University, 5 to 7°C was optimal and more effective than 0°C. Considerable work has been done with many species of commercial fruit trees to determine the minimum period of dormancy. Apples, for example, depending upon the variety, may require 1000 to 1400 hours below 7°C to break their dormancy. Peach varieties have been selected for a shorter chilling requirement, allowing certain cultivars to be grown farther south where the winters are warmer.

The effects of chilling upon the breaking of dormancy are not translocated within the plant but appear to be localized in individual buds. A dormant lilac bush, for example, may be placed in the greenhouse with one branch protruding outside through a small hole in the wall. The branch exposed to the low temperatures of winter leafs out early in the spring, while the rest of the bush inside the greenhouse remains dormant.

This is one example of the plant responses to low temperatures studied by plant physiologists. But there are several of these responses, including the breaking of seed dormancy, a promotion of flowering, induction of tuber formation in potatoes, various responses such as formation of flowers in several bulbs and corms, and the growth promotion by a low (or high) night temperature (thermoperiodism). The same mechanism may possibly be operating within the plant in a number of these different situations. Discovering this mechanism has been especially challenging.

In our biochemical experience low temperatures invariably slow things down. So how can some process be *promoted* by decreasing the temperature? As we delve deeper into the topic, we encounter (in a manner not unfamiliar by now) an increasing number of complications. To begin with, not all dormant buds respond to low temperatures. A midsummer dormancy occurs in some species, especially evergreens, during which the stems cease to elongate for a period of time. This is typically broken by exposure to more long days of summer, even though temperatures remain high. Actually, even winter dormancy can be broken by long days in several species of trees such as beech, birch, larch, yellow poplar, sweetgum, and red oak. Except for beech these species respond to either low temperatures or long days. In a few other trees, to break dormancy the exposure to low temperature must be followed by an exposure to long days. Plant physiologists had begun to feel safe with the generalization that only expanded leaves respond to changing day length and that buds might therefore be expected to respond only to low temperature. But, in the case of the trees just mentioned, it is the buds (probably the immature leaves inside the bud scales) that respond to the long days. Apparently, enough light penetrates the bud scales to account for this response to day length, which has been recognized only during the last two decades.

Further complications become apparent when we consider the induction of bud dormancy in response to short days. To demonstrate this clearly, it is necessary to keep temperatures relatively high (e.g., 21 to 27°C). At lower temperatures plants often cease to grow even under long days. In nature lower temperatures and shorter days typically go hand in hand, so the plant is probably responding to both factors. A few species do not respond to the short days but require the low temperatures.

Development of bud dormancy and appearance of fall color in leaves often occur simultaneously, but the two can be separated. In some plants short days can cause buds to become dormant in midsummer, although leaves remain green and actively photosynthesizing. In the autumn the combination of low temperatures and short days not only results in bud dormancy but also senescence (aging) of the leaves. The photosynthetic mechanism, including the chlorophyll molecules, breaks down, and a number of mineral elements are exported from the leaf back into the stem tissue. Low temperatures, the high light intensities frequently encountered on clear fall days, and perhaps the developing mineral deficiency in the leaf shift metabolic pathways in the direction of both sugar synthesis and ultimately the synthesis of water-soluble pigments called anthocyanins. With the breakdown of the chlorophyll, the orange and red carotenoid pigments are exposed to view, and the developing anthocyanin pigments (also yellow, orange, and red) add to the color of the autumn leaves, so spectacular in our woodland acre. Clearly, we are again discussing a rather complex set of events influenced by a number of environmental factors.

To return to the breaking of winter dormancy, several chemical treatments can be used to break dormancy "artificially." The compound 2-chloroethanol has been successfully used for many years. It is applied in a vapor form to break dormancy of fruit trees. Immersion of the plant or plant part in a warm water bath (40 to 55°C), often for only 15 sec, sometimes results in the breaking of dormancy. In recent years it has been discovered that applied gibberellins break the dormancy of many deciduous plants. As we discussed in Chapter 31, abscisic acid induces dormancy in several species. We have already seen that both gibberellins and abscisic acid are also effective in other plant responses, including some responses to low temperature. Gibberellins break the dormancy of certain seeds and replace the low temperature requirement for flowering in some species.

Just what is the mechanism of the low temperature response in the breaking of winter dormancy? Because of the response to hormonal treatment, it is tempting to think that the low temperature exposure results in changes in the internal hormone levels of the buds (or seeds, bulbs, corms, etc.), that these changes in hormone levels will in turn bring about the activation or repression of certain genes, and that these then account for the growth response. Several workers such as David Walker at Utah State University have observed changes in hormone levels during the breaking of dormancy in buds and seeds.

Seed Dormancy

Seed dormancy proves to be even more complex than bud dormancy. The breaking of seed dormancy is influenced by a host of environmental factors, and recent research has turned up a number of responses to growth regulators as well. We shall begin by examining the response of dormant seeds to low temperature, but we shall also consider some of the other factors known to influence germination of seeds.

Just how long will a seed remain viable; that is, how long can a seed remain alive and able to germinate? A seed is a remarkable object. It lies in its dormant or quiescent condition until conditions are right and germination is appropriate. One often hears about corn or wheat seeds, taken from the tombs of Egypt or the Pueblos of the southwestern United States, that, when planted, produce normal plants—even though the seeds are thousands of years old. Apparently, there is little truth in such stories. Some seeds found in Indian caves in the southwestern United States had recently been placed there by packrats.

A number of experiments have been carried out or at least initiated in an attempt to determine seed life span. Seeds are stored in moist or dry and warm or cold conditions. At intervals (extending over many years) samples are removed for the determination of viability. Some experiments have been in progress for more than half a century, and it is also possible to test even older seeds stored with whole plant specimens in old herbaria (e.g., in Paris). We can now begin to draw some conclusions. It appears that only a few seeds can exceed a maximum life span of about 50 years. Some such as orchids, tea, and certain willow seeds are known to have an extremely short life span, measurable in days. A 10-year life span is common, and a few seeds live up to 50 years. The life spans for wheat and corn are in the neighborhood of 30 to 50 years. Life spans are usually longest under dry, cool conditions, but some species live longest stored under water or in moist soil.

Two authenticated cases of extremely long-lived seeds are now known. One involves the large, very hard seed of the lotus plant in China. Lotus seeds are found deeply embedded in layers of peat, and their age can be determined by their position in the peat bog and by radioactive carbon dating. Such seeds, treated to break the seed coat and then planted, have been known to germinate, even though they were at least 800 and perhaps more than 2000 years old. The other case is that of an arctic lupine found in the central Yukon in a lemming burrow in permanently frozen silt from the Pleistocene age. Viable seeds were at least 10,000 years old!

If seeds, like winter buds, require an extended period at low temperatures to break dormancy, then the lupine seed from the Yukon certainly got it. Actually, many seeds do require a period of cold to break dormancy. A number of rosaceous seeds such as those of the common, temperate zone fruit trees (apple, peach, pear, cherry) provide good examples. This requirement is of obvious survival value, since such seeds will not germinate in the fall but only the following spring. They must be treated with low temperatures (0 to 10°C) while they are in the moist condition, usually for a number of months. In horticultural practice layers of flats containing the moist seed are left outside during the winter or perhaps placed in a cold storage room. This process is called stratification. Various intervals of time are required to bring about stratification, depending upon the species (e.g., 6 weeks for peaches). Temperature requirements are similar to those for breaking of bud dormancy (e.g., 0 to 10°C). Oxygen must be present if the stratification treatment is to be effective.

In many seeds virtually no anatomical changes can be observed during stratification, but in a few such changes have now been described. In cherry seeds a clear-cut elongation of the embryo occurs during stratification (but not in control seeds held in the moist condition at room temperature). Chemical changes also occur. Shifts in the concentrations of nitrogen and phosphorus between the various parts of the seed and changes in amino acids, organic acids, and enzymes have been observed. Growth regulator changes, discussed previously, also take place. For example, it was found in one study that three different promoters in Went's *Avena* coleoptile test increased during stratification, while one inhibitor in this test decreased. All of this supports the tentative conclusion drawn above that low temperature (and other environmental) responses may be mediated through changes in endogenous growth regulators.

In many seeds germination does not occur until the seed coat has been broken. The process of breaking the coat is called scarification. The hard seed coat may serve to keep water or oxygen out of the seed, or in a few cases, it may be so hard that it keeps the embryo in. Many seeds have such coats, including a number of legumes and mallows (such as hollyhocks). Seeds of the black locust (a legume) are sometimes used in classroom demonstrations. Intact seeds will not germinate when planted, but seeds passed around the class in a bottle that is shaken by each student become scarified and germinate. When white sweet clover seeds are shaken, a corky plug is dislodged from the strophiolar cleft, allowing penetration of water.

Scarification may be done in the laboratory by nicking the seed with a knife or a file, by treating it with concentrated sulfuric acid or organic solvents (which apparently dissolve certain fatty substances that otherwise keep the water out), or even by boiling it for a short time. Cotton seeds are soaked for minutes to an hour in concentrated sulfuric acid to promote germination, an exception to the rule that domestic seeds do not have dormancy mechanisms. Most domestic seeds have been selected for easy germination.

In nature the requirement for scarification is of obvious value. If a seed must be scarified, it cannot germinate when it first drops from the mother plant; it must wait until the forces of nature have had time to break the seed coat. Hence germination will not occur in the fall but only during the following spring, after scarification has occurred and when temperature conditions are favorable. Scarification in nature is brought about by a number of forces. In some instances, a bird may accomplish the task by passing seeds through its alimentary canal, and in a few cases (e.g., the frequently burned chaparral of Mediterranean climates), fire is required to break the seed coat, thus ensuring the rapid germination of many seeds as soon as sufficient moisture is available following the fire. Most commonly, scarification is brought about by the normal processes of decay. In one carefully studied instance, a specific fungus was required for scarification of shadscale (*Atriplex confertifolia*) seed in the Utah desert (see Essay 39–2). This appears to be an example of mutualism between two organisms: the seed provided a substrate for growth of the fungus, which in turn allowed germination of the seed. The smoke tree of our southwestern deserts germinates only after the seeds have been washed down the rocky desert gulleys by a cloudburst. When carried too far in the flood, they are ground to pieces. If they are not carried far enough, scarification is incomplete, and they will not germinate. A cloudburst sufficient to carry the seeds just the right distance provides ample moisture in the soil for the subsequent growth of the seedlings.

Frits Went discovered that many desert seeds do not germinate until they have been leached extensively by running water. Again, the value of such a requirement under natural conditions is not difficult to discern. Leaching apparently removes an inhibitor of germination, and a storm heavy enough to remove such an inhibitor wets the ground to a considerable depth. Some seeds require one half inch of rainfall or more before their inhibitors are removed. A lesser amount does not cause germination, nor do a number of short rains separated by intervals of time. Apparently, the level of the inhibitor is reestablished in the seed following a short rain and is maintained until enough rain comes at one time to remove the inhibitor and thoroughly wet the soil.

In a few instances, the chemical nature of the inhibitor has been determined. The shadscale in the Utah desert contains sodium chloride (common table salt) in quantities sufficient to inhibit its germination until it has been leached out by the spring rains. This is in addition to the requirement for scarification by the specific fungus. Organic compounds such as mustard oils, unsaturated lactones, alkaloids, certain organic acids, and molecules breaking down to release ammonia or cyanide provide other examples. The seeds in a tomato, orange, or melon normally germinate readily when planted in the ground. Seeds are certainly subjected to ideal moisture and temperature conditions in these fruits, and one might ask why they fail to germinate. (Most of us have occasionally observed a germinating orange seed in the fruit, but such occurrences are relatively rare.) It can be demonstrated that these seeds remain dormant sometimes in response to chemical substances present in the fruits themselves or sometimes because osmotic potentials of the fruit juices are too negative to permit germination.

In many instances, mature plants produce compounds that inhibit the germination of seeds (see Chapter 40). This is a kind of chemical warfare in which one plant is able to limit the competition (for soil nutrients, moisture, and growing space) of other plants by the production of specific chemical substances. Usually, these substances are produced in the leaves and only reach the soil after the leaves fall to the ground and decay. Decaying organic matter, however, typically produces nitrates, which are known to *promote* germination of many seeds. This would also be of ecological value, since nitrate is often limiting to plant growth, and an ample supply might give the young seedlings an added advantage.

Another dormancy mechanism appears to be the time it takes for the embryo to complete its development after a seed has been separated from the mother plant. That is, some seeds (notably holly, *Ginkgo*, and *Gnetum*) do not germinate when they are first collected from the mother plant simply because they are not yet "ripe." One way to overcome the dormancy of these seeds is to allow sufficient time for the embryo to become mature.

Dormancy mechanisms such as the requirements for cold, scarification or complete embryo development prolong the period before germination, when it is said that the seed is undergoing afterripening. The term is certainly appropriate in relation to seeds with delayed embryo development, but perhaps it is of less value to us now that we are gaining some understanding of the other mechanisms involved. Apparently, afterripening is often concerned (as we have seen) with changes in levels of endogenous growth regulators. Incidentally, it should now be apparent that a given seed may include a number of dormancy mechanisms. A requirement for cold may be in addition to a requirement for scarification or leaching, for example.

Some seeds require special light treatments for germination. A number are promoted by light, but a few are inhibited by it. Some seeds require a certain *quantity* of light (quantity being measured by multiplying intensity $\times$ exposure time). A few other seeds respond to the *duration* of the light treatment; that is, they respond to short days or to long days much as some plants respond to these conditions by flowering (see Chapter 36). Certain birch seeds are long-day seeds, and short-day seeds are also known (e.g., *Ve-*

ronica persica). Seeds having some kind of light requirement for germination make particularly good weeds. They may lie dormant in the ground for years until they are brought into the light by plowing. The study of light requirements for germination of lettuce seeds led to an important discovery about the responses of plants to light—the phytochrome system. We shall discuss it and some of the characteristics of light-controlled germination in the next chapter.

35

Light and the Biological Clock

Consider the events of our woodland ecosystem that occur during a typical growing season and that are dependent upon light. It is late winter in our woodland acre, and little is happening, except for a few spring ephemerals that are growing and turning green under the snow (probably in response to the weak intensities of light penetrating the snow cover). As the snow melts and temperatures go above freezing for at least a few hours each day, these early herbs will flower and flourish for a brief interval. At this same time a few seeds are beginning to germinate—some of them in response to light and warming temperatures. Others are germinating in the soil or beneath the layer of leaf litter where it is dark. The seedlings growing in the dark are quite different from those in the light: they are nearly colorless; their leaves remain quite small; roots aren't developing much; and their stems are elongating rapidly. They are said to be etiolated. Many of the dicots have the ends of their shoots bent into a hook, which does the pushing through the soil and leaf litter, thus protecting the delicate leaves at the end of the stem (the apical bud). Grass leaves are protected by the coleoptile. When these seedlings reach the surface and the light of day, their mode of growth changes markedly: they turn green; leaves begin to expand; roots develop more rapidly; stem internodes and coleoptiles are inhibited in their elongation; and plumular hooks straighten. All of these things occur in response to light. If the tender young stems find themselves under an overhanging rock or log, they bend phototropically toward the light. Exposed roots grow away from the light (probably into the soil).

Light and Its Effects on Plant Growth

35–A Plants grown in the dark are strikingly different from those grown in the light. Leaves are greatly reduced in size and nearly colorless, stems (internodes) are elongated, roots are reduced, and dicots often have a hook at the end of the stem.

With the coming of summer, plants continue to respond in numerous ways to the light environment, but they are often modified in their responses by temperature. Leaves and stems are oriented in relation to the rays of the sun (and, of course, to gravity). Plant growth form is also determined by light intensity, being short and bushy when intensities are high and succulent and elongated under low intensities. Shade leaves are thin, broad, and dark green, while leaves growing in the sunlight have extra layers of palisade tissue or longer palisade cells, are smaller in area, and are a light green. Sun leaves require higher light intensities for photosynthetic saturation.

Leaf positions change throughout the day, being typically rather horizontal at noon and vertical at midnight. If placed under constant controlled conditions, these sleep movements continue with a daily rhythm (usually not quite equal to 24 hr). If their rhythmic periods are not equal to 24 hr, why don't the leaf movements get out of synchronization with the daily cycle of light and darkness? Because the "biological clocks" that control them are reset each day at dawn and sunset. These clocks control numerous other processes, many of which are influenced by light.

Plants even respond to the length of their daily exposure to light. They manifest this in many ways, including their growth form and the time during the year when they form flowers, bulbs, or other storage organs. This ability to detect day length is another manifestation of the biological clock.

Fruit development is often influenced by light in many ways. Formation of the brightly colored pigments in fruit skins is a good example.

Finally, as fall approaches, many changes occur in the members of our ecosystem in response to their light environment. Leaves turn the colors of autumn and then fall from the trees; annuals senesce and die; buds of perennials become dormant; and protoplasm changes so that it can tolerate the freezing temperatures of winter.

In this chapter we shall consider light as an environmental factor affecting plant growth. As in the case of temperatures, we shall again find ourselves dealing with specific plant *processes,* and it will soon become apparent that these are influenced by other environmental factors besides light. Nevertheless, the factor of light serves to pull all these plant functions—including the biological clock—together.

We have already discussed the physics of light in an introductory way (see Chapters 5 and 6). You'll remember that light is propagated with a wave motion: short wavelengths are typical of the ultraviolet and blue parts of the spectrum; long wavelengths occur in the orange, red, and infrared; and intermediate wavelengths occur in the green and yellow parts of the spectrum. At the same time, light has a particulate nature with quanta or photons having individual energy values highest for short wavelengths and lowest for long wavelengths. In a photochemical reaction such as photosynthesis, absorption of a photon depends upon the ability of the absorbing molecule (the pigment) to undergo some energy change exactly equivalent to the energy of the absorbed photon. By undergoing this change, the molecule becomes excited and capable of taking part in a photochemical reaction (or of reemitting a photon in fluorescence or falling back to the ground state and releasing the energy as heat). The First Law of Photochemistry states that light, to be effective, must be absorbed. This leads to the concept of the action spectrum, in which a given response to light is plotted as a function of the wavelength. The resulting curve is then compared with the absorption spectra of pigments thought to take part in the photochemical response. (All of this was discussed in some detail in relation to photosynthesis.)

Kinds of Responses to Light

Actually, there are two kinds of plant responses to light. In the first, light energy is converted directly into chemical bond energy, which is essentially the end of the matter (as far as the light response is concerned). Photosynthesis is the classical example. In the second kind of response, absorption of the light energy acts as a trigger to bring about the change—one of the developmental events discussed in the introduction. Sometimes the growth response triggered by the light requires far more energy than that provided by the light itself; that is, there is an amplification of response. Of the five broad groups of photobiological responses we will discuss in the next five sections, the second, third, fourth, and possibly the fifth are the triggered type, and most are amplified.

The Responses of Chlorophyll or Its Chemical Relatives

As we have seen, photosynthesis is the prime example of a direct conversion of light into chemical bond energy. Furthermore, this chemical bond energy ultimately supplies the entire biosphere. In many angiosperms the synthesis of chlorophyll also depends upon light. A compound known as protochlorophyll is converted to chlorophyll by the action of light. Thus, an angiosperm seedling dug from the dark soil in our woodland acre appears colorless or only slightly greenish-yellow when first brought into the light, but after a few hours, it becomes green. In contrast, a pine seedling germinated in the dark is green when it is first brought into the light. Photochemical reactions are normally temperature-insensitive, so it is not surprising that the direct energy conversion of protochlorophyll to chlorophyll will occur in some angiosperms (e.g., spring ephemerals) under the snow. The surprising thing is that the conversion is as effective as it is under 30 to 40 cm of snow, where light intensities have dropped to very low levels.

Blue Light and Phototropism

In Chapter 30 we discussed the tropisms, including phototropism. It was mentioned that the bending of coleoptiles and stems toward the light is an auxin response to blue light and not to red light. This basic observation was made as early as 1909 by A. H. Blaauw in the Netherlands. Good action spectra were not obtained until the 1930s, and refined action spectra were not obtained until the late 1950s. Figure 30–8 shows a detailed action spectrum for phototropism of oat coleoptiles. The high effectiveness of blue light suggests that a yellow pigment must be the effective photoreceptor in this response. Two common yellow pigments of plants are carotenes and riboflavin. As Figure 30–8 indicates, the action spectrum tends to match the absorption spectrum of a carotene in the blue region, but it matches that of riboflavin in the ultraviolet region. It has recently been possible to reduce considerably the carotene of coleoptiles without reducing either the riboflavin or the phototropic response. This experiment favors riboflavin as the photoreceptor pigment in phototropism.

As in most cases in plant physiology, the more we study, the more complex the picture becomes. Winslow Briggs and others have been studying phototropism during recent years by applying some of the most sophisticated techniques of biophysics. They have looked into the question of reciprocity in phototropism. The Law of Reciprocity says that it is neither the intensity nor the duration of exposure to light that determines a photochemical response; rather, it is the total quantity of light or the number of quanta absorbed.*

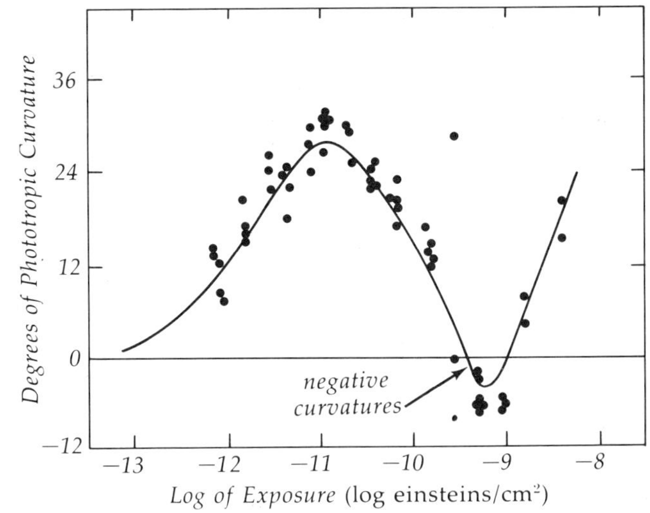

35–1 Phototropic responses (first positive, negative, and second positive curvatures) of oat coleoptiles, caused by increasing energies of blue light at 4,358 A. Intensity was 1.4 × 10^{-12} einsteins/cm²/sec, and exposure times varied to give the total exposures indicated along the abscissa. When the experiment was repeated with different intensities, the left-hand, ascending part of the curve (first positive curvature) remained identical, but the right-hand parts of the curve varied.

That is, a long exposure at low intensities should provide exactly the same response as a short exposure at high intensities, provided both supply the same number of quanta (the same total energy). Briggs studied the degree of phototropic bending as a function of the total energy supplied unilaterally (from a point to one side of the coleoptile). Increasing energy causes increasing phototropic curvature up to a point (Fig. 35–1). Beyond this point, increasing energy decreases the phototropic curvature, until finally no curvature occurs at all. Increasing energy further actually results in a negative curvature (curvature of the coleoptile *away* from the light). Then, with continuing increase in energy, curvature begins to occur again toward the source of the light. It is possible to speak of first positive curvature, first negative curvature, and second positive curvature. Briggs found that reciprocity held nicely for first positive curvature but that it did not hold for second positive curvature. Furthermore, red light given before the blue light greatly reduced the sensitivity to first positive and negative curvature but increased considerably the sensitivity of the system controlling second positive curvature. But just how

*The term reciprocity has the following derivation: If response (R) is proportional to intensity (I) times duration (D), we can write: $R \approx ID$. By a simple algebraic transformation, intensity is seen to be proportional to the *reciprocal* of duration ($I \approx R/D$) or vice versa ($D \approx R/I$).

does absorption of blue or ultraviolet light by carotene or riboflavin result in the movement of auxin toward the dark side of a unilaterally illuminated coleoptile? When our questions are pushed to this level, things can get quite complex, and we remain without answers.

It is important to note that green leafy stems, as well as coleoptiles, respond phototropically. In a sunflower, for example, it has been shown that the leaves apparently supply different quantities of auxin (with most auxin on the dark side), depending upon the direction of illumination.

The Responses Controlled by the Photomorphogenetic Pigment System: Phytochrome

There are a number of plant-growth responses that all exhibit a similar action spectrum. The responses are remarkable because they display a unique property of reversibility. Orange-red light between 620 and 670 nm (with a peak at about 660 nm) causes a given effect, while far-red light from 710 to 750 nm (peak at about 730 nm) nullifies or reverses the effect. Effective intensities are quite low. Figure 35–2 shows a standard action spectrum for these responses, as well as a number of examples of specific responses. The pigment that absorbs the light and controls these phenomena has been termed phytochrome.

This field is an active area of research today. It is especially intriguing to wonder how such varied responses could all be controlled by the same fundamental biochemical system. Furthermore, phytochrome has been extracted and partially characterized. We shall outline a few of the more important responses.

Germination of certain seeds. The red-far-red system was initially observed in the germination of the Grand Rapids variety of lettuce. A large spectroscope was built at the United States Department of Agriculture Plant Industry Laboratory at Beltsville, Maryland. The light source was a powerful carbon arc lamp, making it possible to get relatively high light intensities (even after the beam had been focused and then broken into its component colors by passing it through a prism). Grand Rapids lettuce seeds were known to require light for germination. They were moistened and then placed along the projected spectrum. Seeds in the green part of the spectrum germinated only about 20% — the same as seeds germinated in the dark. Seeds in the orange-red part of the spectrum, however,

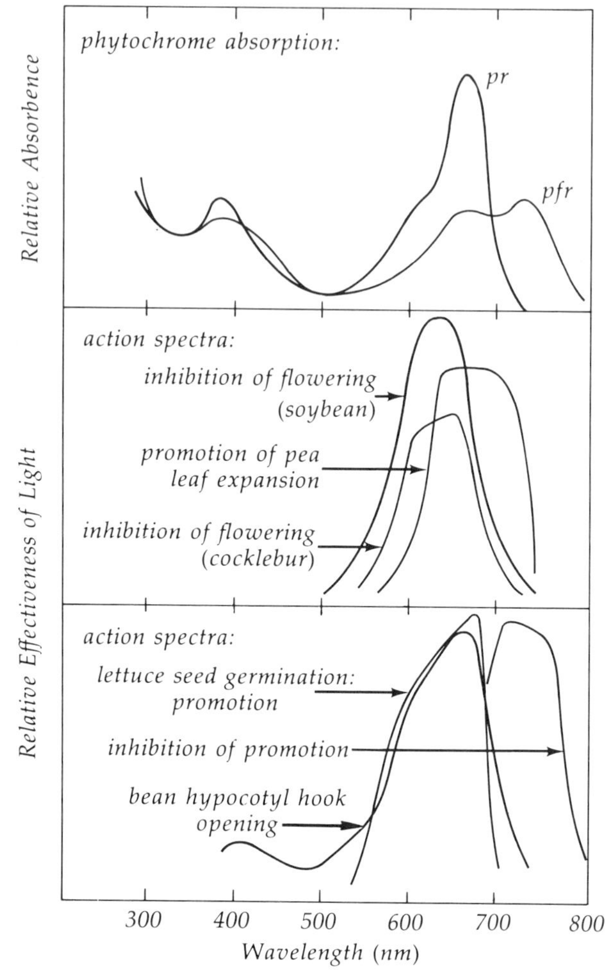

35–2 A comparison of the absorption spectrum of both forms of phytochrome with action spectra for various physiological processes.

germinated nearly 100%. Samples of seeds in the far-red part of the spectrum germinated significantly less than 20%. Hence, germination was promoted by orange-red and inhibited by far-red light.

This observation was first made by L. H. Flint and E. D. McAllister in 1937, but it was not until the beginning of the 1950s that Sterling Hendricks (see Essay 35–1), Harry Borthwick, and their colleagues at Beltsville had the idea that far-red light might reverse the effects of red light. When an experiment was performed, such a reversal did occur. Red light caused nearly 100% germination, but if seeds illuminated with red light were subsequently illuminated with far-red light, germination dropped to less than 20%. If, how-

35–3 Reversal of lettuce seed germination with red and far-red light. Letters under individual petri dishes indicate sequence and number of red and far-red illuminations. Red exposures were for 1 min, and far-red for 4 min. If the last exposure is to red light, seeds germinate; if to far-red, they remain dormant. Temperature during the half hour required to complete the treatments was 7° C; at all other times, it was 19° C.

ever, seeds treated with red and then far-red were subsequently treated with red, germination went back up to nearly 100%. This process was reversible many times (Fig. 35–3). If the last illumination were red light, germination was high; if the last illumination were far-red light, germination was low.

The researchers at Beltsville postulated that a single pigment system was in control of germination. When one form of this pigment system, *Pr,* was illuminated with orange-red light, it shifted into another form, *Pfr,* as illustrated in this reaction.

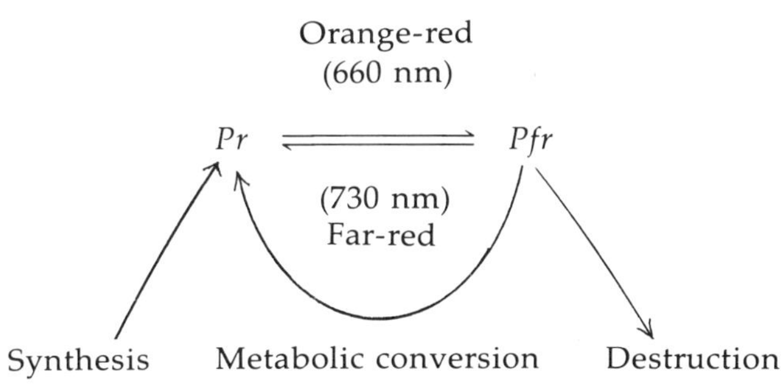

The *Pfr* absorbed far-red light, converting it back to *Pr.* Various lines of reasoning based on these and other experiments indicated that the *Pfr* is converted metabolically within the plant back to *Pr,* even in darkness. But this process is not instantaneous. It requires a certain time interval (many hours in the case of lettuce seeds).

A few years later, still another possibility became apparent. Rather than being converted back to *Pr, Pfr* might be metabolically destroyed and *Pr* synthesized anew. This is apparently what happens in several plants, proving again that nature is often more complicated than we first imagine. Anyway, if *Pfr* were essential for the germination process, then this scheme would explain the experimental results with lettuce seeds.

Since these initial studies, much more has been discovered about the germination response to red light. Many details relating to time and temperatures and various other treatments have been worked out, and many other seeds—even spores of ferns and mosses—that also germinate in response to *Pfr* have been discovered.

Flowering in response to daylength. In Chapter 36 phytochrome is discussed in relation to the flowering process. The effect on flowering was tested at Beltsville, immediately after the effect on lettuce-seed germination was discovered.

Formation of certain plant pigments. Other phytochrome-controlled responses were also studied

Essay 35–1 The Discovery of Phytochrome

The discovery of a new process in the biological world is always an exciting business, and when it proves to be an important one, it may be epoch making. This was true for the discovery of phytochrome. We asked Sterling B. Hendricks to tell us of the events associated with this discovery.

"In 1945, Harry A. Borthwick, Marion W. Parker, and I set out to find something about how plants recognize day lengths. Our method was to note changes in flowering induced by breaking long nights with periods of light of various wavelengths and intensities—or, more exactly, to measure action spectra. A low energy of 660 nm (red) light proved most effective in preventing flowering of the short-day soybean and cocklebur and in inducing flowering of the long-day barley and henbane. These oppositely responding plants had the same action spectrum, in all details, for flowering change. The near infrared, 700 to 900 nm, appeared to be ineffective.

Instead of elaborating the finding about flowering, we (with Frits W. Went) next chose to measure action spectra for inhibition of etiolation (the elongation of stems and restriction of leaf expansion that occur in the dark). Suppression of stem lengthening of barley and enhancement of pea leaf size by light gave the flowering action spectrum. It was exciting to know that such diverse displays were related in initial cause.

A need for light in germination of some seed had been known for about a century. Our first measurements (in 1952 with Eben H. Toole and Vivian K. Toole) were on lettuce seed, which Lewis H. Flint and E. D. McAllister had studied 17 years earlier. Again, 660 nm light was most effective for promoting germination. Seed placed in the 700 to 800 nm region of the spectrum, however, germinated less than the 20% value of those in darkness (controls). Flint and McAllister had found germination to be suppressed in this region when seed were initially potentiated by light to 50% germination. The differences in experimentation had no immediate import to us.

But one day in 1952, during an action spectrum measurement, the thought came that the 700 to 800 nm region had not been correctly tested in flowering controls. If plants had been first exposed to 660 nm light to potentiate the flowering change, instead of being taken directly from darkness, the 700 to 800 nm region might change them back to their dark condition. The 700 to 800 nm region, in truth, might cause a response similar

to darkness rather than being without effect as we had earlier supposed. This region when tested with lettuce seed did indeed suppress potentiated germination from a high to a low value, with maximum effectiveness near 730 nm. The reversal of the potentiated response with short-day plants for flowering also was well enough borne out to support a generalization.

The photoreversibility was the touchstone for deeper understanding of the equivalence of initial action in diverse phenomena. It also led to eventual isolation of the pigment phytochrome. The day in 1952, though, held a further nuance that is not so widely appreciated, nor was it quickly grasped by us. We had done quite a different experiment than had Flint and McAllister. The lettuce seed in our experiments were exposed for short times to low total energy of moderate intensity, while they used high total energy given by continuous exposure at low intensity. The photoreversibility was thus quickly and convincingly evident to us but not to them—as was its implication of determination at the molecular level by quick change in form of a light-absorbing pigment.

In an earlier paragraph, the reversibility of flowering was expressed as '. . . well enough borne out . . . The reversal, in truth, was rather poor. The more we tried to enhance it, by increasing the energy through long exposure, the poorer it became. In fact, 730 nm radiation at low intensity for an hour inhibited cocklebur flowering as effectively as did low-energy exposure at 660 nm. This was recognition of what is now called the 'high-energy reaction.' Flint and McAllister were likely dealing with this reaction in lettuce seed rather than with the simple reversibility of phytochrome. The high-energy reaction, which has many facets in nature, is also thought to arise in part from phytochrome, but in a more complex way than by photoreversibility alone. Finally, our initial objective—how plants recognize day length (or night length)—is somewhat a will-o'-the-wisp. Phytochrome action is involved, but what the action might be and how it ties in with endogenous biological rhythm is still under vigorous debate."

---

at Beltsville. One is the synthesis of certain anthocyanins, such as those produced in autumn leaves, in the skins of apples and turnips, in flower petals, and in various other plant organs. Carotenoids are also synthesized in the skin of tomato fruits in response to red light. In several of these systems, it is possible to float small samples of plant tissue such as the skin of an apple on solutions of sugars and other compounds in petri dishes and then study anthocyanin formation in response to illumination. Biochemical studies are then easy to perform. We now know more about the biochemistry of phytochrome action as it controls synthesis of anthocyanin or carotenoid pigments than we do about its action in any other response.

Etiolation syndrome. As we saw in the introduction, many dark-grown plants exhibit a number of special characteristics that we can refer to collectively as the etiolation syndrome: small leaves, long internodes, reduced roots, pale color, and an epicotyl hook (see Fig. 35–A). For the most part, these features change in response to red light, and the changes are halted by subsequent far-red light. Internode elongation provides a handy tool for phytochrome studies. Red light inhibits elongation of stems and coleoptiles. Typically, the actual amount of elongation depends on the balance between the red and the far-red light used to illuminate the plants for a relatively short period just before they are placed in the dark. Far-red strongly promotes internode elongation. James A. Lockhart, then at the University of Hawaii, found that gibberellic acid treatment overcame the inhibiting effects of red light. He suggested that far-red light resulted in conversion of some precursor to an active gibberellin. Auxin treatments had no effect on response to red and far-red illumination.

The epicotyl hook straightens in response to red light (the *Pfr* of the phytochrome system). Its angle of curvature can be accurately measured, so it is possible to perform detailed quantitative studies. (Bean seedlings have often been used.) Leaf expansion is also promoted by red light (*Pfr*). In the laboratory small discs are cut from unexpanded leaves and floated on various solutions. They may then be illuminated and treated with chemicals, and their subsequent expansion may be accurately measured. Not only is leaf expansion promoted by red light and reversed by far-red light, but it is also promoted by sugars, cobaltous iron, and cytokinins. The significance of these findings is not yet apparent.

Induction and breaking of dormancy. We saw in Chapter 34 that plants of temperate regions, especially trees, go into winter dormancy in response to the short days of autumn, and a few of these come out of dormancy in the spring in response to long days. In a few cases, it was possible to show that an interruption of the night during short-day treatment prevented the short-day response. The interruption was most effective with red light, and the effect could be reversed somewhat with far-red light, thus implicating the phytochrome system. Results were not always clear-cut, however, and the role of phytochrome in the induction and breaking of dormancy remains to be clarified.

Miscellaneous responses. Limited information indicates that the phytochrome system may be involved in epinasty (twisting and bending of stems under certain conditions), leaf abscission, bulb formation (e.g., in onions), production of rhizomes, succulency of leaf tissue, expression of sex (relative numbers of pistils or stamens in certain species), development of roots, formation of plastids within the cell, chromosome breakage due to X-rays, root formation, runner formation (e.g., in strawberries), orientation of chloroplasts, establishing of circadian rhythms, and several other processes as well. It is important to note that the red-far-red system (presumably phytochrome) has been demonstrated in the green algae, the bryophytes, the ferns, gymnosperms, and angiosperms. It does not appear in animals, the fungi, or the other groups of algae.

Purification and characterization of phytochrome. One of the most outstanding scientific achievements in the field of plant physiology occurred in June of 1959, when the phytochrome system was first directly demonstrated in intact plants and then in a plant extract. This was done by the Plant Industry scientists at Beltsville (including K. H. Norris, W. L. Butler, Sterling B. Hendricks, and H. W. Siegelman). Early work with the germination of lettuce seeds and other systems had established a solid foundation of understanding about the nature of this pigment system. Its ability to absorb light had been calculated from experiments with seeds or whole plants. The concentration of the pigment in the plant extract could be estimated accurately from such experiments. The kinetics (time course) of change in *Pfr* had been studied. Other details of action of the pigment were also either fairly well known or at least suspected. Yet all attempts to apply the usual methods of biochemical study to this system had failed, partially because it was present in such small amounts. It was known that the pigment could be recognized by its color (its absorption spectrum) and, most importantly, by its ability to be converted from one color to another by irradiation with far-red or red light.

Isolation of phytochrome became feasible with the development of a sensitive instrument that could detect very slight color changes in extremely dense samples. (For example, the absorption spectrum of an ordinary 4-cm piece of wood could be determined.) It was necessary to determine the difference in absorption of a sample (whole plants or an extract) at 660 nm and at 730 nm. Of course, illumination at 660 nm converts the pigment to the form that absorbs light at 730 nm, so it was necessary to determine the absorption with extreme rapidity and with a relatively small amount of light. Pigment was initially converted to one form or another by illumination with high intensities of red or far-red light; then the difference in absorption between the two peaks was determined 16 times/sec by a special modification of the instrument.

Using this means, researchers first demonstrated the pigment in intact corn seedlings, which proved to be a rich source of this material. The seedlings could

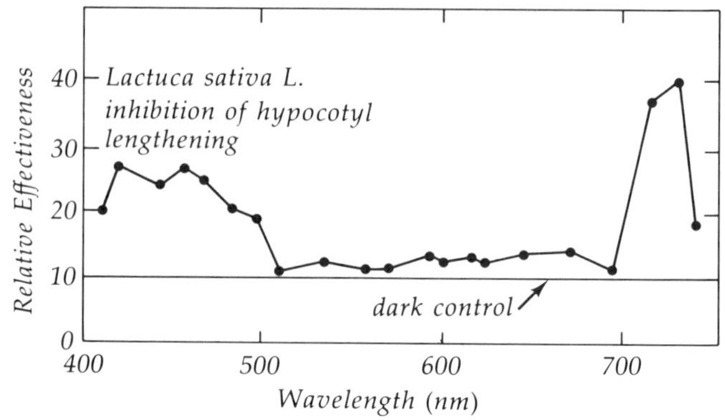

35–4 An action spectrum showing inhibitory effects of blue and far-red light on elongation of lettuce hypocotyls. Exposures were given at high intensities and for long durations. Numerous other responses have also been studied in this way, including most of those known to be controlled by the phytochrome system. Some examples are: inhibition of hypocotyl lengthening, hair formation along the hypocotyl, opening of the hypocotylar ('plumular') hook, formation of plastids in the mesophyll of the cotyledons, synthesis of anthocyanin, increase of RNA synthesis in the cotyledons, and increase or decrease of enzymes.

then be ground up and extracted in various ways, and fractions of the extract could be checked in the instrument. Many other plants have since been studied. It is easier to extract phytochrome from colorless tissues than from those containing chlorophyll, which absorbs in nearly the same part of the spectrum as the phytochrome. Nevertheless, phytochrome has been demonstrated in a few green tissues.

Phytochrome has proven to be a protein, and it has been concentrated to the point where its color becomes apparent to the human eye (*Pr* is blue-green, *Pfr* is light green). The part of the molecule that absorbs the light (the chromaphore) is a phycobilin pigment, similar to the water-soluble pigments that absorb the light used by blue-green and red algae in photosynthesis. The amino-acid sequence of the protein part of phytochrome (which has a molecular weight of about 60,000) may soon be determined, and perhaps this will allow us to understand how the chromophore and the active site combine to account for phytochrome action. There is recent evidence that the molecule with a molecular weight of 60,000 constitutes only a portion of the phytochrome molecule as it exists in the plant. Much recent research activity (using highly sophisticated biophysical methods) has also been concerned with the study and characterization of four intermediate forms that have very short half-lives but that could be important in phytochrome action.

Mode of action of phytochrome. Current research is centered on how the phytochrome acts. Most evidence indicates that *Pfr* is the active form, but what does it do? It is active in small quantities; responses to 40 to 100 *molecules per cell* have been observed, although 10,000 times this amount may be required for full response. This might suggest that phytochrome acts as an enzyme—an idea that agrees with its protein nature. But other modes of action are possible and even likely. Perhaps phytochrome is a gene activator, leading to synthesis of RNA and then protein. Many gene-controlled processes are influenced by phytochrome, and several experimental results implicate gene activation as the mode of phytochrome action.

Other data clearly seem to indicate that phytochrome is located on or in membranes (including the nuclear membranes). For example, red light causes root tips under the right conditions to stick to negatively charged glassware, and this is reversed by far-red light. The root tips adhere to the glass because of a secretion of hydrogen ions that, because of their positive charge, make the root-tip surfaces positive and cause adherence to the negative glass. Effects are observed within 30 sec; apparently, there is no time for gene action.

Mark Jaffe at Ohio University found in the spring of 1970 that acetylcholine was present in plant tissues, especially in active growing regions. (The compound has long been known to be present in animals and to take part in the transmission of nervous impulses.) Jaffe found that red light caused acetylcholine to be secreted by root tips (as used in the experiment just mentioned), and he also found an increase in the acetylcholine within the tissue. Far-red light reduced both the secretion of acetylcholine and the internal increase in concentration. When Jaffe applied acetylcholine to root tips in the dark, he was able to observe an effect exactly comparable to that caused by red light (tips stuck to the negatively charged glassware). Acetylcholine esterase, an enzyme that destroys acetylcholine, inhibited this phenomenon; that is, it acted the same as far-red light. It appears from Jaffe's results (which are certainly preliminary at this point) that phytochrome might act by influencing acetylcholine and that acetylcholine might act in plants as it does in animal systems (i.e., influencing the movement of ions across cell membranes). Perhaps the acetylcholine is a local hormone responding to changes in phytochrome (e.g., changes from *Pr* to *Pfr*), thereby mediating the phytochrome phenomenon.

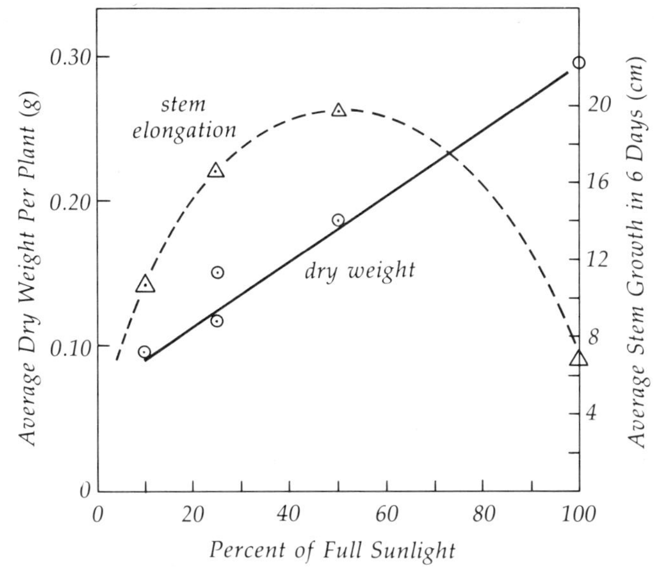

35–5 Stem elongation and dry weight as a function of light intensity. Note that dry weight increases as light intensity increases, up to the intensity of full sunlight. But elongation is maximal at about 50% full sunlight and then becomes inhibited considerably by full intensities.

This most interesting finding does not nullify the suggestions that phytochrome may act as an enzyme by turning genes on or off. It is now our experience that a hormone may act in several ways in a plant system, and phytochrome may also share this characteristic of exhibiting a multiplicity of modes of action.

The High Energy Reactions of Photomorphogenesis

Certain plant growth phenomena (such as the opening of the hook at the top of a lettuce seedling) occur in response to relatively high intensities of blue or far-red light applied for a fairly long time (Fig. 35–4). There are no signs of reversibility. Such responses are said to be under the control of the High Energy Reaction (HER). In some cases, only blue is effective (e.g., in early stages of anthocyanin synthesis), and probably some yet-to-be-characterized pigment other than phytochrome will prove to be responsible. In many other cases, responses to HER are very similar to phytochrome responses. In these instances, we may be dealing with a special phytochrome action. Part of this may be due to the overlapping absorption spectra for *Pr* and *Pfr* (Fig. 35–2). *Pr,* for example, although it absorbs most efficiently in the red part of the spectrum, also absorbs a little far-red (and even more blue) light, thus converting it to *Pfr.* So far-red or blue applied at high intensities and for long intervals of time may act like red—a fact that has been observed. Furthermore, the intermediate forms of phytochrome might be important in HER, and synthesis and destruction of phytochrome may well be influenced by high intensity light.

In any case, study of plant growth responses to high light energies contributes to an area of our understanding that has been neglected because of the emphasis on low-energy phytochrome response. In nature plants are exposed to high-energy radiation, so this study deserves more effort.

James A. Lockhart has studied responses of plants grown in shade compared to full sunlight. Field observations confirm that plants grown under these different conditions differ in significant ways. Shade plants are usually taller and more succulent. In one study (Fig. 35–5), Lockhart studied dry-matter production and stem elongation as a function of light intensity. Dry weight increased linearly from 10% sunlight to full sunlight, but stem elongation was maximum at 50% and minimum at full sunlight. Intensities of 50% sunlight are well above those found in most growth chambers in which artificial light is used to grow plants, so it is not too surprising that plants grown in such units seldom resemble those grown in the field. Of course, light quality in a plant growth chamber is also somewhat different from sunlight. At low intensities Lockhart found that red light inhibits stem elongation more efficiently than does blue light, but at higher intensities the situation is exactly reversed (Fig. 35–6). It is time that studies of this type were emphasized—we need to know more about plants in natural situations!

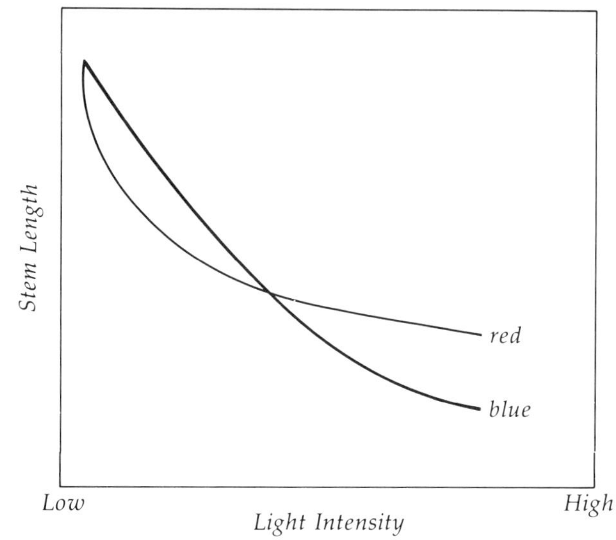

35–6 Both blue and red light are inhibitory to stem extension in some plants. At low intensities red is often more effective, while at higher intensities blue becomes more inhibitory.

The Effects of Ultraviolet Light: Photoreactivation

Ultraviolet light (UV) was discovered in 1801, and the fact that it had powerful photochemical effects (due to the high energy of its photons) was soon noticed. The role of ultraviolet light in sunburn was recognized in 1858. In spite of this early recognition of the importance of UV light (the last of the five photobiological responses we shall discuss), its effects on plants have been rather neglected. This is probably because its effects in nature may be relatively unimportant; yet laboratory studies of UV effects have provided us with insight into some basic biological principles.

Are plants above timberline stunted because of the higher intensities of UV light? The same plants occur north of the arctic circle, where they also appear to be "stunted" (this is their natural growth form), although UV intensities there are even lower than those encountered at sea level near the equator. Martyn Caldwell, working with Dwight Billings at Duke University, studied the importance of UV light for plants of the alpine tundra in the Colorado Rockies. UV intensities at 4350 m above sea level were only 20% higher than at 1670 m—less than most people might have thought. He excluded UV light from some plots in the field and illuminated plants with artificial UV light in the laboratory. Virtually no clear-cut, reproducible effects could be observed (although a clover seemed to be promoted in its flowering by UV light).

What about laboratory effects? Ultraviolet light can have powerful effects upon plant growth when it is applied at relatively high intensities in the laboratory. But most are deleterious. The epidermis of leaves takes on a bronze color, and with sufficient exposure, plants may be killed. There is a wide range of susceptibilities

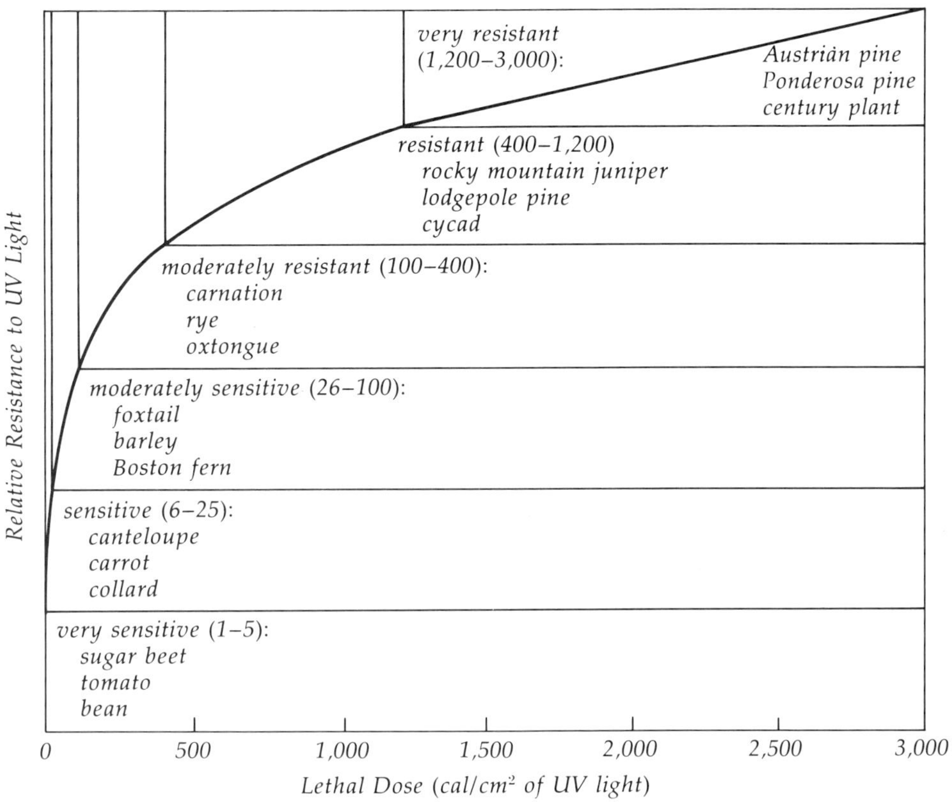

35–7 A representation of species responses to ultraviolet light. Very sensitive species are damaged by low dosages, but resistant species tolerate dosages a thousand times as high.

to UV light. Many plants (e.g., tomatoes, beans, sugar beets, clover, horsetail, cocklebur) are extremely sensitive to UV light, being killed by only 3 or 4 hr of irradiation with intensities equivalent to those occurring on the surface of the planet Mars (Fig. 35–7). (The study was carried out as part of a NASA-supported project designed to investigate the possibilities of life on the various solar planets.) Other species such as rye, wheat, corn, and Engelmann spruce were much more resistant, requiring 60 to 100 hr of irradiation for leaf kill. An Austrian pine survived 635 hr without kill, and a century plant, other pines, and a cactus were also extremely resistant. The Austrian pine could have survived the UV light provided during an entire growing season on Mars.

No single resistance mechanism could be observed in these plants. Conifer leaves seem to be protected by several layers of thick-walled, subepidermal cells, while cactus contains a network of tightly packed, globular crystals in the epidermis that probably reflects and absorbs much of the impinging UV irradiation. The century plant is well fortified with large bundles of fibers just below the epidermis. In each of these cases, the protective mechanisms seem to be a matter of something being present that absorbed the UV irradiation before it could reach the sensitive cellular organelles below.

Can UV effects be reversed? In 1948 it was discovered that the survival of certain soil microorganisms (Actinomycetes) was improved by a factor of several thousand when UV irradiation was followed by exposure to intense visible light. The phenomenon was termed photoreactivation (or photoreversal). It is now known that many other organisms and many different kinds of UV damage exhibit photoreactivation. The damage to higher plants just described is an example, as are certain forms of UV damage in mammals and other higher animals. In certain bacteria, death caused by UV light can apparently be completely reversed by subsequent irradiation with visible light. Blue light (of wavelengths only slightly longer than the UV causing the damage) is most effective in photoreactivation.

How does photoreactivation work? First, we must understand the mechanism of damage by UV light. Although several possible mechanisms could be important, (e.g., a production of damaging peroxides),

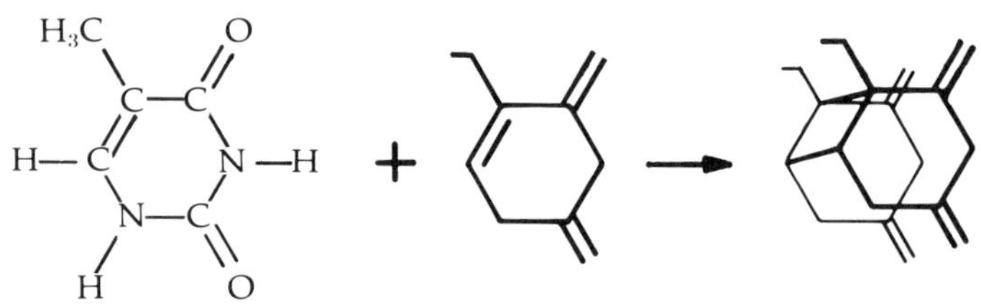

35–8 Formation of thymine dimers by the reaction between two thymine molecules. Reaction occurs across the double bonds in the two rings of the two thymine molecules.

the photoreactivatable forms of UV damage are related to effects upon DNA—specifically, one kind of reaction in the DNA chain. When two molecules of thymine occur next to each other in the chain, it is possible for them to react with each other (through a double bond in the ring) to form a so-called thymine dimer (Fig. 35–8). Other nucleotides can also form dimers but do so much less frequently than thymine. Formation of the dimers prevents DNA replication, RNA transcription, and ultimately the responses dependent upon protein synthesis.

How can blue light reverse these effects? In the bacterium *E. coli* blue light appears to provide energy to activate a DNA repair enzyme. The response is direct and unamplified. Sometimes, other energy sources such as heat will also activate the enzyme. The enzyme seems to "snip out" the damaged portion of the DNA chain, resynthesizing a corrected version by complementary bonding of new nucleotides to the undamaged strand.

Apparently, the enzyme activated by blue light is capable of repairing not only UV damage but other forms of damage to DNA as well (e.g., those caused by such mutagenic agents as nitrogen mustard). Hence the observation could be of broad biological significance. It has been estimated, for example, that the genes in a bacterium may be duplicated as many as 100 million times before there is a 50–50 chance for even one gene to be altered (mutated). Such an extremely accurate copying process certainly seems to require some mechanism for detecting and correcting occasional errors. No modern manufacturing process could mass-produce an object without such mechanisms. Perhaps the enzyme discovered by studies on photoreactivation is the quality control system of living organisms.

The Biological Clock

Just as temperature and light influence such processes as bud and seed dormancy, so does the biological clock. Plants and animals seem to have a built-in timing mechanism, the status of which at any given moment strongly influences many of the morphogenetic events that we have been discussing. At the same time the status of the biological clock is itself influenced by such environmental factors as temperature and light. Manifestations of the biological clock have now been documented and described in considerable detail; yet we know little (if anything) about the clock's mechanism. Understanding this mechanism certainly constitutes one of the most challenging problems of modern biology, but relatively few investigators are engaged in its study. In this section we shall summarize a little of the history relating to the discovery of the biological clock and describe some of its features. In End Paper 35–1 we shall speculate a bit about its fundamental nature.

Man is clearly able to sense time, sometimes with amazing accuracy. For example, most of us are able to wake up at the same time of day with a little practice, and posthypnotic suggestions are sometimes carried out within seconds of the appointed time, even in the absence of time cues such as a clock. Regular patterns of flowering or migration suggest that plants and animals also have the ability to measure time, since day-length might be measured to detect seasons. On the other hand, occurrences at a given time of year could be responses to changes in temperature or some other climatic factor.

Essay 35–2 Discovering and Studying the Biological Clock

One of the great names associated with the biological clock in plants is Erwin Bünning, Professor Emeritus at the University of Tübingen, Germany. One of us (FBS) had opportunities to hear him tell of his early work and wrote to him with the request that he record some of his experiences. This is a translation of the letter he sent to us.

Tübingen, den 2. Juni 1970

Institut für Biologie
der Universität Tübingen
7400 Tübingen
Auf der Morgenstelle 1

Dr. Frank B. Salisbury
Plant Science Department
Utah State University
Logan, Utah 84321, U.S.A.

Dear Frank:

The story went something like this. At the Institute for the Physical Basis of Medicine in Frankfurt, the biophysicist Professor Dessauer (an X-ray specialist) became interested in the effects of the ionic content of the air upon humans. Those were the years when people began to be interested in atmospheric electricity, cosmic rays, etc. Naturally humans could not be used as experimental objects, and so in 1928, Dessauer searched for botanists to work on plants. One that he found was Kurt Stern, who lived in Frankfurt; the other was I, who had just finished my doctoral work in Berlin. So we began in August of 1928 to contemplate the problem. In the process we came upon the work of Rose Stoppel who had been studying the diurnally periodic movements of Phaseolus *(common bean) leaves. In the process she had found, as had several other authors, that under "constant" conditions in the darkroom, most leaves reached the maximum extent of their sinking (their maximum night position) at the same time; namely, between 3:00 and 4:00* A.M. *Her conclusion: Some unknown factor synchronized the movement. Could this be atmospheric ions? We had Rose Stoppel visit us from Hamburg for two or three weeks so that we could become familiar with her techniques. In our group we always called her '*die Stoppelrose' *(stubble rose), and the name was most appropriate. She was energetic and persistent, so persistent that she just died this January in her 96th year. Her results also appeared in our experiments: The night position usually occurred at the indicated time. Then we investigated the effects of air that had been enriched with ions or air from which all the ions had been removed. The result: nothing changed—atmospheric ions are not 'factor* x.*'*

We then decided that our research facilities at the Institute were insufficient. Hence after die Stoppelrose' *had left, Stern and I moved to his potato cellar, where, with the help of a thermostat, we obtained rather constant temperature. Contrary to the practice of Stoppel, who turned on a red 'safe' light to water her plants, we went into the cellar just once a day with a very weak flashlight and felt around with our fingers for the pots and recording apparatus so that we could water the plants and so that we could see if everything was in order with the recorders. The flashlight was weakened with a dark red filter so that one could see only for a few centimeters distance. In those days it was the dogma of all botanical textbooks that red light had absolutely no influence upon plant movements or upon photomorphogenesis. We did one other thing differently from Stoppel. Since Kurt Stern's house was a long way from our laboratory, we didn't make our daily control visit in the morning, rather only in the afternoon. The result: most of the maximal night positions no longer appeared between 3:00 and 4:00* A.M.*, rather between 10:00 and 12:00* A.M. *Hence we concluded: The dogma is false. Red light must synchronize the movements such that a night position always appears about 16 hours after the light's action. That was 'factor x.' When we eliminated this hardly visible red light, we found that the leaf movement period was no longer exactly 24 hr but 25.4 hr.[Ed.: This circadian nature of the clock was the key to understanding the endogenous nature.]*

So that was about how that story went. Naturally I could also tell you the story about how I came upon the significance of the endogenous rhythms for photoperiodism[Ed.: see Chapter 36]. That was in about 1934. Of course, I had already often asked myself how such an endogenous rhythm might ever have any selection value [in evolution], and I had already expressed the opinion in 1932 in a publication (Jahrb. Wiss. Bot. *77:283–320) that some interaction between the internal plant rhythms and the external environmental rhythms must be of significance for plant development. But there were coincidences in the story. As a young scientist, one naturally had to allow himself to be seen by the power-wielding people of his field, so that he might receive invitations for promotion. Hence I traveled in 1934 from Jena to Könisberg in Berlin to introduce myself to the great Professor Kurt Noack. We discussed this and that. He mentioned that discoveries were being made that were so remarkable one simply couldn't believe in them. Such a one, for example, was photoperiodism. He, as a specialist in the field of photosynthesis, must certainly know that it makes no difference whatsoever what program is followed in giving the plant the necessary quantities of light. Then as I was riding back on the train, the idea came to me—Aha, for the plant it does make a difference at which time light is applied, if not exactly in photosynthesis, nevertheless for its development!*

I could present a third story, one from very recent years. I have long felt that the daily leaf movement rhythms had no selection value in themselves. As you know, I have recently changed my mind. The movements could indeed be important in avoiding a disturbance of photoperiodic time measurement by moonlight. Before I had begun to test the idea experimentally or even to think about it, it came very simply one night in a dream. The dream (apparently as a memory of one of my visits to the tropics): tropical midnight, the full moon high above on the zenith, in front of me a field of soybeans, the leaves, however, not sunken in the night position, rather broadly horizontal in the day position. My thoughts (in the dream): how shall these plants know that this is not a long day? They had better hide themselves from the moon if they want to flower.

Okay, see if you can get anything going with these stories.

With kind regards,
Yours, Erwin

Occasional scientific papers before and during the 19th century mention biological time measurement. A persistent rhythm in any plant or animal function that continues even when the organism is placed in unchanging environmental conditions might indicate an ability to measure time. Many rhythms were observed in the 19th century, but seldom was there a clear-cut indication that they were not caused by rhythms in the external environment. In 1918 two researchers at the Plant Industry Station at Beltsville, Maryland, discovered that certain plants flower in response to the length of day and night (see Chapter 36). The scientists, Wightman W. Garner and Harry A. Allard, termed the phenomenon photoperiodism, implying that a response to the *length* of day and night indicates an ability of the organism to measure time. If we can assign a date to the demonstration of biological timing, it probably should be 1920—the year Garner and Allard published the article.

In 1928 Ingeborg Beling discovered that honeybees could be trained to feed at certain times of the day, indicating a biological time measurement in an insect (a time memory). By the 1950s many scientists were engaged in research on this problem, and the term *biological clock* had entered the scientific literature. Perhaps the most striking recent discovery is that certain birds are able to navigate by position of the sun (and probably even the stars). Since these positions change throughout the day and throughout the season, a highly accurate time-measuring ability is clearly implied.

The breakthrough came in the 1920s when a new Institute for the Study of the Physical Basis of Medicine (we would now call it biophysics) was established at the University of Frankfurt in Germany. Two young graduate students at this institute, Erwin Bünning (see Essay 35–2) and Kurt Stern, were looking for a problem to investigate. They encountered the work of Rose Stoppel in Hamburg, a continuation of the studies of the great plant physiologist Wilhelm Pfeffer, who had been recording and studying the sleep movements of the common bean plant for many decades. Although he had convinced himself, Pfeffer had been unable to demonstrate to everyone's satisfaction that the clock controlling the movements was within the plant. Now Stoppel had evidence that it was not. In her experiments a peak was reached in the movements at almost exactly the same time every day (even though the plants were held in continuous darkness and at constant temperature). She reasoned that an internal biological clock could never be so accurate, so some other subtle factor of the environment must be responsible for the timing of the leaf movements.

Bünning and Stern set out to look for Stoppel's "factor *x*," thinking at first that it might be the daily variation in the ionic content of the atmosphere. Ions in the air had no significant effect upon the leaf movements, but Bünning and Stern found Stoppel's factor *x*. It was the red "safe" light that she turned on at the same time each day to water her plants! Because red was ineffective in phototropism, it was thought to be a safe color for studying plants in the dark. Now, of course, we know that its effectiveness in the phytochrome system makes red anything but safe. When Bünning and Stern kept plants in continuous darkness with no safe light, the leaf movement rhythms began to exhibit a period (the time from some recognizable part of the cycle to the same part of the cycle the next time it occurs—Fig. 35–9) that was not equal to 24 hr. In some plants, the period was as much as 28 hr long; in others, it might be as short as 22 hr. It was argued that, if several plants with different periods could be completely out of phase with each other in the same room, then they could hardly all be responding to the same

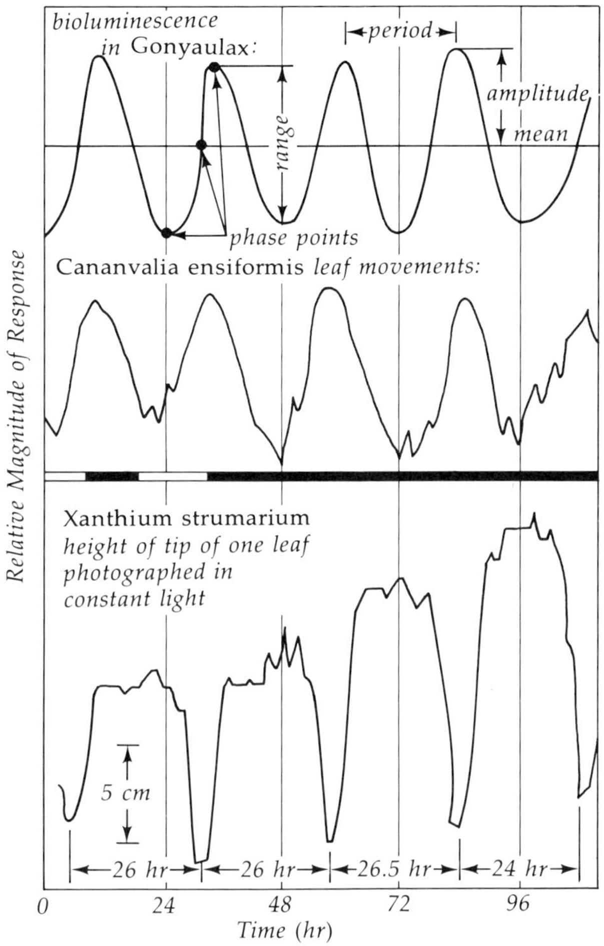

35–9 Some representative data for various circadian rhythms. Top: Bioluminescence in Gonyaulax, *as measured for plants kept under constant conditions of dim light. Characteristics of circadian cycles are indicated (phase points, range, period, amplitude, and mean). Middle: A leaf movement record from* Cananvalia ensiformis. *Light and dark conditions are indicated by the bar. Note gradual shift of the peak as the cycles progress: Bottom: Leaf movements of cocklebur* (Xanthium strumarium), *as recorded by time-lapse photography. Period lengths between the troughs are indicated. Note increase in absolute height of the 'eaves (particularly at the peaks, but also at the troughs). This is largely due to growth of the stem during the course of the experiment, but the increase in range of leaf movement is also very apparent. Light came entirely from fluorescent lamps.*

subtle change in the environment. This experiment and many similar ones conducted during the past half century have been taken as proof that the biological clock is endogenous (i.e., internal or within the plant). As we shall see below, however, the specter of factor *x* has been raised again.

The experiment illustrates some important characteristics of the clock, as well as some terms used to discuss it. When the organism is able to exhibit its free-running rhythm (the rhythm under constant conditions), the period seldom equals exactly 24 hr. Since it *approximates* 24 hr, however, it has been called circadian (circa = approximately, diem = day). If the environment has a rhythmical variation, the organism rhythm can be entrained or synchronized to match the rhythm of the environment, provided that the environment rhythm has a period not differing greatly from a 24-hr cycle (perhaps 16 to 30 hr, depending upon the organism). In Stoppel's experiment, red light was the entraining factor, and light is often the factor most effective in entrainment. Temperature changes may also entrain the cycle, and other synchronizers (or zeitgebers, a German word meaning time-giver) are known. Sparrow songs will entrain the activity rhythms of other sparrows, for example, and a wristwatch can entrain human cycles. (In certain experiments conducted near Munich, Germany, human subjects lived in deep caves for several weeks. A watch that completed its two cycles in 26 instead of 24 hr nicely entrained its wearer.)

Since the pioneering studies of Bünning and Stern, many facets of the problem have been studied in detail, and several other instances of persistent rhythms in plants have been observed, including flower opening and closing, the growth rate of certain seedlings, root pressures, the activities of several enzymes, and discharge of spores in certain fungi. The respiration rates of certain tissues seem to follow a persistent rhythm, as do photosynthesis, mitosis, and volume of the nucleus in some plants. A most interesting and intensively studied system is a marine alga (*Gonyaulax*—a dinoflagellate) that emits light with peaks of intensity separated by about 24 hr. *Gonyaulax* also has rhythms in cell division and the ability to photosynthesize when illuminated with a constant intensity of light at a constant temperature. Persistent rhythms have also been studied intensively in many animals. Indeed, biological timing has been observed in representatives of all organisms that have organized nuclei.

The biological clock is at once an important phenomenon and one that is relatively little understood.

End Paper 35–1 The Nature of the Biological Clock

In this chapter we introduced some of the facts concerning biological clocks in plants and animals. We also saw that circadian rhythms are common and that an organism can be entrained to match the rhythm of the environment.

How effective is light in entrainment? The responses of the biological clock to light have been intensively studied since the initial observations of Bünning and Stern. In view of the discussions in this chapter, for example, a logical question concerns the action spectrum for entrainment by light, but little has been done to answer this question. Clearly, phytochrome is not the photoreceptor pigment in the majority of organisms, and even in plants, experiments are ambiguous. Bean plants grown from seed in the dark show virtually no leaf movements until they have been exposed to light, and then all wavelengths of light prove to be about equally effective in inducing the movements. On the other hand, it was discovered in Bünning's laboratory (by then he was in Tübingen) in the early 1960s that bean leaf movements would continue for a week or more in continuous light from fluorescent lamps that produced no far-red wavelengths, but the rhythms damped out in just a few days when far-red light from incandescent lamps was present (Fig. 35–10). This is highly suggestive that the phytochrome system might be involved in the response of the biological clock to light, but the leads suggested by this experiment have never been pursued.

One of the most intensively studied problems concerns the times of light irradiation in relation to the course of the circadian cycle. Bünning had suggested that cycles were entrained by the change from darkness to light (the dawn signal), but subsequent experiments have shown an interesting relationship between dawn and dusk (the change from light to darkness). Apparently, it is not so much a matter of whether the lights go on or off, but rather when light is given during the cycle. With many organisms (both plants and animals), the rhythm may be observed in continuous darkness for several days after entrainment by a normal light-dark cycle. During this period a brief interruption of light (seconds to an hour or more in duration) may be applied at various times. The subsequent course of the rhythm is then observed. Typically, the phases (characteristic points on the rhythm) are shifted in response to such a flash (Fig. 35–11). The effect is minimal when the flash is given at subjective noon (the phase of the cycle that would occur during noon in

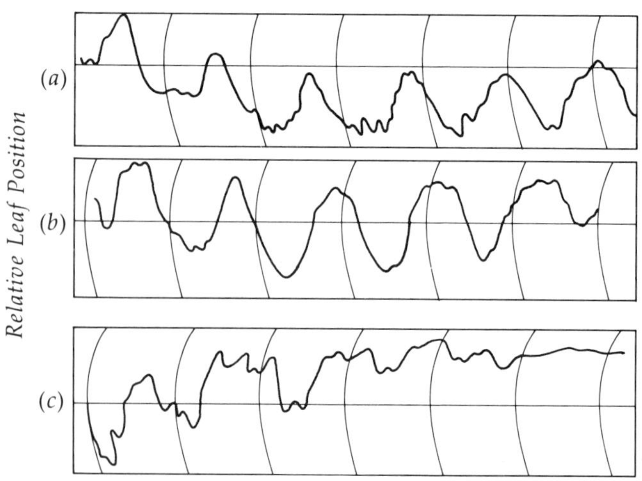

35–10 *Bean-leaf movements (a) in continuous darkness, (b) in continuous red light, and (c) in continuous far-red light.*

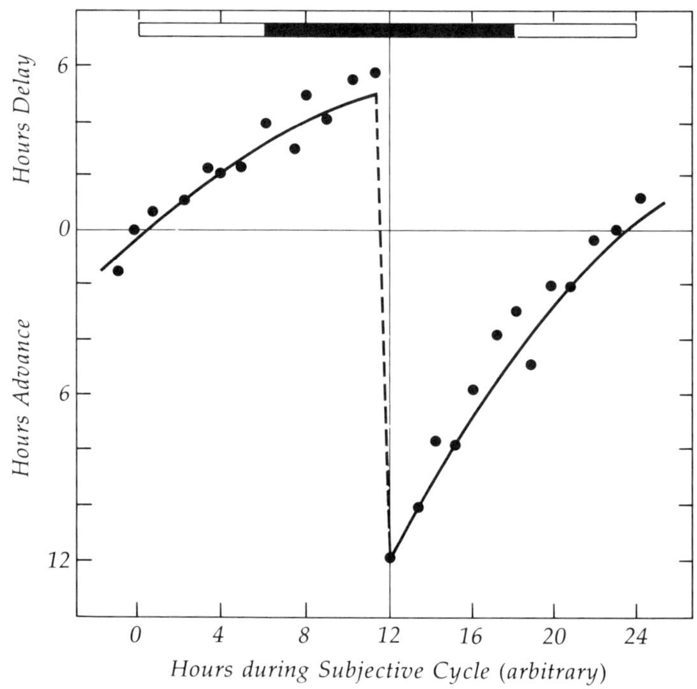

35–11 *Phase shift in petal movements of* Kalanchöe blossfeldiana, *following 2-hour exposures to orange light given at various times during an extended period of continuous darkness. Bar at the top indicates subjective status of the rhythm; that is, the dark part of the bar indicates petal closure or subjective night. As light interruption approaches the middle of subjective night, there is an increasing delay; following the middle of subjective night, there is an advance that decreases as subjective day is approached.*

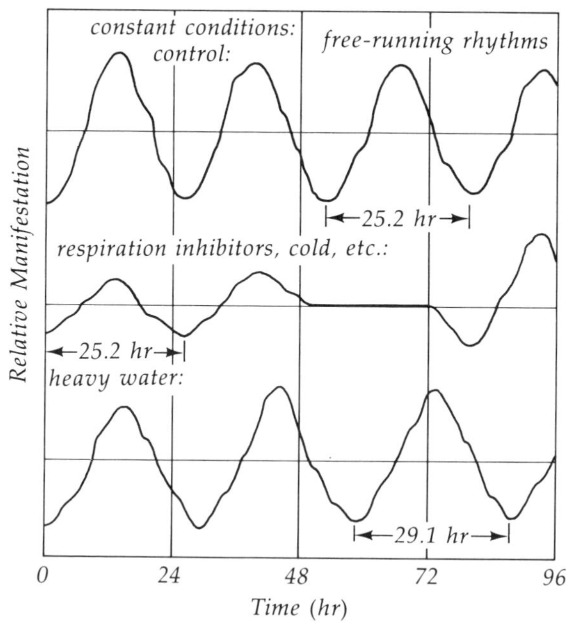

35–12 The effects of various chemicals and other factors upon free-running rhythms. Most compounds that inhibit metabolism reduce the amplitude but do not affect the period. (Compare the center record with the control above.) Heavy water increases the length of the period—29.1 compared with 25.2 hr for eclosion in Drosophila. *(Curves are schematic and do not represent actual data.)*

a normal light-dark environment). As the flash comes later and later during subjective evening, the cycles of the rhythm are increasingly *delayed.* Then, perhaps, as subjective midnight is crossed, there is a sharp shift, after which the phases of the cycle are *advanced* instead of delayed. The flash of light appears to act as a dusk signal up to the time of the shift, after which it then acts as a dawn signal. The extent of the delay or the advance is greatest near the time of the shift.

Colin S. Pittendrigh, an American zoologist who has contributed much to our understanding of the biological clock, showed with his associates at Princeton University that a normal light-dark cycle could be simulated by two flashes of light given during constant darkness, a so-called skeleton photoperiod. When, in this elegant series of experiments, the two flashes were relatively close together, each one repeated once every 24 hr, the first would act as dawn and the second as dusk (as indicated by the relation to the phases of the cycle). As the time of separation increased to about 14 hr, a sudden shift occurred. The flashes seemed to "trade places" in their effects upon the rhythm; the one that had been acting as dawn then acted as dusk and vice versa.

Pittendrigh and his associates also observed with fruitflies (a pupation rhythm) that a phase shift didn't occur suddenly in response to a light flash, but that several days of partial shifting (transients) could be observed before the cycle was stable again. All of these observations must be accounted for in our attempts to understand the mechanism of the biological clock. Incidentally, the transients are not as readily demonstrated with plants as with animals, but in general, most of the features of the biological clock seem to be common to both plants and animals, including man. (The jet age and space travel have made the study of the clocks in man particularly pressing.)

How does the clock work? Is it sensitive to temperature? The approach to understanding how the clock functions has been to discover as many facts about it as possible and then attempt to construct conceptual models that might best account for the observed facts. One of the most difficult facts to account for has been the relative temperature independence of the clock mechanism. The period of persistent rhythms is in many instances only slightly influenced by temperature. Leaf-movement cycles in plants are frequently lengthened somewhat by lower temperatures and shortened by higher temperatures, but the effect is much less even in these cases than one might expect. A temperature change of 10°C commonly influences the leaf movement cycle only by a factor of about 1.3, while chemical reactions are (as we have seen) frequently doubled (or more) by a 10°C temperature increase. Often, the temperature effect on leaf movements can be observed only during the first few cycles, after which there seems to be a compensation so that time is measured about the same as it was at the original temperature. In many instances, the effects of temperature on biological timing are far less; for example, 10°C temperature changes may influence timing by a factor of only 1.02 in the flowering of cocklebur (see Chapter 36).

How can we account for temperature independence? Perhaps the clock is some physical system such as the oscillation of molecules on a membrane. Physical systems are far less sensitive to temperature than chemical ones. Or perhaps we are dealing with temperature-compensated reactions such as those discussed in relation to breaking of bud dormancy. With increasing temperatures, the first reaction might be increased in rate but inhibited by the products of a second reaction that is also increased in rate. No one has been able to gain an entrance (beyond the descriptive level) to the experimental study of these problems.

One of the problems has been that the biological timing phenomenon in plants is remarkably resistant to treatment with many chemicals. A number of compounds that cause harmful effects in plants suppress the *expression* of the rhythm, but few change its *period* (Fig. 35–12). For example, lack of oxygen, excess carbon dioxide, various narcotics, and such respiration inhibitors as dinitrophenol, arsenate, cyanide, and fluoride result in a damping of the leaf movements, but the peaks of the rhythm are still separated by about the

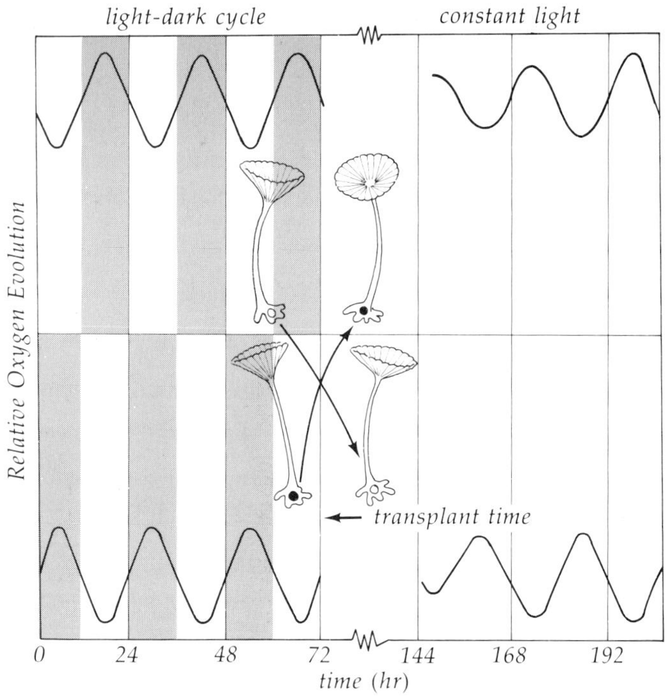

35–13 *The effects of transplantation of the nucleus in* Acetabularia *upon its rhythm of photosynthesis. Following transplantation, the rhythm becomes adjusted to match that of the plant from which the nucleus was obtained, although it cannot be observed for about 3 days.*

same time intervals (approximately 24 hr). These chemicals are apparently inhibiting processes controlled by the clock rather than inhibiting the clock itself. A few chemicals such as colchicine, ethyl carbamate (urethane), and cobaltous iron do change the period, but these chemicals are not effective in all systems. Heavy water (deuterium oxide) clearly lengthens the period in several systems, but no one knows how. At the moment, there is considerable interest in certain tentative lines of evidence indicating that protein synthesis may be a part of the clock.

Where is the clock? Locating the clock might suggest something about its mode of action. Clearly, it has to be within the cell, since many single-celled organisms exhibit it. Plant tissue cultures also show certain persistent rhythms. But where in the cell? So far, we have not been able to locate the clock with certainty, although current evidence seems to indicate that it is in the nucleus. A favorite tool for such studies is *Acetabularia,* a single-celled green alga. The cell is so large (about 2 cm) that it is easy to manipulate and even to operate upon. A holdfast containing the single nucleus anchors a long stalk topped by a "hat." The organism exhibits a rhythm in photosynthesis, and this continues for a few days even after the nucleus is removed, thus suggesting that the clock may not be in the nucleus. But when two organisms are entrained to opposite light-dark cycles and their nuclei are interchanged, the rhythms go with the nuclei. In a cell as special as *Acetabularia,* effects of the nucleus may persist for an unusually long time after it has been removed (first experiment), but the second experiment implies that the clock might well be in the nucleus.

The mode of action of the biological clock remains unknown. Workers go on devising conceptual models such as that of Pittendrigh and Bruce, which suggests that there are two clocks—one sensitive to the environment (e.g., a light flash) and one capable of entraining the other (thus controlling manifestations of the rhythm). If entrainment of the second required a few days, this might account for the transients. Such models have their value, but it would be nice if we could find some specific biophysical or biochemical process that we could clearly relate to the clock.

What good is the clock? It is easy to see how a biological clock might be used to the advantage of the organism in migration, photoperiodism, and a few of its other manifestations. Flowers open according to the schedule controlled by their biological clocks, and bees record this time of opening in their time memories. Some animals are nocturnal, others diurnal—but always in response to the clock. This allows them to coexist in an area without competing for space; they simply use it at different times.

But why should leaves wave on a daily schedule? Pfeffer detected no ill effects when he tied leaves in place so they couldn't move, so he concluded the movements were useless; they were something plants had acquired in their evolutionary history but didn't really need. Maybe so. Others suggested that the movements might have to do with heat transfer. A horizontal leaf at night radiates to the sky, but a vertical one (the usual position) radiates to its neighbors, conserving perhaps a significant quantity of heat. The idea has not been tested. Bünning recently had another idea (see Essay 35–2). Photoperiodically sensitive leaves, provided that they were horizontal instead of vertical, just might be inhibited in their response by light from the full moon. Bünning tested several species, finding that they were sensitive to moonlight when the leaf surface received the full intensity by being at right angles to the beam but not when leaves were parallel to the beam, thus receiving much lower intensity. Perhaps the problem has been solved, although further testing will be required to be sure.

Is factor x really dead? For about 40 years most biologists concerned with biological clocks have been convinced that the clocks are indeed endogenous as Bünning and Stern seemed to show. The goal has been to understand the mechanism of the self-contained clock within the cell. One biologist, Frank A. Brown, Jr., at Northwestern University in Chicago, has not shared this conviction. In the late 1940s he observed rhythms in body color of fiddler crabs with *exactly* 24-hr periods in the dark and at several different but highly constant temperatures. Again the concept of factor x came to mind. Could the organisms be responding to some geophysical factor of the environment other than light and temperature?

To study the problem, Brown and his students have performed a variety of experiments. They are convinced that they have demonstrated (although few other scientists yet agree) that organisms can respond to subtle forces such as the earth's geomagnetic or electromagnetic fields, which are in turn controlled by the primary cosmic radiation (known to have a 24-hr period) striking the earth's upper atmosphere. In 1956 measurements of the respiration of plugs from potato tubers were initiated, and now some 11 years' data have been analyzed. The tissue plugs were held for several months in hermetically sealed recording respirometers (which were serviced once every three to eight days), where light, temperature, pressure, and humidity were held constant within narrow limits (e.g., $\pm$ 0.05°C). This particular tissue does not exhibit the typical circadian periodicity in respiration, but there are three principal peaks in respiration that occur at the same clock times every day: at a little after 6:00 A.M., at noon, and at about 6:00 P.M. The same general observation has now been extended to crabs, mice, carrots, mealworms, and beans. In addition to the daily pattern of potatoes, there is another pattern relating to the phases of the moon, and a strong annual fluctuation with a peak in summer that is as much as 5 times as high as the winter rate. (The daily fluctuation varies only about $\pm$ 5% of the daily mean.) There are also gradual changes occurring from year to year.

All of these rhythms were observed with tissue that is strongly shielded from the obvious changes in the environment to which we imagine organisms to be sensitive. Brown first observed a close correlation between respiration of his potato plugs and changes in barometric pressure that are superimposed upon the much greater random changes due to changing weather. Of course, he knew that the tissue could not be responding to the changes in pressure (it was shielded from these changes), but what other geophysical forces might account for both the pressure and the respiration changes? A close correlation was then observed between measurements made by physicists of primary cosmic radiation and the respiration data. It has long been known that primary cosmic radiation is strongly influenced by the earth's magnetic field. Hence Brown and others began to investigate the effects of magnetism upon various biological systems, finding several clear-cut responses, even when the field strengths were similar to those of our natural geomagnetic field.

Other scientists have tested Brown's early ideas by taking organisms into deep salt mines (no cosmic rays—but the rhythms and magnetic fields remain); by transporting bees from east to west (Paris to New York; rhythms continue on Paris time); and even by observing continuation of rhythms in organisms rotated on 24-hr turntables on the South Pole. (If rotation of the turntable is opposite to that of the earth, the effects of the earth's rotation should be nullified.) Yet none of these tests got at the real questions raised by Brown's data. Perhaps to answer the questions, a space probe will have to be launched away from the earth's influence. Such a probe, the Biopioneer, has been in the planning stages for several years now.

Brown's studies have opened up some highly interesting avenues of research. He presents a theory called autophasing to account for the circadian nature of persistent rhythms, even if timing is exogenous. Basically it suggests that under constant conditions of temperature and light, a regular phase shift occurs each day in relation to the daily input of varying geophysical force. That is, the organism might respond to constant light and temperature by delaying (for example) the phases of its rhythm by one hour each day, just as it may be caused to delay its phases by a single light flash given one hour each day. An electric clock in the United States receives its timing in the 60-cycle current coming from the power station. Of course, such a clock, though its timing is exogenous, could be constructed to run with any desired period. It could run two hours slow every day, for example. (Most clocks have endogenous timers—a pendulum, balance wheel, or vibrating tuning fork.) Clearly, if autophasing is possible, then the circadian nature of the biological clock is not a conclusive argument for its being endogenous.

Brown's theory is challenging. Although it may not be apparent from the brief summary given here, it can account fairly well for virtually all of the facts relating to the biological clock. Furthermore, it accounts better for some of the facts than does the theory of an endogenous clock. The daily peaks in potato respiration in Brown's experiments offer an example, but temperature and drug independence, as well as lunar and annual cycles, are also better explained by an exogenous clock. Clearly, the matter is complex. As Brown suggests, an exogenous timer responding to factor x might exist in the same organism with the endogenous timers postulated by so many other scientists.

One thing is clear: our studies of the biological clock could lead to unexpected and highly significant findings about biological function. These discoveries could rival in interest and importance those relating to the gene and protein synthesis.

we must admit that all such responses are influenced by many factors. The rate of an enzymatically controlled reaction, for example, is highly temperature sensitive, but it is also influenced by such things as water potential, ionic strength, concentrations of reactants and products, and pH—all factors that might in turn be influenced by light, humidity, and so forth. We are dealing with a holistic environment-organism interaction.

In this chapter we shall discuss a single developmental event: conversion from the vegetative to the reproductive state. This is influenced by a number of environmental factors—primarily temperature and light—although other factors are important in certain species. In any case, the flowering process (using the term in a broad sense to cover even those vascular plants that do not produce flowers) is perhaps one of the most complex and intricate morphogenetic events occurring in plants.

36–1 The radish plant (Raphanus sativus) *on the left has been exposed only to short days; the one on the right, only to long days.*

An Historical Perspective

For centuries horticulturists have nurtured their pet theories about how various species might be forced into bloom. By the early part of this century, clear-cut demonstrations of the control of reproductive development by environment were still lacking, although the ancient ideas had been refined somewhat and were taking on an aura of scientific respectability. The flowering of certain trees and shrubs was known to be promoted by severe pruning (virtually any form of mutilation), and it was felt that flowering of other plants could be induced by withholding water. Since flower formation often occurred toward the end of the summer when nitrate levels were low, it seemed that nitrogen levels might be important. The concept of the carbohydrate-nitrogen balance was developed. High carbohydrates (produced by photosynthesis) were supposed to induce flowering, while high nitrogen (provided perhaps as artificial fertilizer) and low carbohydrates were supposed to lead to vegetative growth. The observations relating to mutilation and drought were explained by intricate circumlocution of the carbohydrate-nitrogen theory.

Effects of temperature and light were completely overlooked, even though field observations certainly suggested their importance. Indeed, farmers and agricultural researchers in mid-nineteenth century America had discovered that certain winter cereals could be forced to early flowering by treating their moist seeds before (or shortly after) germination with temperatures close to the freezing point. Although they were published, these observations were forgotten and had no impact upon the course of scientific thought. Other writers in the second half of the nineteenth century had suggested that flowering might be influenced by length of day, which in turn is a function of season and latitude. Facts relating to the times of flowering of various species located at different latitudes from the equator to the arctic clearly suggested this possibility. Yet no one pursued the matter experimentally.

In 1910 and 1918 J. Gustav Gassner published experiments repeating the work of the mid-nineteenth

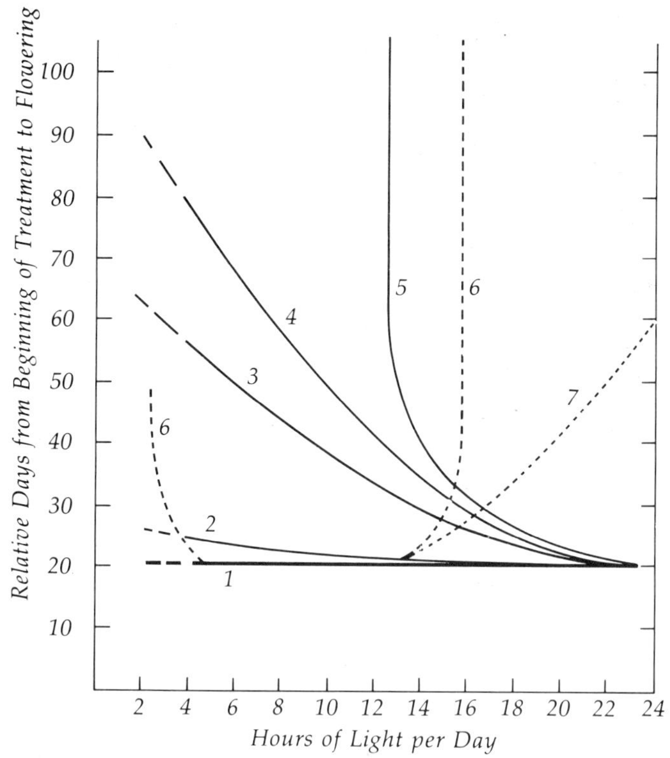

36–2 Representative flowering responses to different daylengths, plotted as time required for appearance of flowers. Curve 1 is for a true day-neutral plant. Curve 2 would also be considered a day-neutral response, although flowering is slightly promoted by long days. Curves 3 and 4 represent quantitative long-day plants, but 4 is more sensitive to daylength than 3. Curve 5 is a true (absolute) long-day plant. Curve 6 is a true (absolute) short-day plant, with its minimum day (about 2.5 hr) indicated on the left and its minimum night (8.5 hr, a 15.5-hr day) on the right. Curve 7 is a quantitative short-day plant.

century American farmers. That is, he clearly demonstrated that certain species would flower sooner following an exposure to cold. Scientists around the world took note, and interest began to develop. Plant physiologists in Russia were particularly interested, and the young scientist, T. D. Lysenko, proposed a term, vernalization, for this promotion of flowering by low temperature.

We have already mentioned that Garner and Allard discovered photoperiodism. They found a variety of tobacco called Maryland Mammoth that grew 10 to 15 ft tall but failed to flower in summer at the latitude of Beltsville, Maryland. In winter, however, this plant flowered profusely in the greenhouse. They also had a number of soybean varieties that all flowered at the same time during late summer, regardless of when they were planted during spring or early summer. In a series of carefully planned and executed studies, these scientists experimented to see whether light intensity, temperature, soil moisture, humidity, or daylength could control flowering of their tobacco or soybean plants. Each of these factors differed in the winter greenhouses from the summer fields. They showed unequivocally that flowering was controlled by daylength (Fig. 36–1) and that other factors had no such effect. Plants placed in cabinets in the summer so that they received short days and long nights flowered profusely, whereas plants held under electric lights in the winter so that they received long days and short nights remained vegetative.

Garner and Allard extended their experiments to a number of other plants and found that most of the species investigated could be grouped into one of three classifications (Fig. 36–2): short-day plants such as Maryland Mammoth tobacco and the various soybean varieties; long-day plants such as spinach, which responded in exactly the opposite way by flowering when the days were long and the nights were short; and day-neutral plants such as tomatoes, which flowered on both long and short days (see examples in Table 36–3). They called this response to daylength photoperiodism.

We should emphasize that short-day plants flower when the day is shorter than some maximum that is specific to the variety. The cocklebur is a short-day plant and flowers when the day is *shorter* than about 15.6 hr, whereas henbane is a long-day plant and flowers when the days are *longer* than about 11 hr. Although one of these plants is a long-day plant and the other is a short-day plant, both flower when the days are between 11 and 15.6 hr long.

Most study of photoperiodism has involved flowering in the angiosperms, but it is now known that there

36–3 Representative plants having long-, short-, and neutral-day requirements for flowering and the critical daylength (light period necessary for flowering) for long- and short-day species.

Long-Day Plants (hr)	Critical Day Length
Barley, winter (*Hordeum vulgare*)	>12
Bent grass (*Agrostis palustris*)	>16
Bromegrass (*Bromus inermis*)	>12.5
Reed canary grass (*Phalaris arundinacea*)	>12.5
Clover, red (*Trifolium pratense*)	>12
Coneflower* (*Rudbeckia bicolor*)	>10
Dill (*Anethum graveolens*)	>11
Henbane* annual (*Hyoscyamus niger*)	>10
Hibiscus (*Hibiscus syriacus*)	>12
Oat (*Avena sativa*)	> 9
Orchard grass (*Dactylis glomerata*)	>12
Ryegrass, early perennial (*Lolium perenne*)	> 9
Ryegrass, Italian (*Lolium italicum*)	>11
Ryegrass, late perennial (*Lolium perenne*)	>13
Sedum (*Sedum spectabile*)	>13
Spinach (*Spinacia oleracea*)	
Timothy, hay (*Phleum pratensis*)	>12
Timothy, pasture (*Phleum nodosum*)	>14.5
Wheatgrass (*Agropyron smithii*)	>10
Wheat, winter (*Triticum aestivum*)	>12
Short-Day Plants	
Bryophyllum (*Bryophyllum pinnatum*)	<12
Chrysanthemum (*Chrysanthemum sp.*)	<15
Cocklebur* (*Xanthium strumarium*)	<15.6
Cosmos, Klondyke (*Cosmus sulphureus*)	<14
Kalanchoë (*Kalanchoë blossfeldiana*)	<12
Lespedeza (*Lespedeza stipylacea*)	<13.5
Poinsettia (*Euphorbia pulcherrima*)	<12.5
Rice, winter (*Oryza sativa*)	<12
Strawberry (*Fragaria chiloensis*)	<10
Tobacco, Maryland Mammoth (*Nicotiana tabacum*)	<14
Violet (*Viola papilionacea*)	<11
Day-Neutral Plants	
Cucumber (*Cucumis sativus*)	
Euphorbia (*Euphorbia peplus*)	
Buckwheat (*Fagopyrum tataricum*)	
Strawberry, everbearing (*Fragaria chiloensis*)	
Cape jasmine (*Gardenia jasminoides*)	
Globe amaranth (*Gomphrena globosa*)	
Cotton, one variety (*Gossypium hirsutum*)	
Artichoke, one variety (*Helianthus tuberosus*)	
English holly (*Ilex aquifolium*)	
Balsam (*Impatiens balsamina*)	
Honesty (*Lunaria annua*)	
Tobacco, one variety (*Nicotiana tabacum*)	
Lima bean, one variety (*Phaseolus lunatus*)	
String bean, one variety (*Phaseolus vulgaris*)	
Bluegrass, annual (*Poa annua*)	
Azalea, coral bell (*Rhododendron sp.*)	
Scrofularia (*Scrofularia peregrina*)	
Senecio (*Senecio vulgaris*)	
Potato, one variety (*Solanum tuberosum*)	
Viburnum (*Viburnum sp.*)	
Corn or maize (*Zea mays*)	

*Weeds, the seeds of which may be collected in the field. Seeds or plants of the other species are available commercially. (>12 hr should be interpreted as 12 hours or more; <12 hr, as 12 hours or less. In many cases, critical daylengths are approximations.)

are many plant and animal responses to daylength. Interest is developing in the induction of reproduction in conifers (see Essay 36–1), and reproduction in some of the nonvascular plants (e.g., mosses) is sometimes controlled by photoperiodism. There are also many vegetative responses. Form of the plant is almost always influenced by daylength, and in strawberry varieties flowering is typically promoted by short days and runner formation by long days. There are also many responses in animals, such as migration and reproduction (e.g., in birds), fur color (snowshoe rabbit, ermine), etc.

The discoveries of Gassner and those of Garner and Allard opened several avenues for investigation of plant morphogenesis in response to environment. As virtually always happens in science, answering one question made it possible to ask a number of other questions more intelligently. The first question was: can environment control flowering? The answer was clearly yes. Low temperatures promote flowering of many species, and specific daylengths promote flowering of many others. Now we could ask many other questions: do temperature and light ever interact in flowering? what are the transducers of temperature and light in the flowering process? how are the received stimuli translated into plant responses? are hormones involved? if so, how are they formed and translocated? how do they act? These and similar questions are fundamental to the broad problem of development.

Vernalization

There is a considerable variety among species in their responses and requirements for low temperature promotion of flowering. Some species have an absolute requirement; these plants remain vegetative indefinitely, unless they are exposed to the required interval of low temperature. Others have only a facultative or quantitative requirement; these flower sooner or more profusely, when they are treated with cold, but they eventually flower anyway, even in the absence of this treatment. Plants also vary in the age at which they can be vernalized. One group of plants (notably the winter cereals such as wheat and rye) can be vernalized as seeds or seedlings. Moist seeds respond, but perfectly dry seeds do not. The requirement is a facultative one in that flowering is usually reduced from about 14 to 7 weeks by the cold treatment. This is provided that plants are exposed to long-day conditions after they have germinated and are growing to maturity. The long-day requirement is also a facultative one. Optimum flowering requires both low-temperature treatment and long days.

Essay 36–1 Induction of Reproduction in Conifers

Many important timber species are conifers, so it is of practical as well as theoretical importance to understand something about how their reproduction is controlled. Yet, until recently, little research has been done with the conifers (or for that matter, with any of the perennial plants). The problem is that of ripeness to flower or juvenility. Who can wait around for 20 or 30 years to obtain the results of his experiment?

The question of juvenility is itself a fascinating one. In some perennials such as ivy or certain species of eucalyptus, the juvenile plants are distinctly different in morphology (especially leaf shape) from the adult forms. In ivy the juvenile form is a vine, and the adult is a shrub. The two forms have even been called different species! In the conifers the principal characteristic of the juvenile form is its inability to form cones (although leaves of small seedlings do differ in shape from those of larger but still vegetative plants).

What is the physiology of the change from the juvenile to the adult plant? *A breakthrough came in 1958 when Japanese researchers observed the formation of cones on young conifers treated with gibberellins. Several scientists have followed this lead, particularly Richard P. Pharis at the University of Calgary in Canada. In his experiments more than 90% of the hundreds to thousands of meristems on a conifer seedling less than a year old developed into male or female cones within 18 to 22 days (longer if light intensities are low), when they are treated continuously with gibberellins (Fig. 36–0). In nature, trees must be 8 to 10 years old (or older) before cones are formed. Pharis has had particular success with Arizona cypress, and redwoods have been induced to cone*

36–0 Arizona cypress seedling, age 16 months, sprayed with gibberellin A_3 for four months. There are 8009 meristems on this plant, 99% of which were induced by gibberellin to become reproductive. Only about 50 of these flowers are female.

with massive doses of gibberellins. But pines, firs, and spruces have proven to be resistant to the known gibberellins.

In nature coning is promoted by long days, provided plants are old enough. On the longest days, male cones are produced, and with shortening days, more and more female cones appear. This control of sexuality by daylength still applies when coning is induced by gibberellins. During short days, large doses of gibberellins induce coning. But most of the cones are females, and a longer time elapses before they appear. The Japanese have suggested that more auxin is produced by terminal buds on short days and that this results in more female cones. They report that applied auxin causes femaleness in conifers (as it is known to do in cucumbers and other angiosperms).

Pharis is testing the hypothesis that juvenile conifers grow so rapidly that their endogenous gibberellins are consumed in the growth process. With age, production may catch up with utilization, resulting in a buildup of gibberellins and a consequent termination of juvenility. In addition, other treatments that inhibit growth (and thus might inhibit gibberellin utilization), such as severe pruning or cambial girdling, may often terminate juvenility. (This is also true in some angiosperms such as apple.) In general, the older the conifer, the less applied gibberellin required for coning. Experimental results also imply that more gibberellin is produced on long days. Pharis and his associates have extracted substances with gibberellin activity from several conifers (in parts/billion to parts/million of dry weight). In Arizona cypress, for example, six or seven active fractions have been obtained; one of these is GA–3 (gibberellic acid). The others remain to be characterized.

Phytochrome is clearly present in conifers, and in germination of several conifer seeds it acts just as it does in lettuce and other seeds (i.e., red light, Pfr, *promotes germination). When a long-day response in conifer seedlings (e.g., production of male cones) is sought by interruption of a long dark period, however, far-red is much more effective than red light. Unexpected responses such as this are being reported with increasing frequency, indicating that we still have much to learn about phytochrome action in particular and the induction of reproduction in general.*

Other species respond to low temperature only after they have reached some minimum size. Such a species may be a seedling (e.g., sugar beet) with only a few leaves, or it may be a somewhat larger plant with a rosette of leaves (e.g., henbane, *Hyoscyamus niger*). Frequently, plants in this category flower in response to the low temperature regardless of whether this is followed by long or short days, but sometimes the low temperature treatment must be followed by long days, if it is to be effective. A henbane variety has an absolute requirement for low temperature followed by an absolute requirement for long days; if either treatment is omitted, plants remain vegetative. In at least one case (*Chrysanthemum*), low temperature must be followed by short days. One variety of *Campanula medium* (the Canterbury bells growing in the forests of Fontainebleau) has an absolute requirement for either low temperatures or short days followed by long days. Short days alone, long days alone, or low temperature followed by short days all result in plants that remain vegetative indefinitely. Many species that respond as mature plants are classified as biennials. These plants normally spend their first year in the vegetative condition; then they flower and die during the second year. This occurs because the low temperatures of the winter following the first year's vegetative growth are required for the induction of flowering. When the requirement is absolute (as in the case of henbane, sugar beet, Canterbury bells, etc.), plants can be maintained in the vegetative condition for literally years by keeping them in a greenhouse at warm temperatures (and in the case of Canterbury bells, by avoiding short days by extending the light period artificially).

The winter annuals (the cereals) and the biennials have been most intensively investigated in studies of vernalization. A number of perennials also require vernalization. Several annual garden vegetables flower somewhat earlier in the season, if they are exposed to a short vernalization treatment.

The temperature optimum for vernalization usually lies between 0 and 6 or 7°C, but often measurable vernalization effects can be observed at temperatures as high as 18°C. Oxygen is required for vernalization, as is an energy substrate such as sucrose. This latter fact can be demonstrated by excising embryos from cereal grains and by placing these on an agar medium, where they may be vernalized. Vernalization is much more effective if the medium contains sucrose or some other sugar.

If plants are taken from the low temperature conditions that induce them to flower and treated immediately with rather high temperatures (e.g., 35°C), the effects of the low-temperature treatment are often nullified. This process is called devernalization. If vernalization is a matter of temperature-compensated reactions (as discussed for winter bud dormancy) in which the product of a reaction with a low temperature coefficient reaction is destroyed by a high temperature coefficient reaction, then the high temperature coefficient reaction would be equal to devernalization. It was the observation of devernalization that suggested the compensated-reaction scheme in the first place. Incidentally, if plants are held at a neutral temperature (one slightly too high for vernalization and slightly too low for devernalization) for three or four days, devernalization then fails. It appears that the product of the vernalization reaction is stabilized, that is, converted in still another reaction to a form not subject to devernalization.

Minimum vernalization times vary from 4 days to 8 weeks for the first observable effect. Saturation times vary from 3 weeks for winter wheat to 3 months for henbane.

It is the meristematic cells in the bud that respond to the cold. This has been demonstrated by cooling different parts of the plant with minute cooling coils and also by grafting different parts of a vernalized plant onto an unvernalized plant. If the meristem of a vernalized plant is grafted onto an unvernalized plant, it will be vernalized when it grows out, but if an unvernalized meristem is grafted to a vernalized plant from which the meristem has been removed, it is vegetative when it grows out. There is even evidence that the meristem cannot respond to the cold, unless its cells are actively dividing or unless at least DNA is being replicated prior to cell division. If this is proven to be true, it will be an interesting observation, since it could imply that a marked change in the morphogenetic program can only be effected during cell division or DNA replication.

Vernalization is one of the best examples known of an inductive response. In such a response, no visible change can be detected during the environmental treatment (in this case, exposure to low temperature), but some time later the response becomes apparent (in this case, the formation of flowers). It is as if a "note" were added to the morphogenetic program stating that flowers must develop after a sufficient interval of vegetative growth.

Although the meristematic cells themselves become vernalized, there is also evidence that a translocatable hormone (called vernalin) forms in response to the cold. In a few cases, a vernalized plant grafted to a nonvernalized plant of the same species causes the latter to flower as though some substance were transmitted across the graft union. In view of these (and other) evidences for a vernalin, numerous attempts have been made to extract such a substance from vernalized plants. In a few cases, these attempts have been moderately successful, but in most instances, the active materials have not been identified.

Are gibberellins the postulated vernalins? In the mid-1950s the discovery was made that gibberellins can often cause cold-requiring plants to flower without a cold treatment. Furthermore, gibberellins tend to build up during the vernalization process (at least in a few species that have been studied). But plant physiologists have been reluctant to accept the conclusion that gibberellins are vernalins. When gibberellins are applied to a cold-requiring rosette plant (a biennial), the first observable response is the elongation of a vegetative shoot. Soon thereafter, the shoot begins to develop flower buds. When plants are induced to flower by exposure to low temperatures, however, flower buds are apparent at the very beginning. Thus gibberellins could be acting in some indirect manner in vernalization. Nevertheless, it is becoming increasingly evident that gibberellins play an important role in normal plant development, and it seems more and more likely that this will also prove to be the case in vernalization. Perhaps the gibberellins that have been tried are not the same as the native ones of the species being tested.

Photoperiodism and the Flowering Process

We have seen how the work of Garner and Allard sets the stage for an understanding of how plants could respond to daylength by flowering. We still have only a foothold on an understanding of photoperiodism—one of the most important problems in plant physiology.

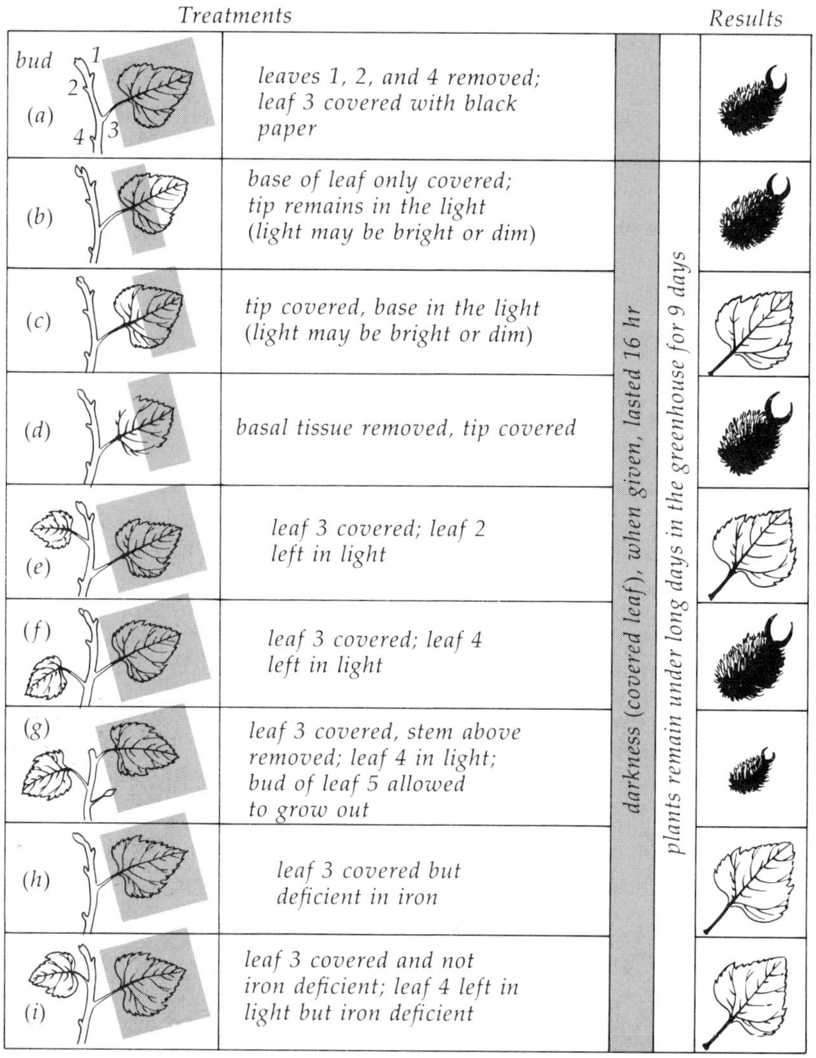

36–4 Experiments demonstrating long-day inhibitory effect in flowering of the short-day Xanthium strumarium *(cocklebur). (a) If all leaves are removed but number 3 (counting number 1 as the first leaf longer than 1 cm) and if this leaf is given a long dark period (covered with black paper), plants flower profusely (large cocklebur symbol). (b) Covering the basal half of the leaf while the tip half remains in continuous light results in flowering almost as high as in the control plant of (a). (c) Covering the tip half, however, while the base remains in the light, results in no flowering at all (leaf symbol). (d) Is the tissue of the tip half of the leaf incapable of responding to the dark period? Not according to this experiment, in which the tip half is covered, while tissue (except for the veins) is cut away from the basal half with a razor blade. (e), (f), and (g) The long-day inhibitory effect can also be observed when one leaf is covered and another is left under continuous light—provided the one left in the light is between the darkened leaf and the bud that will respond. (h) It has been found that leaves deficient in iron could not respond to the long dark period, and this was duplicated in these experiments. (i) If the leaf left on long day is deficient in iron, the long-day inhibitory effect can still be nicely observed; thus the long-night promotion of flowering seems to be chemically different from the long-day inhibition of flowering. It was possible that long-day tissue might interfere with movement of flowering hormone, due to export of sugars from the long-day tissue to the bud. But this seems unlikely because light of low intensities—too low for photosynthesis—is also effective, as shown in (b) and (c). It also seems possible that the long-day tissue might produce an inhibitor that acts against the flowering promoter at the bud. But this, too, appears unlikely because of experiments (b), (c), (e), (f), and (g)—all of which clearly imply that the long-day tissue must be located between the darkened leaf and the responding bud. Perhaps this long-day tissue inactivates or otherwise destroys the flowering hormone as it moves through or nearby the long-day tissue.*

We will now summarize some of the facts obtained and some of the areas of current research. More detailed information is presented in End Paper 36–1.

Which part of the plant responds to daylength? This question can be answered by darkening the leaf, stem, or the tip with black cloth or paper, thus providing short-day conditions for each plant part. Short-day plants treated this way flower only when the leaf is subjected to short days; long-day plants flower only when the leaf remains exposed. Covering the stem or the tip never results in flowering. It is the shoot apical meristem that develops into the flower, but it is the leaf that responds to the environment.

How can events in the leaf control the stem apical meristem? When a plant that has been given a daylength conducive to flowering is grafted to one that has not, the noninduced plant often flowers as though it were receiving a stimulus from the induced plant. This and other experiments support the concept that a hormone is produced in response to daylength in the leaf and transported to the stem tip. The hypothetical hormone was called florigen by the Russian plant physiologist M. Kh. Chailakhyan, who in the 1930s performed grafting and leaf- or bud-covering experiments such as those just described. (Recently, evidence has been found that flower–inhibiting substances are produced under noninductive conditions; Fig. 36–4.)

Can the flowering hormone be extracted? An obvious step is to attempt to make extracts from induced plants and apply them to noninduced plants and then wait for them to flower. This procedure has been tried countless times by many investigators all over the world, and most attempts have failed. During the 1960s an increasing number of successes have been reported, and perhaps in the relatively near future, this problem will finally yield an answer.

Essay 36–2 Discovering the Photoperiodic Sensitivity of the Cocklebur

The accident that results in scientific discovery has always figured prominently in popular accounts of scientific progress. In most cases, such accidents are overemphasized. It is the astuteness of the observer, rather than the accident itself, that causes the discovery. A good example is contained in this account by Karl C. Hamner of the University of California, Los Angeles. Here he discusses the discovery of the cocklebur as the almost ideal plant for research in photoperiodism.

"Upon receiving my Ph.D. from Chicago in 1935, I accepted a position in the U.S. Department of Agriculture at Beltsville, Maryland, to work on the problem of photoperiodism in plants. While there, I worked with Harry Borthwick and Marion Parker, and we spent most of our time investigating the Biloxi soybean. In 1937, I left the USDA to accept a position at the University of Chicago. Upon arrival at Chicago, I felt I should try to find another plant to work with, since Borthwick and Parker were planning to continue research on Biloxi soybean. I arrived in Chicago in the fall and noticed in the greenhouse some cocklebur plants that bore full-sized burs, although the plants were no more than two or three inches tall. Upon inquiry, I learned that Professor Charles Shull had been experimenting with

cocklebur seeds in connection with his studies of water relations of germinating seeds. Some of the seeds had fallen underneath the greenhouse bench and there had germinated. They were exposed to the short days of the fall, and, apparently, flowered promptly upon emerging from the soil and produced flowers and fruits while the plant was still very young. It seemed to me that cocklebur would prove to be a very sensitive plant. I immediately collected some burs from the vacant lots in the vicinity of the greenhouse and initiated some preliminary experiments.

At about this time Edith Niedle expressed the desire to do graduate work under my direction. I suggested that she work with cocklebur and study the effects of mineral nutrient supply on the photoperiodic and flowering responses. She immediately set out several hundred pots of seedlings in nutrient culture. I suggested that she maintain the plants under long-day conditions (using artificial illumination at night) in order to keep the plants vegetative during the early stages of growth. When her plants were several inches high, she came to me one day and protested that all of her plants had small flower buds, even though she was maintaining them under long-day conditions. I looked at her plants and agreed that, indeed, flower buds were present. At this time, I almost discarded the plant as experimental material. However, my earlier experiments had indicated such great prospects for the plant that I was skeptical, and I asked the greenhouse help whether or not the lights had been on every night throughout the experimental period. At first, the greenhouse man and the Buildings and Grounds personnel insisted that the lights had never failed. However, when I appeared so disappointed, one of the men said: 'Well, I think the lights were off for one night a while back; and I can look it up to see when the power failure occurred.' Sure enough, he reported back a little later that the lights had been off all night on one particular occasion. But he insisted that that was the only occasion when the lights were off at night. Edith Niedle and I discussed this for quite a while before we decided that it might be possible that the cocklebur could be induced to flower by only one short day, or more particularly, one long night.

We threw out all of the cocklebur plants and started over again with new seeds. We grew the seedlings under long day (lights on at night until 2 A.M.), and after they had reached about eight inches in height, we gave a group of plants one single short day, or more specifically, a dark period of 16 hr during one night, and returned the plants to long day. Within two weeks, it was apparent that all of these plants had been induced to flower as a result of this single exposure to darkness. Thus we 'discovered' the short-day plant, the cocklebur."

Is it possible to divide the flowering process into component parts? The answer is yes and is based largely on work with cocklebur (Xanthium strumarium), a short-day plant that responds to a single inductive cycle (see Essay 36–2). Several experiments with cocklebur are shown in Figure 36–5.

Cocklebur plants flower when they are exposed to a single dark period longer than about 8 hr and 20 min. But this dark period is not effective unless it has been preceded by light of sufficient intensity and duration. The first part of induction (the steps leading to flowering) is called the high intensity light process. The following reactions of the dark period can be divided into time measurement and florigen synthesis. If the cocklebur is exposed to a longer and longer night, when some critical length is reached, flowering occurs. (Long-day plants respond in an opposite way to increasing daylengths.) As the critical night for short-day plants is exceeded, flowering occurs at an increasing rate. This can be studied with a cocklebur by observing the extent of development of the apical floral bud some days after exposure to a single long dark period (a so-called short-day cycle). The buds are classified according to a series of floral stages (Fig. 36–6), and the results are plotted as a function of the length of the single inductive dark period (Fig. 36–7). The critical night (about 8.3 to 9.7 hr, depending upon conditions) implies the time-measuring part of induction, and the increasing levels of flowering with increasing night lengths beyond the critical length imply the florigen synthesis part. Apparently, more florigen is produced the longer the dark period.

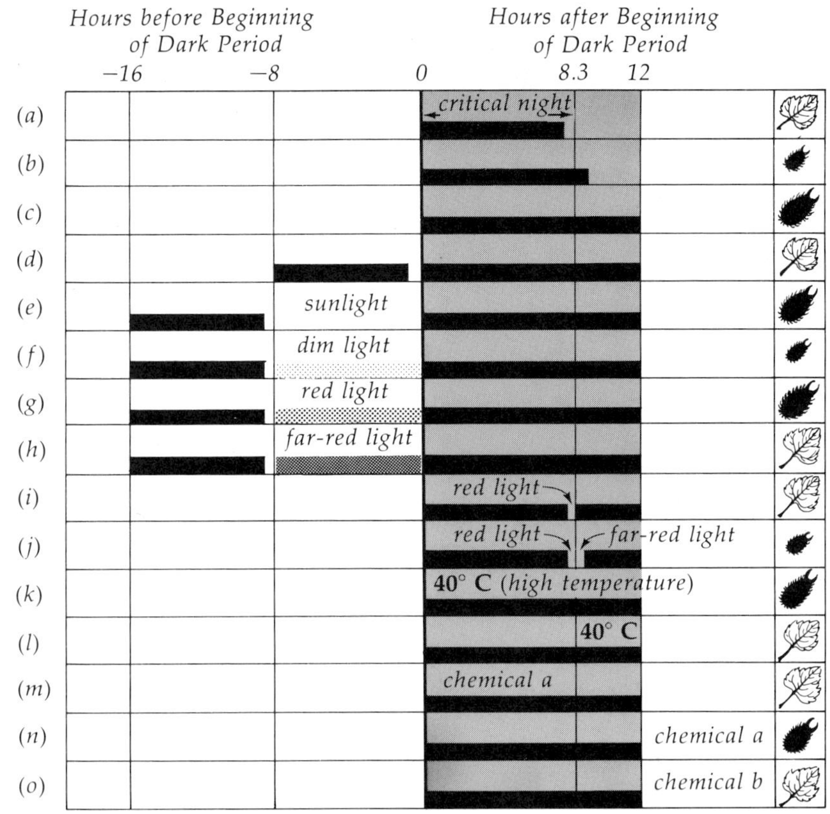

36–5 A number of experiments in which the cocklebur plant is treated with a single, long dark period to induce flowering. (a) Control plants given a dark period shorter than the critical night (shorter than 8.3 hr) fail to flower (leaf symbol). (b) Plants exposed to dark periods slightly longer than 8.3 hr flower at a low level (small bur symbol). (c) Plants given 12 hr flower optimally (large bur symbol). (d) If the 12-hr inductive dark period is preceded by about 7.5 hr of darkness and a brief interval of light, plants remain vegetative; the 12-hr dark period is ineffective in causing flowering. (e) If the 7.5-hr dark period is separated from the 12-hr dark period by high-intensity light for 5 hr or more (8.3 hr shown here), flowering occurs. It is optimal with about 12 hr of intervening light. (f) Flowering is reduced somewhat if the intervening light is of low intensity, but it still occurs. (g) and (h) Red light is most effective, and far-red light is inhibitory when applied between the two dark periods. (i) If the 12-hr inductive dark period is interrupted at about 8 hr with red light, flowering is inhibited—although red light was promotive in (g)! (j) If the red light interruption is followed by exposure to far-red light, flowering is restored. (k) If high temperature is given during the first part of the dark period, flowering is not affected. (l) If the high temperature comes after the critical night, however, flowering is inhibited—as though the flowering hormone were destroyed. (m) A chemical such as dinitrophenol, applied in the middle of the dark period, may inhibit flowering. (n) If the same chemical is applied after the end of the dark period, flowering may not be inhibited—as though this chemical inhibits synthesis of the flowering hormone. (o) Some other chemicals may be applied several hours after the dark period and still inhibit flowering—as though they are inhibiting translocation of the flowering hormone or perhaps the development of the flowering bud.

The next component of the flowering process is florigen translocation. Following induction, florigen is translocated from the leaf to the bud. This can be studied by giving cocklebur the required inductive light, then removing leaves from different groups of plants at various times following induction. Plants are vegetative when leaves are removed immediately after the dark period, but when leaves are removed some hours later, the plants flower nearly as well as controls from which leaves are not removed (Fig. 36–8). Apparently, a certain time interval is necessary for the translocation of florigen from the leaf to the bud.

The final component of the flowering process is the differentiation of the bud. When florigen arrives at the bud, growth is redirected, and the bud differentiates into a flower, as illustrated for the male flower of cocklebur by the floral stages (see Fig. 36–6).

Each of these events has been studied in some detail, and work has proceeded by seeking answers to questions raised by earlier studies.

How is time measured in photoperiodism? This is a perplexing question for which we do not have a definite answer, although we have plentiful speculation (some of which is presented in End Paper 36–1). Using cocklebur, Karl Hamner and James Bonner asked in 1938 whether it was the length of the night or the length of

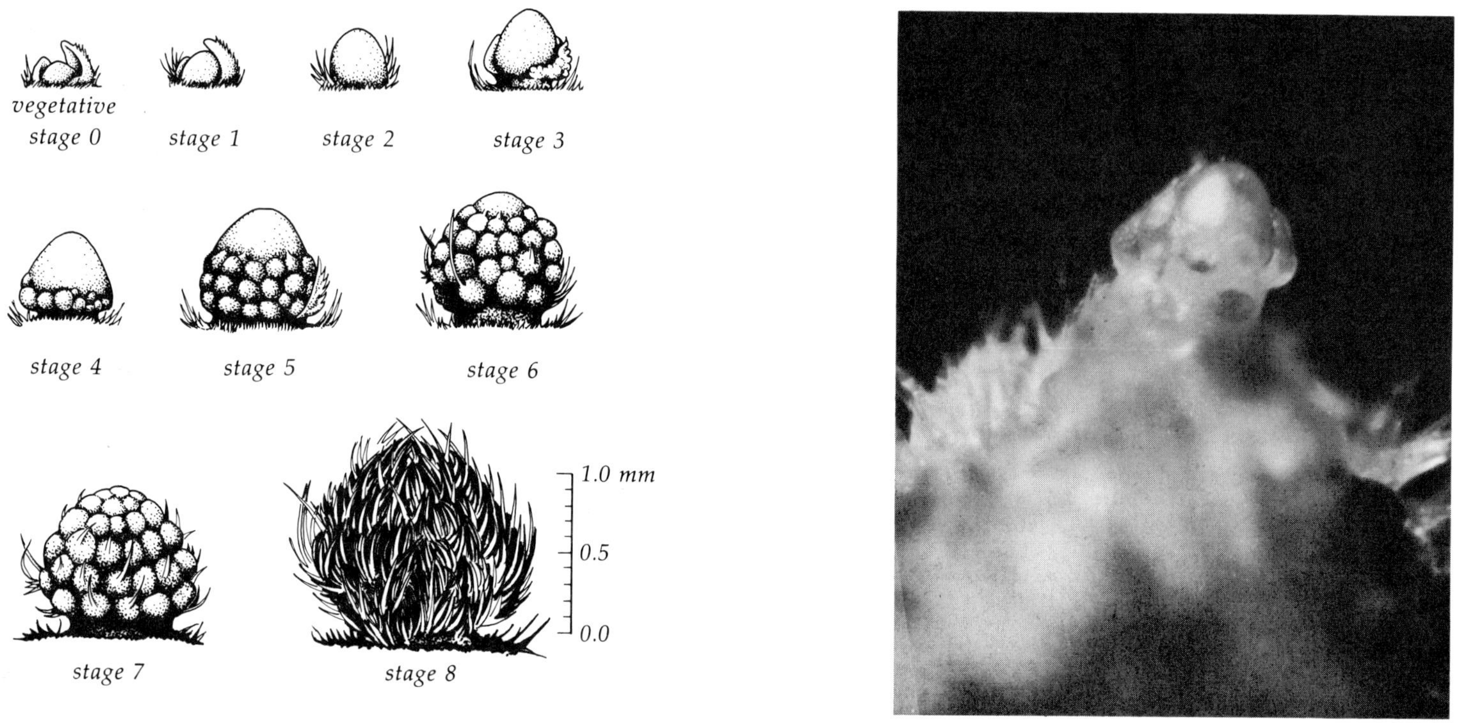

36–6 *Left: Drawings of the developing terminal inflorescence primordium (staminate) of cocklebur, illustrating the system of floral stages. Right: A photograph through a dissecting microscope of a cocklebur inflorescence primordium at stage 3.*

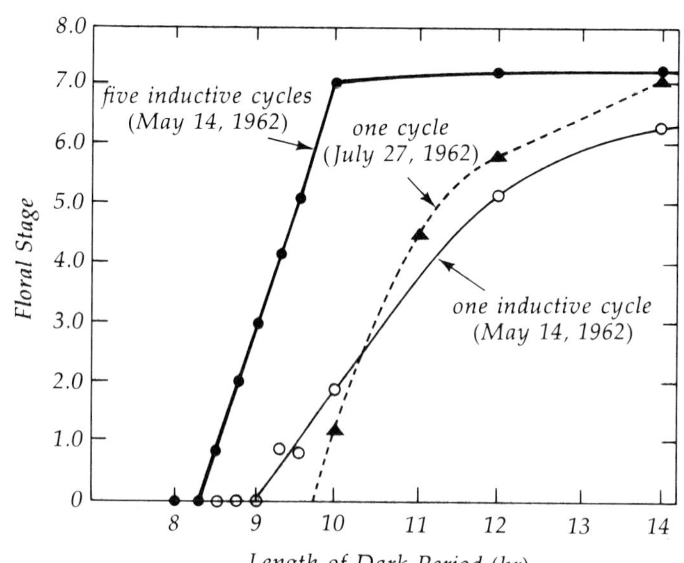

36–7 *Three experiments showing floral stage as a function of night length. The actual length of the critical night depends somewhat on experimental conditions (such as temperature and light intensity before and after the dark period and temperature during the dark period), especially with a single cycle (two curves on the right). With more than one cycle (left curve), the 8.3-hr critical night is always approached very closely.*

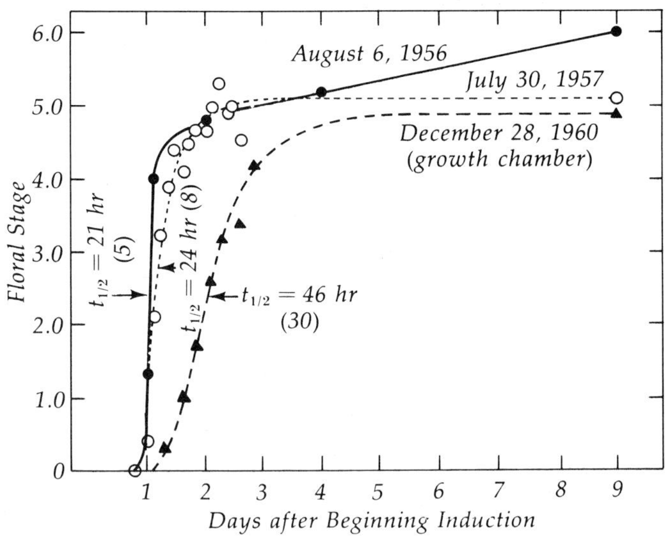

36–8 Three translocation curves obtained by defoliation of plants at various times following a 16-hr inductive dark period. Times of defoliation are shown, with floral stages of all plants being determined on the ninth day. Numbers on the abscissa represent noon of the indicated day, while the bar on the abscissa indicates the inductive dark period. Approximate times when half the stimulus was out of the leaf are indicated by $t_{1/2}$. The figures represent half-times after beginning of the dark period; the figures in parentheses represent half-times after end of the dark period. Dates refer to the day plants were subjected to the dark treatment. Plants represented by the $t_{1/2}$ = 46 hr-curve were kept under 2000 ft-c of fluorescent light in the growth chambers (23° C).

the day that was critical to induction. They found that interrupting the dark period with a light break completely inhibited flowering, but interrupting the light period with an interval of darkness had no effect. This seemed to prove that time was measured only during the dark period, but subsequent experiments have shown that the day is equally important in time measurement. (Days shorter than those used by Hamner and Bonner are effective; see End Paper 36–1.)

In any case, Hamner and Bonner's discovery of the night-break phenomenon allowed other experimenters to probe the question of how time is measured in photoperiodism. The data implicate the phytochrome system discussed in Chapter 35. When plants are first put in the dark for photoperiodic induction, their phytochrome must be predominantly in the *Pfr* form, since this is the result of exposure to daylight. During the dark, *Pfr* decreases and *Pr* builds up. While it seemed reasonable that time measurement could be a matter of the time required for this conversion of phytochrome, accumulating evidence has cast doubt on this concept. It is now clear that time measurement in photoperiodism and time measurement in the circadian rhythms are similar, but research (presented in End Paper 36–1) has shown that the two clocks are probably not identical.

How is florigen synthesized in the leaf? Using chemicals that are known to inhibit florigen synthesis, it has been possible to learn something of the biochemistry of florigen production. Energy in the form of ATP is needed as well as RNA and protein synthesis. One powerful inhibitor of steroid metabolism inhibits flowering, perhaps implying that florigen is a steroid (a complex class of molecules, having 4 six-membered carbon rings as a basic part of their structure, that are effective in certain phases of animal metabolism—several are sex hormones and adrenal cortex hormones).

What happens when florigen arrives at the vegetative bud? The arrival of the florigen causes a shift in the developmental program of the apex. This shift converts the vegetative apex that was producing leaves to a floral apex that produces (in a regular manner) a whole range of flower parts of elaborate design. This conversion is related to the production of new RNA, protein, and enzymes. The number of ribosomes increases, and there are changes in histones and starch. Unfortunately, we know little of the changes that take place. Yet it would appear to be a good place to suspect that certain genes are activated, while others may be deactivated.

Practical Implications of Photoperiodism Research

Research on photoperiodism in plants has been intensively pursued for many years not only because of its basic biological importance but because of the tremendous agricultural implications that it has. Control of flowering in conifers might revolutionize the lumber industry, for example (see Essay 36–1). Many have imagined the advantages that a flowering hormone would give if it could be easily sprayed on a field so that a given number of days later the entire crop would flower and set fruit. Crops might be grown in regions where they would normally not have time to flower before winter set in. Mechanical picking might be possible with plants that cannot be handled this way now, since they flower and fruit over a long period of time. In this case, perhaps, a spray with florigen would cause them all to fruit at the same time (as is done now with one crop, pineapple, where auxin induces flowering). How many such things will be possible when (and if) florigen is available remains to be seen, but it is thoughts like these that sometimes motivate basic research.

End Paper 36–1 Photoperiodism and the Biological Clock

Many facts about flowering in response to daylength have accumulated since Garner and Allard's work in the 1920s. To begin with, a number of additions to Garner and Allard's three basic response types have become apparent. The concept remains the same, but it is now known that many varieties do not *absolutely* require either long or short days to flower—flowering is promoted by one of these treatments. Thus we have the same situation as in vernalization: both absolute and facultative responses, with a few species that are difficult to classify (they fall between these two categories). Furthermore, there are one or two plant species that flower only on intermediate daylengths; that is, when days are neither too short nor too long (see Table 36–3). Examples are even known for the opposite situation: plants that flower when days are either rather long or rather short but not intermediate in length. Several species require long days followed by short days (called long-short-day plants), and a few require short days followed by long days (short-long-day plants). Some plants require or are promoted by a given daylength only at a certain temperature, being day-neutral at other temperatures. Typically, the environmental requirements for flowering are very plastic within a species. Different varieties (or even individuals) may respond to different environmental factors.

A few examples have been reported in which flowering seems to be controlled—sometimes in an absolute way—by factors other than daylength or temperatures. P. Chouard claims, for example, that one species is controlled in its flowering by levels of humidity. Sugarcane is reported to flower profusely when grown on one kind of soil in Hawaii but not when grown on another kind. Here, we seem to have a control by soil-nutrient conditions, which was considered so important during the early decades of flowering study.

Following the discovery of photoperiodism by Garner and Allard, investigators all over the world combined photoperiod experiments with studies on nutrition. It was discovered that nutrition may often influence the level of flowering, but it seldom has an absolute control over flower initiation. High nitrogen levels do indeed inhibit the flowering of several species, as had been proposed in the carbohydrate-nitrogen theory. But flowering of other species (e.g., the cocklebur) is strongly *promoted* by nitrogen fertilizers—a finding exactly contrary to the old ideas. Again, we are faced with the concept of a holistic environment and must realize that many environmental factors interact and could be expected to be influencing. Cocklebur plants deficient in iron, for example, will not respond to short-day induction by flowering, yet such extreme deficiencies as those required for this effect seldom, if ever, occur in nature. We are entitled to be impressed by the absolute controlling effects of light and temperature observable with so many species.

How old must plants be before they will respond to photoperiod? As in vernalization, plants vary considerably in the minimum age that they must achieve before they reach ripeness to flower. None respond as seeds, but some respond as small seedlings. *Chenopodium rubrum* (Lamb's quarters) responds to short days as a tiny seedling only a few days old in a petri dish (Fig. 36–9). The cotyledons of Japanese morning glory are sensitive to a single short day, but most herbaceous plants must be on the order of 10 to 30 days old, and cotyledons typically will not respond at all. Some trees do not flower until they are 5 to 40 years old, depending always upon the species. Often a plant that will not respond is said to be juvenile (see Essay 36–1).

Species also vary in the number of short-day or long-day cycles required to induce flowering (to cause induction). A few known species such as cocklebur, Japanese morning glory, and Lamb's quarters (short-day plants), or perennial ryegrass and white mustard (long-day plants), respond to a

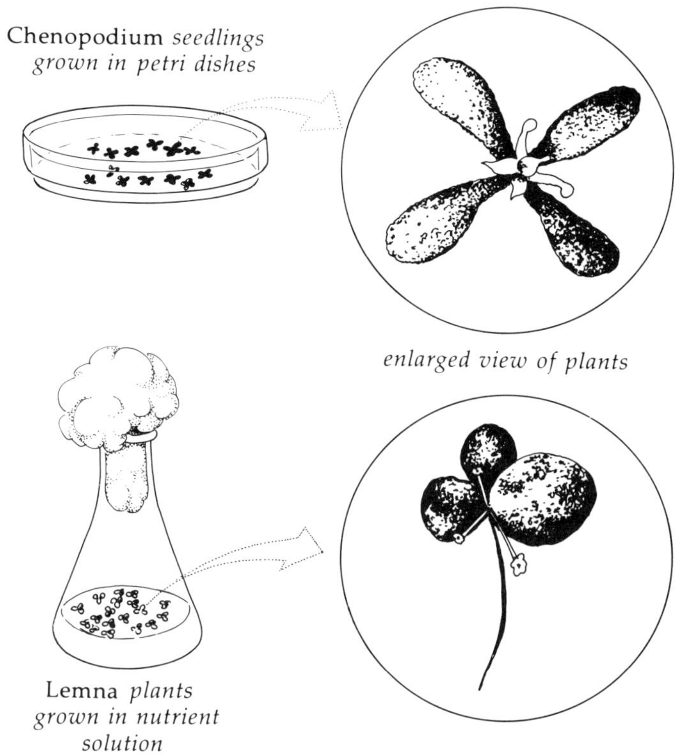

36–9 Chenopodium rubrum *seedlings grown and treated in a petri dish, and* Lemna perpusilla *plants grown and treated in a nutrient solution in a flask. Note minute flowers on the* Chenopodium *seedling and three flowers and a root on the* Lemna *plant.*

single cycle. After this they can be returned to conditions that would not otherwise result in flowering, but flowers continue to develop. This is an example of the induced state, as impressive as that discussed in relation to vernalization. Many other species require several cycles to attain the induced state. Chrysanthemum flowers actually fall off (abort), even when almost completely developed, if the plants are taken from the short days that cause flowering and placed under long-day conditions.

Time and Photoperiodism

In this chapter we asked the question "how is time measured in photoperiodism?" and provided a summary answer. Now, let us ask the same question, but this time we will provide more data for the answer. A start came in Hamner and Bonner's experiment in which they asked the logical question arising from the fundamental observations of Garner and Allard: *which is more important in the induction of flowering, the day or the night?* They combined constant daylengths with varying nightlengths and constant nightlengths with varying daylengths. They found that the critical night remained rather constant regardless of length of day, thus implying the controlling importance of the dark period. To further investigate the matter, they performed the night-break experiment. Light during the night completely inhibited flowering, further supporting the importance of the dark period. They then suggested that we should speak of long-night plants rather than short-day plants and nyctoperiodism rather than photoperiodism. Fortunately, the new terminology was never widely accepted, since the day is important in time measurement after all. The night-break phenomenon has provided a legion of research possibilities.

How much light is required for effectiveness? If the interval is brief (seconds), the intensity must be high (hundreds to thousands of foot-candles); if long (hours), low intensity suffices (1 f-c). In general, we can see that the Law of Reciprocity holds in this case, as it does in other photobiological responses. Actually, when the matter of reciprocity is investigated in detail, it is found that it holds only to an approximation. This is largely because the effectiveness of the night break depends upon when the interruption is given.

What color is most effective in a night break? Red is clearly most inhibitory in short-day plants and most promotive in long-day plants, and far-red at least partially reverses the effects of red. This demonstration of phytochrome implication in cocklebur flowering immediately followed the initial discovery of red-far-red reversibility in germination of lettuce seeds.

Does Pfr inhibit or does Pr promote flowering? In studies by R. J. Downs, a red-light interruption was given in the middle of the dark period, and this was followed by a far-red exposure (allowing various intervals of time to elapse following the red exposure). It was shown with cocklebur and soybean that the far-red exposure was completely ineffective if 30 min were allowed to elapse after the red exposure. This seems to imply that the *Pfr* produced by the red light inhibits flowering and completes its inhibitory act within 30 min after it is formed. After this time, converting it back to *Pr* makes no difference.

As a by-product of the search for answers to these questions, some interesting effects were observed with far-red light. It was found, for example, that when far-red light is applied for a relatively long interval of time, effects are identical to those produced by red light applied for a short interval of time (see Essay 35–1). Apparently, this is because of the overlapping absorption spectra of the two forms of phytochrome. *Pr* absorbs red light most effectively, but it also absorbs a small portion of far-red. Thus when plants are exposed to far-red light, some of the phytochrome is continually being converted to *Pfr* (and rather rapidly converted back to *Pr* by the far-red light). The result is that a small portion of the phytochrome always exists as *Pfr*, and this seems to be enough to cause inhibitory effects in flowering of short-day plants.

When is the light break most effective? This is a difficult question, since the answer depends upon the length of the dark period, the species, etc. With cocklebur, maximum inhibition occurs about 8 hr after the beginning of the dark period for nights from about 10 to 24 hr long. Time measurement, then, seems to be a matter of the plant's measuring the critical night, at which time the flowering process is most sensitive to red light.

What is the nature of time measurement in photoperiodism? When plants are first put in the dark for photoperiodic induction, their phytochrome must be predominantly in the *Pfr* form, since this is the result of exposure to daylight and most artificial light sources. Yet after 8 hr, *red light* inhibits flowering. Apparently, the *Pfr* has been converted to *Pr*. Could time measurement be a matter of the time required for conversion of *Pfr* to *Pr?* This reasonable possibility was accepted for several years as the explanation for time measurement in photoperiodism. But accumulating evidence has cast doubt on this concept. For example, low light intensities (threshold light) that partially inhibit flowering (implying that some *Pfr* is maintained) do not affect the critical night or the time during an inductive dark period, when plants are most sensitive to a red-light interruption. If time measurement is not changed even when some level of *Pfr* is maintained in the leaf, then we can hardly accept the idea that conversion of *Pfr* to *Pr* accounts for time measurement.

What is the relationship between time measurement in photoperiodism and time measurement in the circadian rhythms? Several investigators approached this problem during the 1960s. Hamner and his students and associates at the University of California in Los Angeles were especially active. They found that there is a rhythmical sensitivity to light applied at various times during dark periods extended up to 72 hr (three days). They also found that flowering of soybeans was promoted best when the dark period and the light period added up to 24 hr or multiples of 24 hr.

Salisbury has also investigated this relationship between circadian rhythms and photoperiodic timing. Even the cocklebur goes through a daily cycle in sensitivity to light. In a basic experiment (Fig. 36–10) a dark period (the phasing dark period) too short to cause flowering was first given.

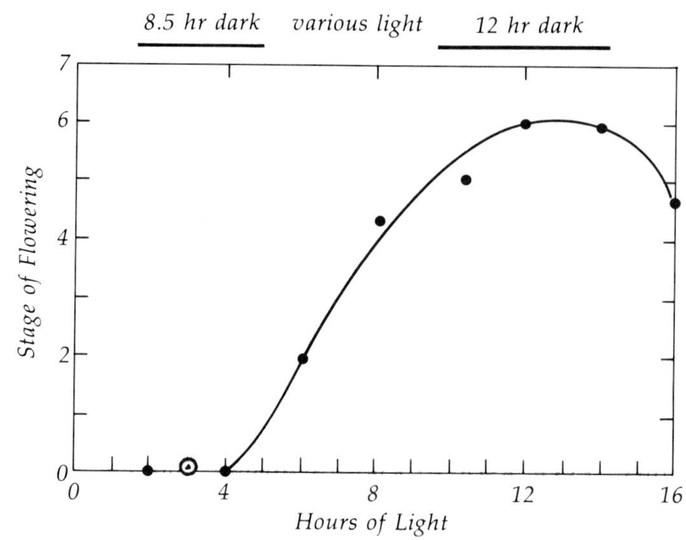

36–10 Flowering response as a function of length of the intervening light period in an experiment symbolized by the bars above the figure.

This was then followed by an intervening light period and a final test dark period, (long enough to normally cause flowering). In such an experimental set-up, it was possible to study the characteristics of the intervening light period. It was found that flowering was inhibited to zero by intervening light periods shorter than five hours (Fig. 36–10). Increasing the light period beyond 5 hr led to an increase in the level of flowering. This demonstrates that the light period is also important in flowering. Indeed, the response is similar to that of a long-day plant, since increasing light duration increases flowering. The importance of the light period was overlooked in the early experiments of Hamner and Bonner because the critical day (in the long-day sense) for cocklebur is only 5 hr. The cocklebur plant may be classified, then, as both short-day and long-day! It was possible to show that red light was highly effective in *promoting* flowering during the intervening light period, and far-red light was *inhibitory*—responses exactly opposite to the ones familiar to us in studies of the night-break phenomenon. Hence plants cycle in their photoperiodic response in sensitivity to red and far-red light. Promotion is therefore obtained by one wavelength at one time, and inhibition is obtained at a later time.

Several workers have shown that photoperiodic time measurement is temperature compensated, another feature common to time measurement in circadian rhythms. This can be shown by studying both the critical dark period and the time of maximum sensitivity to a night break as functions of temperature.

Resetting of the photoperiodic clock was also studied. It was found that night breaks in cocklebur seemed only to delay the clock slightly when given before 4 hr after the beginning of darkness, but when given after 6 hr, a complete shift in the status of the photoperiodic clock could be observed. This again is a feature common to circadian rhythms. Clearly, then, the photoperiodic clock and the circadian rhythm clock have a lot in common.

But are they identical? Salisbury, working with Alice Denney, has recently concluded that they are not. Leaf movement rhythms and photoperiodic flowering have been studied simultaneously in a number of experiments with cocklebur. It has become apparent that the status of the photoperiodic clock cannot be determined by the position of the leaves or by any features of the leaf-movement rhythms (e.g., whether the leaves are moving up or down at a given moment). The tentative hypothesis at present is that the two clocks must be separate but similar in function.

37

Descriptive Ecology

Excerpts from a report:

We, the inhabitants of planet Gamma in the Alpha Centauri star system, are approaching our nearest galactic neighbor, the solar system. I am the expedition's ecological biologist. We have turned off the hyperdrive and are now decelerating to velocities appropriate for exploration of the solar system's planets. We have been measuring their orbits and other features. Considering life as we know it (based upon carbon compounds of DNA and protein), the third planet out from the principal luminary appears to be a promising habitat for life. Its distance from the sun is appropriate for a temperature allowing the existence of liquid water. Indeed, we have spectroscopically detected large quantities of water, oxygen, and nitrogen in an atmosphere around that planet. This is very promising. We have noted that the planet rotates once in 24 hours. This relatively rapid change between day and night will, on the average, not allow conditions to become too hot or too cold. Its equatorial plane is tipped 23° to the plane of its only slightly elliptical orbit. This means that surface conditions will be continually changing as the planet orbits the sun; that is, it will have seasons. The planet has a relatively nearby, fairly large satellite, which should result in tides, if large bodies of water are present.

The chances for life are certainly high. Will life function the same as on our own planet? Will even the ecological principles be the same? The planet is so massive that only a few hydrogen atoms will be able to escape its gravity. Space is relatively empty in its vicinity, so the matter that it sweeps up from space will be quite inconsequential, at least during fairly brief time intervals. Thus its matter exchange can be ignored, but there will be an important energy exchange. It will be receiving large quantities of heat from its sun, approximately 2 cal/cm^2 (normal to the sun's rays) per minute (2.56×10^{18} cal/min for the planet as a

37–A Most of Africa and portions of Europe and Asia can be seen in this spectacular photograph taken on July 17, 1969 from the Apollo 11 spacecraft during its translunar coast toward the moon. Apollo 11, with astronauts Neil A. Armstrong, Michael Collins, and Edwin E. Aldrin, Jr., aboard, was already about 98,000 nautical miles from earth when this picture was made.

whole). Unless the planet is to be continually heating up, all this energy must either be lost again into space or converted to some other form, such as chemical bond energy. The planet, then, is a closed ecological system with respect to matter but an open system with respect to energy.

. . . We are now approaching the planet rather closely. We have just passed its satellite, and we will go into orbit around the planet while we reconnoiter. This is truly a beautiful planet. Brilliant hues of blue and white are the overwhelming colors, although we glimpse traces of greens, reds, and others. It is spectacular when compared to most of the other planets and satellites in this system—although the fourth planet was rather interesting with its dominant reds. Our immediate impression is that this third planet is rather densely covered with clouds, but looking more closely, we begin to discern surface features. Much of the planet appears to be covered by water. The color is right, and broad solar reflections from the calculated points confirm the relatively smooth surface. We have now detected traces of methane in the atmosphere, and these could not exist for long intervals of time in the presence of so much oxygen, unless they were being regenerated by life processes. Indeed, the oxygen itself must be regenerated by life. Its rate of production by ultraviolet breakdown of water in the upper atmosphere would probably be slower than the rate it is tied up by oxidation on a dead planet. Perhaps the vast oceans are teeming with life. Excitement is rising in our group!

. . . The planetary characteristics predicted a rather wide range of temperatures due to day and night, seasons, and changes in latitude. These would lead us to expect a great variety in the living organisms on the planet, which we are now beginning to detect. There seem to be regions, both north and south of the equator, that are relatively cloud-free and that are definitely reddish in aspect. These must be the belts in the planetary atmosphere where air is descending and being warmed. Under such conditions, precipitation will be fairly rare, and deserts should result. The planet's rotation results in large, cyclonic air movements (nicely visible in the cloud patterns), and this will result in prevailing westerly winds over the large continents of the northern hemisphere. Desert regions appear to the east of mountain ranges on these continents. Here, the descending air tends to warm and thus not produce precipitation. In the far north the color suggests a light green, and we might suspect reduced vegetation due to coldness. Clouds are heavy over the coastal areas at moderate latitudes, especially to the east, but surface features that show through appear to be a dark green. Here, moisture must be abundant, resulting in extensive vegetation and perhaps tree forms. Extensive vegetation also occurs in the vicinity of the equator.

We are approaching an earth, then, that is a gem of beauty in the vast reaches of space—one that is essentially a closed ecological system (closed to matter but open to energy) and that has a wide range of climatic conditions resulting in a vast variety of vegetation types. How shall we begin to study these life forms?

If the extraterrestrial visitors from Alpha Centauri are impressed by the variety of the earth's surface as it appears from space, they will certainly be overwhelmed when they land and begin to explore that variety in detail. Earthly ecologists discovered long ago that it is not an easy task to describe something as complex as a biological community. Study of the community structure and function of such groups of organisms is called synecology (in contrast to autecology, the study of individual organisms in relation to their environments).

Yet if we are to understand communities, we must begin by describing them. The *act* of description itself might tell us something about community function. For example, the descriptions carried out during the first four decades of this century had strong philosophical implications. Ecologists of that period were led to the concept of the super organism. Just as the component parts of an organism are its cells, so the components of a discrete, super-organism community were said to be the individual organisms. The ideas of plant succession and climatic climax that developed seemed to imply that a biotic community has a sort of embryogenesis comparable to that of single organisms; that is, the mature adult community is determined by climate and the genetics of all its organisms.

In this chapter we shall investigate some of the methods applied to the study of plant communities. We should note, however, that the term "plant community" might be somewhat artificial and even misleading. All communities are *biological,* containing animals as well as plants. Nevertheless, we can arbitrarily restrict our study to the plants. Finally, we shall consider some of the implications that were drawn from these plant-community investigations.

Methods of Studying Community Structure

Even a rather casual observation reveals some aspects of plant community structure. Most communities have their dominant organisms: the trees of the forest, the grasses of the prairie, the shrubs and succulents of the desert, the herbs of the tundra, and the phytoplankton of the oceans. These various groups are referred to as growth forms, and they may be subdivided well beyond these few categories (see the tables in Chapter 38).

In a forest an extensive vertical structuring is evident. In tropical forests there may be several layers of trees, with a few reaching great heights and some extending above the general canopy (Fig. 37–1). There may be a level of shrubs and, below that, a level of herbs.

Ecologists have investigated horizontal structuring in some detail (Fig. 37–2). As it turns out, plants are rarely, if ever, distributed on the earth's surface in a completely random fashion. Often, plants of a given species are clumped together. Sometimes, as in the desert, individuals or clumps of individuals are distributed in a fairly regular fashion. Mathematical tools have been developed to define rather accurately the horizontal arrangement of plants in a community.

Over long distances, plant distribution on the earth's surface correlates well with climate (the deserts, grasslands, forests, etc.). In mountains there is a belting that is obviously a function of altitude. It was long ago observed that moving up a mountain in elevation is, so far as the vegetation is concerned, much like moving from the equator toward either pole. For example, the tundras of the north country are duplicated on the tops of high mountains, and the taiga (coniferous forests) south of the arctic tundra are duplicated in horizontal belts of trees below the alpine tundra.

The most casual observer also recognizes that in some climates there is a great change in vegetation with the progression of time. This is not only true in forests, where deciduous trees are barren part of the year, but it is also true in grassland and other communities, where one species may be in full flower and conspicuous during a part of the year and other species

coniferous forest profile

deciduous forest profile

tropical rain forest profile

37–1 Stratification in coniferous, deciduous, and tropical rain forests.

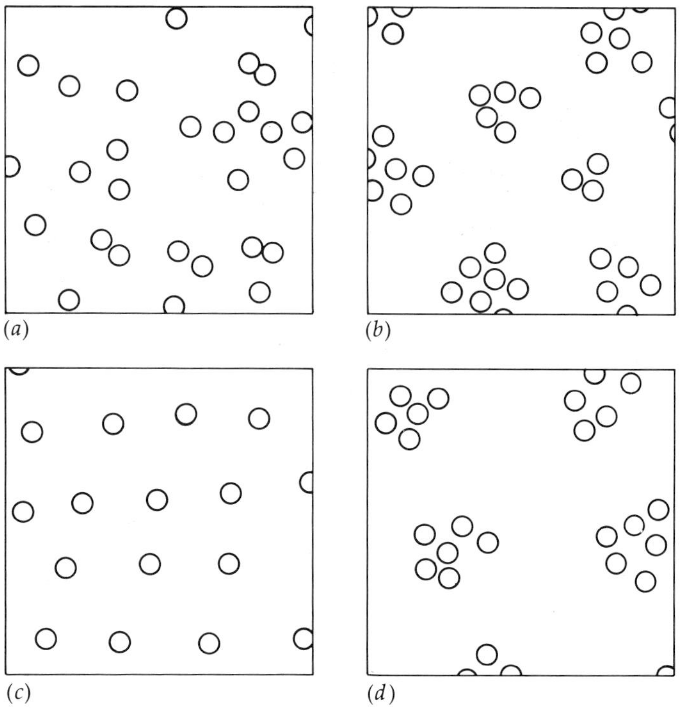

37–2 Horizontal structuring of plants in a community: (a) random distribution, (b) random distribution of clumps, (c) ordered distribution (perfect ordering—as in an orchard—does not occur in nature) and (d) ordered distribution of clumps.

during different parts of the year. The phenomenon of one species being particularly evident at a given time of year (usually due to flowering) is called aspect dominance.

Discrete Associations

During the first few decades of this century, the science of ecology was developed by a number of highly energetic, enthusiastic, original thinkers. The results of these studies were known to the public mostly as natural history, but the individuals involved in the research called themselves ecologists and recognized their responsibilities as developers of a science. To begin with, their conclusions were based largely upon observational studies. Eventually it became evident that quantitative measurements would have to be made to test the hypotheses they were developing. Thus an extensive series of methods were developed to study plant communities. In the meantime, a conclusion about the nature of communities had already begun to form, and it appeared to most ecologists (until as late as about the mid-1950s) that their measurements supported this conclusion.

The individual who might be considered the leader of ecologists (certainly in America) during this period was Frederic E. Clements. He was a brilliant, almost flamboyant naturalist, who occupied a chair of plant physiology, authored books in plant taxonomy, and strongly influenced the development of ecological thought from about 1898 (when he received the Ph.D.) until his death in 1945.*

The conclusion (really a hypothesis to be further tested) developed by Clements and others during this period was that vegetation occurs in discrete associations. The deciduous forest, for example, consists of a maple-beech association, an oak-chestnut association, and an oak-hickory association. Of course, it was recognized that the various associations blend into each other at their boundaries; these regions of overlap were called ecotones. The rationale for the concept of discrete associations was that in plant succession, species react on their environment, changing it so that each member of the succession is better suited to the newly created environment than its predecessor, eventually ending in a climax community that is best able to exist in an environment that it has itself created. The ability of developing successional communities to create their own environments intensified the super-organism concept (proposed as early as 1901 and 1905 by Clements) and seemed to provide a logical basis for a discrete, super taxonomic entity—the specific plant community.

---

*One of the many books written by Clements was entitled *Greek and Latin in Biological Nomenclature* (1902). Clements was clearly intrigued by the use of words in science. His theories of succession and the super-organism community (which he and Shelford later called a biome, implying the participation of both plants and animals) were buttressed with a vast array of technical words that he invented. He even had Greek and Latin names for the ecologist's tools: shovels, picks, and the like. This was to improve the image of the new science!

The theories developed by Clements and others went well beyond the facts that were available at that time. The ecological science of 1938 was at least as intricate and as complex as it is today (see the textbooks authored by Clements and Shelford and Weaver and Clements in that year). Yet significant portions of what was accepted then is now known not to be true.

Is the work of the early ecologists irrelevant today? Of course not. A number of the methods and concepts were those that our present science expanded and built upon. Furthermore, the enthusiasm and excellent writing of the ecologists of that period had a considerable effect (most of it indirect) upon the public—in Europe as well as in the United States. Paul Sears, a student of Cowles, did much excellent writing that strongly influenced American ecology. We are experiencing a rebirth of public interest in ecology at this time, but interest was high in the earlier decades of this century. The dust bowls of the 1930s were known to be the result of poor ecological management, and much of this knowledge and public awareness can be ascribed to the researches and the writings of Frederic Clements and his contemporaries.

Plant Sociology

The basic method used by Clements and others to study vegetation was that of the quadrat. A plot is laid out, and plants contained in this area are studied in various ways. The species may simply be listed; the individuals may be counted; areas covered by plants may be measured or estimated; or an accurate map may be drawn of the vegetation, possibly using a pantograph (a device sometimes used by artists to enlarge or reduce a drawing). In addition to the quadrat, transects are often used in ecological study. A narrow belt will be laid out through the vegetation, and all of the species lying within the belt are measured in various ways. Sometimes a device is used in which a number of pins or pointers are held in a linear arrangement. This is slid along the vegetation, and every plant touched by a pin is listed.

Much investigation has gone into the question of how to sample vegetation. Sizes and shapes of quadrats have been studied. Frequently, in herbaceous vegetation, a quadrat covering 1 m^2 is used. If the plants are small, the quadrat may be even smaller than a square meter. In a forest or other large vegetation, the quadrat must be much larger, sometimes an acre. Shape makes a difference. Fewer 0.25×4.0 m quadrats are needed to get a representative number of species than when 1.0×1.0 m quadrats are used. How many quadrats to use in studying a community may be determined by plotting the number of species as a function of the number of quadrats (Fig. 37–3). Stanley A. Cain suggests that when a 10% increase in quadrat area results in only a 10% increase in species, enough quadrats have probably been used. In order to apply proper statistics, quadrats must be distributed at random. (Clements emphasized that distribution must *not* be random, but rather according to the "judgment" of the investigator. Of course, such judgment is properly called bias.)

The use of quadrats was raised to a high level of sophistication by Josias Braun-Blanquet in France, and other Europeans contributed to the development of these techniques. Braun-Blanquet referred to the science he was developing as plant sociology (or phytosociology). Each species received a certain numerical value that characterized it in certain ways. For example, the frequency of a species in a community was the percentage of quadrats in which it occurred. Of course, this number depends upon quadrat size and shape but is relatively constant when the significant factors are held sufficiently stable. Cover is the percentage of the area covered by a given species. Density is the average number of individuals per unit area. (Range managers in the United States have sometimes used density as cover is defined here.) Braun-Blanquet also suggested several other phytosociological factors such as sociability, vitality, and fidelity.

In the 1950s other methods were developed. In the random-pairs method of studying forest trees, for example, an imaginary grid is laid out in the stand to be studied. (A stand is that real community being investigated.) From each point, the investigator faces the nearest tree 10 cm or more in diameter at breast height. He measures the tree and then finds the nearest tree to the point that is on the other side of an imaginary line drawn at right angles across the point and a line from the first tree to the point (Fig. 37–4). Data are then taken for the second tree, and the distance between the two trees is measured. By applying calculations to these various measurements, information on the number of trees per unit area, the percentage of each tree species, average sizes, etc., can be obtained.

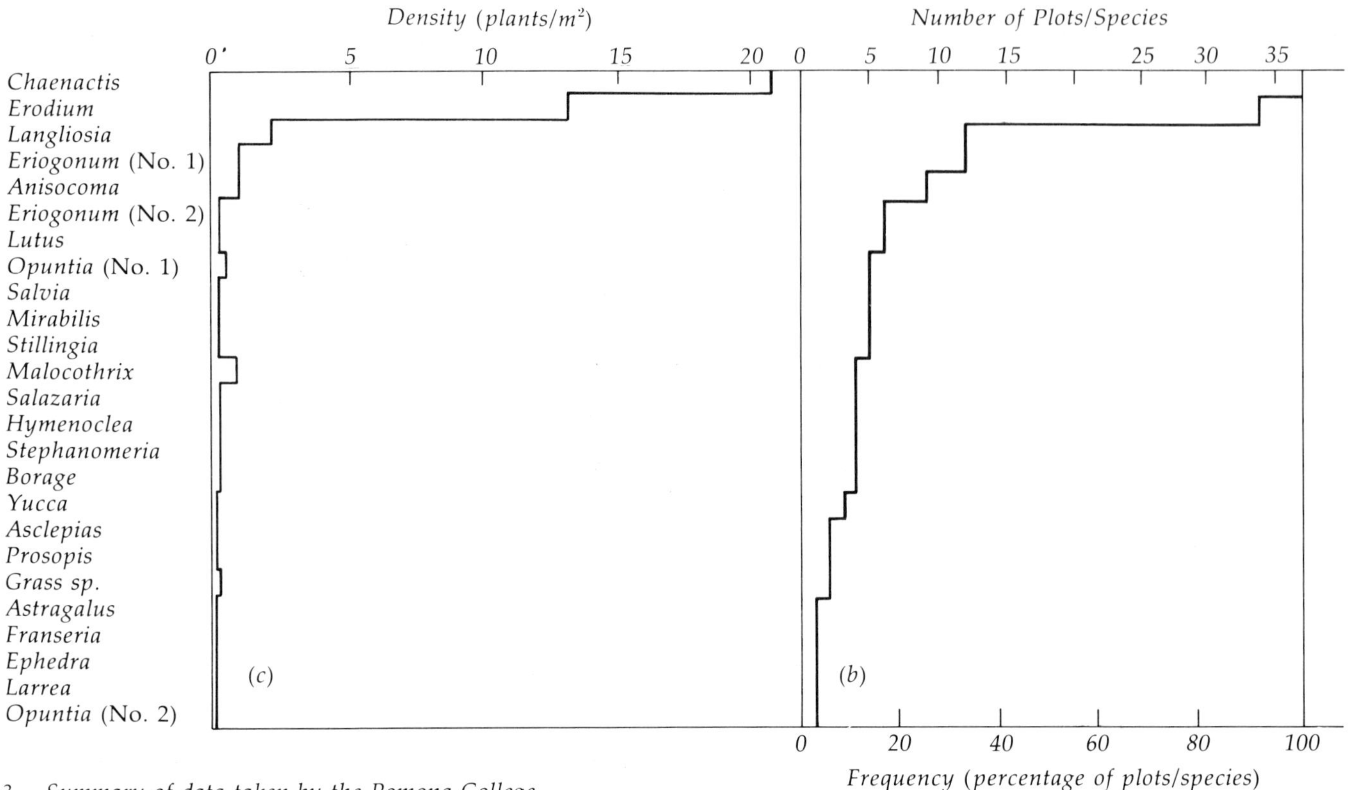

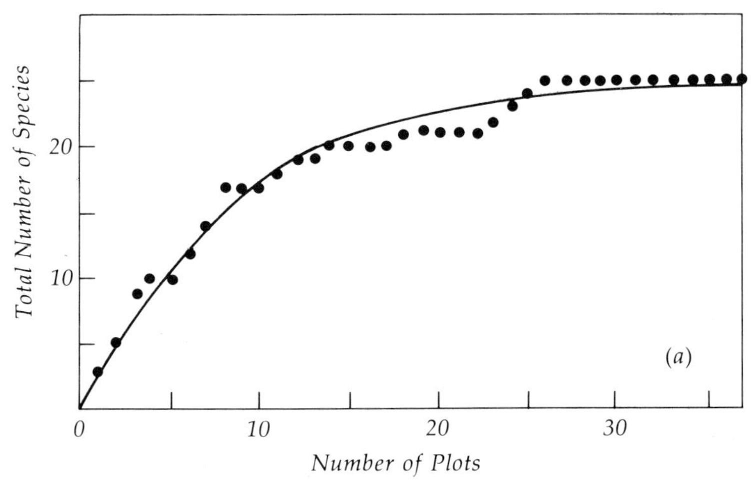

37–3 Summary of data taken by the Pomona College ecology class in Morongo Valley (Joshua Tree National Monument, California) on April 30, 1954. Thirty-seven 1-m² plots (2 × 0.5 m) were laid out at 50-ft intervals along a transect. (a) Total number of species plotted as a function of number of plots along the transect. Most species had been sampled by the 14th plot, although a few others appeared as the transect moved into a slightly different vegetation type (plots 23 to 26). (b) The number of plots containing a given species and percent of the plots containing a given species (frequency). Obviously, Chaenactis *and* Erodium *were the most predominant in this community, although* Opuntia, Larrea, *and two* Yuccas *appeared (because of their size) to dominate the community. (c) Density of each species. This is the number of plants per square meter, and again* Chaenactis *and* Erodium *are the most densely distributed.*

Now it is possible to use sophisticated photographic techniques and to evaluate the results with the help of computers. One example is the continuum index, which may be calculated with a computer from the stand data.

The Continuum

In 1938 Clements published a revised edition of a book with Weaver called *Plant Ecology* and a new book with Shelford in 1939 called *Bio-Ecology*. In these books

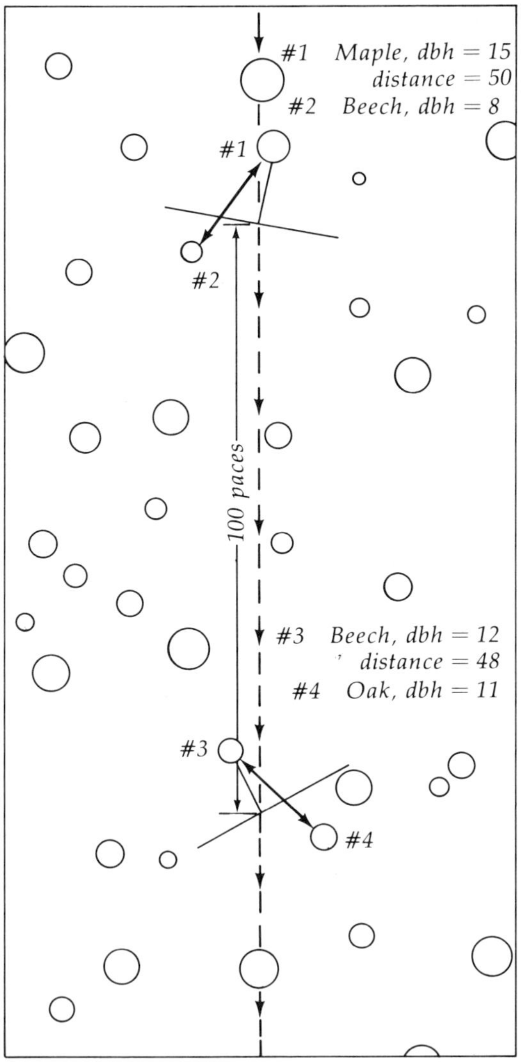

37–4 Method of random pairs used to sample trees in a forest. The observer walks along a surveyed line through the stand of trees. At 100-pace intervals, he stops and faces the tree closest to him. He then raises his arms, forming an imaginary line at right angles to the line between himself and the tree. The second tree of the pair is the nearest one behind this line. Diameters at breast height (4.5 ft) are measured for both trees; so is the distance between them. The species of each tree is recorded. To evaluate the data, distances between all pairs are added, and the sum divided by the number of observation points, thus producing an average of distances. This is multiplied by 0.8 and squared to give the average area occupied by one tree. This may be divided into the area of an acre or other unit to estimate the number of trees per acre or other unit. Percentages for each tree species, average basal area for each species, and other characteristics of the community can also be calculated.

he restated, in eloquent and intricate detail, his concepts of the discrete association and the super organism. But all was not well in the ranks of American ecologists.

As early as 1926, Henry A. Gleason, at the New York Botanical Garden, had taken issue with Clements in a scientific paper.* It was first met with opposition and then ignored. But Gleason published a second paper on the individualistic concept of plant distribution in 1939. This was an extremely well-written paper, presenting a devastating attack on the discrete association and super-organism concepts with a logic as elegant as Clements'. Gleason pointed out that if an association was discrete in the sense that it contained a number of plant species that were essentially restricted to that community, then all of these species ought to have the same environmental requirements, and the entire association should have a uniform environment. This is contrary even to casual experience. He concluded: "Since every community varies in structure, and since no two communities are precisely alike, or have genetic or dynamic connection, a precisely logical classification of communities is not possible." He went on to say that his "individualistic concept is totally at variance with the idea that the association is an organism. . . ." Gleason provided examples, although these were not documented: "Within the state of Michigan, the beech-maple climax forest, always considered to be a definite, well-distinguished association-type, exhibits profound changes from one end of the state to the other." While the philosophy of discrete climatic climax associations and the super organism might be intriguing and stimulating, it is contrary to

*A very similar paper was published in Russia in 1924 by the ecologist L. G. Ramensky. This was followed by field measurements published in 1930. Ramensky really deserves credit for discovering the continuum concept, although his ideas had no impact upon American ecologists until the concept was already well established in America.

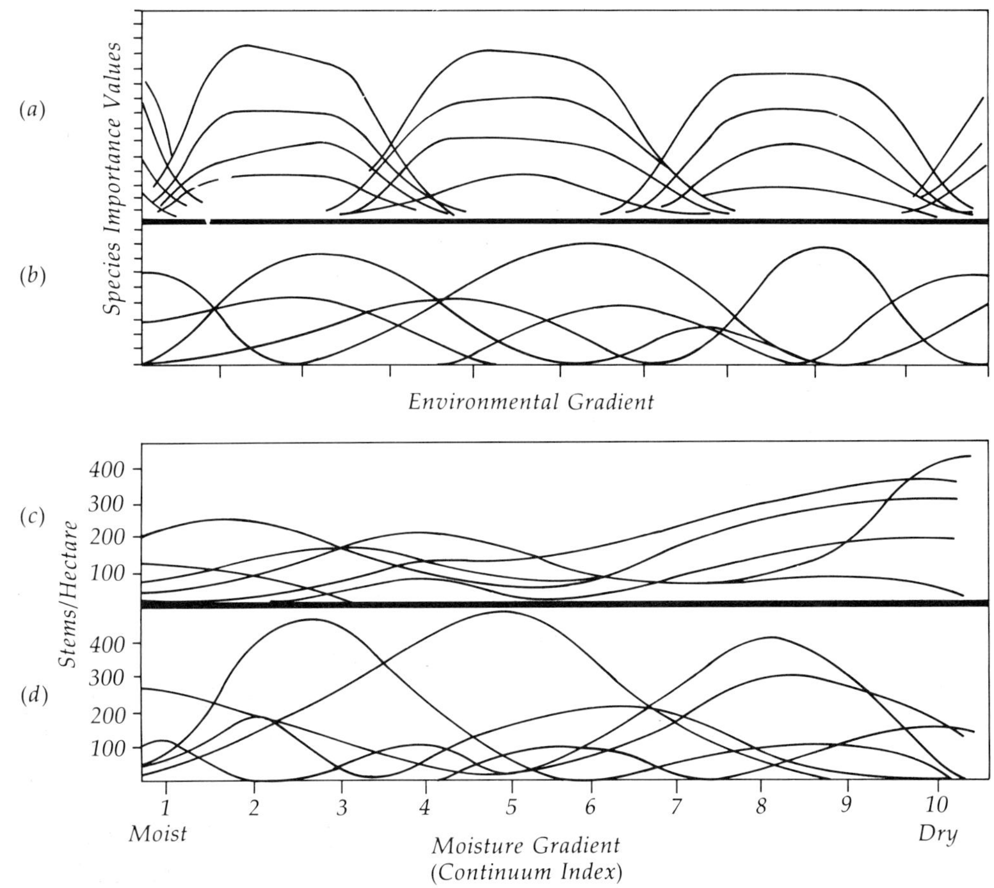

37–5 Expected and observed plant distributions according to the concepts of discrete communities or continua. (a) Curves for discrete communities. Species-importance values are plotted as a function of distance along some environmental gradient. Species occur in discrete units that overlap at their boundaries. (b) Curves for continuum. Each species is distributed according to its own environmental requirements. (c) An actual example. Trees in the forest are plotted as the number larger than 2 cm in diameter per hectare (10^4 m^2) as a function of position along a moisture gradient. The gradient extends from moist ravines to dry southwest-facing slopes in the Siskiyou Mountains (760–1070 m elevation). (d) Similar data from the Santa Catalina Mountains (1830–2140 m elevation).

the facts, Gleason seemed to say. Of course, Clements (and others) had been aware of numerous exceptions like Gleason's, but Clements had rationalized them away.

Things move rather slowly in research areas such as this, but by the early 1950s there was a full-blown controversy developing among American ecologists. Clements died in 1945, but most American universities had ecologists who had been strongly influenced by Clements or by those who shared Clements' views. By then, however, a number of young ecologists had joined Gleason's side of the conflict. One of these was Robert H. Whittaker (see Essay 37–1), then a graduate student at the University of Illinois, and another was John J. Curtis, an ecologist at the University of Wisconsin. Some other ecologists such as A. G. Vestal and Walter Cottam were also impressed. Herbert L. Mason and Stanley A. Cain questioned the logic of the association in separate papers published in 1947, and W. Dwight Billings presented data on the shadscale vegetation zone of Nevada and eastern California that were hard to reconcile with the association concept. He mentioned Gleason, Cain, and Mason but failed to follow through with the needed vegetational analysis. Still, most ecologists accepted the discrete association concept.

Clearly, what was really needed was a large dose of facts and analyses and a much smaller portion of philosophy. These must be collected in a truly objective way (not by placing quadrats according to one's judgment, as Clements had suggested). The most objective approach is to make measurements, obtaining numerical data that represent objective facts that have bearing upon the problem at hand.

In the late 1940s Curtis and Whittaker were obtaining these data, quite unaware of each other's efforts. Measurements were made over long distances along climatic gradients, frequently using the same quadrat techniques that were developed by Clements and others, but often adding modifications to improve objectivity. Stands were measured at intervals of a few miles apart in the deciduous forests of Wisconsin and at different elevations and topographic positions in the Great Smokies. A continuum index was developed (a rather complex procedure), along which the various stands could be located. The importance of each species in a stand could then be plotted as a function of distance along this scale. The discrete association idea predicted that the curves for a number of species should tend to be superimposed upon each other. On the other hand, if communities were not discrete, but each species had its own distribution pattern, then the curves should overlap along the scale (Fig. 37–5). The results were quite clear: as one moves along horizontally on the earth's surface, there is a *continuum* of stands, each somewhat different from the others in species composition; as the environment changes gradually in a continuum, so do the species making up the stands. The discrete association concept had to be discarded.

Essay 37–1 An Hypothesis Rejected: The Natural Distribution of Vegetation

Sometimes graduate study is the most exciting and rewarding interval in a person's life. When the thought processes are stimulated by conflicting but logical views backed by brilliant individuals, when the student has the free rein to develop his own ideas and then to test them objectively—discovering, perhaps, that the unexpected answer proves to be far more exciting than the expected one—when all this has the physical stimulus of long data-collecting hikes under nature's canopy, then a student may achieve an intensity of living that will seldom be duplicated. Robert H. Whittaker, now at Cornell University in New York, tells about such a period in his life.

"The University of Illinois at Urbana in 1946 was a stimulating environment when I came there as a graduate student in ecology. Victor E. Shelford in the Zoology Department, one of the major figures of early American ecology, represented the still strong influence of Frederic E. Clements. Clements had considered that natural communities were something like super organisms and formed well-defined units called associations and formations or biomes. In the Botany Department, Arthur G. Vestal subjected Clementsian ecology to a keen, analytic scrutiny—and rejected it almost wholly in favor of conceptions related to those of Henry A. Gleason. Gleason had argued (in 1926) that plant species distribute themselves 'individualistically,' that is, each according to its own genetics, physiological characteristics, and ways of relating to environment (including the effects of other plants). Species, by this view, would not fit themselves into clearly defined, sharply bounded natural units. With Samuel C. Kendeith, a distinguished ornithologist with ideas on communities developed from the Clementsian tradition, as my thesis advisor and with Vestel as my second advisor, I worked in a cross fire of views on communities. Not only did the Clementsian and Gleasonian viewpoints conflict with

each other, each conflicted with a third viewpoint—the life zones of Charles H. Merriam, with which I had sympathy (based on experience before coming to Urbana).

The topic of community units was thus subject to an exciting confusion. I began groping toward an understanding of my own and was led (with influences of both life zone and biome concepts and of Charles Elton's book Animal Ecology*) to frame an hypothesis. Because of their many interactions, species should tend to evolve toward co-adapted groupings, each representing a favorable, balanced pattern of species interactions adapted to some range of environments. Between these favorable combinations should lie transitions of 'community-level hybridization' with less balanced and more changeable mixtures of species populations. The favorable combinations should appear in some circumstances as life zones; biomes would in some cases correspond to the zones and would in others be groupings of a number of zones of similar vegetation structure. The distributional expression of the hypothesis could be tested. Unless Gleason was right, one could measure species population levels along a continuous environmental gradient and identify groups of coadapted species with parallel distributions, separated by transitions with irregular and rapidly changing population levels.*

Kendeith proposed that several of his graduate students do theses on animal communities in Great Smoky Mountains National Park, Tennessee and North Carolina. I welcomed the opportunity to test my hypothesis in connection with a study of insect communities in a superlative preserve of natural vegetation. In the summer of 1947 I sampled foliage insects in a number of different kinds of plant communities and also in a transect along the elevation gradient from deciduous forests up into spruce-fir forests. Fairly early in the summer it became apparent to me that, first, the latter test of my hypothesis would fail because of the irregularity of insect population levels and, secondly, the former study would lack effectiveness without knowledge of how the plant communities being sampled related to one another and to environments.

Hence I set out to study the vegetation of the Great Smoky Mountains. To test my hypothesis, I took field transects or a series of plant-community samples along particular environmental gradients. But this was not enough, since I was confronted with a range of plant communities the complexity of which I was observing in fascinated but baffled admiration. How to study, how to find order in this complexity? As a broader basis of vegetation study I took 300 samples, largely at random over as much of the surface of the mountain range as I could reach by extensive hiking. This sampling procedure was quite unorthodox at the time. I rejected the usual practice of subjectively choosing samples to represent particular associations, feeling that the natural units I was looking for should emerge from the relations of objectively chosen samples to one another. I assumed also, but without clear plans at the time, that I could fit the many samples together into transects to test my hypothesis. My sample procedure was an intuitive gamble—somehow I could reduce the Smokies vegetation to a comprehensible order.

Back in my student cubicle at Urbana I compiled my data on plant populations in the field transects, then grouped larger numbers of the 300 samples into elevation transects, and then discovered that I could use weighted averages of species composition of the samples to group them into transects of topographic moisture gradients. There emerged in all these transects a characteristic pattern—bell-shaped population distributions of species, with scattered centers and broad overlaps that together formed a population continuum. My hypothesized groups of co-adapted species with parallel distributions were not there and (with certain limited exceptions) the transitions I had been looking for were not in evidence as such, since the many kinds of forest communities intergraded continuously. I spent some time in reluctant astonishment at the overturning of my hypothesis—followed by delight in the feeling of having made a first-order discovery of far-reaching implications for ecological research and understanding.

I then tried plotting my 300 samples, and densities of plant populations in them, on a chart with elevation and the topographic moisture gradient as axes. The diversity of communities that had baffled me fell into place as a pattern of intergrading combinations of species corresponding to the pattern of environments. The vegetation was to be understood, not as a mosaic of associations, but as a coherent design, a subtly wrought tapestry of differently distributed species populations variously combining to form the mantle of plant communities covering the mountains. In my thesis, in 1948, I stated the conception of communities as forming complex and predominantly continuous patterns of populations. I called my approach to the study of such patterns gradient analysis, *a technique for analyzing relations of gradients of environments, species populations, and community characteristics, to each other. The pattern concept led later to the study of vegetation of other, Western mountain ranges and to inquiry into the evolutionary meaning of patterns and their relation to problems of species diversity.*

The discovery should be placed in further context. The scientific stature of the men with whom I worked at Urbana, who both stimulated my thinking and encouraged my research when it departed from their ideas, must be mentioned. It was the ideas of Gleason, caustically rejected by many ecologists when proposed, that my work gave belated vindication. I do not know why Gleason's ideas had been untested for a generation, but the time was ripe for myself and others in 1947–1950. During that period John T. Curtis and his students at the University of Wisconsin came independently to the ideas of species individuality and community continuity. We found at a meeting in 1950 that we were co-discoverers of a new approach to communities. Still later, we learned that Gleason's ideas had been independently advanced in Russia by L. G. Ramensky (1924 and thereafter).

In this account you may note a number of themes—the continuity through generations of scientific developments, the role of readiness of timing and of independent discovery, the close interrelation of new theory and new method, the fortunate choice of object of study, the recognition of unresolved conflict of understanding, and in many cases the essential intuitive jump toward resolution—that are frequent elements in the advance of science."

Why were Clements with his discrete associations and Merriam with his life zones wrong? Clements and Merriam were expressing in detailed scientific language what *seemed* to be rather obvious. If one observes from a distance the San Francisco peaks of northern Arizona (where Merriam worked), the belting in vegetation is clear and undeniable (Fig. 37–6). At the lowest elevations there are the desert plants (the Upper Sonoran vegetation). Then there are the oaks and other deciduous trees (a transition zone). Higher in elevation is an extensive forest of Ponderosa pine (the Canadian forests); spruce-fir forests (the boreal and Hudsonian forests) occur in the mountains themselves; and above tree line is the alpine tundra. So what is so wrong with all this? The life zones of Merriam and the discrete associations of Clements were named upon the basis of the *dominant* vegetation only. The pictures described by these workers applied fairly well to the few species that were large enough to dominate the landscape. When measurements were made, however, and *all* the species were considered objectively in terms of their importance in the community, then it became apparent that life zones and discrete associations were a considerable oversimplification.

If the discrete-association climatic-climax idea had to go, then what about the super organism? Obviously, the concept must lose some of its impact. The super organism is not a taxonomic entity, but an individual organism always is. The individual organism has a portion of the genes constituting the gene pool of the population, but there is nothing comparable about the super organism. Nevertheless, the community or ecosystem does have certain analogies to an organism in the processes of energy flow, cycling or other exchange of elements and compounds, and the developmental sequences of succession. We will examine these functions in Chapter 40.

Actually, our studies on contemporary evolution (carried out particularly in the 1920s and 1930s) should have made us suspicious of the discrete association idea. For example, the niche-exclusion principle states that no two species can occupy *exactly* the same niche. If species cannot occupy the same niche without one or the other tending to dominate, then it would be surprising indeed if a large number of species could be adapted to a single area (thus constituting a discrete community). It is far more reasonable to expect that competition at the ecotones would tend (when climate controls rather than soil) to distribute species along climatic gradients in a continuum. According to observational data, this is exactly what happens.

We have emphasized the question of discrete associations versus a vegetational continuum because this problem is an excellent illustration of how science progresses. But descriptive studies of plant com-

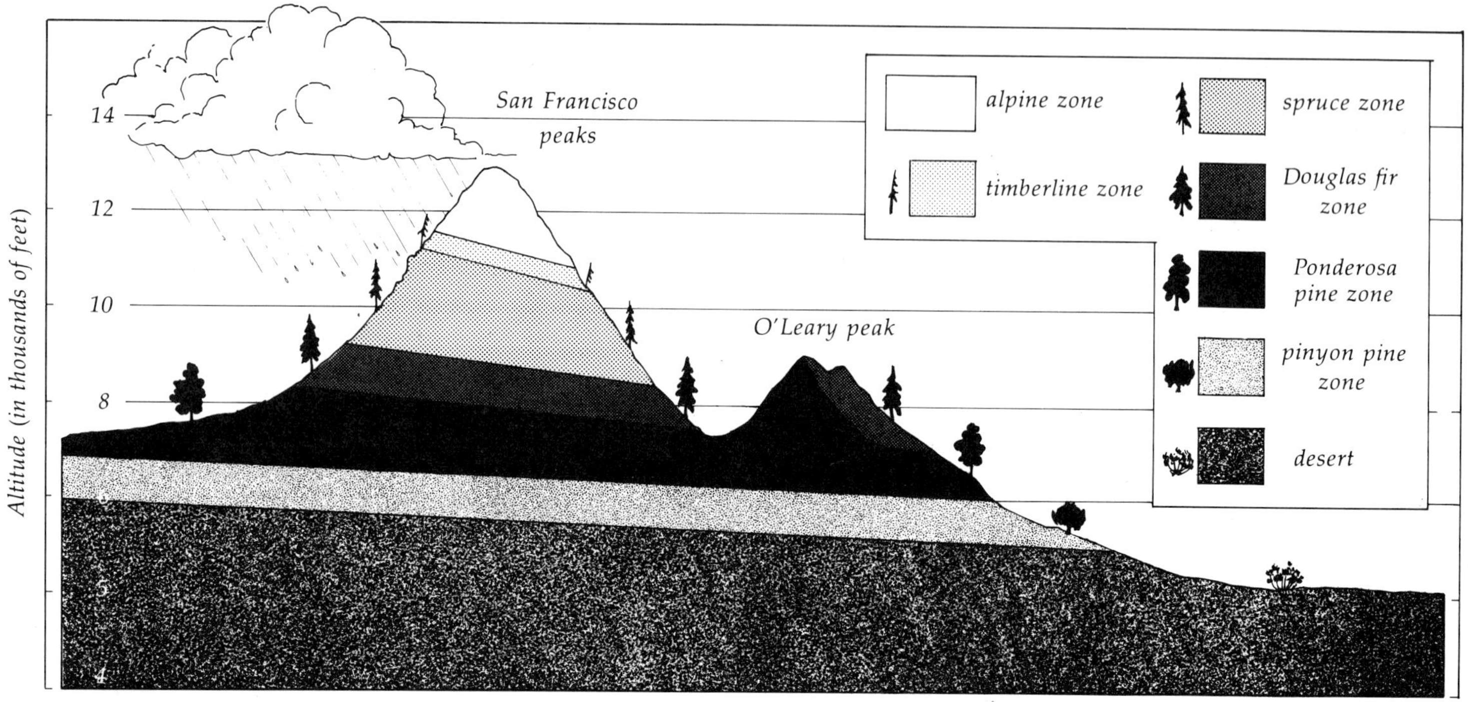

Fig. 37-6 (a) A diagram from Charles Hart Merriam's original publication of 1890 showing vegetational zonation of the San Francisco Peaks in northern Arizona as viewed from the southeast.

munities have led to a number of other conclusions and ideas besides the continuum. For example, when communities are studied on a continental basis, certain generalizations become apparent. The species diversity of plant communities, for instance, increases toward lower elevations and warmer climates (that is, toward warmer temperatures and more available moisture). It is quite apparent that in all of these studies the environmental gradients are important. Continua have been studied as a function of moisture, temperature, and soil gradients. Elevational gradients are often emphasized in mountain areas. These are really gradients in temperature, available moisture, and total atmospheric pressure. (A particularly interesting gradient is found in the tidal zone, discussed in Chapter 18.) In general, such gradient studies imply that *productivity decreases as one progresses from "optimal" to "extreme" environmental conditions.*

Ecotypes

If the distribution of a given species is determined by its genetics, then discovering the environmental limitations on a given species is the key to its distribution—but what if there is genetic diversity within a species? What if, for example, the *Potentilla glandulosa* plants of the high Sierra Nevada have a genetic composition that allows them to get along at relatively low temperatures, while *Potentilla glandulosa* plants at lower elevations have a genetic composition that allows them to do well only at high temperatures? When we think about the principles of evolutionary gene-pool change, we might certainly expect such situations, since the reproductive processes are relatively slow, which makes the rate of gene flow within the gene pool relatively slow. Because the environmental (e.g., climatic) pressures on the population differ depending upon location, we might therefore expect the genetic composition of a population to vary throughout the range of that population.

To what extent is anatomical diversity within any species correlated with geography? For example, the *Potentillas* in the high Sierra Nevada are dwarfed compared to those in the coastal mountains of California. We can imagine two possible explanations: first, the genetic composition of the two *Potentillas* might be alike, but their different appearances might be due to the different climates under which they occur; secondly, the differences could be due to an actual genetic diversity. But merely observing that there are differences does not allow us to distinguish between these two possible explanations. Obviously, the thing to do is to bring the different plants together, growing them either in a uniform garden or in a controlled environment facility (e.g., growth chambers or a phytotron; see Chapter 39).

Experiments of this type were first performed during the 1800s by Alexis Jordan in France. But his gardens were not carefully tended, so his results tended to be ambiguous. In the 1920s Göte Turesson in Sweden developed the uniform-garden approach to a high level of finesse. Since his pioneering studies, many others have also taken this approach. Jens Clausen, William Hiesey, and David Keck of the Carnegie Laboratory at Stanford University in California were the leaders of these studies in America. The results of this work have produced clear implications for our understanding of plant communities—both environment and genetics are important.

The effects of environment upon anatomical differences are common. (It is an area of active research in plant physiology.) Plants exhibiting such differences due to varied natural environments are called ecophenes. Usually in discussions of this type, this is not emphasized because the genetic differences we are about to discuss appear to be the rule. Nevertheless, we should realize that the environment has profound effects upon plant growth and development and that different environments can produce many different ecophenes from any uniform genetic stock.

Turesson and others also found genetic differences in representatives taken from the different parts of a species' distribution. These different genetic representatives of the population are called ecotypes. When the *Potentillas* from the Sierra Nevada and the Coast Range (and from many other locations as well) were brought together in a uniform garden, they continued to exhibit striking and significant anatomical differences. Many species have now been studied (although in terms of the total number of species, their number is few), and it seems obvious that different environments will exert different selection pressures, resulting in different genetic compositions that are directly correlated with geography.

Is ecotypic variation restricted to only the easily observable anatomical characteristics? As might be expected, selection also works on the physiological responses to environment. For example, Charles Olmsted at the University of Chicago, Calvin McMillan at the University of Texas, W. Dwight Billings at Duke University, and others have demonstrated photoperiodic ecotypes in several species. Indeed, Olmsted found both short-day and long-day varieties in the same species of grass (sideoats grama), and this was confirmed for other species by McMillan.

Harold A. Mooney and Billings collected *Oxyria digyna* (alpine sorrel) from several locations in the arctic and also in the alpine areas of the Sierra Nevada and the Rockies. All of these plants flowered in response to long days, but the specimens collected from the far north required much longer days (20 hr or longer) than those collected from the southern Rockies (15 hr). The arctic plants also reached peak photosynthesis rates at lower temperatures than their southern counterparts, but the alpine plants that grow at high elevations and relatively lower carbon-dioxide pressures were more efficient in utilizing carbon dioxide. Any competent taxonomist would classify all of Billings' specimens as the same species, but careful observation revealed a number of morphological differences between the northern and the southern representatives, as well as the physiological differences.

So our hypothetical visitors from the Alpha Centauri star system have a lot to learn about the earth's ecosystems. Perhaps the principles learned on their own planet will apply so well here that they can make short work of the task. It certainly seems reasonable, considering the complexity of any given organism, to imagine that each species will have its own individualistic environmental requirements. Then organisms anywhere will probably be distributed according to a continuum model, even though the dominant ones may at first make their distribution appear to be simpler. And, since the environment will vary even within the range of a species, we can expect ecophenes to appear and gene frequencies to vary, thus producing true ecotypes.

E Where rainfall is great, so is plant growth. The lushness of growth along this stream (the Rio Hicace in Puerto Rico) reflects the fact that the annual rainfall is 375 cm (150 in.).

F High rainfall is also responsible for the tremendous growth of these Douglas firs on Black Oak Creek in the Mount Baker National Forest in Washington. This is an example of a temperate rainforest.

*G Pine trees are often associated with the North, but this forest of longleaf pines (*Pinus palustris*) was photographed near Brunswick, Georgia. This type of southern pine is the source of both turpentine and telephone poles.*

F

E

G

H Coniferous forests cover much of the western United States. This view of Twin Lakes in northern Colorado shows some of the subalpine forests, common to the Sierra Nevada and the Rocky Mountains, that form the counterparts to the boreal forests of Canada (the north woods).

I Perhaps the most awe-inspiring of all the plant communities are the great coast redwoods (Sequoia sempervirens). *They are a never-ending source of wonder and beauty and are also one of the most valuable of all timber trees, producing lumber that is hard to surpass in both beauty and durability. Confined to a narrow coastal zone characterized by dense fog, they are threatened with extinction (except for those protected in state and national parks).*

H

I

J The southwestern mountains, such as the San Gabriel of southern California, are covered with a shrubby growth called chaparral. The shrubs blooming in the foreground are a species of mountain lilac (Ceanothus), *found here on a south-facing slope in the Los Angeles Valley. Smog is partially obscuring the mountains in the background, and smog is also killing thousands of acres of conifers that grow in this area.*

K An open oak forest in which trees are separated by grass is found on the west-sloping foothills of the Sierra Nevada in California. It is an example of a temperate woodland biome. Although this particular photo was taken in 1903, similar woodlands can still be seen today.

L Grasslands make up great areas of the United States and Canada. This is Palouse prairie in eastern Washington, a typical grassland. The black-soil area is one of the richest in the world and was part of a great grassland biome, before it came under cultivation.

K

J

L

At a meeting of the Ecological Society of America in 1959, James Bonner suggested that the problems of ecology could be nicely encompassed by three questions: *Who lives where and why? Who eats whom? What is the physiology of togetherness?*

In Chapter 37 we discussed some of the problems of studying plant distribution—who lives where. Now we shall continue this discussion, considering first the question of how the community changes the environment (especially the soil) and then reviewing in an introductory way the principal biomes of the world. We shall then consider the other two questions in the final chapters.

Climate and Soil

In Chapter 1 we pointed out that climate is the determining factor in establishing biomes, but it is also important to remember that plants affect the environment in which they grow (a point we emphasized in our discussion of plant succession). One aspect of the environment that is strongly influenced by plants is the soil. The study of soils is, of course, important to agriculture and has been carried out for many centuries. During the first part of this century, scientists in Russia began to emphasize the interrelationships between the processes of soil development, climate, and plants that normally occur in a given climate. Through the efforts of these Russian scientists, the principles of soil formation and classification were developed into a fairly extensive science. In this section we shall summarize some of the principal processes of soil formation and their results. Also, Table 38–5 lists the soil types of the various biomes.

What are the three environmental factors that are especially important in the various processes of soil formation? The answer: moisture, temperature, and vegetation. Interestingly enough, the kind of rock from which a soil develops is usually of secondary importance; in a given climate soils often end up the same, regardless of their parent material.

The breakdown of this parent rock requires the presence of moisture, and breakdown rates are strongly influenced by temperature. Another important effect of moisture is to move dissolved minerals in the soil. If precipitation is high relative to evaporation, then minerals are dissolved and leached out through the soil into the groundwater below. Many of the minerals are present in the form of ions—the metallic cations have a positive charge, and the nonmetallic anions have a negative charge. The clay particles in the soil have a negative charge, so the cations are electrically bound to the surfaces of clay particles. The tenacity with which a given cation is held to a clay particle depends, among other things, upon the kind of ion. Hydrogen ions are held most tenaciously of all, and sodium ions are held least tenaciously. The so-called lyotropic series arranges the ions according to how tightly they are held on the clay surfaces.

$$H^+, \quad Ca^{++}, \quad Mg^{++}, \quad K^+ + NH_4^+, \quad Na^+$$

The carbon dioxide dissolved in rainwater becomes carbonic acid.

$$CO_2 + H_2O \rightleftharpoons H_2CO_3$$

Ionization of this acid releases hydrogen ions (H^+), which are acidic.

$$H_2CO \rightleftharpoons H^+ + HCO_3^-$$
$$HCO_3^- \rightleftharpoons H^+ + CO_3^=$$

bicarbonate *carbonate*

Thus, over a period of time, rainfall supplies hydrogen ions to the soil, and since these are held most tenaciously, they tend to displace the other ions, which are then leached away in the groundwater. (We have already discussed cation exchange as it might occur on cell walls in the plant. Anion exchange also occurs on positive surfaces—a few in the soil.) In any case, areas of high rainfall tend to be areas of acidic soil because the hydrogen ions in the rainfall tend to replace many or most of the other ions originally available on the clay particles. If the rainfall is low and evaporation is high, minerals are not leached to the groundwater. Yet the gradual breakdown of minerals in the soil continues, producing soluble and relatively insoluble

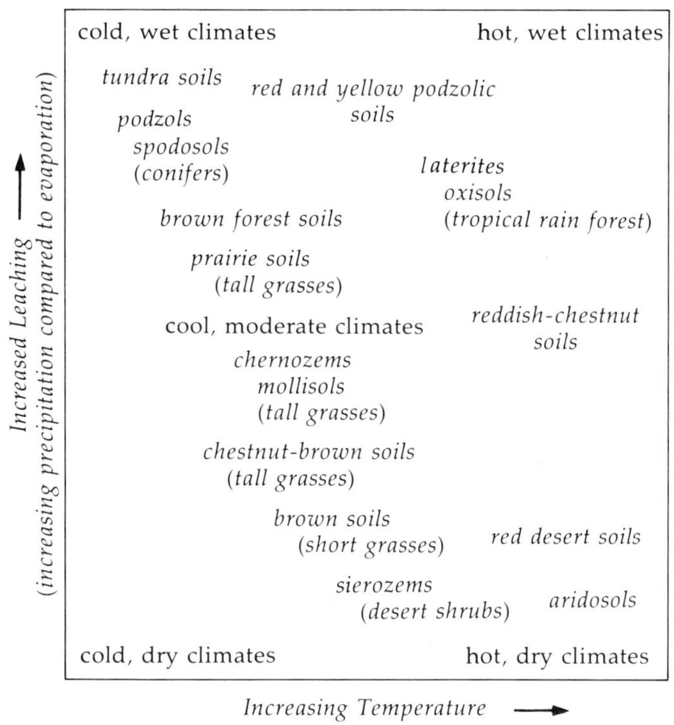

38–1 A diagram illustrating the soils usually formed under different climates. Vegetation is indicated, and soil names (according to the new system of classification) are also given.

salts, which then accumulate in the soils of dry climates (Fig. 38–1).

These powerful effects of moisture (as further influenced by other factors) result in the processes of soil formation. Let us now consider four principal kinds of soil (Fig. 38–2).

Podzols. In cool, moist climates such as those of the taiga (a northern coniferous biome), leaching is extensive, and soils become highly acidic. Here the presence of coniferous vegetation is especially important. The decaying pine, spruce, and fir needles release acids and other substances (chelating agents), which tend to further solubilize some of the minerals and accentuate the process of acidification in the soil. At the relatively low temperatures, the breakdown of litter from trees and other plants is fairly slow, and a deep layer of humus builds up on the surface. Also, at the cool temperatures, silicates break down more slowly than other minerals in the soil, so they tend to accumulate. The result of all these factors is a soil with a characteristic profile. There is a humus layer on the top, then a layer of ash-gray or yellowish soil consisting largely of silicates (the A horizon), and below this a somewhat darker layer (the B horizon), where much of the material from the A horizon has been deposited (Fig. 38–2). This is a podzol profile. (Podzol is a Russian word indicating the ash-gray color.) It is characteristic of the taiga, but in a modified or less extreme form, podzols occur under many other forests. This process of soil formation is called podzolization.

Podzols are poor soils, acidic and relatively infertile due to the extensive leaching. Yet much of Western civilization developed as the forests on these soils were cleared for agriculture. These were mostly cool-temperate deciduous forests instead of coniferous forests, but many of them occurred on modified podzolic soils having several of the characteristics just mentioned. Much lime had to be added to overcome the acidity, and fertilizers were also essential. For years fertilizers were mostly barnyard manure and crop residues; now chemical fertilizers are used. It has been suggested that this challenging environment was a good stimulus for Western man.

Laterites. In tropical rain forests, rainfall is extremely high, resulting in much leaching. Temperatures are also high, and this causes a rapid breakdown of dead plant materials. The consequence is an almost complete lack of humus, even under dense vegetation. At the high temperatures, silicates break down more rapidly than the so-called sesquoxides, which contain large quantities of iron and aluminum and are reddish in color. The results of these processes are red soils called laterites (the process of laterization). Laterites are highly acidic with virtually no organic matter and with A horizons (the zone of leaching) that are very deep, sometimes as much as 15 m. These are well-aerated soils with a loose and friable structure. They are so opposite to many of the sticky muds familiar to us that it is surprising to learn they can sometimes be plowed during a tropical rainstorm. Actually, we should note that the laterites constitute a complex group (as do the podzols).

Chernozems. Under certain conditions, the leaching due to rainfall just balances the production of soluble salts. Hence the salts do not become depleted, and they do not accumulate at the surface in excessive

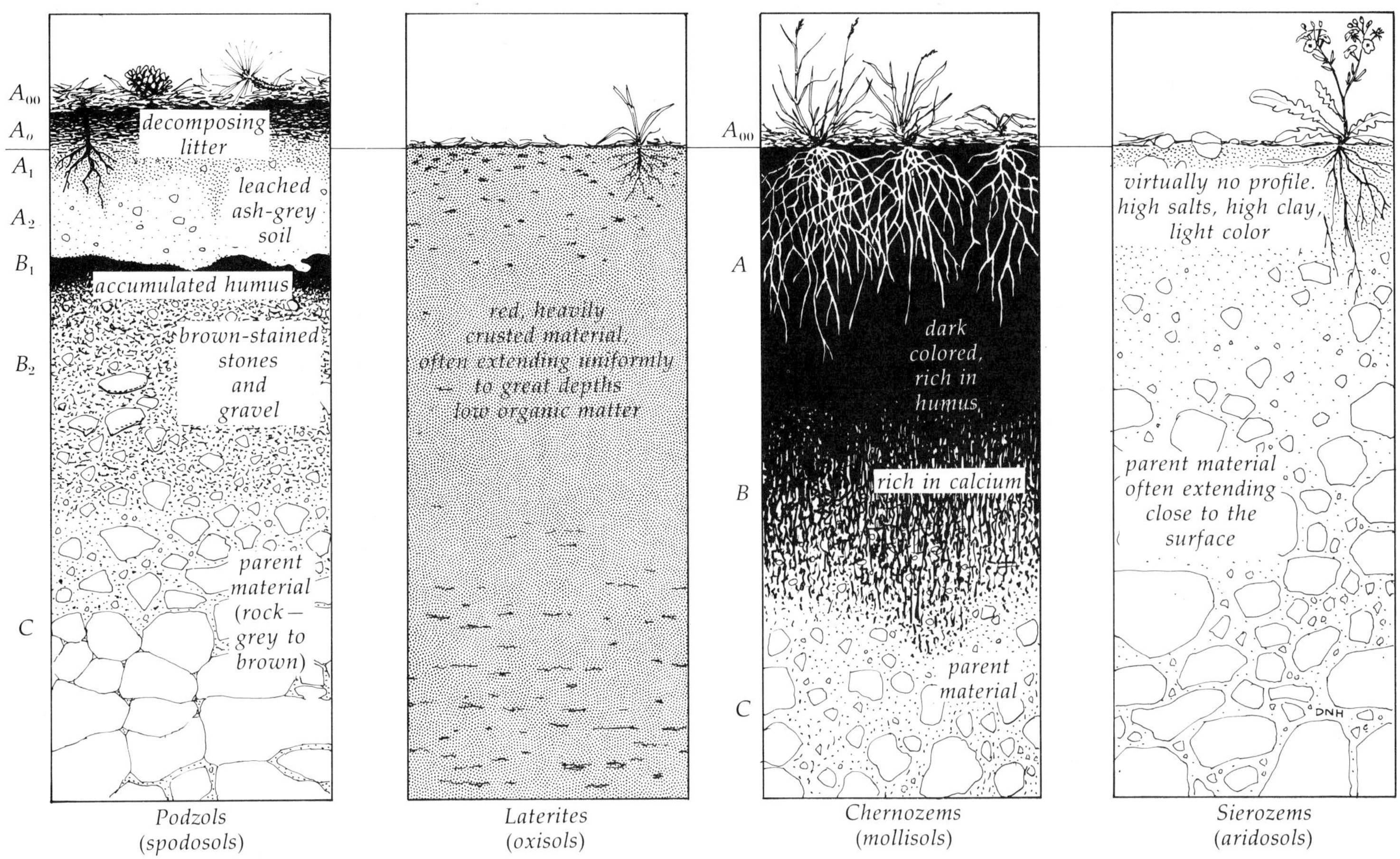

38–2 Representative profiles of the four soil types discussed in the text. Laterites and sierozems typically show little developed profile.

amounts, although a layer of calcium carbonate typically forms below the surface. This process of soil formation is called calcification. Climatic regions where this occurs support the growth of grass, and the resulting prairie soils are called chernozems. The finely divided grass roots penetrate the soil extensively, and as they die, they decay, resulting in a high level of organic matter (humus) incorporated in the soil. These are the black, extremely fertile soils of the Palouse prairie in Washington state, the great valley of California, and much of the Midwestern corn belt. These are rich agricultural lands, although rainfall may be marginal. Most of Western man's food supply now comes from these areas. Again, there are numerous variations of the ideal chernozem.

Sierozems. When conditions are both warm and dry, minerals continue to break down, but the products are not leached out, and the results are the salty soils of the desert. This is calcification carried to an extreme. A typical desert soil is the sierozem. Little organic matter accumulates in the desert, not only because high temperatures result in rapid breakdown when moisture is available but also because little organic matter is produced by the sparse vegetation. With irrigation, the desert soils often become highly productive. When salts are not so high that they are damaging (in which case they must be leached out with excessive irrigation water), they provide ample mineral nutrients for plants, although nitrogen must often be added. The warm temperatures are ideal for many

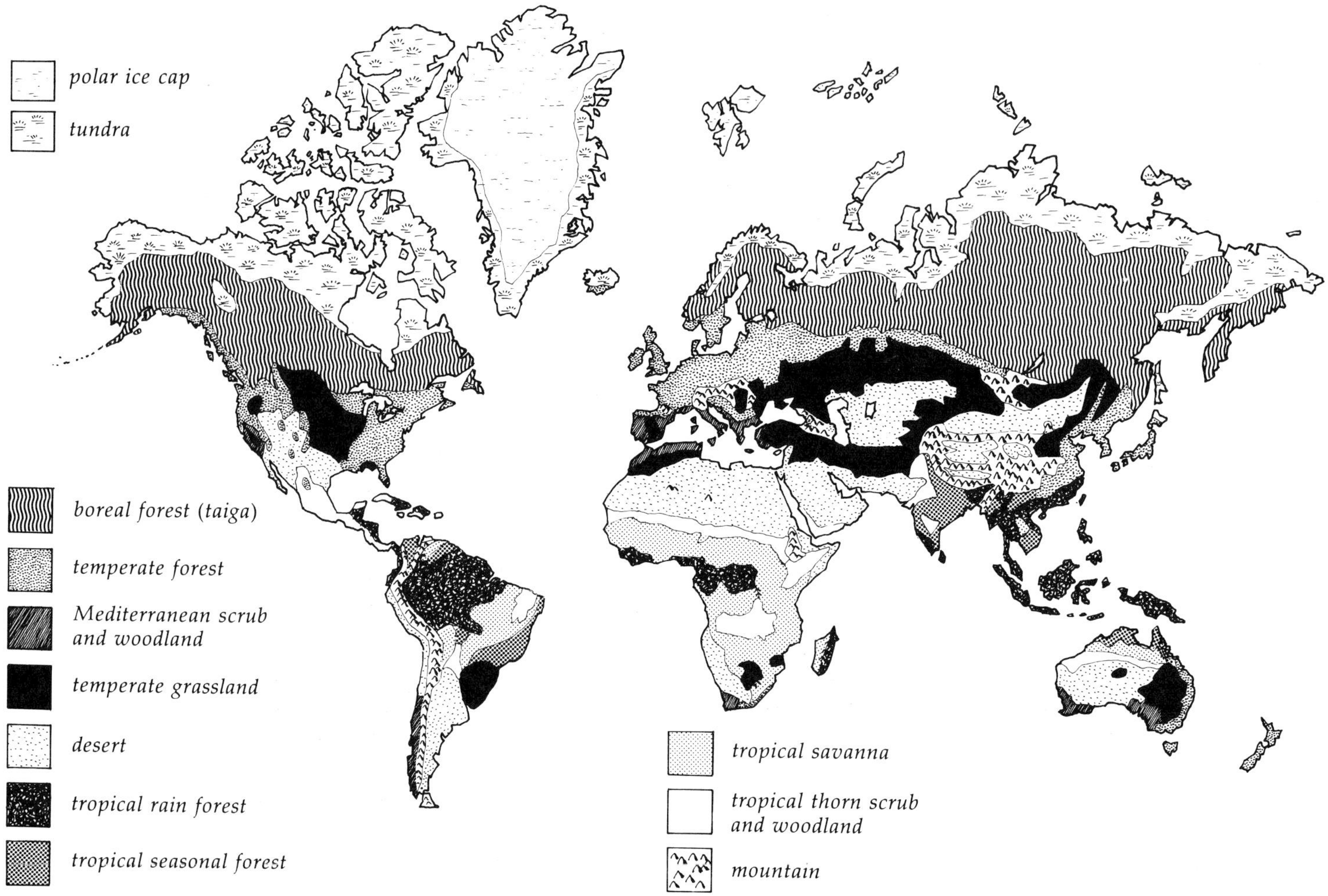

38–3 *Distribution of the principal biomes of the world.*

crops, and much future food will undoubtedly be grown in desert regions.

World Biome Types

A description of the earth's biomes could easily have been provided without such concepts as soil formation, the continuum, or the ecotype—even without the chapters on plant groups. But now with this background you are prepared to realize that the biomes are far more complex than the following brief descriptions might imply. Within each biome there is a continuum of communities, and there will be more or less gradual changes between the biomes. We can discuss the biomes not because they are discrete, but because their dominant vegetation occupies extremely large areas and is itself distinctive enough to allow us to differentiate one biome from another.

Much of the information relating to the biomes is shown in Tables 38–4, 38–5, and 38–6 and in Figure 38–3. You should study these tables and the figure carefully. Then become familiar with biome characteristics relating to growth forms (and perhaps a few principal species), the controlling climates, soils, productivities, and the distribution and importance of the biomes on a worldwide basis.

38–4 Principal biomes with their dominant growth forms and some representative species

Biome	Dominant growth form	Representative plants (mostly northern hemisphere)
Aquatic Systems		
Open oceans	Plankton, floating algae	Diatoms, plankton (dinoflagellates, etc.)
Estuaries and shores	Multicellular algae, grasses	Seaweeds, eelgrass, marsh grass
Lakes and streams	Algae, mosses, higher plants	Plankton algae, filamentous algae, duckweed, water lilies, pondweed, water hyacinth
Swamps, marshes, bogs	Algae, rushes, etc.	Cattails, water plantains, pipeworts, rushes, sedges, sphagnum moss, tamarack, baldcypress, mangrove
Forests		
Tropical rain forests	Trees, broadleaved evergreen	Many species of evergreen, broad-leaved trees (unfamiliar to us), vines, epiphytes (orchids, bromeliads, ferns)
Tropical seasonal forests	Trees, both evergreen and deciduous	Mahogany, rubber tree, papaya, coconut palm
Temperate rain forests	Trees, evergreen	Large coniferous species (Douglas fir, Sitka spruce, coast redwood, western hemlock, white cedar)
Temperate deciduous	Trees, broadleaved deciduous	Maples, beech, oak, hickory, basswood chestnut, elm, sycamore, ash
Temperate evergreen	Trees, needleleaved	Pines, Douglas fir, spruce, fir
Boreal coniferous (taiga)	Trees, needleleaved	Evergreen conifers (spruce, fir, pine), blueberry, oxalis
Reduced Forests—Scrubland		
Chaparral, magius, etc.	Shrubs, sclerophyll evergreen	Live oak, deerbrush, manzanita, buckbrush, chamise
Thorn woodlands	Spinose trees and large shrubs	Acacia, large shrubs
Temperate woodlands	Small evergreen or deciduous trees, grass or shrubs	Pinyon pine, juniper, evergreen oak
Grasslands		
Tropical savanna	Grass (and trees)	Tall grasses, thorny trees, sedges
Temperate grasslands	Grass	Bluestem, Indian grass, grama grass, buffalo grass, bluebunch wheat grass
Tundras		
Arctic	Diverse small plants	Lichens, mosses, dwarf shrubs, grass, sedges, forbs
Alpine	Small herbs (grasslike)	Sedges, grasses, forbs, lichens
Deserts		
Tropical warm	Shrubs, succulents	Spinose shrubs, tall cacti, euphorbias
Temperate warm	Shrubs, succulents	Creosote bush, ocotillo, cacti, Joshua tree, century plant, bur sage (in USA)
Temperate cold	Shrubs	Sagebrush, saltbush, shadscale, winterfat, greasewood (in USA)

38–5 Principal world biomes with climate and soils

Biome	General climate*	Most typical soils*
Aquatic Systems		
Oceans		
Estuaries and shores	Wide range of temperatures from arctic and antarctic cold to tropical warm (with hot springs at one extreme)	Salt or freshwater
Lakes and streams		
Swamp, marsh, and bog	Various, from tropics to arctic	Mud or peat, wet with salt or freshwater
Forests		
Tropical rain forests	125 to 1,250 cm annual rain, no dry period (18 to 35°C)	Mainly reddish laterites
Tropical seasonal forests	Marked dry season, generally lower precipitation	
Temperate rain forests	125 to 900 cm precipitation, nearly even throughout year, some snow (-4 to 21°C)	Podzolic, deep humus
Temperate deciduous	60 to 225 cm ppt., droughts rare, some snow (-30 to 38°C)	Podzolic: gray-brown, red, and yellow
Temperate evergreen	35 to 250 cm ppt., evenly distributed or summer dry season, possibly deep snow (-48 to 27°C)	Various podzolic, often shallow and rocky
Boreal coniferous (taiga)	35 to 600 cm ppt., evenly distributed, much snow (-54 to 21°C)	True podzols
Reduced Forests		
Mediterranean type, broad sclerophyll (chaparral, magius)	25 to 90 cm ppt., nearly all during cool season (2 to 40°C)	Variable, but fairly low carbonate
Thorn woodlands and scrubs	Dry tropical climates, between seasonal forest and desert	
Temperate woodlands, including pigmy conifers, oak woodlands	Temperate climates between forest and grassland or desert	
Grasslands		
Tropical savanna	25 to 90 cm rain, warm season thunderstorms, dry during cool season (13 to 40°C)	Variable, lateritic
Temperate grasslands	30 to 200 cm ppt., evenly distributed or high in summer, snow (-46 to 60°C)	Chernozems and related, lime layer
Tundras (arctic and alpine)	10 to 50 (arctic) and 75 to 200 (alpine) cm ppt., snowdrifts and areas blown free of snow (-57 to 16°C—arctic, -52 to 22°C—alpine)	Rocky, patterned ground, permafrost (especially in arctic)
Deserts		
Warm (tropical or temperate)	0 to 25 cm rain, very irregular, long dry seasons (2 to 57°C with high diurnal fluctuations)	Sandy or rocky, sometimes reddish or saline
Cold (temperate or arctic)	5 to 20 cm ppt., most in winter, some snow, long dry season (-40 to 42°C with diurnal fluctuation)	Gray desert; often sandy, rocky, saline

*Data modified and condensed from W. D. Billings, 1970. *Plants, Man, and the Ecosystem.* Wadsworth, Belmont, California. Used by permission.

38–6 *Areas and productivities of the biomes of the world**

Biome	Area (millions of km^2)	Range of primary productivity (dry g/m^2/yr)	World net production (billions of dry metric tons/yr)
Aquatic Systems			
Oceans	359	2–600	51.0
Estuaries and shores	2	500–4,000	4.0
Lakes and streams	2	100–1,500	1.0
Swamp and marsh (sphagnum bog, tropical and temperate swamps, mangrove swamps, saltmarshes)	2	800–4,000	4.0
Forests			
Tropical forests (rainforests, seasonal forests)	20	1,000–5,000	40.0
Temperate forests (rainforests, deciduous and evergreen forests)	18	600–2,500	23.4
Boreal forests (taiga)	12	400–2,000	9.6
Reduced forests or scrubland (thorn woodlands or scrubs, pigmy conifers, Mediterranean-type broad sclerophyll)	7	200–1,200	4.2
Grasslands			
Tropical savanna	15	200–2,000	10.5
Temperate grasslands	9	150–1,500	4.5
Tundras (alpine and arctic)	8	10–400	1.1
Deserts (warm and cold)	18	10–250	1.3
Extreme deserts, rock and ice	24	0–10	0.07
Agricultural land	14	100–4,000	9.1
Total land	147		108
Total water (aquatic systems)	363		56
Total for earth	510		164

*Data from R. H. Whittaker, 1970. *Communities and Ecosystems.* Macmillan, Toronto, Ontario. Modified slightly and used by permission.

Oceans. Because our discussion here is primarily concerned with terrestrial biomes, we have lumped (in the Tables) all of the open oceans into a single category. Yet the ecology of the deep oceans is certainly not the same as that of the shallower oceans above the continental shelf, and tropical oceans certainly have a different ecology from arctic oceans.

It is important to note that, while the oceans occupy approximately two-thirds of the earth's surface, they account for less than one third of the earth's photosynthetic productivity. We now know that vast areas of the oceans are "nutrient deserts." The organisms in the upper layers of water, where light intensities are high enough to allow photosynthesis, incorporate the available nutrient elements into their bodies. As they die, many sink to the deep waters, taking the nutrients with them. The result is that the upper layers of water become impoverished in such elements as phosphorus and nitrogen—and, finally, in phytoplankton. If some special set of circumstances causes an upwelling of water from the depths, the resultant enrichment of nutrients causes a rapid "flowering" of phytoplankton and the organisms dependent upon them for food. If man is to devise some means of making the deep oceans more productive, he need only find a way to bring the cold waters from the depths to the surface.

Estuaries and shorelines—lakes and streams. These shallow waters tend to be much more productive than the deep oceans, but they make up only a small percentage of the earth's surface, so their contribution to the earth's productivity is relatively small. Algal communities of the ocean shores (especially tidal zones) were discussed in detail in Chapter 18. But there is also a great variety of interacting organisms in lakes and streams. Consider the almost sterile glacial lakes of the high mountains, the extremely saline Great Salt Lake, nutrient-rich (highly polluted) Lake Erie, clear mountain streams, or the silt-laden Mississippi River. In all these situations, plants are reacting to their environments in ways that constitute special ecosystems and communities.

Swamps and marshes. These ecosystems have a rich biota, including many plants with their roots in the water or the mud and with at least parts of their shoots above the water surface. These are some of the most productive systems in the world, but their relative area is also small, so they do not contribute much to the earth's total productivity. Again, this group includes a great variety from sphagnum bogs to temperate and tropical swamps, from mangrove swamps to salt marshes. It also includes, in a sense, the rice paddies that provide most of the food for Eastern man. It is ecologically appropriate that this highly productive habitat should be used for food production.

Tropical rain forests. In the oceans biomass is relatively low, and physical factors are strongly controlling, virtually unmodified by the biota. The forest represents an opposite extreme in both respects. There is a vast standing crop, and modification of the environment has been extreme. Tropical rain forest ecosystems, especially succession ones, represent the extreme in terms of standing crop and modification of the environment. Where temperatures are high and rain comes abundantly and throughout the year (as in the Amazon Basin of Brazil), vegetation proliferates, and productivity is extremely high. The diversity of species far exceeds that of any other ecosystem. In some areas (in Brazil), as many as 300 species of trees may be found within 2 or 3 kl^2. There are literally thousands of species of orchids in the tropics—most of these, along with a number of bromeliads (relatives to the pineapple), grow as epiphytes on the trees. Because of the denseness of the canopy, light becomes limited at the lower levels; thus many plants and animals are adapted to life in the branches of the tall trees. Because of the large biomass of the standing crop, many minerals may be tied up in the biomass and unavailable for future use; thus light and minerals tend to be limited. Minerals are further limited because of the high leaching that is a part of laterization. A climax rain forest may be completely limited in its growth by lack of minerals. Actually, minerals released by decay are rapidly reabsorbed rather than leached (with mycorrhizae playing an important role), so this ecosystem is rather tightly closed with respect to nutrients.

Agriculture is difficult in the tropics. Weeds and plant pathogens grow rampant. Nutrients can be rapidly lost when the forest is cut. Yet potential is clearly high, as the extremely high productivity implies. The goals that could be achieved justify intensive work on the problems.

Tropical seasonal forests. These forests occur in monsoon areas (e.g., India, southern Indochina, northern Australia, southern Brazil) and such other areas as tropical N. America (Panama to southern Mexico), where most of the moisture is concentrated in a certain portion of the year. During the dry season, some of the trees lose their leaves. As might be expected, the composition of these forests is quite different from the nonseasonal rain forests.

Temperate rain forests. There are a few areas in the temperate regions of both hemispheres where rain is extremely abundant throughout the year. This is true for much of the coastal regions of Oregon, Washington, and western Canada, the redwood areas of California, and the *Eucalyptus regnans* forests of Australia. On the Olympic Peninsula of Washington rainfall may reach 500 cm/yr, causing true rain forests. Species diversity is much lower than in tropical rain forests, but the presence of epiphytes is a feature that the two types have in common, although in the temperate rain forests, the epiphytes are largely bryophytes, lichens, and some ferns. Productivities may approach or equal those of tropical rain forests. The lower temperatures allow a considerable buildup of humus in the temperate rain forest compared to the tropical zone. These are the tallest forests in the world.

Temperate deciduous forests. This is the biome where we would find our acre of woodland. The richest and most extensive deciduous forests occurred in eastern North America and eastern Asia. The European deciduous forests had far fewer species; many suffered extinction during the ice-age glaciation. Similar forests still occur in the southern hemisphere. These are the modified podzol areas just discussed. The forests of the eastern United States were destroyed to a great extent with the coming of Western European man, but when agriculture began to develop in the prairie areas, much of the cleared land was allowed to revert to forest. In some areas, such as the Great Smoky Mountains National Park, species diversity is still high. Animals are varied and abundant: deer, bear, rodents, etc. Again, their numbers were reduced during frontier days but have since recovered considerably due to scientific management.

Temperate and boreal coniferous forests. These are the forests of spruces, pines, and firs, which are called taiga by the Russians or the north woods by many of us. Here is where the true podzols develop. Winters are cold and snowy; summers are short. These are the most extreme environmental conditions to which trees have been able to adapt, and only a few species occur. The boreal forests of North America are also represented by belts in the mountains, having somewhat similar temperature and precipitation characteristics, and by related (but far from identical) evergreen, needleleaved forests of somewhat warmer, more southern regions that are not montane. The boreal forest has been little influenced by man—except for fires. The great northern forest could remain relatively unspoiled for many years.

Broad sclerophyll shrubland (chaparral, Mediterranean magius, etc.). In moderately dry, warm-temperate, maritime climates with little or no summer rain, a shrubby vegetation with hard leathery leaves (sclerophylls) tends to develop. This is the dominant vegetation of southern California and other similar areas in the world. Because of the long dry season, fires are frequent, but the plants have adapted so that when aboveground stems are burned off, regrowth occurs rapidly from the surviving roots. An outstanding example of ecological naiveté is the building of expensive homes amongst this type of vegetation—to be burned at intervals by headline-making holocausts. Soils are a rather interesting intermediate between the forest podzolic type (leaching during part of the year) and desert types (no leaching).

Thorn woodlands and thorn scrubs. These are biomes of tropical climates more arid than those of the tropical seasonal forests (e.g., Africa, just south of the Sahara Desert; also Australia and much of Mexico). Spiny acacias and other legumes are widespread as the dominant vegetation. Sometimes, medium-sized trees occur in open growth with grass below, or the understory may be dominated by dense shrubby growth. This may then become a transitional area toward desert climates.

Temperate woodlands. Fairly large areas, typically bordering on temperate desert regions but subject to relatively cold winters, are covered by small trees that are either coniferous, deciduous broadleafed, or sclerophyllous. The understory may be either grass or shrubs. In western North America, this type is represented by the pigmy-conifer woodlands consisting of juniper and pinyon pine trees. Much of the Great Basin, as well as the Colorado Plateau, is occupied by this vegetation type, and its areas are probably expanding as grazing reduces the grasses and shrubs of the cold desert. Various government agencies are now trying to convert some of these pinyon-juniper areas into grasslands.

Tropical savannas. Like thorn woodlands and chaparral, savannas occur in climates with long dry seasons. So fires are common. The savanna type consists of broad expanses of tall grasses with scattered trees or parklike patches—even limited woodlands or forests. In Africa these are the regions of vast herds of hoofed grazing animals, which are preyed upon by large carnivores such as the lion. Savannas also occur in South America and Australia. They are highly varied throughout the world, and man has had a tremendous impact upon them through fire, agriculture, and hunting. The great herds of grazing animals are disappearing rapidly, except where they are protected.

Temperate grasslands. These are the great prairies, plains, and desert grasslands of North America, the steppes of Eurasia, the veldt of Africa, and the pampas of South America. In North America most of these grasslands have been converted to agriculture, but this is somewhat less true in other parts of the

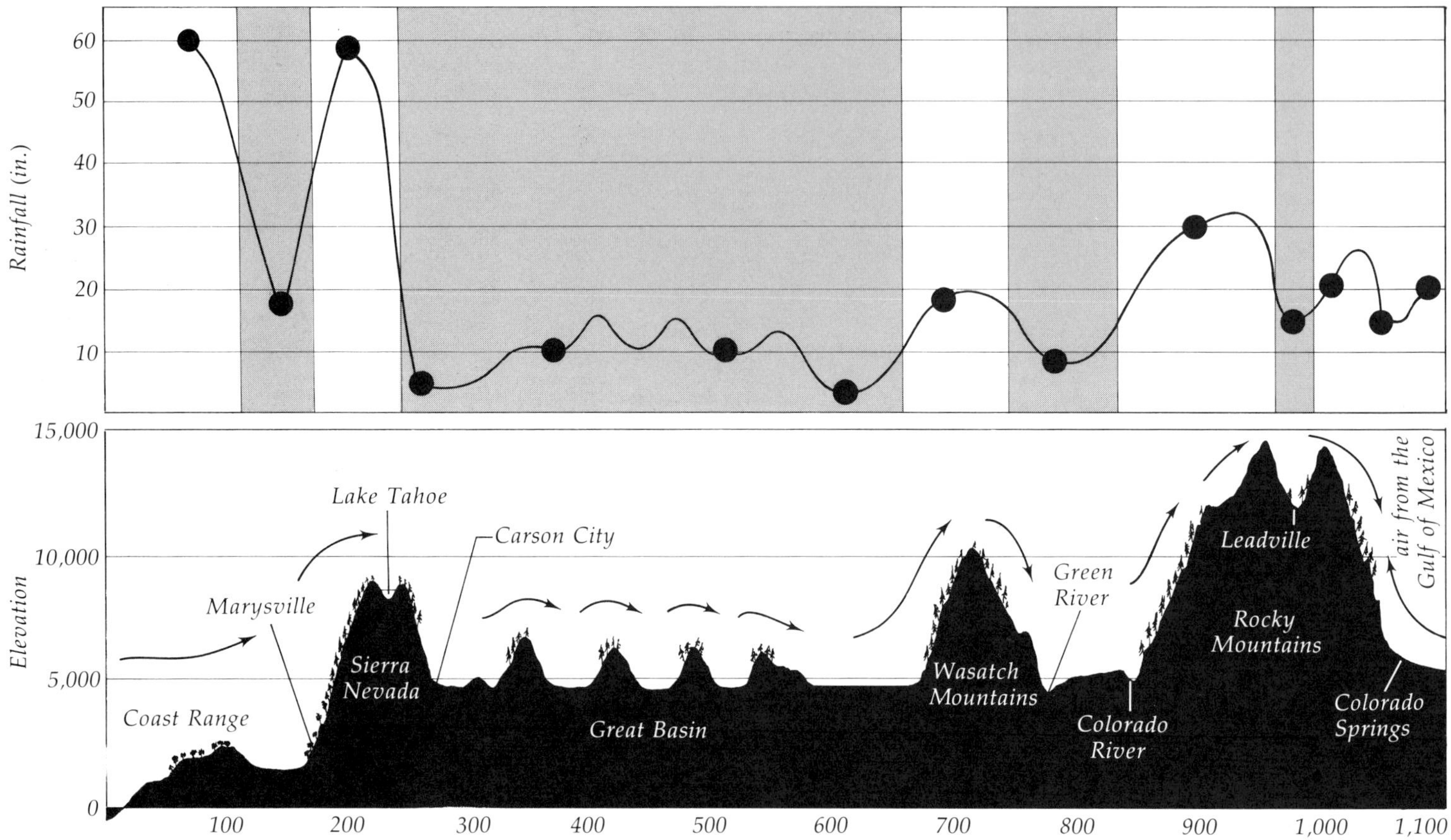

Rainfall (in.)
60
50
40
30
20
10
Elevation
15,000
10,000
5,000
0
Lake Tahoe
Carson City
Marysville
Sierra Nevada
Coast Range
Great Basin
Wasatch Mountains
Green River
Colorado River
Leadville
Rocky Mountains
Colorado Springs
air from the Gulf of Mexico
100
200
300
400
500
600
700
800
900
1,000
1,100
Miles (eastward along the 39th parallel)

(a)

(b)

(c)

world. Where precipitation is highest, the grasses are tall (most are now eliminated), often 1 or 2 m, with many legumes and other plants. In the drier regions the grass is much shorter, producing a patchy sod. Periodic droughts as well as fires tend to keep trees from invading these regions. Here is where the rich chernozems occur, and these are the regions where food is produced for much of the world's population. The only comparable food-producing biome would be the "artificial" marsh—the rice paddy.

Arctic and alpine tundras. Above the treeline on high mountains or in the north, extreme cold reigns supreme throughout all but a few weeks or months of the year, and trees cannot grow. Among vascular plants, only a few dwarfed grasses, sedges, herbs, and shrubs thrive in the tundra regions, but mosses and lichens may be abundant. They must be able to survive frosts at any time during the year and to complete their annual growth in a relatively short time (e.g., 5 to 10 weeks). In the alpine areas gusty winds blow nearly all the time, light intensities are exceptionally high, and carbon-dioxide partial pressures are relatively low. In the arctic regions days are long in summer (24 hr, north of the Arctic Circle), light intensities are not high, and the ground remains frozen close to the surface (permafrost). Thus the arctic and the alpine areas are different in significant ways, although the low temperatures and short growing seasons are common to both. The similarity of the vegetation in these two regions attests to the importance of these latter factors. Areas above timberline on high peaks near the equator (e.g., Kilimanjaro in Africa or the alto plano of Bolivia) or in the southern hemisphere (e.g., New Zealand) also have tundra vegetation, but the species are nearly all different from those of northern-hemisphere tundras. It is significant that growth forms are as similar as they are.

Tropical, temperate warm, and temperate cold deserts. Deserts are regions where plant growth is severely limited by lack of moisture. Tropical deserts

38–7 *(a) Diagram indicating rain shadows produced by mountain ranges on a transect (39th parallel) from the West Coast to the center of Colorado. Air coming in off the ocean is very moist, and rainfall is high as it rises and cools above the Coast Range (even though the Coast Range is at a relatively low elevation). A rain shadow occurs in the Great Valley of California. (Circles on upper part of the graph are actual rainfall measurements; the line connecting them is at best an estimate.) As the air rises over the Sierra Nevada, large quantities of moisture are again precipitated out, resulting in an extreme rain shadow at Carson City and extending throughout the Great Basin. Mountain ranges in the Great Basin cause a little precipitation (no actual data are available). These ranges are typically covered with conifers. The Wasatch Mountains in Utah provide another barrier and another rain shadow in the Green River and Colorado River basins of eastern Utah and western Colorado. Air rising over the Rockies again loses moisture, but a distinct rain shadow occurs at Leadville (although its elevation is above 10,000 ft). The rain shadow east of the Rockies is moderated by moisture arriving from the Gulf of Mexico. As air rises on the west slope of the ranges, it cools, causing a precipitation. Condensation of the moisture releases heat, warming the air; then, as the air descends on the eastern slopes of these ranges, it is further warmed by compression. The result may be an exceptionally warm, dry wind coming over the mountains from the west and blowing into such populated areas as Carson City and Colorado Springs. These winds are called Chinook winds. They melt snow rapidly and may be damaging to vegetation during the growing season. Comparable winds coming over the Alps in Switzerland, Austria, and Germany are referred to as the Föhn, and much superstition is connected with them. (b) A pond in the Tuolumne Meadows near Tioga Pass in Yosemite National Park, California. This is nearly at the top of the Sierras, and rainfall remains high, although temperatures are lower and the growing season is relatively short. (c) Desert vegetation growing near the Ubehebe Crater in Death Valley National Monument. Death Valley lies in the full rain shadow of the Sierras, and its near sea-level elevation (not shown on the chart because it lies south of the transect) contributes to the high temperatures and dry conditions.*

are usually controlled by the global air-circulation patterns in which warm air descends in certain latitudes (the horse latitudes), thus becoming even warmer. Of course, as air is warmed, it becomes capable of holding more moisture and, in a sense, drier. In some situations (e.g., the coast of Peru), this condition is so extreme that rain may not fall for several years. Temperate deserts are often produced by the rain shadow effect (Fig. 38-7). In extreme deserts, when the ground finally is moistened, a number of annuals (called ephemerals) may germinate, grow to sufficient maturity to produce one or more seeds per plant, and then die. These seeds will remain dormant but alive until the next storm occurs. Most deserts are characterized by shrubby growth, which is particularly true in both the warm and the cold deserts of temperate regions. In the warm deserts frost may be rare, but the cold deserts may experience severe winters. In the latter case, plant growth occurs during a short period in the spring. Before this, the temperature is too low for active plant growth; during the summer, water is not available. In the warm temperate deserts, ephemerals may be common along with the shrubs. Some of these are adapted to germinate and grow at relatively low temperatures, and they come up following sporadic winter rains. Warm-temperature germinators appear following summer thunderstorms.

Parallel evolution is often mentioned in relation to desert plants; that is, quite unrelated species may evolve to fill the various available niches in the desert, arriving at similar growth forms. According to the structure of the flowers, the cacti of North America and the *Euphorbia* of Africa are quite unrelated, although their succulent appearances are strikingly similar.

As we mentioned earlier, desert soils may become highly productive when water is made available. Temperatures are ideal for plant growth, and sunlight is abundant due to sparse cloud cover. Thus desert irrigated agriculture often proves to be very successful, especially in the warm deserts. The farms of the Imperial Valley in California, for example, are capable of producing enough food to support a large population, and many cities have grown up in the deserts of the southwestern United States. With a growing world population, irrigation of the deserts appears to be one of the most promising ways to provide food, especially if a cheap way can be found to purify sea water or to efficiently transport water from the north. As to living in these regions, air conditioning is almost as important a development as irrigation. Obviously, the key to all this could be such cheap and unlimited energy as that from atomic or nuclear power plants.

Agricultural land. About 10% of the earth's land area is now covered by agriculture. This is an amazingly high figure when one thinks of the vast areas of desert, forest, mountains, and tundra. Nevertheless, it is also impressive to realize that the tropical savannas of the earth are about equal in area to the agricultural land and that both the temperate forests and the more moderate deserts exceed agricultural land (see Table 38-6). The agricultural lands are close and familiar to most people living on the earth today, so we do ourselves an injustice if we ignore them in an ecological consideration of the earth's biomes.

But it would be a mistake to consider cultivated land as a single agricultural biome. About 650 species of plants have been adapted for agricultural use. A majority of these (over 400) originated in southern Asia, while about 100 originated in the Americas. Most of the world's biome areas include agricultural land, and the crops grown must be suitable to the prevailing climates. In tropical rain forests, for example, cacao and banana can be grown, while banana, sugarcane, pineapple, and coffee may be grown in the tropical savannas. In the subtropics important crops include citrus, olives, grapes, vegetables, sugarcane, cotton, cereal grains (especially rice—the world's second most important crop), and forage crops used to feed cattle and other domestic animals. Cotton and rice are especially important crops of the humid subtropics. Important crops of the intermediate climates are maize (corn), wheat (the world's most important crop), fruits (especially apple), vegetables, soybeans, and forages (including grasses and legumes).

Many agricultural colleges teach courses in crop ecology in which an attempt is made to relate the specific responses of agricultural plants to environmental factors and to derive generalizations about the most economical production of food and fiber. It is important to consider the production of crops in terms of the principles of botany, especially plant physiology and ecology. Actually, knowing about the world's biomes with their associated climates and soils can serve as a basis for study in many areas of botany.

Essay 38–1 Making a Terrarium

You can bring a small ecosystem into your life by making a terrarium for your room or desk. A terrarium is nothing more than a glass container in which you can grow various kinds of plants that reproduce a particular part of nature. You can make terraria that are woodland, desert, or bog—but you cannot mix the plants from these various habitats in the same container.

To make a terrarium, first obtain a glass container such as a wide-mouth gallon jar, a large fish bowl, or an old aquarium tank. On the bottom of the container, place a layer of gravel or small pebbles and on top of this, a layer of coarse sand. These layers are important to maintain good drainage. Next, place a thin layer of crushed charcoal to keep the soil from becoming too acidic. These base materials may be sterilized by heating in an oven at 250°F for one hour. They should fill the terrarium to the depth of about 1/4 in.

If you want to make a desert terrarium, add an inch of sand or sandy soil. Desert plants, such as various types of cacti, aloes, sedums, and succulents, can be placed directly in the sand, or potted plants can be sunk into the sand. The plants should be watered before planting, and then the surface of the sand should be watered gently thereafter. Spray the soil around the plants about once a week, but do not wet the rest of the sand. Place in a sunny location (but not in the direct sun) and keep at 85 to 90°F. You can suspend a 60-watt electric light 1 ft over the top to keep the temperature this high. Cover with a loose-fitting glass plate. If water collects on the sides of the container, remove the plate for a day.

For a woodland terrarium, cover the base with a soil mixture of one part sand and three parts good potting soil or soil obtained directly from a forest floor. Moisten the soil before adding. You can now add such plants as mosses, liverworts, lichens, club mosses, and small ferns. Seedlings of pine, hemlock, spruce, and laurel grow well, and while they are harder to maintain, you can add flowering plants such as violets, hepatica, bloodroot, wild strawberry, Begonia, Croton, Peperonia, and strawberry saxifrage that are usually successful. Oxalis, Selaginella, myrtle, or moneywort make good groundcover plants. Place a woodland terrarium in a north window and keep at a temperature of from 65 to 75°F. Cover with a loose-fitting glass plate that will allow air circulation. Water regularly but lightly, or you will end up culturing fungi. Open the top completely for several hours every few days, particularly if water accumulates on the sides.

A bog terrarium can be made with soil composed of one part sphagnum and two parts potting soil. The bog terrarium should be thoroughly soaked with water; allow the excess water to remain in the gravel layer. Bog terraria are particularly suited to insectivorous plants such as Venus flytrap, pitcher plant, sundew, butterwort, and aquatic bladderwort. Sheep laurel, hemlock, and swamp blueberry seedlings do well. Any light, cool location is suitable for a bog terrarium, although a temperature of 65 to 72°F is best. Water regularly and never let the bog dry out.

39

Physiological Ecology

The Desert and the Alpine Tundra

Deserts and mountains hold a fascination for man that is hardly surprising. There is a great variety in both types of landscape—and many of them are magnificent. Moreover, both of these major biomes represent environmental extremes and present a challenge to human survival. Plants, too, find survival in either the desert or high mountain regions a difficult business, requiring special adaptations.

Water is always limiting in deserts, but due to different temperatures and other factors, there is also a great variety of plants. Hence the general appearance of dominant species often shows striking contrasts from one desert location to another. The desert is fundamentally inhospitable to man, since man needs water, but if he can get water, the desert will blossom with agricultural productivity and support the growth of modern cities.

A This view taken high in the Kofa Mountains of southwestern Arizona (just before sunset) is typical of a warm desert region—a familiar scene to millions of fans of Western movies.

B Not all deserts are hot, since it is the amount of water not the temperature that is the primary limiting factor. This is a view of a relatively cool desert of southwestern Wyoming, where cattle-grazing can be profitable.

A

B

C Blowing, barren sand dunes are associated in people's minds with the desert, and such sand dunes do occur in deserts all over the world. But they are relatively rare, and even then, plants often grow on them. The small dune shown here is in the Mojave Desert of southern California, and the plant is a Creosote bush (Larrea divaricata), one of the most characteristic species of warm deserts in both North and South America.

D Cacti are common to all western deserts and show a great diversity of size and form. The ones shown here are a spineless variety (Opuntia sp.) Spiny opuntia are among the most characteristic of all desert plants.

E This morning dove has built her nest on a cactus branch. Many animals find suitable niches (in relation to plants) even in the inhospitable desert environment. A giant Saguaro cactus, complete with fauna, may be an entire ecosystem itself.

F This striking plant is the ocotillo (Fouquieria splendens). Common in southern deserts, it is a member of the candlewood family, a name derived from the flame-red flowers that appear at the ends of its stems. The roots display succulence for water-storing, and the leaves fall off when water becomes limiting.

G When man has water available, he can literally make the desert bear fruit. These onion fields are being irrigated by water from the Colorado River, which forms the Arizona-California border. Without the addition of water, this land would remain barren and unproductive.

H A desert oasis with fan palms (Washingtonia filifera) that are native to some North American deserts. Where frosts are rare and water is dependable (particularly subsurface groundwater), an oasis appears. The plants growing there may be strikingly different from those in the surrounding areas, as these large fan palms clearly show.

C

D

E

G

F

H

The tundra contrasts sharply with the desert. There is a relative abundance of water, and it is never too hot. Yet temperatures are nearly always too low for optimum growth of most plants, and the growing season may be as short as 3 or 4 weeks. Snow is blown in massive drifts, and the distribution of summer vegetation correlates closely with the distribution of these drifts.

I This cushion of campion (Silene acaulis) *is growing in a form highly typical of plants growing on gravelly, exposed, alpine ridges. Other plants (a grass and* Geum rossii*—avens) have invaded the cushion and are using it as a place to grow.*

J Flag trees at timberline are caused by the winter high-velocity winds, which carry ice crystals that tend to kill the buds on the windward side of the trees near the timberline, thereby allowing growth on the leeward side. Patches of vegetation in the gravelly foreground soil are mountain avens or dryad (Dryas octopetala)*, a species found in alpine tundra of major mountain ranges throughout the northern hemisphere. The insert shows a closer view of the flowers and foliage.*

K The old-man-of-the-mountain (Hymenoxys grandiflora) *is probably the largest flower found in the Rocky Mountain tundra. The flowers, which are 3 to 8 cm in diameter, face the point on the eastern horizon where the sun appears in the morning.*

L Two common flowers that help make the tundra so attractive in the summer are the wallflower (Erysimum nivale, *left*) *and the bistort* (Polygonum bistordoides, *right*).

M Colorado blue columbine (Aquilegia caerulea) *is the state flower of Colorado. This magnificent plant grows in the boreal forest below timberline, but it frequently extends into the tundra, particularly in rocky areas.*

I

J

K

L

M

We are now ready to examine the second part of Bonner's question *who lives where and why?* Why do certain plants grow in our woodland ecosystem? Or, *more generally, why do specific plants grow in any specific environment?* A little reflection reveals that there must be at least three answers to this question: (a) the plant arrived there in the first place; (b) it is capable of survival, growth, and reproduction in that environment; and (c) it hasn't been eliminated by competition, grazing, fire, or any other violent force.

Whether a plant arrived at some location is determined by its means of dissemination (seeds, vegetative propagules, etc.) and the barriers that might prevent this dissemination (oceans, deserts, mountain ranges, etc.). The study of these factors is carried on mostly by evolutionists and taxonomists. The factors of competition, grazing, fire, etc. are now being investigated, but little physiological work has been done in these areas.

But study of the *survival, growth,* and *reproduction* of plants is clearly within the realm of physiology. Emphasis upon these aspects of the why of plant distribution constitute one significant phase of the science of physiological ecology. Here the methods of plant or animal physiology are used to investigate the problems of ecology. This is a true borderline science because we must apply all we know about plant anatomy and function—and even taxonomy and morphology. Yet it is meaningless unless it is in the context of ecology. (An *autecological* investigation—as contrasted to the *synecological* approach used in Chapters 37 and 38—would be mostly physiological and hence would be essentially equivalent to physiological ecology.)

Controlled Environment Research in Physiological Ecology

If physiological ecology is the application of physiological methods to ecological questions, then activities in this discipline should begin in the field. That is, the questions must first be discovered and described where they actually occur; then they may be attacked using physiological methods both in the field and in the laboratory. In recent years a number of investigators have developed field approaches to measurements of photosynthesis, transpiration, and other plant functions. Sometimes, the equipment becomes rather elaborate and cumbersome, but researchers such as W. Dwight Billings at Duke University, Walter Tranquillini in Innsbrück, Austria, and Otto Lange in Würzburg, Germany, have been happy to put up with these inconveniences because of the information gained. Rather elaborate infrared gas analyzers, for example, may be taken into the field to measure CO_2 exchange between a plant and its surrounding atmosphere. An Austrian firm has developed a small unit for measuring photosynthesis that will enclose a plant in the field, thereby duplicating the outside temperature and humidity inside the enclosure. Undoubtedly, this approach will be expanded considerably in the future.

Studies in the laboratory require controlled environmental conditions, where it is possible to maintain virtually all of the environmental factors at a constant level—except for the one in which the investigator is interested. For example, temperature and humidity may be held constant, while light intensity or duration is varied in a controlled manner. This way it becomes possible to study the effect of a single environmental factor upon the growth of an organism.

Only in relatively recent years has such an approach been possible. We have had the engineering capability to control temperature for quite some time. But humidity is somewhat more difficult, and light intensities comparable to those provided by the sun are still difficult to obtain artificially. Nevertheless, it is fairly easy to produce light intensities, high enough for good plant growth. Solar intensities may be as high as 10,000 to 12,000 foot-candles (or even as high as 16,000 foot-candles under suitable conditions on high mountains). With a bank of high-output fluorescent tubes plus a few incandescent bulbs, it is possible to obtain light intensities of 2000 to 5000 ft-c (or sometimes higher). The fluorescent bulbs produce no far-red illumination, and this is why the incandescent bulbs have to be added if the spectral quality of the light is to approximate that of sunlight. If ample funds are available, it is possible to use xenon arc lamps, obtaining light intensities as high as those produced on earth by the sun and with a spectral distribution not unlike that of sunlight. We should remember that stem elongation is sometimes maximum under intensities found in growth chambers and much less in sunlight, while dry matter production is highest in sunlight (see Fig. 35–6). It is still difficult to produce plants under artificial conditions comparable to field-grown plants.

A number of companies now manufacture plant-growth chambers in which temperature, light intensity, and humidity can be artificially controlled. Virtually all of these rely upon fluorescent plus incandescent lamps (xenon lamp units have been custom-built by only a few investigators around the world). The approach of controlling environments in plant studies has become so popular that a number of large facilities now exist at various research centers. The first of these was constructed by Frits Went (see Essay 30–2) in the late 1940s at the California Institute of Technology in Pasadena. Went and Nobel prize-winning physicist Robert T. Millikan had just visited a cyclotron at the University of California in Berkeley. Millikan commented to Went that botanical science should have an equally impressive facility. Referring to this statement, James Bonner and Samuel Wildman jokingly suggested that the new controlled-environment laboratory being built by Went should be christened a *phytotron.* (Its formal name was the Earhart Plant Research Laboratory.) Although the title was strictly a joke to begin with, it has taken hold and now exists in the formal names of comparable environmental laboratories located all over the world. A phytotron, then, is an extensive controlled-environment facility, typically including several rooms and perhaps a number of small growth chambers, air-conditioned greenhouses, and a fairly large staff of technicians and scientists. Such a laboratory built in Wisconsin to study both plants and animals is called a biotron, and in St. Louis, Missouri, Went built a large, dome-shaped greenhouse with temperature gradients called a climatron. These tools for research in physiological ecology are also of considerable value in purely physiological studies or in studies of crop ecology.

But the fundamental ecological problems should not be lost among all the intensive research in phytotrons and other controlled-environment facilities. That is, once the analytical data have been obtained, it is important to synthesize answers to the original field questions.

The Law of Limiting Factors

The environment is complex. As light intensity increases (due to the sun getting higher in the sky), its quality also changes (becomes more ultraviolet for one thing), temperature increases, and relative humidity decreases, although evaporation might increase, thus increasing the absolute humidity. The increasing air temperature, warmed by the warming and drying soil, might cause the air to rise, creating wind. Diffusion rates of such molecules as CO_2 will be increased. In a phytotron we can try to study one factor at a time, but in nature this is extremely difficult to do because virtually all factors are changing continuously. Gravity is probably the most constant factor, and even it changes slightly with the tides.

A breakthrough in approaching the study of plants in complex environments came in 1840 from the research of Justus von Liebig. This German scientist was interested in applying the science of his day to agriculture. It became apparent to him that crops were often limited by a single nutrient element and that adding a combination of nonlimiting nutrients was seldom of much value. He found, for example, that the crops in a given field might not respond to added nitrogen or iron, but addition of phosphorus was followed by a considerable spurt in growth. Thus he formulated his Law of the Minimum: *The growth of a plant is dependent upon the amount of "foodstuff" presented to it in minimum quantities.*

This idea provided us with a wedge into the complexity of natural environments. As a first step, one need not be concerned with all of the environmental factors; rather the research objective is to find that factor that is *limiting.* It was easy to extend the concept beyond mineral nutrients to include temperature, water, light, carbon dioxide, humidity, etc. We have already made implicit use of this principle in several discussions presented in previous chapters. For example, in Chapter 38 we described the various biomes and often had occasion to refer to the environmental factor that seemed to be limiting (e.g., water in the desert and temperature in the arctic).

Unfortunately, we have since had to admit that the situation is seldom as simple as Liebig's law might imply. If we are considering plant distribution, for example, it is seldom, if ever, possible to find a *single* environmental factor determining all the boundaries of a given species. The factor could be temperature during part of the year along part of the boundary or competition with other species along another part. Furthermore, we now know that factors in nature are continually interacting and that more than one (but seldom more than three or four) may be limiting the growth of a plant at a given time. Increasing *any* of these factors may result in a considerable increase in plant growth. Thus the situation remains complex, but we are still entitled to initiate our studies in physiological ecology by searching for the limiting factors.

Essay 39–1 The Art of the Scientific Paper

No matter how brilliant or how important research is, if it remains unpublished, it is the same as if it had never been done. We have tried to show how ideas in science develop and how ideas and data interact to produce an understanding of the way in which nature works. But this growth is dependent on the communication of these data and ideas.

The major medium for this exchange is the scientific journal. These were first started in the 1600s and now number in the thousands in the biological sciences (including several hundred botanical ones). Of these, a few dozen are examined by most botanists, who actually read only a small number of papers per issue. Yet these journals constitute the major means of advancing botanical knowledge.

How does one publish in a journal? Some journals are managed by scientific societies, and you must be a member to submit manuscripts. Other journals have no restrictions and will publish acceptable papers submitted by anyone. This does not mean that all manuscripts are published. In most journals, the editor has the final say in whether a paper will be published. He is assisted by an editorial board and technical reviewers. To see how the process works, let's follow a manuscript from submission to publication.

Say you have done a piece of research that you think is worthy of being shared with your fellow botanists. First, you carefully write the paper and document your findings as fully and clearly as possible. You then send the manuscript to the editor of the journal that you decide is most appropriate for the subject being discussed. The editor reads the manuscript and sends it to two reviewers. These are persons the editor feels are experts in the field. They read the manuscript, looking carefully to see if you have presented enough original, carefully collected data to justify the conclusions you have drawn. They also check to see if the writing is clear and the data are adequately illustrated. Finally, they evaluate the work in terms of its importance. If they are happy with the paper, they will advise the editor to publish it as it stands, and usually he will do so. Normally, however, the reviewers will recommend publication only after changes have been made that they have suggested. Again, the editor will probably agree with these suggestions and may make some himself. In this case, the manuscript is returned to you, and you are asked to make the suggested changes and return the paper to the editor. The reviewers and the editor may feel that the work does not warrant publication for any number of reasons, and the paper is then rejected. There are few scientists that have not had a paper rejected—no matter how famous they may be!

After the revised manuscript is returned to the editor, he sends it to the printer, who sets it in type. A printing is made of this type and sent to you. You read this galley proof for errors and return it to the editor, who returns it to the printer with the corrections indicated. The journal is now assembled in its final form and printed, bound, and mailed out. At this time you can order a special printing of only your paper. These are called reprints and are important in the communication of data. The reason is because few people subscribe to more than a few journals, but they would like to have a copy of many of the papers that appear in a wide range of journals. They will therefore send you a postcard or letter asking for a reprint of your paper. You may also decide to send a reprint to a person you think will be interested in your work. Thus, most scientists have large reprint files consisting of papers in their fields.

Until now there has been no mention of money. But you will receive a bill for the reprints and sometimes an additional charge per page or per plate of figures. These charges vary greatly from journal to journal. But they are usually stated clearly in the journal itself, and you are informed early in your dealings with the editor as to the possible charges. If you cannot pay them, the paper is usually published anyway. In no case are you paid for the publication of a research paper.

Just how is a research paper organized? Research papers have an almost formal organization, and while there are many variations in detail, a general plan is usually followed. The paper starts with an introduction. In the introduction the author states the problem or question to be answered in the paper and then provides the necessary background for the problem. The introduction is followed by a section on materials and methods. Here the author tells you what organisms he will used, how he is grew them, or any other information on materials or methods necessary to understand the research he is presenting. This is followed by a description of the results the author obtained using these organisms and methods. The results are presented with great care and may be illustrated by graphs, tables, *drawings, or photographs. This is the most important part of the paper, since the data here will ultimately be the most useful to other researchers. Following the results is the discussion. This is frequently the most interesting, since the author will tell you what he thinks the data mean. You may accept what he has to say, or you may draw your own conclusions from the data. Finally, there is the summary that presents a brief report of the work and the conclusions presented in the paper. The summary is also called an abstract and is frequently printed at the front of the paper, before the introduction. This is because the summary or abstract is usually the first (frequently the only) part of a paper that is read. Throughout the paper you will see numbers or names followed by a year (Jensen and Salisbury, 1972). These are references to other papers, and a list of these papers can be found at the end of the paper.*

Research papers are usually not easy to read. There are several reasons for this. One is the fact that you are coming into a continuing story that may have many different authors in many parts of the world. Without enough background it is hard to see how this work fits into the overall picture. There is also a formidable vocabulary that has been introduced to ensure precision of statement and communication. Furthermore, the style used by many scientists is not the most readable, but it is tolerated if the meaning is clear. Finally, the research is usually not presented in the way it was actually conducted; rather it is logically organized for the purpose of the paper.

How many people read a given scientific paper? That depends on the paper. Some, like the famous Watson and Crick paper on DNA, have been read by thousands, but they are the exception rather than the rule. Important papers may be read by hundreds, but again the run-of-the-mill paper will probably be read by only a few dozen people around the world. Yet the data are there, and slowly and surely they will accumulate, changing our ideas and concepts. The art of the scientific paper is an important and essential part of the advancement of science.

The Physiology of Response to the Environment

The usual college course in plant autecology emphasizes plant response to each of the pertinent environmental factors. This was the approach taken in the classical book by R. F. Daubenmire, entitled *Plants and Environment* (last published in 1959). We have already discussed many of these responses in earlier chapters. But, as an alternative approach, we might consider plant responses to the environment in a general way and attempt to classify them according to the nature of the response rather than according to the nature of the eliciting environmental factors.

Direct environmental effects. Sometimes, as the environmental factors change, the plant response changes directly and immediately (or in other cases, *almost* immediately). Photosynthesis is an excellent example. When light intensity is limiting, the rate of photosynthesis changes instantly as light intensity changes. Transpiration is a comparable situation, but this is complicated somewhat by the stomatal response, which may be a little less immediate and direct. Many enzyme-controlled reactions respond directly to temperature or, in some cases, to the presence of specific molecules in their environment. This is true, for example, of allosteric enzymes. As the promoting molecule increases in concentration, the enzyme activity becomes proportional.

Triggered or on–off responses. There are a few plant responses to environment (initiated by some environmental change) that may then proceed even if the environment returns to its original state. Frequently, there is a *delay* between the triggering action and the response itself. Often, there is also an *amplification;* that is, the energy supplied to the plant by the environmental change is well below the energy required to bring about the response. The energy for such a response is provided by the plant's metabolism. Seed germination is an excellent example of such a triggered response. In some cases, as moisture increases, seed germination is initiated, and once it has begun, it will continue even if the moisture stops increasing. Actually, examples of triggered responses are not easy to find. Most responses that initially appear to be of the triggered type turn out to belong to the next category.

Modulated or quantitative responses. In this category, the response is not clearly on or off, but the *level* of the response is determined by the level of the environmental factor. Again, there is frequently an amplification, but this amplification is still proportional to the environmental input. There is also typically a delay in such responses, as in the triggered responses. Phototropism is an excellent example of a modulated response. As we discussed earlier (see Fig. 35–2), the degree of bending of the stem is a function of the number of quanta of light absorbed by the stem tip.

If the delay between environmental stimulus and response is long and if the response continues even when the environmental stimulus has been completely removed, then we can speak of *induction.* As we saw, vernalization is an excellent example of induction. Flowering typically occurs long after the plants have been exposed to the low temperature that caused it. Many flowering responses to photoperiodism are equally good examples of induction (e.g., the cocklebur).

Numerous responses in this category are also time-related. In photoperiodism, for example, we saw that the *time* of the light interruption during the dark period was extremely important (see Fig. 36–5). Time of light application is also a controlling factor in circadian rhythms.

Conditioning effects. In this situation, there is not much delay. The change is typically a gradual one occurring in response to a continual application of the environmental factor. Frost and drought-hardening are good examples. The more the plant is exposed to low temperatures, short days, etc., the more frost or drought-hardy it tends to become. Sometimes (e.g., in peach seeds) the germination temperature determines the nature and extent of subsequent seedling growth. This may be an example of a conditioning effect.

Carry-over effects. In the early 1950s Harry Highkin conducted experiments in the phytotron at Caltech in which he discovered that genetically pure, inbred pea plants grew poorly when the day and night temperatures were equal and held constant (10, 17, or 20°C). When the pea plants were grown for several generations under these adverse conditions, each generation (up to about the fifth) grew more poorly than the previous one. When he reversed the situation (germinated seeds from stunted plants under optimum conditions), it required at least three generations to reach the maximum level of growth. Such a carry-over of environmental effects from one generation to the next seems quite contrary to most of our concepts of genetics. (The environment does not influence the genes.) Nevertheless, the phenomenon was real, and it has since been substantiated by other workers. Both Highkin and the others were careful to demonstrate that the effects were not due to selection. Apparently, the developing embryo (or perhaps the stored food material in the cotyledons) is some way conditioned by the environment so that the effect carries over through a number of succeeding generations.

This phenomenon could be of considerable importance. For example, in the study of ecotypes (see Chapter 38), plants are brought to a uniform garden from different locations (where environments may be quite dissimilar). If carry-over effects are playing a part, then the conclusions about ecotypes are not so clear-cut. Regrettably, this phenomenon has been largely ignored by most workers.

In studying these various categories of plant response to environment we must remember that we were originally dealing with questions of ecology, particularly plant distribution. Surely, direct environmental responses are often considered in examining plant distribution (growth responses to water, light, temperature, etc.). Triggered, modulated, and conditioning effects should also be considered—and rather frequently are. Plant distribution, for example, might well be limited by effects on germination or on flower and subsequent seed production. Certainly, Highkin's conditioning effects could be important in plant distribution.

Stress Physiology

If we are going to try to understand plant distribution by physiological ecology studies that are aimed at elucidating the limiting factor or factors, then we will frequently find ourselves dealing with life under extreme conditions. That is, we will be concerned with the response of living organisms to stress conditions and therefore involved in the field of stress physiology. In a broad sense, this will be the case regardless of the biome or ecosystem being studied. But, to make our job easier, we will center our studies on those ecosystems where the stress factors are relatively obvious. In the desert, stress is first imposed due to a shortage of moisture, and in the alpine zone much of the stress is due to low temperatures.

To illustrate of stress physiology and limiting factors, we will now briefly examine the desert and the alpine tundra, and occasionally other situations when they have a strong bearing on stress physiology.

This matter of stress, however, is relative. Desert and alpine plants exist where they do because they have become physiologically adapted to their situations. Since they are adapted, they are not really being stressed. Other plants in those conditions would certainly be stressed, so we are interested in the adaptations that allow some plants to *avoid* stress.

The Desert

Ecologists have long classified plants into three broad groups based upon their water requirements. Hydrophytes are those plants requiring large quantities of water. These are often aquatic, growing either submerged or with roots in continually wet soil. Mesophytes are those plants that require only medium amounts of water (the vast majority of species), and xerophytes are those plants that can grow in the driest regions on the earth's surface.

Two other broad classifications include the glycophytes, which can tolerate only low to moderate concentrations of salt in the soil, and the halophytes, which can tolerate high salt concentrations. Halophytes are often found in deserts, since these soils are often saline (see Chapter 38).

Water as a limiting factor. To the ecologist, xerophytes are all of the plants growing in regions subject to periodic and severe drought—the desert regions (see the Photo Essay). But many of the plants growing in these areas do not directly confront the problem of drought. The desert annuals or ephemerals *escape* drought by existing during the dry season only as seeds. When enough moisture falls to wet the soil to a considerable depth, the seeds germinate, and the plants grow very rapidly. Often these are minute plants, and sometimes each plant produces only one seed. It is not unusual for the seeds to be produced in a period as short as 30 days. How do these tiny plants (called *belly plants* by Frits Went because of the position one must assume to best observe them) "know" when the soil is wet enough to support their rapid growth to maturity? Went showed that their seed coats contain a water-soluble inhibitor and that this is only leached out (thereby allowing germination) by a fairly heavy rainstorm (see Chapter 34). As much as 1.5 cm of precipitation may be required. Smaller rainstorms leach out a little of the inhibitor, but apparently this is replenished between storms, since the effects of storms are not additive.

For many years ecologists and physiologists have studied adaptations of many drought-avoiding perennials. The most successful of these adaptations is a deep-root system, reaching clear to the water table. Mesquite has been found with roots penetrating as deeply as 175 ft and alfalfa with roots at the 129-ft level. Of course, to develop such a system, the plant must be rather special in the first place. Went showed that the smoke tree (*Dalea spinosa*) would extend its roots rapidly when moisture was available and remain in a dormant condition during the drought period, repeating this cycle until the water table was reached—or until death if there were no water table. Seedlings in a tall cylinder of sand survived when given deep watering at intervals as much as 9 or 10 months apart.

Although few direct studies have been made, it is said that other drought-avoiding characteristics include: the sunken stomates that retard transpiration somewhat, small leaves that because of a thin boundary layer tend to keep leaf temperature close to air temperature, and the ability to go dormant by dropping leaves when the drought season approaches (e.g., *Fouquieria splendens,* ocotillo or desert candle). Often, these drought-avoiding characteristics are also present in plants belonging to the next two categories.

The succulent perennials are said to *resist* drought. These include the cacti, the *Euphorbia* of African deserts (similar in appearance to the cacti), the century plant (*Agave*), and several others. Virtually all of these plants resist drought by a combination of two factors. One is a special kind of tissue, displaying succulence, that consists of large thin-walled cells capable of storing water; these cells are embedded below a dense epidermis and a water-impervious waxy cuticle. The second factor is the so-called succulent metabolism. As we saw in Chapter 32, such plants open their stomates at night and close them during the day. Carbon dioxide is fixed at night into organic acids, which are then converted by photosynthetic processes to other metabolic products, including carbohydrates. This neat little adaptation allows the plant to maintain tightly closed stomates during the hot part of the day, when transpiration stresses are extremely high, and to open their stomates only during the cool, more humid night. The cacti and many succulents have one other adaptation important to their survival: a very shallow, wide-spreading root system capable of absorbing the water provided by light rainstorms. Thus the water is captured when it is available and stored in the succulent tissue during drought periods. Incidentally, succulent tissue can also occur in roots (*Fouquieria*), in stems (in the cacti, the spines are modified leaves and the broad, fleshy blades or columns are modified stems), and in leaves (*Agave*).

According to physiologists, the truest xerophytes are the drought-enduring nonsucculent perennials. Their leaf tissue may dry out to levels that would be quite fatal to most species (about 50%), but they still remain alive. The obvious question is what is it about the protoplasm of these plants that confers this hardiness. Much study has been done on this protoplasmic drought-hardiness, but really satisfactory answers have not yet been obtained.

Temperature as a limiting factor. Although the obvious limiting factor in deserts is availability of water, plants living there must often contend with other severe stress factors. One of these is high temperature (not always present in all deserts). Not only do the high air temperatures result in low relative humidity and a high tendency for evaporation, but high temperatures can themselves pose a severe problem for a plant. Protoplasm of the majority of plants is killed if the temperatures exceed about 50 to 54°C. Air temperatures in tropical and subtropical deserts may exceed 40°C, and when radiant energy is high, leaf and stem tissue could exceed the thermal death point, especially if they are in contact with the soil.

As in the case of water, some plants of desert regions *avoid* the problem. The palms growing around an oasis may have large, fan-shaped leaves with a thick boundary layer (see Chapter 32). Since ample water is available, the leaf is cooled by evaporation, and the extensive boundary layer insulates the leaf from the surrounding hot air. But, to manage such an evaporative cooling system, unlimited water must be available. A more typical situation in desert plants is to have small leaves with a thin boundary layer, thus causing the leaf and the air temperatures to remain close to each other. Since the air temperature rarely reaches the lethal point, this is tolerable.

On the other hand, certain tissues (e.g., stem tissue contacting soil) must reach temperatures well above 50°C. These tissues must simply *endure* the high temperatures in some as yet to be understood manner analogous to that in which true xerophytes endure extremely negative water potentials. Survival at high temperature is particularly difficult to understand, since our knowledge of protein chemistry tells us that a majority of enzymes should be denatured at such temperatures. If even one critical enzyme were completely denatured, the plant would die. Therefore how can *all* the enzymes of these tissues manage to survive? At the moment we have essentially no information, but we do have some precedents that are even more extreme than those just mentioned. In the hot springs of Yellowstone National Park, bacteria have been found actively growing close to the boiling point (about 90°C at that elevation). Even DNA "melts" in the laboratory at about 73°C. (The two strands of the double helix separate.) Study has been initiated on these hot-spring organisms (see Chapter 16). It has been noted, for example, that the bacteria are filamentous at 80°C and form short rods below 75°C, but no explanation is yet available.

Salt concentration as a limiting factor. Some desert plants are faced not only with a shortage of water and with high temperatures, but also with high salt concentrations in the soil. (Conversely, the salt may be the primary stress factor, while water may be fairly available, as in a salt marsh.) Many halophytes, such as the mangrove trees that occupy millions of acres of coastal swampland, actually flourish in seawater. A few organisms (mostly bacteria and algae) can even grow in saturated salt solutions such as the Great Salt Lake or the Dead Sea.

Occasionally, there are specific salt effects. Sodium ions may themselves be toxic, for example. More typically, the stress imposed by high salt concentrations in the soil solution or in a liquid medium such as seawater is an osmotic stress. That is, the salt lowers the water potential, making it more difficult for the organism to obtain water. Some halophytes overcome this by being salt accumulators. They absorb a great deal of salt, but for some reason their protoplasm is not damaged by it. (High salt concentrations will denature or precipitate most enzymes.) These, then, are the *endurers*—the "truest" halophytes. But even these accumulators do not take up salt as rapidly as they do water. If they did, they would soon become overloaded with salt, since they are losing water by transpiration. Salt concentrations in the tissues may be high, even though the tissues absorb salt at a rate only one-hundredth as fast as they take up water. Studies on salt enzymes have been initiated. In a few cases, at least, the enzymes are denatured when salt concentrations are lowered.

Some plants are salt regulators. These may expand

their tissues to produce succulence, or they may have special glands on the surfaces of their leaves that excrete salt as small crystals. The metabolism of salt excretion is an interesting and special metabolic adaptation of these halophytes.

In both cases, when plants take up salt (either maintaining it in their tissues or excreting it on the leaf surfaces), they have a profound influence upon the soil in which they grow. Thus, in the deserts of the Great Basin (e.g., the Escalante Desert, where studies were performed), sagebrush plants do not take up salt, and salt concentrations near the soil surface below these plants do not reach particularly high levels. On the other hand, greasewood plants growing in the near vicinity do take up the salt, depositing it on the soil surface when they drop their leaves. The result is a marked increase in salt concentration in the upper layer of soil.

Essay 39–2 Shadscale and Fungus: Desert Partners

E. Dean Vest was working for a master's degree when he first got involved in the physiological ecology of the desert. He found the complex interactions of desert organisms so fascinating that he spent the next decade working with about 30 other scientists on an extensive project at Dugway, Utah, sponsored by the United States Army Chemical Corp. Vest became the director of the project a few years after its inception in 1952. He is now teaching at Southern Utah State College.

"As a graduate student at the University of Utah, seedling establishment in desert plants fascinated me. I soon found that many seeds from different plants would not germinate readily in the laboratory. One desert plant, shadscale (Atriplex confertifolia), *was particularly interesting. The seed was enclosed in two mechanically hard bracts containing stone cells and with calcium oxalate druses (crystals). Seeds inside these bracts never germinated, although many thousands were tested. When the seeds were excised from the bracts and their seed coat scarred with sand paper, however, up to 75% germinated. Interestingly, whole seeds (with bracts) when soaked in five times their weight of water produced a water-soluble extract that would inhibit scarred shadscale seeds, lettuce seeds, and radish seeds—even when diluted five times with distilled water. Inhibitory effects could still be observed in dilutions up to 20 times (Table 39–1).*

Leaching the inhibitor from the whole fruit still did not permit embryo germination through the hard bracts enclosing the seed, indicating that a mechanical inhibitor was present as well as the chemical one. Cracking the water-leached bracts resulted in immediate germination of the enclosed seeds, although cracking unleached bracts did not allow germination. Therefore, even desert storms capable of leaching the water-soluble inhibitor would not be sufficient to germinate seeds inside these hard bracts. A further solution had to be sought to explain natural germination.

40

Functional Ecology: The Ecosystem

Time and Again—The Ecology of Plant Succession

Secondary plant succession occurs following the removal of vegetation—often by a natural or man-made catastrophe such as fire. The first plants to move in ("pioneers") will be those that can tolerate high intensities of sunlight and burned or barren soil. The "climax plant community" will ultimately be capable of reproducing itself; that is, its seedlings can grow in the shade of their parents. The steps followed in such a succession depend upon several factors, not the least of which is the variety of species in the area. These photographs show a few stages in a succession, which often occurs in the northern Rocky Mountains, from a recently burned area through lodgepole pine to spruce-hemlock forests.

A A burned stump in a forest of lodgepole pine (Pinus contorta) *in Grand Teton National Park. The fire occurred nearly 100 years ago, but careful searching reveals a few remaining charred stumps.*

B A mature lodgepole pine forest in Yellowstone National Park. Note the many small pioneer lodgepole seedlings (lower right), where road building has exposed the soil. These seedlings require high light intensities.

A

B

C Young spruce trees (Picea engelmannii) *growing in the shade of mature lodgepole pine trees.*

D A mature climax stand of spruce and hemlock (Tsuga heterophylla). *No lodgepole trees occur here, but young spruce and hemlock trees are reproducing themselves in the deep shade of their elders.*

C

D

Now our acre of woodland becomes even more intricate. We have defined ecosystems as highly dynamic, integrated, complex, functioning communities of plants, animals, and their environments. But just how do they function? Are their functions analogous, as Clements suggested, to an organism? Or do they function in ways that are unique and not analogous to anything? In this final chapter, we shall consider the activities of ecosystems, discovering that most functions are indeed unique to the ecosystems themselves.

Energy transfer is central to ecosystem functions—at least in terms of the intensity of many modern studies. Transfer of matter virtually always accompanies energy transfer, and in some ways, the transfer of matter is more distinctive and important to ecosystem function. We shall also have occasion to investigate some other interesting ecosystem activities, such as competition, symbiosis, homeostasis, and plant succession.

All of this is beginning to be of the most profound importance to man, who is perturbing most of the earth's functioning ecosystems in an almost infinite variety of ways. He has established his own agricultural and horticultural (urban) ecosystems at the expense of the natural ones, but his very survival now depends upon the artificial ecosystems—as well as the natural ones. Man is the rational animal. Can he apply his reason to an understanding of ecosystems, solve his problems, and ensure his survival and joy of living?

Energy Transfer in the Ecosystem

In recent years ecologists have been active in applying the techniques developed by physicists and physiologists to measure energy transfer. This has proven to be a profitable research field. Many parameters (environmental factors) can be measured quite accurately, and others can at least be estimated. For example, radiometers measure radiant energy transfer. In one kind of radiometer, a small black wafer at the apex inside a highly mirrored glass cone comes to radiant equilibrium with its target; that is, by exchanging radiant energy with its target area (the sky, the ground, a single leaf, etc.), it changes its temperature, which is measured by a sensitive thermocouple or thermister. After the instrument has been calibrated, it can indicate the calories per square centimeter per minute ($cal/cm^2/min$) being transferred to or from some surface by the process of radiation. Recording thermometers placed in the soil, in the air, etc., may allow calculations of convective or conductive heat flow. Leaves, plants, or even groups of plants may be enclosed in transparent containers (cuvettes), so that carbon dioxide and oxygen exchange between the plant and the atmosphere can be measured. Photosynthesis and respiration rates can be calculated from these measurements. Respiration is usually measured in the dark and assumed not to be affected by light. But this is not true for those plants capable of photorespiration, so other corrections must be made. Harvesting sample plots at intervals throughout the growing season allows net productivity to be measured (the carbon fixed by photosynthesis minus that released by respiration). In the past, many of these harvest measurements gave faulty impressions because roots were not taken into account.

Physical energy transfer in an ecosystem is essentially an enlarged version of physical energy transfer between a leaf and its environment, as discussed in Chapter 32. The solar energy impinging on the outer edge of the earth's atmosphere is equal to about 2.0 $cal/cm^2/min$ (normal to the sun's rays). Some of this is reflected back into space by the atmosphere (especially clouds) or by the plants and animals in an ecosystem. Some of the energy is transmitted through leaves, and some is absorbed. Most of that which is absorbed is converted to heat, but a small portion (an important portion) drives the process of photosynthesis. All objects above absolute zero, including plants and animals, radiate energy. Much of the energy absorbed by an ecosystem is thus reradiated to the atmosphere and to space. Some of the energy absorbed by plants or animals warms the surrounding air or soil by conduction and by convection, and some of this energy is used to evaporate water in transpiration from plants or evaporation from animals. (These processes are summarized in Figure 40–1.) The energy trapped by photosynthesis may be stored in the ecosystem, and when organic matter moves into or out of an ecosystem, chemical energy moves with it. This could be a highly significant aspect of the energy exchange in a section of a stream, for example.

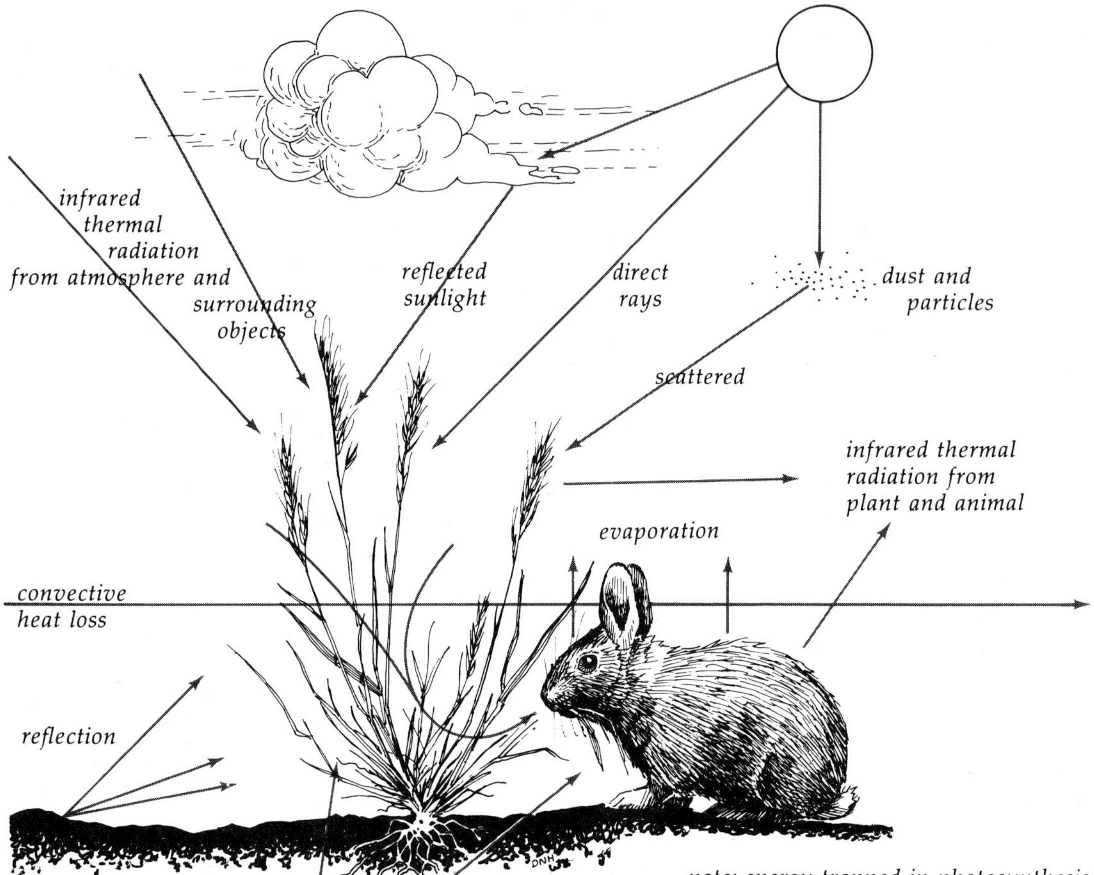

40–1 Some of the principal physical energy transfers in the ecosystem.

Ecosystem Productivity

Although the light energy used in photosynthesis is only a small portion of the total, it nevertheless accounts for all of the primary productivity of most ecosystems (except for those supplied with organic matter from outside, as in the stream). The primary productivity of an ecosystem is, therefore, the productivity of its autotrophic plants (green plants, capable of living only on minerals, water, carbon dioxide, and light energy).

What limits the primary productivity—photosynthesis—of an ecosystem? In arid climates water is clearly the limiting factor. Probably water is limiting to a certain extent in all climates receiving less than 100 cm of precipitation per year. Table 38–5 shows precipitation figures for various biomes, and Table 38–6 includes figures of primary productivity. The maximum productivity of deserts is only about 250 dry grams of organic matter/m^2/year. Grasslands receive a little more moisture and have maximum productivity values of 1500 g/m^2/yr, while temperate forests (even more moist) reach 2500 g/m^2/yr. Tropical rain forests supplied with much moisture have figures that go as high as 5000 g/m^2/yr, and swamps and estuaries may produce 4000 g/m^2/yr. Properly managed agricultural lands can approach the productivity values of tropical forests and swamps (4000 g/m^2/yr). But most temperate cereal crops are in a lower range (300 to 1200 g/m^2/yr), and marginal agricultural lands may have productivities comparable only to those of deserts (100 $g/m^2/yr$). In each of these cases, the range of productivities shown in the table is great. We can imagine that at least the upper productivity of each biome is limited by moisture (at least in the drier biomes), while the lower productivities may be due to some other limiting factor.

In some situations, temperature is obviously limiting. This becomes clear when we compare the arctic or alpine tundra with the tropics. In areas where the tundra is amply supplied with water (bogs, etc.), productivity still never approaches that possible in the tropics, even when a correction is made for the much shorter growing season.

Many agricultural ecosystems respond markedly when fertilized with certain nutrient elements, particularly nitrogen, phosphorus, or potassium. Depending upon the local situation, potassium is probably never limiting in deserts or semidesert regions, while nitrogen may often be limiting. The limitations of nutrients on agricultural land are largely due to their removal by continual cropping. We nevertheless have every reason to believe that nutrients may also be limiting to natural ecosystems, although this has been studied only superficially. Many natural soils are highly infertile for crops, even immediately following removal of the natural ecosystem. Furthermore, in a few cases, natural ecosystems have responded to application of nutrients.

Light intensities are often limiting, at least in portions of many natural ecosystems. This is at first surprising because studies of photosynthesis in some agricultural plants have shown that the process is typically saturated by light intensities of 1000 to 2000 ft-c, and these intensities may exist even under fairly heavy cloud cover. Of course, many known species are not saturated, even at intensities of 10,000 ft-c, and these intensities are about the maximum encountered in most natural environments. Furthermore, only a portion of the leaves in a natural ecosystem are exposed to full sunlight, while many others are shaded. It is interesting to study the placement of leaves on plants, since this placement often ensures that a maximum number of leaves will be exposed to the direct rays of the sun (see the Photo Essay for Chapter 31). Nevertheless, in forests especially, where several strata may exist, only the upper stratum will receive full sunlight. On the typical forest floor, intensities may drop to only a few hundred ft-c, (even on a clear day) and can be less than 100 ft-c in a dense forest.

The total light received during the day will influence total photosynthesis. Although light intensities are relatively low in the arctic because the sun is low in the sky, days are so long (up to 24 hr) that the total light energy available for photosynthesis in the arctic summer may exceed that available even in the tropics. In any case, although light intensities may limit the photosynthesis of certain portions of an ecosystem, in ecosystems where vegetation is dense (warm temperatures and ample moisture), the utilization of the light energy falling on the ecosystem as a whole may be highly efficient.

Studies have also shown that light intensities may sometimes be limiting in agricultural situations. This is true in regions where heavy clouds are the rule, and it is certainly true in winter greenhouses. In these situations, the limitations would probably not be apparent if agriculture were not carried on so intensively, that is, if such factors as water, nutrients, and temperature were not artificially raised to optimal levels.

When these factors are optimal (high light intensity, warm temperatures, excess moisture), the carbon dioxide concentration of the atmosphere may limit photosynthesis. This has been shown to be important above a mature corn field or an apple orchard, for example, but most ecologists feel that it is seldom important in natural ecosystems. Yet the productivity of tropical successional forests and several other ecosystems equals that of the best agricultural situation, so carbon dioxide might be expected to be limiting. In a climax forest, decay of organic matter produces carbon dioxide as rapidly as photosynthesis uses it. In any case, laboratory studies indicate that increasing carbon dioxide may often increase photosynthesis. If this proves to be true in nature, then we might expect an increased productivity of many of the earth's ecosystems in response to the gradually increasing carbon dioxide of the atmosphere* brought about by the burning of fossil fuels in man's industries (see Essay 40–1). In practice, however, this may be more than counterbalanced by the increases in concentration of toxic atmospheric components such as ozone and sulfur dioxide.

---

*Actually, the increase in carbon dioxide is much slower than one might expect, since most of the released carbon dioxide is absorbed by the oceans, which act as an efficient buffer, maintaining rather constant concentrations in the atmosphere.

Essay 40–1 Man, Energy, and Ecosystems

The only energy transfer of importance between most organisms is the consumption of one organism by another. All consumers, including many backward peoples, obtain virtually all of the energy they use in the form of food calories. At some point primitive man discovered the use of fire and beasts of burden and thus broke the pattern.

When man invented agriculture, he learned to exercise conscious control over the biological production of his food calories. This was a great improvement. No longer did his problems deal with fang and claw, and his average life span could be extended—from about 25 to 30 years. Yet he exchanged his old problems for some new ones. Now he had to worry about drought and pestilence. Yet he could sometimes produce more than he actually needed; that is, he had wealth. *He became a social organism living in cities. There was not much wealth, and a few clever members of the species typically managed to corner most of it, while the rest lived at a subsistence level.*

About 300 years ago, man found fossil fuels, which greatly extended his energy supply. This brought with it the invention of machines and the industrial revolution. It became possible for a large number of individuals to share the wealth (specifically in Europe, North America, and a few other places). Quite a few now have luxury. With ample free time, a few could utilize man's special abilities—creative thinking and the scientific method. Medicines were invented, and life spans increased greatly. This was true not only for the affluent societies but also for many others in the world, since medicines, once invented, were relatively inexpensive.

But why don't all segments of mankind share in the affluence enjoyed by a few? The ultimate limitation still seems to be the ability of man to produce food. Ample food is easy to produce on really good land, where temperatures are relatively high, rains are reliable, and fertility is maintained by frequent flooding. Unless conditions are as ideal as these, however, much effort is required to make the land produce abundantly. And there is not much good land on the earth's surface. All of the cereal grains—the staples for nearly all mankind—are produced on about 7% of the earth's surface, for example. As it turns out, there are about 3 billion acres of farmable land on our planet—about 1 acre for each person living today.

The basic missing ingredient that keeps all peoples on the earth from having an adequate food supply is energy. *To make the land produce well one needs energy requiring tractors, fertilizers, insecticides, special seeds, powered wells for irrigation, and other devices of modern agriculture. For example, the input of solar energy is about 21,000 kilowatt-hours/acre/day. High-producing grains capture about 80 kwh/a/d in a form suitable for human consumption (counting only the grain—no roots, stems, or leaves). This is an efficiency of about 0.4%. To obtain this efficiency, some 60 kwh/a/d must be expended with tractors, fertilizers, etc. In the United States this energy would cost only about 80¢ per acre per day, and if this acre feeds 20 people, then the cost is only about 4¢ per person. In India, however, the same energy produced by cattle costs about 30 times as much or $24 per acre per day ($1.20 per person per day).*

Energy is cheap in rich countries and expensive in poor countries—probably rich countries are rich because they have been able to make energy cheap! The average energy expenditure for the world as a whole is about 36 kwh/d/person. In the United States, however, each person utilizes about 240 kwh/d (in Western Europe, about 70 kwh/d), and in the most backward countries, individuals utilize only about 3 kwh/d. Energy is required not just to build and fuel industries but to produce an abundance of food. Without this, industries can never develop.

Within recent years we have become aware of some serious problems. Will energy production destroy us by the pollutants it dumps into the environment? Or will the fossil fuels upon which we are so dependent run out? Or will we run out of raw materials? Or freshwater?

The fossil fuels will be gone in 300 to 1000 years. Some of them (oil) are already becoming difficult to obtain. But there are two potential sources of energy that are essentially unlimited. The deuterium in the seas might be used in controlled fusion reactions. The energy available from a pint of seawater would be approximately equal to the energy produced by burning 60 pounds of coal. In addition, the lithium in rocks can be used to produce tritium, a fuel for fusion reactors, and the uranium and thorium can be used in breeder reactors in which nonfissionable elements are converted to fissionable fuel, so that as the reactor functions, an excess of fuel is produced. Fusion reactors are not yet feasible, but they may become so in the near future. The breeder reactors are perfectly feasible, and some are being constructed.

There are enough rich ores containing lithium, thorium, and uranium to last many centuries. Actually, energy obtained from uranium and thorium in oil shales might exceed the energy obtainable from the oil by 10 to 100 times! The granites are far more extensive than the shales, and uranium and thorium can be extracted from them for less than 1% of the energy produced. (Many metals and other important elements are also obtained as by-products.) The ores, shales, and granites would supply us with vast quantities of energy for millions of years. Adding the deuterium in the ocean, our energy supply might last as long as the solar system itself—perhaps another 10 billion years.

The other raw materials required by affluent man don't present a clear picture. Yet only two things are required to have ample materials: ingenuity and energy. With energy we can produce huge quantities of freshwater from seawater. Water and air (rather than the natural gas now being used) can be used to produce ammonia for fertilizer. Plastics and even gasoline could be produced from limestone. Energy can be used to recycle the products of man's technology. Rare metals difficult to recycle can be obtained from low-grade ores or even from seawater—if enough energy is available. Actually, the granites contain not only the energy but virtually all the raw materials that we might need.

What about pollution? Obviously, we must tackle our problems promptly and solve them. With ample energy this should be possible, but the heat produced provides a special problem because if it reaches high enough levels, our climate might be influenced. Warming of the climate could melt the polar ice, raising the levels of the oceans and inundating most of our coastal cities. Increased carbon dioxide in the atmosphere might have the same effect, since it traps the infrared radiation being emitted from the surface of the earth. On the other hand, our solid pollutants increase the reflectivity of the earth's atmosphere so that less heat strikes the surface, and this could lead to a cooling trend. Problems of heat and climate could ultimately be the most serious ones we have to face.

Energy is almost certain to be the key to our solution of the problems of solid and gaseous wastes. To begin with, technology is not the only producer of waste. Consider the medieval or Asian cities with their vast quantities of garbage, sewage, and smoke. In India a pall of smoke covers the cities at mealtime when dung is burned for fuel. By using energy, Western cities have a much higher living standard with little increase in pollution. The immediate problem in the energy-rich countries is that burning of fossil fuels produces smoke, sulfur dioxide, and other pollutants. Perhaps we are lucky that we are running out of fossil fuels.

So far as pollution goes, atomic energy is much easier to handle than fossil fuels. Fission reactors have now been operating for several years, and we know how to handle their waste products. At present the radioactive wastes are converted to dry ash, sealed in metal containers, and buried in the earth (preferably in abandoned salt mines, which are always dry). If New York City were provided energy exclusively from nuclear reactors, the wastes for one year would occupy only about 1 m^3. When fusion reactors can be used to produce energy, the problems will be even less serious, since they produce no radioactive wastes at all. Furthermore, they will pose no threat of accidental explosions.

So the promise is there! Study groups have investigated proposals for coastal cities fueled entirely by atomic reactors, producing large quantities of freshwater from the sea, which is then used to produce ample nutritious food for everyone. Basic minerals such as clay are used to produce metals and other raw materials for technology and commerce. With clean atomic energy—cheaper even than fossil fuels—there seems to be no limit to the possibilities. With a population of controlled size (no one knows how large this might be), ample food could be produced on less land per person than is presently required. Clean, silent, efficient, electrically powered transport would be available. Wastes would be managed, and scarce elements would be recycled. There would be ample work for man, but this would not be the kind of work presently performed by animals or machines. It would be work that only a man, with his brain, can accomplish.

Will we build this utopia, or will we lose our opportunity thanks to nuclear war, greedy industries that overwhelm the environment with wastes, or other weaknesses of human nature? It is up to us. But there is room for optimism. We have just begun to learn how to utilize energy. Consider the history of the earth by compressing each year into a second. In the words of R. Philip Hammond: "Time began for our part of the universe only 300 years ago, and life appeared on earth less than 100 years ago. Three years ago dinosaurs roamed the land, and mammals began to predominate only one year ago. Mankind appeared 11 days ago, and a week ago he invented tools. Three hours ago agriculture was invented, and written history began 90 minutes ago. Five minutes ago we began our first experiments with energy, coal, and steam, and only two minutes ago did we really get going with autos, machines, and mass production. Around 20 seconds ago we began to notice that energy use had some effects called smog, and about three seconds ago it became clear we had to do something about taking care of our wastes. Fossil fuels will be exhausted in another five or ten minutes, but the nuclear fuels will last 300 years."

Carbon dioxide can be depleted in greenhouses, particularly in winter when these are tightly closed to retain heat. In such situations, it may be economically profitable to fertilize with carbon dioxide. This is done by allowing a block of dry ice to melt (sublime) in the greenhouse, by adding carbon dioxide from cylinders of liquid, or by burning highly pure propane or some other gas, to produce carbon dioxide. The latter method is efficient and widely used but slightly dangerous because if combustion is not complete, carbon monoxide is produced. (Incidentally, the human workers in a greenhouse release a surprising amount of carbon dioxide, enough to supply many plants.)

Productivity is also limited in an ecosystem sense by the number of plants per unit area capable of photosynthesis. Sometimes attempts have been made to measure this factor by measuring the quantity of chlorophyll per unit land or water-surface area, but since not all chlorophyll is equally efficient, this method is not wholly satisfactory. In any case, it is obvious that the desert will be less productive than the tropics, simply because there are fewer plants per unit area—a limitation imposed by the lack of water in the desert. Studies of the productivity of the various stages in a plant successional sequence demonstrate that the early stages, just before the climax, exhibit maximum productivity. All stages before the climax produce more than they use, so the total biomass of the ecosystem increases. The climax, on the other hand, is at an equilibrium where productivity equals consumption.

Man has efficiently applied his knowledge of the factors limiting productivity to agriculture. To increase productivity, he adds water by irrigation, increases temperatures in greenhouses (and, on a much smaller scale, darkens the soil surface with asphalt mulches to increase temperature by absorption of more radiant energy), fertilizes with the major and the minor nutrient elements, grows plants under artificial light (on a very limited scale), removes competing plants that would shade his crops, fertilizes with carbon dioxide in his greenhouses, and attempts to maintain in his artificial agricultural ecosystem stages that are comparable in their productivity to natural subclimaxes. Quite an impressive achievement.

What is secondary productivity? Primary productivity is due to the autotrophs, but secondary productivity is a function of the heterotrophs—those organisms that must obtain their energy supply and organic nutrients from other organisms. Secondary productivity is a measure of the efficiency with which the heterotrophs can convert the organic matter they consume into the organic matter constituting their own bodies. An oft-quoted figure for the efficiency with which herbivores, carnivores, and decomposers can assimilate the food they consume is 10%, although this is complicated by many factors. In modern agriculture efficiency has been increased considerably in the raising of chickens, turkeys, cattle, and swine. In turkey production, for example, the figure may be as high as 45% (2.2 lb of feed, mostly cereals, producing 1 lb of turkey). Unfortunately, little is known of the productivity efficiencies of plant heterotrophs (e.g., the fungi).

The Ecosystem Pyramids

One way of stating the Second Law of Thermodynamics is that energy transfers can never take place without some of the energy being converted to heat—in other words, energy transfers are never 100% efficient. Thus primary productivity will always be larger than secondary productivity. If the productivities of the various trophic levels in the ecosystem are stacked one upon the other with the autotrophs at the bottom, then a pyramid results—the so-called energy pyramid

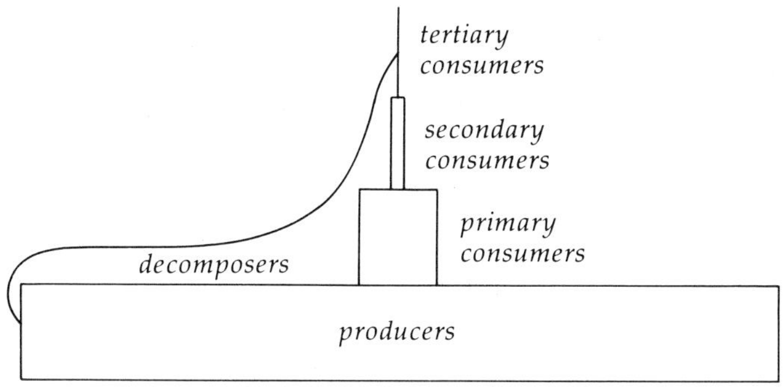

An Energy Pyramid

40–2 A representative energy pyramid, each trophic level containing 10% of the energy of the one below it.

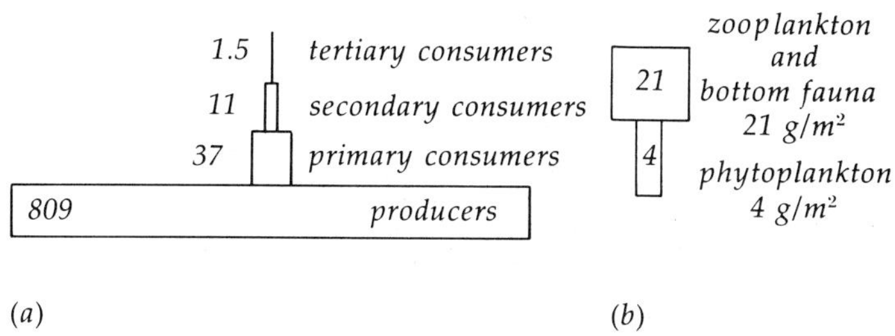

40–3 Pyramids of biomass. (a) A pyramid of biomass in the aquatic ecosystem of Silver Springs, Florida. Figures represent grams of dry biomass per square meter. (b) An aquatic community in which the biomass of the consumers is five times larger than that of the producers.

of ecosystem productivities (Fig. 40–2). Such a pyramid is simple in concept but not always easy to construct. The herbivores fit nicely upon the autotrophs, and the carnivores that eat nothing but herbivores would fit nicely. But what about omnivorous man (or grizzly bear), who is part herbivore and part carnivore? To complicate matters further, there may be secondary levels of carnivores that eat carnivores. One of the worst problems concerns the placing of the decomposers. In one sense, all but the autotrophs are decomposers, but we usually think of the decomposers as microorganisms. This being the case, they would lie over the entire pyramidal structure, since they decompose not only the uppermost carnivores but the herbivores and autotrophs as well. If each trophic level in the ecosystem energy pyramid represents at most only about 10% of the energy included in the level below it, then the upper levels become quite small. It is said in India that a hill can support only one tiger. This is partly a statement of the habits of tigers, which occupy a given territory to the exclusion of competitors, but it is also a recognition of the energy pyramid principle.

Since man is omnivorous, he can occupy not only parts of two trophic levels, but he can also occupy one or the other. Clearly, if he eats nothing but meat and animal products (eggs, cheese, milk, etc.), his numbers will ultimately be only 10 to 45% of what they might be if he occupied an exclusively herbivore level, living only on plant products (e.g., wheat, rice, maize, soybeans, and cassava).

Ecosystems also have pyramids of biomass (Fig. 40–3). In a forest system, for example, the primary producers often have an extremely large biomass, although much of it consists of the dead cells of wood. In the ocean the biomass of consumers (whales and other plankton-eating fish) at any moment may exceed that of the primary producers (the plankton) so that the "pyramid" is inverted or spindle-shaped. But the total phytoplankton biomass that lives during the course of the year exceeds that of their consumers.

Ecosystems also have pyramids of numbers (Fig. 40–4). Sometimes (e.g., a grassland biome), these figures may have the typical pyramid shape. There are other cases, however, where the number of consumers may greatly exceed that of the producers, as for example, the hundreds or thousands of aphids living on a single plant.

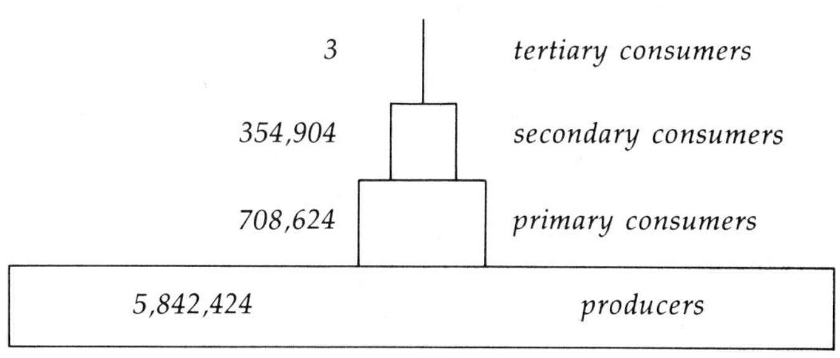

40–4 A pyramid of numbers in a bluegrass field.

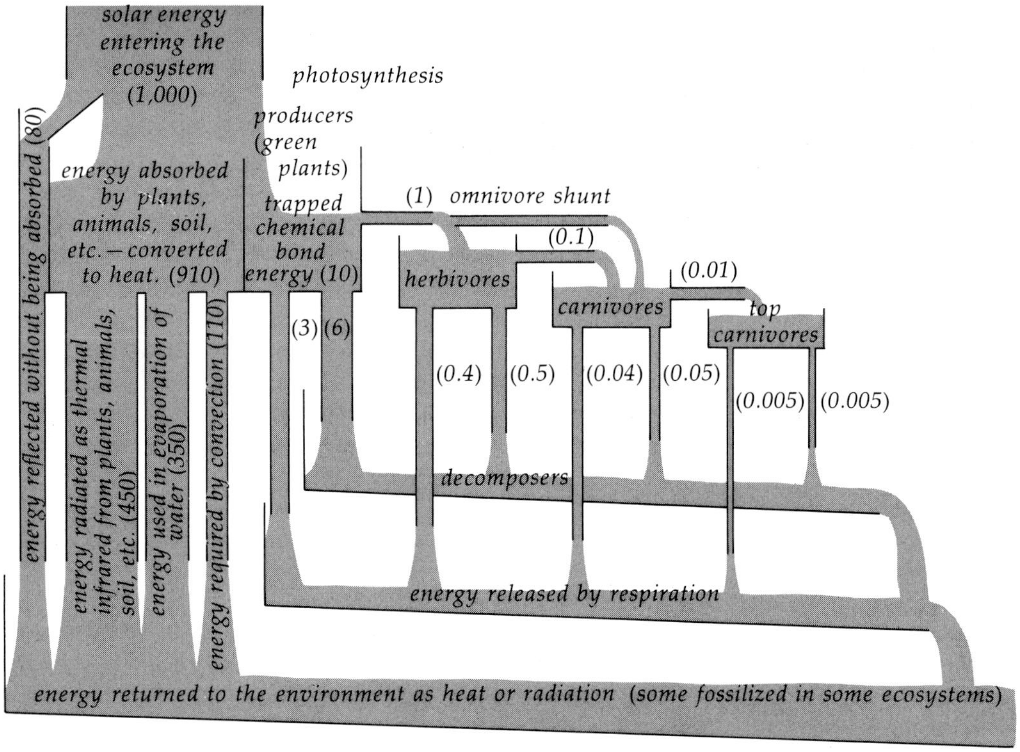

40–5 A generalized diagram for energy flow through an ecosystem. The numbers are arbitrary but representative.

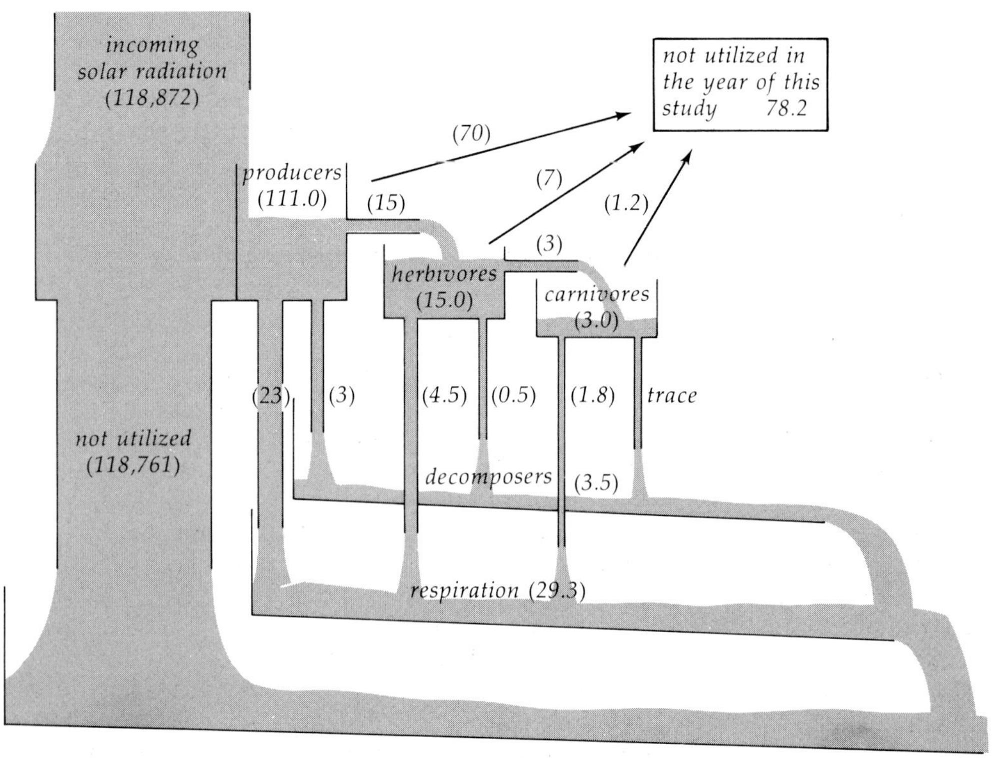

40–6 Energy flow through Cedar Bog Lake, Minnesota, in g-cal/cm²/year. Some items indicated in Figure 40–5 were not measured in this study, and a large portion of the energy was not utilized during the year of the study (box in upper right).

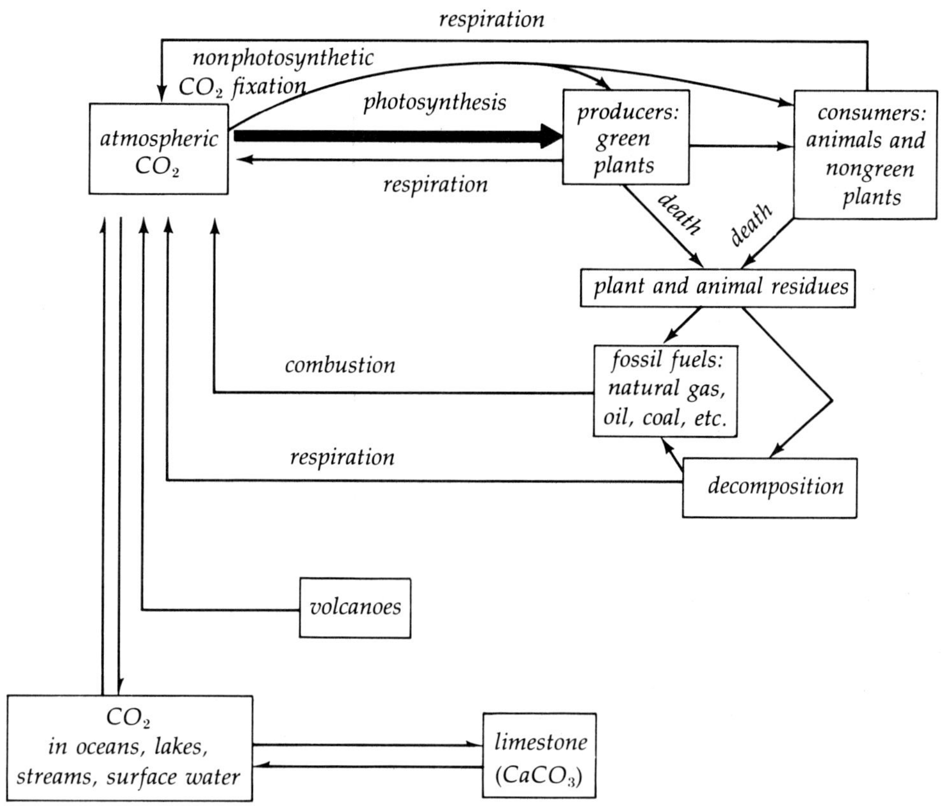

40–7 *The carbon cycle.*

Energy-Flow Diagram

Figure 40–5 emphasizes the physical aspects of energy flow in an ecosystem. The energy pyramid (Fig. 40–4) implies something about the biological exchanges of energy within an ecosystem, but such a figure is rather static. A more dynamic approach to energy-flow diagrams has been developed by Eugene P. Odum for several ecosystems that he has studied in detail. Biomass can be indicated simultaneously with energy-flow rates for different parts of the ecosystem. Such diagrams can be drawn to scale, indicating the biomass of certain components as a solid volume. Figure 40–6 shows an energy-flow diagram for a specific ecosystem. In detail the diagrams always depend strongly upon the ecosystem with which one is dealing. For example, the energy capture could be photosynthetic (as it usually is) or it could be an organic matter input (as found in a stream, lake, cave, or the ocean floor). Obtaining the data, formulating ecosystem energy-flow diagrams, and interpreting their meaning for community function and agricultural harvests are important objectives of modern ecological research.

Transfer of Matter in the Ecosystem

The carbon cycle. Matter as well as energy is transferred in ecosystems. Much of this transfer has been studied for at least a hundred years. Textbooks written for the early grades discuss the carbon cycle in nature (Fig. 40–7). There are several physical transfers of carbon on our planet (e.g., the equilibrium between carbon dioxide in the atmosphere and carbonic acid dissolved in the ocean or the carbonate held in limestone rocks). Carbon is outgased through volcanoes, fumaroles, etc., from the interior of the earth into the atmosphere. Much carbon is now released into the atmosphere by the burning of fossil fuels. Indeed, if all the carbon held in fossil fuels were returned to the atmosphere (and none were dissolved in the ocean),

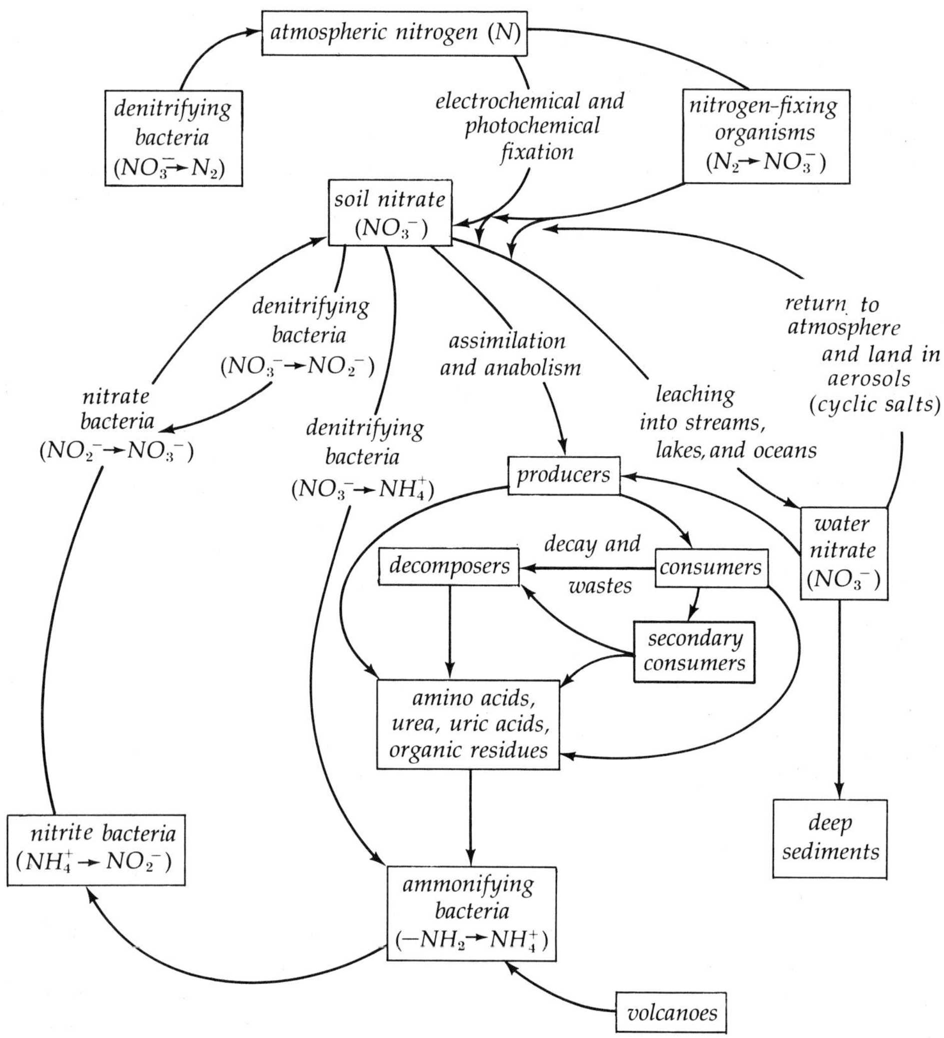

40–8 *The nitrogen cycle.*

our atmosphere would consist primarily of carbon dioxide, as does the dense atmosphere of Venus. The only important movement of carbon into the biosphere is via the process of photosynthesis. Once in the biosphere, it moves through food chains and is released back into the atmosphere at each stage by respiratory processes. Some of the carbon may never be returned; it may be fossilized, as in contemporary peat bogs or the ancient swamps that formed coal and oil.

The nitrogen cycle. The nitrogen cycle is also often discussed (see Chapter 16), and in some ways it is more complex than the carbon cycle (Fig. 40–8). Certain organisms (nitrogen fixers) are capable of utilizing molecular nitrogen, but most organisms are not. Plants convert the nitrogen to amino acids (protein), chlorophyll, nucleotides, and many other products, most of which may be used almost directly by herbivores. Animals convert much of this nitrogen to urea, and microorganisms convert the urea and many other nitrogen-containing compounds to molecular nitrogen, thus completing at least a part of the cycle.

Other biogeochemical cycles. Phosphate is also cycled to some extent in nature (Fig. 40–9). Its primary origin is the mineral apatite occurring in rocks

40–9 *The nutrient cycle in a forest ecosystem.*

and soils. It is transferred through an ecosystem along with other matter and may ultimately be released as phosphate by decomposers. In this form it may be reabsorbed or leached (although not readily) into the water table and ultimately into the oceans. There it is consumed by plankton, some of which die and settle to the bottom, taking the phosphate out of the ecosystem—a fate that awaits many mineral nutrients. We are particularly concerned now because the addition of phosphates from detergents seems to be upsetting many of our aquatic ecosystems.

Of course, the biogeochemical cycles have been worked out for numerous elements besides carbon, nitrogen, and phosphorus. The movement of oxygen is particularly important. It was the buildup of oxygen (and its product, ozone) in the earth's atmosphere that filtered out the ultraviolet rays from the sun and allowed evolution on land to proceed. Even chlorine (as chloride) cycles in the biosphere. As winds whip the surface of the ocean into whitecaps, minute droplets of ocean water (aerosols) go into suspension in the atmosphere. The water rapidly evaporates, and the salts (including chloride and all of those present in the ocean) move with the atmosphere's circulation—some over the land. These minute particles then form the condensation nuclei around which raindrops develop, when temperature and humidity conditions are suitable. They are then returned to the land as the rain falls. (Chloride has been emphasized in studies of these cycles because it is easy to measure, and its ultimate source must be the ocean.)

Transfers of energy compounds. The flow of elements through the earth's ecosystems is relatively easy to study, but we must not lose sight of the equally important flow of compounds. First are the energy compounds—the foodstuffs that move from one trophic level to the level above. Primary food compounds are carbohydrates, fats, and proteins. These exist in a great variety, and there are many individual species differences with regard to their ability to digest, decompose, and assimilate these compounds. Cellulose, for example, is a carbohydrate that cannot be digested by most animals. Certain microorganisms, including fungi, bacteria, ascomycetes, etc., do digest cellulose. In the stomachs of ruminant animals or termites, cellulose-digesting bacteria make this carbohydrate available as sugar for their own nutrition and that of their host. Lignin is an especially resistant product, but there are bacteria that can decompose it. Insects are adapted to digest certain kinds of plant products, and birds and lizards are adapted to digest insects. Certainly no organism (except perhaps an organic chemist) can produce an organic molecule that cannot be digested by some other organism.

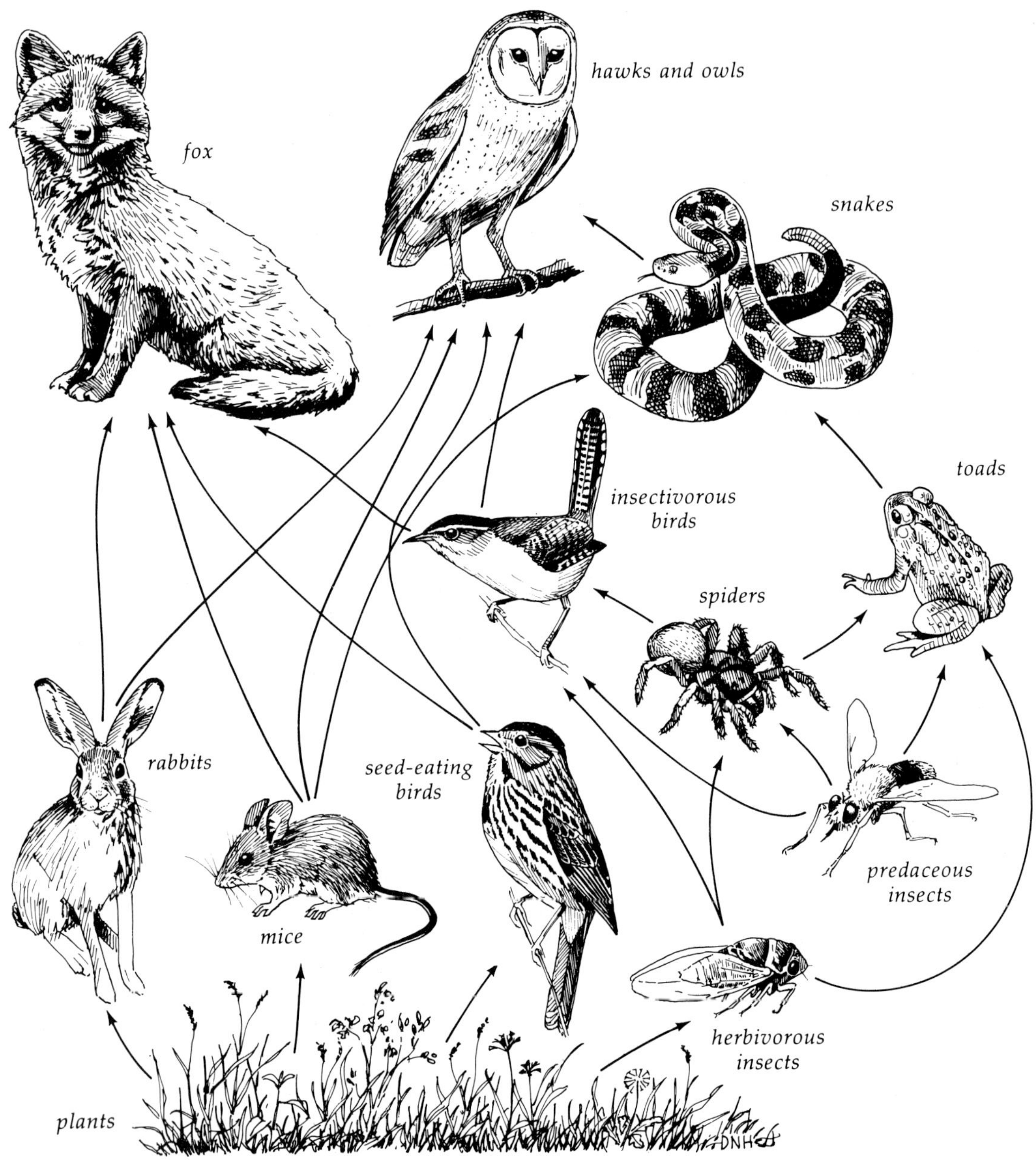

40–10 A diagram of a hypothetical food web. No real food web would be as simple as this one.

Some parasites are physiologically adapted to live efficiently upon the life fluids (e.g., blood, contents of the digestive tract, etc.) of their hosts. Typically, the biochemical relationship between a parasite and its host is a very intricate and carefully balanced one. Yet it is obvious that all organisms must be physiologically adapted to utilize the food they consume. Herbivores have a different digestive system from carnivores, for example. The herbivore stomach has four compartments, and the presence of suitable microorganisms is essential to its proper function. But carnivores have a simple stomach. Thus enzymes, acidity, and other factors are distinctly different in the two cases but are always adapted to the diet.

"Food webs," often displayed in animal ecology texts (Fig. 40–10), are a further elaboration of the energy-flow diagrams. Obviously, since a given organism may consume a wide variety of other organisms, these food webs, if worked out in ultimate detail, would be extremely complex.

Allelopathics. Not only are energy compounds transferred through the energy-flow system of a community, but many other compounds that are often capable of powerful physiological effects go with these energy substances. An allelopathic is defined as a substance produced by one organism that is harmful to another (this harm is usually beneficial to the first). Fungi produce allelopathics, called antibiotics, that are effective against bacteria (and that man uses in medicine), and higher plants also produce allelopathics that are effective against other species. The walnut produces juglone, and soft chaparral produces terpenes that inhibit the germination of other plants. Rarely will they act directly against a mature plant. In any case, when the plant produces these compounds (possibly in leaves—released upon their decay—or in roots), it assures for itself a sort of "territory" in which it will not be bothered by competition from other species.

Many plants produce narcotics that may not result in death but do exert a profound influence upon the physiology or even psychology of the organism consuming them. When they result in death, we call them poisons. These have been shown to be defense mechanisms for the plants that produce them. Grazing animals avoid eating such plants. Most of the narcotics are alkaloids, and they fit the definition of the allelopathics, being produced by one organism and harmful toward another. Only man is perverse enough to cultivate tobacco, tea, coffee, and other narcotics producers. Properly used, some of these narcotics can be beneficial to man; we are also aware of their harmful effects. Alcohol (produced by yeast) was not evolved to repel anything. Hence it is not an allelopathic, although its effects are similar to many narcotics.

Vitamins and other required compounds. Having emphasized the negative aspects of molecular transfer within ecosystems, we might tend to overlook a more important and positive side to these processes. The vast majority of animals require not only carbohydrates, fats, and proteins in their diets but also a number of other highly specific compounds that are essential to some phase of the animal's metabolism. These are the vitamins. An ecosystem must be properly balanced so that the consumers can obtain a suitable supply of vitamins from the organisms they consume. Consumers also require specific amino acids (obtained in the protein), specific fatty acids, and, undoubtedly, specific carbohydrates. When sugars that have no counterpart in nature are synthesized in the laboratory, few organisms are able to utilize them. Usually, we do not emphasize these latter compounds in discussing nutrition, since virtually all of the essential amino acids, fatty acids, and carbohydrates are present in almost all plants. The essential amino acid lysine is present only at low concentrations in many plants, thus leading to severe nutrition problems in several countries. The recent development of a high-lysine corn is having profound effects in these areas.

Material transfers and the super organism. The transfer of elements and compounds through ecosystems has been compared to the metabolic processes going on within an organism. This brings us back to the concept of the super organism. Perhaps the analogy of an ecosystem to an organism has some value, but if we are not careful, it could be misleading. Metabolism in an organism is the chemistry that occurs in the presence of enzymes. *Ecosystem metabolism* is merely a transfer of compounds from one organism to another, perhaps in a manner analogous to the transfer of compounds from the leaf to the root or from the gut to the bloodstream. Furthermore, metabolism in an organism is under rigid genetic control (modified somewhat by environment), but the transfer of compounds in ecosystems varies greatly from community to community, forming a continuum of types as the communities themselves form continua. One thing is clear: whether transfer of materials in ecosystems has anything to do with a super organism or not, it is certainly an important integrating function of these systems. Furthermore, it is a function that has only been studied superficially. It definitely deserves more attention.

Man and material transfers. Certainly, the transfer of materials in ecosystems could be of paramount importance in our attempts to understand man in his environment. To begin with, man must understand and control his own nutrition and intake of narcotics (e.g., nicotine from tobacco, caffeine from tea and coffee, and theobromine from cocoa), alcohol, hallucinogens and other mind-affecting drugs (e.g., mescaline from peyote, tetrahydrocannabinol from marijuana, opium and its derivatives from the poppy), and other drugs that occur naturally in the ecosystem (e.g., such tranquilizers as reserpine from *Rauwolfia* and salicylic acid—aspirin—from willow bark). Furthermore, man is no longer dependent upon the metabolic synthesis of drugs by the plants and animals that he consumes. Thanks to the development of organic chemistry, man has become a prolific drug synthesizer himself. Probably the number of compounds he has synthesized artificially now rivals the number of compounds produced in nature. Obviously, he is faced with some grave problems in utilizing these.

Many of these synthetic compounds have been produced specifically for use against other members of man's ecosystems; in a sense, therefore, they are allelopathics. We have a host of insecticides, fungicides, and herbicides that are used primarily in agriculture. Like the allelopathic-producing organism, our "friends," the agricultural food-producing plants, gain a competitive advantage with these substances. Although this has allowed us to stave off world famine in spite of increasing human population, we are becoming aware that some of these compounds have evaded our strict control. DDT is building up in the world's ecosystems to levels that many scientists consider to be dangerous—it has now been banned in many states.

Some (probably to be thought of as extremists) even suggest that our very survival is in doubt because of the effects of these compounds and our waste products upon our ecosystems. This seems unlikely, if failing to survive means that *all* members of the human race would die off in response to pollution. Surely, if large numbers began to die off, our technology, industry, and economy would collapse. The survivors might gradually assume a position in the earth ecosystem not too far removed from that occupied before the industrial revolution. But, regardless of whether we are actually dealing with the question of survival, we must recognize that we are now faced with extremely severe problems because of the manner in which we must tamper with ecosystems to maintain our present population and standard of living. Our kind of civilization may well be in grave danger.

Some Other Community Interactions

Competition is often alluded to in discussions of ecology. There is intraspecific competition, between the members of a population, and interspecific competition, between members of different populations. In either case, two or more competitors are placing an over-demand on resources. In spite of the frequent reference to this important ecological principle, much of our knowledge about it exists at a rather superficial, observational level. Much research needs to be done in this area.

Competition for food was one of Darwin's primary mechanisms essential to the operation of natural selection. Yet the few studies that have been carried out seem to indicate that the competition so often discussed by the early Malthusian evolutionists seldom occurs in nature. Before populations can reach a level where they are limited by the food supply, other mechanisms become operative that limit the population. Guppies eat their young, for example, and in many species of birds, the female is sexually unresponsive to males without a territory. Nevertheless, much competition does occur in natural communities. When an area of a forest floor is surrounded by a trench (cutting the roots coming from nearby trees), other plants in the plot begin to grow much more vigorously, indicating that they had been retarded by root competition. Competition for water and light can also be observed. But such competition often results in a *reduced state of growth* rather than complete elimination. Thus, if seeds are sown very densely, most will germinate, and most may survive. But the resulting plants will be much smaller than comparable plants growing from seeds sown much more sparsely.

Neo-Darwinian evolution emphasizes that competition does not have to be severe to bring about changes in the gene frequencies of a population. Even a slight competitive advantage leads to a drastic shift in gene frequencies in a relatively small number of generations. The principle that states that two populations occupying the same niche will compete with each other until one has been completely eliminated. This was demonstrated with populations of *Drosophila*, grown on a nutrient medium. At one temperature one species won out to the elimination of the other, while at a lower temperature the other species won out. Yet it has recently been shown that when temperatures are held at an exactly intermediate level, both species coexist together indefinitely—again indicating the complexity of competition and the need for future research.

The original definition of the word symbiosis was *living together*, and it was used to describe all situations

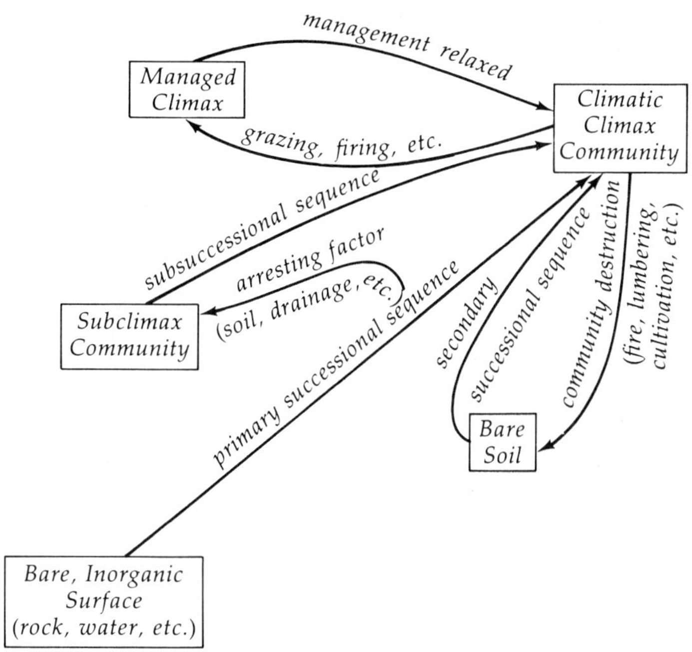

40–11 *Some of the kinds of ecological succession.*

in which two organisms lived in close association. Using this approach, we can recognize three distinct examples of symbiosis. In commensalism species *A* is benefited, while species *B* is affected neither positively nor negatively. In mutualism, both species benefit (this is now often called symbiosis). In parasitism, species *A* benefits, while species *B* is harmed. If organisms are not closely associated, but a predator consumes its prey, we are dealing with predation. It is important to realize that in parasitism and predation, a careful balance must be maintained between parasite and host or between predator and prey. If the predator consumes all its prey, then it will starve to death, as will a parasite if it kills all its hosts. In the broad sense, therefore, symbiosis and predation are other functions of ecosystems.

We have referred to ecological succession in Figs. 40–11 and 40–12. In a sense, this dynamic function of community ecosystems pulls together all the ecosystem activities discussed in this book. As the ecosystem functions, the community is not only being strongly influenced by the environment, but it is also changing the environment in a holistic way. The changes that take place will be a summation response to all of the primary, the secondary, and the tertiary productivities—the energy flows, the food webs, and everything else that is happening in the ecosystem. When the climax has been achieved, all of these processes will have reached an equilibrium or a near equilibrium (fluctuations may occur around a mean). The conditions produced by the ecosystem will be the ideal conditions for the members of the ecosystem, considering their numbers and the niches that they occupy. This being the case, the ecologist can hardly be more challenged than when he is attempting to understand ecological succession.

Feedback in Ecosystems

The functions of ecosystems tend to result in a balanced, steady-state condition. We have often mentioned the holistic approach to understanding environments. This says that communities must be considered as integrators of numerous environmental factors impinging upon them and that ecosystems themselves generate and influence important parameters. Of course, natural ecosystems are typically unstable, if long time intervals are considered. Fire may destroy an ecosystem so that succession must begin anew. Flood, glaciation, or other natural catastrophes may do the same. But when environments are relatively stable, we have good reason to believe that ecosystems also achieve considerable stability. Apparently, the more complex the ecosystem (the more species in it), the more stable it may become. A perturbation upsetting one of its components may be easily absorbed by all the others.

This is an example of the tendency of ecosystems toward homeostasis—a tendency for ecosystem components and functions to maintain constant levels once these have been achieved. Often, this comes about because of negative feedback processes. In a negative feedback arrangement, when some environmental variable is altered, there will be an automatic tendency to readjust back to the original (e.g., the home thermostat). In ecosystems, if a predator becomes too populous, it will be too successful in catching its prey, which then becomes in short supply and automatically brings the predator population back down. A parasite and its host act the same way.

But positive feedback can also occur when an ecosystem is perturbed. In a situation such as this, when

(*a*)

(*b*)

40–12 *(a) Ponderosa pine* (Pinus ponderosa) *trees and manzanita bushes* (Arctostaphylos *sp.*) *growing on hydrothermally altered material near Monroe, Utah. The light-colored material represents the eroded cross section of an ancient hot spring. The hot waters contained numerous dissolved minerals. Many sulfides were deposited, and as the surface eroded away exposing the altered rocks, they oxidized producing sulfuric acid. The present material has a pH between 2.5 and 3.5 and is rich in iron, aluminum, and calcium. Certain species, notably ponderosa pine, manzanita, limber pine* (Pinus flexilis), *and a few others, are able to grow under these extreme soil conditions. Gradually, as they deposit plant debris, the pH becomes rather basic (e.g., 8.5), and species more typical of this semi-desert region invade. Under the low moisture conditions, the original pioneers cannot compete with the desert species, and they are crowded out. (b) The Big Rock Candy Mountain south of Richfield, Utah. This is another hydrothermally altered area with ponderosa pine, particularly at the base. The desert species (sagebrush—* Artemisia tridentata*—and others) at the top of the mountain are growing on rather typical desert soil that has developed from hydrothermally altered material.*

the level is changed, there will be an automatic tendency to change the level even more. A thermostat might be built that, when the temperature began to get hot, turned on the furnace so that the temperature got even hotter. Positive feedback accounts for the squeal that is sometimes heard with a public address system. If the volume is set too high, the microphone begins to pick up the sound from the loudspeaker, rebroadcasting it at an even louder, reinforced level. (Turn down the volume or cover the microphone with your hands!) Ultimately, of course, a positive-feedback system is almost always limited in one way or another. Raising the temperature of the room might eventually set it on fire, causing even more positive feedback and raising the temperature even more. But eventually the house burns down, and the temperature returns to normal.

The logarithmic part of a population growth curve is a positive-feedback system. The more organisms there are, the more they can reproduce to produce more of their own species. This happened for a long time when rabbits were introduced into Australia, and the end result was a series of severe ecosystem changes. As nutrients are poured into a lake (e.g., via sewage or seepage from surrounding cesspools and septic tanks), some of the organisms in the lake begin to multiply at an exponential rate, while others suffer in response to this. Perhaps "trash fish" multiply at the expense of desirable trout. Often the positive-feedback explosion is finally halted by a triggered decline. As a population increases at its logarithmic rate, stresses build up (limitation of food or other stresses) until a breaking point is reached. A little more increase in the stress then results in a sudden and rapid decline—often quite unanticipated.

Exobiology

Are the principles of ecosystem function of universal application? Say there is life on Mars. Might we expect the principles of productivity, nutrient cycling, competition, and even ecological succession to apply? If they do, then perhaps the ecologist is finally beginning to formulate general principles somewhat comparable to the laws of physics and chemistry. Of course, life on Mars (or Venus or any planet in the universe) might be different from our own—perhaps in ways that would upset the ecological principles we have just summarized. But at the moment, many of us are brash enough to imagine that the principles will apply to extraterrestrial ecosystems. But we must wait to see.

Man in the Ecosystem

It is especially important at this time for us to be aware of the relationship between the principles of ecology and man's place in the ecosystems of his planet. Decisions made now and acted upon will certainly determine our future. Man is now such an important part of the earth's ecosystems that his collective actions can no longer be absorbed by the negative-feedback, stabilizing features of ecosystems.

Is it too late? Man has already brought about profound and irreversible changes in many ecosystems. Right now he is engaged in highly significant positive-feedback perturbations of his natural ecosystems. His population is exploding, and he is pushing up his standard of living at a logarithmic rate. The larger his population and the better he lives, the greater the quantities of pollutants introduced into his environment. Particularly noxious are the industrial contaminants that destroy the purity of his water and his atmosphere, but he is also releasing pesticides, radioactive wastes, and heat from atomic power plants. The latter are being released on a relatively limited scale so far, but their potential for harm is most frightening. In any case, although the release of pollutants follows the population positive-feedback curve, eventually they must act as a negative feedback to limit the rate of population increase. And this may result in a triggered decline of important ecosystem populations.

We frequently blame man's industries for pollution and overlook the sizable impact of waste disposal by individuals. The garbage produced by a single person is considerable, and often at least a portion of this garbage is disposed of as litter scattered around our natural and agricultural ecosystems. In the teeming cities of Asia, for example, individual wastes constitute the principal problem.

Man has never had much social conscience. When he has had the power, he has desecrated and destroyed his environments. Witness the hills of the Middle East that were grazed to death or the destroyed forests of medieval China, Europe, and imperial England. Then, however, he could always expand into some yet unruined part of the world and begin anew. Now we are approaching the limits. We can not much longer destroy faster than we can rebuild.

Man, then, is faced with some extremely severe problems because of the effects he is having upon the earth's ecosystems. To exist with his present numbers and standard of living, he must produce prodigious quantities of food. This he does by creating his own agricultural ecosystems. But, although we know that the more complex a community, the greater its potential stability, man continues to strive for the simplest of ecosystems: thousands of acres covered by a single species such as wheat or corn. Such a system is extremely vulnerable. A particularly virulent plant disease can eliminate enough of a crop to cause extensive starvation. Of course, man has no choice but to develop pure-stand crops, although he can diversify by growing several kinds of crops in a given large area.

To add to the problems created by man's agriculture, he contaminates his environment with vast quantities of waste materials and highly active chemicals. This he must do to continue his industry and support his standard of living. But where will it all lead? Will the house finally burn down so that man's current endeavors collapse with a mighty roar and the natural negative feedbacks of ecosystems take over? Or is there a way out?

The Challenge of Our Times

The writer of Genesis stated that man was to "have dominion over . . . all the earth . . ." Surely, this is man's inclination. Under suitable conditions all organisms might tend to expand until they "have dominion," but in the long run, this dominion must be temporary. If the dominion means the elimination of the predator's prey, then the predator must also disappear.

Cursed is the ground for thy sake . . . Behold, the man is become as one of us, to know good and evil . . . (Genesis 3:17, 22). Man's dominion could be different. It does not have to be one of overwhelming all other organisms on the face of the earth. Rather, his dominion *could* be one of understanding the ecosystems of his planet—for his own sake. The utopia envisioned by the ecologist is one in which man lives happily with his fellow man and as part of his planet's ecosystems. They serve him as they serve all other herbivores and carnivores, but he is grateful for their service and protects them rather than destroying them.

What philosophy shall we adopt in striving for utopia? Some advocate that our present troubles came about because of science and technology. Hence science and technology are bad, and they must go. We should return to some kind of agrarian society in which most of us extract our living directly from the land, as our forefathers did. In short, we act as other animals, struggling for individual and national survival until the raw forces of the ecosystem destroy us or force us to occupy our niche.

But this is a hopeless dream. A few of us may voluntarily give up the blessings of science and technology, but the vast majority will not. They are unintrigued by the thought of returning to an ecosystem in which man's population is limited by the natural forces: famine, pestilence, and war. A life expectancy of 33 is not exciting to modern man, even though it might allow him to come to a better balance with nature. Most of us prefer to retain immunization programs, surgical procedures, nutritious and balanced diets, and even rapid transportation and electronic entertainment.

So, the application of reason—science—is just about our only hope for the future. Of course, extraterrestrial intelligent beings might visit our planet and give us the solutions to all of our problems. But, even if this were likely, we would be foolish stewards indeed to abdicate our own responsibilities and count upon such an intervention. The important thing is for us to apply all of the intelligence and all of the science and technology at our disposal in solving the problems that beset us. We must be wise—not just smart. We must try to *act* now that we know good and evil. Above all, we must apply all the powers of our reason to create a world in which we occupy a logical place in a functioning earth ecosystem that provides each of us with an opportunity to grow and to have joy in doing so. Is this an impossible goal? Perhaps, yet we have no choice but to strive for it.

Index

Abscisic acid, *588*, 590–592, 594–596, 638
Abscission, 567, 591, 596, 635, 651
Absolute requirement, 666
Absorption spectrum, 84–85, 566, 616, 646, 652; of algae and bacteria, *269*; overlapping, 654; of phytochrome, *648*
Acacia, 707
Acalypha, 589
Accessory fruits, 518
Accium, *370*
Accumulation, 58
Acer, 2, 519, *534*
Acetabularia, 114–115, *662*
Acetaldehyde, *44*
Acetanilides, *599*
Acetate buffer solution, *77*
Acetic acid, *44*, 268
Acetone, *44*
Acetonitrile, *44*
Acetylcholine, 653
Acetyl Coenzyme A, 107, *109*–110
Acetylene, 42, *44*, 568
Acetyl glucosamine, 41
Achene, 519, *521*
Achlya ambisexualis, 344–345
Acid group, *62*
Acidic soil, 699
Acidity and proteins, 75–76
Acrasin, 367–*368*
Acrasiomycete, 365, 367
Actaea, 519
Actinomorphy, 564
Actinomycetes, 655
Action spectrum, 84–*85*, 646, *648*, 650, 653; for entrainment, 660; for phototropism, 565–*566*
Activated amino acid, 117
Active: site, 24, 57–58, *75*–76, 108
Actomyosin, 366
Acuminate tip, *534*
Acute tip, *534*
Adaptation, 164, 501
Adaptive radiation, 213
Addicott, F. T., 590
Adenine, *118*–123, 572, 588
Adenosine diphosphate. *See* ADP
Adenosine triphosphate. *See* ATP
Adhesion, 623–624
Adiabatic cooling, 726
ADP, 87, 88, 105, 107, 110
Adventitious: bud, *600*; root, *552*, 587
Aeciospores, 369–*370*
Aerosol, 741
Aesculus hippocastanum, 419, *550*
African violet, 588, 636–637
After ripening, 641
Agar carrageenan, 231
Agardhiella, 323
Agaricus, *350*
Agave, 723–724
Aggregate fruits, 518, *520*
Aggregation, 218
Aging, 638
Agrarian society, 748
Agricultural: ecosystem, 733, 748; land, 710, 732
Agriculture, 16–17, 27, 522, 562, 596, 598, 612, 675, 699–700, 707, 710, 719, 727, 734, 736, 744; artificial, 609; in the tropics, 706
Agrobacterium tumefaciens, 255–*256*, 583
Ahmadjian, Vernon, 376
Ailanthus, 517
Air: conditioning, 710; ionic content of, 657–658; layering, *600*; pollution, 377
Alachlor, *599*
Alanine, *46*, 63, 110
Albino, 185
Albugo candida, 339
Alcohol, 33, 106–107, 110–111, 743
Aldehyde, *95*
Alder, 259, 542, 545
Aleuria, *349*
Aleurone layer, 77, 595
Alexopoulos, C., 365
Alfalfa, 191, 723
Algae, *84*, 226, *237*, 239, *300–301*, 306, 311–312, 314, *318*, 320; bloom, 279, 287; blue-green, 11, *78*, 86, 114, 117, 213, 219–220, 226, 230, *237*, 239, 257–258, *271*, *274*, 276, 279–282, 288, 290, 306, 315, 321, 362, 367, 464; brown, 12, 14, 65, *230*, *239*; coralline, 314; in ecosystem, 280; encrusted, *311*; as epiphytes, 290; *Euglena*, *230*; as food, 309; golden, 228, *230*, *237*, 239, 242; green, 11, 14–15, 80, 229–*230*, *237*, *239*, 274, *276*, 279–282, 289, 293–295, *298*, 384; in lichens, 375–378; marine, 298–324; red, 12, 15, *78*, 86, 95, 229–*230*, *237*, *239*, *300–301*, 306, 313–314, 321–*322*; snow, 290; soil, 279; succession of, 286; yellow-green, 229–*230*, *237*, *239*
Algal: cells, *375–377*; communities, 706; culture, 97
Alginic acid, 231
Aliphatic amino acids, *63*
Alisma, 519
Alismales, 542, 549
Alkaloid, 111, 357, 640, 743
Allard, H. A., 658, 665–666, 669, 676–677
Allele, 183, 190
Allelopathic, 743–744
Allfrey, V. G., 127
Allomyces, *328*, 334–*335*, 336, 344–347
Allopatric: population, 199; species, 203
Allosteric enzyme, 76, 159, 595
Alpha amylase, 77
Alpine: environment, 726; meadows, 538; sorrel, 693, 727; tundra, 614
Alternaria, 726
Alternation of generations, 14, 228, 230; isomorphic, 306, 334; in lower plants, 337
Altitude, 683
Altmann, R., 102
Aluminum, 605, 700
Amanita phalloides, 362
Amaranthos, 92
Amaryllis, 543
Amber, 487
Amentiferae, 541–542, 545–546, 548
American laurel, 538
Amici, G., 515
Amide, 62–*63*
Amino acid, 23, 25, 58–*62*, *63*, 71–72, 90–94, 109–111, 117, 126, 153, 159, *253*, 573, 581, 606, 626; in bacterial walls, *264*; basic, 63; essential, 743; in origin of life, 217–218; sequence, 71–72, 123, 126; sequence of phytochrome, 652; synthesis, 95; with tRNA, 124
Amino group, *62*, *76*
Aminotriazole, *599*
Ammonia, 111, 217, 258, *259*, 640, 735
Amo, 571
Amoeba, 59, 366
Amphibian, 192, 390
Amphoteric 76
Amplification, 646, 721–722
Amplitude, *659*
Amylase, 595–596, 608, 612
Amylopectin, 47, *49*
Amyloplast, 94, 389, 424, 566
Amylose, 47, *49*
Anabaena, *276*, 281, 289, 464
Analysis of growth, 557
Ananas comosus, 210
Anaphase, 141, *144*, *151*
Anatomy, 22, 585
Anderson, Edgar, 531
Anemone, 545
Angel of Death mushroom, 362
Angiosperm, *210*, *216*, 224, 234–235, *237* *239*, 389, *441*, 497–549, 647, 651, 663; evolution of, 485, 498, 533–544; families of, 537; seedling, 646; shoot apex, 419; success of, 499; taxonomy, 585
Angstrom, 29
Animal, 11, 14, 27, 39, 101, 142, 202, 257, 390, 497, 505, 521, 651, 655, 659
Animalism, 218
Anion exchange, 699
Anise, 540
Anisogamic, 283
Annual, 500; fluctuation, 663; ring, *439*–*440*; *442*
Annulus, 467–468
Antarctica, 376
Anther, 174, 176, 187, *502*–502, 505
Antheridial, 345, *347*
Antheridial initials, 344
Antheridium, 234, *238*, 354–355, *453–454*, 456, 469–470
Anthesis, 544
Anthocyanin, 608, 635, 638, 649
Anthophyta, 548
Anthrax, 262
Antibiotic, 254, 743
Anticlinal, 420
Anti-gibberellin compound, 571
Antimetabolites, 103
Antipodal cells, *510*, 511
Antithamnion, 321–*323*
Apatite, 740
Apex: root, *406*; shoot, 584
Aphid, 248, 629–*630*
Apiaceae, 540
Apical: cell, 321, *421*, 456, 464–*466*, 470; dominance, 588; meristem, *410*, *412*, *417*, 418–423, *421*
Apomixis, 182, 186, 207–288, 544
Apoplast, 434–435, 611, 625, 628–*629*, 632, 633
Apothecium, *355*
Apple, 191, 518–521, *520*, 568, 637, 649, 733
Apricot, *520*
Aquatic: community, *737*; fern, 465; fungi, 284–285; system, 15, *703*–705
Aquilegia, 519, 520, 716
Araucaria, 235, *486*
Archaeopteris, 451, 479, 494–495
Archegonium, 234, 453–456, *454*, *455*, 469–470
Arctic, 464, 654, 733
Arctostaphylos, *746*
Arginine, *63*
Aridosols, *700–701*
Aristotle, 211, 515, 562
Arizona cypress seedling, 667–668
Armillaria mellea, 362, *364*
Arnold, C., 495
Aromatic: amino acids, *63*; compounds, 40, 111
Arrowhead, 542
Arsenate, 76
Arsenault, 345
Arsenicals, *599*
Artemisia tridentata, *746*
Arthropod, 41
Arum maculatum, *506*
Ascent of sap, 623–627
Asclepias, 519, *520*, *522*
Ascocarp, 354, *356*
Ascogonium, 354–*355*
Ascomycete, 229, *232*, *237*, *239*, 332–333, *348*, 354, *355*, 363, 375, 380–382; life history of, *355*; sexual reproduction in, 354
Ascospore, 232, 354–*355*, 376
Ascus, 354–*356*
Ash, 518–519, 534
Asparagine, *63*
Asparagus, 519
Aspartic acid, *63–64*, 93, 111
Aspartyglycyl, *64*
Aspect dominance, 684
Asphalt, 624
Aspirin, 744
Assimilates, 626
Association, 689
Assymetric carbon, 46
Aster, 541
Asteraceae, 541
Asteridae, *548*
Asterionella formosa, 334
Athlete's foot, 363
Atmosphere, 29, 636, 657, 681
Atom, 38, 42–*43*, 54; energy, 735; explosion, 70
ATP, 59, 87–88, 91, 101–111, 156, 608, 675
Atrazine, *599*
Atriplex confertifolia, 92, 640, 725
Atropine, 539
Austrian pine, *655*
Autecology, 682, 718
Autocatalytic reaction, 557–558
Autophasing, 663
Autoradiography, 134, 148, 632
Autosome, 192
Autotroph, 288, 606, 736
Autumn, 635, 646; colors, 96, 638, 651; overturn in lake, 287
Auxin, 110, 159, 194, 561–563, 564–568, 583–585, 587–588, 590–591, 594–596, 648, 651, 668, 676
Avena: coleoptile, *567*; *sativa*, *554*, *559*
Avens, *716*
Avery, A. G., 438
Avocado, 96
Axelrod, D. I., 498
Axil, 436
Axillary bud, 436, 567, 587–588, 590
Azalea, 538
Azolla, *461*, 464, 473

Bacillariophyceae, 292
Bacilli (rods), 261
Bacillus: *pestis*, 254; subtilis, 267
Bacon, Francis, 561
Baconian viewpoint, 562
Bacteria, 11, 75, 101, 114–117, 126, 160, 219–220, 226, 237, 239, *247*, 250–273, 655–656; cell division in, *265–266*; cell, interior of, 264; colony, *257*; denitrifying, 258–*259*; and the ecosystem, 256–261; in evolution experiments, 200; filamentous, 261; generalized, *265*; green, *269*; nitrogen-fixing, 258–*259*; nutrition, 261; photosynthetic, *270*; purple, 268–*269*, *271*; size, 261; spores in, 266; structure, 261; sulfur, *250*, *269*, 271; thermophilic, 257; wall, 263, *265*
Bacteriophage, 120–*121*, 244–*245*, *246*, 261
Baeria charysostoma, 637
Bailey, I. W., 546, 585
Bajer, A., 151
Baker, R. R., *606*
Ball, E., 585–587
Balsam, 431
Baltimore, D. 244
Bamboo, *544*
Banana, *521*, 568, 575, 710
Baneberry, 519
Bangio-phycidae, 321–*322*
Banks, H. P., 251, 390, 449, 494
Barberry, 368–369, *370*, 519, 533, 538
Bark, 417, 440–443
Barksdale, A., 345
Barley, 77, 188, 544, 595, 650, *655*, 666
Barnacles, 170
Barometric pressure, 625, 663
Barriers in reproductive isolation, 203
Basal body, *227*

Basalt, *225*
Base (alkali), *62*
Basidiocarp, 359
Basidiomycete, 229, *232, 237, 239,* 331–332, *348,* 357–363, 375–377, 380–381; cell structure in, 358; fruiting body in, *360;* sexual reproduction in, 359
Basidium and basidiospore, 232, 360–*361, 362,* 369–370
Basswood, 519
Bat, 507
Bates, H. W., 170
Bateson, W., 183
Beadle, G., 24, 116, 132, 185
Bean, *85,* 145, *552–553,* 572, *632, 648,* 651, *655,* 657, 658, *660,* 663, 666
Becherer, A., 212
Beck, C., 479, 493
Bee, 204, 505, 662. *See also* Honeybee
Beech, 2, 542, 545, 639
Beefwood, 533, 541, 546
Beer, Egyptian, 354
Beggar tick, 165
Begonia semperflorens, 429, *600*
Beling, I. 658
Belladonna, 539
Bellflower, 540
Benchuga bug, 169
Bennettitales, *216, 484,* 498–499, 546
Benson, A., 89–91
Benson, L., 500, *530,* 533, *536,* 546, 548
Benson's orders of flowering plants, *537,* 538
Bentham, G., 545
Benzene, *44;* ring structures, 40–*41,* 62, 111
Benzoics, *599*
Benzyladenine, 573
Berberis, 519
Berry, 519, 520
Bertalanffy, L. von, 558
Bertsch, W., 549
Bessey, C. E., 533, 537–548
Bessey's cactus, 545–*546,* 548
Betula papyrifera, 534
Biddulph, O., 631
Bidens, 165–*166*
Bidirectional movement, 631–632
Biemann, 345
Biennials, 668–669
Bierhorst, D. W., 472–473
Bilateral, 504
Billings, W. D., 654, 688, 693, *704,* 718
Biloxi soybean, 671
Bindweed, 571
Binomial system, 13, 208
Binucleate cells, 232, *358*
Binz, A., 212
Bioassay, 346–347, 564, 575
Biochemistry, 22, 24, 34, 87
Biogeochemical cycle, 740–741
Biological: clock, 111, 646, 653–663, 676; community, 682; species concept, 202
Biologist, 561–562
Bioluminescence, *659*
Biomass, 736–737
Biome, 15, 17, 18, 598, 684, 690; growth forms, 703; of the world, *694–698, 702–710*
Biophysics, 561, 613, 647
Biosphere, 710
Biosystematics, 527
Biotic potential, 610
Biotin, 608
Biotron, 719
Birch, *534,* 542, 545, 637, 641
Birds, 193, 202, 316, 473, 505, 666; navigation, 658; pollinating flower, 507
Bird's nest fungus, *351*
Bistort, *716*
Blaauw, A. H., 562, 647
Black body, 71
Black bread mold, 342
Black Death, 254
Black locust, 639
Black oak, *204*
Black pine, 597
Black stem rust, 368, *370*
Blade, 306, 438, 462–*463*
Blakeslee, A. F., 438, 575
Blastocladiella emersonii, 337
Blender, *66*
Blood, *622–623*
Blood type, human, 184
Bluebell, 540
Blueberries, 538
Blue camas lily, 542
Bluegrass, 208, *737*
Blue gum, *540*
Blue light, 647, *654*–655
Blue oak, 530, 531
Boat and chair hexoses, 46–*48*
Bog, 16, *703*–705; terrarium, 711
Bold, H. C., 238
Boletus, 362
Bolting, 570, 594
Bond: double, 42–*43;* single, 42–43
Bonner, J., 128, 157, 571, 673, 675, 677, 699, 718–719
Bonsai, *596–597*
Borage, 538–540
Bordeaux mixture, 340
Boreal forest, 691, *694,* 697, *702, 703*–705, 707
Borlaug, N. E., 191
Boron, 575, 599, 607–608
Borthwick, H., 648, 650, 671
Boston fern, 655
Botanical Congress, 13, 211
Botanical gardens, 528
Botanist, 12, 336
Botrychium, 465
Botulism, 273
Bouillenne, 589
Boundary layer, 613, *618,* 724
Bower, F. O., 473
Box elder, *534*
Boysen-Jensen, P., 560–561, 563
Brachet, J., 119, 132
Bracken fern, 465
Bracket fungi, 360
Bract, 541, *725*
Bragg, W. L., 129–130
Braun, A. C., 581, 584
Braun-Blanquet, J., 685
Brazil nuts, 540
Breeder reactor, 734
Breeding lines, 187
Briggs, L., 625
Briggs, W., 647
Bristlecone pine, 418
Broad sclerophyll shrubland, 707
Brock, T., 290
Bromacil, *599*
Bromeliaceae, 210
Bromeliad, *211,* 543, *706*
5-Bromouracil, 187
Brown, F. A., 663
Brown, R., 31, 64
Browne, H. T., 615
Brownian movement, 64
Bruce, V., 662
Bryonia, 194
Bryophyllum, 600
Bryophyta, 233, 235, *237–239*
Bryopsis, 295, *297*
Buchner brothers, 22
Buckwheat, 519
Bud, 435, 590, 669; adventitious, 588; formation of, 575; yeast, *356*
Buellia, 373
Buffalo, 257–*258*
Buffer, 77
Buffon, Comte de, 167
Bulb formation, 651
Bulk flow, 627–632
Bundle, vascular, *153, 412,* 436–438, *437,* 488–*489,* 500–*501,* 587
Bunning, E., 656–660, 662
Bur, *673*
Burweeds, 545
Burr, G. O., 91
Butane, *44*
Butler, W. L., 652
Butter, 64, 268
Buttercup, *410, 434,* 519, 533, 545; aquatic, 593
Butterflies, 193
Button stage, *360*
Byrum, 99

Cacao, 710
Cactus, *164, 539,* 655, 710, *714,* 723–724
Caenothus, *698*
Caesalpinus, A., 209–210
Caffeine, 743
Cain, S. A., 685, 688
Calamites, 457
Calbovista subsculpta, 351
Calcification, 701
Calcium, 65, 133, 508, 605, 632
Calcium oxalate, 59
Caldwell, M., 654
California live oak, 542
California poppy, *537*
Callixylon, 479, 494–495
Callophyllis, 323
Callose, *428,* 504
Callus, *429,* 586; culture, 588; growth, 583; lilac, *584*
Calorie, 71
Calvin, M., 89–91
Calvin-Benson Cycle, *91–92,* 106
Cambial cell, *429*
Cambium, *412, 439–441,* 465, 490, 500, 585–586, 600
Cambrian, 220
Campanula medium, 668
Campanulales, 540
Campion, 193, *716*
Canadian thistle, 600
Cananvalia ensiformis, 659
Cancer, 126, 189, 243, 583
Candlewood, *714*
Candolle, Augustin Pyrame de, 545
Cannabis sativa, 431
Canteloupe, 655
Cantino, E., *336*
Cap, 359
Capillarity, 623
Capillary z-tubes, 625
Capsella, 113, 114, 511, 517, 529
Capsomere, 243–*244*–245
Capsule, *265,* 519–*520;* bacterial, *121*
Caraway, 540
Carbamates, *599*
Carbohydrate, 38–39, 42–49, 92, 94, 741; essential, 743; general formula for, 88; nitrogen balance, 664, 676
Carbon, 14, 42–*43,* 88–89, 605; chain, 62, 82; cycle, *739;* fixation, *89*
Carbon dioxide, 76, 88–89, 91, 101, 103, *105,* 107, 111, 585, 604–605, *632,* 636, 735–736; concentration in atmosphere, 733; dissolved in rainwater, 699; evolution or uptake, *86;* fixation, *89, 90–91,* 101, 151 in a lake, 286; in origin of life, 218; penetration of the cell, 95; in photosynthesis, 610
Carboniferous period, *216,* 222, 233, 235, 390, 449, 451, 455, 457, 461, 473, 480–486, 490, 498
Carbon monoxide, 311, 321
Carboxydismutase, 91–*92*
Carboxyl, 61–*62,* 76
Cardamine hirsuta, 523
Cardinal temperatures, 636
Carlquist, S., *165*
Carnation, 538, *655*
Carnivore, 26, 736, 738, 742
Carotene, 96, 228, 305, 565–*566,* 647
Carotenoid, 82–84, 94, 108, *228, 292,* 315, 324, 635, 638, 649
Carpels-gynoecium, *502,* 507, 518
Carrier substance, 58–60
Carrot, 22, 540, *574,* 581, *582,* 663
Carrying capacity, 610
Carry-over effect, 722
Carya, 419
Caryophyallaes, 538
Caryopsis, 519, 544
Casein, 573
Casparian strip, *434*–436, *435*
Casperssen, T. O., 119, 132
Castilleja, 522
Castor bean, 514, 518, 537, *552*
Casuarinales, 541
Catalpa, 519, *522*
Catellaria, *372*
Catenaria anguillulae, 143
Cation, 606, 699; exchange, 41, 65, 699
Catkins, 541, 545
Cattails, 533, 546
Cattle, 257, *712*
Cavitation, 468, 624–625
CCC, 571, *592*
C_4-dicarboxylic acid pathway, *93*
Cedar, 488
Cedrus deodora, 489
Celery, 540
Cell, 29, 31; algal, 51; apical, 420; bacterial, 261; biosynthesis, 153–161; crystal containing, *430;* division, 418, 558, 568, 570–571, *580,* 583, 659; division in bacteria, *265–266;* duplication, 156–159; elongation, 570–571; enlargement, *580;* epidermal, *20;* in origin of life, 218; meristematic, 423; motile, 142, 335; multinucleate, 334, 453; plate, 141–*142,* 440; red algal, *320;* red blood, 52
Cellular respiration, 103
Cellular slime molds, 367
Cellulose, 39–41, 44, *49, 57,* 65, 91, 133, 231–232, 257–*258,* 281, 292, 295, 305–306, *333–334,* 337, 368, *388,* 424, 431, 741
Cell wall, 22, 34, 36–37, 38–39, 51, 57, 60, 65, 132, *142,* 156–157, 231–232, *253,* 288; in bacteria, *250,* 264–265; components, 110; formation, 572; in green algae, 295
Cenozoic era, *216,* 220, 224–225, 233, 461, 482, *498*
Centaurea, 522
Centigrade or celsius, 29
Central: cell, 511; pore, 354
Centrifugal force, 302, 625
Centrifuge, 35, 66, 77
Centrioles, 142, *227,* 321
Centromere, 140–141, 144, 147, *150*–151, 157, 291, 293, 324, 532
Century plant, *655,* 723
Cephalotaxaceae, *486*
Cereal, 668, 732, 734
Ceres wheat, 371
Cesium chloride, 158
Chaenactis, 686
Chagas' disease, 169
Chance, 179, 199
Chandra, Ram, 595
Chaparral, *698, 703*–705, 707
Charcoal, 711
Chargaff, E., 121, 132
Charophyceae, 295
Check valves, 625
Cheese, 355
Chelate, 606, 700
Chemical: in biological timing, 661–662; bond, 70, 87, 99, 646; inducing mutations, 188–189; potential, 55, 58, 610; warfare, 641
Chemistry, 22; of the cell, 42–49; organic, 34, 42; of photosynthetic pigments, 96; in taxonomy, 532
Chemosynthesis, 216, 268
Chenopodium rubrum, 676–677
Chernozems, *700–701,* 704, 709
Cherry, *497,* 519, 522, 639
Chestnut, 542
Chiasmata, 143–147
Chinese fan palm, *543*
Chitin, 41, 232, 333–334, 337, 354
Chlamydomonas, 80, *276,* 280–283, 282, 290, *308,* 295
Chloramphenicol, 595
Chlorates, *599*
Chlorella, 80, 89–90, *269*
Chlorenchyma, 424
Chlorine, 42–*43,* 607–608, 632, 741
Chlorobium, 269–270
2-chloroethanol, 638
Chlorogenic acid, 572
Chloro-IPC, *599*
Chlorophenoxy herbicides, *599*
Chlorophyceae, 295
Chlorophyll, 84, 86–87, *95*–96, 99, 105–111, 228, 231, 238, 280, 288, 292, 305–306, 315, 324, *389,* 575, 607–608, 635, 647, 736; a, b, and d, *85, 95;* and cytokinins, 575; in origin of life, 218; responses of, 646
Chlorophyta, 229, *230, 237,* 238, *239,* 274, 280, 293–296, 305, 391
Chloroplast, *20,* 22, 36–37, 52–53, 59, *79*–86, 102, 153–156, 160, 242, *276, 291, 294,* 317, *408,* 424, 429; in algae, 281; dark-grown, *80;* destruction, 248; development, 84; dividing, *83;* fragments, *82;* of green algae, 293; membranes, 61; orientation of, 651; in various plants, 83–85
Chlorosis, 608
Chokecherry, *2*
Cholera, 262–263, 254
Cholodny, N., 562
Chouard, P., 676
Chromatid, *140*–146, 149–150, 156
Chromatin, 36, 38, *113,* 158
Chromatium, 270
Chromatography, *67, 90,* 532, 631–632
Chromium, 607
Chromonema, *140*
Chromophore, *652*
Chromoplast, 36
Chromosome, 28, 34, 36, 38, 140–151, 156–158, 179, 192, 317, 324, 581; aberrations, 146–147; behavior, *205;* breakage, 651; changes in, *197;* duplication, 147, *149;* homologous, 146; metaphase, *140, 150;* movement, *151;* number of, 14; pairing, 205–207; sex, 192; size, 140; structure, *146, 149,* 180; in taxonomy, 530
Chroococcus, 292
Chrysanthemum, 600, 666, 668, 677
Chrysophyceae, 292
Chrysophyta, 228, *230, 239,* 289, 292 291
Chymotrypsin, 24
Chytrid, 232, *334,* 382–*383*
Chytridomycetes, 229, *232, 237,* 239, 333–336, 382
Cilia, *227*
Cinnamon fern, 473
Circadian rhythm, 651, *659*–662, 675, 678–679; characteristics of, *659*
Circaeaster agrestis, 419
Circulation in plants, 632–633
Cisternae, 133
Cistron, 124
Citric acid, 107, *109*
Citrus, 710
Citrus fruits, 59
Civilization, 610
Cladophora, 289, 297, *304,* 306
Clamp connection, *358,* 358
Class, 13, 210, 212
Classification, 12, 22, 208–213, 235–239; artificial, 211; natural, 13, 212, 544
Clausen, J., 693
Clavaria aurea, 350
Claviceps purpura, 357
Clay particles, 57, 699
Clematis, 522
Clements, E., 545, 547
Clements, F. E., 537, 545, 547, 684, 686–689, 691, 731
Clements' chart of flowering plants, *547*
Climate, 683, 699–702, *704*
Climatron, 561, 719
Climax community, 27, 684, 728, *730,* 733
Clone, 600
Closed ecological system, 682
Clostridium botulinum, 273

Clouds, 682
Clover, *260*, 654
Club mosses, 214, 216, 221, 224, 449
Coastal barrier, *450*
Cobalt, 607, 632, 651, 662
Cocci (spheres), 261
Cocklebur, 541, 583, 588, 600, *603*, 614, 636, 648, 650–651, 655, 659, 661, 665–666, 670–*674*, 676, 678, 722; experiments with, *673*
Cockroaches, *216*
Coco-de-mer, 515
Coconut, 515, 518–519, 543; milk, 573, 581–*582*
Codium, 295, *297*
Codon, 25, 123
Coenzyme, 24, 74, 571–572, 594, 608
Coenzyme A, 107, *109*
Coffee, 710, 743
Cohesion, 468, 623–626
Cohn, Ferdinand, 262–263
Colacium, *273*
Colchicine, 192, 197, 438, 662
Cole, Fernley, 493
Coleoptile, *564*, *567*, 645, 648
Coleus, *578*, 585
Collard, *655*
Collenchyma, *406*, 422, *424*, 434, 436
Colloid, 64–*65*, 566
Colonies, 230
Color, *85*, 98
Colpodexylon deatsii, 452
Columbine, 519, *716*
Comatricha typhoides, *353*
Commensalism, 27, *745*
Common garden, 529
Community, 12, 15, 18, 27, 682, 744
Companion cell, *410*, *428*, 428–*429*
Compensation intensity, 86
Competing ion groups, 59
Competition, 454, 718, 744
Competitive inhibitors, 76, 103
Complementary bonding, 25
Compositae, 519, 541
Composite, 504, 537; inflorescence, *503*
Compost heaps, 383
Compound leaf, 438, *534*
Compression, 240
Compression wood, 566
Computer, 532, 580, 686
Concentration, 54–55, *59*, 628; ionic, 58; water, 57
Conceptacles, 307
Condensation nuclei, 741
Conditioned medium 581
Conditioning effect, 722
Conducting cells in algae, 313
Conduction, 616
Cone, 235, 451, 456–*457*, *476*, 480, 487–488, *491*; of a cycad, *481*
Conidia, *355*, *365*; spore, 339, *339*, *343*
Conidium, *339*, *343*
Conifer, 9, 97, *216*, 237, 239, 378, 485–490, 498–500, 666, *700*; distribution, *486*; induction of reproduction, 667–668; leaf size and shape, 487
Coniferal alcohol, 40–*41*
Coniferophyta, 234–*237*, *239*
Coniferous forests, 374, *683*, *697*, *703*–705, 707
Conjugation, *268*, 268, 284–*285*
Conklin, M. E., 575
Conservation of energy, 71
Constance, L., *528*
Consumer, 26, 737; cell, 628, *629*
Continental drift, *224*–226
Continuum, 686–691, 702; index, 686, *688*
Controlled environment research, 718
Control mechanisms, 158–159
Convection, 613, 616–617, 619
Convergent evolution, *164*, 533
Conversion of PFr to Pr, 678
Convolvulus, 571
Cooksonia, 386, 390, 449
Cooling: of the earth, 168; by transpiration, 613
Cooper, B., 589
Copernicus, N., 183
Copper, 108, 340, 607–608
CoQ, 108, *110*
Coral, 314; atoll, 170, 549
Corallina, *304*
Coralline red algae, 313
Coral mushroom, *350*
Cordaitales, 235, 485, 489–490, *493*
Cork, 22, 441, 443; cambium, 442–443, 487, 501, 572
Corn, 16, 92, *133*, *414*, *416*, *432*, 518, 544, *552*, *556*, 572, 575, 636, 639, 655, 666; blight, 191, 383; coleoptile, *564*; dwarf, 190; field, 196, 733; high lysine, 191, 743; hybrid, 187; oil and protein in, *196*
Cornus florida, 2
Corpus, *421*–422
Correns, K., 183
Cortex, *378*, *410*, 420, *422*–*423*, *434*, *436*, 442
Cory, R., 631
Corylus, 519
Cosmic radiation, 663
Cotoneasters, 597
Cottam, W., 688
Cotton, 44, 112; fiber, 431; industry, 596; seeds, 640
Cottonwood, *534*, 545
Cotyledon, *483*, 489, *492*, 500–501, 514–515, *517*, 552
Coumarin, 567, *588*, 592
Cousteau, J., 302
Covalent bond, 42–*43*, 47
Cover, 685
Cowles, H. C., 685
Crab, 663
Crabgrass, 624
Cranberries, 538
Crassulaceae, 540
Creighton, H., 190
Creosote bush, 207, *714*
Cretaceous, 224–225, 240, 472, 482, 484, 498–499
Crick, F. H. C., 24, 119, 121, 129, 132
Cristae, *102*
Critical night, 672–*674*
Cronartium ribicola, 371
Cronquist, A., 533, 548
Cronquist's chart of flowering plants, *548*
Crop, 97, 259, 719
Cross-breeding population, 199
Crossing, 187
Crossing-over, 175, 190
Crosswalls, hyphae, 232
Crowfoot, 519
Crown, gall, 255–256, 583
Crown of thorns, *537*
Crucibulum levis, *351*
Cruciferae, 519, 537
Crustose lichens, *372*–*373*
Cryptomeria, 597
Cryptomonad, 228, 239
Cryptophyceae, 324
Crystal, 21, 36, 129, 430
Cucumbers, 194, 520, 594, 636
Cucumis sativus, 520
Cucurbita, *552*, *580*
Cue, environmental, 663
Culms, 544
Cultivar, 13
Cumagloia, *323*
Cuneate, *534*
Cupressaceae, *486*
Curlytop, 191
Currant, 522
Curtis, John J., 687–688, 691
Cuticle, 95, *386*–*389*, *388*, 429, *443*, 611, 723
Cutin, *388*, 429
Cutting, 581, 588, *600*
Cuvette, 731
Cyanide, 640, 661
Cyanophyta, 230, 237, 239, 274, 279, 290–*292*
Cycad, *216*, 234, *237*, *239*, 419, *474*, 481, 485, 498, *655*
Cycadophyta, 234, *237*, *239*
Cycas, 419, 482
Cyclic AMP, 367–*368*, 595
Cycling of elements, 26
Cyclohexamide, 568
Cyclonic air movements, 682
Cydonia, 519
Cypress, *694*
Cypripedium calceolus, 6
Cypsela, 519
Cysteine, *63*
Cystine, 62
Cytochrome, 73, 96, 111, 607–608; oxidase, 73, 108; system, 107–*110*
Cytokinesis, 141
Cytokinin, 159, 570, 572–575, 586, 588, 590–591, 594, 651; effect on translocation, *575*
Cytology, 22, 532
Cytoplasm, 22, 34–37, *113*, 116–*117*, 123, 147, 156, 240, 242, *265*, 424–425, 428
Cytoplasmic membrane, 36
Cytoplasmic streaming, 36, 129, 157, 434, 611, 627–631, *628*
Cytosine, *118*–123

2,4-D, 564–*565*, *599*, 631
Dacthal, *599*
Dactylella bembicodes, *365*
Dalapon, 598–*599*
Dalea spinosa, 723
Dandelion, 207, 541
Dandelion tree, 541
Dark period, 674
Dark reaction, 87–88
Darwin, C., 13, 28, 132, 169, *170*, 175, 181, 187, 194–195, 202, 559–560
Darwin, E., 167
Darwin, R., *170*
Date, 519
Datura, 539, 575, 581
Daubenmire, R. F., 721
Davidson, J. N., 119
Dawn redwood, *482*
Dawn signal, 660–661
Dawson, J. W., 449
Day length, 646, 650–651, 665
Day-neutral plant, 665
DDT, 99, 200, 316–317, 744
Deats, W., 452
de Barry, H., 365
Decay, 11, 111, 257, 380
Deciduous forest, 2, 365, *683*–684, 707
Decomposer, 26, 736–738
Deduction, 12, 562
Deficiencies, symptoms of, 608
Defoliation, 675
Degenerate code, 123
Degeneria, 537
Dehiscent fruits, 519–*520*
Dehydration, 33
Delayed, 661
Deletion, 147
Delevoryas, T., 238
Delphinium, 519, *520*
Deltoid, 534
Delwiche, C. C., 258
Denaturation, 66, 72, 75
Denney, A., 679
Density, 685–*686*
Deoxy-D-ribose, *45*
Deoxyribonucleic acid. *See* DNA
Deoxyribose, *118*–123
Derbesia, 295, *297*
Dermatogen, *517*
Desert, 705, 709, *712*–*715*, 723–726; as environment for ferns, 464; terrarium, 711
Desert candle, 723
Desert fungi, 381
Desert saltbush, 92
Desert scrub oak, 530–531
Desiccation, 727
Desmarestia, 320
Desmids, 269, 289
Desmococcus, 280, 289, *296*
Dessauer, F., 657
Detergent enzyme, 267
Determinate, 504
Deuterium, 62, 662, 734–735
Deuteromycete, 229, 332, 363
Development, 25, 69, 157, 186, 576–601, 666; of roots, 651
Developmental pathway, 157
Devernalization, *669*
Devonian Period, *214*, *216*, 221, 233, 235, 386, 390, 449, 451–452, 455, 461, 472, 479, 494–495, 498
DeVries, H., 164, 179, 182, 191
Dewey, W. G., *172*, 186
Diakinesis, *144*
Dialysis, 67, 74, 77
Diatom, 228, *230*, *237*, *239*, 250, 289, 292, 293, *295*, 315, 334
Diatomaceous earth, 240, 292
Dicamba, *599*
2,4-dichlorophenoxyacetic acid. *See* 2,4-D
Dichotomous, 451, 532, *534*
Dicksonia, 465
Dicot, 2, 10, 210, *500*–501, 533, 542, 545, 548
Dicotyledoneae, 210, *234*, *237*, *500*–501
Dictiyota, 308
Dictyosome, 36, 38, *101*, 130–*132*, *133*, 141, 157, *276*, *293*–*294*, 320, 432, 510–511; vesicle, 156, 426, *427*
Dictyostele, 464–*465*
Dictyostelium discoideum, 367–*368*
Dictyota, 320
Dictyuchus, *326*
Differentiation, 418, 558, 579; of bud, 673; of vascular elements, 584
Diffraction, 98, 129
Diffusion, 51, 53–55, 71, 434, 627; pressure, 55; through small pores, 611
Dihybrid, 179
Dihydroxyacetone phosphate, *106*
Dikaryotic, 232, 354, 359, *361*, 369; hypha, *355*; mycelium, *358*, *361*
Dill, 540
Dillenius, J. J., 209
6-(γ,γ-dimethylallyl) amino purine, *573*
Dinitrophenol, 661, 673
Dinoflagellate, 228, *230*, *237*, *239*, 289, 314–315, 321, 659
Dinophyceae, 321, 324
Dinosaurs, *216*, 224, 484, 498
Dioecious, 545
Dioon calocoma, 419
Dipeptide, *64*
Diphenamid, *599*
1,3-Diphenylurea, 574
Diphtheria, 263
Diplococcus pneumoniae, *250*
Diploid, 14, 140, 143, 192, 306
Diplonema, *144*, *145*
Diquat, *599*
Direct environmental effect, 721
Directional selection, 202
Disaccharide, 47, 91
Disc, flower, *503*–504, 541
Discrete associations, 684–686, 688, 691
Disc-shaped chloroplasts, 321
Disease-resistant native plants, 383
Disruptive selection, 202
Dissolved substances, 55
Distribution of plants, *684*
Disulfide bonds, 72
Diversity, 12, 13, 21, 25, 528, 555
Division, 13, 210
Dixon, H., 623, 625
DNA, 24, 28, 34, 38–39, 69, 78, 84, 111, *115*, 119–*129*, *122*, *127*, 132, 140, 148, 153, 158–160, 175, 184, 188, *272*, 317, 572, 669; bacterial, *246*–*247*, 250, 264–265, 324; in blue-green algae, *274*–*276*, 291; and histones, in algae, 295; in nucleus, 148; in phage, *245*; polymerases, 122; protein fibers, 149; repair enzyme, 656; replication, 123, 149, 159, *266*; strand, *189*, structure of, 121–122, *149*, *268*; synthesis, 147–*148*, 244; viral, 243, *246*–*247*
Dobzhansky, T., 202, 208
Dodder, 248
Dogbanes, 538
Dogtooth violet, *20*
Dogwood, 2
Dolk, F., 589
Dolk, H., 565, 589
Dominance, 176, 178, 181
Dominant organisms, 683
Dominion, 748
Dormancy, 588, 591, 637, 651; bud, 590, 637–638
Dorsiventral symmetry, 504
Double helix, 24, 38, *72*, 122, 124, 129, 130, 132
Double ions, 62
Douglas fir, 235, 626, *696*
Downs, R. J., 677
Downy mildew, 338–340
Doyle, W. T., 238
Dragon tree, 544
Drake, B. G., 615
Draper, 161
Driving Force, 610, 623–624
Drones, 193
Drosophila, 126, 188, 192–193, 744; eclosion in, *661*
Drought, 375, *723*, 734; avoiding perennial, 723; enduring nonsucculent perennials, 724; resistance, 192
Drug, 743–744
Drupe, 519, *520*
Druse, 725
Dryas octopetala, *716*
Duckweed, *6*, 11
Duct, 436
Dung, 341, 343, 735; fungus, 361, *342*, 356, 361, *381*
DuPaw, E., 126, 149
Duplications, 147
Dusk signal, 660–661
Dutrochet, H., 31
Dwarf plant, 570
Dysentery, 263

Eames, A. J., 452
Early Wood, *441*
Earth, *303*, 681–682; photosynthetic productivity, 705
East, E. M., 196
Echallium elaterium, *523*
Ecological field studies, *686*
Ecological principles, 681
Ecological succession, *745*
Ecology, 9, 12, 76, 97, 561–563, 689, 718, 722; and energy exchange, 614; in taxonomy, 529
Ecophenes, 693
Ecosystem, 15, 17, 27, 95, 111, 273, 383, 598, 603, 663, 711, 731; aquatic, 229; metabolism, 743; seasonal progression of, 663
Ecotone, 684, 691
Ecotype, 529, 692–693, 722
Ectocarpus, 320
Ectothiorhodospira mobilis, *253*
Eelworm, *365*
Efficiency of respiration, 110
Efficiency, photosynthetic, 97
Egg, 14, 178–179, 283, 307, 416, 454, *469*, 511, 513, 515, 517; in algae, 283; cell, *510*; chicken, 273; in fern, 470–*471*; in pine, *491*
Eggplant, 539
Egregia, *304*, 311–312
Eichler, A. W., 235
Einstein, A., 69, 98
Ejected electron, 87
Elaters, 456–*457*
Electrical currents in plants, 560, 563
Electrochemical transformations, 70
Electromagnetic spectrum, 70
Electron, 42–*43*, 45, 70, 98, 108
Electron-probe analysis, 612
Electron pump, 88, 109
Electron transfer system, 102, 107–*110*
Electrophoresis, 68
Elementary particles, 102

Elements, 26, 428, 604; essential, *607*
Elephant head, 539
Elliptic, 534
Elm, 518–519, 534, 597
Elodea, 194
Elton, C., 689
Elymus glaucus, 198–199
Emasculation, 187
Embedding, 33, 35
Embryo, *77*, 234–235, 389, 416, 450, 481, *483*, 489, 513–514, 529, 581, 639; *Datura*, 575; development delayed, 641; development in *Capsella*, 517; heart-shaped, 514; sac, 511–515
Embryoid, 582
Embryophyta, 238
Emerson, R., 337, 346
Emission spectrum, *618*
Emulsion, 64
Encasement, 240
Endocarp, *518*–518, *521*
Endodermis, *434–436*, *442*, 466, 623; in ferns, 464–465
Endogenous rhythm, 651, 657, 659
Endophytic fungus, 453–*454*
Endoplasmic reticulum. *See* ER
Endosperm, 77, 514–515; liquid, 518, 573, 575; nucleus, 513
Endospore, *261*
Endothal, 599
End plate projection, *245*
Energetics of respiration, 110–*115*
Energy, 56, 69–70, 87, 101, 663; of activation, *75*; absorbed by leaf, 616; barrier, 22, 74–*75*; compounds in ecosystem, 741–744; diagram, *75*; electrical, 70; exchanges of, 15, 70, 613, 615–619; flow through an ecosystem, *738–739*; forms of, 70; importance of for man, 734–736; kinetic and potential, 69; level, 74, 99; mitosis, 147; of a photon, 98; pyramid, 26, 736–*737*; radiant, 70; solar, 718, 734; transfer, 26–27, 69–70, 86, 731–739
Engelmann, T. W., 84–*85*
Engler, A., 533, 538–539, 541, 545–546, 548
Enteromorpha, *298*, *304*, 306
Entire leaf, 534
Entomophthora muscae, *343*
Entrainment, 659
Entropy, 71, 558
Environment, 304–305; effects on anatomical differences, 693; and gene expression, 185; response to, 21, 26, *27*; in selection, 163–164
Enzyme, 22–26, 28, 40–41, 51, 53, 58–59, 65–66, 103, 116, 126, 156, 175, 185, 217, 272, 329, 532, 608, 652–653, 675; allosteric, 24; bacterial, 267–273; characteristics, 72–77, 124; controlled reaction, 721; controlling respiration, 107; destruction, 636, 724; digestive, 432; inhibitors, 103; in Krebs cycle, 130; purification, 68, 77; reactions, temperature of, *76*; regulation, 159; replicating RNA, 244, substrate complex, *46*, 74, *76*, synthesis, 145; synthesizing starch and cellulose, 47
Enzymology, 66–68, 76
Ephedra, 234, *478*, 493, 545
Ephemeral, 710, 723
Epicotyl, 559; hook, 651
Epidemic, 383
Epidermal cell, *30*, 389–*390*, *408*
Epidermal hair, *433*
Epidermis, 389, *410*, *414*, 422–423, *425*, 429, 434, 438, *443*, 450, *456*, 488, 514, 723
Epigynous, *503*, 545–546
Epilobium angustifolium, 540
Epinasty, 568, 651
Epiphyte, 210–211, 231, 233, 257, 305, 313, *460*, 464, 543, 706
Episome, 268
Epistatic gene, 185
Epithelium, 432
Epoch, 220
Epstein, E., 58
Equatorial plate, 141–144, *142*, 150
Equilibrium, 55, 628
Equisetum, 6, 41, 233, 456–*457*, 541, 545; stem, *456*
ER, 36–*37*, 60, 94, *101*, *113*, *117*, 130–131, 141, 156, 158, 294, *320*, 330, *427–429*, 510–511
Era, 220
Ergastic substance, 36
Ergot, *232*, 354, 357
Erharta erecta, 197
Ericaceae, 538
Erodium, 686
Erwinia amylouara, 254–*255*
Erxleben, H., 590
Erysimum nivale, *716*
Erythrocladia, 321–322
Erythrocyte, 52–53
Erythronium grandiflorum, 20
Erythrose, *45*
Erythrotrichia, 321–*322*
Escherichia coli, 21, 84, 158, 244, *250*, *253*, 261, 265–268, 656
Escombe, F., 615
Essential mineral elements, 74
Ester, *44*
Esterification, *61*, 96
Estuary, *703*–706, 732
Ethane, 42, *44*
Ethyl alcohol, *44*
Ethyl carbamate, 662
Ethylene, 42, *44*, 568, 591, 594
Ethyl mercaptan, 44
Etiolation syndrome, 645, 651
Eucalyptus, 540, 667, 706
Eucaryotic algae, 220
Eucaryotic cell, 160, 230, 274, 321
Eucaryotic organisms, 127–128
Euglena, 281, 289
Euglenophyta, 229, *230*, *237*, *239*, 281, 291–293; nuclear division in, 291
Eumycophyta, 235, 322
Eumycota, 229, 237, *239*
Euphorbia, *164*, *537*, 710, 723
Eutrophication, 286–287, 606
Evans, H. A., 605
Evans, J. O., *68*
Evaporation, 387, *603*, 610, 618, 699
Evaporative cooling, 614
Evening primrose, 187, 195, *540*
Evolution, 29, 92, 161–171, 175, 212, 528, 558; algal, 317–324; in angiosperms, 498; convergent, 493, 710; early, 218–219; in green algae, 296–297; mechanisms, 163, 585; plant, 220; theory of, 529
Evolutionary reduction, 542
Evolutionary sorting of characters, 531
Excited state, 99, 646
Exine, *504*
Exobiology, 747
Exocarp, 518
Exoskeleton, 41
Experiment, 12, 562
Expressivity, 189
Extension, 595
Extinction, 213
Eyepiece, 32
Eyespot, *80*, *278*, 280, *293*

Factor limiting productivity, 736
Factor *x*, 657–658, 662
Facultative: flowering requirement, 666; heterotroph, *271*; parasite, 256; saprophyte, 256
FAD, 107, 110
Fagales, 542
Fagopyrum, 519
Fagus grandifolia, 2
Fairy ring, *359*
Fallout, radioactive, 188
Family, 13, 210, 533
Famine, 97, 229, 332, 748
Fan palm, 544, 714
Farinales, 210
Far-red light, 648, 654, 660, *673*, 677
Fast green, 34
Fat, 36, 38, 60–*61*, 94, 231, 515, 741; metabolism of, 38
Fatty acid, 60–61, 107, 110, 156, 217; essential, 743
Feedback, 27, 107–108, 159, 581; in ecosystems, 745–747
Fennel, 540
Fermentation, 105–107, 218, *258*
Fern, *214*, *216*, 221, 234, *237*–239, 433, 449, *452*, 458–473, 485, 587, 651, 663; archegonium, 583; bracken, *6*, 10; embryo, *472*: eusporangiate, 466, 473; evolution of, 472–473; gametophyte, 583; leptosporangiate, 466, 473; life cycle of, 467; prothalli, 586; root, *466*; sporophyte, 583: tree, *460*, 464–465; water, *461*, 464, 473
Fernandez-Moran, H., 102
Ferredoxin, 73, 608
Fertilization, *14*, 142, 389, 512–515; in angiosperms, 515; double, 513, 515; in ferns, 471; tube, *338*
Fertilizer, 258–259, 605–606, 700
Fiber, 150, *406*, 424–*425*, 427, *441*
Fibonacci series, *593*
Fibril, *227*
Fibrous root, 501
Fiddlehead, *458*, 462
Fiddler crab, 663
Field research, 529, 718
Fig, 518, *521*
Filament, *322*, *502*; fungal, 11
Filamentous algae, 289
Filial generation, first, 176
Filicales, 473
Filiform apparatus, *514*
Filtration, *66*
Fire, 728, 745
Fireblight, 254–*255*
Fireweed, 540
First positive and negative curvature, 647
Fish, 191, 221, 287, 316, 382
Fisher, R. A., 200
Fission reactor, 735
Fittonia argyroneura, *579*
Fixing tissue, 33
Flagellum, 142, *227*, 231–232, 238, *253*, 263, *265*, 278, 280, *291*, 292, 315, 328, 338, *470*; tinsel, 320, 337; whiplash, 320, 333–335, 337, 389
Flag tree, *716*
Flax, 425, 429
Flexibacterium, 257
Flightless insects and birds, 164–165
Flint, L. H., 648, 650–651
Flora, 532
Floral: characteristics, 212; morphology, 503; stages, 672, *674*; stages and night length, 674; tube, 521, 539
Floret, *172*, 544
Florideophycidae, 321, *323*
Florigen, 593–594, 671, 676; synthesis, 672, 675; translocation, 673
Flower, 235, 436, 502–514, 662; bud initiation, 593, 727; complete, 502; inhibiting substances, 671; parts, 500, 502; pollination mechanisms, *506*, *507*, *508*
Flowering, 591, *648*, 650
Flowering plants. *See* Angiosperm
Flowering process, 664–679
Flowering quince, 597
Flowering responses, 651, *665*
Flow rate, 629
Fluorescence, 86–87, 99, 646
Fluoride, 76, 661
Fly, 343, 361, 505–*506*; fungus, 337
Föhn, 709
Foliar organs, 419
Foliose lichens, *372*
Folke Skoog, 572–573, 575
Follicle, *519–520*
Food, 741; additives, 273; calories, 734; poisoning, 254, 273; preservation, 273; web, *742*
Foolish seedling disease of rice, 568, 570
Forages, 191
Forest, 9, 16, 376, *700*, *703*–705, 732–733, 737; coniferous, 16; deciduous, 9; floor, 95; fungi, 378–380; reduced, 16; seasonal, *702*, 706; temperate, *702*
Formaldehyde, 42–*43*, 88
Formic acid, *44*
Forskal, 210
Fossil, 214, 225, 386, 390, 472, *477*, 479, 494, 528–529; fuel, 733–736, 739; making of, 240; oldest, 288
Foster, A. S., 32, 418–*419*
Fouquieria splendens, *714*, 723–724
Four-o'clock, 538
Foxtail, *655*
Fractional precipitation, 66
Fragaria, *2*, *520*
Fragmentation, 282, 291, 376
Frame shift, 124
Fraxinus, 519, *534*
Free cell suspension, 582
Free energy, 55
Free-running rhythm, 659
Freeze etching, 35, *50*, 80, *291*
Frequency, 98, 685–*686*
Freshwater, 231; from seawater, 735
Fries, N., 316
Fritschiella, 295, *297*, 389–*390*
Frond, fern, *458*, *463*
Frost, 192, 727
Fructose, *45*, 49, *106*
Fruit, 436, 514–515, 518–523, 567; airborne, 522; classification of, *519*–520; development, 568, 571, 594, 627, 646; simple, 518; tree, 637–638, 639
Fruitfly, 128, 183, 187, 200, 661
Fruiting body, *352*, *361*, *366*
Fruticose lichens, *362*
Fucoidin, 231
Fucoxanthin, 85, 231, 305
Fucus, *298*, *304*, 307–309, 312, 314
Functional group, 46
Fungal hyphae, 375–377, 379
Fungi, *8*, 10, 41, 101, 229, *237–239*, 326–343, 390, 521, 640, 651, 726; aquatic, *326*; endophytic, 450; forest, 380; imperfecti, 229, 332, 363, 365, 375, 380, 382; in lichens, 365–368; marine, 382; pathogenic, 569; sex hormones in, 334; spores of, 329; true, 229, *237*, *239*, 332
Fungicide, 340, 744
Funiculus, *510*, 512
Fur, 522, 666
Fusiform initial, 440–*441*
Fusion of petals, 545
Fusion reactor, 734

Galactose, *45*
Galactoronic acid, 40
Galeopsis tetrahct, 197
Gall, 248, 583
Gallium, 607
Galton, F., 784
Gametangium, 334, 338, 340, *469*
Gamete, 14, 140, 142–143, 178–179, 230, *283*, 334, *338*, 389; biflagellate, 282; male, *321*
Gametophyte, *14*, 15, 140, 142, 192, 228, 233–234, 389, *454–457*, 469, 514; female, *455*, *510*; in fern, 470–471; male, 505; subterranean, 453
Gap: leaf, *437*; phase, 148
Garbage, 606, 747
Garden geranium, 538
Garner, W., 658, 665–666, 669, 676–677
Garrya elliptica, *505*
Gartner, 182
Gas, 52, 54, 389, 611–616
Gasoline, 51, 735
Gassner, J. G., 664, 666
Gates, D., 615
Gehring, P., 494
Gel, electrophoresis, 68, 77
Gelatinous sheath, 280
Gammae, 471
Gene, 25–26, 34, 145–146, 157–158 160, 175, 177–178, 179–181, 183, 193, 532, 570, 579, 581, 638, 656; activation, 652, 675; and growth regulator, 595; deactivation, 595; defective, 184; evolution, 195–215; flow, 692; frequency, 200, 529, 693, 744; and growth regulator, 595; interaction, 185; linked, 190; pool, 199, 692; replication, 120; repressor, 127
Generative cell, *483*, *503–504*, *509*
Generative nucleus, 505
Genetic, 24, 28, 529; analyses, 116; chance event in, 177; code, 122–*125*–127; diversity, 200, 692; drift, 200; material, 120, 581; recombination, *197*, 199; tumor, 583
Genetum, 234, *237*, *239*, 493, 641
Genic disharmony, 205
Genome, 186, 581
Genotype, 179, 190, 195
Gentian, 538–541
Gentian violet, 263
Genus, 13, 208
Geologic clock, *221*
Geologic period, *452*
Geology, 240
Geophysical factor, 663
Geotropism 384, 553, 565, 567
Geranium, 519, *523*, 537
Germ plasm, 192
Germination, 77, 416, 571, 591, 639–640, 721; light in, *649*, 650
Germlings, 314
Geum rossii, *716*, 727
Gibberellic acid, *569–570*
Gibberellin, 77, 111, 190, 193, 566, 568–*569*, 570, 571, 592, 594, 596, 638, 651, 667–669; molecular structure, *569*
Gibby, D. D., *600*
Gigartina, 313
Gill fungal, 359–*362*
Ginkgo biloba, 13, 213, 216, 234, 419, *477*, 481, 485, 490–493, 498, 641
Ginkgophyta, 234, *243–245*
Girdle, 571, *627*
Gleason, H. A., 687–691
Gleditsia triancanthos, *534*
Globular embryo, *517*
Gloeocapsa, 78–79, *274*, 279–280, 289, 292
Glossopteris flora, 225–226
Glucans, 232, 237
Glucose, *39*, 41, *45–47*, *49*, 77, 104–107, 111, 268, 334, 507, 584
Glucose amine, 334
Glucose-1-phosphate, 106
Glucose-6-phosphate, *105–106*
Glume, 544
Glutamic acid, *63*, 111
Glutamine, *63*
Glyceraldehyde, *45*
Glycerine, 217
Glycerol, *45*, 60–*61*, 106, 110
Glycine, *44*, *63–64*
Glycogen, 111, 231, 291
Glycolic acid, 94
Glycolipids, 61
Glycolysis, 104–108, *106*, 110
Glycophyte, 723
Glyoxysome, 94
Gnetales, 493, 499
Gnetophyta, 234, *237*, *239*
Gnotobiotic tent, *606*
Gondwana, *224*–225
Goniotrichum, *322*
Gonium, 280, 296
Gonorrhea, 254, 263
Gonyaulax, *659*
Gooseberry, 522
Gradient, 55, 621, 690–691
Grafting, *600*
Gram stain, 263
Gramineae, 519, 544
Grana, *79*, 82–83, 293
Grand phase, 555, 557–558
Granite, 735
Grape, 339, 519–*520*, 571, 594, 596, 710

Grass, 41, 198, 216, 505, 507, 537, *700*–701, 707, 709–710; ribbon, 507; tropical, 92–93
Grasshopper, 193
Grassland, 16, 259, *698*–705, 732; fungi, 380–381; temperate, 707
Gravity, 69, 302, 388, 416, 635
Gray, Asa, 167, 171
Grazing, 27, 498; animal, 707, 743
Greasewood, 17, 725
Green, D., 102
Greenhouse, 719, 733
Green revolution, 174, 191
Griffith, J., 149
Groniotrichum, 321
Groundwater, 699
Growing season, short, 709
Growth, 21, 69, 111, 158, 555
Growth chamber, 654, 719
Growth curve, *556*
Growth form, 683, *703*–705
Growth rate, 555–*556*
Growth regulator, 564, *592*, 594, 596, 632, 639–641
Growth retardant, 570–571, 598
Gryza sativa, 243
Guanine, *118*–123
Guard cell, 387–*388*, *408*, *414*, 430, 438, 488, 611–612; development, *430*; mother, 430; solutes in, 611
Guignard, L., 515
Gum, 45
Gymnosperm, 15, *216*, 221, 224, 234, *237*–239, 389, *441*, 449, *452*, 498, 651, 663, *727*; shoot apices, 418
Gynura avrantiaca, *578*

Haagen-Smit, A. J., 575, 590
Habenaria unalascensis, 543
Haberlandt, G., 572
Haeckel, E., 171
Haemmerling, J., 115
Hair, *431*, 435; cell, 38, *408*
Hales, S., 610, 627
Halley, 183
Hallucinogen, 743
Halophytes, 57, 723–725
Hamamelis virginiana, *523*
Hammond, R. P., 735
Hammer, K. C., *671*, 673, 675, 677–678
Hapalosiphon, *292*
Haploid, 14, 140, 143, 192, 306
Hardening, 722, 724
Hardin, G., 181
Hardy, G. H., 199
Hardy-Weinburg law, *199*, 200
Harvey, W., 633
Hashish, 432
Hasselquist, 210
Hatch, M. D., 91
Hatch–Slack pathway, 91–*93*–95
Haustorium, 70, 331, *331*–*332*, 334, 339
Hawk, 317
Hawk moth, *506*
Hawkweed, 182
Haystack, 383
Hazelnut, 519, 542
Heart, 621–*622*–623; stage, *517*, *582*
Heartwood, 380, 440
Heat, 70–71, 99, 110, 726, 735, 738; denaturation, 77; transfer, 613, 615–619, 662
Heath, *538*
Heavy water, 662
Hedychium, 99
Heeger, G., 529
Height, 555
Helianthus, *576*
Helium, 42–43, 217
Helmont, J. B. von, 604
Hemicellulose, 40–41, 45, 133, 157, 424
Hemlock, 235, *730*
Hemoglobin, 73, 97
Hemophilia, 194
Henbane, 656, 665–666, 668
Hendricks, S., 648, *650*, 652
Henle, J., 262–263
Henslow, J. S., 168–169, 170
Hepatophyta, 237–238
Heptose, *45*
Herb, 9
Herbarium, 12, *524*, *526*, 527–529, 548, 639
Herbert, 182
Herbicide, 598–*599*, 744
Herbivore, 736, 738, 742
Hereditary information, 145
Hermaphrodite, 346
Heterocyclic amino acids, *63*
Heterocysts, *276*
Heterokaryotic, 354
Heterosis, 190
Heterosporous, 453–454, 470
Heterotroph, 736
Heterozygosity theory, 190
Heterozygous, 179, 186
Hevea brasiliensis, 433, 537
Hexokinase, 73
Hexose, *45*, 91
Hexose phosphate, 110
Hickory, 545
Hieracium, 182
Hiesey, W., 693
High energy reaction (HER) 651, *653*–*654*
Highkin, Harry, 722
Hildebrandt, A. C., 581
Hill reaction, 82, 87
Hill, R., 82
Himalayan rabbit, 185
Histidine, *63*
Histochemistry, 34
Histogen concept, 419
Histone protein, 128, 140, 148, 156–159, 317, 675
Hofmann, A., 357
Hofmeister, W., 515
Holdfast, 230, 281, *300*–*301*, 306, 389
Holley, R., 124, 132
Holly, 641
Hollyhock, 639
Holmgren, A., *526*
Holbistic environment, 664, 676, 719, 745; growth, 580
Holy fire (St. Anthony's fire), 357
Homeostasis, 27, 557–558, 745
Homogenize, 35, *66*
Homokaryotic, 354
Homologous pairs, 143
Homosporous, 453, 470
Homozygous, *179*, 186, 190
Honeybee, 182, 193, 658
Honeydew, *630*
Honey locust, *534*
Honeysuckle, 541, 546
Hooke, R., 22, 31–32, 132, 262
Hooker, D., 545
Hooker, J., 161, 167, 183
Hook opening, 648, 653
Hop tree, 519
Horizontal structuring, 683–684
Hormone, 157–158, 345, 560, 564, 568, 571–572, 626, 638, 669; flowering. *See* florigen
Horneophyton, 451
Horse chestnut, *550*, 575
Horsemint, 539
Horsetail, 43, 213–214, 221, 224, 233, *237*, 239, 449, *452*, 541, 545, 587, *655*; tree, 216, *541*
Host, 745
Hot spring, 257
Huang, R., 127, 157
Huckleberries, 538
Hudsonian forest, 691
Humidity, 624, 676; controlled, 718
Hummingbird, 204
Humulus, 194
Humus, 378, 606, 700–*701*, 706
Hutchinson, J., 548
Hutton, J., 167
Huxley, T., 161–*162*, 171
Hybrid, 13, 190, 198, 204–205; corn, 186; swarm, 530–531; vigor, 190–194
Hybridization, 199
Hydration, 57, 64–65, 624
Hydrocyanic acid, 191
Hydrodictyon, *278*, *297*
Hydrogen, 42–*43*, 45, 217, 605; bond, 42–*43*, 47; ion, 62, 108, 699
Hydrogen peroxide, 188
Hydrogen sulfide, 268–271
Hydrolysis, 77
Hydrolytic enzymes, 38
Hydrophilic, 52, 54, 65, 67; surface, 624
Hydrophobic, 52
Hydrophyte, 723
Hydroponic, *604*, 609
Hydrothermally altered material, *746*
Hydroxyl (alcohol), 45, 62
Hydroxylated amino acids, *63*
Hydroxyproline, 62
Hymenoxys grandiflora, *716*
Hyoscyamus niger, 668
Hyphae (fungal), 11, 232, 324–331, 334, 338–340, 343, 354
Hypochaeris, 522
Hypocotyl, *65*, *492*
Hypodermal cells, 488
Hypogynous, 503, 537, 545

Ice crystal, 727
Ideal angle, 593
Idioblast, 419, 425, 430
Illumination, 98
Imbibition, 65, 624
Immunity protein, 246
Impatiens aurella, *523*
Incomplete dominance, 183
Incubation time, 248
Indian paint brush, 539
Indian pipe, 538
Individualistic concept of plant distribution, 687
Induleacetic acid (IAA), 255, 378, 564–*565*, 568–587, 590, 608
Indolebutyric acid (IBA), 564–*565*, *592*, 596, 600
Indole compounds, 110, *565*
Indole ethanol, *565*
Induction, 12, 672, 674, 677, 722; of bud dormancy, 638
Inductive approach, 562
Inductive dark period, 672
Inductive response, 669
Indusium, 467–*468*
Industrial: contaminant, 747; revolution, 610, 734, 744; waste, 76, 287, 289, 376, 382
Industry, 734
Infection, 583
Inferior ovary, 503, 546
Inflorescence, *503*, 504, *506*
Information theory, 73
Infrared gas analyzer, 718
Infrared photography, *616*
Inheritance, 164; of acquired characteristics, 28, 167
Inhibitor of germination, 640–641, 725–726
Inorganic phosphate (Pi), 87–88
Inositol, 574
Insect, 11, 26, 41, 192, 200, 221, 224, 243, 248, 330, 361, 380, 390, 432, 473, 505, 658; communities, 690
Insecticide, 744
Insectivorous plant, 432
Insulin, 71
Integument, 481, *510*
Intensity, 98
Intercellular space, *423*
Interface, 64
Intergradations of phenotypes, 529
Intermediate, compound, 103
Intermediate inheritance 183
Intermembrane space, *102*
Internode, 435, 651; elongation, 651
Intervening light period, *678*
Intine, *504*
Inversion, 147
Iodine, 47, 607; reagent, 77
Ion, 42–43, 699; absorption, *59*; uptake, 58
Ionic bend, 42–*43*
Ionizing radiation, 128, 188
IPC, *599*
Iridaea, *301*, *304*, 313
Iris, 542–543
Iron, 96, 605, 607–608, 609, 632, 676, 700; containing porphyrins, 108; deficiency in, 670
Irregular corolla, 545
Irrigation, 701
Irritability, 21
Island, 165
Isodiametric, 423, 568
Isoetes, 451
Isogametes, 282
Isolating mechanism, 203–204, 529
Isolation in selection, 164
Isoleucine, *63*
Isomorphic, 306, 308
Isopentenyl adenine, *573*
Isoprenoid, 96, 110
Isotope, 42, 88, 92, 103
Isozyme, 74
Ivy, 667

Jack-in-the-pulpit, 543
Jacob, F., 123, 126–128, 132
Jacobs, W. P., 585
Jaffe, M., 653
Jansen, H., 32
Jansen, Z., 32
Jimson weed, 581
Johannsen, W., 195
Johnson, R. M., *596*
Jolly, J., 623
Jordan, A., 692
Jordan, D. S., 196
Juglone, 743
Juniper, 235, *655*
Juniperus chinensis, *597*
Jupiter, 219
Jurassic, *216*, 225 484, 492
Jussieu, A., 545
Juvenility, 667–668, 676

Kalanchöc blossfeldiana, *660*
Kalm, 216
Karyokinesis, 141
Keck, D., 693
Keel, 178
Kelvin, W. T., 168
Kendeith, S. C., 689–690
Kerb, *599*
Ketoglutaric acid, 109–111
Key for classification, 532
Khorana, H. G., 123, 132
Kigelia, *507*
Kinetic activity, 23
Kinetic energy, 99
Kinetic theory, 54, 64
Kinetics, 652
Kinetin, 572–*573*, *575*, 587–588
Kingdom, 13; multiple, 238
Kingdonia uniflora, 419
Kinin, 573
Knopf, J., 605–606, 609
Knopf's solution, *605*–606
Koch, R., 262
Koch's postulates, 263
Koepfli, J., 590
Kögl, F., 564, 589–590
Kölliker, 182
Kölreuter, J. G., 182, 190
Korosawa, E., 568
Krebs cycle, 102, 107–110, *109*
Krill, 316
Kronberg, A., 122, 132

Labiatae, 519, 539
Labyrinthulomycete, 365, *367*
Lactic acid, 106–107, 110, 268
Lactobacillus casei, 270
Lactones, unsaturated, 592
Lactuca sativa L., *653*
Lady's slipper, *6*, 543
Lake, 270, 382, *703*–706; algae, 285–288
Lamarck, J. B., 169
Lamb's quarters, 676
Lamellar membrane, 608
Lamiales, 539
Laminaria, 231, *300*, *304*, 311–314, 320
Laminariales, 312
Lamp shell, 12
Land plant, 229, 233–235, 386–389, 498; origin of, 389–390, 449
Lange, O., 718
Lanolin, *583*
Larch, 235, 637
Larix lyallii, 207
Larkspur, 519, 537
Larrea divaracata, 207, 686, *714*
Late blight, of potato, 331, 338–339, 383
Late riser, 498
Late wood, *441*
Lateral movement from sieve tube, 631–632
Laterization, 700–*701*, *704*
Latex, 433, 537
Laticifer, 433, 436
Latin name, 13, 208–211
Latitude, 303
Latosol, *700*
Laurasia, *224*–225
Laurel, 533
Lauric acid, *61*
Law of the minimum, 719
Law of reciprocity, 647, *677*
Layering, *600*
Lead, 76
Leaf, 234, 387, 389, 437–439, 463; attachment, 436; broad, 499; expansion, *438*, 651; fall, *635*; of fern, 462, *468*; formation, 593; gap, 437, 462, 464; of grass, 439; movement rhythm, 658–*659*; phyllotaxy, 593; position, 646; primordium, *421*, 435; resistance, 613–614; shape, 438–439, 503, 534, 613–614, 618; spectrum, *617*; temperature, 614, 616; trace, 437, 451
Leafhopper, 248
Leaflet, 438, *534*
Leathesia, 320
Lecog, 182
Lederberg, J., 200–201, 266
Leeuwenhoek, A. von, 356
Legume, 258–259, 519, *520*, 537, 707, 710
Leguminosae, 178, 539–540
Lemna perpusilla, 544, *677*
Lemon, 589
Lens, 32
Leoanora, *373*
Lescarpus fragilis, *352*–*353*
Lepidium root, *567*
Lepidodendron, *455*
Leptonema, *144*, *145*
Lethal mutations, 188
Lettuce, 648, *649*, *651*–*653*
Leucine, 63
Leucoplast, 36
Lichen, 8, 10, 97, 306, 372, 375, 377, *727*; acid, 376; ascomycete, 37; in ecosystem, 375–377; section, *375*
Licorice, 540
Liebig, J. von, 719
Life, 17, 23, 69, 75, 681, 722; characteristics of, 21; cycle, 13, *14*, 416; origins of, 28; on other planet, 219; zones, 684, 689–691
Light, 305, 635; blue, 470, 565, 653; break, 675; effect on fucus development, 308; in entrainment, 659–660; in flowering, 678; infrared, 86; intensity of, 10, *86*, 95, 638, 653–654, *672*, 673, 718, 727, 733; physics of, 98–99, 646; quality, 95, 653; reaction, 86–89; threshold, 678; treatment for germination, 641; ultraviolet, *655*
Lignification, *427*
Lignin, 40, 41, 133, *387*, 389, 424, 741
Lilac, *534*, 637; apex, *584*
Liliopsida, *548*
Lilium martagon, *506*
Lily, 508–*509*, 512–*513*–514, 542
Limber pine, *474*, 746

Lime, 340, 735
Limiting factor, 719, 722, 732
Linde, 263
Linden, *534*
Linen, 425
Linkage groups, 190
Linnaeus, C. (E. von Linné) 13, 22, 167, 208, *209*, 212, 544
Linnaeus' artificial system, 212
Linolenic acid, *61*
Linum usitatissimum, 425
Lep cell, 468
Lipid, *54*, 60–61, 115, 156; droplet, *101*; material, 52; matrix with watery holes, 52
Lipophilic, 52
Lipopolysaccharide, 263
Liquid, 54
Liriodendron tulipifera, 210, *441*, 534
Lister, 263
Lithium, 734–735
Liverwort, 192, *237–239*, 521
Lobed leaves, *534*
Lockhart, J. A., 651, 654
Locust, *520*
Lodgepole pine, *553*, *655*, *728*, *730*
Lodging, 191
Log phase, 555, 557
Long day, 651, 667; inhibitory effect, *670*; plant, 664, 665, *665–666*, 670, 672, 676, 678
Longitudinal section, *420*
Longleaf pine, *696*
Long-night plant, 677
Long-short-day plant, 676
Lotus seed, 639
LSD, *357*
Lucas, 182
Lumen, *423*
Lunaria annoa, *520*
Lupine, 585, 587, 590, *636*, 639
Lwoff, A., 123, 132
Lycopersicum esculentum, *520*
Lycopod, *237*, *239*, 452, *455*
Lycopodium, 433, 437, 451–452, *454*
Lycopsid, 449, 451, 454, 461
Lycopsida, 233, *237–239*, 451
Lyall, Charles, 167–169
Lyngbya, *292*
Lycotropic series, 699
Lysenko, T. D., 665
D-lysergic acid diethylamide, *357*
Lysimeter, 611
Lysine, *63*, 191, 743
Lysis, 57, 245
Lysogenic phage, 246

McAllister, E. D., 648, 650, 651
Machine, 734
Machlis, L., 345–*346*
McMillan, C., 693
McMorris, 345
Macrocystis, 311, 320
Magius, *703–705*, 707
Magnesium, 96, 108, 605, 607–608, 632
Magnetism, 663
Magnifying power, 32
Magnol, P., 210
Magnolia, *80*; 204, 210, *533*, *534*, 545
Magnoliopsida, *548*
Mahonia, 519
Major nutrient element, 605
Malate ion, 470
Male sterile line, 191
Maleic hydrazide, *592*, 598
Malia moschatus, *520*
Malic acid, *73*, *93*
Malphigi, M., 627
Malthus, T., 28, 169–170
Malting process, 596
Maltose, 47, *49*
Malus, 519
Malvaceae, 519
Mammal, 192, 224, 316, 473, 498
Man, 12, 27, 192, 224, 522, 523; in the ecosystem, 743–744, 747–748; sensing time, 656
Manganese, 607–608
Mangrove, 16, 57, *521*, 540, 724
Manihot, 537
Mannan, 231, 295
Mannitol, 574
Mannose, 46
Manzanita, 538, *746*
Maple, 2, 208, 518–519, *534*, *591*
Marattiales, 473
Margulis, L., 160
Marigold, 541
Marijuana, 431, 744
Marquis wheat, 371
Mars, 82, 655, 747
Marsh, 16, *703–706*
Marsilea, 473
Martin, G., 365
Mascarhanus, J., 508
Mason, H. L., 688
Mass flow, *628*–633
Mastodon, *224*
Mathematics, 178, 181, 532
Mating, selective, 200
Matonia pentinata, *465*
Matrix, 39, 60, 102
Matter, 69–70
Maximum, 76, 555
Maxwell-Boltzmann distribution, *55*
Mayr, E., 202
Mealworm, 663
Mechanical harvesting, 192
Megagametophyte, 453, *480*–481, 489, *491*, 492, 502, 507, 510–511, 513
Megaphyll, 438, 462, *462*, 463, 486, 501
Megasporangia, 481, *491*
Megaspore, 453, *455*, 481, *483*, 502
Megaspore mother cell, 481, *510*, *513*
Megasporocyte, 481, 510
Megasporophyll, 481
Meiosis, *14*, 142–*144*–*146*, 190, 204, 338, 450
Melandrium, 194
Meljubow, D., 568
Membrane, 37, *51*–60, 101, 156, 608, 652; chloroplast, *54*, *252*; holes, 53; in origin of life, 226; properties and structure, 52–53; stability, 290; stimulated by auxin, 595; unit, 53–*54*
Mendel, G., 24, *175*–184, 193, 195, 262
Mercury, 76, 97, 271, 316–317
Mericarp, 519
Merismopedia, *292*
Meristem, *417*–420, 555; diattente, 421; development, 587
Meristematic cell, 669
Meristematic tissue, 423
Merriam, C. H., 689–692
Mertensia, 540
Mescaline, 744
Mesocarp, *518*–518, *521*
Mesophyll, 422, 438, 488
Mesophyte, 723
Mesosome, 264–*265*, *266*
Mesozoic, *216*, 220, 224–225, 234–235, 461, 473
Messenger-RNA (m-RNA), 25, 123–*124*, *127–128*; 156–158, 272, 508, 595
Metabolism, 21, 69, 71, 158; in bacteria, 268–271; in translocation, 630
Metabolite, toxic, 380
Metal activator, *74*
Metaphase, 141, *143*, *144*, *149*, *513*
Metasequoia glyptostroides, 482
Metaxylem, 456
Methane, 42–*44*, *43*, 217, 219, 682
Methionine, 63
Methyl group, *44*
Methyl mercury, 271
Metric measurement, 28–29,
Mice, 663
Michelsen-Morley experiment, 98
Micrasterias, *278*
Microbody, 36, 38, 94, *99*, 130
Microcladia, *304*, 313
Microcolony, *257*
Microcycas calocoma, 419
Microcystis, 275, 281, *291*
Microfibril, 39, 41, 157
Microgametophyte, 453, *483*
Micrographia, 32
Microphyll, 437–438, 451, 456, *462*, *463*
Microphyllophyta, 238
Micropyle, 510–512
Microscope, electron, 22, 31, 32, 34, *35*–36, 250; light, 250; phase-contrast, 36, 38, 151
Microscopic morphology, 529
Microspecies, 198
Microspectrophotometer, 147
Microsporangia, 481
Microspore, *453*, 453, 502; mother cell, 481, *503*, 505
Microsporophyll, 481
Microtechnique, 32
Microtome, 33
Microtubule, 28, *142*, *150*, 157, *227*, 291, *293*, 324, 426–*427*
Middle lamella, 37, 39–41, 65, 423, 608
Migration, *197*, 199–200, 662, 666
Mildew, 38, 337
Milk, 270
Milkweed, 519, 543
Milky sap, 537
Millardet, P., 340
Miller, C., 572, 575
Miller, S. L., 218
Millikan, R. T., 719
Mimosa, 633
Mimulus: cardinalis, 204; *lewisii*, 204
Mineral, 51, 379; dissolved in soil, 699; nutrition of plant, 316, 604, 608
Miniscus, 623
Mint, 538–539, 604
Mirror image, 23, 45, 75
Mirsky, A. E., 127
Mistletoe, 10, *27*
Mist propagatron, *600*
Mitochondrion, 20, 36–38, 37, *51*–53, 59, 101–103, *102*, 117, 130, 153, 156–158, 160, *276*, 293–*294*, *320*, *330*, *332*, *470*, 511
Mitosis, 14, 141–142, 147–151, *148*, 230, 324; in diatoms, 292; in euglenophyta, 291; in green algae, 293
Mizutani, S., 244
Moccasin flower, 543
Model building, 557, 580
Modern synthetic theory of evolution, 196–206
Modifier, 185
Modulated response, 721–722
Moisture, 699–700; in atmosphere, *619*
Molecular: biology, 24, 25; sieve hypothesis, 52; velocity, *55*
Molecule, 21, 38, 39, 42–*43*, 54; phytochrome, 652
Mole fraction, 56
Molisch, H., 568
Mollisols, *701*
Molybdenum, 607–608
Monera, 238
Monkey flower 204, 539
Monkshood, 537
Monocot, *2*, 9–10, 210, *500*–501, 533, 542, 545, 548
Monocotyledoneae, *239*, 500–501
Monod, J., 123, 126–128, 132
Monoecious, 542, 545
Monograph, 532
Monohybrid, 179
Monokaryotic, 354, *358*, *361*
Monosaccharide, 44, *45*
Monostroma, 295–*296*
Monsoon area, 706
Monterey pine, *497*
Moon, *303*, 658, 662–663
Mooney, H. A., 693
Morchella conica, *350*
Morel, *350*
Morgan, T. H., 187, 194
Morning dove, *714*
Morning glory, 538, 676
Morphactin, 598
Morphine, 433
Morphogenesis, 555, 558, 579–580, 593
Morphological species concept, 202
Morphological study, 527–529
Morphology, 22
Morus alba, 520
Moss, 233, *237*, *239*, 587, 663, 666
Moth, 193, 505, 540
Motile cell, 321
Mougeotia, 281
Moulton, F. R., 168
Mountain: ash, 519; avens, 716, 727; lilac, *698*; pine, *576*
Movement, 21
Mucilage, 432; duct, 320
Mucopeptide, 231, 263, 279
Mucor mucedo, 340
Muir, W. H., 581
Mulberry, *520*
Müller, H. J., 188
Multiflagellate sperm, 470–*471*
Multinucleate cell, 114, 141, 230, 232, 572; endosperm, 514
Multiple allele, 184
Multiple fruits, 518
Münch, E., 628, 631; hypothesis of, 632; model of, 628–629
Müntzing, A., 197
Muscle, 111, 366
Mushroom, *232*, 348, 624, 663; life history of, 358–*361*–363; poisonous, 362
Mustard: family, 537; oil, 640
Mutation, 24, 28, 69, 116, 128, 145–146, 188–*189*, 197, 199, 217, 254, 570, 656; theory, 195
Mutualism, 27, 745
Mutualistic symbiosis, 27, 375, 363
Mycelia, fusion of, *358*
Mycelium, 232, 329–*330*, *332*, 359–360, 369
Mycology, 229, 337
Mycoplasm, 11, *272*, 272
Mycorrhiza, 11, 27, *377*, 380
Mycorrhizal: association, 488; root, *378*
Myristica, 519
Myrtle order, 540
Myxamoebae, 366–367
Myxomycete, *352*, 380
Myxomycophyta, *232*, 332, 365–368
Myxomycota, 229, *237*, *239*

NAA, 564, 565, *583*, 592, 600
NADH, 106–111
NADP, 87, 104
NADPH, 87–88, 91, 101, 104–107, 111, 156
Nägeli, K. W. von, 181–184
α-Naphthalene acetic acid, *see* NAA, 583, 592
Narcotic, 661, 743
Naringenin, *588*, 592
Nason, A., 605
Nasturtium, 585
Natural gas, 735
Natural history, 684
Natural selection, 28, 146, 163, 165, 169, 195–197, 199, 201, 612, 744; and population, 200–208
Navel orange, 418
Navicula pelliculosa, *292*
Nawaschin, S., 515
Neap tide, *303*–303
Necrosis, 608
Nectar, 432, 507
Needle, of conifer, 438–439, 488–*489*
Negative: charge, 41; curvature, *647*; feedback, 747; staining, 102
Neisseria gonorrhoeae, 254
Nemalion, *323*
Nematode, *365*
Nemec, B., 566
Neo-Darwinian evolution, 744
Neptune, 219
Nereocystis, *304*, 311–*312*–314
Nerve cell, 61
Nervous stimulus, 111, 560, 653
Net productivity, 731
Nettler, 431
Neurospora erassa, 24, 116, 185, 355
Newton, I., 183
Niche-exclusion principle, 691
Nicholas of Cusa, 604
Nickel, 607
Nicotiana, *142*
Nicotine, 539, 743
Nicotinic acid, 571
Niedle, E., 672
Night-break phenomenon, 675, 677
Nightshade, 538–539
Nilsson-Ehle, H., 196
Nirenburg, M., 123, 132
Nitrate, 258, 609–610, 641; reduction, 608
Nitrite, 258
Nitrogen, 42, 44, 62, 217, 219, 259, 605, 608; base, 217; compound, 632; cycle, 257, *259*, *740*; fixation, 226, 288, 290, 611, 664, 676, 740; mustard, 188, 656
Number, 140
Noack, K., 651
Nobel, A., 131
Noctiluca, 315
Node, 435, 437, 559
Nodule, 258–259, *260*
Nonsense codon, 123
Norimaki-Sushi, 309–310
Norris, K. H., 652
North woods, *694*, *697*, *707*
Noxious weed, 600
Nucellus, 510, 513
Nuclear: cap, *328*; explosion, 188; fusion, 515; membrane (envelope), 36–58, *115*, 141, 144, 150–151, 157–158, 230, 291, 293, 295, 317, 652; protein, 217; reactor, 735; sap, 36, 115
Nucleic acid, 21, 38, 45, 117–*118*, 122, *240*, 608; in virus, 243
Nucleolus, 34, 36, 38, 115, 143, 153, 244, *291*, *328*, 428
Nucleoplasm, *250*–*264*–*265*
Nucleotide, 24, 91, 111, 122, 125, 608, 656
Nucleus, 20, 22, 36, 37, 38, 84, *112*–*113*–117, 154, 158, 276, 293–294, *320*, *328*, *330*, *343*, *410*, 424–425, 428–*429*, *470*, 510, 662; polar, *511*–512; transportation of, *662*; vegetative, *503*–*504*, 505
Numbers, pyramid of, *737*
Nut, 519, 522
Nutation, 559
Nutmeg, 519
Nutrient: on agricultural land, 733; cycle, 382, *741*; desert, element, 97; 705; level, 314; in a lake, 286
Nutrition, 217; of algae in oceans, 303–304
Nyctoperiodism, 677

Oak, 522, 531, *534*, 542, 545; Californian, 530; forest, *698*; woodland, 530
Oasis, 714
Oat, 77, 544, 666; coleoptile, *554*, *559–560*; seedling, *559*, *563*
Obligate parasite, 256
Ocean, 15, 694, *703–705*, 737; as environment, 302; photosynthesis and respiration in, 97; of time, 549
Ochoa, S., 122
Ocotillo, *714*, 723
Ocular, 34
Odum, E. P., 739
Oedogonium, *276*, 281–283, 289; reproduction in, *284*
Oenothera, 540
Oil, 36–37, 94, 110, 192, 231, 515, 734; in ocean, 97; gland, 521; shale, 735
Old-man-of-the-mountain, *716*
Olea, 519
Oligosaccharide, *77*
Oligotropic, 286
Olive, 519, 538, 710
Oliver, D., 484
Olmsted, C., 693

One-gene–one-enzyme concept, 24, 116
Onion, *112–113, 406, 410*; field, *714*; root, 422
Oogamy, 283
Oogonial initial, 344
Oogonium, *338*
Oomycete, 227, *232, 237, 239*, 333, 337–340, 344, 382
Oosphere, *338*
Open system, 417, 557, 603
Operon, 126–*127*, 128
Opium, 744; poppy, 433
Opuntia, 686, *714*
Orange, *521*, 640
Orchid, *6*, 10, 27, *195*, 537, 543, 639, 706; growing, 248; seed, 378–*379*, 518
Order, 13, 210, 533
Organ, 22, 571, 590
Organelle, 153; membrane-bound, 160, 264, 279, 280
Organic: acid, 90–91, 94, 611, 614, 640; gardening, 605–606; matter, 606
Organism, 12, 22; procaryotic, 226; thermophilic, 290
Origin: of life, 212, 217; of plastid, 83–84
Orotic acid, 74
Orotidine-5′-phosphate pyrophosphorylase, *74*
Osborn, H. F., 196
Oscillatoria, *274*, 279, 289, 292, 315
Osmium, 102
Osmometer, *56*, 435, 623, 628
Osmosis, 51, 53, 55–56, 82
Osmotic: gradient, 629; potential, 641; stress, 724
Osmundaceae, 473
Ostiole, *356*
Ostracoderm, 451
Ovary, *173*, 235, *502*–502, 507, 510, 515, 545
Overbeck, J. van, 573–*575*, 581
Overdominance theory, 190
Overproduction, 163–164
Overton, C. E., 52
Orule, *502*, 507, 510, 512, 515
Oxaloacetic acid, 91, *93*, 107, *109*, 111
Oxidation, 70; of chlorophyll, 85
Oxidation-reduction potential, 108
Oxisols, *701*
Oxtongue, *655*
Oxygen, 42, *43*, 44, 45, 82, 84, 86, 88, 94, 101, 106–108, 111, 217, 585, 605, 610, 639, 669, 681–682; in atmosphere, 97; cycle, 741; dissolved, 382; in a lake, 287; revolution, 219
Oxyria digyna, 693, 727
Oxysome, 102
Ozone, 219, 741

Paal, A., 560–561, 563
Pachynema, *144–145*
Paleobotany, 390, 451, 479, 529
Paleontologist, 240
Paleozoic, 220–221, 224, 234, 240, 498
Palisade parenchyma, *30*, 424, 438
Palm, *216*, 437, 543, 614, 629
Palmetto, 544
Palouse prairie, *698*
Pampas, 707
Pandanales, 545
Pandorina, 280
Pangaea, *224–225*
Panicum, *2*
Pantograph, 685
Papain, *75*
Papaver, 520, 537; somniferum, 433
Paper chromatography, *90*
Papilionaceae, 519
Pappus, 541
Paraffin wax, 33
Parallel or convergent evolution, 533, 543
Paramylon, 231
Paraphysis, *355*
Paraquat, *599*
Parasite, 10, 11, 27, 218, 226, 229, 331, 375, 383, 742, 745
Paratyphoid, 254
Parenchyma, *406, 410, 412, 414*, 422, *423*, *441*, 450, 487, *488*, 501
Parent material, 699, 701
Park, D., 380
Parker, M. W., 650, 671
Parsley, 540
Parthenocarpy, 515, 594
Particles, 54, *65*
Passive uptake, 58
Pasteur, L., 262–263, 357
Pathology, 562
Pathways of respiration, 103–111, *104*
Patten, W., 451
Pauling, L., 72, 121, 129, 132
P-coumaryl alcohol, *41*
PCPA, *592*
Pea, 24, 158, 190, 540, *552*, 568, 571, 589, 722; Alaska, *556*; Family, 539; garden, 175–*176*, 178–179, 182; leaf expansion, *648*; stem segments, 590; Swartbekkie, *556*
Peach, 519, 637, 639
Peale, J., 332
Pear, 519
Pectic substance, 40, 133, 157, 254, 260, 289, 295, 424
Pectin, 41, 65, 231
Pectinase, 40
Peduncle, 502
Pelargonium, 538
Pelvetia, *300*, 308
Penetrance, 185
Penicillin, 254, 263, 288
Penicillium, *232*, 354–357
Pentstemon, *539*
Pentose phosphate pathway (PPP), 103–108, *105*
Pentoses, *45*
Peony, 519, 545
PEP carboxylase, *93*
Peperomia argyreia, 600
Peptide bonds, 62, *64*, 71–72
Peptidoglycans, 263–*264*
Peracetic acid, 606
Percursaria, 295, *296*
Perennating bud, 727
Perennial ryegrass, 676
Perforation plate, 427
Perianth, 502, 521, 543
Pericarp, *518*–519, 521
Periclinal, 420
Pericycle, *434, 435*, 436, 440, *442*, 466, 587
Periderm, 443, 501
Perigynous, *503*–503, 545
Period, 658, *639*; of rhythm, 639, 658, 661
Perithecium, *356*
Permafrost, 709
Permeability, 52, 55
Permian, 233, 455, 461, 482, 484, 498
Peronospora parasitica, 339
Peroxides, 655
Peroxisome, 36, 94
Persistent rhythm, 658–659, 662
Pesticide, 76, 747
Pestilence, 734, 748
Petal (corolla), *502*–502, 541, 545, 649
Petal movement, *660*
Petiole, 416, 436, 438
Petry, L. C., 452
Petunia, 508, 539
Peyote, 744
Pfeffer, W., 658, 662
PGA, 91, 92–93, 106
PGAldehyde, 91, 92, 106
Phaeophyta, 228, *237, 238*, 239, 305, 311, 317
Phage, 244–*246–247*
Phallus impudicus, 361, *363*
Pharis, R. P., 667–668
Phase, *659–660*
Phaseolus vulgaris, *552*, 657
Phasing dark period, 678
Phelloderm, 443
Phellogen, 442
Phenol, 51, 110, 572, 588
Phenotype, 179, 184, 195
Phenoxyalkancarboxylic acid, 598
β-phenyl-isothiocyanate, *588*
Phenylalanine, *63*, 595
Phenylmercuric acetate, 612
Phloem, 233, 238, 244, 388–*389, 410, 412, 414*, 420, 422, *423*, 427–428, *434–435, 437*, 440–442, 450, *456, 501, 574*, 584–587, 621, 623, 629–630, 631–*632*; accessory, 540; exudate, 629; in ferns, 464–465; fibers, *443*; internal to xylem, 538; parenchyma, *428*; primary, *439*; sucking insects, 629
Phlogiston theory, 562
Phlox, 538–539, 540; *divaricata*, *2*
Phosfon, 571
Phosphate, 274, 287, 740–741
Phosphatidic acid, 62
Phosphoenol: pyruvate carboxylase, 2, *91*, 93–94; pyruvic acid, 91, *93, 106*, 110
Phosphogluconic acid, *105*
Phosphoglyceraldehyde. *See* PGAldehyde
Phosphoglyceric acid. *See* PGA
Phospholipids, 61–*62*, 110, 608
Phosphorescence, 99, 315
5-Phosphoribosyl-1-pyrophosphate, *74*
Phosphoric acid, *118*
Phosphorus, 605, 608, 632
Phosphorylated sugars, 105
Phosphotungstic acid, 102
Photochemistry, 84, 98–99, 646
Photoelectric effect, 99
Photomorphogenesis, 645–657
Photomorphogenetic pigment system, 648–653
Photon, 87, 88, 94, 98–99, 646
Photo oxidation, 85
Photoperiodism, 314, 646–648, 657–658, 662, 665, 669–679, 693, 722
Photoreactivation, 654–656, 655
Photorespiration, 38, 92, 94–95
Photoreversibility, 650–651, 655
Photosynthesis, 37, 44–45, 82–97, 101, 107–109, 218–219, 250, 387, 605, 611, 616, 646, 721, 740; in alpine plants, *727*; bacterial, 268–269; rhythm of, *662*; temperature range for, 97
Photosynthetic unit, 87
Phototropism, 381, 565, *647*, 722
Phragmoplast, 141
Phycobilin pigments, 86, 95, 97, 228, 279, 290, 306, 321, 324, *652*
Phycocolloid, 305–309
Phycocyanin, *86*, 228, 239
Phycoerythrin, 86, 95, 228, 231
Phycomycetes, 332
Phyllotaxy, 593
Phylogenetic classification, 235, 212, 532
Phylum, 216
Physarium polycephalum, 365–*366, 367*
Physcia stellaris, 274
Physiological ecology, 561, 718–727
Physiology, 22, 561
Phytochrome, 470, 648–653, 658, 660, 675–678; absorption, *648*; in conifers, 668; discovery, 650; transformations of, 649, 652
Phytol, *95*–96, 111
Phytophthora infestans, 331, 338–340
Phytoplankton, 231, 281, 285, 289, 302, *315*, 317, 705; in oxygen balance, *97*
Phytosociology, 685
Phytotron, 561, 563, 692, 719
Picea engelmannii, *730*
Picloram, 598–*599*
Picolinic acid, 598–*599*
Pigment, 37, 78, 84, 228, 231, 647–649; photosynthetic, 96, 288, 389
Pili, *253*, 263, *265*
Pilobolus, 340–343, 356, 361, 381
Pilophorus, *374*
Pine, 235, 378, 426, 429, 452, *486*–490, 597, 668; life history of, 490; seedling, 647
Pineapple, 211–*212*, 518, 676, 710
Pink bread mold, 24, 116, 183, 232, 355
Pink order, 538
Pinnae, 462
Pinnately lobed leaf, *534*
Pinocytosis, 59, *60*, 366
Pinus, *383*, 487–490; *contorta*, *553, 728*; *flexilis*, 746; *mugo*, *576*; *palustris*, *696*; *parviflora*, 597; *thunbergii*, 597
Pinyon juniper, 707
Pipsissewa, 538
Pistil, 176
Pisum sativum, 176, 178, 182, *552*
Pit, bordered, *423*, 425, *426*, 487, 625; simple, *426*
Pith, 412, 420, 422–*423*, 436, 440, 501
Pittendrigh, C. S., 661–662
Placenta, *521*
Plagiotropic growth, 559, 566, 568
Plague, 254
Planck, M., 98
Planes of cell division, *420*
Plankton, 286–287, 293, 334, 382
Plant, *703*–705; breeding, 172–174, 190–194; cell, *56*, 423; community, 15, 682; day-length requirements for Flowering, *666*; diseases, 40, 248, 254, 748; Kingdom, *239*; growth regulator, 564, *588*; hormones, 550–575, 564; press, *524*; sociology, 685–686; succession, 684, *728–730*
Plantlet, *600*
Plasma membrane, 22, 37–38, *50*, 53–*54*, 59–*60*, 156–157, *227, 250, 253, 265*, 330, 510
Plasmalemma, 22, 36–37
Plasmodesmata, 36–38, 60, *242*, 244, *423*, 428–*429*, *434*, 611, 632
Plasmodium, 232, 365–*366–367*
Plasmolysis, 57
Plasmopara viticola, 339–340
Plastic, 257, 735
Plastid, 36–37, *80*, 82, 94, 102, 157–158, *293*, 428–*429*, 510, 651; DNA, RNA, and ribosomes in, 84; origin of, 83–84, 160
Platanaceae, 540
Platanus acerifolia, *occidentalis*, and *orientalis*, *207–208*
Pleiotropy, 185
Pleodorina, *296*
Pleuropneumonia-like organism (PPLO), 272
Plum, 519
Pneumococcus, *120–121*
Pneumonia, 263
Poa sandbergi, 208
Podocarpaceae, *486*
Podzolization, 700
Podzols, 699–*701, 704*, 707
Poinsettia, 537, 666
Poison, 76, 743; hemlock, 540; ivy, *6*
Polar: auxin movement, 562, 564; nuclei, 513–514
Polarized light, 45
Polemoniales, 538–540
Poles, 141
Pollen: grain, 15, *172, 174*, 176, 240, 382–*383*, 388, *476*, 481, 488, *491*, 502–507, 515; fossil, 498; wall of, *504*, 508; mother cells, 530; tube, 508, 510–512, *514*–515, 594; viability, 192
Pollination, 187, 204, 481, 488, 505–508, 515
Pollinium, 543
Pollister, A., 147
Pollution, 28, 284, 735, 744, 747; atmospheric, 383, 635; water, 287, 289, 382
Polyacrilamide gel, 68
Polydopus, *351*
Polyembryogony, 489
Polygonum bistordoides, *716*
Polymerase enzyme, 158
Polyneme concept, 149
Polyneura, *322*
Polypeptide, 71, 116
Polyphosphate, 264, 291
Polyploidy, 192–193, 205–*206*, 295, 584
Polyribosome, 38, 117, *117*, 156
Polysaccharide, 40, 41, 47, 91, 217
Polystichum munitum, 461
Polystyrene, *240*
Pome, 519, *520*
Pomegranate, 540
Pomoideae, 519
Pond, 382; algae, 280–282
Ponderosa pine, *655, 746*
Poplar, 519, *534*, 545, 588
Poppy, 537, 744
Population, 27, 97, 199; growth curve, 97, 747; interbreeding, 146; limitation, 744; of controlled size, 735; variety in a, 28
Population geneticist, 200
Pore, 53, 60, *387*; nuclear, *113*
Pore fungus, *362*
Porphyra, 298, *304*, 309, 314, 321, *323*
Porphyridium, *279*, 321
Porphyrin, 86, 96, 111, *219*
Positive: curvature, *647*; feedback, 745, 747; geotropic response, 559
Postelsia palmiformis, *301*, 304, 311–*313*
Posthypnotic suggestions, 656
Potassium, 605, 608, 612, 632; permanganate, 102
Potato, 519, 539, 572, 666; disease, 331; tissue, 575
Potentilla, 519, 692–693
Powerhouses, cellular, 103
Power rule, 73
PPLO, 272
PPP, 107
Prantl, K., 533, 538–539, 541, 545–546, 548
Prairie, 16, 701, 707
Precambrian, 220
Precipitation, 699
Predation, 27, *745*–745
Prefern, 472, 485
Pressure, 55–56, 628; bomb, *626*; flow theory, *628*–633, *629*; gradient, 629
Prey, 745
Primary: cell wall, 37, 39–40–41, 44; endosperm nucleus, 514; meristem, 420; productivity, 732; protein structure, 71; tissue, 418, 420
Primitive plants, 212
Primordium, root, *436*
Procambium, 420, *439*–440–*442*, 427
Procaryotic cell, 128, 160, 220, 230, 274, 279, 288, 324
Procellular life, 246
Producer, 26, 628–*629*, 737–*738*
Product, 74–*75*
Productivity, 26, *705*–706, 732–736
Profile, soil, 700
Program, developmental, 26, 580–581, 598 669, 675
Progymnosperm, *216*, 235, 449, *452*, 479–480, 485, 489, 494; discovery of, 493
Prolamellar body, *80*, 84
Promeristem, *421*
Propagation, asexual, 418, *600*
Prophase, 141, 150
Proplastid, 36, 83
Prop roots, 540
Prosthetic group, *74*
Protein, 21–24, 37, 41, 53–54, 57, 66, 71–77, 83, 112, 115, 117, 120, 130, 140, 156, 158, 175, 184, 196, 217, 231, *240, 244*, 515, 608, 627, 741; into amino acids, 258; bodies, 36, 428–*429*; breakdown, 591; carbohydrate, 133; enzyme, 38; molecule, *65*, 126; precipitant, 35; synthesis, 38, 116, 123–128, 145, 156, 159–160, 662; variety, 73; in virus, 243
Prothallus, 470
Protista, 238
Protochlorophyll, 646–647
Protoderm, 420, *422*
Proton, 42, 43, *62*, 108
Protophloem, *422*
Protoplasm, 37, 57, 65, 366

Protoplast, 36–37, 54, 57, 60; multinucleate, 340; naked, 292
Protostele, 464–*465*
Protoviruses, 218
Protozoa, 238, 257–*258*
Pruning, 664
Prunus, 519, *520; viriniana, 2*
Pseudoplasmodium, 367–*368*
Psilophyton, 449, *452*
Psilopsida, 221, 233, *237, 239,* 449–450, 452, 472–473
Psilotum nudum, 433, 437, 450, 452–*453,* 472
Ptelea, 519
Pteridium, 6, 473
Pteridosperm, 234–236, *484*
Pteropsida, 234, *237, 239,* 473
Pucinia graminis, 368–*370*
Puffball, *351,* 360, 381
Punnett R. C., 179
Punnett square, 179
Pure-stand crop, 748
Puromycin, 595
Purple plant, *578*
Pycnia, 370
Pyramid: ecosystem, 736–737; of life, 316
Pyrenoid, 78, *80,* 94, *276,* 280, 321
Pyridoxine, 577
Pyrole, 96
Pyrrophyta, 228, *230, 237, 239,* 315, 321–*324*
Pyrus, 519, *520*
Pyruvic acid, *45,* 93, *106*–107, *109*–110

Quackgrass, 600
Quadrat, 685
Quadrose, 44–*45*
Quality control system, 656
Quantitative inheritance 185, 196
Quantitative requirement, 666
Quantum, 87, 98–99, 646
Quaternary, 220; protein structure, 72
Quercus, 519, *534; coccinea, 204; Douglasii,* 531; *turbinella,* 531; *velutina, 204*
Quiescent center, 420–*422*
Quince, 519

Race, 13, 202
Radial, 420
Radial symmetry, 504
Radiation, 217, 219, 288, 613–614, 616, 619
Radioactive: carbon dating, 639; disintegration, 88; element, 90, 377; fallout, 378; water, 631
Radioactivity, 168
Radiometer, 71, 731
Radish, *664*
Ragweed, 541
Rainfall, 640, *704, 708*
Rain: forest, 166, 224, 378, 696, *700, 702–703*–705, 706, 710, 732; shadow, *709*–710; water, 604
Ralfsia, 320
Ramensky, L. G., 687, 691
Ranales, *533,* 537–542, 545–547; primitive character of, 537
Randomness, 558
Random pair, 685, *687*
Random reproduction, 200
Ranunculus, 410, 519; aquatilis, 593, 186
Raper, J., *341*
Raphanus sativus, 664
Rapoport, H., 346
Raspail, F., 34
Raspberry, 518, 522
Rate: of diffusion, 55; of growth, *556;* of assimilate flow, 629–630
Rattlesnake plantain, 543
Rauwolfia, 744
Ray, *441, 443,* 440; flower, *503*–504, 541
Ray, J., 209, 544
Reactant, *75*
Reaction: center, 87; of the dark period, 672; rate, 74; wood, 566–*567*
Receptacle, 502
Reciprocity, 647, 677
Recycle, 378, 735
Red: light, *654, 673;* oak, 638; pepper, 539; "safe" light, 657–658
Red-far-red system, 470
Reduction, 70; in flower parts, 545
Redwood, 235, 418, 667, *697,* 706
Refraction, index of, 36
Refrigerated chamber, *67*
Regal fern, 473
Regeneration, 580
Regular flower parts, 537
Reindeer, 377; moss, 375, *370*
Relative humidity, 10, 57, *619*
Rendle, A. B., 546
Repressor, *128*
Reproduction, 21, 69, 217; in algae, 314; in cycads, 482, *488;* in gymnosperms, 480, 488–490; in lichens, 376; sexual, 13
Reproductive: fitness, 165; isolation, 199, 202, 207–208; state, 417; structure in pine, *491*
Reptile, 221, 224, 390
Reserpine, 744
Resin duct, 487–488, 432
Resistant variety, 191
Respiration, 37, 44, 45, *86,* 92, 94, 101–111, 158, 330; aerobic, 219; in blue-green algae, *291;* energetics, 110–*111;* inhibitor, *661;* of potato tuber, 663
Restorer gene, 191
Rhizobium, 258, *260*
Rhizoid, *334,* 387, 389, 433, 307, *453,* 455
Rhizome, 450, 456, 461, 463–464, 501, 651
Rhizomorph, 362, *364*
Rhizophydium planktonicum, 334
Rhizophydium sphaerothica, 383
Rhizopus stolonifer, 340–*341,* 589–590
Rhododendron, 538
Rhodoglossum, 323
Rhodophyta, 229, *239, 237, 239,* 305–306, 313, 321–*322*
Rhodospeudomonas vannielii, 250
Rhodospirillum rubrium, 252
Rhus toxicodendron, 6; typhina, *534*
Rhynia, 449–452
Rhynic chert, 449
Riboflavin, 565–*566,* 647
Ribonoclease, 24
Ribonucleic acid. *See* RNA
Ribose, *45, 118*–119, 123
Ribosome, 25, 36–37, 60, 78, 84, *101,* 115–117, *127,* 130, 153, 160, *250, 272,* 276, 288, *293,* 328, *330,* 608, 675; in blue-green algae, 291; in pollen grain, 508, 510
Ribulose, *45*
Ribulose-diphosphate, 88, 91–*92,* 104
Ribulose-diphosphate carboxylase, 91
Ribulose-5-phosphate, *105*
Rice, 77, 191, 243, 544, 666; paddy, 288, 706, 709
Ricinus communis, 552
Rickettsiae, 11
Rier, J. P., 584
Rift, 225, *225*
Riker, A. J., 579
Ring: alpha (α) and beta (β), 46; fungal, *360;* structures, *46*
RNA, 25, 38, 84, 111, 115, 117, 119–129, 158, 175, *240,* 244, 591, 652; on DNA template, 595; low molecular-weight, 128; polymerase, 158; replicase, 244; replication of viral, 244; in virus, 243
Roach, A. W., 609
Robinia pseudoacacia, 520
Rockweed, *304,* 307
Rodent, 254
Root, 416–417, 433–436, 387, *389, 567,* 590, 608; adventitious, 435, 588, 466; apex, *389, 410,* 652; cap, 132, *410, 422, 436, 389;* exchange problem, 610; excised, 418, 571, 587; fern, *466;* formation, 651; geotropic response, 566; hair, *260, 389, 430*–431; hormone, 571, 589, 600; nodule, *260;* pressure, 435, 623; stock, *600;* system, 500
Rooting medium, *600*
Rose, 539–540
Roseleaf mountain ash, *534*
Rosen, W., 508
Rosette plants, 570, 614, 669
Rosidae, 548
Rosin, 432
Rothwell, K., 590
Rubber tree, 433
Rubiales, 541
Rubidium chloride, *59*
Rumen, *258*
Rumex, 194; *acetosa,* 192; *acetosella,* 192
Runner formation, 651
Rust, 191, 368
Rye, 77, 544, *655,* 667

Saccharomyces cerevisiae, 356
Sacchorhiza polyschides, 317
Sac fungi, *349*
Sachs, J. von, 422, 590, 604–605, 609
Safranin, 34
Sagebrush, 539, 544, 725, *746*
Saguaro cactus, 714
Salicylic acid, 744
Salisbury, F. B., 678–679
Salkowski test, 590
Salmonella typhosa, 253–254, 273
Salt, 65, 700, 723; accumulator, 724; concentration as limiting factor, 724; excretion, 725; mine, 663; regulator, 724; uptake, 595; water, 16
Salvia pratensis, 506
Salvinia, 461, 464, 473
Samara, 519
Sample collector, 66
Sand dune fungi, 381
Sanger, F., 71
Saponin, 191
Saprolegnia, 326, 337–338, 344, 382
Saprophyte, 10–11, 218, 226, 229, 330, 543; fungi as, 329
Sapwood, 380, 440
Sarcoscypha coccinea, 349
Sargasso sea, 15
Sargent juniper, 597
Satellite, *140*
Satina, S., 438
Saturated fatty acids, 61
Saturation: curve, 58; intensity, 86–89; reaction, 74
Saturn, 219
Sausage tree, *507*
Savanna, 16, *702–703*–705, 707, 710
Saxifragaceae, 540
Saxifrage, 521, 540
Scagel, R. F., 238
Scarification, 522, 639–640–641
Scarlet: cup, *349;* gilia, *538;* oak, 204
Schizocarp, 519
Schizophyta, *237, 239*
Schleiden, M., 31, 515
Scholander, P. F., 626
Schwann, T., 31, 515
Scientific: method, 12, 212–213; name, 13; paper, 720–721
Scion, *600*
Sclereid, *406,* 419, 424–*425*
Sclerenchyma, *406,* 422, 424–*425,* 434, 436–437, *456,* 466
Sclerophylls, 707
Sclerotium, 357, *357,* 366
Scopoletin, 567, *588,* 592
Scott, H., 484
Scouring rush, *6,* 233, 541
Screw pine, 545
Scrophulariales, 539, 541
Scrub, *701*
Scrubland, 16, *703*–705
Sea: lettuce, *304;* palm, 301, 311; urchin, 150
Seabird, 168
Sears, P., 685
Season, 681
Sebastiana, 537
Secondary: productivity, 736; protein structure, 72; root, 435, *436;* wall, 37, 39–*40,* 44
Second positive curvature, 647
Secretory hair, *408,* 431–432
Sedoheptulose, *45*
Seed, 15, 72, 130, 176, 234, 388–389, 417, 454, *480,* 482, 489, 514–515, 518–521–523; coat, 515, 639; dispersal, 500, 518–*523;* dormancy, 638–641; fern, 234, 239, 461, *484,* 485; life span, 639; of monocot and dicot, *518;* and production of duxin, 594; size, 515
Seedling, *416–417,* 492
Selaginella, 433, 437, 451, 453, *455*
Self-pollination, 178
Senescence, 555, 557–558, 575, 591, 638
Sensitive plant, *633*
Sepal, calyx, *502*–502
Sequoia, 140; *sempervirens, 697*
Serenin, 345
Serine, *63*
Sesquoxides, 700
Sewage, 269, 287, 289, 382
Sex: determination, 192–194; expression of, 651, hormone in fungi, 344–347
Sex-influenced gene, 194
Sex-linked inheritance, 194
Sexes, separation of, 545
Sexual reproduction: in bacteria, 246, 266–*268;* in bread mold, 340; in diatoms, 292; in land plants, 388
Shade leaf, *591,* 646; plant, 654
Shadowed virus, *240*
Shadscale, 640, 688, 725–726
Sheath, 501
Shelf fungi, *351,* 360–361
Shelford, V. E., 684–686, 689
Shepherd's purse, *517*
Ship, wooden sailing, 331
Shizocarp, 520
Shoot, 387, *388,* 416, 437; apex, *388, 412;* tip, 248
Shore, *703*–705, 706
Short-day plant, 651, *663,* 665–*666,* 670, 672, 677, 650
Shrub, *2,* 9, 707, 710
Shull, G. H., 190, 529, 671
Sibling species, 199, 202
Sideoats grama, 693
Siegelman, H. W., 652
Sierozem, 700–*701*
Sieve: plate, *410, 428;* pore, 429; tube, 242, 320, *410,* 427–428, *429,* 499–500, 629–630
Silene, 192; *acaulis, 716; dioica,* 194
Silica, 41, 240, 281, *286,* 292, 329, 456, 607, 700; cell, 430
Silique, 519
Silurian, 220–221, 381
Sink, 627
Sinnott, E. W., 580
Siphonostele, 464–*465*
Siphonous line, 295, *297*
Sirenin, 334, 346–347
Sitanion jubatum, 198–199
Siu, R. G. H., 575
Skeleton photoperiod, 661
Skin color, 185
Slack, C. R., 91
Sleep movement, 646, 658
Slime, 263, 428–*429,* 630; mold, 11, 64, 114, 229, *232, 235, 237,* 334, *352,* 365
Smith, W., 167
Smog, *698,* 736
Smoke, 735; tree, 640, 723
Smut, 191, *232*
Snapdragon, 184, 508, 538–539, 546
Sneezeweed, 541
Snow: accumulation, *727;* algae, 86; as limiting factor, 726
Soap, 65
Social conscience, 747
Socrates, 540
Sod, 709
Sodium, 42–43, 51, 65, 273, 607, 632, 640
Soil, 16, 27, 57, *257,* 606, 699–702, *700, 704;* formation, 699–700; nutrient condition, 676; sterilant, *599;* temperature, *727*
Solanaceae, 539
Solander, 210
Solanum, 519
Solar spectrum, 617
Solar system, 681
Sol-gel system, 64–65
Solute, 37, 52, 54, 55, 56–58
Solution, 39, *56*
Solvent, 37, *90*
Somniferum, 520
Sorbus, 519, *534*
Sordaria, 356, 381
Soredia, 376–*377*
Sorghum, 191, 544
Sori, 467–*468*
Soulangean, M., 210
Sound, 98
Soybean, 575, 648, 650, 658, 665, 677, 710
Space probe, 663
Spanish moss, 211–*212*
Spathe, 543
Spatial spearation, 203
Special creation, 169
Speciation, 197, *203,* 208, 529
Species: 214–221, 208, 706; definition of, 202; designation, 13
Specificity, 72
Spectrophotometer, 77, 84
Spectroscope, 648
Spectrum, 84–*85,* 616, *618,* 646, 648
Sperm, 14, 176, 178–179, *283,* 307, 416, 457, 469–*470, 504*–505, 508, *509,* 513, 515, 347; flagellated, 453; multiflagellate, 234, 450, 456, 481–482, *483,* 492; swimming, 388–389
Spermagonia, 369
Spermatophyta, 235–236
Sphacelaria, 320
Sphaerobolus, 361–362, *363,* 381
Sphaerocarpus, 192
Sphagnum moss, 600
Sphenopsid, 233, *237, 239,* 449, 455, 456, 461
Spherosome, *38, 130*–131
Spiderwort, 543
Spikelet, *174,* 544
Spinach, 194, 665–666
Spindle, 38, 141, 142, *143,* 144, 150–151, 230; fiber, 141, *150,* 156–157, 291, 293; microtubule, 317
Spirilli, 261
Spirogyra, 86–*87,* 281, 283, *285,* 289
Spodosol, *701*
Spomer, G. G., 626
Sponge fungus, *350*
Spongy parenchyma, 424, 438
Sporangiophore, 340–342
Sporangium, 326, 334, 339–342, *367,* 450–451, *453,* 456, *457, 491;* in fern, 466–470
Spore, 10, 14, 232, 234, 309, 320, *330*–331, 360, 366, 376, 381, 388–389, *453, 455–456*–457, 480, 649; asexual, 291; in bacteria, *266;* dark-colored, 340; discharge, *363;* in fern, 466–470; motile, 335
Sporocarp, 367
Sporophyll, 537
Sporophyte, *14*–15, *140,* 192, 228, 233, 235, 389, *416,* 450, *453–454,* 457, 514–515; fern, 471; of pine, 490
Sporopollenin, *504*
Spring: ephemeral, 645; hot, 746; overturn, 286; tide, *303*
Spruce, 235, 259, 668, 730
Squash, 194, *552;* fruit growth, *580*
Squid, 96, 316
S-shaped curve, 555, 557
Stability, 213
Stadler, L. J., 188
Staghorn fern, *460*
Stahl, F. W., 188
Staining method, 31
Stalked: cell, *261;* particle, 102
Stamen, *172–173, 502*–502, 505, *521*
Standing crop, 706
Stanley, W., 243
Staphylococci, 263, 273
Starch, 45, 47, *49,* 77–78, 91, 94, 110, 231, 238, 280, 293, 335–336, 389, *406,* 423–424, *429,* 436, 515, 595, 675, 612; grain, 36–37, 566

Starfish, 150
Stark, N., 378
Statistics, 532
Statolith, *566*
Steady-state, 616, 745
Stearic acid, *61*
Stebbins, G. L., 196, 197–198, 203
Stele, *410*, 434, 450, 464–*465*
Stem, 387, *389*, 417, 435–437, 697; cutting, *600*; elongation, *653–654*; of fern, 464
Stemonitis fusca, *352*
Stephan-Boltzmann constant, 617
Stephan's law, 617
Steppes, 707
Stereoisomerism, 45–46, 62
Sterigmata, 360
Sterile cells, jacket of, 389
Stern, K., 657–660
Steroid, 111, 675
Steward, F. C., 573, 581; experiment of, *582*
Stigeoclonium, 289, *297*
Stigma, *173*, 176, 502, 507
Stigmatic: crest, 537; surface, *172*
Stinging hair, 431
Stipe, *300*, 306, 462–*463*
Stomach, four-chambered, *258*
Stomate, 95, *339*, 369, 387–*388*, 389, *414*, 430, 438, 450, *456*, 488, 610–613, 619; function, *331*, 608; sunken, 723
Stone, 240; cell, *425*
Stonecrop, 540
Stoppel, R., 657–659
Storage tissue, 627
Strata, 240
Stratification, 639
Strawberry, 2, 518, *520–521*, 522, 651, 666
Stream, 694, *703–705*, 706
Streptococcus thermophilus, 270
Streptomycin, 200–202, 263
Stress physiology, 722–727
Strobilus, 451, 545
Stroma, 83, 91, 103
Strophiolar cleft, 640
Structural gene, *128*
Structure and function, 41, 82, 109, 586
Struggle for existence, 26, 163, 169, 512, 562
Sturtevant, A. H., 184
Style, *502*, 507, 508–509, 512
Stylet, 248, 629–*630*
Subalpine: forest, *697*; larch, 207
Subclimax, *745*
Suberin, 435, 443
Subspecies, 13, 202
Substituted benzoic acid, 598
Substrate, 159, 557; enzyme, 23, 73–*75*, 103
Succession, 27, 381, 736, 745–746
Succinic acid, 110
Succinyl CoA, 111
Succulent, 611–612, 614, 723
Sucrose, 47, *49*, 91, 94, 110, 571, 584, 586, 626, 629, 669
Sugar, 45, 47, 156, 217, 584, 586, 626, 631; beet, 191, 655, 668; derivative in bacterial wall, *254*; phosphate, 90–91, 608; secretion in phloem, 630; sugarcane, *30*, 91–92, 153, 544, 632, 676, 710; synthesis, 638
Sulfur, 42, 44, 62, 63, *250*, 271, 605, 608, 632; dioxide, 735; radioactive, *632*
Sulfuric acid, 271, 640
Sumac, *534*
Summary: equation for photosynthesis, 94; of cell structure and function, *154*
Summer dormancy, 638
Sun, 26, 217, *303*, 681; leaf, 95, *591*, 646
Sunburn, 654
Sunflower, *503*, *541*, *576*, 648
Sunlight, 10, 69, 71, 82, 654
Super organism, 27, 682, 684, 687, 689, 691
Surface, 64; of protein, 72; volume relationship, 65
Survival of the fittest, 28, 163, 563
Suspensor, *483*, 514, *517*
Swamp, *694*, *703–705*, 732
Swarm cell, 366–*367*
Sweetgum, 638
Sweet william, *2*
Swift, H., 148
Swordfern, 461
Sycamore, 206–*207*, 540
Symbiosis, 27–28, 160, 744–745
Symmetrical, 504, 576, 580
Sympatric species, 203
Symplast, *434*–435, 611, 626–*629*, 631, 633
Synapsis, 143
Synaptonemal complex, 144, *147*
Synchronizer, 659
Synechococcus, *269*
Synecology, 682, 718
Synergid, *510–511*, 511, 513–*514*, 515
Synergism, 574
Syngamy, 142, 232, *285*, *338*
Synopsis, 532
Synthesis, 87, 218
Syphilis, 254, 263
Syringa vulgaris, *534*
Systema Naturae, 209
Systematics, 12, 206–213, 528

2,4,5-T, 598–*599*
Taiga, 16, 683, 700, *702*, *703*–705, 707
Takhtajian, A., 548
Tamarix pentandra, 611
Tangential, 420
Tannins, 36, *406*
Taproot, 501
Tapetum, *503*, 505
Tartaric acid, 45
Tatum, E. L., 24, 116, 132, 185, 266
Tautomeric state, 189
Taxaceae, *486*
Taximetrics, 532
Taxedium distichum, *694*
Taxonomist, 207, 527, 532; role in biology, 528
Taxonomy, 12, 13, 208–213, 527, 561
Taylor, H. T., 149
TCNB, *592*
Tea, 545, 639, 743
Technology, 735, 748
Teleology, 612
Telescope, 32
Teliospore, 363, 369–*370*
Telome theory, 462
Telophase, 141, 144, 151
Temin, H., 244
Temperate phage, 246
Temperate zone, 16
Temperature, 10, 54, 55, 57, 314, 613, 617, 624, 635–639, *673*, 682, 699, 704, 709; compensation, 661, 669, 678; controlled, 718; cycle of a lake, *287*; and enzymes, *75*; and gene expression, 185; and growth, *636*; leaf, 619, 723; as a limiting factor, 724; and photosynthesis, *86*; for vernalization, 668
Template, 158
Tensile strength, 47
Tension, 57, 566, *626*
Terminal plate, *227*
Terpene, 110, 743
Terrarium, *711*
Tertiary period, *216*
Tertiary protein structure, 72
Test cross, 179
Test dark period, 678
Tetanus, 263
Tetrahedron, *46*
Tetrahydrocannabinol, 431, 744
Tetraploid, 140
Tetrasaccharide, 47
Tetrasporineline, 295
Thallophyta, 235–236
Thallus, *305*
Theobromine, 743
Theophrastus, 211, 234, 515, 544–545
Thermocouple, 731
Thermodynamics: First Law of, 71; Second Law of, 71, 74, 168, 558, 562, 736
Thermonuclear reaction, 217
Thermoperiodism, 636
Thermophile, 290
Thermophilic fungi, 382
Thermoplasma acidophila, 272
Thermostat, 581, 747
Thiamine, 571, 608
Thimann, K. V., 566, 589
Thiocapsa, *250*
Thiopedia, *250*
Thiospirillum jenense, *250*
Thistle, 519, 541
Thorium, 734–735
Thorne, R., 548
Thorn woodland, 16, 707
Threonine, *63*, *65*
Thuja occidentalis, *441*
Thylakoid, 93, *230*, *276*, 290, 292, 293, 317, *320*; membrane, 78, *80*, 82–84, 86, 91, 274
Thyme, 539
Thymidine, 123, 148–*149*
Thymine, *118*–122; dimer, *656*–658
Thyroxin, 607
Tidal zone, 15, *450*, 691, 706
Tide, 302–*303*–*304*, 681
Tilia, 519, 534
Timberline, 654, 727
Time: measurement, 658, 672, 675, 678; memory, 658, 662; scale, 169; and vegetation changes; 683
Time lapse microcinematography, 151
Timing mechanism, 656
Tinsel flagella, *227*–228
Tip growth, 431, 433
Tippo, O., 235
Tissue, 22, 422; culture, 572, 584, 587; somatic, 188; staining, 36; system, *423*, *436*; transmitting, *509*; types, 422–433
Tmesipteris, 450
TMV, *240*, *242*–244, 248
Toadstool, *348*
Tobacco, *79–80*, *83*, *142*, 242–243, 539, 572, 581, 743; Maryland mammoth, 665–666; mosaic virus (TMV), *240*, *242*–243, *244*, 248; pith tissue culture, 587
Tomato, 192, 519, *520*, *521*, *539*, *583*, 594, 609, *636*, 640, 651, *655*, 665
Tongue graft, *600*
Tonoplast, 22, 36–37, 53, 59
Toole, E. H., 650
Toole, V. K., 650
Toothed leaf, 534
Torden, 598–*599*
Torpedo embryo, *517*, *582*
Torrey, J. G., 585
Torus, 426
Totipotent, 424, 581, 587
Tournefort, J. de, 210
Toxin, 362
2,4,5-TP, *592*
Trace, leaf, *437*
Tracer, isotopic, 69, 103; radioactive, 631
Tracheid, *409*, 425–*426*, 438, *441*, 487, 500–501
Tracheophyta, 233, *237–239*
Trailing arbutus, 538
Tranquilizer, 595, 744
Tranquillini, W., 718
Transcription, 123, 158–159
Transducer, 580–581, 637
Transduction, 246–*247*
Transect, 685, 690
Transfer of compounds, 26, 739–744
Transfer RNA (t-RNA), 25, 124, *126–127*, 154, 156, 575
Transfer wall, 37
Transformation of matter, 69
Transfusion tissue, 488
Transgressive segregate, 187
Transient, 661–662
Transition zone, 691
Transtation, 25, 126, 159, 580
Translocation, 147, 388, 626–633, 673, *675*
Transpiration, 57, *603*, 610–616, 618, 623, 721; stream, 613, 632
Transport mechanism, 621–624
Traube, M., 52
Traumatic acid, 572
Tree: form, 378, 454; in miniature, 597
Tree of heaven, 519
Tremella, *350*
Trench, 222–*225*
Treponema pallidum, 254
Triassic, 225, 490, 492, 498
Triazine, 598–*599*
Trichia montana, *352*
Trichloroacetic acid, 598
Trichomanes reniforme, *465*
Trichome, 430, 529
Trichophyton interdigitale, 363
Trident maple, 597
Trifluralin, *599*
Trigger, 646, 721
Triggered decline, 747
Trihybrid, 179
Triose, 44–*45*
Triose phosphate, 88
Triple bond, 42
Triplet, 123–124
Triploid, 140
Trisaccharide, 47
Trisporic acid, 340
Tritium, 62, 734
t-RNA, *See* Transfer RNA
Trophic level, 736
Tropic: forest as fern environment, 464; rain forest, 16, *683*, 706; soil, 259
Tropics, 16, 658, 733
Tropism, 558, 647
Truffle, 356
Tryon, R., 463
Tryptophane, *63*, 110, 590
Tschermak, E., 183
Tsuga heterophylla, *730*
Tuber aestirum, 356
Tuberculosis, 262–263
Tuber formation, 591
Tubules, 142–143
Tulip tree, 210, *441*, *534*, 545
Tumor, 572, 575, 581, 583; inducing principle, 255, 583
Tuna, 316
Tundra, 16, 17, 375, 683, *703*–705, 709, 726–727, 733; alpine, 691, *716–717*; soil, *699*
Tunica, *421*, 422
Tunica-corpus concept, 419, 421, 438
Turesson, G., 693
Turgidity, 37, 57
Turnip, 22, 649
Two-carbon fragment, 91
Type specie, 208
Typhoid, 254, 263
Tyrosine, *63*

UDP-glucose pyrophosphorylase, *68*
Ulmus, 519, 534
Ulothrix, 281, 283, *296*
Ultracentrifuge, 35
Ultramicroscope, 32
Ultramicrotome, 34–35
Ultraviolet light, 128, 188, 246, 654–656, 741
Ulva, 14, 295–296, 298, *304*, 306–*307*, 309
Umbelliferae, 519, 540–541
Uniform garden, 692–693
Uniformitarianism, 169
Uniformity, 21, 25, 628
Unineme concept, 149
Uninucleate, 232
Unit, determiner, 175
Unit inheritance concept, 178
Unit membrane, 60, 82, 130, 272
Unity, 528
Unsaturated: fatty acid, 61; lactone, 567, 640
Upwelling, 304, 316
Uracil, *118*, 119
Uranium, 734–735
Uranus, 219
Urea, 22, 44, 72
Urediospore, 369–*370*
Uredium, *370*
Urethane, 662
Uridine, 123
Urospora, 297
Urtica, 431
Usnea Florida, *374*
Ustilago maydis, 363
Utopia, 735, 748

Vacuolar membrane, 36–37
Vacuole, 22, 36–37, *51*, 59–60, 264, *320*, *330*, 424, *427*, *429*, 611; gas, *291*
Vacuum boils, 625
Valine, *63*
Vallisneria, 507–*508*
Valonia, 295–*297*
Vanadium, 607
Vapor pressure, 613, 618
Variability, genetic, 198
Variation, genetic, 165–167, 175, 195
Variegated, 188
Variety, dwarf, 191
Varner, J. E., 595
Vascular: bundle. *See* Bundle, vascular; cambium 439–440–*443*, 480, 501; canal, 456; cylinder, 420, *437*; plants, 233–235, 295, 389, *499*, 555; tissue, 233, 389, *409–410*, *412*, *414*, 436–437, 462, 487, 586
Vector, 243, 248
Vegetation, 15, 690, 699
Veil, 359–*360*
Vein, *414*, 425, 438, 500, *621*
Veinlet, 438
Velocity of particles, 54
Venation, dichotomous, 420; pattern, 419
Venereal disease, 254
Venus flytrap, *433*, 506
Vernalin, 669
Vernalization, 665, 666–669, 722
Veronica persica, 641
Vertebrate, 11, 390
Vertical structuring, 683
Vesicle, 38, 84, *133*, *142*, 157
Vessel, 254, 425, 441, 499, 501; differentiation, *427*; element, 426–427
Vest, E. D., *725*
Vestal, A. G., 687, 689
Viable, 639
Vibrio cholerae, 254
Vicia gigantea, *523*
Virus, 226, 242–248, 631; duplication, 244; origin of, 246; particle, 627; structure, *245*
Vitalism, 22
Vitamin, 27, 74, 96, 571–572, 581, 594, 596, 605, 607, 743
Vitis, 519, *520*
Vogel, Orville A., 191
Volatile: oil, 431; organic molecule, 610
Volcano, 739
Voltage, 70
Volume, *65*, 555; to surface ratio, 387
Volva, *360*
Volvocine line, *294*, 293
Volvox, *278*, 280–281, 289, *296*

Wain, R. L., 590
Walker, D. R., 592, 638
Walkingshaw, C., 606
Walkingstick stage, *517*
Wall, *406*; diatom, *294*; hyphal, *330*; primary, *423*
Wallace, A. R., 28, 167, 170, 171
Wallflower, *716*
Walnut, 545, 575, 743
War, 748
Wareing, P. F., 588
Warm desert, *712*
Washingtonia filifera, 614, *714*
Waste, solid and gaseous, 735
Water, 42–*43*, 219, 280, 681–682; availability of, 97; columns of, 625; concentration, 56; deficit, 612, 613; density of, 286; distribution, 386, and land as environments, *387*; as a limiting factor, 723, 732; pollution, 382; potential, 55–57, 82, 435, 558, 618, *622*, 624, 633; softener, 65; uptake of, 610; vapor, 610, 612
Water butterup, 186